INTRODUCTION TO
FLUID MECHANICS

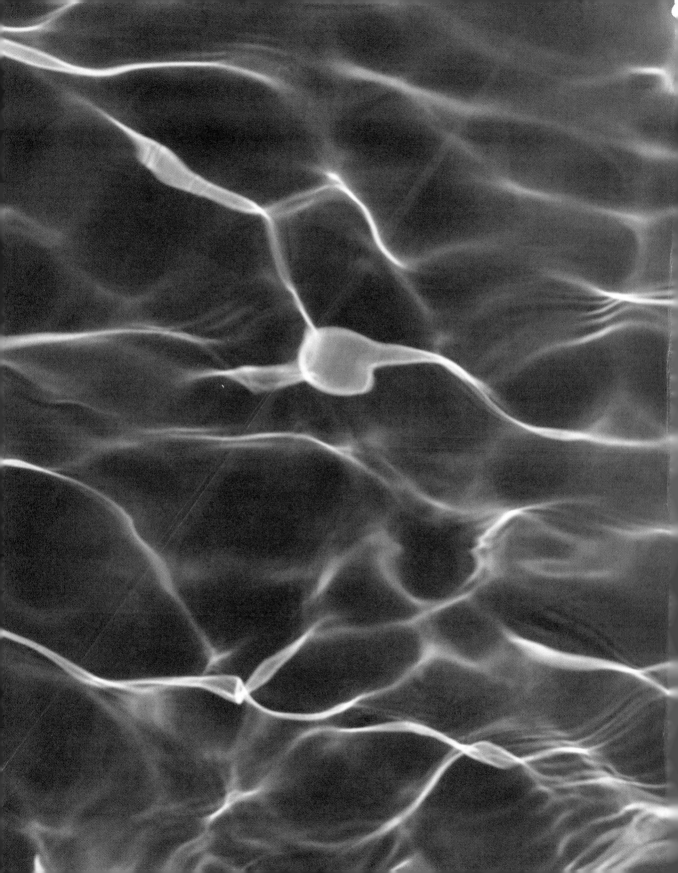

INTRODUCTION TO
FLUID MECHANICS

Edward J. Shaughnessy, Jr.
Duke University

Ira M. Katz
Lafayette College

James P. Schaffer
Lafayette College

New York   Oxford
OXFORD UNIVERSITY PRESS
2005

This book is dedicated to Nana and Pops, who inspired me, Meg, who loved me, and Mike, Erin, and Kerry, who cheered me on. —EJS

Oxford University Press

Oxford   New York
Auckland Bangkok Buenos Aires Cape Town Chennai
Dar es Salaam Delhi Hong Kong Istanbul Karachi Kolkata
Kuala Lumpur Madrid Melbourne Mexico City Mumbai Nairobi
São Paulo Shanghai Taipei Tokyo Toronto

Copyright © 2005 by Oxford University Press, Inc.

Published by Oxford University Press, Inc.
198 Madison Avenue, New York, New York 10016
www.oup.com

Oxford is a registered trademark of Oxford University Press

All rights reserved. No part of this publication may be reproduced,
stored in a retrieval system, or transmitted, in any form or by any means,
electronic, mechanical, photocopying, recording, or otherwise,
without the prior permission of Oxford University Press.

Library of Congress Cataloging-in-Publication Data

Shaughnessy, Edward J.
  Introduction to fluid mechanics/Edward J. Shaughnessy, Ira M. Katz, James P. Schaffer.
    p. cm.
  Includes index.
  ISBN-13: 978-0-195154-51-1
  ISBN 0-19-515451-7 (cloth: acid-free paper)
  1. Fluid mechanics.   I. Katz, Ira M., 1958- II. Schaffer, James P. III. Title.

QA911.S452 2004
620.1'06—dc22
                                                                2004044802

Printing number: 9 8 7 6 5 4 3 2 1

Printed in the United States of America
on acid-free paper

CONTENTS

PREFACE xiii

1 FUNDAMENTALS

CHAPTER 1
Fundamental Concepts

   1.1 Introduction 3
   1.2 Gases, Liquids, and Solids 14
   1.3 Methods of Description 22
       1.3.1 Continuum Hypothesis 23
       1.3.2 Continuum and Noncontinuum Descriptions 24
       1.3.3 Molecular Description 26
       1.3.4 Lagrangian Description 27
       1.3.5 Eulerian Description 27
       1.3.6 Choice of Description 28
   1.4 Dimensions and Unit Systems 29
       1.4.1 $\{MLtT\}$ Systems 30
       1.4.2 $\{FLtT\}$ Systems 31
       1.4.3 $\{FMLtT\}$ Systems 31
       1.4.4 Preferred Unit Systems 32
       1.4.5 Unit Conversions 33
   1.5 Problem Solving 34
   1.6 Summary 35
   Problems 36

CHAPTER 2
Fluid Properties

   2.1 Introduction 43

2.2 Mass, Weight, and Density 43
    2.2.1 Specific Weight 48
    2.2.2 Specific Gravity 49
2.3 Pressure 51
    2.3.1 Pressure Variation in a Stationary Fluid 54
    2.3.2 Manometer Readings 57
    2.3.3 Buoyancy and Archimedes' Principle 58
    2.3.4 Pressure Variation in a Moving Fluid 60
2.4 Temperature and Other Thermal Properties 64
    2.4.1 Specific Heat 65
    2.4.2 Coefficient of Thermal Expansion 67
2.5 The Perfect Gas Law 70
    2.5.1 Internal Energy, Enthalpy, and Specific Heats of a Perfect Gas 70
    2.5.2 Limits of Applicability 71
2.6 Bulk Compressibility Modulus 73
    2.6.1 Speed of Sound 76
2.7 Viscosity 80
    2.7.1 Viscous Dissipation 83
    2.7.2 Bulk Viscosity 83
2.8 Surface Tension 85
    2.8.1 Pressure Jump Across a Curved Interface 87
    2.8.2 Contact Angle and Wetting 90
    2.8.3 Capillary Action 90
2.9 Fluid Energy 93
    2.9.1 Internal Energy 93
    2.9.2 Kinetic Energy 94
    2.9.3 Potential Energy 95
    2.9.4 Total Energy 95
2.10 Summary 97
Problems 99

CHAPTER 3
Case Studies in Fluid Mechanics

3.1 Introduction 103
3.2 Common Dimensionless Groups in Fluid Mechanics 105
3.3 Case Studies 114
    3.3.1 Flow in a Round Pipe 115
    3.3.2 Flow Through Area Change 120
    3.3.3 Pump and Fan Laws 124
    3.3.4 Flat Plate Boundary Layer 128
    3.3.5 Drag on Cylinders and Spheres 132
    3.3.6 Lift and Drag on Airfoils 137

3.4 Summary **140**
Problems **141**

CHAPTER 4
Fluid Forces

4.1 Introduction **146**
4.2 Classification of Fluid Forces **148**
4.3 The Origins of Body and Surface Forces **149**
4.4 Body Forces **152**
4.5 Surface Forces **160**
    4.5.1 Flow Over a Flat Plate **171**
    4.5.2 Flow Through a Round Pipe **173**
    4.5.3 Lift and Drag **175**
4.6 Stress in a Fluid **178**
4.7 Force Balance in a Fluid **187**
4.8 Summary **190**
Problems **191**

CHAPTER 5
Fluid Statics

5.1 Introduction **197**
5.2 Hydrostatic Stress **199**
5.3 Hydrostatic Equation **201**
    5.3.1 Integral Hydrostatic Equation **202**
    5.3.2 Differential Hydrostatic Equation **205**
5.4 Hydrostatic Pressure Distribution **210**
    5.4.1 Constant Density Fluid in a Gravity Field **211**
    5.4.2 Variable Density Fluid in a Gravity Field **218**
    5.4.3 Constant Density Fluid in Rigid Rotation **222**
    5.4.4 Constant Density Fluid in Rectilinear Acceleration **229**
5.5 Hydrostatic Force **233**
    5.5.1 Planar Aligned Surface **234**
    5.5.2 Planar Nonaligned Surface **238**
    5.5.3 Curved Surface **248**
5.6 Hydrostatic Moment **252**
    5.6.1 Planar Aligned Surface **256**
    5.6.2 Planar Nonaligned Surface **260**
5.7 Resultant Force and Point of Application **267**
5.8 Buoyancy and Archimedes' Principle **269**
5.9 Equilibrium and Stability of Immersed Bodies **275**
5.10 Summary **278**
Problems **280**

CHAPTER 6
The Velocity Field and Fluid Transport

- 6.1 Introduction 299
- 6.2 The Fluid Velocity Field 300
- 6.3 Fluid Acceleration 312
- 6.4 The Substantial Derivative 319
- 6.5 Classification of Flows 320
  - 6.5.1 One-, Two-, and Three-Dimensional Flow 321
  - 6.5.2 Uniform, Axisymmetric, and Spatially Periodic Flow 327
  - 6.5.3 Fully Developed Flow 331
  - 6.5.4 Steady Flow, Steady Process, and Temporally Periodic Flow 332
- 6.6 No-Slip, No-Penetration Boundary Conditions 336
- 6.7 Fluid Transport 337
  - 6.7.1 Convective Transport 340
  - 6.7.2 Diffusive Transport 348
  - 6.7.3 Total Transport 352
- 6.8 Average Velocity and Flowrate 358
- 6.9 Summary 363
- Problems 365

CHAPTER 7
Control Volume Analysis

- 7.1 Introduction 375
- 7.2 Basic Concepts: System and Control Volume 376
- 7.3 System and Control Volume Analysis 377
- 7.4 Reynolds Transport Theorem for a System 381
- 7.5 Reynolds Transport Theorem for a Control Volume 382
- 7.6 Control Volume Analysis 385
  - 7.6.1 Mass Balance 386
  - 7.6.2 Momentum Balance 397
  - 7.6.3 Energy Balance 420
  - 7.6.4 Angular Momentum Balance 439
- 7.7 Summary 450
- Problems 452

CHAPTER 8
Flow of an Inviscid Fluid: the Bernoulli Equation

- 8.1 Introduction 474
- 8.2 Frictionless Flow Along a Streamline 475

8.3 Bernoulli Equation 477
    8.3.1 Bernoulli Equation for an Incompressible Fluid 479
    8.3.2 Cavitation 482
    8.3.3 Bernoulli Equation for a Compressible Fluid 487
8.4 Static, Dynamic, Stagnation, and Total Pressure 490
8.5 Applications of the Bernoulli Equation 496
    8.5.1 Pitot Tube 496
    8.5.2 Siphon 503
    8.5.3 Sluice Gate 509
    8.5.4 Flow through Area Change 511
    8.5.5 Draining of a Tank 518
8.6 Relationship to the Energy Equation 521
8.7 Summary 524
Problems 526

CHAPTER 9

Dimensional Analysis and Similitude

9.1 Introduction 534
9.2 Buckingham Pi Theorem 536
9.3 Repeating Variable Method 540
9.4 Similitude and Model Development 549
9.5 Correlation of Experimental Data 554
9.6 Application to Case Studies 557
    9.6.1 DA of Flow in a Round Pipe 557
    9.6.2 DA of Flow through Area Change 558
    9.6.3 DA of Pump and Fan Laws 559
    9.6.4 DA of Flat Plate Boundary Layer 561
    9.6.5 DA of Drag on Cylinders and Spheres 562
    9.6.6 DA of Lift and Drag on Airfoils 562
9.7 Summary 563
Problems 564

2 DIFFERENTIAL ANALYSIS OF FLOW

CHAPTER 10

Elements of Flow Visualization and Flow Structure

10.1 Introduction 573
10.2 Lagrangian Kinematics 578
    10.2.1 Particle Path, Velocity, Acceleration 578
    10.2.2 Lagrangian Fluid Properties 589

10.3   The Eulerian–Lagrangian Connection   590
10.4   Material Lines, Surfaces, and Volumes   592
10.5   Pathlines and Streaklines   597
10.6   Streamlines and Streamtubes   603
10.7   Motion and Deformation   607
10.8   Velocity Gradient   612
10.9   Rate of Rotation   619
    10.9.1   Vorticity   622
    10.9.2   Circulation   628
    10.9.3   Irrotational Flow and Velocity Potential   632
10.10   Rate of Expansion   635
    10.10.1   Dilation   636
    10.10.2   Incompressible Fluid and Incompressible Flow   638
    10.10.3   Streamfunction   643
10.11   Rate of Shear Deformation   650
10.12   Summary   653
Problems   654

CHAPTER 11
Governing Equations of Fluid Dynamics

11.1   Introduction   659
11.2   Continuity Equation   660
11.3   Momentum Equation   666
11.4   Constitutive Model for a Newtonian Fluid   671
11.5   Navier–Stokes Equations   678
11.6   Euler Equations   683
    11.6.1   Streamline Coordinates   689
    11.6.2   Derivation of the Bernoulli Equation   692
11.7   The Energy Equation   699
11.8   Discussion   702
    11.8.1   Initial and Boundary Conditions   702
    11.8.2   Nondimensionalization   703
    11.8.3   Computational Fluid Dynamics (CFD)   706
11.9   Summary   708
Problems   709

CHAPTER 12
Analysis of Incompressible Flow

12.1   Introduction   713
12.2   Steady Viscous Flow   718
    12.2.1   Plane Couette Flow   720

  12.2.2 Circular Couette Flow **723**
  12.2.3 Poiseuille Flow Between Parallel Plates **732**
  12.2.4 Poiseuille Flow in a Pipe **737**
  12.2.5 Flow over a Cylinder (CFD) **741**
12.3 Unsteady Viscous Flow **744**
  12.3.1 Startup of Plane Couette Flow **749**
  12.3.2 Unsteady Flow over a Cylinder (CFD) **752**
12.4 Turbulent Flow **754**
  12.4.1 Reynolds Equations **756**
  12.4.2 Steady Turbulent Flow Between Parallel Plates (CFD) **757**
12.5 Inviscid Irrotational Flow **760**
  12.5.1 Plane Potential Flow **761**
  12.5.2 Elementary Plane Potential Flows **769**
  12.5.3 Superposition of Elementary Plane Potential Flows **772**
  12.5.4 Flow over a Cylinder with Circulation **777**
12.6 Summary **780**
Problems **782**

3 APPLICATIONS

CHAPTER 13
Flow in Pipes and Ducts

13.1 Introduction **791**
13.2 Steady, Fully Developed Flow in a Pipe or Duct **793**
  13.2.1 Major Head Loss **799**
  13.2.2 Friction Factor **801**
  13.2.3 Friction Factors in Laminar Flow **805**
  13.2.4 Friction Factors in Turbulent Flow **812**
13.3 Analysis of Flow in Single Path Pipe and Duct Systems **817**
  13.3.1 Minor Head Loss **824**
  13.3.2 Pump and Turbine Head **835**
  13.3.3 Examples **838**
13.4 Analysis of Flow in Multiple Path Pipe and Duct Systems **846**
13.5 Elements of Pipe and Duct System Design **851**
  13.5.1 Pump and Fan Selection **853**
13.6 Summary **864**
Problems **867**

CHAPTER 14
External Flow

14.1 Introduction **882**

14.2 Boundary Layers: Basic Concepts 884
    14.2.1 Laminar Boundary Layer on a Flat Plate 887
    14.2.2 Turbulent Boundary Layer on a Flat Plate 894
    14.2.3 Boundary Layer on an Airfoil or Other Body 898
14.3 Drag: Basic Concepts 902
14.4 Drag Coefficients 905
    14.4.1 Low Reynolds Number Flow 905
    14.4.2 Cylinders 908
    14.4.3 Spheres 913
    14.4.4 Bluff Bodies 916
14.5 Lift and Drag of Airfoils 926
14.6 Summary 933
Problems 935

CHAPTER 15
Open Channel Flow

15.1 Introduction 942
15.2 Basic Concepts in Open Channel Flow 945
15.3 The Importance of the Froude Number 952
    15.3.1 Flow over a Bump or Depression 953
    15.3.2 Flow in a Horizontal Channel of Varying Width 961
    15.3.3 Propagation of Surface Waves 965
    15.3.4 Hydraulic Jump 972
15.4 Energy Conservation in Open Channel Flow 978
    15.4.1 Specific Energy 981
    15.4.2 Specific Energy Diagrams 986
15.5 Flow in a Channel of Uniform Depth 989
    15.5.1 Uniform Flow Examples 994
    15.5.2 Optimum Channel Cross Section 999
15.6 Flow in a Channel with Gradually Varying Depth 1003
15.7 Flow Under a Sluice Gate 1003
15.8 Flow Over a Weir 1009
15.9 Summary 1012
Problems 1014

Appendixes

Appendix A Fluid Property Data for Various Fluids A-1
Appendix B Properties of the U.S. Standard Atmosphere B-1
Appendix C Unit Conversion Factors C-1

CREDITS D-1

INDEX I-1

PREFACE

This book is intended primarily for use in a one-quarter or one-semester introductory fluid mechanics course. Our goal is to provide both a balanced introduction to all the tools used for solving fluid mechanics problems today and a foundation for further study of this important and exciting field. By learning about analytical, empirical (existing experimental data and accepted engineering practice), experimental (new experimental data, which will need to be obtained), and computational tools, students learn that an engineering problem can be approached in many different ways and on several different levels. This distinction of approach is especially important in fluid mechanics, where all these tools are used extensively. Although the traditional methodology of engineering fluid mechanics is thoroughly covered, this text also includes elements of differential analysis presented at a level appropriate for the target student audience. We also make use of outputs from commercially available computational fluid dynamics codes to help illustrate the phenomena of interest. It is not expected that students will perform any computational fluid mechanics simulations. However, with computational solutions becoming routine, economical, and accessible to engineers with bachelor's degrees, it is important that students be familiar with the use of this type of information. Therefore, computationally produced figures are used in the text for expository purposes. Throughout the text, *CFD* icons indicate when the subject matter directly, or indirectly, relates to computational methods. There are also a large number of figures, photographs, and solved problems to give students an understanding of the many exciting problems in fluid mechanics and the tools used to solve them. The visual approach to understanding fluid mechanics is highlighted with the use of *visual* icons that point students to resources

| | CD/Kinematics/Compressibility/Compressible and Incompressible Fluids |

 such as websites, books, and especially, the excellent CD *Multi-Media Fluid Mechanics.** A third icon, called *FE*, is used to note material that is covered in the Fundamentals of Engineering exam.

We have organized the text in three parts to give to each instructor the flexibility needed to meet the needs of his or her students and course(s).

Part 1, Fundamentals, contains the first nine chapters and covers the traditional body of introductory material. Our emphasis here is on developing an understanding of fundamental ideas. The judicious use of software packages to perform routine mathematical and graphical operations is intended to allow the student to concentrate on *ideas* rather than *mathematics*. Here, and elsewhere in the text, we employ a visual presentation of results to enhance student learning and to encourage students to do the same in their own problem solving. An important feature of Part 1 is the introduction to empirical methods in Chapter 3, rather than covering this much later, as is the case in other texts. Chapter 3 includes simple but effective case studies on pipe flow,

---

*© 2000, 2004 by Stanford University and its licensors, published by Cambridge University Press.

drag on spheres and cylinders, lift and drag on airfoils, and other topics. The student is thus empowered to solve important and interesting fluid mechanics problems in these areas without being forced to wait until the end of the course for the "good stuff." The early exposure to these topics in the lecture also serves to broach these topics early in the traditional laboratory portion of an introductory course, which also helps to build student interest. As Part 1 unfolds, the student learns more and more about the source of the empirical rules presented in the case studies. The text revisits the case studies not only in Part 1 but also in Parts 2 and 3 of the text to show the student how advanced methods contribute to a deeper understanding of a flow than can be gained from empirical methods alone.

Part 2, Differential Analysis of Flow, consists of three chapters and represents the core of our added emphasis on differential analysis and a visual presentation of fluid dynamics. It is important to note that we have written the chapters in Part 2 in a modular fashion. That is, instructors can select to cover as much or as little of this material as they see fit without losing the ability to continue on into the third part of the text on applications. This section begins with Chapter 10, Elements of Flow Visualization and Flow Structure, where we introduce classic kinematic concepts from both the Lagrangian and Eulerian descriptions. This chapter demonstrates the importance of flow visualization in the context of modern experimental approaches to flow measurement, as well as in flow simulation. An additional feature of our coverage in this and subsequent chapters is that the student begins to appreciate the wealth of information available from skillful postprocessing of CFD simulations. This chapter discusses flow structure in preparation for a discussion of the governing equations of fluid dynamics. Chapter 12 allows instructors to expose their students to one or more of the classic exact solutions to the Navier–Stokes equations.

The remaining chapters of the text constitute *Part 3, Applications*. It is here that students see how analytical, empirical, experimental and computational methods come together to solve engineering problems. Chapters on traditional topics such as Flow in Pipes and Ducts, External Flow and Open Channel Flow extend students' understanding of the breadth of fluid mechanics and its applications.

In writing the text, we are well aware of the needs of different instructors. Someone faced with selecting material for a one-quarter course may elect to cover *Part 1, Fundamentals,* and feel comfortable in going straight to *Part 3, Applications,* to continue with one or more of these chapters. Those who have a semester course may elect to cover *Part 1, Fundamentals,* followed by some of *Part 2, Differential Analysis of Flow,* and then finish the course by discussing one or more of the chapters in *Part 3, Applications*. There is enough material in the text for a second, intermediate fluid mechanics course in either the quarter or semester system. In this respect we have found that current introductory texts fail to provide sufficient material on differential analysis, while advanced texts place a reliance on mathematics that is far too heavy for the intermediate student. Ample problems at the end of each chapter are designed to meet the needs of instructors and students alike. Today's students are accustomed to, and thrive in, a visual learning environment. These students have already integrated the use of computers into many daily tasks. We believe a fluid mechanics textbook that provides the same visual, computer-oriented environment will be an extremely effective aid in learning. Fluid mechanics is a notoriously challenging subject, but one that lends itself to a visual learning process and the use of computers to accomplish routine tasks and examine results.

The text is designed to accommodate different disciplines: mechanical, civil, aerospace, and chemical engineering. It also permits instructors to select the level of treatment appropriate for the course and setting, without relinquishing an opportunity to employ a visual text with early exposure to interesting fluid mechanics problems and an effective use of today's computational tools.

We wish to extend our thanks to our Editor at Oxford University Press, Danielle Christensen. We also thank Barbara Brown, Editorial Assistant, for pulling together the photo program for the book, and thanks to Karen Shapiro, Managing Editor, for guiding the production of our book.

PART 1
FUNDAMENTALS

1 FUNDAMENTAL CONCEPTS

1.1 Introduction
1.2 Gases, Liquids, and Solids
1.3 Methods of Description
   1.3.1 Continuum Hypothesis
   1.3.2 Continuum and Noncontinuum Descriptions
   1.3.3 Molecular Description
   1.3.4 Lagrangian Description
   1.3.5 Eulerian Description
   1.3.6 Choice of Description
1.4 Dimensions and Unit Systems
   1.4.1 $\{MLtT\}$ Systems
   1.4.2 $\{FLtT\}$ Systems
   1.4.3 $\{FMLtT\}$ Systems
   1.4.4 Preferred Unit Systems
   1.4.5 Unit Conversions
1.5 Problem Solving
1.6 Summary
Problems

1.1 INTRODUCTION

The fluid-covered Earth from space.

Fluid mechanics is concerned with understanding, predicting, and controlling the behavior of a fluid. Since we live in a dense gas atmosphere on a planet mostly covered by liquid, a rudimentary grasp of fluid mechanics is part of everyday life. For an engineer, fluid mechanics is an important field of the applied sciences with many practical and exciting applications. If you examine municipal water, sewage, and electrical systems, you will notice a heavy dependence on fluid machinery. Pumps and steam turbines are obvious components of these systems, as are the valves and piping found in your home, under your city streets, in the Alaska oil pipeline, and in the

1 FUNDAMENTAL CONCEPTS

HISTORY BOX 1-1

The history of science and technology, including that of engineering fields like fluid mechanics, is often ignored in engineering courses. Yet a sound historical perspective can help engineers avoid the mistakes of the past while creatively building on the achievements of their predecessors. By learning something about the insights of the pioneers of fluid mechanics, you will sharpen your own insight into the subject, and perhaps develop new ideas in areas such as turbulent flow, where progress is very slow.

Fluid mechanics has been important to virtually all societies because people need water to drink, irrigation for agriculture, and the economic advantages of waterborne transportation. The earliest known hydraulic engineering was accomplished in the river valleys of Mesopotamia and Egypt. Mechanical devices and canals were used for distributing water for agriculture. Perhaps the greatest engineering feat of antiquity was the Roman water system, with aqueducts that could supply fresh water miles from the source.

Figure HB.1 The Roman aqueducts are one of the great engineering achievements of the ancient world.

natural gas pipelines that crisscross the country. Moreover, aircraft, automobiles, ships, spacecraft, and virtually all other vehicles involve interactions with fluid of one type or another, both externally and internally, within an engine or as part of a hydraulic control system.

 CD/Gallery of Flows

Learning more about fluid mechanics also allows us to better understand our bodies and many interesting features of our environment. The heart and lungs, for example, are wonderfully designed pumps that operate intermittently rather than steadily as most man-made pumps do. Yet the heart moves blood efficiently through the branching network of arteries, capillaries, and veins, and the lungs cycle air quite effectively through the branching pulmonary passages, thereby keeping the cells of our bodies alive and functioning. Many other sophisticated fluid handling devices are found throughout the biological world in living creatures of all types, sizes, and degree of complexity.

 CD/Video Library/Flow in the Lungs

The environment is another source of complex and interesting fluid mechanics problems. These range from the prediction of weather, hurricanes, and tornadoes to the spread and control of air and water pollution. Add to this list the flow of rivers and streams, the movement of groundwater, the jet stream and great ocean currents, and the tidal flows in estuaries. The lava flows of volcanoes and the movements of molten rock within the earth also lie within the domain of fluid mechanics. Looking beyond Earth, stellar processes and interstellar events are striking examples of fluids in motion on a grand scale. Knowledge of fluid mechanics is also the key to understanding and sometimes controlling other interesting, if not vital phenomena, such as the curving flight of a tennis, golf, or soccer ball, and the many different pitches in baseball.

HISTORY BOX 1-2

The most illustrious name in ancient Greek engineering is Archimedes (287–212 B.C.). Among this man's accomplishments was the determination, by means of a clever application of the principles of fluid statics, of the percentage of gold in the crown of the king of Syracuse. For thousands of years fluid mechanics depended on principles deduced by trial and error. During the Italian Renaissance Leonardo da Vinci (1452–1519) used his acute powers of observation to describe fluid flows and to imagine fluid machines.

 CD/History/Leonardo Da Vinci

It was not until the seventeenth century, however, that the history of modern fluid mechanics began. A disciple of Galileo (1564–1642), Evangeliston Torricelli (1608–1647) invented the barometer, a device for measuring atmospheric pressure variations caused by weather. The principles of the barometer were clarified by the noted scientist and philosopher Blaise Pascal (1623–1662). This work laid the foundation for our understanding of fluid statics. For good reason, then, the units of pressure called the torr and pascal, respectively, were named in their honor.

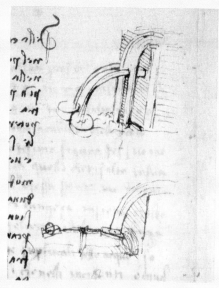

(A) (B)

Figure HB.2 (A) Leonardo da Vinci. (B) Sketch of falling water by Leonardo.

 CD/Special Features/Demonstrations/Sports Balls

Figure 1.1 (A) Hoover Dam. (B) The space shuttle at takeoff.

The field of fluid mechanics has historically been divided into two branches, fluid statics and fluid dynamics. Fluid statics, or hydrostatics, is concerned with the behavior of a fluid at rest or nearly so. Fluid dynamics involves the study of a fluid in motion. Consider the engineering systems illustrated in Figure 1.1. Which branch of fluid mechanics is applicable in each case? Do any of these applications involve both branches?

Modern engineering science is rooted in the ability to create and solve mathematical models of physical systems. Students often view fluid mechanics as a challenging subject, primarily because the underlying mathematical model appears to be complex and difficult to apply. We will show that the governing equation of fluid statics, called the hydrostatic equation, is actually relatively simple and may always be solved to find the pressure distribution in the fluid. On the other hand, the governing equation of fluid dynamics, called the Navier–Stokes equation, would never be described as simple. The inherent difficulties of fluid mechanics have been recognized for centuries, yet engineers have demonstrated great ingenuity in developing a number of different approaches to solving specific fluid flow problems. The common theme is to simplify the mathematical or experimental model used to describe the flow without sacrificing the relevant physical phenomena. For example, a standard approximation in the prediction of low speed airflow is to neglect the compressibility of air. This assumption is accurate at vehicle speeds as high as 250 mph. Thus in the flows over automobiles and light aircraft, the compressibility of air can be ignored. By the way, do you consider 250 mph to be a low speed flow? It definitely is in fluid dynamics applications!

It is often said that there is both an art and science to the practice of fluid mechanics. One learns the science of fluid mechanics in a class or from self-study, with a textbook like this one serving as a guide. The art of fluid mechanics, however, is developed primarily through experience, both your own and that of others. This art consists in knowing when it is safe to neglect the effects of physical phenomena that are judged to have little impact on the flow. Once we decide to neglect certain physical phenomena, we drop the corresponding terms in the governing equations, thereby decreasing the difficulty in obtaining a solution. Such fundamental topics as boundary layer theory, the Bernoulli equation, potential flow, and even fluid statics can be considered to be part of the art of fluid mechanics in this sense. Learning about these historical approximations

HISTORY BOX 1-3

In 1687 Sir Isaac Newton's *Mathematical Principles of Natural Philosophy* was published. The second book of this work was devoted to fluid mechanics. It was in the *Principia* that Newton (1642–1727) established on a rational basis the relationship between the mass of a body, its acceleration, and the forces acting upon it. Although he tried, Newton was unable to properly apply these concepts to a moving fluid. Even though Newton's second law (**F** = m**a**) applies to a fluid in motion, the dynamics of fluid flow are inherently more difficult to understand than the dynamics of solid bodies. For example, what mass of water should be included in the analysis of a ship sailing on the ocean? How do forces such as friction behave between a fluid and a solid surface? The answers to these and other fundamental questions eluded Newton.

 CD/History/Sir Isaac Newton

In the eighteenth century, mathematicians built on the foundations of mechanics and calculus Newton had laid. The leading lights of this era were Daniel Bernoulli (1700–1782), Jean le Rond d'Alembert (1717–1783), and Leonhard Euler (1707–1783). The fundamental equations of fluid mechanics relating the conservation of mass, momentum, and energy were being developed. An equation for the conservation of mass was first appropriately applied by d'Alembert for plane and axisymmetric flows in 1749. Euler first published the generalized form of the equation for

Figure HB.3 Daniel Bernoulli.

History/Jean le Rond d'Alembert

mass conservation 8 years later. At this time he also published the proper form of fluid acceleration in the momentum equation; however, the role of friction was not understood. The well-known Bernoulli equation can be considered to be a form of the energy equation under special conditions. The proper statement for the conservation of energy in a fluid could not be fully derived until the nature of heat was understood in the nineteenth century.

 CD/History/Leonhard Euler

HISTORY BOX 1-4

The complete mathematical statement for the conservation of momentum, including the role of friction, was derived independently by the Frenchman C. L. M. H. Navier (1785–1836) and the Englishman Sir George Stokes (1819–1903). These equations, called the Navier–Stokes equations, are the fundamental mathematical model for fluid mechanics, and as such, are the basis for all the analytical solutions presented in this text. Because the equations were intractable, mathematical solutions could be obtained only for "ideal" fluids, meaning fluids of zero viscosity. An ideal fluid does not exhibit friction as it passes along a surface because shear stress is completely absent. While the elegant methods developed for ideal fluid flow could produce the pressure distribution about bodies moving through a fluid, they could not provide answers for the important practical problem of the drag force that was exerted by the fluid. This difficulty was known as d'Alembert's paradox.

The Navier–Stokes equations are formidable nonlinear partial differential equations, and extremely difficult to solve. For this reason engineers have relied heavily on experiments to answer their questions. Among the experimentalists active during this period were Henri Pitot

 CD/History/C. L. M. H. Navier

 CD/History/Sir George Stokes

(1695–1771) and J. L. M. Poiseuille (1799–1869), who developed a simple device for measuring fluid velocity and measured the relationship between pressure drop and flow in pipes, respectively.

 CD/History/J. L. M. Poiseuille

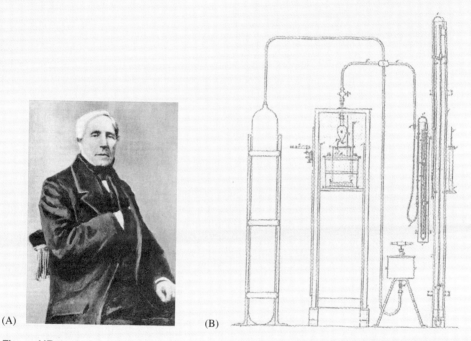

Figure HB.4 (A) J. L. M. Poiseuille. (B) The apparatus Poiseuille used to study pressure drop in pipes.

and others like them is an essential goal of a first course in fluid mechanics, and engineers working with fluids should continue to seek to attain it.

Once the analysis of a fluid mechanics problem has been cast in the form of an appropriate mathematical model, a solution method must be chosen. For example, one might employ an analytical solution method that results in a representation of the flow variables as functions of space and time. Figure 1.2 compares an analytical solution and a visual representation of the low speed flow of fluid between two flat plates. An analytical solution is a highly compact and useful form of solution that should always be acquired if possible. Be aware, however, that an analytical solution of the governing equations of fluid dynamics is usually not possible. Complex engineering geometries and a natural tendency for fluid flows to become unstable ensure that analytical solutions will remain elusive. Nevertheless, it is wise to consult the engineering literature to determine what has been accomplished in treating the same or related flow problems. If you find that an

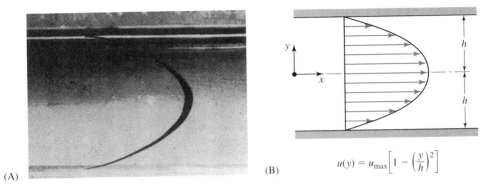

(A) (B) $u(y) = u_{max}\left[1 - \left(\frac{y}{h}\right)^2\right]$

Figure 1.2 Comparison of the parabolic velocity profiles in low speed fluid flow between parallel flat plates: (A) photograph from a flow visualization experiment and (B) plot of the analytical solution given by the equation $u(y) = u_{max}[1 - (y/h)^2]$.

approximate analytical solution to a problem of current interest is available, you may be able to use it as the starting point for your analysis.

Today, an engineer will increasingly choose to employ computational methods to solve the equations of fluid motion. These methods include finite difference, finite element, finite volume, and other computational approaches in which digital computers are used to supply numerical solutions of approximate versions of the governing equations. These solutions are discrete, meaning that the flow variables are known only at specific spatial locations in the flow field. Computational tools of all kinds, ranging from commercially available computational fluid dynamic codes to visualization packages and symbolic mathematics codes, are among the most important aids in the modern practice of fluid mechanics. An image produced by a computational fluid dynamics (CFD) code is shown in Figure 1.3.

One of our motivations in writing this book is to integrate these modern computational aids into a first course in fluid mechanics. The symbolic mathematics codes Mathematica, MATHCAD, and others like them are superb aids in learning fluid mechanics. We recommend their use to simplify calculations and to visualize the mathematics.

Figure 1.3 The flow field around an automobile as simulated by CFD.

Many of you are already skilled in the use of these mathematics packages and need only learn how to employ them effectively in solving fluid mechanics problems. For those who have yet to acquire this engineering skill, an introductory fluid mechanics course offers an excellent vehicle in which to start learning to do high level mathematics and visualization on a personal computer or workstation.

Regardless of the source of the information about a particular fluid flow (i.e., a commercially available or proprietary CFD code, an exact or approximate analytic solution), we believe that a picture is worth more than the proverbial thousand words when it comes to understanding what is happening in a fluid flow. Consider the flow field shown in Figure 1.3, and imagine trying to describe this flow with words alone.

Our emphasis on the use of numerical computation and visualization in fluid mechanics does not imply that these can replace experimental methods. In fact, it is foolish to think that all, or perhaps even most flows can be completely simulated on a computer. The simulation of even relatively simple flows can tax the capability of today's most powerful workstations. A numerical simulation of a complex flow like that shown in Figure 1.3 generally requires a supercomputer, and even then, the resolution of the fine-scale structure in the flow field may be lacking. For these reasons, and despite working in this computer age, engineers must be knowledgeable about using experiments to guide their design and problem-solving efforts in fluid mechanics.

 CD/Special Features/Flow Visualizations

Experimental methods employ a wide range of sophisticated equipment to obtain numerical data describing the velocity, pressure, and other properties of a fluid flow. Flow visualization techniques provide a visual picture of the flow by making portions of a normally transparent fluid visible. For example, Figure 1.2A was generated with a flow visualization system.

Internal flows, meaning those that occur within confining walls, are often studied in the laboratory by reduced scale models of the physical device. For external flows (typically a flow over a body immersed in fluid), a wind or water tunnel is often employed to expose a scale model to the flowing fluid. These tunnels range in size from tabletop to major installations. With speeds ranging from an imperceptible breeze to hypersonic flow, wind tunnels like that in Figure 1.4 provide opportunities to explore flows by using a variety of sophisticated sensors. Many external flows simply cannot be adequately

Figure 1.4 Model of F-18E in a 30 ft × 60 ft wind tunnel at the NASA Langley Research Center.

HISTORY BOX 1-5

Great strides in the development of fluid mechanics were stimulated by the dream of human flight. The Wright brothers brought their mechanical ingenuity to bear on aerodynamic experiments at the turn of the nineteenth century. Wilbur (1867–1912) wrote that "having set out with absolute faith in the existing scientific data, we were driven to doubt one thing after another, until finally after two years of experiment, we cast it all aside, and decided to rely entirely upon our own investigations." They built a wind tunnel in which they tested small models.

The field of aerodynamics, and fluid mechanics in general, was dominated through the first half of the twentieth century by Ludwig Prandtl (1875–1953) (Figure HB.6A). His work included thin airfoil theory, finite wing theory, supersonic shock wave and expansion wave theory, compressibility corrections, and his most important contribution to fluid mechanics, the boundary layer concept. Beyond his own contribution, his students, such as H. Blasius (1883–1970) and T. von Karman (1881–1963), have also had an enormous impact on the field.

Prandtl's boundary layer concept resulted from his willingness to combine theory and experiment. Examples of his flow visualization experiments are shown in Figure HB.6B. Prandtl compared his experimental data with the results of the inviscid, or frictionless, theoretical calculations. What he found was that theory and experiment were in good agreement except for the velocity profile in a thin layer of fluid located near the solid–fluid boundary. Prandtl's experimental efforts revealed the problem—a breakdown in the inviscid assumption in the boundary layer. His theoretical efforts solved the problem—he developed a novel solution to the Navier–Stokes equations for the thin layer of fluid adjacent to the solid boundary. The result of linking Prandtl's solution within the boundary layer to the inviscid solution outside the boundary layer resolved d'Alembert's paradox.

 CD/History/Ludwig Prandtl

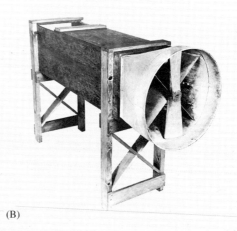

(A) (B)

Figure HB.5 (A) The Wright brothers' flying at Kitty Hawk, North Carolina. (B) The Wright brothers' wind tunnel, which they used to test airfoils.

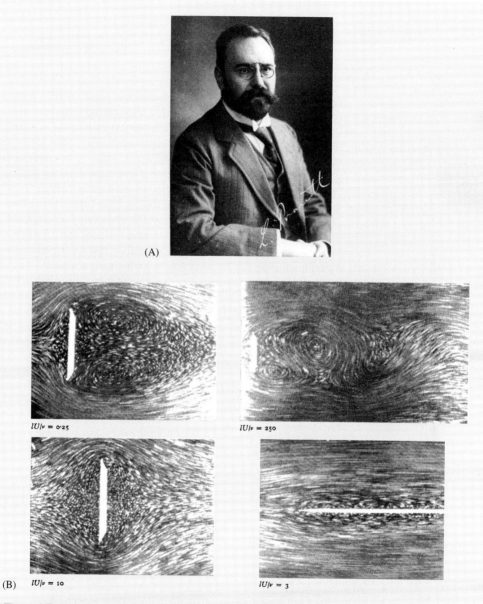

Figure HB.6 (A) Ludwig Prandtl. (B) Flow visualizations by Prandtl.

simulated in any other way. Water tunnels are also used to simulate external flows, particularly for marine applications. In the foreseeable future, experimentation will remain an important means of solving flow problems.

Empirical results, in the form of experimental data correlations, are used extensively in the design of piping systems, pumps, turbines, engines and many other

HISTORY BOX 1-6

To put the history of fluid mechanics into perspective, the early pioneers of mathematical physics, from Newton to Navier and Stokes, developed a mathematical model for fluid mechanics. Prandtl, and researchers since his day, have been applying this mathematical model. The development of digital computers has made a tremendous impact in this respect. With computational fluid dynamics (CFD), it is becoming possible to obtain solutions for an arbitrary flow geometry.

well-understood classes of fluid machinery. A number of important empirical results are introduced early in this text in the form of case studies. These case studies, which expose you to simple design formulas, allow you to immediately begin to solve a number of practical engineering problems involving fluid flow.

The use of empirical results in the analysis of a flow problem requires engineering judgment and a level of experience similar to that mentioned in connection with mathematical approaches. Thus your application of the case study results must initially be cautious and generally guided by your instructor. The case studies in Chapter 3 are revisited in subsequent chapters as we develop the theoretical tools of fluid mechanics, thus allowing you to appreciate the underlying assumptions, methodologies, and flow characteristics that lie behind the otherwise simple formulas. Just as engineers must understand the assumptions made in constructing a mathematical model, they must also understand the process required to design an experiment that allows data to be applied to a particular flow problem. The ability to design a fluid flow experiment and interpret the outcome requires an understanding of dimensional analysis and similitude. You will learn about these powerful tools, which allow us to relate results achieved with a scale model to those that occur with the full-scale prototype, and see how they underlie all the case studies.

CD/Special Features/Virtual Labs/Dynamic Similarity

1.2 GASES, LIQUIDS, AND SOLIDS

In everyday language we casually use the terms fluid, liquid, gas, and solid. Many people mistakenly assume that "fluid" is a synonym for a liquid and that a solid is a rigid material that is incapable of deformation. Neither of these popular conceptions is precise enough for engineering work, so in this section we formally define each of these terms.

CD/Kinematics/Compressibility/Compressible and Incompressible Fluids

The fundamental difference between a fluid and a solid lies in the response to a shear stress of the respective materials. Suppose you glue a brick to your desk with epoxy, and also carefully pour a small quantity of water next to it, making a puddle. Imagine placing the palm of your left hand on the brick and your right palm on the surface of the water. Now apply a small but equal force on each material in a direction parallel to the surface of the desk. What happens? The response of a solid such as the brick to a small shear stress (defined as tangential force divided by area) is a static deformation. Although your eyes cannot always detect it, all solids change shape upon the application of a shear stress. A solid quickly returns to rest, however, and retains the new shape for as long as the (constant) shear stress is maintained.

Unlike the brick, the water in our experiment proves unable to withstand the shear stress. Your hand continues to move because the water is set in motion by the applied shear stress. Instead of coming to rest in a deformed state, the water continues to deform as long as the shear force is applied. This continuing deformation in shear is characteristic of all fluids. We therefore define a fluid as any material that is unable to prevent the deformation caused by a shear stress. A related conclusion is that in a fluid at rest all shear stresses must be absent.

Let us formalize our examination of differences between solids and fluids by imagining that we have placed solid and fluid samples into identical unidirectional shear testing devices as shown in Figure 1.5A. Each material is placed in contact with a rigid upper plate free to move and with a stationary rigid lower plate. A unidirectional shear force is transmitted to the material by applying a tangential force to the upper plate. The applied shear stress, τ, is defined as the ratio of the tangential force, F, to the area of the upper plate, A, i.e., $\tau = F/A$.

As shown in Figure 1.5B, for a shear stress below the elastic limit, the relationship between shear stress τ, and shear strain γ in a solid is:

$$\tau \propto \gamma \tag{1.1a}$$

The shear strain is defined as the displacement in the direction of the applied force, Δx, normalized by the height of the solid (perpendicular to the applied force), Δy, i.e., $\gamma = \Delta x/\Delta y$. The proportionality between τ and γ can be converted to an equality by

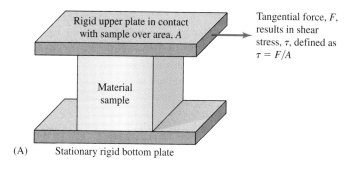

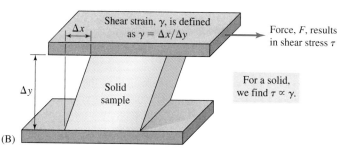

Figure 1.5 Illustration of a parallel plate testing device for determining the response of materials to an applied shear stress. (A) The sample in its original form just prior to the application of the tangential force, F. (B) The response of a solid sample to the application of the shear stress, τ ($= F/A$). The solid deforms by an amount Δx almost immediately upon the application of the shear stress and then remains stationary in that position throughout the duration of the applied stress.

EXAMPLE 1.1

Consider 0.5 in. thick steel and aluminum plates. If each metal plate is subjected to a shear stress of 2×10^5 psi, what is the magnitude of the displacement in the direction of the applied force for the two materials?

SOLUTION

This exercise can be solved by using Eq. 1.1b, and the definition of shear strain. From the appropriate shear moduli for steel and aluminum given earlier, the result is that the steel sheet experiences a displacement of 8.3×10^{-3} in. while the aluminum sheet experiences a displacement of 2.6×10^{-2} in.

inserting a constant, G, known as the shear modulus of the solid. Thus we can write the relationship as:

$$\tau = G\gamma \tag{1.1b}$$

Since the shear modulus of steel is about three times that of aluminum (12×10^6 vs 3.8×10^6 psi), Eq. 1.1b tells us that if similar blocks of steel and aluminum were subjected to the same shear stress, the displacement in the aluminum block would be more than three times as large as that in the steel block. That is, steel is stiffer (experiences less deflection per unit stress) than aluminum.

As shown in Figure 1.6A, if a fluid sample is placed in our imaginary shear testing device, the fluid will continue to deform no matter how small the applied shear stress. The upper plate will move faster if the tangential force is increased. A fluid resists being sheared, but its underlying molecular structure does not allow it to prevent the resulting deformation. Thus, the fluid will be set in motion, and it is found that there is a linear velocity distribution in the gap between the plates as shown in Figure 1.6B. Instead of having a proportionality between shear stress and shear strain, a fluid exhibits a relationship between shear stress and the shear strain rate, $d\gamma/dt$.

The relationship between the shear stress and the shear strain rate for a fluid is

$$\tau \propto \frac{d\gamma}{dt} \tag{1.2a}$$

For many common liquids, and all gases, this relationship is linear and can be written as

$$\tau = \mu \frac{d\gamma}{dt} \tag{1.2b}$$

where the proportionality constant, μ, is a fluid property known as the absolute or dynamic viscosity. Equation 1.2b is known as Newton's law of viscosity, and a fluid that obeys this equation is termed a Newtonian fluid. If you subjected two Newtonian fluids to the same shear stress, the one with the higher viscosity would exhibit a lower shear strain rate. If two Newtonian fluids are subject to the same strain rate, the one with the higher viscosity will have a higher shear stress.

1.2 GASES, LIQUIDS, AND SOLIDS

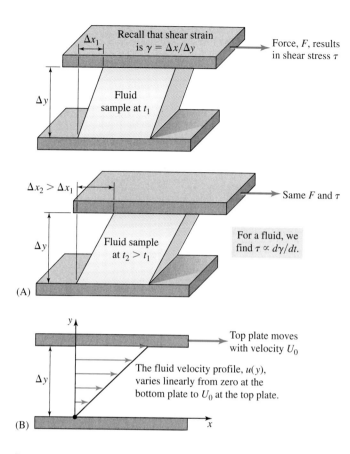

Figure 1.6 (A) When a fluid is tested in the device described in Figure 1.5A the displacement Δx and the corresponding shear strain γ increase linearly with time. For a fluid, the relationship between shear stress and shear strain is $\tau \propto d\gamma/dt$. (B) In this situation, the fluid velocity in the x direction, u, is a function of the y coordinate. That is, $u(y)$ varies linearly from 0 at the bottom plate to U_0 at the top plate. Note that $u(y) = d(\Delta x)/dt$. This result is used in the text to show that $d\gamma/dt = du/dy$; i.e., the shear rate and velocity gradient are equal.

Have you noticed that a playing card can glide across a smooth surface for quite a distance, "riding on a cushion of air"? The air is sheared in the thin gap between the card and the surface, and although the frictional resistance to this shear is small, it is definitely nonzero.

An examination of Figure 1.6B should help convince you that the shear rate $d\gamma/dt$, in a fluid sheared between parallel plates, is related to the transverse velocity gradient du/dy, where u is the velocity of the fluid in the x direction and y is the spatial coordinate perpendicular to the parallel plates. This result is obtained by noting that the shear strain γ is defined as $\gamma = \Delta x/\Delta y$, so the strain rate $d\gamma/dt$ is given by $d\gamma/dt = d/dt(\Delta x/\Delta y)$. Since Δy is constant, and the time rate of change of the displacement Δx is u, the fluid velocity in the x direction, we obtain $d\gamma/dt = du/dy$. Thus the shear rate and velocity gradient are equal, and Newton's law of viscosity (Eq. 1.2b) is usually expressed in fluid mechanics as:

$$\tau = \mu \frac{du}{dy} \qquad (1.2c)$$

Because the velocity profile is linear across the gap in the shear flow between parallel plates, the velocity gradient du/dy is constant, and Eq. 1.2c tells us that the shear stress is uniform in the region between the plates.

We have seen that a solid differs from a fluid in its response to an applied shear stress. Since a fluid may be a liquid or a gas, on what basis do we distinguish liquids and

EXAMPLE 1.2

A sample of motor oil is tested in a parallel plate shearing device like the one shown in Figure 1.6 with the following results: $\tau = 1.54 \, \text{lb}_f/\text{ft}^2$, the plate separation distance is 0.5 in., and the top plate velocity is 10 ft/s. Determine the viscosity of the fluid and shear rate.

SOLUTION

This problem can be solved by manipulating Eq. 1.2c to isolate the fluid viscosity. Since $\tau = \mu(du/dy)$, we have $\mu = \tau/(du/dy)$. Since the velocity gradient is constant, we can write $du/dy = \Delta u/\Delta y$ and substitute the values $\tau = 1.54 \, \text{lb}_f/\text{ft}^2$, $\Delta y = 0.5 \, \text{in.} = (0.5/12) \, \text{ft}$ and $\Delta u = 10 \, \text{ft/s}$ into the expression for viscosity:

$$\mu = \tau \frac{dy}{du} = \tau \frac{\Delta u}{\Delta y} = (1.54 \, \text{lb}_f/\text{ft}^2) \frac{0.5/12 \, \text{ft}}{10 \, \text{ft/s}} = 6.4 \times 10^{-3} (\text{lb}_f\text{-s})/\text{ft}^2$$

The shear rate is given by

$$\frac{d\gamma}{dt} = \frac{du}{dy} = \frac{10 \, \text{ft/s}}{(0.5/12) \, \text{ft}} = 240 \, \text{s}^{-1}$$

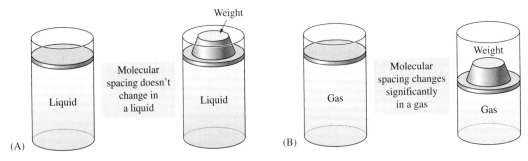

Figure 1.7 Differences in the compressibility of liquids and gases. (A) The molecular spacing between liquid molecules is not changed appreciably when a weight is applied to the piston. (B) In contrast, the same weight on the piston will cause a significant change (decrease) in the spacing between gas molecules.

gases from each other? It is tempting to use the widely different densities of liquids and gases, or to determine how each fills a container; but the more important difference between these two fluids lies in their vastly different response to a compressive stress. Consider an experiment in which we attempt to use two identical piston–cylinder devices to compress a liquid and a gas, as shown in Figure 1.7. The intermolecular spacing in liquids is essentially constant, so a fixed amount of liquid occupies nearly the same volume under most conditions. A liquid is therefore difficult to compress, and even a large weight on the piston will result in only a small volume change. In contrast, the distance between molecules in a gas is much greater and highly variable. It takes a relatively small force on the piston to significantly decrease the intermolecular distance in a

EXAMPLE 1.3

Imagine applying a shear stress to a puddle of water with your hand. Suppose the water puddle is initially 2 mm thick and that the viscosity of water is 0.001 kg/(m-s). If your hand is applying a shear stress of magnitude 0.05 kg/(m-s^2), calculate the shear strain rate and velocity gradient in the fluid, and the speed u at which your hand is moving.

SOLUTION

The exercise asks us to calculate $d\gamma/dt$, du/dy, and u at the top surface of the water. We are given $\tau = 0.05$ kg/(m-s^2) (= 0.05 N/m^2) and $\mu = 0.001$ kg/(m-s). A sketch of the system would look like the fluid testing device in Figure 1.6. This problem can be solved using Newton's law of viscosity, which is given in Eq. 1.2b as $\tau = \mu(d\gamma/dt)$. Assuming that water is a Newtonian fluid (a very safe assumption), the equation can be solved for the desired quantity:

$$\frac{d\gamma}{dt} = \frac{\tau}{\mu}$$

Substituting the given values for τ and μ yields the constant shear strain rate in the water:

$$\frac{d\gamma}{dt} = \frac{0.05 \text{ kg/(m-s}^2)}{0.001 \text{ kg/(m-s)}} = 50 \text{ s}^{-1}$$

Since the shear rate and velocity gradient are equal, we conclude that $du/dy = 50$ s^{-1}. This value is also uniform across the gap. To find the velocity of the top plate we write

$$\frac{du}{dy} = \frac{\Delta u}{\Delta y} = \frac{(u_{\text{Top}} - u_{\text{Bottom}})}{0.002 \text{ m}} = 50 \text{ s}^{-1}$$

Since the bottom plate is not moving, $u_{\text{Bottom}} = 0$. Solving for the velocity of the top plate p, which is the speed at which our hand is moving, we get

$$u_{\text{Top}} = (50 \text{ s}^{-1})(0.002 \text{ m}) = 0.1 \text{ m/s} = 10 \text{ cm/s}$$

Note that since the velocity gradient is uniform in the gap, the velocity profile in the direction of the applied force is a linear function of the vertical distance above the table given by

$$u(y) = \left(\frac{du}{dt}\right)y = (50 \text{ s}^{-1})y$$

The velocity at the top plate could also be obtained by evaluating this function at $y = 0.002$ m to obtain $u = (50 \text{ s}^{-1})(0.002 \text{ m}) = 0.1$ m/s = 10 cm/s.

gas, so the same weight on the piston compresses the gas into a substantially smaller volume than that occupied by the liquid under similar conditions. We conclude that all gases are far easier to compress than any liquid.

What would happen if we attempted to use our piston–cylinder device to put a liquid or gas into tension? Perhaps you intuitively feel that it is not possible to put a fluid in tension. If so, you are correct. The molecular structure of a gas, with its distant molecules, ensures that a gas cannot support a tensile stress. The gas molecules are too far apart on average for intermolecular forces to maintain cohesion. The molecular structure of a liquid is quite different, however. There are always nearby molecules exerting significant intermolecular forces. A liquid should in principle be able to support a tensile stress because of these strong intermolecular forces. However, under normal circumstances, a liquid boils as the pressure on the liquid decreases to a value known as the vapor pressure. The presence of vapor pockets in the boiling liquid prevents the liquid from being placed in tension. Thus under normal circumstances liquids and gases cannot be put in tension.

We may also compare the behavior of a liquid and a gas in another way. Consider what happens when a sample of each type of fluid is transferred from a small rectangular container to a larger cylindrical container. As shown in Figure 1.8, the liquid occupies the same volume in each container but changes its shape to conform to that of the container, forming an interface with liquid vapor above. The gas expands to completely fill the larger container—its volume is not fixed, and no interface is present. You have undoubtedly noticed that liquids are able to form a stable, lasting interface that is not associated with container boundaries, something gases cannot do. Such an interface may also occur between two immiscible liquids, or between a liquid and a gas. The interface between a liquid and a gas is referred to as a free surface. A summary of the important differences between liquids, gases and solids is contained in Table 1.1.

We conclude our discussion of fluids and solids with a word of caution. While there is little doubt about the classification of a particular substance as a gas, many common materials exhibit characteristics of both liquids and solids. For example, what is toothpaste? Shearing a small amount of toothpaste between your fingers should convince you that toothpaste is not a solid. Yet if you carefully squeeze toothpaste into a straw, and

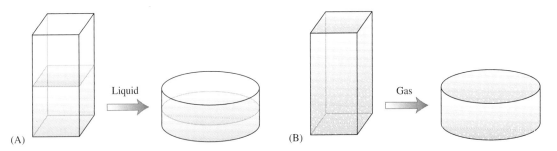

Figure 1.8 Differences in characteristics of liquids and gases. (A) A liquid takes the shape of its container but retains a constant volume. Note that a liquid is capable of forming an interface with its vapor that is not associated with a container boundary. (B) In contrast, a gas not only takes the shape of its container but expands or contracts to completely fill a container of any shape. Gases do not form interfaces other than those associated with the container boundaries.

1.2 GASES, LIQUIDS, AND SOLIDS

TABLE 1.1 Comparison of Solids, Liquids, and Gases

Characteristic	Solids (crystalline)	Fluids	
		Liquids	Gases
Response to a shear stress, τ	$\tau = G\gamma$ (resists deformation)	$\tau = \mu(d\gamma/dt) = \mu(du/dy)$ (resists rate of deformation)	
Distance between adjacent molecules	Smallest	Small	Large
Molecular arrangement	Ordered	Semiordered (short-range order only)	Random
Strength of molecular interaction	Strong	Intermediate	Weak
Ability to conform to the shape of a container	No	Yes	Yes
Capacity to expand without limit	No	No	Yes
Able to exhibit a free surface	Yes	Yes	No
Able to resist a small tensile stress	Yes	Theoretically yes, practically no	No
Compressibility	Essentially zero	Virtually incompressible	Highly compressible

hold the straw vertically with both ends open, the toothpaste will stay in the straw despite the effects of gravity. Water, which obeys Eq. 1.2c, would never behave this way.

Many other materials exhibit behavior that seems to fall somewhere between a fluid and a solid. Paint behaves in much the same way as toothpaste. It does not flow off the brush owing to the force of gravity alone, but it will flow onto the wall under the application of larger shear stress when brushed or rolled. Toothpaste and paint are examples of non-Newtonian fluids. A fluid of this type does not obey Eq. 1.2c; rather, it exhibits a more complex relationship between the shear stress and velocity gradient. The behavior of a non-Newtonian fluid varies greatly, but is often fluidlike, with some features of a solid at low shear stress levels. Other examples of non-Newtonian fluids are polymer solutions (i.e., macromolecules dissolved in a solvent), colloidal suspensions (e.g., milk or mayonnaise), slurries (e.g., nuclear fuel particles or paper pulp in water), and clay suspensions. Although the behavior of non-Newtonian fluids is important in polymer and food processing, and in chemical and pharmaceutical manufacturing, this book deals primarily with Newtonian fluids.

EXAMPLE 1.4

In which of the following situations would you feel comfortable using Newton's law of viscosity? Justify your decisions.

  A. Modeling the flow of caulk out of a caulking gun.
  B. Modeling the flow of water out of a squirt gun.
  C. Modeling the flow of hot gases out of an aircraft turbojet engine.

SOLUTION

We are asked to determine which of the situations listed can be adequately modeled by Newton's law of viscosity. We will assume that Newton's law is appropriate if the fluid is a gas, a liquid with a simple molecular structure and no additives, or a material that we have observed behaving like a Newtonian fluid under the conditions of interest.

A. The vinyl-based caulk used in most home improvement projects is a high molecular weight polymer and does not flow out unless substantial pressure is applied with the plunger of the caulking gun. We conclude that caulk is not a Newtonian fluid.

B. Water has a simple molecular structure, and Newton's law of viscosity is applicable. Thus water is a Newtonian fluid.

C. A turbojet exhaust stream is a mixture of hot gases, primarily air. There is no reason to think that the mixture will behave differently from any other gas. We conclude that the exhaust mixture is a Newtonian fluid and that Newton's law of viscosity is applicable.

 CD/Kinematics/Fields, Particles, and Reference Frames

1.3 METHODS OF DESCRIPTION

To construct a theory of fluid mechanics, it is necessary to represent basic laws describing conservation of mass, momentum, and energy in a form suitable for mathematical analysis. This requires the selection of a description, or a conceptual framework in which to work. The breadth of fluid dynamic problems extends from the description of star formation within interstellar gas clouds to the water flow in the wake of a diving pelican, from the exchange of gases with the blood flowing in the capillaries of our bodies to the spreading of a drop of epoxy glue on a surface to the drag on a modern communications satellite. Given this breadth of topics, is it reasonable to expect to find a unique method of description that works well in all cases? Before we answer this question, it is important to realize that even though a fluid has a molecular structure, it is not always necessary to incorporate this molecular structure into a model. The deliberate omission of the fluid's molecular structure in a mathematical model of fluid flow is known as the continuum hypothesis.

 CD/Dynamics/Newton's Second Law of Motion/A Fluid as a Continuum

1.3.1 Continuum Hypothesis

To place the need for the continuum hypothesis in context, suppose you are given the assignment to predict the fluid velocity distribution created by stirring a glass of water with a spoon. How would you propose to model the fluid in this situation? One approach might be to attack the problem on a molecular level by using your understanding of dynamics and modern physics to model the water molecules as interacting hard spheres with a known intermolecular force field. Good luck! A typical glass of water contains on the order of 10^{26} water molecules. Even with the help of a powerful computer, tracking such a large number of objects is impossible, at least in the foreseeable future. An alternate approach to modeling the water in the glass, or any fluid in general, is to consider all macroscopic properties at a point in a fluid as averages of molecular characteristics in a small region about that point. In this approach we assume that a fluid may be treated as a continuous substance or continuum, rather than as a group of discrete molecules. Therefore, this concept is called the continuum hypothesis.

 CD/Video Library/Stirring

In employing the continuum hypothesis, the underlying molecular structure of a fluid is conveniently ignored and replaced by a limited set of fluid properties, defined at each point in the fluid at every instant. Mathematically speaking, the continuum hypothesis allows the use of differential calculus in the modeling and solution of fluid mechanics problems. Each fluid property is considered to be a continuous function of position and time. There may be discontinuous jumps in a property value such as density at fluid–fluid or fluid–solid interfaces, but the continuum model generally assumes that all properties are described by continuous functions.

How do we know if a continuum model is valid for a specific application? As with most engineering theories that attempt to explain or model physical phenomenon, the proof is in how well the theory describes reality. Continuum models of fluid mechanics have been applied to an extraordinarily wide range of problems with excellent results, so there is little doubt of the general validity of this approach. Does the molecular structure of a fluid ever become important and perhaps cause a breakdown in the continuum theory? The answer to this question is yes: the continuum theory is in jeopardy when the length scale of a physical phenomenon or object exposed to a fluid is of the same order as molecular dimensions. In a problem involving a gas, for example, the largest molecular dimension of practical importance is the mean free path of gas molecules, which is inversely proportional to the density of the gas. The mean free path in air is approximately 10^{-7} m (0.1 μm) at standard conditions. Is a continuum model appropriate to describe the flow of gas in 0.5 μm pores of a filter media? Since the pore diameter is approximately five times the mean free path, we should be cautious in using results based on the continuum hypothesis in this application.

There is another category of flow problems in which the physical scale is not small but the mean free path is much larger than normal. NASA engineers have to deal with the fact that in the upper atmosphere, the mean free path approaches the length scale of a vehicle or satellite. The drag on the vehicle in a high orbit will not be accurately predicted

EXAMPLE 1.5

The pressure in a vacuum system is 10^{-6} atm. Estimate the mean free path λ of the air inside the system. Assume that the air can be modeled using the perfect gas law, i.e., at constant temperature the air density is proportional to its pressure.

SOLUTION

This problem can be solved by remembering that the mean free path in a gas is inversely proportional to its density. If we combine this fact with the information provided in the problem statement (at constant temperature the air density ρ is proportional to its pressure p), we can obtain the relationship: $\lambda_2/\lambda_1 = \rho_1/\rho_2 = p_1/p_2$ for the ratio of the mean free path at two different conditions. We know that at atmospheric pressure the mean free path in air is 0.1 μm and that the air inside the vacuum system is at a pressure of 10^{-6} atm. Thus, if we let $p_1 = 1$ atm, $p_2 = 10^{-6}$ atm, and $\lambda_1 = 10^{-7}$ m, then λ_2 is calculated as

$$\lambda_2 = \lambda_1 \frac{p_1}{p_2} = (10^{-7}\,\text{m}) \frac{1\,\text{atm}}{10^{-6}\,\text{atm}} = 0.1\,\text{m} = 10\,\text{cm}$$

by formulas based on continuum theory. Industrial coating processes in which gases are metered or flow through vacuum chambers are also examples of situations for which the continuum hypothesis may break down. Note, however, that a large mean free path in and of itself does not always mean that the continuum hypothesis is invalid. The mean free path in the vicinity of a stellar nebula is enormous, but the scale of the astronomical flow structure is larger still, so the continuum hypothesis is valid. Hydrodynamic models of nebulae based on continuum fluid mechanics appear to work quite well in describing the observed characteristics of these fascinating astronomical structures.

The continuum hypothesis is nearly always applicable in liquids because the distance between liquid molecules is small and relatively constant. Nevertheless the continuum hypothesis is known to fail for very thin lubricant films whose thickness approaches the molecular dimensions of the lubricant molecules. Under these conditions the lubricant may exhibit distinctly different behavior under shear than it will in a thicker film. Fluid mechanics problems like these must be approached by using a model that correctly accounts for the molecular structure of the fluid.

The continuum hypothesis is valid for the problems addressed in this book. It is well to be alert, however, for situations in which the hypothesis is invalid. We recommend the habit of checking all of your assumptions before proceeding with a formulaic approach to problem solving.

1.3.2 Continuum and Noncontinuum Descriptions

We have seen that an engineer has the option of including the fluid's molecular structure in the description of a fluid or leaving it out. Once the choice of a continuum or

EXAMPLE 1.6

In each of the following situations, would you feel comfortable using the continuum hypothesis in your analysis of the fluid flow? For any doubtful cases, list the additional information you would like to obtain before feeling confident of your decision.

A. Flow of gas at very low pressure through an orifice.
B. Calculating the shear stress in a lubricant layer 1 nm thick.
C. Modeling the flow of blood in the smallest capillaries.
D. Airflow over a passenger plane at normal altitude.
E. Respiration of insects through tubes connected to pores on their bodies.

SOLUTION

To determine which of the situations listed can be adequately modeled with a continuum theory we will compare a length scale in the problem to an appropriate molecular dimension such as the mean free path in a gas.

A. Since the flow is occurring at very low pressures, the mean free path of the gas molecules could be comparable to the dimensions of the orifice (see Example 1.5). If so, the continuum hypothesis may be invalid. We would need to know the gas pressure to estimate the mean free path and compare it to the dimensions of the orifice before being able to justify the use of the continuum hypothesis for this flow.

B. Since atoms typically have radii on the order of 0.1 nm, the size of the lubricant molecules is similar to the thickness of the lubricating film. Therefore, it is not valid to use the continuum hypothesis to predict the shear stress in this problem.

C. We should determine the diameter of the smallest capillaries and compare it to the characteristic molecular dimension for blood (on the order of nanometers) to see if the continuum hypothesis is valid for this fluid system. In this case the continuum hypothesis does apply. However, we should also be aware that blood is a complex liquid made up of cells of several types immersed in plasma. The diameter of the smallest capillaries is approximately the same size as red blood cells. This creates an unusual flow as the red blood cells squeeze through them in single file.

D. Since the mean free path of gas molecules at ordinary pressures is much less than the characteristic length of an airplane, we should feel comfortable using the continuum hypothesis in this case. By obtaining a relationship between gas pressure and mean free path, and a relationship between altitude and pressure, one could determine the range of altitudes for which the continuum hypothesis is useful.

E. We must determine the characteristic dimension for insect respiration tubes and pores as well as the mean free path of gas molecules at atmospheric pressure before being able to justify the use of the continuum hypothesis.

noncontinuum approach has been made, the next task is to select an appropriate method of describing the fluid and its behavior. Engineers have developed three principal methods, each of which has advantages and disadvantages. These are known as the molecular, Lagrangian, and Eulerian descriptions. The first, as the name implies, is a molecular level treatment that provides the basis for a noncontinuum theory of fluid mechanics. The Lagrangian and Eulerian approaches are both continuum treatments.

These three descriptions may be usefully distinguished from one another by the way in which the fundamental entity, or basic structural unit characteristic of the fluid, is defined. We think of a bulk fluid as being divisible into a large number of these entities, with the properties of the fluid ultimately residing in, and derived from, characteristics of the fundamental entity. The fundamental entity is always assumed to be small enough to preclude property variation within an individual entity. In the next three sections we discuss the three descriptions and their corresponding fundamental entities, beginning with the molecular description.

1.3.3 Molecular Description

In the molecular description the smallest identifiable element or fundamental entity is a single fluid molecule. Thus, as noted earlier, the molecular description of a fluid is inherently a noncontinuum model. The advantage of this approach is that we have a very good understanding of the molecular structure of common fluid molecules, so we can expect to get excellent results for a broad range of problems without having to tune our model in any significant way. The major disadvantage, however, is that we must track an enormously large number of molecules in most circumstances. For example, in modeling the flow of a breath of air traveling into our lungs through the trachea we must track on the order of 10^{21} air molecules. The storage requirements for the spatial coordinates alone exceed one million billion megabytes! This problem of sheer size has always been recognized and thought insurmountable. Recent computational efforts using molecular dynamics are nevertheless proving to be both exciting and fruitful. Researchers are dealing with the problem of tracking a large number of molecules by considering only a very small region of fluid for actual calculation. For example, in the spreading of a liquid the interesting physical phenomena are controlled by what is happening in the vicinity of the moving contact line where the liquid is advancing over the substrate. A molecular dynamics simulation of this problem may require tracking only on the order of 10^4 molecules. This is possible on many of today's fast workstations. So although molecular dynamics is unlikely to become a more widely used method of describing large-scale fluid mechanics problems, it has a promising future for small-scale problems. With its foundation in the underlying molecular structure of a fluid, molecular dynamics is a powerful and exciting tool for answering critical questions that cannot be approached at the continuum level.

 CD/Special Features/Simulations/A Molecular Dynamics Simulation

 CD/Special Features/Demonstrations/Euler vs. Lagrange: What is Steady and Unsteady?

1.3.4 Lagrangian Description

In the Lagrangian description the fundamental entity is a fluid particle. The conceptual model here assumes that the fluid consists of a continuous distribution of small discrete particles of fluid, each of which has a fixed mass but is otherwise shapeless. The particles are tracked to arrive at the motion of the bulk fluid. The fluid consists of a large enough number of fluid particles to permit it to be treated as if it were infinitely divisible. A fluid particle is composed of an extremely large, but finite, number of molecules whose average behavior defines the properties of the fluid particle. Consequently there will be many orders of magnitude fewer fluid particles in the Lagrangian description than molecules in the molecular description. This reduction in the number of fundamental entities in the Lagrangian description is a significant advantage over the molecular description. Since the Lagrangian description is a continuum description, the continuum hypothesis must be valid for the Lagrangian description to succeed in describing fluid behavior.

To illustrate the Lagrangian concept, imagine standing on a bridge and spraying a shower of small droplets of a red dye into a river. As your eyes follow the droplets downstream you are tracking a number of marked (dyed) fluid particles on the river surface. The movement of the water at the surface is revealed by the movement of fluid particles made visible by the dye. If droplets of dye are also introduced below the river's surface, a Lagrangian description of the flow is made visible, at least for the small number of dyed fluid particles.

Engineering students are usually comfortable with the Lagrangian description because it is the standard description of ordinary dynamics. Consider the problem of describing the motion of a point mass. Tracking a point mass through time by means of its position vector is a natural way to describe mathematically what we observe with our senses. For example, if a ball is thrown through the air, our eyes naturally track the ball (Lagrangian). Perhaps you have noticed the swiveling heads of spectators at a tennis match as they follow the flight of the ball. Although describing the motion of a cloud of fluid particles is more complex than describing the motion of a point mass, many of the mathematical concepts are similar. Since the Lagrangian description is applied only for relatively rare, specialized applications, we will not derive or solve the corresponding governing equations in this introductory textbook. However, kinematic concepts from the Lagrangian description are essential in analyzing and understanding fluid flow and form the basis for flow visualization. Several useful Lagrangian kinematic concepts are discussed later in the text.

1.3.5 Eulerian Description

In the Eulerian description of continuum fluid mechanics, we describe physical space by means of a coordinate system that serves as a backdrop for fluid motion. This is the principal method of description in fluid mechanics today, and the one used in this textbook. To gain an immediate sense of what the Eulerian description is like, imagine returning to our hypothetical bridge over a river. Instead of using dye to mark the movement of fluid particles, suppose we use a fixed array of flow meters to measure a simultaneous set of fluid velocity vectors at different points in the river. In this experiment we are not focusing on fluid particles and watching them move down stream. Instead, we are focusing on a set of specific locations in the flow field, and observing the velocity vectors at those points.

In describing fluid motion via the Eulerian description, we associate a complete set of fluid property values with every point in the spatial grid defined by a selected coordinate system. When we say that the fluid velocity vector has a value at a certain point in space, that value is the velocity vector of the fluid particle that happens to be at that spatial point at that instant. Similarly, if we refer to the fluid pressure at a certain point at a certain instant, we are speaking of the pressure associated with the fluid particle that coincides with that spatial point at that time.

The Eulerian approach constitutes a field description of a fluid in contrast to the particle description of the molecular and Lagrangian approaches. The idea of a field description should be familiar from calculus. In fact the field description of calculus and the Eulerian description of a fluid property are identical. Every property of a fluid has a corresponding functional dependence on position and time when represented in the Eulerian description. There is a velocity field, acceleration field, density field, pressure field, temperature field, and so on. Some properties, like the fluid temperature, require a scalar field description; others, like the fluid velocity, require a vector field description. You will find that many of the concepts of fluid mechanics in this description are simply applications of vector calculus to describe physical phenomena in fluids.

1.3.6 Choice of Description

In the last three sections we have introduced the molecular, Lagrangian, and Eulerian descriptions of a fluid. Although the concepts behind the three descriptions are very different, we expect all the approaches to yield identical answers to a given question. This assumes, of course, that the underlying assumptions of the chosen description are valid. For example, in the case of air flowing over the wing of an aircraft, we may want to know the lift and drag acting on that wing. As engineers, we expect to get the same values for the lift and drag regardless of the description in which we choose to work. For an aircraft at normal altitudes, all three descriptions should work in principle. Our selection of one description over the other two is therefore likely to be based on practical considerations. Since the incalculable number of molecules involved in describing the flow over a wing rules out a molecular description, we are limited to the other two descriptions. In this case, we would undoubtedly chose the Eulerian description because it is the standard description of fluid mechanics. On the other hand, for an aircraft at a very high altitude, the continuum hypothesis is violated and the Eulerian and Lagrangian descriptions are invalid. The high altitude case must therefore be tackled in the molecular description.

In problems for which the continuum hypothesis holds, some questions that are naturally asked and answered in one description are totally foreign to another. For example, asking where a certain volume of fluid (perhaps a volume of polluted air or water) will go in a flow is appropriately answered in the Lagrangian description, since this description tracks specific fluid particles. Likewise, asking about what is happening in a certain region of space (maybe a wind shear zone over an airport) is naturally answered in the Eulerian description because this description describes a flow in terms of spatial location.

EXAMPLE 1.7

Which method of description, molecular, Lagrangian or Eulerian, would you recommend for use in each of the following flow situations?

A. It is possible to use semiconductor fabrication techniques to create gas flow valves that can be embedded directly into a silicon wafer. Consider a device with a characteristic dimension in the micrometer range and a gas pressures of about 10^{-4} atm. You are asked to predict the flow through these microvalves.

B. You are asked to conduct an analysis of the wind patterns at Wrigley Field to determine why so many home runs are hit on certain days.

C. As the engineer on an emergency rescue team, you are responsible for predicting the spread of toxic gas released into the subway system during a terrorist attack on a major city.

SOLUTION

We are asked to determine whether the molecular, Lagrangian, or Eulerian description should be used to analyze the flow situations listed. The molecular approach is appropriate when the validity of the continuum hypothesis is in question. If the continuum hypothesis is valid, the Lagrangian approach focuses attention on individual fluid particles, while the Eulerian approach concentrates on flow through a region of space.

A. The combination of small dimensions and low gas pressures suggests that the continuum hypothesis is not valid. The molecular description is necessary.

B. To study the wind characteristics at Wrigley Field we need to know the velocity field in the stadium as a function of time and position. The continuum hypothesis is satisfied, so the Eulerian description is appropriate.

C. The continuum hypothesis is satisfied, so either the Lagrangian or Eulerian description is applicable. To predict the path of a specific volume of contaminated fluid we would use the Lagrangian description.

1.4 DIMENSIONS AND UNIT SYSTEMS

Thus far we have used familiar fluid properties such as pressure and density without definition. Before defining these and other fluid properties more precisely in the next chapter, it is worthwhile to review their dimensions and unit systems. Fluid mechanics embodies a wealth of fluid properties, many with distinctive units, and in a global economy it is important for an engineer to be able to work confidently in any customer's preferred unit system.

A dimension is a physical variable used to specify some characteristic of a system. Examples include mass, length, time, and temperature. In contrast, a unit is a particular amount of a physical quantity or dimension. For example, a length can be measured in units of inches, centimeters, feet, meters, miles, furlongs, and so on. Consistent units for a variety of physical quantities can be grouped together to form a unit system.

The dimensions length $\{L\}$, time $\{t\}$, and temperature $\{T\}$ are considered to be base dimensions from which other dimensions are derived. Examples of supplementary dimensions are velocity, $\{Lt^{-1}\}$, and acceleration, $\{Lt^{-2}\}$. The set of base dimensions $\{LtT\}$ is not complete, for how would you define density (mass per unit volume) using only $\{LtT\}$? If we add mass, M, we get a base dimension set $\{MLtT\}$ that can describe most problems in fluid mechanics. Exceptions include the description of the effects of electric fields on fluid flow, which requires the additional base dimensions of charge or current. The $\{MLtT\}$ set is not, however, the only set of valid base dimensions. Force can be substituted for mass to yield an $\{FLtT\}$ system, or both force and mass can be included to form an $\{FMLtT\}$ system.

Although a specific base dimension set may be freely chosen, other aspects of a valid dimensional system are restricted by the laws of physics. Every valid physical law can be cast in the form of a dimensionally homogeneous equation; i.e., the dimensions of the left side of the equation must be identical to those of the right side. Consider Newton's second law in the form

$$\mathbf{F} \propto M\mathbf{a} \qquad (1.3a)$$

The proportionality symbol is required here because there is no a priori reason to assume that the units in which force is measured in a given experiment will automatically be consistent with those of the mass–acceleration product. More generally, Newton's second law is written as

$$\mathbf{F} = \frac{M\mathbf{a}}{g_c} \qquad (1.3b)$$

where $1/g_c$ is a proportionality constant inserted to account for units. The units for g_c depend on the unit system under consideration. The key observation is that the units for only three of the four terms in Eq. (1.3b) can be arbitrarily defined. In the discussion that follows we examine three common engineering unit systems.

1.4.1 $\{MLtT\}$ Systems

Since acceleration has dimensions $\{Lt^{-2}\}$, Eq. (1.3b) describes a restriction on the relationship between the constant g_c and the dimensions $\{F\}$, $\{M\}$ and $\{Lt^{-2}\}$. Since only $\{M\}$ and $\{Lt^{-2}\}$ have been defined, one can arbitrarily define the dimensions for either g_c or $\{F\}$. If $\{F\}$ is defined, an $\{FMLtT\}$ system results. The other option, which is adopted by all $\{MLtT\}$ systems, is to define g_c as a dimensionless constant of unit magnitude. The advantage to this approach is that g_c no longer needs to appear in Newton's second law. Let us now examine the most common $\{MLtT\}$ system in use today.

Système International d'Unités (SI): In this system the base units for mass, length, time, and temperature are respectively the kilogram (kg), meter (m), second (s), and Kelvin (K). The magnitude of each unit is specifically defined (e.g., a meter is defined as 1,650,763.73 wavelengths of the radiation corresponding to the transition between the electronic energy levels $2p^{10}$ and $5d^5$ of the krypton-86 atom in a vacuum). Since force

EXAMPLE 1.8

What are the units of viscosity and pressure in SI?

SOLUTION

From Eq. 1.2c, the dimensions of viscosity are $\{Ft/L^2\}$. Thus in SI, viscosity has units of newton-seconds per square meter, (N-s)/m², or in terms of the base units $\{\mu\}$ = kilograms per meter-second, kg/(m-s). Pressure is a force per unit area, so the units of pressure in SI are newtons per square meter (N/m²), or in terms of the base units $\{p\}$ = kg/(m-s²).

is a supplemental dimension, its unit, called the newton (N), is defined by Eq. (1.3b) as:

$$1\,N = 1\,(kg\text{-}m)/s^2 \tag{1.4}$$

The SI system has been structured so that g_c and all other proportionality constants, including those in Fourier's law (a basic heat transfer equation) and Gauss's law (a basic electromagnetic equation) are dimensionless constants of unit magnitude. This is one of the major advantages of SI.

1.4.2 {FLtT} Systems

We have shown that Newton's second law describes a restriction on the relationship among g_c and the dimensions $\{F\}$, $\{M\}$, and $\{Lt^{-2}\}$. In an $\{FLtT\}$ system, g_c is made dimensionless with unit magnitude and M is a supplemental dimension. A commonly employed unit system of this type is the British gravitational unit system.

British Gravitational Unit System (BG): In this system the base units are the pound force (lb$_f$), foot (ft), second (s), and degree Rankine (°R). If g_c is to vanish, a mass unit known as a slug must be defined:

$$1\,\text{slug} = 1\,(lb_f\text{-}s^2)/ft \tag{1.5}$$

The weight of one slug is then 32.2 lb$_f$. Newton's second law applies in the usual form $F = Ma$. The British gravitational system of units is used in the United States in aerospace engineering, mechanical engineering, and several other engineering fields. See if you can convince yourself that the units for viscosity and pressure in BG are (lb$_f$-s)/ft² and lb$_f$/ft², respectively.

1.4.3 {FMLtT} Systems

In an $\{FMLtT\}$ system, base units are defined for force, mass, length, time, and temperature. The units and magnitude of the constant g_c are completely defined by

On September 23, 1999, NASA's *Mars Climate Orbiter* was lost as it attempted to enter orbit around the red planet. One of the major factors contributing to the failed mission was confusion over metric and English units. This technical disaster serves to remind us of the importance of checking units for internal consistency in every calculation we perform.

Newton's second law. A common $\{FMLtT\}$ system in use today is the English engineering unit system.

English Engineering Unit System (EE): The base units for mass, length, time and temperature are the pound mass (lb_m), foot (ft), second (s), and degree Rankine (°R). The force unit, pound force (lb_f), is arbitrarily defined so that 1 lb_m "weighs" 1 lb_f when acted on by gravity at sea level. Solving $\mathbf{F} = M\mathbf{a}/g_c$ for g_c, and substituting the appropriate values including the local gravitational acceleration of 32.2 ft/s² yields:

$$g_c = \frac{M\mathbf{a}}{\mathbf{F}} = \frac{(1\,lb_m)(32.2\,ft/s^2)}{1\,lb_f} = 32.2\,(lb_m\text{-}ft)/(lb_f\text{-}s^2) \qquad (1.6)$$

This is the customary unit system in the United States. It suffers from the fact that g_c does not vanish, a result that has caused significant confusion among generations of students and even some practicing engineers. See if you can convince yourself that the units for viscosity and pressure in EE are $(lb_f\text{-}s)/ft^2$ and lb_f/ft^2, respectively.

1.4.4 Preferred Unit Systems

There are three major reasons for the continued use by engineers of several unit systems. The first is psychological: old habits and customs are hard to change. The second is based on economics. Tooling, standards, educational systems, instructional materials, and research in the existing unit systems entail massive capital investments. The third is more subtle and perhaps even sensible: certain classes of problems are easier to solve in specific unit systems. With respect to the last point, all $\{MLtT\}$ and $\{FLtT\}$ systems are defined such that g_c is dimensionless and of unit magnitude, including SI and BG. This is frequently an advantage because it often simplifies calculations. In a few situations, however, an $\{FMLtT\}$ system, such as the EE system, offers significant advantages.

One could argue that our decision to expose you to three unit systems in this text, (SI, BG, and EE) is confusing. We recognize this argument, but students must realize that in the real world there are industries that commonly use these different systems. Many industries have customary units that are unique and not part of any standard system. For example, most pumps in the United States are sized in gallons per minute. The bottom line is that anyone who works in fluid mechanics must be able to perform unit conversion, a skill that will be explained in the following section. What we recommend is that you solve problems symbolically, i.e., work a problem all the way to the final result using only symbols and no data. Your result is a relationship among symbols that can then be checked for dimensional consistency. Then insert the data into this relationship to arrive at a numerical value that includes the appropriate units. This method reduces significantly the number of errors associated with incorrect units and will give you an understanding of the influence of different fluid variables on the result that does not occur when numbers alone are manipulated.

1.4.5 Unit Conversions

One method for converting a value from one unit system to another is to use the concept of multiplication by unity. Suppose we must convert the base unit of length in SI to the corresponding unit in the BG system. Appendix C gives the relationship 1 ft = 0.3048 m, and simple algebra yields: 1 ft/0.3048 m = 1 = 0.3048 m/1 ft. Clearly we can multiply any quantity by a fraction equal to one without changing the value of the quantity. Thus, we convert a meter into an equivalent number of feet as follows:

$$1 \text{ m} \times (1) = 1 \text{ m} \times \frac{1 \text{ ft}}{0.3048 \text{ m}}$$
$$= 3.281 \text{ ft}$$

Note that identical units are canceled algebraically. Thus the process treats a unit algebraically like any other mathematical symbol.

Suppose the only SI/English length conversion you remember is 1 in. = 2.54 cm. The preceding conversion can still be accomplished by multiplying by a string of 1s as follows:

$$1 \text{ m} \times \frac{100 \text{ cm}}{1 \text{ m}} \times \frac{1 \text{ in.}}{2.54 \text{ cm}} \times \frac{1 \text{ ft}}{12 \text{ in.}} = 3.281 \text{ ft}$$

EXAMPLE 1.9

As a sales representative for a U.S. pump manufacturer, you have a request from a European customer for the price of a pump with a capacity of at least 1.5 liters per second (L/s). Your company makes pumps with capacities ranging from 5 to 100 gallons per minute in 5 gal/min increments. Your client also wants your opinion on how a 1.5 L/s pump compares in capacity to those typically used by gas stations to dispense fuel. One of your office mates remembers that there are 28.316 liters in a cubic foot and another knows there are 231 cubic inches in a gallon. How are you going to answer your customer's questions?

SOLUTION

We are to convert a pump capacity of 1.5 L/s into an equivalent capacity in gallons per minute and then compare the size of such a pump to the ones typically used in service stations. The information obtained, which we should not assume to be correct without checking, is that (1 ft^3 = 28.316 L) and (1 gal = 231 in.^{3}). Using the procedure described in Section 1.4.4, we write these conversion factors in the form of 1s as

$$\frac{1 \text{ ft}^3}{28.316 \text{ L}} = 1 = \frac{28.316 \text{ L}}{1 \text{ ft}^3} \quad \text{and} \quad \frac{1 \text{ gal}}{231 \text{ in.}^3} = 1 = \frac{231 \text{ in.}^3}{1 \text{ gal}}$$

We will also need to use the facts that 12 in. = 1 ft and 1 min = 60 s in the forms:

$$\frac{1 \text{ ft}}{12 \text{ in.}} = 1 = \frac{12 \text{ in.}}{1 \text{ ft}} \quad \text{and} \quad \frac{1 \text{ min}}{60 \text{ s}} = 1 = \frac{60 \text{ s}}{1 \text{ min}}$$

We can now convert 1.5 L/s to an equivalent number of gallons per minute by multiplication by a string of 1s:

$$1.5 \left(\frac{\text{L}}{\text{s}}\right) \times \left(\frac{1 \text{ ft}^3}{28.316 \text{ L}}\right) \times \left(\frac{12 \text{ in.}}{1 \text{ ft}}\right)^3 \times \left(\frac{1 \text{ gal}}{231 \text{ in.}^3}\right) \times \left(\frac{60 \text{ s}}{1 \text{ min}}\right) = 23.8 \left(\frac{\text{gal}}{\text{min}}\right)$$

Since our customer requires a pump that can handle at least 23.8 gal/min, the smallest pump we can offer is our 25 gal/min model. How does this pump compare in capacity to those typically encountered at a gas station? An experiment found that it took just over two minutes to put 10 gallons of gas in a car. This corresponds to a pump capacity of roughly 5 gal/min, which is about 20% of the capacity of the pump we will recommend to our customer. (Many symbolic manipulator software packages such as Mathematica and MATHCAD contain unit conversion subroutines that you may find helpful.)

1.5 PROBLEM SOLVING

As we explore different approaches to solving fluid mechanics problems in the rest of this book, you will be exposed to a variety of problem types and solution techniques. You will gradually fill your "tool bag" with a powerful array of analytical, empirical, and computational tools. No one tool is inherently superior to another, only more appropriate. Each is ideally suited for certain jobs. One of the most important skills for you to develop is the ability to recognize the critical aspects of a problem so that you can choose the optimal tool(s) for its solution. For example, you will learn to recognize the similarities and differences among fluid mechanics problems. Recognizing similarities will save you time by allowing you to reuse parts of solutions you have obtained previously. Recognizing differences will save you time, embarrassment, or worse, since a failure to adjust the problem-solving approach can result in errors.

In previous courses you have probably developed certain problem-solving habits. A good problem-solving procedure is helpful in avoiding mistakes. You will also find that when you have no idea how to proceed, a standard procedure allows your mind to work subconsciously as you carry out the routine initial phases of the process. The approach outlined in Table 1.2 is offered as a suggestion for solving a fluid mechanics problem. Faithfully carrying out the different steps has proven helpful to many engineering students. As your skills develope, you will become adept at classifying problems.

Many of the example problems in this book will be solved by using the procedure outlined in Table 1.2. The major exception is that solving the dimensional analysis problems of Chapter 9 calls for a special procedure explained in that chapter. Although we recognize that the recommended procedure is not always necessary for routine calculations, it can be helpful in solving unfamiliar and more complex problems.

One of our goals in this textbook is to give you an early start on learning fluid mechanics in an engineering context. Thus, in the next chapter you will find familiar fluid

TABLE 1.2 Problem-Solving Procedure

Step 1	State the problem and draw a schematic diagram of the physical system (properly labeled and dimensioned).
Step 2	List the information provided in the problem statement and the basic laws you think you need to solve the problem.
Step 3	List any simplifying assumptions and limits of the analysis.
Step 4	Solve the problem. The following ideas may be helpful:
	Complete the analysis algebraically *before* substituting numerical values.
	Check the answer for dimensional consistency, then substitute numerical values (using a consistent set of units) to obtain a numerical answer.
	Check the numerical answer for physical reasonableness. Review the assumptions made in the solution to make sure they are still reasonable in light of the answer.
Step 5	Comment on, or suggest, alternative solutions or techniques that might be appropriate.

properties such as density and pressure discussed in the context of well-known formulas derived from an application of the principles of fluid mechanics to a practical problem. We continue this approach in Chapter 3, where you will be introduced to a variety of interesting case studies drawn from engineering practice. These include flow in pipes and ducts, flows with area changes (e.g., in a nozzle or venturi), pump and fan laws, drag on cylinders and spheres, and lift and drag on airfoils. The important problem characteristics, i.e., fluid properties, service conditions, and flow geometry, are summarized, and the problem is solved by using well-accepted design equations. In subsequent chapters we will show you the source of these equations. Some are derived from first principles, while others are the result of fitting a curve through experimental data. By using different tools to revisit case studies in subsequent chapters, you will see that the accuracy and complexity of a solution changes depending on the approach taken. In some cases the more sophisticated tools yield superior results. In other cases, the use of a high-powered computational solution of a complex theoretical model offers only marginal improvement, at a significant increment in cost, time, and difficulty over a perfectly satisfactory back-of-the-envelope calculation.

Examples and case studies are the glue that holds this book together. They are intended to show you that the material in each chapter is related to that in all the other chapters. They also demonstrate that engineering fluid mechanics problems are generally open ended—they can be solved on several different levels using a variety of tools. Remember that you will be practicing two related skills: the ability to select the right tool and the ability to use each tool correctly.

1.6 SUMMARY

This chapter serves as an introduction to the field of fluid mechanics. We study fluid mechanics because we want to understand the world around us: biological flow (air and blood), environmental flows (air and water pollution), and design mechanical devices (pumps and turbines), transportation systems (planes and boats), athletic equipment (golf balls and parachutes), and many other important devices and systems that improve lives.

Fluids are unable to resist a shear stress and will continue to deform no matter how small the applied stress happens to be. Thus there is no shear stress in a fluid at rest. Liquids are nearly incompressible, while gases are highly compressible.

The continuum hypothesis states that under certain conditions a fluid may be treated as a continuous substance or continuum rather than as a group of discrete molecules. This concept is valid as long as the dimensions of the physical system are large in comparison to the mean free path (in gases) or molecular dimensions (in liquids).

The three main approaches for describing a fluid are the molecular, Lagrangian, and Eulerian descriptions. The methods differ from one another in the choice of the fundamental fluid entity in each. In the molecular description the fundamental entity is a single fluid molecule, in the Lagrangian description the fundamental entity is a fluid particle, and in the Eulerian description the fundamental entity is a point in space. The Eulerian description is the customary description of engineering fluid mechanics.

A dimension is a physical variable used to specify some characteristic of a system while a unit is a particular amount of a physical quantity or dimension. The engineering systems commonly designated as SI, British gravitational, and English engineering are all used to some extent in engineering practice. Thus, anyone who works in fluid mechanics must be able to perform unit conversion that can be accomplished using the concept of multiplication by 1 in the form of a ratio of equivalent units.

In Table 1.2 we have provided a recommended procedure for use in solving fluid mechanics problems. We also suggest that before selecting your tools and begining to work, you think about the type of solution required and the resources required to achieve that solution.

PROBLEMS

Section 1.1

1.1 List five fluid systems you encounter every day. Include at least one biological system, one transportation system, and one natural (non-man-made) system. Describe the fluid involved and function of the system.

1.2 What is the difference between an analytical and a computational solution in fluid dynamics?

1.3 Look through a magazine and find a picture of a moving fluid. Write a paragraph that could be used to replace the picture in the magazine. Which representation of the flow do you prefer, the visual or the verbal? Why?

1.4 Describe the difference between internal and external flows, and give examples of each.

1.5 Describe the difference between fluid statics and fluid dynamics, and give three examples of typical problems from each subfield.

1.6 Describe the contributions of each of the following people to the field of fluid mechanics.
(*a*) Archimedes
(*b*) d'Alembert
(*c*) Wright brothers

1.7 Describe the contributions of each of the following people to the field of fluid mechanics.
(*a*) Bernoulli
(*b*) Poiseuille
(*c*) Prandtl

Section 1.2

1.8 Determine whether each of the substances listed is a gas, liquid, or solid under

ambient conditions. Be sure to state any assumptions and justify your decisions.
(a) Aluminum
(b) Water
(c) Nitrogen
(d) Blood
(e) Jello

1.9 Each of the substances listed can be modeled as either a fluid or a solid under certain circumstances. Explain this observation and provide examples.
(a) Sand at room temperature
(b) Nitrogen
(c) Paint at room temperature
(d) Eggs

1.10 Each of the substances listed can be modeled as either a fluid or a solid under certain circumstances. Explain this observation and provide examples.
(a) Toothpaste at room temperature
(b) Window glass
(c) Wax
(d) Marbles at room temperature

1.11 What is the difference between a Newtonian and a non-Newtonian fluid? Give three examples of each.

1.12 In which of the following situations would you be comfortable using Newton's law of viscosity? Justify your decisions.
(a) The flow of water through a straw
(b) The flow of a milkshake through a straw
(c) The flow of air through a straw

1.13 In which of the following situations would you be comfortable using Newton's law of viscosity? Justify your decisions.
(a) The flow of air through a runner's hair
(b) The flow of water through a swimmer's hair
(c) The flow of styling gel through a model's hair

1.14 Use the information in Table 1.1 to determine whether each of the substances listed is a liquid, solid, or gas.
(a) This material has the ability to conform to the shape of its container and can exhibit a free surface.
(b) This material has strong intermolecular bonds and experiences limited deformation under the application of a shear stress.
(c) This material is highly compressible.
(d) This material continues to deform under the application of a shear stress but is unable to take the form of its container.

1.15 Use the information in Table 1.1 to determine whether each of the substances listed is a liquid, solid, or gas.
(a) This material has the ability to conform to the shape of its container and is virtually incompressible.
(b) This material does not exhibit a free surface.
(c) This materials is able to resist a tensile stress.
(d) This material has the ability to expand without limit and resist the application of a tensile stress.

1.16 Consider steel and aluminum sheets of 6 mm thickness; each metal sheet is subjected to a shear stress of 2×10^5 psi. Compare the magnitudes of the displacements in the direction of the applied force for the two materials.

1.17 Consider steel and aluminum sheets of 5 mm thickness. If the steel sheet is subjected to a shear stress of 8×10^5 psi, calculate the magnitude of the shear stress that must be applied to the aluminum sheet so that both materials experience the same displacement in the direction of the applied force.

1.18 A sample of motor oil has been tested in a flat plate shearing device like the one shown in Figure 1.5 with the following results: $\tau = 7.7$ lb_f/ft^2 for a plate separation distance of 0.25 in. and a top plate velocity of 25 ft/s. Determine the viscosity of the fluid.

1.19 A sample of motor oil has been tested in a flat plate shearing device like the one shown in Figure 1.5A with the following results: $\tau = 4.0$ lb$_f$/ft^2, for a plate separation distance of 0.05 in. and a fluid viscosity of $\mu = 6.5 \times 10^{-3}$ (lb$_f$-s)/ft^2. Determine the top plate velocity.

1.20 Revisit the thought experiment in which we applied a shear stress to a puddle of water with our hand. Suppose the water puddle is initially 1.5 mm thick and that the viscosity of water is 0.001 kg/(m-s). If your hand is applying a shear stress of magnitude 0.10 kg/(m-s^2), calculate the shear strain rate, $d\gamma/dt$, in the fluid, the velocity gradient, du/dy, in the fluid, and the speed u at which your hand is moving.

1.21 The laminar flow of fluid between parallel flat plates is illustrated in Figure 1.2b. The velocity distribution for this flow is $u/u_{max} = 1 - (y/h)^2$. The fluid is water with a viscosity of 0.001 kg/(m-s), $u_{max} = 0.4$ m/s, and $h = 3.75$ mm.
(a) Calculate the shear stress on the top plate.
(b) Calculate the shear stress on the bottom plate.
(c) What is the magnitude of the total force experienced by a 5 m^2 section of either plate?

1.22 The laminar flow of fluid between parallel flat plates is illustrated in Figure 1.2b. The velocity distribution for this flow is $u/u_{max} = 1 - (y/h)^2$. The fluid is water with a viscosity of 0.001 kg/(m-s), $u_{max} = 0.4$ m/s, and $h = 2.5$ mm.
(a) Calculate the shear stress on the top plate and give its direction.
(b) Determine the shear strain rate, $d\gamma/dt$, as a function of position for this flow.

1.23 A child is sledding down an ice-covered hill. Each of the two runners on the sled has dimensions of 1.1 m by 1.3 cm. The sled is supported on a thin film of water [$h = 0.15$ mm, $\mu = 0.001$ kg/(m-s)] created by a local melting of the ice due to the pressure exerted by the sled. If the child is cruising at a velocity of 8 m/s, calculate:
(a) The shear strain rate, $d\gamma/dt$, as a function of position for this flow

(b) The total shear stress on the runners of the sled
(c) The magnitude of the force on one of the runners of the sled

1.24 A child is sledding down an ice-covered hill. Each of the two runners on the sled has dimensions of 1.1 m by 1.3 cm. The sled is supported on a thin film of water [$\Delta y = 0.15$ mm, $\mu = 0.001$ kg/(m-s)] created by a local melting of the ice due to the pressure exerted by the sled. If the shear stress on the runners is 70 Pa, calculate:
(a) The velocity of the sled
(b) The shear strain rate, $d\gamma/dt$, for this flow
(c) The magnitude of the force on one of the runners of the sled

1.25 The laminar flow of fluid over a stationary fixed plate is illustrated in Figure P1.1. The velocity distribution, defined for $y \leq h$, for this flow is $u/u_{max} = \frac{3}{2}(y/h) - \frac{1}{2}(y/h)^3$. The fluid is glycerin with $\mu = 1.5$ kg/(m-s), $u_{max} = 2$ cm/s, and $h = 6$ mm. Calculate the shear stress on the stationary plate and give its direction.

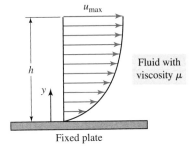

Figure P1.1

1.26 The laminar flow of fluid over a stationary fixed plate is illustrated in Figure P1.1. The velocity distribution, defined for $y \leq h$, for this flow is $u/u_{max} = \frac{3}{2}(y/h) - \frac{1}{2}(y/h)^3$. The fluid is glycerin with $\mu = 1.5$ kg/(m-s) and $u_{max} = 4$ cm/s. If the shear stress on the plate is 20 N/m^2, calculate the value of h for this flow.

1.27 The apparatus shown in Figure P1.2 is used to coat a thin sheet of steel with oil. The fluid is SAE 10W-30 oil, which has a viscosity of 0.02 kg/(m·s) at the temperature of interest. If $h = 4$ mm, and the total oil–steel contact area is 20 m², calculate the force necessary to move the steel at a velocity of 5 m/s. Assume that the velocity profile in the fluid is linear.

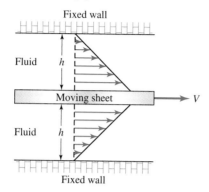

Figure P1.2

1.28 The apparatus shown in Figure P1.2 is used to coat a thin sheet of steel with oil. The fluid is SAE 10W-30 oil, which has a viscosity of 0.01 kg/(m·s) at the temperature of interest. If $h = 3$ mm, the total oil–steel contact area is 50 m², and the applied force is 1600 N, calculate the velocity at which the steel is moving through the system. Assume that the velocity profile in the fluid is linear.

Section 1.3

1.29 Explain how an engineer can decide whether the continuum hypothesis is valid for a particular engineering application.

1.30 The pressure in a vacuum system is 10^{-7} atm. Estimate the mean free path in the gas inside the system. Assume the gas can be modeled by using the perfect gas law so that the gas density is proportional to its pressure.

1.31 The pressure in a vacuum system is 0.1 Pa. Estimate the mean free path in the gas inside the system. Assume the gas can be modeled by using the perfect gas law so that the gas density is proportional to its pressure.

1.32 In each of the following situations, would you feel comfortable using the continuum hypothesis in your analysis of the fluid flow? For any doubtful cases, list the additional information you would like to obtain before feeling confident of your decision.
(a) The flow of interstellar gas at very low pressures
(b) The movement of an aircraft carrier through the ocean
(c) The flow of blood through an elephant
(d) The flow of blood through a mosquito

1.33 In each of the following situations, would you feel comfortable using the continuum hypothesis in your analysis of the fluid flow? For any doubtful cases, list the additional information you would like to obtain before feeling confident of your decision.
(a) The flow of gaseous nitrogen through a valve
(b) The flight of a space shuttle through Earth's atmosphere
(c) The flight of a vehicle halfway through its journey to Mars
(d) The flow of liquid nitrogen through a valve

1.34 Describe the fundamental fluid entity for the molecular, Eulerian, and Lagrangian descriptions.

1.35 Which, if any, of the three common descriptions (molecular, Eulerian, Lagrangian) are continuum descriptions? Justify your answer.

1.36 Explain why engineers frequently use the density of a fluid rather than its mass when solving fluid mechanics problems.

1.37 Which, if any, of the three common descriptions (molecular, Eulerian, Lagrangian) are field descriptions?

1.38 Which method of description, molecular, Lagrangian or Eulerian, would you recommend for use in each of the following flow situations?
(a) An analysis of wind patterns at an airport
(b) The motion of a manned balloon in a trip around the globe
(c) Rarefied gas flow

1.39 Which method of description, molecular, Lagrangian or Eulerian, would you recommend for use in each of the following flow situations?
(a) A study of monolayers of fluids acting as lubricants
(b) An investigations of tomorrow's weather patterns around the globe
(c) A study of the movement of mold spores through an air-conditioning system

Section 1.4

1.40 What is the difference between a dimension and a unit?

1.41 List the base dimensions for each of the following unit systems:
(a) SI
(b) English absolute
(c) British gravitational

1.42 Newton's second law can be written in the form $\mathbf{F} = M\mathbf{a}/g_c$.
(a) In which unit systems is g_c a dimensionless constant of magnitude 1?
(b) Give the magnitude and units for g_c in each of the units systems not included in your answer to part a.

1.43 The column headings in the following table name the three basic types of unit system. The row headings are a series of important fluid or flow properties. Fill in the remaining blocks in the table by providing the base dimensions for each of the various fluid/flow properties. Note that your solutions should contain only a combination of base dimensions contained in the column heading. That is, don't include M as a dimension in an entry in the $\{FLtT\}$ column.

Property	$\{FLtT\}$	$\{MLtT\}$	$\{FMLtT\}$
Velocity			
Mass			
Stress			
Kinematic viscosity			

1.44 The column headings in the following table names the three basic types of unit system. The row headings are a series of important fluid or flow properties. Fill in the remaining blocks in the table by providing the base dimensions for each of the various fluid/flow properties. Note that your solutions should only contain a combination of base dimensions contained in the column heading. That is, don't include M as a dimension in an entry in the $\{FLtT\}$ column.

Property	$\{FLtT\}$	$\{MLtT\}$	$\{FMLtT\}$
Acceleration			
Density			
Pressure			
Viscosity			

1.45 The column headings in the following table name three important unit systems. The row headings are a series of important fluid or flow properties. Fill in the remaining blocks in the table by providing the appropriate units for each of the various fluid/flow properties for each of the unit systems.

Property	SI	BG	EE
Velocity			
Mass			
Stress			
Kinematic viscosity			

1.46 The column headings in the following table name three important unit systems. The row headings are a series of important fluid or flow properties. Fill in the remaining blocks in the table by providing the appropriate units for each of the various fluid/flow properties.

Property	SI	BG	EE
Acceleration			
Density			
Pressure			
Viscosity			

1.47 Let P represent a force and x, y, and z represent distances. Determine the dimensions for each of the quantities listed in each of the indicated dimensional systems.

Quantity	$\{FLtT\}$	$\{MLtT\}$
$\dfrac{dP}{dx}$		
$\int P\,dx$		
$\dfrac{d^2 P}{dx\,dy}$		
$\int P\,dx\,dy$		

1.48 Let V represent a velocity and let x, y, and z represent distances. Determine the dimensions for each of the quantities listed in each of the indicated dimensional systems.

Quantity	$\{FLtT\}$	$\{MLtT\}$
$\dfrac{dV}{dx}$		
$\int V\,dx\,dy$		

1.49 Determine the SI units for each of the groups of dimensions shown. Also give the common name for this group of units. For example, the dimensional group $\{Lt^{-1}\}$ has SI units of meters per second and is commonly known as velocity.
(a) $\{ML^{-3}\}$
(b) $\{MLt^{-2}\}$
(c) $\{ML^{-1}t^{-2}\}$
(d) $\{ML^2t^{-2}\}$

1.50 Determine the BG units for each of the groups of dimensions shown. Also give the common name for this group of units. For example, the dimensional group $\{Lt^{-1}\}$ has BG units of feet per second and is commonly known as velocity.
(a) $\{FL^{-2}\}$
(b) $\{FL^{-1}t^2\}$
(c) $\{FL\}$
(d) $\{FLt^{-1}\}$

1.51 Determine the English engineering units for each of the groups of dimensions shown. Also give the common name for this group of units. For example, the dimensional group $\{Lt^{-1}\}$ has EE units of feet per second and is commonly known as velocity.
(a) $\{ML^{-3}\}$
(b) $\{FL^{-2}\}$
(c) $\{ML^2t^{-2}\}$
(d) $\{FtL^{-2}\}$

1.52 If M is a mass, determine the dimensions of A, B, and C in the dimensionally homogeneous equation $M = A(B - 1) + B/C$.

1.53 If V is a velocity and x is a distance, determine the dimensions of A, B, and C in the dimensionally homogeneous equations $A = V + \int B\,dx + dC/dx$.

1.54 If x is a distance and t is time, determine the dimensions of A, B, and C in the dimensionally homogeneous equation $Ax + B(dx/dt) + C(d^2x/dt^2) = 0$.

1.55 Use the data in Appendix C to express each of the following quantities in SI units:
(a) 60 mph
(b) 10 slugs
(c) 7.8 g/cm³
(d) 75°F
(e) 14.7 lb_f/in.²

1.56 Use the data in Appendix C to express each of the following quantities in BG units:
(a) 60 m/s
(b) 10 kg
(c) 7.8 g/cm³
(d) 75°C
(e) 101,000 N/m²

1.57 Use the data in Appendix C to express each of the following quantities in EE units:
(a) 25 m/s
(b) 20 slugs
(c) 2.7 g/cm³
(d) 196 K
(e) 101,000 N/m²

1.58 As a sales representative for a European pump manufacturer, you have a request from an American customer for the price of a pump with a capacity of at least 12 gal/min. Your company makes pumps with capacities ranging from 0.25 to 6.0 L/s in increments of 0.25 L/s. Which of your company's pumps will you recommend? Why?

1.59 A certain garden slug has a mass of 125 g. What is the mass of this garden slug in the BG unit system?

1.60 A rectangular swimming pool has a surface area of 15 ft × 30 ft and a uniform depth of 4 ft. If the density of the water in the pool is 1.0 g/cm³, determine the mass and the weight of the water in the pool. Report your answer in SI, BG, and EE units.

2 FLUID PROPERTIES

2.1 Introduction
2.2 Mass, Weight, and Density
    2.2.1 Specific Weight
    2.2.2 Specific Gravity
2.3 Pressure
    2.3.1 Pressure Variation in a Stationary Fluid
    2.3.2 Manometer Readings
    2.3.3 Buoyancy and Archimedes' Principle
    2.3.4 Pressure Variation in a Moving Fluid
2.4 Temperature and Other Thermal Properties
    2.4.1 Specific Heat
    2.4.2 Coefficient of Thermal Expansion
2.5 The Perfect Gas Law
    2.5.1 Internal Energy, Enthalpy, and Specific Heats
    2.5.2 Limits of Applicability
2.6 Bulk Compressibility Modulus
    2.6.1 Speed of Sound
2.7 Viscosity
    2.7.1 Viscous Dissipation
    2.7.2 Bulk Viscosity
2.8 Surface Tension
    2.8.1 Pressure Jump Across a Curved Surface
    2.8.2 Contact Angle and Wetting
    2.8.3 Capillary Action
2.9 Fluid Energy
    2.9.1 Internal Energy
    2.9.2 Kinetic Energy
    2.9.3 Potential Energy
    2.9.4 Total Energy
2.10 Summary
Problems

2.1 INTRODUCTION

To address the interesting flow problems in biology, the environment, mechanical devices, transportation systems, athletic equipment, and the many other important devices and systems, our first step must be to agree on a common language and set of symbols to use in defining fundamental fluid properties and the role of these properties in fluid mechanics. In this chapter we will distinguish a fluid property, defined to be a characteristic of the material structure of the fluid, from a flow property, whose value is determined in part by how the fluid is moving. The color of a fluid is purely a fluid property, while the velocity of a fluid is purely a flow property. The density, pressure, temperature, and viscosity of a fluid, which you may have thought of as fluid properties, are actually flow properties whose precise values depend on the nature of the fluid and type of flow. In reading this chapter, keep in mind then that all fluid properties are variables rather than constants, and assumed to be functions of position and time.

The discussion of fluid and flow properties in this chapter will enable you to begin to solve some simple but interesting fluid mechanics problems. It will also prepare you to read the interesting set of case studies drawn from engineering practice contained in the next chapter. These case studies include flow in pipes and ducts, flow in a nozzle, pump and fan laws, drag on cylinders and spheres, and lift and drag on airfoils. These case studies are revisited throughout the remainder of the text as we develop additional tools to help you solve more complicated fluid mechanics problems.

As you read through this chapter and the rest of the textbook, you'll see many standard fluid and flow properties along with their corresponding symbols, dimensions and units in the major unit systems. Note carefully that in fluid mechanics a few symbols are used to represent more than one property. This is unavoidable but usually creates no confusion because the context of a problem makes it clear to which property the symbol refers. Fluid property data for various fluids are contained in Appendix A and in tables throughout this chapter.

2.2 MASS, WEIGHT, AND DENSITY

The mass of an object is defined as a measure of its resistance to acceleration, i.e., resistance to a change in velocity. For an object of fixed mass, the relationship between acceleration, **a**, force, **F**, and mass, M, is given by Newton's second law:

$$\mathbf{F} = M\mathbf{a} \tag{2.1}$$

Mass is most often measured in units of kilograms (kg) or pounds mass (lb_m), but engineers working in the aerospace and related fields also use the slug.

The weight of an object, W, is the magnitude of the force acting on the object due to Earth's gravity field. Thus, weight is defined by the previous equation. If the acceleration produced by gravity is g, the weight of a mass M is:

$$W = Mg \tag{2.2}$$

Weight has the same dimensions as force so is expressed in units of newtons (N) or pounds force (lb_f).

Mass and weight depend on the amount of material in an object or system. In the case of a fluid, the weight obviously depends on whether you are describing the amount of water in a swimming pool or a teacup, or the air in an automobile tire or a hot air balloon. Properties of a fluid that depend on the amount of fluid in a system are termed extensive properties. The weight of a volume of fluid is an extensive property. If an extensive property is divided by the total mass in a system, the result is an intensive property, i.e., a property per unit mass. Extensive and intensive properties are both common in thermodynamics. In fluid mechanics, where the total mass of fluid is usually not relevant, we normally use intensive properties defined per unit volume.

In this text we generally use capital letters (both Roman and Greek) to represent extensive properties and lowercase letters to represent intensive properties. Thus, capital letters represent properties that depend on the extent of a fluid system, and lowercase letters represent properties that do not depend on the extent of a fluid system.

Fluids differ in their resistance to acceleration in proportion to their density. The density of a fluid, ρ, may be determined by dividing the mass of a sample of that fluid by its volume V:

$$\rho = \frac{M}{V} \qquad (2.3)$$

Density has dimensions of $\{ML^{-3}\}$ and units abbreviated as kg/m^3, lb$_m$/ft^3, or slugs/ft^3. The reciprocal of density is called specific volume, v:

$$v = \frac{1}{\rho} \qquad (2.4)$$

As a thermodynamic property of a fluid, density has a certain numerical value defined by the pressure and temperature of the fluid. This relationship is expressed by an equation of state. Since pressure and temperature are generally functions of position and time in a flow, density must also be a function of position and time. Thus density, pressure, and temperature are flow properties. In liquids, density normally remains nearly

EXAMPLE 2.1

A 40-gallon barrel of transformer oil is found to have a mass of 303 lb$_m$. What is the density of the oil?

SOLUTION

This exercise can be solved by using the definition of density given in Eq. 2.3. We will also require the unit conversion factor for gallons to cubic feet, which can be found in Appendix C to be 7.48 gal = 1 ft^3. The oil density is then found to be

$$\rho = \frac{M}{V} = \frac{303 \text{ lb}_m}{(40 \text{ gal})(1 \text{ ft}^3/7.48 \text{ gal})} = 56.7 \text{ lb}_m/\text{ft}^3$$

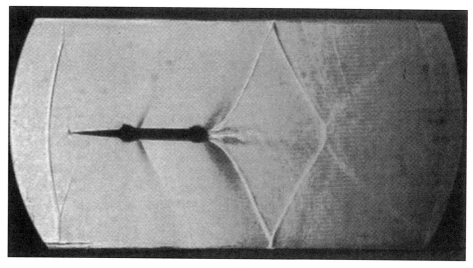

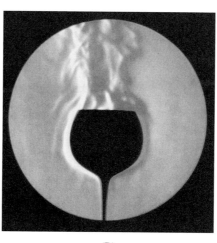

Figure 2.1 (A) Transonic flow past a thin wedged plate visualized with the Schlieren method. The lines preceding and trailing the flow are shock waves, which are a discontinuity in flow variables such as pressure, density, and pressure. (B) The heated glass creates a plume of low density fluid that rises through the relatively colder surrounding air.

constant throughout a region of flow. In gases, however, density often varies significantly (see Figure 2.1).

The accepted values of density and other properties of air at sea level are listed in Table 2.1. The variation of temperature with elevation for the U.S. standard atmosphere is shown in Figure 2.2 and similar data for pressure and density profiles in the atmosphere are given in Appendix B. These are, of course, average values. The atmosphere

46 | **2 FLUID PROPERTIES**

TABLE 2.1 Properties of U.S. Standard Atmosphere at Sea Level

Property	Symbol	SI Units	BG Units	EE Units
Temperature	T	288.15 K (15°C)	518.67°R (59°F)	518.67°R (59°F)
Pressure	p	101.33 kPa (abs)	2116.2 lb_f/ft^2 (14.696 psia)	2116.2 lb_f/ft^2 (14.696 psia)
Density	ρ	1.225 kg/m^3	0.002377 $slug/ft^3$	0.076539 lb_m/ft^3
Specific weight	γ	12.014 N/m^3	0.07647 lb_f/ft^3	0.07647 lb_f/ft^3
Absolute viscosity	μ	1.781×10^{-5} kg/(m-s) or Pa-s or $(N\text{-}s)/m^2$	3.737×10^{-7} $(lb_f\text{-}s)/ft^2$	3.737×10^{-7} $(lb_f\text{-}s)/ft^2$

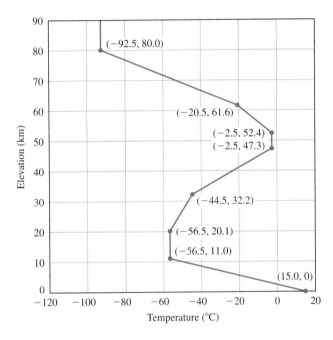

Figure 2.2 Temperature as a function of elevation for the U.S. Standard Atmosphere.

changes daily and seasonally. You can use Appendix B to verify that the pressure in the atmosphere at 30,000 ft is 4.373 psia.

Before concluding this discussion, it is important to clear up a potential point of confusion. On many occasions we will use the term "standard temperature and pressure." Although this phrase sounds similar to "standard atmosphere" the two terms are not synonymous. Standard temperature and pressure, referred to as STP, are defined to be (20°C = 68°F) and (1 atm = 101,300 N/m^2 = 2116 lb_f/ft^2 = 14.7 $lb_f/in.^2$).

The difference between mass and density is something most students understand. You will find, however, that imprecise language can occasionally cause confusion. Suppose you ask your neighbor, Which is heavier, gasoline or water? The answer may depend on what he is doing at the time. If he is walking toward his lawn mower with a cold glass of water in one hand and a 2.5-gallon container of gasoline in the other, he is likely to tell you that the gasoline is heavier (the extensive answer). If you ask the

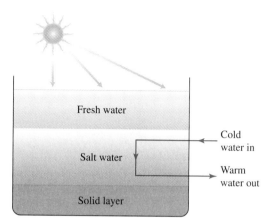

Figure 2.3 In a solar pond, most of the thermal energy from the sun passes through the water and is absorbed by the solid layer below. The warm solid raises the temperature of the adjacent salt water, and the fresh water insulates the heated salt water from the atmosphere. The thermal energy trapped in the salt water can be extracted via a piping system and used for space heating.

question when he is hosing out his garage to clean up a gasoline spill, he may tell you that since the gasoline is floating on the water, the water must be heavier (the intensive answer). The confusion arises because the word "heavier" is open to interpretation—it could mean "has a greater weight" or "has a greater density."

In fluid mechanics, "heavy" or "light" serves to compare the density of one fluid with respect to another. Mercury is heavier than water, water is heavier than oil, seawater is heavier than fresh water, cold air is heavier than hot air. When two immiscible liquids, such as oil and vinegar, are poured simultaneously into the same container, the lighter liquid will eventually settle on top of the heavier liquid. A practical application of this concept is a solar pond. As shown in Figure 2.3, a solar pond consists of a layer of fresh water on top of a layer of salt water. Since the water layers are thin, most of the thermal energy from the sun passes through the water and is absorbed by the solid material below. This raises the temperature of the solid, which in turn heats the salt water layer adjacent to it. The presence of the fresh water surface layer insulates the heated salt water from the atmosphere. The energy in the warm salt water is extracted for space heating.

The tendency of heavy flammable gases such as propane to settle in low places creates an explosion hazard. Hydrogen and methane, being light, tend to rise and mix with the surrounding heavier air, eventually decreasing in concentration below the limits of flammability. In the early moments after an accidental release of these two light gases, both will burn catastrophically if ignited (see Figure 2.4).

Is there a characteristic of a gas that determines its density? If you thought to answer "the molecular weight," you are correct. As shown in Table 2.2, the density of a gas at a given pressure and temperature is proportional to its molecular weight. Examination of Table 2.2 shows, however, that the molecular weight of liquids is not a reliable indicator of density. Other factors such as molecular polarity and shape, which are not important at the large intermolecular separation distance characteristic of gases, become important in the more closely packed structure of liquids.

The density of fluids at STP varies from 9×10^{-5} g/cm^3 for hydrogen gas to 13.6 g/cm^3 for liquid mercury. This is a difference of more than five orders of magnitude! Earlier we stated that the fluid density is related to pressure and temperature through a state equation. Since flow conditions influence the pressure and temperature

Figure 2.4 The Texas City explosion of April 1947 is an example of the potential destructive power of explosions.

TABLE 2.2 Influence of Molecular Weight on Fluid Density at STP

Fluid	Molecular Weight (amu)	Density (kg/m^3)	Density (slug/ft^3)
Gases			
Hydrogen (H$_2$)	2	0.09	1.63×10^{-4}
Helium (He)	4	0.18	3.23×10^{-4}
Methane (CH$_4$)	16	0.72	1.29×10^{-3}
Air	29	1.23	2.38×10^{-3}
Propane (C$_3$H$_8$)	44	2.01	
Liquids			
Water	18	1.00×10^3 (= 1.00 g/cm^3)	
Acetone	58	7.9×10^2 (= 0.79 g/cm^3)	
Glycerin	92 (92.11)	1.26×10^3 (= 1.26 g/cm^3)	
Toluene	92 (92.15)	8.7×10^2 (= 0.87 g/cm^3)	
Carbon tetrachloride	154	1.59×10^3 (= 1.59 g/cm^3)	
Mercury	201	1.359×10^4 (= 13.59 g/cm^3)	

in a fluid, they must also influence the fluid density through the state equation. This is often not important for liquids, so the state equation for a liquid reduces to the statement that density is constant. For gases, changes in density induced by temperature and pressure can be significant and must be accounted for to correctly predict the behavior of a gas. We investigate the most common state equation for gases, the ideal gas law, in Section 2.5.

2.2.1 Specific Weight

The specific weight γ of a fluid is defined as the weight of a fluid per unit volume in Earth's gravity field. For a volume of fluid V of mass M, the weight of the fluid is

2.2 MASS, WEIGHT, AND DENSITY

EXAMPLE 2.2

Estimate the density of krypton gas at STP given that its molecular weight is 83.8 amu.

SOLUTION

Examination of Table 2.2 shows that for gases, density is roughly proportional to molecular weight (MW). Thus, since krypton has an MW of 83.8 amu and helium has an MW of 4 amu we can estimate the density of krypton as

$$\rho(\text{Kr}) = \frac{\text{MW(Kr)}}{\text{MW(He)}} \rho(\text{He}) = \frac{83.8 \text{ amu}}{4 \text{ amu}} (0.18 \text{ kg/m}^3) = 3.77 \text{ kg/m}^3$$

This estimate compares favorably with the literature value for krypton's density of 3.74 kg/m³.

$W = Mg$, which we can also write as $W = \rho \forall g$. Thus the specific weight of a fluid is given by any of the following formulas

$$\gamma = \frac{W}{\forall}, \quad \gamma = \frac{Mg}{\forall}, \quad \text{or} \quad \gamma = \rho g \quad \quad (2.5\text{a–c})$$

Specific weight has dimensions of $\{F/L^3\}$ and typical units abbreviated as N/m³ or lb$_\text{f}$/ft³. The value of a fluid's specific weight depends on the specific physical conditions to which the fluid is exposed.

2.2.2 Specific Gravity

If you work for an international corporation and your product list contains dozens of different fluids, you will have a significant number of density values at STP to deal with, and these may be given in a number of different units. Placed in this situation, you

EXAMPLE 2.3

What is the specific weight of hydrogen at STP?

SOLUTION

Using Eq. 2.5c with the data in Table 2.2 and the gravitational constant $g = 9.81 \text{ m/s}^2$, we find:

$$\gamma = \rho g = (0.09 \text{ kg/m}^3)(9.81 \text{ m/s}^2) = 0.883 \text{ N/m}^3$$

EXAMPLE 2.4

Use the information provided for the specific weight of air under standard atmosphere conditions in Table 2.1 to determine the implied value of the gravitational constant at sea level.

SOLUTION

Solving Eq. 2.5c for the gravitational constant and substituting data from Table 2.1 in SI units yields:

$$g = \frac{\gamma}{\rho} = \frac{12.014 \text{ (N/m}^3)}{1.225 \text{ (kg/m}^3)} = 9.807 \text{ N/kg} \times \frac{1 \text{ (kg-m)/s}^2}{1 \text{ N}} = 9.807 \text{ m/s}^2$$

If we repeat the process for BG units, we find

$$g = \frac{\gamma}{\rho} = \frac{0.07647 \text{ (lb}_f/\text{ft}^3)}{0.002377 \text{ (slug/ft}^3)} = 32.17 \text{ lb}_f/\text{slug} \times \frac{1 \text{ slug}}{1 \text{ (lb}_f\text{-s}^2)/\text{ft}} = 32.17 \text{ ft/s}^2$$

By using specific gravity, petroleum engineers in Kuwait, the North Sea, Alaska, and Venezuela can easily compare the density of their respective crude oils regardless of a counterpart's preferred unit system. Density variations of crude oil occur because crude is a complex mixture of hydrocarbons, but the practical importance of knowing the precise density of a crude lies in its sale on a volumetric basis (barrels). The oil industry has developed a density scale that takes temperature effects into consideration.

might want to use a dimensionless, and therefore unitless, measure of density. This fluid property, known as specific gravity, sg, is defined as the ratio of a fluid's density to that of a standard reference fluid (water for liquids, air for gases) at STP. Thus, for gases, we define the specific gravity as

$$\text{sg(gas)} = \frac{\rho_{gas}}{\rho_{air}} \quad (2.6a)$$

while for liquids, we write

$$\text{sg(liquid)} = \frac{\rho_{liquid}}{\rho_{water}} \quad (2.6b)$$

EXAMPLE 2.5

What is the specific gravity of methane at STP?

SOLUTION

Using Eq. 2.6a with data from Table 2.2, we find:

$$\text{sg(methane)} = \frac{\rho_{methane}}{\rho_{air}} = \frac{0.72 \text{ kg/m}^3}{1.23 \text{ kg/m}^3} = 0.59$$

EXAMPLE 2.6

Although most modern thermometers contain alcohol, older ones often contained mercury (Hg). During fabrication of a certain thermometer, the Hg was inserted under standard conditions so that it filled a reservoir of volume 50 mm³. Given that Hg has a specific gravity of 13.6 and the density of water is 1 g/cm³, calculate the weight and mass of the Hg in the thermometer.

SOLUTION

We are asked to calculate the weight and mass of mercury in a thermometer. This can be done without the aid of a sketch. We are given sg(Hg) = 13.6, $\rho(H_2O) = 1$ g/cm³, and V (Hg) = 50 mm³. This problem can be solved by noting that the weight of a volume of fluid is $W = \rho g V$, while the mass of the same volume is $M = \rho V$. We are given the specific gravity of the mercury so we can calculate the density of the mercury using Eq. 2.6b: sg(liquid) = $\rho_{liquid}/\rho_{water}$, to write ρ_{Hg} = SG(Hg) × ρ_{water} = (13.6)(1 g/cm³) = 13.6 g/cm³. Thus the weight of the mercury in the thermometer is

$$W = [(13.6 \text{ g/cm}^3)(1 \text{ kg}/10^3 \text{ g})](9.81 \text{ m/s}^2)[(50 \text{ mm}^3)(1 \text{ cm}/10 \text{ mm})^3]$$
$$= 6.67 \times 10^{-3} \text{ (kg-m)/s}^2 = 6.67 \times 10^{-3} \text{ N}$$

The mass of the mercury in the thermometer is

$$M = \rho V = (13.6 \text{ g/cm}^3)[(50 \text{ mm}^3)(1 \text{ cm}/10 \text{ mm})^3] = 0.680 \text{ g}$$

Notice the use of the unit conversion factors. Since most of us do not have an intuitive feeling for the magnitude of a Newton (0.2248 lb$_f$), it may be more useful to examine the calculated value of mass to decide if our answer is reasonable. Recalling that a gram of water occupies 1000 mm³, we see that the mass of 50 mm³ of water is 0.05 g. The same volume of mercury must therefore have a mass 13.6 times larger, which agrees with what we found.

2.3 PRESSURE

All fluids are composed of energetic molecules in motion. When these molecules collide with a surface, they exert a normal and tangential force on the surface due to the change in momentum of colliding molecules. Since the resulting surface forces are of critical importance in all fluid mechanics problems, it is important for engineers to be able to quantify them. In this section we focus on the normal force exerted by a fluid on a surface. This normal force exists in fluids at rest and in motion, whereas tangential (shearing) forces exist only for fluids in motion.

As noted in Section 1.2, liquids and gases are unable to exert tensile stresses. Thus the normal force exerted by a fluid on a surface is always compressive; i.e., it is directed into the surface. It is easiest to envision this force as being applied by a fluid to a real physical surface, but it is also important to realize that a normal force is applied by a fluid on every surface it contacts, as illustrated in Figure 2.5. Note that on the imaginary

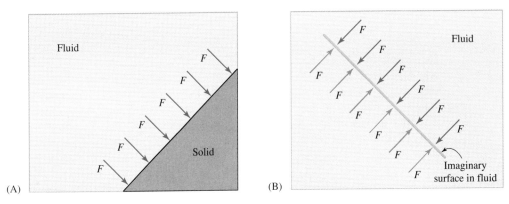

Figure 2.5 Fluids exert a compressive force on any surface they contact. (A) A fluid exerting a force normal to a solid surface. (B) A fluid exerting a compressive force on an imaginary interior surface. Note that the force exerted on the fluid above and to the right of the imaginary surface is equal and opposite to the force exerted by the fluid below and to the left of that same surface.

interior surface shown, the fluid on one side of the surface exerts a normal force on the fluid on the other side and vice versa.

The pressure p on a planar surface is defined as the compressive normal force applied by the fluid to the surface, F_N, divided by the area of that surface, A. Thus we write

$$p = \frac{F_N}{A} \qquad (2.7)$$

and apply the convention that pressure is a positive quantity. Note that Eq. 2.7 defines an average pressure acting on the surface. The dimensions for pressure are $\{FL^{-2}\}$ and the corresponding common units are newtons per square meter [N/m² also known as a pascal (Pa), lb$_f$/ft², or lb$_f$/in.² (abbreviated psi)].

EXAMPLE 2.7

Robert Fulton's steamboat the *Clermont* was powered by a steam-driven piston engine. The net steam pressure acting on the piston was about 10 psi and the piston surface was circular, with a diameter of 27 in. What was the magnitude of the force applied by the steam to the piston in this historically important engineering system?

SOLUTION

According to Eq. 2.7 the normal force is the product of the pressure in the fluid and the area of contact, thus we calculate the force in this case as

$$F_N = pA = p\left(\frac{\pi D^2}{4}\right) = 10\,\frac{\text{lb}_f}{\text{in.}^2}\left(\frac{\pi(27\,\text{in.})^2}{4}\right) \approx 5730\,\text{lb}_f$$

The pressure defined in Eq. 2.7 is an absolute pressure, which may take any positive value. The use of the label "psia" signifies that the numerical value for the pressure is an absolute pressure, measured in the units abbreviated $lb_f/in.^2$. In a vacuum, there are no fluid molecules to exert a force on a surface. Thus the absolute pressure in a perfect vacuum is zero. As mentioned previously, standard atmospheric pressure at sea level is 1 atm = 101,300 Pa = 2116 lb_f/ft^2 = 14.7 psia. Other common units used for pressure in discussing the atmosphere are the torr (760 torr = 1 atm) and the bar (1 bar = 100,000 Pa). High pressures are not uncommon. A water jet cutter, like the one shown in Figure 2.6, operates at pressures of ~50,000 psia, an order of magnitude higher than pressures seen in hydraulic power and control systems.

Absolute pressure is defined and measured in reference to a perfect vacuum. The pressure employed in an equation of state is always an absolute pressure. Most pressure measurements, however, are made by comparing the unknown pressure to an ambient pressure. This results in a reading of what is referred to as the gage pressure. The gage then responds to the difference in the two pressures. A pressure reading made this way gives zero when the unknown pressure is equal to atmospheric pressure, and −14.7 psi when the unknown pressure is a perfect vacuum. A gage pressure is therefore specified with reference to atmospheric or ambient pressure and can range from minus atmospheric pressure to large positive values. Figure 2.7 shows a common pressure gage. Note that the pressure indicated by the device represents the difference between the unknown pressure source and the ambient pressure.

Figure 2.6 A water jet cutter can cut a variety of metallic or nonmetallic materials such as stainless steel, felt, rubber, graphite, fiberglass, and titanium.

54 | 2 FLUID PROPERTIES

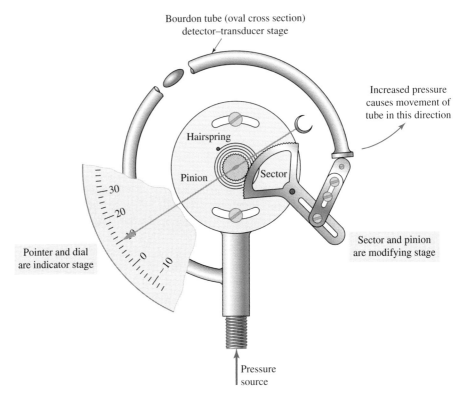

Figure 2.7 Schematic of Bourdon tube pressure gage.

Would you agree that the only kind of pressure gage that gives a reading of −20 psi is a broken one?

The relationship between gage and absolute pressure is given by

$$p(\text{gage}) = p(\text{absolute}) - p(\text{ambient}) \quad (2.8)$$

Thus, the pressure in a tire might be measured to be either 32 psig (gage) or 46.7 psia (absolute), and the pressure in a vacuum pump may be −10 psig or 4.7 psia. Note that a pressure reported in units of psi is likely to be a gage pressure (as was the case in Example 2.7), but the circumstances should always be examined carefully before a definitive conclusion is determined.

2.3.1 Pressure Variation in a Stationary Fluid

Anyone who has flown in an airplane, climbed a mountain, or been underwater knows that the pressure in a stationary fluid in Earth's gravity field is a function of height. The deeper you swim or dive, the greater the water pressure, and attendant ear pain. The pressurized cabin of an aircraft maintains only a certain minimum pressure at altitude. Since this pressure is much less than the sea level value, your ears may "pop" during the pressure changes characteristic of takeoff and landing. The pressure in a stationary fluid,

EXAMPLE 2.8

You have been asked to recommend an appropriate compressor for inflating automobile tires to a Swiss customer who imports American automobiles. If the tires are to be inflated to 32 psi, what is the corresponding pressure in SI units? Suppose the customer inflates the tires in a car to the "correct" SI pressure and finds that the tires look underinflated. What could be wrong?

SOLUTION

We are asked to calculate the pressure in SI units that corresponds to 32 psi and then to comment on the observation that tires inflated to this pressure appear to be underinflated. No illustration is required to solve this problem. Since the recommended tire pressure is 32 psi, we will require the unit conversion factor from BG/U.S. pressure units of psi to SI pressure units of pascals. This unit conversion factor can be found in Appendix C or in the text, where standard pressure is defined to be 101,300 Pa = 14.7 psi. Thus, the unit conversion factor is:

$$\frac{14.7 \text{ psi}}{101{,}300 \text{ Pa}} = 1 = \frac{101{,}300 \text{ Pa}}{14.7 \text{ psi}}$$

No symbolic manipulation is required, and the unit conversion is simply executed as

$$32 \text{ psi} \times \frac{101{,}300 \text{ Pa}}{14.7 \text{ psi}} = 2.21 \times 10^5 \text{ Pa} = 221 \text{ kPa}$$

What about the underinflated appearance of the tire? The problem is that we do not know whether the recommended 32 psi value is a gage pressure or an absolute pressure. If we filled the Swiss tire to 221 kPa(absolute), the equivalent of 32 psia, while the tire in the United States was filled to 32 psig (= 46.7 psia), then clearly the Swiss tire will appear underinflated. If, however, we fill the Swiss tire to 221 kPa(gage) it should appear identical to its properly inflated sister tire. As you probably know, tire pressures are always given as gage pressures, so this problem should not arise.

---

referred to as the hydrostatic pressure, varies with height. Hydro is an outdated reference to water, and static means at rest.

Using Figure 2.8 and a simple force balance, we can demonstrate that the pressure at a point within a stationary fluid is directly related to the weight of the column of fluid above that point. Note that this means that a sea level atmospheric pressure of 14.7 psia (= 2115 lb$_f$/ft^2 = 101,300 Pa) is equal to the weight of the entire column of air in the atmosphere above a square inch of surface. Using a similar force balance on a column of liquid or other constant density fluid at rest or nearly so, we may calculate the change in pressure with depth using the formula

$$\Delta p = \rho g d \tag{2.9}$$

where Δp is the pressure increase at the bottom of a fluid column of depth d.

2 FLUID PROPERTIES

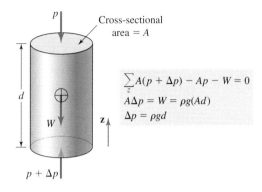

Figure 2.8 Force balance in the vertical direction for a cylindrical column of constant density fluid. The key result is that $\Delta p = \rho g d$.

$$\sum_z A(p + \Delta p) - Ap - W = 0$$
$$A\Delta p = W = \rho g(Ad)$$
$$\Delta p = \rho g d$$

EXAMPLE 2.9

Estimate in meters the depth below the surface of a lake at which the pressure is equal to twice atmospheric pressure.

SOLUTION

The pressure at the top of the lake is atmospheric and twice this value at the bottom. Thus the pressure change is one atmosphere. We solve Eq. 2.9 for the height of the water column d and then substitute the known values $\Delta p = 101{,}300\,\text{Pa}$ [$= 101{,}300\,\text{kg/(m-s}^2)$], $\rho = 1000\,\text{kg/m}^3$, and $g = 9.81\,\text{m/s}^2$ to find:

$$d = \frac{\Delta p}{\rho g} = \frac{101{,}300\,[\text{kg/(m-s}^2)]}{1000\,(\text{kg/m}^3)9.81\,(\text{m/s}^2)} = 10.3\,\text{m}$$

EXAMPLE 2.10

What is the change in air pressure experienced in riding an elevator to the top of a 10-story building? Assume the air density remains constant over this change in height.

SOLUTION

This exercise can be solved by using Eq. 2.9 and Table 2.1. The density and pressure of air at standard atmosphere conditions are given as $0.076539\,\text{lb}_m/\text{ft}^3$ and $2116.2\,\text{lb}_f/\text{ft}^2$, respectively. Recognizing that the gravitational constant is $32.2\,\text{ft/s}^2$, estimating the distance between successive floors in the building as 10 ft, and substituting the appropriate values into Eq. 2.9 gives

$$\Delta p = \rho g d = (0.076539\,\text{lb}_m/\text{ft}^3)(32.2\,\text{ft/s}^2)(100\,\text{ft}) = 246.5\,\text{lb}_m/(\text{ft-s}^2)$$

The units for Δp are not the expected lb_f/ft^2. Why? Since we are using the English engineering unit system, we have to deal with the pesky g_c factor. Equation 1.7 can be treated as a unit conversion factor of the form 32.2 (lb$_m$-ft) = 1 (s^2-lb$_f$). We can then use this "unit conversion" to find that

$$\Delta p = p_2 - p_1 = (246.5 \text{ lb}_m/(\text{ft-s}^2)) \left(\frac{1 \text{ s}^2\text{-lb}_f}{32.2 \text{ lb}_m\text{-ft}}\right) = 7.65 \text{ lb}_f/\text{ft}^2$$

Recognizing that the air pressure at the top of the building is less than that at the bottom and using $p_2 = 2116.2$ lb$_f$/ft^2, we solve for p_1 to find:

$$p_1 = (2116.2 - 7.65)(\text{lb}_f/\text{ft}^2) = 2108.6 (\text{lb}_f/\text{ft}^2)$$

This represents a less than 0.4% reduction in pressure.

The pressure–depth relationship in a fluid of variable density, such as the air in Earth's atmosphere, is more complex than the linear relation in Eq. 2.9. For example, look at the Standard Atmosphere data for pressure in Appendix B. We will show how to calculate the pressure distribution in a variable density fluid in Chapter 5, but note that Eq. 2.9 may be used if the density is interpreted as an average value over the whole column of variable density fluid.

2.3.2 Manometer Readings

A manometer is a liquid-filled device used to measure pressure. As shown in Figure 2.9, in a simple water filled U-tube manometer, an unknown pressure can be compared to atmospheric pressure by using Eq. 2.9 to estimate the pressure difference. By changing the liquid from water to mercury, a manometer of reasonable size may be used to read higher pressures. To illustrate the importance of replacing water with a denser fluid such as mercury, recall that a pressure difference of 1 atm requires a water column 10.3 m high. How high would the corresponding column of mercury have to be? Examination

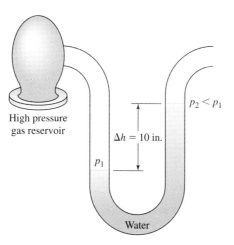

Figure 2.9 A U-tube manometer can be used to measure pressure differences by recording the difference in the height of the fluid levels in the two legs of the device. In this example $\Delta h = 10$ in.

EXAMPLE 2.11

A blower operates with a pressure difference of 10 in. of water as measured by a manometer. What is the pressure difference in psia and Pa? Use $\rho = 0.0361$ lb_m/in^3.

SOLUTION

Equation 2.9 tells us that $\Delta p = \rho g d$. In this example the height $d = \Delta h$ is 10 in. and the manometer fluid is water. From the problem statement we know that the fluid density is $\rho = 0.0361$ lb_m/in^3. Thus, using $g = 32.2$ $ft/s^2 = 386.4$ $in./s^2$, Δp is calculated as

$$\Delta p = \rho g d = \rho g \Delta h = (0.0361\ lb_m/in.^3)(386.4\ in./s^2)(10\ in.) = 139.5\ lb_m/(in.\text{-}s^2)$$

As was the case in Example 2.9, we must make use of Eq. 1.7 as a unit conversion factor of the form 32.2 lb_m-ft = 1 s^2-lb_f. In this case we find:

$$\Delta p = 139.5\ lb_m/(in.\text{-}s^2) \left(\frac{1\ s^2\text{-}lb_f}{32.2\ lb_m\text{-}ft} \right) \left(\frac{1\ ft}{12\ in.} \right) = 0.36\ lb_f/in.^2 = 0.36\ \text{psia}$$

The equivalent pressure difference in SI units can be found directly by using the unit conversion factor (1 psia = 6.8947×10^3 Pa) found in Appendix C. The result is $\Delta p = 2489$ Pa.

How would you estimate the total weight of air in Earth's atmosphere given that the diameter of the earth is ~12,750 km? Since the surface area of a sphere is $4\pi r^2$, Earth's surface area is about $A = 4\pi r^2 = 4\pi(6375\ m)^2 = 5.1 \times 10^8\ m^2$. The sea level atmospheric pressure of 101,300 Pa represents the weight of the column of air above a square meter of Earth's surface. Thus the total weight of air in the atmosphere may be estimated by multiplying sea level pressure by Earth's surface area. The result is a weight of 5.17×10^{13} N or equivalently 1.16×10^{13} lb_f. The mass of the atmosphere may be calculated as 5.27×10^{12} kg. Note that this calculation is based on the fact that the atmosphere is very thin in comparison to the radius of Earth.

of Eq. 2.9 shows that the height of the fluid column scales linearly with density. Since the density of mercury is roughly 13.6 times that of water (see Table 2.2) the column of mercury in the manometer would be only 0.76 m high.

2.3.3 Buoyancy and Archimedes' Principle

The increase in hydrostatic pressure with depth in a fluid creates a net force on an immersed object. The net vertical force acting on an object due to hydrostatic pressure is called the buoyancy force. Consider a force balance on an object as shown in Figure 2.10. Archimedes' principle states that a buoyancy force, $F_{buoyancy}$, acts in the direction opposite to that of the gravitational force, $F_{gravity}$, and has a magnitude equal to the weight of the displaced fluid. Figure 2.10 indicates that the buoyancy force arises because the increase in hydrostatic pressure with depth creates a net upward hydrostatic force on the surface of the object. Since the gravitational force on the object is equal to its weight, the net force F_{net} on a submerged object is the difference between its weight W_{obj} and

2.3 PRESSURE

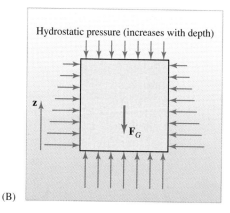

(A)   (B)

Figure 2.10 Hot air balloon: (A) photograph and (B) schematic, indicating that the net force on the balloon is calculated as the difference between the gravitational force and the buoyancy force, i.e., $F_N = F_G - F_B$.

the weight of the displaced fluid W_{fluid}. That is, we can write the net force as

$$F_{\text{net}} = F_{\text{gravity}} - F_{\text{buoyancy}} = W_{\text{obj}} - W_{\text{fluid}} \qquad (2.10)$$

Note that a positive value of F_{net} represents a net force in the direction of the gravitational field. A negative value implies that the object is being pushed upward in the opposite direction. This principle may be used to estimate the buoyancy force acting on a heated volume of fluid immersed in similar fluid. In this case, the buoyancy force is the difference in the weight of the volume of hot fluid and the weight of an equal volume of the surrounding colder fluid.

EXAMPLE 2.12

Suppose a thin plastic bag containing 200 cm³ of hot water at 90°C is held submerged in a tank of cold water at 20°C. Calculate the net force acting on the bag of fluid. Is the force pushing the bag up or down? What is the buoyancy force acting on the bag?

SOLUTION

We are asked to determine the magnitude and direction of the net force acting on a bag of warm water immersed in a bucket of cold water, and the buoyancy force. Figure 2.10 is a reasonable representation of the physical situation. The bag contains 200 cm³ ($= 2 \times 10^{-4}$ m³) of water at 90°C and the temperature of the cold water is 20°C. The relevant equation is 2.10:

$$F_{\text{net}} = F_{\text{gravity}} - F_{\text{buoyancy}} = W_{\text{obj}} - W_{\text{fluid}}$$

We also require the definitions of weight and density from Eq. 2.2 ($W = Mg$), and Eq. 2.3 ($\rho = M/\forall$). We will assume that the weight and volume of the plastic bag may be neglected. Since we have one fluid immersed in another, W_{obj} here represents the weight of the hot fluid, and W_{fluid} represents the weight of an equal volume of displaced cold fluid. The weight of a volume of fluid is given by $W = Mg = \rho \forall g$. Since the volumes of cold and hot water are identical, Eq. 2.10 can be written

$$F_{net} = F_{gravity} - F_{buoyancy} = W_{obj} - W_{fluid} = \rho_H \forall g - \rho_C \forall g = (\rho_H - \rho_C) \forall g$$

where the subscripts H and C stand for hot and cold, respectively. The volume is known, and the densities of water at 90 and 20°C are found in Appendix A to be $\rho_H = 965.3$ kg/m^3, and $\rho_C = 998.2$ kg/m^3. Substituting these values into the preceding equation yields

$$F_{net} = (\rho_H - \rho_C) \forall g$$
$$= (965.3 - 998.2) \text{ (kg/m}^3\text{)}(2 \times 10^{-4} \text{ m}^3)(9.81 \text{ m/s}^2) = -0.0645 \text{ N}$$

The negative sign indicates that the net force is up. Therefore, if the bag is released it will tend to rise through the cooler surrounding fluid. The buoyancy force is easily calculated as

$$F_{buoyancy} = \rho_C \forall g = (998.2) \text{ (kg/m}^3\text{)}(2 \times 10^{-4} \text{ m}^3)(9.81 \text{ m/s}^2) = 1.958 \text{ N}$$

The buoyancy force points up (opposite the gravitational force), since the pressure in the surrounding water increases with depth.

2.3.4 Pressure Variation in a Moving Fluid

In the preceding section we learned about the pressure–depth relationship in a stationary fluid. In this section we investigate the pressure–velocity relationship in a steadily moving fluid in the absence of any frictional effects. For a constant density fluid in motion at the same elevation, there is an inverse relationship between the square of the fluid speed and the pressure. Including the effect of elevation, the relationship between pressure, speed, and elevation at two points along the path of a fluid particle is given by

$$\left[p + \tfrac{1}{2}\rho V^2 + \rho g h\right]_1 = \left[p + \tfrac{1}{2}\rho V^2 + \rho g h\right]_2 \tag{2.11}$$

where h is the height of the point above a datum level. Equation 2.11 is a version of the famous Bernoulli equation. The derivation of this equation and the limits of its applicability are discussed further in Chapter 8, but it is sufficient here to recall from elementary physics that the equation may be thought of as a statement of energy conservation for a particle of fluid moving in the absence of any losses due to frictional effects.

Equation 2.11 can be used to understand the operation of a venturi, a smoothly shaped constriction through which fluid flows. At the position of minimum area in a venturi, called the throat, the fluid experiences a higher speed and lower pressure. Therefore, a venturi may be used to provide suction, as in the carburetor of a gasoline engine (Figure 2.11). The flow of air into the engine pulls fuel into the airstream due to the venturi effect.

2.3 PRESSURE

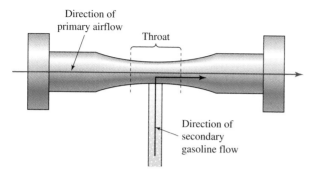

Figure 2.11 The flow through a venturi tube can be used to induce the flow of a second fluid. A common example is a carburetor, in which the primary airflow draws gasoline into the engine. As the air moves through the throat (the region of reduced cross-sectional area), its velocity increases and its pressure decreases. The drop in pressure provides suction, which draws the gasoline into the throat and out through the exit of the venturi.

EXAMPLE 2.13

Consider a venturi tube of the type illustrated in Figure 2.11 with water flowing through it at 20°C. If the pressure at the throat is 2 atm below that at the inlet and the inlet velocity is 0.6 m/s, estimate the velocity of the water at the throat of the venturi tube.

SOLUTION

We are asked to calculate the velocity of water at the throat of a venturi tube as illustrated in Figure 2.11. The pressure drop is 2 atm (= 202,600 Pa), and the upstream velocity is 0.6 m/s. To solve this problem we will apply Eq. 2.11 between the inlet and the throat along the path shown. Since the elevation is the same for these points, we have

$$\left[p + \tfrac{1}{2}\rho V^2\right]_1 = \left[p + \tfrac{1}{2}\rho V^2\right]_2$$

We have assumed that the flow is horizontal, incompressible, and frictionless, so that this form of the Bernoulli equation is valid. If we let point 1 correspond to the inlet of the venturi tube and point 2 correspond to the throat, then the variable to be isolated is V_2. Manipulation of the equation yields:

$$V_2 = \sqrt{\frac{2}{\rho}(p_1 - p_2) + V_1^2}$$

From Appendix A, the density of water at 20°C is 998.2 kg/m³. Substituting this value and the other values given in the problem statement [$(p_1 - p_2) = 202{,}600$ Pa and $V_1 = 0.6$ m/s] into the preceding equation gives

$$V_2 = \sqrt{\frac{2}{\rho}(p_1 - p_2) + V_1^2} = \sqrt{\frac{2}{998.2 \text{ kg/m}^3}(202{,}600 \text{ Pa}) + (0.6 \text{ m/s})^2} = 20.2 \text{ m/s}$$

Note that the pressure at the inlet must be greater than 2 atm. We will revisit this problem in a future chapter as we develop additional methods of analyzing a flow.

Try putting your hand out an automobile window at different speeds, holding your palm perpendicular to the flow. The pressure on the front of your hand can be roughly estimated as $\frac{1}{2}\rho V^2$ higher than ambient, and somewhat lower than ambient on the back. Now suppose you dipped your hand into the water while traveling at a similar speed. Why would the force on your hand be nearly three orders of magnitude greater in water than in air at the same velocity?

Consider what happens when high velocity fluid at speed V impacts a solid surface as shown in Figure 2.12. The fluid cannot penetrate the surface and must therefore come to rest on the surface. In this situation, Eq. 2.11 predicts that the local pressure on the surface will be increased $\frac{1}{2}\rho V^2$ above the ambient pressure upstream in the moving fluid.

There are many other fluid flow situations for which changes in pressure and velocity are important. As shown in Figure 2.13A, a parachute takes advantage of a higher pressure under the canopy than the ambient value above to develop a net vertical force to slow the

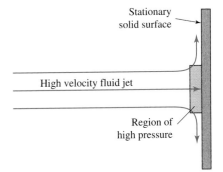

Figure 2.12 When a high speed fluid stream with initial velocity V impacts a solid surface, its velocity is reduced to zero. As a result, its pressure must increase to a value of magnitude $\frac{1}{2}\rho V^2$ above the ambient pressure.

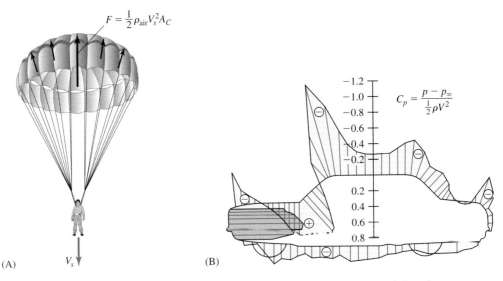

Figure 2.13 (A) A parachute slows the fall of a jumper because the higher pressure below the canopy produces a force in the vertical direction that opposes the force of gravity. The magnitude of the net vertical force on the canopy is $F = \frac{1}{2}\rho_{air} V_s^2 A_C$, where V_s is the sinking speed of the jumper and A_C is the area of the open end of the canopy. (B) Pressure distribution along the centerline of an automobile.

EXAMPLE 2.14

A jet of water with a cross-sectional area of 2 cm² strikes a surface at a speed of 50 m/s as shown in Figure 2.12. Estimate the force applied to the surface by the jet. What force is applied by a jet of air at this speed?

SOLUTION

Based on Eq. 2.11, the pressure will increase above the ambient on the surface where the jet comes to rest. Since the force on the surface is given by the pressure times the area, the force generated is given by $F = (\Delta p)A = \frac{1}{2}\rho A V^2$. Knowing that the density of water is ~ 1000 kg/m³ and that the water velocity is 50 m/s, we write

$$F_{H_2O} = \frac{1}{2}\rho A V^2 = \frac{1}{2}(1000 \text{ kg/m}^3)\left[(2 \text{ cm}^2)\left(\frac{1 \text{ m}}{100 \text{ cm}}\right)^2\right](50 \text{ m/s})^2$$
$$= 250 \text{ (kg-m)/s}^2 = 250 \text{ N}$$

The force applied by air is found by using the density of air (1.225 kg/m³) to obtain

$$F_{air} = \frac{1}{2}\rho A V^2 = \frac{1}{2}(1.225 \text{ kg/m}^3)(2 \text{ cm}^2)\left(\frac{1 \text{ m}}{100 \text{ cm}}\right)^2 (50 \text{ m/s})^2$$
$$= 0.306 \text{ (kg-m)/s}^2 = 0.306 \text{ N}$$

fall of the jumper. The force developed may be estimated as $\frac{1}{2}\rho_{air} V_S^2 A_C$, where A_C is the area of the open end of the canopy, and V_S is the sinking speed. Similarly, at higher speeds a significant fraction of the drag of an automobile is due to pressure differences over the front and rear of the vehicle (see Figure 2.13B). The drag force can be shown to be proportional to the product of the vehicle's frontal area and $\frac{1}{2}\rho V^2$.

EXAMPLE 2.15

Estimate the increase in the aerodynamic drag due to pressure forces if the speed of an automobile increases from 55 mph to 70 mph.

SOLUTION

Assuming that the pressure distribution over the vehicle behaves according to our simplified form of Bernoulli's equation, the increase in pressure will be proportional to velocity squared, which turns out to be a factor of $(70/55)^2 = 1.62$. Thus we can calculate an increase of 62% in the aerodynamic drag. This must cause a decrease in the fuel economy of the automobile. However, the decrease in fuel economy is less than 62% because aerodynamic drag is only a fraction of the total drag on an automobile at these speeds.

Just as a pressure variation may cause a fluid to flow or be caused by a fluid flow, temperature variations in a fluid may create a fluid flow or be a consequence of a fluid flow. Perhaps you have experienced the low temperature resulting from venting a pressurized gas, or have felt a temperature change with depth while swimming in a lake. Soaring birds and sailplanes take advantage of rising air masses called thermals, which are warm parcels of air pushed up by buoyancy forces through the surrounding cooler fluid.

Thus far our discussion of the pressure variation in a moving fluid has focused on pressure differences resulting from fluid flow. One must also recognize that pressure differences can drive fluid flow. For example, wind is created by relatively small pressure differences acting over large distances. Water squirts out of a hose because the pressure at the faucet is greater than the (atmospheric) pressure at the open end of the hose. Pumps are designed to increase the pressure of a fluid, enabling it to move through a piping system to a location at lower pressure. In most flow situations the magnitude of a pressure difference is more important than the actual value of the pressure itself. The flow characteristics of water moving from a supply at 50 psig to a region at 0 psig will be identical to those of water moving from a supply at 150 psig to a region at 100 psig.

2.4 TEMPERATURE AND OTHER THERMAL PROPERTIES

The temperature T of a fluid is a thermodynamic state variable that provides a measure of the internal energy of the fluid. For a fluid in equilibrium, the temperature is proportional to the mean kinetic energy of the random motion of the molecules. Temperature is a base dimension in all unit systems and is often expressed in units of °C or °F. The relationships between the Celsius and Fahrenheit temperature scales and their corresponding absolute temperature scales are:

$$K = °C + 273.16 \quad \text{and} \quad °R = °F + 459.69 \quad (2.12a,b)$$

where K (no degree symbol) is the temperature unit in the Kelvin scale, and °R is the temperature unit in the Rankine scale. Remember that the temperature used in an equation of state and all other thermodynamic relationships is always an absolute temperature.

As mentioned earlier, the density, pressure, and temperature of a fluid are related to one another by an equation of state. For many gases, the state equation is well approximated by the perfect gas law to be described in Section 2.5. Although all liquids also exhibit some dependence of density on pressure and temperature, the state equation of a liquid is usually replaced by the assumption that the liquid density is constant. This approximation should be examined before being applied to a specific flow problem. For example, if a constant density approximation is applied to a problem involving a liquid with an imposed temperature variation, the analysis cannot predict the tendency of warm, less dense liquid to rise in response to buoyancy forces. Sound waves are also absent in a constant density model, since these waves inherently involve fluid density changes. Despite these failings, we will see that the constant density assumption is appropriate for many of the liquid flows encountered in fluid mechanics, and also for gas flows at low speeds.

2.4 TEMPERATURE AND OTHER THERMAL PROPERTIES

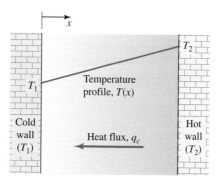

Figure 2.14 When a fluid is located between closely spaced parallel walls held at different temperatures, a linear temperature profile develops. For the specified coordinate system, T increases with increasing x so that the temperature gradient is positive. Thus, since the thermal conductivity of the fluid is defined to be positive, the heat flux q_c, defined as $q_c = -k(dT/dx)$, is in the negative x direction for this situation (this is the expected result, since we know that heat flows from the hot wall to the cold wall).

Temperature differences in a fluid are always accompanied by the flow of heat by molecular conduction. Consider a one-dimensional flow of heat in a fluid as shown in Figure 2.14. Let q represent the heat flux, or rate at which heat is crossing a surface parallel to the walls per unit area per unit time. It is found empirically that the heat flux is proportional to the temperature gradient existing in the fluid at the location of the surface, i.e., $q_c \propto -dT/dx$. A minus sign is necessary because heat flows from hot to cold regions. The missing constant of proportionality in this expression is called the thermal conductivity of the fluid, k. Thus we model the relationship between the heat flux and temperature gradient as

$$q_c = -k\frac{dT}{dx} \tag{2.13}$$

where k has dimensions of $\{Ft^{-1}T^{-1}\}$ with corresponding SI units of newtons per second-kelvin [N/(s-K)]. However, it is customary to use the definition of a joule [1 J = 1 N-m] to express k in units of J/(m-s-K). Equation 2.13 is known as Fourier's law of heat conduction. Values for thermal conductivity for several fluids and solids are contained in Table 2.3. The thermal conductivity of a fluid is a function of temperature and pressure.

2.4.1 Specific Heat

The specific heat capacity of a fluid c is defined to be the amount of heat energy required to raise the temperature of a unit mass of fluid by one degree. This is expressed by writing

$$c = \frac{dq}{dT} \tag{2.14}$$

where dq is the amount of heat required to produce a temperature change dT per unit mass of material. The dimensions for heat capacity are $\{FLM^{-1}T^{-1}\}$ and the common units are J/(kg-K) or (ft-lb$_f$)/(lb$_m$-°F). Note carefully that the q in this equation is not the same as the q_c in Fourier's law. In this case q has dimensions of $\{FLM^{-1}\}$ and common units of J/kg or (ft-lb$_f$)/lb$_m$.

The value of c depends both on temperature and the process by which heat is added. In a thermodynamic sense there are many possible processes, but in fluid mechanics

2 FLUID PROPERTIES

TABLE 2.3 Thermal Conductivity and Specific Heat Values for Various Substances at 1 atm and 25°C

Substance	Thermal Conductivity, k [J/(m-s-K)]	Specific Heat, c_p [kJ/(kg-K)]
Gases		
Hydrogen	0.183	14.6
Helium	0.151	5.23
Nitrogen	0.0259	1.09
Oxygen	0.0266	0.921
Air	0.0262	1.05
Liquids		
Gasoline	0.135	2.22
Acetone	0.176	2.15
Glycerin	0.284	2.41
Water	0.616	4.816
Mercury	8.36	0.139
Solids		
Copper	398	0.386
Aluminum	273	0.900
Steel (1020)	52	0.486
Pyrex	1.1	0.75
Window glass	0.80	0.840
Brick	0.72	0.835
Nylon	0.30	1.67

EXAMPLE 2.16

What is the conductive heat flux through a 1 cm wide water gap if the temperature difference is 50°C? Repeat for the same gap filled with air.

SOLUTION

This exercise is solved by using Eq. 2.13: $q_c = -k(dT/dx)$. From the given information we can determine that the temperature gradient is:

$$\frac{dT}{dx} \approx \frac{T_2 - T_1}{x_2 - x_1} = \frac{50°C}{0.01 \text{ m}} = 5000°C/m = 5000 \text{ K/m}$$

Note that we have assumed that the temperature is increasing in the direction of increasing x as shown in Figure 2.14. That is, we have assumed a positive temperature gradient. From Table 2.3 we find: $k_{H_2O} = 0.616$ J/(m-s-K) and $k_{air} = 0.0262$ J/(m-s-K). Substituting these values in the expression for the heat flux gives:

$$q_c^{H_2O} = -k\frac{dT}{dx} = -[0.616 \text{ J/(m-s-K)}](5000 \text{ K/m}) = -3080 \text{ J/(m}^2\text{-s)}$$

$$q_c^{air} = -k\frac{dT}{dx} = -[0.0262 \text{ J/(m-s-K)}](5000 \text{ K/m}) = -131 \text{ J/(m}^2\text{-s)}$$

2.4 TEMPERATURE AND OTHER THERMAL PROPERTIES

> The fact that the heat flux is negative indicates that heat is being transported in the negative x direction which is what we would expect for the case of T increasing as x increases. If we had assumed that T decreased as x increased, then we would have obtained a negative thermal gradient and a positive heat flux. In either case, the magnitude of the heat flux is the same.

It is useful to note the similarities between Newton's law of viscosity and Fourier's law of heat conduction. The two equations are $\tau = \mu(du/dy)$ and $q_c = -k(dT/dx)$. Fourier's law relates the heat flux to the temperature gradient through the thermal conductivity k. This equation is a model of thermal transport. In direct analogy, Newton's law of viscosity can be considered to be a model for momentum transport. Since momentum is the product of mass and velocity, it has the dimension $\{MLt^{-1}\}$ or equivalently $\{Ft\}$. The rate of tangential momentum transfer per unit area, or momentum flux, is the counterpart of the heat flux. Momentum flux has dimensions of $\{Ft/(tL^2)\}$ or simply $\{FL^{-2}\}$, which of course are also the dimensions of shear stress. This analysis justifies an interpretation of shear stress as a momentum transfer rate. To complete the analogy, the roles of the temperature gradient and thermal conductivity in Fourier's law are played, respectively, by the velocity gradient and the shear viscosity in Newton's law. Thus, we can interpret shear viscosity as the proportionality constant that relates the velocity gradient to the momentum transfer rate. Note that in both heat and momentum transfer, something flows downhill—heat flows from regions of high temperature to regions of low temperature and momentum flows from regions of high tangential velocity to regions of low tangential velocity.

only two processes need to be considered: heat added at constant pressure and heat added at constant volume. For liquids these two specific heats are virtually identical, and a liquid is said to have a specific heat c with no further consideration of the process. For gases the specific heat capacity at constant pressure c_p is greater than the specific heat capacity at constant volume c_v because extra energy is required for the expansion of a gas at constant pressure.

For incompressible liquids the change in internal energy of the fluid is directly related to the specific heat of the liquid through the relationship $\Delta u = c_p \Delta T$. The corresponding relationship for gases is more complex and is discussed in Section 2.5.1.

Specific heat values for some common liquids and gases are shown in Table 2.3 in SI units, along with values for several solids for comparison (corresponding values in BG units are given in Appendix A). Specific heat values at constant pressure range from 14.6 kJ/(kg-K) for hydrogen gas to 0.139 kJ/(kg-K) for liquid mercury. Air and water have intermediate values of 1.05 and 4.816 kJ/(kg-K), respectively. It may seem counterintuitive for air and water to have specific heat values that differ by less than a factor of 5, but remember that these values are based on a unit mass rather than a unit volume. On a per-unit-volume basis, it takes much less energy to heat air than to heat water because the density of air is nearly a thousand times smaller.

2.4.2 Coefficient of Thermal Expansion

Gases and liquids expand when heated, and density decreases, as a result of the enhanced kinetic energy of the individual molecules that make up the fluid. The relationship between temperature and density change at constant pressure is described by β, the coefficient of

EXAMPLE 2.17

Calculate the amount of thermal energy required to heat each of the following quantities of fluid from 20°C to 30°C at standard pressure: 1 g of air, 1 g of water, 1 cm³ of air, and 1 cm³ of water. Assume that the density of air and water remain constant over this temperature range at values of 1.225 and 997 kg/m³, respectively.

SOLUTION

We are to determine the amount of thermal energy required to raise the temperatures of various quantities of matter from 20°C to 30°C. No sketch is required. The densities for air and water are given as 1.225 and 997 kg/m³, and specific heat values are available in Table 2.3. Since pressure is constant, we use the specific heat at constant pressure c_p. Rearranging Eq. 2.14 we have $dq = c_p\,dT$. We will assume that the specific heats of air and water do not change appreciably over the temperature range 20–30°C so that $q = c_p \Delta T$. The total thermal energy Q required to raise the temperature of a fixed mass of fluid M by an amount ΔT when the specific heat is constant is $Q = Mq = Mc_p\Delta T$. When the fluid mass is known we can use this relationship directly. If, however, we are given the volume, V, of fluid, then we must use the definition of density to obtain $Q = \rho V c_p \Delta T$. The thermal energy required to raise the temperature of the various quantities of matter from 20°C to 30°C is calculated as follows.

For 1 g (= 0.001 kg) of air [$c_p = 1050$ J/(kg-K)]:

$$Q = Mc_p\Delta T = 1050 \text{ J/(kg-K)} \times 0.001 \text{ kg} \times 10 \text{ K} = 10.5 \text{ J}$$

For 1 g (= 0.001 kg) of water [$c_p = c = 4816$ J/(kg-K)]:

$$Q = Mc\Delta T = 4816 \text{ J/(kg-K)} \times 0.001 \text{ kg} \times 10 \text{ K} = 48.16 \text{ J}$$

For 1 cm³ (= 10^{-6} m³) of air [$c_p = 1050$ J/(kg-K) and $\rho = 1.225$ kg/m³]:

$$Q = \rho V c_p \Delta T = 1050 \text{ J/(kg-K)} \times 10^{-6} \text{ m}^3 \times 1.225 \text{ kg/m}^3 \times 10 \text{ K} = 0.0129 \text{ J}$$

For 1 cm³ (= 10^{-6} m³) of water [$c_p = c = 4816$ J/(kg-K) and $\rho = 997$ kg/m³]:

$$Q = \rho V c \Delta T = 4816 \text{ J/(kg-K)} \times 10^{-6} \text{ m}^3 \times 997 \text{ kg/m}^3 \times 10 \text{ K} = 48.02 \text{ J}$$

These answers seem reasonable. Since the density of water is approximately 1 g/cm³, the similarity of the two values for water is not surprising.

---

thermal expansion of the fluid. This coefficient is defined in terms of specific volume and density as

$$\beta = \frac{1}{v}\left(\frac{\partial v}{\partial T}\right)_p \qquad (2.15a)$$

$$\beta = -\frac{1}{\rho}\left(\frac{\partial \rho}{\partial T}\right)_p \qquad (2.15b)$$

2.4 TEMPERATURE AND OTHER THERMAL PROPERTIES

Values of the coefficient of thermal expansion for several common liquids can be found in Appendix A. Note that water at 4°C is an exception to the general rule of expansion upon heating. Under ordinary conditions, the coefficient of thermal expansion for gases is given by

$$\beta = -\frac{1}{\rho}\left(\frac{\partial \rho}{\partial T}\right)_p = \frac{1}{T} \tag{2.16}$$

where T is an absolute temperature. This relationship may be derived from the perfect gas law.

EXAMPLE 2.18

Calculate the volume change if 1 ft³ of air is heated from 80°F to 100°F at constant pressure. Repeat the calculation for 1 ft³ of water. The coefficient of thermal expansion for water at 90°F is approximately $1.67 \times 10^{-4}\,°\text{R}^{-1}$.

SOLUTION

We are asked to determine the change in volume of 1 ft³ of air and water as a result of a temperature increase from 80°F to 100°F at constant pressure. No sketch is required to solve this problem. We are told that β for water at 90°F is $1.67 \times 10^{-4}\,°\text{R}^{-1}$. The relevant Eq. is 2.15a. Since we are considering a fixed fluid mass, we can write $1/v(\partial v/\partial T)_p = 1/V(dV/dT)$ and use this equation to write

$$\frac{1}{V}\frac{dV}{dT} = \beta \tag{A}$$

We assume that air can be treated as an ideal gas in this situation, using Eq. 2.16 to write $\beta = 1/T$. We will also assume that the value of β for water does not change appreciably over the temperature range 80–100°F and use the value of β at 90°F. Separating the variables in (A), we obtain $\int (dV/V) = \int \beta\, dT$. For air, we use $\beta = 1/T$. After integration we have:

$$V_2/V_1 = T_2/T_1 \tag{B}$$

For water, β is assumed to be constant and we get

$$V_2/V_1 = e^{\beta(T_2 - T_1)} \tag{C}$$

Thus when 1 ft³ of air is heated from 80°F to 100°F (539°R to 559°R) (B) predicts the following volume at 100°F:

$$V_2 = V_1\left(\frac{T_2}{T_1}\right) = (1\text{ ft}^3)\left(\frac{559\,°\text{R}}{539\,°\text{R}}\right) = 1.037\text{ ft}^3$$

The corresponding result for water is found using (C) to be

$$V_2 = V_1\, e^{\beta(T_2 - T_1)} = (1\text{ ft}^3)\exp[(1.67 \times 10^{-4}\,°\text{R}^{-1})(20\,°\text{R})] = 1.0034\text{ ft}^3$$

The volume increase is 0.037 ft³ (3.7%) for air and 0.0033 ft³ (0.33%) for water. The observation that the volume increase for air is much greater than that for water is consistent with experience.

2.5 THE PERFECT GAS LAW

Under a broad range of conditions, the equation of state for a gas is well modeled by the perfect (or ideal) gas law. In extensive form this law is

$$p\forall = nR_u T \tag{2.17}$$

where R_u is the universal gas constant, $\forall$ is the volume of the system, and n is the number of moles of gas in the system. The value of the universal gas constant is

$$R_u = 8314 \text{ (N-m)/(kgmol-K)} = 1545 \text{ (ft-lb}_f\text{)/(lbmol-}°\text{R)} \tag{2.18}$$

The specific gas constant R is found by dividing the universal gas constant by the molecular weight M_w of the gas:

$$R = \frac{R_u}{M_w} \tag{2.19}$$

The dimensions of the specific gas constant are $\{FLM^{-1}T^{-1}\}$ and the units are (N-m)/(kg-K) or (ft-lb$_f$)/(lb$_m$-°R). For example, the value of R for air is 287 (N-m)/(kg-K) or 53.3 (ft-lb$_f$)/(lb$_m$-°R).

By using Eq. 2.19, and noting that nM_w equals the total mass M of the gas, the perfect gas law can also be written as

$$p\forall = MRT \tag{2.20a}$$

In fluid mechanics, we employ the intensive (per-unit-volume) form of the perfect gas law, namely

$$p = \rho RT \tag{2.20b}$$

2.5.1 Internal Energy, Enthalpy, and Specific Heats of a Perfect Gas

In the model known as a calorically perfect gas, which we use throughout this text, the specific heats are assumed to be constants. In this model the internal energy change, $u_2 - u_1$, and the enthalpy change, $h_2 - h_1$, are related to temperature change $(T_2 - T_1)$ by the equations

$$u_2 - u_1 = c_V(T_2 - T_1) \tag{2.21a}$$

$$h_2 - h_1 = c_P(T_2 - T_1) \tag{2.21b}$$

The ratio of specific heats occurs so often in gas flow problems that it is given a special symbol:

$$\frac{c_P}{c_V} = \gamma \tag{2.22}$$

Since the specific heats are constants for a calorically perfect gas, the ratio of specific heats is also a constant. From thermodynamics, the following relationships can be

2.5 THE PERFECT GAS LAW

EXAMPLE 2.19

A volume of air originally at STP is heated 10°C at constant pressure. What is the new density?

SOLUTION

This exercise is solved by using Eq. 2.20b. We have $p_0 = \rho_0 R T_0$ at STP and $p_f = \rho_f R T_f$ in the final state. Dividing the second equation by the first and noting that R is constant gives $p_f/p_0 = \rho_f T_f / \rho_0 T_0$. Rearranging we have $\rho_f / \rho_0 = p_f T_0 / p_0 T_f$. Since $p_0 = p_f$ in this case, we have $\rho_f = \rho_0 (T_0/T_f)$. Substituting the values for air at STP, $T_0 = 20°C = 293$ K and $\rho_0 = 1.20$ kg/m³ (Appendix A), gives the new density as:

$$\rho_f = \rho_0 \left(\frac{T_0}{T_f}\right) = 1.20 \text{ kg/m}^3 \times \left(\frac{293 \text{ K}}{303 \text{ K}}\right) = 1.16 \text{ kg/m}^3$$

As an example of the utility of the perfect gas law, consider the problem of estimating the density of air at standard conditions. In SI units, standard pressure and temperature are 101,300 Pa and 20°C, respectively. Rearranging Eq. 2.20b to solve for density gives $\rho = p/RT$. Recognizing that R for air was given in the text as 287 (N-m)/(kg-K), and substituting the appropriate values into the perfect gas law with the unit conversion factor (1 Pa = 1 N/m²) gives

$$\rho = \frac{101,300 \text{ Pa}}{287 \left[(\text{N-m})/(\text{kg-K})\right](293 \text{ K})} \times \frac{1 \text{ N/m}^2}{1 \text{ Pa}}$$
$$= 1.205 \text{ kg/m}^3$$

shown to hold for a calorically perfect gas:

$$c_p - c_V = R, \quad c_p = \frac{\gamma R}{\gamma - 1}, \quad \text{and} \quad c_V = \frac{R}{\gamma - 1}$$
(2.23a–c)

The classical kinetic theory of gases suggests that for monatomic and diatomic gases the following relationships are excellent approximations:

$$c_p = \frac{n+2}{2} R, \quad c_v = \frac{n}{2} R, \quad \text{and} \quad \gamma = \frac{n+2}{n}$$
(2.24a–c)

where $n = 3$ for monatomic gases and $n = 5$ for several diatomic gases including oxygen and nitrogen (at normal temperatures and pressures). These results can be seen to be consistent with the perfect gas relationships given in Eqs. 2.22 and 2.23, and provide a means of estimating the specific heats and specific heat ratio of a gas from its molecular structure. Values of the specific heats and other perfect gas parameters for a number of common gases are listed in Table 2.4. The reported c_p and c_v values were calculated from the given values of R and γ using Eqs. 2.23b and 2.23c.

2.5.2 Limits of Applicability

The perfect gas law is applicable for a wide range of engineering problems involving air and other gases. It should be used with caution, however, if the gas pressure or

TABLE 2.4 Perfect Gas Parameters for Several Common Gases

Gas	MW	γ	In SI Units: J/(kg-K)			In BG Units: (ft-lb$_f$)/(slug-°R)		
			R	c_p	c_v	R	c_p	c_v
Air	29	1.40	287	1,004	717	1,716	6,006	4,290
CO_2	44	1.30	189	819	630	1,130	4,897	3,767
He	4	1.66	2077	5,224	3,147	12,420	31,238	18,818
H_2	2	1.41	4124	14,180	10,060	24,660	84,806	60,146
Methane	16	1.31	518	2,190	1,672	3,099	13,096	9,997
N_2	28	1.40	297	1,039	742	1,775	6,213	4,438
O_2	32	1.40	260	909	650	1,130	5,439	3,885

EXAMPLE 2.20

What are the internal energy and enthalpy changes for nitrogen in a temperature change of 100°C? Assume that nitrogen is a perfect gas.

SOLUTION

Since we are assuming that nitrogen can be modeled as a calorically perfect gas, this exercise can be solved by using Eqs. 2.21 and 2.23. We have:

$$u_2 - u_1 = c_V(T_2 - T_1) \quad \text{and} \quad h_2 - h_1 = c_P(T_2 - T_1)$$

From Table 2.4, the specific gas constant for nitrogen is 297 (N-m)/(kg-K) and the specific heat ratio is 1.4. Thus we calculate

$$u_2 - u_1 = c_V \Delta T = \frac{R}{\gamma - 1}(T_2 - T_1) = \left(\frac{297 \text{ (N-m)/(kg-K)}}{1.4 - 1}\right)(100 \text{ K})$$

$$= 74{,}250 \text{ (N-m)/kg} = 74{,}250 \text{ J/kg}$$

$$h_2 - h_1 = c_P \Delta T = \frac{1.4R}{1.4 - 1}(T_2 - T_1) = \left(\frac{1.4}{0.4}\right) 297 \text{ (N-m)/(kg-K)}(100 \text{ K})$$

$$= 103{,}950 \text{ (N-m)/kg} = 103{,}950 \text{ J/kg}$$

Note that nitrogen is a diatomic gas. Thus, even if we did not have Table 2.4 available we could use Eq. 2.19 to calculate that the gas constant is 297 (N-m)/(kg-K). We would then insert $n = 5$ into Eqs. 2.24a and 2.24b to obtain estimates for the two specific heat values. You may wish to demonstrate to yourself that this procedure results in the same values for the change in internal energy and enthalpy for the conditions specified in this example.

temperature is extremely high. The perfect gas law breaks down at high pressure because as gas molecules are forced closer together, molecular interactions occur that are not accounted for in the perfect gas model. At high temperatures, some of the intermolecular collisions are so violent that polyatomic gas molecules are ionized, another process not accounted for in the perfect gas law.

What is the range of temperatures and pressures over which it is "safe" to use the perfect gas law? For the pressures encountered in engineering, the perfect gas law can be used as long as the temperature is below the dissociation temperature of the polyatomic molecules in the gas (e.g., 3000 K for O_2 and 6000 K for N_2). At the low end of the temperature spectrum, perfect gas behavior can be assumed if temperatures are above the critical temperature ($T_C = $ 155 K for O_2 and 126 K for N_2) up to pressures of at least 70 atm. For values outside the range specified, one can use advanced state equations, such as the van der Waals, Beattie–Bridgeman, or Sutherland equations, or use a table of thermodynamic properties or charts for the specific fluid of interest.

2.6 BULK COMPRESSIBILITY MODULUS

When a fluid is subjected to a pressure increase, the volume decreases, and the density increases. For many fluids the pressure–volume relationship is linear and may be characterized by a proportionality constant called the bulk compressibility modulus, E_V. The relationship between a change in pressure, dp, and the corresponding fractional change in specific volume, dv/v, is written in terms of E_V as:

$$dp = -E_V \frac{dv}{v} \tag{2.25}$$

The minus sign is necessary because a positive change in pressure (a pressure increase) results in a negative change in volume (a volume decrease). Since $-dv/v = d\rho/\rho$, Eq. 2.25 may be written as

$$dp = E_V \frac{d\rho}{\rho} \tag{2.26}$$

This equation can be rearranged to define the bulk modulus as

$$E_V = \frac{dp}{d\rho/\rho} \tag{2.27}$$

Since $d\rho/\rho$ is a dimensionless ratio, the dimensions and common units for E_V are the same as those for pressure, $\{FL^{-2}\}$, and thus pascals or psi. Values of the bulk modulus for several liquids may be found in Table 2.5 (see also Appendix A).

TABLE 2.5 Bulk Modulus Values E_V, for Several Common Liquids at 20°C

	Bulk Modulus, E_V	
Liquid	SI (GPa)	BG (lb$_f$/ft²)
Gasoline	0.958	2.00 × 10⁷
Mercury	25.5	5.32 × 10⁸
Methanol	0.83	1.73 × 10⁷
SAE 30W oil	1.38	2.88 × 10⁷
Water	2.19	4.57 × 10⁷
Seawater (30% salinity)	2.33	4.86 × 10⁷

2 FLUID PROPERTIES

For gases, the precise value of the bulk modulus depends on the type of compression process which occurs. For a perfect gas undergoing an isothermal process $dp = RT_0 d\rho$. Thus the isothermal bulk modulus of a perfect gas can be written as

$$E_V = \frac{dp}{d\rho/\rho} = \frac{RT_0 d\rho}{d\rho/\rho} = \rho RT_0 = p$$

and we can write the isothermal bulk modulus of a perfect gas in two ways:

$$E_V = \rho RT_0 \qquad (2.28a)$$

$$E_V = p \qquad (2.28b)$$

If a perfect gas undergoes an isentropic (frictionless and without heat transfer to the surroundings) process, then it can be shown that the isentropic bulk modulus is given by

$$E_V = \gamma p \qquad (2.29)$$

EXAMPLE 2.21

The pressure in the deepest part of the ocean is ~110 MPa. For seawater at this depth, use a constant bulk modulus to estimate the density increase over the sea level value.

SOLUTION

This exercise is solved by using Eq. 2.26, $dp = E_V(d\rho/\rho)$. Solving for the normalized density change $d\rho/\rho$ and integrating yields:

$$\int \frac{d\rho}{\rho} = \int \frac{dp}{E_V}$$

$$\ln\left(\frac{\rho_2}{\rho_1}\right) = \frac{p_2 - p_1}{E_V}$$

$$\frac{\rho_2}{\rho_1} = \exp\left(\frac{p_2 - p_1}{E_V}\right)$$

For seawater $E_V = 2.33$ GPa (Table 2.5), and since atmospheric pressure is about 0.1 MPa, we find:

$$\Delta p = 110 \text{ MPa} - 0.1 \text{ MPa} = 109.9 \text{ MPa} = 0.1099 \text{ GPa}$$

Substituting the appropriate values into the expression for the density change gives:

$$\frac{\rho_2}{\rho_1} = \exp\left(\frac{\Delta p}{E_V}\right) = \exp\left(\frac{0.1099 \text{ GPa}}{2.33 \text{ GPa}}\right) = 1.048$$

Thus, the density increases about 4.8% at this depth over its sea level value. Note that if we anticipated that the density change here is small, then we could do an approximate calculation by writing $dp = E_V(d\rho/\rho)$ in the form $p_2 - p_1 = E_V(\rho_2 - \rho_1)/\rho_1 = E_V[(\rho_2/\rho_1) - 1]$. Solving for the density ratio and inserting the data gives $\rho_2/\rho_1 = 1 + (p_2 - p_1)/E_V = 1 + (0.1099 \text{ GPa})/2.33 \text{ GPa} = 1.047$, which is essentially the same as the preceding answer.

2.6 BULK COMPRESSIBILITY MODULUS

A material with a large value of the bulk modulus, i.e., a liquid or solid, undergoes a negligible change in density when exposed to the highest pressures normally encountered in engineering. Such materials are incompressible. For example, a pressure increase of over 3100 psi is required to increase the density of water by 1%. Since liquids are virtually impossible to compress by using reasonable pressures, we will assume that a liquid has a constant density unless otherwise noted.

EXAMPLE 2.22

Suppose you have a compressed-air shock absorber on your car with dimensions shown in Figure 2.15. After filling your shocks to the recommended pressure of 80 psig, you load the car for a trip and observe that the shock absorber is compressed an additional 3 in. What force must be applied to the shock absorber to compress it this amount? Assume that the original temperature of the air in the shock absorber is 30°C.

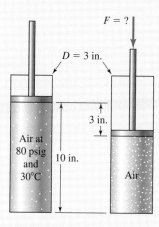

Figure 2.15 Cylindrical compressed-air shock absorber with a diameter of 3 in. In its unloaded state the air column is 10 in. long and the air is at a pressure of 80 psig and a temperature of 30°C. When the unknown load is applied, the length of the air column decreases by 3 in.

SOLUTION

We are to calculate the force applied to a shock absorber when a car is loaded. Figure 2.15 is a sketch of the situation. The air column in the shock absorber has a diameter of 3 in. and length of 10 in. at a pressure of 80 psig. After the load is applied, the length of the column of air is reduced by 3 in. This problem is solved using Eq. 2.26: $dp = E_V(d\rho/\rho)$. We will assume that the air in the shock may be approximated as a perfect gas undergoing an isothermal process (the compression occurs rather slowly as the car is loaded) and use Eq. 2.28b to define the isothermal bulk modulus as $E_V = p$. Combining the equations just listed, we get: $dp/p = d\rho/\rho$. Separating variables and integrating gives $p_2/p_1 = \rho_2/\rho_1$. Since the mass of gas in the shock absorber is fixed, for an air column of constant cross section, we have $\rho_2 L_2 = \rho_1 L_1$, and $\rho_2/\rho_1 = L_1/L_2$. Since $p_2/p_1 = \rho_2/\rho_1$, we get $p_2/p_1 = L_1/L_2$. Since the initial length is 10 in., and a reduction in length of 3 in. corresponds to a final length of 7 in., we know $p_2/p_1 = 10/7$. Since we are using the ideal gas law, we must use absolute pressure.

Substituting $p_1 = 80 \text{ psig} + 14.7 \text{ psi} = 94.7 \text{ psia}$ gives

$$p_2 = p_1 \left(\frac{L_1}{L_2}\right) = (94.7 \text{ psia})\left(\frac{10 \text{ in.}}{7 \text{ in.}}\right) = 135 \text{ psia} = 120 \text{ psig}$$

To calculate the increase in the corresponding force applied to the shock absorber, we recognize that $\Delta F = \Delta p A$. Since the pressure increased from 80 psig to 120 psig, we use $\Delta p = 40$ psi. The cross-sectional area of the air column is $A = \pi d^2/4 = \pi(3 \text{ in.})^2/4 = 7.07 \text{ in.}^2$. Thus, the additional force is found to be $\Delta F = \Delta p(A) = 40 \text{ psi } (7.07 \text{ in.}^2) = 280 \text{ lb}_f$.

Assuming four identical shocks on this car, the total load is 1120 lb$_f$, which is likely to be well above the manufacturer's recommendation for maximum load. Note that in this case of a slow loading, the isothermal model is appropriate. As the vehicle goes over a bump, the compression process is more likely to be isentropic.

2.6.1 Speed of Sound

Consider the stereo speaker shown in Figure 2.16. The speaker cone moves in response to the output signal from the amplifier, acting on the adjacent air like a moving piston without an enclosing cylinder. When the cone moves toward the air, the air is slightly compressed and is forced to move in the direction of travel of the cone. When the cone retreats from the air, the air is slightly expanded and follows the retreating cone. At the molecular level, air molecules colliding with the moving speaker surface experience an

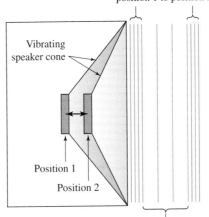

Figure 2.16 Vibrating stereo speaker cone. When the cone moves from position 1 to position 2, the air in front of the moving cone is slightly compressed as it is forced to move in the direction of travel of the cone. When the cone moves in the opposite direction, the air molecules adjacent to the cone follow its retreat and create a region of slightly expanded air to the right of the moving cone.

increase in their normal component of momentum. At the macroscopic level this is seen as a pressure increase at the speaker surface. Through molecular collisions, the increase in normal momentum propagates away from the speaker surface into the undisturbed air at a rate proportional to the mean speed of the random motion of the air molecules. This speed depends on the air temperature, since the average random kinetic energy of the molecules is proportional to the temperature. The overall effect is to cause a pressure wave, i.e., a sound wave to propagate away from the speaker at a fixed speed, expanding and compressing the air in its passage.

  Kinematics/Compressibility/Sound Waves

It takes time for a sound wave to propagate, or equivalently, for information to be transferred over a distance through a fluid. We see lightning before we hear thunder, and music sounds "unbalanced" if one sits much closer to one of the stereo speakers. In each case, pressure waves are traveling with a specific speed relative to the fluid itself, called the speed of sound, c. The speed of sound in a fluid is found to be proportional to the compressibility of the fluid as measured by the isentropic bulk modulus. The exact relationship can be shown to be

$$c = \left(\frac{dp}{d\rho}\right)^{1/2} = \left(\frac{E_V}{\rho}\right)^{1/2} \tag{2.30}$$

The sound speed has dimensions of $\{Lt^{-1}\}$ and units of meters or feet per second. Values for the speed of sound in several common fluids can be found in Appendix A.

EXAMPLE 2.23

Estimate the speed of sound in water and in mercury.

SOLUTION

This exercise is solved by using Eq. 2.30. We require bulk modulus values from Table 2.5 (for water $E_V = 2.19$ GPa and for mercury $E_V = 25.5$ GPa) and density values from Table 2.2 (for water $\rho = 1000$ kg/m^3 and for mercury $\rho = 13{,}590$ kg/m^3). Thus, the speed of sound in water is estimated as

$$c = \sqrt{\frac{E_V}{\rho}} = \sqrt{\frac{2.19 \text{ GPa}}{1000 \text{ kg/m}^3}} = \sqrt{\frac{2.19 \times 10^9 \text{ (N/m}^2)}{1000 \text{ (kg/m}^3)}} = 1480 \text{ m/s}$$

Similarly, the speed of sound in mercury is estimated as:

$$c = \sqrt{\frac{E_V}{\rho}} = \sqrt{\frac{25.5 \text{ GPa}}{13{,}590 \text{ kg/m}^3}} = \sqrt{\frac{25.5 \times 10^9 \text{ (N/m}^2)}{13{,}590 \text{ (kg/m}^3)}} = 1370 \text{ m/s}$$

What is the speed of sound in a perfectly incompressible material? Since an incompressible material has a bulk modulus approaching infinity, Eq. 2.30 shows that the sound speed is infinite. Are there actually any perfectly incompressible materials? Sound wave propagation in a liquid is clear evidence that all liquids are actually compressible fluids. Nevertheless, in many engineering applications we model the liquid as an incompressible fluid. Sound waves travel so rapidly in liquids (at about 1500 m/s) that the sound speed is effectively infinite. This means that all parts of a liquid begin to move almost immediately if a boundary is moved.

In gases, experiments show that the sound propagation process is isentropic, so it is appropriate to use the isentropic bulk modulus to compute the sound speed for a perfect gas. Using $E_V = \gamma p$ in Eq. 2.30, we have

$$c = \left(\frac{\gamma p}{\rho}\right)^{1/2} = \sqrt{\gamma RT} \qquad (2.31)$$

Remember the connection between temperature and the average speed of the random motion of gas molecules? According to Eq. 2.31, a disturbance in a gas propagates a bit more slowly than the average speed of the molecules. The speed of sound in air at room temperature is 343 m/s. Notice that this is slow compared with liquids, in which the sound speed is roughly 1500 m/s.

EXAMPLE 2.24

Estimate the distance to a thunderstorm if you see the lightning 8 s before you hear the thunder on a summer evening when the temperature is 20°C. How far would sound travel through water in the same time at the same temperature?

SOLUTION

We are asked to estimate the distance to a thunderstorm given the time lag between the sight of lightning and the sound of thunder. In addition, we are to calculate the distance sound would travel through water in the same time at the same temperature. No sketch is required to solve this problem. The temperature is given as 20°C and the time lag is 8 s. This problem is solved using Eqs. 2.30 for water and 2.31 for air. These equations are:

$$c = \sqrt{\frac{E_V}{\rho}} \quad \text{and} \quad c = \sqrt{\gamma RT}$$

Since velocity is defined as distance divided by time, the distance, L, traveled in air and water can be calculated as

$$L_{\text{air}} = ct = t\sqrt{\gamma RT} \quad \text{and} \quad L_{\text{water}} = ct = t\sqrt{\frac{E_V}{\rho}}$$

For air at 20°C (= 293 K), with $t = 8$ s, $\gamma = 1.4$, and $R = 287$ (N-m)/(kg-K), we find

$$L_{\text{air}} = t\sqrt{\gamma RT} = (8\,\text{s})\sqrt{1.4[(287\,\text{(N-m)/kg-K})] \times (293\,\text{K})}$$
$$= (8\,\text{s})(343.1\,\text{m/s}) = 2745\,\text{m}$$

2.6 BULK COMPRESSIBILITY MODULUS

Similarly, for water with $E_V = 2.19$ GPa (Table 2.5) and $\rho = 998.2$ kg/m³ (Appendix A), we find

$$L_{\text{water}} = ct = t\sqrt{\frac{E_V}{\rho}} = 8\text{ s}\sqrt{\frac{2.19 \times 10^9 \text{ N/m}^2}{998.2 \text{ kg/m}^3}}$$
$$= (8 \text{ s})(1481 \text{ m/s}) = 11{,}850 \text{ m}$$

The answers seem reasonable. Note that the speed of sound in water is 4.3 times that of the speed of sound in air under the conditions specified in the problem. You may be familiar with the rule of thumb that the distance to a thunderstorm can be crudely estimated in miles by dividing the time lag (in seconds) by 5 or in kilometers by dividing the time lag by 3.

The ratio of the speed of a moving fluid to the speed of sound in that fluid is known as the Mach number, M:

$$M = \frac{V}{c} \qquad (2.32)$$

A vehicle is also said to move at a certain Mach number, M. In that case the Mach number is calculated as the ratio of the vehicle speed to the speed of sound in the surrounding fluid.

EXAMPLE 2.25

What is the Mach number of an airplane flying at 500 mph at an altitude of 30,000 ft?

SOLUTION

First use Eq. 2.31, $c = \sqrt{\gamma RT}$, to find the speed of sound under these conditions, then use Eq. 2.32, $M = V/c$, to find the Mach number. The specific gas constant for air is given in the text (or Appendix A) as 287 (N-m)/(kg-K) and $\gamma = 1.4$ for air. The air temperature at 30,000 ft is found from Appendix B to be $-47.83°$F ($= 228.8$ K). Substituting these values into the equation for the speed of sound gives:

$$c = \sqrt{\gamma RT} = \sqrt{1.4[287 \text{ (N-m)/(kg-K)}](228.8 \text{ K})} = 303 \text{ m/s}$$

The unit conversion factor for m/s to mph is found in Appendix C to be 1 mph $=$ 0.44704 m/s. Therefore, the speed of sound under these conditions is $c = (303$ m/s$)$ (1 mph/0.44704 m/s) $= 678$ mph. Thus, the Mach number of this airplane is: $M = V/c = 500$ mph/678 mph $= 0.74$.

Density changes are very small in flows at low Mach numbers, defined as those for which Mach number is less than ~0.3. In flows of liquids, speeds are typically very low in comparison to the sound speed, so virtually all liquid flows are at a Mach number near zero. Thus density changes in a liquid flow are truly negligible, and a constant density model for a liquid is appropriate. A gas flow at low Mach number may also sometimes be modeled by constant density. For air at standard conditions, a Mach number of 0.3 corresponds to a velocity of about 100 m/s (~225 mph). Below this speed, pressure changes induced by the motion are very small and temperature changes are negligible. The resulting density change is also small enough to ignore, even though air is highly compressible. Thus in many flows, including those associated with surface vehicles, light aircraft, fans and blowers, and hurricanes, air behaves like an incompressible fluid. A low Mach number flow of gas is therefore referred to as an incompressible flow.

2.7 VISCOSITY

In our discussion in Chapter 1 of the basic characteristics of fluids and solids, the viscosity of a fluid, μ, was defined as the constant of proportionality between shear stress and the transverse velocity gradient. In fluid mechanics the viscosity μ is referred to as the absolute or dynamic viscosity, but a more descriptive name is shear viscosity.

Figure 2.17 shows a thin layer of fluid being sheared between two closely spaced parallel flat plates. This geometry models the lubricant-filled space between a piston and cylinder wall in an engine, in a journal bearing, or between your foot and the wet floor in your bathroom. Note the linear velocity profile within the fluid layer, and note also that at both fluid–solid interfaces the fluid velocity matches the velocity of the solid surface. The latter condition, which is virtually always satisfied, is known as the no-slip condition. The no-slip condition reflects the fact that there is a thin layer of fluid adsorbed (held by molecular forces) to a solid surface.

The first consequence of fluid viscosity is a resistance to shear in accordance with Newton's law of viscosity (Eq. 1.2c): $\tau = \mu(du/dy)$. The shear viscosity of a fluid is a strong function of temperature but only a weak function of pressure. The temperature dependence of viscosity differs for liquids and gases. Shear viscosity increases with temperature for gases but decreases with temperature for liquids.

To understand the temperature dependence of viscosity, we must consider the mechanisms by which momentum is transferred in fluids. In liquids the viscosity, or ability to transfer momentum, is a result of the intermolecular attractive forces between adjacent molecules. As the temperature increases, the strength of this cohesive force decreases, and the average separation distance between liquid molecules increases. The liquid transfers momentum less effectively, so viscosity decreases. The temperature

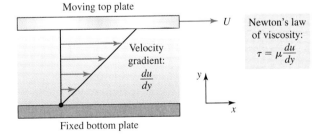

Figure 2.17 A fluid in shear between two parallel flat plates. The bottom plate is stationary and the top plate is moving to the right ($+x$ direction) with a velocity U. The velocity gradient, du/dy, results in a shear stress τ, as given by Newton's law of viscosity: $\tau = \mu(du/dy)$.

dependence of shear viscosity for a liquid is modeled by the exponential relation

$$\mu = Ae^{B/T} \quad (2.33a)$$

where A and B are constants for a given liquid.

In gases, molecules are not close enough together for intermolecular forces to be important. The mechanism responsible for momentum transfer is fundamentally different. Gas molecules are far more mobile than those in a liquid. Some of these energetic gas molecules move in a direction with a component perpendicular to the motion of the top plate. The molecules moving from the "fast" fluid layer to the "slower" moving layer below act to pull the "slower" layer forward. In contrast, the molecules moving from the

EXAMPLE 2.26

Calculate the temperature at which the shear viscosity of water is equal to 40% of its value at 20°C, given $\mu(20°C) = 0.001$ (N-s)/m^2 and $\mu(0°C) = 0.0018$ (N-s)/m^2.

SOLUTION

This exercise is solved by using Eq. 2.33a: $\mu = Ae^{B/T}$. We will use the information given to obtain a value for the constant B and then use that value to determine the temperature at which the viscosity of water is 40% of its room temperature value. The ratio of the viscosity of water at any two temperatures is given by:

$$\frac{\mu_1}{\mu_2} = \frac{A\exp(B/T_1)}{A\exp(B/T_2)} = \exp\left[B\left(\frac{1}{T_1} - \frac{1}{T_2}\right)\right]$$

Solving this expression for B and substituting the values given in the problem statement (with the subscript 1 referring to 20°C and the subscript 2 referring to 0°C) yields:

$$B = \frac{\ln(\mu_1/\mu_2)}{(1/T_1 - 1/T_2)}$$

$$= \frac{\ln(0.001[\text{N-s/m}]/0.0018\,[\text{N-s/m}])}{[(1/293.16\,\text{K}) - (1/273.16\,\text{K})]} = 2353\,\text{K}$$

Now we solve the viscosity ratio equation for the temperature T_2:

$$T_2 = \left[\frac{1}{T_1} - \frac{\ln(\mu_1/\mu_2)}{B}\right]^{-1}$$

If subscript 1 continues to refer to 20°C and subscript 2 now refers to the unknown temperature at which the viscosity 40% of its room temperature value (i.e., $\mu_2/\mu_1 = 0.4$), we find:

$$T_2 = \left[\frac{1}{293.16\,\text{K}} - \frac{\ln(1/0.4)}{2353\,\text{K}}\right]^{-1} = 330.94\,\text{K} = 57.78°\text{C}$$

"slower" layer to the "faster" layer exert a drag on the "faster" layer. Since a rise in temperature increases the random molecular motion in all directions (including the perpendicular direction), the shear viscosity of a gas increases with rising temperature. The appropriate form of the viscosity–temperature relationship is

$$\frac{\mu}{\mu_0} = \left(\frac{T}{T_0}\right)^n \tag{2.33b}$$

where μ_0 is the known viscosity at the reference temperature T_0 (often 273 K) and n is a constant for a specific gas (0.7 for air).

Suppose you are asked to predict the temperature of minimum shear viscosity for water in the temperature range 50–150°C. To do so, you must recognize that water undergoes a liquid–gas phase change in this temperature range. First consider the temperature dependence of viscosity for the liquid phase. As shown in Eq. 2.33a, viscosity decreases with increasing temperature in a liquid, so the minimum viscosity for the liquid phase occurs at a temperature just below the boiling point. In contrast, by Eq. 2.33b, the viscosity of a gas increases with increasing temperature, so the minimum viscosity for the gas phase occurs just above the boiling temperature. The viscosity of steam is always less than that of water, so the minimum viscosity for H_2O over the range 50–150°C must occur in the vapor phase just above the boiling temperature.

A different way to express the shear viscosity is to divide it by the density. This normalized form, called the kinematic viscosity, ν, is defined as:

$$\nu = \mu/\rho \tag{2.34}$$

The dimensions of kinematic viscosity are $\{L^2 t^{-1}\}$ and it is expressed in units of m^2/s, or ft^2/s.

EXAMPLE 2.27

Calculate the temperature at which the shear viscosity of air is equal to twice its value at 20°C, given that $\mu(20°C) = 0.0001813$ (N-s)/m^2 and $n = 0.7$.

SOLUTION

Solving Eq. 2.33b for temperature we get

$$T = T_0 \left(\frac{\mu}{\mu_0}\right)^{1/n}$$

Substituting the data provided in the problem statement gives us

$$T = T_0 \left(\frac{\mu}{\mu_0}\right)^{1/n} = (293\text{ K})(2)^{1/0.7} = 789\text{ K} = 516°C$$

All fluids, with the curious exception of liquid helium in the temperature range at which it exhibits superfluidity, exhibit shear viscosity. An important historic approximation in fluid mechanics treats a fluid as if it were inviscid, meaning that its shear viscosity is zero. An inviscid, constant density fluid is also referred to as an ideal fluid. A closely related approximation, called inviscid flow, treats a fluid in motion as if it were free of all shear stress. We mention these approximations here because there are many exact solutions known for inviscid flow, and we will be examining a number of these in later chapters.

The kinematic viscosity often occurs in modeling the dynamics of fluid flow. Notice that this is true in Eq. 2.36, which defines the rate of viscous dissipation of energy in flow between parallel plates.

In keeping with our belief in exposure to different unit systems, it is necessary to pause here and note that engineers frequently encounter viscosity values in handbooks with cgs units. In this system the shear viscosity μ has a unit called poise P, defined as

$$1\,P = 1\,\text{dyn-s/cm}^2 \qquad (2.35a)$$

The corresponding unit of kinematic viscosity ν is the Stoke (St) defined as

$$1\,St = 1\,\text{cm}^2/\text{s} \qquad (2.35b)$$

Kinematic viscosity data for various fluid can be found in Appendix A.

2.7.1 Viscous Dissipation

An important consequence of the existence of shear viscosity is a loss of energy when fluid is sheared. This frictional energy loss is referred to as viscous dissipation. The general action of viscosity in a fluid flow is a tendency to convert the useful energy content of the fluid into heat. The useful energy lost appears as an increase in the internal energy of the fluid, corresponding to a rise in temperature.

The rate of dissipation of energy per unit mass of fluid by the shear viscosity is given by the viscous dissipation, Φ. Although the general formula for the viscous dissipation will be given later, it is instructive to note that for the specific situation depicted in Figure 2.17, the viscous dissipation rate at any point in the flow is given by:

$$\Phi = 2\frac{\mu}{\rho}\left(\frac{du}{dy}\right)^2 \qquad (2.36)$$

Viscous dissipation has the dimension $\{L^2 t^{-3}\}$ and is usually expressed in units of power per unit mass, i.e., J/(s-kg) or W/kg in SI, or Btu/(lb$_m$-s) or hp/lb$_m$ in EE. Equation 2.36 can be used to estimate the work required to shear a fluid in this physical arrangement, since all the work done on the fluid by the moving plate is dissipated by viscosity.

2.7.2 Bulk Viscosity

Thus far our discussion has been about shear viscosity. When fluids with a complex molecular structure undergo compression or expansion, they may exhibit a second, quite different type of viscosity. This second viscous property of a fluid, called the bulk or expansion viscosity, is represented by κ. Like shear viscosity, expansion viscosity has dimensions of $\{FtL^{-2}\}$ and units of (N-s)/m^2 [= kg/(m-s)] or (lb$_f$-s)/ft^2.

Just as the shear viscosity of a fluid causes an irreversible conversion of mechanical energy into heat when a fluid is sheared, the bulk viscosity of a fluid is responsible for a viscous loss of energy when certain fluids are compressed or expanded very rapidly. The

EXAMPLE 2.28

The planar configuration in Figure 2.17 has been used to model the flow of liquid lubricants in cylindrical journal bearings (see Figure 2.18). Osborne Reynolds first made the simplifying assumption that the fluid film was so thin in comparison to the bearing radius that the curvature could be neglected. What is the viscous dissipation in the lubricant oil SAE 30W between the journal bearing and the rotating shaft shown in Figure 2.18? The operating temperature is 40°C, at which $\mu = 0.075$ (N-s)/m².

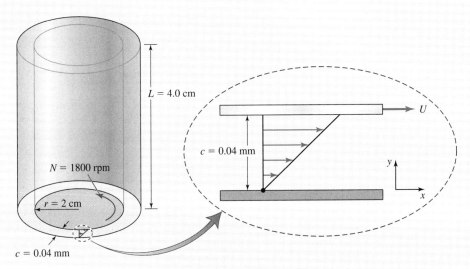

Figure 2.18 Schematic illustration of a journal bearing. Notice the enlarged drawing of the region between the inner rotating shaft and the stationary outer surface. Since the thickness of the fluid layer is small compared with the shaft radius, the curvature can be neglected and we make use of the parallel plate model developed earlier.

SOLUTION

We are asked to find the viscous dissipation of the lubricant SAE 30W oil in the journal bearing. The enlarged sketch of critical region in Figure 2.18 shows the x-y coordinate system of the planar model. The bottom plane represents the journal bearing and has zero velocity. The shaft surface is the upper plane moving at velocity $u = 2\pi r N$. We are given that the shaft is rotating at $N = 1800$ rpm, the radius and length of the bearing are $r = 2.0$ cm, and $L = 4.0$ cm, respectively, and the gap between the bearing and the shaft is $c = 0.04$ mm.

To solve this problem, use Eq. 2.36, $\Phi = 2(\mu/\rho)(du/dy)^2$. The key assumption is that curvature can be neglected (see inset, Figure 2.18). The velocity distribution of the oil in the gap is $u = (2\pi r N/c)y$, so the velocity gradient is constant and equal to $du/dy = 2\pi r N/c$ everywhere in the space. In this problem we will determine the viscous dissipation per unit volume, which is simply the viscous dissipation per unit mass given in Eq. 2.36 multiplied by ρ:

$$\rho\Phi = 2\mu\left(\frac{du}{dy}\right)^2 = 2\mu\left(\frac{2\pi r N}{c}\right)^2$$

Because the dissipation rate is uniform, the total dissipation is the product of the dissipation rate per unit volume and the volume $\mathcal{V}$,

$$\rho\Phi\mathcal{V} = 2\mu\left(\frac{2\pi r N}{c}\right)^2 \mathcal{V} = 2\mu\left(\frac{2\pi r N}{c}\right)^2 L[\pi(r+c)^2 - \pi r^2]$$

Substituting numerical values into the foregoing result and noting that the viscosity of SAE 30W oil is $\mu(40°C) = 0.075$ (N-s)/m² gives:

$$\rho\Phi\mathcal{V} = 2[0.075 \text{ kg/(m-s)}]\left[\frac{2\pi(2.0 \text{ cm})(1800 \text{ rpm})(\text{min/60 s})}{(0.04 \text{ mm})(\text{cm/10 mm})}\right]^2$$

$$\times (4.0 \text{ cm})[\pi(2.0 + 0.004)^2 - \pi(2.0 \text{ cm})^2](10^{-6} \text{ m}^3/\text{cm}^3)$$

$$\rho\Phi V = 268 \text{ (kg-m}^2)/\text{s}^3 = 268 \text{ (N-m)/s} = 268 \text{ W}\left(\frac{1 \text{ hp}}{745.7 \text{ W}}\right) = 0.36 \text{ hp}$$

Note that we have used the unit conversion factor 1 hp = 745.7 W and recognize that the result is in units of power, as expected. A shaft of this size is typical for a 60 hp electric motor, which has losses of about 3 hp. Bearing losses of 10% of this total are reasonable. While this estimate of the dissipation in the bearing is reasonable, an engineer would normally apply lubrication theory, a branch of fluid dynamics concerned with this type of problem to calculate the flow in the bearing.

rate of dissipation of energy per unit mass due to the bulk viscosity is given by Φ_κ:

$$\Phi_\kappa = \frac{\kappa}{\rho^3}\left(\frac{d\rho}{dt}\right)^2 \tag{2.37}$$

where $d\rho/dt$ is the time rate of change of density occurring in the fluid. Large values of $d\rho/dt$ are necessary for dissipation by the bulk viscosity to be significant. For example, in gases this occurs in the interior of a shock wave. Since large density changes are difficult to create in a liquid, the only easy way to achieve a large value of $d\rho/dt$ in a liquid is with ultrasound, an acoustic process at frequencies in the megahertz range (i.e., by making dt small). Practical devices in which a large value of $d\rho/dt$ may occur include ultrasonic cleaners for jewelry, sonic disrupters used in molecular biology to rupture cell walls, and ultrasound imaging devices in medicine.

2.8 SURFACE TENSION

Perhaps you have wondered why a sponge soaks up water so readily, whereas the treated fabric of a raincoat is impervious to water. Water forms a thin film on a clean windshield, but it beads on the same surface after the use of a windshield treatment. Water also beads

Figure 2.19 A liquid–gas interface. The liquid molecules (represented by solid colored spheres) are packed in a semiorderly fashion (atoms in a crystalline solid are packed in a highly ordered fashion). The molecules in the interior of the liquid, including molecule I, generally have six nearest molecular neighbors in the plane of the paper. In contrast, liquid molecules at the liquid surface, such as molecule S have only four nearest molecular neighbors in the plane of the paper. As discussed in the text, this difference in number of nearest neighbors result in a surface tension or equivalently a surface energy at any fluid interface.

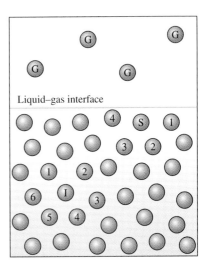

on a waxed or oily automobile surface, but when a little detergent is added, it will spread into a thin film. A jet of water tends to break up into a spray of small spherical droplets. Shake a bottle of Italian dressing, and the oil is dispersed into fine drops suspended in the water and vinegar. Liquids may either advance up a small tube against the force of gravity or stubbornly refuse to enter the tube at all. These are all examples of surface tension phenomena. What is responsible at the molecular level for the variety of behaviors associated with the presence of an interface between two fluids?

As shown in Figure 2.19, molecules below the surface of a liquid have a characteristic number of nearest neighbors. The total energy of the fluid system is minimized when each molecule has the "correct" number of neighbors (determined by the bonding characteristics of the atoms in the fluid). Molecules at the liquid surface, however, have a different number of nearest neighbors. There are two ways to interpret this observation. First, since the surface molecules have the "wrong" number of neighbors, they will be at a higher energy state. The excess energy associated with the molecules at the surface is known as surface energy. Surface energy, which is represented by the symbol σ, has dimensions of energy per unit area, $\{FL/L^2\}$, or equivalently, $\{F/L\}$, and is expressed in units of ergs/cm^2, J/m^2, or (ft-lb$_f$)/ft^2.

A different but equivalent model for a liquid–vapor interface recognizes that since the surface molecules of liquid do not have identical molecules above them, they will be more strongly attracted to their neighbors below and in the plane of the interface. The result is a layer of surface molecules that at the macroscopic level behaves as an elastic membrane. The corresponding net force on a molecule in the interface acts in the plane of the surface in all directions and is referred to as the surface tension (see Figure 2.20). Surface tension has dimensions of force/length, which is dimensionally equivalent to surface energy. The units for surface tension are N/m, dyn/cm, or lb$_f$/ft. We will use the concept of surface tension in the remainder of this section, but note that both approaches (surface tension and energy) are valid ways to investigate interfacial phenomena.

2.8 SURFACE TENSION

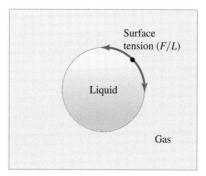

Figure 2.20 Schematic illustration of a spherical liquid bubble surrounded by a gas. The surface tension acts on a liquid molecule located at the interface in the plane of the liquid–gas interface in all directions. In this two-dimensional view, the surface tension acts on the colored "molecule" in the direction of the indicated arrows.

TABLE 2.6 Surface Tension Values for Various Fluid Systems at Room Temperature

Liquid	Surface Tension, σ (N/m)	
	When in Contact with Air	When in Contact with Water
Benzene	0.029	0.035
Carbon tetrachloride	0.027	0.045
Glycerin	0.063	—
Hexane	0.018	0.051
Mercury	0.484	0.375
Methanol	0.023	0.023
Octane	0.022	0.051
Water	0.073	—

The magnitude of the surface tension is a function of the fluids on both sides of the interface. Thus, the surface tension at an air–water interface is different from that at an oil–water interface. The influence of both fluids on the surface tension is a reflection of the fact that atoms in the second fluid are serving as substitute nearest neighbors for the surface atoms in the first fluid and vice versa. Surface tension values for several fluid systems are given in Table 2.6.

Since it is the near neighbors of a molecule at an interface that affect the resulting surface tension, foreign molecules adsorbed onto the interface will change the surface tension. For example, soaps and detergents reduce the surface tension of an air–water interface and thus allow the water to spread out on a surface.

2.8.1 Pressure Jump Across a Curved Interface

To see how the concept of surface tension is used to analyze a practical problem, consider a small spherical drop of one fluid at rest in another, such as a tiny fuel droplet in the vicinity of a fuel injection nozzle. Since the drop is small, we can neglect the slight hydrostatic pressure variation with height and assume a uniform but different pressure

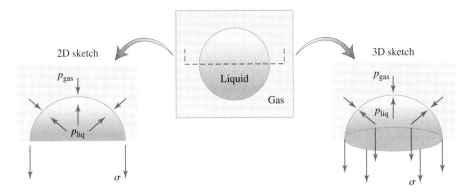

Figure 2.21 Schematic illustration of the vertical force balance on a hemispherical portion of a liquid bubble suspended in gas. The sum of the surface tension, σ, and the uniform pressure, p, acting on the outside of the bubble must be balanced by the higher pressure, $p + \Delta p$, acting on the inside of the bubble.

inside and outside the drop. Imagine cutting the drop in half as shown in Figure 2.21 and performing a force balance in the vertical direction on the hemispherical interface. If the drop is at rest, then Newton's second law tells us that the sum of the forces acting on the interface in any direction must be zero. The only forces acting on the interface are those due to pressure and surface tension. The interface itself is infinitely thin and massless. Equilibrium requires that the net force up due to pressure inside, plus the net force down due to pressure outside, plus the surface tension force pulling down on the edge of the hemispherical interface, add to zero.

The net action of a uniform pressure inside a hemisphere is as if the same pressure acted on the equatorial plane of area πr^2. Surface tension acts on the circumference of the hemisphere, $2\pi r$. If we let Δp be the pressure difference, inside minus outside, then the force equilibrium condition just stated becomes

$$\Delta p(\pi r^2) = \sigma(2\pi r)$$

Solving for the pressure difference (jump) across the interface yields

$$\Delta p = \frac{2\sigma}{r} \qquad (2.38a)$$

This result shows that surface tension causes the pressure inside a drop or bubble to be greater than the pressure outside. Smaller drops and bubbles experience larger pressure differences, and for a given drop or bubble size, a higher surface tension value produces a larger pressure difference.

Perhaps you have enjoyed using a wand and a bubble solution to create soap bubbles. A soap bubble has air inside and out, separated by a very thin film of soap

Equation 2.38a is a special case of a more general result for the difference in pressure inside and outside a point on a curved interface:

$$P_{in} - P_{out} = \sigma \left(\frac{1}{R_1} + \frac{1}{R_2} \right) \qquad (2.38b)$$

where R_1 and R_2 are the two orthogonal radii of curvature at the point on the interface, and a radius of curvature is given a positive sign if the center of curvature lies on the inside and a negative sign if the center of curvature is outside. Do you see that when applied to spherical interface, the general result Eq. 2.38b simplifies to 2.38a?

EXAMPLE 2.29

Suppose the small air bubbles in a glass of tap water may be on the order of 50 μm in diameter. What is the pressure inside these bubbles?

SOLUTION

Surface tension causes the pressure inside a bubble to be higher than that outside. Using Eq. 2.38a and the data in Table 2.6, we can calculate a pressure difference of about:

$$\Delta p = \frac{2\sigma}{r} = \frac{2(0.073 \text{ N/m})}{25 \times 10^{-6} \text{ m}} = 5840 \text{ Pa}$$

Thus, the pressure inside the bubble is about 6% higher than the atmospheric pressure outside.

---

solution having a surface tension σ. Can you explain why the pressure difference across a spherical soap bubble is

$$\Delta p = \frac{4\sigma}{r} \tag{2.39}$$

that is, twice the value calculated for a fluid drop under the same conditions? [*Hint:* How many liquid–gas interfaces are present in this case?] The formula suggests that the pressure inside large soap bubbles must not be very different from that outside. The tension in the skin of large soap bubbles is so small that they are easily deformed by the slightest breeze.

EXAMPLE 2.30

Estimate the pressure difference in a 1 m diameter soap bubble in air. Use a surface tension value of half that of an air–water interface in your calculation.

SOLUTION

We use Eq. 2.39, and the value of 0.073 N/m for the air–water surface tension found in Table 2.6 to find

$$\Delta p = \frac{4\sigma}{r} = \frac{4(0.5)(0.073 \text{ N/m})}{0.5 \text{ m}} = 0.292 \text{ Pa}$$

Gasoline cannot form a drop on water in the presence of air. If a drop of gasoline falls on water, surface tension acts to pull the contact line away from the initial body of gasoline until the gasoline forms a thin film. This also occurs with oils, although the process is much slower. An extraordinarily thin gasoline or oil film on water is responsible for the refraction of light into a rainbow of colors. It can be shown that the film is only a few molecules thick.

2.8.2 Contact Angle and Wetting

When a liquid contacts a solid surface, the line at which liquid, gas, and solid meet is called the contact line (see Figure 2.22A). The effect of surface tension is evident in the contact angle θ_c, defined to be the angle in the liquid between the solid surface and the interface at the contact line. The net surface tension acting on a contact line depends on all three materials—liquid, gas, and solid. A force balance on the contact line shows that

$$\sigma_{SG} - \sigma_{SL} = \sigma \cos(\theta_c) \qquad (2.40)$$

where σ_{SG} is the surface tension of the gas–solid interface, σ_{SL} is the surface tension of the solid–liquid interface, and σ is the surface tension of the gas–liquid interface. Experimental observations show that the contact angle for an air–water–glass interface is $\sim 0°$, while the contact angle for a air–mercury–glass interface is $\sim 140°$. If the contact angle is less than $90°$ (Figure 2.22B), the surface is said to be wetted by the liquid. Perfect wetting occurs if the contact angle is $\sim 0°$. If the contact angle is greater than $90°$ (Figure 2.22C), the surface is not wetted by the liquid.

2.8.3 Capillary Action

Now consider what happens if a thin glass tube is inserted into a liquid. If the liquid wets the glass, it will enter the tube by capillary action. If the liquid does not wet the glass, it will be prevented from entering the tube. Both effects are due to surface tension. Let us analyze the case in which a liquid wets a round tube with a contact angle θ_c as shown in

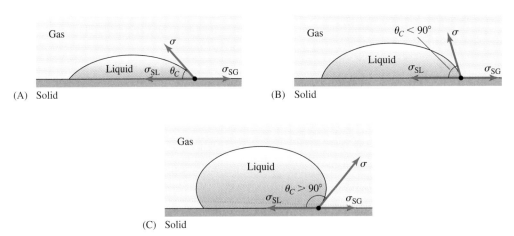

Figure 2.22 Schematic diagrams of a liquid in contact with a solid and a gas. (A) The relevant force balance at the contact line. (B) Example of a liquid wetting a solid as defined by a contact angle $\theta_c < 90°$. (C) In contrast, the liquid does not wet the solid, since $\theta_c > 90°$. The air–water–glass system forms a contact angle of $\sim 0°$ so that water wets glass. In contrast, the air–mercury–glass system forms a contact angle of $\sim 140°$ so that Hg does not wet the glass.

2.8 SURFACE TENSION

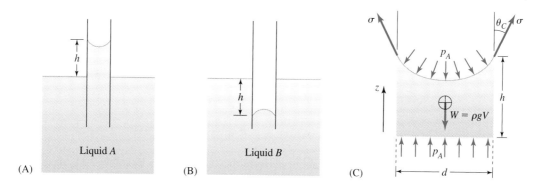

Figure 2.23 The capillary action in a solid tube depends on the contact angle associated with the corresponding gas–liquid–solid system. (A) When the liquid wets the solid ($\theta_c < 90°$), the liquid level within the tube will be above the liquid–gas interface outside the tube. (B) When the liquid does not wet the solid ($\theta_c > 90°$), the liquid level within the tube will be below the general liquid–gas interface. (C) Enlarged view of the case in (A) including the terms associated with the force balance on the liquid within the capillary tube. See text for discussion of this force balance.

Figure 2.23C. The liquid will be drawn up a tube of diameter d to a height h. We will perform a force balance in the z direction on the column of liquid shown in Figure 2.23C, taking into account surface tension, gravity, and pressure. The column is at rest, so the sum of all forces acting on the fluid column is zero. The pressure acting on the top of the meniscus formed in the tube is atmospheric. The pressure acting at the bottom of the liquid column inside the tube is also atmospheric, because lines of constant pressure in a stationary fluid in a gravity field are horizontal, and the tube is open. We are neglecting the tiny change in atmospheric pressure over the height of the liquid column.

Since atmospheric pressure acts over an area equal to the cross section of the tube at each end, the net effect of pressure on the liquid column is zero. The surface tension

EXAMPLE 2.31

Calculate the height of capillary rise for water in a glass tube of diameter 1 mm.

SOLUTION

This exercise can be solved using Eq. 2.41. The air–water surface tension is found from Table 2.6 to be 0.073 N/m. Earlier we noted that the contact angle for the air–water–glass system is $\sim 0°$. The glass tube has $d = 0.001$ m and the density of water is ~ 1000 kg/m³. Using the known value of 9.81 m/s² for the gravitational constant gives

$$h = \frac{4\sigma \cos \theta_c}{d\rho g} = \frac{4(0.073 \text{ N/m})(\cos(0°))}{(0.001 \text{ m})(1000 \text{ kg/m}^3)(9.81 \text{ m/s}^2)} = 0.0298 \text{ m} = 2.98 \text{ cm}$$

force acting up on the contact line, plus the force of gravity on the liquid column acting down must therefore add to zero. From the geometry at the contact line we find

$$\sigma \pi d \cos \theta_c - \pi \frac{d^2}{4} \rho g h = 0$$

Solving for the height of capillary rise, we have

$$h = \frac{4\sigma \cos \theta_c}{d\rho g} \quad (2.41)$$

Suppose we attempt to perform the calculation in Example 2.31 for the case of a glass tube ($d = 1$ mm) in liquid mercury. What happens? The contact angle for air–mercury–glass is given in the text as $\sim 140°$, the value of σ is 0.484 N/m (Table 2.6), and the density of Hg is 13,550 kg/m³ (Appendix A). Substituting these values into Eq. 2.41 yields $h = -1.12$ cm. The physical interpretation of this result is shown in Figure 2.23B. You may be wondering if Eq. 2.41 is in fact valid for $\theta_c > 90°$. The short answer is yes, although the details of the force balance leading to the result are slightly different from those employed for the $\theta_c < 90°$ case.

Capillary action is often cited as causing the movement of fluid from the roots of a plant to its crown where the leaves are present. In Sequoia redwood trees, the capillary rise would have to exceed 200 ft. This seems unlikely. Alternate theories argue that the minute water columns in the trees' tissues are in a state of tension (otherwise normally prohibited) caused by evaporation at the leaf surface.

Capillary action also explains the tendency of liquids to penetrate cracks even when there is no differential pressure acting to drive the liquid into the cracks. A two-dimensional model for the rise of a liquid in a crack of width w is shown in Figure 2.24. A force balance indicates that the height of the rise of a liquid in a crack which it wets is given by

$$h = \frac{2\sigma \cos \theta_c}{w \rho g} \quad (2.42)$$

Notice that the height to which a liquid rises in a round tube of diameter d is twice the height to which the same liquid rises in a crack of width $w = d$.

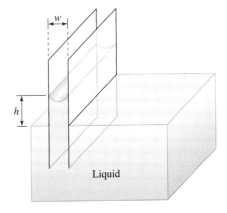

Figure 2.24 Schematic diagram of the rise of a liquid into a crack due to capillary action. The height of the rise is given by h.

Try using a garden hose to force water through the fabric of your raincoat, or a piece of treated tent material. It does not take much pressure differential to defeat the effects of a surface tension treatment.

For a crack or tube inclined at an angle ϕ from vertical, we replace g by $g\cos\phi$ in the appropriate formula. Thus, as ϕ approaches $\pi/2$, meaning the crack or tube is horizontal, capillary action causes a liquid to penetrate deeply. If we wish to prevent a liquid from entering a crack, and there is minimal differential pressure acting to drive the liquid through the crack, we can employ surface tension forces to exclude the liquid. This is accomplished by applying a surface treatment to ensure that the liquid does not wet the solid. Fabric waterproofing treatments coat the fabric surface with a thin layer of material that water does not wet. Unfortunately this approach will not ensure a dry basement, because water in saturated soil is under several feet of hydrostatic pressure in comparison to the air on the other side of the (cracked) wall. This pressure differential overcomes the surface tension force which might resist the advance of water in larger cracks.

2.9 FLUID ENERGY

A fluid possesses energy in various forms. When applied to a fluid, the first law of thermodynamics relates the change in the internal, kinetic, and potential energies of a mass of fluid to the work done on that fluid plus the heat added to the fluid. Changes in the energy content of a fluid are important in many applications. In some applications a fluid does work (e.g., turbines, windmills, waterwheels), in other applications work is done on the fluid (e.g., pumps, fans, compressors). In either case it is important to remember that viscous effects result in an irreversible increase in the internal energy of a fluid at the expense of a decrease in the kinetic and potential energy. While a complete discussion of fluid energy must await the development of the energy balance in Chapter 7, we think it is helpful for you to begin to learn about the types of fluid energy here.

2.9.1 Internal Energy

The internal energy U of a mass M of fluid is a macroscopic measure of microscopic (molecular, atomic, and subatomic) energy content. The internal energy of a fluid is primarily a function of its temperature. In fluid mechanics we are interested in the change in internal energy rather than its absolute value. For applications involving liquids with no external heat transfer, the internal energy change is negligible and normally neglected. For a gas it is appropriate to employ the perfect gas law. We can use Eq. 2.21a to relate the change in internal energy in a mass M of gas to the change in temperature by

$$U_2 - U_1 = Mc_V(T_2 - T_1) \tag{2.43a}$$

The change in internal energy per unit mass is

$$u_2 - u_1 = c_V(T_2 - T_1) \tag{2.43b}$$

which may also be written on a per-unit-volume basis as

$$\rho(u_2 - u_1) = \rho c_V(T_2 - T_1) \tag{2.43c}$$

Internal energy has dimensions of $\{ML^2t^{-2}\}$ or equivalently $\{FL\}$. The dimensions of internal energy are the same as those for any other energy quantity, but each type of energy tends to be expressed in a traditionally preferred unit. The preferred units for internal energy are Btu or joule, while u is most often expressed in Btu/lb$_m$ or J/kg.

2.9.2 Kinetic Energy

The energy associated with fluid in motion is called kinetic energy, E_K. This energy is proportional to the mass of fluid in the system, and to the square of the fluid speed, V. For a mass of fluid M, the total kinetic energy is given by

$$E_K = \tfrac{1}{2}MV^2 \qquad (2.44\text{a})$$

The corresponding kinetic energy per unit mass is

$$e_K = \tfrac{1}{2}V^2 \qquad (2.44\text{b})$$

while kinetic energy per unit volume is given by:

$$\rho e_K = \tfrac{1}{2}\rho V^2 \qquad (2.44\text{c})$$

EXAMPLE 2.32

Water flows from a garden hose at 10 ft/s. What is the kinetic energy content of the water per unit mass? What is the kinetic energy content per unit volume?

SOLUTION

This exercise can be solved by using Eqs. 2.44b and 2.44c. Water has a density of ~ 62.4 lb$_m$/ft^3, and the speed of the water is 10 ft/s. On a per-unit-mass basis,

$$e_K = \tfrac{1}{2}V^2 = \tfrac{1}{2}(10\text{ ft/s})^2 = 50\text{ ft}^2/\text{s}^2$$

Using g_c in the form of a unit conversion factor gives

$$e_K = \tfrac{1}{2}V^2 = 50\text{ ft}^2/\text{s}^2 \left(\frac{1\text{ lb}_f\text{-s}^2}{32.2\text{ lb}_m\text{-ft}}\right) = 1.55\text{ (ft-lb}_f)/\text{lb}_m$$

On a per-unit-volume basis,

$$\rho e_K = \tfrac{1}{2}\rho V^2 = \tfrac{1}{2}(62.4\text{ lb}_m/\text{ft}^3)(10\text{ ft/s})^2$$

$$= (3120\text{ lb}_m/(\text{ft-s}^2))\left(\frac{1\text{ lb}_f\text{-s}^2}{32.2\text{ lb}_m\text{-ft}}\right) = 96.9\text{ (ft-lb}_f)/\text{ft}^3$$

Kinetic energy has the usual energy dimensions of $\{ML^2t^{-2}\}$ or equivalently $\{FL\}$. The preferred units of extensive kinetic energy are mechanical: $(kg\text{-}m^2)/s^2 = N\text{-}m$ or $(lb_m\text{-}ft^2)/s^2 = lb_f\text{-}ft$, with appropriately normalized values for the two intensive forms.

2.9.3 Potential Energy

A change in the gravitational potential energy of a fluid occurs whenever the fluid moves with, or against, the force of gravity. Suppose we chose a coordinate system with the z axis vertical. Then the gravitational potential energy, E_G, of a small volume of fluid at height z, relative to the potential energy the volume of fluid has at the origin, is given by

$$E_G = Mgz \quad (2.45a)$$

The potential energy per unit mass is given by

$$e_G = gz \quad (2.45b)$$

and the potential energy per unit volume is

$$\rho e_G = \rho gz \quad (2.45c)$$

The dimensions and preferred units for gravitational potential energy are identical to those for kinetic energy.

A second form of potential energy is related to the ability of fluid at high pressure to perform mechanical work. The pressure potential energy E_P of a mass M of fluid is given by

$$E_P = M\frac{p}{\rho} = p\forall \quad (2.46a)$$

where $\forall$ is the volume of the fluid, and pressure is measured with respect to the ambient pressure at which spent fluid is exhausted. The per-unit-mass form of this potential energy is

$$e_P = \frac{p}{\rho} \quad (2.46b)$$

with the per-unit-volume counterpart being

$$\rho e_P = p \quad (2.46c)$$

The dimensions and units for pressure potential energy are the same as those for gravitational potential energy.

2.9.4 Total Energy

The total energy E of a mass of fluid M can be represented as the sum of its internal, kinetic, and two types of potential energy as

$$E = U + \tfrac{1}{2}MV^2 + Mgz + M\frac{p}{\rho} \quad (2.47a)$$

In thermodynamics, the enthalpy h, is defined as the sum $h = u + pv$. In fluid mechanics this is expressed in terms of density as:

$$h = u + \frac{p}{\rho} \quad (2.48)$$

Strictly speaking, enthalpy is not a form of fluid energy in the sense we have been using. It does, however, have the dimensions and units of energy, i.e., Btu or joules. For a perfect gas, we can use Eq. 2.21b to write the change enthalpy in terms of the change in temperature, as $(h_2 - h_1) = c_p(T_2 - T_1)$.

In engineering applications it is the change in total energy that is important rather than an absolute value. On a per-unit-mass basis, total energy is given by

$$e = u + \frac{1}{2}V^2 + gz + \frac{p}{\rho} \quad (2.47b)$$

and on a per-unit-volume basis we have

$$\rho e = \rho\left(u + \frac{1}{2}V^2 + gz + \frac{p}{\rho}\right) \quad (2.47c)$$

The sum $\frac{1}{2}V^2 + gz + p/\rho$ is usually referred to as the mechanical energy per unit mass, and it is this energy that is turned into internal energy by the action of viscosity and lost.

EXAMPLE 2.33

An airplane is flying at an altitude of 10,000 ft at a speed of 620 mph (909 ft/s). The ambient conditions are $T = 430°R$ and $p = 11.8$ psi. Air exits the engine at 2070 mph (3040 ft/s) relative to the plane with $T = 1025°R$. Calculate the total energy change per unit mass for the air moving through this jet engine.

SOLUTION

We will use Eq. 2.47b to determine the total energy change, calculating the energy content of the air relative to the engine. As shown in Figure 2.25, air approaches the engine at 909 ft/s, and leaves the engine at 3040 ft/s. The energy change is calculated from Eq. 2.47b as

$$e_2 - e_1 = (u_2 - u_1) + \left(\tfrac{1}{2}V_2^2 - \tfrac{1}{2}V_1^2\right) + g(z_2 - z_1) + \left(\frac{p_2}{\rho_2} - \frac{p_1}{\rho_1}\right)$$

Next we assume the inlet and exit of the engine are at the same elevation (i.e., $\Delta z = 0$) and introduce enthalpy by using Eq. 2.48 to obtain

$$e_2 - e_1 = (h_2 - h_1) + \left(\tfrac{1}{2}V_2^2 - \tfrac{1}{2}V_1^2\right)$$

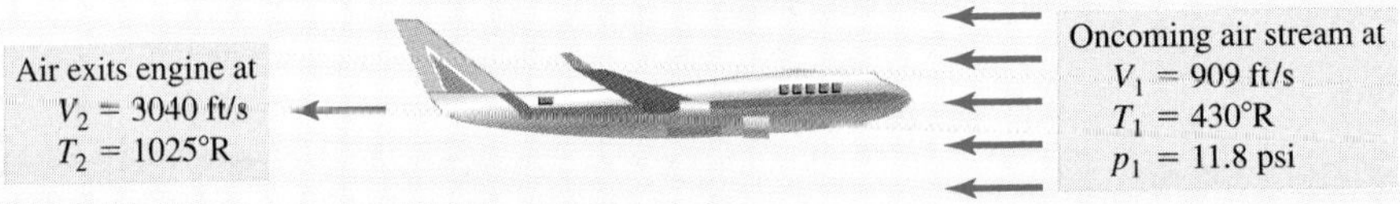

Air exits engine at
$V_2 = 3040$ ft/s
$T_2 = 1025°R$

Oncoming air stream at
$V_1 = 909$ ft/s
$T_1 = 430°R$
$p_1 = 11.8$ psi

Figure 2.25 Schematic illustration of an airplane flying at a constant velocity of 909 ft/s (620 mph) at an altitude of 10,000 ft ($T = 430°R$, $p = 1.8$ psi). The air leaves the jet engine with $V = 3040$ ft/s (2070 mph) and $T = 1025°R$. Notice that we have selected a reference frame attached to the moving plane.

Finally we assume air behaves as a perfect gas and use Eq. 2.21b $(h_2 - h_1) = c_P(T_2 - T_1)$ to relate enthalpy to temperature:

$$e_2 - e_1 = c_P(T_2 - T_1) + \tfrac{1}{2}(V_2^2 - V_1^2)$$

Using values from Table 2.4 for air, we find that the change in total energy of the air passing through the engine is

$$e_2 - e_1 = [6{,}006 \text{ (ft-lb}_\text{f})/(\text{slug-}°\text{R})](1025°\text{R} - 430°\text{R})$$
$$+ \frac{1}{2}[(3040^2 - 909^2) \text{ ft}^2/\text{s}^2]\left[\frac{1 \text{ lb}_\text{f}}{1 \text{ (slug-ft)/s}^2}\right]$$
$$= (3.57 \times 10^6 + 4.21 \times 10^6) \text{ (ft-lb}_\text{f})/\text{slug}$$

The necessary energy is supplied by burning jet fuel.

2.10 SUMMARY

This chapter serves as an introduction to the fluid and flow properties needed to specify the dynamic and thermodynamic state of a fluid.

The mass of an object is defined as a measure of its resistance to acceleration, and its weight is the magnitude of the force acting on the object due to gravity. Properties that depend on the amount of fluid in a system are termed extensive properties. If an extensive property is divided by the total mass or the total volume of the fluid, the result is an intensive property. For example, we will make frequent use of the fluid density defined as the mass of a fluid divided by its volume. The local density is defined by the conditions of pressure and temperature existing in the fluid at that point. Specific gravity is defined as the ratio of a fluid's density to that of a standard reference fluid (usually water or air) at STP.

Pressure is the compressive normal component of the force applied by a fluid to a surface, divided by the area of that surface. Absolute pressure is defined in reference to a perfect vacuum, while gage pressure is specified with reference to atmospheric or ambient pressure. Pressure differences can cause a fluid to flow or be caused by a fluid flow.

Archimedes' principle defines the magnitude and direction of the buoyancy force acting on an object immersed in a stationary fluid. The buoyancy force acts in the direction opposite to that of the gravitational force and has a magnitude equal to the weight of the displaced fluid.

The temperature of a fluid is a thermodynamic state variable that provides a measure of the internal energy stored in the fluid. In a fluid in equilibrium, the temperature is proportional to the mean kinetic energy of the random motion of the molecules. Temperature differences may create a fluid flow, or they may be a consequence of a fluid flow. Temperature differences in a fluid are always accompanied by the flow of heat by molecular conduction. The relationship between heat flux, temperature gradient, and thermal conductivity is given by Fourier's law of heat conduction.

The specific heat capacity of a fluid represents the amount of heat energy required to raise the temperature of a unit mass of fluid by one degree. The value of the specific heat depends on whether the heat is added at constant pressure or at constant volume. For liquids these two specific heats are virtually identical, but for gases the specific heat capacity at constant pressure is greater than the specific heat capacity at constant volume. Most fluids expand when heated. The effects of temperature on density are described by the coefficient of thermal expansion of the fluid.

Under a broad range of conditions, the equation of state for a gas is well modeled by the perfect gas law. In fluid mechanics we normally employ the per-unit-volume form of the perfect gas law, $p = \rho RT$. As an engineering rule of thumb, the perfect gas law can be used as long as the temperature is below the dissociation temperature of the polyatomic molecules in the gas (3000 K for O_2 or 6000 K for N_2). In addition, perfect gas behavior can be assumed at temperatures that are above twice the critical temperature ($T_C = 155$ K for O_2 and 126 K for N_2) up to pressures of at least 70 atm. If the service conditions fall outside the range just specified, one can either use advanced state equations or a table of thermodynamic properties for the fluid of interest.

When a volume of any material is subjected to a pressure rise, the volume decreases, and the density increases. For many fluids the pressure–volume relationship is linear and characterized by a constant of proportionality called the bulk compressibility modulus. Fluids of large compressibility modulus (liquids) undergo negligible changes in density when subjected to even the largest pressures encountered in engineering, making them effectively incompressible. The speed of sound in a fluid, c, is found to be proportional to the compressibility of the fluid as measured by the isentropic bulk modulus. For an ideal gas, $c \propto \sqrt{T}$. The ratio of the speed of a moving fluid to the speed of sound in that fluid is known as the Mach number.

The absolute or shear viscosity of a fluid is defined as the constant of proportionality between an applied shear stress and the resulting transverse velocity gradient. The shear viscosity can also be interpreted as the constant of proportionality between the velocity gradient and the momentum transfer rate. The shear viscosity of a fluid is a strong function of temperature but a weak function of pressure. It increases with temperature for gases but decreases with temperature for liquids. An important consequence of the existence of a shear viscosity in a fluid is that of viscous dissipation or loss of energy when a fluid is sheared.

The extra energy associated with the molecules at the surface of a fluid is known as the surface energy. The corresponding net force on a molecule in the surface layer acts in the plane of the surface in all directions and is referred to as the surface tension. Surface tension causes the pressure inside a fluid drop to be greater than the pressure outside, and since the Δp is inversely related to the size of the bubble, smaller bubbles experience larger pressure differences. Surface tension is an important factor in a variety of fluid phenomena including capillary action, the waterproofing of fabric, and the cleaning action of soaps and detergents.

A fluid contains internal, kinetic, and potential energy. Significant changes in the energy content of a fluid are important in many applications including those in which the fluid performs work and those in which work is done on the fluid. The various forms of energy in a fluid can be considered to be extensive quantities or on a per-unit-mass or per-unit-volume basis.

PROBLEMS

Section 2.1

2.1 What is the difference between a fluid property and a flow property? Give an example of each kind of property.

2.2 Browse through the text and appendices to find three different symbols that are used to represent more than one fluid or flow property in this text and list the multiple properties that correspond to each symbol. Select one of these symbols and describe how you will "know" which property the symbol is meant to represent in a given application. Use examples to illustrate your answer.

Section 2.2

2.3 Are mass, weight, and density intrinsic or extrinsic properties? Justify your answer.

2.4 A certain type of jet fuel has a specific gravity of 0.85.
(a) What is the density of this fuel (in cgs units)?
(b) What is the specific weight of this fuel (in SI units)?
(c) What is the mass of a 55-gallon drum of this fuel (in lb_m)?
(d) What is the weight of a 55-gallon drum of this fuel (in lb_f)?

2.5 Estimate the mass of the air in your bedroom? State any assumptions.

2.6 What volume of air has the same mass as 12 cm^3 of toluene?

2.7 Are the terms "standard temperature and pressure" and "standard atmosphere" equivalent? Why or why not?

2.8 How is the molecular weight of a gas related to its density? Does the same relationship hold for liquids? Why or why not?

2.9 Estimate the density at STP of argon gas, having an atomic weight of 39.95 amu.

2.10 Determine the specific weight of each of the following fluids at STP. Report your answers in SI units.
(a) Helium
(b) Propane
(c) Mercury
(d) Acetone

2.11 Determine the specific gravity of each of the following fluids at STP.
(a) Helium
(b) Propane
(c) Mercury
(d) Acetone

Section 2.3

2.12 The text suggested that "the only kind of pressure gage that gives a reading of -20 psig (or psia) is a broken one." Explain this statement.

2.13 A Newcomen steam engine, like those used to drain water from coal mines in England in the 1770s, operated with a pressure of about 7.5 psi acting on a circular piston head with a diameter of about 50 in. What was the magnitude of the force applied by the fluid to the piston?

2.14 In Example 2.7 we calculated the force developed by the steam on the piston in Robert Fulton's *Clermont*. Oliver Evans built a steamboat called the *Aetna* for use between Philadelphia and Wilmington, Delaware. He used a higher steam pressure so that he could make his engine smaller and lighter. If the *Aetna*'s engine used steam at ~150 psi, what size piston would have been required to produce the same force as that obtained in the low pressure steam engine of the *Clermont*? [*Note:* One of the boilers in the *Aetna* burst in 1824, killing several passengers.]

2.15 Estimate the depth below the surface of a column of acetone at which the pressure is equal to 1.5 times atmospheric pressure.

2.16 A 10 cm diameter piston experiences a force of 5000 N. What is the corresponding fluid pressure responsible for generating this force? Give your answer in both SI and EE units.

2.17 Suppose a thin plastic bag containing 150 cm³ of cold water at 20°C is submerged in a tank of hot water at 80°C. Calculate the buoyancy and net forces acting on the bag of fluid. Is the force pushing the bag up or down?

2.18 The density of gold is 19.28 g/cm³. What is the volume of a gold crown that weighs 10,000 N when the crown is fully submersed in water at STP.

2.19 The pressure difference between two points at the same elevation in an airflow is 100 Pa. What is the difference in velocity? Assume that the assumptions of Eq. 2.11 are valid for this flow.

2.20 Estimate the pressure increase generated at the moment of impact during a belly flop from a 5 m high dive.

2.21 An air blower is capable of producing a pressure change of 15 in. of water. Estimate the maximum air velocity this blower might produce.

2.22 How does the aerodynamic drag and fuel economy of an automobile change if the speed increases from 55 mph to 65 mph?

2.23 A strong gust of wind will create a large, sudden increase in pressure on fixed structures, and the pressure increase may cause damage. Estimate the pressure force on a 30 m² traffic sign exposed to a 100 km/h wind gust.

2.24 Recall our discussion of the force your hand encounters when you hold it perpendicular to the airflow outside a moving car or perpendicular to the water flow outside a jet ski. Calculate the wind speed necessary to provide the same pressure force on your hand as it experiences in the water on a jet ski traveling at 10 mph.

2.25 Find the ratio of the speed of descent in water and air for a parachute suspending a known mass.

2.26 Provide a physical interpretation of Eq. 2.11 and list the assumptions that must be satisfied for this form of Bernoulli's equation to be valid.

Section 2.4

2.27 The four common temperature scales are Celsius, Fahrenheit, Kelvin, and Rankine. Convert each of the temperature listed to the other three temperature scales.
(*a*) 10°C
(*b*) 10°F
(*c*) 600 K
(*d*) 600°R

2.28 Explain why there is a minus sign on the right-hand side of Fourier's law of heat conduction, Eq. 2.13.

2.29 What is the heat flux through a 1 in. wide air gap if the temperature difference is 70°F? Repeat for the same gap filled with water.

2.30 The heat flux through a 5 mm wide air gap is measured to be 500 J/(m²-s). What is the temperature change across this air gap?

2.31 Which is generally larger, the specific heat capacity at constant pressure c_p or the specific heat capacity at constant volume c_v? Why?

2.32 Calculate the amount of thermal energy required to heat each of the following quantities of fluid from 50°F to 80°F at standard pressure: 1 lb$_m$ of hydrogen and 1 in.³ of acetone. State any assumptions.

2.33 Calculate the volume change if 1 m³ of hydrogen is heated from 10 to 30°C at standard pressure. Repeat the calculation for 1 m³ of glycerin.

2.34 What temperature change is necessary to cause a 5% increase in the volume of a fixed mass of air if the reference temperature is 50°C. Does your answer change if the reference temperature is 100°C? Why or why not?

Section 2.5

2.35 What is the difference between the universal gas constant and a specific gas constant? Determine the specific gas constant for hydrogen.

2.36 Air at STP is heated 20°F at constant pressure. What is the new density?

2.37 Describe the conditions under which it is "safe" to use the perfect gas law.

2.38 What is the internal energy and enthalpy change for oxygen if the temperature change is 50°C?

2.39 An oxygen tank for medical use (modeled as a cylinder of radius r and height h) contains 12 kg of pure oxygen at a pressure of 15 MPa at a temperature of 30°C. If the radius of the tank is 10 cm, what is its height?

2.40 A tire having a volume of 0.85 m³ contains air at a gage pressure of 180 kPa at 21°C. Determine the density and weight of the air in the tire.

2.41 Nitrogen is compressed to a density of 4 kg/m³ at a temperature of 70°C. What is the pressure of the gas?

Section 2.6

2.42 Use the definition of E_V and perfect gas law to show that at constant temperature $E_V = p$.

2.43 A cubic meter of oxygen at an absolute pressure of 100 kPa is compressed isentropically to two-thirds of its original volume. What is the final pressure?

2.44 Use the bulk modulus to estimate the speed of sound in gasoline and seawater.

2.45 Estimate the speed of sound in hydrogen and helium.

2.46 Estimate the distance between you and a thunderstorm if you see the lightning 5 seconds before you hear the thunder on a summer evening when the temperature is 80°F. How far would sound travel through water in the same time at the same temperature?

2.47 What is the Mach number of an airplane flying at 800 km/h at an altitude of 10,000 m?

Section 2.7

2.48 How does the shear viscosity of a gas change with temperature? How does the shear viscosity of a liquid change with temperature?

2.49 Calculate the temperature at which the shear viscosity of water is equal to half its value at 20°C given that $\mu(20°C) = 0.010019$ (N-s)/m² and $\mu(0°C) = 0.01787$ (N-s)/m². Repeat the calculation for air given that $\mu(20°C) = 0.0001813$ (N-s)/m² and $n = 0.7$.

2.50 A certain fluid is known to have a viscosity of 1.240 cP at 100°C and a viscosity of 0.950 cP at 300°C. Is this fluid likely to be a liquid or a gas? Why? Estimate the viscosity of this fluid at 200°C.

2.51 Olive oil has a viscosity of 138 cP at 10°C and a viscosity of 12.4 cP at 70°C. Estimate its viscosity at 40°C.

2.52 Determine the kinematic viscosity of mercury at STP.

Section 2.8

2.53 Cleaning processes involve the use of soaps and detergents to alter surface tension so that dirty surfaces are brought into contact with water, the universal solvent. Do you think soaps increase or decrease the surface tension? Why?

2.54 An air bubble in glycerin has a diameter of 2000 μm. What is the pressure difference across the surface of this bubble due to surface tension?

2.55 Calculate the height of capillary rise for water in a glass tube with $D = 0.5$ mm.

2.56. When a glass tube is inserted into liquid mercury, the depression is found to be 4 cm. Estimate the diameter of the tube.

2.57 What is the rise of water in a vertical crack formed by two glass plates 20 μm apart?

2.58 Calculate the capillary rise of water in a pair of the glass plates separated by 2 mm if the plates are inclined at 75° from the vertical.

Section 2.9

2.59 Water flows from a pipe at 6 m/s. What is the kinetic energy per unit mass of the water? What is the kinetic energy per unit volume of water?

2.60 A windmill extracts energy from air moving at 40 km/h. What is the total kinetic energy per unit volume of moving air?

2.61 A hydroelectric plant will employ a total elevation change of 350 ft. What is the gravitational potential energy change per unit volume of water?

2.62 What is the pressure potential energy stored in 75 L of water at 20 MPa?

3 CASE STUDIES IN FLUID MECHANICS

3.1 Introduction
3.2 Common Dimensionless Groups in Fluid Mechanics
3.3 Case Studies
    3.3.1 Flow in a Round Pipe
    3.3.2 Flow Through Area Change
    3.3.3 Pump and Fan Laws
    3.3.4 Flat Plate Boundary Layer
    3.3.5 Drag on Cylinders and Spheres
    3.3.6 Lift and Drag on Airfoils
3.4 Summary
Problems

3.1 INTRODUCTION

In Chapter 2 you learned how to combine your understanding of fluid and flow properties and a force balance based on Newton's second law to solve simple fluid mechanics problems. Here we focus our attention on some of the results that have been obtained by engineers for more complex fluid mechanics problems. We have selected some of these results to form the basis of a number of interesting case studies. In each case study, you will find a brief description of the flow field of interest, and one or more design formulas that can be used to calculate important quantities of engineering and design interest. These formulas rely primarily on results obtained by using experimental methods, and in particular on the dimensional analysis and modeling tools to be presented in Chapter 9. In some cases, the formulas can be developed or otherwise explained by means of the more sophisticated analysis tools you will also learn about in later chapters. In any case, the amount of information given in a case study is not unlike what you might find in an engineering handbook, and applying the material should not be difficult.

3 CASE STUDIES IN FLUID MECHANICS

A twofold goal of this chapter is to expose you to interesting flow fields early in the text and to allow you to calculate some engineering characteristics of these flows at an early stage in the learning process. As we revisit these case studies in later chapters, our hope is that you will progress from a cautious first application of the case study results to a fuller understanding of the underlying flow fields. Furthermore, these results may help you better comprehend your laboratory course work.

 CD/Video Library/Laminar and Turbulent Flow on a Flat Plate

At this point you might be wondering: Why do we need to rely on experimental results in fluid mechanics? Why not just use a better analytical model or a bigger computer to solve a flow problem? An answer to these questions lies in recognizing the difference between laminar and turbulent fluid flow. As the name implies, laminar flow involves the movement of fluid in "layers." As shown by the dye in the top of Figure 3.1, the motion of a fluid in laminar flow is orderly, often slow and steady, and generally amenable to observation, measurement, and prediction. Analytical and computational solutions to laminar flow problems are both feasible and common, and the need for experiments is often minimal. However, laminar flows are relatively rare both in nature and in engineering practice. This is because a laminar flow undergoes a transition (middle of Figure 3.1) and eventually becomes turbulent as flow speeds increase. Turbulent flow, as illustrated at the bottom of Figure 3.1, is encountered in almost all flows in nature and engineering practice. This type of flow consists of a chaotic, disordered, and unsteady motion of fluid that is generally difficult to visualize, measure, and predict. There are no analytical solutions for turbulent flow, and computational models of turbulence are limited in their applicability. Thus experimental results are necessary for engineering designs involving turbulent flows.

Although the future of fluid mechanics will undoubtedly be marked by an increasing dependence on computational solutions for both laminar and turbulent flows, models of turbulence and other physical processes of interest in fluid mechanics will continue to require calibration and verification by well chosen experiments.

In the case studies that follow, you will find frequent references to dimensionless groups. Examples of these groups include the Reynolds and Mach numbers. Simply put, a dimensionless group is an algebraic combination of the parameters describing a particular flow that proves to be both dimensionless as a whole and significant in terms of understanding the flow field. In fluid mechanics, the most important dimensionless group is called the Reynolds number. The Reynolds number of a flow, written as $Re = \rho V L/\mu$, is the product of density ρ, a fluid velocity scale V, and a length scale L, all divided by

Figure 3.1 Dye injected into a pipe flow indicates laminar flow (top), transitional flow (middle), and turbulent flow (bottom).

viscosity, μ. In a given unit system Re is dimensionless, which we can demonstrate by writing the dimensions of each quantity in the Reynolds number to obtain

$$\{Re\} = \frac{\{\rho\}\{V\}\{L\}}{\{\mu\}} = \frac{[(M/L^3)(L/t)(L)]}{M/Lt} = \frac{ML^{-1}t^{-1}}{ML^{-1}t^{-1}} = 1$$

Flows with large Reynolds numbers are usually turbulent, an important consideration in understanding how a flow will behave.

3.2 COMMON DIMENSIONLESS GROUPS IN FLUID MECHANICS

As you learn more about fluid mechanics you will discover that some dimensionless groups occur repeatedly in analyses of fluid mechanics problems. Most dimensionless groups have been given names in honor of their discovers or other prominent individuals in the study of fluid mechanics. It is important to become familiar with the common dimensionless groups to ensure that you present the results of your analysis in the form other engineers expect. Also, the numerical values of these traditional dimensionless groups are used in the classification of a particular fluid mechanics problem, in the selection of efficient solution techniques, and to compare results with those obtained by investigations of similar flows. Let us take a look at some of the more important dimensionless groups in fluid mechanics and learn about their relationship to various physical phenomena.

Reynolds Number: As discussed earlier, the Reynolds number, the most important dimensionless group in fluid mechanics, is defined to be

$$Re = \frac{\rho V L}{\mu} \tag{3.1}$$

where ρ is the fluid density, V is a fluid velocity scale, L is a length scale, and μ is the fluid viscosity. This dimensionless group is named in honor of Osborne Reynolds (1842–1912), a noted pioneer in the study of pipe flow and turbulence. The velocity and length scales involved in its definition are illustrated for internal and external flows in Figure 3.2.

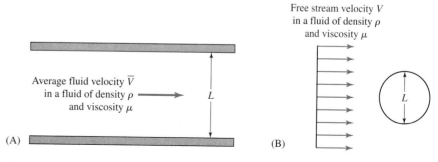

Figure 3.2 Velocity and length scales used in defining Re for examples of (A) internal flow and (B) external flow.

CD/History/Osborne Reynolds

It is important for you to have an understanding of the physical significance of the Reynolds number. One way to interpret Re is to think of it as a ratio of inertial to viscous forces in a fluid flow. An inertial force can be written using Newton's second law as $\mathbf{F} = M\mathbf{a}$. If we recognize that mass is equal to the product of density and volume and write the equation in terms of dimensions we find:

$$\{\mathbf{F}_I\} = \{M\}\{\mathbf{a}\} = \{\rho L^3\}\{Vt^{-1}\} = \{\rho L^3 V t^{-1}\} = \{\rho V^2 L^2\} \qquad (3.2)$$

where we have made use of the fact that the dimensions for velocity are $\{Lt^{-1}\}$. To generate a similar expression for the viscous force, we begin with Newton's law of viscosity, $\tau = \mu(du/dy)$, in dimensional form:

$$\{\tau\} = \{\mu\}\{VL^{-1}\} \qquad (3.3)$$

But we require an expression for the viscous force, which is equal to the shear stress multiplied by the area over which that stress acts. Thus,

$$\{\mathbf{F}_V\} = \{\tau A\} = \{\mu\}\{VL^{-1}\}\{L^2\} = \{\mu VL\} \qquad (3.4)$$

If we divide Eq. 3.2 by Eq. 3.4 we obtain:

$$\frac{\{\mathbf{F}_I\}}{\{\mathbf{F}_V\}} = \frac{\{\rho V^2 L^2\}}{\{\mu VL\}} = \left\{\frac{\rho VL}{\mu}\right\} \qquad (3.5)$$

Since the right-hand side of this equation is equivalent to the Reynolds number, we are justified in interpreting Re as a ratio of inertial to viscous forces.

Except within a thin boundary layer near solid surfaces, high Re flows are dominated by inertial forces and are usually turbulent. Low Re flows, or creeping flows, are highly viscous in character and laminar. Flows at intermediate Re are often laminar, with inertial and viscous forces both playing significant roles in determining flow structure throughout the flow field.

CD/Video Library/Flow Past a Cylinder

The effect of Re on flow structure for flow over a cylinder is illustrated in Figure 3.3. At very low values, $Re = 0.038$ (Figure 3.3A), the inertia is so small that fluid particles easily flow around the cylinder while remaining in their laminar layers. At $Re = 19$ (Figure 3.3B) the inertia has increased to the point that some fluid particles cannot "make the turn," like Formula 1 racecar drivers who spin out going too fast through a curve. This phenomenon is called flow separation. As Re increases to 55 (Figure 3.3C), the separation bubble is pushed downstream. Thus Re indicates the presence of structural changes in the flow field. In Chapters 12 and 14 we will discuss in greater detail the flow over a cylinder and the interesting results that occur at higher Reynolds numbers.

Before we continue with an example, let us sound a note of caution concerning the interpretation of Re. It would be a gross simplification to consider Re to be only the ratio

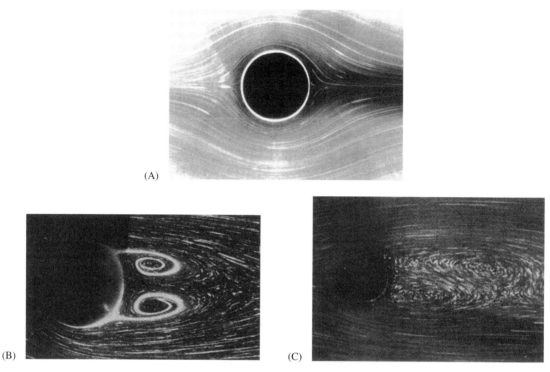

Figure 3.3 Flow field over a cylinder at (A) $Re = 0.038$, (B) $Re = 19$, and (C) $Re = 55$.

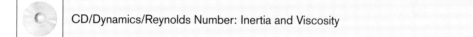
CD/Dynamics/Reynolds Number: Inertia and Viscosity

of inertial to viscous forces. For example, $Re = 1$ should not be interpreted as inertial and viscous forces being equal. The choice of length and velocity scales used in Re have most often been chosen for convenience, not physical significance. Thus Re should be compared and interpreted for a single flow field only, not between flow fields. Consider the critical Re_{cr}, where the transition of a laminar flow to turbulent flow is an important application of the Reynolds number: Re_{cr} can differ by several orders of magnitude between an internal flow and an external flow. Thus the physical meaning cannot be precisely the same.

Mach Number: The Mach number, named in honor of Ernst Mach (1838–1916), a pioneer in the study of high speed flow, was introduced in Section 2.6.1 and is defined to be the ratio of fluid velocity V to c, the speed of sound in the fluid. Thus the Mach number is given by

$$M = \frac{V}{c} \tag{3.6}$$

EXAMPLE 3.1

A good serve from a professional tennis player may reach 190 km/h. If the diameter of a tennis ball is approximately 6.5 cm, what is the Reynolds number for the flow over the ball?

SOLUTION

The Reynolds number for a tennis ball is found using Eq. 3.1: $Re = \rho V L/\mu$. The characteristic velocity is $V = 190$ km/h, and we will use the diameter of the tennis ball as the characteristic length scale so that $L = 6.5$ cm. The density and viscosity of air at STP are found in Appendix A to be $\rho = 1.204$ kg/m^3 and $\mu = 1.82 \times 10^{-5}$ (N-s)/m^2. Substituting these values into the expression for Re and using the appropriate unit conversion factors found in Appendix C yields:

$$Re = \frac{\rho V L}{\mu}$$

$$= \frac{1.204 \text{ kg/m}^3 \left[(190 \text{ km/h})\left(\frac{1 \text{ h}}{3600 \text{ s}}\right)\left(\frac{1000 \text{ m}}{1 \text{ km}}\right)\right]\left[(6.5 \text{ cm})\left(\frac{1 \text{ m}}{100 \text{ cm}}\right)\right]}{1.82 \times 10^{-5}[(\text{N-s})/\text{m}^2]\left[\frac{1 \text{ (kg-m)/s}^2}{1 \text{ N}}\right]}$$

$$Re = 2.27 \times 10^5$$

This is a high value of Re (we will define "high value" later in the context of specific types of flows); thus for the movement of the tennis ball through the air, inertial forces are significant and viscous forces will be important only in the boundary layer.

The Mach number provides a measure of the effects of compressibility on a flow. An incompressible fluid, i.e., a liquid, has $M \approx 0$ because the sound speed is very large in comparison to a typical liquid flow speed. Gases tend to flow much faster than liquids relative to their sound speeds, hence Mach number is of great interest in classifying the flow of a gas such as air. When air flows with a small Mach number, nominally $M < 0.3$, the air behaves like an incompressible fluid. Thus a flow with $M < 0.3$ is called an incompressible flow. A flow with a Mach number greater than this is termed a compressible flow, since variations in the density of the air must be accounted for. We further classify compressible flows according to Mach number as subsonic if $M < 1$ and supersonic if $M > 1$. Flows near the sonic velocity have unique characteristics such that $0.9 < M < 1.2$ flows are classified as transonic. Flows at very high velocity, $M > 5$, are termed hypersonic.

CD/Video Library/Shock Waves

3.2 COMMON DIMENSIONLESS GROUPS IN FLUID MECHANICS

Figure 3.4 A ship's wake, photograph from the space shuttle. The wake trails several miles behind the ship.

Froude Number: The Froude number is defined to be the ratio

$$Fr = \frac{V}{\sqrt{gL}} \tag{3.7}$$

where V is a fluid velocity scale, L is a length scale, and g is the acceleration of gravity. This dimensionless group is named in honor of William Froude (1810–1879), who used models to perform pioneering studies of the drag on ships due to wave making (Figure 3.4).

The Froude number can be interpreted as the ratio of inertial forces to gravitational forces. From Eq. 3.2 we know that the dimensions for the inertial force can be written as $\{\mathbf{F}_I\} = \{M\}\{a\} = \{\rho L^3\}\{Vt^{-1}\} = \{\rho L^3 V t^{-1}\} = \{\rho V^2 L^2\}$. Similarly, the dimensions for the gravitational force are:

$$\{\mathbf{F}_G\} = \{M\}\{g\} = \{\rho L^3\}\{g\} \tag{3.8}$$

Taking the ratio of the inertial force to the gravitational force yields:

$$\frac{\{\mathbf{F}_I\}}{\{\mathbf{F}_G\}} = \frac{\{\rho V^2 L^2\}}{\{\rho L^3 g\}} = \left\{\frac{V^2}{gL}\right\} \tag{3.9}$$

Since this ratio is clearly dimensionless (units of force in the numerator and denominator), the square root of the ratio is also dimensionless, and we see that the Froude number can in fact be interpreted as a ratio of inertial to gravitational forces.

The Froude number is important in ship hydrodynamics, in the study of water waves, and in the classification of free surface flows, which do not involve a moving body. In such cases the length scale is often taken to be the liquid depth. Free surface flows are of interest to civil engineers involved in large-scale projects such as canals, weirs, spillways, and waterways of all kinds.

 CD/Video Library/River Flow

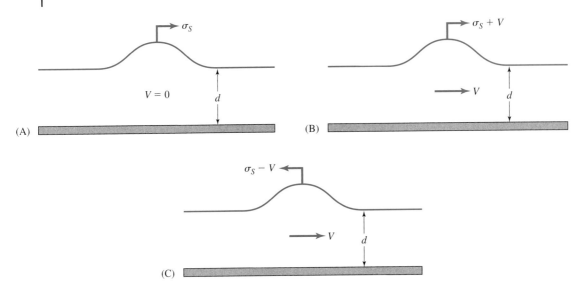

Figure 3.5 Infinitesimal wave moves (A) to the right on stationary fluid (B) to the right on fluid moving to the right and (C) to the left on fluid moving to the right. To an observer moving with the fluid, the wave speed is σ_S in both cases. For an observer on the shore, the wave speed is σ_S for (A), $\sigma_S + V$ for (B), and $\sigma_S - V$ for (C).

Free surface flows with $Fr < 1$ are said to be subcritical; those with $Fr > 1$ are supercritical, and a flow at $Fr = 1$ is said to be critical. An understanding of the physical phenomenon behind the use of the adjective critical in free surface flows can be gained by noting that the wave propagation speed of an infinitesimal wave in stationary water of depth d is

$$\sigma_S = \sqrt{gd} \qquad (3.10a)$$

Here σ_S is the speed at which the wave moves relative to the water (see Figure 3.5A). The Froude number in a problem involving wave propagation in water moving at speed V is

$$Fr = \frac{V}{\sqrt{gd}} = \frac{V}{\sigma_S} \qquad (3.10b)$$

If the water is moving at a velocity V to the right, then, as shown in Figure 3.5B, a wave moving to the right (in the flow direction) travels at a velocity $\sigma_S + V$, and to the left at a velocity $\sigma_S - V$ (Figure 3.5C). If the water is moving at a velocity $V = \sigma_S$, then a wave cannot propagate upstream. This is the critical water speed for a free surface flow of depth d, and Eq. 3.10b shows that this speed corresponds to $Fr = 1$. In a subcritical flow, $Fr < 1$ and $V < \sigma_S$, so waves may travel in both directions. In a supercritical flow, $Fr > 1$ and $V > \sigma_S$, so waves can travel downstream only.

Weber Number: The Weber number is an important dimensionless group in flow problems involving surface tension. It is named after Moritz Weber (1871–1951), who

3.2 COMMON DIMENSIONLESS GROUPS IN FLUID MECHANICS

EXAMPLE 3.2

The flow in a wide tidal channel separating a back bay from the ocean may approach 0.75 m/s. If the tidal channel is 6 m deep, what are the Reynolds and Froude numbers for the flow?

SOLUTION

The Reynolds and Froude numbers for this flow are found using Eqs. 3.1 and 3.10b: $Re = \rho V L/\mu$ and $Fr = V/\sqrt{gd} = V/\sigma_S$. The characteristic velocity is $V = 0.75$ m/s and the depth of the tidal channel serves as the characteristic length scale such that $L = d = 6$ m. We assume conditions at 20°C for water and use Appendix A to find: $\rho = 998$ kg/m³ and $\mu = 1 \times 10^{-3}$ (N-s)/m². Also note that $g = 9.81$ m/s². Substituting these values into the expressions for Re and Fr and using the definition of a newton as a unit conversion factor, we have:

$$Re = \frac{\rho V L}{\mu} = \frac{(998 \text{ kg/m}^3)(0.75 \text{ m/s})(6 \text{ m})}{1 \times 10^{-3}[(\text{N-s})/\text{m}^2]} \left[\frac{1 \text{ (kg-m)/s}^2}{1 \text{ N}}\right]$$

$$= 4.49 \times 10^6$$

$$Fr = \frac{V}{\sqrt{gd}} = \frac{0.75 \text{ m/s}}{\sqrt{(9.81 \text{ m/s}^2)(6 \text{ m})}} = 0.098$$

A Reynolds number of this magnitude would result in turbulent flow in the channel, and since the Froude number is less than one, we can conclude that the flow is subcritical.

worked on problems involving capillary effects. In a problem involving a moving liquid, the Weber number is defined by

$$We = \frac{\rho V^2 L}{\sigma} \qquad (3.11a)$$

where σ is the surface tension, and V and L are velocity and length scales, respectively. The Weber number in a moving liquid can be thought of as the ratio of inertial force to surface tension (or equivalently a ratio of kinetic energy to surface energy). In a problem involving liquid at rest in a gravitation field g, the importance of surface tension can be characterized by defining the Weber number as

$$We = \frac{\rho g L^2}{\sigma} \qquad (3.11b)$$

In this case We may be viewed as the ratio of gravitational forces to surface tension (or equivalently, gravitational potential energy to surface energy). Surface tension effects are only important when $We \ll 1$. Otherwise the effects of surface tension can be safely

EXAMPLE 3.3

Water flows from a 1 mm diameter orifice at 4 m/s. Is it likely that surface tension effects will be important in this application?

SOLUTION

The Weber number for this flow is found by using Eq. 3.11a: $We = \rho V^2 L / \sigma$. The characteristic velocity is $V = 4$ m/s, and the diameter of the orifice serves as the characteristic length scale so that $L = 1$ mm $= 0.001$ m. Assuming conditions at 20°C for water, we use Appendix A to find $\rho = 998$ kg/m³ and $\sigma = 0.073$ N/m. Note that we have used the surface tension for a water–air interface because we are assuming that water exits the orifice into air. Substituting the appropriate values into the expressions for We and using the definition of a Newton as a unit conversion factor yields:

$$We = \frac{\rho V^2 L}{\sigma} = \frac{(998 \text{ kg/m}^3)(4 \text{ m/s})^2(0.001 \text{ m})}{(0.073 \text{ N/m})} \left[\frac{1 \text{ (kg-m)/s}^2}{1 \text{ N}}\right] = 219$$

Since $We \gg 1$, we can safely neglect surface tension effects in this application.

The Bond number $B = [g(\rho_1 - \rho_2)D^2]/\sigma$ is used to characterize problems involving fluid droplets or bubbles of density ρ_1 immersed in another fluid of density ρ_2. It can be thought of as a measure of the ratio of buoyancy force to surface tension force.

ignored. Can you confirm that the Weber number for the capillary rise of water ($\sigma = 0.073$ N/m) in a round glass tube 1 mm in diameter is $We = 0.134$? How large would the diameter of the tube need to be for you to predict that capillary rise would be negligible?

Euler Number: The Euler number is defined to be

$$Eu = \frac{p - p_0}{\frac{1}{2}\rho V^2} \tag{3.12a}$$

where $p - p_0$ is the difference between a local value of pressure and that at some reference location. Leonhard Euler (1707–1783) was a great mathematician who first derived many fundamentals of fluid mechanics. The Euler number can be interpreted as a measure of the ratio of pressure force to inertial force. A number of variations on the Euler number appear in fluid mechanics. In aerodynamics, the pressure difference in the Euler number refers to the upstream static pressure p_∞, and the Euler number then becomes the pressure coefficient

$$C_p = \frac{p - p_\infty}{\frac{1}{2}\rho V^2} \tag{3.12b}$$

The Euler number does not have the great physical significance of the Mach or Froude numbers; however, as with all dimensionless groups, it allows for the compact communication of data.

3.2 COMMON DIMENSIONLESS GROUPS IN FLUID MECHANICS

Figure 3.6 The Karman vortex street in the wake of a cylinder.

Slender structures such as suspended power transmissions lines, struts on small airplanes, and smokestacks are known to have natural vibration frequencies that can be calculated by using techniques from structural mechanics. When the natural vibrational frequency of a structure coincides with the frequency of the flow-induced Karman vortices, a condition known as resonance develops. During resonance, the amplitude of the structural vibrations can increase significantly. The Karman vortices are implicated in the wind-induced failure of the Tacoma Narrows suspension bridge in 1940 (Figure 3.7); however, there is still disagreement about the precise cause of the disaster.

Figure 3.7 The Tacoma Narrows Bridge shortly before its collapse in 1940. The sidewalk to the right is over 28 ft above the one to the left.

Strouhal Number: The Strouhal number, which is defined as

$$St = \frac{\omega L}{V} \qquad (3.13)$$

is important in problems involving flow oscillations in which the frequency of the oscillations is ω. The Strouhal number can be interpreted as the ratio of vibrational velocity to translational velocity. Many flows over bluff bodies develop oscillations. The most well known is the generation of Karman vortices that are shed periodically from the wake of a cylinder (see Figure 3.6). In this case it is known that over a range of Reynolds numbers $10^2 \leq Re \leq 10^7$, the Strouhal number is approximately 0.21 if the frequency of vortex shedding is measured in radians per second. Thus St can be used to predict the expected frequency of vortex shedding. Vincenz Strouhal (1850–1922) did pioneering work on the vibration or "singing" of wires due to this effect.

 CD/Video Library/Tacoma Narrows Bridge Disaster

Prandtl Number: Fluid mechanics is integrally related to the field of convective heat transfer, which is the study of heat transport processes in fluid flows. In fact, an important dimensionless number used in convective heat transfer is the Prandtl number, named after Ludwig Prandtl (1875–1953), one of the giants of twentieth-century fluid mechanics. The Prandtl number

$$Pr = \frac{\nu}{\alpha} \qquad (3.14)$$

EXAMPLE 3.4

A smokestack at a power plant is 9 ft in diameter. The natural vibrational frequency for this structure is known to be 7 rad/s. Calculate the wind velocity that would induce Karman vortex shedding at a frequency of 7 rad/s and comment on the likelihood of wind-induced resonance leading to structural failure.

SOLUTION

The Strouhal number for this flow is found by using Eq. 3.13: $St = \omega L/V$. Resonance occurs when the wind-induced vortex shedding frequency calculated in this way corresponds to the natural frequency of the structure. Thus, we must determine the wind velocity at which the vortex shedding frequency is $\omega = 7$ rad/s. The diameter of the smokestack serves as the characteristic length scale, so that is $L = 9$ ft. Assuming that the critical Strouhal number is 0.21 and substituting the appropriate values into Eq. 3.7 after solving for V yields:

$$V = \frac{\omega L}{St} = \frac{(7 \text{ rad/s})(9 \text{ ft})}{0.21} = 300 \text{ ft/s}$$

$$V = (300 \text{ ft/s})\left(\frac{1 \text{ mph}}{1.467 \text{ ft/s}}\right) = 204 \text{ mph}$$

Since it is unlikely that the smokestack will experience wind speeds in excess of 200 mph, one need not be concerned about vortex shedding leading to structural failure.

---

is the ratio of kinematic viscosity ν to thermal diffusivity α. Heat transfer between a solid surface and a fluid that is in motion due to external means (e.g., a fan or pump) is called forced convection. In cases of forced convection the heat transfer rate depends on the Prandtl and Reynolds numbers.

Other Dimensionless Groups: There are many more named dimensionless groups in fluid mechanics as well as some that are simply physically descriptive and not named after a particular historical figure. For example, the dimensionless group known as the relative roughness e/D occurs in pipe flow. This group is defined as the ratio of the average height of the pipe wall roughness e to inside diameter of the pipe D.

3.3 CASE STUDIES

The following case studies represent a varied selection of the type of information available to engineers. Our emphasis in selecting these particular studies is their broad applicability in engineering design. Engineers use results like these to successfully practice design after a single course in fluid mechanics. Each of these flow problems has been investigated theoretically, but the majority of useful results have been obtained empirically. If you are careful to apply the formulas developed in a case study in the original context, your analysis will provide answers to design questions within the range of normal engineering accuracy.

The empirical results presented here give the impression of simplicity because they involve only global characteristics of the flow field. Notice that nothing is said in any of the case studies about local details of the flow field. Actually, all these flow fields are quite complex. Our purpose in designating these problems as case studies is to give you an early introduction to the design aspects of engineering fluid mechanics and to emphasize the relevance of the subsequent theoretical chapters to developing a better understanding of the fluid mechanics of engineering problems. We do this by revisiting these same flow problems in later chapters and using the new tools we have developed to better understand the sources of the case study formulas.

3.3.1 Flow in a Round Pipe

Pumping a fluid through a pipe or duct is a common, and arguably the most important, application of fluid mechanics. Society could not function without the water, steam, air, natural gas, oil, and other hydrocarbons transported via piping systems. Our homes and workplaces depend on central heating, ventilation, and air conditioning. Indeed, social historians in the United States have commented that the migration of people to the southern states after World War II would not have occurred without the universal availability of air conditioning. Virtually all engines require delivery of fuel, lubricant, and coolant through a pipe or hose. Can you think of other important technical applications of these systems? Does pipe flow also occur in biological systems?

In this first case study we consider steady, fully developed incompressible flow in a straight, horizontal, round pipe as shown in Figure 3.8. The adjective "steady" implies that the flow is unchanging in time, and "fully developed" implies that the flow is the same at every location along the pipe. "Incompressible" here implies that the fluid density is constant. This type of flow commonly occurs in the movement of liquid through relatively long pipes subjected to a continuous pumping action. In later chapters we show how to handle a rectangular, square, or other shape for the pipe or duct, as well as flows that are not steady or fully developed. Low speed gas flow occurs at constant density, so the techniques developed in this case study may also be used to analyze flow of air in heating, ventilating, and air-conditioning systems.

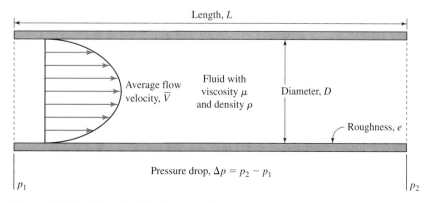

Figure 3.8 Variables for fully developed flow in a horizontal pipe.

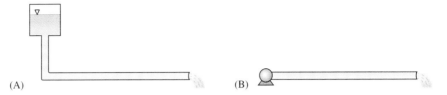

Figure 3.9 Flows can be driven by elevation changes such as a head tank (A) or mechanically driven such as by a pump (B).

As shown in Figure 3.9, a liquid can be caused to flow through a pipe by placing a reservoir at a higher elevation (see the head tank of Figure 3.9A). A liquid or gas can also be caused to flow by simply using a pump, fan, or blower (Figure 3.9B). A significant portion of the total energy delivered to a fluid by a head tank, pump, fan or blower is needed to overcome the frictional pressure drop in straight sections of a pipe or duct system. In the case of the pump, fan, or blower, hereafter simply referred to as a pump, we are interested in predicting the pressure increase and power a pump must provide to overcome this pressure drop.

We begin our analysis with the experimental observation suggesting that the pressure drop Δp may depend upon the pipe length L, diameter D, wall roughness e, average velocity $\bar{V}$, fluid density ρ, and fluid viscosity μ. The wall roughness, defined as the average height of random protuberances, depends on the type of pipe and how long it has been in service. How did we decide on this list of parameters? We know that Δp is related to frictional losses (viscous dissipation). Since we expect frictional losses to increase with an increase in average velocity, viscosity, pipe length, and/or pipe wall roughness, it makes sense to include $\bar{V}$, μ, L, and e in our model. The inclusion of fluid density and pipe diameter should also seem reasonable to you as parameters that might affect the frictional pressure drop.

After writing the proposed functional relationship mathematically as $\Delta p = f(L, D, e, \bar{V}, \rho, \mu)$, it is possible to use dimensional analysis (explained further in Chapter 9) to express the dependence of the pressure drop on the problem parameters as

$$\Delta p = \rho f \frac{L}{D} \frac{\bar{V}^2}{2} \qquad (3.15)$$

where the quantity f, known as the friction factor is an (unknown) function of the relative roughness e/D and Reynolds number $\rho \bar{V} D/\mu$. That is, we can write

$$f = f\left(\frac{e}{D}, \frac{\rho \bar{V} D}{\mu}\right)$$

It should now be evident that a major factor in analyzing pipe flow and specifying a pump is determining the dependence of friction factor on relative roughness and Reynolds number. If we know this functional relationship, we can calculate a value for the friction factor and analyze or otherwise design a straight segment in a pipe or duct system.

In a later chapter we will show that for a flow of liquid or constant density gas, the relationship between volume flowrate Q (defined as volume of fluid per unit time flowing through a cross section of the pipe), average velocity, and the cross-sectional area of a pipe A is

$$Q = \bar{V} A \qquad (3.16)$$

Although we are not deriving any of the earlier formulas here, note that they are dimensionally consistent and intuitively reasonable. For example, the formula for volume flowrate follows from thinking about the volume of fluid that moves through a pipe in one second if all the fluid moves with the average velocity.

and the power P required to pump the fluid through a pipe whose pressure drop is Δp is

$$P = Q\Delta p \qquad (3.17)$$

Note that this expression for the power does not include an allowance for the pump efficiency. Thus Eq. 3.17 provides a value that is somewhat less than the actual pump power needed.

 CD/Video Library/The Reynolds Transition Experiment

It is known that steady flow in a pipe may be laminar or turbulent depending on the value of the Reynolds number. From observation it has been found that a smoothly flowing laminar flow of fluid occurs at low Re (when viscous forces dominate). This flow becomes unstable as the Reynolds number is increased to approximately 2300. This value marks a transition from laminar to turbulent flow. Pipe flows at $Re < 2300$ are almost always laminar. At higher Reynolds numbers (when inertial forces dominate) the flow is likely to be turbulent. Since, however, under carefully controlled conditions laminar flows in a pipe can persist at a Reynolds number greater than 2300, it may be necessary to verify the presence of turbulent flow before selecting one of the two methods outlined shortly for determining the friction factor. As mentioned earlier, laminar flow is characterized by highly ordered smooth layers, or laminae, of fluid. Turbulent flow is an unsteady, disordered flow. Observation shows that in pipe flow the friction factor f is strongly dependent on whether the flow is laminar or turbulent.

Laminar Flow

The governing equations for steady, fully developed laminar flow in a round pipe may be solved analytically. The solution, which is discussed in detail in Chapter 13, reveals that the friction factor in laminar flow is given by

$$f = \frac{64}{Re} \qquad (3.18)$$

One of the interesting characteristics of laminar flow is that for the normal range of roughness encountered in pipes, neither the friction factor nor the pressure drop depends on the relative roughness.

Turbulent Flow

In turbulent flow the friction factor and pressure drop are functions of the relative roughness and the Reynolds number. There are no analytical solutions to turbulent flow, so in this case we must rely on the large body of empirical observations. The data were correlated by Colebrook, resulting in the following expression for the friction factor:

$$\frac{1}{\sqrt{f}} = -2.0 \log\left(\frac{e/D}{3.7} + \frac{2.51}{\sqrt{f}\, Re}\right) \qquad (3.19a)$$

Note that this is a transcendental equation and will require iteration to determine the friction factor for known values of relative roughness and Reynolds number. We may also determine the friction factor by means of the Chen equation, another empirically based relationship:

$$f = \left\{-2.0\log\left[\frac{e/D}{3.7065} - \frac{5.0452}{Re}\log\left(\frac{(e/D)^{1.1098}}{2.8257} + \frac{5.8506}{Re^{0.8981}}\right)\right]\right\}^{-2} \quad (3.19b)$$

EXAMPLE 3.5

Normal saline solution flows with an average velocity of $\bar{V} = 0.5$ mm/s in a 2 m length of polymer tubing before entering a patient's arm intravenously. If the inside diameter of tubing is $D = 2$ mm, determine the friction factor, volume flowrate, and pressure drop in the tubing. Assume that the saline solution has the same properties as water, and that the IV line is horizontal.

SOLUTION

We are asked to determine f, Q, and Δp for a flow of saline through a horizontal tube with $L = 2$ m and $D = 2$ mm. This problem can be solved without the aid of a sketch. We are given $\bar{V} = 0.5$ mm/s and assume conditions at 20°C for water. From Appendix A we find $\rho = 998$ kg/m³ and $\mu = 1 \times 10^{-3}$ (N-s)/m². The problem is solved by using Eq. 3.15 $[\Delta p = \rho f(L/D)(\bar{V}^2/2)]$ to find Δp and Eq. 3.16 to find Q. We begin, however, by returning to Eq. 3.1 ($Re = \rho\bar{V}L/\mu$) to determine Re and then using either Eq. 3.18 or 3.19 to determine the appropriate friction factor for the calculated value of Re.

Substituting the foregoing values into the expression for Re yields:

$$Re = \frac{\rho\bar{V}D}{\mu} = \frac{(998 \text{ kg/m}^3)(0.5 \times 10^{-3} \text{ m/s})(2 \times 10^{-3} \text{ m})}{[1 \times 10^{-3} \text{ (N-s)/m}^2]}\left[\frac{1 \text{ (kg-m)/s}^2}{1 \text{ N}}\right] = 1$$

Since $Re < 2300$, the flow is laminar, and we can use Eq. 3.18 to find the friction factor: $f = 64/Re = \frac{64}{1} = 64$. Next, use Eq. 3.15 to solve for the pressure drop:

$$\Delta p = \rho f \frac{L}{D} \frac{\bar{V}^2}{2} = (998 \text{ kg/m}^3)(64)\left(\frac{2 \text{ m}}{0.002 \text{ m}}\right)\left[\frac{(0.5 \times 10^{-3} \text{ m/s})^2}{2}\right]$$

$$= 7.98 \text{ N/m}^2 = 7.98 \text{ Pa}$$

Finally we use Eq. 3.16 to find the volume flowrate:

$$Q = \bar{V}A = \bar{V}\frac{\pi D^2}{4} = (0.5 \times 10^{-3} \text{ m/s})\left[\frac{\pi(2 \times 10^{-3} \text{ m})^2}{4}\right] = 1.57 \times 10^{-9} \text{ m}^3/\text{s}$$

A very small volume flowrate like this can also be expressed in milliliters per minute:

$$Q = (1.57 \times 10^{-9} \text{ m}^3/\text{s})\left(\frac{10^2 \text{ cm}}{\text{m}}\right)^3\left(\frac{1 \text{ mL}}{1 \text{ cm}^3}\right)\left(\frac{60 \text{ s}}{1 \text{ min}}\right) = 0.09 \text{ mL/min}$$

EXAMPLE 3.6

Gasoline flows with an average velocity of $\bar{V} = 4$ ft/s in a horizontal steel pipe of length $L = 100$ ft with an inside diameter $D = 1$ in. The pipe connects the bulk storage tank to the pump at a gas station as shown in Figure 3.10. Determine the friction factor, volume flowrate, pressure drop, and pump horsepower required for this flow if the relative roughness of the pipe is $e/D = 0.001$. The fluid properties are $\rho = 42.45$ lb$_m$/ft^3 and $\mu = 1.96 \times 10^{-4}$ lb$_m$/(ft-s).

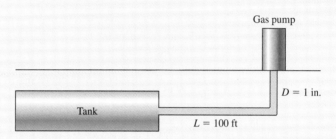

Figure 3.10 Schematic of gas station for Example 3.6.

SOLUTION

We are asked to determine f, Q, Δp, and P required for a flow of gasoline through a horizontal pipe with $L = 100$ ft and $D = 1$ in. Figure 3.10 is an adequate sketch of the flow situation. We are given $\bar{V} = 4$ ft/s, $\rho = 42.45$ lb$_m$/ft^3, and $\mu = 1.96 \times 10^{-4}$ lb$_m$/(ft-s). The problem is solved by using Eq. 3.15 [$\Delta p = \rho f (L/D)(\bar{V}^2/2)$] to find Δp, Eq. 3.16 ($Q = \bar{V}A$) to find Q, and Eq. 3.17 ($P = Q\Delta p$) to find P. We begin of course by using Eq. 3.1 ($Re = \rho \bar{V} L/\mu$) to determine Re and then use either Eq. 3.18 or 3.19 to determine the friction factor.

Substituting appropriate values into the expression for Re yields:

$$Re = \frac{\rho \bar{V} D}{\mu} = \frac{(42.45 \text{ lb}_m/\text{ft}^3)(4 \text{ ft/s})(1 \text{ in.})(1 \text{ ft}/12 \text{ in.})}{1.96 \times 10^{-4} \text{ lb}_m/(\text{ft-s})} = 7.22 \times 10^4$$

Since $Re > 2300$, the flow is turbulent and we must use Eq. 3.19a or 3.19b to find the friction factor. Choosing Eq. 3.19a and substituting $e/D = 0.001$ and $Re = 72{,}200$ gives:

$$\frac{1}{\sqrt{f}} = -2.0 \log \left(2.7 \times 10^{-4} + \frac{3.48 \times 10^{-5}}{\sqrt{f}} \right)$$

Using repeated hand calculations (painful), a spreadsheet (still time-consuming), or a symbolic manipulator program (good idea) we find $f = 0.023$.

Next, use Eq. 3.15 to solve for the pressure drop.

$$\Delta p = \rho f \frac{L}{D} \frac{\bar{V}^2}{2} = (42.45 \text{ lb}_m/\text{ft}^3)(0.023)\left[\frac{100 \text{ ft}}{(1 \text{ in.})(1 \text{ ft}/12 \text{ in.})}\right]\left[\frac{(4 \text{ ft/s})^2}{2}\right]$$
$$= 9373 \text{ lb}_m/(\text{ft-s}^2)$$

To obtain Δp in common units, we use the definition of g_c as a unit conversion factor:

$$\Delta p = 9373 \text{ lb}_m/(\text{ft-s}^2)\left(\frac{1 \text{ lb}_f\text{-s}^2}{32.2 \text{ lb}_m\text{-ft}}\right) = 291 \text{ lb}_f/\text{ft}^2$$

$$\Delta p = 291 \text{ lb}_f/\text{ft}^2 \left(\frac{1 \text{ ft}}{12 \text{ in.}}\right)^2 = 2.0 \text{ psi}$$

Next we use Eq. 3.16 to find the volume flowrate:

$$Q = \bar{V}A = \bar{V}\frac{\pi D^2}{4} = (4 \text{ ft/s})\left[\frac{\pi(1 \text{ in.})^2(1 \text{ ft}/12 \text{ in.})^2}{4}\right] = 2.18 \times 10^{-2} \text{ ft}^3/\text{s}$$

$$Q = (2.18 \times 10^{-2} \text{ ft}^3/\text{s})\left(\frac{1 \text{ gal}}{0.13368 \text{ ft}^3}\right)\left(\frac{60 \text{ s}}{1 \text{ min}}\right) = 9.8 \text{ gal/min}$$

Finally, use Eq. 3.17 to find the pump horsepower:

$$P = Q\Delta p = (2.18 \times 10^{-2} \text{ ft}^3/\text{s})(291 \text{ lb}_f/\text{ft}^2)\left[\frac{1 \text{ hp}}{550 \text{ (ft-lb}_f)/\text{s}}\right] = 0.01 \text{ hp}$$

The volume flowrate of $\sim$10 gal/min seems reasonable for a gasoline pump at a service station. The power required to operate the pump is small because we are considering only the pressure drop needed to overcome friction. In most cases the pump must also produce enough pressure to overcome the hydrostatic pressure variation due to elevation change as well as losses due to valves and other fittings in the pipe network. We will discuss these additional aspects of piping system design in Chapter 13. This problem can also be solved by using the Chen equation, Eq. 3.19b, to estimate the friction factor.

3.3.2 Flow Through Area Change

If you examine a pipe or duct system in a building, it is evident that changes in the cross-sectional area of a flow passage are quite common (Figure 3.11). The area change is often abrupt owing to space limitations, and turbulent flow is the norm in these systems. In this section we provide a method for using a loss coefficient to estimate the frictional pressure drop in steady incompressible turbulent flow through a sudden area change. Frictional pressure drops also occur when flow passes through nozzles, diffusers, bends, valves, entrances, exits, and other features of a pipe or duct system. Methods to compute the pressure drop through these elements will also be described later (see Chapter 13).

In examining the flow through an area change it is critical to realize that even in the absence of frictional effects, there is always a pressure change due to the change in the speed of the flow as it passes through the area change. (This is the change in pressure predicted by the Bernoulli equation as discussed in Chapter 2.) The total change in pressure as a flow passes through an area change may therefore be thought of as the sum of a pressure change associated with the change in average flow velocity (which may be either positive or negative depending on whether the flow slows down or speeds up) and a frictional pressure drop (a negative pressure change). We model this effect in turbulent flow as

$$p_2 - p_1 = \left[\tfrac{1}{2}\rho(\bar{V}_1^2 - \bar{V}_2^2)\right] - \Delta p_F \qquad (3.20)$$

> Did you recognize that if the frictional pressure drop is set to zero in Eq. 3.20, this equation becomes identical to Bernoulli's equation (Eq. 2.11) for a flow along a horizontal path? Notice also how the empirical model here (Eq. 3.20) builds on an earlier ideal result by adding a term to account for friction.

where p_2 is the downstream pressure, p_1 is the upstream pressure, and Δp_F is the frictional pressure loss. The velocities $\bar{V}_1$ and $\bar{V}_2$ in this formula are the average velocities in the upstream and downstream sections. We can calculate Δp_F by using empirical results. Note from Eq. 3.16 that since the same volume flowrate passes through each section the average velocities are related by

$$\bar{V}_1 A_1 = \bar{V}_2 A_2 \qquad (3.21)$$

Now consider what happens in the idealized case of a frictionless flow through an area decrease. Since the frictional pressure loss Δp_F is assumed to be zero, Eqs. 3.20 and 3.21 show that the value of the pressure downstream is less than that upstream because the area decrease causes the flow to speed up. Conversely, for an increase in area the value of the pressure downstream is greater than that upstream because the flow slows down in the larger area downstream. Equation 3.20 shows that the effect of friction is to cause a lower pressure downstream than the ideal result irrespective of the area change.

The four basic types of cross-sectional area change are shown in Figure 3.12. As noted earlier, flows in systems of engineering interest usually have high Reynolds numbers and are turbulent. Because the section of a pipe or duct in which area change occurs is often relatively short, the portion of the frictional pressure loss due to viscous effects at the walls is negligible in comparison to the loss caused by turbulence. Thus, fluid viscosity is not an important parameter in these flows. Observation suggests that for

Figure 3.11 Ductwork system with several area changes.

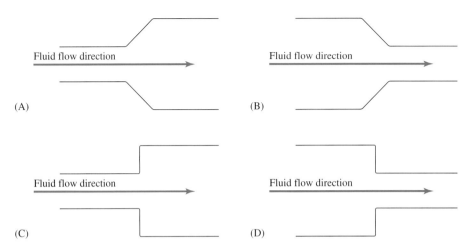

Figure 3.12 Schematics of area changes: (A) enlargement, (B) gradual contraction, (C) sudden expansion, and (D) sudden contraction.

gradual enlargements or contractions, the pressure loss in turbulent flow is a function of the inlet and outlet areas, fluid density, average velocity through the section, and an angle defining the geometry of the area change. For a sudden area change, however, there is no angle to consider, hence the pressure change depends only upon the remaining variables.

 CD/Video Library/Flow Past a Back Step

Sudden Expansion

Suppose we analyze the case of an abrupt enlargement of a round pipe as shown in Figure 3.12C. We assume that in a turbulent flow through a sudden area change the frictional pressure loss Δp_F is described by a functional relationship of the form $\Delta p_F = f(A_1, A_2, \rho, \bar{V}_1)$, where A_1 is the inlet area, A_2 is the outlet area, ρ is the fluid density, and $\bar{V}_1$ is the average inlet velocity. Note that we do not have to include $\bar{V}_2$ in our analysis since Eq. 3.16 makes its inclusion redundant. By using dimensional analysis, we can write the frictional pressure drop as

$$\Delta p_F = K_E \tfrac{1}{2} \rho \bar{V}_1^2 \qquad (3.22)$$

where conventional engineering practice introduces a dimensionless loss coefficient K_E for the enlargement. Note that we can think of $\tfrac{1}{2} \rho \bar{V}_1^2$ as representing the kinetic energy per unit volume in the upstream flow. Thus, the result suggests that the frictional pressure drop may be represented as some fraction of the upstream kinetic energy content of the fluid. From the available experimental data we can also deduce that K_E is a function of the area ratio of the enlargement. The problem reduces to finding the enlargement loss coefficient, since when K_E is known, the frictional pressure drop can be calculated from Eq. 3.22.

The enlargement loss coefficient for high Reynolds number turbulent flow is shown in Figure 3.13. Note that the enlargement loss coefficient is always positive. If the inlet

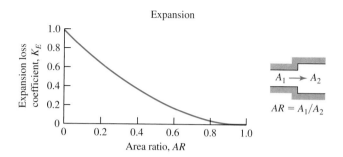

Figure 3.13 Loss coefficients for flow through a sudden expansion.

and outlet areas are equal, there is no frictional pressure loss, and the loss coefficient must be zero. If the ratio of the outlet area to inlet area is very large, the loss coefficient should approach unity because all the kinetic energy in the incoming flow is dissipated.

EXAMPLE 3.7

What is the frictional pressure drop in air flowing in a round duct due to a sudden change in diameter from 0.4 m to 0.6 m? The flowrate in the duct is 0.5 m³/s. What is the total pressure change across this enlargement?

SOLUTION

We are asked to find the frictional pressure drop and total pressure change across a sudden enlargement in a pipe. Figure 3.13 will serve as a sketch of the geometry of the enlargement. The first part of this problem is solved by using Eq. 3.22 ($\Delta p_F = K_E \frac{1}{2}\rho \bar{V}_1^2$). The area ratio is found to be

$$\frac{A_1}{A_2} = \frac{\pi D_1^2/4}{\pi D_2^2/4} = \left(\frac{D_1}{D_2}\right)^2 = \left(\frac{0.4 \text{ m}}{0.6 \text{ m}}\right)^2 = 0.444$$

By using Figure 3.13, we find a loss coefficient of $K_E \approx 0.3$. Next we determine the upstream average velocity, $\bar{V}_1$, using the definition of volume flowrate given in Eq. 3.16 ($Q = \bar{V}A$). Solving this expression for $\bar{V}_1$ and substituting known values gives

$$\bar{V}_1 = \frac{Q}{A_1} = \frac{Q}{\pi D_1^2/4} = \frac{0.5 \text{ m}^3/\text{s}}{\pi(0.4 \text{ m})^2/4} = 3.98 \text{ m/s}$$

Next use Eq. 3.22, along with the density of air at 20°C (Appendix A) $\rho = 1.204 \text{ kg/m}^3$, to find the frictional pressure loss:

$$\Delta p_F = K_E \tfrac{1}{2}\rho \bar{V}_1^2 = 0.3 \left(\tfrac{1}{2}\right)(1.204 \text{ kg/m}^3)(3.98 \text{ m/s})^2 = 2.86 \text{ Pa}$$

Next find $\bar{V}_2$ using the volume flowrate:

$$\bar{V}_2 = \frac{Q}{A_2} = \frac{Q}{\pi D_2^2/4} = \frac{0.5 \text{ m}^3/\text{s}}{\pi (0.6 \text{ m})^2/4} = 1.77 \text{ m/s}$$

Finally, use Eq. 3.20 to find the total pressure change across the enlargement.

$$p_2 - p_1 = \left[\tfrac{1}{2}\rho(\bar{V}_1^2 - \bar{V}_2^2)\right] - \Delta p_F$$
$$= \{\tfrac{1}{2}(1.204 \text{ kg/m}^3)[(3.98 \text{ m/s})^2 - (1.77 \text{ m/s})^2]\} - 2.86 \text{ Pa}$$

$$p_2 - p_1 = 7.65 \text{ Pa} - 2.86 \text{ Pa} = 4.79 \text{ Pa}.$$

CD/Video Library/Forward Facing Step

Sudden Contraction

A similar analysis of the turbulent flow in a sudden contraction as shown in Figure 3.14 leads to the introduction of the contraction loss coefficient K_C and the following formula for calculating the pressure drop

$$\Delta p_F = K_C \tfrac{1}{2}\rho \bar{V}_2^2 \qquad (3.23)$$

Note carefully that the contraction loss coefficient is defined in terms of the kinetic energy in the higher speed outlet flow. The value of the contraction loss coefficient can be found in Figure 3.14. If the inlet and outlet areas are equal, there is no pressure loss, and the contraction loss coefficient must be zero. For very small ratios of outlet area to inlet area, the loss coefficient has been found to approach 0.5.

3.3.3 Pump and Fan Laws

The preceding case studies have dealt with calculating the frictional pressure drop in a section of a pipe or in a sudden area change. We now consider the problem of choosing a pump or fan with the performance needed to move fluid through a system once the total pressure drop at the desired flowrate has been determined. It is beyond the scope of this section to address the question of what type of pump or fan should be selected. For

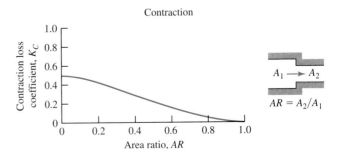

Figure 3.14 Loss coefficients for flow through a sudden contraction.

example, in air-handling applications one can choose a centrifugal or vane–axial fan (Figure 3.15A). Similar choices exist for pumps (Figure 3.15B). When a certain type of device has been chosen, and the manufacturer selected, it is necessary to pick the appropriate size machine from the manufacturer's family of geometrically similar equipment. In fact, the overall process of choosing a pump or fan is called sizing. The pump and fan laws developed next will allow you to use information provided by the manufacturer to predict the characteristics of geometrically similar, differently sized devices. They will also give you the ability to predict the performance of a specific device under different operating conditions.

In our earlier analysis of flow in a pipe or duct system the focus was on the frictional pressure drop. There are other contributions to the total pressure drop in a system, for example, a change in elevation. It is customary to use a parameter called total head, H, in the design of pipe and duct systems. This total head, with dimensions of energy per unit mass (or equivalently $\{L^2 t^{-2}\}$), is a measure of the total load seen by a pump or fan moving fluid through the system. The power, P, required by the pump or fan is also an important parameter in the design of these systems. Thus, in analyzing the performance of a pump or fan, both the head and power are considered to be important dependent variables.

We begin our analysis with the observation that for geometrically similar machines of a given type, only one length scale is required to specify the machine geometry. This length scale is conveniently taken to be the diameter D of the impeller or other rotating element. We assume that the head and power of a fan or pump depends on ω, the angular speed of the impeller, the volume flowrate, and the density and viscosity of the fluid. Thus, we postulate that the head and power are functions of these variables:

$$H = f_1(D, Q, \omega, \rho, \mu) \quad \text{and} \quad P = f_2(D, Q, \omega, \rho, \mu)$$

A dimensional analysis (to be performed in Chapter 9) would show that the dimensionless head can be expressed as follows

$$\frac{H}{\omega^2 D^2} = g_1\left(\frac{Q}{\omega D^3}, \frac{\rho D^2 \omega}{\mu}\right) \tag{3.24a}$$

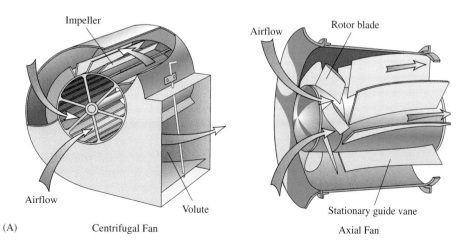

Figure 3.15 Schematics of common designs of (A) fans and (B) pumps. The three-lobe, gear, and sliding-vane devices are all rotary pumps.

126 | 3 CASE STUDIES IN FLUID MECHANICS

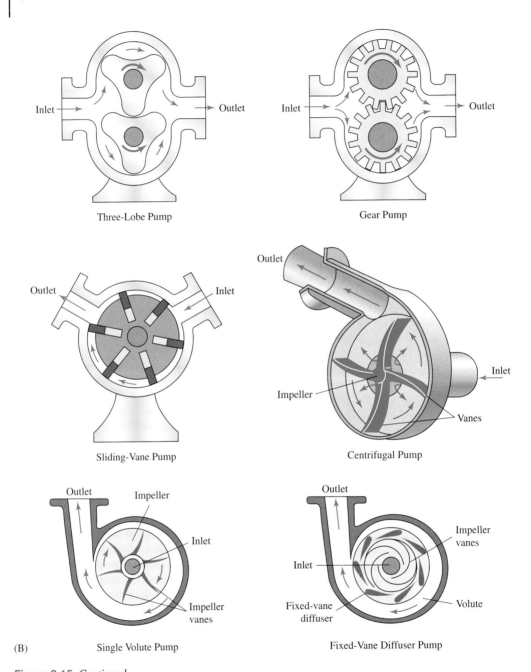

(B)

Figure 3.15 Continued.

while the dimensionless power may be written as

$$\frac{P}{\rho\omega^3 D^5} = g_2\left(\frac{Q}{\omega D^3}, \frac{\rho D^2 \omega}{\mu}\right) \qquad (3.24b)$$

In pump and fan engineering, the dependent dimensionless groups $H/\omega^2 D^2$ and $P/\rho\omega^3 D^5$ are known as the head and power coefficients, respectively. The independent dimensionless group $Q/\omega D^3$ is known as the flow coefficient, while the group $\rho D^2 \omega/\mu$ can be considered to be a form of the Reynolds number because the product $D\omega$ has the dimensions of velocity.

In considering the scaling of two geometrically similar systems, the principle of similitude, (also discussed in Chapter 9) tells us that all independent dimensionless groups must be the same for each system. However, in dealing with pumps and fans of reasonable size, it is found that the performance is independent of Re as defined earlier. Thus the appropriate scaling law for comparing two pumps or fans in the same family is

$$\frac{Q_1}{\omega_1 D_1^3} = \frac{Q_2}{\omega_2 D_2^3} \qquad (3.25a)$$

If the flow coefficient of two machines are equal, then the head and power coefficients are also equal:

$$\frac{H_1}{\omega_1^2 D_1^2} = \frac{H_2}{\omega_2^2 D_2^2} \quad \text{and} \quad \frac{P_1}{\rho\omega_1^3 D_1^5} = \frac{P_2}{\rho\omega_2^3 D_2^5} \qquad (3.25b)$$

These equations are known as the pump laws or fan laws. Not only do they relate the performance of two differently sized machines in the same family, but they also allow us to determine how a given machine will operate under a new set of operating conditions.

EXAMPLE 3.8

To upgrade a ventilation system it is required that the flowrate be increased from 5000 ft³/min to 8000 ft³/min. This is to be accomplished by increasing the angular velocity of the ventilation fan. If the current system operates at a fan rpm of 1000, what fan rpm is required for the upgrade? What will be the power increase for the upgrade?

SOLUTION

Use the fan law, Eq. 3.25a, to determine the new angular velocity (noting that the fan is the same so the characteristic dimension D is constant).

$$\left(\frac{Q}{D^3\omega}\right)_{\text{upgrade}} = \left(\frac{Q}{D^3\omega}\right)_{\text{existing}}$$

$$\omega_{\text{upgrade}} = Q_{\text{upgrade}}\left(\frac{\omega}{Q}\right)_{\text{existing}} = (8000 \text{ ft}^3/\text{min})\left(\frac{1000 \text{ rpm}}{5000 \text{ ft}^3/\text{min}}\right) = 1600 \text{ rpm}$$

Now use Eq. 3.25b to find the increase in power (with fluid density constant).

$$\left(\frac{P}{\omega^3 D^5 \rho}\right)_{\text{upgrade}} = \left(\frac{P}{\omega^3 D^5 \rho}\right)_{\text{existing}}$$

$$\frac{P_{\text{upgrade}}}{P_{\text{existing}}} = \frac{\omega^3_{\text{upgrade}}}{\omega^3_{\text{existing}}} = \frac{(1600 \text{ rpm})^3}{(1000 \text{ rpm})^3} = 4.096$$

Thus, the increase in rpm of 60% results in a power increase of over 300%.

CD/Boundary Layers

3.3.4 Flat Plate Boundary Layer

The case studies thus far have involved internal flow. A flow is classified as internal if the fluid moves within an interior space defined by a number of bounding walls. Pipe flow is obviously of this type, as is the flow in a pump. Engineers also deal with many important external flows, i.e., flows in which a fluid moves around an object. An external flow also occurs whenever a body such as a vehicle moves through a fluid. The next three case studies deal with external flows.

Consider what happens when flow occurs over a flat plate. As shown in Figure 3.16, the fluid at the plate surface does not move relative to the plate. A short distance away from the plate, however, the fluid is moving at the free stream velocity. The effect of viscosity is to create a boundary layer near the plate in which the velocity changes smoothly and continuously from zero on the plate to the free stream value.[1] The boundary layer thickness increases downstream of the leading edge, and the flow in the boundary layer eventually changes from laminar to turbulent (see Figure 3.17). Because there is a transverse velocity gradient at the plate surface, the fluid exerts a shear stress on the plate that results in a drag force (recall that Newton's law of viscosity relates the shear stress to the velocity gradient via the fluid viscosity).

CD/History/Ludwig Prandtl

A quantity of great interest in the flat plate boundary layer is the wall shear stress. If we know how the wall shear stress varies along the plate, we can calculate the

---

[1] Given that the fluid velocity and viscous effects are likely to be important, which dimensionless group do you expect to see play a major role in the model for the flat plate boundary layer?

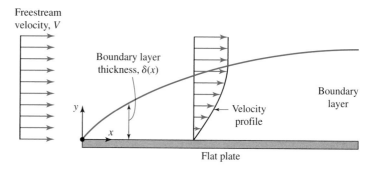

Figure 3.16 Development of the boundary layer on a flat plate.

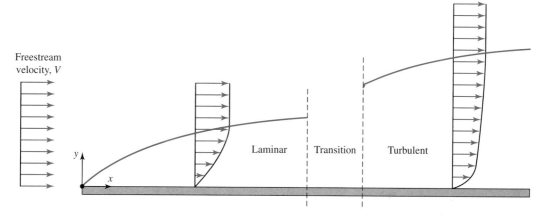

Figure 3.17 Laminar-to-turbulent transition of the boundary layer on a flat plate.

The concept of a boundary layer was conceived by Ludwig Prandtl, who reasoned that in a high Reynolds number flow over a body, viscous effects would be significant only within the boundary layer. His boundary layer theory was one of the most important contributions to fluid mechanics in the twentieth century.

frictional force applied by the fluid to the plate. The flat plate boundary layer may be used to model flow over relatively flat surfaces such as ship hulls and the walls of various structures, and as a crude approximation to the more complex boundary layers on airplane wings, fuselages, and similar surfaces.

Observations suggest that in an incompressible flow at high Re the shear stress τ_W on the wall in a flat plate boundary layer (Figure 3.18) depends on the distance from the leading edge x, the freestream velocity V, and the fluid density and viscosity. Thus we propose a relationship between these variables of the form:

$$\tau_W = f(x, V, \rho, \mu)$$

Dimensional analysis reveals that this relationship can be expressed as

$$\frac{\tau_W}{\frac{1}{2}\rho V^2} = g\left(\frac{\rho V x}{\mu}\right) \tag{3.26}$$

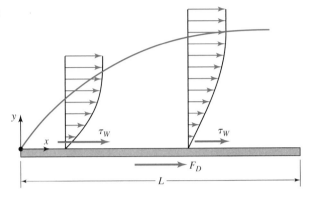

Figure 3.18 Shear stress due to flow over a flat plate.

It is customary in boundary layer analysis to define the skin friction coefficient C_f as

$$C_f = \frac{\tau_W}{\frac{1}{2}\rho V^2} \tag{3.27}$$

and to define a Reynolds number based on the distance x from the leading edge as

$$Re_x = \frac{\rho V x}{\mu} \tag{3.28}$$

From the dimensional analysis we can also conclude that there is a relationship between the skin friction coefficient and the Reynolds number of the form

$$C_f = C_f(Re_x) \tag{3.29}$$

The force exerted by the shear stress on one side of a plate of width w and length L shown in Figure 3.18 is found by integrating the (variable) shear stress along the length of the plate. This frictional force, or drag (since it acts in the flow direction), is given by

$$F_D = w \int_0^L \tau_W(x)\, dx$$

and can also be written in terms of the skin friction coefficient as

$$F_D = w \int_0^L \frac{1}{2}\rho V^2 C_f(x)\, dx \tag{3.30}$$

We can calculate the drag on a flat plate due to a laminar or turbulent boundary layer by using Eq. 3.30, provided we have an expression for the appropriate skin friction coefficient.

 CD/Special Features/Blasius Boundary Layer Growth

Laminar Boundary Layer: H. Blasius, a student of Prandtl's, developed an approximate solution for the laminar flat plate boundary layer that gave the following

A laminar boundary layer usually transitions to turbulence very close to the leading edge of a plate, so close that in calculating the drag, the laminar portion of the boundary layer can often be ignored and the whole boundary layer treated as if it were turbulent from the leading edge.

expression for the skin friction coefficient:

$$C_f = \frac{0.664}{\sqrt{Re_x}} \quad (3.31)$$

From empirical observation we know that the transition to turbulence occurs at about $Re_x = 5 \times 10^5$, so Eq. 3.31 is limited to $Re_x < 5 \times 10^5$.

Turbulent Boundary Layer: An approximate model of the velocity distribution in turbulent flow (see Figure 3.17) yields an expression for the skin friction coefficient of the form:

$$C_f = \frac{0.0594}{(Re_x)^{1/5}} \quad (3.32)$$

EXAMPLE 3.9

A cruise missile 5 m long and 1 m in diameter is cruising at 200 m/s at an altitude of 500 m. If the boundary layer on the missile skin is modeled as that over a flat plate, what is the drag force on the missile due to skin friction?

SOLUTION

From Appendix A we find for air at 500 m, $\rho = 1.17$ kg/m^3 and $\mu = 1.77 \times 10^{-5}$ (N-s)/m^2. First we use the critical Reynolds number of 5×10^5 to locate the transition to turbulence:

$$x_{cr} = Re_{cr}\frac{\mu}{\rho V} = 5 \times 10^5 \frac{[1.77 \times 10^{-5}\ (\text{N-s})/\text{m}^2]}{(1.17\ \text{kg/m}^3)(200\ \text{m/s})} = 0.04\ \text{m}$$

The laminar region is small enough to be neglected. We will use Eq. 3.32 for the skin friction coefficient and calculate the drag force on the wetted surface by using Eq. 3.30.

$$F_D = w\frac{1}{2}\rho V^2 \int_0^L C_f\,dx = w\frac{1}{2}\rho V^2 \int_0^L \frac{0.0594}{(Re_x)^{1/5}}\,dx = w\frac{1}{2}\rho V^2 \int_0^L \frac{0.0594}{(\rho V x/\mu)^{1/5}}\,dx$$

$$F_D = w\frac{1}{2}\rho V^2 \left(\frac{0.0595}{(\rho V/\mu)^{1/5}}\right)\int_0^L x^{-1/5}\,dx = w\frac{1}{2}\rho V^2 \left[\frac{0.0595}{(\rho V/\mu)^{1/5}}\right]\left(\frac{x^{4/5}}{4/5}\bigg|_0^L\right)$$

In this case the "width" is the circumference of the missile, πD. Substituting appropriate numerical values yields:

$$F_D = \pi(1\ \text{m})\frac{1}{2}(1.17\ \text{kg/m}^3)(200\ \text{m/s})^2 \frac{0.0594}{\left\{\left[\frac{(1.17\ \text{kg/m}^3)(200\ \text{m/s})}{1.77 \times 10^{-5}\ (\text{N-s})/\text{m}^2}\right]\left[\frac{1\ \text{N}}{1\ (\text{kg-m})/\text{s}^2}\right]\right\}^{1/5}} \frac{(5\ \text{m})^{4/5}}{4/5}$$

$$= 744\ \text{N}$$

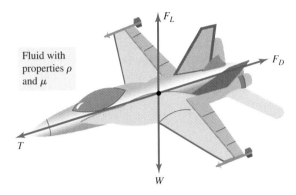

Figure 3.19 Free body diagram of a nonaccelerating body showing that the drag and thrust are equal.

3.3.5 Drag on Cylinders and Spheres

One of the most important problems in fluid mechanics is to determine the drag on a body immersed in a moving fluid. Drag is the component of the total retarding force acting on the body in the direction of the oncoming stream. A bit of thought shows that drag can be due to unbalanced pressures on the fore and aft surfaces of a body as well as to skin friction in the form of shear stress on the wetted surface. Applying Newton's second law to the body shown in Figure 3.19 shows that the thrust and drag forces acting on a nonaccelerating body are equal and opposite. Thus, estimating the force (thrust) needed to move a body through a stationary fluid at constant velocity requires estimating the drag. The power required to move the body through the fluid is the product of the magnitude of the thrust (or drag) and the speed of the body.

The ability to calculate drag is a critical element in the design of virtually all modern modes of transportation. Historically, problems of this type have been investigated

---

EXAMPLE 3.10

What is the power required to fly the cruise missile in Example 3.9? Assume that the drag is primarily due to skin friction.

SOLUTION

The power required is the product of the thrust and the flight speed. Since the missile is at constant velocity the thrust is equal to the drag 720 N and the flight speed is 200 m/s, the power required is

$$P = (720 \text{ N})(200 \text{ m/s})\left(\frac{1 \text{ J}}{1 \text{ N-m}}\right)\left(\frac{1 \text{ W}}{1 \text{ J/s}}\right) = 144{,}000 \text{ W} = 144 \text{ kW}$$

Another approach to this problem would be to find a drag coefficient that includes both the effects of skin friction and the pressure distribution as discussed shortly.

experimentally by using a wind tunnel to provide a flow over a scale model, with the results presented in terms of the drag coefficients. Analytical results are available to estimate the drag force in a very few cases, but generally engineers rely on a large body of empirical results. In this section we discuss the drag in steady, incompressible flow for two very simple geometries: an infinitely long circular cylinder and a sphere. For simplicity, the flow approaching the cylinder is required to be perpendicular to the axis of the cylinder, and neither the cylinder nor the sphere is rotating.

 CD/Video Library/Flow Past a Cylinder

Cylinder

The circular cylinder is a common structural shape. Examples include bridge cables, chimney pipes, wing struts, and flagpoles. Although the geometry of a circular cylinder is simple, the wake of a cylinder can be quite complex (see Figure 3.3).

Now consider the steady flow over a cylinder. We are interested in the drag force F_D on a cylinder of diameter D and length L. The drag will depend on these two geometric parameters as well as on the velocity, density and viscosity of the fluid. We summarize the proposed relationship mathematically as $F_D = f(D, L, V, \rho, \mu)$. Dimensional analysis (details to be provided in Chapter 9) then shows that the relationship between these groups is

$$\frac{F_D}{\frac{1}{2}\rho V^2 DL} = g\left(Re, \frac{L}{D}\right) \quad (3.33)$$

where the Re is based on the cylinder diameter. The standard way to present this result is to write:

$$F_D = C_D \frac{1}{2}\rho V^2 DL \quad (3.34)$$

where the drag coefficient for a cylinder is defined as

$$C_D = \frac{F_D}{\frac{1}{2}\rho V^2 DL} \quad (3.35)$$

Note that from Eq. 3.33 the drag coefficient is $C_D = g(Re, L/D)$, or simply

$$C_D = C_D\left(Re, \frac{L}{D}\right) \quad (3.36)$$

From Eq. 3.36 we conclude that the drag on a cylindrical body depends on the Reynolds number and on the aspect ratio of the cylinder. As the length of the cylindrical body approaches infinity, the flow over the cylinder anywhere along its length must become independent of position. In this limiting case of long cylinders, the drag coefficient for a cylinder is only a function of Re:

$$C_D = C_D(Re) \quad (3.37)$$

CD/Video Library/Flow Past a Sphere

Sphere

The drag on a sphere is needed to predict the behavior of spherical objects of all sizes including pollen and the particles in mists and smoke, as well as the balls used in golf, soccer, and baseball. The drag force F_D on a smooth sphere of diameter D will depend on this single geometric parameter as well as on the fluid velocity, density, and viscosity.

In this case we postulate the relationship as

$$F_D = f(D, V, \rho, \mu)$$

and find that the drag is given by

$$F_D = C_D \frac{1}{2} \rho V^2 \frac{\pi D^2}{4} \tag{3.38}$$

The drag coefficient is defined for a sphere by

$$C_D = \frac{F_D}{\frac{1}{2}\rho V^2 (\pi D^2/4)} \tag{3.39}$$

Since there is only one length scale in the flow, namely the sphere diameter, the drag coefficient for a sphere depends only on the Reynolds number:

$$C_D = C_D(Re) \tag{3.40}$$

where Re is based on the sphere diameter.

Drag Coefficient

At this point the problem of calculating the drag on a cylinder or sphere is reduced to finding information on the variation of the drag coefficient with Re. Reynolds numbers of interest may range from near zero to 10^8 or even larger, depending on the application: contrast the Re for wind flow over a strand of a spiderweb with that for a guide wire on an early biplane in flight at 90 mph.

Flows for which $Re \ll 1$ are called creeping flows. A creeping flow is dominated by viscous forces. There are analytical results for creeping flows over cylinders and spheres, and we can take advantage of these to deduce the drag coefficients for $Re \ll 1$. An approximate solution due to Oseen for creeping flow over a very long cylinder gives the following formula for the drag coefficient:

$$C_D = \frac{4\pi}{Re\left[\ln\left(2\frac{L}{D}\right) - 0.72\right]} \quad \text{or} \quad C_D = \frac{8\pi}{Re\left[\log_{10}\left(\frac{7.4}{Re}\right)\right]} \tag{3.41}$$

The exact solution for creeping flow over a sphere was derived by Stokes. This solution gives the drag coefficient for a sphere in the creeping flow regime as

$$C_D = \frac{24}{Re} \qquad (3.42)$$

For higher Reynolds numbers we can take advantage of empirical data and read the drag coefficients for flow over a sphere or cylinder from Figure 3.20. The interesting variations in drag coefficient with increasing Reynolds number reflect changes in the flow structure. These changes will be discussed in more detail in Chapter 14 on external flow.

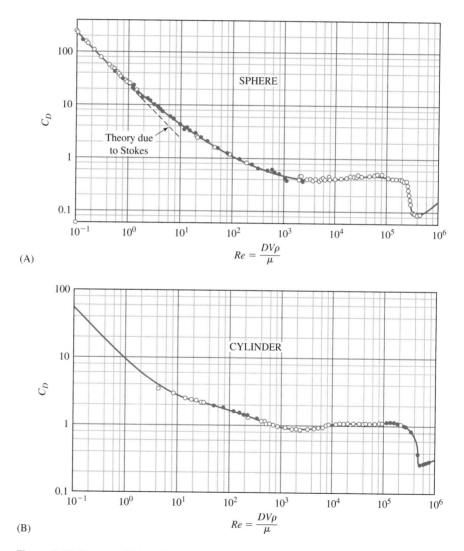

Figure 3.20 Drag coefficient for (A) a smooth sphere and (B) an infinite cylinder as a function of Reynolds number.

EXAMPLE 3.11

What is the drag force on a spherical particle 1 μm in diameter settling in air at $V = 0.1$ m/s?

SOLUTION

From Appendix A we find for air at 20°C: $\rho = 1.204$ kg/m³ and $\mu = 1.82 \times 10^{-5}$ (N-s)/m². The first step is to calculate the Reynolds number:

$$Re = \frac{\rho V D}{\mu} = \frac{(1.204 \text{ kg/m}^3)(0.1 \text{ m/s})(1 \times 10^{-6} \text{ m})}{1.82 \times 10^{-5} \text{ (N-s)/m}^2} = 6.6 \times 10^{-3}$$

Since $Re \ll 1$, we can use Eq. 3.42, $C_D = 24/Re$, for the drag coefficient. Finally, we use Eq. 3.38 to calculate the drag force:

$$F_D = C_D \frac{1}{2} \rho V^2 \frac{\pi D^2}{4} = \left(\frac{24}{Re}\right) \frac{1}{2} \rho V^2 \frac{\pi D^2}{4}$$

$$F_D = \left(\frac{24}{6.6 \times 10^{-3}}\right) \frac{1}{2} (1.204 \text{ kg/m}^3)(0.1 \text{ m/s})^2 \frac{\pi (1 \times 10^{-6} \text{ m})^2}{4} = 1.7 \times 10^{-11} \text{ N}$$

EXAMPLE 3.12

A radio transmission tower is 1000 ft tall and employs 0.5 in. diameter wire cables to stabilize and strengthen the structure as shown in Figure 3.21. What is the normal force on a cable in the highest expected wind of 100 mph (146.7 ft/s)?

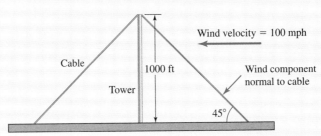

Figure 3.21 Schematic of radio transmission tower for Example 3.12.

SOLUTION

From the geometry, the length of the longest cable is 1414 ft, and the wind velocity is 146.7 ft/s. The component of wind velocity normal to the cable is

(146.7 ft/s cos 45°) = 103.7 ft/s. From Appendix A we find for air at 70°F: $\rho = 0.002329$ slug/ft^3 and $\mu = 3.82 \times 10^{-7}$ (lb$_f$-s)/ft^2. Next calculate Reynolds number as

$$Re = \frac{\rho V D}{\mu} = \frac{(0.002329 \text{ slug/ft}^3)(103.7 \text{ ft/s})(0.5/12 \text{ ft})}{3.82 \times 10^{-7} \text{ (lb}_f\text{-s)/ft}^2} = 2.63 \times 10^4$$

From Figure 3.20b we read a drag coefficient for a cylinder of ~1.2. Next we compute the force acting normal to the cable with Eq. 3.34:

$$F_D = C_D \tfrac{1}{2}\rho V^2 DL = (1.2)(0.5)(0.002329 \text{ slug/ft}^3)(103.7 \text{ ft/s})^2(0.5/12 \text{ ft})(1000 \text{ ft}) = 626 \text{ lb}_f$$

The aspect ratio of the cable is over 3×10^4, so the assumption, implicit in using Figure 3.20b, that it is an infinite cylinder is appropriate.

CD/Video Library/Flow Past an Airfoil

3.3.6 Lift and Drag on Airfoils

A wing is a specially shaped body designed to produce lift when exposed to a stream of fluid. Lift is defined to be the component of fluid force acting on a body at a right angle to the oncoming stream. Thus, lift is a vertical force for a vehicle or object in level flight and may be thought of as being created by unbalanced pressures acting on the top and bottom of the object. The pressure on a wing, for example, is much higher on the bottom surface than on the top surface. The total lift developed by a wing supports the weight of an aircraft.

Many factors influence the design of a wing. The cross section at any given point along a wing has the form known as an airfoil. This airfoil shape is carefully designed to maximize lift and minimize drag. There are many different airfoil shapes for different applications such as airplane wings, propellers, and impeller blades in turbomachines. Example airfoil shapes are shown in Figure 3.22. In this section we discuss the problem of calculating the total lift and drag produced by a wing with a constant airfoil shape all along its length under the assumption that the wing is effectively infinitely long. Real wings of finite length are subject to end effects, which lower their performance. Airfoils are discussed in more detail in Chapter 14 on external flow.

The standard nomenclature for airfoil geometry is illustrated in Figure 3.23. In steady subsonic flow the lift and drag forces, F_L and F_D, respectively, are each found to depend on the thickness t, span b, chord length c, and angle of attack α. They also depend on the freestream velocity V, and on the fluid density and viscosity. If we postulate the dependence of lift and drag on the physical parameters as

$$F_L = f(t, b, c, V, \rho, \mu) \quad \text{and} \quad F_D = f(t, b, c, V, \rho, \mu)$$

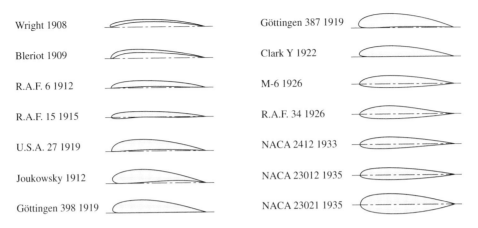

Figure 3.22 Important airfoil shapes in the history of aerodynamics.

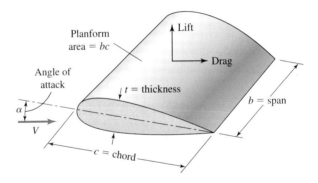

Figure 3.23 Airfoil nomenclature.

dimensional analysis leads to the following standard relationships among dimensionless groups:

$$\frac{F_L}{\frac{1}{2}\rho V^2 bc} = g_1\left(Re_c, \frac{t}{c}, \frac{b}{c}, \alpha\right) \quad \text{and} \quad \frac{F_D}{\frac{1}{2}\rho V^2 bc} = g_2\left(Re_c, \frac{t}{c}, \frac{b}{c}, \alpha\right)$$

where Re_c is the Reynolds number based on chord length, i.e., $Re_c = \rho V c/\mu$. The lift and drag coefficients for an airfoil section are defined as

$$C_L = C_L\left[Re_c, \frac{t}{c}, \frac{b}{c}, \alpha\right] \quad \text{and} \quad C_D = C_D\left[Re_c, \frac{t}{c}, \frac{b}{c}, \alpha\right]$$

thus the lift and drag are given by

$$F_L = C_L \tfrac{1}{2}\rho V^2 bc \quad \text{and} \quad F_D = C_D \tfrac{1}{2}\rho V^2 bc \quad \text{(3.43a, b)}$$

where the product bc is called the planform area. For an infinitely long wing, the ratio of span to chord, b/c, disappears from the expressions for C_L and C_D, and we conclude

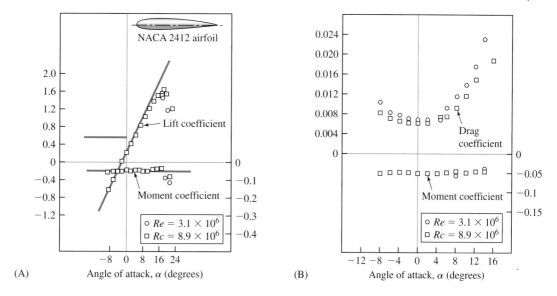

Figure 3.24 Experimental (A) lift and (B) drag coefficients as a function of angle of attack for a NACA 2412 airfoil.

that the lift and drag coefficients of a long wing are a function only of Reynolds number, the geometry of the airfoil as expressed by the ratio of thickness to chord, and the angle of attack.

Lift and drag data were made available for a large number of airfoils by the predecessor to NASA, the National Advisory Committee for Aeronautics (NACA). Figure 3.24 shows lift and drag coefficients for a typical airfoil shape, NACA 2412.

EXAMPLE 3.13

Calculate the lift force on a Cessna 150 wing cruising at an airspeed of 120 mph at an altitude of 5000 ft. The wing is constructed of a NACA 2412 airfoil at an angle of attack of 2°. Its span is 32 ft., 8 in., and the wing planform area is 157 ft².

SOLUTION

From Appendix B we find for air at 5000 ft ($T = 41°F$): $\rho = 2.048 \times 10^{-3}$ slug/ft³ and $\mu = 3.637 \times 10^{-7}$ (lb$_f$-s)/ft². The lift coefficient for the NACA 2412 at 2° angle of attack is ~ 0.3 from Figure 3.24. Next, use Eq. 3.43a to calculate the lift as

$$F_L = C_L \tfrac{1}{2} \rho V^2 bc$$

$$= (0.3)\tfrac{1}{2}(2.048 \times 10^{-3} \text{ slug/ft}^3)\left[120 \text{ mph}\left(1.4667 \frac{\text{ft/s}}{\text{mph}}\right)\right]^2 157 \text{ ft}^2 = 1.5 \times 10^3 \text{ lb}_f$$

Note that the Reynolds number based on the chord length (calculated as area divided by length) is

$$Re_c = \frac{\rho V c}{\mu}$$

$$= \frac{(2.048 \times 10^{-3} \text{ slug/ft}^3)(120 \text{ mph})\left(1.4667 \, \frac{\text{ft/s}}{\text{mph}}\right)(157 \text{ ft}^2/32.667 \text{ ft})}{3.637 \times 10^{-7} (\text{lb}_f\text{-s})/\text{ft}^2}$$

$$= 4.8 \times 10^6$$

which is within the range of the experimental data given in Figure 3.24.

3.4 SUMMARY

In this chapter several case studies were introduced. Each case study had two parts, a brief description of the flow field of interest and the introduction of design formulas used to calculate important quantities of engineering interest. These formulas rely primarily on results obtained using experimental methods, and in particular on the dimensional analysis and modeling tools, which you will learn about eventually, in Chapter 9. The amount of information given in a case study is not unlike what you might find in an engineering handbook.

The case studies included frequent references to dimensionless groups. A dimensionless group is an algebraic combination of the parameters describing a particular flow problem that proves to be both dimensionless as a whole and significant in terms of understanding the flow field. The use of dimensionless groups allows an engineer to classify a fluid mechanics problem, relate it to work by others, and select an effective solution method. Although a large number of dimensionless groups occur in fluid mechanics, only a limited number of them are used on a regular basis. We list five examples.

1. Reynolds number, $Re = \rho V L/\mu$, is the most common dimensionless group in fluid mechanics. It can be interpreted as the ratio of inertial forces to viscous forces. If the Re is small, viscous forces dominate the flow and inertial forces can be neglected. Conversely, if Re is large, inertial forces dominate outside of boundary layers.
2. Euler number, $Eu = \Delta p/\rho V^2$, is the ratio of pressure forces to inertial forces.
3. Froude number, $Fr = V^2/gL$, is the ratio of inertial forces to gravity forces. It is important in the classification of free surface flows.
4. Mach number, $M = V/c$, is the ratio of the velocity scale to the speed of sound in the fluid. The Mach number is important in compressible fluid mechanics and is used to determine when compressible effects must be considered.
5. Weber number, $We = \rho V^2 L/\sigma$, is the ratio of inertial forces to surface tension forces. The Weber number is important in a limited number of instances such as capillary flows.

This chapter concludes with six important case studies: fully developed flow in pipes and ducts, flow through sudden area change, pump and fan laws, flat plate boundary layer, drag on cylinders and spheres, and lift and drag on airfoils. These case studies represent a substantial amount of the material with which an engineer could practice design after a single course in fluid mechanics. Each of these problems may be studied theoretically, but the majority of useful results have been obtained empirically. In this chapter we have introduced each flow, indicated the important dimensionless groups that can be used to understand the flow, and provided formulaic solutions to each flow. These problems were designated as case studies to emphasize the relevance of the subsequent theoretical chapters to everyday engineering problems. We do this by revisiting these problems throughout the book.

PROBLEMS

Section 3.2

3.1 For each of the common dimensionless groups listed, demonstrate that the group is, in fact, dimensionless. In addition, offer a physical interpretation of each dimensionless group.
(a) Reynolds number, Re
(b) Froude number, Fr
(c) Euler number, Eu
(d) Prandtl number, Pr

3.2 For each of the common dimensionless groups listed, demonstrate that the group is, in fact, dimensionless. In addition, offer a physical interpretation of each dimensionless group.
(a) Reynolds number, Re
(b) Mach number, M
(c) Weber number, We
(d) Strouhal number, St

3.3 Air initially at STP is flowing over an airplane wing with a chord of 2 m. If the air velocity is 300 km/h, determine the Reynolds number and Mach number for this flow. At what speed must the plane fly in a standard atmosphere at an altitude of 3000 m for the flow over the wing to have the same value of Re?

3.4 A sphere of diameter 2 mm is moving through glycerin at a velocity of 5 mm/s.
(a) Calculate the Reynolds number for this flow.
(b) Would you characterize this flow as turbulent, laminar, or creeping flow?
(c) Do you think viscous effects are important in this flow?
(d) Do you think inertial effects are important in the previous flow?

3.5 As mentioned in Chapter 2, an engineer must consider compressibility effects when the Mach number exceeds 0.3. What is the flight speed for an aircraft flying in a standard atmosphere at 10,000 ft necessary to achieve this Mach number?

3.6 As mentioned in Chapter 2, an engineer must consider compressibility effects when the Mach number exceeds 0.3. For water at STP, what velocity must the fluid reach to achieve this Mach number? If the water is moving through a 1 in. diameter pipe at $M = 0.3$, what is the corresponding Re number?

3.7 A thin film of SAE 30W oil is experiencing a velocity of 0.75 m/s at a depth of 1.5 mm below its free surface. Calculate the Froude number and the Weber number for this flow. What is the significance of the relative values of Fr and We in this flow?

3.8 In open channel flows the characteristic dimension in the Froude number is the depth of the fluid. What is the minimum fluid velocity necessary to achieve supercritical flow in a channel that is 100 ft deep? What does it mean for a flow of this type to be termed supercritical?

3.9 A thin film ($d = 0.1$ in.) of glycerin is at rest at room temperature. Calculate the Weber number for this situation. Are surface tension effects important in this situation?

3.10 Water is flowing through a piping system at 40°F at atmospheric pressure. The vapor pressure of water at 40°F is 0.122 psia. Use the concept of the Euler number to calculate the minimum velocity for which cavitation damage might become important.

3.11 The pressure coefficient for flow over a cylinder as predicted by inviscid flow theory is $C_p = 1 - (4 \sin^2 \theta)$. If the cylinder is moving through air at 100 mph at sea level, what is the lowest pressure predicted by the theory? Does it seem reasonable?

3.12 A fluid is flowing through a cylindrical pipe of diameter 0.5 m with an oscillating inlet velocity given by $u = U_{max} \cos \omega t$. If $U_{max} = 5$ m/s, and $\omega = 0.2$ s^{-1}, calculate the Strouhal number for this flow.

3.13 A fluid is flowing over a sphere of diameter 1.5 in. at a velocity of 96 mph. If the Strouhal number for this flow is 0.2, estimate the frequency of oscillations in the flow.

3.14 For wind tunnel testing of prototype golf ball dimple patterns, it is known that Re and St must be the same for both the model and actual prototype balls. Experiments with a representative sampling of members of the PGA tour indicate prototype parameters of $V = 250$ ft/s and $\omega = 900$ s^{-1}. The diameter of a golf ball is 1.68 in. The company president wants to use a video clip from your tests in a commercial and has told you to use a geometric factor of 4 : 1 (the model "ball" will be larger than the prototype). Determine the required model fluid velocity and model angular velocity.

3.15 The pressure drop through an oil (SAE 30W at 15°C) supply line at a service station is to be modeled by using the flow of water at 15°C through an identical length of tubing. The required oil velocity in the prototype is known to be 35 cm/s. The Re and Eu of the model and the actual flow must be the same.
(a) Use Reynolds number to find the velocity of the model fluid.
(b) Use Euler number to predict the pressure drop given that the pressure drop in the prototype was 0.05 psi.

3.16 An airplane wing with a 2 m chord is designed to fly through standard atmosphere at an elevation of 8 km and a velocity of 850 km/h. The wing is to be tested in a water tank at STP. What fluid velocity is necessary in the model to ensure that the Reynolds numbers are the same? What other dimensionless groups might be important in this flow?

3.17 You are in charge of testing a 1 : 50 scale model of a flat bottom barge. The prototype fluid is seawater at 15°C, and from your analysis you know that you must obtain equal Reynolds and Froude numbers. What characteristics must your model fluid possess? What fluid would you recommend using for your model?

3.18 At 20°C the thermal diffusivity of water and air are 0.00142 and 0.208 cm^2/s, respectively. What are the Prandtl numbers for each case?

Section 3.3

Section 3.3.1

3.19 Gasoline flows with an average velocity of $U = 1$ ft/s in a $D = 2$ in horizontal pipe. The e/D ratio for this pipe is 0.001. What is the pressure drop in a 50 ft length of pipe?

3.20 Air at STP flows with an average velocity of $U = 1$ cm/s in a $D = 1$ cm horizontal pipe. The e/D ratio for this pipe is 0.001. What is the pressure drop in a 20 m length of pipe?

3.21 Air at STP flows with an average velocity of $U = 100$ cm/s in a $D = 1$ m horizontal duct. The e/D ratio for this pipe is 0.001. What is the pressure drop in a 10 m length of pipe?

3.22 Glycerin flows with an average velocity of $U = 0.5$ in./s in a $D = 0.25$ in. horizontal tube. The e/D ratio for this pipe is 0.0001. What is the pressure drop in a 5 ft length of tube?

3.23 Glycerin flows with an average velocity of $U = 3$ cm/s in a $D = 1$ cm horizontal tube. The e/D ratio for this pipe is 0.0001, the viscosity of glycerin in the flow is 2.0 (N-s)/m², and the density is 1260 kg/m³. The pressure drop in a 1 m length of pipe is 12.8 kPa. Estimate the fluid velocity through the tube.

3.24 Oil flows through a horizontal section of the Alaska Pipeline at the rate of 1.5 million barrels (42 gallons per barrel) per day. At the pumping conditions of 140°F SG = 0.93 and $\mu = 3.5 \times 10^{-4}$ (lb$_f$-s)/ft², the pipe has $e/D = 0.00012$ and $D = 48$ in. What is the pressure drop per mile of pipeline?

3.25 For the design in problem 3.19 it is suggested that smoother pipe ($e/D = 0.00001$) be used to reduce the operating cost. What would now be the pressure drop in the pipe?

Section 3.3.2

3.26 What is the pressure change for air in a round duct due to a sudden change in diameter from 0.6 to 0.4 m? The flowrate in the duct is 0.5 m³/s.

3.27 What is the pressure change for air in a round duct due to a sudden change in diameter from 1.0 ft to 2.0 ft? The flowrate in the duct is 15 ft³/s.

3.28 What is the pressure change for water in a round duct due to a sudden change in diameter from 1.0 ft to 2.0 ft? The flowrate in the duct is 1.5 ft³/s.

3.29 What is the pressure change for water in a round duct due to a sudden change in diameter from 0.4 m to 0.6 m? The flowrate in the duct is 0.05 m³/s.

3.30 What is the pressure change for water in a round duct due to a sudden change in diameter from 0.6 m to 0.4 m? The flowrate in the duct is 0.05 m³/s.

3.31 Two installations for a 4 ft heat exchanger are being considered for a 2 ft diameter wind tunnel. Air flows at 5000 ft³/min. The first design, a sudden expansion and contraction, costs $1000. The second design is a gradual expansion and contraction that has 95% less frictional pressure loss than the first design but costs $5000. The electric power required is estimated as $1.25 Q \Delta p_F$ and costs $0.07/kW-h. Based on the costs described, which design would you choose? Use properties of air at standard conditions for your analysis. What factors besides cost might be important?

3.32 The manufacturer of a room air conditioner for motel rooms has added a sudden contraction in the round internal duct of 8 in. to 6 in. Each unit provides 200 ft³/min of supply air. The electric operating power this modification requires is estimated as $1.25 Q \Delta p_F$ and costs $0.07/kW-h. How much will this design modification cost a 100-room motel each year? Assume air at standard conditions and that each unit operates 8 h/day.

Section 3.3.3

3.33 To upgrade a ventilation system it is required that the air flowrate be increased from 200 m³/min to 300 m³/min. This is to be accomplished by increasing the angular velocity of the ventilation fan. If the current rpm = 800, what rpm is required for the upgrade? What will be the power increase for the upgrade?

3.34 To upgrade a pumping system, it is required that the water flowrate be increased from 50 ft³/min to 80 ft³/min. This is to be accomplished by increasing the angular velocity of the pump impeller. If the current rpm = 100, what rpm is required for the upgrade? What will be the power increase for the upgrade?

3.35 A chemical purifier must be added to the wastewater discharge at a manufacturing plant. It is determined that the purifier will increase the head the pump must overcome by 50%; thus the angular speed must be increased. By what percentage will the power required increase?

3.36 In the situation described in Problem 3.35 the existing pump has a 20 hp motor. What size motor will be required for the new conditions? If the pump operates 5000 h/yr and energy costs $0.07/kW-h, how much more will it cost to operate the pump each year?

3.37 In the installation of duct work for heating and air-conditioning systems, interferences occur that require changes in the designed duct layout. For each 1% increase in head required by the fan, how much more angular speed is required? How much more power is required?

3.38 An exhaust fan for a mine shaft needs to be specified. The choice has been narrowed down to a 48 in. or a 60 in. impeller. Compare the power required to run each impeller.

Section 3.3.4

3.39 A missile 12 ft long and 3 ft in diameter is cruising at 400 mph at an altitude of 1500 ft. If the flow is modeled as that over a flat plate, what is the drag force on the missile?

3.40 A torpedo 2 m long and 0.5 m in diameter is moving through water at STP at 20 m/s. If the flow is modeled as that over a flat plate, what is the drag force on the torpedo?

3.41 A probe 10 cm long and 1 cm in diameter is moving through glycerin at STP at a velocity of 0.3 cm/s. If the flow is modeled as that over a flat plate, what is the drag force on the probe?

3.42 Find the drag force on a flat plate, 1.0×1.0 m, towed at 1.0 m/s through (a) air, and (b) water.

3.43 An 8.5 in. $\times$ 11 in. piece of paper with mass of 1×10^{-4} slugs has a coefficient of static friction with the ground of 0.2. What velocity of air over the paper will cause it to start to move? Assume the boundary layer starts to format at the leading edge of the paper.

3.44 A V-shaped ship's hull 50 m long is submerged to a depth such that the total wetted hull area is 1000 m^2. If the ship is moving at 20 knots and each half of the hull is modeled as a flat plate, what is the drag force on the hull?

Section 3.3.5

3.45 What is the drag force on a cylinder of dimensions $D = 0.04$ in. and $L = 1$ in. falling in air at $V = 3$ in./s? Assume the relative airflow is perpendicular to the cylinder axis.

3.46 What is the drag force on a $D = 2$ mm particle settling in glycerin at STP at $V = 0.03$ m/s?

3.47 What is the drag force on a cylinder of dimensions $D = 1$ mm and $L = 5$ mm falling in glycerin at STP at $V = 2$ cm/s?

3.48 A fiber is settling in air at the terminal velocity where $W = F_D$. How fast is the fiber moving if $D = 0.5$ mm, $L = 15$ mm, and $\rho_{fiber} = 10$ kg/m^3? Assume creeping flow.

3.49 Hotwire anemometers have been used extensively for fluid velocity measurements by taking advantage of the "wind chill effect." The probe shown in Figure P3.1 is heated by electrical current. The device is calibrated by recording the amount of current required to achieve a known temperature for a fluid at a known velocity. If the wire has $D = 1$ mm and $L = 10$ mm, what is the drag force when used in (a) water and (b) air when $V = 10$ m/s?

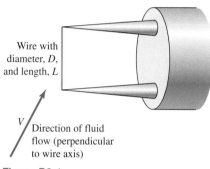

Figure P3.1

Section 3.3.6

3.50 Calculate the lift force on the wing of an airplane cruising at an airspeed of 190 km/h at 1500 m altitude. The wing is constructed of a NACA 2412 airfoil at an angle of attack of 3°. Its span is 10 m and the wing planform area is 17.5 m².

3.51 A streamlined support strut has thickness $t = 1$ in. and has $C_D = 0.02$ where the drag force is given by $F_D = C_D \frac{\rho V^2}{2} t L$. L is the length of the strut. Calculate the force per unit length of strut if it is moving at $V = 10$ mph in water. Compare this result to the force per unit length of a 1 in. diameter circular strut.

3.52 What is the downward force and drag created by the spoiler on a race car traveling at 150 mph? The spoiler consists of a 3 ft × 1 ft NACA 2412 airfoil at −3° angle of attack.

4 FLUID FORCES

- 4.1 Introduction
- 4.2 Classification of Fluid Forces
- 4.3 The Origins of Body and Surface Forces
- 4.4 Body Forces
- 4.5 Surface Forces
    - 4.5.1 Flow Over a Flat Plate
    - 4.5.2 Flow Through a Round Pipe
    - 4.5.3 Lift and Drag
- 4.6 Stress in a Fluid
- 4.7 Force Balance in a Fluid
- 4.8 Summary
- Problems

4.1 INTRODUCTION

 CD/Video Library/Everyday Flows

This chapter establishes a basis for understanding the fluid forces encountered in nature and technology. You will learn about the fluid forces of different types and about their origins. We will show that we must consider both the forces fluids exert on objects they contact and the forces exerted on fluids by their surroundings. An understanding of forces exerted by and on fluids is the key to harnessing fluid forces in the design of vehicles, devices, and structures.

Have you ever seen trees downed by a hurricane or volcanic blast, buildings blown apart by a tornado, or beach cottages destroyed by a storm surge (Figure 4.1)? How are

Figure 4.1 The blast wave from a volcano can also be enormous. These trees were blown down in the aftermath of the eruption of Mount St. Helens in 1980.

the enormous forces necessary to cause such damage generated by air and water in motion? If you have access to a fan you can conduct a simple experiment to illustrate the basic characteristics of the force exerted on a surface by a moving fluid. Take a piece of cardboard and hold it at different angles in the airstream to gain a sense of the dependence on orientation of the force applied to the cardboard by the air. Notice that with the cardboard held normal to the airstream the drag is much greater than if you hold the cardboard parallel to the flow, and at intermediate angles you will detect both a lift and a drag force. Now change the fan speed, and notice that for a given orientation the lift and drag forces increase dramatically with an increase in air speed. You can also readily observe that these forces are proportional to the surface area of the cardboard. This experiment demonstrates that the force applied by a moving fluid to a surface depends on orientation, speed, and surface area.

A fluid at rest is also capable of applying a substantial force on a surface. It does so through its hydrostatic pressure field, the nature of which was introduced in Section 2.3 and will be discussed in more detail in Chapter 5. You may have experienced the effect of hydrostatic pressure after diving into a swimming pool. The sense of discomfort in your ears at a depth of a meter or more is the direct result of the hydrostatic pressure force on the outside of the eardrum being temporarily larger than that on the inside. This hydrostatic force depends on the area of surface involved; in contrast to a moving fluid, however, it does not depend on the orientation of the surface. Can you reduce the pressure in your ears by reorienting your head underwater? The fact that you cannot do so indicates that the hydrostatic force exerted on a surface is independent of the orientation of that surface. This is quite different from the force exerted by a fluid in motion, for which orientation effects are pronounced.

The preceding examples have focused on forces exerted by fluids on their surroundings. It is equally important to consider forces applied to fluids. If you have enjoyed running river rapids in a canoe or raft, or spent a pleasant afternoon sailing or surfing, you might have wondered about the specific forces that causes a river to flow, the wind to blow, and waves to form and break. Understanding the origins and nature of these forces is also of interest to us in predicting and describing the fluid behavior.

As you read this chapter it will be helpful if you keep in mind the case studies introduced in Chapter 3. It should be clear that the values of lift and drag coefficients associated with flow over flat plates, spheres, cylinders, and airfoils are determined by the fluid forces acting on these immersed objects. The pressure drop in flow in a pipe is also determined by the frictional force of the fluid on the wall, as is the drag of the flat plate boundary layer. As we revisit these case studies in this chapter, you should begin to appreciate the important role of fluid forces in practical applications of fluid mechanics.

In the remainder of this chapter we will describe (1) the two basic types of force in fluid mechanics and their origins, (2) the mathematical representation or model used to describe and calculate each type of fluid force, (3) the state of stress in a fluid, and (4) the balance of forces in a fluid. We begin our discussion by placing fluid forces in the broader context of other forces you have encountered in physics and dynamics.

 CD/Dynamics/Newton's 2nd Law of Motion

4.2 CLASSIFICATION OF FLUID FORCES

Inertial force, gravitational force, pressure force, electromagnetic force, and centrifugal force are some of the many different forces in nature that are encountered in engineering and physics. Forces that are evident to an observer in an inertial (nonaccelerating) reference frame are usually termed real forces. Examples include gravitational and electromagnetic forces as well as forces associated with pressure and shear stress.[1] When Newton's second law is written in its inertial form as $\Sigma \mathbf{F} - m\mathbf{a} = 0$, the product $-m\mathbf{a}$ is called the inertial force because it has the dimensions of force, $\{MLt^{-2}\}$. Since this inertial force is evident to an observer in an inertial reference frame, it too should be classified as a real force.

A second category, called apparent forces, are forces that appear to be present to an observer for a noninertial (accelerating) reference frame. They appear in Newton's second law when written in a noninertial reference frame as additional acceleration terms. In a noninertial reference frame the fluid behaves as if forces are acting due to the acceleration. Centrifugal and Coriolis forces are familiar examples of apparent forces.

All forces in fluid mechanics, real or apparent, are divided into two distinctive types: body and surface forces. Body forces, of which gravity and electromagnetic forces are the familiar examples, are long-range forces that act on a small fluid element in such a way that the magnitude of the body force is proportional to the element's mass. Since the element's mass is defined to be density times volume, the magnitude of a body force is also proportional to the volume of a fluid element. Thus, body forces are expressed on a per-unit-volume basis in units such as newtons per cubic meter (N/m³) or pound-force per cubic foot (lb_f/ft³), and on a per-unit-mass basis with units of acceleration. In fluid mechanics, body forces are considered to be external forces; i.e., they are thought of as acting on a fluid, but not as forces applied by a fluid. Body forces exert

---

[1]Recall that both pressure (which is a form of a normal stress) and shear stress have units of force per unit area, so that p or τ must be multiplied by the applied area to obtain the force.

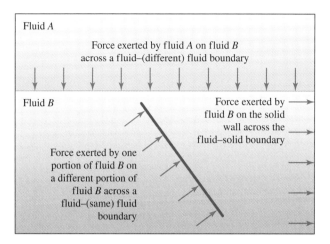

Figure 4.2 Examples of surfaces forces acting on fluid boundaries.

their influence on a fluid at rest or in motion without the need for physical contact between the external source of the body force and the fluid. The apparent forces described above also act in this way. Thus, centrifugal and Coriolis forces are categorized as body forces.

Surface forces, such as those exerted by pressure or shear stress, are short-range forces that act on a fluid element through physical contact between the element and its surroundings. As shown in Figure 4.2, a surface force is exerted across every boundary or interface between a fluid and another material. The second material may be a solid, another portion of the same fluid, or a different fluid. Surface forces are also frequently referred to as contact forces. In fluid mechanics, surface forces are thought of as acting on a fluid, and also as applied by a fluid to its surroundings. These forces exist at every interface wetted by a fluid and are present irrespective of whether the fluid is at rest (pressure only) or in motion (pressure and shear stress). Since the magnitude of a surface force is proportional to the contact area between the fluid and its surroundings, surface forces are expressed in units of force per unit area (e.g., N/m² or lb_f/ft^2).

4.3 THE ORIGINS OF BODY AND SURFACE FORCES

Body and surface forces arise from the underlying structure of matter and the four fundamental forces in the universe: gravitational, electromagnetic, and strong and weak intermolecular. The cause of a body force such as gravity is the presence of matter itself. In physics you learned that the gravitational force $\mathbf{F}_G$ exerted by a particle of mass M_1 on a second particle of mass M_2 is given by

$$\mathbf{F}_G = \left(\frac{GM_1M_2}{r^2}\right)\mathbf{r}$$

where G is the universal gravitational constant, r is the distance between the particles, and $\mathbf{r}$ is a unit vector pointing from the second particle toward the first. Although it is true that one portion of a fluid exerts a gravitational attraction for other portions of the

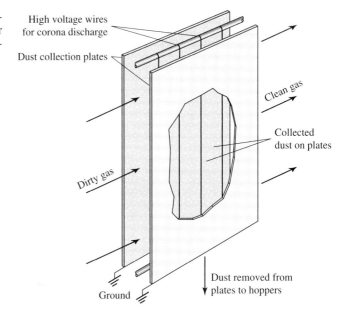

Figure 4.3 Schematic of an electrostatic precipitator, which is used to clean particulate matter from gas flows such as the exhaust from a coal-fired boiler.

same fluid, this effect is negligible unless the problem involves a mass of fluid the size of a planet or star. In that case the fluid is referred to as self-gravitating. In normal engineering practice, Earth's gravity field is assumed to be due to the mass of the planet and unaffected by the presence of fluid. The variation in Earth's gravitational force with distance is so small that all portions of a fluid volume of normal dimensions experience a gravitational force of equal magnitude. Thus, the gravitational force clearly has the characteristic of a body force: it is an external force that acts on the mass or volume of fluid.

Electromagnetic forces are also body forces and have the same long-range character. You are unlikely to encounter electromagnetic forces in fluid mechanics unless you work on specialized applications involving intense electric and magnetic fields. Examples include electrostatic precipitation (see Figure 4.3) and magnetohydrodynamics. In any case it should be evident that body forces arise from the fundamental gravitational and electromagnetic forces in nature.

Now let us look at the origins and characteristics of surface forces in more detail. Consider the surface force acting on the interface between a fluid and a solid piston in the system shown in Figure 4.4A. When an external force is applied to the piston, the piston moves and fluid in the cylinder is compressed. What is actually happening at the fluid–piston interface? We will consider this question from the macroscopic perspective first, then from the molecular perspective. Figure 4.4B shows a macroscopic force balance on this infinitely thin, and therefore massless interface. Surface forces are applied to this interface by the piston and by the fluid. Since the interface is massless, it can have no momentum; Newton's second law shows that the two forces are equal in magnitude and opposite in direction. This equality holds regardless of whether the interface, piston, and fluid are moving. Thus our first conclusion about surface forces is that the force exerted by a fluid on a solid is always equal and opposite to the force exerted by a solid on a fluid.

4.3 THE ORIGINS OF BODY AND SURFACE FORCES

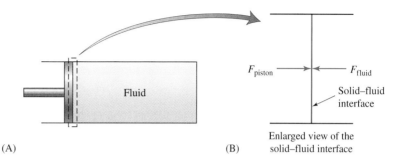

Figure 4.4 (A) Interface between a fluid and solid. (B) Enlargement of the massless interface.

Now consider the fluid–piston interface from a molecular perspective. The surface forces applied to this interface by the fluid are generated through molecular momentum transfer. For simplicity, we will assume that the fluid in our cylinder is a gas. Gas molecules have a mean and random thermal component to their motion. The mean motion defines the macroscopic gas motion, and the random component causes the molecules to continuously collide with one another. As gas molecules collide, there is on average a net momentum transfer from higher momentum regions to lower momentum regions. This momentum transfer process is the mechanism by which surface forces are transmitted to the interior of a gas. Molecular momentum transfer also occurs in liquids but differs in the details of the interaction process.

How does the gas exert a surface force on the stationary piston in our example? We may think of the force as being transmitted at the molecular level through collisions of moving gas molecules with the molecules of the solid piston. A moving fluid molecule possesses translational momentum. In its encounter with a solid molecule, momentum transfer takes place and is felt as a force on the solid molecule and an equal and opposite force on the gas molecule. After the collision with the solid molecule, the gas molecule rebounds away and continues moving until it undergoes its next collision. This collision is likely to be with a neighboring gas molecule approaching the piston, at which time another momentum transfer occurs but this time within the gas. This latter process is responsible for the existence of stress in the gas or fluid and will be discussed further in a moment.

We have shown that surface forces are the macroscopic consequence of molecular momentum transfer. The transfer process is through collisions, and this idea of a collision is simply another way of talking about the action of short range intermolecular forces. Thus we can think of surface forces as arising from fundamental intermolecular forces. By recognizing the underlying mechanism of surface forces, we see that this force occurs regardless of the volume of fluid or solid involved provided only that the thickness of either material is more than a few molecular diameters in a direction normal to the interface. A surface force therefore depends on the contact area between a fluid and a second material but not upon the volume of either material.

Now that you know about the two types of fluid force and their origins, it is time to introduce the mathematical models that describe body and surface forces.

4.4 BODY FORCES

The body forces encountered in fluid mechanics are vector functions of position and time. We therefore write the body force per unit mass **f** in general as

$$\mathbf{f} = \mathbf{f}(\mathbf{x}, t) \tag{4.1}$$

where $\mathbf{f}(\mathbf{x}, t)$ represents the net long-range force acting on a fluid particle or infinitesimal volume of fluid located at position **x** at time t. In this text we do not discuss body forces that depend on time, so we will write Eq. 4.1 in a Cartesian coordinate system in the form

$$\mathbf{f} = f_x(x, y, z)\mathbf{i} + f_y(x, y, z)\mathbf{j} + f_z(x, y, z)\mathbf{k} \tag{4.2}$$

where (f_x, f_y, f_z) are the three components of the body force, each of which may depend on position.

It is often possible and convenient to select the coordinate system so that the body force vector is aligned with one of the coordinate axes. For example, if we are modeling Earth's gravitational body force in Cartesian coordinates, we can elect to align the z axis vertically upward, parallel to the direction of gravity. The force of gravity, therefore, acts downward in this coordinate system. The resulting expression for the gravitational body force per unit mass is

$$\mathbf{f} = -g\mathbf{k} \tag{4.3}$$

where gravity, g, has dimensions of force per unit mass (or acceleration). We shall always treat gravity as a constant, but you should be alert to situations in which the body force acting on a fluid is not spatially constant. Comparing Eq. 4.3 with the general form of Eq. 4.2, we see that it makes sense to align the coordinate system with the body force if possible.

EXAMPLE 4.1

Consider the case of a liquid film flowing down a tilted flat plate under the action of gravity as shown in Figure 4.5. Write the body force in the coordinate system illustrated in the figure.

Figure 4.5 Schematic of liquid flowing down an incline.

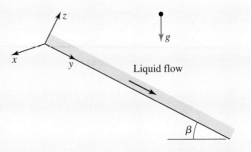

SOLUTION

The coordinate system shown does not have an axis aligned with the gravitational body force. Thus we must use Eq. 4.2 and the geometry to determine the appropriate expressions for the components of the body force. By inspection of Figure 4.5 and the use of Eq. 4.2 we see that the body force is given by

$$\mathbf{f} = 0\mathbf{i} + g\sin\beta\mathbf{j} - g\cos\beta\mathbf{k}$$

In a problem like this we are free to align the coordinates any way we like, but it is usually better to align one of the coordinates in the flow direction. In this case, one coordinate is aligned parallel to the plate and causes the body force to have two components in these coordinates.

Body forces act on individual fluid particles and on entire volumes of fluid. Consider a fluid particle of volume δV in a body force field per unit mass given by $\mathbf{f}$. The long-range nature of a body force ensures that over the scale of a fluid particle, any variation in the body force is negligible. By definition the density ρ does not vary over the scale of a fluid particle, so the mass of the particle is $\rho\delta V$, and the body force $\delta\mathbf{F}_B$ acting on the fluid particle is simply given by the product of the body force per unit mass and the particle mass:

$$\delta\mathbf{F}_B = \rho\mathbf{f}\delta V \tag{4.4}$$

Since the density and body force per unit mass may vary with the position of the fluid particle in the fluid, they must generally be evaluated at the location of the fluid particle.

For a fluid particle in Earth's gravity field with no other body forces acting, we can use Eq. 4.3, $\mathbf{f} = -g\mathbf{k}$, to write Eq. 4.4 in Cartesian coordinates with z upward as

$$\delta\mathbf{F}_B = -\rho g\delta V\mathbf{k} \tag{4.5}$$

The mass of this particle is $M = \rho\delta V$, and the product $\rho g\delta V$ is simply the weight of the fluid particle δW. Thus the body force acting on a fluid particle in this case can also be written as

$$\delta\mathbf{F}_B = -\delta W\mathbf{k} \tag{4.6}$$

EXAMPLE 4.2

The density of the liquid in the rotating cylinder shown in Figure 4.6 is $\rho = 865$ kg/m^3. Because of gravity and rotation this liquid is subjected to a spatially varying body force field of the form $\mathbf{f} = r\Omega^2\mathbf{e}_r - g\mathbf{k}$. The angular velocity is $\Omega = 60$ rpm. Compare the body force acting on 1 cm^3 of fluid located at positions A and B in Figure 4.6.

Figure 4.6 Schematic for Example 4.2.

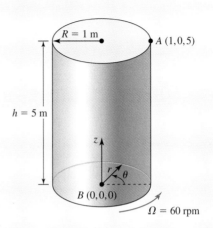

SOLUTION

Applying Eq. 4.4 with $\mathbf{f} = r\Omega^2 \mathbf{e}_r - g\mathbf{k}$ we have

$$\delta \mathbf{F}_B = \rho \mathbf{f} \delta V = \rho(r\Omega^2 \mathbf{e}_r - g\mathbf{k})\delta V$$

The density of a liquid is constant, so for point A, which is located at $(1, 0, 5)$, and with $\delta V = 1 \text{ cm}^3 = 10^{-6} \text{ m}^3$, we find:

$$\delta \mathbf{F}_B = 865 \text{ kg/m}^3 \left[(1 \text{ m}) \left[(60 \text{ rpm}) \left(\frac{2\pi \text{ rad}}{\text{rev}} \right) \left(\frac{1 \text{ min}}{60 \text{ s}} \right) \right]^2 \mathbf{e}_r - 9.81 \text{ m/s}^2 \, \mathbf{k} \right] (10^{-6} \text{ m}^3)$$

$$\delta \mathbf{F}_B = (0.034 \, \mathbf{e}_r - 0.0085 \mathbf{k}) \text{ N}$$

For point B, which is located at $(0, 0, 0)$ we find:

$$\delta \mathbf{F}_B = 865 \text{ kg/m}^3 (0 \mathbf{e}_r - 9.81 \text{ m/s}^2 \, \mathbf{k})(10^{-6} \text{ m}^3) = (-0.0085 \mathbf{k}) \text{ N}$$

In both locations the body force vector has a vertical component due to gravity. In addition, a fluid particle located off the rotation axis also experiences a centrifugal force in the radial direction because of the rotation. Note that we have assumed that 1 cm³ of fluid can be considered to be a fluid particle.

The preceding discussion enables us to represent the body force acting on a single fluid particle. Suppose we have an arbitrary volume of fluid of variable density $\rho(\mathbf{x}, t)$, as shown in Figure 4.7. If the net body force per unit mass from all sources is given by $\mathbf{f}(\mathbf{x})$, how do we calculate the total body force $\mathbf{F}_B$ acting on this finite volume of fluid? The answer is found by realizing that a finite volume of fluid contains a large number of fluid particles. To find the total body force acting on the entire volume, we must add up the body force contributions on each fluid particle using a volume integral.

To derive this volume integral, we first subdivide the total volume of interest into infinitesimal volume elements of size dV as shown in Figure 4.7. The density and body

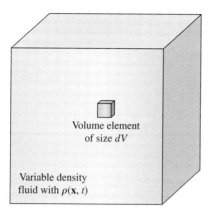

Figure 4.7 A finite volume of fluid with a variable density field.

force will then be spatially constant within each of these infinitesimal volume elements, although they may still vary with the position of the element. Since each of the infinitesimal volume elements may be thought of as a fluid particle, we can apply the results of the last section to calculate the body force on each individual element. Noting that the mass of each volume element is $\rho\, dV$, and the body force per unit mass is **f**, the body force acting on an individual volume element is

$$d\mathbf{F}_B = \rho \mathbf{f}\, dV$$

Finally, to find the total body force $\mathbf{F}_B$ acting on the entire volume of fluid, we use a volume integral to sum the contributions from every infinitesimal volume element:

$$\mathbf{F}_B = \int \rho \mathbf{f}\, dV \tag{4.7}$$

In most problems of interest, setting up and evaluating the volume integral in Eq. 4.7 is straightforward. For example, for a constant density fluid in Earth's gravity field, we can take both the density and body force outside the integral in Eq. 4.7 and obtain

$$\mathbf{F}_B = \int \rho \mathbf{f}\, dV = -\rho g \mathbf{k} \int dV$$

The remaining integral is now equal to the volume of fluid $\mathcal{V}$ so we have

$$\mathbf{F}_B = -\rho g \mathcal{V} \mathbf{k} \tag{4.8a}$$

Since the density is constant, the mass of this volume of fluid is $M = \int \rho\, dV = \rho \mathcal{V}$. The product $\rho g \mathcal{V}$ is simply the weight of the fluid in the volume W. Thus we can also write this result as

$$\mathbf{F}_B = -W \mathbf{k} \tag{4.8b}$$

As you would expect, the gravitational body force acting on a volume of constant density fluid is equal to the weight of the fluid.

If we encounter a variable density fluid in earth's gravity field, we can evaluate the general expression Eq. 4.7 by taking the constant body force $\mathbf{f} = -g\mathbf{k}$ outside the integral while leaving the variable density inside and writing $\mathbf{F}_B = \int \rho \mathbf{f}\, dV = -g\mathbf{k} \int \rho\, dV$. Defining an average density as $\bar{\rho} = (1/\mathcal{V}) \int \rho\, dV$, we can write the gravitational body

EXAMPLE 4.3

Calculate the total body force acting on the water in a cylindrical tank with a height 34 in. and a radius of 11 in. Express your answer in newtons and pounds-force.

SOLUTION

The geometry for this problem is similar to that shown in Figure 4.6 and we will make use of the same coordinate system illustrated there. Gravity is the only body force acting in this case, thus from Eq. 4.8a, we have $\mathbf{F}_B = -\rho g V\mathbf{k}$, where the volume of the tank is easily seen to be $V = \pi R^2 H$. The gravitational body force on the volume of water is thus $\mathbf{F}_B = -\rho g \pi R^2 H \mathbf{k}$. Inserting the given data we find:

$$\mathbf{F}_B = -(998 \text{ kg/m}^3)(9.81 \text{ m/s}^2)(\pi)(11 \text{ in.})^2(34 \text{ in.}) \left(\frac{2.54 \times 10^{-2} \text{ m}}{1 \text{ in.}}\right)^3 \mathbf{k}$$

$$= -2074 \text{ N } \mathbf{k}$$

Converting this to pounds-force, we find: $\mathbf{F}_B = -2074 \text{ N } \mathbf{k} = -466 \text{ lb}_f \mathbf{k}$. You might have noticed that this tank is approximately the size of a 55-gallon drum. In fact the volume is

$$V = \pi R^2 h = \pi (11 \text{ in.})^2 (34 \text{ in.}) \left(\frac{1 \text{ gallon}}{231 \text{ in.}^3}\right) = 55.95 \text{ gallons}$$

If you have ever tried to move a 55-gallon drum filled with liquid, you know it is heavy. Note that to solve this problem by using the volume integral in Eq. 4.7, we would write

$$\mathbf{F}_B = \int \rho \mathbf{f} \, dV = \int_0^H \int_0^{2\pi} \int_0^R \rho(-g\mathbf{k}) r \, dr \, d\theta \, dz$$

and obtain the same result.

---

force as

$$\mathbf{F}_B = -\bar{\rho} g V \mathbf{k} \qquad (4.9\text{a})$$

In this case $M = \int \rho \, dV = \bar{\rho} V$ is the mass of the volume of fluid of interest, so the weight of this fluid is $W = Mg = \bar{\rho} g V$. Thus, as was the case for a constant density fluid,

$$\mathbf{F}_B = -W\mathbf{k} \qquad (4.9\text{b})$$

That is, the gravitational body force acting on a volume of variable density fluid is also equal to the weight of the fluid. In using any of the preceding equations, it is important to recognize that the total body force $\mathbf{F}_B$ is a vector. Thus we have been careful to include the appropriate unit vector in our expressions. It is always possible to write $\mathbf{F}_B$ in terms of its three components by equating components on each side of the corresponding equation.

In calculating the total body force on a volume of fluid, it is important to recognize that if either the density or the body force per unit mass is spatially variable, we may need to evaluate the volume integral in Eq. 4.7. The following example illustrates this process.

EXAMPLE 4.4

The liquid in the cylinder illustrated in Figure 4.8 is subjected to a temperature variation along its axis. This results in a linear density profile in the fluid given by $\rho(z) = \rho_0 - \alpha z$, where $\rho_0 = 1020$ kg/m^3. It is known that the density at the top of the tank is 40 kg/m^3 less than that at the bottom of the tank. The cylinder dimensions are $R = 1$ m and $H = 10$ m. What is the total body force acting on the liquid?

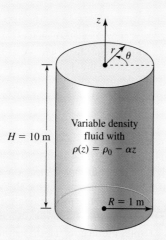

Figure 4.8 Schematic of a cylinder filled with a variable density fluid.

SOLUTION

We are asked to determine the total body force acting on a volume of liquid. The geometry is illustrated in Figure 4.8. The liquid density is variable, its volume is defined in the figure, and the only body force acting is gravity. The constant α can be deduced from the problem statement to be $\alpha = -d\rho/dz = 40$ kg/m^3/10 m $= 4$ kg/m^4. This problem can be solved by using Eq. 4.7: $\mathbf{F}_B = \int \rho \mathbf{f} \, dV$. Although the body force is constant, we must account for the fact that ρ varies significantly over the volume of fluid under investigation. This fluid volume is conveniently described by using cylindrical coordinates. In this case, Eq. 4.7 is

$$\mathbf{F}_B = \iiint \rho \mathbf{f} r \, dr \, d\theta \, dz = \iiint (\rho_0 - \alpha z)(-g\mathbf{k}) r \, dr \, d\theta \, dz$$

The limits of integration are 0 to R for r and $-H$ to 0 for z. The theta integration can be performed by inspection, yielding 2π. Thus we have

$$\mathbf{F}_B = 2\pi(-g\mathbf{k}) \int_{-H}^{0} \int_{0}^{R} (\rho_0 - \alpha z) r \, dr \, dz = \pi R^2 (-g\mathbf{k}) \int_{-H}^{0} (\rho_0 - \alpha z) \, dz$$

Completing the integration we have:

$$\mathbf{F}_B = \pi R^2 (-g\mathbf{k}) \left(\rho_0 H + \alpha \frac{H^2}{2} \right)$$

We can rearrange this result slightly to yield:

$$\mathbf{F}_B = \left(\rho_0 + \alpha \frac{H}{2}\right)(-g\mathbf{k})\pi R^2 H$$

Inserting the data, the total body force is found to be:

$$\mathbf{F}_B = [1020 \text{ kg/m}^3 + 4 \text{ kg/m}^4 \,(10 \text{ m}/2)](-9.81 \text{ m/s}^2 \,\mathbf{k})\pi(1 \text{ m})^2(10 \text{ m})$$
$$= -320{,}191 \text{ N}\,\mathbf{k}$$

Since the only body force is gravity, we could have used Eq. 4.8a ($\mathbf{F}_B = -\bar{\rho} g \mathbf{\Psi} \mathbf{k}$), observing that the average density of the liquid in this case is the value $\bar{\rho} = \rho_0 - \alpha(-H/2)$ at the midheight of the tank, since the density profile is linear. This gives the same result and would have saved a lot of effort.

In Example 4.4 the fluid was at rest. Body forces also act on a volume of fluid in motion. How do we calculate the total body force on a volume filled with moving fluid? To accomplish this task, we must calculate the total body force on a given volume of fluid at a specified instant of time, for in the next instant of time the value may change, perhaps owing to a change in density. Once a specific instant of time has been chosen, we consider the volume to be fixed for the calculation and ignore the motion of the fluid. Mathematically speaking, the limits of integration are fixed, so the volume integral defining the total body force is carried out as usual.

In some applications, the effect of apparent forces must be included in the calculation of the total body force. If this is the case, the total body force is calculated as usual by using Eq. 4.7, but it is necessary to include the applicable apparent body force (i.e., Coriolis, centrifugal, or inertial) in addition to any real body forces present. For example, in Example 4.2, which involved gravity and rotation, this led to writing the body force per unit mass acting on the liquid in a rotating cylinder as $\mathbf{f} = r\Omega^2 \mathbf{e}_r - g\mathbf{k}$. The total body force acting on the liquid in the cylinder in that example may be calculated by using Eq. 4.7 to write

$$\mathbf{F}_B = \int \rho \mathbf{f}\, dV = \int_0^H \int_0^{2\pi} \int_0^R \rho(r\Omega^2 \mathbf{e}_r - g\mathbf{k})r\, dr\, d\theta\, dz$$

You probably realize that the effect of gravity here is a contribution $-\rho g \pi R^2 H \mathbf{k}$, i.e., the usual weight of the liquid term. The effect of the centrifugal force is zero. This may not be intuitively clear unless you realize that the symmetry prevents a net body force in the radial direction.[2] In certain cases an apparent body force may result in a net body force on a volume of fluid. This situation is illustrated next.

---

[2] To evaluate the integral correctly and confirm this statement, you must recognize that the unit vector in the radial direction is actually a function of θ given by $\mathbf{e}_r = \cos\theta \mathbf{i} + \sin\theta \mathbf{j}$. Substitute this expression into the body force integral above and see if you can confirm that the centrifugal body force gives a zero contribution to the total body force.

4.4 BODY FORCES

EXAMPLE 4.5

A tanker truck is accelerating at 1 mph/s as shown in Figure 4.9. The cylindrical tank is filled with gasoline. What is the body force per unit mass acting on the gasoline due to the truck's acceleration and gravity? What is the total body force acting on the gasoline?

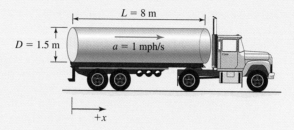

Figure 4.9 Schematic of an accelerating truck.

SOLUTION

We can write the absolute acceleration of any moving body as $\mathbf{a} = a_x\mathbf{i} + a_y\mathbf{j} + a_z\mathbf{k}$. The inertial force per unit mass corresponding to this acceleration is then $-\mathbf{a}$. The sum of the real body force per unit mass and this (apparent) inertial force is therefore $\mathbf{f} - \mathbf{a}$. We can check this conclusion by asking what the body force per unit mass acting on a volume of fluid in free fall in Earth's gravity field would be? In that case, we have $\mathbf{a} = -g\mathbf{k}$, since a volume in free fall will accelerate downward with the acceleration of gravity. The net body force is $\mathbf{f} - \mathbf{a} = [(-g\mathbf{k}) - (-g\mathbf{k})] = 0$, which is correct because free fall simulates zero gravity. For the tanker truck in this example, we find

$$\mathbf{a} = a_x\mathbf{i} + a_y\mathbf{j} + a_z\mathbf{k} = 1 \text{ mph/s } \mathbf{i} = 1.47 \text{ ft/s}^2 \mathbf{i}$$

The total body force per unit mass acting on the gasoline is $\mathbf{f} - \mathbf{a} = -32 \text{ ft/s}^2 \mathbf{k} - 1.47 \text{ ft/s}^2 \mathbf{i}$. Thus, the acceleration of the truck in the $+x$ direction creates an apparent body force in the $-x$ direction. Since the density and body force per unit mass are constant, the total body force acting on the gasoline can be calculated by using Eq. 4.7: $\mathbf{F}_B = \int \rho \mathbf{f} \, dV = \rho \mathbf{f} \int dV = \rho \mathbf{f} \mathcal{V}$. We can interpret this as showing that the truck's acceleration modifies the apparent "gravity" insofar as the fluid is concerned. In this case we know $\mathbf{f} - \mathbf{a} = -32 \text{ ft/s}^2 \mathbf{k} - 1.47 \text{ ft/s}^2 \mathbf{i}$, which shows that the acceleration acts like a sideways inertial force. If the acceleration happened to occur in the vertical direction, the inertial force would modify the felt "gravity." To complete the calculation we would use the known capacity of the tanker truck and obtain the density of gasoline from Appendix A.

4.5 SURFACE FORCES

From our earlier discussion of surfaces forces, we know that in general the force exerted by a fluid on a solid surface depends on the area and orientation of the surface and whether the fluid is or is not moving. This is also true of the surface force exerted by a fluid on an interior surface (see Figure 4.2). In the case of a solid surface, the surface force is thought of as being exerted by the fluid on the solid. In the case of an interior surface, the surface force can be thought of as being exerted by one part of a fluid on another. In either case, the magnitude and direction of the surface force may vary on different parts of the selected surface. To account for these observed features we will employ the concept of a stress vector Σ, and explain the features of the stress vector by using an interior surface lying in the fluid.

Consider an infinitesimal planar element of surface of area dS and outward unit normal $\mathbf{n}$ as shown in Figure 4.10. The infinitesimal element is so small that there is no spatial variation in the surface force acting on the element. We can now define the stress vector Σ as the surface force per unit area acting on this infinitesimal surface element of area dS. Since the stress vector is uniform, the surface force, $d\mathbf{F}_S$ acting on this element is given by the product of the stress vector and the area of the element, or

$$d\mathbf{F}_S = \Sigma dS \qquad (4.10)$$

Alternately, we may write that the stress vector is given by

$$\Sigma = \frac{d\mathbf{F}_S}{dS} \qquad (4.11)$$

The convention regarding the stress vector is that Σ represents the surface force per unit area applied to a surface by the fluid on the side of the surface toward which the unit normal points.

Note that the dimensions of the stress vector are $\{FL^{-2}\}$. Any quantity with these dimensions may be termed a stress. For example, pressure is a normal force per unit area or normal stress, and you are also familiar with the concept of a shear stress. In moving fluids, the surface force per unit area has both normal and tangential components. Thus, it is necessary to represent the surface force per unit area as a vector, which is why we employ the stress vector.

Observation suggests that the stress vector acting on an infinitesimal planar surface element in contact with fluid may be dependent on the position of the element $\mathbf{x}$, on the time t, and on the orientation of the surface element as given by $\mathbf{n}$. To account for these observed features, the stress vector Σ is represented mathematically as

$$\Sigma = \Sigma(\mathbf{x}, t, \mathbf{n}) \qquad (4.12)$$

Figure 4.10 Surface element dS with outward unit normal $\mathbf{n}$ and stress vector Σ.

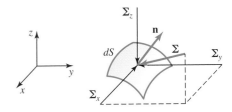

EXAMPLE 4.6

Find the surface force acting on a square element of surface 1 mm on a side located on a surface in the interior of a fluid. The stress vector acting on the front side of this element is given by $\boldsymbol{\Sigma} = -10^5 \text{ N/m}^2\,\mathbf{i} + 2.5 \times 10^{-3} \text{ N/m}^2\,\mathbf{j} + 0\mathbf{k}$. What is the stress vector and surface force acting on the back side of this element?

SOLUTION

Assuming that an element of this size may be considered to be infinitesimal, and using Eq. 4.10, $d\mathbf{F}_S = \boldsymbol{\Sigma}\,dS$, the surface force on the element's front side is the product of the corresponding stress and the element area, $dS = 10^{-6} \text{ m}^2$. Thus we find $d\mathbf{F}_S = -0.1 \text{ N}\,\mathbf{i} + 2.5 \times 10^{-9} \text{ N/m}^2\,\mathbf{j} + 0\mathbf{k}$. On the opposite side of the element the normal is in the opposite direction; thus by Eq. 4.13, the stress vector on the back side has the opposite sign. The stress vector and surface force on the back side of the element are found to be $\boldsymbol{\Sigma} = 10^5 \text{ N/m}^2\,\mathbf{i} - 2.5 \times 10^{-3} \text{ N/m}^2\,\mathbf{j} + 0\mathbf{k}$ and $d\mathbf{F}_S = 0.1 \text{ N}\,\mathbf{i} - 2.5 \times 10^{-9} \text{ N/m}^2\,\mathbf{j} + 0\mathbf{k}$. As expected, the surface force on the back side is equal and opposite to that acting on the front side.

From the force balance argument for a massless interface given earlier, the stress vector acting on the back side of this same element must be equal and opposite to the stress vector acting on the front side. This is expressed by writing

$$\boldsymbol{\Sigma}(\mathbf{x}, t, -\mathbf{n}) = -\boldsymbol{\Sigma}(\mathbf{x}, t, \mathbf{n}) \tag{4.13}$$

The dependence of the stress vector on the orientation of the surface on which it acts is a highly significant feature, which you should keep in mind as we proceed. In Cartesian coordinates the stress vector may be written as

$$\boldsymbol{\Sigma} = \Sigma_x \mathbf{i} + \Sigma_y \mathbf{j} + \Sigma_z \mathbf{k} \tag{4.14}$$

where the three components of the stress vector are given in the usual way by

$$\Sigma_x = \boldsymbol{\Sigma}\cdot\mathbf{i}, \quad \Sigma_y = \boldsymbol{\Sigma}\cdot\mathbf{j}, \quad \text{and} \quad \Sigma_z = \boldsymbol{\Sigma}\cdot\mathbf{k} \tag{4.15}$$

In fluid mechanics, the orientation of any surface is specified by the outward unit normal $\mathbf{n}$ of the surface. Since a surface may have an arbitrary shape and orientation, the outward unit normal vector may vary at different positions on the surface. Compare the unit normal on the flat surface in Figure 4.11A to that on the curved surfaces in Figure 4.11C. If the shape and orientation of a surface are specified, the outward unit normal is known, and vice versa. Thus, the outward unit normal to a surface plays a key role in the description of surface forces.

Notice that the outward unit normal on the surface in Figure 4.11A is pointing from the solid toward the fluid. Why is $\mathbf{n}$ drawn in this direction rather than in the opposite direction? The answer is that the direction of $\mathbf{n}$ is set by a convention. For closed surfaces, like those shown in Figure 4.11C and 4.11D, $\mathbf{n}$ always points away from the interior of

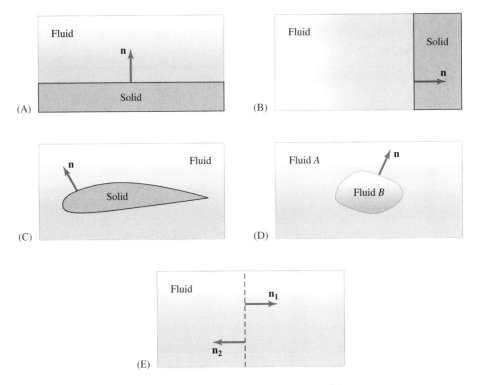

Figure 4.11 Outward unit normal for fluid mechanics problems of five types.

the region bounded by the closed surface. For an open surface, like those shown in Figure 4.11A, 4.11B, and 4.11E, the convention is that **n** points from the surface toward the agent responsible for the surface force. If we are considering the force applied to a solid by a fluid, then **n** points at the fluid (Figure 4.11A). If we are interested in the force applied to the fluid by the solid wall, **n** points at the solid (Figure 4.11B). In what direction does **n** point for the force on an interface within a fluid (Figure 4.11E)? The direction of **n** on this open surface obeys the standard rule: it points toward the fluid on the side of the interface from which the surface force of interest originates. Thus, **n** can point in either direction, and our analysis must identify the agent responsible for the surface force.

To continue our discussion of the stress vector, recall that a fluid in motion exerts normal and tangential stresses on a surface. Thus, as shown in Figure 4.12, we can

Figure 4.12 Relationship between **n**, σ_n, τ, and Σ.

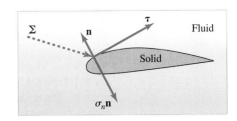

always write the stress vector at any point on the surface as the following sum

$$\Sigma = \sigma_N \mathbf{n} + \boldsymbol{\tau} \tag{4.16}$$

where $\sigma_N \mathbf{n}$ is a vector pointing in the normal direction and $\boldsymbol{\tau}$ is a vector tangent to the surface. From this equation it is evident that σ_N is defined by the dot product

$$\sigma_N = \Sigma \cdot \mathbf{n} \tag{4.17}$$

thus it is appropriate to refer to σ_N as the normal stress. The normal stress σ_N also has dimensions $\{FL^{-2}\}$ and thus has typical units of N/m² (Pa), lb_f/ft^2, or psi. From Eq. 4.17 and our convention for $\mathbf{n}$, we conclude that a positive value for σ_N represents a tensile surface force applied by the fluid to the surface at this point. Since we know that a fluid cannot exert a tensile force under ordinary circumstances, the normal stress σ_N is always negative, and the vector $\sigma_N \mathbf{n}$ must always point into the surface.

Examination of Figure 4.12 and Eq. 4.16 shows that the vector $\boldsymbol{\tau}$ defines the surface force acting in the tangential direction at this point on the surface. If we represent the tangential or shear stress as τ, assumed positive as a convention, then it is clear that τ is the magnitude of the vector $\boldsymbol{\tau}$ in Eq. 4.16, and also has dimensions $\{FL^{-2}\}$ and typical units of N/m² (Pa), lb_f/ft^2, or psi. Keep in mind that in most flows, the normal stress is actually very large compared with the shear stress, often thousands of times larger. This means that in Figure 4.12 we have had to exaggerate the scale of the vector $\boldsymbol{\tau}$; otherwise it would not be visible on the same plot as the vector $\sigma_N \mathbf{n}$. To give you a rough idea of some numbers, a flow may have a shear stress of 100 Pa, and a normal stress of roughly atmospheric pressure or 101 kPa.

Unlike the normal stress, the direction of the shear stress, i.e., the direction of the vector $\boldsymbol{\tau}$, is not known a priori from the surface geometry. Although the line of action of the normal stress is uniquely determined solely by the outward unit normal, the line of action of the tangential stress is known only to lie in the plane tangent to the surface. The direction of the tangential stress within the tangent plane is determined by the flow itself. As you gain experience with fluid mechanics, you will learn to recognize the likely direction of the shear stress on a surface from the direction of the nearby flow. Recall, for example, the case study on the flat plate boundary layer and the alignment of the shear stress on the plate with the freestream flow direction (Figure 3.18). In most cases the shear stress points downstream. However, in the event of flow reversal, which is illustrated for the flow over a cylinder in Figure 3.3, there is also a reversal of the direction of the shear stress. As a matter of guidance, we can say that we expect the shear stress on a surface to generally act in the same direction as the nearby flow.

In most cases, we have a good idea of the direction of the shear stress. If the stress vector and unit normal are given or known, we don't need to worry about the direction of the tangential stress at all, because we can rearrange Eq. 4.16 to get

$$\boldsymbol{\tau} = \Sigma - \sigma_N \mathbf{n} \tag{4.18}$$

and use this equation to determine the vector $\boldsymbol{\tau}$. We do this by first using the known stress vector and unit normal to evaluate Eq. 4.17, $\sigma_N = \Sigma \cdot \mathbf{n}$, thereby obtaining the normal stress, then use the normal stress and unit normal to evaluate Eq. 4.18 and find the vector $\boldsymbol{\tau}$. Once we have this vector, we can determine both the shear stress τ (the magnitude of $\boldsymbol{\tau}$), and its direction.

EXAMPLE 4.7

Find the normal and tangential stresses at the point (1, 2, 1) on the plane defined by $x - y + 1 = 0$, if the stress vector at this point is $\mathbf{\Sigma} = 4\mathbf{i} - 10\mathbf{j} + 2\mathbf{k}$, where the units for stress are *psia*. Choose the side of the plane by selecting the unit normal so that it points away from the xz plane. The relevant geometry is shown in Figure 4.13.

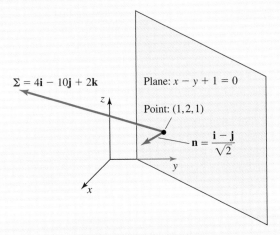

Figure 4.13 Schematic for Example 4.7.

SOLUTION

The normal stress is found first by using Eq. 4.17, $\sigma_N = \mathbf{\Sigma} \cdot \mathbf{n}$. For a surface defined by an equation of the form $f(x, y, z) = 0$, we can use the gradient to construct the normal: $\mathbf{n} = \nabla f / |\nabla f|$. Note that there are two possible normals to the surface, one for each side. In this case, we have $f(x, y, z) = x - y + 1$ so with $\nabla f = \mathbf{i} - \mathbf{j}$ and $|\nabla f| = \sqrt{2}$, we find $\mathbf{n} = (\mathbf{i} - \mathbf{j})/\sqrt{2}$. The opposite normal is $\mathbf{n} = (-\mathbf{i} + \mathbf{j})/\sqrt{2}$, which is the normal we want. To find the normal stress, we now use Eq. 4.17:

$$\sigma_N = \mathbf{\Sigma} \cdot \mathbf{n} = (4\mathbf{i} - 10\mathbf{j} + 2\mathbf{k}) \cdot \frac{-\mathbf{i} + \mathbf{j}}{\sqrt{2}} = \left(\frac{-4}{\sqrt{2}} - \frac{10}{\sqrt{2}} + 0\right)$$

$$= \frac{-14}{\sqrt{2}} = -7\sqrt{2} \text{ psia}$$

To find the tangential stress, we use Eq. 4.18 to write

$$\boldsymbol{\tau} = \mathbf{\Sigma} - \sigma_N \mathbf{n} = (4\mathbf{i} - 10\mathbf{j} + 2\mathbf{k}) - (-7\sqrt{2}) \frac{-\mathbf{i} + \mathbf{j}}{\sqrt{2}} = -3\mathbf{i} - 3\mathbf{j} + 2\mathbf{k}$$

The tangential stress is the magnitude of this vector or

$$\tau = |\boldsymbol{\tau}| = \sqrt{(-3)^2 + (-3)^2 + (-2)^2} = \sqrt{22} \text{ psia}$$

Note that the unit normal vector may also be found by inspection or by using the direction numbers to the plane. In this example the normal and shear stress are of the same order of magnitude.

In a moving fluid the normal stress acting at a point on a surface and the pressure in the fluid at that point are not necessarily identical. There can be a viscous contribution to the normal stress that is not accounted for by the pressure. Nevertheless, in fluid mechanics, the terms "normal stress" and "pressure" are often regarded as synonymous because the difference between these two quantities (the viscous contribution) is almost always negligible in flows of engineering interest. Thus, we can usually assume $\sigma_N = -p$ and write the stress vector for a fluid in motion as

$$\boldsymbol{\Sigma} = -p\mathbf{n} + \boldsymbol{\tau} \qquad (4.19)$$

Equation 4.19 suggests that in thinking about the force exerted by a moving fluid on a surface, we consider the action of pressure and shear stress.

In the next chapter, on fluid statics, we show that the normal stress and pressure are exactly the same in a fluid at rest; i.e., writing $\sigma_N = -p$ is not an approximation of any kind. We also know that in a fluid at rest the tangential stress is identically zero. Thus we can write the stress vector in a fluid at rest as

$$\boldsymbol{\Sigma} = -p\mathbf{n} \qquad (4.20)$$

We conclude that the stress vector in a fluid at rest acts normal to a surface at every point. This type of stress field, in which tangential components are absent, is called a hydrostatic state of stress. If we have information about the pressure and shear stress acting on a surface, we can use the preceding results to calculate the stress vector acting on the surface.

EXAMPLE 4.8

Consider pressure-driven viscous flow through the parallel plate channel shown in Figure 4.14A. It is known that the shear stress distribution on the top and bottom wall is given by $\tau = (h/2)(p_1 - p_2)/L$, and the pressure distribution on each wall is $p(x) = p_1 - [(p_1 - p_2)/L]x$. Find the stress vector acting on each wall.

166 | 4 FLUID FORCES

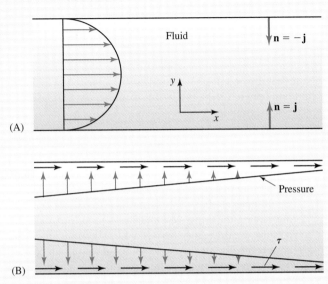

Figure 4.14 Schematic for Example 4.8. (A) Velocity distribution for pressure-driven flow through parallel plates. (B) Stress distribution.

SOLUTION

We can use Eq. 4.19, $\boldsymbol{\Sigma} = -p\mathbf{n} + \boldsymbol{\tau}$, to determine the stress vector. For flow from left to right as shown in Figure 4.14A, the shear stress vector $\boldsymbol{\tau}$ on each wall must point downstream and must be aligned with the flow direction, i.e., $\boldsymbol{\tau} = \tau\mathbf{i}$. The unit normal on each wall points into the fluid. Thus the stress vector on the bottom plate, for which $\mathbf{n} = \mathbf{j}$ is given by $\boldsymbol{\Sigma} = -p\mathbf{n} + \boldsymbol{\tau} = -p(x)\mathbf{j} + \tau\mathbf{i}$. There is no stress acting in the z direction, so $\boldsymbol{\Sigma} = \tau\mathbf{i} - p(x)\mathbf{j} + 0\mathbf{k}$. The constant wall shear stress is $\tau = (h/2)(p_1 - p_2)/L$, and the pressure distribution on the wall is $p(x) = p_1 - [(p_1 - p_2)/L]x$. On the top plate with $\mathbf{n} = -\mathbf{j}$ we find $\boldsymbol{\Sigma} = -p\mathbf{n} + \boldsymbol{\tau} = -p(x)(-\mathbf{j}) + \tau\mathbf{i} = p(x)\mathbf{j} + \tau\mathbf{i}$, or $\boldsymbol{\Sigma} = \tau\mathbf{i} + p(x)\mathbf{j} + 0\mathbf{k}$, with τ and $p(x)$ given by the same functions. The stress vectors (which are the vector sums of the normal and tangential stress vectors) vary with position along each wall as shown in Figure 4.14B.

EXAMPLE 4.9

For the liquid at rest shown in Figure 4.15A the hydrostatic pressure distribution is $p(z) = p_A - \rho g(z - h)$, where p_A is the ambient pressure at the liquid surface. What is the stress vector on the bottom surface A and the side surface B?

4.5 SURFACE FORCES

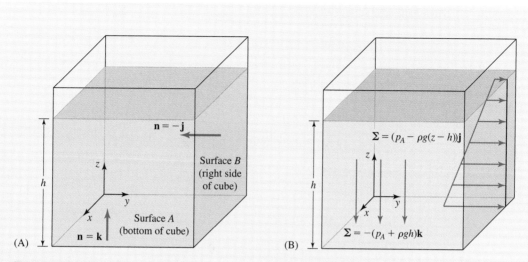

Figure 4.15 Schematic for Example 4.9. (A) surface definition. (B) Stress vectors.

SOLUTION

Since the fluid is at rest, the stress vector can be found by using Eq. 4.20: $\boldsymbol{\Sigma} = -p\mathbf{n}$. On surface A, located at $z = 0$, the unit normal vector is $\mathbf{n} = \mathbf{k}$, so on this surface the stress vector is given by $\boldsymbol{\Sigma} = -p\mathbf{n} = -p\mathbf{k}$. To find the pressure on this surface, we evaluate the pressure distribution at $z = 0$ (which defines the surface) and obtain $p_{z=0} = (p_A - \rho g(0 - h)) = (p_A + \rho g h)$. Thus the stress vector on surface A is $\boldsymbol{\Sigma} = -(p_A + \rho g h)\mathbf{k}$. The stress vector is a constant on surface A and points into the surface (compression).

On surface B, $\mathbf{n} = -\mathbf{j}$, and the pressure at a height z on this surface is $p(z) = p_A - \rho g(z - h)$. Thus the stress vector varies on this surface and is a function of z given by

$$\boldsymbol{\Sigma} = -p\mathbf{n} = -[p_A - \rho g(z - h)](-\mathbf{j}) = [p_A - \rho g(z - h)]\mathbf{j}$$

This vector also points into the surface. These stress vectors are shown in Figure 4.15B.

To calculate the total surface force $\mathbf{F}_S$ acting on an entire surface in contact with fluid, we must account for the fact that the magnitude and direction of the stress vector may vary on the surface. To do this we divide the surface into infinitesimal oriented surface elements, each of area dS, with an outward unit normal $\mathbf{n}$ as shown in Figure 4.16. The differential surface force $d\mathbf{F}_S$ acting on any one of these elements is given by Eq. 4.10 as $d\mathbf{F}_S = \boldsymbol{\Sigma}\, dS$, and the total force on the entire surface may be found by using a surface integral to sum the individual contributions from each surface element.

4 FLUID FORCES

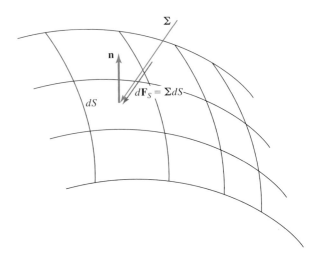

Figure 4.16 An infinitesimal surface element.

Thus the total surface force acting on any surface is given by

$$\mathbf{F}_S = \int_S \mathbf{\Sigma}\, dS \qquad (4.21)$$

The total surface force $\mathbf{F}_S$ is a vector quantity whose components may be resolved in any desired direction. We often wish to write the total force in Cartesian coordinates. Recalling that Eq. 4.14 gives the stress vector in these coordinates as $\mathbf{\Sigma} = \Sigma_x \mathbf{i} + \Sigma_y \mathbf{j} + \Sigma_z \mathbf{k}$, we can substitute this definition into Eq. 4.21 to obtain

$$\mathbf{F}_S = \int_S \mathbf{\Sigma}\, dS = \int_S (\Sigma_x \mathbf{i} + \Sigma_y \mathbf{j} + \Sigma_z \mathbf{k})\, dS$$

The Cartesian components of the total force ($F_{S_x}, F_{S_y}, F_{S_z}$) are related to the three Cartesian components of the stress vector by the integrals:

$$F_{S_x} = \int_S \Sigma_x\, dS, \quad F_{S_y} = \int_S \Sigma_y\, dS, \quad \text{and} \quad F_{S_z} = \int_S \Sigma_z\, dS \qquad (4.22\text{a–c})$$

We have shown that the total surface force $\mathbf{F}_S$ acting on any surface in contact with fluid is found by integrating the stress vector over the surface. In a fluid in motion, we can use Eq. 4.19 to write the stress vector as $\mathbf{\Sigma} = -p\mathbf{n} + \boldsymbol{\tau}$, and substitute this into Eq. 4.21: $\mathbf{F}_S = \int_S \mathbf{\Sigma}\, dS$. The result is

$$\mathbf{F}_S = \int_S -p\mathbf{n}\, dS + \int_S \boldsymbol{\tau}\, dS \qquad (4.23)$$

We see that in a fluid in motion the total surface force may be found by integrating the pressure and shear stress over the surface and that it is the net action of pressure and shear stress acting on each element of a surface that produces the resulting surface force. To evaluate the integrals, we must know the pressure and shear stress at each point on the surface. This is often accomplished by using a CFD code to solve the governing equations of fluid mechanics.

4.5 SURFACE FORCES

In the special case of a fluid at rest, the shear stress is zero, and we can use Eq. 4.20, $\Sigma = -p\mathbf{n}$, in Eq. 4.21. Thus in a fluid at rest the total surface force is given by

$$\mathbf{F}_S = \int_S -p\mathbf{n}\,dS \tag{4.24}$$

Of course this result can also be obtained by simply putting $\tau = 0$ in Eq. 4.23. In either case, Eq. 4.24 shows that the total surface force in a fluid at rest is found by integrating the (hydrostatic) pressure over the surface. Methods for determining the pressure distribution in a fluid at rest are described in the next chapter on fluid statics.

The next two exercises demonstrate how the integrals defining the total surface force are evaluated in a simple case. These integrals can be tedious for a curved surface, and although modern CFD programs evaluate these integrals at the click of the mouse, it is important to know where they come from and what they represent.

EXAMPLE 4.10

Consider the liquid at rest in the tank as described in Example 4.9 and shown here in Figure 4.17. The hydrostatic pressure distribution is $p(z) = p_A - \rho g(z - h)$, where p_A is the ambient pressure at the liquid surface. What is the force applied by the liquid to the side of the tank marked B in the figure?

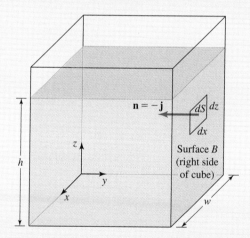

Figure 4.17 Schematic for Example 4.10.

SOLUTION

Since the fluid is at rest, we can use Eq. 4.24, $\mathbf{F}_S = \int_S -p\mathbf{n}\,dS$. The pressure distribution in the tank is given by $p(z) = p_A - \rho g(z - h)$, and the unit normal to the wall is $\mathbf{n} = -\mathbf{j}$ (see Figure 4.17). An infinitesimal element of area on this side of the tank may be written as $dS = dx\,dz$. Thus, the integral defining the total surface force becomes

$$\mathbf{F}_S = \int_S -p\mathbf{n}\,dS = \int_0^h \int_{-w/2}^{w/2} [-(p_A - \rho g(z-h))][-\mathbf{j}]\,dx\,dz$$

where the limits have been chosen for a rectangular side of height h and width w. Completing the integration yields the answer: $\mathbf{F}_S = wh(p_A + \rho gh/2)\mathbf{j}$. Alternately, the stress vector on the side of the tank was found to be given by $\mathbf{\Sigma} = (p_A - \rho g(z-h))\mathbf{j}$ in Example 4.9. Since we have the stress vector, we can calculate the total surface force using Eq. 4.21, $\mathbf{F}_S = \int_S \mathbf{\Sigma}\,dS$, employing the same area element and limits as previously. Thus, the integral defining the total surface force becomes

$$\mathbf{F}_S = \int_S \mathbf{\Sigma}\,dS = \int_0^h \int_{-w/2}^{w/2} (p_A - \rho g(z-h))\mathbf{j}\,dx\,dz$$

and we obtain the same result. We could also obtain the answer by starting with Eq. 4.22a–c or Eq. 4.23. Since the stress vector has only a $\mathbf{j}$ component in this case, we would write Eq. 4.22b as $F_{S_y} = \int_S \Sigma_y\,dS = \int_S [p_A - \rho g(z-h)]\,dS$ and proceed to evaluate the integral as shown earlier. If we decided to use Eq. 4.23, $\mathbf{F}_S = \int_S -p\mathbf{n}\,dS + \int_S \boldsymbol{\tau}\,dS$, as a starting point, we would set the shear stress to zero and obtain the same answer. It is always a good idea to check a result by using your intuition and experience. The pressure distribution on the side of the tank is shown in Figure 4.15B. Can you confirm by inspection that the calculated force is consistent with the action of this pressure distribution in both magnitude and direction?

EXAMPLE 4.11

Reconsider the pressure-driven flow through the parallel plate channel introduced in Example 4.8 and illustrated in Figure 4.14. What is the total surface force on the bottom plate?

SOLUTION

The viscous flow creates a pressure and shear stress on the channel walls. The fluid is in motion, so we will use Eq. 4.23, $\mathbf{F}_S = \int_S -p\mathbf{n}\,dS + \int_S \boldsymbol{\tau}\,dS$, to calculate the total surface force. We know $\mathbf{n} = \mathbf{j}$ and that the direction of the shear stress is in the flow direction. The total surface force on the bottom plate can thus be written as $\mathbf{F}_S = \int_S -p(x)\mathbf{j}\,dS + \int_S \tau\mathbf{i}\,dS$. Inserting the known functions for the pressure and shear stress, and writing the integrals for a unit width w of the channel into the plane of the paper, we can write the three components of the total surface force as

$$F_{S_x} = w \int_0^L \tau\, dx = wL\tau = \frac{wh}{2}(p_1 - p_2)$$

$$F_{S_y} = \int_S -p(x)\, dS = -w \int_0^L \left(p_1 - \frac{p_1 - p_2}{L} x\right) dx = -\frac{wL}{2}(p_1 - p_2)$$

$$F_{S_z} = \int_S 0\, dS = 0$$

The force on the bottom plate acts down and to the right as expected, with the force in the flow direction being $F_{S_x} = (wh/2)(p_1 - p_2)$. This exercise can also be solved by using the stress vector found in Example 4.8 and evaluating the total surface force either with Eq. 4.21, $\mathbf{F}_S = \int_S \mathbf{\Sigma}\, dS$, or with Eq. 4.22a–4.22c.

CD/Video Library/Laminar and Turbulent Flow on a Flat Plate

4.5.1 Flow Over a Flat Plate

Let us apply what we have just learned about surface forces to the case study of Section 3.3.4, the boundary layer on a flat plate. Our goal is to find an expression for the total surface force on the plate (see Figure 4.18A). Since the shear stress clearly must point in the x (flow) direction on both sides of the plate, we can apply Eq. 4.23, $\mathbf{F}_S = \int_S -p\mathbf{n}\, dS + \int_S \boldsymbol{\tau}\, dS$, to the upper and lower plate surfaces. Noting the direction of $\mathbf{n}$ on each surface, and that the shear stress points in the x direction on both surfaces, we can write $\boldsymbol{\tau} = \tau \mathbf{i}$, and the total surface force is given by

$$\mathbf{F}_S = \int_{\text{upper}} -p_U(\mathbf{j})\, dS + \int_{\text{upper}} \tau_U(\mathbf{i})\, dS + \int_{\text{lower}} -p_L(-\mathbf{j})\, dS + \int_{\text{lower}} \tau_L(\mathbf{i})\, dS$$

where we have labeled the pressures and shear stresses on the upper and lower surfaces.

If the plate is at zero angle of attack, the pressure distribution on the top of the plate will be the same as that on the bottom. Therefore, the two pressure integrals cancel. Symmetry also suggests that the shear stress distributions on the top and bottom of the plate are identical as shown in Figure 4.18B. Thus we can write the total force as $\mathbf{F}_S = 2\int_{\text{upper}} \tau(x)\mathbf{i}\, dS$ and anticipate that the development of the boundary layer will cause variation in the shear stress along the plate. For a plate of width w (into the paper) we can write $F_{S_x} = 2w\int_0^L \tau(x)\, dx$. By using the friction coefficient, Eq. 3.27, and identifying F_{S_x} as the drag, we can calculate $F_D = F_{S_x} = 2w\int_0^L \frac{1}{2}\rho U^2 C_f(x)\, dx$, which is the same as Eq. 3.30 (where $U = V$) in the case study (with the factor of 2 accounting for the two sides of the plate).

Figure 4.18 Flow over a flat plate: (A) geometry, (B) stress distribution at zero angle of attack, and (C) stress distribution at a small angle of attack.

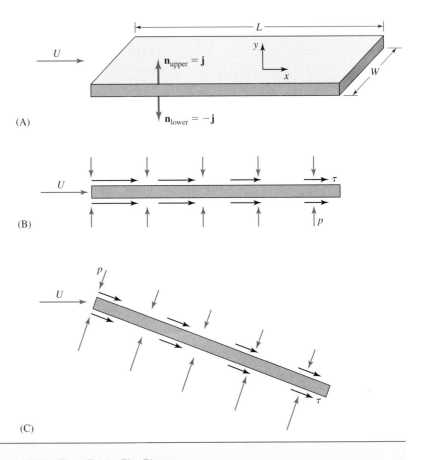

CD/Video Library/Flow Past a Flat Plate

If the plate is at a small angle of attack, it seems reasonable to suppose that the pressure and shear stress distributions on the top and bottom of the plate are different, but that the shear stress will still be directed along the plate in the flow direction. This is illustrated in Figure 4.18C. In this case we again apply Eq. 4.23 to each side of the plate and write the shear stress as $\boldsymbol{\tau} = \tau \mathbf{s}$, where $\mathbf{s}$ is a unit vector tangent to the plate and pointing downstream. This gives

$$\mathbf{F}_S = \int_{\text{upper}} -p_U(\mathbf{n}_U)\,dS + \int_{\text{upper}} \boldsymbol{\tau}_U(\mathbf{s}_U)\,dS + \int_{\text{lower}} -p_L(\mathbf{n}_L)\,dS + \int_{\text{lower}} \boldsymbol{\tau}_L(\mathbf{s}_L)\,dS$$

where the unit normals $\mathbf{n}_U$ and $\mathbf{n}_L$ and unit tangents $\mathbf{s}_U$ and $\mathbf{s}_L$ (which point downstream along the plate) depend on the angle of attack. Although we are not given enough information about the pressure and stress distributions to complete the integrations, it seems likely that the pressure difference on the two surfaces will produce lift on the plate and as well as a contribution to the drag. The shear stress can be seen to produce drag and negative lift.

4.5.2 Flow Through a Round Pipe

Next consider the steady, fully developed flow of a liquid in a round pipe as introduced in the case study of Section 3.3.1 and shown in Figure 4.19A. If the effects of gravity are neglected (i.e., ignoring the weight of the liquid), what is the force applied by the liquid to the wall of the pipe? Computing the total surface force requires that we set up and evaluate a surface integral on a curved surface. We will begin our analysis by using a force balance on the volume of fluid shown in Figure 4.19B to find a relationship between the wall shear stress and the pressure drop. In steady, fully developed flow this volume is not accelerating, so the sum of the body and surface forces on this volume must be zero. Since gravity is neglected, the body force is zero and, therefore, the total surface force on this volume is also zero. The total surface force on this cylindrical volume of liquid may be found using Eq. 4.23, $\mathbf{F}_S = \int_S -p\mathbf{n}\,dS + \int_S \boldsymbol{\tau}\,dS$. Dividing the surface into parts we can write

$$\mathbf{F}_S = \int_{\text{left}} -p\mathbf{n}\,dS + \int_{\text{right}} -p\mathbf{n}\,dS + \int_{\text{outer}} -p\mathbf{n}\,dS + \int_{\text{left}} \boldsymbol{\tau}\,dS$$

$$+ \int_{\text{right}} \boldsymbol{\tau}\,dS + \int_{\text{outer}} \boldsymbol{\tau}\,dS = 0$$

Consider each of the six integrals in order from left to right. We must retain the first two integrals because there is a uniform but different pressure acting on these two end

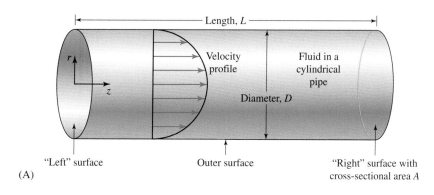

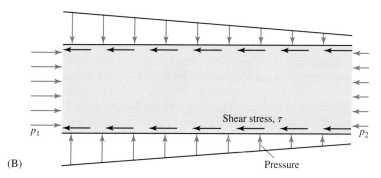

Figure 4.19 Flow in a cylindrical pipe: (A) geometry and velocity distribution and (B) fluid volume and stress distribution acting on the fluid.

surfaces of area $A = \pi D^2/4$. With respect to the third integral, the pressure on the outer surface results in a zero net force due to the symmetry about the axis. Since there is no reason to expect a shear stress to act on the left or right surfaces because there is no tangential flow there, the fourth and fifth integrals are zero. To evaluate the last integral, note that the shear stress exerted by the liquid on the pipe wall clearly points in the flow direction. This means that the shear stress applied by the wall to the fluid in our volume must point opposite to the flow direction, so we write $\boldsymbol{\tau} = \tau(-\mathbf{k})$ at all points on the outer curved surface. Thus, the total surface force on the entire volume of liquid is given by

$$\mathbf{F}_S = \int_{\text{left}} -p_1\mathbf{n}_1\, dS + \int_{\text{right}} -p_2\mathbf{n}_2\, dS + \int_{\text{outer}} \tau(-\mathbf{k})\, dS = 0$$

Since the pressure is uniform (but different) on the end surfaces, the integrals reduce to the product of the integrand and the area of each surface. Using the known unit normals, we have:

$$\int_{\text{left}} -p_1\mathbf{n}_1\, dS = -p_1\mathbf{n}_1 A = -p_1(-\mathbf{k})A = p_1 A\mathbf{k}$$

$$\int_{\text{right}} -p_2\mathbf{n}_2\, dS = -p_2\mathbf{n}_2 A = -p_2(\mathbf{k})A = -p_2 A\mathbf{k}$$

The shear stress on the surface of the volume must be uniform in fully developed flow, so the remaining integral gives

$$\int_{\text{outer}} \tau(-\mathbf{k})\, dS = \tau A_W(-\mathbf{k}) = -\tau A_W \mathbf{k}$$

The total surface force on this volume of fluid is therefore given by

$$\mathbf{F}_S = [(p_1 - p_2)A - \tau A_W]\mathbf{k} = 0$$

Solving for the shear stress applied by the wall on the curved surface of the fluid volume we obtain

$$\tau = (p_1 - p_2)\frac{A}{A_W} = \frac{D(p_1 - p_2)}{4L}$$

We see that the frictional drop in pressure in fully developed pipe flow is caused by the viscous shear stress applied to the liquid by the wall of the pipe.

To determine the total force exerted by the moving fluid on the pipe wall, note that the shear stress just calculated is that applied by the pipe wall to the liquid and points to the left. The wall shear stress, τ_W, which is applied by the moving liquid to the wall, has the same magnitude but points in the flow direction. Thus the wall shear stress is given by $\tau_W = D(p_1 - p_2)/4L$. The total surface force applied by the fluid on the wall in the flow direction is the product of the uniform wall shear stress and the contact area, i.e.,

$$F = \tau_W A_W = \frac{D(p_1 - p_2)}{4L}(\pi DL) = (p_1 - p_2)\frac{\pi D^2}{4}$$

4.5.3 Lift and Drag

As discussed in the case studies of Sections 3.3.5 and 3.3.6, it is possible to compute the lift and drag on an object from the lift and drag coefficients. Empirical results for C_L and C_D for a typical airfoil shape, as well as for cylinders and spheres, were given in Section 3.3.6. Our goal here is to discuss the relationship between the stress vector, total surface force, and lift and drag forces applied to a body by a moving fluid.

For an aircraft in level flight, positive lift is the component of total surface force acting upward. Lift is necessary to support the weight of an aircraft in flight and to provide the force needed for vertical acceleration. Since lift is a component of the total surface force $\mathbf{F}_S$, in a specific direction, the lift F_L is given by the dot product

$$F_L = \mathbf{F}_S \cdot \mathbf{n}_L \tag{4.25a}$$

where $\mathbf{n}_L$ is a unit vector pointing in the lift direction. Combining Eq. 4.23, which defines the total surface force as $\mathbf{F}_S = \int_S -p\mathbf{n}\,dS + \int_S \boldsymbol{\tau}\,dS$, and Eq. 4.25a yields $F_L = [\int_S -p\mathbf{n}\,dS + \int_S \boldsymbol{\tau}\,dS] \cdot \mathbf{n}_L$. Since $\mathbf{n}_L$ is a constant vector that does not depend on the orientation of any surface element, we can move the dot product inside each surface integral and combine the integrals to obtain

$$F_L = \int_S (-p\mathbf{n} + \boldsymbol{\tau}) \cdot \mathbf{n}_L\,dS \tag{4.25b}$$

The integrand here is simply the component of the stress vector in the lift direction, i.e., $\boldsymbol{\Sigma} \cdot \mathbf{n}_L$. Equation 4.25b shows that to produce lift, the stress vector on a surface must have a component in the lift direction. Notice the stresses acting on the two sides of the flat plate at a small angle of attack as shown in Figure 4.20A. There is a net lift on this plate from the asymmetric normal stress (pressure) distribution. At zero angle of attack, Figure 4.20B, there is no lift because the ambient pressure distribution is symmetric.

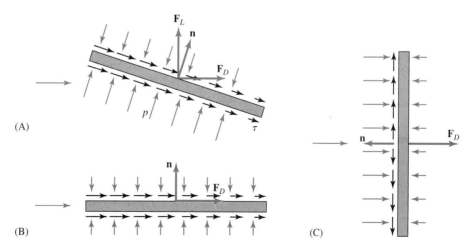

Figure 4.20 Stress distribution, lift, drag, and outward unit normal for flow over a flat plate: (A) small angle of attack, (B) horizontal, and (C) vertical.

Figure 4.21 Pressure distribution on an airfoil in terms of departure from ambient.

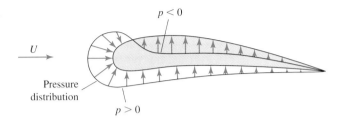

Equation 4.25b shows that in principle, pressure and shear stresses may each contribute to lift. For example, in Figure 4.20A, the tangential stress must clearly reduce the lift on a flat plate at an angle of attack. Nevertheless, since the normal stress (pressure) is usually much larger than the shear stress, lift tends to be correctly thought of as produced almost entirely by the pressure distribution on a body. An airfoil shape is carefully optimized to develop a large lift as a result of a higher pressure on the bottom of the wing than on the top. An example of the departure of the pressure distribution from the ambient value on an airfoil is shown in Figure 4.21. Note that the pressure is not uniform, but varies continuously along the wing on both the upper and lower surfaces. On an airfoil, the contribution from the tangential stress to lift may be safely neglected.

 CD/Boundary Layers/Boundary Layer Concepts/Origins of Drag Forces

---

The drag F_D is the component of total surface force acting in the direction of the oncoming fluid stream. For an aircraft in level flight, thrust is needed to counteract the effect of drag and keep the speed of the aircraft constant. Reducing the drag of an aircraft is an important element in achieving good performance. The drag is defined by the dot product of the total surface force with a unit vector pointing in the drag direction. Let $\mathbf{n}_\infty$ be a unit vector in the direction of the upstream fluid flow. Then the drag F_D is given by

$$F_D = \mathbf{F}_S \cdot \mathbf{n}_\infty \tag{4.26a}$$

Recalling again that Eq. 4.23 defines the total surface force as $\mathbf{F}_S = \int_S -p\mathbf{n}\,dS + \int_S \boldsymbol{\tau}\,dS$, we can rewrite Eq. 4.26a as $F_D = [\int_S -p\mathbf{n}\,dS + \int_S \boldsymbol{\tau}\,dS] \cdot \mathbf{n}_\infty$. Since $\mathbf{n}_\infty$ is also a constant vector that does not depend on the orientation of any surface element, we can move the dot product inside each surface integral and combine the integrals, obtaining

$$F_D = \int_S (-p\mathbf{n} + \boldsymbol{\tau}) \cdot \mathbf{n}_\infty\,dS \tag{4.26b}$$

The integrand here is the component of the stress vector in the drag direction, $\boldsymbol{\Sigma} \cdot \mathbf{n}_\infty$. We see that to produce drag, the stress vector on a surface must have a component in the drag direction. Notice the stresses acting on the two sides of the flat plate at different angles

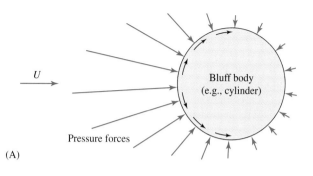

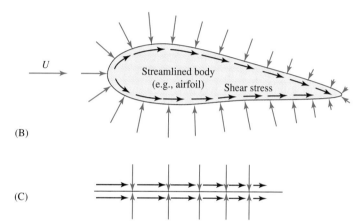

Figure 4.22 Stress distribution on (A) a bluff body, (B) a streamlined body, and (C) a thin flat plate.

of attack as shown in Figure 4.20A and 4.20B. There is a net drag on this plate in both cases.

Equation 4.26b shows that both normal and tangential stresses may contribute to drag. To drive this point home, reconsider the flat plate at zero angle of attack in Figure 4.20B, for which the drag is due exclusively to the shear stress. However, if the flat plate is turned normal to the oncoming stream (Figure 4.20C), the drag is due exclusively to the effects of normal stress, i.e., pressure.

Pressure differences over the front and rear of an object are important contributors to drag, particularly if the object is a bluff body (see Figure 4.22A). The drag on a bluff body is primarily due to this effect and is called pressure or form drag. The drag on a streamlined body (Figure 4.22B) is due to both form drag and skin friction drag. The relative sizes of the two contributions depend on the shape of the streamlined body. The drag on a long, thin body is primarily skin friction drag. The infinitely thin aligned plate in Figure 4.20C illustrates this idea in the extreme. Drag on an airfoil arises from both pressure and shear stress, i.e., both form and skin friction drag.

Now that you understand the concept of total surface force and how to calculate it, you can perhaps begin to appreciate the complexity of the mathematics of fluid mechanics and why the empirically based approaches illustrated in the case studies are so valuable. Which method do you prefer: calculating lift and drag using surface integrals, or using lift and drag coefficients?

4.6 STRESS IN A FLUID

When body and surface forces act on a fluid, the fluid is said to be under stress. That is, there is a state of stress in the fluid caused by these forces, and the fluid transmits force from one point in its interior to another. We can demonstrate this transmission of forces through a fluid using the example of a column of fluid loaded by a piston as shown in Figure 4.23. The fluid is at rest, so there are no shear stresses acting on it. The load applied by the piston to the fluid (a surface force) plus the force of gravity acting on the fluid (a body force) are transmitted to the bottom of the fluid column and must be balanced there by a force applied to the fluid by the bottom of the container (a surface force). The transmission of forces through the interior of the fluid is indicated by saying that the fluid is in a state of stress.

In general, the state of stress in a fluid is influenced by body forces, by surface forces acting at the interface between the fluid and its surroundings, by fluid motion, and by the fluid's physical and thermodynamic properties. The state of stress in a fluid is responsible for the surface force exerted by the fluid on the materials it contacts. For a fluid at rest, as in the case of a fluid column loaded by a piston, the state of stress is purely compressive and is characterized by a hydrostatic pressure distribution. For a fluid in motion, the state of stress is more complex, as we shall describe more fully in this section. An understanding of the concept of fluid forces must include an awareness of body and surface forces, and of the presence of a state of stress in the fluid.

Although the stress vector may be used to calculate the surface force, it is not used to describe the state of stress in the interior of a fluid, and it does not appear in the governing equations of fluid dynamics. Why is this the case? Recall that the stress vector, $\mathbf{\Sigma} = \mathbf{\Sigma}(\mathbf{x}, t, \mathbf{n})$, is a function of the location of the point of interest within the fluid, $\mathbf{x}$, the time, t, and the normal to the corresponding surface, $\mathbf{n}$. For an interior point, there are an infinite number of possible surface orientations that contain that point. Thus, there are potentially an infinite number of valid stress vectors. Since the stress vector is not a unique entity at a point within a fluid, it cannot be used to describe the state of stress at a point. Instead, we use an orientation-free representation of stress called the stress tensor, $\boldsymbol{\sigma}(\mathbf{x}, t)$, which represents the state of stress in the fluid at point $\mathbf{x}$ at time t.

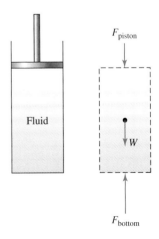

Figure 4.23 Transmission of forces through a column of fluid.

4.6 STRESS IN A FLUID

The number of values provided by scalar, vector, and tensor fields is given by the formula 3^n, where n represents the number of "directions" associated with each quantity. A scalar field has no directions associated with it, so $n = 0$, and the number of values provided by the scalar field at each point in space is $3^0 = 1$. For a vector field there are $3^1 = 3$ values provided at each point in space, one in each principal direction. For a tensor field there are $3^2 = 9$ values provided at each point, one for each pairwise combination of directions.

To place the concept of a stress tensor in a physical context, consider the hierarchy of mathematical functions in fluid mechanics. Some properties, such as density, pressure, and temperature, are described by scalar (magnitude-only) functions. Other quantities, including velocity, force and, of course, $\boldsymbol{\Sigma}$, are described by vectors, which convey information about both magnitude and direction. Now consider the concept of stress, which is defined to be a force divided by the area over which that force acts. To adequately describe the stress tensor, we must specify not only the magnitude and direction of the relevant force, but also the orientation of the corresponding surface. The surface orientation is a second directional quantity typically described by the normal to that surface. Thus, the description of stress at a point within a fluid requires that the stress tensor have a doubly directional character. In fact, we can write the stress tensor in vector notation using the usual unit vectors as

$$\boldsymbol{\sigma} = \sigma_{xx}\mathbf{ii} + \sigma_{xy}\mathbf{ij} + \sigma_{xz}\mathbf{ik} + \sigma_{yx}\mathbf{ji} + \sigma_{yy}\mathbf{jj} + \sigma_{yz}\mathbf{jk} + \sigma_{zx}\mathbf{ki} + \sigma_{zy}\mathbf{kj} + \sigma_{zz}\mathbf{kk} \quad (4.27)$$

There are nine terms in this equation, each of which is the product of a scalar stress component and two unit vectors. The nine scalar stress components are simply referred to as stresses. These nine stresses are functions of position and time, and the pairs of unit vectors give the tensor its doubly directional characteristic.

To investigate the relationship between the stress vector and this new concept of a stress tensor, consider the forces acting on an element of fluid in the shape of a small tetrahedron located at a point in a fluid as shown in Figure 4.24A. Body forces act on the

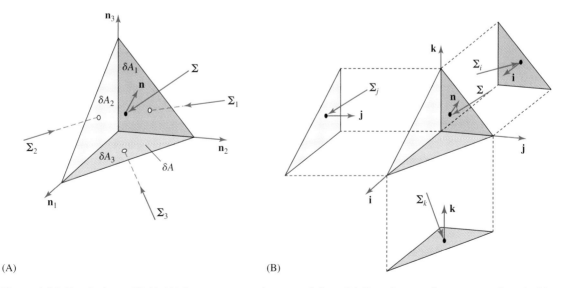

Figure 4.24 Tetrahedron of fluid. (A) Stress vectors acting on each face. (B) Cartesian coordinate system aligned with orthogonal planes.

volume of fluid inside the tetrahedron, and surface forces act on this fluid across the four facets of the tetrahedron through the action of the stress vector on each facet. By Newton's second law we can equate the sum of the body and surfaces forces acting on the fluid in this tetrahedron to the product of the mass of the fluid and its instantaneous acceleration to obtain:

$$\rho \mathbf{f} \delta V + \mathbf{\Sigma} \delta A + \mathbf{\Sigma}_1 \delta A_1 + \mathbf{\Sigma}_2 \delta A_2 + \mathbf{\Sigma}_3 \delta A_3 = \rho \mathbf{a} \delta V$$

This result should hold for any size tetrahedron. Taking the limit as the size of the tetrahedron approaches zero causes the body and inertial forces to vanish, leaving only a balance of surface forces:

$$\mathbf{\Sigma} \delta A + \mathbf{\Sigma}_1 \delta A_1 + \mathbf{\Sigma}_2 \delta A_2 + \mathbf{\Sigma}_3 \delta A_3 = 0$$

To understand this conclusion, note that the body and inertial forces are proportional to the linear dimension of the tetrahedron cubed, while the surface forces are proportional to the dimension squared. As the linear dimension approaches zero, cubic terms vanish first.

Using a dot product to relate each of the smaller areas to δA, i.e., $\delta A_1 = (\mathbf{n} \cdot \mathbf{n}_1) \delta A$, $\delta A_2 = (\mathbf{n} \cdot \mathbf{n}_2) \delta A$, and $\delta A_3 = (\mathbf{n} \cdot \mathbf{n}_3) \delta A$, we find

$$\mathbf{\Sigma} \delta A + \mathbf{\Sigma}_1 (\mathbf{n} \cdot \mathbf{n}_1) \delta A + \mathbf{\Sigma}_2 (\mathbf{n} \cdot \mathbf{n}_2) \delta A + \mathbf{\Sigma}_3 (\mathbf{n} \cdot \mathbf{n}_3) \delta A = 0$$

Solving this expression for the stress vector yields:

$$\mathbf{\Sigma} = -\mathbf{n} \cdot (\mathbf{n}_1 \mathbf{\Sigma}_1 + \mathbf{n}_2 \mathbf{\Sigma}_2 + \mathbf{n}_3 \mathbf{\Sigma}_3)$$

This very important result, which follows directly from Newton's law, shows that the stress vector on a surface of arbitrary orientation $\mathbf{n}$ is related by a dot product to the three stress vectors acting on any three orthogonal planes passing through the same point.

Since we are free to choose the orientation of three orthogonal planes, it is convenient to align them with the relevant Cartesian coordinate system as shown in Figure 4.24B. In this case, the outward unit normals on the three orthogonal faces become $\mathbf{n}_1 = \mathbf{i}$, $\mathbf{n}_2 = \mathbf{j}$, and $\mathbf{n}_3 = \mathbf{k}$. Similarly, the three stress vectors acting on the orthogonal faces become $\mathbf{\Sigma}_1 = -\mathbf{\Sigma}_i$, $\mathbf{\Sigma}_2 = -\mathbf{\Sigma}_j$, and $\mathbf{\Sigma}_3 = -\mathbf{\Sigma}_k$. Thus, the preceding analysis shows that we can write

$$\mathbf{\Sigma} = \mathbf{n} \cdot (\mathbf{i} \mathbf{\Sigma}_i + \mathbf{j} \mathbf{\Sigma}_j + \mathbf{k} \mathbf{\Sigma}_k) \qquad (4.28)$$

As shown in Figure 4.25A, the stress vector acting on each of the three orthogonal planes can be resolved as usual into a normal and tangential stress. We can then take any tangential stress and further resolve it into components along the coordinate axes. Notice for example, that Figure 4.25B, showing the stress vector acting on the yz plane, shows that the tangential stress has been further resolved into a pair of components aligned with the y and z directions. Considering each plane in turn, and resolving the stress vector into a normal and two tangential components, we can write each of the stress vectors, $\mathbf{\Sigma}_i$, $\mathbf{\Sigma}_j$, and $\mathbf{\Sigma}_k$ in terms of their three Cartesian components. For convenience, as shown in Figure 4.25C, we can label the three components of $\mathbf{\Sigma}_i$ as $(\sigma_{xx}, \sigma_{xy}, \sigma_{xz})$. Similarly, we can label the three components of $\mathbf{\Sigma}_j$ as $(\sigma_{yx}, \sigma_{yy}, \sigma_{yz})$, and those of $\mathbf{\Sigma}_k$ as $(\sigma_{zx}, \sigma_{zy}, \sigma_{zz})$. Then the three stress vectors are

$$\mathbf{\Sigma}_i = \sigma_{xx}\mathbf{i} + \sigma_{xy}\mathbf{j} + \sigma_{xz}\mathbf{k}, \quad \mathbf{\Sigma}_j = \sigma_{yx}\mathbf{i} + \sigma_{yy}\mathbf{j} + \sigma_{yz}\mathbf{k}, \quad \text{and} \quad \mathbf{\Sigma}_k = \sigma_{zx}\mathbf{i} + \sigma_{zy}\mathbf{j} + \sigma_{zz}\mathbf{k}$$

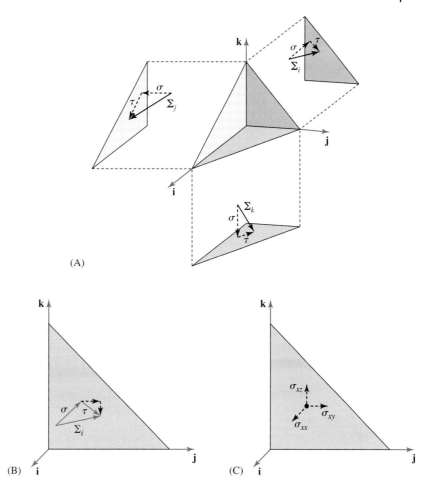

Figure 4.25 (A) Stress vectors resolved into normal and tangential components. (B) Tangential stress components in the yz plane. (C) Notation for stress components in the yz plane.

If we now substitute these definitions of Σ_i, Σ_j, and Σ_k into Eq. 4.28 and rearrange, we find

$$\Sigma = \mathbf{n} \cdot [\sigma_{xx}\mathbf{ii} + \sigma_{xy}\mathbf{ij} + \sigma_{xz}\mathbf{ik} + \sigma_{yx}\mathbf{ji} + \sigma_{yy}\mathbf{jj} + \sigma_{yz}\mathbf{jk} + \sigma_{zx}\mathbf{ki} + \sigma_{zy}\mathbf{kj} + \sigma_{zz}\mathbf{kk}]$$

The quantity inside the square brackets is the stress tensor

$$\boldsymbol{\sigma} = \sigma_{xx}\mathbf{ii} + \sigma_{xy}\mathbf{ij} + \sigma_{xz}\mathbf{ik} + \sigma_{yx}\mathbf{ji} + \sigma_{yy}\mathbf{jj} + \sigma_{yz}\mathbf{jk} + \sigma_{zx}\mathbf{ki} + \sigma_{zy}\mathbf{kj} + \sigma_{zz}\mathbf{kk}$$

as given earlier by Eq. 4.27. Thus in vector notation our analysis shows that

$$\Sigma = \mathbf{n} \cdot \boldsymbol{\sigma} \qquad (4.29)$$

We see that by applying Newton's law to a small tetrahedron of fluid, we have shown that the stress vector acting on a surface of arbitrary orientation $\mathbf{n}$ located at a point in a fluid is related by a dot product to the stress tensor at this point. This key result allows us to employ the orientation-free stress tensor in our analysis and then recover the oriented stress vector in a specific calculation involving a surface.

The preceding analysis also provides the basis for the subscript system used to label individual stresses. To understand this system, note that while we can write the stress tensor in vector notation as

$$\boldsymbol{\sigma} = \sigma_{xx}\mathbf{ii} + \sigma_{xy}\mathbf{ij} + \sigma_{xz}\mathbf{ik} + \sigma_{yx}\mathbf{ji} + \sigma_{yy}\mathbf{jj} + \sigma_{yz}\mathbf{jk} + \sigma_{zx}\mathbf{ki} + \sigma_{zy}\mathbf{kj} + \sigma_{zz}\mathbf{kk}$$

it is customary in fluid mechanics to represent the stress tensor in a matrix form as

$$\boldsymbol{\sigma} = \begin{pmatrix} \sigma_{xx} & \sigma_{xy} & \sigma_{xz} \\ \sigma_{yx} & \sigma_{yy} & \sigma_{yz} \\ \sigma_{zx} & \sigma_{zy} & \sigma_{zz} \end{pmatrix} \qquad (4.30)$$

Regardless of which notation we use, the subscripts on each stress component serve to identify both the plane on which the stress acts, and the direction of the stress. The first subscript on a stress identifies the plane on which it acts. The subscript x in σ_{xz}, for example, indicates a plane normal to the x axis, meaning a plane parallel to the zy coordinate plane. The second subscript on a stress indicates the direction of the force on this plane. Thus σ_{xz} is a force per unit area acting in the z direction on a plane normal to the x direction.

Figure 4.26 labels the stresses on the visible sides of an infinitesimal cube of fluid according to this system. A stress component with the two identical subscripts represents a normal stress. Notice that in Figure 4.26 the three normal stresses are σ_{xx}, σ_{yy}, and σ_{zz}. The remaining six stress components with different subscripts are shear stresses, since they act in the plane of the surface on which they occur.

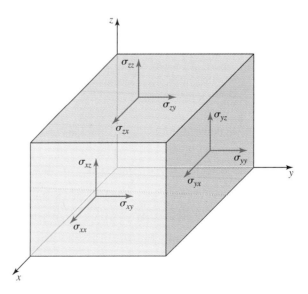

Figure 4.26 Stresses acting on the visible sides of an infinitesimal cube of fluid.

4.6 STRESS IN A FLUID

By considering a moment balance on the cube of fluid shown in Figure 4.26, and again considering the limit as the size of this cube approaches zero, it can be shown that if there are no externally imposed body moments acting on a fluid, the stress tensor $\boldsymbol{\sigma}(\mathbf{x}, t)$ is symmetric. Since very few engineering applications involve external body moments, we will assume throughout the remainder of this text that the shear stresses are related to each other by

$$\sigma_{xy} = \sigma_{yx}, \qquad \sigma_{xz} = \sigma_{zx} \quad \text{and} \quad \sigma_{yz} = \sigma_{zy}$$

Thus there are only six independent stress components in the stress tensor: the three normal stresses and three shear stresses.

Since a stress component may be positive or negative, it is necessary to employ a sign convention. The convention used in this text is based on the directions of the corresponding force and unit normal. If a force points in the positive x, y, or z direction, then the force is considered to be "positive." Similarly, if the unit normal to the plane on which the force acts points in the positive x, y, or z direction then the unit normal is "positive." The sign of a stress component is defined by "multiplying" the sign of the force and the sign of the normal. Thus, a positive stress component either has a positive force and a positive unit normal or a negative force and a negative unit normal. A negative stress component involves a force and unit normal with opposite signs.

For example, consider Figure 4.27A. Suppose a stress component $\sigma_{yy} = -10$ psia acts on a plane of area 1 in.^{2} with a unit normal in the $+y$ direction. This is a normal stress on this plane, a compressive force that pushes the plane in the direction opposite the unit normal. The corresponding force on the plane of 10 lb$_f$ acts in the $-y$ direction. Similarly, if there is a stress $\sigma_{yz} = 5$ psia at the same point in the fluid, we conclude that the direction of the applied force on the indicated plane must be in the $+z$ direction, since the normal to the plane is in the $+y$ direction. This positive shear stress is shown in Figure 4.27B.

Now that you understand the concepts of the stress vector $\boldsymbol{\Sigma}$ acting on a surface, and the stress tensor $\boldsymbol{\sigma}$ acting at a point in a fluid, we can give you a better sense of how a problem in fluid mechanics is solved. Suppose we are asked to solve a flow problem and then calculate the stress vector and force applied by the fluid to a surface. The general approach would be to solve the appropriate governing equations of fluid mechanics. This will provide the stress tensor. Next, we use the relationship between the stress

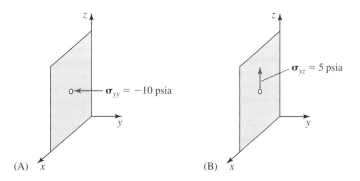

Figure 4.27 Relationship of (A) normal and (B) tangential stress to force on a surface.

EXAMPLE 4.12

Sketch the two possible orientations of a negative normal stress on a plane parallel to the xy plane. Also show the two possible orientations for a positive σ_{xz}.

SOLUTION

A plane parallel to the xy plane will have a unit normal of either $+\mathbf{k}$ or $-\mathbf{k}$. A negative normal stress, $\sigma_{zz} < 0$, on the $+\mathbf{k}$ plane acts in the $-\mathbf{k}$ direction. On the $-\mathbf{k}$ plane the normal stress acts in the $+\mathbf{k}$ direction. These two cases are shown in Figure 4.28A. Since the first subscript indicates the plane, the shear stress σ_{xz} acts on a plane whose unit normal is either $+\mathbf{i}$ or $-\mathbf{i}$. This is a plane parallel to the yz plane. By convention, a positive shear stress σ_{xz} points in the $+\mathbf{k}$ direction on the $+\mathbf{i}$ plane, and in the $-\mathbf{k}$ direction on the $+\mathbf{i}$ plane. These two cases are shown in Figure 4.28B. Since the stress tensor is symmetric we have $\sigma_{xz} = \sigma_{zx}$. Therefore, we can identify the directions of the shear stresses on a plane parallel to the xy plane as shown in Figure 4.28C.

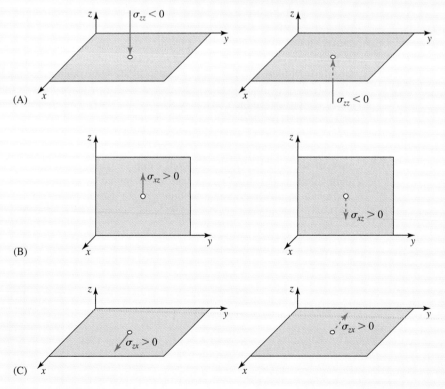

Figure 4.28 Schematic for Example 4.12. (A) Normal stresses on the xy plane, (B) shear stresses on the yz plane, (C) shear stresses on the xy plane.

vector and stress tensor as given by Eq. 4.29, $\mathbf{\Sigma} = \mathbf{n} \cdot \boldsymbol{\sigma}$, to find the stress vector. We can evaluate the dot product in two ways. In vector notation, with the stress tensor given by Eq. 4.27, the dot product $\mathbf{\Sigma} = \mathbf{n} \cdot \boldsymbol{\sigma}$ is

$$\mathbf{\Sigma} = (n_x\mathbf{i} + n_y\mathbf{j} + n_z\mathbf{k}) \cdot (\sigma_{xx}\mathbf{ii} + \sigma_{xy}\mathbf{ij} + \sigma_{xz}\mathbf{ik} + \sigma_{yx}\mathbf{ji} + \sigma_{yy}\mathbf{jj} \\ + \sigma_{yz}\mathbf{jk} + \sigma_{zx}\mathbf{ki} + \sigma_{zy}\mathbf{kj} + \sigma_{zz}\mathbf{kk})$$
(4.31a)

If the stress tensor is represented in matrix form, Eq. 4.30, then we can evaluate the dot product defining the stress vector by using the matrix product of a row vector $\mathbf{n}$ with the matrix $\boldsymbol{\sigma}$. The dot product is then found by evaluating

$$\mathbf{\Sigma} = \mathbf{n} \cdot \boldsymbol{\sigma} = (n_x, n_y, n_z) \begin{pmatrix} \sigma_{xx} & \sigma_{xy} & \sigma_{xz} \\ \sigma_{yx} & \sigma_{yy} & \sigma_{yz} \\ \sigma_{zx} & \sigma_{zy} & \sigma_{zz} \end{pmatrix}$$
(4.31b)

You should be able to see by inspection that this produces the same result as using Eq. 4.31a. The final step in the analysis is to calculate the total surface force. We can also

EXAMPLE 4.13

If the stress tensor (in units of psia) in a flow is given by $\boldsymbol{\sigma} = -5\mathbf{ii} + 3\mathbf{ij} - 1\mathbf{ik} + 3\mathbf{ji} - 4\mathbf{jj} + 4\mathbf{jk} - 1\mathbf{ki} + 4\mathbf{kj} - 2\mathbf{kk}$, find the stress vector acting on the surface characterized by the unit normal $\mathbf{n} = (1/2)\mathbf{i} + (1/\sqrt{2})\mathbf{j} - (1/2)\mathbf{k}$. Next write this stress tensor in matrix form and calculate the stress vector on the same surface.

SOLUTION

We can use Eq. 4.31a to solve the first part of this problem. We have

$$\mathbf{\Sigma} = \left(\frac{1}{2}\mathbf{i} + \frac{1}{\sqrt{2}}\mathbf{j} - \frac{1}{2}\mathbf{k}\right) \cdot (-5\mathbf{ii} + 3\mathbf{ij} - 1\mathbf{ik} + 3\mathbf{ji} - 4\mathbf{jj} + 4\mathbf{jk} - 1\mathbf{ki} + 4\mathbf{kj} - 2\mathbf{kk})$$

$$= \left(\frac{-5}{2} + \frac{3}{\sqrt{2}} + \frac{1}{2}\right)\mathbf{i} + \left(\frac{3}{2} - \frac{4}{\sqrt{2}} - \frac{4}{2}\right)\mathbf{j} + \left(\frac{-1}{2} + \frac{4}{\sqrt{2}} + \frac{2}{2}\right)\mathbf{k}$$

Thus the stress vector on this surface is

$$\mathbf{\Sigma} = \left(\frac{3\sqrt{2}-4}{2}\right)\mathbf{i} - \left(\frac{4\sqrt{2}+1}{2}\right)\mathbf{j} + \left(\frac{4\sqrt{2}+1}{2}\right)\mathbf{k}$$

Expressing the stress tensor in matrix notation we have:

$$\boldsymbol{\sigma} = \begin{pmatrix} -5 & 3 & -1 \\ 3 & -4 & 4 \\ -1 & 4 & -2 \end{pmatrix}$$

Evaluating the stress vector using the matrix product, Eq. 4.31b, gives

$$\Sigma = \left(\frac{1}{2}, \frac{1}{\sqrt{2}}, -\frac{1}{2}\right) \begin{pmatrix} -5 & 3 & -1 \\ 3 & -4 & 4 \\ -1 & 4 & -2 \end{pmatrix}$$

$$= \left[\left(\frac{-5}{2} + \frac{3}{\sqrt{2}} + \frac{1}{2}\right), \left(\frac{3}{2} - \frac{4}{\sqrt{2}} - \frac{4}{2}\right), \left(\frac{-1}{2} + \frac{4}{\sqrt{2}} + \frac{2}{2}\right)\right]$$

$$= \left(\frac{3\sqrt{2} - 4}{2}\right)\mathbf{i} - \left(\frac{4\sqrt{2} + 1}{2}\right)\mathbf{j} + \left(\frac{4\sqrt{2} + 1}{2}\right)\mathbf{k}$$

which can be seen to be identical to the foregoing result. Which approach do you prefer?

do this in two different ways. We can use the stress vector to calculate the total surface force from Eq. 4.21:

$$\mathbf{F}_S = \int_S \Sigma \, dS$$

or we can substitute for the stress vector, using Eq. 4.29, $\Sigma = \mathbf{n} \cdot \boldsymbol{\sigma}$, and write the total surface force in terms of the stress tensor as

$$\mathbf{F}_S = \int_S (\mathbf{n} \cdot \boldsymbol{\sigma}) \, dS \tag{4.32}$$

EXAMPLE 4.14

Write an expression for the total surface force on a plane with normal $\mathbf{n} = (0, 1, 0)$, i.e., a plane parallel to the xz plane.

SOLUTION

We can use either Eq. 4.21, $\mathbf{F}_S = \int_S \Sigma \, dS$, or Eq. 4.32, $\mathbf{F}_S = \int_S (\mathbf{n} \cdot \boldsymbol{\sigma}) \, dS$, to write the desired expression. Using Eq. 4.31b with the given unit normal yields:

$$\Sigma = \mathbf{n} \cdot \boldsymbol{\sigma} = (n_x, n_y, n_z) \begin{pmatrix} \sigma_{xx} & \sigma_{xy} & \sigma_{xz} \\ \sigma_{yx} & \sigma_{yy} & \sigma_{yz} \\ \sigma_{zx} & \sigma_{zy} & \sigma_{zz} \end{pmatrix} = (0, 1, 0) \begin{pmatrix} \sigma_{xx} & \sigma_{xy} & \sigma_{xz} \\ \sigma_{yx} & \sigma_{yy} & \sigma_{yz} \\ \sigma_{zx} & \sigma_{zy} & \sigma_{zz} \end{pmatrix}$$

$$= (\sigma_{yx}, \sigma_{yy}, \sigma_{yz})$$

The total surface force on this plate is given by either Eq. 4.21 or Eq. 4.32 as

$$\mathbf{F}_S = \int_S (\sigma_{yx}\mathbf{i} + \sigma_{yy}\mathbf{j} + \sigma_{yz}\mathbf{k}) \, dS$$

If we write this in terms of the components of the total surface force, we have

$$F_{S_x} = \int_S \sigma_{yx}\,dS, \quad F_{S_y} = \int_S \sigma_{yy}\,dS, \quad \text{and} \quad F_{S_z} = \int_S \sigma_{yz}\,dS$$

Are you able to convince yourself that these surface integrals account for the force applied by the fluid to this plane? Notice that the component of force parallel to the plane normal ($\mathbf{n} = \mathbf{j}$) is related to a normal stress, while the components of force perpendicular to the plane normal are related to shear stresses, as expected. It may also be helpful to recall that we can use Eq. 4.23 to write the total surface force as $\mathbf{F}_S = \int_S -p\mathbf{n}\,dS + \int_S \boldsymbol{\tau}\,dS$.

4.7 FORCE BALANCE IN A FLUID

We may draw important conclusions about the state of stress in a fluid and the role of body and surface forces in creating fluid motion if we consider a force balance on an infinitesimal cube of fluid of sides dx, dy, dz centered at (x_0, y_0, z_0) as shown in Figure 4.29A. The volume of this cube is $\delta V = dx\,dy\,dz$. Applying Newton's second law to the cube, we see that the sum of the body and surface forces acting on the cube must equal the time rate of change of linear momentum, i.e., the inertial force. If the density of the fluid at the center of the cube is ρ, and the acceleration of the fluid is $\mathbf{a} = (a_x, a_y, a_z)$, then Newton's law can be written as $\rho \mathbf{a}\,\delta V = \mathbf{F}_B + \mathbf{F}_S$. Since the cube is infinitesimal, we can write the total body force acting on the cube as $\mathbf{F}_B = \rho \mathbf{f}\,\delta V$, where $\mathbf{f} = (f_x, f_y, f_z)$ is the body force per unit mass. Thus the force balance on the cube is $\rho \mathbf{a}\,\delta V = \rho \mathbf{f}\,\delta V + \mathbf{F}_S$, and the three components of this vector equation are

$$\rho a_x\,\delta V = \rho f_x\,\delta V + F_{Sx} \tag{4.33a}$$
$$\rho a_y\,\delta V = \rho f_y\,\delta V + F_{Sy} \tag{4.33b}$$
$$\rho a_z\,\delta V = \rho f_z\,\delta V + F_{Sz} \tag{4.33c}$$

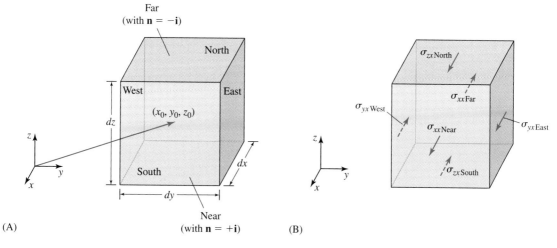

Figure 4.29 Infinitesimal cube of fluid: (A) nomenclature and (B) stresses.

To complete the force balance, we must write expressions for the surface forces acting on the six cube faces. Consider the x component of the force balance, Eq. 4.33a. We can calculate the component of the surface force acting in the x direction by using the information in Figure 4.29B. First note that from the subscript convention for the stresses, or by direct observation, the only stresses acting on any cube face in the x direction are σ_{xx}, σ_{yx}, and σ_{zx}. Since the dimensions of this cube are small, the surface force produced by a stress on a face is equal to the product of that stress at the center of the face and the area of the face. If we label the cube faces in Figure 4.29 as North, South, East, West, Near, and Far, then we can immediately write the x component of the total surface force on each face as the product of the individual stresses and the area of each face. For example, on the Far face, the surface force acting in the x direction is

$$F_{S_x} = (-\sigma_{xx_{Far}})\, dy\, dz$$

Adding the contributions from all six faces of the cube, we obtain the x component of the total surface force on the cube as

$$F_{S_x} = (\sigma_{xx_{Near}} - \sigma_{xx_{Far}})\, dy\, dz + (\sigma_{yx_{East}} - \sigma_{yx_{West}})\, dx\, dz \\ + (\sigma_{zx_{North}} - \sigma_{zx_{South}})\, dx\, dy \tag{4.34}$$

We can use a two-term Taylor series to relate the stress on any of the cube faces to the value of the same stress at the center of the cube. A two-term approximation is sufficient because the cube is small, and higher order terms in the Taylor series are negligible. For example, on the Near and Far faces we can write

$$\sigma_{xx_{Near}} = \sigma_{xx_{Center}} + \frac{\partial \sigma_{xx}}{\partial x}\frac{dx}{2} \quad \text{and} \quad \sigma_{xx_{Far}} = \sigma_{xx_{Center}} - \frac{\partial \sigma_{xx}}{\partial x}\frac{dx}{2}$$

The difference in these two terms is

$$\sigma_{xx_{Near}} - \sigma_{xx_{Far}} = \frac{\partial \sigma_{xx}}{\partial x}\, dx$$

Similarly, we can write Taylor series for the other two terms and show that the x component of the total surface force on the cube is

$$F_{S_x} = \left(\frac{\partial \sigma_{xx}}{\partial x} + \frac{\partial \sigma_{yx}}{\partial y} + \frac{\partial \sigma_{zx}}{\partial z}\right) dx\, dy\, dz$$

Since $\delta V = dx\, dy\, dz$, we can write this as

$$F_{S_x} = \left(\frac{\partial \sigma_{xx}}{\partial x} + \frac{\partial \sigma_{yx}}{\partial y} + \frac{\partial \sigma_{zx}}{\partial z}\right) \delta V$$

Considering the y and z components of the surface force in the same way, we find

$$F_{S_y} = \left(\frac{\partial \sigma_{xy}}{\partial x} + \frac{\partial \sigma_{yy}}{\partial y} + \frac{\partial \sigma_{zy}}{\partial z}\right) \delta V \quad \text{and} \quad F_{S_z} = \left(\frac{\partial \sigma_{xz}}{\partial x} + \frac{\partial \sigma_{yz}}{\partial y} + \frac{\partial \sigma_{zz}}{\partial z}\right) \delta V$$

Upon substituting into Eq. 4.33, and dividing by $\delta \Psi$, the three components of the force balance on the cube are

$$\rho a_x = \rho f_x + \left(\frac{\partial \sigma_{xx}}{\partial x} + \frac{\partial \sigma_{yx}}{\partial y} + \frac{\partial \sigma_{zx}}{\partial z} \right) \tag{4.35a}$$

$$\rho a_y = \rho f_y + \left(\frac{\partial \sigma_{xy}}{\partial x} + \frac{\partial \sigma_{yy}}{\partial y} + \frac{\partial \sigma_{zy}}{\partial z} \right) \tag{4.35b}$$

$$\rho a_z = \rho f_z + \left(\frac{\partial \sigma_{xz}}{\partial x} + \frac{\partial \sigma_{yz}}{\partial y} + \frac{\partial \sigma_{zz}}{\partial z} \right) \tag{4.35c}$$

Although we have referred to these three equations as the force balance in a fluid, they are also the equations of motion, since they relate the acceleration of the fluid to the body and surface forces.

Notice that each of the terms in these equations has the dimensions of a force per unit volume. For example, the three components of the inertial force per unit volume $\rho \mathbf{a}$ are $(\rho a_x, \rho a_y, \rho a_z)$, and the three components of the body force per unit volume $\rho \mathbf{f}$ are given by $(\rho f_x, \rho f_y, \rho f_z)$. The three components of the surface force per unit volume acting at a point in a fluid are seen to be

$$\frac{F_{S_x}}{\delta \Psi} = \frac{\partial \sigma_{xx}}{\partial x} + \frac{\partial \sigma_{yx}}{\partial y} + \frac{\partial \sigma_{zx}}{\partial z} \tag{4.36a}$$

$$\frac{F_{S_y}}{\delta \Psi} = \frac{\partial \sigma_{xy}}{\partial x} + \frac{\partial \sigma_{yy}}{\partial y} + \frac{\partial \sigma_{zy}}{\partial z} \tag{4.36b}$$

$$\frac{F_{S_z}}{\delta \Psi} = \frac{\partial \sigma_{xz}}{\partial x} + \frac{\partial \sigma_{yz}}{\partial y} + \frac{\partial \sigma_{zz}}{\partial z} \tag{4.36c}$$

If all the stresses are spatially uniform, the net surface force acting on an infinitesimal volume of fluid is zero, and the stresses do not contribute to an acceleration. Thus to have a net surface force on an infinitesimal volume of fluid, there must be a spatial variation in one or more components of the stress tensor describing the state of stress in the fluid.

In vector notation the right-hand side of Eq. 4.36a–c can be expressed as

$$\nabla \cdot \boldsymbol{\sigma} \tag{4.37}$$

This dot product, known as the stress divergence, involves the del operator

$$\nabla = \left(\frac{\partial}{\partial x} \mathbf{i} + \frac{\partial}{\partial y} \mathbf{j} + \frac{\partial}{\partial z} \mathbf{k} \right)$$

and can be evaluated in vector or matrix notation. For example, if we use the matrix product of the del operator with the stress tensor we get

$$\nabla \cdot \boldsymbol{\sigma} = \left(\frac{\partial}{\partial x} \mathbf{i} + \frac{\partial}{\partial y} \mathbf{j} + \frac{\partial}{\partial z} \mathbf{k} \right) \begin{pmatrix} \sigma_{xx} & \sigma_{xy} & \sigma_{xz} \\ \sigma_{yx} & \sigma_{yy} & \sigma_{yz} \\ \sigma_{zx} & \sigma_{zy} & \sigma_{zz} \end{pmatrix} \tag{4.38}$$

Accordingly, the three Cartesian components of the stress divergence, are

$$(\nabla \cdot \boldsymbol{\sigma})_x = \frac{\partial \sigma_{xx}}{\partial x} + \frac{\partial \sigma_{yx}}{\partial y} + \frac{\partial \sigma_{zx}}{\partial z} \qquad (4.39a)$$

$$(\nabla \cdot \boldsymbol{\sigma})_y = \frac{\partial \sigma_{xy}}{\partial x} + \frac{\partial \sigma_{yy}}{\partial y} + \frac{\partial \sigma_{zy}}{\partial z} \qquad (4.39b)$$

$$(\nabla \cdot \boldsymbol{\sigma})_z = \frac{\partial \sigma_{xz}}{\partial x} + \frac{\partial \sigma_{yz}}{\partial y} + \frac{\partial \sigma_{zz}}{\partial z} \qquad (4.39c)$$

Writing the force balance Eq. 4.35a–4.35c in vector notation, we have

$$\rho \mathbf{a} = \rho \mathbf{f} + \nabla \cdot \boldsymbol{\sigma} \qquad (4.40)$$

This deceptively simple equation governs all aspects of the behavior of a fluid. From it we can conclude that if the body and surface forces are not in balance, a fluid must accelerate. It follows then that in the absence of acceleration, the body and surface forces are in an exact balance at every point in the fluid.

4.8 SUMMARY

In this chapter we have discussed body and surface forces in fluid mechanics, the origins of these forces, and the mathematical representation or model used to describe and calculate fluid forces.

Forces in fluid mechanics that have their origins in the structure and behavior of matter are termed real forces. Examples include gravity and pressure forces. Apparent forces appear to be present because a fluid is observed in a noninertial reference frame. Examples include centrifugal and Coriolis forces and the force created when a volume of fluid is given a uniform rectilinear acceleration. All the forces in fluid mechanics, real and apparent, are divided into two types: body forces and surface forces. Body forces, of which gravity and electromagnetic forces are examples, are long-range forces that act on a volume of fluid in such a way that the magnitude of the body force is proportional to the mass or volume of the fluid element. Body forces act on a fluid but are not applied by a fluid. They exert their influence on fluids at rest and in motion without the need for physical contact between the fluid and the external source of the body force. In general, a body force must be represented as a function of both position and time. The gravitational body force is most conveniently represented on a per-unit-mass basis.

Surface forces, such as those exerted by pressure or shear stress, are short-range forces that act on an element of fluid through physical contact. A surface force is created as a result of physical contact between the molecules of a fluid and the molecules of a bounding material. Surface forces not only act on a fluid but are applied by a fluid to its surroundings. These forces exist at every interface involving a fluid, both in a fluid at rest and in a fluid in motion.

The total surface force depends on the size, shape, and orientation of the surface on which the force acts, as well as on characteristics of the fluid and its motion. In developing a model for the surface force, we may distinguish three different situations: the surface force exerted by a fluid on a structure, the surface force applied to a fluid by a structure, and the state of stress in a fluid. The orientation of any surface is specified by the outward unit normal of the surface.

In a fluid in motion, the surface force per unit area acting on an infinitesimal surface generally has both normal and tangential components. Thus, stress in fluid dynamics is a vector quantity because it has a magnitude expressed in dimensions of $\{FL^{-2}\}$ and a direction. The surface force acting on a plane surface element depends on position, and on the size and orientation of the element. The total surface force acting on any surface in contact with fluid is found by using a surface integral to sum the individual contributions from each infinitesimal surface element. The total surface force is a vector quantity whose components may be resolved in any desired direction.

The components of total surface force on an aircraft, vehicle, or other object are given distinctive names. The component of total surface force in the direction of the oncoming fluid stream is called the drag. The component of total surface force that acts orthogonally to the oncoming fluid stream is called the lift.

The state of stress in a fluid is represented by a stress tensor. The stress tensor is conveniently represented in matrix form as

$$\boldsymbol{\sigma} = \begin{pmatrix} \sigma_{xx} & \sigma_{xy} & \sigma_{xz} \\ \sigma_{yx} & \sigma_{yy} & \sigma_{yz} \\ \sigma_{zx} & \sigma_{zy} & \sigma_{zz} \end{pmatrix}$$

with a pair of unit vectors associated with each component of the stress tensor by the position of the component in the matrix representation. If there are no externally imposed body moments acting on a fluid, as is normally the case, the stress tensor is symmetric. For a symmetric stress tensor there are six independent stress components. These are the three normal stresses and three shear stresses. We shall see later that the governing equations of fluid mechanics are written in terms of spatial derivatives of the nine components of the stress tensor, each of which is a function of position and time.

The total surface force acting on an arbitrary volume of fluid may be expressed as a volume integral of the dot product of the del operator and the stress tensor. The dot product, $\boldsymbol{\nabla} \cdot \boldsymbol{\sigma}$, is the stress divergence. The stress divergence may be interpreted as the surface force per unit volume. To have a net surface force on a volume of fluid, there must be a spatial variation in one or more components of the stress tensor describing the state of stress in the fluid. If the stresses are uniform, the stress divergence is zero at every point in a fluid, and the net surface force acting on a volume of fluid is also zero.

PROBLEMS

Section 4.1

4.1 Hold out your hand with your palm facing down. Estimate the net force on your palm as a result of atmospheric pressure (use units of Pa). Now change the orientation of your hand so that your palm is facing away from you, i.e., 90° away from the original orientation. Does the magnitude of the force applied to your palm by the atmosphere change?

4.2 Suppose you are supporting a sheet of 8.5 in. × 11 in. paper by placing your open palm underneath the paper so that the paper is parallel to the ground. Estimate the force applied to the top side of the paper by the atmosphere (use units of lb$_f$). Why doesn't the paper "feel" heavy?

Section 4.2

4.3 Describe the similarities and differences between "real" and "apparent" forces. Give an example of each kind of force. Is an inertial force a real force or a fictitious force?

4.4 Describe the similarities and differences between body and surface forces. Give an example of each kind of force. Which of these two types of force can act on a fluid? Which of these types can be applied by a fluid?

4.5 Determine whether each of the forces listed is a real or a fictitious force. Also decide whether it is a body or a surface force.
(*a*) Gravity force
(*b*) Pressure force
(*c*) Coriolis force
(*d*) Inertial force
(*e*) Electromagnetic force
(*f*) Centrifugal force

4.6 List and describe each of the various kinds of force that contribute to the motion of air (wind) in the atmosphere.

4.7 Provide an example of a fluid flow that is driven exclusively by body forces. Also give an example of a fluid flow that is dominated by surface forces.

Section 4.3

4.8 The gravitational body force is a direct result of the presence of matter. Therefore, one portion of a fluid exerts a gravitational attraction for other portions of the same fluid. Under most engineering circumstances, however, this effect is negligible. Provide an example of an exception to this rule.

4.9 From physics we know that the magnitude of the gravitational force is a function of position. In fact, gravity obeys an inverse square law. Why is it that we can model the gravitational force on a fluid as constant?

4.10 Which type of force, body or surface, is able to act over large distances? Why is the other kind of force restricted to action over small distances?

4.11 Provide at least two engineering applications in which body forces other than gravity are important.

4.12 Explain the mechanism by which a gas can transmit force through its interior. Is the mechanism by which a gas transmits a surface force to a solid object it contacts the same or different?

4.13 Could you reduce friction in a pipe flow by coating the pipe wall with Teflon? Why or why not?

4.14 How does the concept of surface tension fit into a discussion of fluid forces?

Section 4.4

4.15 Consider the fluid flow shown in Figure 4.5. At what angle β are the magnitudes of the components of the gravitational body force acting in the flow direction and normal to the flow direction equal? At what angle β is the magnitude of the component of the gravitational body force acting in the flow direction three times as great as that of the component of the body force acting normal to the flow direction?

4.16 What volume of air at STP experiences the same gravitational body force per unit volume as does a 1 in.^{3} volume of water at STP?

4.17 What volume of mercury at STP experiences the same gravitational body force per unit volume as does a slug of oxygen at STP?

4.18 Air at STP has a uniform free space electrical charge per unit volume of 6×10^9 C/m^3. The air is subjected to a gravitational body force in the $-z$ direction and a constant electric field of unknown strength in the $+z$ direction. Given that the body force exerted on a fluid by an electric field is described by the equation.

$$F_B = \int Eq\,dV$$

where E is the electric field strength (in units of V/m or equivalently N/C) and q is the electric charge per unit volume (in units of C/m^3). If the total body force acting on a fluid particle is zero, estimate the magnitude of the electric field.

4.19 Express the centrifugal body force on a per-unit-volume basis and plot its value as a function of distance from the rotation axis for a water particle in a horizontal centrifuge. Assume a rotation rate of 10,000 rpm and that the outside radius of the centrifuge is 100 mm.

4.20 The liquid in the cylinder illustrated in Figure 4.8 is subjected to a temperature variation along its axis. This results in a linear density profile in the fluid given by $\rho(z) = \rho_0 - \alpha z$, where $\rho_0 = 1020$ kg/m^3. The cylinder dimensions are $R = 1$ m and $H = 10$ m. It is known that the total body force acting on the tank is -302 kN k. Determine the density at the top of the tank.

4.21 Calculate the total body force acting on a cup of coffee. State your assumptions.

4.22 Calculate the total body force on a room full of air. The room 12 ft × 16 ft and it has an 8 ft ceiling. Assume that 10% of the room volume is occupied by furniture. State any additional assumptions.

4.23 The tank from Example 4.3 is filled with a mixture of liquids. After settling, it is found that the mixture is affected by temperature and gravity such that the density distribution is $\rho = A + Br + Cz$, where $A = 500$ kg/m^3, $B = 5$ kg/m^2, and $C = -10$ kg/m^2. What is the total body force on the fluid in the tank?

4.24 Is the total body force acting on a volume of fluid an intrinsic or an extrinsic quantity? Why?

4.25 A fish tank has dimensions $L = 75$ cm, $W = 36$ cm, and $H = 30$ cm. Calculate the total gravitational body force on the contents of the tank under each of the following conditions:
(a) The tank is filled with water at STP. (Use a volume integral.)
(b) The tank is filled with water at STP. (Don't use any integrals.)
(c) The lower 90% of the tank is filled with water and the top 10% is filled with air at STP. (Use a volume integral.)
(d) The lower 90% of the tank is filled with water and the top 10% is filled with air at STP. (Don't use any integrals.)

4.26 The fish tank described in Problem 4.25 is being transported in the back of your friend's pickup truck. The tank is half full of water, and the truck is accelerating at a rate of 1.2 km/h/s. Calculate the total apparent body force acting on the water in the tank due to the truck's acceleration.

4.27 A 55-gallon drum (see Example 4.3 for dimensions) is two-thirds full of oil. Calculate the total gravitational body force on the fluid in the drum under the following conditions:
(a) The remaining third of the drum is filled with air at STP.
(b) The remaining third of the drum is filled with water at STP. Assume that the fluid is stratified; i.e., one horizontal layer of fluid is sitting directly on top of the other layer (which layer is on top?).
(c) The remaining third of the drum is filled with water at STP. Assume that the two fluids are completely mixed (how this is done is not obvious).

4.28 The 55-gallon drum described in part a of Problem 4.27 is being transported in an airplane that is undergoing a horizontal acceleration of 25 mph/s. What is the total apparent body force acting on the fluid in the drum?

4.29 A rectangular swimming pool has a surface area 15 m × 8 m. Two-thirds of the pool has a uniform depth of 1.2 m, and the diving well has a uniform depth of 3 m. Calculate the total gravitational body force acting on the water in the pool under the following conditions:
(a) It is assumed that the water has a constant density of 1000 kg/m³.
(b) It is assumed that temperature variations cause the density of the water in the pool to obey the functional form: $\rho(z) = $ (998.9 kg/m³) $- [(1.0$ kg/m²$)z]$, where gravity acts in the $-z$ direction, $z = 0$ corresponds to the water surface, and z is measured in meters.

Section 4.5

4.30 Describe in your own words the convention used to identify outward unit normal vectors on surfaces in fluid mechanics.

4.31 Sketch a fluid system with which you are familiar. Identify and correctly label at least four outward unit normal vectors associated with surface forces in this fluid system.

4.32 In a certain fluid flow situation the stress vector on a surface is known to have the form

$$\Sigma = \alpha \mathbf{i} + \beta x \mathbf{j} + \gamma y^2 \mathbf{k}$$

where $\alpha = 30$ kPa, $\beta = 20$ kN/m³, and $\gamma = 10$ kN/m⁴.
(a) Determine the stress vector acting on a flat surface area of magnitude 1 mm² centered at the position (5 m, 0, 3 m) with an outward unit normal $\mathbf{n} = \mathbf{i}$.
(b) Determine the force acting on the surface described in part a.
(c) Repeat the calculation in part a but do it for the back side of the same surface; i.e., this time the outward unit normal is $\mathbf{n} = -\mathbf{i}$.

4.33 For the surface $f(x, y, z) = x + 2y - 3z$, find the normal and tangential stresses at the point (1, 0, 1) if the stress vector at that point is $\Sigma = \mathbf{i} + 2\mathbf{j} - 3\mathbf{k}$.

4.34 For the surface $f(x, y, z) = x^3 + 2/y + 8z^2$, find the normal and tangential stresses at each of the indicated points assuming the stress vector is $\Sigma = 8\mathbf{i} - 5\mathbf{j} + 10\mathbf{k}$.
(a) (0, 1, 0)
(b) (1, 1, 2)

4.35 For the surface $f(x, y, z) = x + 2y^2 + 1/z - 5$, find the normal and tangential stresses at each of the indicated points assuming the stress vector is $\Sigma = x\mathbf{i} - y\mathbf{j} + \mathbf{k}$.
(a) (0, 0, 0)
(b) (0, 3, −3)

4.36 A stress vector is given by $\Sigma = 2\mathbf{i} + z\mathbf{j} + \mathbf{k}$. Calculate the total surface force acting on the portion of a vertical surface parallel to the xz plane and intercepting the y axis at $y = 2$, defined by $1 < x < 3$ and $0 < z < 5$.

4.37 Calculate the surface force on a flat plate (1 × 1 m²) due to air flowing at $U = 5$ m/s, at 20°C. The plate is horizontal to the airstream, so the shear stress is the same on the upper and lower surfaces and is equal to

$$\tau = 0.0332 \left(\frac{\rho\mu}{x}\right)^{1/2} U^{3/2}$$

where x is along the plate in the direction of the flow. The pressure is atmospheric everywhere along the plate.

4.38 If the flow over the plate in Problem 4.37 is water, what will be the surface force?

4.39 If the pressure drop in a pipe having an inner diameter of 1 in. is 1.0 psig over 10 ft of length, what is the wall shear stress?

4.40 If the wall shear stress in a 50 mm diameter tube is 50 N/m², what is the pressure drop per meter of length?

4.41 The maximum specified weight for a Boeing 747SP (Special Performance) aircraft is $W = 3115$ kN. While cruising, it requires a thrust $T = 600$ kN. What is the lift F_L developed by the wing? What is the drag F_D of the aircraft?

4.42 Define, using both words and equations, each of the following terms:
(a) Lift
(b) Drag

Section 4.6

4.43 Sketch the two possible orientations of a negative normal (compressive) stress on a plane parallel to the yz plane. Also show the two possible orientations for the (positive) shear stress, σ_{yz}.

4.44 Sketch the two possible orientations of a positive normal (tensile) stress on a plane parallel to the xy plane. Also show the two possible orientations for the (negative) shear stress, σ_{xy}.

4.45 Sketch the two possible orientations of a negative normal (compressive) stress on a plane parallel to the zx plane. Also show the two possible orientations for the (negative) shear stress, σ_{zx}.

4.46 Find the stress vector acting on the surface characterized by the normal

$$\mathbf{n} = (1/\sqrt{3})\mathbf{i} + (1/\sqrt{3})\mathbf{j} - (1/\sqrt{3})\mathbf{k}$$

arising from the stress tensor

$$\boldsymbol{\sigma} = -5\mathbf{ii} + 3\mathbf{ij} - 1\mathbf{ik} + 3\mathbf{ji} - 4\mathbf{jj} + 4\mathbf{jk} - 1\mathbf{ki} + 4\mathbf{kj} - 2\mathbf{kk}$$

4.47 Find the stress vector acting on the surface characterized by the normal

$$\mathbf{n} = (1/3)\mathbf{i} - (2/3)\mathbf{j} + (2/3)\mathbf{k}$$

arising from the stress tensor

$$\boldsymbol{\sigma} = -25\mathbf{ii} + 18\mathbf{ij} + 7\mathbf{ik} + 30\mathbf{ji} + 16\mathbf{jj} - 4\mathbf{jk} - 10\mathbf{ki} - 45\mathbf{kj} + 10\mathbf{kk}$$

4.48 Find the stress vector acting on the surface characterized by the normal

$$\mathbf{n} = (1/\sqrt{3})[-\mathbf{i} - \mathbf{j} + \mathbf{k}]$$

arising from the stress tensor

$$\boldsymbol{\sigma} = \begin{pmatrix} \sigma_{xx} & \sigma_{xy} & \sigma_{xz} \\ \sigma_{yx} & \sigma_{yy} & \sigma_{yz} \\ \sigma_{zx} & \sigma_{zy} & \sigma_{zz} \end{pmatrix}$$

$$= \begin{pmatrix} 10 & -20 & 30 \\ -20 & 0 & 5 \\ 30 & 5 & -10 \end{pmatrix}$$

4.49 Assuming that there are no externally imposed body moments acting on a fluid, find the stress vector acting on the surface characterized by the normal

$$\mathbf{n} = (1/3)[-2\mathbf{i} + \mathbf{j} + 2\mathbf{k}]$$

arising from the stress tensor

$$\boldsymbol{\sigma} = \begin{pmatrix} \sigma_{xx} & \sigma_{xy} & \sigma_{xz} \\ \sigma_{yx} & \sigma_{yy} & \sigma_{yz} \\ \sigma_{zx} & \sigma_{zy} & \sigma_{zz} \end{pmatrix}$$

$$= \begin{pmatrix} -10 & -50 & ? \\ ? & -10 & ? \\ 25 & -5 & -10 \end{pmatrix}$$

4.50 After reviewing Example 4.14, write the expression for the total surface force on a plate with normal $\mathbf{n} = (0, 0, 1)$.

4.51 After reviewing Example 4.14, write the expression for the total surface force on a plate with normal $\mathbf{n} = (1/\sqrt{3})[-\mathbf{i} - \mathbf{j} + \mathbf{k}]$.

Section 4.7

4.52 For the stress tensor given, calculate the following quantities:

$$\boldsymbol{\sigma} = \begin{pmatrix} \sigma_{xx} & \sigma_{xy} & \sigma_{xz} \\ \sigma_{yx} & \sigma_{yy} & \sigma_{yz} \\ \sigma_{zx} & \sigma_{zy} & \sigma_{zz} \end{pmatrix}$$

$$= \begin{pmatrix} -Az & 0 & Bx^2 - Cy^3 \\ 0 & -Az & 0 \\ Bx^2 - Cy^3 & 0 & -Az \end{pmatrix}$$

(a) The stress divergence
(b) The component of the surface force acting on a cube of edge length α in the x direction
(c) The total surface force acting on a cube of edge length α

4.53 For the stress tensor given, calculate the following quantities:

$$\boldsymbol{\sigma} = \begin{pmatrix} \sigma_{xx} & \sigma_{xy} & \sigma_{xz} \\ \sigma_{yx} & \sigma_{yy} & \sigma_{yz} \\ \sigma_{zx} & \sigma_{zy} & \sigma_{zz} \end{pmatrix}$$

$$= \begin{pmatrix} -2Az & 0 & Ax \\ 0 & -2Az & Ay \\ Ax & Ay & -2Az \end{pmatrix}$$

(a) The stress divergence
(b) The component of the surface force acting on a cube of edge length α in the y direction
(c) The total surface force acting on a cube of edge length α
(d) The appropriate dimensions for the constant A

5 FLUID STATICS

5.1 Introduction
5.2 Hydrostatic Stress
5.3 Hydrostatic Equation
    5.3.1 Integral Hydrostatic Equation
    5.3.2 Differential Hydrostatic Equation
5.4 Hydrostatic Pressure Distribution
    5.4.1 Constant Density Fluid in a Gravity Field
    5.4.2 Variable Density Fluid in a Gravity Field
    5.4.3 Constant Density Fluid in Rigid Rotation
    5.4.4 Constant Density Fluid in Rectilinear Acceleration
5.5 Hydrostatic Force
    5.5.1 Planar Aligned Surface
    5.5.2 Planar Nonaligned Surface
    5.5.3 Curved Surface
5.6 Hydrostatic Moment
    5.6.1 Planar Aligned Surface
    5.6.2 Planar Nonaligned Surface
5.7 Resultant Force and Point of Application
5.8 Buoyancy and Archimedes' Principle
5.9 Equilibrium and Stability of Immersed Bodies
5.10 Summary
Problems

5.1 INTRODUCTION

In this chapter we discuss fluid statics, the branch of fluid mechanics associated with the behavior of fluid at rest. Fluids at rest exert significant surface forces. Since shear stresses are completely absent in a fluid at rest, these surface forces arise solely from the action of normal stress, i.e., hydrostatic pressure. Thus, the total surface force applied by a fluid at rest is due to the pressure distribution on the surface. The key problem in fluid statics is to determine the pressure distribution in the fluid. In this chapter we will use a force balance to derive the governing equation of fluid statics from Newton's second law. We then show how to solve this equation to determine the hydrostatic pressure distribution for situations of engineering interest. The concepts of fluid forces and stresses developed in Chapter 4 provide the foundation on which your understanding of hydrostatics will be constructed.

As you will discover, all problems in fluid statics can be solved exactly, meaning in analytical form, with no approximations needed to simplify the mathematics. The ability to solve a problem analytically is rare in fluid mechanics, and while this fact alone might be sufficient motivation for studying hydrostatics, our interest in the subject is far more practical. There are many engineering applications involving a fluid at rest or nearly so. For example, fluid statics is used to analyze problems involving Earth's atmosphere, oceans, lakes, and rivers. Since the fluid in liquid and gas storage vessels is often at rest, the safe design of such vessels relies on information gained from hydrostatics. The fluid in hydraulic systems in industrial equipment and automobile braking systems is at rest or moving slowly, so analysis of pressures and forces in these devices also falls within the realm of hydrostatics.

The references to "a fluid at rest or nearly so" and to "at rest or moving slowly" in the preceding paragraph are quite deliberate. Earth's atmosphere is never completely at rest, so how is it possible to analyze this and similar situations with fluid statics? Consider our discussion in Chapter 2 of the relationship between pressure and velocity in a moving fluid. At modest velocities, Bernoulli's equation (Eq. 2.11) shows that the change in pressure due to the fluid motion is small and may be neglected. As our first example illustrates, it is not necessary for a fluid to be absolutely at rest to conclude that a fluid statics analysis will accurately provide the pressure distribution.

EXAMPLE 5.1

The water near the intake pipe of a dam is moving at 1 m/s as shown in Figure 5.1. If the intake pipe is located 10 m below the surface, is the pressure distribution in the vicinity of the intake significantly changed by the flow?

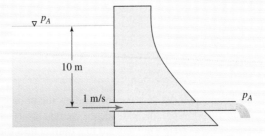

Figure 5.1 Schematic for Example 5.1.

> **SOLUTION**
>
> If the water density is approximated as 1000 kg/m^3, then the pressure at this depth for still water may be calculated by using Eq. 2.9 as $\Delta p = \rho_{H_2O}gd = 98.1$ kPa above atmospheric pressure. Thus the absolute pressure 10 m below the surface is about 199 kPa or 2 atm. From Bernoulli's equation we know that the pressure will be lower than this for water that is moving. The decrease in pressure due to the flow may be estimated by using Eq. 2.11, applied between a point near the intake pipe (where the speed is 1 m/s) and a second located at the same depth but well away from the intake pipe (where the speed is 0 m/s). The pressure difference between these points is $\Delta p = \frac{1}{2}\rho_{H_2O}V^2 = 500$ Pa. This represents a pressure decrease of $\sim 0.25\%$ below the background hydrostatic pressure of 199 kPa. Clearly the pressure distribution near the intake pipe is very close to that predicted by fluid statics.

In Section 5.2 we explore the state of stress in a fluid at rest, develop the governing equation of fluid statics, and demonstrate how to solve this equation to determine the hydrostatic pressure distribution. Finally, we show how to calculate the forces and moments exerted by a fluid at rest on a surface.

5.2 HYDROSTATIC STRESS

Since shear stresses must be completely absent in a fluid at rest, the stress vector given by Eq. 4.16 as $\mathbf{\Sigma} = \sigma_N\mathbf{n} + \boldsymbol{\tau}$, reduces to $\mathbf{\Sigma} = \sigma_N\mathbf{n}$, where the normal stress σ_N is the magnitude of the stress vector. Since we know from Chapter 1 that fluids are unable to exert a tensile stress, the hydrostatic stress vector must be compressive in its action, i.e., $\sigma_N < 0$.

Now consider the forces acting on a small tetrahedron of fluid located at some point x within the fluid, as shown in Figure 5.2. We have selected a set of Cartesian coordinates (X, Y, Z) with its origin located at the vertex of the tetrahedron and unit vectors $(\mathbf{I}, \mathbf{J}, \mathbf{K})$ aligned with the edges of the tetrahedron. The tetrahedron is of height h, has volume $V = h^3/6$, and has three faces labeled A_1, A_2, and A_3, each of area $h^2/2$. The outward unit normals $\mathbf{n}_1$, $\mathbf{n}_2$, and $\mathbf{n}_3$ to these three faces are easily found by inspection. The larger face labeled A has area $(\sqrt{3}/2)h^2$ and an outward unit normal $\mathbf{n} = (1/\sqrt{3})\mathbf{I} + (1/\sqrt{3})\mathbf{J} + (1/\sqrt{3})\mathbf{K}$.

The fluid outside the tetrahedron applies a normal stress on each face. Let the normal stresses on each orthogonal face be given by σ_{N1}, σ_{N2}, and σ_{N3}, respectively, with σ_N as the normal stress on the largest face. The surface force acting on a face of the tetrahedron is found by combining Eqs. 4.21 and $\mathbf{\Sigma} = \sigma_N\mathbf{n}$ to obtain, $\mathbf{F}_S = \int_S \mathbf{\Sigma}\,dS = \int_S \sigma_N\mathbf{n}\,dS$. Since the faces are planar, the unit normal on each face is a constant vector. The normal stress on each face is also constant, since the tetrahedron is small. Thus, the surface force on a face is simply given by $\mathbf{F}_S = \sigma_N A\mathbf{n}$. The surface forces acting on the three orthogonal faces are $\sigma_{N1}A_1(-\mathbf{I})$, $\sigma_{N2}A_2(-\mathbf{J})$, and $\sigma_{N3}A_3(-\mathbf{K})$. The surface force on the larger slanted face is $\sigma_N A[(\mathbf{I}/\sqrt{3}) + (\mathbf{J}/\sqrt{3}) + (\mathbf{K}/\sqrt{3})]$.

5 FLUID STATICS

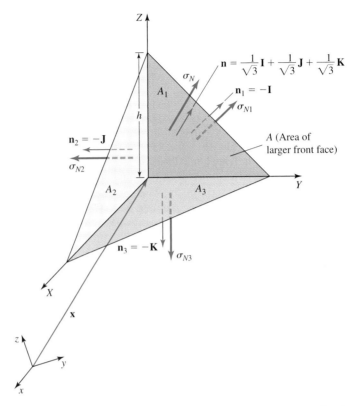

Figure 5.2 Tetrahedron of fluid with its associated stresses and unit normals.

Next consider the body force $\mathbf{f} = \mathbf{f}(\mathbf{x}, t)$ acting on the fluid inside the tetrahedron. For a sufficiently small tetrahedron, both the fluid density and the body force per unit mass are constant and equal to their value at location x. The total body force acting on the small tetrahedron filled with fluid is then given by Eq. 4.7 as $\mathbf{F}_B = \int \rho \mathbf{f}\, dV = \rho \mathbf{f}\, V$, and the components of the body force can be written as $\mathbf{F}_B = (\rho f_X V)\mathbf{I} + (\rho f_Y V)\mathbf{J} + (\rho f_Z V)\mathbf{K}$.

We now apply Newton's second law to the tetrahedron by equating the sum of the body and surface forces to zero. Consider first the forces acting in the X direction. There are surface force contributions from face 1 and the slanted face, plus the X component of the total body force acting on the fluid inside the tetrahedron. Equating the sum of these forces to zero gives

$$-\sigma_{N1} A_1 + \frac{\sigma_N A}{\sqrt{3}} + \rho f_X V = 0$$

Inserting the known areas and volume into this expression yields

$$-\sigma_{N1}\left(\frac{h^2}{2}\right) + \sigma_N \frac{1}{\sqrt{3}}\left(\frac{\sqrt{3}}{2}h^2\right) + \rho f_X \left(\frac{h^3}{6}\right) = 0$$

To be consistent, this equation must hold as the tetrahedron shrinks, a process equivalent to taking the limit of this expression as $h \to 0$. After we first divide by h^2, we discover that the X component of the total body force goes to zero faster in the limit than the X component of the total surface force. Thus we obtain, $\sigma_{N1} = \sigma_N$. Writing similar force balances in the Y and Z directions leads to the additional relationships $\sigma_{N2} = \sigma_N$, and $\sigma_{N3} = \sigma_N$. These results show that the normal stress acting on the slanted face of the tetrahedron is identical to that acting on each of the three orthogonal faces of the same tetrahedron. We have proven an important characteristic of the normal stress in a fluid at rest: it acts equally in all directions. This explains why tilting your head underwater does not change the sense of pressure on your eardrums.

It is relatively easy to show that a state of hydrostatic stress also occurs if a fluid is in motion at constant velocity as a whole, in linear acceleration as a whole, or in a solid body rotation as a whole. In these cases, the tetrahedron analysis may be repeated in a reference frame fixed to the moving volume of fluid. In this moving frame the fluid is at rest, but acted upon by the appropriate noninertial body force needed to account for the acceleration of the noninertial reference frame. Since this noninertial body force may be absorbed into the actual body force, it has no effect on the outcome of the tetrahedron analysis, and we conclude that the fluid in these situations is also in a state of hydrostatic stress.

How is the normal stress σ_N in a fluid at rest related to the pressure? The answer is that the pressure p, called the static pressure, is defined to be the negative of the normal stress acting on any surface in a fluid at rest. Thus, for a fluid at rest we have

$$p = -\sigma_N \tag{5.1a}$$

and the stress vector in a fluid at rest, or hydrostatic stress, is given by

$$\mathbf{\Sigma} = -p\mathbf{n} \tag{5.1b}$$

Note that this result is identical to Eq. 4.20, given previously without proof. The foregoing definition of the static pressure is consistent with our earlier description of pressure in Section 2.3 in terms of a compressive surface force. Since the normal stress acting on any surface is $\sigma_N = -p$ and all shear stresses are zero, the stress tensor in a fluid at rest has the form

$$\boldsymbol{\sigma} = \begin{pmatrix} \sigma_{xx} & \sigma_{xy} & \sigma_{xz} \\ \sigma_{yx} & \sigma_{yy} & \sigma_{yz} \\ \sigma_{zx} & \sigma_{zy} & \sigma_{zz} \end{pmatrix} = \begin{pmatrix} -p & 0 & 0 \\ 0 & -p & 0 \\ 0 & 0 & -p \end{pmatrix} \tag{5.1c}$$

To complete the discussion of the hydrostatic stress distribution, we need only determine the pressure distribution in the fluid.

5.3 HYDROSTATIC EQUATION

In fluid mechanics pressure is represented by a scalar field, $p = p(\mathbf{x}, t)$. For a fluid at rest, both the pressure distribution and the body forces that create it are normally time independent, so in this chapter we consider the pressure and body force to be functions of position alone. To learn more about the pressure field in a fluid at rest, we must develop a governing equation for fluid statics. This governing equation, called the hydrostatic equation, is an expression of Newton's second law for a fluid at rest. The hydrostatic equation, in integral and differential form, applies to every problem in fluid statics. However, we do not necessarily need to solve this equation repeatedly, since the general characteristics of the hydrostatic pressure distribution are known for every common engineering situation.

5.3.1 Integral Hydrostatic Equation

Consider an arbitrary volume of fluid at rest. The total body force acting on this fluid volume is given by Eq. 4.7 as $\mathbf{F}_B = \int \rho \mathbf{f} \, dV$, and the surface force applied by the fluid outside the volume on the fluid inside the volume is given in general by Eq. 4.21 as $\mathbf{F}_S = \int_S \mathbf{\Sigma} \, dS$. Substituting for the stress vector in this equation by using Eq. 5.1b, we obtain $\mathbf{F}_S = \int_S -p\mathbf{n} \, dS$. Newton's second law states that the sum of the body and surface forces acting on this arbitrary volume of fluid at rest is zero, thus we can write

$$\int_S -p\mathbf{n} \, dS + \int \rho \mathbf{f} \, dV = 0 \tag{5.2}$$

Equation 5.2 is the integral hydrostatic equation. It tells us that the total force exerted by the pressure on the surface of a volume of fluid at rest is exactly balanced by the total body force exerted on the fluid within the volume. In applying this equation it is important to realize that the density and body force may be functions of position.

To see how this equation might be employed to analyze a problem in fluid statics, consider the free body diagram for a stationary cylindrical column of water of diameter D and height H shown in Figure 5.3. Gravity is the only body force acting, and the top of the container is open to the atmosphere. What can the integral hydrostatic equation tell us

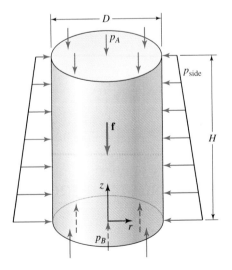

Figure 5.3 Free body diagram of a cylindrical column of water.

about this situation? The first step is the judicious selection of a volume of fluid for analysis. In this case, we choose a fluid volume consisting of the entire water column, for reasons that should become clear as we proceed. To evaluate the volume integral in Eq. 5.2, we note that gravity is the only body force acting, and the water density is constant. Thus the integral gives $\int \rho \mathbf{f}\, dV = -\rho_{H_2O} g \pi (D^2/4) H \mathbf{k}$. The magnitude of the total body force is equal to the weight of water in the volume, and the force acts downward.

To evaluate the pressure integral in Eq. 5.2, note that the surface of interest has three parts, a top, a bottom, and a cylindrical wall. Consider the top and bottom integrals first. In Chapter 2 we stated that the hydrostatic pressure in a fluid at rest in a gravity field is uniform on horizontal planes. Thus, the pressure on the top and bottom surface is uniform. On the top surface the pressure is the ambient atmospheric pressure p_A, the outward unit normal is $\mathbf{n} = \mathbf{k}$, and the area is $(\pi D^2/4)$. Thus the pressure integral on the top surface is $\int_S -p\mathbf{n}\, dS = -p_A \mathbf{k}(\pi D^2/4)$. On the bottom surface the outward unit normal is $\mathbf{n} = -\mathbf{k}$, and if we let p_B represent the unknown pressure at the bottom of the column, then the pressure integral yields: $\int_S -p\mathbf{n}\, dS = -p_B(-\mathbf{k})(\pi D^2/4)$. These two integrals contribute a net force of $(p_B - p_A)(\pi D^2/4)\mathbf{k}$. On the cylindrical wall, the pressure varies in the z direction, but because of symmetry, the total force applied to the wall by the fluid is zero.

Substituting the expressions for the body and surface forces into the integral hydrostatic equation yields $(p_B - p_A)(\pi D^2/4)\mathbf{k} - \rho_{H_2O} g(\pi D^2/4) H \mathbf{k} = 0$, which is simply a force balance on the chosen volume of fluid. The only unknown is the pressure at the bottom of the fluid column, p_B. Solving for this pressure we have $p_B = p_A + \rho_{H_2O} g H$. By applying the integral hydrostatic equation, we have learned that the pressure at the bottom of a stationary column of water is greater than that at the top, by the amount $\rho_{H_2O} g H$.

A general statement of this result is that the pressure at the bottom of a layer of constant density fluid of thickness H in Earth's gravity field exposed to atmospheric pressure at the top is given by

$$p_B = p_A + \rho g H \qquad (5.3)$$

Thus, the surface force per unit area at the bottom of a fluid column, p_B, is equal to the surface force per unit area at the top of the column, p_A, plus the weight per unit area of the fluid in between, $\rho g H$. Note that this equation is equivalent to Eq. 2.9, which was stated without proof in our discussion of pressure in Chapter 2.

Equation 5.3 applies only to a constant density fluid. What can the integral hydrostatic equation tells us about a layer of variable density fluid? Suppose a fluid has an arbitrary but known density distribution given by $\rho(z)$. Applying Eq. 5.2 to a cylindrical column of this fluid, of height H, the body force integral gives

$$\int \rho \mathbf{f}\, dV = -(\pi D^2/4)g\mathbf{k} \int_0^H \rho(z)\, dz = -(\pi D^2/4)\bar{\rho} g H \mathbf{k}$$

where the average density $\bar{\rho}$ of the fluid in the column is defined by

$$\bar{\rho} = \frac{1}{H}\int_0^H \rho(z)\, dz \qquad (5.4)$$

EXAMPLE 5.2

What is the pressure increase per foot of water depth in a lake? What is the pressure decrease per foot of height in the atmosphere at sea level.

SOLUTION

Examination of Eq. 5.3 shows that the pressure increase in a constant density fluid over H feet of depth is $\rho g H$. Therefore, the pressure increase per foot of water depth is simply ρg. Using a nominal density of water of 62.4 lb$_m$/ft^3, we find:

$$\rho g = (62.4 \text{ lb}_m/\text{ft}^3)(32.2 \text{ ft/s}^2) = 2.01 \times 10^3 \text{ lb}_m/(\text{ft}^2\text{-s}^2)$$

As is frequently the case when one is working with EE units, we need the g_c constant to make sense of our solution. Recall that the unit conversion form of g_c is 32.2 ft-lb$_m$ = 1 lb$_f$-s^2. Therefore the preceding result becomes:

$$\rho g = [2.01 \times 10^3 \text{ lb}_m/(\text{ft}^2\text{-s}^2)] \left(\frac{1 \text{ s}^2\text{-lb}_f}{32.2 \text{ lb}_m\text{-ft}} \right) = 62.4 \text{ lb}_f/\text{ft}^3$$

Thus, the pressure change per foot of water depth is 62.4 (lb$_f$/ft^2)/ft or 0.433 psia/ft.

Assuming that the density of the air in the atmosphere is constant and equal to the sea level value, the pressure decrease in air at sea level is also ρg. Using the Standard Atmosphere data of Appendix B, the pressure decrease per foot of elevation in Earth's atmosphere is

$$\rho g = (0.002377 \text{ slug/ft}^3)(32.2 \text{ ft/s}^2) \left(\frac{32.2 \text{ lb}_m}{1 \text{ slug}} \right) \left(\frac{1 \text{ lb}_f\text{-s}^2}{32.2 \text{ lb}_m\text{-ft}} \right)$$

$$= 0.0765 \text{ lb}_f/\text{ft}^3 \left(\frac{1 \text{ ft}}{12 \text{ in.}} \right)^2 = 5.3 \times 10^{-4} \text{ psia/ft}$$

This change in pressure with height in the atmosphere is often small enough to be ignored.

Notice that the total body force is equal in magnitude to the weight of the variable density fluid, given by the product of the weight per unit area of the fluid $\bar{\rho} g H$ and the column area ($\pi D^2/4$). The pressure integrals on the top, bottom, and cylindrical walls are exactly the same as those calculated earlier for the column of water, so the integral hydrostatic equation in this case gives:

$$(p_B - p_A) \left(\frac{\pi D^2}{4} \right) \mathbf{k} - \bar{\rho} g \left(\frac{\pi D^2}{4} \right) H \mathbf{k} = 0$$

Solving for the pressure at the bottom of the column of variable density fluid we find

$$p_B = p_A + \bar{\rho} g H \tag{5.5}$$

Equation 5.5 tells us that the surface force per unit area at the bottom of a column of variable density fluid, p_B, is equal to the surface force per unit area at the top of the column, p_A, plus the weight per unit area of the fluid in between. The latter can be expressed using the average density as $\bar{\rho} g H$.

Equation 5.5 applies to any fluid, including a constant density fluid for which $\bar{\rho} = \rho$; thus it is useful in many engineering applications. For example, suppose a column of water has a temperature gradient so that the water density is slightly higher at the bottom than at the top, and the resulting linear density distribution is given by

$$\rho(z) = \rho_H - (\rho_B - \rho_H)\frac{(z - H)}{H} \tag{5.6}$$

where ρ_H and ρ_B are the densities at the top (i.e., at height H) and bottom, respectively. What is the pressure at the bottom of the column? By inspection, the average density is $\bar{\rho} = (\rho_H + \rho_B)/2$, a result you can confirm by using Eq. 5.4. By Eq. 5.5, the pressure at the bottom of the column is

$$p_B = p_A + \left(\frac{\rho_B + \rho_H}{2}\right) g H \tag{5.7}$$

which, as expected, is the pressure at the top plus the weight per unit area of the variable density fluid.

5.3.2 Differential Hydrostatic Equation

A differential form of the hydrostatic equation can be derived directly from the integral hydrostatic equation with the help of Gauss's theorem. The advantage of a differential governing equation is that it allows us to find the pressure distribution at any point in the fluid. To derive the differential equation, we start by using the scalar form of Gauss's theorem to convert the (surface) pressure integral in Eq. 5.2 to a volume integral. The result is

$$\int_S -p\mathbf{n}\, dS = \int -\nabla p\, dV \tag{5.8}$$

where ∇p is the gradient of the pressure field. Using concepts from vector calculus it can be shown that the negative of the pressure gradient may be interpreted as the pressure force per unit volume.

EXAMPLE 5.3

A mixture of crude oil (SG = 0.87) and water is pumped into a settling tank that is 10 m high and vented at the top (see Figure 5.4). After a period of time, it is found that a layer of crude oil 2 m deep has accumulated on top of 7.5 m of water. What is the average density of the liquid column? What is the pressure at the bottom of the tank?

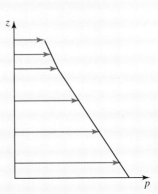

Figure 5.4 Schematic for Example 5.3.

SOLUTION

To determine the average density, we use Eq. 5.4:

$$\bar{\rho} = \frac{1}{H}\int_0^H \rho(z)\,dz = \frac{1}{H}\left[\left(\int_0^{h_{water}} \rho_{water}\,dz\right) + \left(\int_{h_{water}}^{h_{water}+h_{oil}} \rho_{oil}\,dz\right)\right]$$

where the integration has been separated into water and oil parts. Completing the integral, we have

$$\bar{\rho} = \frac{1}{H}[\rho_{water}h_{water} + \rho_{oil}h_{oil}] = \frac{\rho_{water}}{H}\left[h_{water} + \left(\frac{\rho_{oil}}{\rho_{water}}\right)h_{oil}\right] \qquad (A)$$

where h_{water} is the height of the water column, and h_{oil} is the height of the oil column. Evaluating this expression yields:

$$\bar{\rho} = \frac{998 \text{ kg/m}^3}{9.5 \text{ m}}[7.5 \text{ m} + (0.87)(2 \text{ m})] = 971 \text{ kg/m}^3$$

The pressure at the bottom of the tank is calculated by using Eq. 5.5, $p_B = p_A + \bar{\rho}gH$, i.e., the pressure is $\bar{\rho}gH$ above that at the free surface. Therefore,

$$p_B = 1.013 \times 10^5 \text{ N/m}^2 + (971 \text{ kg/m}^3)(9.81 \text{ m/s}^2)(9.5 \text{ m})\frac{1\,(\text{N-s}^2)}{1\,(\text{kg-m})}$$
$$= 1.92 \times 10^5 \text{ N/m}^2$$

The pressure at the bottom of the tank is almost 2 atm. An alternate way to think about this problem is to realize that the pressure on the bottom is higher than at the free surface owing to the weight per unit area of the two fluids: oil and water. These weights per unit area are $\rho_{oil}gh_{oil}$ and $\rho_{water}gh_{water}$, respectively. Notice that (A) can be written as $\bar{\rho}gH = \rho_{water}gh_{water} + \rho_{oil}gh_{oil}$, which confirms the validity of this approach.

EXAMPLE 5.4

Figure 5.5A shows a hemispherical ($R = 5$ m) undersea habitat located at a depth of $H = 20$ m in the ocean. What is the total force applied by the seawater on the structure? Assume that the specific gravity for seawater is 1.025.

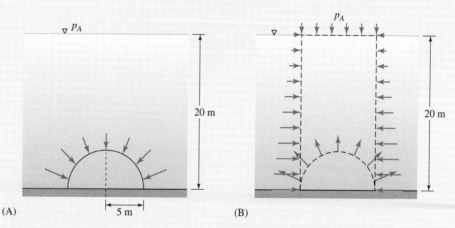

Figure 5.5 Schematic for Example 5.4. (A) force on hemispherical structure. (B) forces on corresponding fluid control volume.

SOLUTION

We are to find the force applied by the seawater on the undersea habitat. The hemispherical ($R = 5$ m) structure is located 20 m below the surface. We will assume constant density and negligible motion of the seawater. The key to this problem is to realize that the force $\mathbf{F}_{dome}$ applied by the seawater to the dome is equal and opposite to the force $\mathbf{F}$ of the dome on the bottom of the selected fluid volume shown in Figure 5.5B, which extends to the surface. Having selected this volume, we can apply the integral hydrostatic equation (Eq. 5.2), writing it as

$$\int_{top} -p\mathbf{n}\,dS + \int_{bottom} -p\mathbf{n}\,dS + \int_{cylinder} -p\mathbf{n}\,dS + \int \rho \mathbf{f}\,dV = 0$$

Since the force $\mathbf{F}_{dome}$ applied by the seawater to the dome is equal and opposite to the force $\mathbf{F}$ of the dome on the bottom of the selected fluid volume, we have $\mathbf{F} = -\mathbf{F}_{dome} = \int_{bottom} -p\mathbf{n}\,dS$. The surface integral on the cylindrical surface is zero owing to symmetry, so the equation becomes $\int_{top} -p\mathbf{n}\,dS - \mathbf{F}_{dome} + \int \rho \mathbf{f}\,dV = 0$. Solving for $\mathbf{F}_{dome}$, we find $\mathbf{F}_{dome} = \int_{top} -p\mathbf{n}\,dS + \int \rho \mathbf{f}\,dV$. The pressure at the ocean surface is atmospheric, the outward normal points up, and the area is known. Thus we find

$$\mathbf{F}_{\text{dome}} = \int_{\text{top}} -p\mathbf{n}\,dS + \int \rho \mathbf{f}\,dV = -p_A \pi R^2 \mathbf{k} + \int \rho_{\text{sw}}(-g)\mathbf{k}\,dV$$
$$= -p_A \pi R^2 \mathbf{k} - \rho_{\text{sw}} g \mathbf{k} \pi \left(R^2 H - \tfrac{2}{3}R^3\right)$$

Note that the volume is that of the cylinder from the surface to the ocean bottom less the volume of the hemisphere. A dimension check of the two terms yields force, as expected. With an ambient pressure of $p_A = 101{,}300\text{ N/m}^2$, the pressure term contributes a force of -7.96×10^6 N $\mathbf{k}$. Since $\rho_{\text{sw}} = 1.025 \rho_{\text{H}_2\text{O}} = 1023\text{ kg/m}^3$, the body force term contributes a force of

$$-\rho_{\text{sw}} g \mathbf{k} \pi \left(R^2 H - \tfrac{2}{3}R^3\right) = -(1023\text{ kg/m}^3)(9.81\text{ m/s}^2)\mathbf{k}(\pi)\left[(5\text{ m})^2(20\text{ m}) - \tfrac{2}{3}(5\text{ m})^3\right]$$
$$= -1.31 \times 10^7\text{ N }\mathbf{k}.$$

The force applied by the seawater to the dome is -2.11×10^7 N $\mathbf{k}$, which is the sum of a force of -7.96×10^6 N $\mathbf{k}$ due to atmospheric pressure, representing the weight of air above the dome, and a force of -1.31×10^7 N $\mathbf{k}$, due to the weight of water above the dome.

A force balance on the dome would include the force applied by the seawater plus the structural weight of the dome and the force applied by the air pressure inside the dome. The net force applied by the two fluids involved, air and seawater, depends on the air pressure on the inside of the dome. If the habitat air supply is maintained at atmospheric pressure, the result would be an upward force on the dome that would exactly balance the force of -7.96×10^6 N $\mathbf{k}$ due to the atmospheric pressure at the top of the water column above the dome, leaving the net force on the structure due to the two fluids equal to the weight of the water column above the habitat.

Substituting Eq. 5.8 into the integral hydrostatic equation gives

$$\int -\nabla p\,dV + \int \rho \mathbf{f}\,dV = 0$$

Since these two integrals refer to the same volume of fluid, we may combine them to obtain

$$\int (-\nabla p + \rho \mathbf{f})\,dV = 0$$

This equation applies to an arbitrary volume of fluid; thus the integrand must be zero, allowing us to conclude that the pressure and body forces are related by

$$\nabla p = \rho \mathbf{f} \tag{5.9}$$

The hydrostatic equation can also be obtained from the force balance in a fluid as discussed in Chapter 4. Recall that the balance is given by Eq. 4.40 as $\rho \mathbf{a} = \rho \mathbf{f} + \nabla \cdot \boldsymbol{\sigma}$. Since the fluid is at rest, the acceleration is zero and the force balance can be written as $-\nabla \cdot \boldsymbol{\sigma} = \rho \mathbf{f}$. Evaluating the stress divergence using the form of the stress tensor in a fluid at rest, Eq. 5.1c, gives the hydrostatic equation.

Equation 5.9 is the differential form of the hydrostatic equation. This equation is customarily referred to as the hydrostatic equation. It tells us that the pressure gradient at a point in a fluid at rest points in the direction of the body force and has a magnitude equal to the product of density and body force. Note that ∇p is the surface force per unit volume acting on the fluid and $\rho \mathbf{f}$ is the body force per unit volume. Thus at each point in a fluid at rest the pressure and body forces per unit volume are in balance. As was the case with the integral hydrostatic equation, the density and body force in Eq. 5.9 may be functions of position.

In Cartesian coordinates, the three components of the hydrostatic equation are

$$\frac{\partial p}{\partial x} = \rho f_x, \qquad \frac{\partial p}{\partial y} = \rho f_y, \qquad \text{and} \qquad \frac{\partial p}{\partial z} = \rho f_z \qquad (5.10\text{a–c})$$

These three partial differential equations describe the variation in pressure in each coordinate direction at any point in the fluid. In cylindrical coordinates, the three components of the hydrostatic equation are

$$\frac{\partial p}{\partial r} = \rho f_r, \qquad \frac{1}{r}\frac{\partial p}{\partial \theta} = \rho f_\theta, \qquad \text{and} \qquad \frac{\partial p}{\partial z} = \rho f_z \qquad (5.11\text{a–c})$$

EXAMPLE 5.5

Figure 5.6 shows a constant density fluid in a cylindrical container rotating at angular velocity Ω. The fluid is at rest (relative to the container) in the rotating container. Find the body force acting on the fluid, and determine the pressure gradient.

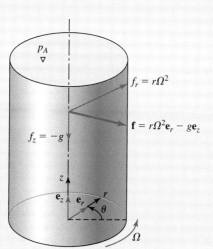

Figure 5.6 Schematic for Example 5.5.

SOLUTION

We must determine the three components of the body force acting on the fluid. Cylindrical coordinates fixed to the rotating container appear to be appropriate given the geometry. Since this a rotating reference frame, we must be careful to account for noninertial body forces. In the r direction there is a noninertial body force due to the centrifugal acceleration, given by $f_r = r\Omega^2$. There is no body force acting in the θ direction so, $f_\theta = 0$, but gravity acts in the vertical direction as usual, so $f_z = -g$. Thus the body force is given by $\mathbf{f} = r\Omega^2 \mathbf{e}_r - g\mathbf{e}_z$. The hydrostatic equation relates the pressure gradient in a fluid at rest to the body force. Thus from Eqs. 5.11, the pressure gradient is given by

$$\frac{\partial p}{\partial r} = \rho f_r = \rho r \Omega^2, \qquad \frac{1}{r}\frac{\partial p}{\partial \theta} = \rho f_\theta = 0, \quad \text{and} \quad \frac{\partial p}{\partial z} = \rho f_z = -\rho g$$

which in vector notation is $\nabla p = \rho r \Omega^2 \mathbf{e}_r - \rho g \mathbf{e}_z$. We see that, as expected, the pressure increases in the r direction owing to the centrifugal body force and increases in the negative z direction owing to the gravitational body force.

5.4 HYDROSTATIC PRESSURE DISTRIBUTION

The solution of the hydrostatic equation provides the pressure distribution in a fluid at rest. To solve this partial differential equation, the density and body force must be known. There can be a number of different body forces acting in a hydrostatics problem, either alone or in combination. In addition, the fluid may be a liquid or gas, and the density may be constant or variable. Although an engineer must be prepared to analyze any problem involving a fluid at rest, gravity is the most common body force. Thus, in the remainder of this chapter we devote most of the discussion to problems in which a fluid is at rest in a uniform gravity field.

The governing equation for the pressure distribution, applicable to a constant or variable density fluid at rest in Earth's gravity field, is developed from the hydrostatic equation. In Cartesian coordinates with the z axis upward, the gravitational body force per unit mass is $\mathbf{f} = -g\mathbf{k}$. Substituting $f_x = 0$, $f_y = 0$, and $f_z = -g$ into Eqs. 5.10, we obtain:

$$\frac{\partial p}{\partial x} = 0, \qquad \frac{\partial p}{\partial y} = 0, \quad \text{and} \quad \frac{\partial p}{\partial z} = -\rho g$$

These three equations reveal two interesting features of the pressure and density variations in a fluid at rest. First, we conclude that the pressure is only a function of height z. This is the basis for the statement in Chapter 2 that in Earth's gravity field the hydrostatic pressure is uniform on horizontal planes. By separately differentiating the last equation with respect to x and y, interchanging the resulting derivatives of the pressure, and using the fact that $(\partial/\partial z)(\partial p/\partial x) = (\partial/\partial x)(\partial p/\partial z)$, we discover that $\partial \rho/\partial x = \partial \rho/\partial y = 0$. This shows that the density of a fluid at rest in Earth's gravity

5.4 HYDROSTATIC PRESSURE DISTRIBUTION

field is either a constant, or a function of height z. The density cannot vary on a horizontal plane.

The pressure distribution in any fluid at rest in Earth's gravity field therefore satisfies

$$\frac{dp}{dz} = -\rho(z)g \tag{5.12}$$

where $\rho(z)$ is the density distribution. This is a first-order ordinary differential equation, so we must specify one boundary condition, normally a known value, p_0, for the pressure at a certain elevation z_0. The solution to this equation is obtained by separating variables and integrating from z_0 to z, $\int_{p_0}^{p} dp = \int_{z_0}^{z} -\rho(z)g\, dz$, where p is the pressure at elevation z. Completing the integration of the left-hand side yields:

$$p(z) - p_0 = \int_{z_0}^{z} -\rho(z)g\, dz \tag{5.13}$$

Since the density occurs inside the integral in Eq. 5.13, the nature of the fluid involved in a hydrostatics problem is important. It is not simply a question of whether the fluid is a liquid or a gas. Rather, it is whether the density of the liquid or gas is constant, or varies with height z.

5.4.1 Constant Density Fluid in a Gravity Field

Engineers are often called upon to examine structural issues related to the pressure distribution in liquids in storage or confinement. Similar issues arise in applications involving oceans, lakes, and rivers. The motion of the liquid involved is often negligible. Thus the pressure distribution in the liquid is the hydrostatic pressure distribution for a constant density fluid at rest in Earth's gravity field. In this case, the integral in Eq. 5.13 reduces to $p(z) = p_0 - \rho g \int_{z_0}^{z} dz$, which upon integration yields

$$p(z) = p_0 - \rho g(z - z_0) \tag{5.14}$$

Equation 5.14 is the hydrostatic pressure distribution in a constant density fluid. The shape of this pressure distribution is shown in Figure 5.7, along with the locations of the datum pressure p_0, and datum elevation z_0. Notice that the pressure begins at the specified value p_0 at a specified elevation z_0 and increases linearly with depth at a rate given by ρg.

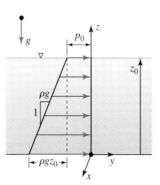

Figure 5.7 Linear pressure distribution in a constant density fluid.

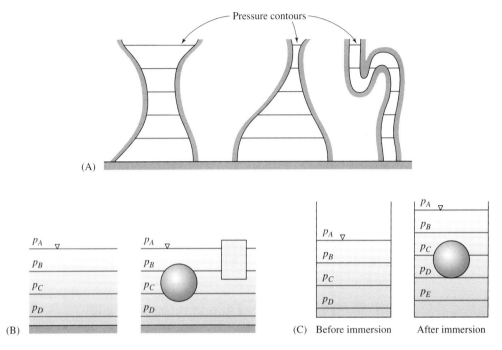

Figure 5.8 (A) Pressure contours in several containers filled with fluid and open to the environment. The shape of a container does not affect the pressure distribution. (B) A large body of contstant density fluid, in which the presence of an object either partially or fully immersed in the fluid does not affect the pressure distribution. (C) Object immersed in a container that does affect the pressure distribution.

An interesting feature of the hydrostatic pressure distribution given by Eq. 5.14 is that the distribution does not depend on the shape of the container. This feature is illustrated in Figure 5.8A, showing the pressure contours in several different containers, all of which are open at the top and exposed to the ambient pressure. The hydrostatic pressure distribution in a large body of constant density fluid is usually not changed by the presence of an object either partially or fully immersed in the fluid (Figure 5.8B). Note however, that as shown in Figure 5.8C, if the volume of an immersed object is large relative to the total volume of a container, there will be an overall increase in fluid depth and the pressure distribution must change to reflect this increase.

In cases involving a more general form of the body force, a hydrostatic pressure distribution may be represented in Cartesian coordinates by $p = p(x, y, z)$. In many applications it is of interest to know exactly what a surface of constant hydrostatic pressure looks like, and to be able to predict at what locations a certain pressure value occurs. Surfaces of constant hydrostatic pressure are found by setting the pressure distribution $p(x, y, z)$ equal to a constant:

$$p(x, y, z) = C \qquad (5.15)$$

For example, in problems in which gravity is the only body force, the pressure distribution is given by Eq. 5.14. We can reference the actual pressure to the datum pressure in this case and write

$$p(z) - p_0 = C = -\rho g(z - z_0)$$

5.4 HYDROSTATIC PRESSURE DISTRIBUTION

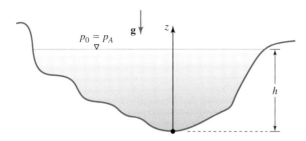

Figure 5.9 Parameters necessary to describe the pressure distribution in a lake.

The surfaces of constant pressure, as measured relative to the datum pressure, are then given by

$$z = z_0 - \frac{C}{\rho g} \quad (5.16)$$

which are a family of horizontal planes, in agreement with earlier statements that the hydrostatic pressure in a constant density fluid is uniform on horizontal planes.

Suppose we wish to use the parameters shown in Figure 5.9 to write an analytical expression for the pressure distribution in a lake. At a point near the middle of the lake, the water depth is h. Note that we have elected to place the origin of our coordinate system at the lake bottom and that the pressure at the lake surface is known to be atmospheric. A point on the lake surface is therefore an excellent location to define the datum pressure and elevation as $p_0 = p_A$ and $z_0 = h$. From Eq. 5.14, and referring to the coordinates shown in Figure 5.9, the pressure distribution in the lake is

$$p(z) = p_A - \rho_{H_2O} g(z - h) \quad (5.17)$$

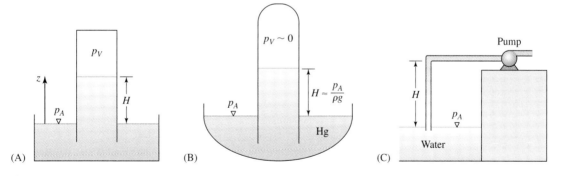

Figure 5.10 (A) Schematic for Example 5.6. (B) Schematic of barometer. (C) Schematic of pump intake with pressure less than atmospheric.

EXAMPLE 5.6

An evacuated tube is immersed in a liquid as shown in Figure 5.10A. How high will the liquid rise into the tube? Calculate this height for water and mercury. You can neglect surface tension in the solution of this problem.

SOLUTION

We begin by recognizing that the pressure at the liquid surface just outside the tube at height $z = 0$ must be atmospheric pressure. That is, $p(0) = p_A$. Since the tube is open to the liquid, we can conclude that the pressure inside the tube at $z = 0$ is the same as that outside the tube at the same elevation (horizontal surfaces in open fluids have equal pressures). Since this is a constant density fluid, we know that the pressure decreases linearly with height and we can conclude that at the fluid interface inside the tube $(z = H)$, we have $p(H) = p(0) - \rho g H = p_A - \rho g H$. To finish the problem, we must know the pressure within the evacuated space at the top of the tube. If a perfect vacuum existed above the liquid, we would have $p(H) = 0$. However liquid molecules leave and rejoin the surface through evaporation and condensation, and at equilibrium a vapor pressure p_V will exist above a particular liquid at a particular temperature. Thus $p(H) = p_V$, and we can write $p_V = p_A - \rho g H$ and solve for H to obtain $H = (p_A - p_V)/\rho g$.

From Appendix A, we find that the vapor pressures for water and mercury (at 20°C) are 2.34×10^3 and 1.1×10^{-3} N/m², respectively, and the corresponding densities are 998 and 13,550 kg/m³. Substituting appropriate values into our equation for H yields:

For water:

$$H = \frac{p_A - p_v}{\rho g} = \frac{1.01 \times 10^5 \text{ N/m}^2 - 2.34 \times 10^3 \text{ N/m}^2}{(998 \text{ kg/m}^3)(9.81 \text{ m/s}^2)} = 10.1 \text{ m}$$

For mercury (Hg):

$$H = \frac{p_A - p_v}{\rho g} = \frac{1.01 \times 10^5 \text{ N/m}^2 - 1.1 \times 10^{-3} \text{ N/m}^2}{(13{,}550 \text{ kg/m}^3)(9.81 \text{ m/s}^2)} = 0.76 \text{ m}$$

Since these heights are very large in comparison to a capillary rise in a reasonably sized tube, we are justified in ignoring the effects of surface tension. The rise in mercury of 760 mm is the basis for the unit called the torr: 1 atm = 760 torr = 760 mm Hg.

The formula obtained in the preceding example, namely

$$H = \frac{p_A - p_v}{\rho g} \tag{5.18a}$$

is applicable to several engineering situations. For example, it is the basis for the mercury filled device shown in Figure 5.10B, long used to report the barometric pressure in Earth's atmosphere. Mercury's vapor pressure is so low that Eq. 5.18a becomes $H = (p_A - p_v)/\rho g \approx p_A/\rho g$, which means that by measuring the height of the mercury column, we effectively measure the absolute pressure in the atmosphere, rather than the difference between the absolute pressure and the vapor pressure of mercury. Equation 5.18a also explains why a pump cannot function if the liquid level is too far below the pump intake. Consider the arrangement of Figure 5.10C. If the water level

5.4 HYDROSTATIC PRESSURE DISTRIBUTION

drops and approaches a value equal to H, the water boils near the pump inlet and the flow through the pump ceases.

The pressure-measuring device shown in Figure 5.11 is called a U-tube manometer. This device is of historical and practical importance in fluid mechanics. Let us use our knowledge of fluid statics to interpret the reading of a U-tube manometer. Consider a procedure similar to the one we used to solve Example 5.6 (and, later, Example 5.7). Starting at point 1, the pressure is p_1. If we move down the tube to the U and back up to point 1', the pressure must be p_1 no matter how far we have traveled. Continuing to ascend to height H up the right side of the tube to the surface, the pressure decreases by the amount $\rho g H$ and equals p_2, which completes the analysis. Thus, we find $p_1 - \rho g H = p_2$. Rearranging this expression to isolate the pressure difference yields the manometer formula

$$p_1 - p_2 = \rho g H \tag{5.18b}$$

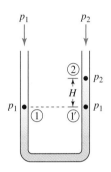

Figure 5.11 U-tube manometer. The pressure change in each air column is negligible.

A manometer pressure reading is usually reported as the value of H along with the manometer liquid. For example, a reading may be given in inches of water or millimeters of mercury. The range of pressures that can be measured depends on the density of the liquid in the manometer. A dense liquid such as mercury greatly extends the range of readable pressures. On the other hand, sensitivity, the change in height per unit change in pressure, increases as fluid density decreases.

The precision of a pressure measurement with a U-tube manometer depends to some extent on the type of liquid used and the effects of surface tension in causing a meniscus to form in the tube at the free surface. We can improve the precision greatly by using the inclined manometer shown in Figure 5.12. Repeating the analysis for the inclined manometer shows that the pressure difference is still given by Eq. 5.18b. However, the measuring distance L is related to H by $H = L \sin\theta$, so we can rewrite the formula as

$$p_1 - p_2 = \rho g H = \rho g L \sin\theta \tag{5.18c}$$

Equation 5.18c demonstrates that for an inclined manometer, a pressure difference will give a larger value of L and corresponding improvement in sensitivity as the angle θ decreases.

Manometers are used less frequently today owing to the development of low cost and accurate electronic transducers, but they remain important experimental tools. It is still common for mechanical and electronic pressure gages to report measured pressures in terms of inches of water or millimeters of mercury above or below some ambient or reference value. To find the corresponding value for the pressure difference in a unit of

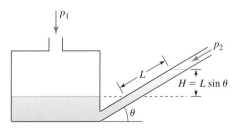

Figure 5.12 Inclined manometer.

force per unit area, the height value is multiplied by the specific weight of the fluid. Can you show that a reading of 1 inch of water corresponds to a pressure of 0.036 psia, and a reading of 1 mm of mercury corresponds to 132.9 Pa?

A pressure difference of Δp results in a reading of $H = \Delta p/\rho g$ on a U-tube manometer. If we select $\Delta p = 1$ atm, we can calculate the corresponding height in water at room temperature to be

$$H = \frac{\Delta p}{\rho_{H_2O} g} = \frac{(14.696\ \text{lb}_f/\text{in.}^2)(144\ \text{in.}^2/\text{ft}^2)}{(1.936\ \text{slugs/ft}^3)(32.2\ \text{ft/s}^2)}$$
$$= 33.95\ \text{ft} = 407.4\ \text{in.}$$

and in mercury we find:

$$H = \frac{\Delta p}{\rho_{Hg} g} = \frac{101{,}300\ \text{N/m}^2}{(13{,}550\ \text{kg/m}^3)(9.81\ \text{m/s}^2)}$$
$$= 0.762\ \text{m} = 762\ \text{mm}$$

These values give some indication of the maximum pressures that can be read by a U-tube manometer of reasonable size: a fraction of an atmosphere with water, and up to 2 atm with mercury.

In Example 5.2 we showed that for modest elevation changes in air, the change in hydrostatic pressure with height is negligible. This is the basis for the simplest approximation or model for the atmosphere, namely that the pressure in the atmosphere is

EXAMPLE 5.7

A water-filled manometer like the one shown in Figure 5.11 is connected on one side to a duct through which pressurized air is flowing and is open to the atmosphere on the other side. If the height H is found to be 18 cm, determine the air pressure in the duct.

SOLUTION

This exercise is solved by using Eq. 5.18b, $p_1 - p_2 = \rho g H$. In this case $H = 18\ \text{cm} = 0.18\ \text{m}$, $p_2 = p_A = 101{,}300\ \text{N/m}^2$, $\rho = 998\ \text{kg/m}^3$ (at 20°C), and we are interested in finding the pressure at the duct wall. A simple substitution of the relevant values into the equation yields:

$$p_1 - p_A = \rho_{H_2O}\, gH = (998\ \text{kg/m}^3)(9.81\ \text{m/s}^2)(0.18\ \text{m})\frac{1\ (\text{N-s}^2)}{1\ (\text{kg-m})} = 1762\ \text{N/m}^2$$

This is a gage pressure. The corresponding absolute pressure is

$$p_1 = 1762\ \text{N/m}^2 + 101{,}300\ \text{N/m}^2 = 103{,}062\ \text{N/m}^2$$

5.4 HYDROSTATIC PRESSURE DISTRIBUTION

constant and equal to the appropriate ambient pressure for the given elevation. It is customary in engineering to use this constant pressure model in most problems. Note that the air density is effectively assumed to be zero in this model.

An improved model of Earth's atmosphere results if we assume that the air density is constant and equal to its ambient value. From Eq. 5.14, the pressure distribution in Earth's atmosphere in this constant density model is given by the linear distribution

$$p(z) = p_0 - \rho_{air} g(z - z_0)$$

Since the pressure at Earth's surface is known to be atmospheric or ambient, we may choose a point on the surface to define the datum pressure and elevation as $p_0 = p_A$ and $z_0 = 0$. The absolute pressure distribution in the atmosphere is then given by

$$p(z) = p_A - \rho_{air} g z \tag{5.19a}$$

and the gage pressure distribution is given by

$$p_{gage}(z) = \rho_{air} g z \tag{5.19b}$$

Equation 5.19b can be used to show that the air pressure at the level of the Skydeck of the Sears Tower in Chicago, which is 1353 ft above the ground, is approximately 0.7 psia less than at the ground.

EXAMPLE 5.8

Find the pressure distribution in the tank containing crude oil and water discussed in Example 5.3 and illustrated in Figure 5.4.

SOLUTION

The density is constant in each layer of fluid; therefore, Eq. 5.14 can be applied in a piecewise fashion to obtain the pressure distribution. Starting with the oil layer, we know at the surface the pressure is atmospheric because the tank is vented; $p(z_0 = 9.5 \text{ m}) = p_A$. Thus, in the oil layer, the pressure distribution is given by

$$p(z) = p_0 - \rho g(z - z_0) = p_A - \rho_{oil} g(z - 9.5 \text{ m}) \qquad \{7.5 \text{ m} \leq z \leq 9.5 \text{ m}\}$$

From this we conclude that the pressure at the oil–water interface is

$$p(7.5 \text{ m}) = p_A - \rho_{oil} g(7.5 \text{ m} - 9.5 \text{ m}) = p_A + \rho_{oil} g(2 \text{ m})$$

This pressure is used as the datum value p_0 in the pressure distribution in the water layer:

$$p(z) = p_0 - \rho g(z - z_0) = p(7.5 \text{ m}) - \rho_{water} g(z - 7.5 \text{ m})$$
$$= p_A + \rho_{oil} g(2 \text{ m}) - \rho_{water} g(z - 7.5 \text{ m})$$

Thus in the water layer we have

$$p(z) = p_A + \rho_{oil}g(2\text{ m}) - \rho_{water}g(z - 7.5\text{ m}) \qquad \{0 \leq z \leq 7.5\text{ m}\}$$

From Example 5.3 the densities of the water and oil are 998 kg/m^3 and 868 kg/m^3, respectively, and atmospheric pressure may be taken as 1.013×10^5 N/m^2. Thus, the pressure in the oil layer is

$$p(z) = 1.823 \times 10^5 \text{ N/m}^2 - (8.52 \times 10^3 \text{ N/m}^3)z \qquad \{7.5\text{ m} \leq z \leq 9.5\text{ m}\}$$

and in the water layer it is

$$p(z) = 1.918 \times 10^5 \text{ N/m}^2 - (9.79 \times 10^3 \text{ N/m}^3)z \qquad \{0 \leq z \leq 7.5\text{ m}\}$$

This exercise illustrates the procedure for matching the pressure at the interface between two immiscible fluids. Note that the pressure at the bottom of the water layer is

$$p(0) = 1.918 \times 10^5 \text{ N/m}^2 - (9.79 \times 10^3 \text{ N/m}^3)(0) = 1.918 \times 10^5 \text{ N/m}^2$$

which agrees with the value found via a different method in Example 5.3.

5.4.2 Variable Density Fluid in a Gravity Field

In Section 5.4 we showed that the pressure distribution in any fluid at rest in Earth's gravity field is given by Eq. 5.12, which states that $dp/dz = -\rho(z)g$. For a constant density fluid, the solution to this equation is a linear pressure distribution, in which pressure increases with depth at a rate defined by the specific weight of the fluid, ρg. In a variable density fluid, usually referred to in fluid mechanics as a stratified fluid, the pressure distribution is nonlinear.

In theory, any vertical density variation is consistent with the foregoing equation, and the corresponding pressure distribution may be found by integrating this equation after inserting the known density distribution. As a practical matter, however, certain density distributions are unstable. The most common unstable stratified fluid situation is a layer of heavy fluid on top of a layer of light fluid. Given a sufficient disturbance, unstable stratified fluid will overturn. In liquids and gases, colder fluid is normally heavier, so a temperature decrease with increasing height is unstable. The creation of an unstable density profile in a fluid is a challenging task, but one which nature has mastered. For example, consider the type of atmospheric inversion that often occurs in the Los Angeles Basin when a layer of smog is trapped below a layer of colder, denser air.

The overturning of an unstable stratified fluid is a complex process, and the onset of motion is not easily predicted. This phenomenon has been observed in nature, both in the atmosphere and in lakes. Overturning in a water reservoir due to temperature stratification in the fall sometimes results in a brief period of bad-tasting, discolored drinking water. On rare occasions, overturning in nature results in tragedy. A massive release of CO_2 in 1986 from the gas-saturated waters of Lake Nyos, a caldera lake in Cameroon, killed numerous people. The release was attributed to an unstable density profile in the deepest water. As the lake waters overturned, massive amounts of CO_2 that had been dissolved in the deep waters were released, and the water behaved like a shaken carbonated drink.

To find the hydrostatic pressure distribution in a variable density fluid, we must solve Eq. 5.12, $dp/dz = -\rho(z)g$, with the known density distribution. We do this by integrating Eq. 5.13 $[p(z) - p_0 = \int_{z_0}^{z} -\rho(z)g\,dz]$, after inserting the known density distribution. For example, consider a variable density water column exposed to the atmosphere with a linear density distribution as given by Eq. 5.6, $\rho(z) = \rho_H - (\rho_B - \rho_H)(z - H)/H$. To find the corresponding pressure distribution, we insert this density distribution into the integral in Eq. 5.13 and integrate from the top of the water column where the pressure is atmospheric down to some elevation z. Thus, we have $p_0 = p_A$, and $z_0 = H$ and Eq. 5.13 becomes

$$p(z) - p_A = \int_{H}^{z} -\left[\rho_H - (\rho_B - \rho_H)\frac{z-H}{H}\right]g\,dz$$

Completing the integration gives

$$p(z) = p_A - \rho_H g(z - H) + g(\rho_B - \rho_H)\frac{(z-H)^2}{2H} \quad (5.20)$$

This linear density distribution and the corresponding nonlinear pressure distribution are shown in Figure 5.13. Convince yourself that $p(H) = p_A$, $p_B = p(0) = p_A + [(\rho_B + \rho_H)/2]gH$, and note that this result for $p(0)$ is consistent with that obtained earlier as Eq. 5.7. Can you confirm by inspection of Eq. 5.20 that the surfaces of constant pressure in this stratified water column are horizontal planes?

The hydrostatic pressure variation in a variable density gas may be found by solving Eq. 5.12, $dp/dz = -\rho(z)g$, with the known density distribution, or by completing the integration of Eq. 5.13: $p(z) - p_0 = \int_{z_0}^{z} -\rho(z)g\,dz$. It is important to recognize that although the density of a liquid is only a function of temperature, the density of a gas is a function of both pressure and temperature. This function must be known to proceed. For example, we can use the perfect gas law to write a relationship between pressure and density for a gas at a constant temperature T_0 as $p = \rho R T_0$, and solve for the density yielding

$$\rho = \frac{p}{RT_0} \quad (5.21)$$

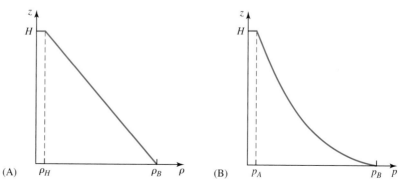

Figure 5.13 (A) Linear density distribution and (B) the resulting nonlinear pressure distribution.

EXAMPLE 5.9

The hot water tank in Figure 5.14 is 50°F hotter at the top than at the bottom. If the thermostat maintains the water at the bottom of the tank at 140°F, and the tank is 5 ft tall and vented at the top, find the pressure distribution in the tank as a function of height. How does the pressure at the bottom of the tank compare with that which would occur in the same tank with all of the water at 140°F? Assume $\beta = 2.3 \times 10^{-4}$°F^{-1}.

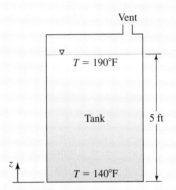

Figure 5.14 Schematic for Example 5.9.

SOLUTION

We are to find the pressure distribution in a hot water tank. Figure 5.14 serves as a sketch of the system. We know that $H = 5$ ft and the temperatures at the top and bottom of the water column are $T_T = 190$°F and $T_B = 140$°F. The key equation is 5.13, $p(z) - p_0 = \int_{z_0}^{z} -\rho(z) g \, dz$. We will assume that the water is at rest with a linear temperature distribution, $T(z) = T_B + (T_T - T_B)(z/H)$. Next we must determine the density distribution in the tank. A first-order approximation for the density as a function of temperature uses the coefficient of thermal expansion β to write $\rho(T) = \rho_B[1 - \beta(T - T_B)]$, where ρ_B is the density at temperature T_B. Substituting the linear temperature distribution for $T(z)$ into this expression gives us $\rho(z) = \rho_B[1 - \beta(T_T - T_B)(z/H)]$. Now substitute this result into Eq. 5.13 and integrate to find the pressure distribution:

$$p(z) - p_A = \int_H^z -\rho_B \left[1 - \beta(T_T - T_B)\frac{z}{H}\right] g \, dz = \rho_B g \left[-z + \frac{z^2}{2H}\beta(T_T - T_B)\right]\Big|_H^z$$

which gives the pressure distribution as

$$p(z) = p_A + \rho_B g(H - z)\left[1 - \frac{H + z}{2H}\beta(T_T - T_B)\right] \quad \text{(A)}$$

The pressure at the bottom of the tank is

$$p(0) = p_A + \rho_B g H \left[1 - \frac{\beta(T_T - T_B)}{2}\right] \quad \text{(B)}$$

If the water had been at a uniform temperature of 140°F the pressure distribution would be

$$p(z) = p_A + \rho_B g(H - z) \quad \text{(C)}$$

and the pressure at the bottom of the tank would be

$$p(0) = p_A + \rho_B g H \quad \text{(D)}$$

Comparing the pressures at the bottom for each density distribution as given by (B) and (D), we see that in the water column that is hotter at the top, the pressure on the bottom is less by an amount $\rho_B g H[\beta(T_T - T_B)]/2$. This makes sense, since the hotter water is less dense, and the weight of the fluid column is therefore less.

To calculate this pressure difference, we use Table A.3 to find that for water at 140°F $\rho_B = 1.908$ slugs/ft³. Finally, we substitute appropriate values into the $p(z)$ expression to obtain the pressure difference as

$$\Delta p = -\rho_B g H \frac{\beta(T_T - T_B)}{2}$$

$$= (1.908 \text{ slugs/ft}^3)(32.2 \text{ ft/s}^2)(5 \text{ ft}) \frac{(2.3 \times 10^{-4}\,°\text{F}^{-1})(190°\text{F} - 140°\text{F})}{2}$$

$$\Delta p = 1.77 \text{ lb}_f/\text{ft}^2 \left(\frac{1 \text{ ft}}{12 \text{ in.}}\right)^2 = 0.012 \text{ psi}$$

The answers seem reasonable, but note that the linear temperature and density distribution assumed here represent approximations. When the heating element is turned on, it will create convection currents in the water, which change the temperature and density distribution.

To find the pressure distribution in an isothermal perfect gas, we rearrange Eq. 5.12 to obtain $dp/dz = -\rho(z)g$, and substitute for the density to find $dp = -\rho(z)g\,dz = -(p/RT_0)g\,dz$. Separating variables, and integrating from elevation z_0 where the pressure is p_0 to elevation z where the pressure is p, we have $\int_{p_0}^{p} dp/p = -\int_{z_0}^{z}(1/RT_0)g\,dz$. After this integration has been performed, the pressure distribution in an isothermal gas at temperature T_0 is found to be

$$p(z) = p_0 \exp\left[-\frac{g(z - z_0)}{RT_0}\right] \quad (5.22)$$

This exponential pressure–height relationship is illustrated in Figure 5.15A.

A more general model of the pressure–density relationship in a perfect gas is provided by the polytropic law:

$$\frac{p}{p_0} = \left(\frac{\rho}{\rho_0}\right)^n \quad (5.23)$$

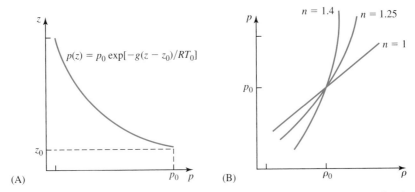

Figure 5.15 (A) Pressure distribution for an isothermal gas and (B) pressure–density relationships for a perfect gas with the polytropic exponents $n = 1$ (isothermal), 1.25, and 1.4 (isentropic).

Earlier we discussed two models for Earth's atmosphere, both of which apply to any gas at rest in Earth's gravity field. The first simply assumed the pressure in the atmosphere to be constant, which is equivalent to assuming that the air density is zero. The second assumed constant air density and resulted in a linear pressure distribution. We can construct an additional model for the atmosphere by treating it as a perfect gas at rest with a uniform temperature T_0. This model is discussed in Example 5.10 and compared with the U.S. Standard Atmosphere, an empirical model adopted by a committee convened by the federal government (see Figure 2.2). It is the job of the engineer to decide which, if any, of these models is appropriate for the problem at hand.

where n is the polytropic exponent, and p_0 and ρ_0 are the pressure and density of the gas at some reference condition. If n is set equal to unity, the gas is isothermal. If n is set equal to γ, the specific heat ratio, the pressure–density relationship is isentropic, meaning that the process involved is reversible and adiabatic. Intermediate values of n may be used to model a variety of practical situations. The pressure–density relationships for isothermal, polytropic, and isentropic gases are shown in Figure 5.15B.

5.4.3 Constant Density Fluid in Rigid Rotation

An important class of fluid statics problems occurs when a container filled with fluid is placed in steady rotation. When a container first begins to rotate, the fluid tends to remain at rest, but a thin layer of fluid near the wall is dragged around with the wall as a result of the effects of viscosity and the no-slip condition. You can see this for yourself if you hold a cup of opaque liquid in your hand while turning in a circle. Notice that the liquid does not turn with you. As the container continues to rotate, however, more and more fluid gradually begins to rotate. After a sufficient time, viscous effects cause the entire volume of fluid inside the container to rotate along with it as a rigid body. When a rigid body rotation is achieved in a fluid, all shear stresses vanish, and there is no deformation of fluid elements. The fluid is at rest relative to the rotating container with a hydrostatic pressure distribution.

An understanding of this class of problems may be gained by writing the hydrostatic equation for a fluid in solid body rotation in a closed cylindrical container as shown

EXAMPLE 5.10

If Earth's atmosphere is modeled as an isothermal perfect gas at 20°C, find the pressure predicted by this model at an altitude of 3 km and compare it with both the pressure predicted by a constant density model and that given by the U.S. Standard Atmosphere.

SOLUTION

To apply the isothermal perfect gas model for the atmosphere, we use Eq. 5.22, $p(z) = p_0 \exp(-gz/RT_0)$, with $z_0 = 0$, $p_0 = 101{,}300$ N/m², $T_0 = 20 + 273 = 293$ K, and $R = 287$ (N-m)/(kg-K) for air. The pressure at $z = 3$ km $= 3000$ m is then

$$p(3000) = (101{,}300 \text{ N/m}^2) \exp\left\{-\frac{(9.81 \text{ m/s}^2)(3000 \text{ m})}{[287 \text{ (N-m)/(kg-K)}](293 \text{ K})}\right\} = 71.4 \text{ kPa}$$

The isothermal pressure distribution in general is

$$p(z) = (101{,}300 \text{ N/m}^2) \exp\left\{-\frac{(9.81 \text{ m/s}^2)(z)}{[287 \text{ (N-m)/(kg-K)}](293 \text{ K})}\right\}$$

which simplifies to

$$p(z) = (101.3 \text{ kN/m}^2) \exp[-(1.17 \times 10^{-4} \text{ m}^{-1})(z)]$$

The pressure distribution for the constant density model is $p(z) = p_0 - \rho_0 g z$, thus the pressure at 3000 m with this model is

$$p(3000) = 101{,}300 \text{ N/m}^2 - (1.204 \text{ kg/m}^3)(9.81 \text{ m/s}^2)(3000 \text{ m}) = 65.9 \text{ kPa}$$

From Appendix B, the U.S. Standard Atmosphere has $p(3000 \text{ m}) = 70.1$ kPa. The isothermal and constant density pressure distributions just derived are compared with the U.S. Standard Atmosphere in Figure 5.16.

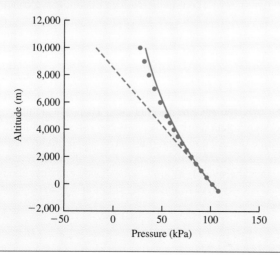

Figure 5.16 From Example 5.10, the atmospheric pressure distribution based on the constant density model (dashes), the isothermal perfect gas model (solid), and the U.S. Standard Atmosphere (circles).

Figure 5.17 Schematic of solid body rotation in a closed cylindrical container.

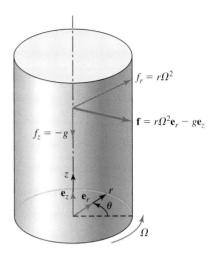

in Figure 5.17. Choose cylindrical coordinates (r, θ, z) fixed to the rotating container as shown, with $(\mathbf{e}_r, \mathbf{e}_\theta, \mathbf{e}_z)$ the corresponding unit vectors. The fluid is at rest in this noninertial reference frame. There is a centrifugal body force in the radial direction and the force of gravity in the z direction. In Example 5.5, we showed that the total body force is given by $\mathbf{f} = r\Omega^2 \mathbf{e}_r - g\mathbf{e}_z$, and that the three components of the differential hydrostatic equation are given by

$$\frac{\partial p}{\partial r} = +\rho r \Omega^2, \quad \frac{1}{r}\frac{\partial p}{\partial \theta} = 0, \quad \text{and} \quad \frac{\partial p}{\partial z} = -\rho g \quad (5.24\text{a--c})$$

Equation 5.24a shows that the pressure gradient in the r direction balances the centrifugal force due to the rotation of the fluid. The positive sign in this equation indicates that the pressure increases monotonically in the radial direction, reaching a maximum at the container wall. Since there is no θ component of the body force, the pressure gradient in the θ direction is zero. From the remaining equation, we conclude that the pressure gradient in the z direction balances the gravitational body force as usual.

Since the pressure gradient in the θ direction is zero, taking a derivative of Eq. 5.24a or 5.24c with respect to θ, interchanging derivatives, and noting that $\partial p/\partial \theta = 0$ shows that $\partial \rho / \partial \theta = 0$. This means that a density variation in the θ direction is not allowed in a fluid in rigid rotation. A density variation in the r and z directions is permitted. Thus the density of a fluid in a rotating container does not need to be constant for the fluid to be at rest relative to the container. There are stability issues for a stratified fluid in rigid body rotation just as there are for a stratified fluid in a gravity field. A stable density profile in a rotating fluid must have heavier fluid "beneath" lighter fluid, but this statement must now be interpreted in terms of the direction of the total body force vector, which is the sum of the gravitational and centrifugal body force.

For a constant density fluid in rigid body rotation, a general solution of Eqs. 5.24a and 5.24c, is given by

$$p(r, z) = p_0 - \rho g(z - z_0) + \tfrac{1}{2}\rho \Omega^2 \left(r^2 - r_0^2\right) \quad (5.25)$$

EXAMPLE 5.11

A water-filled cylindrical container, 20 cm in diameter and 50 cm tall, rotates about its axis at 1000 rpm. If the cylinder axis is vertical, as shown in Figure 5.17, what are the maximum and minimum radial and vertical pressure gradients, and where do they occur?

SOLUTION

Equations 5.24 describe the three components of the pressure gradient in this case. The radial pressure gradient $\partial p/\partial r = +\rho r \Omega^2$ is linear with a minimum of zero on the axis of rotation. The maximum is at the wall, where $r = 10$ cm. Inserting the data, we have

$$\frac{\partial p}{\partial r} = \rho r \Omega^2 = (998 \text{ kg/m}^3)(0.1 \text{ m}) \left[(1000 \text{ rpm}) \left(\frac{2\pi \text{ rad}}{1 \text{ rev}} \right) \left(\frac{1 \text{ min}}{60 \text{ s}} \right) \right]^2$$
$$= 1.09 \times 10^6 \text{ N/m}^3$$

Note that the radial pressure gradient can also be expressed as 1.09×10^6 Pa/m.
The vertical pressure gradient given by Eq. 5.24c is found to be

$$\frac{\partial p}{\partial z} = -\rho g = -(998 \text{ kg/m}^3)(9.81 \text{ m/s}^2) = -9.79 \times 10^3 \text{ N/m}^3$$
$$= -9.79 \times 10^3 \text{ Pa/m},$$

which is spatially constant and about 100 times smaller than the radial pressure gradient. This result illustrates the strong pressure gradient that can be generated by centrifugal acceleration. In this problem the maximum centrifugal acceleration is greater than $100g$.

where p_0 is the pressure at location (r_0, z_0). The pressure in a rotating fluid depends on both z and r, increasing linearly with depth as usual, but quadratically with radius. The equation describing surfaces of constant pressure can be found by dividing Eq. 5.25 by ρg, and gathering constant terms to obtain

$$\frac{p - p_0}{\rho g} - z_0 + \frac{1}{2}\frac{\Omega^2 r_0^2}{g} = C = -z + \frac{1}{2}\frac{\Omega^2 r^2}{g}$$

Solving for z gives:

$$z(r) = \frac{1}{2}\frac{\Omega^2 r^2}{g} - C \qquad (5.26)$$

These constant pressure surfaces are parabolic as shown in Figure 5.18A.

Figure 5.18 Pressure distribution in a rotating container filled with water. (A) In Earth's gravitational field, and (B) in a zero gravity environment.

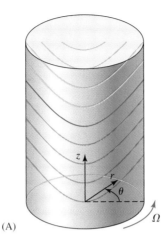
(A)

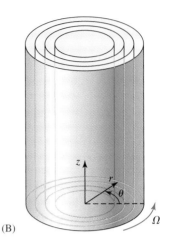
(B)

EXAMPLE 5.12

Find the pressure distribution in the rotating water-filled container described in Example 5.11. Plot surfaces of constant pressure. If this container were rotating inside a spacecraft in Earth orbit, what would the pressure distribution be? What would the surfaces of constant pressure look like?

SOLUTION

This exercise is solved by using Eq. 5.25 with the origin of the coordinate system on the cylinder axis at the bottom of the container. The pressure there is $p(r_0, z_0)$, and substituting this value along with $r_0 = 0$ and $z_0 = 0$ into Eq. 5.25, we find $p(r, z) = p(r_0, z_0) - \rho g z + \frac{1}{2}\rho \Omega^2 r^2$. Surfaces of constant pressure are parabolas (see Figure 5.18A). In orbit, the gravitational body force is zero. Using the pressure distribution just given and substituting $g = 0$, we get $p(r, z) = p(r_0, z_0) + \frac{1}{2}\rho \Omega^2 r^2$. The surfaces of constant pressure are concentric cylinders (see Figure 5.18B). Note that in both cases further information is needed to establish the datum value of pressure, i.e., $p(r_0, z_0)$.

In our discussion of a fluid in rigid body rotation, we assumed that the fluid completely fills the container. This is always true for gases, but a rotating container may be only partially filled with a liquid. As shown in Figure 5.19, the equilibrium configuration of a rotating liquid depends on the volume of liquid in the container and the rotation rate. In a partially filled rotating container there is a free surface between the liquid and the gas or vapor above it. In the absence of surface tension effects, the pressure on this free surface is atmospheric or ambient. We can use the results of the foregoing analysis to predict the shape of this free surface. For example, consider a partially filled cylindrical container of water rotating about its axis with a free surface as shown in Figure 5.19A. The container is open, and the pressure above the free surface is ambient. To find the

5.4 HYDROSTATIC PRESSURE DISTRIBUTION

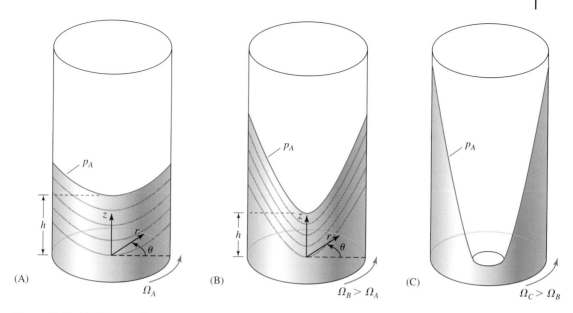

Figure 5.19 (A) Pressure distribution in a container partially filled with water and rotating at speed Ω_A. (B) As speed is increased to Ω_B, the depression deepens. (C) Eventually the container rotates fast enough at Ω_C that a dry spot can develop on the bottom of the container.

The parabolic shape of the free surface in a liquid rotating about a vertical axis has been cleverly exploited in a telescope design in which a rotating mercury surface forms the reflector of the telescope. A photograph of such a telescope is shown in Figure 5.20.

Figure 5.20 Parabolic liquid mercury mirror formed of the surface of the rotating vessel.

pressure distribution in the water, choose a datum location on the free surface at its intersection with the cylinder axis. Then the parameters appearing in Eq. 5.25 are $r_0 = 0$, $z_0 = h$ and $p_0 = p_A$. Inserting these values into Eq. 5.25, we can determine the pressure distribution in the water to be:

$$p(r, z) = p_A - \rho_{H_2O} g(z - h) + \tfrac{1}{2}\rho_{H_2O}\Omega^2 r^2$$

(5.27a)

The constant pressure contours corresponding to this solution are plotted in Figure 5.19A. Note that these surfaces are parabolic, as expected.

How do we find the shape of the free surface? The answer lies in our knowledge that the entire free surface is at atmospheric or ambient pressure. In the absence of surface tension, the water just inside the free surface is also at atmospheric pressure; thus the constant pressure contour for a value of the pressure equal to atmospheric pressure must coincide with the free surface. By setting $p(r, z) = p_A$ in Eq. 5.27a, we can express the shape of the free surface as

$$z(r) = h + \frac{1}{2}\frac{\Omega^2 r^2}{g}$$

(5.27b)

Since a free surface is a surface of constant pressure, Eq. 5.27b could have also been obtained by evaluating the constant in Eq. 5.26. From the geometry and datum values already chosen, you should be able to show that $C = -h$.

Now suppose you have been given values for the dimensions of the rotating water-filled cylinder, as well as its rotation rate. Curiously, you will find that you cannot plot the shape of the free surface by using Eq. 5.27b, or calculate the rise of the free surface at the wall. The parameter h, which is the free surface elevation at the axis, is unknown! What is missing in our analysis? The answer is that the volume of liquid in a rotating container is a critical part of the specification of this type of problem. Once a volume of liquid has been specified, h is known, since the volume is given by the integral $V = 2\pi \int_0^R r z(r)\, dr$, which in this case gives

$$V = \pi R^2 h + \pi \frac{\Omega^2 R^4}{4g} \tag{5.27c}$$

EXAMPLE 5.13

A 15 in. diameter cylinder, 30 in. tall, rotates about its axis at 60 rpm. If the cylinder contains 1 ft³ of water and its axis is vertical, find the pressure distribution in the water and the shape of the free surface. What is the pressure at the bottom of the cylinder at the axis and at the wall? What is the free surface elevation at the axis and the wall?

SOLUTION

We are to determine the pressure distribution, shape of the free surface, pressure at the points $(0, 0)$ and $(R, 0)$, and the elevation at the axis and wall in a rotating cylinder partially filled with water. Figure 5.19 is an appropriate sketch for this problem. The cylinder has $D = 15$ in. $= 1.25$ ft and rotates with $\Omega = 60$ rpm $= 2\pi$ rad/s. The water volume is $V = 1$ ft³, and we will assume a water density of $\rho = 62.4$ lb$_m$/ft³. This problem is solved using Eqs. 5.27a–5.27c. We begin by solving Eq. 5.27c for h and then substituting the appropriate values to obtain the elevation of the free surface at the axis as

$$h = \frac{V}{\pi R^2} - \frac{\Omega^2 R^2}{4g} = \left[\frac{1\text{ ft}^3}{\pi (0.625\text{ ft})^2}\right] - \left[\frac{(2\pi\text{ s}^{-1})^2 (0.625\text{ ft})^2}{4(32.2\text{ ft/s}^2)}\right] = 0.695\text{ ft}$$

Equation 5.27a gives the pressure distribution within the rotating cylinder as

$$p(r, z) = p_A - \rho_{H_2O} g(z - h) + \tfrac{1}{2}\rho_{H_2O}\Omega^2 r^2$$

The pressure at the bottom of the cylinder at the axis is found to be

$$p(0, 0) = p_A - \rho_{H_2O} g(0 - h) + \tfrac{1}{2}\rho_{H_2O}\Omega^2 (0)^2 = p_A + \rho_{H_2O} gh$$

and the pressure at the bottom of the cylinder at the wall is

$$p(R, 0) = p_A - \rho_{H_2O} g(0 - h) + \tfrac{1}{2}\rho_{H_2O}\Omega^2 R^2 = p_A + \rho_{H_2O} gh + \tfrac{1}{2}\rho_{H_2O}\Omega^2 R^2$$

Inserting the data, recognizing that $p_A = 2116 \text{ lb}_f/\text{ft}^2$, and using g_c as a unit conversion factor, we can calculate the pressure at $(0, 0)$ as

$$p(0,0) = 2116 \text{ lb}_f/\text{ft}^2 + (62.4 \text{ lb}_m/\text{ft}^3)(32.2 \text{ ft/s}^2)\left(\frac{1 \text{ lb}_f\text{-s}^2}{32.2 \text{ ft-lb}_m}\right)(0.695 \text{ ft}) = 2159.4 \text{ lb}_f/\text{ft}^2$$

while that at $(R, 0)$ the pressure is

$$p(R,0) = 2116 \text{ lb}_f/\text{ft}^2 + (62.4 \text{ lb}_m/\text{ft}^3)(32.2 \text{ ft/s}^2)\left(\frac{1 \text{ lb}_f\text{-s}^2}{32.2 \text{ ft-lb}_m}\right)(0.695 \text{ ft})$$

$$+ \frac{1}{2}(62.4 \text{ lb}_m/\text{ft}^3)(2\pi \text{ s}^{-1})^2(0.625 \text{ ft})^2 \left(\frac{1 \text{ lb}_f\text{-s}^2}{32.2 \text{ ft-lb}_m}\right)$$

$$p(R,0) = 2174.3 \text{ lb}_f/\text{ft}^2$$

The increment in pressure at the wall due to the rotation is small at this low rpm.

Next we use Eq. 5.27b with our known values of h and Ω to obtain the expression for the free surface of the liquid in the rotating cylinder:

$$z(r) = h + \frac{1}{2}\frac{\Omega^2 r^2}{g} = (0.695 \text{ ft}) + \frac{1}{2}\frac{(2\pi \text{ s}^{-1})^2 r^2}{(32.2 \text{ ft/s}^2)} = 0.695 \text{ ft} + (0.613 \text{ ft}^{-1})r^2$$

The elevation at the axis is found by substituting $r = 0$ into this expression or more simply by noting that this is the definition of h. In either case we find $z(0) = 0.695$ ft. The elevation at the wall is found to be $z(0.625 \text{ ft}) = 0.695 \text{ ft} + (0.613 \text{ ft}^{-1})(0.625 \text{ ft})^2 = 0.934$ ft.

Sloshing of fuel in a vehicle fuel tank is also an important problem in structural design and vibration analysis, particularly in aerospace vehicles. Despite its importance, the dynamics of sloshing is beyond the scope of this text. You actually have considerable practical experience with the problem caused by a tilted free surface if you have ever had a drink spill in your car during hard acceleration or braking. Was it due to a tilted free surface or did the drink spill because of sloshing? A question like this reminds us that real-world problems can be difficult to categorize.

5.4.4 Constant Density Fluid in Rectilinear Acceleration

The structural design of the fuel tanks of a liquid-fueled rocket must take into account the pressure forces developed during acceleration. Similar considerations apply to storage tanks for fuel and lubricants in automobiles and aircraft, where acceleration may cause the free surface of a liquid to tilt, with the inlet of a pump then momentarily running dry. Although it is unusual to maintain a constant rectilinear acceleration for any length of time, the hydrostatic pressure distribution caused by acceleration is established almost instantly and adjusts itself rapidly to a changing acceleration. Thus, in this section we will assume that the liquid is at rest with respect to its container under the imposed acceleration, and focus on calculating the shape of the free surface.

Consider a container completely filled with a fluid that undergoes a constant linear acceleration $\mathbf{a} = (a_x, a_y, a_z)$ as shown in Figure 5.21. Choosing a noninertial reference

Figure 5.21 A completely filled container of fluid that undergoes constant linear acceleration, $\mathbf{a} = (a_x, a_y, a_z)$.

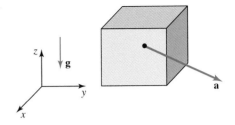

frame attached to the container, we conclude that for a fluid at rest relative to the container, the total body force per unit mass acting on the fluid is the sum of an noninertial body force per unit mass in the amount $\mathbf{f} = -\mathbf{a}$, and the gravitational body force $\mathbf{f} = -g\mathbf{k}$. The total body force is therefore

$$\mathbf{f} = (-g\mathbf{k}) - \mathbf{a} \tag{5.28}$$

To find the pressure distribution in a constant density fluid undergoing a constant translational acceleration, we must insert the appropriate expression for the total body force into the hydrostatic equation and solve the resulting system of equations. Reconsider a container completely filled with constant density fluid as shown in Figure 5.21. In Cartesian coordinates fixed to the container, the three components of Eqs. 5.10a–5.10c acting on fluid at rest under a constant acceleration $\mathbf{a} = (a_x, a_y, a_z)$ are

$$\frac{\partial p}{\partial x} = \rho(f_x - a_x), \quad \frac{\partial p}{\partial y} = \rho(f_y - a_y), \quad \text{and} \quad \frac{\partial p}{\partial z} = \rho(f_z - a_z)$$

If gravity is the only body force, and with z upward as usual, these equations become

$$\frac{\partial p}{\partial x} = -\rho a_x, \quad \frac{\partial p}{\partial y} = -\rho a_y, \quad \text{and} \quad \frac{\partial p}{\partial z} = -\rho(g + a_z) \tag{5.29a–c}$$

EXAMPLE 5.14

A container of liquid has been dropped and is in free fall. What is the total body force per unit mass acting on the liquid in the moment after its release?

SOLUTION

Since the container falls with the acceleration of gravity, $\mathbf{a} = (0, 0, -g)$. The total body force per unit mass can be found by using Eq. 5.28: $\mathbf{f} = (-g\mathbf{k}) - \mathbf{a} = (-g\mathbf{k}) - (-g\mathbf{k}) = 0$. The total body force is zero. This *zero-g* effect is utilized in the production of round lead shot, zero-gravity research, and many amusement park rides.

5.4 HYDROSTATIC PRESSURE DISTRIBUTION

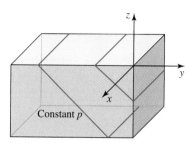

Figure 5.22 Constant pressure planes that result from linear acceleration.

The general solution of these equations can be written immediately as

$$p(x, y, z) = p_0 - \rho a_x(x - x_0) - \rho a_y(y - y_0) - \rho(g + a_z)(z - z_0) \quad (5.30)$$

where p_0 is the datum pressure at location (x_0, y_0, z_0). As shown in Figure 5.22, surfaces of constant pressure are planes whose precise orientation in space; i.e., tilt, depends on the values of the three acceleration components. If a free surface is present in a problem involving rectilinear acceleration, the free surface will coincide with the corresponding pressure contour at the atmospheric or ambient pressure.

An interesting effect of acceleration on a fluid at rest is illustrated in the following example. Suppose a liquid-fueled rocket accelerates upward at $\mathbf{a} = a_0\mathbf{k}$. What is the pressure distribution in the fuel tank if it is completely filled with fuel? What is the pressure at the bottom of the fuel tank? We can answer these questions by recognizing that the only component of the hydrostatic equation in this case is given by Eq. 5.29c:

$$\frac{dp}{dz} = -\rho_{\text{fuel}}(g + a_z)$$

The pressure distribution may be found by inserting the known constant acceleration of $a_z = a_0$, and integrating this equation from an arbitrary elevation z to the top of the tank at $z = H$, where the pressure is assumed to be known. The result after rearrangement is

$$p(z) = p_{\text{top}} - \rho_{\text{fuel}}(g + a_0)(z - H)$$

This is a linear pressure distribution in which the pressure increases with depth at a rate of $\rho_{\text{fuel}}(g + a_0)$. This pressure distribution may also be obtained from the general solution, Eq. 5.30, by substituting the same datum values for pressure and location, and appropriate values for each component of acceleration. Because there is acceleration in the z direction only, the free surface remains horizontal. The pressure at the bottom of the tank is found to be

$$p_{\text{bottom}} = p_{\text{top}} + \rho_{\text{fuel}}(g + a_0)H$$

and if the tank is tall and the acceleration is several times the force of gravity, the pressure on the bottom may be very large.

When more than one component of acceleration is present, the free surface will tilt and take up a position so that the free surface lies with its normal parallel to the total body force. The calculation of the equilibrium position of a free surface under acceleration, and the pressure distribution in the liquid, is illustrated by the following example.

EXAMPLE 5.15

A truck engine with a sump reservoir ($L = 2$ ft, $W = 0.5$ ft, $H = 1$ ft) is half-filled with SAE 30W oil as shown in Figure 5.23A; the sump has an emissions control breather pipe located as shown. Find the slope of the free surface if the vehicle undergoes a lateral acceleration of $a_y = 0.25\,g$. At what acceleration will the oil reach the breather inlet?

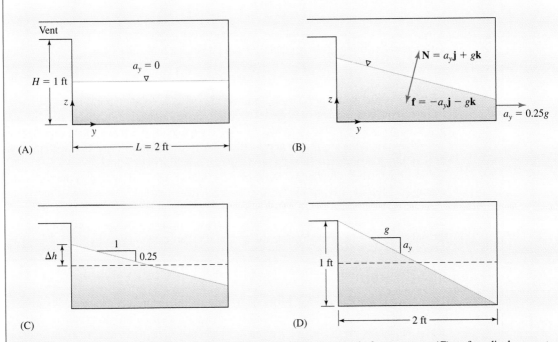

Figure 5.23 (A) Schematic for Example 5.15. (B) surface normal and body force vectors. (C) surface displacement. (D) surface displacement into the breather.

SOLUTION

We are asked to determine the slope of the free surface for a lateral acceleration of $a_y = 0.25g$. We must also calculate the acceleration for which oil reaches the breather pipe. Figure 5.23A will serve as the schematic and we will assume that the oil is at rest under constant acceleration. The sump dimensions are $L = 2$ ft, $W = 0.5$ ft, and $H = 1$ ft. The sump is half-filled with SAE 30W oil and is accelerating at $a_y = 0.25g$. The volume of oil is easily calculated to be $V = \frac{1}{2}LWH = \frac{1}{2}(2\,\text{ft})(0.5\,\text{ft})(1\,\text{ft}) = 0.5\,\text{ft}^3$. Equation 5.30 defines the relevant pressure distribution as $p(x, y, z) = p_0 - \rho a_x(x - x_0) - \rho a_y(y - y_0) - \rho(g + a_z)(z - z_0)$. In this problem $a_x = a_z = 0$, so the pressure distribution in the oil is

$$p(y, z) = p_0 - \rho a_y(y - y_0) - \rho g(z - z_0) \tag{A}$$

where the datum pressure p_0 and location y_0, z_0 are as yet unspecified. Setting $p(y, z) = p_A$, the equation for the free surface is

$$z(y) = -\frac{a_y}{g}y + \left(z_0 + \frac{p_0 - p_A}{\rho g} + \frac{a_y}{g}y_0\right) \tag{B}$$

This is a linear distribution, indicating that the free surface is a plane. We see that the slope of the free surface is $dz/dy = -a_y/g$, which for $a_y = 0.25g$ is $dz/dy = -a_y/g = -0.25g/g = -0.25$. Note that the normal to this surface lies along the line whose slope is the negative reciprocal, i.e., $dz/dy = g/a_y$. Thus by inspection of Figure 5.23B, we can write the (nonunit) normal to the free surface in general as $\mathbf{N} = a_y\mathbf{j} + g\mathbf{k}$. According to Eq. 5.28, the total body force is $\mathbf{f} = (-g\mathbf{k}) - \mathbf{a} = -a_y\mathbf{j} - g\mathbf{k}$. We see that these vectors are antiparallel, which confirms our earlier statement that the free surface arranges itself perpendicular to the total body force.

Since the volume of oil, $\forall = \frac{1}{2}LWH$, is fixed, the free surface must arrange itself so the area under the free surface shown in Figure 5.23A is always $A = \frac{1}{2}LH$. Together with the slope, this is sufficient to allow us to graph the free surface for $a_y = 0.25g$ as shown in Figure 5.23C. From Figure 5.23D, it can be seen that the slope of a free surface that just reaches the breather is -0.5. Setting $dz/dy = -a_y/g = -0.5$ shows that this slope occurs for an acceleration $a_y = 0.5g$. This horizontal acceleration corresponds to going from 0 to 60 mph in 5.5 s, which is far beyond the capability of a heavy vehicle, but within reach of high performance automobiles and motorcycles.

The selection of a datum location at which the pressure is known in a free surface problem can be tricky because the location of the free surface is unknown until the problem has been solved. In the preceding example it was helpful to establish the slope of the free surface first, then recognize the constraint of a fixed fluid volume. Note that for accelerations in the range $0 \leq a_y \leq 0.5g$, the geometry shown in Figure 5.23C indicates that $g/a_y = (L/2)/\Delta h$, so the oil height at the left wall is $\Delta h + H/2 = La_y/2g + H/2$. We can then pick the datum location as $y_0 = 0$, $z_0 = La_y/2g + H/2$, where we know the pressure is $p_0 = p_A$, and write (B) for the free surface as $z(y) = (H/2) - (a_y/g) \times (y - L/2)$. The pressure distribution in the oil given by Eq. (A) is $p(y, z) = p_A - \rho a_y y - \rho g(z - H/2 - La_y/2g)$. Another

5.5 HYDROSTATIC FORCE

In the majority of applications of fluid statics, an engineer is concerned with the forces and moments applied by a fluid to a structure. These surface forces and moments were discussed in a general way in Chapter 4. In the next two sections we discuss them in greater detail for the specific case of a fluid at rest.

Since a solution to a fluid statics problem results in an analytical expression for the hydrostatic pressure distribution, the hydrostatic stress vector is immediately known from Eq. 5.1b to be $\boldsymbol{\Sigma} = -p\mathbf{n}$. Substituting this into Eq. 4.21, which defines the total surface force as $\mathbf{F}_S = \int_S \boldsymbol{\Sigma}\, dS$, we see that the total hydrostatic force exerted by the pressure is given by the integral

$$\mathbf{F}_S = \int_S -p\mathbf{n}\, dS \tag{5.31}$$

The difficulty, if any, in calculating the total surface force in a fluid statics problem arises from the requirement to

good datum location for this case is $y_0 = L/2$, $z_0 = H/2$, since the pressure is also atmospheric at this location for this range of accelerations.

calculate this integral on surfaces of various degrees of complexity.

The three common types of surface encountered in fluid statics are illustrated in Figure 5.24. The first, a planar aligned surface, is one that is flat and aligned with one of the Cartesian coordinate planes (see Figure 5.24A). The second, a planar nonaligned surface, is not aligned with one of the coordinate planes but may be aligned by using a new set of X, Y, Z coordinates as shown in Figure 5.24B and 5.24C. As shown in Figure 5.24D, the third type of surface is curved.

Now consider the general problem of evaluating the integral in Eq. 5.31 for each surface type. Each term in the integrand must be given the appropriate value on the surface. The pressure p is the value of the pressure on the surface as determined from the hydrostatic pressure distribution. The outward unit normal vector **n** is usually obtained from the mathematical description of the surface. It is a constant on a planar surface but varies with the position of each infinitesimal element on a curved surface. Finally, the description of the scalar area element dS depends on the surface involved, and on the coordinates in which the calculation is carried out. It is usually straightforward to write down the description of dS in Cartesian coordinates for a planar surface, and in cylindrical or spherical coordinates for a cylindrical or spherical surface, respectively. On a curved surface the description of dS is a more challenging task.

5.5.1 Planar Aligned Surface

Let us now focus on the simplest type of surface, namely the planar aligned surface. Consider the three planar surfaces shown in Figure 5.25. Each surface is parallel to one

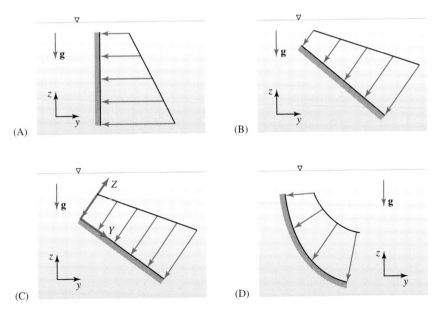

Figure 5.24 Surface types encountered in fluid statics: (A) planar aligned, (B) planar nonaligned, (C) planar nonaligned with aligned coordinates, and (D) curved.

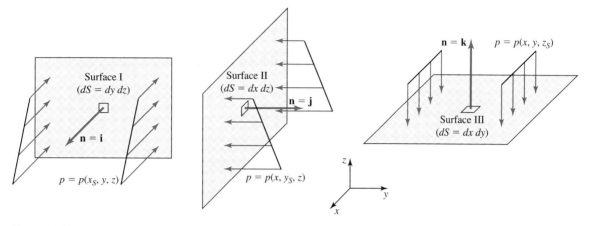

Figure 5.25 Planar surfaces aligned with each coordinate plane (see text).

of the coordinate planes (aligned), but not necessarily coincident. We assume that these surfaces are immersed in a fluid at rest and that we are to use Eq. 5.31 to calculate the total surface force on the near face of each surface. To allow for any type of body force, we assume that the hydrostatic pressure distribution in the fluid is given by $p = p(x, y, z)$.

Consider surface I, parallel to the (y, z) coordinate plane. Although the pressure distribution in the fluid may vary in all three directions, the pressure distribution on surface I does not. The pressure on this surface can be a function of the y and z coordinates only. It cannot depend on x, since all points on the surface have exactly the same value for the x coordinate, say $x = x_S$. Thus the pressure on this surface may be written as $p = p(x_S, y, z)$. An infinitesimal element of area dS on surface I may be represented in Cartesian coordinates by a rectangle with sides of length dy and dz, thus the area of the element is $dS = dy\,dz$. By inspection, the outward unit normal for an infinitesimal element on this surface is constant and given by $\mathbf{n} = \mathbf{i}$ (the normal points at the fluid).

Using these results to construct the integrand of Eq. 5.31, we obtain

$$-p\mathbf{n}\,dS = -p(x_S, y, z)(\mathbf{i})\,dy\,dz$$

valid for any planar surface aligned with the (y, z) coordinate plane with the fluid on the positive x side as shown. The hydrostatic force is therefore given by the double integral

$$\mathbf{F}_S = \iint -p(x_S, y, z)(\mathbf{i})\,dy\,dz \qquad (5.32a)$$

where the limits of integration are chosen to describe the surface of interest. For a fluid on the other side of this surface, the sign of the unit normal is negative in order to point at the fluid, and we use

$$\mathbf{F}_S = \iint -p(x_S, y, z)(-\mathbf{i})\,dy\,dz \qquad (5.32b)$$

A similar analysis for surface II shows that the pressure is given by $p = p(x, y_S, z)$, where y_S is the location of this surface along the y axis. The outward unit normal and area element are given by $\mathbf{n} = \mathbf{j}$ and $dS = dx\,dz$. The integrand on this surface is $-p\mathbf{n}\,dS = -p(x, y_S, z)(\mathbf{j})\,dx\,dz$, valid for any planar surface aligned with the (x, z) plane with the fluid on the positive y side. The hydrostatic force is now given by

$$\mathbf{F}_S = \iint -p(x, y_S, z)(\mathbf{j})\,dx\,dz \tag{5.33a}$$

with appropriate limits. For a fluid on the negative y side of this surface, we use

$$\mathbf{F}_S = \iint -p(x, y_S, z)(-\mathbf{j})\,dx\,dz \tag{5.33b}$$

For surface III we find $-p\mathbf{n}\,dS = -p(x, y, z_S)(\mathbf{k})\,dx\,dy$, and the hydrostatic force for a fluid on the positive z side of the surface is given by

$$\mathbf{F}_S = \iint -p(x, y, z_S)(\mathbf{k})\,dx\,dy \tag{5.34a}$$

with appropriate limits. For a fluid on the negative z side of this surface, we use

$$\mathbf{F}_S = \iint -p(x, y, z_S)(-\mathbf{k})\,dx\,dy \tag{5.34b}$$

In Example 5.17, we show that the resultant force on the hatch differs from the hydrostatic force applied by the water because of the presence of air on the opposite side of the hatch. In engineering applications, the net hydrostatic force is also of interest. Assuming that air at atmospheric pressure is in contact with the opposite side of a surface, the net hydrostatic force can be obtained by using gage pressure rather than absolute pressure in calculating the hydrostatic force with Eq. 5.31. Alternately, we can subtract the atmospheric pressure term from the answer obtained by using the absolute pressure.

EXAMPLE 5.16

Consider a rectangular tank of water open at the top, as shown in Figure 5.26. What is the force applied by the water to the hatch shown in the bottom of the tank? What is the net or resultant force on the hatch from all fluids?

SOLUTION

The hatch is a planar aligned surface lying parallel to the (x, y) plane with the fluid on the positive z side. To use Eq. 5.34a, we need to find the pressure $p(x, y, z_S)$ acting on this surface. From Eq. 5.14 the pressure distribution in the water is $p(z) = p_A - \rho_{H_2O}g(z - h)$. On the hatch $z_S = 0$, so the pressure on this surface is $p(x, y, z_S) = p(x, y, 0) = p_A + \rho_{H_2O}gh$, a constant. Since the integrand of Eq. 5.34a is a constant on this surface, the integration is easy, and the force of the water is found to be given by $\mathbf{F}_S = -(p_A + \rho_{H_2O}gh)A\mathbf{k}$, where $A = LW$ is the area of the hatch. We see that the hydrostatic force acts downward and is equal in magnitude to the area of the hatch multiplied by the sum of the air pressure and weight of water per unit area above the hatch. The same result may be obtained by selecting the volume of fluid shown in Figure 5.26B and employing the integral hydrostatic equation.

To find the net or resultant force of all fluids on the hatch, we must also consider the force applied on the hatch by the air below. The outward unit normal on this surface is $\mathbf{n} = -\mathbf{k}$, and the pressure is constant and equal to p_A (see Figure 5.26C). Thus the force of the air on the hatch is found to be $\mathbf{F}_S = p_A A \mathbf{k}$. The net force is, therefore, $\mathbf{F}_S = -\rho_{H_2O} g h A \mathbf{k}$.

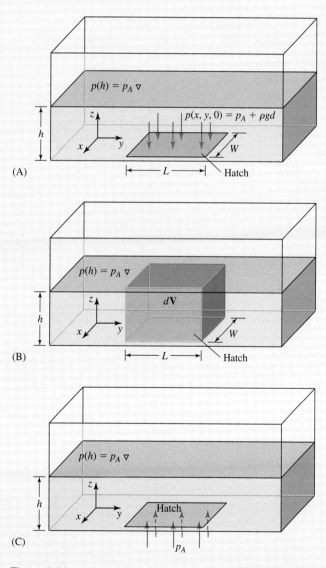

Figure 5.26 (A) schematic, (B) fluid volume, and (C) force acting on hatch due to air below.

EXAMPLE 5.17

A vertical gate of width W controls the water level in the intake reservoir of a power plant as shown in Figure 5.27. What is the force of the water on the gate? What is the net hydrostatic force on the gate?

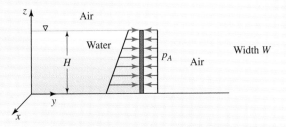

Figure 5.27 Schematic for Example 5.17.

SOLUTION

The gate is aligned parallel to the (x, z) plane with the water on the left side, so we will use Eq. 5.33b. Thus the force of the water on the gate is given by $\mathbf{F}_S = \iint -p(x, y_S, z)(-\mathbf{j})\,dx\,dz$. The pressure distribution along the gate is $p(z) = p_A - \rho g(z - H)$. Substituting, and evaluating the integral with appropriate limits, gives

$$\mathbf{F}_{\text{water}} = \int_0^H \int_{-W/2}^{W/2} -[p_A - \rho g(z-H)](-\mathbf{j})\,dx\,dz = WH\left(p_A + \frac{\rho g H}{2}\right)\mathbf{j}$$

This is the force of the water on the gate. Note that it can be thought of as an average pressure $p_A + \rho g H/2$ times an area WH pushing the gate to the right. The net force on the gate is found by using gage pressure, $p_{\text{gage}}(z) = -\rho g(z - H)$, in Eq. 5.33b to find:

$$\mathbf{F}_{\text{net}} = \int_0^H \int_{-W/2}^{W/2} -[-\rho g(z-H)](-\mathbf{j})\,dx\,dz = WH\left(\frac{\rho g H}{2}\right)\mathbf{j}$$

This net force is, of course, the sum of the force of the water $WH[p_A + \rho g H/2]\mathbf{j}$, and the force of the air on the opposite side of the gate, $(-p_A)WH\,\mathbf{j}$, and could also have been obtained by simply subtracting the atmospheric pressure term from the force of the water calculated earlier.

5.5.2 Planar Nonaligned Surface

Now consider the problem of calculating the hydrostatic force on a planar nonaligned surface. We will investigate two approaches to solving this problem. The first, a direct approach, involves setting up and evaluating the surface integral in Eq. 5.31. Since the majority of engineering applications involve a constant density fluid in Earth's gravity

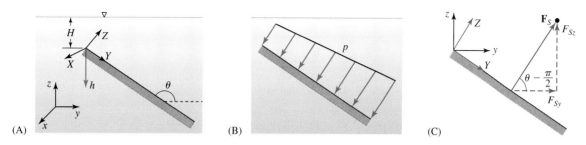

Figure 5.28 Planar inclined surface: (A) geometry, (B) hydrostatic pressure distribution, and (C) force components.

field, we will limit our discussion of the direct approach to this class of problems. A second, indirect approach, employs the integral hydrostatic equation to replace the surface integral on the inclined surface with integrals on the horizontal and vertical projections of the inclined surface plus a volume integral.

To illustrate the direct approach, we will find the hydrostatic force due to the fluid on the right-hand side of the inclined surface shown in Figure 5.28A. We do this by evaluating Eq. 5.31 in a new set of coordinates X, Y, Z, which are aligned with the surface, then expressing the result in the usual x, y, z Cartesian coordinates. The Z axis is perpendicular to the surface of interest and pointing at the fluid, with Y along the surface as shown. The origin has been placed at the uppermost point of the surface for convenience. A new variable h measures the distance below the origin. Consider the hydrostatic force on the inclined surface in Figure 5.28A due to the fluid on the right side of this surface. This force is given by Eq. 5.31 as $\mathbf{F}_S = \int_S -p\mathbf{n}\, dS$. We will follow the procedure of Section 5.5.1 to evaluate this integral, but use the X, Y, Z coordinate system.

From the geometry of Figure 5.28A, the outward unit normal to the inclined surface is

$$\mathbf{n} = +\mathbf{K} \tag{5.35a}$$

and an infinitesimal surface element on the inclined surface is given by

$$dS = dX\, dY \tag{5.35b}$$

By inspection, the pressure distribution on the inclined surface can be seen to be a function only of the Y coordinate of a point on the surface. (Since there is no component of gravity in the X direction, the pressure cannot depend on X; and since Z is a constant on the surface, the pressure cannot depend on Z.) We need to find the pressure distribution $p(Y)$ to proceed. From earlier discussions, we know that the pressure at any depth h is equal to $(p_0 + \rho g h)$ where p_0 is the pressure at the $h = 0$. From the geometry in Figure 5.28A we find $h = Y \sin\theta$; thus we can write the pressure distribution on this surface as $p(Y) = p_0 + \rho g(Y \sin\theta)$, or

$$p(Y) = p_0 + \rho(g \sin\theta)Y \tag{5.36}$$

We see that the pressure increases along the surface due to the component of gravity, $g \sin\theta$, in this direction. This pressure distribution is shown in Figure 5.28B. The hydrostatic force on the inclined surface may now be found by combining Eqs. 5.35a,

5.35b, and 5.36 to express Eq. 5.31 as

$$\mathbf{F}_S = \int_S -p\mathbf{n}\,dS = \iint -(p_0 + \rho(g\sin\theta)Y)(+\mathbf{K})\,dX\,dY \qquad (5.37)$$

with the limits of integration chosen to describe the surface. Notice that the force acts in the $-\mathbf{K}$ direction, which is normal to the surface as expected.

In most engineering applications we will want to express the hydrostatic force on a surface in terms of its components in the usual lower case Cartesian coordinate system shown in Figure 5.28C. We may obtain the horizontal and vertical components, F_{Sy} and F_{Sz}, of this force by simple geometry. The results are:

$$F_{Sy} = -|\mathbf{F}_S|\sin\theta \qquad \text{and} \qquad F_{Sz} = |\mathbf{F}_S|\cos\theta \qquad (5.38\text{a,b})$$

Note that the preceding results apply for a fluid on the right-hand side of the inclined surface shown, for any angle θ between 0 and π. The use of these formulas is illustrated in Example 5.18.

EXAMPLE 5.18

An inclined gate of length $L = 3$ m and width $W = 1.5$ m holds back water in an irrigation channel as shown in Figure 5.29. What is the force of the water on the gate when the channel is filled to height of $H = 1$ m? What is the net force on the gate? Express the forces in the x, y, z coordinate system.

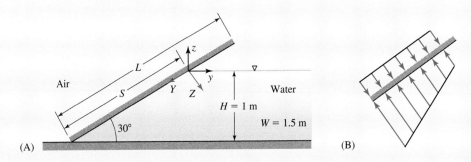

Figure 5.29 (A) Schematic and (B) pressure distribution for Example 5.18.

SOLUTION

The hydrostatic force on this inclined gate may be calculated by using Eq. 5.37, since the water is on the right side, and the geometry is otherwise the same. The coordinate origin is conveniently placed at the free surface where $p_0 = p_A$. From the geometry, the length of gate wetted by the water is $S = H/\sin\theta$, and the pressure distribution is

$p(Y) = p_A + \rho(g\sin\theta)Y$ as shown in Figure 5.29B. The force of the water on the gate is therefore given by

$$\mathbf{F}_S = \int_0^S \int_{-W/2}^{W/2} -(p_A + \rho(g\sin\theta)Y)(+\mathbf{K})\,dX\,dY$$

$$= W\int_0^S -(p_A + \rho(g\sin\theta)Y)(+\mathbf{K})\,dY$$

Completing the integration gives the force of the water on the gate as

$$\mathbf{F}_S = -WS\left[p_A + \frac{(\rho g\sin\theta)S}{2}\right]\mathbf{K}$$

The horizontal and vertical components of this force are found by using Eqs. 5.38a and 5.38b (or by inspection) as

$$F_{Sy} = -|\mathbf{F}_S|\sin\theta = -WS\left[p_A + \frac{(\rho g\sin\theta)S}{2}\right]\sin\theta$$

$$F_{Sz} = |\mathbf{F}_S|\cos\theta = WS\left[p_A + \frac{(\rho g\sin\theta)S}{2}\right]\cos\theta$$

The corresponding results for the net hydrostatic force, which includes the effect of the air behind the gate, are obtained easily by subtracting the atmospheric pressure contribution to obtain

$$\mathbf{F}_{\text{net}} = -WS\left[\frac{(\rho g\sin\theta)S}{2}\right]\mathbf{K}$$

$$F_{\text{net}\,y} = -|\mathbf{F}_{\text{net}}|\sin\theta = -WS\left[\frac{(\rho g\sin\theta)S}{2}\right]\sin\theta \quad\text{and}\quad F_{\text{net}\,z} = |\mathbf{F}_{\text{net}}|\cos\theta = WS\left[\frac{(\rho g\sin\theta)S}{2}\right]\cos\theta$$

For $\theta = 30°$, $S = H/\sin\theta = 2$ m. Also $W = 1.5$ m, and we have $p_A = 101{,}300$ N/m^2, and $\rho = 998$ kg/m^3. Substituting these data gives the magnitude of the force of the water as

$$|\mathbf{F}_S| = (1.5\text{ m})(2\text{ m})\left[101{,}300\text{ N/m}^2 + 998\text{ kg/m}^3\,(9.81\text{ m/s}^2)(\sin 30°)\left(\frac{2\text{ m}}{2}\right)\left(\frac{1\text{ N}}{1\text{ (kg-m)/s}^2}\right)\right]$$

$$= 318{,}586\text{ N}$$

The two components of this force are

$$F_{Sy} = -|\mathbf{F}_S|\sin\theta = -(318{,}586\text{ N})\sin 30° = -159{,}293\text{ N}$$

$$F_{Sz} = |\mathbf{F}_S|\cos\theta = (318{,}586\text{ N})\cos 30° = 275{,}904\text{ N}$$

The magnitude of the net force is easily found to be $|\mathbf{F}_{\text{net}}| = 14{,}686$ N, and the components of this force are $F_{\text{net}\,y} = -7343$ N and $F_{\text{net}\,z} = 12{,}718$ N.

A word of caution here. Equations 5.35 through 5.38 are valid only for the specific geometry shown in Figure 5.29 with the fluid on the right. If the fluid is on the left of the surface, or with an inclined surface whose geometry is different from that just discussed, you must develop new formulas by defining the X, Y, Z coordinates so that the surface is aligned, then applying Eq. 5.31, $\mathbf{F}_S = \int_S -p\mathbf{n}\, dS$, by assigning the unit normal and surface element dS. Next find the pressure distribution on the surface in the X, Y, Z coordinates, and evaluate the resulting surface integral. Don't forget to use the geometry to find new expressions for the components of the hydrostatic force in the usual lowercase Cartesian coordinate system. The general complexity of using this approach to evaluate the hydrostatic force on an inclined surface is further reason to consider the use of the indirect method discussed next.

There is a powerful indirect approach to finding the hydrostatic force on a planar nonaligned surface that uses the integral hydrostatic equation. The value of this approach is that it is applicable in every type of fluid statics problem, including variable density and nongravitational body forces, and is much easier than the direct method in many cases. To learn how to apply this method, consider the inclined planar surface S in contact with fluid at rest as shown in Figure 5.30A. Suppose we are asked to calculate the hydrostatic force on this surface due to the fluid on the right. We can find this force by applying the integral hydrostatic equation, choosing the volume of fluid V shown in Figure 5.30B. This volume is defined in part by a decal surface, S_D, just inside the fluid but otherwise identical to surface S as shown in Figure 5.30B. The volume is further defined by S_H, the horizontal projection of surface S, and S_V, the vertical projection of surface S. Two additional near and far vertical surfaces, S_N and S_F as shown, complete the boundary of this closed volume. Take note of the direction of the unit normals in Figure 5.30B.

The chosen volume of fluid is at rest, so the integral hydrostatic equation, Eq. 5.2, applies, giving $\int_S -p\mathbf{n}\, dS + \int \rho \mathbf{f}\, dV = 0$. This equation tells us that the hydrostatic force applied by the pressure field to the various surfaces of this volume is balanced by the total body force acting on the fluid in the volume. Writing the surface integral as the sum of the integrals on each piece of the surface, we obtain

$$\int_{S_D} -p\mathbf{n}\, dS + \int_{S_H} -p\mathbf{n}\, dS + \int_{S_V} -p\mathbf{n}\, dS + \int_{S_N} -p\mathbf{n}\, dS + \int_{S_F} -p\mathbf{n}\, dS + \int_V \rho \mathbf{f}\, dV = 0$$

(5.39)

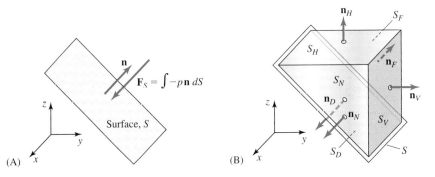

Figure 5.30 (A) Inclined planar surface. (B) Fluid volume for indirect approach to finding the hydrostatic force on a planar nonaligned surface.

5.5 HYDROSTATIC FORCE

Each of these surface integrals represents a surface force.

Now consider the first integral over the decal surface in Eq. 5.39. The pressure acting on the decal surface S_D is the same as that acting on the original surface of interest S. The surface element dS on the decal surface is the same as that for the original surface, but the outward unit normal on the two surfaces are opposite (see Figure 5.30B). Thus we may write the first integral in Eq. 5.39 as

$$\int_{S_D} -p\mathbf{n}\, dS = \int_S +p\mathbf{n}\, dS = -\mathbf{F}_S \tag{5.40}$$

where $\mathbf{F}_S$ is the desired force on the original surface. Using this to replace the first integral in Eq. 5.39 and rearranging, we obtain

$$\mathbf{F}_S = \int_{S_H} -p\mathbf{n}\, dS + \int_{S_V} -p\mathbf{n}\, dS + \int_{S_N} -p\mathbf{n}\, dS + \int_{S_F} -p\mathbf{n}\, dS + \int_V \rho \mathbf{f}\, dV \tag{5.41a}$$

This useful and completely general result shows that the hydrostatic force on a planar nonaligned surface may be found indirectly as the sum of surface integrals on the horizontal and three vertical projections of the surface, plus a volume integral. Each of the surface integrals refers to a planar aligned surface and represents the hydrostatic force of the fluid on that surface. The volume integral represents the vector weight $\mathbf{W}$ of the fluid in the volume. This "weight" is due to the action of all body forces, rather than just gravity. Thus we can write Eq. 5.41a in symbolic form as

$$\mathbf{F}_S = \mathbf{F}_H + \mathbf{F}_V + \mathbf{F}_N + \mathbf{F}_F + \mathbf{W} \tag{5.41b}$$

where each of the forces is defined by the corresponding integral in Eq. 5.41a.

The indirect method of calculating the surface force on an inclined surface consists of choosing an appropriate volume and applying Eq. 5.41a in the normal Cartesian coordinate system. The key steps in the method are as follows: (1) identify the decal surface, (2) define a volume within the fluid adjacent to the decal surface by selecting the horizontal and vertical projections of this surface, and (3) close out the volume with additional horizontal or vertical surfaces as needed.

In most engineering applications of fluid statics, namely, in problems involving constant density fluid in Earth's gravity field, the pressure varies only in the vertical direction. Thus, with reference to Figure 5.30B, the forces applied by the fluid on the near and far vertical surfaces cancel in this case because they are equal and opposite. The equation to be employed in applying the indirect method in problems of a constant density fluid in Earth's gravity field is

$$\mathbf{F}_S = \int_{S_H} -p\mathbf{n}\, dS + \int_{S_V} -p\mathbf{n}\, dS - \rho g \mathcal{V} \mathbf{k} \tag{5.42a}$$

where the z axis is up as usual. This may also be written symbolically as

$$\mathbf{F}_S = \mathbf{F}_H + \mathbf{F}_V - W\mathbf{k} \tag{5.42b}$$

where the weight of the fluid in the volume is $W = \rho g \mathcal{V}$.

EXAMPLE 5.19

Apply the indirect method based on the integral hydrostatic equation to calculate the force of the water on the inclined gate in Example 5.18.

SOLUTION

We are to determine the force of the water on the inclined gate in Example 5.18. Figure 5.31A is the appropriate sketch. The pressure in the water can be written as $p(z) = p_A - \rho g(z - H)$. This problem is solved by using Eq. 5.42a and selecting a volume adjacent to the surface as shown in Figure 5.31B. The force on the horizontal surface can be found by using Eq. 5.34b: $\mathbf{F}_S = \iint -p(x, y, z_S)(-\mathbf{k})\,dx\,dy$. This surface is exposed to a constant pressure of magnitude $p_A + \rho g H$, so we find $\mathbf{F}_H = (p_A + \rho g H)A\mathbf{k}$. From the geometry, the area of this surface is $A = WS\cos\theta$, thus $\mathbf{F}_H = WS\cos\theta(p_A + \rho g H)\mathbf{k}$. The force on the vertical surface is found using Eq. 5.33a, $\mathbf{F}_S = \iint -p(x, y_S, z)(\mathbf{j})\,dx\,dz$. The pressure on this surface is $p(z) = p_A - \rho g(z - H)$, and we find

$$\mathbf{F}_V = \int_0^H \int_{-W/2}^{W/2} -[p_A - \rho g(z - H)](\mathbf{j})\,dx\,dz = -WH\left(p_A + \frac{\rho g H}{2}\right)\mathbf{j}$$

The remaining term in Eq. 5.42a is the vector weight of the fluid in the volume, $-\rho g \mathcal{V}\mathbf{k}$. In this case, from the geometry we find $-\rho g \mathcal{V}\mathbf{k} = -W\mathbf{k} = -\rho g(\tfrac{1}{2}HWS\cos\theta)\mathbf{k}$. Adding these three terms together we have

$$\mathbf{F}_S = \mathbf{F}_H + \mathbf{F}_V - W\mathbf{k} = WS\cos\theta(p_A + \rho g H)\mathbf{k} - WH\left[p_A + \frac{\rho g H}{2}\right]\mathbf{j} - \rho g\left(\frac{1}{2}HWS\cos\theta\right)\mathbf{k}$$

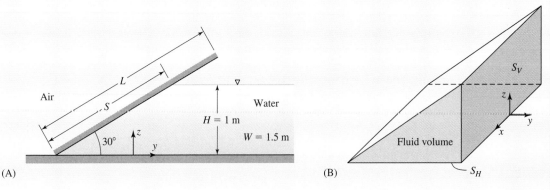

Figure 5.31 (A) Schematic and (B) fluid volume for indirect approach to finding the hydrostatic force on the gate for Example 5.19.

Simplifying this, we find

$$\mathbf{F}_S = -WH\left[p_A + \frac{\rho g H}{2}\right]\mathbf{j} + WS\left[p_A + \frac{\rho g H}{2}\right]\cos\theta\,\mathbf{k}$$

Since $H = S\sin\theta$, we can also write this as

$$\mathbf{F}_S = -WS\left[p_A + \frac{(\rho g \sin\theta)S}{2}\right]\sin\theta\,\mathbf{j} + WS\left[p_A + \frac{(\rho g \sin\theta)S}{2}\right]\cos\theta\,\mathbf{k}$$

The two components of this force are

$$F_{S_y} = -WS\left[p_A + \frac{(\rho g \sin\theta)S}{2}\right]\sin\theta \quad \text{and} \quad F_{S_z} = WS\left[p_A + \frac{(\rho g \sin\theta)S}{2}\right]\cos\theta$$

Comparing these results with those obtained in Example 5.18, we see that they are identical.

Many engineering applications in hydrostatics involve determining the forces acting on an object at rest in contact with one or more stationary fluids, usually air and water. The normal procedure in analyzing this type of problem is to write a force balance on the object, which includes body and surface forces plus any external or structural force that might be present. Example 5.20 illustrates how a calculation of the hydrostatic force enters this type of problem.

EXAMPLE 5.20

Consider the stationary barge shown in Figure 5.32. Write a force balance on the barge. What is the force applied by the water on the barge? What is the net force on the barge due to all fluids? What is the weight of the barge and its contents?

SOLUTION

We define the barge and its contents as the object of our force balance. Since barge, air, and water are at rest, the sum of the forces on the barge is zero:

$$\mathbf{F}_{body} + \mathbf{F}_{air} + \mathbf{F}_{water} + \mathbf{F}_{ext} = 0$$

Here $\mathbf{F}_{body}$ is the force due to all body forces acting on the barge and its contents, $\mathbf{F}_{air}$ is the hydrostatic force of the air, $\mathbf{F}_{water}$ is the hydrostatic force of the water, and $\mathbf{F}_{ext}$ is any external force on the object. In this case there is no external force so the force balance is

$$\mathbf{F}_{body} + \mathbf{F}_{air} + \mathbf{F}_{water} = 0 \tag{A}$$

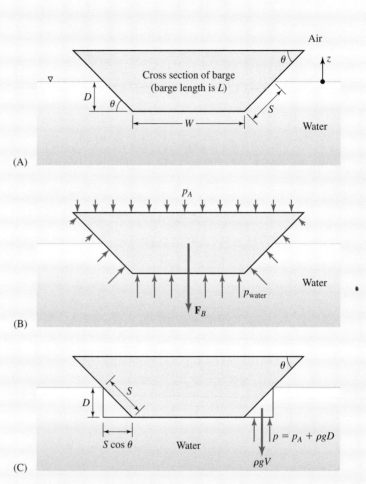

Figure 5.32 (A) Schematic, (B) pressure distribution, and (C) fluid volumes used for indirect approach for Example 5.20.

The forces of the air and water are applied to the external surface of the barge through the pressure on these surfaces, as shown in Figure 5.32B. The only body force acting here is gravity, so we can write $\mathbf{F}_{body} = -W_{barge}\mathbf{k}$, where W_{barge} is the weight of the barge and its contents. The force of the air and water on the barge may be calculated by applying Eq. 5.31, considering each fluid and the surfaces it contacts separately. To find the force of the water on the barge, we first note that this force is exerted on the bottom of the barge and on the wetted parts of the two sides. By inspection, the force of the water on the bottom of the barge of length L into the paper, is

$$\mathbf{F}_{water/bottom} = WL(p_A + \rho g D)\mathbf{k} \qquad (B)$$

We will use the indirect method on the sides of the barge because these are planar, non-aligned surfaces, choosing the two volumes shown in Figure 5.32C. We will apply

Eq. 5.42a to each volume: $\mathbf{F}_S = \int_{S_H} -p\mathbf{n}\,dS + \int_{S_V} -p\mathbf{n}\,dS - \rho g \forall \mathbf{k}$. Thus the force of the water on the two sides is

$$\mathbf{F}_{\text{water/sides}} = \left(\int_{S_H} -p\mathbf{n}\,dS + \int_{S_V} -p\mathbf{n}\,dS - \rho g \forall \mathbf{k}\right)_{\text{left}} + \left(\int_{S_H} -p\mathbf{n}\,dS + \int_{S_V} -p\mathbf{n}\,dS - \rho g \forall \mathbf{k}\right)_{\text{right}}$$

From the symmetry (see Figure 5.32C), we see that the two integrals over the vertical surfaces cancel, the two integrals over the horizontal surfaces are the same, and the weight terms are the same. Thus we have $\mathbf{F}_{\text{water/sides}} = 2[(\int_{S_H} -p\mathbf{n}\,dS) - \rho g \forall \mathbf{k}]$, and we can use either volume to evaluate these terms. By inspection, the force of the water on the horizontal surface of one the selected volumes is $\int_{S_H} -p\mathbf{n}\,dS = SL\cos\theta[p_A + \rho g D]\mathbf{k}$, where we have made use of the fact that the width of the horizontal side is $S\cos\theta$ and its length is L. The vector weight of the fluid in either volume is $-\rho g \forall \mathbf{k} = -\rho g(\frac{1}{2}SD\cos\theta)L\,\mathbf{k} = -\frac{1}{2}\rho g SDL\cos\theta\,\mathbf{k}$. Thus we have

$$\mathbf{F}_{\text{water/sides}} = 2\left(\int_{S_H} -p\mathbf{n}\,dS - \rho g \forall \mathbf{k}\right) = 2SL\cos\theta[p_A + \rho g D]\mathbf{k} - \rho g SDL\cos\theta\,\mathbf{k}$$

$$\mathbf{F}_{\text{water/sides}} = 2SL\cos\theta\left[p_A + \frac{\rho g D}{2}\right]\mathbf{k} \tag{C}$$

The force of the water on the barge is therefore the sum of (B) and (C), or

$$\mathbf{F}_{\text{water}} = WL[p_A + \rho g D]\mathbf{k} + 2SL\cos\theta\left(p_A + \frac{\rho g D}{2}\right)\mathbf{k} \tag{D}$$

The net force of all fluids on the barge is the sum of the force applied by the water and the force applied by the air. The force applied by the air is given by $\mathbf{F}_{\text{air}} = \int_{S_{\text{air}}} -p_A \mathbf{n}\,dS$, where the integral is taken over the external barge surface above the waterline. Rather than go through another long process to calculate the force of the air on the barge, we can employ a trick. Note that if atmospheric pressure acted on all sides of the barge, the net force applied by the air would be zero. We can write this as $\int_{S_{\text{air}}} -p_A \mathbf{n}\,dS + \int_{S_{\text{water}}} -p_A \mathbf{n}\,dS = 0$, where the first integral refers to surfaces wetted by air and the second to surfaces actually wetted by water. Rearranging, we have $\int_{S_{\text{air}}} -p_A \mathbf{n}\,dS = -\int_{S_{\text{water}}} -p_A \mathbf{n}\,dS$. Thus we have proven that the force of the air is equal and opposite to the effect of atmospheric pressure on the surfaces wetted by water. When we calculated the force due to water earlier, we found the necessary contributions of atmospheric pressure on the surfaces wetted by water. From (D) we have

$$\int_{S_{\text{water}}} -p_A \mathbf{n}\,dS = WL p_A \mathbf{k} + 2SL\cos\theta\, p_A \mathbf{k}$$

Thus the force of the air is

$$\int_{S_{\text{air}}} -p_A \mathbf{n}\,dS = -\int_{S_{\text{water}}} -p_A \mathbf{n}\,dS = -WL\, p_A \mathbf{k} - 2SL\cos\theta\, p_A \mathbf{k} \tag{E}$$

The net fluid force is then the sum of the force of the water, (D), and the force of the air, (E), or

$$\mathbf{F}_{net} = WL\rho g D\mathbf{k} + 2SL\cos\theta\left(\frac{\rho g D}{2}\right)\mathbf{k} = \rho g[WLD + (S\cos\theta)LD]\mathbf{k} \quad (F)$$

Note that the net force acts upward as expected and can also be obtained by using gage pressure to calculate the force of the water. If you examine (F), which defines the net force, you will notice that the magnitude of this force is equal to the weight of the water displaced by the barge. We will discuss this result in more detail later in this chapter (see Section 5.8, Buoyancy and Archimedes' Principle).

Using the force balance (A), $\mathbf{F}_{body} + \mathbf{F}_{air} + \mathbf{F}_{water} = 0$, we can find the weight of the barge and its contents from $\mathbf{F}_{body} = -W_{barge}\mathbf{k} = -\mathbf{F}_{air} - \mathbf{F}_{water} = -\mathbf{F}_{net}$. By using (F) we find

$$-W_{barge}\mathbf{k} = -\mathbf{F}_{net} = -\rho g[WLD + (S\cos\theta)LD]\mathbf{k} \quad (G)$$

5.5.3 Curved Surface

The hydrostatic force on a curved surface in contact with fluid at rest is also given by Eq. 5.31, $\mathbf{F}_S = \int_S -p\mathbf{n}\,dS$. Calculating this integral for a curved surface can be challenging, however, because it is necessary to represent the surface element and unit normal on the curved surface. It is usually easier to apply the indirect approach based on the integral hydrostatic equation. In the preceding section we showed how to apply the indirect method to a planar nonaligned surface. In this section we demonstrate how one can apply the indirect method to a curved surface.

Suppose you are asked to calculate the hydrostatic force on the curved surface S shown in Figure 5.33. The idea of the indirect approach is to replace the difficult

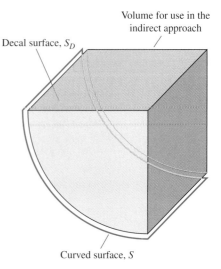

Figure 5.33 Fluid volume used for indirect calculation of the hydrostatic force on a curved surface.

integral over the curved surface with an integral over a decal surface S_D just inside the fluid. We then judiciously choose a fluid volume bounded by the decal surface and as many other planar and aligned surfaces as needed to define a closed volume, and apply Eq. 5.41a. If the problem involves constant density fluid in Earth's gravity field, we use Eq. 5.42a. This process is illustrated in Example 5.21.

EXAMPLE 5.21

The gate illustrated in Figure 5.34 is in the shape of a quarter-cylinder with a radius of 3 ft and a width of 10 ft. Develop a formula for the force exerted by the water on the gate and the net force applied by all fluids on the gate. Find the net force.

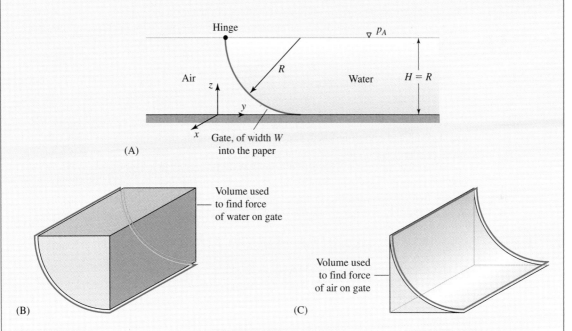

Figure 5.34 (A) Schematic, (B) fluid volume used for indirect calculation of the hydrostatic force due to the water on the gate, and (C) volume used for indirect calculation of the force due to the air on the gate for Example 5.21.

SOLUTION

We are to determine the force exerted by the water and the net force applied by all fluids on a cylindrical gate with $W = 10$ ft and $R = 3$ ft. Figure 5.34 is the appropriate sketch. We will apply Eq. 5.42a, $\mathbf{F}_S = \int_{S_H} -p\mathbf{n}\,dS + \int_{S_V} -p\mathbf{n}\,dS - \rho g \forall \mathbf{k}$, to solve this problem, and analyze the volume of water shown in Figure 5.34B. The horizontal surface is exposed to atmospheric pressure, and the area of this surface is RW, thus by inspection $\mathbf{F}_H = -p_A RW\mathbf{k}$. The vertical surface is aligned with the (x, z) plane, so we use the method leading up to Eq. 5.33a to evaluate the surface integral with $p(z) = p_A - \rho g(z - R)$, yielding

$$\mathbf{F}_V = \int_{S_V} -p\mathbf{n}\, dS = \int_0^R \int_{-W/2}^{W/2} -[p_A - \rho g(z - R)](\mathbf{j})\,dx\,dz = -WR\left[p_A + \frac{\rho g R}{2}\right]\mathbf{j}$$

The vector weight of the fluid in the volume is found by inspection as $-\rho g \mathcal{V}\mathbf{k} = -\rho g(\frac{1}{4}\pi R^2 W)\mathbf{k}$. Thus the force of the water on the curved surface is

$$\mathbf{F}_{\text{water}} = -p_A RW\mathbf{k} - WR\left(p_A + \frac{\rho g R}{2}\right)\mathbf{j} - \rho g\left(\frac{\pi R^2 W}{4}\right)\mathbf{k} \tag{A}$$

The net force on this surface includes the effect of the air on the back of the gate. We can get the net force by removing the atmospheric pressure terms from the force of the water to obtain

$$\mathbf{F}_{\text{net}} = -WR\left(\frac{\rho g R}{2}\right)\mathbf{j} - \rho g\left(\frac{\pi R^2 W}{4}\right)\mathbf{k} \tag{B}$$

We can check this result by computing the force of the air on the gate by using the indirect method and applying Eq. 5.42a to the volume of air shown in Figure 5.34C. By inspection, the force on the horizontal surface is $\mathbf{F}_H = -p_A RW(-\mathbf{k}) = p_A RW\mathbf{k}$, while on the vertical surface we get $\mathbf{F}_V = -p_A RW(-\mathbf{j}) = p_A RW\mathbf{j}$. The weight of air in this volume is easily calculated, but to be consistent with the constant atmospheric pressure model (which assumes air to have zero density), this tiny contribution must be neglected. The force of the air on the gate is therefore calculated as

$$\mathbf{F}_{\text{air}} = p_A RW\mathbf{k} + p_A WR\mathbf{j} \tag{C}$$

It can be seen that adding (C) to (A) does produce the net force (B).

The net force on this gate can be calculated from (B) by using $\rho = 62.4$ lb$_m$/ft^3, $p_A = 2116$ lb$_f$/ft^2, $W = 10$ ft, and $R = 3$ ft. Applying (B), we have

$$\mathbf{F}_{\text{net}} = -WR\left(\frac{\rho g R}{2}\right)\mathbf{j} - \rho g\left(\frac{\pi R^2 W}{4}\right)\mathbf{k}$$

$$= -(10\,\text{ft})(3\,\text{ft})\left[\frac{1}{2}(62.4\,\text{lb}_m/\text{ft}^3)(32.2\,\text{ft/s}^2)(3\,\text{ft})\right]\left(\frac{1\,\text{lb}_f\text{-s}^2}{32.2\,\text{ft-lb}_m}\right)\mathbf{j}$$

$$- (62.4\,\text{lb}_m/\text{ft}^3)(32.2\,\text{ft/s}^2)\left[\frac{\pi}{4}(3\,\text{ft})^2(10\,\text{ft})\right]\left(\frac{1\,\text{lb}_f\text{-s}^2}{32.2\,\text{ft-lb}_m}\right)\mathbf{k}$$

$$\mathbf{F}_{\text{net}} = -2808\,\text{lb}_f\,\mathbf{j} - 4410\,\text{lb}_f\,\mathbf{k}$$

We see that the net force acts down and to the left, as expected.

EXAMPLE 5.22

A thin-walled pipe is tested hydrostatically by pressurizing it with water as shown in Figure 5.35. Water is used in this application because a failure of the pipe wall results in a leak rather than the explosion that occurs with a pressurized gas. Calculate the tensile stress in the pipe wall in the axial and circumferential direction. Assume the pressure in the water is uniform.

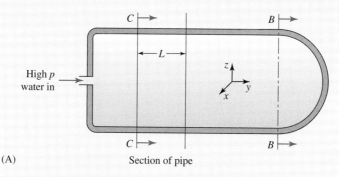

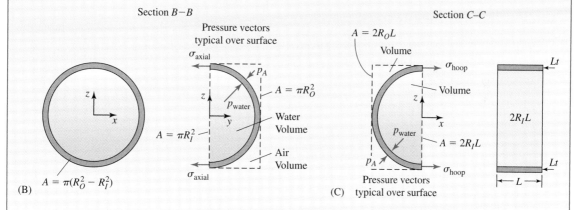

Figure 5.35 (A) Schematic, (B) section B–B of end cap, and (C) section C–C showing length of pipe L for Example 5.22.

SOLUTION

Consider a force balance in the horizontal y direction on the section of pipe with hemispherical cap shown in Figure 5.35B. We will neglect the effect of the weight of the pipe on the axial stress in the pipe wall. The force of the pressurized water acting on the inside of the cap, plus the force of the air on the outside, must be balanced by the tensile force developed in the pipe wall. By using the indirect volumes shown, and neglecting the weight of the water and air in the volumes, we find that the net force of the two fluids is $\mathbf{F}_{\text{net}} = [p_W(\pi R_I^2) - p_A(\pi R_O^2)]\mathbf{j}$, where p_W is the water pressure inside the pipe and R_I and R_O are the inner and outer radii of the pipe. The tensile force in the pipe wall is $\mathbf{F}_{\text{tensile}} = -\sigma_{\text{axial}}[\pi(R_O^2 - R_I^2)]\mathbf{j}$, where σ_{axial} is the average axial stress in the wall. Thus we have

$$[p_W(\pi R_I^2) - p_A(\pi R_O^2)]\mathbf{j} - \sigma_{\text{axial}}[\pi(R_O^2 - R_I^2)]\mathbf{j} = 0$$

Solving for the axial stress we find

$$\sigma_{\text{axial}} = p_W\left(\frac{R_I^2}{R_O^2 - R_I^2}\right) - p_A\left(\frac{R_O^2}{R_O^2 - R_I^2}\right) \tag{A}$$

A similar force balance in the x direction, found by means of the indirect method and the volumes shown in Figure 5.35C for the section of pipe of length L, gives

$$[p_W(2R_IL) - p_A(2R_OL)](-\mathbf{i}) + \sigma_{\text{hoop}}[2Lt]\mathbf{i} = 0$$

where we have again neglected the weight of the water and air in the volumes, and σ_{hoop} is the average circumferential or hoop stress in the wall. Since $t = R_O - R_I$, we have $R_O = R_I + t$. Solving for the hoop stress in this case gives

$$\sigma_{\text{hoop}} = \frac{p_W R_I}{t} - \frac{p_A R_O}{t} \tag{B}$$

These results are valid for any pipe wall thickness. We can simplify the results further by noting that $R_O^2 = R_I^2 + 2R_It + t^2$, and $R_O^2 - R_I^2 = 2R_It + t^2$. For a thin-walled pipe with $t/R_I \ll 1$, we get the following approximations: $R_O \approx R_I$, $R_O^2 = R_I^2(1 + 2t/R_I + t^2/R_I^2) \approx R_I^2$ and $R_O^2 - R_I^2 = 2R_It(1 + t/2R_I) \approx 2R_It$. Thus we can write $R_I^2/(R_O^2 - R_I^2) \approx R_I^2/2R_It = R_I/2t$ and $R_O^2/(R_O^2 - R_I^2) \approx R_I^2/2R_It = R_I/2t$. The preceding expressions for the stress then become

$$\sigma_{\text{axial}} = p_W\left(\frac{R_I^2}{R_O^2 - R_I^2}\right) - p_A\left(\frac{R_O^2}{R_O^2 - R_I^2}\right) \approx (p_W - p_A)\frac{R_I}{2t} \tag{C}$$

$$\sigma_{\text{hoop}} = \frac{p_W R_I}{t} - \frac{p_A R_O}{t} \approx (p_W - p_A)\frac{R_I}{t} \tag{D}$$

We see that the average hoop stress is twice the average axial stress.

5.6 HYDROSTATIC MOMENT

As you have seen in the preceding examples, the hydrostatic pressure acting on a surface creates a significant surface force. This hydrostatic force must be considered in the design of a structure exposed to fluid at rest. The hydrostatic moment created by the hydrostatic pressure is also important in practical applications. For example, consider the problem of tethering a storage tank to the floor of the ocean as shown in Figure 5.36.

5.6 HYDROSTATIC MOMENT

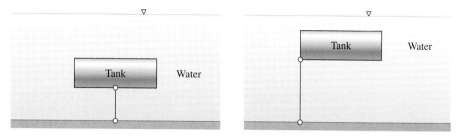

Figure 5.36 Two possible attachment points for a cylindrical tank of oil submerged in water, tethered.

Suppose the cylindrical tank is full of oil but light enough to float to the surface if not tethered. Which of the two proposed points of attachment would you recommend to keep the tank horizontal? Will an attempt to tether the tank at the end result in the tank orienting itself vertically rather than in the desired horizontal position? How do we determine the proper way to restrain a structure immersed in a stationary fluid in a more complicated situation where our intuition fails?

Body and surface forces, as well as an external force like that produced by the tether on the tank in Figure 5.36, produce moments that must be considered in analyzing the behavior of a stationary object immersed in fluid at rest. To calculate the moment due to a body force, consider the arbitrary volume of fluid (or other material) shown in Figure 5.37A. To compute the total body moment $\mathbf{M}_B$ acting on this volume, we must sum the contributions from each infinitesimal volume element dV. The body moment on any

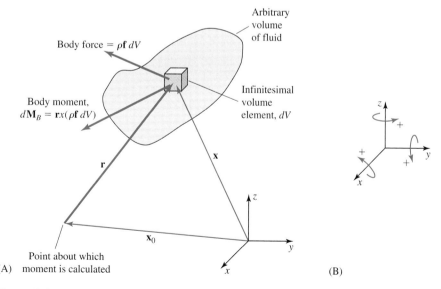

Figure 5.37 (A) Fluid volume experiencing a body moment. (B) The sign convention for moments.

volume element is given by the cross product of the moment arm $\mathbf{r}$ and the body force $(\rho \mathbf{f}\, dV)$:

$$d\mathbf{M}_B = \mathbf{r} \times (\rho \mathbf{f}\, dV)$$

where ρ is the density of the material. To calculate the total body moment, we use a volume integral to sum all the contributions from throughout the volume:

$$\mathbf{M}_B = \int \mathbf{r} \times (\rho \mathbf{f}\, dV)$$

Since $\mathbf{r} = \mathbf{x} - \mathbf{x}_0$, we can write this integral for a moment about the point $\mathbf{x}_0$ as

$$\mathbf{M}_B = \int (\mathbf{x} - \mathbf{x}_0) \times (\rho \mathbf{f}\, dV)$$

We normally calculate the moment in Cartesian coordinates. The components of the moment arm are given by

$$r_x = x - x_0, \quad r_y = y - y_0, \quad \text{and} \quad r_z = z - z_0 \tag{5.43}$$

where the point about which the moment is to be taken is $\mathbf{x}_0 = (x_0, y_0, z_0)$. Evaluating the cross product and writing the integral in terms of the three Cartesian components, we have

$$M_{B_x} = \int [(y - y_0)\rho f_z - (z - z_0)\rho f_y]\, dV \tag{5.44a}$$

$$M_{B_y} = \int [(z - z_0)\rho f_x - (x - x_0)\rho f_z]\, dV \tag{5.44b}$$

$$M_{B_z} = \int [(x - x_0)\rho f_y - (y - y_0)\rho f_x]\, dV \tag{5.44c}$$

The convention regarding the components of a moment vector is shown in Figure 5.37B. For example, a positive value for the $\mathbf{i}$ component of a moment tends to cause a counterclockwise rotation about the $\mathbf{i}$ axis when seen from the end of the axis looking in toward the origin. This convention applies to all moments in this book.

Several of the terms in Eqs. 5.44a–5.44c disappear for the important case of gravity, and often the moments can be determined by inspection. This is illustrated in Example 5.23.

EXAMPLE 5.23

The rectangular tank shown in Figure 5.38 is full of wet concrete with a density of 1900 kg/m³. The tank dimensions are $w = 1.5$ m, $d = 1$ m, and $L = 2$ m. Find the total body moment due to gravity acting on the concrete about the indicated moment origin location.

SOLUTION

We will apply Eqs. 5.44a–5.44c, noting that the body force due to gravity is $\mathbf{f} = (0, 0, -g)$. Thus, the only terms remaining in these equations are those containing

5.6 HYDROSTATIC MOMENT 255

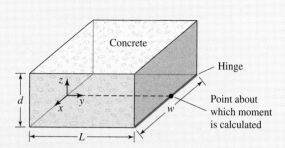

Figure 5.38 Schematic for Example 5.23.

the z component of the body force. The only component of interest of the moment in Figure 5.38 is the **i** component, since the symmetry ensures that the other two components are zero. The **i** component represents the tendency for tipping about the indicated moment origin located at $x_o = 0$, $y_o = L$, $z_o = 0$. Thus, we use Eq. 5.44a to write

$$M_{B_x} = \int [(y - y_0)\rho f_z - (z - z_0)\rho f_y] dV = \int [-\rho g(y - y_0)] dV = \int_{-w/2}^{w/2} \int_0^L \int_0^d [-\rho g(y - L)] dx\, dy\, dz$$

The integration yields $M_{B_x} = -\rho g w\, dL(L/2 - L) = \rho g(wdL)(L/2)$, and inserting the data gives

$$M_{B_x} = 1900 \text{ kg/m}^3 (9.81 \text{ m/s}^2)(1.5 \text{ m})(1 \text{ m})(2 \text{ m})(1 \text{ m}) = 5.59 \times 10^4 \text{ N-m}$$

We see that the moment is positive and equal to the product of the weight of the wet concrete times an effective moment arm of half the length of the tank. This is correct, since the weight of the concrete can be thought of as concentrated at the center of gravity.

As shown in Figure 5.39, the moment due to hydrostatic pressure, or hydrostatic moment, may be calculated by recognizing that the moment created about a point $\mathbf{x}_0$ by the surface force acting on an infinitesimal surface element is the cross product of the moment arm, $\mathbf{r} = \mathbf{x} - \mathbf{x}_0$, and the force $\mathbf{\Sigma}\, dS$ acting on this surface element. Thus the surface moment on this element is given by

$$d\mathbf{M}_S = \mathbf{r} \times \mathbf{\Sigma}\, dS$$

The moment applied by the fluid to the entire surface is found by summing the individual contributions by means of a surface integral. Thus the total surface moment $\mathbf{M}_S$ is given by

$$\mathbf{M}_S = \int_S [(\mathbf{x} - \mathbf{x}_0) \times \mathbf{\Sigma}]\, dS$$

We can write this for a fluid at rest as

$$\mathbf{M}_S = \int_S [\mathbf{r} \times (-p\mathbf{n})]\, dS$$

In Cartesian coordinates, the three components of the hydrostatic moment are given by

$$M_{S_x} = \int_S [r_y(-pn_z) - r_z(-pn_y)]\, dS \qquad (5.45a)$$

5 FLUID STATICS

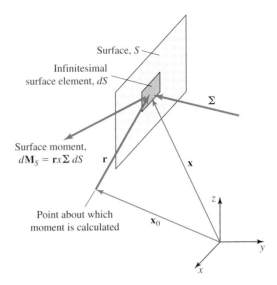

Figure 5.39 Hydrostatic moment about a point due to the surface force.

$$M_{S_y} = \int_S [r_z(-pn_x) - r_x(-pn_z)]\,dS \qquad (5.45b)$$

$$M_{S_z} = \int_S [r_x(-pn_y) - r_y(-pn_x)]\,dS \qquad (5.45c)$$

with $r_x = x - x_0, r_y = y - y_0, r_z = z - z_0$, and the usual sign convention.

The calculation of the hydrostatic moment by means of a surface integral involves the same mathematical procedures discussed earlier for the hydrostatic force. Setting up the integrals and calculating the hydrostatic moment on the two types of planar surface (aligned and nonaligned) will be discussed in the next two sections. We do not discuss the calculation of these integrals on curved surfaces in this textbook.

5.6.1 Planar Aligned Surface

We can compute the hydrostatic moment about a point x_0 for the planar aligned surface shown in Figure 5.40A by using Eqs. 5.45a–5.45c, since these equations are valid for any surface. The surface shown is parallel to the yz plane, normal to the x axis, and it has fluid on the positive x side. We can express the unit normal, pressure, and surface element for the surface as $\mathbf{n} = \mathbf{i} = (1, 0, 0)$, $p = p(x_S, y, z)$, and $dS = dy\,dz$, where x_S is the location of the surface along the x axis. Note that as usual, the normal points at the fluid. Since $n_y = n_z = 0$, many of the terms in the integrands of Eqs. 5.45a–5.45c are identically zero. Upon substituting the unit normal, pressure, and surface element into Eqs. 5.45a–5.45c, we see that the three components of the hydrostatic moment on any planar surface parallel to the yz plane are given by

$$M_{S_x} = 0 \qquad (5.46a)$$

5.6 HYDROSTATIC MOMENT 257

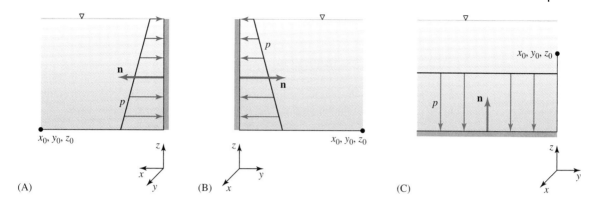

Figure 5.40 Hydrostatic moment on an aligned planar surface in (A) the yz plane, (B) the xz plane, and (C) the xy plane.

$$M_{S_y} = \int_S -(z - z_0) p(x_S, y, z)\, dy\, dz \qquad (5.46b)$$

$$M_{S_z} = \int_S (y - y_0) p(x_S, y, z)\, dy\, dz \qquad (5.46c)$$

where the limits are selected to cover the entire surface.

The first equation confirms our intuition that pressure acting normal to the surface shown cannot create a moment tending to twist the surface about the x axis. The remaining two equations give the values of the y and z components of the moment created by the hydrostatic pressure. Note that these equations apply only to the $+x$ side of the surface shown in Figure 5.40A.

To calculate the moment on the $-x$ side of this surface, we must account for the fact that the unit normal on the back side is pointing in the opposite direction. The resulting equations are

$$M_{S_x} = 0 \qquad (5.46e)$$

$$M_{S_y} = \int_S (z - z_0)(p(x_S, y, z))\, dy\, dz \qquad (5.46f)$$

$$M_{S_z} = \int_S -(y - y_0)(p(x_S, y, z))\, dy\, dz \qquad (5.46g)$$

In a typical engineering application involving constant density fluid in Earth's gravity field, with air at atmospheric pressure on the far side of the surface, the net hydrostatic force and net hydrostatic moment are of interest. These are calculated by using gage pressure in the preceding equations or by subtracting the atmospheric pressure contribution obtained in considering the effect of the liquid alone. (See example 5.24).

To continue our discussion of how to calculate the hydrostatic moment on a planar aligned surface, consider next a surface parallel to the xz plane, normal to the y axis,

EXAMPLE 5.24

A proposed design for a container for transporting wet concrete is shown in Figure 5.41A. Find the net hydrostatic force acting on the gate and the net hydrostatic moment on the gate about the hinge pin. The density of wet concrete is 2400 kg/m³.

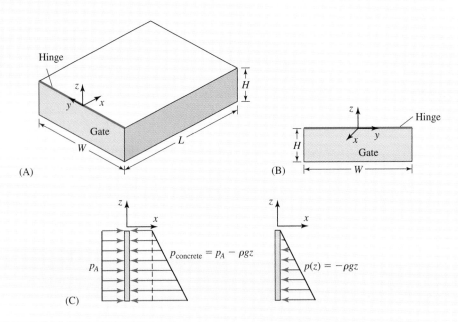

Figure 5.41 Schematics for Example 5.24: (A) geometry, (B) detail of gate and (C) pressure distribution on two sides of gate, gage pressure distribution on side.

SOLUTION

We are to determine the net hydrostatic force and moment acting on the gate of a concrete container. Before proceeding, note that it simplifies the calculation to place the coordinate origin at x_0 whenever possible. Figures 5.41B and 5.41C will serve as the schematics. The container has dimensions $H = 1$ m, $W = 2$ m, and $L = 2$ m and is open at the top. The gate is aligned with the yz plane, with the fluid (wet concrete) on the $+x$ side. We will use Eqs. 5.32a and 5-46a–c for the force and moment and choose the coordinate origin as shown in the figure. With gage pressure to account for the air behind the gate, the pressure distribution in the concrete and on this surface can be written as $p(z) = -\rho g z$. The normal is $\mathbf{n} = \mathbf{i}$. From Eq. 5.32a, the net hydrostatic force on the gate is

$$\mathbf{F}_S = \int_{-H}^{0} \int_{-W/2}^{W/2} (\rho g z)(\mathbf{i}) \, dy \, dz = -WH \frac{\rho g H}{2} \mathbf{i} \qquad \text{(A)}$$

The force acts in the $-x$ direction as expected and has the form of an average gage pressure times the area of the gate. Note the limit on the z integration in this case. As convention demands, the limit is from the most negative value of z on the surface to most positive value.

To compute the moment on the gate we apply Eqs. 5.46a–5.46c, noting that $M_{S_x} = 0$. In selecting the origin of the coordinates, we recognized that the component M_{S_y} of the moment is responsible for rotation about the axis of the hinge pin. By selecting the coordinate origin and point about which the moment is to be computed on the hinge pin and at its center, we ensure that the symmetry works to our advantage and causes the z component of the moment to be zero. Thus we have $\mathbf{x}_0 = (0, 0, 0)$, and the only component we need to evaluate here is given by Eq. 5.46b:

$$M_{S_y} = \int_S -(z - z_0) p(x_S, y, z)\, dy\, dz$$

Setting up the integral we have

$$M_{S_y} = \int_{-H}^{0} \int_{-W/2}^{W/2} (-z)(-\rho g z)\, dy\, dz = \rho g W \frac{H^3}{3}$$

Thus the net hydrostatic moment is

$$\mathbf{M}_S = M_{S_y} \mathbf{j} = \rho g W \frac{H^3}{3} \mathbf{j} \tag{B}$$

These answers seem reasonable. As you will recall from our discussion of the sign convention for a moment, a positive value for the $\mathbf{j}$ component of a moment tends to cause a counterclockwise rotation about the $\mathbf{j}$ axis when seen from the end of the axis looking in toward the origin. When looking from the right toward the origin along the y axis as shown in Figure 5.41B, the hydrostatic pressure tends to rotate the gate in a counterclockwise direction around the hinge pin. Also note that the magnitude of the moment can be written as $M_{S_y} = \rho g W (H^3/3) = \rho g W (H^2/2)(2H/3)$, which is in the form of the product of the magnitude of the force due to the linear gage pressure distribution shown in Figure 5.41C times a moment arm from a point of application of this force, which is two-thirds of the way down the gate.

To complete the problem we insert the data [$\rho = 2400$ kg/m^3 for concrete, $H = 1$ m, $W = 2$ m] into (A) and (B) to yield

$$\mathbf{F}_S = -\rho g W \frac{H^2}{2} \mathbf{i} = -(2400 \text{ kg/m}^3)(9.81 \text{ m/s}^2)(2 \text{ m}) \frac{(1 \text{ m})^2}{2} \mathbf{i} = -23.5 \text{ kN } \mathbf{i}$$

$$\mathbf{M}_S = M_{S_y} \mathbf{j} = \rho g W \frac{H^3}{3} \mathbf{j} = (2400 \text{ kg/m}^3)(9.81 \text{ m/s}^2)(2 \text{ m}) \frac{(1 \text{ m})^3}{3} \mathbf{j} = (15.7 \text{ kN-m}) \mathbf{j}$$

with fluid on the $+y$ side, as shown in Figure 5.40B. For this surface we have $\mathbf{n} = \mathbf{j} = (0, 1, 0)$, $p = p(x, y_S, z)$, and $dS = dx\,dz$. Since the only component of the unit normal is $n_y = 1$, the components of the hydrostatic moment on any planar surface parallel to the xz plane with fluid on the $+y$ side are

$$M_{S_x} = \int_S (z - z_0) p(x, y_S, z)\,dx\,dz \tag{5.47a}$$

$$M_{S_y} = 0 \tag{5.47b}$$

$$M_{S_z} = \int_S -(x - x_0) p(x, y_S, z)\,dx\,dz \tag{5.47c}$$

For fluid on the $-y$ side of this plane, the normal points in the opposite direction, so the relevant equations for this case can be obtained by placing a negative sign in front of the right-hand sides of Eqs. 5.47a and 5.47c.

Finally, consider the surface in Figure 5.40C parallel to the xy plane, normal to the z axis, with fluid on the $+z$ side of the surface. By inspection, we find $\mathbf{n} = \mathbf{k} = (0, 0, 1)$, $p = p(x, y, z_S)$, and $dS = dx\,dy$. For any planar surface parallel to the xy plane with fluid on the $+z$ side the components of the hydrostatic moment are

$$M_{S_x} = \int_S -(y - y_0) p(x, y, z_S)\,dx\,dy \tag{5.48a}$$

$$M_{S_y} = \int_S (x - x_0) p(x, y, z_S)\,dx\,dy \tag{5.48b}$$

$$M_{S_z} = 0 \tag{5.48c}$$

For fluid on the $-z$ side of this plane, the normal points in the opposite direction so the appropriate equations will have a negative sign in front of the right-hand side in Eqs. 5.48a and 5.48b.

In our discussion of the hydrostatic force, we mentioned that engineering applications in hydrostatics often involve determining the forces acting on an object at rest in contact with one or more stationary fluids, usually air and water. A force balance on the object includes the effects of body, surface, and external or structural forces. Each of these forces exerts a moment on the object. Thus the normal procedure in analyzing this type of problem is to write both a force and a moment balance on the object. A calculation of the hydrostatic moment therefore usually occurs in the context of a broader analysis that includes the effects of body, surface, and structural forces and moments. A comprehensive analysis of this type is illustrated in Example 5.25.

5.6.2 Planar Nonaligned Surface

The calculation of the hydrostatic moment on a planar nonaligned surface follows the same procedure used to calculate the hydrostatic force on a planar nonaligned surface. That is, we employ a new uppercase set of Cartesian coordinates in which the surface is aligned, carry out the required integration of the appropriate equations to determine the moment, then find the components of the moment in the normal lowercase Cartesian

5.6 HYDROSTATIC MOMENT

EXAMPLE 5.25

A rectangular hatch ($L = 0.5$ m and $W = 0.5$ m) in the bottom of the large cylindrical kerosene storage tank shown in Figure 5.42a and 5.42b is hinged on one edge and held closed by a latch. Find the force on each of the two hinges and on the latch when the hatch is closed and the tank is full of kerosene to a depth of $H = 5$ m. The mass of the hatch is 50 kg.

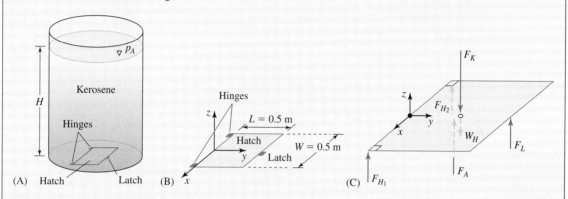

Figure 5.42 Schematics for Example 5.25: (A) geometry, (B) dimensions of hatch, and (C) free body diagram of hatch showing the pressure force and reaction forces.

SOLUTION

To find the forces on each hinge and the latch, we will make use of the fact that the force applied to each of these objects by the hatch is equal and opposite to the force applied by each object to the hatch. Thus we will analyze the forces on the hatch. The hatch is a stationary object exposed to two fluids at rest, with external forces applied to it by the hinges and the latch. A gravitational body force also acts on the hatch in the amount $-Mg\mathbf{k}$, where M is the mass of the hatch. The point of application of this force is the center of gravity of the hatch, which by symmetry is located at the center of the hatch.

We will construct a free body diagram on the hatch to isolate the various forces on it as shown in Figure 5.42c. Writing a force balance on the hatch we have

$$\mathbf{F}_{H1} + \mathbf{F}_{H2} + \mathbf{F}_L - W_H\mathbf{k} + \mathbf{F}_A + \mathbf{F}_K = 0$$

where $\mathbf{F}_{H1}$ and $\mathbf{F}_{H2}$ are the forces applied by each hinge to the hatch, $\mathbf{F}_L$ is the force applied by the latch to the hatch, $-W_H\mathbf{k}$ is the body force acting on the hatch due to gravity, $\mathbf{F}_A$ is the force of the air on the hatch, and $\mathbf{F}_K$ is the force of the kerosene on the hatch. By inspection, the force of the air on the hatch is seen to be $\mathbf{F}_A = p_A L W \mathbf{k}$, while the force of the kerosene is $\mathbf{F}_K = -(p_A + \rho_K g H) L W \mathbf{k}$. From the symmetry, the force of each hinge on the hatch is the same, and we can combine the force of the air and kerosene as a net hydrostatic force acting at the center of the hatch. The force balance now becomes

$$2\mathbf{F}_H + \mathbf{F}_L - W_H\mathbf{k} + \mathbf{F}_{\text{net}} = 0 \quad\quad\quad (A)$$

where $\mathbf{F}_{net} = -(\rho_K gH)LW\mathbf{k}$. Although we are given the weight of the hatch and have calculated the net hydrostatic force, we see that a force balance can only give us the sum of the forces on the hatch due to the hinges and latch, not each force separately. However, from the physical arrangement we anticipate that the forces applied to the hatch by the each hinge and the latch will have only a z component. [Force balances in x and y directions by inspection give this result.]

Now consider a moment balance on the hatch, taking moments about the (judiciously chosen) coordinate origin shown in Figure 5.42B. The moment of a force is of the form $\mathbf{R} \times \mathbf{F}$, where $\mathbf{R}$ is the moment arm, in this case a vector drawn from the origin to the location at which the force acts. The moment balance in this case is

$$(\mathbf{R}_{H1} \times \mathbf{F}_{H1}) + (\mathbf{R}_{H2} \times \mathbf{F}_{H2}) + (\mathbf{R}_L \times \mathbf{F}_L) + (\mathbf{R}_W \times (-W_H\mathbf{k})) + \mathbf{M}_{net} = 0$$

where, $\mathbf{M}_{net}$ is the net hydrostatic moment. By inspection $\mathbf{R}_{H1} = (W/2)\mathbf{i}$, $\mathbf{R}_{H2} = -(W/2)\mathbf{i}$, $\mathbf{R}_L = L\mathbf{j}$, and $\mathbf{R}_W = (L/2)\mathbf{j}$. To find the net hydrostatic moment, we can use Eqs. 5.48a–5.48c, with the constant gage pressure on the hatch given by $\rho_K gH$. The result is

$$M_{S_x} = \int_0^L \int_{-W/2}^{W/2} -y(\rho_K gH)\,dx\,dy = -W(\rho_K gH)\frac{L^2}{2}$$

$$M_{S_y} = \int_0^L \int_{-W/2}^{W/2} x(\rho_K gH)\,dx\,dy = 0$$

$$M_{S_z} = 0$$

Thus we have $\mathbf{M}_{net} = -W(\rho_K gH)(L^2/2)\mathbf{i}$. Note that the same result is obtained by taking the moment of the net hydrostatic force $\mathbf{R}_{net} \times \mathbf{F}_{net}$, where by inspection $\mathbf{R}_{net} = (L/2)\mathbf{j}$ because the pressure is constant on the hatch. Then $\mathbf{R}_{net} \times \mathbf{F}_{net} = (L/2)\mathbf{j} \times [-(\rho_K gH)LW\mathbf{k}] = -W(\rho_K gH)(L^2/2)\mathbf{i}$, which agrees with the result obtained by calculating the moment with the surface integrals. Completing the moment balance, we have

$$\left(-\frac{W}{2}\mathbf{i} \times \mathbf{F}_H\right) + \left(\frac{W}{2}\mathbf{i} \times \mathbf{F}_H\right) + (L\mathbf{j} \times \mathbf{F}_L) + \left[\frac{L}{2}\mathbf{j} \times (-W_H\mathbf{k})\right] + \left[-W(\rho_K gH)\frac{L^2}{2}\mathbf{i}\right] = 0$$

where we have used the fact that the forces applied by each hinge to the hatch are the same. The first two terms in this moment balance can be seen to cancel. This is due to the symmetry and our selection of the coordinate origin and coincident point about which the moment is taken. When we evaluate the cross product involving the weight of the hatch, the moment balance becomes

$$(L\mathbf{j} \times \mathbf{F}_L) - \frac{L}{2}W_H\mathbf{i} - W(\rho_K gH)\frac{L^2}{2}\mathbf{i} = 0 \qquad (B)$$

This is a vector equation that can be solved by writing $\mathbf{F}_L = F_x\mathbf{i} + F_y\mathbf{j} + F_z\mathbf{k}$ in general, then using our understanding of the physical arrangement to conclude, as noted earlier, that the force of the latch on the hatch must only have a z component. Thus we get $\mathbf{F}_L = F_z\mathbf{k}$. The moment of this force is $L\mathbf{j} \times (\mathbf{F}_L) = L\mathbf{j} \times F_z\mathbf{k} = LF_z\mathbf{i}$. Substituting this into (B) and solving, we find

$$LF_z\mathbf{i} = \frac{L}{2}W_H\mathbf{i} + W(\rho_K gH)\frac{L^2}{2}\mathbf{i} \quad \text{and} \quad F_z = \frac{W_H}{2} + \frac{WL(\rho_K gH)}{2}$$

Thus we find

$$\mathbf{F}_L = \tfrac{1}{2}[W_H + (\rho_K gH)WL]\mathbf{k} \qquad (C)$$

which tells us that the latch bears half the weight of the hatch and half the net force of the kerosene and air on the hatch. We can now complete the analysis by using (A) to write

$$2\mathbf{F}_H = -\mathbf{F}_L + W_H\mathbf{k} - \mathbf{F}_{\text{net}} = -\tfrac{1}{2}[W_H + WL(\rho_K gH)]\mathbf{k} + [W_H + (\rho_K gH)LW]\mathbf{k}$$
$$2\mathbf{F}_H = \tfrac{1}{2}[W_H + WL(\rho_K gH)]\mathbf{k}$$

We now see that the two hinges together support the other half of the load on the hatch. The force on the hatch from each hinge is

$$\mathbf{F}_H = \mathbf{F}_{H1} = \mathbf{F}_{H2} = \tfrac{1}{4}[W_H + WL(\rho_K gH)]\mathbf{k} \qquad (D)$$

The problem statement asked us to find the forces on each hinge and the latch. These forces are equal and opposite to (D) and (C), respectively. Substituting the given dimensions and using $\rho = 804$ kg/m³ for kerosene (from Appendix A), we find:

$$-\mathbf{F}_L = -\tfrac{1}{2}[W_H + (\rho_K gH)WL]\mathbf{k}$$
$$= -\tfrac{1}{2}[(50 \text{ kg})(9.81 \text{ m/s}^2) + (804 \text{ kg/m}^3)(9.81 \text{ m/s}^2)(5 \text{ m})(0.5 \text{ m})(0.5 \text{ m})]\mathbf{k}$$
$$-\mathbf{F}_L = -5175 \text{ N}\,\mathbf{k}$$

The force on each hinge is half this value, so $-\mathbf{F}_{H1} = -\mathbf{F}_{H2} = -2587.5$ N k. We can check this result by noting that the weight of the hatch is 490.5 N and the weight of the column of kerosene above the hatch is 9859 N. The air pressure below the hatch cancels the air pressure above the kerosene, so the net load on the hatch is 10,350 N. It is reasonable, given the symmetric arrangement of the hinges and latch, to find half the load borne by the hinges and half by the latch. The direction of the force on each of these objects also appears to be correct.

coordinates. To illustrate this procedure, consider the inclined surface shown in Figure 5.43A, which we used in Section 5.5.2 to discuss the calculation of the hydrostatic force on a planar nonaligned surface.

Using the uppercase Cartesian coordinate system introduced in Section 5.5.2, we see that the surface is parallel to the XY plane with fluid on the positive Z side. For a

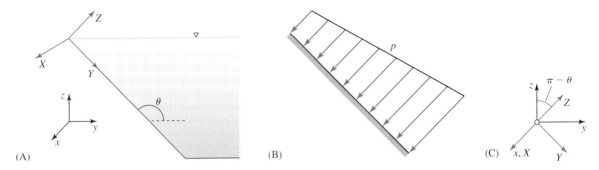

Figure 5.43 Planar nonaligned surface: (A) geometry, (B) pressure distribution, and (C) coordinate system.

moment about a point $\mathbf{X}_0 = (X_0, Y_0, Z_0)$, the moment arm $\mathbf{R}$ to any surface element is given by

$$R_x = X - X_0, \quad R_y = Y - Y_0, \quad \text{and} \quad R_z = Z - Z_0 \quad \text{(5.49a–c)}$$

Thus we will apply Eqs. 5.48a–5.48c, writing these equations in the uppercase coordinates as

$$M_{S_X} = \int_S -(Y - Y_0) p(X, Y, Z_S)\, dX\, dY \quad \text{(5.50a)}$$

$$M_{S_Y} = \int_S (X - X_0) p(X, Y, Z_S)\, dX\, dY \quad \text{(5.50b)}$$

$$M_{S_Z} = 0 \quad \text{(5.50c)}$$

We will again limit our discussion to a constant density fluid in Earth's gravity field. From the geometry of Figure 5.43A, the outward unit normal to the inclined surface is $\mathbf{n} = +\mathbf{K}$, and an infinitesimal surface element on the inclined surface is given by $dS = dX\, dY$. As discussed in Section 5.5.2, the pressure distribution on the inclined surface is a function only of the Y coordinate of a point on the surface, and can be written as $p(Y) = p_o + \rho(g \sin \theta) Y$. Inserting these values into Eqs. 5.50a–5.50c, we obtain

$$M_{S_X} = \int_S -(Y - Y_0)\, p(Y)\, dX\, dY,$$

$$M_{S_Y} = \int_S (X - X_0)\, p(Y)\, dX\, dY \quad \text{(5.51a–c)}$$

$$M_{S_Z} = 0$$

Note that any problem is simplified by placing the coordinate origin at the point about which moments are taken. This is a good practice to follow in all problems involving moments. In this case, for a surface of unit width into the paper, there is no Y component of the moment if we select an origin and coincident moment center at the midwidth of the surface. This choice causes the Y component of the moment of the pressure on each half of the surface to cancel.

5.6 HYDROSTATIC MOMENT

The remaining step in any moment analysis on a planar nonaligned surface is to transform the hydrostatic moment calculated in the uppercase coordinates

$$\mathbf{M}_S = M_{S_X}\mathbf{I} + M_{S_Y}\mathbf{J} + M_{S_Z}\mathbf{K} \tag{5.52}$$

into the lowercase coordinates, so that we have the desired components of the hydrostatic moment about the original axes:

$$\mathbf{M}_S = M_{S_x}\mathbf{i} + M_{S_y}\mathbf{j} + M_{S_z}\mathbf{k} \tag{5.53}$$

This is accomplished by using the known geometric relationships between the unit vectors of the uppercase and lowercase coordinate systems.

From Figure 5.43C, we can illustrate this process in the present case by writing the following three relationships between the unit vectors of the two-coordinate systems:

$$\mathbf{I} = \mathbf{i}, \quad \mathbf{J} = -\cos\theta\,\mathbf{j} - \sin\theta\,\mathbf{k}, \quad \text{and} \quad \mathbf{K} = \sin\theta\,\mathbf{j} - \cos\theta\,\mathbf{k} \tag{5.54a–c}$$

We now substitute these relations into Eq. 5.52 to get

$$\begin{aligned}\mathbf{M}_S &= M_{S_X}\mathbf{I} + M_{S_Y}\mathbf{J} + M_{S_Z}\mathbf{K} = M_{S_X}(\mathbf{i}) + M_{S_Y}(-\cos\theta\,\mathbf{j} - \sin\theta\,\mathbf{k}) \\ &\quad + M_{S_Z}(\sin\theta\,\mathbf{j} - \cos\theta\,\mathbf{k}) \\ &= M_{S_X}\mathbf{i} + (-M_{S_Y}\cos\theta + M_{S_Z}\sin\theta)\mathbf{j} + (-M_{S_Y}\sin\theta - M_{S_Z}\cos\theta)\mathbf{k}\end{aligned}$$

Thus the components of the moment in the lowercase coordinates are

$$M_{S_x} = M_{S_X} \tag{5.55a}$$

$$M_{S_y} = -M_{S_Y}\cos\theta + M_{S_Z}\sin\theta \tag{5.55b}$$

$$M_{S_z} = -M_{S_Y}\sin\theta - M_{S_Z}\cos\theta \tag{5.55c}$$

Since $M_{S_Z} = 0$ in the present case, and we can make $M_{S_Y} = 0$ by proper choice of the coordinate origin and moment center as explained earlier, these relationships can be further simplified.

The preceding formulas apply to the calculation of the hydrostatic moment on a planar surface parallel to the XY plane with fluid on the positive Z side. If the fluid is on the negative Z side, the normal changes sign, which leads to a corresponding sign change in Eqs. 5.55a–5.55c. For a surface parallel to one of the other uppercase coordinate planes, you must develop an appropriate set of formulas, paying attention to the alignment of the plane in the uppercase coordinates and to the side of the plane in contact with the fluid. Example 5.26 illustrates the calculation of the hydrostatic moment on a planar nonaligned surface.

EXAMPLE 5.26

Consider the inclined gate of length $L = 3$ m and width $W = 1.5$ m, which holds back water in an irrigation channel as shown earlier (Figure 5.29) and investigated in Example 5.18. Calculate the net hydrostatic moment exerted by all fluids on the gate with respect to the hinge line at the end of the gate when the channel is filled to height of $H = 1$ m. Express the results in the x, y, z coordinate system.

SOLUTION

Figure 5.44A serves as a sketch for this problem. In Example 5.18, we calculated the net hydrostatic force on the gate to be $\mathbf{F}_{net} = -WS\,[(\rho g \sin\theta)S/2]\,\mathbf{K}$. The net force acts normal to the surface, and we can see from Figure 5.44B that there will be only an I component of the net hydrostatic moment. We will use the same uppercase coordinate system to find the moment in the $\mathbf{I}$ direction about the hinge with Eq. 5.51a, but select the moment center on the hinge line so that $\mathbf{X}_0 = (X_0, Y_0, Z_0) = (0, Y_0, 0)$, where $Y_0 = -(L-S) = -1\,\text{m}$ as shown in Figure 5.44A. Recall that the wetted length of gate is $S = H/\sin\theta$. The coordinate origin remains at the free surface where $p_0 = p_A$, thus the pressure distribution on the surface is $p(Y) = p_A + \rho(g\sin\theta)Y$, and the gage pressure distribution on the surface is $p(Y)_{\text{gage}} = \rho(g\sin\theta)Y$. Using Eq. 5.51a, we find

$$M_{S_X} = \int_0^S \int_{-W/2}^{(W/2)} -(Y-Y_0)\rho(g\sin\theta)\,Y\,dX\,dY = -W\rho g \sin\theta \left[\frac{S^3}{3} - \frac{Y_0 S^2}{2}\right] \quad (A)$$

The negative sign indicates a tendency to rotate the gate clockwise about the hinge, as expected. Recognizing that this moment is due to the net hydrostatic force, we can rewrite this result as

$$M_{S_X} = -W\rho g \sin\theta \left(\frac{S^3}{3} - \frac{Y_0 S^2}{2}\right) = -WS \left(\frac{\rho g \sin\theta \, S}{2}\right)\left(\frac{2}{3}S - Y_0\right) \quad (B)$$

The net hydrostatic moment can also be thought of as the cross product $\mathbf{R} \times \mathbf{F}$, where the moment arm, as shown in Figure 5.44B, is seen to be $\mathbf{R} = \frac{2}{3}S\mathbf{J} + (-Y_0\mathbf{J}) = (\frac{2}{3}S - Y_0)\mathbf{J}$, and $\mathbf{F}$ is the net hydrostatic force, $\mathbf{F}_{net} = -WS\,[(\rho g \sin\theta)S/2]\,\mathbf{K}$. Thus the net hydrostatic moment in the uppercase coordinates is $\mathbf{M}_S = -WS\,[\rho g \sin\theta \, S/2]\,(\frac{2}{3}S - Y_0)\mathbf{I}$, which agrees with (A) and (B). Since $\mathbf{I} = \mathbf{i}$ in this problem, the moment in the normal lowercase coordinates is

$$\mathbf{M}_S = -WS \left(\frac{\rho g \sin\theta \, S}{2}\right)\left(\frac{2}{3}S - Y_0\right)\mathbf{i} \quad (C)$$

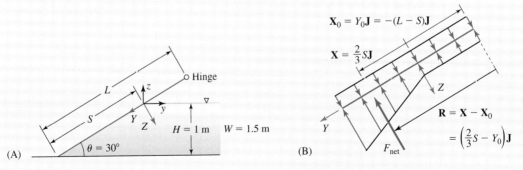

Figure 5.44 Schematics for Example 5.26: (A) geometry and (B) pressure distribution.

In Example 5.18 we found $|\mathbf{F}_{\text{net}}| = WS[\rho g \sin\theta S/2] = 14{,}686\,\text{N}$, and from the data we find $(\tfrac{2}{3}S - Y_0) = \tfrac{2}{3}(2\,\text{m}) - (-1\,\text{m}) = \tfrac{7}{3}\,\text{m}$, thus the moment is

$$\mathbf{M}_S = -WS\left(\frac{\rho g \sin\theta S}{2}\right)\left(\frac{2}{3}S - Y_0\right)\mathbf{i} = -(14{,}686\,\text{N})\left(\frac{7}{3}\,\text{m}\right)\mathbf{i} = -34{,}267\,(\text{N-m})\mathbf{i}$$

Also note that the moment arm $\tfrac{2}{3}S - Y_0 = \tfrac{2}{3}S - (-(L-S)) = L - \tfrac{1}{3}S$ as expected.

5.7 RESULTANT FORCE AND POINT OF APPLICATION

We have seen that the effect of hydrostatic pressure on a surface is to exert a force and moment on the surface. In engineering applications, particularly those that involve fluid–structure interactions, it is traditional to think of the net effect of surface forces as a resultant force $\mathbf{F}_R$ and a point of application (POA) for this force, which lies on the surface. The resultant force is defined to be the net hydrostatic force on the surface from all fluids. Hence from Eq. 5.31, the resultant force is given by the surface integral

$$\mathbf{F}_R = \mathbf{F}_S = \int_S -p\mathbf{n}\,dS \qquad (5.56)$$

where we must consider all the fluids and surfaces involved. The POA of the resultant force is defined by requiring that the moment of the resultant force when this force is acting at the POA equals the net hydrostatic moment on the surface from all fluids. If the POA is located relative to the moment origin by its moment arm $\Re$, then by definition we have

$$\Re \times \mathbf{F}_S = \mathbf{M}_S \qquad (5.57\text{a})$$

and using Eq. 5.45a–5.45c in vector notation, we obtain

$$\Re \times \mathbf{F}_S = \int_S [\mathbf{r} \times (-p\mathbf{n})]\,dS \qquad (5.57\text{b})$$

That is, the moment of the resultant force is equal to the moment due to the surface force.

The vector Eq. 5.57b can be expressed in the lowercase Cartesian x, y, z coordinates as the following three algebraic equations from which the three unknown components of $\Re = (\Re_x, \Re_y, \Re_z)$ can be found:

$$\Re_y F_{S_z} - \Re_z F_{S_y} = M_{S_x}, \quad \Re_z F_{S_x} - \Re_x F_{S_z} = M_{S_y}, \quad \Re_x F_{S_y} - \Re_y F_{S_x} = M_{S_z}$$
$$(5.58\text{a–c})$$

or in the uppercase Cartesian coordinates X, Y, Z, with $\Re = (\Re_X, \Re_Y, \Re_Z)$, from

$$\Re_Y F_{S_Z} - \Re_Z F_{S_Y} = M_{S_X}, \quad \Re_Z F_{S_X} - \Re_X F_{S_Z} = M_{S_Y}, \quad \Re_X F_{S_Y} - \Re_Y F_{S_X} = M_{S_Z}$$
$$(5.59\text{a–c})$$

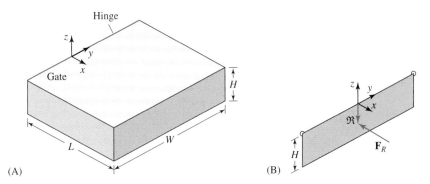

Figure 5.45 Schematics of concrete container and gate: (A) geometry and (B) resultant force and point of application.

In most problems, the fact that the POA must lie on the surface, and that symmetry considerations cause one or more components of the net moment to be zero, allow us to determine some of the components of the moment arm, $\mathfrak{R} = (\mathfrak{R}_x, \mathfrak{R}_y, \mathfrak{R}_z)$, defining the POA by inspection.

We can illustrate the process of determining the resultant force and POA by revisiting Example 5.24. In that example we calculated the net hydrostatic force and moment about the hinge pin on the gate of a container of wet concrete (see Figure 5.45). The net hydrostatic force was given by (A) as $\mathbf{F}_S = \int_{-H}^{0} \int_{-(W/2)}^{(W/2)} (\rho g z)(\mathbf{i}) dy\, dz = -WH(\rho g H/2)\mathbf{i}$, and the moment was given by (B) as $\mathbf{M}_S = M_{Sy}\mathbf{j} = \rho g W (H^3/3)\mathbf{j}$. The resultant force is therefore $\mathbf{F}_R = \mathbf{F}_S = -WH(\rho g H/2)\mathbf{i}$. To find $\mathfrak{R} = (\mathfrak{R}_x, \mathfrak{R}_y, \mathfrak{R}_z)$, we note that since the POA must be on the surface, $\mathfrak{R}_x = 0$. Since the force acts solely along the x axis, and there is no moment about the z axis, we conclude that $\mathfrak{R}_y = 0$. We can determine the remaining component of $\mathfrak{R} = (\mathfrak{R}_x, \mathfrak{R}_y, \mathfrak{R}_z)$ from Eq. 5.58b in the form $\mathfrak{R}_z F_{S_x} - \mathfrak{R}_x F_{S_z} = M_{S_y} = \rho g W (H^3/3)$. Inserting the known components of $\mathbf{F}_S$ and using the fact that $\mathfrak{R}_x = 0$, we find $\mathfrak{R}_z[-WH(\rho g H/2)] = \rho g W (H^3/3)$, which we solve to find $\mathfrak{R}_z = -\frac{2}{3}H$. Thus the location of the POA is defined by the moment arm $\mathfrak{R} = (0, 0, -\frac{2}{3}H)$, which is shown in Figure 5.45B. The POA is on the centerline of the gate two-thirds of the way down the gate. This seems reasonable for the linear gage pressure distribution on the surface.

The procedure for finding the resultant force and POA for a planar nonaligned surface is the same as that just described, but the process is carried out in the uppercase X, Y, Z coordinate system. Once the resultant force and POA are determined in the uppercase coordinates, it is straightforward to write them in the usual lowercase Cartesian coordinates. Consider the inclined gate used in Examples 5.18 and 5.26, and illustrated in Figure 5.44. Note carefully that in this problem the coordinate origin and point about which moments are calculated do not coincide. The moment arm of the POA by inspection must be of the form $\mathfrak{R} = (0, \mathfrak{R}_Y, 0)$. The resultant force was calculated in Example 5.18 as $\mathbf{F}_{\text{net}} = -WS[(\rho g \sin\theta)S/2]\mathbf{K}$, and the moment was found in Example 5.26 to be $\mathbf{M}_S = -WS[\rho g \sin\theta S/2](\frac{2}{3}S - Y_0)\mathbf{I}$. Applying Eq. 5.59a, we find $\mathfrak{R}_Y F_{S_Z} = M_{S_X}$, which gives $\mathfrak{R}_Y = M_{S_X}/F_{S_Z} = \frac{2}{3}S - Y_0$. Thus the POA has a moment arm $= (0, \frac{2}{3}S - Y_0, 0)$. To verify this result we note that when the origin and center of moment do not coincide, we have $\mathfrak{R} = \mathbf{X}_{\text{POA}} - \mathbf{X}_0$, where $\mathbf{X}_{\text{POA}}$ is the position of the POA, and

5.8 BUOYANCY AND ARCHIMEDES' PRINCIPLE

$\mathbf{X}_0$ is position of the center about which moments are taken. Thus in this case we write $\Re_Y = Y_{\text{POA}} - Y_0 = \frac{2}{3}S - Y_0$ and solve to obtain $Y_{\text{POA}} = \frac{2}{3}S$. We see that the POA is located two-thirds of the way down the inclined gate as expected, and at a distance $\frac{2}{3}S - Y_0 = L - \frac{1}{3}S$ from the moment origin.

5.8 BUOYANCY AND ARCHIMEDES' PRINCIPLE

The concepts of a buoyancy force and Archimedes' principle were introduced in Section 2.3.3 in the context of our discussion of fluid pressure. Recall that the buoyancy force is defined as the net vertical force acting on an immersed object due to the variation in hydrostatic pressure with height (see Figure 2.10). From this definition we conclude that the buoyancy force is given by the surface force acting on the object. Thus using Eq. 5.31 we can write

$$\mathbf{F}_{\text{buoyancy}} = \int_S -p\mathbf{n}\, dS \qquad (5.60)$$

where the integration is taken over the entire surface of the object. Archimedes' principle defines the magnitude and direction of this force by stating that a buoyancy force acts in the direction opposite to that of gravity and has a magnitude equal to the weight of the displaced fluid. We will now use our newfound understanding of fluid statics to prove this statement.

An immersed object may be either free to move or restrained by a structural element. If an object is free to move, it is acted on by gravity and buoyancy forces only, as shown in Figure 5.46A and 5.46B. If an object is restrained, the structural connections transmit an external force to the object that must be accounted for in an analysis. Thus, a restrained object is acted upon by gravity, buoyancy, and external forces, as shown in Figures 5.46C. In both cases, an application of Newton's second law to the stationary object requires that the sum of the forces be zero.

An immersed object free to move will remain in place if the buoyancy force is exactly equal to its weight. If the buoyancy force is greater than its weight, the object will

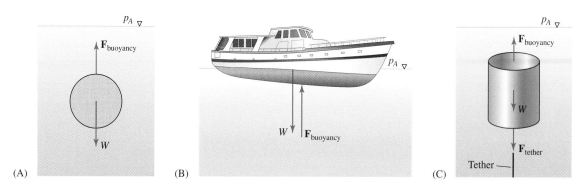

Figure 5.46 Force balance on (A) a fully immersed object, (B) a ship, and (C) a tethered object.

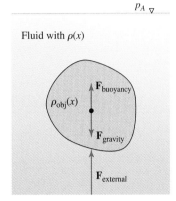

Figure 5.47 Forces on an arbitrary, tethered, fully immersed object.

rise. If the buoyancy force is less than its weight, the object sinks, stopping in its downward trajectory only if some additional force, such as that created by contact with the bottom, occurs. If an object is tethered or otherwise held in place by an external force created by a structural element of some type, as in Figure 5.46C, then the difference between the buoyancy force and the weight of the object must be balanced by the external restraining force.

Consider an object wholly immersed in a fluid at rest and subject to gravity, buoyancy, and external forces as shown in Figure 5.47. To account for the nonuniform physical structure of real materials, we suppose the object has a density distribution $\rho_{obj}(\mathbf{x})$. For further generality, let the fluid have a variable density $\rho(\mathbf{x})$. If the object is at rest, then by Newton's second law, the sum of the forces acting on the object is zero. As shown in Figure 5.47, these forces include an external force $\mathbf{F}_{ext}$, a gravitational body force given by the integral $\mathbf{F}_{gravity} = \int_V \rho_{obj}(\mathbf{x})(-g\mathbf{k})\,dV$, and a surface (buoyancy) force given by Eq. 5.60, $\mathbf{F}_{buoyancy} = \int_S -p\mathbf{n}\,dS$.

For an object at rest, applying Newton's second law gives

$$\mathbf{F}_{ext} + \int_V \rho_{obj}(\mathbf{x})(-g\mathbf{k})\,dV + \int_S -p\mathbf{n}\,dS = 0 \tag{5.61a}$$

where the volume and surface integrals refer to the volume and surface of the object. We can also write this balance of forces symbolically as

$$\mathbf{F}_{ext} + \mathbf{F}_{gravity} + \mathbf{F}_{buoyancy} = 0 \tag{5.61b}$$

If the sum of the gravity and buoyancy force is zero, the object is said to be neutrally buoyant, since it will remain in place indefinitely with no external force applied to it.

Since the volume integral in Eq. 5.61a defines the weight of the object, we can also write this integral as $\int_{V_B} \rho_{obj}(\mathbf{x})(-g\mathbf{k})\,dV = -W_{obj}\mathbf{k}$, where W_{obj} is the weight of the object, and express the balance of forces as

$$\mathbf{F}_{ext} + (-W_{obj}\mathbf{k}) + \int_S -p\mathbf{n}\,dS = 0 \tag{5.62}$$

5.8 BUOYANCY AND ARCHIMEDES' PRINCIPLE

EXAMPLE 5.27

A cylindrical instrument container of height $H = 1$ m and radius $R = 0.25$ m has been designed so that it is neutrally buoyant when partially immersed to a depth $H/2$ in water, as shown in Figure 5.48. Find the weight of the cylinder and its contents.

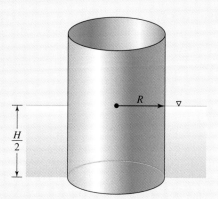

Figure 5.48 Schematic for Example 5.27.

SOLUTION

There is no external force present, and the only body force is gravity. According to Eq. 5.62, which is now a force balance on this instrument, the condition for neutral buoyancy (zero external restraining force) is

$$-W_{\text{obj}}\mathbf{k} + \int_S -p\mathbf{n}\, dS = 0$$

The first term contains the unknown weight of the object. The pressure integral is equal to the buoyancy force acting on the container. The integral over the entire surface of the object can be separated into integrals over the top, bottom, and the side (immersed and dry). On the top the pressure is uniform and equal to $p = p_A$ so the surface force is

$$\mathbf{F}_S = -pA\mathbf{k} = -p_A\pi R^2 \mathbf{k}$$

On the bottom the pressure is also uniform and equal to $p = p_A + \rho g H/2$, so the surface force there is

$$\mathbf{F}_S = -pA(-\mathbf{k}) = \left(p_A + \frac{\rho g H}{2}\right)\pi R^2 \mathbf{k}$$

The resultant surface force on the side of the cylinder is zero because of the symmetry of the object and pressure distribution. The sum of forces in the z direction is

$$-W_{\text{obj}} - p_A\pi R^2 + \left(p_A + \frac{\rho g H}{2}\right)\pi R^2 = 0$$

and it can be seen that the net surface force, i.e., buoyancy force, is $\rho g \pi R^2(H/2)\mathbf{k}$. Solving for the weight of the cylinder and its contents we find

$$W_{\text{obj}} = \left(\frac{\rho g H}{2}\right)\pi R^2 = (998 \text{ kg/m}^3)(9.81 \text{ m/s}^2)(1.0 \text{ m/2})(\pi)(0.25 \text{ m})^2 = 961 \text{ N}$$

where the density of water is taken from Appendix A. The mass of the cylinder and its contents is therefore 98 kg, which seems reasonable. It can be very challenging to adjust the weight of a container so that it is exactly neutrally buoyant. Try to do it with a soda bottle, using sand or gravel to adjust the weight of the bottle so that it floats fully immersed but stationary.

Equations 5.61a, 5.61b, and 5.62 also apply to a partially immersed object. For such an object, the surface integral defining the buoyancy force has several parts, one for each fluid in contact with the object. However, the weight refers to the entire object.

Equations 5.61a, 5.61b, and 5.62 are general expressions of Newton's second law for any stationary object in a fluid at rest in Earth's gravity; they apply to every situation an engineer might encounter. However, it is possible to replace the surface integral in Eq. 5.62 by a certain volume integral by taking advantage of the fact that the pressure distribution in a fluid at rest is generally unaffected by the presence of an object. Consider the immersed object shown in Figure 5.49A. Suppose the object is no longer present so that the volume in space formerly occupied by the object is now filled with fluid, as shown in Figure 5.49B. This volume is referred to as the volume of displaced fluid, and we suppose that the fluid filling the displaced volume has the density distribution of the surrounding undisturbed fluid. If this is the case, the pressure distribution on the surface of the displaced volume is identical to that on the object, and we conclude that the hydrostatic force on the displaced volume is identical to that on the object.

Suppose we now apply Eq. 5.62 to the displaced volume of fluid, noting that there is no external force on the displaced volume and that the density in the volume integral

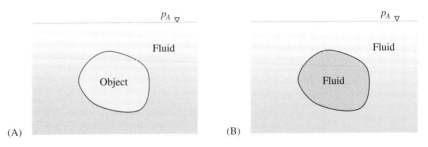

Figure 5.49 (A) An immersed object and (B) the equivalent volume of fluid.

5.8 BUOYANCY AND ARCHIMEDES' PRINCIPLE

HISTORY BOX 5-1

"Eureka, Eureka!" meaning "I found it, I found it!", was reportedly Archimedes' cry one day when he overflowed his bath. What he had found was a method of determining whether the golden crown of the king of Syracuse, Hero II, was made of pure gold or an alloy of gold and silver. Archimedes placed in a bowl a block of pure gold of the same weight as the crown and carefully filled the bowl to the brim with water. After removing the block of pure gold, he placed the crown in the bowl. Because silver is less dense than gold, a crown containing silver would cause the bowl to overflow. Does this method seem practical? Does it use the principle named after Archimedes? Can you think of any way to solve the problem stated in Example 5.27 by using Archimedes' principle?

defining the body force is now that of the fluid. Rearranging the resulting equation gives

$$\int_S -p\mathbf{n}\, dS = -\int_V \rho(\mathbf{x})(-g\mathbf{k})\, dV$$

where the integrals refer to the displaced volume. Since the surface integral here also represents the hydrostatic force on the object, i.e., buoyancy force on the object, we conclude that

$$\mathbf{F}_{\text{buoyancy}} = \int_S -p\mathbf{n}\, dS$$
$$= -\int_V \rho(\mathbf{x})(-g\mathbf{k})\, dV \quad (5.63a)$$

a result that can also be written as

$$\mathbf{F}_{\text{buoyancy}} = W_{\text{fluid}}\mathbf{k} \quad (5.63b)$$

where W_{fluid} is the weight of the displaced fluid. Note that this result applies to any density distribution in the fluid, and it tells us that the buoyancy force is equal in magnitude to the weight of the displaced fluid and acts vertically upward, a result said to have been known to Archimedes. Thus, Eq. 5.63b is called Archimedes' principle. Equation 5.63a is simply a generalization of this principle to allow the calculation of the weight of a displaced volume of fluid of variable density. For a constant density fluid, we can write Archimedes' principle as

$$\mathbf{F}_{\text{buoyancy}} = \rho_{\text{fluid}} g V \mathbf{k} \quad (5.64)$$

Now consider the balance of forces on an immersed object as given by Eq. 5.61a:

$$\mathbf{F}_{\text{ext}} + \int_V \rho_{\text{obj}}(\mathbf{x})(-g\mathbf{k})\, dV + \int_S -p\mathbf{n}\, dS = 0$$

where the pressure integral is the buoyancy force. If we use Eq. 5.63a to replace the integral pressure and rearrange terms, the force balance on a stationary immersed object becomes

$$\mathbf{F}_{\text{ext}} = \int_V \rho_{\text{fluid}}(\mathbf{x})(-g\mathbf{k})\, dV - \int_V \rho_{\text{object}}(\mathbf{x})(-g\mathbf{k})\, dV \quad (5.65a)$$

Finally we can evaluate the volume integrals and write

$$\mathbf{F}_{\text{ext}} = (W_{\text{object}} - W_{\text{fluid}})\mathbf{k} \quad (5.65b)$$

which states that the external restraining force required to keep an immersed object stationary is equal to the difference between the weight of the object and the weight of the displaced fluid. Notice that this result is in perfect agreement with our earlier discussion of the behavior of partially and fully immersed objects.

EXAMPLE 5.28

What is the buoyancy force acting on the cylindrical instrument container shown in Figure 5.48?

SOLUTION

According to Eq. 5.64, the buoyancy force contributed by the water is

$$\rho g \mathcal{V} \mathbf{k} = \rho_{H_2O} g \pi R^2 \frac{H}{2} \mathbf{k} = 961 \text{ N} \mathbf{k}$$

The air also contributes a buoyancy force of

$$\rho g \mathcal{V} \mathbf{k} = \rho_{air} g \pi R^2 \frac{H}{2} \mathbf{k} = 1.18 \text{ N} \mathbf{k}$$

The buoyancy force due to the air is only about 0.1% the force due to the water. However, in the constant pressure model of the atmosphere, the air density is zero. To be consistent, we must neglect the buoyancy force due to air, and thus the result here is the same as found earlier in Example 5.27.

EXAMPLE 5.29

A spherical ($D = 3$ ft) sea mine weighing 500 lb$_f$ is chained to the bottom of a harbor as shown in Figure 5.50. What external force must the chain provide to keep the mine from floating to the surface?

Figure 5.50 Schematic for Example 5.29.

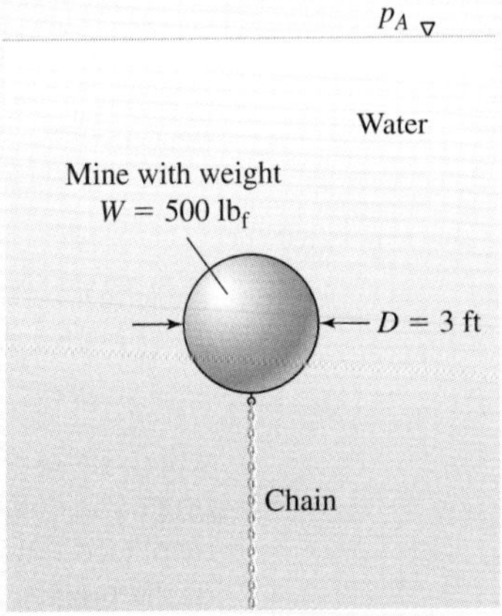

SOLUTION

A force balance on the sea mine shows that Eq. 5.65b, $\mathbf{F}_{ext} = (W_{object} - W_{fluid})\mathbf{k}$, is applicable. Using the density of seawater (Appendix A) $\rho_{sw} = 1025$ kg/m^3 = 1.989 slugs/ft^3, we have

$$\mathbf{F}_{ext} = \left[500 \text{ lb}_f - (1.989 \text{ slugs/ft}^3)(32.2 \text{ ft/s}^2)\left(\frac{4}{3}\pi\right)\left(\frac{3}{2}\text{ ft}\right)^3\left(\frac{1 \text{ lb}_f\text{-s}^2}{1 \text{ slug-ft}}\right)\right]\mathbf{k} = -405 \text{ lb}_f\, \mathbf{k}$$

The mine will float to the surface unless restrained.

Our analysis of the forces on an immersed object was limited to a gravitational body force; however, the general conclusion equating the buoyancy force to the body force acting on the displaced fluid is true for any body force. The force balance on an immersed object is given in general by

$$\mathbf{F}_{ext} = \int_V \rho_{fluid}(\mathbf{x})\mathbf{f}\, dV - \int_V \rho_{object}(\mathbf{x})\mathbf{f}\, dV \quad (5.66)$$

where $\mathbf{f}$ is the body force per unit mass. For example, in a fluid in rigid body rotation, the solution of Example 5.5 showed that the body force in cylindrical coordinates is given by $\mathbf{f} = r\Omega^2 \mathbf{e}_r - g\mathbf{e}_z$. The external force needed to hold an object immersed at some location in this case must balance the effects of gravity and the centrifugal force simultaneously.

5.9 EQUILIBRIUM AND STABILITY OF IMMERSED BODIES

The occasional report of the capsizing of an overloaded passenger ferry or other vessel (Figure 5.51A) suggests that it is not sufficient to consider only the balance of forces acting on a partially submerged or even fully submerged object. In fact it is also necessary to consider the moments acting on the object, since an unbalanced moment will result in a rolling motion that is not easily resisted. Although a complete treatment of this subject is beyond the scope of this text, we can observe that a necessary condition for an object immersed in fluid to be stationary is that the sum of the forces and the sum of the moments be zero. When the sum of the forces and moments is zero, the object is in a state of equilibrium, but it is also necessary to consider the stability of this state.

For the object shown in Figure 5.51B, the sum of the forces is given by Eq. 5.61a and 5.61b or its equivalent, Eqs. 5.65a and 5.65b. Setting the sum of the moments acting on this object about a moment origin at $\mathbf{x}_0$ to zero gives

$$\mathbf{R}_{ext} \times \mathbf{F}_{ext} + \int_V [(\mathbf{x} - \mathbf{x}_0) \times (\rho_{object}(\mathbf{x})(-g\mathbf{k}))]\, dV + \int_S [(\mathbf{x} - \mathbf{x}_0) \times (-p\mathbf{n})]\, dS = 0 \quad (5.67a)$$

where $\mathbf{R}_{ext}$ defines the moment arm of the external force. The remaining two integrals in this equation represent the moment due to body forces acting on the object and the moment due to the buoyancy force. The calculation of the moments acting on a three-dimensional object is very tedious. However, some insight can be gained if we introduce the concept of a resultant gravitational body force $\mathbf{F}_{gravity}$ and its POA, and a resultant

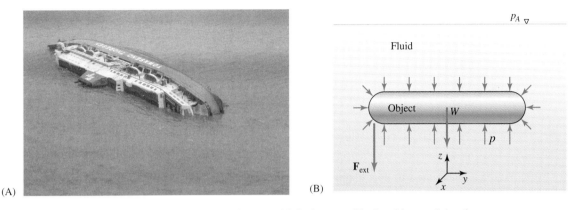

Figure 5.51 (A) Capsized vessel. (B) Free body diagram of fully immersed body with restraining force.

buoyancy force $\mathbf{F}_{\text{buoyancy}}$ and its POA. This allows us to write the condition for moment equilibrium as

$$\mathbf{R}_{\text{ext}} \times \mathbf{F}_{\text{ext}} + \mathbf{R}_{\text{gravity}} \times \mathbf{F}_{\text{gravity}} + \mathbf{R}_{\text{buoyancy}} \times \mathbf{F}_{\text{buoyancy}} = 0 \quad (5.67b)$$

Using the fact that $\mathbf{F}_{\text{gravity}} = -W_{\text{obj}}\mathbf{k}$, $\mathbf{F}_{\text{buoyancy}} = W_{\text{fluid}}\mathbf{k}$, we can also write this condition for moment equilibrium, symbolically, as

$$\mathbf{R}_{\text{ext}} \times \mathbf{F}_{\text{ext}} + \mathbf{R}_{\text{gravity}} \times (-W_{\text{object}}\mathbf{k}) + \mathbf{R}_{\text{buoyancy}} \times (W_{\text{fluid}}\mathbf{k}) = 0 \quad (5.67c)$$

We can now analyze a problem by considering the locations of the center of gravity and center of buoyancy. Using the familiar idea of the displaced fluid, it is possible to show that the moment due to the buoyancy force is equal and opposite to the moment due to the gravitational body force acting on the displaced fluid. Thus the center of buoyancy is the same as the center of gravity for the displaced fluid and, of course, the same as the center of pressure.

We can illustrate this approach to analyzing the requirement for equilibrium by examining the horizontal capsule shown in Figure 5.52. The extra weight on one end of the object causes its center of gravity to be to the right of the midpoint, and the gravitational body force acting on the object (its weight) acts down at this location. The symmetry of the body causes the center of gravity of the displaced fluid to be at the midpoint, and the

Figure 5.52 Relative locations of center of buoyancy and gravity.

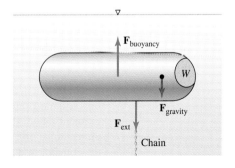

5.9 EQUILIBRIUM AND STABILITY OF IMMERSED BODIES

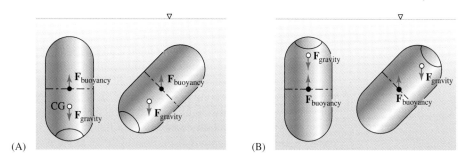

Figure 5.53 Neutrally buoyant capsule in (A) stable equilibrium with restoring couple, and (B) unstable equilibrium with overturning couple.

buoyancy force acts upward at this point. Notice that these two forces contribute a clockwise moment on the capsule. For equilibrium, external force must act down at the attachment point, and it must have a magnitude that exactly balances both the net force and net moment contributed by the body and buoyancy forces.

In the absence of a restraining force, equilibrium for a neutrally buoyant body is analyzed by examining the locations of the center of gravity and center of buoyancy. Consider the two situations shown in Figures 5.53A and 5.53B. Equilibrium exists in both cases, but if it occurred to you that the arrangement in Figure 5.53B is unstable, you are correct. The general rule for a fully immersed object is that a position of equilibrium with the center of gravity above the center of buoyancy is unstable. This rule can be established rigorously by examining the balance of forces and moments when an

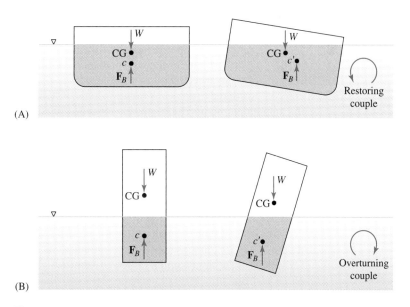

Figure 5.54 Partially submerged object in (A) stable and (B) unstable conditions: c, centroid of original displaced volume; c', centroid of new displaced volume.

immersed body is perturbed from its proposed equilibrium position. If the resulting moment is nonzero and tends to return the body to its former orientation, the equilibrium position is stable. If the resulting moment tends to cause the body to move away from its former position of equilibrium, the equilibrium position is unstable.

Equilibrium and stability for a partially immersed (floating) object is analyzed in the same manner. A floating body, however, may be in a position of stable equilibrium with its center of gravity slightly above its center of buoyancy, as shown in Figure 5.54. A key feature of the stability analysis for a floating body is that when its orientation is changed, the center of gravity of a floating body remains fixed, but the center of buoyancy may shift significantly. Stability depends on the magnitude and direction of the change in the center of buoyancy, and whether a righting or overturning moment is created. Thus one can encounter a configuration that is stable to any perturbation as illustrated in Figure 5.54A or a configuration that is stable for small roll angles but unstable if the roll is large enough (Figure 5.54B).

5.10 SUMMARY

Fluid statics, or hydrostatics, is the branch of fluid mechanics concerned with the behavior of fluid at rest. In a fluid at rest, all shear stresses are completely absent. The stress vector acting on any surface in a fluid under these conditions is given by $\mathbf{\Sigma} = \sigma \mathbf{n}$, where σ is the magnitude of the stress vector. Since fluids are unable to exert a tensile stress, the hydrostatic stress is compressive, and the normal stress σ in a fluid at rest is a negative quantity. The pressure p, called the static pressure, is defined to be the negative of the normal stress acting on any surface in a fluid at rest: $p = -\sigma_N$. The corresponding hydrostatic stress is $\mathbf{\Sigma} = -p\mathbf{n}$, and this stress acts equally in all directions at a point in the fluid. In the Eulerian description, pressure is represented by a scalar field $p = p(\mathbf{x}, t)$. Under normal circumstances, both the pressure distribution and the body forces that create the pressure field are time independent.

A state of hydrostatic stress also occurs if a fluid is in motion at constant velocity as a whole, in linear acceleration as a whole, or in a solid body rotation as a whole. In a reference frame fixed to an accelerating volume of fluid, the fluid is at rest but acted on by the appropriate noninertial body force needed to account for the noninertial reference frame. This apparent body force may be treated as if it were an additional body force.

The governing equation for fluid statics is the hydrostatic equation, which in integral and differential form, applies to every problem in fluid statics. The integral form of the hydrostatic equation is $\int_S -p\mathbf{n}\, dS + \int \rho \mathbf{f}\, dV = 0$. This equation tells us that for any volume of fluid at rest, the total force exerted on the surface of that volume by the pressure is exactly balanced by the total body force exerted on the fluid within that volume. We can use this equation to show that the pressure at the bottom of a column of constant density fluid is greater than the pressure at the top by the amount $\rho g H$. The product $\rho g H$ has dimensions of force per unit area, and we can interpret this product as the weight of the fluid column divided by the area of its footprint.

The differential form of the hydrostatic equation states that $\nabla p = \rho \mathbf{f}$ at every point in the fluid. That is, the pressure force per unit volume is balanced by the body force per unit volume. To solve this partial differential equation for the pressure, we must know the density and body force. The solution to this equation, for a known distribution of body forces, completely describes the state of stress (pressure) at every point in the fluid, as well as the stress vector on every surface the fluid contacts. Armed with this knowledge, an engineer can calculate the total surface force and moment on any surface of interest.

Many important problems in fluid statics involve a constant density fluid at rest in Earth's gravity field. The pressure distribution in this case, $p(z) = p_0 - \rho g(z - z_0)$, describes a linear increase in pressure with depth. Surfaces of constant pressure are found by setting the pressure distribution equal to a constant. For this pressure distribution, these surfaces are a family of horizontal planes. In a variable density fluid, usually referred to as a stratified fluid, the pressure distribution is nonlinear and may be quite complex.

In a second class of problems in fluid statics, a container completely filled with fluid is placed in steady rotation. Once this rigid body rotation has been achieved in a fluid, the fluid is at rest relative to the rotating container with a hydrostatic pressure distribution. This distribution can be described by the hydrostatic equation with respect to coordinates fixed to the moving container, with the centrifugal body force included in the total body force. The resulting pressure gradient in the r direction balances the centrifugal force caused by the rotation of the fluid. The pressure increases monotonically in the radial direction, reaching a maximum at the container wall. The pressure gradient in the θ direction is zero and the pressure gradient in the z direction balances the gravitational body force as usual.

To find the pressure distribution in a constant density fluid undergoing a constant rectilinear acceleration, we must insert the appropriate expression for the total body force into the hydrostatic equation and solve the resulting system of equations. For a container filled with fluid in constant linear acceleration $\mathbf{a} = (a_x, a_y, a_z)$, and assuming the only body force is gravity, the pressure distribution is given by $p(x, y, z) = p_0 - \rho a_x(x - x_0) - \rho a_y(y - y_0) - \rho(g + a_z)(z - z_0)$, where p_0 is the datum pressure at location (x_0, y_0, z_0). Surfaces of constant pressure are planes whose orientation in space depends on the values of the various acceleration components.

The hydrostatic force exerted by a fluid on a structure is given by the surface integral $\mathbf{F}_S = \int_S -p\mathbf{n}\, dS$. The difficulty in calculating the surface force in fluid statics arises from calculating this integral on various surfaces. We showed how to evaluate this integral for surfaces of three types: planar aligned (a flat surface aligned with a Cartesian coordinate plane); planar nonaligned (a flat surface not aligned with a coordinate plane); and curved. An indirect method of calculating the surface force, which is recommended for planar nonaligned and curved surfaces, uses the force on the horizontal and vertical projections of the surface plus the weight of the fluid in a selected volume to replace the direct calculation of the hydrostatic force by means of a surface integral on the original surface.

The hydrostatic moment created by the pressure on a surface is also important in practical applications. For a fluid at rest the hydrostatic moment may be calculated as $\mathbf{M}_S = \int_S [\mathbf{r} \times (-p\mathbf{n})]\, dS$. The moment arm is $\mathbf{r} = \mathbf{x} - \mathbf{x}_0$, where the moment is defined with respect to position $\mathbf{x}_0$. The use of this surface integral to calculate the hydrostatic moment involves the same mathematical procedures used for the hydrostatic force.

In some cases of interest, particularly those that involve rigid structures, it is traditional to think of the net effect of surface forces as a resultant force $\mathbf{F}_R$, and a point of application for this force defined by the moment arm $\mathfrak{R}$. This resultant force is the net pressure force on the surface from all fluids. The point of application of the resultant force is defined by requiring that the net hydrostatic moment on the surface from all fluids is equal to the moment of the resultant force when this force is acting at the point of application. That is, $\mathfrak{R} \times \mathbf{F}_S = \int_S [\mathbf{r} \times (-p\mathbf{n})] \, dS$.

There are many situations in which an object is at rest but partially or fully immersed in fluid. If the fluid is at rest or moving slowly, the pressure field is hydrostatic, and the hydrostatic force on the object is identical to the total surface force applied by the fluid to the immersed body. The force resulting from hydrostatic pressure on a stationary object under these conditions is called a buoyancy force. An object free to move will float if the buoyancy force is equal to its weight. If the buoyancy force is greater than its weight, the object will rise unless restrained. If the buoyancy force is less than an object's weight, the object sinks. If an object is held in place by a restraining force created by a structural element of some type, then the difference between the buoyancy force and the weight of the object must be balanced by a restraining force.

The buoyancy force on a submerged object is equal to the body force acting on the displaced volume of fluid. For the case of constant density fluid in Earth's gravity field, the buoyancy force is equal in magnitude to the weight of the displaced fluid and acts vertically. This result is called Archimedes' principle.

In the absence of a restraining force, equilibrium for a neutrally buoyant body is analyzed by examining the locations of the center of gravity and center of buoyancy. The general rule for a fully immersed object is that a position of equilibrium with the center of gravity above the center of buoyancy is unstable. A floating body, however, may be in a position of stable equilibrium, with its center of gravity slightly above its center of buoyancy. Thus one can encounter a configuration that is stable for small roll angles, but unstable if the roll is large enough.

PROBLEMS

Section 5.1

5.1 Describe the similarities and differences between fluid statics and fluid dynamics. Give an example of an engineering problem that falls into each category.

5.2 Is it ever possible for the equations of fluid statics to apply to a moving fluid? If so, give an example. If not, explain why not.

5.3 In the text it was stated that the air in a hurricane or tornado may move at speeds of 150 mph or more. If the ambient hydrostatic pressure in the atmosphere is approximately 14.7 psia, verify that the change in pressure from this value due to a flow at 150 mph may be estimated using ideas from Section 2.3 as $\Delta p = 0.398$ psia.

5.4 In Example 5.1 we showed that if the water near the intake pipe of a dam is moving at 1 m/s and the intake pipe is located 10 m below the surface, the corresponding decrease in pressure is ~0.5%. How fast would the water have to be moving for the corresponding pressure decrease to be ~10%?

Section 5.2

5.5 Explain why the stress vector acting on a surface in contact with a stationary fluid must be purely compressive in its action.

5.6 Use a procedure similar to that demonstrated in the text to show that $\sigma_{N3} = \sigma_N$.

5.7 For a fluid at rest, describe the relationship between the normal stress in the fluid, the stress vector, and the static pressure.

Section 5.3

5.8 Provide a physical interpretation of Eq. 5.2. Which terms in this equation, if any, may be functions of time? Which terms in this equation, if any, may be functions of position?

5.9 What is the pressure increase per meter of depth in a gasoline storage tank?

5.10 Compare the heights of columns of water, carbon dioxide, and mercury required to generate a pressure of 1.1 atm at the base of the column. Assume the top of the column is exposed to air at a pressure of 1 atm.

5.11 A bottle of a certain type of salad dressing can be considered to be a mixture of olive oil and vinegar. The oil has a density of 0.92 g/cm^3 and the vinegar has a density of 1.01 g/cm^3. If an unopened and unshaken bottle of dressing contains 3 in. of oil on top of 5 in. of vinegar, estimate the pressure acting on the bottom surface of the bottle. What is the pressure acting on the side of the bottle at the line between the oil and the vinegar? What is the pressure acting on the bottom surface of the bottle after the bottle is shaken so that the oil and vinegar are (temporarily) thoroughly mixed?

5.12 Consider the linear density distribution given in Eq. 5.6. Use Eq. 5.5 to verify that $\bar{\rho} = (\rho_H + \rho_B)/2$. Is the pressure at a depth of $H/2$ greater than or less than $(p_B - p_A)/2$?

5.13 What is the average density of a layer of fluid of thickness H, characterized by a density distribution of the form $\rho(z) = \rho_H - (\rho_B - \rho_H)[(z/H)^2 - 1]$? What is the corresponding pressure at the bottom of this layer of fluid?

5.14 Consider two columns of fluid of equal dimensions. Both fluid columns contain the same fluid and experience a linear temperature profile with the same high and low boundary temperatures. The only difference is that the first column has the hot fluid on top (a stable situation) and the second column has the cold fluid on top (at best a metastable situation). Which column has a higher pressure at its base? Why?

5.15 Figure P5.1 shows an undersea structure with a shape that can be approximated as one half of a cylinder. If the ocean floor is 40 m below the surface, estimate the total surface force acting on this structure.

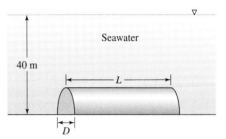

Figure P5.1

5.16 Provide a physical interpretation of Eq. 5.9.

5.17 A variable density fluid whose linear density distribution is given by Eq. 5.6 is at rest in Earth's gravity field. Find the pressure gradient acting in the fluid at a depth of $H/3$.

5.18 A fluid whose density distribution is given by $\rho(z) = \rho_H - (\rho_B - \rho_H)[(z/H)^2 - 1]$ is at rest in Earth's gravity field. Find the pressure gradient acting in the fluid at a depth of $2H/3$.

5.19 The cylinder described in Example 5.5 is filled with water and has a radius of 6 ft and a length of 10 ft. If the angular velocity is 0.5 rad/s, determine the pressure gradient at the following points:
(a) $r = 0, z = 0$
(b) $r = 0, z = 5$ ft
(c) $r = 3$ ft, $z = 5$ ft

Section 5.4

5.20 Use the procedure outlined in the text to verify that $\partial p/\partial x = \partial p/\partial y = 0$ for a fluid at rest in Earth's gravity field (with z upward). State any assumptions.

5.21 Consider a Cartesian coordinate system defined so that the z points up (opposite to the direction of the gravitational body force). Is it possible for the hydrostatic pressure to be a function of the *x* coordinate? If so, how? If not, why not?

Section 5.4.1

5.22 Find the pressure distribution in the salad dressing bottle described in Problem 5.11 before it is shaken. What is the pressure distribution after the bottle is shaken?

5.23 Use Eq. 5.14 to determine the pressure distribution in Earth's atmosphere. Let the reference pressure be atmospheric pressure at zero elevation. What is the rate of decrease in pressure with elevation in air?

5.24 Find an expression relating the height of the liquid level of the piezometer shown in Figure P5.2 to the pressure at the connection to the pipe.

5.25 A closed tank is partially filled with glycerin. If the air pressure above the glycerin is 300 kPa, what is the pressure in the glycerin at a depth of 0.75 m below the glycerin–air surface?

5.26 The underground gasoline storage tank illustrated in Figure P5.3 has developed a leak such that water has entered the tank as shown. For the dimensions given, determine the hydrostatic pressure at the following points:
(a) The water–gasoline interface
(b) The base of the tank
(c) The free surface of the gasoline

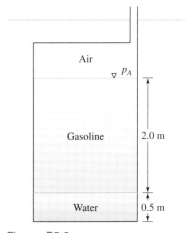

Figure P5.3

5.27 The device shown in Figure P5.4 is filled with hydraulic fluid ($\rho = 0.88$ g/cm^3). If the diameter of the larger cylinder is 8 cm, determine the weight of the unloaded plate if the corresponding height *h* is 3 cm. If a 10 kg mass is placed on the plate, calculate the corresponding fluid height *H*.

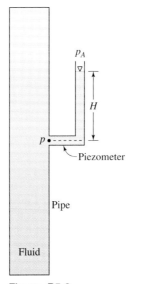

Figure P5.2

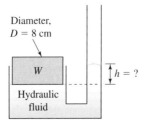

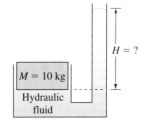

Figure P5.4

5.28 The hydraulic jack shown in Figure P5.5 is filled with fluid of density 0.88 g/cm³. What magnitude force, **F**, is required to lift a car that weighs 2200 lb$_f$? State any assumptions.

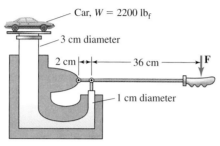

Figure P5.5

5.29 As shown in Figure P5.6, a U-tube manometer filled with mercury is connected to a water supply line. Use the information given in the figure to determine the water pressure at the manometer location. State any assumptions.

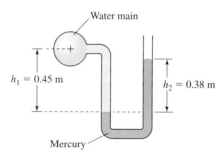

Figure P5.6

5.30 Consider the geometry shown in Figure P5.7. Assuming the water has a density of 1.0 g/cm³, estimate the density of the unknown fluid.

5.31 Determine the unknown height H in Figure P5.8.

5.32 Determine the pressure difference between points A and B in Figure P5.9. What might cause such a pressure difference in a horizontal tube?

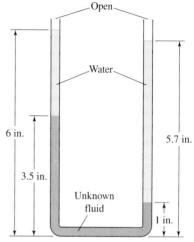

Figure P5.7

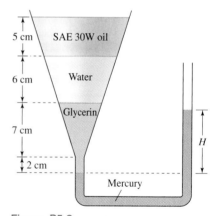

Figure P5.8

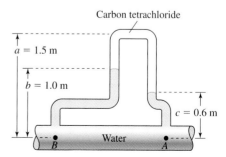

Figure P5.9

5.33 Consider the inclined manometer shown in Figure 5.12. If the working fluid is mercury and the inclination angle is 20°, calculate the pressure resolution of the instrument (i.e., the smallest change in pressure that can be measured accurately) if the scale L is marked in $\frac{1}{8}$ in. intervals. How does the resolution change if the inclination angle is increased to 45°?

5.34 In Figure P5.10 all the fluids are at room temperature and the pressure in reservoir A is 120 kPa. The specific gravity of Meriam Red manometer oil is 0.827, and the density of the brine is 1.18 g/cm^3. Calculate the pressure in reservoir B.

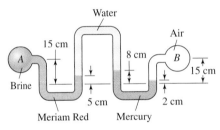

Figure P5.10

5.35 Figure P5.11 shows a micromanometer that can be used to measure small pressure differences in gases. Notice that the device has two working fluids, which are selected so that the density of fluid two is slightly greater than that for fluid one. Derive an expression

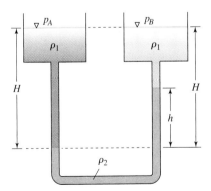

Figure P5.11

for the relationship between h and $\Delta p = p_A - p_B$. Assume that the reservoirs are large enough that H remains constant. What two working fluids would you recommend for this application? Why?

Section 5.4.2

5.36 Find the pressure distribution for the fluid layer described in Problem 5.13. What do the surfaces of constant pressure look like in this situation?

5.37 Find the pressure distributions for the two columns of fluid described in Problem 5.14. What do the surfaces of constant pressure look like in this situation?

5.38 The variation of density with pressure in water can be approximated by $(p + B)/(p_{\text{Atm}} + B) = (\rho/\rho_0)^N$, where the constant $N = 7$, $B = 3000$ atmospheres, the pressure p is in atmospheres, and ρ_0 is the density of water at 1 atmosphere. Find the pressure distribution in water using this relationship and compare it to the constant and linear density profiles. Calculate the pressure 10,000 m beneath the surface and compare the result with that obtained from the constant density model.

5.39 In the text we discussed the isothermal model for the pressure distribution of a stationary gas under the action of gravity. A more complicated model of the pressure–density relationship in a gas is provided by the polytropic law, $p/p_0 = (\rho/\rho_0)^n$, where n is the polytropic exponent. If n is set equal to γ, the specific heat ratio (1.4 for air), the pressure–density relationship is isentropic. Find the pressure–height relationship in a polytropic gas, and plot the pressure distribution as a function of height with $n = 1.25$. Use the polytropic model with $n = 1.25$ to determine the pressure in the Earth's atmosphere at an altitude of 2 km and compare this result with the corresponding predictions from the isentropic model, the isothermal model, and the standard atmosphere.

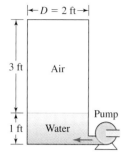

Figure P5.12

5.40 The cylindrical tank shown in Figure P5.12 is being filled with water by means of a pump that can generate a maximum exit pressure of 36 psig. Just before the pump is turned on, the air above the water is at STP and the height of the water is 1 ft. Assuming the air undergoes isothermal compression, estimate the height of the water in the tank when the pump stops, i.e., when the pump can no longer raise the water pressure.

5.41 Airplane passengers experience ear "popping" as a result of cabin pressure changes. The same phenomenon can occur as you drive through a mountain range. Suppose that you have begin at a sea level and drive up the side of a mountain until your ears "pop." You note that the first "pop" occurs at 250 ft and have been told that sequential "pops" will occur for constant pressure changes. Estimate the elevation at which the second "pop" occurs if:
(a) The atmosphere has a constant density.
(b) The atmosphere is adiabatic.

5.42 Assume that the atmosphere on Mars is composed of CO_2 and that the surface temperature averages 200 K. If the density of the atmosphere at zero elevation is 0.015 kg/m^3 and the gravitational constant on Mars is about 40% of that on Earth, estimate:
(a) The value of atmospheric pressure on the surface of Mars.
(b) The atmospheric pressure at an elevation of 10 km above the surface of Mars. State any assumptions.

Section 5.4.3

5.43 Show that the pressure distribution given by Eq. 5.25 satisfies the hydrostatic equation.

5.44 Consider a partially filled cylindrical container of water rotating about its axis with a free surface as shown in Figure 5.19. The container is open and the pressure above the free surface is ambient. Use Eqs. 5.25 and 5.26 to find an equation for the free surface. Compare your result with Eq. 5.27. [*Hint:* Choose a datum location on the free surface at its intersection with the cylinder axis and note that the pressure on the surface of interest is atmospheric pressure.]

5.45 At a garage sale you purchase an ancient piece of stereo equipment known as a turntable. When the turntable is installed in your living room, you decide to place a partially filled glass of water ($h = 8$ in., $d = 3$ in.) on the center of the rotating surface.
(a) If an angular velocity of $33\frac{1}{3}$ rpm raises the level of the water to the rim of the cup as shown in Figure P5.13, what was the height of the water in the cup when the fluid was at rest?
(b) Determine the location and magnitude of the maximum pressure in this rotating fluid.

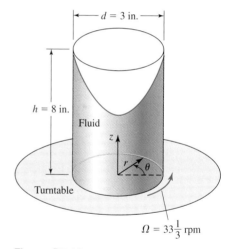

Figure P5.13

5.46 A U-shaped glass tube filled with mercury at STP is rotated around a vertical axis as shown in Figure P5.14. If the angular velocity is 3.5 rev/s, calculate
(a) The angle of the mercury–air interface
(b) The location and magnitude of the maximum pressure in this rotating fluid
(c) The location and magnitude of the minimum pressure in this rotating fluid

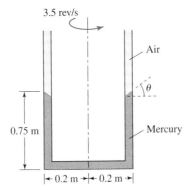

Figure P5.14

5.47 A 55-gallon drum initially full of SAE 30W oil is rotated around its vertical axis at an angular velocity Ω given in units of rpm. What value of Ω is required such that enough oil spills out of the drum so that the center of the bottom is just barely exposed?

5.48 A cylindrical bucket of height 30 cm and diameter 15 cm is half-filled with water. A string is tied to the handle of the bucket so that the distance from the end of the string to the free surface of the water is 0.8 m. You hold the end of the string and swing the bucket in a circular path in a horizontal plane; determine the pressure at the bottom of the bucket.

5.49 The bucket described in Problem 5.48 is now swung in a circular path in a vertical plane. Calculate the pressure at the bottom of the bucket when:
(a) The bucket is at its highest point
(b) The bucket is at its lowest point
(c) The string is horizontal

5.50 Reconsider the U-shaped tube illustrated in Figure P5.14. Suppose that this tube is rotated around one of its vertical legs. At what rotational velocity will the difference in height between the mercury in the two legs be 20 cm?

5.51 A long transparent cylindrical tube of diameter of 25 cm contains 10 cm of water resting on top of 15 cm of mercury. If the cylinder is rotating at 20 rad/s, calculate:
(a) The shape of the water–mercury interface.
(b) The shape of the water–air interface.
(c) The location and magnitude of the maximum pressure in the rotating fluid.

5.52 A plane in level flight at a speed of 700 km/h begins a turn to the left along a circular path of radius 35 km. A young passenger is holding a balloon filled with helium. Will the balloon string remain vertical, tilt to the left (toward the center of the turn), or tilt to the right (away from the center of the turn)? Why?

Section 5.4.4

5.53 A container of fluid is placed in an elevator and accelerated upward at 1 g. What is the total body force per unit mass acting on the fluid?

5.54 It is proposed to use a U-shaped glass tube filled with mercury as a simple accelerometer as shown in Figure P5.15. What range of acceleration can be measured?

5.55 A 55-gallon drum of SAE 30W oil (diameter = 22.5 in.) rests on the floor of an elevator. If the elevator is has an upward acceleration of 5 ft/s², calculate:
(a) The total body force per unit mass acting on the fluid
(b) The fluid pressure at the bottom of the drum

5.56 A cubical container of water with an edge dimension of 2 m rests on the floor of an elevator.
(a) If the upward acceleration is 4 m/s², calculate the absolute and gauge pressure at the bottom of the tank.

(b) What acceleration is required to give a gage pressure at the bottom of the tank of 0 kPa?

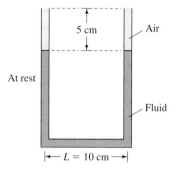

Figure P5.15

5.57 A gasoline reservoir towed by a certain brand of tanker truck can be modeled as a horizontal cylinder of length 7.5 m and diameter 2.5 m, which is completely filled with gasoline.
(a) Determine the pressure difference between the front surface of the tank and the back surface of the tank (along the centerline) when the truck is undergoing a horizontal acceleration of 1.5 m/s².
(b) Determine the pressure difference between the front surface of the tank and the back surface of the tank (along the centerline) when the truck is undergoing a horizontal deceleration of 2.5 m/s².

5.58 The tank of fluid shown in Figure P5.16 is experiencing a constant acceleration to the right.

(a) Calculate the magnitude of a_x if the fluid in the tank is glycerin.
(b) For the acceleration found in part a, determine the location and magnitude of the maximum pressure in the tank.
(c) Calculate the magnitude of a_x if the fluid in the tank is water.

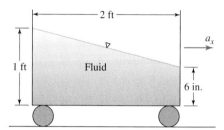

Figure P5.16

5.59 The test driver for the Tar Foot drag racing team rests his coffee cup on the dashboard of his dragster during time trials. The cup is 4 in. tall and has a diameter of 3 in. When the dragster reaches its maximum (constant) acceleration of 3 ft/s², determine the angle θ associated with the free surface and the height h (see Figure P5.17). Assume that the density of coffee is 63.1 lb$_m$/ft³ and that the cup is $\frac{3}{4}$ full at the start of the race.

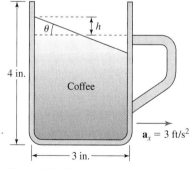

Figure P5.17

5.60 The test driver for the Blue Devil racing team has a coffee cup in her dragster

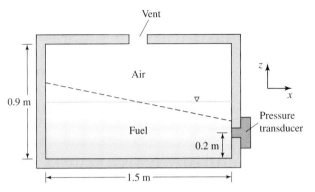

Figure P5.18

that is identical to the one described in Problem 5.59 (except it is restrained in an appropriate cup holder so it doesn't slide off the dash and land on the driver's lap). During the period of constant acceleration, the coffee in the cup just reaches the top of the mug (i.e., $h = 1$ in.). Determine the acceleration of the car.

5.61 The gas tank for a light truck is illustrated schematically in Figure P5.18. The pressure transducer serves as the active component for the fuel gage.
(a) When the tank is half-full, determine the equation that relates the pressure at the gage to the constant horizontal acceleration of the truck.
(b) How does the relationship change if the tank is three-quarters full?
(c) If the tank is half-full, at what acceleration will the gasoline level drop below the gage port?
(d) Can you suggest a better location for the location of the pressure gage? Justify your selection.

5.62 Your mom has talked you into moving her friend's prize-winning goldfish for her. The fishtank is 30 cm high and has a rectangular base 25 cm × 65 cm. You are going to transport the tank in the trunk of your car, and you know that during acceleration the water surface will tilt in a manner similar to that illustrated for the gas tank in Problem 5.61. Because you don't want water all over your trunk, you have removed enough water to bring the level at rest to 3 cm below the top of the tank.
(a) Is it better to orient the tank so that the long dimension is parallel or perpendicular to the direction of motion? Why?
(b) For the orientation you recommend in part a, calculate the maximum acceleration for safe (dry) transport.

5.63 A half-filled tank of water is moving up an incline as shown in Figure P5.19.
(a) Find the magnitude and direction of the total body force acting on this fluid.
(b) Find the maximum acceleration in the direction of motion such that no fluid spills over the back wall.
(c) Find the location and magnitude of the maximum pressure in the fluid for the maximum acceleration calculated in part b.

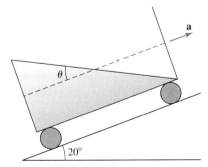

Figure P5.19

5.64 A half-filled tank of glycerin is moving down an incline under the action of gravity as shown in Figure P5.20. Find the angle θ associated with the free surface of the fluid. Explain your result.

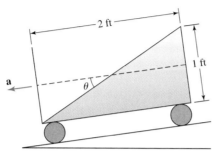

Figure P5.20

5.65 A child sitting in an airplane is holding a helium-filled balloon by a string. During takeoff (acceleration in the $+x$ direction), will the balloon string remain vertical, tilt forward, or tilt backward? Why?

5.66 A luggage strap is hanging out of an overhead compartment on an airplane. During the deceleration associated with landing, will the luggage strap hang straight down, tilt toward the front of the plane, or tilt toward the back of the plane? Why?

Section 5.5

Section 5.5.1

5.67 Suppose you are asked to use Eq. 5.31 to find the hydrostatic force on the bottom of the engine sump in Example 5.15.
(a) What type of surface is this, and what is the value of the pressure in the integrand of the surface integral?
(b) Find the outward unit normal on the bottom of the engine sump.

5.68 Suppose a flattened cylindrical rubber stopper of diameter 3 cm covers the drain hole in your bathtub. If the water depth is 10 cm and the water temperature is 40°C, estimate the total hydrostatic force on the stopper.

5.69 Concrete walls are often formed by pouring the concrete between temporary forms. One specific wall is to be 8 ft high, 1 ft thick, and 20 feet long. Thus, two forms, each 8 ft × 20 ft, will be used. If the wet concrete has a density of 150 lb_m/ft^3, calculate the total hydrostatic force on each vertical form.

5.70 The concrete forms described in Problem 5.69 must contain access ports.
(a) In one form the access port is a square with an edge length of 15 in. The top edge of the port is located 2 ft below the free surface of the concrete. What is the total hydrostatic force acting on this port?
(b) The other form contains a circular access port of diameter 15 in. The top of this port is also located 2 ft below the free surface of the concrete. What is the total hydrostatic force acting on this port?

5.71 Concrete is poured into the forms shown in Figure P5.21 to create a stand for the winners of swimming ribbons at the local community pool. After the concrete has been poured, two sandbags are placed as shown to stop the form from lifting off the ground. The concrete has a density of 2400 kg/m³; estimate the minimum total weight of the two sandbags.

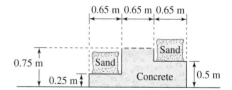

Figure P5.21

5.72 The hot water heater shown in Figure P5.22 is fitted with a drain tube as shown. A circular plug is used to seal the drain. Calculate the total hydrostatic force on the plug, assuming that the water is at room temperature. Will the force on the plug be greater or less than this value when the water is heated?

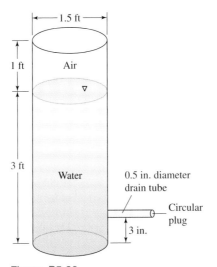

Figure P5.22

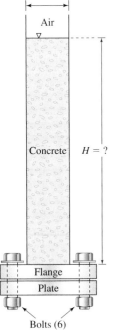

Figure P5.23

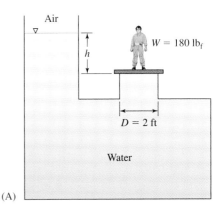

(A)

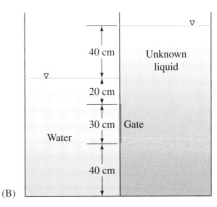

(B)

Figure P5.24

5.73 An open rectangular tank 2 m in width and 4 m in length contains 16 m³ of gasoline and 28 m³ of water at 20°C. The fluid is arranged in two stable layers. Determine the total hydrostatic stress on:
(a) The bottom of the tank
(b) One of the two smaller vertical sides of the tank
(c) One of the two larger vertical sides of the tank

5.74 Concrete columns are being formed by using the arrangement shown in Figure P5.23. The base plate is attached to the vertical structure with 6 bolts, each of which is $\frac{3}{8}$ in. in diameter with a failure strength of 40,000 lb$_f$/in.². If the density of the concrete is 150 lb$_m$/ft³, what is the tallest column that can be constructed using this method?

5.75 Your friend (weight = 180 lb$_f$) is standing on the device shown in Figure P5.24A. The gate on which he is standing covers a circular pipe of diameter 2 ft. What is the maximum height, h, of water above the level of his feet necessary to cause water to leak out of the pipe?

5.76 Determine the density of the fluid on the right in Figure P5.24B such that the net hydrostatic force on the rectangular gate is zero.

5.77 Repeat Problem P5.76, but this time let the gate have a circular shape.

5.78 If the "unknown fluid" in Problem P5.76 is SAE 30W oil, determine the magnitude and direction of the net hydrostatic force acting on the circular gate.

5.79 Consider the pressurized tank shown in Figure P5.25. The manometer is filled with Meriam Red manometer oil with a specific gravity of 0.827. Calculate the total hydrostatic force on the rectangular gate shown if its width is 4 ft.

5.81 For the tank shown in Figure P5.26, if the right side is filled with gasoline and $H = 4$ m, calculate the total hydrostatic force on gates C and D. Assume the structure has a width of 4 m.

5.82 For the tank shown in Figure P5.26, for what height of water H will the magnitude of the total hydrostatic force on all four gates A–D be equal? Assume the structure has a width of 4 m.

5.83 For the tank shown in Figure P5.26, if it is known that the magnitude of the total hydrostatic force on all four gates A–D is the same and the height of the fluid in the right side of the tank is $H = 6$ m, determine the density of the fluid in the right side of the tank.

5.84 Consider the tank shown in Figure P5.27. Calculate the total hydrostatic force on the gate for each of the following gate shapes:
(a) A rectangular gate with width into the paper of 2 ft
(b) A circular gate

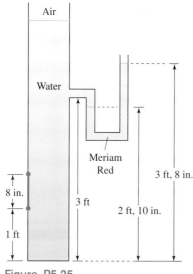

Figure P5.25

Section 5.5.2

5.80 Consider the fluid container shown in Figure P5.26. Determine the total hydrostatic force on gate A. What is the total hydrostatic force on gate B? Assume the structure has a width of 2 m.

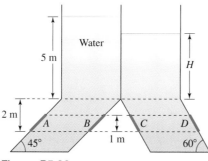

Figure P5.26

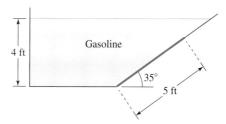

Figure P5.27

5.85 A pressurized tank is shown in Figure P5.28. The gas pressure is 30 psig and the liquid is nitric acid with a specific gravity of 1.5. Determine the total hydrostatic force on the plug (which is in the shape of a truncated cone) shown in the figure.

5.86 A concrete dam with a triangular cross section is shown in Figure P5.29. Calculate the total hydrostatic force on the dam due to the water if it is 800 ft wide.

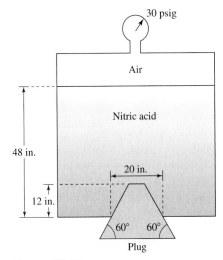

Figure P5.28

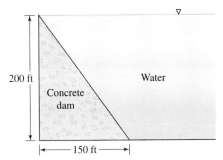

Figure P5.29

5.87 Consider the gate shown in Figure P5.30. For what water height H is the gate

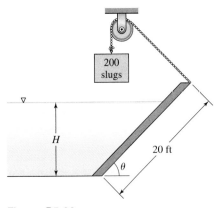

Figure P5.30

in equilibrium? Assume the gate has negligible mass, a width of 1 ft, and that $\theta = 45°$.

5.88 Consider the gate shown in Figure P5.30. The gate is 5 ft wide normal to the plane of the paper, the mass of the gate is 50 slugs, and $\theta = 30°$. For what water height is the gate in equilibrium?

5.89 A steel gate of density 7.8 g/cm³ and thickness 1 cm is shown in Figure P5.31. What is the maximum water depth, H, that can be held back by the gate if $\theta = 45°$?

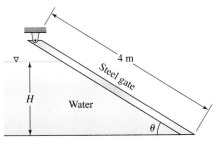

Figure P5.31

5.90 For the steel gate of density 7.8 g/cm³ and thickness 1 cm shown in Figure P5.31, calculate the angle θ if the gate is in equilibrium when the water depth is 2.5 m.

5.91 Consider the submarine shown in Figure P5.32.
(a) Calculate the total hydrostatic force on the hatch for the geometry shown.

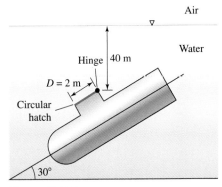

Figure P5.32

(b) What depth would the submarine have to attain in a perfectly horizontal orientation for the total hydrostatic force on the hatch to be the same as that calculated in part a?

Section 5.5.3

5.92 Consider the fluid tank shown in Figure P5.33. Calculate the total hydrostatic force per unit width on gate A. Do you think the corresponding total force on gate B will be higher or lower than that on gate A? Why?

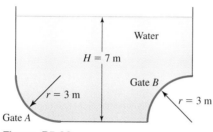

Figure P5.33

5.93 Consider the fluid tank shown in Figure P5.33. Calculate the total hydrostatic force per unit width on gate B. Do you think the corresponding total force on gate A will be higher or lower than that on gate B? Why?

5.94 Consider the fluid tank shown in Figure P5.34. Calculate the total hydrostatic force per unit width on gate A if $a = 0.3 \text{ ft}^{-1}$. Do you think the corresponding total force on gate B will be higher or lower than that on gate A? Why?

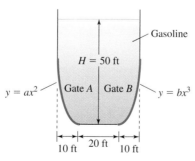

Figure P5.34

5.95 Consider the fluid tank shown in Figure P5.34. Calculate the total hydrostatic force per unit width on gate B if $b = 0.03 \text{ ft}^{-2}$. Do you think the corresponding total force on gate A will be higher or lower than that on gate B? Why?

5.96 A water heater company produces tanks with the designs shown in Figure P5.35. Calculate the total hydrostatic force on the base of tank A given that the water temperature is 40°C and the radius of the hemispherical base is 25 cm. Will the total hydrostatic force on the base of tank B be higher or lower than that for tank A? Why?

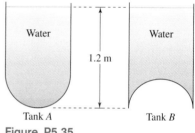

Figure P5.35

5.97 For hot water tanks with the designs shown in Figure P5.35, calculate the total hydrostatic force on the base of tank B given that the water temperature is 50°C and the radius of the hemispherical base is 25 cm. Will the total hydrostatic force on the base of tank A be higher or lower than that for tank B? Why?

5.98 Suppose the gasoline tanker truck described in Problem 5.57 has hemispherical end caps. Calculate the total hydrostatic force on one of the end caps when the truck is at rest.

5.99 As shown in Figure P5.36, a Tainter gate has the shape of a partial cylinder. Calculate the total hydrostatic force per unit width on the gate for this geometry. Note that the water height corresponds to the height of the gate pivot point.

5.100 Calculate the total hydrostatic force per unit width on the gate shown in Figure P5.37. The gate is composed of a lower

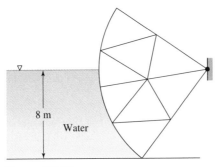

Figure P5.36

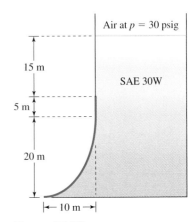

Figure P5.38

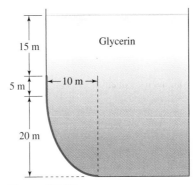

Figure P5.37

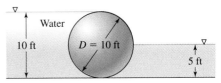

Figure P5.39

curved section and an upper vertical section. The curved section is parabolic and is described by the equation $y = cx^2$, where $c = 0.2\,\text{m}^{-1}$.

5.101 Calculate the total hydrostatic force per unit width on the gate shown in Figure P5.38. The gate is composed of a lower curved section and an upper vertical section. The curved section is cubic in shape and described by the equation $y = dx^2$, where $d = 0.2\,\text{m}^{-1}$.

5.102 Consider the geometry shown in Figure P5.39. If the fluid on the left is water, what density must the fluid on the right have if the total hydrostatic force on the cylinder (length = 50 ft) is zero?

5.103 A 5 ft diameter tank of length 10 ft is fabricated from two half-cylinders fastened together by 12 bolts as shown in Figure P5.40. Neglecting the weight of the container itself, calculate the force supported by each bolt when the tank is full of gasoline.

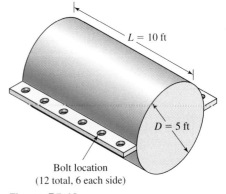

Figure P5.40

Section 5.6

Section 5.6.1

5.104 The forms described in Problem 5.69 were built to construct a concrete wall 8 ft high × 20 ft long × 1 ft thick. The wet concrete has a density of $150\,\text{lb}_m/\text{ft}^3$. Calculate the hydrostatic moment on one of the form walls about the top edge of the wall. Will the magnitude of the moment about the bottom edge of the same wall be greater or less than that about the top edge? Why?

5.105 The concrete forms described in Problem 5.104 must contain access ports. In one form the access port is a square with an edge length of 15 in. The top edge of the port is located 2 ft below the free surface of the concrete. If the port is hinged along its top edge, calculate the total hydrostatic moment acting on this port?

5.106 In Problem 5.73 we investigated an open rectangular tank of dimensions 2 m in width and 4 m in length contains $16\,\text{m}^3$ of gasoline and $28\,\text{m}^3$ of water at 20°C. The fluid is arranged in two stable layers. Determine the total hydrostatic moment on:
(a) The bottom of the tank if the bottom is hinged along one of its shorter ends
(b) One of the two smaller vertical sides of the tank if it is hinged along its lower edge
(c) One of the two larger vertical sides of the tank if it is hinged along its upper edge

5.107 Reconsider the tank shown in Figure P5.24b. If the "unknown fluid" is gasoline, determine the total hydrostatic moment on the rectangular gate assuming that it is hinged along its top edge and that it has a width of 50 cm.

5.108 Reconsider the tank shown in Figure P5.24B. If the "unknown fluid is SAE 30W oil, determine the total hydrostatic moment on the rectangular gate assuming that it is hinged along its bottom edge and that has a width of 50 cm.

5.109 Reconsider the pressurized tank illustrated in Figure P5.25. Calculate the total hydrostatic moment on the gate assuming that it is hinged along its bottom edge and that it has a width of 4 ft.

Section 5.6.2

5.110 Reconsider the fluid container illustrated in Figure P5.26. Recall that the structure has a width of 2 m.
(a) Calculate the total hydrostatic moment on gate A if it is hinged along its upper edge.
(b) Calculate the total hydrostatic moment on gate B if it is hinged along its lower edge.

5.111 Reconsider the fluid tank shown in Figure P5.26. Let the fluid on the right side be gasoline with height 4 m.
(a) Calculate the total hydrostatic moment on gate C if it is hinged along its lower edge.
(b) Calculate the total hydrostatic moment on gate D if it is hinged along its upper edge.

5.112 Reconsider the tank shown in Figure P5.27. Calculate the total hydrostatic moment on the gate for each of the conditions described.
(a) A rectangular gate hinged along its top edge with width into the paper of 2 ft
(b) A square gate hinged along its bottom edge

5.113 Reconsider the concrete dam shown in Figure P5.29. Calculate the total hydrostatic moment about the lower edge of the dam that is in contact with the water. Assume that the only surface of the dam in contact with fluid is the one touching the water.

5.114 Repeat Problem 5.113 but this time calculate the total hydrostatic moment about the lower edge of the dam that is not in contact with the water.

Section 5.7

5.115 Determine the resultant force and point of application for the concrete form (hinged at the top) described in Problem 5.104.

5.116 Determine the resultant force and point of application for the access port described in Problem 5.105.

5.117 Determine the resultant force and point of application for the tank sides described in Problem 5.106.

5.118 Determine the resultant force and point of application for the gate described in Problem 5.107.

5.119 Determine the resultant force and point of application for the gate described in Problem 5.108.

5.120 Determine the resultant force and point of application for the gate described in Problem 5.109.

5.121 Determine the resultant force and point of application for gate A described in Problem 5.110.

5.122 Determine the resultant force and point of application for gate D described in Problem 5.111.

5.123 Determine the resultant force and point of application for the rectangular gate described in Problem 5.112.

5.124 Determine the resultant force and point of application for the dam described in Problem 5.113.

5.125 Determine the resultant force and point of application for the dam described in Problem 5.114.

Section 5.8

5.126 Consider the geometry shown in Figure P5.41. The cube center is level with the surface of the liquid.
(a) If the mass M is 300 kg, calculate the density of the cube.
(b) For the density calculated in part a, will the cube float or sink if the mass is removed? If it floats, is it stable? Why or why not?
(c) Repeat part a, but this time let the partially submerged object be spherical, with its center level with the surface of the liquid.
(d) Repeat part b for the spherical object investigated in part c.
(e) Repeat part a, but this time let the centerline of the cube be located 3 m below the surface of the liquid.

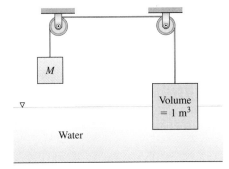

Figure P5.41

5.127 A hydrometer is a simple device for measuring the specific gravity, SG, of a fluid. As shown in Figure P5.42, SG is indicated by the level at which the free surface of the fluid intersects the floating hydrometer. For example, when the device is floating in pure water, the line marked 1.0 is even with the fluid surface. The diameter of this hydrometer is $\frac{3}{8}$ in. and its mass 1×10^{-6} slugs. Calculate the following.
(a) The relative position of the line marked 1.0 (i.e., distance above or below the fluid surface) when the hydrometer is immersed in ethylene glycol at room temperature.

(b) The density of the fluid in which the hydrometer floats with the 1.0 line 1 in., above the fluid level.

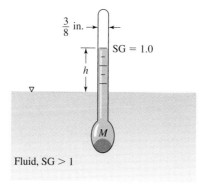

Figure P5.42

5.128 As shown in Figure P5.43, a simple spar buoy can be fabricated from a weighted piece of wood. A particular oak (SG = 0.71) spar buoy of dimensions 10 cm × 10 cm × 4 m is floating in seawater (SG = 1.02) such that $h = 60$ cm.
(a) Calculate the weight of steel (SG = 7.87) attached to the bottom of the buoy.
(b) Determine the value of h if an identical spar buoy is used in a pristine inland lake with SG = 1.00.

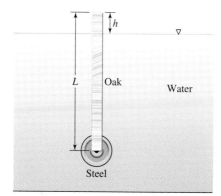

Figure P5.43

5.129 What fraction of the fatal iceberg would a passenger on the *Titanic* have seen above the waterline? Assume sg(ice) = 0.89 and sg(seawater) = 1.025.

5.130 Balloonists must decide whether to use hot air, hydrogen, or helium to provide lifting power. Calculate the lifting power (in lb_f/ft^3) for each of the options listed. In addition, determine the balloon volume required to lift a payload of 750 lb_f. Assume that the ambient air temperature is 70°F. State any additional assumptions.
(a) Hot air at 225°F
(b) Hot air at 325°F
(c) Helium (unheated)
(d) Hydrogen (unheated)

5.131 A spherical weather balloon of weight 16 N and radius 0.9 m is filled at room temperature with He at a pressure of 120 kPa. Determine the altitude in a standard atmosphere for which this balloon will be neutrally buoyant? How does the answer change if the balloon must carry a payload of 150 kg? State any assumptions.

5.132 A toy submarine is fabricated out of a sheet of aluminum of thickness 1 mm and density 2.71 g/cm³. Assume that the shape of the sub can be adequately modeled as a cylinder of diameter 8 cm. Determine the appropriate length so that the toy is neutrally buoyant when submersed in the bathtub. State any assumptions.

5.133 A toy soldier is cast form zinc ($\rho_{sol} = 7.13$ g/cm³) and has a volume of 7 cm³. The soldier is equipped with a foamed plastic life vest ($\rho_{vest} = 0.40$ g/cm³). Calculate the volume of the life vest if the toy is to be neutrally buoyant in water. State any assumptions.

5.134 A solid cube of material with a density of 1.04 g/cm³ and a volume of 1 cm³ is stable at the water–glycerin interface shown in Figure P5.44. Determine the distance between the top of the block and the water–glycerin interface.

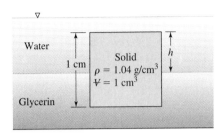

Figure P5.44

Section 5.9

5.135 A long cylindrical rod is a composite of two meterials. Half (0.5 m) of the rod is made of material A ($\rho_A = 500$ kg/m^3) and the other half is made of material B ($\rho_B = 1{,}500$ kg/m^3). Will the rod be neutrally buoyant in water? In what submerged orientation will the rod be stable?

5.136 Determine the locations of the center of gravity and center of buoyancy for the "point up" and "point down" orientations shown in Figure P5.45. Comment on the stability of each orientation.

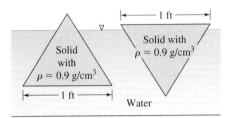

Figure P5.45

6 THE VELOCITY FIELD AND FLUID TRANSPORT

6.1 Introduction
6.2 The Fluid Velocity Field
6.3 Fluid Acceleration
6.4 The Substantial Derivative
6.5 Classification of Flows
    6.5.1 One-, Two-, and Three-Dimensional Flow
    6.5.2 Uniform, Axisymmetric, and Spatially Periodic Flow
    6.5.3 Fully Developed Flow
    6.5.4 Steady Flow, Steady Process, and Temporally Periodic Flow
6.6 No-Slip, No-Penetration Boundary Conditions
6.7 Fluid Transport
    6.7.1 Convective Transport
    6.7.2 Diffusive Transport
    6.7.3 Total Transport
6.8 Average Velocity and Flowrate
6.9 Summary
Problems

6.1 INTRODUCTION

In this chapter we begin the study of a fluid in motion, focusing primarily on the velocity field in the standard Eulerian description. After introducing the language and mathematics used to describe the velocity field, we show that acceleration is defined by a special derivative, called the substantial derivative. Next we show how engineers classify flow problems based on their geometric and temporal characteristics, and introduce the concept of the no-slip, no-penetration boundary condition. We conclude this chapter with a discussion of fluid transport. The transport of mass, momentum, energy, heat, and other substances by a moving fluid is of great interest to engineers. Thus we develop a general method for calculating the rate of transport of any property or substance across a surface by a moving fluid. We use this fluid transport model to define the concept of an average velocity, and to provide a foundation for control volume analysis, a powerful tool for analyzing flow problems without solving the governing differential equations.

6.2 THE FLUID VELOCITY FIELD

In the Eulerian description, a fluid is not thought of as consisting of molecules or fluid particles, but rather as an indivisible, continuous material that moves through space under the action of external and internal forces. The visual perspective of this description is that of watching a flow of material past a background coordinate grid as shown in Figure 6.1. Strictly speaking, there are no fluid particles in the Eulerian description because the fluid is indivisible. Nevertheless, the concept of a fluid particle remains useful, and it is customary to speak of a point in the fluid and a fluid particle interchangeably. That is, at every point in a fluid-filled space there is a fluid particle. We therefore define the value of velocity or any other fluid property at a point in the fluid as that of the fluid particle that happens to be at that point.

 CD/Kinematics/Field, Particles, and Reference Frames.

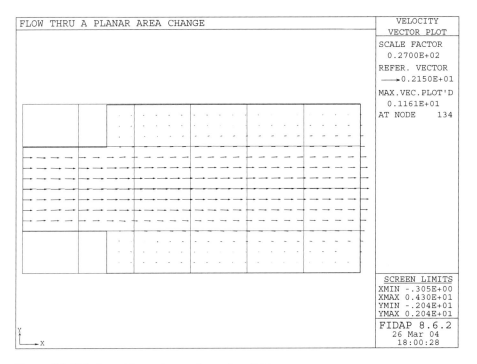

Figure 6.1 Eulerian grid superimposed on a flow field.

6.2 THE FLUID VELOCITY FIELD

Since the Eulerian description is a continuum model, both a Eulerian spatial point and the fluid particle that resides there are orders of magnitude larger than a molecule, but so small that there is no variation in any fluid or flow property on this scale. Thus, the mathematical representation of fluid and flow properties in this description is identical to the field description of vector calculus. A solution to a flow problem consists of knowing the appropriate scalar and vector functions describing various fluid and flow properties as a function of space and time.

The fundamental variable in fluid dynamics is the velocity field. This field is represented by a vector function

$$\mathbf{u} = \mathbf{u}(\mathbf{x}, t) \tag{6.1}$$

We say that $\mathbf{u}$ is the velocity at position $\mathbf{x}$ at time t. The presence of the independent variables $\mathbf{x}$ and t indicates that the velocity may vary throughout space at a given time, and over time at a given point. Therefore, in general, the velocity field is a three-dimensional, time-dependent vector field.

It is customary in Cartesian coordinates to use (u, v, w) for the three components of the velocity vector, with position represented by (x, y, z). Thus, as illustrated in Figure 6.2A, the velocity field $\mathbf{u} = \mathbf{u}(\mathbf{x}, t)$ is represented in Cartesian coordinates as

$$\mathbf{u} = u(x, y, z, t)\mathbf{i} + v(x, y, z, t)\mathbf{j} + w(x, y, z, t)\mathbf{k} \tag{6.2}$$

Cylindrical and polar coordinates are also frequently used in fluid mechanics. The velocity field is represented in cylindrical coordinates (r, θ, z) as

$$\mathbf{u} = v_r(r, \theta, z, t)\mathbf{e}_r + v_\theta(r, \theta, z, t)\mathbf{e}_\theta + v_z(r, \theta, z, t)\mathbf{e}_z \tag{6.3}$$

where (v_r, v_θ, v_z) are the three cylindrical velocity components as shown in Figure 6.2B. In polar coordinates we simply drop the third component from Eq. 6.3 and represent the velocity field by

$$\mathbf{u} = v_r(r, \theta, z, t)\mathbf{e}_r + v_\theta(r, \theta, z, t)\mathbf{e}_\theta \tag{6.4}$$

where the two velocity components are shown in Figure 6.2C.

There are a number of software packages that provide a graphical representation of a Cartesian vector field. These packages draw an arrow originating at each point in a certain region for which the corresponding vector is known. The length and direction of

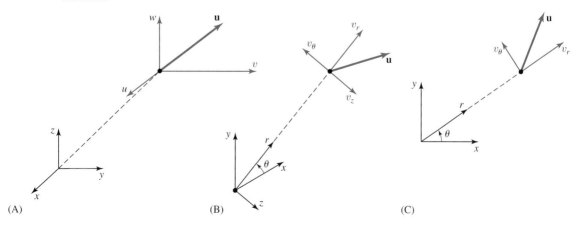

Figure 6.2 Velocity vectors in (A) Cartesian, (B) cylindrical, and (C) spherical coordinate systems.

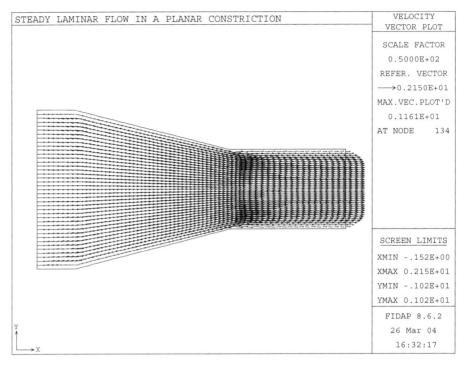

Figure 6.3 CFD solution of velocity vector field for flow through a constriction.

the arrow represent the magnitude and direction of the vector, respectively. An arrow color is often available along with the arrow length to indicate the vector magnitude. This representation, called a vector plot, is a standard tool in fluid mechanics. A vector velocity plot provides an easily grasped visual representation of a flow field as can be seen in Figure 6.3.

EXAMPLE 6.1

The velocity field for the pressure driven flow in the gap between the large parallel plates shown in Figure 6.4A is described in Cartesian coordinates by

$$u(y) = \left(\frac{h^2(p_1 - p_2)}{2\mu L}\right)\left[1 - \left(\frac{y}{h}\right)^2\right], \quad v = 0, \quad \text{and} \quad w = 0$$

where μ is the viscosity, $2h$ is the channel height, and the pressures are measured a distance L apart as shown. Determine the velocity of the fluid at the midpoint of the gap and on the plates. Plot the velocity vectors, and velocity profile in the gap.

SOLUTION

This exercise is solved by substituting the y coordinates for the points of interest into the velocity components.

At the top plate, $y = +h$, so that

$$u(h) = \left[\frac{h^2(p_1 - p_2)}{2\mu L}\right]\left[1 - \left(\frac{h}{h}\right)^2\right] = 0, \quad v = 0, \quad \text{and} \quad w = 0$$

At the midpoint, $y = 0$, so that

$$u(0) = \left[\frac{h^2(p_1 - p_2)}{2\mu L}\right]\left[1 - \left(\frac{0}{h}\right)^2\right] = \left[\frac{h^2(p_1 - p_2)}{2\mu L}\right], \quad v = 0, \quad \text{and} \quad w = 0$$

At the bottom plate, $y = -h$, so that

$$u(h) = \left[\frac{h^2(p_1 - p_2)}{2\mu L}\right]\left[1 - \left(\frac{-h}{h}\right)^2\right] = 0, \quad v = 0, \quad \text{and} \quad w = 0$$

Notice that the velocity at both stationary solid boundaries is zero, and that the maximum speed $U_{max} = [h^2(p_1 - p_2)]/2\mu L$ occurs at the centerline. Figure 6.4A shows a plot of the velocity profile $u(y) = U_{max}[1 - (y/h)^2]$.

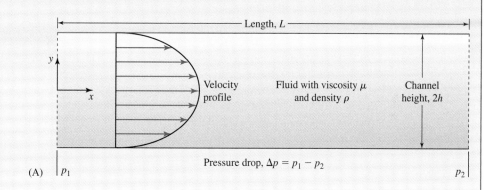

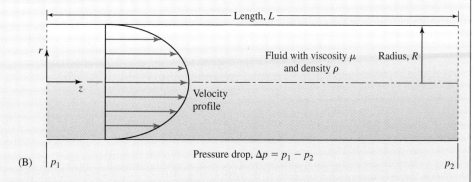

Figure 6.4 Coordinate systems and velocity distributions for fully developed flow (A) between parallel plates and (B) in a round pipe.

EXAMPLE 6.2

The pressure-driven flow in the round pipe shown in Figure 6.4B is described in cylindrical coordinates by $\mathbf{u} = 0\mathbf{e}_r + 0\mathbf{e}_\theta + v_z(r)\mathbf{e}_z$, with

$$v_z(r) = \frac{(p_1 - p_2)R^2}{4\mu L}\left[1 - \left(\frac{r}{R}\right)^2\right]$$

where μ is the viscosity, R is the pipe radius, and the pressures are measured a distance L apart as shown. Find the velocity of the fluid at the axis of the pipe and on the wall, and plot the velocity profile.

SOLUTION

This exercise is solved by substituting the r coordinates for the points of interest into the velocity components. At the wall, $r = +R$ so that,

$$v_z(R) = \frac{(p_1 - p_2)R^2}{4\mu L}\left[1 - \left(\frac{R}{R}\right)^2\right] = 0$$

On the axis, $r = 0$, so that

$$v_z(0) = \frac{(p_1 - p_2)R^2}{4\mu L}\left[1 - \left(\frac{0}{R}\right)^2\right] = \frac{(p_1 - p_2)R^2}{4\mu L}$$

Note that the fluid velocity at the stationary pipe wall is zero, and that the maximum speed is $U_{\max} = [(p_1 - p_2)R^2]/4\mu L$ at the centerline. Figure 6.4B shows a plot of the velocity profile $v_z(r) = U_{\max}[1 - (r/R)^2]$ for this flow. This pressure-driven flow through a round pipe is known as Poiseuille flow (see History Box 1.4).

It is often important to know the flow speed, i.e., the magnitude of the velocity, in a specific region. Contour plots of the speed are helpful in locating high and low velocity regions and in optimizing the design of a flow passage. It is also possible to plot contours of speed of individual velocity components. The contour plotting capabilities of modern software packages are illustrated by the speed and individual velocity component contours for flow through a constriction in Figure 6.5A–6.5C.

 CD/Kinematics/Streamlines and Streamfunctions.

An important tool for visualizing a flow is the concept of a streamline. As shown in Figure 6.6A, a streamline is defined to be a line in space that is everywhere tangent to the local instantaneous velocity vector. As a result the arrows of a vector velocity plot

6.2 THE FLUID VELOCITY FIELD   305

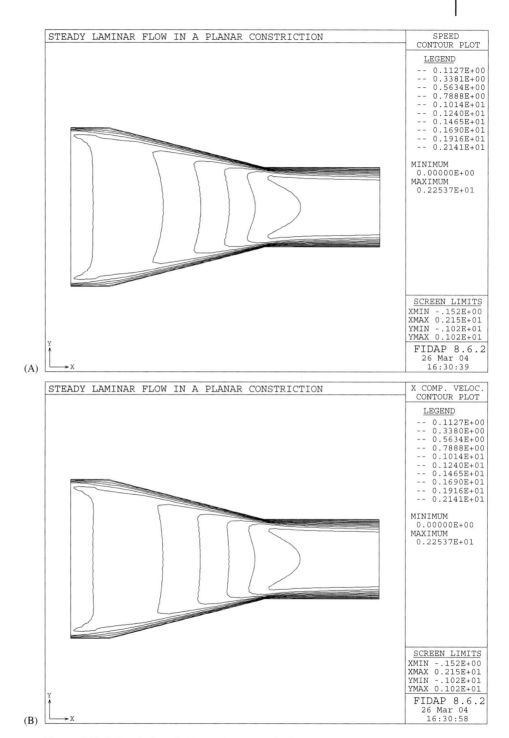

Figure 6.5 CFD solution of (A) speed contours for flow through a constriction and contours of (B) x and (C) y velocity components for the same flow.

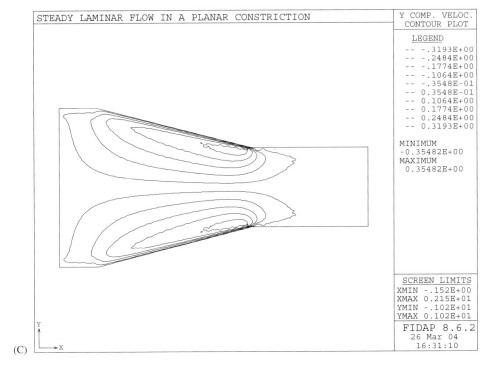

(C)

Figure 6.5 (*Continued*)

CASE STUDY

Consider the flow field over a cylinder at two Reynolds numbers in Figure 6.6B and 6.6C. Recall the case study on the drag of cylinders and spheres in Section 3.3.5. The creeping flow solution for $Re = 0.1$ in Figure 6.6B does not include the regions of flow reversal shown in Figure 6.6C for the flow at $Re = 50$. Visualization of flow over a cylinder highlights the complexity of this common flow and its dependence on Reynolds number.

are tangent to the streamlines at different locations. Plotting an entire family of streamlines is another effective way to visualize a fluid velocity field. Looking at Figures 6.6B and 6.6C, do you find the streamline plot helpful in understanding the flow over a cylinder?

Before concluding this section we note that a full description of a fluid in motion requires that we be able to specify such other fluid and flow properties as density, pressure, and temperature. These properties have magnitude only, hence are modeled by scalar functions of **x** and t. This representation is called a scalar field. For example, fluid density is represented in Cartesian coordinates by the scalar field $\rho = \rho(x, y, z, t)$. If we know the density field as an analytical function, then inserting the values of x, y, z, t into the function gives us the value of the fluid density at that location at that instant of time. Similarly, the pressure field and temperature field in a fluid are represented by scalar fields of the form $p = p(x, y, z, t)$, and $T = T(x, y, z, t)$. In the Eulerian description, every fluid and flow property is defined at each point in space occupied by the fluid.

 CD/Kinematics/Fields, Particles, and Reference Frames/Eulerian Representation/ Scalar Fields

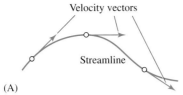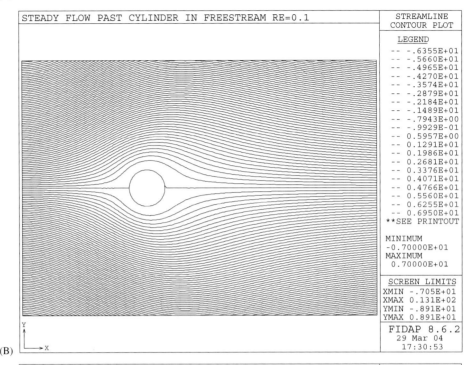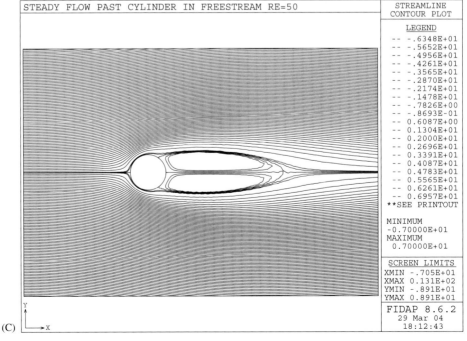

Figure 6.6 (A) Velocity vectors along a streamline. Streamlines for flow over a cylinder at (B) Re = 0.1 and (C) Re = 50.

6 THE VELOCITY FIELD AND FLUID TRANSPORT

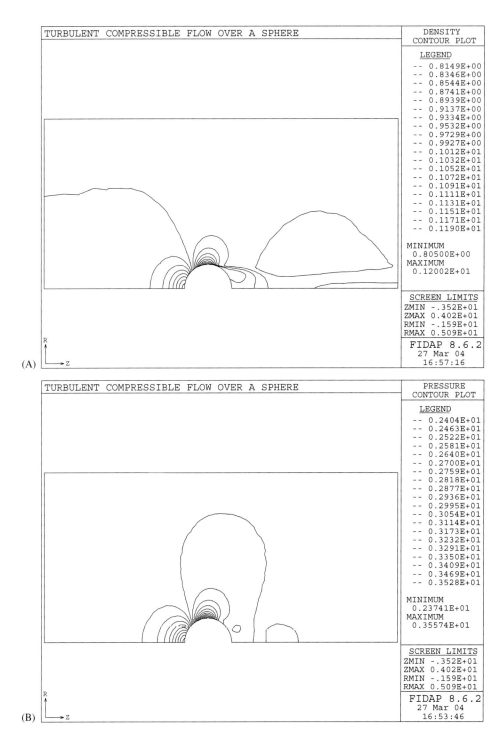

Figure 6.7 CFD contour plots of (A) density, (B) pressure, and (C) temperature, for turbulent compressible flow over a sphere.

6.2 THE FLUID VELOCITY FIELD 309

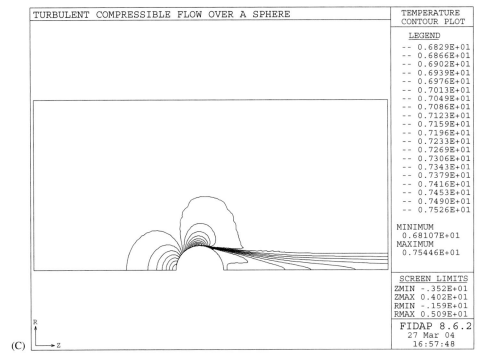

Figure 6.7 (*Continued*)

 General-purpose graphics packages and computational fluid dynamics codes provide contour plotting capabilities for visualizing the distribution of any Cartesian scalar field. The graphical output takes the form of contour lines and surfaces of constant property value. For example, contour plots of density, pressure, and temperature for a high speed gas flow over a sphere are shown in Figure 6.7.

EXAMPLE 6.3

Figure 6.8A shows the velocity field and streamlines for inviscid (frictionless) two-dimensional flow over a cylinder. The pressure distribution in this flow is given by

$$p(x, y) = p_\infty + \frac{1}{2}\rho U_\infty^2 \left[\left(\frac{2R^2}{r^4}\right)(x^2 - y^2) - \left(\frac{R^4}{r^4}\right) \right]$$

where p_∞ and U_∞ are the pressure and speed in the freestream far upstream, x and y are the usual Cartesian coordinates, R is the radius of the cylinder, and $r^2 = x^2 + y^2$. Plot pressure contours, and find the pressure distribution on the surface of the cylinder. Plot this pressure distribution, and find the locations and values of the maximum and minimum pressure on the surface. What is the force on a length L of the cylinder due to this pressure distribution?

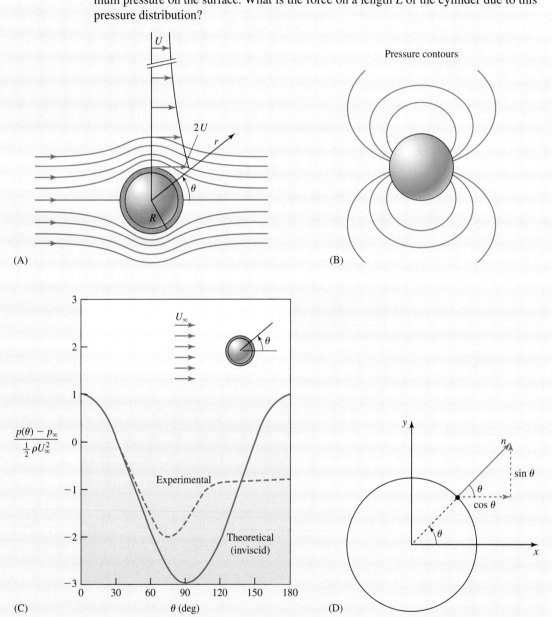

Figure 6.8 For flow over a cylinder the (A) streamlines and velocity profile, (B) pressure contours, (C) pressure distribution along the surface, and (D) surface normal vector.

SOLUTION

We are asked to determine the pressure distribution on the surface of a cylinder and to plot pressure contours for this flow. In addition, we must determine the locations of the pressure extremes on the surface and calculate the force acting on a cylinder of length L due to this pressure distribution. This flow is illustrated in Figure 6.8A, and we are told to assume that the flow is inviscid. Since $r^2 = x^2 + y^2$, we can substitute this into the pressure distribution to obtain the pressure contours in Cartesian coordinates as shown in Figure 6.8B.

The pressure distribution at a point (x_S, y_S) on the surface of the cylinder is obtained by noting that on the surface $x = x_S$, $y = y_S$, $r = R$, and $R^2 = x_S^2 + y_S^2$. Substituting this into the pressure distribution yields

$$p(x_S, y_S) = p_\infty + \frac{1}{2}\rho U_\infty^2 \left[\left(\frac{2R^2}{R^4}\right)(x_S^2 - y_S^2) - \left(\frac{R^4}{R^4}\right) \right]$$

Thus the pressure on the surface is given in Cartesian coordinates by

$$p(x_S, y_S) = p_\infty + \frac{1}{2}\rho U_\infty^2 \left[\frac{2}{R^2}(x_S^2 - y_S^2) - 1 \right] \tag{A}$$

Alternately, we might have recognized that the geometry suggests the use of cylindrical coordinates. The original pressure distribution may be written in cylindrical coordinates by making the substitution $x = r\cos\theta$, $y = r\sin\theta$, to obtain

$$p(r, \theta) = p_\infty + \frac{1}{2}\rho U_\infty^2 \left[\frac{2R^2}{r^4}(r^2\cos^2\theta - r^2\sin^2\theta) - \frac{R^4}{r^4} \right]$$

$$= p_\infty + \frac{1}{2}\rho U_\infty^2 \left[\frac{2R^2}{r^2}(\cos^2\theta - \sin^2\theta) - \frac{R^4}{r^4} \right]$$

which simplifies to

$$p(r, \theta) = p_\infty + \frac{1}{2}\rho U_\infty^2 \left[\frac{2R^2}{r^2}(1 - 2\sin^2\theta) - \frac{R^4}{r^4} \right] \tag{B}$$

On the surface of the cylinder $r = R$, and the surface pressure distribution is

$$p(R, \theta) = p(\theta) = p_\infty + \frac{1}{2}\rho U_\infty^2 \left[\frac{2R^2}{R^2}(1 - 2\sin^2\theta) - \frac{R^4}{R^4} \right]$$

$$= p_\infty + \frac{1}{2}\rho U_\infty^2 (1 - 4\sin^2\theta) \tag{C}$$

To plot this distribution, it is helpful to rearrange and normalize (C) as follows

$$\frac{p(\theta) - p_\infty}{\frac{1}{2}(\rho U_\infty^2)} = (1 - 4\sin^2\theta) \tag{D}$$

The left-hand side of (D) is in the form of the Euler number, given in general by Eq. 3.12a as $Eu = (p - p_0)/\frac{1}{2}(\rho V^2)$. Thus we could also write the pressure distribution as $Eu = (1 - 4\sin^2\theta)$.

It is evident from a plot of the pressure on the surface (Figure 6.8C) or from (D) that the maximum pressure occurs at $\theta = 0, \pi$, and the minimum pressure at $\theta = \pm \pi/2$. Checking the value of the maximum pressure, we find

$$p(\pi) = p_\infty + \tfrac{1}{2}\rho U_\infty^2 (1 - 4\sin^2 0) = p_\infty + \tfrac{1}{2}\rho U_\infty^2$$

which is the same as that predicted by Eq. 2.11, Bernoulli's equation, applied along the horizontal streamline from the far left and ending on the surface of the cylinder where the speed is zero. The minimum pressure is

$$p\left(\pm\frac{\pi}{2}\right) = p_\infty + \frac{1}{2}\rho U_\infty^2 \left[1 - 4\sin^2\left(\pm\frac{\pi}{2}\right)\right] = p_\infty - \frac{3}{2}\rho U_\infty^2$$

The force on a cylinder of length L due to this pressure distribution is found by using Eq. 4.23, $\mathbf{F}_S = \int_S -p\mathbf{n}\,dS + \int_S \boldsymbol{\tau}\,dS$ and recognizing that in this inviscid (frictionless) flow the tangential stress is zero. Thus we will calculate the force with $\mathbf{F}_S = \int_S -p\mathbf{n}\,dS$. From the symmetry in the pressure distribution, it is clear that there can be no net force on the cylinder in any direction. Is this confirmed by direct calculation? To set up the integral, we note that from Figure 6.8D the surface geometry is described by $\mathbf{n} = \mathbf{e}_r = \cos\theta\mathbf{i} + \sin\theta\mathbf{j}$, and $dS = R\,d\theta\,dz$. Thus we find

$$\begin{aligned}\mathbf{F}_S &= \int_S -p\mathbf{n}\,dS \\ &= \int_{-L/2}^{L/2}\int_0^{2\pi} -\left(p_\infty + \frac{1}{2}\rho U_\infty^2(1 - 4\sin^2\theta)\right)(\cos\theta\mathbf{i} + \sin\theta\mathbf{j})\,R\,d\theta\,dz\end{aligned} \quad (E)$$

Integrating this result gives zero, as expected.

6.3 FLUID ACCELERATION

Acceleration in a flow has a profound influence on the pressure distribution and other flow characteristics. Recall the case studies on flows in Chapter 3: flow through area change, the performance of pumps and fans, the flat plate boundary layer, and the forces on cylinders, spheres, and airfoils. Acceleration is part of the underlying nature of all these flows. This is evident in the flow over a cylinder in Figure 6.6. Notice the change in the direction and magnitude of the velocity vectors in this figure as the fluid passes around the cylinder. In this section we continue our analysis of the velocity field by discussing fluid acceleration. Consider the following questions: What is the relationship

6.3 FLUID ACCELERATION

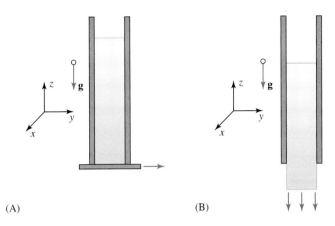

Figure 6.9 Vertical tube with (A) stationary fluid and (B) slug flow, which occurs immediately after the bottom end of the tube is opened.

between the acceleration $\mathbf{a}(\mathbf{x}, t)$ and the velocity field $\mathbf{u}(\mathbf{x}, t)$? How is this relationship expressed mathematically?

It is tempting to hypothesize that the mathematical connection between velocity and acceleration is given by the partial time derivative of velocity, as is true in dynamics. To test this theory, consider the experiment shown in Figure 6.9A in which a flow is created by opening the bottom end of a long vertical tube full of liquid. Gravity causes the liquid in the tube to accelerate downward; hence the velocity field is time dependent. In the first few moments after the flow starts, the liquid may be assumed to have a nearly spatially uniform velocity $\mathbf{u} = -U(t)\mathbf{k}$, as shown in Figure 6.9B. Since it appears as if the entire volume of fluid is moving together as a solid slug of material, a flow of this type is referred to as slug flow. In a slug flow, every fluid particle has the same velocity. Thus, the acceleration at each location and of each fluid particle must be the same. We can calculate the acceleration from the velocity field by using $\mathbf{a} = \partial \mathbf{u}/\partial t = -(dU/dt)\mathbf{k}$, which suggests that in this case the fluid acceleration is given by the partial time derivative of velocity as we hypothesized.

However, suppose we now consider the steady, turbulent flow of liquid through a short contraction as shown in Figure 6.10. Upstream of the contraction, turbulent mixing causes the liquid to have a nearly uniform velocity $\mathbf{u} = U_1\mathbf{i}$. The same is true downstream, where there is a uniform velocity $\mathbf{u} = U_2\mathbf{i}$. Suppose the areas at the upstream and downstream cross sections are A_1 and A_2, respectively. For mass to be conserved, the volume flowrate at the upstream section $U_1 A_1$ must equal the volume flowrate $U_2 A_2$ at the downstream section, i.e., $U_1 A_1 = U_2 A_2$. Since A_2 is less than A_1, U_2 must be greater than U_1. The liquid is moving faster downstream; thus the fluid must be accelerating through the contraction in response to the area change. Is the acceleration in this

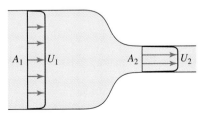

Figure 6.10 Velocity profiles for flow through a contraction.

flow given by the time derivative of velocity? A bit of thought should convince you that the velocity at any *particular point* in this flow does not change with time. If we use the time derivative of the velocity field to calculate the acceleration, we obtain $\mathbf{a} = \partial \mathbf{u}/\partial t = 0$. How can this be?

CD/Kinematics/Fields, Particles and Reference Frames/Euler vs. Lagrange; What's Steady and Unsteady

We can resolve this difficulty by recognizing that the Eulerian acceleration $\mathbf{a}(\mathbf{x}, t)$ at point $\mathbf{x}$ at time t must be identical to the acceleration of the fluid particle located at this point. Thus the acceleration should be calculated by following the motion of this fluid particle, as shown in Figure 6.11. To do this we use the definition of a derivative in calculus to compute the acceleration by taking the limit as Δt approaches zero of the difference in the new and old velocities of the fluid particle of interest, i.e., $\mathbf{a} = \lim_{\Delta t \to 0}(\mathbf{u}_{\text{new}} - \mathbf{u}_{\text{old}})/\Delta t$. During the interval Δt, the fluid particle moves from position (x, y, z), where its velocity is $u(x, y, z, t)$, to a new location $(x + \Delta x, y + \Delta y, z + \Delta z)$, where its velocity is $u(x + \Delta x, y + \Delta y, z + \Delta z, t)$. Thus, the expression for the acceleration vector becomes

$$\mathbf{a} = \lim_{\Delta t \to 0} \frac{\mathbf{u}(x + \Delta x, y + \Delta y, z + \Delta z, t + \Delta t) - \mathbf{u}(x, y, z, t)}{\Delta t} \tag{6.5}$$

Equation 6.5 provides a means for calculating the three components of acceleration (a_x, a_y, a_z) from the three components of velocity (u, v, w). For example, the x component of acceleration a_x, which involves only the x component of velocity u, is given by

$$a_x = \lim_{\Delta t \to 0} \frac{u(x + \Delta x, y + \Delta y, z + \Delta z, t + \Delta t) - u(x, y, z, t)}{\Delta t} \tag{6.6}$$

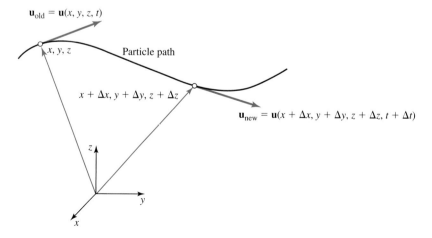

Figure 6.11 Velocity change for a fluid particle and its pathline.

6.3 FLUID ACCELERATION

To evaluate the limit, we use a Taylor series expansion to write $u(x + \Delta x, y + \Delta y, z + \Delta z, t + \Delta t)$ in terms of the nearby value $u(x, y, z, t)$. For a continuous function $u(x, y, z, t)$ of several variables, the Taylor series expansion is

$$u(x + \Delta x, y + \Delta y, z + \Delta z, t + \Delta t)$$
$$= u(x, y, z, t) + \left(\frac{\partial u}{\partial x} \Delta x + \frac{\partial u}{\partial y} \Delta y + \frac{\partial u}{\partial z} \Delta z + \frac{\partial u}{\partial t} \Delta t \right)$$

plus higher order terms. The distance Δx is the x component of the displacement vector defining the change in position of the fluid particle in time Δt. This distance is approximately $u \Delta t$, with the estimate becoming exact in the limit as Δt approaches zero. Similarly, we find $\Delta y = v \Delta t$, and $\Delta z = w \Delta t$, so the Taylor series can be written as

$$u(x + \Delta x, y + \Delta y, z + \Delta z, t + \Delta t)$$
$$= u(x, y, z, t) + \left(\frac{\partial u}{\partial x} u \Delta t + \frac{\partial u}{\partial y} v \Delta t + \frac{\partial u}{\partial z} w \Delta t + \frac{\partial u}{\partial t} \Delta t \right) \quad (6.7)$$

plus higher order terms. Substituting Eq. 6.7 into Eq. 6.6, subtracting $u(x, y, z, t)$, dividing by Δt, and taking the limit provides the x component of acceleration. Using this approach, we discover that the three components of the acceleration in Cartesian coordinates are

$$a_x = \frac{\partial u}{\partial t} + u \frac{\partial u}{\partial x} + v \frac{\partial u}{\partial y} + w \frac{\partial u}{\partial z} \quad (6.8a)$$

$$a_y = \frac{\partial v}{\partial t} + u \frac{\partial v}{\partial x} + v \frac{\partial v}{\partial y} + w \frac{\partial v}{\partial z} \quad (6.8b)$$

$$a_z = \frac{\partial w}{\partial t} + u \frac{\partial w}{\partial x} + v \frac{\partial w}{\partial y} + w \frac{\partial w}{\partial z} \quad (6.8c)$$

We have shown that a component of acceleration is given by the sum of one temporal and three spatial derivative terms involving the corresponding velocity component. This suggests that fluid acceleration is present in virtually all flows of engineering interest.

It is customary in fluid mechanics to use the symbol $D(\ )/Dt$ to represent the following combination of derivatives:

$$\frac{D(\ )}{Dt} = \frac{\partial (\ )}{\partial t} + (\mathbf{u} \cdot \nabla)(\ ) \quad (6.9)$$

This combination of a time derivative and three spatial derivatives involving the velocity components is referred to by a special name: the substantial, material, or co-moving derivative. The substantial derivative can be expressed in Cartesian coordinates as

$$\frac{D(\ )}{Dt} = \frac{\partial (\ )}{\partial t} + u \frac{\partial (\ )}{\partial x} + v \frac{\partial (\ )}{\partial y} + w \frac{\partial (\ )}{\partial z} \quad (6.10)$$

and used to write the three Cartesian components of acceleration as:

$$a_x = \frac{Du}{Dt}, \quad a_y = \frac{Dv}{Dt}, \quad \text{and} \quad a_z = \frac{Dw}{Dt} \qquad (6.11\text{a–c})$$

or in a compact form as a vector equation:

$$\mathbf{a} = \frac{D\mathbf{u}}{Dt} \qquad (6.12)$$

By using the definition of the substantial derivative in Eq. 6.9, we can expand Eq. 6.12 to find

$$\mathbf{a} = \frac{\partial \mathbf{u}}{\partial t} + (\mathbf{u} \cdot \nabla)\mathbf{u} \qquad (6.13)$$

Note that the leading term $\partial \mathbf{u}/\partial t$ is a partial time derivative of velocity, and the remaining term $(\mathbf{u} \cdot \nabla)\mathbf{u}$, called the convective derivative of velocity, involves partial space derivatives.

CD/Kinematic/The Material Derivative

From Eq. 6.13 we see that the total acceleration at a point in a fluid can be interpreted as the sum of two distinct types of acceleration called the local and convective accelerations. The local acceleration $\mathbf{a}_L$ is defined by the partial time derivative of velocity, or

$$\mathbf{a}_L = \frac{\partial \mathbf{u}}{\partial t} \qquad (6.14\text{a})$$

The convective acceleration $\mathbf{a}_C$ is defined by the convective derivative of velocity, or

$$\mathbf{a}_C = (\mathbf{u} \cdot \nabla)\mathbf{u} \qquad (6.14\text{b})$$

The local acceleration is a contribution to the total acceleration at a particular point in a flow due to a temporal change in the velocity. Clearly the velocity field must be a function of time for this contribution to exist. In the slug flow of a liquid falling under gravity in a long vertical tube (Figure 6.9), the only type of acceleration that is present is local acceleration. In the steady flow through a contraction (Figure 6.10), the local acceleration is zero.

The convective acceleration is a contribution to the total acceleration at a point in a flow from the convection (movement) of fluid along the instantaneous streamline through the point. A flow must be spatially nonuniform for this contribution to exist. For the flow in a contraction (Figure 6.10), the convective derivative of velocity is nonzero everywhere in the flow, since an observer moving along any streamline from point 1 to point 2 experiences a continuous change in velocity. Thus in this flow acceleration is given by the convective derivative. In the gravity-driven slug flow in a vertical tube (Figure 6.9), the velocity field is given by $\mathbf{u}(\mathbf{x}, t) = (0, 0, -U(t))$. This flow is accelerating, but the acceleration calculated by using the convective derivative is zero because all the spatial derivatives in the convective derivative are zero.

6.3 FLUID ACCELERATION

In cylindrical coordinates, the three components of acceleration (a_r, a_θ, a_z) are given in terms of the three components of velocity (v_r, v_θ, v_z) as

$$a_r = \frac{\partial v_r}{\partial t} + v_r \frac{\partial v_r}{\partial r} + \frac{v_\theta}{r} \frac{\partial v_r}{\partial \theta} + v_z \frac{\partial v_r}{\partial z} - \frac{v_\theta^2}{r} \quad (6.15a)$$

$$a_\theta = \frac{\partial v_\theta}{\partial t} + v_r \frac{\partial v_\theta}{\partial r} + \frac{v_\theta}{r} \frac{\partial v_\theta}{\partial \theta} + v_z \frac{\partial v_\theta}{\partial z} + \frac{v_r v_\theta}{r} \quad (6.15b)$$

$$a_z = \frac{\partial v_z}{\partial t} + v_r \frac{\partial v_z}{\partial r} + \frac{v_\theta}{r} \frac{\partial v_z}{\partial \theta} + v_z \frac{\partial v_z}{\partial z} \quad (6.15c)$$

EXAMPLE 6.4

A velocity field is given by $\mathbf{u} = \alpha x \mathbf{i} + \beta y \mathbf{j} - (\alpha + \beta) z \mathbf{k}$, where α and β are constants with dimension $\{t^{-1}\}$. What is the acceleration in this flow?

SOLUTION

Substituting the velocity field in Cartesian coordinates into Eqs. 6.8a–c to determine the acceleration of the flow yields

$$a_x = \frac{\partial u}{\partial t} + u \frac{\partial u}{\partial x} + v \frac{\partial u}{\partial y} + w \frac{\partial u}{\partial z}$$
$$= \frac{\partial (\alpha x)}{\partial t} + \alpha x \frac{\partial (\alpha x)}{\partial x} + \beta y \frac{\partial (\alpha x)}{\partial y} - (\alpha + \beta) z \frac{\partial (\alpha x)}{\partial z}$$
$$= \alpha^2 x$$

$$a_y = \frac{\partial v}{\partial t} + u \frac{\partial v}{\partial x} + v \frac{\partial v}{\partial y} + w \frac{\partial v}{\partial z}$$
$$= \frac{\partial (\beta y)}{\partial t} + \alpha x \frac{\partial (\beta y)}{\partial x} + \beta y \frac{\partial (\beta y)}{\partial y} - (\alpha + \beta) z \frac{\partial (\beta y)}{\partial z}$$
$$= \beta^2 y$$

$$a_z = \frac{\partial w}{\partial t} + u \frac{\partial w}{\partial x} + v \frac{\partial w}{\partial y} + w \frac{\partial w}{\partial z}$$
$$= \frac{\partial (-(\alpha + \beta) z)}{\partial t} + \alpha x \frac{\partial (-(\alpha + \beta) z)}{\partial x} + \beta y \frac{\partial (-(\alpha + \beta) z)}{\partial y} - (\alpha + \beta) z \frac{\partial (-(\alpha + \beta) z)}{\partial z}$$
$$= (\alpha + \beta)^2 z$$

In this case the local acceleration is zero, so the total acceleration is made up of the convective contributions only. The final result is $\mathbf{a} = \alpha^2 x \mathbf{i} + \beta^2 y \mathbf{j} + (\alpha + \beta)^2 z \mathbf{k}$.

EXAMPLE 6.5

The velocity distribution for a certain type of wave motion in a rotating fluid is given by $\mathbf{u} = A\cos(\alpha x - 2\Omega t)\mathbf{j} - A\sin(\alpha x - 2\Omega t)\mathbf{k}$, where A, α, and Ω are constants with dimension of $\{Lt^{-1}\}$, $\{L^{-1}\}$, and $\{t^{-1}\}$, respectively. What is the acceleration in this flow?

SOLUTION

Substituting the velocity field in Cartesian coordinates into Eqs. 6.8a–6.8c to determine the acceleration of the flow we find:

$$a_x = \frac{\partial u}{\partial t} + u\frac{\partial u}{\partial x} + v\frac{\partial u}{\partial y} + w\frac{\partial u}{\partial z} = \frac{\partial(0)}{\partial t} + (0)\frac{\partial(0)}{\partial x} + v\frac{\partial(0)}{\partial y} + w\frac{\partial(0)}{\partial z} = 0$$

$$a_y = \frac{\partial v}{\partial t} + u\frac{\partial v}{\partial x} + v\frac{\partial v}{\partial y} + w\frac{\partial v}{\partial z}$$

$$= \frac{\partial}{\partial t}[A\cos(\alpha x - 2\Omega t)] + (0)\frac{\partial[A\cos(\alpha x - 2\Omega t)]}{\partial x}$$

$$+ [A\cos(\alpha x - 2\Omega t)]\frac{\partial[A\cos(\alpha x - 2\Omega t)]}{\partial y}$$

$$+ [-A\sin(\alpha x - 2\Omega t)]\frac{\partial[A\cos(\alpha x - 2\Omega t)]}{\partial z}$$

$$a_y = A2\Omega\sin(\alpha x - 2\Omega t) + 0 + [A\cos(\alpha x - 2\Omega t)](0)$$

$$+ [-A\sin(\alpha x - 2\Omega t)](0)$$

$$a_y = A2\Omega\sin(\alpha x - 2\Omega t)$$

$$a_z = \frac{\partial w}{\partial t} + u\frac{\partial w}{\partial x} + v\frac{\partial w}{\partial y} + w\frac{\partial w}{\partial z}$$

$$= \frac{\partial}{\partial t}[-A\sin(\alpha x - 2\Omega t)] + (0)\frac{\partial[-A\sin(\alpha x - 2\Omega t)]}{\partial x}$$

$$+ [A\cos(\alpha x - 2\Omega t)]\frac{\partial[-A\sin(\alpha x - 2\Omega t)]}{\partial y}$$

$$+ [-A\sin(\alpha x - 2\Omega t)]\frac{\partial[-A\sin(\alpha x - 2\Omega t)]}{\partial z}$$

$$a_z = -A2\Omega\cos(\alpha x - 2\Omega t) + 0$$

$$+ [A\cos(\alpha x - 2\Omega t)](0) + [-A\sin(\alpha x - 2\Omega t)](0)$$

$$a_z = -A2\Omega\cos(\alpha x - 2\Omega t)$$

The convective acceleration is zero, so the total acceleration is made up of the local acceleration components in the $\mathbf{j}$ and $\mathbf{k}$ directions. The result is

$$\mathbf{a} = 0\mathbf{i} + 2\Omega A\sin(\alpha x - 2\Omega t)\mathbf{j} - 2\Omega A\cos(\alpha x - 2\Omega t)\mathbf{k}$$

If you apply Eqs. 6.15a–6.15c to the Poiseuille flow described in Example 6.2, you will find that the acceleration is zero. Another simple flow called solid body rotation is described in cylindrical coordinates by $\mathbf{u} = 0\mathbf{e}_r + v_\theta(r)\mathbf{e}_\theta + 0\mathbf{e}_z$, with $v_r = 0$, $v_\theta = r\Omega$, and $v_z = 0$. This is the velocity field that occurs in a container full of fluid rotating about an axis at a constant angular velocity with the fluid at rest relative to the rotating container. See if you can confirm that the acceleration in this flow is $\mathbf{a} = (-r\Omega^2)\mathbf{e}_r + 0\mathbf{e}_\theta + 0\mathbf{e}_z$. Does the radial acceleration here look familiar? It should, since it is the centrifugal acceleration acting on a particle of fluid located at radius r.

Note the last term in equations 6.15a and 6.15b. These extra terms arise from the curvilinear nature of cylindrical coordinates. In cylindrical coordinates the substantial derivative is written as

$$\frac{D(\ )}{Dt} = \frac{\partial(\ )}{\partial t} + v_r \frac{\partial(\ )}{\partial r} + \frac{v_\theta}{r}\frac{\partial(\ )}{\partial \theta} + v_z \frac{\partial(\ )}{\partial z} \quad (6.16)$$

By using the substantial derivative, we may write the acceleration in cylindrical coordinates as

$$a_r = \frac{Dv_r}{Dt} - \frac{v_\theta^2}{r}, \quad a_\theta = \frac{Dv_\theta}{Dt} + \frac{v_r v_\theta}{r}, \quad \text{and} \quad a_z = \frac{Dv_z}{Dt} \quad (6.17)$$

Expressions for acceleration in spherical and other coordinate systems may be found in many advanced textbooks.

6.4 THE SUBSTANTIAL DERIVATIVE

In the last section we showed that fluid acceleration is defined by the substantial derivative of velocity. If the Taylor series approach used to calculate acceleration is applied to calculate the rate of change of an arbitrary fluid or flow property ε following the motion of a fluid particle, it is straightforward to show that the rate of change is given by $D\varepsilon/Dt$, i.e., by the substantial derivative of the property. According to Eq. 6.9, the substantial derivative is given by, $D(\ )/Dt = \partial(\ )/\partial t + (\mathbf{u} \cdot \nabla)(\ )$, thus the rate of change of a fluid property ε following the motion of a fluid particle is given by

$$\frac{D\varepsilon}{Dt} = \frac{\partial \varepsilon}{\partial t} + (\mathbf{u} \cdot \nabla)\varepsilon \quad (6.18)$$

We see that the rate of change consists of a local rate of change given by the time derivative and a convective rate of change given by the convective derivative.

The substantial derivative of a fluid property appears in each of the governing equations of fluid dynamics to be derived in Chapter 11. There are also a number of other important uses of the substantial derivative in fluid mechanics. For example, the substantial derivative is used in problems involving wave motion to derive the boundary condition known as the kinematic condition, and also to interpret measurements made with moving sensors in applications such as environmental sampling in the oceans and atmosphere.

As an example of somewhat unusual but illuminating use of Eq. 6.18, recall that the velocity at a point in a fluid is defined to be that of the fluid particle which happens to be located there. The Eulerian velocity $\mathbf{u}$ is therefore equal to the rate of change in particle position following the motion of the particle. Consider a fluid particle located at position $\mathbf{x} = x\mathbf{i} + y\mathbf{j} + z\mathbf{k}$. We should be able to apply Eq. 6.18 with $\varepsilon = \mathbf{x}$ to calculate the velocity of this particle from

$$\mathbf{u} = \frac{D\mathbf{x}}{Dt} \qquad (6.19)$$

Let us see what this operation produces. Inserting $\mathbf{x} = x\mathbf{i} + y\mathbf{j} + z\mathbf{k}$, using the product rule, and noting that the unit vectors are constant, we get

$$\mathbf{u} = \frac{D\mathbf{x}}{Dt} = \frac{Dx}{Dt}\mathbf{i} + \frac{Dy}{Dt}\mathbf{j} + \frac{Dz}{Dt}\mathbf{k} \qquad (6.20)$$

Upon using Eq. 6.10 to evaluate the substantial derivative in Cartesian coordinates, and realizing that a given spatial coordinate is not a function of time or the remaining two coordinates, we find

$$\frac{Dx}{Dt} = \frac{\partial x}{\partial t} + u\frac{\partial x}{\partial x} + v\frac{\partial x}{\partial y} + w\frac{\partial x}{\partial z} = u$$

$$\frac{Dy}{Dt} = \frac{\partial y}{\partial t} + u\frac{\partial y}{\partial x} + v\frac{\partial y}{\partial y} + w\frac{\partial y}{\partial z} = v$$

$$\frac{Dx}{Dt} = \frac{\partial z}{\partial t} + u\frac{\partial z}{\partial x} + v\frac{\partial z}{\partial y} + w\frac{\partial z}{\partial z} = w$$

Substituting these results into Eq. 6.20 gives the expected result:

$$\mathbf{u} = \frac{Dx}{Dt}\mathbf{i} + \frac{Dy}{Dt}\mathbf{j} + \frac{Dz}{Dt}\mathbf{k} = u\mathbf{i} + v\mathbf{j} + w\mathbf{k}$$

We see that in the Eulerian description, position, velocity, and acceleration are related by the substantial derivative, i.e., by

$$\mathbf{u} = \frac{D\mathbf{x}}{Dt} \quad \text{and} \quad \mathbf{a} = \frac{D\mathbf{u}}{Dt} \qquad (6.21)$$

6.5 CLASSIFICATION OF FLOWS

An engineer may encounter a wide variety of fluid flows. In most cases it will be necessary to consult databases, journals, textbooks, and monographs, and to conduct online searches of reference materials to find out what is known about a particular flow of interest. To do this effectively, it is necessary to know how flows are classified and to learn the key technical terms used to describe broad classes of flows. In this section we introduce some of the vocabulary used to describe and classify fluid flows.

CD/Kinematics/Two Dimensional Flow and Vorticity/2D Flow: Definition

A coordinate system for a flow problem is usually chosen to fit the geometry of the problem. For example, it seems sensible to describe channel flow between parallel plates in Cartesian coordinates aligned with the plates, and flow in a round pipe in cylindrical coordinates with the z axis along the pipe centerline. This leads to the simplest mathematical description of the boundaries and often reduces the number of velocity components as well. Thus picking a set of coordinates on the basis of geometry is a good initial step in trying to reduce the number of velocity components in problem.

6.5.1 One-, Two-, and Three-Dimensional Flow

Earlier we described the Eulerian velocity field in general as three dimensional. A flow is characterized as having one, two, or three dimensions depending on the corresponding number of components needed to describe its Eulerian velocity field. A 1D flow is also referred to as a unidirectional flow. Note that we are defining a characteristic of a flow here by reference to its velocity field alone. We have chosen to define flow dimensionality in terms of the number of vector components needed to describe the velocity field. Thus, in Cartesian coordinates, if u, v, w are each nonzero, then the flow is 3D. If u, v are nonzero but $w = 0$ at all points in the velocity field, then the flow is 2D. Similarly, in a cylindrical coordinate system, if v_z is nonzero but v_r, v_θ are zero, the flow field is 1D.

The same dimensionality terminology is also used to describe scalar property fields. In this usage, however, 3D refers to the dependence of the scalar field on three spatial coordinates. It obviously cannot refer to three vector components, since these do not exist for scalar properties like density, pressure, and temperature. Thus, if a temperature field in a flow problem is said to be 3D, the temperature function must depend on (x, y, z) in Cartesian coordinates, or (r, θ, z) in cylindrical coordinates. Figure 6.12 shows the temperature contours and velocity vectors calculated in a computational simulation of one chip in an array on an air-cooled electronic circuit board. It can be seen that both the velocity and temperature fields are in fact 3D in this case.

An engineer often encounters a fluid flow that is constrained in some way by the physical geometry. For example, constraints can be imposed in an internal flow by the wall configuration, or in external flow by the shape of the object over which the fluid flows. The velocity field in constrained flows may not need to be represented by all three vector components, or by all three spatial variables. One can often determine how many velocity components are likely to occur in a flow by inspection of the geometry and boundary conditions, as illustrated in Examples 6.6, 6.7, and 6.8.

Part of the art of solving a fluid dynamics problem is choosing a coordinate system in which the velocity field is described by the minimum number of components. To accomplish this goal, it may be necessary to rotate the original Cartesian coordinates or change to cylindrical or spherical coordinates to take advantage of spatial symmetry. For example, the 2D flow in Figure 6.14A becomes a 1D flow in the coordinates shown in Figure 6.14B. Describing the flow in the long rotating pipe in Cartesian coordinates requires all three velocity components since the flow about the axis occurs in the xy plane (see Figure 6.15A). Thus this problem is clearly simplified by using cylindrical coordinates since there is no radial component of velocity. Similarly, the analysis of flow over a stationary sphere shown in Figure 6.15B benefits from the selection of a spherical coordinate system, since this will reduce the number of required velocity components from three in Cartesian coordinates to two in spherical coordinates.

A 1D or 2D flow requires fewer governing equations than are needed for a 3D flow. Thus 1D and 2D flows are much easier to model, solve, measure, and interpret. This is

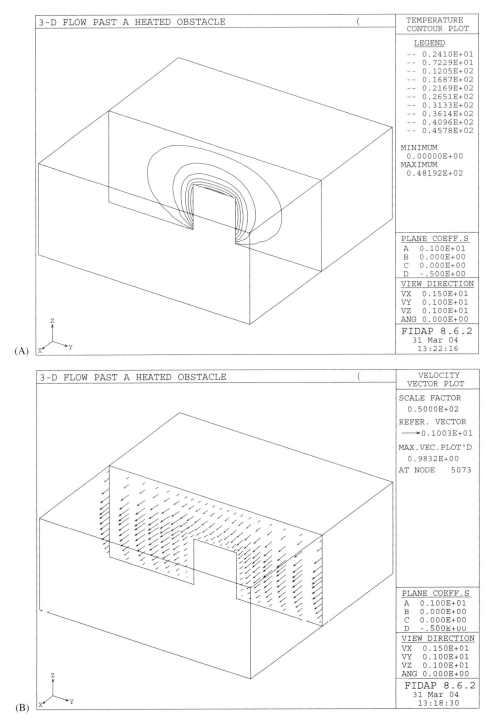

Figure 6.12 Temperature contours ((A) and (C)) and velocity vectors ((B) and (D)) for flow past a heated obstacle. Generated by CFD.

6.5 CLASSIFICATION OF FLOWS

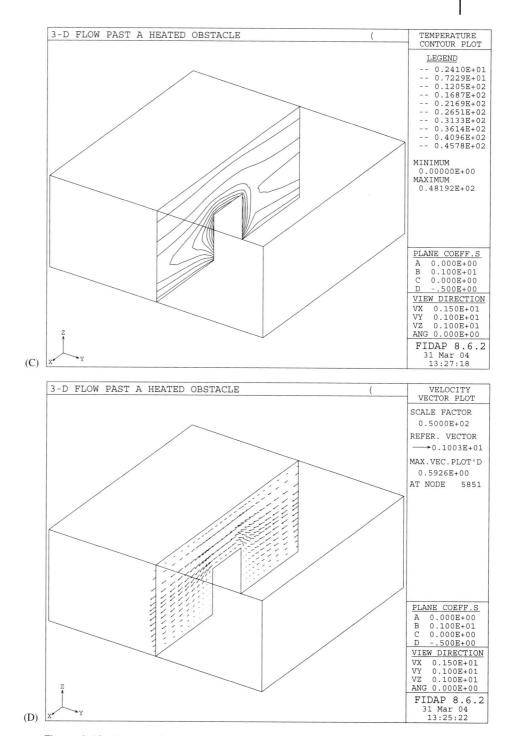

Figure 6.12 (*Continued*)

EXAMPLE 6.6

Consider the three flows illustrated in Figure 6.13. Determine the number of velocity components and spatial coordinates necessary to describe each velocity field.

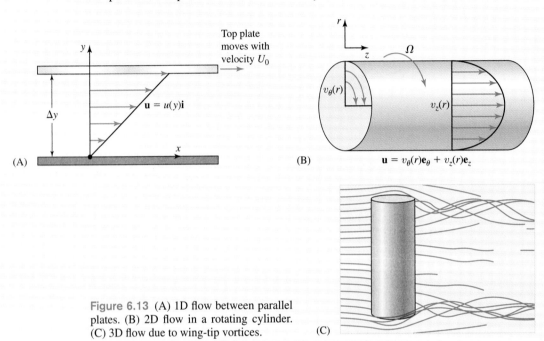

Figure 6.13 (A) 1D flow between parallel plates. (B) 2D flow in a rotating cylinder. (C) 3D flow due to wing-tip vortices.

SOLUTION

Flow A. Shear flow between large parallel plates: The motion of the upper plate and the fact that we are in the middle of very large plates suggest that the only nonzero velocity component is in the x direction. That is, $v = w = 0$, and only one velocity component is necessary to describe the velocity field: $\mathbf{u} = u(x, y, z)\mathbf{i}$. To determine the number of spatial coordinates necessary to describe the velocity field, we examine the functional dependence of the nonzero velocity component on the spatial coordinates x, y, and z. Since the plates are large, there is no reason to think that the velocity in the x direction would depend on x or z. Since, however, the velocity must vary in the y direction to match the motion of the plates, we conclude that $\mathbf{u} = u(y)\mathbf{i}$.

Flow B. Flow through a long rotating pipe: This flow is conveniently described by means of cylindrical coordinates. Examination of the flow shows that there must be a velocity component along the pipe axis and also in the θ direction, since the pipe wall is rotating. There does not appear to be anything that would cause a flow in the radial direction. We conclude that two components are required to describe the velocity field: $\mathbf{u} = v_\theta(r, \theta, z)\mathbf{e}_\theta + v_z(r, \theta, z)\mathbf{e}_z$. Since the pipe is long and axisymmetric, we further assume that the two velocity components are functions of only the radial coordinate.

Thus we feel safe in assuming that the velocity field in the absence of time dependence takes the form $\mathbf{u} = v_\theta(r)\mathbf{e}_\theta + v_z(r)\mathbf{e}_z$.

Flow C. Flow over a tapered wing: The finite length and taper of the wing suggest that all three velocity components will be needed to describe the flow. The taper and curvature of the wing surface suggest that the velocity components may depend on all three spatial coordinates. Thus, we conclude that $\mathbf{u} = u(x, y, z)\mathbf{i} + v(x, y, z)\mathbf{j} + w(x, y, z)\mathbf{k}$. [*Note:* For a long straight wing, the flow over the midpoint of the wing is described by $\mathbf{u} = u(x, y)\mathbf{i} + v(x, y)\mathbf{j}$, since there is no reason for flow along the wing or dependence on position along the wing.]

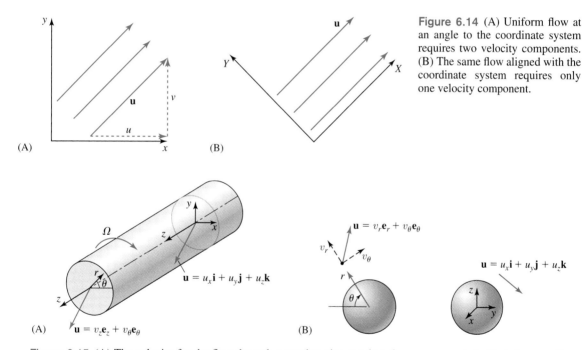

Figure 6.14 (A) Uniform flow at an angle to the coordinate system requires two velocity components. (B) The same flow aligned with the coordinate system requires only one velocity component.

Figure 6.15 (A) The velocity for the flow through a rotating pipe requires three components in Cartesian coordinates but only two in cylindrical coordinates. (B) Similarly, flow over a sphere is described by three velocity components in Cartesian coordinates but only two in spherical coordinates.

true regardless of how the problem is approached: analytically, experimentally, or numerically. Until recently, few engineers had access to computers powerful enough to model 3D flow. Even today, a numerical solution for 3D flow is often crudely resolved because of the memory and CPU times such flow problems demand. On the other hand, engineers now have access to desktop computers that can readily solve virtually all 1D and 2D flow problems. Known analytical solutions tend to be limited to 1D flow and 2D flow rather than to 3D flow. Thus it is important for an engineer to be able to select and utilize a coordinate system that minimizes the dimensionality of a velocity field.

EXAMPLE 6.7

What is the dimensionality of the three flows in Example 6.6?

SOLUTION

The shear flow has only one velocity component; thus this flow is 1D. The flow in the rotating pipe has two velocity components, so this flow is 2D. The flow over the tapered wing is a 3D flow since all three velocity components are nonzero.

EXAMPLE 6.8

What is the dimensionality of each of the following velocity fields? (Each field is investigated in one of the problems at the end of this chapter.)

Flow A:

$$\mathbf{u} = \left(\frac{\rho g R_P^2}{4\mu}\right)\left[1 - \left(\frac{r}{R_P}\right)^2 + 2\alpha^2 \ln\left(\frac{r}{R_P}\right)\right]\mathbf{k}$$

Flow B:

$$\mathbf{u} = C_1 t x \mathbf{i} + 2C_2 x y \mathbf{j} - C_3 t^2 z \mathbf{k}$$

Flow C:

$$\mathbf{u} = \left\{\Omega_0 R\left[\frac{(\kappa R/r - r/\kappa R)}{(\kappa - 1/\kappa)}\right]\right\}\mathbf{e}_\theta$$

Flow D:

$$\mathbf{u} = W_0\left[\left(\frac{h-x}{h}\right)^{1/n}\right]\mathbf{k}$$

Flow E:

$$\mathbf{u} = C[x\mathbf{i} - y\mathbf{j}]$$

Flow F:

$$\mathbf{u} = V_\infty \cos\theta\left[1 - \frac{3}{2}\left(\frac{R}{r}\right) + \frac{1}{2}\left(\frac{R}{r}\right)^3\right]\mathbf{e}_r - V_\infty \sin\theta\left[1 - \frac{3}{4}\left(\frac{R}{r}\right) + \frac{1}{4}\left(\frac{R}{r}\right)^3\right]\mathbf{e}_\theta$$

6.5 CLASSIFICATION OF FLOWS

SOLUTION

Flow A has only one velocity vector component (in the z direction) so it is a 1D, or unidirectional, flow. Flow B is a 3D flow because it has velocity vector components in the x, y, and z directions. Flow C is a 1D, or unidirectional, flow because its only velocity vector component is in the θ direction. Flow D is also a 1D flow; its only velocity vector component is in the z direction. Flow E is an example of a 2D flow, since it has velocity vector components in the x and y directions. Finally, flow F is an example of a 2D flow because it has nonzero velocity vector components in the r and θ directions.

6.5.2 Uniform, Axisymmetric, and Spatially Periodic Flow

A uniform flow in a region has velocity vectors that are constant in magnitude and direction throughout that region. For the pilot of a vehicle flying through a still atmosphere at a constant speed, all the air appears to be approaching the vehicle with a constant velocity (Figure 6.16A). The upstream flow field is therefore an example of a uniform flow from the pilot's perspective. In a well-designed wind tunnel (Figure 6.16B), the sharp contraction upstream of the test section and a slight divergence in the test section walls create a highly uniform flow over the entire test section with $\mathbf{u} = U\mathbf{i}$ to simulate flight conditions. Of course no real flow is ever perfectly uniform: all flow fields exhibit at least some minor spatial variation. The occasional bumpy aircraft flight demonstrates that the upstream flow field is not always spatially uniform.

Many flows cannot be considered uniform throughout a region. For example, consider the planar nozzle shown in Figure 6.17a. In this figure we see that the flow at the inlet plane is nearly uniform. As the fluid moves through the nozzle it accelerates, moving

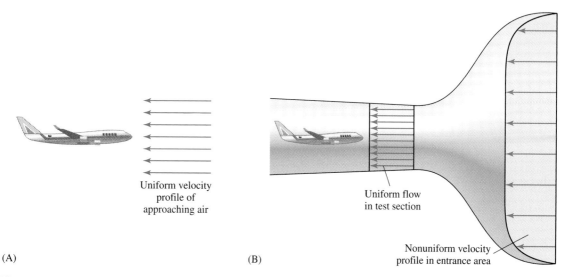

(A)   (B)

Figure 6.16 Uniform flows: (A) aircraft flying in still air and (B) wind tunnel test section.

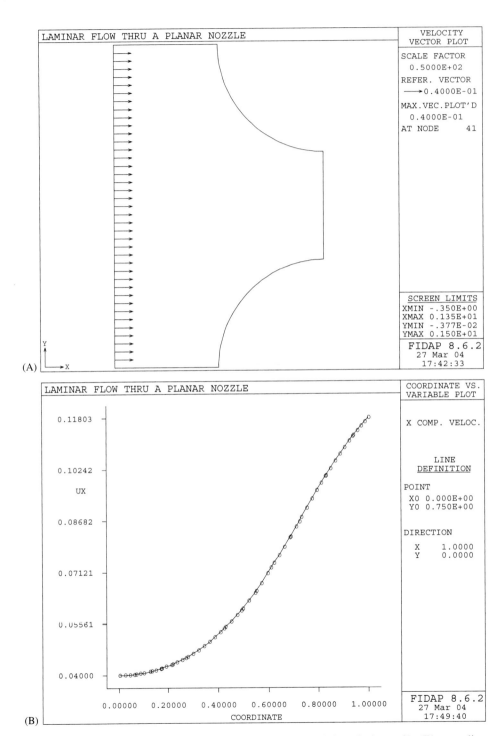

Figure 6.17 CFD plots of (A) geometry of a planar nozzle and inlet velocity profile, (B) center line velocity as a function of axial coordinate, (C) velocity vectors for this flow, and (D) x component of velocity across the exit plane. Notice that the velocity at the exit plane is, in fact, nearly uniform.

6.5 CLASSIFICATION OF FLOWS 329

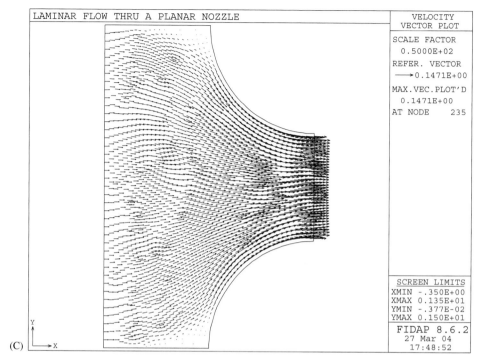

(C)

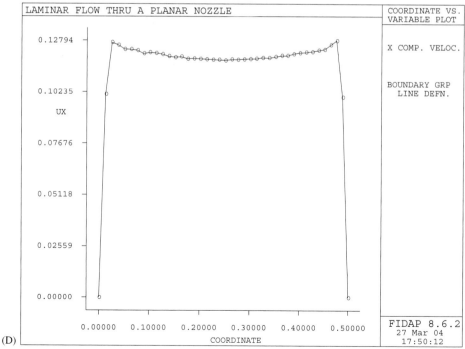

(D)

Figure 6.17 (*Continued*)

faster where the area is smaller as shown by the centerline x-component velocity plot in Figure 6.17b. An examination of Figure 6.17c, the velocity vector plot for this flow, shows that the velocity is non-uniform inside the nozzle. However, for a nozzle like this one with a short, sharp contraction, it can be seen from both Figure 6.17c and 6.17d that the velocity vector is approximately the same everywhere on the nozzle exit plane. The flow at the nozzle exit may be modeled as a uniform flow at a cross section, meaning that on the nozzle exit plane the velocity vector may be approximated as a spatially constant vector. We will often be able to assume a uniform flow at a cross section as an engineering approximation. Later in this chapter we will show that the assumption of uniform flow at a cross section greatly simplifies a calculation of the mass or volume flowrate through the cross section. Figure 6.18 illustrates several geometries through which fluid flows. Can you identify the uniform flow cross section in each example?

As mentioned earlier, when fluid is flowing between rigid boundaries or in a region with some type of symmetry, the velocity field may not depend on all three spatial variables. An axisymmetric flow is described in cylindrical coordinates by a velocity field that does not depend on the angular coordinate. If a swirl component of velocity v_θ is present in an axisymmetric pipe flow, as shown in Figure 6.19A, the velocity field in cylindrical coordinates, $(0, v_\theta, v_z)$, is 2D. If swirl is absent, the velocity field is 1D, as shown in Figure 6.19B. If there is a bend in a pipe, the flow upstream may be axisymmetric and without swirl, but the flow downstream will be nonaxisymmetric and swirling due to the turn. Is the flow in a long rotating pipe axisymmetric?

It is typical for flows to exhibit a complex dependence on one or more spatial coordinates. In some cases, however, a flow may exhibit a spatial periodicity along a certain direction. For a spatially periodic flow, the velocity field may be represented by

$$\mathbf{u}(\mathbf{x} + \mathbf{x}_0, t) = \mathbf{u}(\mathbf{x}, t) \tag{6.22}$$

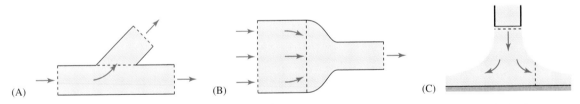

Figure 6.18 Schematic illustration of fluid flow through (A) a branch, (B) a contraction, and (C) a vertical exit.

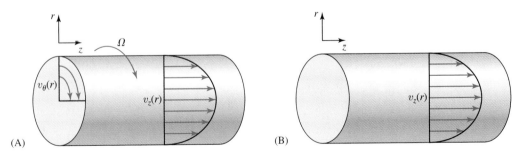

Figure 6.19 Pipe flow (A) with swirl due to the rotation and (B) without swirl.

6.5 CLASSIFICATION OF FLOWS

Equation 6.22 shows that at two points in a spatially periodic flow a certain distance apart the velocity is the same. Examples of spatially periodic flow are shown in Figure 6.20.

 CD/Video Library/Fully Developed Flow

6.5.3 Fully Developed Flow

The term "fully developed flow" is used to describe a velocity field that does not change in the flow direction. The velocity vector is therefore independent of the coordinate along that direction. In the axisymmetric flow in a long constant area pipe, there is always a region near the entrance to the pipe (Figure 6.21A) where the velocity field in polar coordinates is 2D because of the developing boundary layer. The velocity depends on both r and z near the entrance. Further into the pipe, however, the velocity field becomes 1D and independent of the z coordinate, and it may be represented by a velocity vector of the form $(0, 0, v_z(r))$. The flow well inside the pipe is thus fully developed because the velocity field is not changing in the axial direction.

The term "fully developed" should never be assumed to apply to other fluid or flow properties in a fully developed flow in the absence of evidence to support this conclusion. In the fully developed flow far from the entrance of the long pipe in Figure 6.21,

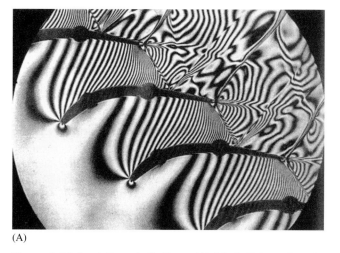
(A)

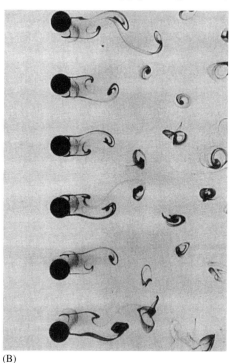
(B)

Figure 6.20 Spatially periodic flows: (A) Mach–Zehnder interferometry image of flow through a turbine blade cascade and (B) flow through a bank of circular cylinders.

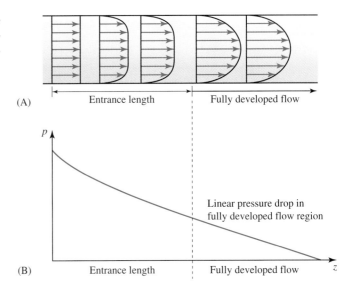

Figure 6.21 Flow in a round pipe: (A) velocity profiles through entrance length to fully developed region and (B) pressure change in the pipe.

CASE STUDY

You should be familiar with the concept of fully developed flow in a pipe because it was presented in Chapter 3, Section 3.3.1, in the first case study on flow in a round pipe. The determination of the pressure drop in fully developed pipe flow is the most frequently computed quantity in engineering fluid mechanics.

the velocity field does not depend on the z coordinate, but viscous friction is causing the pressure in the fluid to fall continuously through the pipe. Thus, the pressure field is not fully developed, as can be seen in Figure 6.21B.

6.5.4 Steady Flow, Steady Process, and Temporally Periodic Flow

A flow is said to be steady if the Eulerian velocity field is independent of time. Thus, a steady flow is described by

$$\mathbf{u} = \mathbf{u}(\mathbf{x}) \qquad (6.23)$$

If a velocity field is known in functional form, the absence of time in the function representing the velocity field confirms that the flow is steady.

Since many engineering devices and systems are designed to operate in a steady state mode, steady flows are common. For example, after several minutes of operation, a wind tunnel reaches a stable operating condition and the speed control will maintain a constant air speed in the test section unless the controls are deliberately changed. The output of a velocity sensor positioned to record airspeed in the test section of a wind tunnel will typically exhibit little if any variation over time. The flow in the test section may therefore be modeled as a steady flow.

Figure 6.22 shows typical velocity signals recorded by a velocity sensor in a steady laminar flow (Figure 6.22A), an unsteady laminar flow (Figure 6.22B), and a turbulent flow (Figure 6.22C). Notice that there is a small amount of fluctuation in the steady

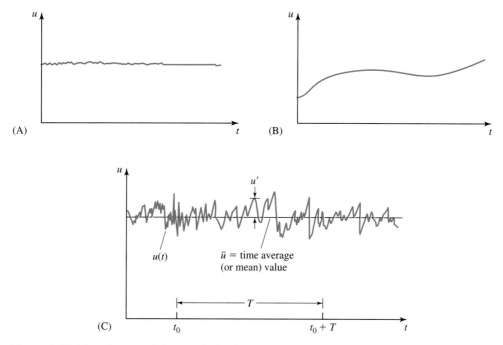

Figure 6.22 Time history of (A) steady laminar flow, (B) unsteady laminar flow, and (C) steady turbulent flow.

laminar flow. This is unavoidable. For an experimentally measured velocity field, a determination that the flow is steady is always a matter of engineering judgment. A flow may exhibit temporal fluctuations, but if these are small in comparison to the time-averaged value of velocity, the flow may be considered to be steady. Otherwise the flow is unsteady (Figure 6.22B). The velocity signal from a turbulent flow is always unsteady, as shown in Figure 6.22C. If this signal is averaged over a short time interval, and the average velocity is time invariant, the turbulent flow is described as steady. Thus in a turbulent flow, "steady" is defined in terms of the time-averaged velocity and not the instantaneous velocity, as is the case in laminar flow.

The definition of steady flow in terms of the time dependence of the velocity field is a traditional one. Some fluid dynamists prefer to define steady flow on the basis of the time independence of all fluid and flow properties rather than just the velocity field. Since these two definitions are not the same, we shall refer to a flow in which all fluid and flow properties are independent of time as a steady process. A steady flow is therefore described by a time-independent velocity field, and a steady process by time-independent velocity and property fields.

It is important to realize that identifying a flow as steady does not imply the absence of acceleration. In a steady flow the local acceleration, $\partial \mathbf{u}/\partial t$, is zero but the convective acceleration, $(\mathbf{u} \cdot \nabla)\mathbf{u}$, need not be zero. To illustrate this point, recall our discussion of flow through a contraction (see Figure 6.10). This flow is steady, yet the liquid accelerates as it moves through the area change. If measured, the value of fluid velocity at every point in the venturi will show no significant temporal variation, yet the fluid is accelerating.

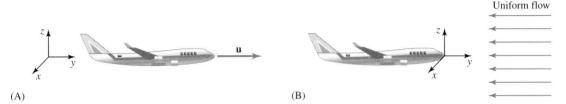

Figure 6.23 (A) Aircraft with an inertial coordinate system results in unsteady velocity field. (B) With coordinate system moving with the aircraft, the velocity field is steady.

The classification of a flow as steady results in a significant simplification in the governing equations of fluid motion. For this reason you should be alert to the possibility of transforming an unsteady flow into a steady flow by a change to a different reference frame in which the velocity field is time independent. The new frame is typically either translating at constant velocity or rotating at constant angular velocity. For example, the flow about an aircraft moving at a constant velocity through a still atmosphere shown in Figure 6.23A is an unsteady flow when described in a reference frame fixed to Earth. To see this, pick a point upstream of the aircraft and in its path and ask yourself whether the velocity at this fixed point will change as the aircraft comes closer. Suppose, however, that we choose to describe the flow in a frame of reference moving with the aircraft, as shown in Figure 6.23B. In this frame the flow is steady. The pilot in the aircraft frame of reference sees a spatially uniform flow of air approaching at constant velocity and observes a velocity field about the aircraft that is not time dependent.

As mentioned earlier, all flows exhibit slight temporal variations in velocity even when the driving forces themselves are steady. These slight variations do not affect a decision to model the flow as steady. On the other hand, because of natural flow instabilities, many flows are unsteady even when the driving forces are steady. The Karman vortex sheet behind a cylinder (Figure 6.24) is a good example of an unsteady flow of

Figure 6.24 Karman vortex sheet behind a cylinder.

6.5 CLASSIFICATION OF FLOWS

Figure 6.25 Velocity waveform in the descending thoracic aorta of a dog obtained with a hot-film anemometer.

this type; recall our discussion of vortex shedding in the context of the Strouhal number in Section 3.2. Temporally periodic driving forces will also create a temporally periodic (pulsatile) flow that exhibits a repetitive behavior over time with a period τ. This is evident in the velocity record shown in Figure 6.25. The velocity vector for a temporally periodic flow behaves according to

$$\mathbf{u}(\mathbf{x}, t + \tau) = \mathbf{u}(\mathbf{x}, t) \tag{6.24}$$

indicating that the same velocity vector occurs at a given spatial point every τ seconds. Fluid machinery that operates cyclically must create a temporally periodic velocity field, and many biological flows are of this type. Given the many possible types of flow behavior, an engineer must judge whether a flow is accurately modeled as a steady flow, a steady process, or a temporally periodic flow.

 CD/Dynamics/Boundary Conditions

EXAMPLE 6.9

Water flows through a flume (a liquid version of a wind tunnel) with a velocity field far from the walls given by $\mathbf{u}(\mathbf{x}, t) = [5 \text{ m/s} + (0.005 \text{ m/s})(\sin t/\tau)]\mathbf{i}$. Characterize the time dependence of this flow. Could this flow be modeled as a steady flow?

SOLUTION

This single component velocity field is a function of time, so this is a 1D unsteady flow. Further inspection reveals that the flow is temporally periodic. However, the deviation from a constant velocity of 5 m/s is minimal, about 0.1%. Engineering judgment suggests that it is valid to ignore the unsteady term and approximate the velocity field as $\mathbf{u}(\mathbf{x}, t) = [5 \text{ m/s}]\mathbf{i}$.

One important exception to the no-slip condition occurs if the fluid is modeled as inviscid, meaning that it has zero viscosity, or if the flow is modeled as an inviscid flow, meaning that the effects of viscosity are neglected. In either case the flow is frictionless, there are no shear forces acting in the fluid, and the fluid is assumed to have the ability to freely slip in any tangential direction along the solid surface.

6.6 NO-SLIP, NO-PENETRATION BOUNDARY CONDITIONS

In using the governing equations to solve a flow problem, it is necessary to specify conditions that model the behavior of the fluid and flow properties at boundaries of various types. The boundary conditions that apply specifically to the velocity field are referred to as the no-slip and no-penetration conditions. In nearly all flows encountered in engineering, observation shows that a fluid does not move relative to a solid surface in the tangential direction. Rather, the fluid sticks to the surface, a phenomenon referred to as no-slip. From this we conclude that the tangential component of velocity u_T is equal to the tangential component of boundary velocity U_T. This boundary condition

$$u_T = U_T \tag{6.25}$$

is called the no-slip condition in fluid dynamics. If a solid surface is not moving, $U_T = 0$, and the no-slip condition becomes $u_T = 0$. The no-slip condition is normally invoked for every solid–fluid interface.

The no-penetration condition arises because bulk fluid cannot penetrate an impermeable boundary. Thus, the normal component of fluid velocity u_N must match the normal boundary velocity U_N. This boundary condition

$$u_N = U_N \tag{6.26}$$

is called the no-penetration condition and applies to any solid or otherwise impermeable surface in contact with a fluid. If a solid or impermeable surface is not moving in a direction normal to its own surface, $U_N = 0$, and the fluid in contact is also not moving in the normal direction, so $u_N = 0$.

 CD/History/D'Alembert

For a viscous fluid in contact with a solid boundary, the complete no-slip, no-penetration boundary condition may be summarized by saying that the velocity vector of the fluid on the boundary is always equal to the velocity of the boundary at the same point. Thus, a convenient way of representing the complete no-slip, no-penetration boundary condition is to write

$$\mathbf{u} = \mathbf{U}_B \tag{6.27}$$

where $\mathbf{U}_B$ is the boundary velocity vector. It is interesting to consider the implications of this statement. Equation 6.27 indicates that the surface of a ship that has sailed from New York to South Hampton, England, is still coated with water from New York Harbor. This improbable result is true insofar as the continuum hypothesis is concerned, but as you probably suspect, water molecules originally on the surface will have long since been replaced by others owing to molecular diffusion. This thought experiment reminds

EXAMPLE 6.10

The piston in the tube shown in Figure 6.26 is advancing at speed U. What is the no-penetration condition on the piston face and tube wall?

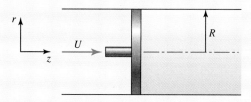

Figure 6.26 Schematic for Example 6.10.

SOLUTION

The no-penetration condition on the piston face is expressed in cylindrical coordinates as $v_z = U$, since the piston is moving at speed U in the z direction and the normal component of fluid velocity, v_z, must match the normal component of boundary velocity, U. The tube wall is stationary; thus the no-penetration condition on the wall is $v_r = 0$ (at $r = R$). That is, the normal component of fluid velocity, v_r, matches the normal component of boundary velocity, 0.

HISTORY BOX 6.1

The historical arguments related to the fluid–solid boundary interaction are valuable in that they highlight the dangers of not having a complete model of fluid flow. In 1744 d'Alembert formulated his paradox, noting that while the inviscid pressure distribution about a cylinder could be calculated accurately (see Example 6.3), the inviscid flow theory predicted zero drag, in contradiction to empirical observations. This theory did not include the effects of viscosity, hence imposed a boundary condition that allowed fluid to slip along the surface of the cylinder. It was not until Prandtl's contribution of boundary layer theory 150 years later that d'Alembert's paradox was fully understood.

us that in employing the continuum hypothesis, some questions involving molecular level events and phenomena cannot be answered.

In general, applying the no-slip, no-penetration boundary condition involves specifying conditions for each component of velocity. This process is illustrated in Examples 6.10 and 6.11.

6.7 FLUID TRANSPORT

An understanding of fluid transport is fundamental to many engineering fields, including the performance of mechanical devices such as engines, pumps, and transportation systems; the operation of chemical engineering processes ranging from the refining of oil to the production of the basic chemicals of a modern economy; the biological systems studied by biomedical engineers; and the movement of substances within local and global ecosystems studied by civil and environmental engineers. From oil refining and bioreactors to

EXAMPLE 6.11

Write the no-slip, no-penetration boundary conditions for the two flow situations illustrated in Figure 6.27.

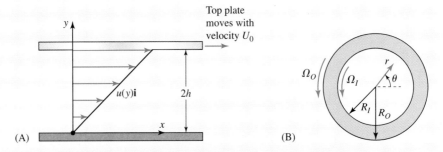

Figure 6.27 Schematic for Example 6.11.

SOLUTION

A. This flow is traditionally described in Cartesian coordinates with x along the plates and y up. Since the bottom plate is stationary, the no-slip, no-penetration boundary condition at this location is $\mathbf{u}(x, 0, z, t) = 0$. The top plate is moving in the $+x$ direction with constant velocity U_0 so the no-slip, no-penetration boundary condition is $\mathbf{u}(x, 2h, z, t) = U_0 \mathbf{i}$. On a component basis, we have $u = v = w = 0$ on the bottom plate, and $u = U_0, v = 0, w = 0$ on the top plate.

B. Cylindrical coordinates are appropriate here. The inside cylinder is rotating with an angular velocity Ω_I so the no-slip, no-penetration boundary condition is $\mathbf{u}(R_I, \theta, z, t) = R_I \Omega_I \mathbf{e}_\theta$. Similarly, since the outer cylinder is rotating with an angular velocity Ω_O, the boundary condition is $\mathbf{u}(R_O, \theta, z, t) = R_O \Omega_O \mathbf{e}_\theta$. On a component basis we have $v_r = v_z = 0, v_\theta = R_I \Omega_I$ on the inner cylinder and $v_r = v_z = 0, v_\theta = R_O \Omega_O$ on the outer cylinder.

artificial hearts and drug delivery, from tracking pollutants to predicting the growth of the ozone hole in the atmosphere, fluid transport plays a fundamental role.

To calculate the transport of mass, momentum, energy, and other properties and substances across a surface, it is helpful to recognize that fluid motion occurs on two different length scales—the macroscopic and microscopic. In the continuum Eulerian description, the macroscopic, or bulk, motion of fluid molecules is modeled by the fluid velocity field. The transport resulting from this visible movement (i.e., convection) of fluid particles across a surface is called convective transport. The transport resulting from the microscopic, or molecular-scale, motion of fluid molecules, is called diffusive transport. Convective transport can occur only in a moving fluid. Diffusive transport may occur in fluids at rest or in motion but is nonzero only in the presence of a spatial

6.7 FLUID TRANSPORT

To appreciate the importance of fluid transport in your life, consider the circulatory system in your body. You may think of this system in mechanical terms, with your heart functioning as a pump that develops a (blood) pressure driving the flow of fluid through a system of pipes in the form of arteries and veins. While this mechanical view is certainly valid, we should also consider the fluid transport characteristics of this system. Your blood transports dissolved oxygen to all your organs and then carries away the waste carbon dioxide. In addition, a host of other compounds, including sodium bicarbonate, which controls the blood pH, must be transported in carefully controlled amounts.

variation in a fluid property. For example, a spatial variation in temperature, i.e. a nonzero thermal gradient, results in diffusive transport of heat (also known as thermal conduction). Similarly, a concentration gradient of a certain chemical species within a fluid (e.g., dissolved oxygen in blood) will result in diffusive transport of that chemical species. If a spatial variation in temperature or concentration is absent, diffusive transport does not occur.

Convective and diffusive transport are equally important topics in the study of fluid mechanics, but not equally important modes of transport in every problem you encounter. In a fluid at rest, diffusive transport may be present, but convective transport is impossible. In a moving fluid, both modes of transport are possible, but it is common for the convective transport to far exceed the diffusive transport. In a turbulent flow the convective transport of mass, momentum, and energy can be surprisingly large, causing, for example, rapid evaporation of water by the wind and dangerous hypothermia due to wind chill.

Before discussing how to model convective and diffusive transport across a surface, it is helpful to describe the characteristics of different surfaces encountered in fluid mechanics. We may usefully distinguish between internal and bounding surfaces, and between permeable and impermeable surfaces. An internal surface is an imaginary surface wholly contained within a fluid. A bounding surface is located at the interface between the fluid and a solid, or between two fluids. A surface is permeable if bulk motion of fluid through the surface is possible. All internal surfaces are permeable. Bounding surfaces may be permeable or impermeable depending on whether they allow macroscopic quantities of the fluid to pass through them. An engineer must select and classify the surfaces in a problem as permeable or impermeable and apply the appropriate methods described in this section to compute the convective and diffusive transport.

Next consider the relationship between the surface type (permeable or impermeable) and the convective and diffusive transport across that surface. Since convective transport involves the bulk movement of fluid across a surface, it is possible only if the surface is permeable. The distinction between a permeable and an impermeable surface is not relevant to diffusive transport. For example, hydrogen can diffuse into steel, yet steel is impermeable to the bulk motion of gases and liquids. Thus diffusive transport is possible across surfaces of both types if an appropriate spatial gradient of a species or property is present. Table 6.1 summarizes these important results.

TABLE 6.1 Relationship Between Surface Type and Permissible Transport Mechanisms

	Type of Surface	
Transport Mechanism	**Permeable**	**Impermeable**
Convection	Possible	Impossible
Diffusion	Possible	Possible

Notice that in our discussion of convective transport no restriction was placed on the velocity or property fields. Thus the results apply to both steady and unsteady velocity and property fields. If a flow is unsteady, the convective transport defined by Eq. 6.28 provides the corresponding value at a certain instant of time.

6.7.1 Convective Transport

To describe convective transport, consider a permeable surface exposed to a fluid in motion as shown in Figure 6.28. The convective transport of any substance across this surface can be explained as follows. The velocity at a point on the surface describes the motion of the fluid particle located at that point at a given instant of time. The fluid particle possesses a certain amount of the transportable substance, and if this particle crosses the surface, the substance is transported across the surface along with it. Only the normal component of velocity is involved in the transport, since the component of velocity tangential to the surface does not cause the particle to cross the surface. Thus, the transport is due to the normal component of velocity and the amount of the substance carried by the fluid particle. Since different points on the surface may have different velocity and property values, a mathematical representation of convective transport must account for the potential variation of velocity and property fields on a surface.

The convective transport, represented by the symbol Γ_C, is defined to be the net rate at which a property is crossing a given surface at an instant of time due to the movement of fluid. If we let ε represent a transportable fluid property of interest per unit mass of fluid, then $\rho\varepsilon$ is the amount of this fluid property on a unit volume basis. At any point in a flow, the instantaneous rate at which this property is being transported by the velocity field in the direction of the local velocity vector is given by the convective flux vector $\rho\varepsilon\mathbf{u}$. Thus at a certain point on a permeable surface, the rate at which a property is transported across a surface element of area dS with unit normal $\mathbf{n}$ is

$$\delta \Gamma_C = \rho\varepsilon(\mathbf{u} \cdot \mathbf{n})\,dS$$

where $(\mathbf{u} \cdot \mathbf{n})$ produces the required normal component of velocity. In general there may be a different convective transport rate at each point of a surface due to local variations in the values of the convective flux vector. The convective transport for an entire surface, Γ_C, is found by summing the contributions from each surface element with a surface integral. That is,

$$\Gamma_C = \int_S \rho\varepsilon(\mathbf{u} \cdot \mathbf{n})\,dS \qquad (6.28)$$

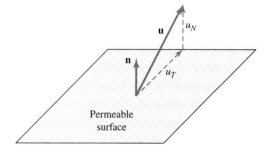

Figure 6.28 Velocity along a permeable surface can have convective transport only if there is a normal velocity component.

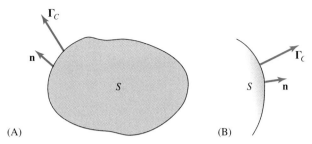

Figure 6.29 (A) Closed surface with positive **n** pointing outward and positive flux outward. (B) Open surface with positive flux in the direction of **n**.

A surface integral of this type is often referred to as a convective flux integral, or simply a flux integral. Note carefully however, that the result of the integration, the convective transport Γ_C, is a rate (property per unit time) rather than a flux (property per unit area per unit time).

Because of the presence of the outward unit normal in the integrand of Eq. 6.28, when a flux integral is evaluated for a closed surface, as shown in Figure 6.29A, it gives the net rate of transport of the property or substance out of the closed volume bounded by the surface. A positive value for Γ_C is therefore an outflow. For the open surface shown in Figure 6.29B, a positive value of Γ_C represents the rate of transport across the surface in the direction of the unit normal. Table 6.2 gives the appropriate form of the integral for various types of convective transport.

To illustrate these ideas, it is instructive to work through an example in which we calculate the convective transport of mass and volume, i.e., the mass and volume flowrates. Consider the steady, 1D channel flow of a constant density fluid between

TABLE 6.2 Selected Properties and Their Corresponding Convective Flux Integrals

Property	Property per unit mass, ε	Property per unit volume, $\lambda = \rho\varepsilon$	Convective flux integral, $\Gamma_C = \int_S \rho\varepsilon\,(\mathbf{u}\cdot\mathbf{n})\,dS$
Mass	1	ρ	$\int_S \rho(\mathbf{u}\cdot\mathbf{n})\,dS$
Volume	ρ^{-1}	1	$\int_S (\mathbf{u}\cdot\mathbf{n})\,dS$
Momentum	**u**	$\rho\mathbf{u}$	$\int_S \rho\mathbf{u}(\mathbf{u}\cdot\mathbf{n})\,dS$
Angular momentum	**r** × **u**	**r** × $\rho\mathbf{u}$	$\int_S (\mathbf{r}\times\rho\mathbf{u})(\mathbf{u}\cdot\mathbf{n})\,dS$
Heat	$c_p T$	$\rho c_p T$	$\int_S \rho c_p T(\mathbf{u}\cdot\mathbf{n})\,dS$
Concentration*	c	ρc	$\int_S \rho c\,(\mathbf{u}\cdot\mathbf{n})\,dS$
Internal energy*	u	ρu	$\int_S \rho u(\mathbf{u}\cdot\mathbf{n})\,dS$
Kinetic energy	$\frac{1}{2}(\mathbf{u}\cdot\mathbf{u})$	$\rho\frac{1}{2}(\mathbf{u}\cdot\mathbf{u})$	$\int_S \rho\frac{1}{2}(\mathbf{u}\cdot\mathbf{u})(\mathbf{u}\cdot\mathbf{n})\,dS$

*Lowercase u and c represent internal energy and concentration per unit mass, respectively.

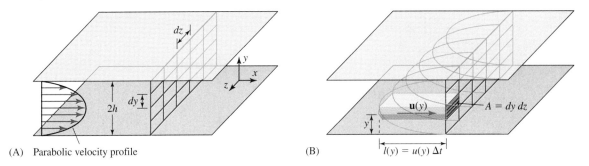

(A) Parabolic velocity profile (B)

Figure 6.30 (A) Schematic of flow between parallel plates and surface for flux calculation. (B) Elemental flow channel for the flux calculation.

parallel plates as shown in Figure 6.30A. This flow has a parabolic velocity distribution, with an x component of velocity given by

$$u(y) = U_{max}\left[1 - \left(\frac{y}{h}\right)^2\right] \qquad (6.29)$$

The maximum velocity, U_{max}, occurs at the center of the gap between the plates. If we choose a surface of width w at a right angle to the plates and spanning the entire gap between them, it is evident that fluid particles at a height y cross this surface at a right angle at a speed $u(y)$ as given by Eq. 6.29. Since each fluid particle has mass, and fluid particles at every height move from left to right across this surface, there must be a net convective transport of mass across this surface.

It is evident that the convective mass transport through the channel also passes through the selected surface. To calculate this transport, note that the gap between the plates may be divided into a large number of elemental flow channels each of height dy and width dz as shown in Figure 6.30A. The cross-sectional area of each of these elemental flow channels is $dz\,dy$. In a time interval Δt, all the fluid shown to the right of the parabolic lines shown in Figure 6.30B and to the left of the surface passes through the surface. The equation describing one of these parabolic lines is given by $x(y) = u(y)\Delta t$, where $x(y)$ represents the position of a fluid particle in the elemental flow channel at elevation y at a time Δt prior to the present instant. As you can see, during the selected time interval, a greater volume of fluid passes through the surface by way of the channel in the center of the gap than in any other channel because the fluid speed is a maximum there.

Consider the highlighted channel at elevation y. The length of the fluid column in this channel is $l(y) = u(y)\Delta t$. All the fluid in this column is destined to cross the surface during the time interval Δt, contributing to the convective transport of various fluid properties. The incremental fluid volume $\Delta V(y)$ in the highlighted channel of length $l(y)$ and area $dz\,dy$ is given by $\Delta V(y) = l(y)\,dz\,dy = u(y)\,\Delta t\,dz\,dy$.

The fluid mass in the selected channel is found by multiplying the incremental volume, $\Delta V(y)$, by the fluid density ρ. Thus, the total mass of fluid in the selected channel is

$$\rho \Delta V(y) = \rho u(y)\,dz\,dy\,\Delta t$$

The quantity $\rho u(y)\,dz\,dy\,\Delta t$ also represents the mass transported through the area element $dz\,dy$ during the time interval Δt. The corresponding mass flowrate is found by dividing the preceding result by the time interval to obtain $\rho u(y)\,dz\,dy$. The total mass flowrate across the entire surface, $\dot{M}$, is the sum of the contributions occurring in each elemental flow channel. For this 1D flow, this sum may be calculated by the surface integral

$$\dot{M} = \iint \rho u(y)\,dz\,dy \tag{6.30}$$

with the limits of integration chosen to describe the entire surface. Note that the integrand, which is the convective mass transport per unit area per unit time for a surface element of area $dz\,dy$, is proportional to the product of the density and the normal velocity of the fluid approaching the surface. This integrand, which has dimensions of mass per unit area per unit time, is referred to as the convective mass flux.

We can also use the same approach to calculate the volume flowrate (i.e., volume of fluid per unit time crossing the surface). Since the volume of fluid in the selected channel is

$$\Delta \mathcal{V}(y) = u(y)\,dz\,dy\,\Delta t$$

the total volume flowrate Q is given by

$$Q = \iint u(y)\,dz\,dy \tag{6.31}$$

The integrand $u(y)$, which has dimensions of volume per unit area per unit time, is the volume flux.

We can easily confirm these results by using Eq. 6.28 or the appropriate entries in Table 6.2. Taking Eq. 6.28, $\Gamma_C = \int_S \rho\varepsilon(\mathbf{u}\cdot\mathbf{n})dS$, and noting that x component of velocity is given by Eq. 6.29 as $u(y) = U_{\max}[1 - (y/h)^2]$ (with the other two velocity components being zero), we have $\mathbf{u} = u(y)\mathbf{i} + 0\mathbf{j} + 0\mathbf{k}$. Taking $\mathbf{n} = \mathbf{i}$ to find the transport downstream, we find: $\mathbf{u}\cdot\mathbf{n} = u(y)$. Thus the convective transport is given in general by Eq. 6.28 as

$$\Gamma_C = \int_S \rho\varepsilon(\mathbf{u}\cdot\mathbf{n})\,dS = \iint \rho\varepsilon u(y)\,dz\,dy$$

The integrand is the product of density, property per unit mass, and normal velocity on the surface as expected.

To obtain the mass flowrate, we use $\varepsilon = 1$ in this integral, or use Table 6.2, to write $\Gamma_C = \int_S \rho(\mathbf{u}\cdot\mathbf{n})\,dS = \iint \rho u(y)\,dz\,dy$. Note that this is identical to Eq. 6.30, obtained by using elemental flow channels. Similarly, by taking $\varepsilon = \rho^{-1}$ in this integral, or using Table 6.2, we find the following volume flowrate or volume transport

$$\Gamma_C = \int_S \rho(\rho)^{-1}(\mathbf{u}\cdot\mathbf{n})dS = \int_S (\mathbf{u}\cdot\mathbf{n})\,dS = \iint u(y)\,dz\,dy$$

which is identical to Eq. 6.31. The use of these formulas to calculate the mass and volume flowrate in channel flow is demonstrated in Example 6.12.

EXAMPLE 6.12

Derive expressions for the mass and volume flowrates across the surface shown in Figure 6.30 for the channel flow of a liquid.

SOLUTION

The velocity profile for this flow is given by Eq. 6.29, $u(y) = U_{max}[1 - (y/h)^2]$. Substituting this expression into the mass flowrate formula, Eq. 6.30, we find:

$$\dot{M} = \iint \rho u(y)\, dz\, dy = \iint \rho U_{max}\left[1 - \left(\frac{y}{h}\right)^2\right] dz\, dy$$

To evaluate this integral, the limits must be determined. The limits for y are from $-h$ to h, and we can consider the surface to have width w. Density is constant, and we know that $u = u(y)$ only. Therefore the z integration will yield the width, w, leaving us with $\dot{M} = \rho U_{max} w \int_{-h}^{h}[1 - (y/h)^2]\,dy$. Completing the integration, we have

$$\dot{M} = \rho U_{max} w \left[y - \left(\frac{y^3}{3h^2}\right)\right]_{-h}^{h} = \frac{4\rho U_{max} wh}{3}$$

The dimensions of this result are $\{Mt^{-1}\}$, as expected. To find the volume flowrate, we can use Eq. 6.31, or simply observe that for constant density, $Q = \iint u(y)\, dz\, dy = \dot{M}/\rho$. Thus, we have

$$Q = \frac{4 U_{max} wh}{3},$$

which has dimensions of $\{L^3 t^{-1}\}$, as expected.

We can also use Eq. 6.28 or the appropriate entries in Table 6.2 to develop expressions for the mass and volume flowrates across the surface in the pipe flow illustrated in Figure 6.31. The integrals giving the mass and volume flowrates in this case are written in cylindrical coordinates as

$$\dot{M} = \iint \rho v_z(r) r\, dr\, d\theta \quad (6.32)$$

$$Q = \iint v_z(r) r\, dr\, d\theta \quad (6.33)$$

Note the integrands in these expressions. Once again we see that the product of density and normal velocity defines the mass flux, and the normal velocity alone defines the volume flux.

The preceding exercises illustrate how to calculate the convective transport of mass and volume in a simple flow. In more complicated situations, students often find it

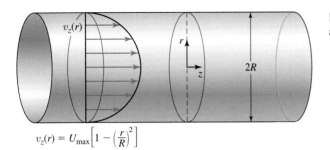

Figure 6.31 Schematic of pipe flow to determine flux across the surface.

helpful to use a step-by-step method to calculate one or more of the flux integrals contained in Table 6.2. We recommend the method outlined.

General Method to Evaluate a Flux Integral

1. Select a coordinate system that provides the simplest possible representation of the unit normal and surface element (e.g., align one axis with the unit normal).
2. Use the selected coordinates to write the unit normal $\mathbf{n}$ and scalar area element dS.
3. Write the velocity vector and evaluate the dot product $\mathbf{u} \cdot \mathbf{n}$.
4. Write the integrand $\rho\varepsilon(\mathbf{u} \cdot \mathbf{n})\, dS$.
5. Assign appropriate limits to the convective flux integral $\Gamma_C = \int_S \rho\varepsilon(\mathbf{u} \cdot \mathbf{n})\, dS$ and complete the integration.

EXAMPLE 6.13

Derive expressions for the mass and volume flowrates across the surface shown in Figure 6.31 for Poiseuille flow in a round pipe. The velocity profile is

$$v_z(r) = U_{\max}\left[1 - \left(\frac{r}{R}\right)^2\right]$$

SOLUTION

The mass flowrate is given by Eq. 6.32, $\dot{M} = \iint \rho U_{\max}[1 - (r/R)^2]\, r\, dr\, d\theta$. The limits on θ are from 0 to 2π, while r goes from 0 to R. Since density is constant, we have

$$\dot{M} = 2\pi\rho U_{\max} \int_0^R \left[1 - \left(\frac{r}{R}\right)^2\right] r\, dr$$

Evaluating this integral gives

$$\dot{M} = 2\pi\rho\, U_{\max} \left[\left(\frac{r^2}{2}\right) - \left(\frac{r^4}{4R^2}\right)\right]_0^R = \frac{1}{2}\rho\, U_{\max} \pi R^2$$

The volume flowrate can be found by using Eq. 6.33, or more simply for this constant density fluid, by dividing the mass flowrate by the density to obtain $Q = \frac{1}{2} U_{\max} \pi R^2$.

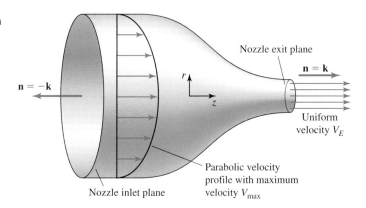

Figure 6.32 Schematic of flow through a constriction.

One of the most important tasks in fluid transport analysis is to calculate momentum transport. This calculation is required to determine the force exerted by a fluid on a moving object or vehicle, and in the calculation of the thrust produced by a rocket or air-breathing engine. From Table 6.2, the momentum transport across a surface is given by:

$$\Gamma_C = \int (\rho \mathbf{u})(\mathbf{u} \cdot \mathbf{n})\, dS$$

This integral can be particularly challenging to evaluate correctly. To illustrate the use of the method just outlined, consider the calculation of the momentum flux at the inlet and exit of the round nozzle shown in Figure 6.32.

We begin by noting that the momentum flux integral $\int_S \rho \mathbf{u}(\mathbf{u} \cdot \mathbf{n})\, dS$ is to be evaluated on the surfaces shown at the inlet and exit of the nozzle. For simplicity we assume steady flow of a constant density fluid. The geometry suggests the use of cylindrical coordinates, with the z axis normal to both surfaces. Consider first the nozzle exit, where fluid is leaving with a uniform velocity. To complete the second step in the method we must write the unit normal to the desired surface. In this case $\mathbf{n}$ is aligned with the z axis, so we have $\mathbf{n} = \mathbf{k}$. The exit surface lies in the (r, θ) plane so the surface element is given in cylindrical coordinates by $dS = r\, dr\, d\theta$. The velocity vector at the exit is normal to the surface; consequently the v_r and v_θ components of velocity are both zero, and the v_z component is uniform at the exit. We can therefore write the uniform fluid velocity vector on the surface of the exit as $\mathbf{u} = (0, 0, V_E \mathbf{k})$, where V_E is a positive constant. Having established the outward unit normal and velocity vector at the exit, we can complete the third step by writing the dot product as $(\mathbf{u} \cdot \mathbf{n}) = (V_E \mathbf{k}) \cdot (\mathbf{k}) = V_E$. The magnitude and sign of the normal velocity $(\mathbf{u} \cdot \mathbf{n})$ can be determined automatically by using the correct expressions for the velocity vector and outward unit normal. The positive normal velocity at the exit indicates that the transport of momentum is out of the nozzle in the direction of the outward unit normal. The integrand for a momentum flux is $\rho \varepsilon = \rho \mathbf{u} = \rho(V_E \mathbf{k})$; thus the integrand is written in the fourth step as $(\rho \mathbf{u})(\mathbf{u} \cdot \mathbf{n})\, dS = \rho(V_E \mathbf{k}) V_E r\, dr\, d\theta$. The final step in evaluating the momentum flux integral is to insert the integrand and write the appropriate limits,

$$\int_{\text{exit}} (\rho \mathbf{u})(\mathbf{u} \cdot \mathbf{n})\, dS = \int_0^{2\pi} \int_0^{R_E} \rho(V_E \mathbf{k})(V_E) r\, dr\, d\theta$$

The integrand $\rho(V_E^2 \mathbf{k})$ is constant, so it may come out of the integral. What is left is an integral defining the area of the exit, so the momentum transport at the nozzle exit is

$$\int_0^{2\pi}\int_0^R \rho(V_E\mathbf{k})(V_E)r\,dr\,d\theta = \rho\left(V_E^2\mathbf{k}\right)\int_0^{2\pi}\int_0^{R_E} r\,dr\,d\theta = \rho A_E\left(V_E^2\mathbf{k}\right)$$

where $A_E = \pi R_E^2$ is the area of the nozzle exit. The momentum transport at the exit has only a $\mathbf{k}$ component because the velocity vector at the exit port has only a $\mathbf{k}$ component. The transport is positive because a positive $\mathbf{k}$ component of momentum is leaving the nozzle. Note that the overall sign of a momentum transport depends on whether $(\rho\mathbf{u})$ is positive or negative, and on whether the normal velocity $(\mathbf{u}\cdot\mathbf{n})$ is positive or negative. The integrand contains the product of two signs, one from $(\rho\mathbf{u})$ and the other from $(\mathbf{u}\cdot\mathbf{n})$.

Now let us consider the momentum transport at the nozzle inlet where the flow is nonuniform. The general procedure is the same. For the momentum flux integral here, we write the outward unit normal to the surface as $\mathbf{n} = -\mathbf{k}$ and again write the surface element in cylindrical coordinates as $dS = r\,dr\,d\theta$. The velocity profile at the inlet port is axisymmetric and parabolic, with a maximum velocity $V_{\max}$ on the axis of the nozzle. Thus the nonuniform fluid velocity vector entering the nozzle is given by $\mathbf{u} = (v_r, v_\theta, v_z) = v_z(r)\mathbf{k}$, where $v_z(r)$ is the axial velocity component at the inlet surface, and the other two velocity components are zero by inspection of Figure 6.32. Since the velocity varies in the radial direction, it is represented by a function $v_z(r)$, which specifies the value of v_z at each spatial location on the inlet port surface. The sign of v_z is part of its specification, and there is no z dependence because the integrand is being written on a surface where z is constant.

Proceeding step by step to formulate the momentum flux integral, we write

$$(\mathbf{u}\cdot\mathbf{n}) = (v_z(r)\mathbf{k})\cdot(-\mathbf{k}) = -v_z(r)$$
$$\rho\varepsilon = \rho\mathbf{u} = \rho v_z(r)\mathbf{k}$$
$$(\rho\mathbf{u})(\mathbf{u}\cdot\mathbf{n})\,dS = \rho v_z(r)(-v_z(r)\mathbf{k})r\,dr\,d\theta = -\rho[v_z(r)]^2\mathbf{k}\,r\,dr\,d\theta$$

The magnitude and sign of the normal velocity are determined by the directions of the velocity vector and outward unit normal. The negative sign shows that the normal component of the velocity vector is pointing opposite to the outward unit normal, indicating momentum transport into the nozzle.

To complete the final step, we note that $v_z(r) = V_{\max}[1 - (r^2/R_I^2)]$. Thus the integrand is

$$(\rho\mathbf{u})(\mathbf{u}\cdot\mathbf{n})\,dS = -\rho[v_z(r)]^2\mathbf{k}\,r\,dr\,d\theta = -\rho V_{\max}^2\left[1 - \left(\frac{r^2}{R_I^2}\right)\right]^2\mathbf{k}$$

After the appropriate limits of integration have been applied, the momentum transport is given by

$$\int_{\text{inlet}}(\rho\mathbf{u})(\mathbf{u}\cdot\mathbf{n})\,dS = \int_0^{2\pi}\int_0^{R_I}\left(-\rho V_{\max}^2\left(1 - \frac{r^2}{R_I^2}\right)^2\mathbf{k}\right)r\,dr\,d\theta = -\frac{1}{6}\rho V_{\max}^2 A_I\mathbf{k}$$

where $A_I = \pi R_I^2$ is the area of the nozzle inlet. The momentum transport at the nozzle inlet has only a $\mathbf{k}$ component because the velocity vector at the inlet has only a $\mathbf{k}$

As you gain experience in calculating convective transport, you may find that it is not necessary to carry out every step in the recommended procedure for calculating a flux integral. Many additional examples of the use of this method to calculate convective transport can be found in the next chapter on control volume analysis.

component. The momentum transport is negative at this surface because a positive **k** component of momentum is crossing the surface in the direction opposite to the selected unit normal.

Our discussion of convective transport has been based on the use of ε, a property per unit mass. Although every fluid property may be defined on a unit mass basis, some properties, such as the density, are more naturally defined on a per-unit-volume basis, i.e., as the mass per unit volume. Engineers may encounter situations in which a convective transport calculation is required for a property defined on a unit volume basis. The necessary change in the preceding formulas becomes evident if we note that if we let the property per unit volume be represented by λ, then λ and ε are related by

$$\lambda = \rho \varepsilon \tag{6.34}$$

Substituting this relation into Eq. 6.28, which defines convective transport for a property per unit mass, allows us to write the integral defining the convective transport of any property per unit volume λ as

$$\Gamma_C = \int_S \lambda \, (\mathbf{u} \cdot \mathbf{n}) \, dS \tag{6.35}$$

The calculation of convective transport for a property defined on a unit volume basis may be done by means of the same five-step method discussed earlier for a property defined on a unit mass basis. The next example illustrates this procedure when the property involved is the concentration of a substance mixed in air.

CD/Boundary Layer Instability, Transition, and Turbulence/Turbulent Mixing and Diffusion

6.7.2 Diffusive Transport

Diffusive transport arises from the microscopic motion of fluid molecules. If there is a larger value of a property on one side of a surface than on the other, a net transport of this property will occur as fluid molecules diffuse back and forth across the surface. You might be wondering whether mass, momentum, and energy are diffused in a fluid. The answer to this question comes in three parts. First, in the continuum model of a fluid such as air and water, diffusive transport of mass occurs only for a chemical species mixed in the fluid. There is no diffusive transport of density, which describes the mass per unit volume of the carrier fluid. Second, momentum transport arises both from the macroscopic, or bulk, motion of fluid as represented by the velocity field and from the microscopic, or molecular-scale, motion of fluid molecules. The transport of momentum by the bulk motion is modeled as a convective transport. However, the molecular scale transport of momentum is not modeled as a diffusive transport. Instead the continuum model postulates the existence of a state of stress in the fluid characterized by the

EXAMPLE 6.14

Air containing 0.5% ammonia (NH_3) by volume flows from an absorber column through a round exit pipe with $D = 10$ cm. Suppose the velocity field in the pipe is given in cylindrical coordinates by $\mathbf{u} = U_{max}[1 - (r/R)^2]\mathbf{k}$, where $U_{max} = 0.5$ m/s. If the density of ammonia at the pressure and temperature of the air leaving the column is $\rho_{NH_3} = 0.649$ kg/m³, what is the rate at which ammonia leaves the absorber column because of convective transport? Use the concept of a property per unit volume in your calculation, and assume the ammonia is well mixed.

SOLUTION

We are asked to calculate the rate at which ammonia leaves the absorber column due to convective transport. Figure 6.33 serves as a sketch for this problem. Since the ammonia that leaves the absorber column passes through the exit pipe, the convective transport can be calculated over the surface of the exit pipe. First we note that to obtain the mass flowrate of ammonia, we would use, $\varepsilon = c_{NH_3}$, where c_{NH_3} is the concentration of ammonia in dimensions of mass of ammonia per mass of air. The ammonia is well mixed, so this concentration is spatially uniform. The mass of ammonia per unit volume of air is $\lambda_{NH_3} = \rho_{air} c_{NH_3}$, which we can write as

$$\lambda_{NH_3} = \rho_{air} c_{NH_3} \equiv \frac{\text{mass of air}}{\text{volume of air}} \times \frac{\text{mass of } NH_3}{\text{mass of air}} \equiv \frac{\text{mass of } NH_3}{\text{volume of air}}$$

If we knew the mass concentration c_{NH_3} and the density of the air ρ_{air} leaving the column, we could easily calculate λ_{NH_3} and proceed with the five-step method to

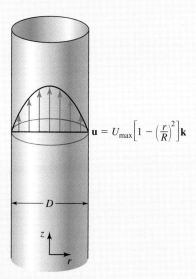

Figure 6.33 Schematic for Example 6.14.

calculate the convective transport. We are given that the concentration of ammonia in the air is 0.5% ammonia by volume. If we write this information as

$$\frac{0.5\% \text{ volume of NH}_3}{100\% \text{ volume of air}} = 0.005 \text{ m}^3\text{NH}_3/\text{m}^3\text{air}$$

the units suggest that we multiply this value by the density of ammonia at the flow condition to find

$$(0.005 \text{ m}^3\text{NH}_3/\text{m}^3\text{air})(0.649 \text{ kg NH}_3/\text{m}^3 \text{ NH}_3) = 3.25 \times 10^{-3} \text{ kg NH}_3/\text{m}^3 \text{ air}$$

which is the mass of ammonia per unit volume of air. This is the desired value of λ_{NH_3}.

The convective transport is now found using Eq. 6.35, $\Gamma_C = \int_S \lambda (\mathbf{u} \cdot \mathbf{n}) \, dS$, and following the five-step method. In cylindrical coordinates, aligned as shown in Figure 6.33, the unit normal is $\mathbf{n} = \mathbf{k}$, and the dot product is

$$\mathbf{u} \cdot \mathbf{n} = U_{\max}\left[1 - \left(\frac{r}{R}\right)^2\right]\mathbf{k} \cdot \mathbf{k} = U_{\max}\left[1 - \left(\frac{r}{R}\right)^2\right]$$

The surface element for the pipe exit is $dS = r \, dr \, d\theta$. Applying the limits of integration from 0 to R and 0 to 2π, and noting that λ_{NH_3} is spatially uniform because the ammonia is well mixed, we can evaluate the integral as follows

$$\Gamma_C = \int_S \lambda (\mathbf{u} \cdot \mathbf{n}) \, dS = \lambda_{\text{NH}_3} \int_0^{2\pi} \int_0^R U_{\max}\left[1 - \left(\frac{r}{R}\right)^2\right] r \, dr \, d\theta$$

$$\Gamma_C = 2\pi \lambda_{\text{NH}_3} U_{\max}\left[\left(\frac{r^2}{2}\right) - \left(\frac{r^4}{4R^2}\right)\right]_0^R = \lambda_{\text{NH}_3} \frac{U_{\max}}{2} \pi R^2$$

which we recognize as the product of the concentration of ammonia per unit volume of air times the volume flowrate of air in the pipe. Inserting the data, we have

$$\Gamma_C = \lambda_{\text{NH}_3} \frac{U_{\max}}{2} \pi R^2 = (3.25 \times 10^{-3} \text{ kg NH}_3/\text{m}^3 \text{ air})\left(\frac{0.5 \text{ m/s}}{2}\right)(\pi)(0.05 \text{ m})^2$$

The rate at which ammonia is leaving the absorber is therefore $\Gamma_C = 6.38 \times 10^{-6}$ kg NH$_3$/s. The dimensions for the solution are $\{Mt^{-1}\}$, as expected.

presence of short-range forces or stresses acting on surfaces within the fluid and on boundary surfaces the fluid contacts. The presence of these internal stresses in the fluid, as represented by the stress tensor, fully accounts for the molecular-scale momentum transport in the model of the fluid. Thus, there is no diffusive momentum transport in the continuum model. You may wish to review the discussion of the state of stress and stress tensor in Section 4.6 to add to your understanding of this point. The third and final answer to our original question about diffusive transport is that diffusion of energy (in the form know as heat) does occur and is modeled by the concept of heat conduction. Thus, to summarize, in a continuum model we need consider only the diffusive transport of chemical species, often simply referred to as diffusion, and the diffusive transport of heat, normally referred to as heat conduction.

The diffusive transport across a surface at a specific point in a fluid depends on the orientation and shape of the surface as well as on the spatial distribution of the property at this point. It proves useful to define the diffusive transport rate at a given point on a surface in terms of a diffusive flux vector $\mathbf{q}_D$, which has dimensions of the transportable property per unit area per unit time. For fluids such as air and water, the relationship between $\mathbf{q}_D$ and the gradient of a transportable scalar property is modeled with a linear relationship, which proves to be sufficiently accurate for engineering applications. Consider first the transport of heat across a surface, i.e., heat conduction. According to Fourier's law, the diffusive flux vector for heat conduction is

$$\mathbf{q}_D = -k\nabla T \qquad (6.36)$$

where k is the thermal conductivity of the fluid and T is the temperature. Similarly, according to Fick's law, the diffusive flux vector for the mass transfer of a chemical species or substance is

$$\mathbf{q}_D = -k_C \nabla C \qquad (6.37)$$

where C is the concentration of a substance dissolved in a fluid and k_C is the diffusion coefficient. Values for the thermal conductivity of various fluids and diffusivities of several substances are contained in many engineering handbooks.

Equations 6.36 and 6.37 show that the diffusive flux vector always points in the direction opposite the property gradient, since property flows down the gradient from the higher value of the property to the lower value. However the flux vector is not necessarily aligned with the normal to a given surface element. This is illustrated in Figure 6.34. It can be seen that the component of the flux vector normal to the surface, i.e., $\mathbf{q}_D \cdot \mathbf{n}$, is the only component that contributes to the net rate at which heat or mass crosses the surface. The tangential component of the flux vector represents diffusion along the surface in a tangential direction; hence it makes no contribution to the diffusive transport across the surface.

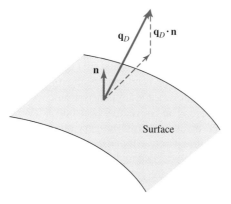

Figure 6.34 The flux vector and the surface normal are not necessarily aligned.

The diffusive transport, represented by the symbol Γ_D, is defined to be the net rate at which a property is crossing a given surface at an instant of time as a result of diffusion. If the surface is divided into infinitesimal surface elements as usual, the rate at which heat or mass is transported across a surface element of area dS and unit normal $\mathbf{n}$ by molecular diffusion is given by

$$\delta\Gamma_D = \mathbf{q}_D \cdot \mathbf{n}\, dS$$

The diffusive transport Γ_D for the entire surface is then given by the surface integral

$$\Gamma_D = \int_S (\mathbf{q}_D \cdot \mathbf{n})\, dS$$

From Eqs. 6.36 and 6.37, we see that the heat conduction across this surface is given by

$$\Gamma_D = \int_S -k(\nabla T \cdot \mathbf{n})\, dS \tag{6.38}$$

and the diffusion (of a chemical species) is given by

$$\Gamma_D = \int_S -k_C(\nabla C \cdot \mathbf{n})\, dS \tag{6.39}$$

These integrals define the net rate of transport of heat or mass of a chemical species across a surface in the direction of $\mathbf{n}$ due to diffusion (the molecular-scale motion of fluid molecules). The interpretation of the sign of Γ_D follows the same rule given earlier for convective transport: for a closed surface, a positive value for Γ_D is an outflow; on an open surface, a positive value represents the diffusive transport across the surface in the direction of the outward unit normal. Examples 6.15 and 6.16 illustrate the use of Eqs. 6.38 and 6.39 to calculate diffusive transport.

6.7.3 Total Transport

In most fluid flows convective and diffusive transport occur simultaneously. The total transport rate of a property across a surface, Γ, is the sum of the convective transport rate Γ_C and the diffusive transport rate Γ_D:

$$\Gamma = \Gamma_C + \Gamma_D \tag{6.40}$$

We can use the models for convective and diffusive transport developed in the preceding two sections to calculate the total transport rate Γ of any quantity across a surface within a fluid or across a surface defining the boundary between a fluid and another material. In a given situation, an engineer should first consider the nature of the surface involved to see if the convective flux is nonzero (Table 6.1). At an impermeable boundary between a fluid and a solid at rest, the no-penetration condition, $\mathbf{u} \cdot \mathbf{n}$, ensures that the convective transport is zero. The convective transport is also zero for an impermeable surface in motion, since in that case the no-slip, no-penetration condition is $\mathbf{u} = \mathbf{u}_S$, the relative velocity is then $\mathbf{u}_R = \mathbf{u} - \mathbf{u}_S = 0$, and therefore $\mathbf{u}_R \cdot \mathbf{n} = 0$ on the surface. The type of substance transported should be considered next to decide whether both convective and diffusive fluxes are present. Finally, using the appropriate integrals to calculate the convective and diffusive fluxes (if present) on the specified surface will

EXAMPLE 6.15

The steady heat flow between vertical parallel plates a distance $2h$ apart arising from the unequal plate temperatures is shown in Figure 6.35. A linear temperature profile develops in the fluid as given by $T(y) = \frac{1}{2}(T_{\text{hot}} + T_{\text{cold}}) - \frac{1}{2}(T_{\text{hot}} - T_{\text{cold}})(y/h)$. The fluid has a thermal conductivity of $k = 0.65\,\text{W/(m-K)}$, and $(T_{\text{hot}} - T_{\text{cold}})/2h = 100\,°\text{C/m}$. Find the heat flux vector in the fluid and the diffusive heat transport rate to a square meter of each plate.

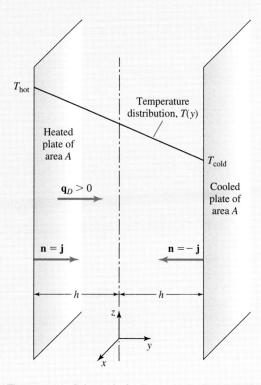

Figure 6.35 Schematic for Example 6.15.

SOLUTION

We are asked to determine the diffusive heat flux vector and the diffusive heat transport to a square meter of each plate in the parallel plate geometry shown in Figure 6.35 (which serves as the sketch for this problem). The diffusive heat flux vector is given by Eq. 6.36 as $\mathbf{q}_D = -k\nabla T$. Since $\nabla T = (\partial T/\partial x)\mathbf{i} + (\partial T/\partial y)\mathbf{j} + (\partial T/\partial z)\mathbf{k}$, and only $\partial T/\partial y$ is nonzero, the heat flux vector in this problem is

$$\mathbf{q}_D = -k\nabla T = -k\frac{\partial T}{\partial y}\mathbf{j} = -k\left(-\frac{T_{\text{hot}} - T_{\text{cold}}}{2h}\right)\mathbf{j} = k\left(\frac{T_{\text{hot}} - T_{\text{cold}}}{2h}\right)\mathbf{j}$$

The heat flux vector is a constant and points to the right.

The diffusive heat transport Γ_D to either of the plates is given by Eq. 6.38 as $\Gamma_D = \int_S -k(\nabla T \cdot \mathbf{n})\, dS$. The outward unit normal to each plate is needed to evaluate this integral. This normal points at the fluid. Thus for the left plate we have $\mathbf{n} = \mathbf{j}$, while for the right plate $\mathbf{n} = -\mathbf{j}$. For the left plate the diffusive heat transport is

$$\Gamma_D = \int_S -k(\nabla T \cdot \mathbf{n})\, dS = \int_S \left(k\frac{T_{\text{hot}} - T_{\text{cold}}}{2h}\right) \mathbf{j} \cdot \mathbf{j}\, dS = k\left(\frac{T_{\text{hot}} - T_{\text{cold}}}{2h}\right) A$$

where A is the plate area. The positive sign indicates heat is flowing from the hot plate to the fluid. On the right plate the diffusive transport is

$$\Gamma_D = \int_S -k(\nabla T \cdot \mathbf{n})\, dS = \int_S \left(k\frac{T_{\text{hot}} - T_{\text{cold}}}{2h}\right) \mathbf{j} \cdot (-\mathbf{j})\, dS = -\left(k\frac{T_{\text{hot}} - T_{\text{cold}}}{2h}\right) A$$

The negative sign indicates heat is flowing to the cold plate from the fluid.

Since we know that $(T_{\text{hot}} - T_{\text{cold}})/2h = 100°\text{C}/\text{m}$, and $k = 0.65\,\text{W}/(\text{m-K})$, the heat flux vector is

$$\mathbf{q}_D = k\left(\frac{T_{\text{hot}} - T_{\text{cold}}}{2h}\right)\mathbf{j} = [0.65\,\text{W}/(\text{m-K})](100\,°\text{C}/\text{m})\left(\frac{1\text{K}}{1°\text{C}}\right)\mathbf{j} = 65\,\text{W}/\text{m}^2\,\mathbf{j}$$

and the heat transfer to a square meter of each plate is $k[(T_{\text{hot}} - T_{\text{cold}})/2h]A = (65\,\text{W}/\text{m}^2)(1\,\text{m}^2) = 65\,\text{W}$. The conduction of heat is from the hot plate into the fluid and to the cold plate.

EXAMPLE 6.16

A laboratory experiment involves the flow of aerated water on both sides of a polymer sheet as illustrated in Figure 6.36. The process involves the diffusion of O_2 through a polyethylene terephthalate membrane of thickness $h = 100\,\mu\text{m}$. The diffusion coefficient for oxygen through this membrane is $k_C = 3.6 \times 10^{-9}\,\text{cm}^2/\text{s}$. You may assume that water molecules have a negligible rate of diffusion through this polymer. The oxygen concentrations on the two sides of the membrane at steady state are 8.9×10^{-8} and $21.2 \times 10^{-8}\,\text{mol/cm}^3$, and it is known that there is a linear concentration gradient through the membrane. If the device produces an hourly flow of 1.9×10^{-5} moles of O_2 through the membrane, estimate the total membrane surface area.

SOLUTION

We are asked to determine the surface area of a membrane through which oxygen is diffusing. We will accomplish this task by determining the diffusive mass transport through

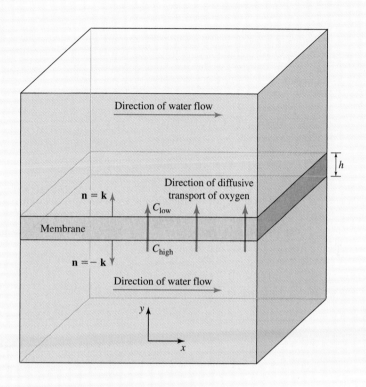

Figure 6.36 Schematic for Example 6.16 showing the surface for oxygen diffusion through a membrane.

the membrane. Figure 6.36 serves as the sketch for this problem. The mass flux vector is given by Eq. 6.37 as $\mathbf{q}_D = -k_C \nabla C$. The concentration gradient is found by inspection to be

$$\nabla C = \left(\frac{C_{\text{low}} - C_{\text{high}}}{h}\right)\mathbf{j}$$

The diffusive mass transport Γ_D through the membrane is given by Eq. 6.39, the surface integral $\Gamma_D = \int_S -k_C(\nabla C \cdot \mathbf{n})\,dS$. The outward unit normal to the membrane is needed to evaluate this integral. This normal points at the fluid. Thus on the top side of the membrane (the side with the lower O_2 concentration), we have $\mathbf{n} = \mathbf{j}$. Thus the diffusive mass transport is

$$\Gamma_D = \int_S -k_C(\nabla C \cdot \mathbf{n})\,dS$$

$$= \int_S -\left(k_C \frac{C_{\text{low}} - C_{\text{high}}}{h}\mathbf{j}\right) \cdot \mathbf{j}\,dS = -k_C\left(\frac{C_{\text{low}} - C_{\text{high}}}{h}\right)A$$

where A is the area of the membrane, and we have made use of the fact that none of the terms in the integrand are functions of position on the membrane surface. Substituting the known values for the two concentrations and k_C into the preceding expression yields:

$$\Gamma_D = -(3.6 \times 10^{-9} \text{cm}^2/\text{s}) \left[\frac{(8.9 \times 10^{-8} - 21.2 \times 10^{-8})\text{mol/cm}^3}{0.01 \text{ cm}} \right]$$

$$= [4.43 \times 10^{-14} \text{ mol/(cm}^2\text{-s)}]A$$

Finally, to solve for the membrane area, we must recognize that the value of the diffusive mass transport through the membrane was specified in the problem statement to be 1.9×10^{-5} mol O_2 per hour. Solving the preceding expression for the membrane area and inserting the known value for Γ_D yields:

$$A = \left[\frac{1.9 \times 10^{-5} \text{ mol/h}}{4.43 \times 10^{-14} \text{ mol/(cm}^2\text{-s)}} \right] \left(\frac{1 \text{ h}}{3600 \text{ s}} \right) = 1.2 \times 10^5 \text{ cm}^2 \left(\frac{1 \text{ m}}{100 \text{ cm}} \right)^2 = 1.2 \text{ m}^2$$

EXAMPLE 6.17

Consider the convective flow between vertical parallel plates shown in Figure 6.37. The velocity field used to model this flow is given by $\mathbf{u} = 0\mathbf{i} + 0\mathbf{j} + w(y)\mathbf{k}$ with

$$w(y) = \frac{\rho \beta g h^2 (T_{\text{hot}} - T_{\text{cold}})}{12\mu} \left[\left(\frac{y}{h} \right)^3 - \frac{y}{h} \right]$$

where ρ and β are the density and coefficient of thermal expansion of the fluid evaluated at the mean temperature $\frac{1}{2}(T_{\text{hot}} + T_{\text{cold}})$, g is the acceleration of gravity, and $2h$ is the spacing between the plates. Recall that the temperature profile in the fluid is $T(y) = \frac{1}{2}(T_{\text{hot}} + T_{\text{cold}}) - \frac{1}{2}(T_{\text{hot}} - T_{\text{cold}})(y/h)$. Find the total heat transport across the surface shown of width b into the paper.

SOLUTION

We are asked to determine the total heat transport across the surface indicated in Figure 6.37, given the corresponding velocity and temperature profiles for this flow. This problem can be solved by using Eq. 6.40, which gives the total transport as $\Gamma = \Gamma_C + \Gamma_D$. The selected surface is permeable; thus in general we must consider

6.7 FLUID TRANSPORT

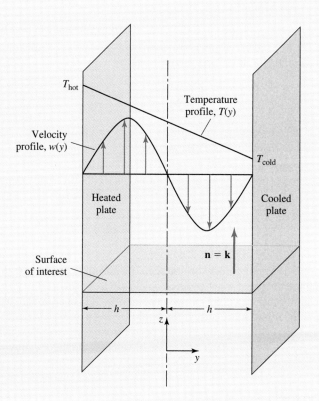

Figure 6.37 Schematic for Example 6.17.

the possibility of both convective and diffusive transport of heat across this surface. The transport integrals are determined by consulting Table 6.2 and Eq. 6.38. The result is

$$\Gamma = \Gamma_C + \Gamma_D = \int_S \rho c_p T\,(\mathbf{u}\cdot\mathbf{n})\,dS + \int_S -k(\nabla T \cdot \mathbf{n})\,dS$$

which can also be represented as a single integral

$$\Gamma = \int_S [\rho c_p T\,(\mathbf{u}\cdot\mathbf{n}) - k(\nabla T \cdot \mathbf{n})]\,dS$$

In Example 6.15 we found that the heat flux vector is $\mathbf{q}_D = -k\nabla T = k[(T_{\text{hot}} - T_{\text{cold}})/2h]\,\mathbf{j}$, and we know the temperature and velocity fields. By inspection the unit normal to this surface is $\mathbf{n} = \mathbf{k}$, so

$$\mathbf{u}\cdot\mathbf{n} = w(y)(\mathbf{k}\cdot\mathbf{k}) = w(y),\quad \rho c_p T\,(\mathbf{u}\cdot\mathbf{n}) = \rho c_p T(y)\,w(y)$$

$$-k(\nabla T\cdot\mathbf{n}) = k\left(\frac{T_{\text{hot}} - T_{\text{cold}}}{2h}\right)(\mathbf{j}\cdot\mathbf{k}) = 0 \quad\text{and}\quad \Gamma = \int_S \rho c_p T(y)\,w(y)\,dS$$

The diffusive transport of heat across this surface is zero because the heat flux vector is parallel to the surface. The total transport is therefore due to convection alone and can be found by evaluating the following double integral:

$$\Gamma = \int_{-h}^{h} \int_{-b/2}^{b/2} \rho c_p T(y)\, w(y)\, dx\, dy$$

Substituting for the temperature and velocity profiles and evaluating the resulting integral with a symbolic mathematics code, we find:

$$\Gamma = \rho c_p b h \left(\frac{2}{15}\right) \left(-\frac{T_{hot} - T_{cold}}{2}\right) \left(\frac{\rho \beta g h^2 (T_{hot} - T_{cold})}{12\mu}\right)$$

give the total transport across the surface. The Example 6.17 illustrates the process of calculating the total heat transfer across a surface in a flow.

We can summarize all this in two general rules: (1) one should consider convective transport of the relevant property if the fluid is moving and the surface of interest is permeable, and (2) one should consider the possibility of diffusive transport of heat or concentration across every surface regardless of whether the fluid is moving or at rest.

6.8 AVERAGE VELOCITY AND FLOWRATE

The concept of an average velocity $\bar{V}$ and its relationship to mass and volume flowrate are fundamental to any discussion of fluid transport. In our case study on flow in a round pipe, Section 3.3.1, we stated that in a steady flow of constant density fluid, the relationship between the volume flowrate Q, average velocity $\bar{V}$, and the cross-sectional area of a pipe A is given by Eq. 3.16, as

$$Q = \bar{V} A \qquad (6.41a)$$

Since we will now show that this relationship holds for any permeable surface, we relist it here as Eq. 6.41a. The corresponding mass flowrate $\dot{M}$ crossing the surface is defined by

$$\dot{M} = \rho A \bar{V} \qquad (6.41b)$$

Combining these two equations, we see that for a flow of constant density fluid, the mass flowrate and volume flowrate across a surface are related by

$$\dot{M} = \rho Q \qquad (6.41c)$$

The dimensions of Q are volume per unit time, $\{L^3 t^{-1}\}$. Volume flowrate is specified for liquids in units of gallons per minute (gal/min) or cubic feet or cubic meters per

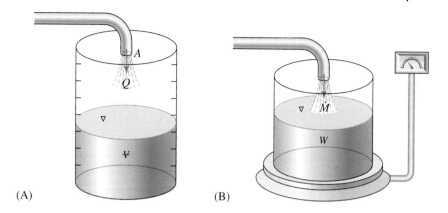

Figure 6.38 Schematics of (A) volume flowrate measurement and (B) mass flowrate measurement.

second (ft³/s, m³/s). Units of cubic feet per minute are often used in gas flows. The dimensions of $\dot{M}$ are mass per unit time, $\{Mt^{-1}\}$. Typical units are kg/s or lb_m/s.

Now imagine that we are measuring a steady flow of liquid through a pipe by catching the discharge in a container and measuring the volume of liquid that accumulates over an interval of time, as shown in Figure 6.38A. In a given time interval, we calculate the volume flowrate Q from the measured discharge volume divided by the time interval. All the volume flow passes across the exit plane normal to the pipe axis of area A. Thus by Eq. 6.41a, the average velocity $\bar{V}$ on the exit plane may be calculated as the ratio of the measured volume flowrate to the area or

$$\bar{V} = \frac{Q}{A} \tag{6.41d}$$

Another common flow measurement technique is to measure the mass of liquid discharged in a time interval by placing the discharge container on a scale as shown in Figure 6.38B. We can use Eq. 6.41b to solve for the average velocity, obtaining

$$\bar{V} = \frac{\dot{M}}{\rho A} \tag{6.42a}$$

Since, we actually measure the weight of liquid W discharged in a measured interval of time, we must still calculate the mass flowrate. The mass flowrate is given in this case by dividing the measured weight by the product of the time interval and acceleration of gravity:

$$\dot{M} = \frac{W}{g \Delta t} \tag{6.42b}$$

Combining this result with Eq. 6.42a gives the following formula for average velocity:

$$\bar{V} = \frac{\dot{M}}{\rho A} = \frac{W}{\rho A g \Delta t} \tag{6.42c}$$

EXAMPLE 6.18

Suppose a pipe of 5 cm inside diameter is used to fill a 55-gallon drum with water in 3 minutes. Calculate the volume flowrate and average velocity across the exit plane of the pipe. What is the mass flowrate?

SOLUTION

This exercise can be solved by means of Eq. 6.41d. The volume flowrate is just the volume of the drum divided by the time required to fill it. Making use of the unit conversion factor found in Appendix C, the volume flowrate Q can be found to be

$$Q = \left(\frac{55 \text{ gal}}{180 \text{ s}}\right)\left(\frac{3.786 \times 10^{-3} \text{m}^3}{1 \text{ gal}}\right) = 1.157 \times 10^{-3} \frac{\text{m}^3}{\text{s}}$$

The pipe cross section has area $A = \pi D^2/4 = \pi (0.05 \text{ m})^2/4 = 1.96 \times 10^{-3} \text{ m}^2$. If we divide the flowrate by the cross-sectional area of the pipe, we find the average velocity is:

$$\bar{V} = \frac{Q}{A} = \frac{1.157 \times 10^{-3} \text{m}^3/\text{s}}{1.96 \times 10^{-3} \text{m}^2} = 0.59 \text{ m/s}$$

The mass flowrate is $\dot{M} = \rho Q$. Thus for a nominal density of water, we have

$$\dot{M} = \rho Q = 998 \text{ kg/m}^3 (1.157 \times 10^{-3} \text{m}^3/\text{s}) = 1.15 \text{ kg/s}$$

Equation 6.42c can be used, for example, to calculate how long it would take to load 30,000 pounds of jet fuel into a commercial aircraft if the density of the fuel is known, along with the fuel hose diameter and the average velocity through the hose.

For a liquid, density is constant, so there is no variation in density on a surface on which a mass flowrate or average velocity is to be calculated. Significant density variations may occur in gas flows, particularly at high speed. Often, however, the gas density on a surface of interest may be considered to be uniform, with a value that depends on the pressure and temperature on the surface. In a gas flow under these circumstances, the foregoing relationships apply to define mass and volume flowrate and average velocity, provided we use a value of gas density corresponding to the pressure and temperature on the surface. Of course, at other locations in the flow field there may be different values of density, pressure, and temperature.

The preceding formulas involving the average velocity were essentially stated without proof. We can use our understanding of convective transport to give a more precise definition of the concept of average velocity in fluid dynamics and show where these formulas come from.

Consider the problem of defining an average velocity for a surface through which a constant density fluid is flowing. Since there is usually a fluid velocity distribution on

6.8 AVERAGE VELOCITY AND FLOWRATE

EXAMPLE 6.19

An empty (nonelastic) helium balloon is filled through a 1/4 in. diameter tube. Helium exits the tube at $T = 50°F$ and $p = 1$ atm. If the average gas velocity across the exit plane of the tube is 500 ft/s, calculate the time required to fill a balloon with a volume of 1.5 ft^3. At the same gas velocity, how long would it take to fill a similar balloon with 3×10^{-3} lb$_m$ of He? The density of He at 50°F is 1.10×10^{-2} lb$_m$/ft^3.

SOLUTION

This exercise can be solved by using the equations and concepts discussed in Section 6.7. The tube cross-sectional area is:

$$A = \frac{\pi d^2}{4} = \frac{\pi (0.25 \text{ in.})^2}{4} \left(\frac{1 \text{ ft}}{12 \text{ in.}}\right)^2 = 3.41 \times 10^{-4} \text{ ft}^2$$

Multiplying the area by the velocity gives the volume flowrate

$$Q = \bar{V} A = (500 \text{ ft/s})(3.41 \times 10^{-4} \text{ ft}^2) = 1.70 \times 10^{-1} \text{ ft}^3/\text{s}$$

The time required to fill the balloon is simply the balloon volume divided by the volume flowrate:

$$t = 1.5 \text{ ft}^3 / 0.170 \text{ ft}^3/\text{s} = 8.8 \text{ s}$$

The second part of the exercise requires the use of the density of He at 50°F and $p = 1$ atm which is given in the problem statement as 1.10×10^{-2} lb$_m$/ft^3. The mass flowrate can then be calculated using Eq. 6.44a as

$$\dot{M} = \bar{\rho} A \bar{V} = (1.10 \times 10^{-2} \text{ lb}_m/\text{ft}^3)(3.41 \times 10^{-4} \text{ ft}^2)(500 \text{ ft/s}) = 1.88 \times 10^{-3} \text{ lb}_m/\text{s}$$

The time required to fill the second balloon is found by dividing the mass of He required to fill the balloon by the mass flowrate:

$$t = \frac{3 \times 10^{-3} \text{ lb}_m}{1.88 \times 10^{-3} \text{ lb}_m/\text{s}} = 1.6 \text{ s}$$

the surface, a definition of an average velocity on this surface should reflect this fact by including contributions from the velocity values on different parts of the surface. This requirement is met if the average velocity $\bar{V}$ is defined by the value of the normal velocity component that reproduces the mass flowrate through the surface when multiplied by the product of density and surface area. Given the appropriate convective flux integral (see Table 6.2) to define the mass flowrate, the foregoing definition implies

$$\dot{M} = \rho A \bar{V} = \int_S \rho(\mathbf{u} \cdot \mathbf{n}) \, dS \tag{6.43a}$$

EXAMPLE 6.20

Calculate the average velocity for channel flow of a liquid between parallel plates and Poiseuille flow through a round pipe. The velocity fields of these flows are given in Examples 6.1 and 6.2, and illustrated in Figure 6.4.

SOLUTION

Since density is constant, we can use Eq. 6.43b, $\bar{V} = (1/A)\int_S (\mathbf{u}\cdot\mathbf{n})\,dS$ in both cases. In Example 6.1, the velocity field for channel flow in the gap between the large parallel plates shown in Figure 6.4A is described in Cartesian coordinates by

$$u = \left[\frac{h^2(p_1-p_2)}{2\mu L}\right]\left[1-\left(\frac{y}{h}\right)^2\right], \quad v=0, \quad \text{and} \quad w=0$$

We choose a surface spanning the channel of width b with a unit normal in the flow direction. Then

$$\mathbf{u}\cdot\mathbf{n} = u(y) = \left[\frac{h^2(p_1-p_2)}{2\mu L}\right]\left[1-\left(\frac{y}{h}\right)^2\right]$$

the surface area is $2bh$, and the average velocity is

$$\bar{V} = \frac{1}{A}\int_S (\mathbf{u}\cdot\mathbf{n})\,dS = \frac{1}{2bh}\int_{-h}^{h}\int_{-b/2}^{b/2}\left[\frac{h^2(p_1-p_2)}{2\mu L}\right]\left[1-\left(\frac{y}{h}\right)^2\right]dz\,dy$$

Using a symbolic mathematics package to evaluate the integral, we find $\bar{V} = \frac{2}{3}[h^2(p_1-p_2)/2\mu L]$. Since $U_{\max} = h^2(p_1-p_2)/2\mu L$, this result is normally written as $\bar{V} = \frac{2}{3}U_{\max}$. In channel flow, the maximum velocity is 1.5 times the average velocity.

In Example 6.2, the velocity field for Poiseuille flow is given in cylindrical coordinates by $v_r = v_\theta = 0$, and $v_z(r) = [(p_1-p_2)R^2/4\mu L][1-(r/R)^2]$. Choosing a surface spanning the interior of the pipe and perpendicular to the flow direction, the area is πR^2, and we can write the dot product as $\mathbf{u}\cdot\mathbf{n} = v_z(r)\mathbf{e}_z\cdot\mathbf{e}_z = [(p_1-p_2)R^2/4\mu L][1-(r/R)^2]$. The integral defining the average velocity is

$$\bar{V} = \frac{1}{\pi R^2}\int_0^{2\pi}\int_0^R \frac{(p_1-p_2)R^2}{4\mu L}\left[1-\left(\frac{r}{R}\right)^2\right]r\,dr\,d\theta$$

and evaluating this integral with a symbolic mathematics code gives $\bar{V} = \frac{1}{2}[(p_1-p_2)R^2/4\mu L]$. In Example 6.2, we showed that $U_{\max} = (p_1-p_2)R^2/4\mu L$, so that this result is normally written as $\bar{V} = U_{\max}/2$. Thus, in Poiseuille flow the maximum velocity is twice the average velocity.

Thus the average velocity on a surface may be calculated from the integral

$$\bar{V} = \frac{1}{\rho A} \int_S \rho(\mathbf{u} \cdot \mathbf{n}) \, dS$$

Since density is constant, we can also write the preceding equation as

$$\bar{V} = \frac{1}{A} \int_S (\mathbf{u} \cdot \mathbf{n}) \, dS \qquad (6.43b)$$

The integral here can be recognized as the volume flowrate. Thus, our formal definition of average velocity and its relationship to mass and volume flowrate is consistent with Eqs. 6.41 and 6.42.

For a fluid whose density is not constant, an average velocity may be defined as the value of velocity that reproduces the mass flowrate through the surface when multiplied by $\bar{\rho}$, the average fluid density on the surface, and surface area. Using the appropriate convective flux integral for mass flowrate (see Table 6.2) we now have

$$\dot{M} = \bar{\rho} A \bar{V} = \int_S \rho(\mathbf{u} \cdot \mathbf{n}) \, dS \qquad (6.44a)$$

The average velocity is therefore defined by the integral

$$\bar{V} = \frac{1}{\bar{\rho} A} \int_S \rho(\mathbf{u} \cdot \mathbf{n}) \, dS \qquad (6.44b)$$

where the average density $\bar{\rho}$ is defined by

$$\bar{\rho} = \frac{1}{A} \int_S \rho \, dS \qquad (6.44c)$$

In engineering problems involving the use of average velocity in compressible flow, the density on the surface of interest is often uniform. In that case Eq. 6.44b reduces to equation 6.43b because the density inside the integral is constant and equal to the average density $\bar{\rho}$.

6.9 SUMMARY

In the standard Eulerian description of fluid mechanics, a fluid is thought of as an indivisible, continuous material that occupies and moves through space under the action of external and internal forces. The mathematical representation of fluid and flow properties in this description is identical to the field description of vector calculus. A solution to a flow problem in the Eulerian description consists of knowing the scalar and vector functions describing all the various fluid and flow properties as a function of space and time. The primary variable in a discussion of fluid motion in the Eulerian description is the fluid velocity $\mathbf{u} = \mathbf{u}(\mathbf{x}, t)$. In general, the Eulerian velocity field is a 3D, unsteady vector field.

In the Eulerian model, fluid acceleration is given by the equation $\mathbf{a} = \partial \mathbf{u}/\partial t + (\mathbf{u} \cdot \nabla)\mathbf{u}$. The leading term is a partial time derivative of velocity, known as the local acceleration, and the remaining term is the convective derivative of velocity, known as the convective acceleration. The local acceleration is a contribution to the total acceleration at a particular point in a flow resulting from a temporal change in the velocity. For this contribution to exist, the velocity field must be a function of time. The convective acceleration is a contribution to the total acceleration at a point in a flow from the convection (movement) of fluid along the instantaneous streamline through the point. A flow must be spatially nonuniform for this contribution to exist. The substantial derivative, $D(\ )/Dt$, is defined as $D(\ )/Dt = \partial(\ )/\partial t + (\mathbf{u} \cdot \nabla)(\ )$. Thus, acceleration is conveniently represented as the substantial derivative of velocity, $\mathbf{a} = D\mathbf{u}/Dt$.

A flow is characterized as 1D, 2D, or 3D according to the number of components needed to describe its velocity field. To determine the minimum number of components needed to describe a velocity field, it may be necessary to rotate the original Cartesian coordinates or change to cylindrical coordinates to take advantage of spatial symmetry. A uniform flow in a region has a velocity vector that is constant in magnitude and direction throughout that region. An axisymmetric flow is described in cylindrical coordinates by a velocity field that does not depend on the angular coordinate. A spatially periodic flow exhibits repetitive behavior over a space and is characterized by a velocity field of the form $\mathbf{u}(\mathbf{x} + \mathbf{x}_0, t) = \mathbf{u}(\mathbf{x}, t)$.

The term "fully developed flow" implies that the velocity field is not changing in the flow direction; that is, the velocity vector is independent of the coordinate along the flow direction. A fluid flow is said to be steady if the velocity field is independent of time. A flow in which the velocity and fluid and flow properties are independent of time is defined as a steady process. Temporally, periodic or pulsatile flow exhibits repetitive behavior over time with a period τ such that $\mathbf{u}(\mathbf{x}, t + \tau) = \mathbf{u}(\mathbf{x}, t)$.

The boundary conditions that apply to the velocity field are referred to as the no-slip and no-penetration conditions. In nearly all flows encountered in engineering, fluid does not move relative to a solid surface in the tangential direction. Rather, the fluid sticks to the surface, a phenomenon referred to as no-slip. From this we conclude that the tangential component of velocity u_T is equal to the tangential component of boundary velocity U_T. Since bulk fluid cannot penetrate an impermeable boundary, the normal component of fluid velocity u_N must match the normal boundary velocity U_N. For a viscous fluid in contact with a solid boundary, the complete no-slip, no-penetration boundary condition may be summarized by saying that the velocity vector of the fluid on the boundary is always equal to the velocity of the boundary at the same point.

The total rate at which a property crosses a surface in a flow is the sum of the convective and diffusive transport. The convective transport is defined with the aid of a convective flux vector $\mathbf{q}_C$. If ε represents the fluid property of interest per unit mass, ρ is the fluid density, and $\mathbf{u}$ is the fluid velocity vector, then the convective flux vector is $\mathbf{q}_C = \rho \varepsilon \mathbf{u}$. The convective transport Γ_C is given by the surface integral $\Gamma_C = \int_S \rho \varepsilon (\mathbf{u} \cdot \mathbf{n}) \, dS$. A surface integral of this type is referred to as a flux integral. Because of the presence of the outward unit normal in the integrand, when a flux integral is evaluated for a closed surface it gives the net rate of transport of the property or substance out of the closed volume bounded by the surface. A positive value for Γ_C is therefore an outflow. For the open surface, a positive value of Γ_C represents the rate of transport across the surface in the direction of the outward unit normal.

Diffusive transport arises from the molecular-scale motion of fluid molecules. The diffusive rate of transport at a point in a fluid is represented by a diffusive flux vector $\mathbf{q}_D$. If D represents a scalar property field, the diffusive flux vector is written as $\mathbf{q}_D = k_D \nabla D$, where k_D is the coefficient of diffusion for the specific property. In a continuum model, the two quantities transported by diffusion are heat and mass (of a chemical species other than the host fluid, or other substance mixed in the fluid). The diffusive transport Γ_D for a surface is given by the surface integral $\Gamma_D = \int_S k_D (\nabla D \cdot \mathbf{n}) \, dS$. This integral defines the net rate of transport of a property across the surface in the direction of $\mathbf{n}$ due to the random motion of fluid molecules.

Although there may be significant variations in fluid and flow properties on a surface, it is often useful (and always possible) to define an average value of the normal component of velocity on any surface. The average value is often used to describe the convective transport of mass, momentum, and energy across the surface. For a surface of area A, the mass flowrate created by fluid of constant density ρ moving across the surface at average velocity $\bar{V}$ is defined to be $\dot{M} = \rho A \bar{V}$. The volume flowrate Q across this same surface is defined by $Q = A\bar{V}$. If the velocity field is known, the average velocity can be computed from a surface integral. For example, in a constant density flow the average velocity is defined by $\bar{V} = (1/A) \int_S (\mathbf{u} \cdot \mathbf{n}) \, dS$.

PROBLEMS

Section 6.2

6.1 The Eulerian velocity field can be visualized using velocity vector plots, velocity contour plots, or streamline plots. Define each type of plot. Which type do you prefer? Why?

6.2 A certain Eulerian velocity field is found to be three dimensional and time dependent. Offer an interpretation of this observation. Can you think of an example of a velocity field that may not be three dimensional? How about an example of a velocity field that may not be time dependent?

6.3 Consider the flow of fluid through the cylindrical annulus as shown in Figure P6.1. In cylindrical coordinates the velocity field for this flow is

$$\mathbf{u} = W_0 \left[1 - \left(\frac{r}{R_P} \right)^2 + \frac{1 - \kappa^2}{\ln(1/\kappa)} \ln\left(\frac{r}{R_P} \right) \right] \mathbf{e}_z$$

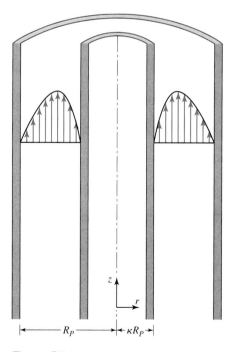

Figure P6.1

(a) Determine the location at which the maximum velocity occurs.
(b) Determine the magnitude of the maximum fluid velocity in this field.
(c) Sketch the velocity profile as a function of the radial coordinate.

6.4 The Eulerian velocity field for upward flow through a simple cylindrical pipe of radius R_P is given by

$$\mathbf{u} = W_0 \left[1 - \left(\frac{r}{R_P} \right)^2 \right] \mathbf{e}_z$$

where W_0 is a constant.
(a) Determine the location of the maximum fluid velocity in this flow field.
(b) Determine the position of the minimum fluid velocity in this flow field.
(c) Sketch (by hand or with a computer) the velocity profile as a function of the radial coordinate r (at constant z) for this flow field.

6.5 Fully developed turbulent flow in a pipe can be approximated by

$$\mathbf{u} = W_0 \left(1 - \frac{r}{R} \right)^{1/n} \mathbf{e}_z$$

where n is a function of Re and W_0 is a constant. Given that at $Re \sim 1 \times 10^5$, $n = 7$:
(a) Determine the location of the maximum fluid velocity in this flow field.
(b) Determine the position of the minimum fluid velocity in this flow field.
(c) Sketch (by hand or with a computer) the velocity profile as a function of the radial coordinate r (at constant z) for this flow field.

6.6 The laminar flow of a falling film on a flat surface is illustrated in Figure P6.2. The Eulerian velocity field for this flow is given in Cartesian coordinates as

$$\mathbf{u} = \left(\frac{\rho \delta^2 g \cos \beta}{2\mu} \right) \left[1 - \left(\frac{x}{\delta} \right)^2 \right] \mathbf{k}$$

(a) Determine the location of the maximum fluid velocity in this flow field.
(b) Determine the position of the minimum fluid velocity in this flow field.

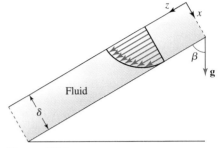

Figure P6.2

(c) Sketch (by hand or with a computer) the velocity profile as a function of the coordinate x (at constant z) for this flow field.

6.7 The adjacent flow of two immiscible liquids between parallel horizontal flat plates at low Reynolds numbers is illustrated in Figure P6.3. The Eulerian velocity field for this flow is given in Cartesian coordinates as

$$\mathbf{u}_1 = \left(\frac{C}{\mu_1} \right) \left[\left(\frac{2\mu_1}{\mu_1 + \mu_2} \right) \right.$$
$$\left. + \left(\frac{\mu_1 - \mu_2}{\mu_1 + \mu_2} \right) \left(\frac{x}{h} \right) - \left(\frac{x}{h} \right)^2 \right] \mathbf{k}$$

$$\mathbf{u}_2 = \left(\frac{C}{\mu_2} \right) \left[\left(\frac{2\mu_2}{\mu_1 + \mu_2} \right) \right.$$
$$\left. + \left(\frac{\mu_1 - \mu_2}{\mu_1 + \mu_2} \right) \left(\frac{x}{h} \right) - \left(\frac{x}{h} \right)^2 \right] \mathbf{k}$$

where $\mathbf{u}_1$ represents the velocity in fluid 1 $\mathbf{u}_2$ represents the velocity in fluid 2, and C is a constant.
(a) Determine the location of the maximum fluid velocity in this flow field. Assume that the less dense fluid is also less viscous.

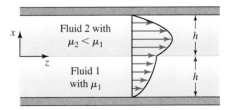

Figure P6.3

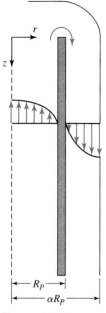

Figure P6.4

(b) Determine the position of the minimum fluid velocity in this flow field.
(c) Sketch (by hand or with a computer) the velocity profile as a function of the coordinate x (at constant z) for this flow field.

6.8 The flow of a falling film on the outside of a vertical cylindrical tube at low Reynolds numbers is illustrated in Figure P6.4. The Eulerian velocity field for this flow is given in Cartesian coordinates as

$$\mathbf{u} = \left(\frac{\rho g R_P^2}{4\mu}\right)\left[1 - \left(\frac{r}{R_P}\right)^2 + 2\alpha^2 \ln\left(\frac{r}{R_P}\right)\right]\mathbf{k}$$

(a) Determine the location of the maximum fluid velocity in this flow field.
(b) Determine the position of the minimum fluid velocity in this flow field.
(c) Sketch (by hand or with a computer) the velocity profile as a function of the coordinate x (at constant z) for this flow field.

6.9 Figure P6.5 illustrates horizontal annular flow with the inner cylinder moving axially with velocity V. The Eulerian velocity field for this flow at low Reynolds numbers is given in cylindrical coordinates as

$$\mathbf{u} = V\left[\frac{\ln(r/R)}{\ln \kappa}\right]\mathbf{e}_z$$

(a) Determine the location of the maximum fluid velocity in this flow field.
(b) Determine the position of the minimum fluid velocity in this flow field.
(c) Sketch (by hand or with a computer) the velocity profile as a function of the coordinate x (at constant z) for this flow field.

6.10 Figure P6.6 illustrates pressure-driven flow between horizontal parallel plates when the top plate is moving parallel to the bottom plate with velocity V. The Eulerian velocity field for this flow at low Reynolds numbers is given in Cartesian coordinates as:

$$\mathbf{u} = \left[C(x^2 - hx) + V\left(\frac{x}{h} - 1\right)\right]\mathbf{k}$$

where C is a constant.
(a) Determine the location of the maximum fluid velocity in this flow field.
(b) Determine the position of the minimum fluid velocity in this flow field.
(c) Sketch (by hand or with a computer) the velocity profile as a function of the coordinate x (at constant z) for this flow field.

6.11 Flow between two vertical concentric cylinders, with the outer cylinder rotating, is illustrated in Figure P6.7. At low Reynolds numbers the Eulerian velocity field is given by

$$\mathbf{u} = \left\{\Omega_O R\left[\frac{(\kappa R/r - r/\kappa R)}{(\kappa - 1/\kappa)}\right]\right\}\mathbf{e}_\theta$$

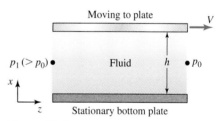

Figure P6.6

(a) Determine the location of the maximum fluid velocity in this flow field.
(b) Determine the position of the minimum fluid velocity in this flow field.

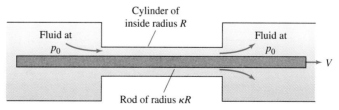

Figure P6.5

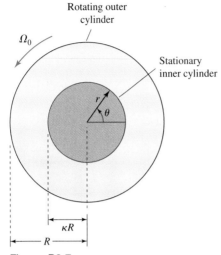

Figure P6.7

(c) Sketch (by hand or with a computer) the velocity profile as a function of the coordinate r (at constant θ) for this flow field.

6.12 Flow around a horizontal cylinder is illustrated in Figure P6.8. In cylindrical coordinates the Eulerian velocity field is given by

$$\mathbf{u} = V_\infty \cos\theta \left[1 - \left(\frac{R}{r}\right)^2\right] \mathbf{e}_r$$

$$- V_\infty \sin\theta \left[1 + \left(\frac{R}{r}\right)^2\right] \mathbf{e}_\theta$$

(a) Determine the location of the maximum fluid velocity in this flow field.
(b) Determine the position of the minimum fluid velocity in this flow field.

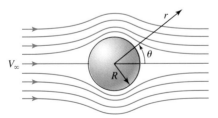

Figure P6.8

(c) Sketch (by hand or with a computer) the velocity profile as a function of the coordinate r at $\theta = 90°$ for this flow field.

6.13 Creeping (very low Re) flow around a sphere is illustrated in Figure P6.9. In spherical coordinates the Eulerian velocity field is given by

$$\mathbf{u} = V_\infty \cos\theta \left[1 - \frac{3}{2}\left(\frac{R}{r}\right) + \frac{1}{2}\left(\frac{R}{r}\right)^3\right] \mathbf{e}_r$$

$$- V_\infty \sin\theta \left[1 - \frac{3}{4}\left(\frac{R}{r}\right) + \frac{1}{4}\left(\frac{R}{r}\right)^3\right] \mathbf{e}_\theta$$

(a) Determine the location of the maximum fluid velocity in this flow field.
(b) Determine the position of the minimum fluid velocity in this flow field.
(c) Sketch (by hand or with a computer) the velocity profile as a function of the coordinate r (at $\theta = \phi = 0$) for this flow field.

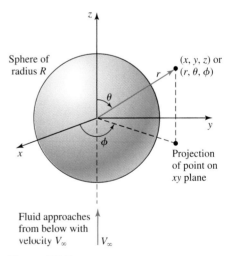

Figure P6.9

6.14 The Eulerian velocity field $\mathbf{u} = C[x\mathbf{i} - y\mathbf{j}]$ is characteristic of what is known as "flow in a corner." Generate a visual representation of this velocity field (in the first quadrant) that can be used to justify this name.

6.15 The flow between parallel disks, one of which is rotating, is illustrated in Figure P6.10. The Eulerian velocity field for this flow is given by:

$$\mathbf{u} = \left(\frac{rz\Omega}{h}\right)\mathbf{e}_\theta$$

(a) Determine the location(s) of the maximum fluid velocity in this flow field.
(b) Determine the position(s) of the minimum fluid velocity in this flow field.
(c) Sketch (by hand or with a computer) the velocity profile as a function of the coordinate r (constant z) for this flow field.
(d) Sketch (by hand or with a computer) the velocity profile as a function of the coordinate z (constant r) for this flow field.

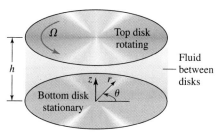

Figure P6.10

6.16 The Eulerian velocity field for a tornado can be approximated as

$$\mathbf{u} = \left(\frac{-C_1}{r}\right)\mathbf{e}_r + \left(\frac{-C_2}{r}\right)\mathbf{e}_\theta$$

where C_1 and C_2 are constant.
(a) Determine the units for the two constants.
(b) Generate a plot that you feel adequately represents this velocity field.

Section 6.3

6.17 For the following velocity field:
(a) Determine the acceleration. Note the local and convective components.

$$\mathbf{u} = W_0\left[1 - \left(\frac{r}{R_P}\right)^2\right]\mathbf{e}_z$$

(b) Determine the location of the maximum acceleration.

6.18 For the following velocity field:
(a) Determine the acceleration. Note the local and convective components.

$$\mathbf{u} = W_0\left[\left(\frac{h-x}{h}\right)^{1/n}\right]\mathbf{k}$$

(b) Determine the location of the maximum acceleration.

6.19 For the following velocity field:
(a) Determine the acceleration. Note the local and convective components.

$$\mathbf{u} = (C\sin\omega t)[x\mathbf{i} - y\mathbf{j}]$$

(b) Determine the location of the maximum acceleration.

6.20 For the following velocity field:
(a) Determine the acceleration. Note the local and convective components.

$$\mathbf{u} = \left[\Omega_0 R\left(\frac{\kappa R/r - r/\kappa R}{\kappa - 1/\kappa}\right)\right]\mathbf{e}_\theta$$

(b) Determine the location of the maximum acceleration.

Section 6.5

6.21 Reconsider the velocity field given in Problems 6.13. For this velocity field determine all the following.
(a) 1D, 2D, or 3D flow? (Justify your response.)
(b) Steady or unsteady flow?
(c) Fully developed flow?
(d) Uniform flow?
(e) Axisymmetric flow?

6.22 Reconsider the velocity field given in Problem 6.6. For this velocity field determine all the following.
(a) 1D, 2D, or 3D flow? (Justify your response.)
(b) Steady or unsteady flow?
(c) Fully developed flow?
(d) Uniform flow?
(e) Axisymmetric flow?

6.23 Reconsider the velocity field given in Problem 6.15. For this velocity field determine all the following.
(a) 1D, 2D, or 3D flow? (Justify your response.)
(b) Steady or unsteady flow?
(c) Fully developed flow?
(d) Uniform flow?
(e) Axisymmetric flow?

6.24 Reconsider the velocity field given in Problem 6.18. For this velocity field determine all the following.
(a) 1D, 2D, or 3D flow? (Justify your response.)
(b) Steady or unsteady flow?
(c) Fully developed flow?
(d) Uniform flow?
(e) Axisymmetric flow?

6.25 Reconsider the velocity fields given in Problems 6.4 and 6.5. In which case is the assumption of uniform flow more reasonable?

6.26 Consider the Eulerian velocity field given by

$$\mathbf{u} = C_1 r \mathbf{e}_\theta + C_2\left[1 - \left(\frac{r}{R}\right)^2\right]\mathbf{e}_z$$

Use the concepts discussed in Section 6.5 of the text to fully classify this flow. That is, is it 1D, 2D, or 3D? Steady or unsteady? Periodic? Axisymmetric? ... In addition, offer a physical interpretation of this velocity field. What sort of flow does it represent?

6.27 Consider the Eulerian velocity field given by

$$\mathbf{u} = Ctx\mathbf{i} + 0\mathbf{j} + Czt\mathbf{k}$$

Use the concepts discussed in Section 6.5 of the text to fully classify this flow. That is, is it 1D, 2D, or 3D? Steady or unsteady? Periodic? Axisymmetric?

6.28 Consider the Eulerian velocity field given by

$$\mathbf{u} = C_1 \sin\left[\omega\left(t - \frac{y}{C_2}\right)\right]\mathbf{i} + C_2\mathbf{j}$$

(a) Use the concepts discussed in Section 6.5 of the text to fully classify this flow. That is, is it 1D, 2D, or 3D? Steady or unsteady? Periodic? Axisymmetric?
(b) What are the dimensions for the constants C_1 and C_2?

6.29 Consider the Eulerian velocity field given by

$$\mathbf{u} = C_1 tx\mathbf{i} + 2C_2 xy\mathbf{j} - C_3 t^2 z\mathbf{k}$$

Use the concepts discussed in Section 6.5 of the text to fully classify this flow. That is, is it 1D, 2D, or 3D? Steady or unsteady? Periodic? Axisymmetric?

6.30 The flow of water through a cylindrical pipe is controlled by a valve. When the valve is open, the Eulerian velocity field in the pipe can be approximated by

$$\mathbf{u} = C_1(1 - e^{-t/C_2})\mathbf{k}$$

Use the concepts discussed in Section 6.5 of the text to fully classify this flow. That is, is it 1D, 2D, or 3D? Steady or unsteady? Periodic? Axisymmetric?

6.31 An analysis of the flow of gases through the exhaust pipe of a car shows that the corresponding Eulerian velocity field is approximately described by the equation

$$\mathbf{u} = C_1[1 + C_2 e^{-C_3 t}\sin(C_4 t)]\mathbf{k}$$

(a) Use the concepts discussed in Section 6.5 of the text to fully classify this flow. That is, is it 1D, 2D, or 3D? Steady or unsteady? Periodic? Axisymmetric?
(b) What are the dimensions for the constants C_1, C_2, C_3, and C_4?

Section 6.6

6.32 Explain the physical basis for the no-slip and no-penetration boundary conditions.

6.33 Given the validity of the no-slip and no-penetration boundary conditions, what can you say about the relationship between the fluid velocity vector on a solid surface and the velocity vector of that solid surface at the same point?

6.34 List several flow situations, either real or simplified models, for which the no-slip or no-penetration boundary conditions would not be satisfied.

6.35 Verify that the no-slip and no-penetration boundary conditions are satisfied for the velocity fields in both
(a) Poblem 6.5
(b) Problem 6.14

6.36 Verify that the no-slip and no-penetration boundary conditions are satisfied for the velocity fields in both
(a) Problem 6.6
(b) Problem 6.15

6.37 Verify that the no-slip and no-penetration boundary conditions are satisfied for the velocity fields in both
(a) Problem 6.10
(b) Problem 6.16

6.38 Verify that the no-slip and no-penetration boundary conditions are satisfied for the velocity fields in both
(a) Problem 6.13
(b) Problem 6.18

6.39 Comment on the likely locations of the physical boundaries associated with the velocity field

$$\mathbf{u} = C_1 r \mathbf{e}_\theta + C_2 \left[1 - \left(\frac{r}{R}\right)^2\right] \mathbf{e}_z$$

6.40 Comment on the likely locations of the physical boundaries associated with the velocity field $\mathbf{u} = Ctx\mathbf{i} + 0\mathbf{j} + Czt\mathbf{k}$.

6.41 Comment on the likely locations of the physical boundaries associated with the velocity field

$$\mathbf{u} = C_1 \sin\left[\omega\left(t - \frac{y}{C_2}\right)\right]\mathbf{i} + C_2 \mathbf{j}$$

6.42 Is it possible for the Eulerian velocity field $\mathbf{u} = C_1[1 + C_2 e^{-C_3 t} \sin(C_4 t)]\mathbf{k}$ to be a complete representation of the flow field in an actual exhaust pipe? Why or why not?

Section 6.7

6.43 Calculate the mass flux through a cylindrical annulus using the velocity profile given in Problem 6.3.

6.44 Calculate the mass flux for the laminar flow of a falling film using the velocity profile given in Problem 6.6.

6.45 Calculate the mass flux for the laminar flow of a falling film on the outside of a vertical cylindrical tube using the velocity profile given in Problem 6.8.

6.46 Calculate the mass flux for laminar flow between parallel flat plates with the top plate moving using the velocity profile given in Problem 6.10.

6.47 Determine expressions for the mass, volume, and momentum convective flux vectors for the flow described in Problem 6.4. Also calculate the total volume flux through a cross section of the pipe.

6.48 Determine expressions for the mass, volume, and momentum convective flux vectors for the flow described in Problem 6.6. Also calculate the total momentum flux through a cross section of the film.

6.49 Determine expressions for the mass, volume, and momentum convective flux vectors for the flow described in Problem 6.8. Also calculate the total volume flux through a cross section of the film.

6.50 Determine expressions for the mass, volume, and momentum convective flux vectors for the flow described in Problem 6.10. Also calculate the total momentum flux through a cross section of the channel.

6.51 What is the relationship between ε and λ? Give examples of problems for which you would prefer the use of each property type.

6.52 Describe the procedure you would use to calculate the total mass flux through a moving surface. What changes will be necessary if you are calculating a total momentum flux through a moving surface?

6.53 Describe the mechanisms associated with convective and diffusive transport in fluids.

6.54 The Tar Foot Consulting Company has applied for a patent for a device that measures the magnitude and direction of convective transport of any fluid property through an impermeable boundary. Would you like to purchase the rights for this device? Why or why not?

6.55 In an attempt to recoup loses from an unsuccessful venture, the Tar Foot Consulting Company has applied for a patent for a device that measures the magnitude and direction of diffusive mass and momentum transport through either a permeable or impermeable boundary. Would you like to purchase the rights for this device? Why or why not?

6.56 The Blue Devil Consulting Company has applied for a patent for a microscopic device that measures the magnitude and direction of the diffusive transport of oxygen through a permeable membrane. Would you like to purchase the rights for this device? Why or why not?

6.57 The laminar free convection flow between hot and cold vertical plates is illustrated in Figure P6.11. The temperature field is given by the equation:

$$T = 0.5(T_H + T_C) - \left[0.5(T_H - T_C)\frac{x}{W}\right]$$

(a) Sketch the temperature profile as a function of x.
(b) Do the fluid temperature values at $x = +W$ and $x = -W$ agree with your intuition? Why or why not?
(c) Calculate the diffusive heat flux per unit area to the hot plate.

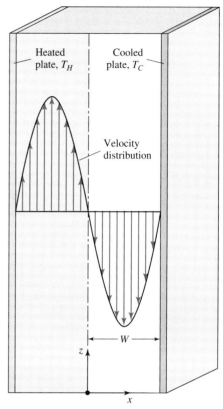

Figure P6.11

6.58 In the text we investigated the parabolic velocity profile associated with the laminar flow of fluid through a circular pipe. If the pipe wall is heated in such a way the wall heat flux is independent of the axial position (e.g., via a resistance heating element wrapped around the pipe at constant pitch), the temperature profile within the fluid is given by

$$T = T_s - C_0\left(\frac{3R_p^2}{16} - \frac{r^2}{4} + \frac{r^4}{16R_p^2}\right)$$

where T_s is the temperature of the pipe surface and C_0 is a constant.
(a) What are the dimensions for C_0?
(b) Sketch the temperature profile as a function of r for constant axial position.
(c) Compare and discuss the shapes of the temperature and velocity profiles.

(d) Calculate the diffusive heat flux per unit area to the pipe wall.

6.59 Reconsider the flow field illustrated in Figure P6.2. Suppose that the free surface of the fluid is maintained at constant temperature T_0 and the surface at $x = \delta$ is maintained at a higher constant temperature T_δ. (The velocity profile will be different from that previously described for a constant temperature film but that change is not relevant for this problem.) In this case the Eulerian temperature field is given by:

$$T = T_0 + (T_L - T_0)\left(\frac{x}{\delta}\right)$$

The corresponding viscosity field is given by

$$\mu = \mu_0 \left(\frac{\mu_L}{\mu_0}\right)^{x/\delta}$$

where μ_0 is the fluid viscosity at $x = 0$ and μ_L is the fluid viscosity at $x = \delta$.
(a) Determine the locations of the minimum and maximum temperature in this flow.
(b) Determine the locations of the minimum and maximum viscosity in this flow.
(c) Compare and discuss the shapes of the temperature and viscosity profiles.
(d) Calculate the diffusive heat flux per unit area of the solid surface.

6.60 The flow of chilled air between two concentric porous spherical shells is illustrated in Figure P6.12. The outer surface of the inner shell is maintained at a constant low temperature T_L, while the inner surface of the outer shell has a constant temperature of T_H. The radii of the inner and outer shells are receptively κR and R. Air is supplied to the interior of the inner shell and flows radially outward through the porous inner shell toward the outer shell and then out of the system through the outer shell. The Eulerian temperature profile for this flow is given by

$$\frac{T - T_H}{T_L - T_H} = \frac{e^{-C/r} - e^{-C/R}}{e^{-C/\kappa R} - e^{C/R}}$$

where C is a constant that reflects the thermal properties of the gas and the mass flowrate. If

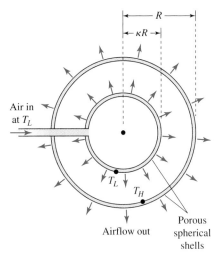

Figure P6.12

C is small, the temperature field can be approximated by the expression:

$$\frac{T - T_H}{T_L - T_H} = \frac{1/r - 1/R}{1/\kappa R - 1/R}$$

Calculate the diffusive heat flux per unit area at the outer shell walls.

Section 6.8

6.61 Determine the mass flowrate, volume flowrate, and average velocity for the laminar flow of a falling film on a flat surface in Problem 6.6.

6.62 Determine the mass flowrate, volume flowrate, and average velocity for the adjacent flow of two immiscible liquids between parallel horizontal flat plates at low Reynolds number in Problem 6.7.

6.63 Determine the mass flowrate, volume flowrate, and average velocity for the laminar flow of a falling film on the outside of a vertical cylindrical tube described in Problem 6.8.

6.64 Determine the mass flowrate, volume flowrate, and average velocity for the horizontal annular flow with the inner cylinder moving described in Problem 6.9.

6.65 Determine the mass flowrate, volume flowrate, and average velocity for flow between parallel flat plates with the top plate moving described in Problem 6.10.

6.66 Water is flowing through a cylindrical pipe of diameter 10 in. at the rate of 4.5 slugs/s. Determine the average fluid velocity. How long would it take to fill a 55-gallon drum with water from this pipe?

6.67 Water is flowing through a rectangular open channel at a velocity of 0.5 m/s. If the depth of the channel is 0.4 m, calculate the mass flowrate per unit width of channel.

6.68 Estimate the mass and volume flowrates and the average fluid velocity for gasoline moving through a typical gas pump hose. State all your assumptions.

6.69 Estimate the mass and volume flowrates and the average fluid velocity for water moving through a typical garden hose. Now estimate the time required to fill a rectangular swimming pool of dimensions 5 m by 15 m to a depth of 1.5 m using this hose. State all your assumptions.

6.70 Estimate the average velocity through the opening of a typical plastic gallon milk jug if the jugs are filled in 7.5 s at the dairy. Would it take more or less than half this time to fill a half-gallon cardboard container? Why?

6.71 What is the relationship between the average fluid velocity and the maximum fluid velocity in the flow described in Problem 6.3?

6.72 What is the relationship between the average fluid velocity and the maximum fluid velocity in the flow described in Problem 6.5?

6.73 What is the relationship between the average fluid velocity and the maximum fluid velocity in the flow described in Problem 6.7?

6.74 What is the relationship between the average fluid velocity and the maximum fluid velocity in the flow described in Problem 6.10?

7 CONTROL VOLUME ANALYSIS

7.1 Introduction
7.2 Basic Concepts: System and Control Volume
7.3 System and Control Volume Analysis
7.4 Reynolds Transport Theorem for a System
7.5 Reynolds Transport Theorem for a Control Volume
7.6 Control Volume Analysis
    7.6.1 Mass Balance
    7.6.2 Momentum Balance
    7.6.3 Energy Balance
    7.6.4 Angular Momentum Balance
7.7 Summary
Problems

7.1 INTRODUCTION

Control volume analysis is a tool for analyzing flow problems that applies the fundamental laws governing the behavior of a fluid to a region of space. The first of these laws, conservation of mass, states that in a moving fluid, mass is neither created nor destroyed. The second, conservation of momentum, is established by applying Newton's second law to a volume of fluid. The result is that the time rate of change of linear momentum of a volume of fluid is equal to the sum of the body and surface forces acting on the fluid. Conservation of energy, which is the result of applying the laws of thermodynamics to a fluid, states that the time rate of change in the internal plus kinetic energy of a volume of fluid is equal to the rate at which heat is added plus the rate at which work is done on the fluid by body and surface forces. The fourth law, conservation of angular momentum, equates the time rate of change of the angular momentum of a volume of fluid to the sum of the torques acting on the fluid.

7 CONTROL VOLUME ANALYSIS

Control volume analysis has a long history of effective use as one of the most powerful and frequently used tools in the preliminary engineering analysis of flow problems. It provides insight into the global characteristics of a flow and is one of the most important analytical methods in fluid mechanics. Since control volume analysis is accomplished with pencil and paper, it is economical and fast. Most engineers would agree that mastery of control volume analysis is a basic skill every engineer working in fluid mechanics should posses. As you will discover, control volume analysis is often the source of empirical results of the type discussed in the case studies of Chapter 3.

A region of space through which fluid flows is known as a control volume. Applying the four conservation laws to a control volume results in integral equations known as the mass, momentum, energy, and angular momentum balances. These equations are derived by using the Reynolds transport theorem to relate the time rate of change of the total amount of some property in a fluid system to a coincident control volume. We begin our discussion of control volume analysis in Section 7.2 by describing the differences between a fluid system and a control volume; then we derive the mass, momentum, energy, and angular momentum balances and illustrate their application to problems of engineering interest.

7.2 BASIC CONCEPTS: SYSTEM AND CONTROL VOLUME

Most engineers encounter the concepts of system and control volume in a thermodynamics course, where a system is defined to be a fixed identifiable quantity of matter. This body of matter has a boundary surface that separates its constituent particles from everything external. Mass may not cross this boundary, but other interactions of the system and its surroundings are permitted. A fluid system is a fixed identifiable volume of fluid; thus a fluid system always contains the same fluid particles. This volume is able to move and deform as illustrated in Figure 7.1, but in doing so, the original fluid particles must remain within the volume. It is important to realize that a fluid system is assumed to contain spatially and temporally variable velocity and property fields. Thus, the momentum and kinetic energy of different fluid particles in a fluid system will vary, as will all other properties such as density, pressure, and temperature.

When we use the term "system" in this book, we are referring to a fluid system: a fixed identifiable volume of fluid with spatially and temporally varying properties, one that may move and deform. Since by definition fluid particles do not leave a system, there can be no convective transport across the system boundary. However, other types of transport and interaction across the system boundary are allowed. For example, there

Figure 7.1 A system deforming as it passes through a sudden expansion.

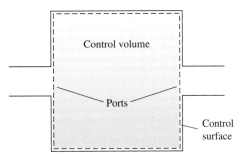

Figure 7.2 A control volume and its control surface made up of ports and sections adjacent to solid boundaries.

may be a transport of energy across a system boundary by heat conduction, and work may be done on the fluid within a system by body and surface forces.

A control volume (abbreviated CV throughout this chapter) is a region in space of any size or shape through which fluid flows. As shown in Figure 7.2, the extent of the CV is defined by its boundary, the control surface. A section of the control surface at which fluid enters or leaves a CV is called a port. By definition then, convective transport of a fluid property can occur only at a port. However, diffusive transport may occur at a port or elsewhere on a control surface. Knowledge of the velocity profile at a port is an important part of CV analysis. It is standard practice to select the port of a CV at a right angle to the flow direction to simplify the calculation of convective transport there (see Figure 7.2). An engineer must make use of information gathered from experiment, numerical simulation, or experience to model the velocity distribution at a port. In a turbulent flow, it is usually appropriate to model the velocity profile at a port as a uniform flow at a cross section. In laminar flow, a uniform flow may be unsatisfactory, and an approximate or exact velocity profile should be specified. Knowledge of the profiles of pressure, stress, density, temperature, and other fluid properties at ports also play a role in CV analysis. In the absence of detailed information, uniform profiles of velocity and other flow properties at ports are often assumed for both laminar and turbulent flows.

In some cases an engineer will select a CV completely filled with fluid. It is also possible, however, to define a mixed CV containing fluid as well as all or part of a physical device that is in contact with the fluid. These two CV types are illustrated in Figure 7.3A and 7.3B. If a CV is fixed in space, meaning that it is stationary, it is called a fixed CV. By definition, neither the volume nor shape of a fixed CV can change. In contrast, a moving CV is one that is moving through space as a whole relative to the reference frame of the observer (Figure 7.3C). The size and shape of a moving CV do not change; thus the velocity of each point on a moving control surface is the same and generally known from the problem statement. A moving CV is useful, for example, when one is analyzing the flow associated with an aircraft flying at constant velocity through stationary air. In that case it is customary to surround the aircraft with a CV that moves along with it.

7.3 SYSTEM AND CONTROL VOLUME ANALYSIS

Although the system and control volume approaches are equally valid choices for analyzing the behavior of a fluid, they are based on different conceptual models. This difference is evident if you remember that fluid is allowed to flow across the boundary of a

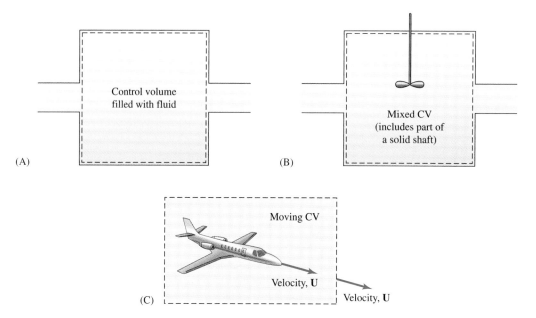

Figure 7.3 Control volume examples: (A) fluid only, (B) mixed fluid and solid, and (C) moving.

CV but not across the boundary of a system. How do we choose between system and control volume analysis in a given problem? Before we answer this question, keep in mind that the outcome of a fluid dynamic analysis must logically be independent of the choice of a system or CV approach. Thus, we expect the two approaches to yield identical results. The choice between system and CV approaches is therefore made for reasons of convenience and economy of effort. Making this choice for a specific problem is guided by an understanding of the basic characteristics of system and CV analysis, and experience.

To illustrate the issues involved in choosing a type of analysis, consider the process of compressing a gas by means of a single stage piston compressor like those sold in most hardware and auto-parts stores (see Figure 7.4A). A piston compressor operates as follows. As the piston is retracted, air is drawn into the cylinder through a valve, causing the cylinder to fill with air at ambient pressure. As the piston reverses its motion, the inlet valve closes. The piston then moves forward, and the air in the cylinder is compressed to a higher pressure and passes through a check valve into a storage tank. Suppose we are interested in the compression process from the moment the inlet valve closes until the check valve opens. How might we select an appropriate system or CV in an engineering problem like this and perform the two types of analysis?

Let us consider the problem of choosing an appropriate system and CV first, then illustrate the analysis using each. Suppose we select a system consisting of all the air in the cylinder, as shown in Figure 7.4B. During the compression, the system deforms (i.e., changes shape and volume), but the deformation is limited by the solid boundaries formed by the piston surface, the walls of the cylinder, and the cylinder head. This deformation is not difficult to specify, for if we know the speed of the piston we know the

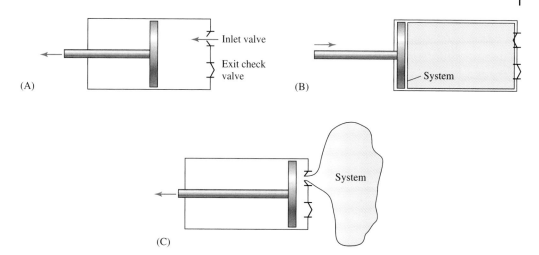

Figure 7.4 (A) Schematic of piston-type compressor. (B) System for analysis of the isolated compression process. (C) System before entering the cylinder has uncertain definition.

shape and volume of the system at all times in this interval. Therefore, the use of a system analysis to describe the fluid dynamics and thermodynamics of this part of the compression process appears to be a viable option. However, notice that a system analysis of the single stage piston compressor through an entire cycle must include more than just what happens to the air during the isolated compression process. Certainly frictional losses will be incurred as the air travels through the inlet manifold, inlet valve, and check valve. If we define a set of air particles (a system filled with air) just before it enters the inlet manifold as shown in Figure 7.4C, it is evident that the deformation of this system is so severe that we are unable to specify it. That is, we do not know the shape or size of the system for all times in the interval of interest. If we cannot specify the deformation of a system as a function of time, we cannot use the system approach. Thus, while the system approach appears to be viable for analyzing the isolated compression process, it is unsuited to a broader analysis of the flow of air through the compressor.

Can we select a fixed CV to analyze the overall compression process? One CV choice is shown in Figure 7.5. The fixed CV includes the entire machine and cuts across the intake port as well as across the check valve port. Since this CV encloses more than just fluid, it is an example of a mixed CV. Air crosses the two ports, but these are the only sections of the control surface at which a convective mass transport of air occurs. It also appears that an analysis of the isolated compression process using a mixed CV is feasible.

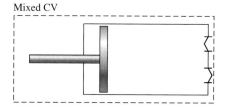

Figure 7.5 Mixed CV used to analyze the overall compression process.

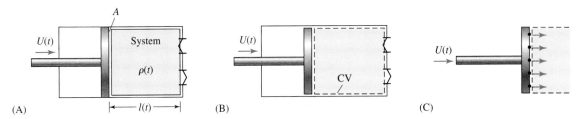

Figure 7.6 Schematic of a piston-type compressor: (A) system, (B) coincident CV, and (C) fluid particles crossing CS in front of piston face.

In this application, both the system and CV approaches appear to be feasible for analyzing the isolated compression; but because of the severe deformation experienced by a system, the mixed CV looks like a better choice for the full cycle analysis.

Let us now analyze mass conservation in the isolated compression process, using the system and a fixed CV containing fluid only. Consider first the system shown in Figure 7.6A. The mass M of a system is constant; thus the law of mass conservation for any system is $M = $ constant, which can also be expressed in differential form as $dM/dt = 0$. Assuming that the gas density at any moment in time is spatially uniform in the cylinder, the total mass of gas in the cylinder is

$$M(t) = \rho(t)l(t)A \tag{7.1}$$

where $\rho(t)$ is the gas density in the cylinder, $l(t)$ is the distance between the piston and the cylinder head, and A is the cylinder cross-sectional area. Since $M =$ constant for a system, the product $\rho(t)l(t)A$ is a constant, so the law of mass conservation in this case can be written as $\rho(t)l(t)A = $ constant. Alternately, since $dM/dt = 0$, applying the time derivative directly to Eq. 7.1 gives $l(t)(d\rho/dt) + \rho(t)(dl/dt) = 0$, and after rearranging, we obtain

$$\frac{1}{\rho(t)}\frac{d\rho}{dt} = -\frac{1}{l(t)}\frac{dl}{dt} \tag{7.2}$$

This form of the mass conservation equation relates the fractional time rate of change in density to the fractional time rate of change of the distance $l(t)$. The word "fractional" here indicates that on each side of this equation, the time rate of change of the variable is divided by the value of the variable, giving a result with the dimension of inverse time.

Additional insight into the compression process can be gained from further manipulation of Eq. 7.2. Since $l(t)$ defines the position of the piston, the time derivative of $l(t)$ is $dl/dt = -U(t)$, where $U(t)$ is the speed of the piston. Noting that the volume of the system is given by $V(t) = l(t)A$, we can now write Eq. 7.2 in terms of the piston speed and volume as

$$V(t)\frac{d\rho}{dt} = \rho(t)AU(t) \tag{7.3}$$

This version of the mass conservation equation is equivalent to Eq. 7.2, but it provides a relationship between the time rate of change of gas density in the cylinder and the physical parameters defining the cylinder geometry, piston motion, and air density. This completes our system approach to mass conservation analysis. Notice that the system geometry, movement, and deformation in this example is simple enough to allow us to execute a system analysis easily.

Let us now use a CV to perform a mass conservation analysis of the same problem. At a given time, we chose a CV as shown by the dashed line in Figure 7.6B. The control surface encloses all the gas in the cylinder and passes just in front of the moving piston and just inside the cylinder walls. It can be seen that this is a fixed CV containing fluid only, which coincides with the system defined earlier. Since the piston is displacing fluid across an adjacent section of the stationary control surface, there is a convective mass transport occurring on the port just in front of the advancing piston, (see Figure 7.6C). The fact that the piston will eventually cross the control surface does not concern us because the CV analysis is performed at a single instant of time. Assuming a uniform flow across this port, and a spatially uniform gas density in the cylinder, we can use the appropriate integral in Table 6.2 to determine that the convective mass transport rate across the port is $\Gamma_C = -\rho(t)AU(t)$. There is no mass transport on the remaining control surfaces because they are located at stationary solid boundaries, and the normal component of fluid velocity is zero on these surfaces by the no-penetration condition. We see that the piston motion enters the CV analysis by creating a flow of air across the section of the control surface adjacent to it.

At this point we must temporarily suspend our CV analysis because we have yet to formulate a statement of mass conservation for a fixed CV. Although the magnitude of the convective mass transport across the entire control surface, $\rho(t)AU(t)$, appears in the mass conservation equation for a system, Eq. 7.3, how does the other term in this equation, i.e., $V(t)(d\rho/dt)$, enter a CV analysis? At this point we do not know because all four conservation laws are stated only for a system. What form will the laws take for a CV? To answer this question, you must understand the Reynolds transport theorem. There are two forms of this theorem, one for a system and one for a CV. Together they allow us to write the conservation laws for a control volume that coincides with a system, thus providing a link between system and CV analysis. In the next section we discuss the theorem in system form.

7.4 REYNOLDS TRANSPORT THEOREM FOR A SYSTEM

Consider a system to be analyzed by an observer at rest in the reference frame of the coordinates. For this observer, the fluid velocity field $\mathbf{u}(\mathbf{x}, t)$ describes the velocity of the fluid inside and on the boundary of the system. The system moves and deforms because of this velocity field, but it always contains the same fluid particles. Let E_{sys} represent any extensive fluid property, defined to be the total amount of the property in the system at a certain instant of time. We can calculate E_{sys} by dividing the system into an infinite number of volume elements of size dV. Then the amount of property in a volume element located at position $\mathbf{x}$ is given by $\rho(\mathbf{x}, t)\varepsilon(\mathbf{x}, t)\,dV$, where ε is the appropriate intensive (per unit mass) variable representing the property. The total amount of the property in the system is then given by $E_{sys} = \int_{R(t)} \rho\varepsilon\,dV$, where $R(t)$ is the region occupied by the system.

Now suppose we want to write the time rate of change of the total amount of some property in a system at a certain instant of time. This time rate of change is given by

$$\frac{dE_{sys}}{dt} = \frac{d}{dt}\int_{R(t)} \rho\varepsilon\, dV$$

The direct evaluation of the time derivative of this integral over a system is often difficult because the system is moving, its boundary is changing shape, and the integrand is also changing in time.

The Reynolds transport theorem allows us to evaluate this time rate of change in a way that avoids the difficulties just mentioned. The theorem states that the time rate of change in the total amount of a property in a system is given by

$$\frac{dE_{sys}}{dt} = \int_{R(t)} \frac{\partial}{\partial t}(\rho\varepsilon)\, dV + \int_{S(t)} (\rho\varepsilon)(\mathbf{u}\cdot\mathbf{n})\, dS \tag{7.4}$$

where the labels $R(t)$ and $S(t)$ on the integral signs indicate that the integrations are to be performed over the volume of the system and over its surface. The theorem shows that the time rate of change in the total amount of a property in a system may be calculated as the sum of two integrals, each of which is evaluated with the system frozen in the volume and shape it has at the instant of time in question. The fluid velocity and property fields refer to the values seen by an observer at rest with respect to the system. The volume integral in Eq. 7.4 accounts for the instantaneous change in the total amount of property within the system due to temporal variations in fluid properties. The surface integral accounts for the contribution from the deformation of the system due to the instantaneous motion of its boundaries. Equation 7.4 the Reynolds transport theorem for a system, allows us to calculate the time rate of change in the total amount of a property in a system even when the system deformation is complex. However, the more important version of the theorem is the control volume form, discussed next.

7.5 REYNOLDS TRANSPORT THEOREM FOR A CONTROL VOLUME

Consider an arbitrary system indicated by the solid line in Figure 7.7, and suppose there is a fixed CV that instantaneously coincides with this system as indicated by the dashed line. The CV and the observer are at rest with respect to the coordinates. If we calculate the time derivative of some total property in this system by using the Reynolds transport theorem for a system, then according to Eq. 7.4 we obtain

$$\frac{dE_{sys}}{dt} = \int_{R(t)} \frac{\partial}{\partial t}(\rho\varepsilon)\, dV + \int_{S(t)} (\rho\varepsilon)(\mathbf{u}\cdot\mathbf{n})\, dS$$

Since the system and the fixed CV coincide at this specific instant in time, the value of a fluid property at any point in the system is the same as the value of that property at the same point within the CV. Thus, the two integrals over the system may be interpreted as applying to the coincident CV. We conclude that the Reynolds transport theorem for a system also provides the following relationship between a time rate of change for a

7.5 REYNOLDS TRANSPORT THEOREM FOR A CONTROL VOLUME

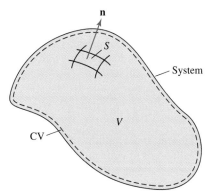

Figure 7.7 An arbitrary system and its coincident CV.

The derivation of the Reynolds transport theorem for a CV provided here relies on the system version of the theorem. It is also possible to derive the CV version directly.

system and two integrals over a fixed coincident CV:

$$\frac{dE_{sys}}{dt} = \int_{CV} \frac{\partial}{\partial t}(\rho\varepsilon)\,dV + \int_{CS} (\rho\varepsilon)(\mathbf{u}\cdot\mathbf{n})\,dS \tag{7.5}$$

This result is the Reynolds transport theorem for a CV, one of the most important equations in fluid mechanics. The labels CV and CS on the integral signs indicate that the integrations are to be performed over the control volume and control surface, with the fluid velocity and property fields referring to the values seen by an observer at rest with respect to the fixed CV.

The volume and surface integrals in the Reynolds transport theorem for a CV have a physical interpretation different from the corresponding integrals in the system formulation of the theorem. To understand the meaning of the volume integral in Eq. 7.5, let E_{CV} represent the total amount of a fluid property in a fixed CV at a certain time. Then E_{CV} is defined by the volume integral

$$E_{CV} = \int_{CV} (\rho\varepsilon)\,dV \tag{7.6}$$

and the time rate of change of the total amount of this property in the control volume is given by

$$\frac{dE_{CV}}{dt} = \frac{d}{dt}\int_{CV} (\rho\varepsilon)\,dV$$

Since the CV is a fixed region in space, the time derivative may be put inside the integral sign, allowing us to write this as

$$\frac{dE_{CV}}{dt} = \int_{CV} \frac{\partial}{\partial t}(\rho\varepsilon)\,dV \tag{7.7}$$

Comparing Eqs. 7.7 and 7.5, we conclude that the volume integral is an accumulation term that represents the rate at which the total amount of a fluid property within the fixed CV is changing with time. A positive value for the accumulation term represents an increase of the selected property within the CV.

Next consider the surface integral in Eq. 7.5, $\int_{CS} (\rho\varepsilon)(\mathbf{u}\cdot\mathbf{n})\,dS$. From our discussion of convective transport in Section 6.7.1, we recognize that this integral defines the net outward convective transport of the fluid property represented by ε across the fixed control surface. Representing the outward convective transport rate over the entire control surface with the symbol Γ_{CV}, we can write

$$\Gamma_{CV} = \int_{CS} (\rho\varepsilon)(\mathbf{u}\cdot\mathbf{n})\,dS \qquad (7.8)$$

A positive value for Γ_{CV} represents a net outflow of property from the CV. Substituting Eqs. 7.7 and 7.8 into Eq. 7.5 allows us to write the Reynolds transport theorem for a control volume as

$$\frac{dE_{sys}}{dt} = \frac{dE_{CV}}{dt} + \Gamma_{CV} \qquad (7.9)$$

We see that the rate of increase in the total amount of fluid property in a system is equal to the rate of increase in the total amount of this property in the coincident CV (accumulation), plus the instantaneous outflow of the fluid property through the control surface (convective transport). The mass, momentum, energy, and angular momentum balances for a CV developed later in this chapter are all based on Eq. 7.5; hence each balance will include an accumulation term and a transport term.

Because of the prevalence of steady state operation of fluid-handling devices of all types, as well as operation in a periodic or cyclic mode for which the time-averaged fluid properties within a CV are constant, we may consider the flow in such devices to be a steady process. By definition, the time rate of change of the amount of a fluid property or other transportable substance in a fixed CV in a steady process is zero. Thus in a steady process, we will always assume that

$$\frac{dE_{CV}}{dt} = \int_{CV} \frac{\partial}{\partial t}(\rho\varepsilon)\,dV = 0 \qquad (7.10)$$

EXAMPLE 7.1

Starting with Eq. 7.5, derive a form of Reynolds transport theorem for a CV that applies to a fluid property λ defined on a per-unit-volume basis.

SOLUTION

From Eq. 6.34, the relationship between a property per unit volume and the same property per unit mass is $\lambda = \rho\varepsilon$. Thus, the equivalent to Eq. 7.5 for a property per unit volume is

$$\frac{dE_{sys}}{dt} = \int_{CV} \frac{\partial(\lambda)}{\partial t}\,dV + \int_{CS} (\lambda)(\mathbf{u}\cdot\mathbf{n})\,dS$$

Control volume analysis may also be applied to moving and accelerating control volumes, but the latter topic is not discussed in this text.

indicating that there is no accumulation of any type of fluid property in a fixed CV. The assumption of a steady process is normally made in a CV analysis of any fluid machine that operates steadily or cyclically. Thus, a compressor driven by an electric motor would normally qualify as a steady process, as would an internal combustion engine operating at a constant rpm.

7.6 CONTROL VOLUME ANALYSIS

In the early stages of the engineering design process, the engineer often needs global values of selected fluid properties and process parameters rather than the detailed information provided by a solution to the governing differential equations. For example, in a nozzle design, we may initially be interested only in the volume flowrate through the nozzle, its relationship to the pressure difference across the inlet and outlet of the nozzle, and the nozzle geometry. Later, the flow in the nozzle may be calculated in more detail by using a computational fluid dynamics code to optimize performance and refine the geometry. Although preliminary design data can be obtained from a computational solution to the governing differential equations, a CV analysis will provide the desired information quickly and with far less effort and expense.

Control volume laws are derived from the equivalent system laws using the Reynolds transport theorem for a CV to provide the necessary connection between the system and its coincident CV. Thus in CV analysis, the conservation laws are applied in integral form, rather than in the form of differential equations. In Sections 7.6.1 through 7.6.4 we employ the Reynolds transport theorem for a CV to develop integral conservation laws for a CV that is at rest in an inertial reference frame. These integral relationships are the mass, momentum, energy, and angular momentum balances referred to earlier. Although our derivation of these balances is based on a fixed CV containing fluid only, the results also apply to problems involving a mixed CV that contains fluid and solid.

The judicious selection of a control volume plays a critical role in CV analysis. There is no simple rule about how to choose a CV for a specific problem, and experience plays a role in making the most efficient choice. After reading our recommendations for CV selection and studying the examples in this chapter, you should feel comfortable in making your own informed choice of a CV, modifying the rules given here as necessary.

Control volume selection
1. Choose a fixed control volume at rest in an inertial reference frame.
2. Place sections of control surface at all fluid–solid interfaces to take advantage of the no-slip, no-penetration boundary condition.
3. Place sections of control surface at all ports with the control surface oriented so that the normal to the surface is aligned with the velocity vector at the port.
4. Place sections of control surface at locations where information is requested.
5. Place sections of control surface at locations where information is provided.
6. Place additional control surface segments to form a closed control volume.

We employ these rules in the examples of CV analysis in the remainder of this chapter. In some cases we will choose a CV containing only fluid, in others a mixed CV that contains both fluid and solid. As you gain experience with CV analysis, you will learn that the use of a mixed CV is preferable in problems of certain types.

7.6.1 Mass Balance

The derivation of an integral mass conservation law for a fixed CV begins with the Reynolds transport theorem for a control volume, Eq. 7.5:

$$\frac{dE_{sys}}{dt} = \int_{CV} \frac{\partial}{\partial t}(\rho\varepsilon)\,dV + \int_{CS}(\rho\varepsilon)(\mathbf{u}\cdot\mathbf{n})\,dS$$

Although our interest here is on what is happening inside and on the surface of a CV, remember that the left-hand side of the theorem refers to the coincident system. To develop a mass conservation statement, E_{sys} is chosen as M_{sys}, the total mass in the system. The intensive counterpart of the total mass is the mass per unit mass, so we take $\varepsilon = 1$. The transport theorem then states:

$$\frac{dM_{sys}}{dt} = \int_{CV} \frac{\partial \rho}{\partial t}\,dV + \int_{CS}\rho(\mathbf{u}\cdot\mathbf{n})\,dS$$

Since mass is conserved for a system, we write $dM_{sys}/dt = 0$ and rearrange the equation to obtain:

$$\int_{CV} \frac{\partial \rho}{\partial t}\,dV + \int_{CS}\rho(\mathbf{u}\cdot\mathbf{n})\,dS = 0 \qquad (7.11)$$

This is the integral mass conservation equation, or mass balance, for a fixed CV. In deriving this equation it was not necessary to say anything about the type of fluid; thus the equation applies to all fluids, Newtonian or non-Newtonian, liquid or gas, and under all circumstances.

The meaning of each term in Eq. 7.11 is revealed if we recognize that the rate at which mass accumulates inside the control volume is $dM_{CV}/dt = \int_{CV}(\partial\rho/\partial t)\,dV$, while the convective mass transport rate out of the control volume is $\Gamma_{CV} = \int_{CS}\rho(\mathbf{u}\cdot\mathbf{n})\,dS$. Thus we can write the integral mass conservation equation for a fixed control volume as $dM_{CV}/dt = -\Gamma_{CV}$, which states that the rate of increase in the total amount of mass in a fixed control volume is equal to the instantaneous convective transport of mass into the control volume.

To illustrate the use of Eq. 7.11, suppose we reconsider the compression of gas in a cylinder as discussed earlier in connection with the derivation of Eq. 7.3. Using the recommended procedure for choosing a fixed CV, we choose a CV containing all the gas in the cylinder as shown in Figure 7.8. The two integrals in Eq. 7.11 are to be evaluated instantaneously at time t. Assuming a spatially uniform gas density within the CV, the volume integral gives

$$\int_{CV}\frac{\partial\rho}{\partial t}\,dV = \frac{d\rho}{dt}\,\forall(t)$$

7.6 CONTROL VOLUME ANALYSIS

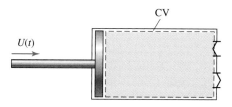

Figure 7.8 The fluid-only CV for the compressor.

where $\mathcal{V}(t)$ is the volume of the CV at time t. There is no mass transport across the control surfaces adjacent to the cylinder walls, but there is a mass transport across the control surface near the piston as discussed earlier. The mass transport across this control surface is given by

$$\int_{CS} \rho(\mathbf{u} \cdot \mathbf{n})\, dS = -\rho(t) A U(t)$$

since the velocity of the gas on the control surface just in front of the advancing piston is $U(t)\mathbf{i}$ at time t, the outward unit normal is $\mathbf{n} = -\mathbf{i}$, the piston area is A, and the density is spatially uniform on the surface. Thus a mass balance by means of Eq. 7.11 gives

$$\frac{d\rho}{dt}\mathcal{V}(t) - \rho(t)AU(t) = 0$$

which upon rearrangement is identical to the earlier result, Eq. 7.3.

EXAMPLE 7.2

Consider the charging of a gas bottle from a high pressure line as shown in Figure 7.9. What is the time rate of change of density in the bottle at the instant shown?

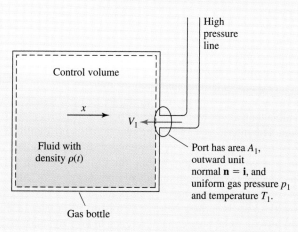

Figure 7.9 Schematic and control volume definition for Example 7.2.

SOLUTION

We can solve this problem by picking a fixed CV and performing a mass balance via Eq. 7.11, $\int_{CV} (\partial \rho/\partial t)\, dV + \int_{CS} \rho(\mathbf{u} \cdot \mathbf{n})\, dS = 0$. By means of the recommended procedure, we choose a fixed CV containing all the gas in the bottle as shown in Figure 7.9. Assuming a spatially uniform density within the bottle, the volume integral gives $\int_{CV} (\partial \rho/\partial t)\, dV = (\partial \rho/\partial t) \mathcal{V}(t)$.

There is no transport across the control surfaces adjacent to the walls of the bottle, but there is mass transport across the inlet surface. The mass transport across this control surface is calculated by realizing that the uniform velocity of the gas on the inlet surface is $-V_1(t)\mathbf{i}$, and the outward unit normal is $\mathbf{n} = \mathbf{i}$. Thus we have $\mathbf{u} \cdot \mathbf{n} = (-V_1(t)\mathbf{i}) \cdot \mathbf{i} = -V_1(t)$. Assuming that the density ρ_1 is uniform on the inlet surface, the integrand of the surface integral is constant, so we find $\int_{CS} \rho(\mathbf{u} \cdot \mathbf{n})\, dS = -\rho_1 A_1 V_1(t)$, where A_1 is the inlet area. Thus, from Eq. 7.11 the mass balance is $(\partial \rho/\partial t)\mathcal{V}(t) - \rho_1 A_1 V_1(t) = 0$, which upon rearrangement give: $\partial \rho/\partial t = \rho_1 A_1 V_1(t)/\mathcal{V}(t)$.

This answer is dimensionally correct, but we have not been given the density of the gas entering the inlet to the bottle. However, since we know the pressure and temperature of the gas entering the bottle, we can use the perfect gas law to write the density as $\rho_1 = p_1/RT_1$, where R is the specific gas constant of the gas involved. Thus our final expression for the time rate of change of the density of the gas inside the bottle at the instant of time shown is $\partial \rho/\partial t = p_1 A_1 V_1(t)/RT_1 \mathcal{V}(t)$.

The previous analysis requires a few comments. Recall that in Section 7.3 we suspended our CV analysis of the gas compression problem after computing the convective mass transport crossing the control surface in front of the advancing piston. At that point we argued that it was not intuitively clear how to complete the analysis. The integral mass conservation equation, Eq. 7.11, shows us that it is always necessary to account for both the rate at which mass is accumulating within a fixed control volume and the mass transport across the control surface. This is illustrated in Example 7.2.

To decide whether mass is accumulating within a CV, we must take into account the compressibility of the fluid and the type of process the fluid is undergoing. The type of fixed CV you select (i.e., containing fluid only or mixed), also affects the value of the accumulation term in a mass balance. For example, given a steady flow of water displacing gasoline in a tank as shown in Figure 7.10, let us perform a mass balance on the mixed CV shown. The interface between the two fluids is not stationary, and the total mass in the CV is increasing as the more dense water displaces the less dense gasoline. Thus there is an unsteady density field within this CV even though the fluids entering and leaving at a constant rate are both liquids of constant density. Although the flow is steady, the displacement of the gasoline by the water is not a steady process. You will have an opportunity to apply a mass balance to this problem later (see Problem 7.18).

In the following example, notice the difference in how the mass balance is applied for a fixed and mixed control volume.

7.6 CONTROL VOLUME ANALYSIS

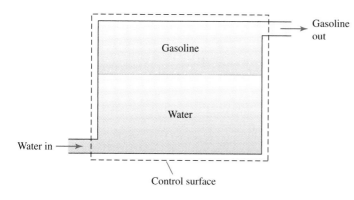

Figure 7.10 Schematic and control volume for water filling a tank partially filled with gasoline.

EXAMPLE 7.3

A rod enters a cylinder of glycerin as shown in Figure 7.11 at speed $V_{\text{rod}}(t)$. Find an expression for the average velocity in the annular space between the rod and the cylinder. If the rod is moving at a fixed speed of 1 in./s into the cylinder, determine how fast the glycerin–air interface is moving.

SOLUTION

We are asked to determine the velocity of the glycerin in the annulus. Figure 7.11B serves as a sketch of an appropriate fixed CV containing fluid only. Notice that we have placed control surfaces at all fluid–solid interfaces, ports, and locations at which information is given or requested. At this instant the CV contains only fluid. We know that $R_{\text{rod}} = 0.5$ in., $R_{\text{cyl}} = 1.5$ in., $z_1 = 3$ in., $z_2 = 10$ in., and the rod moves at speed $V_{\text{rod}}(t)$. The problem is solved by applying Eq. 7.11, $\int_{\text{CV}} (\partial \rho / \partial t) \, dV + \int_{\text{CS}} \rho (\mathbf{u} \cdot \mathbf{n}) \, dS = 0$ to

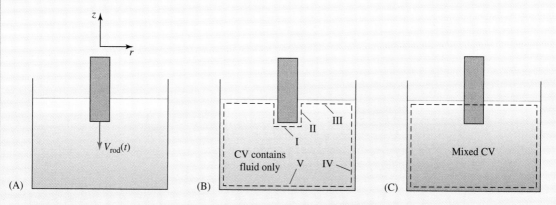

Figure 7.11 Schematic for Example 7.3, (A) geometry, (B) fluid only CV, (C) mixed CV, (D) CFD solution.

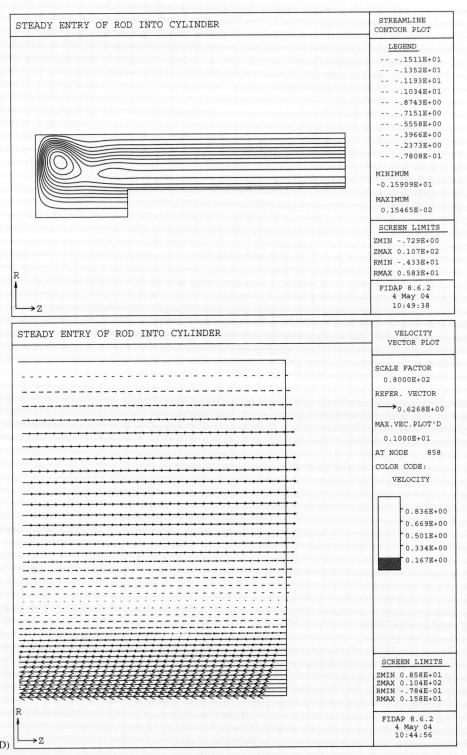

Figure 7.11 (*Continued*)

the CV at this instant of time. The CV is fixed and contains glycerin, a liquid. Thus the density is constant, $\partial \rho / \partial t = 0$, and the volume integral representing mass accumulation is zero. The key to this problem is the evaluation of the surface integrals where fluid is crossing the control surface. In Figure 7.11B the control surface is broken into five parts. To apply Eq. 7.11 we must know **u** and **n** for each surface, as well as the differential area and limits of each integral. The following table summarizes this information.

Surface	Description	u	n	dS	Limits
I	Rod face	$V_{rod}(-\mathbf{k})$	$\mathbf{k}$	$2\pi r\, dr$	0 to R_{rod}
II	Rod side	$V_{rod}(-\mathbf{k})$	$-\mathbf{e}_r$	$2\pi R_{rod}\, dz$	z_1 to z_2
III	Exit	$\mathbf{u}$	$\mathbf{k}$	$2\pi r\, dr$	R_{rod} to R_{cyl}
IV	Cylinder side wall	0	$\mathbf{e}_r$	$2\pi R_{cyl}\, dz$	0 to z_2
V	Cylinder end wall	0	$\mathbf{k}$	$2\pi r\, dr$	0 to R_{cyl}

The integrand for surface II is zero because on this surface **u** and **n** are perpendicular, so that $\mathbf{u} \cdot \mathbf{n} = 0$. As a result of the no-penetration condition, $\mathbf{u} \cdot \mathbf{n} = 0$ on surfaces IV and V. On surface I, just in front of the rod face, the integral is $\int_0^{R_{rod}} \rho V_{rod}(-\mathbf{k} \cdot \mathbf{k}) 2\pi r\, dr = -\rho_{gly} V_{rod}(t) \pi R_{rod}^2$. At the exit, surface III, the velocity profile is unknown. We use the concept of an average velocity and anticipate a mass transport out of the CV to write the integral: $\int_{R_{rod}}^{R_{cyl}} \rho(\mathbf{u} \cdot \mathbf{k}) 2\pi r\, dr = \rho_{gly} \bar{V}_{III}(t) \pi (R_{cyl}^2 - R_{rod}^2)$. Thus, the mass balance becomes $-\rho_{gly} V_{rod}(t) \pi R_{rod}^2 + \rho_{gly} \bar{V}_{III}(t) \pi (R_{cyl}^2 - R_{rod}^2) = 0$. Solving for $\bar{V}_{III}$ while noting that density of glycerin is constant gives the average velocity in the annulus

$$\bar{V}_{III}(t) = \frac{V_{rod}(t) R_{rod}^2}{R_{cyl}^2 - R_{rod}^2} \tag{A}$$

The result has proper dimensions for velocity and is positive, indicating an outflow of mass at the exit. Notice also that the answer is valid for a rod velocity that is a function of time. For a rod moving at a constant speed $V_{rod} = 1$ in./s, substituting the known quantities yields

$$\bar{V}_{III} = \frac{V_{rod} R_{rod}^2}{R_{cyl}^2 - R_{rod}^2} = \frac{(1 \text{ in./s})(0.5 \text{ in.})^2}{(1.5 \text{ in.})^2 - (0.5 \text{ in.})^2} = 0.125 \text{ in./s}$$

This is the average speed at which the glycerin–air interface is moving.

We can also solve this problem by using the mixed CV shown in Figure 7.11C. Applying Eq. 7.11 to this mixed CV we have $\int_{CV} (\partial \rho / \partial t)\, dV + \int_{CS} \rho(\mathbf{u} \cdot \mathbf{n})\, dS = 0$. In this case, the volume integral is nonzero because mass is accumulating within the CV. Since the mass in the CV at any time is $M(t) = \rho_{rod} \mathcal{V}_{rod} + \rho_{gly} \mathcal{V}_{gly}$, the mass accumulation rate is $dM(t)/dt = \rho_{rod}(d\mathcal{V}_{rod}/dt) + \rho_{gly}(d\mathcal{V}_{gly}/dt)$. By inspection we can calculate the rate of accumulation of volume of rod as $d\mathcal{V}_{rod}/dt = V_{rod} \pi R_{rod}^2$, since this is the additional amount of rod volume that enters the CV in the next instant of time. Since

EXAMPLE 7.4

A flow of constant density fluid enters a round pipe at low speed as shown in Figure 7.12. The uniform velocity field at the inlet can be represented by $v_r = 0$, $v_\theta = 0$, and $v_z(r, \theta, z) = U_0$. The flow gradually changes until at some distance downstream it becomes a fully developed Poiseuille flow described by $v_r = 0$, $v_\theta = 0$, and $v_z = U_{max}[1 - (r/R)^2]$. Find the ratio of U_{max} to U_0.

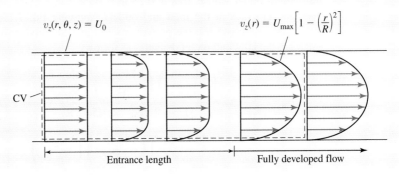

Figure 7.12 Schematic for Example 7.4, velocity profiles at the entrance to a pipe.

SOLUTION

We can solve this problem with a mass balance, using the CV shown in Figure 7.12. Since ρ is constant, Eq. 7.14 takes the form: $\int_{cs} (\mathbf{u} \cdot \mathbf{n}) \, dS = \int_{inlet} (\mathbf{u} \cdot \mathbf{n}) \, dS + \int_{exit} (\mathbf{u} \cdot \mathbf{n}) \, dS = 0$. By inspection, the value of the integral on the inlet surface is $\int_{inlet} (\mathbf{u} \cdot \mathbf{n}) \, dS = -U_0 A$, where the pipe area is $A = \pi(D^2/4)$. The integral at the exit is

$$\int_{exit} (\mathbf{u} \cdot \mathbf{n}) \, dS = \int_0^{2\pi} \int_0^R U_{max} \left[1 - \left(\frac{r}{R}\right)^2\right] r \, dr \, d\theta = \frac{1}{2} U_{max} \pi R^2$$

Note that we can write the volume flowrate at the exit as $\frac{1}{2} U_{max} \pi R^2 = \frac{1}{2} U_{max} A$. Thus the mass balance gives us a relationship $-U_0 A + \frac{1}{2} U_{max} A = 0$, and we discover $U_{max}/U_0 = 2$.

An alternate approach to solving this problem is to use Eq. 7.15 to write $\bar{V}_{in} A_{in} = \bar{V}_{exit} A_{exit}$, then cancel the areas and use the fact that the average velocity at the inlet is U_0 to write $U_0 = \bar{V}_{exit}$. The remaining step is to recall that in Example 6.20 we showed that for a Poiseuille velocity profile, the maximum velocity is twice the average velocity. Thus $\bar{V}_{exit} = \frac{1}{2} U_{max}$, and we have $U_0 = \bar{V}_{exit} = \frac{1}{2} U_{max}$, which is the same result.

This is simply a statement that the net transport of fluid volume out of the control volume is zero. Under these conditions, Eq. 7.13 becomes

$$\sum A_{in} \bar{V}_{in} = \sum A_{out} \bar{V}_{out} \tag{7.15}$$

or equivalently

$$\sum Q_{in} = \sum Q_{out} \tag{7.16}$$

where Q_{in} and Q_{out} are volume flowrates in and out at the ports of a control volume. With a constant density fluid, the fluid volume coming into the CV per unit time must always equal the volume going out per unit time. This is because the fluid inside the CV cannot be compressed.

The examples in this section give some indication of the variety of problems that can be addressed by means of a mass balance. In each case note how the CV and equation expressing the mass balance are selected, and pay particular attention to why terms are retained or dropped.

EXAMPLE 7.5

Air flows steadily at low speed through the sudden expansion in a round air-conditioning duct as shown in Figure 7.13A. Derive an expression relating the velocities and duct diameters at the inlet and exit of the expansion.

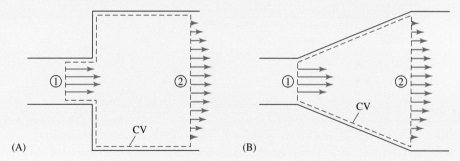

Figure 7.13 Control volume for (A) sudden expansion and (B) gradual area change.

SOLUTION

We can solve this problem with a mass balance, and we assume that the density of air flowing at low speed in an air conditioning duct is constant. We will choose the CV shown in Figure 7.13A, by using the recommended method. Applying Eq. 7.15 to this CV, and noting that there is one inlet port and one exit port, we find

$$A_1 \bar{V}_1 = A_2 \bar{V}_2 \tag{A}$$

This is identical to Eq. 3.21 in case study 3.3.2 (flow through area change). When we used this formula earlier, we did not yet have the tools to derive it. Now we see that a simple application of a mass balance is all that is needed.

For a round duct, the areas are given by the usual formulas. Thus we can also write the following relationship between average velocities

$$\frac{\bar{V}_1}{\bar{V}_2} = \frac{A_2}{A_1} = \frac{D_2^2}{D_1^2} \tag{B}$$

These formulas also apply to flow through a gradual area change as shown in Figure 7.13B, as you can easily confirm by using the CV shown in that figure. Also note that the result, (A), is stated in terms of average velocity and area, hence is not dependent on the precise nature of the velocity profile and applies to a duct or pipe of any cross section.

EXAMPLE 7.6

Derive an expression for the density of the mixture of oil and vinegar leaving the mixer shown in Figure 7.14. Assume the free surface height remains fixed.

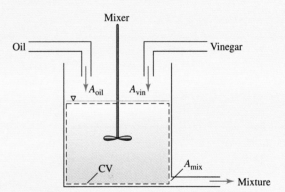

Figure 7.14 Schematic for Example 7.6.

SOLUTION

This appears to be a steady process involving liquids, so we suspect there is no mass accumulation. Since two fluids are involved, however, we must be cautious in using Eq. 7.13:

$$\sum \rho_{in} A_{in} \bar{V}_{in} = \sum \rho_{out} A_{out} \bar{V}_{out}$$

The CV ports are indicated in Figure 7.14. In this case there are two input streams and one output stream so that, $\rho_{oil} A_{oil} \bar{V}_{oil} + \rho_{vin} A_{vin} \bar{V}_{vin} = \rho_{mix} A_{mix} \bar{V}_{mix}$, where all variables are positive numbers. The mixture density here is that of the oil and vinegar emulsion, and the velocities are averages across the ports (since there is no information given on the velocity distributions). Solving for the mixture density, we obtain:

$$\rho_{mix} = \frac{\rho_{oil} A_{oil} \bar{V}_{oil} + \rho_{vin} A_{vin} \bar{V}_{vin}}{A_{mix} \bar{V}_{mix}}$$

It is common in industrial processes involving the mixture of liquids to control the volume flowrates of input streams to be able to control the volume flowrate and composition of the output stream.

Students often experience difficulty in deciding whether a mass balance is the appropriate way to solve a problem. If a question involves mass or volume flowrates, the rate of change of density within a control volume or device, or the values of average velocity, density, or area at a port, a mass balance is clearly indicated. The momentum and energy balances discussed in the next two sections are always accompanied by a mass balance, so it is a good idea to routinely perform a mass balance as part of every type of CV analysis.

7.6.2 Momentum Balance

According to Newton's second law, the time rate of change of the total linear momentum of a system is equal to the sum of the body and surface forces acting on the fluid within the system. We express this by writing $d\mathbf{L}_{\text{sys}}/dt = \mathbf{F}_B + \mathbf{F}_S$, where $\mathbf{L}_{\text{sys}}$ is the total linear momentum in the system, and $\mathbf{F}_B$ and $\mathbf{F}_S$ are the body and surface forces, respectively. Now consider a fixed CV that coincides with the system. According to the Reynolds transport theorem for a control volume, Eq. 7.5, we can also write

$$\frac{d\mathbf{L}_{\text{sys}}}{dt} = \int_{\text{CV}} \frac{\partial}{\partial t}(\rho \varepsilon)\, dV + \int_{\text{CS}} (\rho \varepsilon)(\mathbf{u} \cdot \mathbf{n})\, dS$$

where we have replaced E_{sys} by $\mathbf{L}_{\text{sys}}$, the total linear momentum in the system. Since the linear momentum per unit volume is $\varepsilon = \rho\mathbf{u}$, this becomes

$$\frac{d\mathbf{L}_{\text{sys}}}{dt} = \int_{\text{CV}} \frac{\partial}{\partial t}(\rho \mathbf{u})\, dV + \int_{\text{CS}} (\rho \mathbf{u})(\mathbf{u} \cdot \mathbf{n})\, dS$$

The system and fixed CV coincide, so the total body and surface forces acting on the system must also act on the fixed CV. Thus since $d\mathbf{L}_{\text{sys}}/dt = \mathbf{F}_B + \mathbf{F}_S$, the preceding expression becomes

$$\int_{\text{CV}} \frac{\partial}{\partial t}(\rho \mathbf{u})\, dV + \int_{\text{CS}} (\rho \mathbf{u})(\mathbf{u} \cdot \mathbf{n})\, dS = \mathbf{F}_B + \mathbf{F}_S \qquad (7.17)$$

This is the integral momentum conservation equation, or momentum balance, for a fixed CV. Since it was not necessary to say anything about the type of fluid in deriving this equation, it is valid for all fluids, under all conditions.

We can provide a physical interpretation for this equation by writing it symbolically as

$$\frac{d\mathbf{L}_{\text{CV}}}{dt} = -\mathbf{\Gamma}_{\text{CV}} + \mathbf{F}_B + \mathbf{F}_S$$

where we have made use of the facts that the rate of accumulation of momentum inside the CV is given by $d\mathbf{L}_{\text{CV}}/dt = \int_{\text{CV}}(\partial/\partial t)(\rho\mathbf{u})\, dV$, and the net convective transport of momentum into the CV is $-\mathbf{\Gamma}_C = -\int_{\text{CS}}(\rho\mathbf{u})(\mathbf{u}\cdot\mathbf{n})\, dS$. Thus, the momentum balance for a fixed CV states that the rate of accumulation of momentum within the CV is equal to the sum of the net inflow of momentum into the CV and the rate at which momentum is created by the sum of the body and surface forces.

A clearer understanding of the role of body and surface forces in the momentum balance can be gained by representing these forces by their corresponding integral expressions. In Chapter 4, we showed that the total body force is given by a volume integral, Eq. 4.7. Writing this integral over the CV gives $\mathbf{F}_B = \int_{\text{CV}} \rho \mathbf{f}\, dV$. Similarly, the total surface force is given by a surface integral, Eq. 4.21, which when written over the control surface is $\mathbf{F}_S = \int_{\text{CS}} \mathbf{\Sigma}\, dS$. We can use these integrals to write the momentum balance as

$$\int_{\text{CV}} \frac{\partial}{\partial t}(\rho \mathbf{u})\, dV + \int_{\text{CS}} (\rho \mathbf{u})(\mathbf{u} \cdot \mathbf{n})\, dS = \int_{\text{CV}} \rho \mathbf{f}\, dV + \int_{\text{CS}} \mathbf{\Sigma}\, dS \qquad (7.18)$$

A momentum balance is a powerful tool in engineering analysis. It is universally applicable, meaning that it applies to the unsteady and steady flow of compressible or incompressible fluids at any speed. Thus a momentum balance can be applied to both laminar and turbulent flow. You may be surprised to learn that a momentum balance also applies to a fluid at rest. In a fluid at rest, the velocity is zero, so Eq. 7.18 becomes $\int_{CV} \rho \mathbf{f}\, dV + \int_{CS} \mathbf{\Sigma}\, dS = 0$. Since the stress vector in a fluid at rest is given by Eq. 5.1b as $\mathbf{\Sigma} = -p\mathbf{n}$, the momentum balance in a fluid at rest is $\int_{CV} \rho \mathbf{f}\, dV + \int_{CS} -p\mathbf{n}\, dS = 0$. Do you recognize this as Eq. 5.2, the integral hydrostatic equation?

This form explicitly shows the integrals representing the body and surface forces acting on the fluid inside the CV.

Now consider the force terms on the right-hand side of the momentum balance, Eq. 7.18. The total body force is generated by the body force acting on each element of fluid inside the CV. Consider the role of the most common body force: gravity. The magnitude of the total body force acting on a CV due to gravity is W_{CV}, the weight of the CV contents. Thus with the z axis in the vertical direction as usual, the body force will appear in a momentum balance as $-W_{CV}\mathbf{k}$, and account for the weight of the fluid and any other material inside the CV.

The total surface force is generated by the stress acting on the surface of the CV. Each section of a control surface has acting on it a stress vector that represents the surface force per unit area applied to the fluid inside the CV by the fluid or other agent outside the CV. At ports, where fluid enters or leaves a CV, the surface force is primarily due to the pressure in the fluid acting on the control surface. On control surfaces located at the interface between a fluid and a solid, the effects of both pressure and shear stress are important. We will see that the execution of a momentum balance requires a good understanding of surface forces, their magnitude, and the direction in which they act on a given control surface.

The first term on the left-hand side of Eq. 7.18 represents the accumulation of momentum within the CV. Most flows of engineering interest involve a steady process in which the accumulation of momentum in a properly chosen CV is zero. Thus for simplicity we will limit ourselves to problems of this type in this book. Dropping the momentum accumulation term in Eqs. 7.17 and 7.18, we can write the two equivalent forms of the steady process momentum balance as

$$\int_{CS} (\rho \mathbf{u})(\mathbf{u} \cdot \mathbf{n})\, dS = \mathbf{F}_B + \mathbf{F}_S \tag{7.19a}$$

and

$$\int_{CS} (\rho \mathbf{u})(\mathbf{u} \cdot \mathbf{n})\, dS = \int_{CV} \rho \mathbf{f}\, dV + \int_{CS} \mathbf{\Sigma}\, dS \tag{7.19b}$$

It is important to keep in mind that the momentum balance is a vector equation. We can therefore write the preceding equations in terms of their three Cartesian components as

$$\int_{CS} (\rho u)(\mathbf{u} \cdot \mathbf{n})\, dS = F_{B_x} + F_{S_x} \tag{7.20a}$$

$$\int_{CS} (\rho v)(\mathbf{u} \cdot \mathbf{n})\, dS = F_{B_y} + F_{S_y} \tag{7.20b}$$

$$\int_{CS} (\rho w)(\mathbf{u} \cdot \mathbf{n})\, dS = F_{B_z} + F_{S_z} \tag{7.20c}$$

and

$$\int_{CS}(\rho u)(\mathbf{u}\cdot\mathbf{n})\,dS = \left[\int_{CV}\rho\mathbf{f}\,dV\right]_x + \left[\int_{CS}\mathbf{\Sigma}\,dS\right]_x \qquad (7.21\text{a})$$

$$\int_{CS}(\rho v)(\mathbf{u}\cdot\mathbf{n})\,dS = \left[\int_{CV}\rho\mathbf{f}\,dV\right]_y + \left[\int_{CS}\mathbf{\Sigma}\,dS\right]_y \qquad (7.21\text{b})$$

$$\int_{CS}(\rho w)(\mathbf{u}\cdot\mathbf{n})\,dS = \left[\int_{CV}\rho\mathbf{f}\,dV\right]_z + \left[\int_{CS}\mathbf{\Sigma}\,dS\right]_z \qquad (7.21\text{c})$$

Writing the momentum balance in terms of its Cartesian components encourages us to look carefully at all parts of the CV and control surfaces to get a sense of the magnitudes and directions of the momentum transport, body forces, and surface forces. In some cases it is evident that there is only a single nonzero component of momentum transport or surface force. This suggests that we need evaluate only the corresponding component of the momentum balance. Although this approach minimizes the amount of calculation, there is no harm in using one of the vector forms of the momentum balance and calculating each term as a vector. In fact, we recommend this to you as a standard approach to start with and illustrate it in many of the examples in this section.

A momentum balance is helpful in understanding the distributions of pressure and shear stress in a flow, and it serves as a foundation for the empirical analysis of pressure drop, lift and drag forces, and other quantities of engineering interest. You will find connections to the case studies in a number of the examples that follow. The manner of selecting the CV is an important element in the successful use of a momentum balance. Study the next two examples and note how the CV selection allows us to isolate the effects of pressure and shear stress.

EXAMPLE 7.7

Use a momentum balance to analyze the steady, fully developed flow of constant density fluid in a round pipe. Find an expression relating the pressure drop down the pipe to the wall shear stress. Consider laminar and turbulent flow as shown in Figure 7.15, and use your results to investigate the friction factor.

SOLUTION

The laminar flow velocity profile in the pipe is parabolic, while the turbulent flow velocity profile is nearly uniform (Figure 7.15A). We will analyze both cases by assuming that the fully developed axisymmetric velocity field is given in cylindrical coordinates by $v_r = 0$, $v_\theta = 0$, and $v_z(r)$, and insert the appropriate function $v_z(r)$ for laminar and turbulent flow later. To focus on the pressure drop and wall shear stress in the two flows, we choose a CV consisting of all the fluid in a section of pipe of length L (see Figure 7.15B). The control surfaces consist of an inlet, an exit, and a surface along the wall

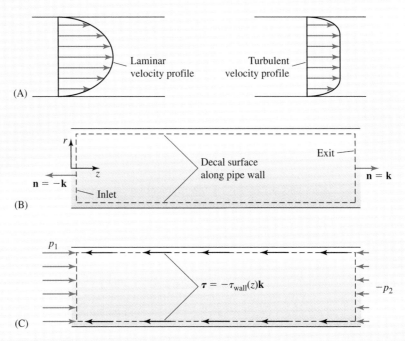

Figure 7.15 Schematic for Example 7.7: (A) velocity profiles for laminar and turbulent flow in a round pipe, (B) CV, and (C) stress distribution on the fluid in the CV.

of the pipe. We will refer to the latter as a decal surface to emphasize that this surface is directly adjacent to the solid wall. Note that we have defined the CV by placing control surfaces at the locations at which the pressures and shear stress act.

Since this is a steady constant density flow, we will apply Eq. 7.19b

$$\int_{CS} (\rho \mathbf{u})(\mathbf{u} \cdot \mathbf{n})\, dS = \int_{CV} \rho \mathbf{f}\, dV + \int_{CS} \mathbf{\Sigma}\, dS$$

to this CV. Since gravity does not effect the pressure drop down the pipe, we will drop the body force term. The flow is fully developed; thus the velocity fields at the inlet and exit to the CV are the same. On the inlet we have $\mathbf{u} = v_z(r)\mathbf{k}$, $\mathbf{n} = -\mathbf{k}$; thus $\mathbf{u} \cdot \mathbf{n} = -v_z(r)$. On the exit, $\mathbf{u} = v_z(r)\mathbf{k}$, $\mathbf{n} = \mathbf{k}$, and $\mathbf{u} \cdot \mathbf{n} = v_z(r)$. The momentum transport into the CV is seen to be

$$\int_{\text{inlet}} (\rho \mathbf{u})(\mathbf{u} \cdot \mathbf{n})\, dS = \int_0^{2\pi} \int_0^R (\rho v_z(r)\mathbf{k})(-v_z(r)) r\, dr\, d\theta = -\int_0^{2\pi} \int_0^R \rho v_z(r)^2 \mathbf{k}\, r\, dr\, d\theta$$

while the transport out of the CV is

$$\int_{\text{exit}} (\rho \mathbf{u})(\mathbf{u} \cdot \mathbf{n})\, dS = \int_0^{2\pi} \int_0^R (\rho v_z(r)\mathbf{k})(v_z(r)) r\, dr\, d\theta = +\int_0^{2\pi} \int_0^R \rho v_z(r)^2 \mathbf{k}\, r\, dr\, d\theta$$

Because the unit normal to the inlet and exit ports point in opposite directions, and the velocity profiles are the same, these terms add to zero irrespective of the exact nature of the velocity profile. This is true of all fully developed flows: the net momentum transport in the flow direction is zero. There is also no momentum transport across the decal surface, since this surface is adjacent to the solid wall of the pipe. Thus the momentum balance in this case reduces to $\int_{CS} \Sigma \, dS = 0$, which simply states that the net surface force acting on the fluid inside the CV is zero. That is,

$$\int_{\text{inlet}} \Sigma \, dS + \int_{\text{exit}} \Sigma \, dS + \int_{\text{decal}} \Sigma \, dS = 0$$

where the terms refer to the inlet and exit ports, and the decal surface adjacent to the pipe wall. We use Eq. 4.19 to write the stress vector in terms of its normal and tangential components as

$$\Sigma = -p\mathbf{n} + \boldsymbol{\tau}$$

On the inlet and exit ports, the tangential stress is negligible; thus on these surfaces we write the stress vector as $\Sigma = -p\mathbf{n}$. On the decal surface, the stress consists of both pressure and shear stress; thus on this surface we write $\Sigma = -p\mathbf{n} + \boldsymbol{\tau}$. The momentum balance becomes

$$\int_{\text{inlet}} -p\mathbf{n} \, dS + \int_{\text{exit}} -p\mathbf{n} \, dS + \int_{\text{decal}} (-p\mathbf{n} + \boldsymbol{\tau}) \, dS = 0$$

The inlet and exit ports both have area A. Assuming uniform but different pressures on the ports, and noting that the effect of pressure on the decal surface cancels owing to axisymmetry, we have

$$p_1 A \mathbf{k} - p_2 A \mathbf{k} + \int_{\text{decal}} \boldsymbol{\tau} \, dS = 0$$

This result applies to both laminar and turbulent fully developed flow in a round pipe. It shows that the force applied to fluid inside the CV by the pressure is balanced by the frictional shear force exerted by the wall on the fluid.

A mass balance for this same CV confirms that the average velocity is the same at the inlet and exit surfaces. Thus in a fully developed, steady, constant density flow, viscous friction does not slow the fluid down. Instead it makes itself felt in a pressure drop and a loss of the fluid energy associated with pressure, as discussed in Section 2.9.3. We see that the use of control surface segments at the inlet and exit, along with a decal surface adjacent to the pipe wall, introduces the pressure and shear stress into the momentum balance via the surface force term.

To evaluate the remaining integral containing the shear stress, we note that by the law of action–reaction, the shear force exerted by the wall on the fluid is equal and opposite to the shear force exerted by the fluid on the wall. Thus we can write

$$\int_{\text{decal}} \boldsymbol{\tau} \, dS = -\bar{\tau}_{\text{wall}} A_{\text{wall}} \mathbf{k}$$

where $\bar{\tau}_{wall}$ is the average shear stress on the wall, and A_{wall} is the area of the wall in contact with the fluid. (In this case the flow is fully developed and axisymmetric, so the shear stress $\tau_{wall}(z)$ acting on the wall is uniform (i.e., $\tau_{wall}(z) = \bar{\tau}_{wall} = \tau_{wall}$).) The momentum balance thus gives

$$p_1 A \mathbf{k} - p_2 A \mathbf{k} - \tau_{wall} A_{wall} \mathbf{k} = 0$$

Figure 7.15C provides a visual representation of this equation. Solving for the pressure drop yields:

$$p_1 - p_2 = \frac{\tau_{wall} A_{wall}}{A} \quad \text{(A)}$$

Noting that $A_{wall} = \pi D L$ and $A = \pi D^2/4$, we can write the pressure drop as

$$\Delta p = p_1 - p_2 = 4\tau_{wall}\frac{L}{D} \quad \text{(B)}$$

Thus the wall shear stress can be determined by measuring the pressure drop.

To investigate the friction factor recall that in case study 3.3.1 (flow in a round pipe), the pressure drop is given in terms of the friction factor by Eq. 3.15 as $\Delta p = \rho f (L/D)(\bar{V}^2/2)$. Equating this to the pressure drop predicted by the momentum balance in (B), we find

$$\Delta p = \rho f \frac{L}{D}\frac{\bar{V}^2}{2} = 4\tau_{wall}\frac{L}{D}$$

Solving for the friction factor we have

$$f = \frac{4\tau_{wall}}{\frac{1}{2}\rho \bar{V}^2} \quad \text{(C)}$$

We see that after using the methods outlined in case study 3.3.1 to calculate the friction factor, we can also calculate the wall shear stress.

EXAMPLE 7.8

The steady turbulent flow of a constant density fluid through a sudden expansion in a round pipe is shown in Figure 7.16A. Find an expression for the pressure change across the expansion, and use it to determine the loss coefficient. Assume a uniform flow at the inlet and exit stations as shown, and neglect the effects of gravity.

SOLUTION

In Example 7.7 we saw that pressure and shear stress appear in the surface force terms of a momentum balance. Since we are asked to derive an expression for the pressure

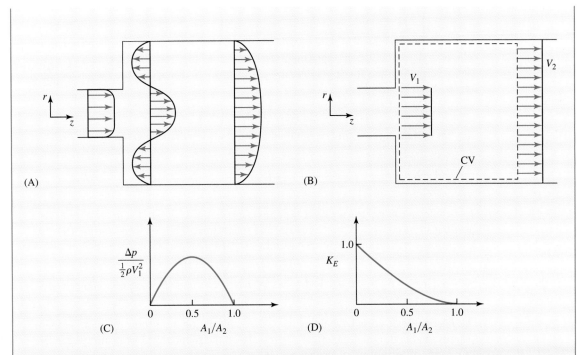

Figure 7.16 Schematic for Example 7.8: (A) velocity distribution through the sudden expansion, (B) uniform velocity distribution approximation and control volume, (C) normalized pressure change as a function of area ratio, and (D) loss coefficient as a function of area ratio.

change across the sudden expansion, we will perform a steady flow momentum balance and choose a CV as shown in Figure 7.16B. Note the placement of control surfaces at the inlet and exit of the expansion and two decal surfaces adjacent to the walls of the expansion. The latter are a wall decal surface adjacent to the round walls and a washer decal surface adjacent to the washer-shaped shoulder. Upon applying Eq. 7.19, neglecting the body force term, we have

$$\int_{CS} (\rho \mathbf{u})(\mathbf{u} \cdot \mathbf{n}) \, dS = \int_{CS} \mathbf{\Sigma} \, dS$$

Momentum transport occurs only on the inlet and exit ports. We will use cylindrical coordinates. On the inlet we have $\mathbf{u} = V_1 \mathbf{k}$, $\mathbf{n} = -\mathbf{k}$, and $\mathbf{u} \cdot \mathbf{n} = -V_1$. On the exit, $\mathbf{u} = V_2 \mathbf{k}$, $\mathbf{n} = \mathbf{k}$, and $\mathbf{u} \cdot \mathbf{n} = V_2$. The momentum transport at the inlet port is given by

$$\int_{\text{inlet}} (\rho \mathbf{u})(\mathbf{u} \cdot \mathbf{n}) \, dS = \int_0^{2\pi} \int_0^{R_1} (\rho V_1 \mathbf{k})(-V_1) r \, dr \, d\theta = -\rho V_1^2 A_1 \mathbf{k}$$

where $A_1 = \pi R_1^2$. On the exit port, where $A_2 = \pi R_2^2$, we get

$$\int_{\text{exit}} (\rho \mathbf{u})(\mathbf{u} \cdot \mathbf{n}) \, dS = \int_0^{2\pi} \int_0^{R_2} (\rho V_2 \mathbf{k})(V_2) r \, dr \, d\theta = \rho V_2^2 A_2 \mathbf{k}$$

A mass balance on this CV shows that $\dot{M} = \rho A_1 V_1 = \rho A_2 V_2$. Since the density is constant and $A_2 > A_1$ for an expansion, we know from the mass balance that $V_1 > V_2$. Thus, the momentum transport into and out of this CV differ, and the net momentum transport can be written as

$$\left(\rho V_2^2 A_2 - \rho V_1^2 A_1\right) \mathbf{k} = \dot{M}(V_2 - V_1)\mathbf{k}$$

The surface force acting on the fluid in the control volume is given by

$$\int_{CS} \mathbf{\Sigma}\, dS = \int_{\text{inlet}} \mathbf{\Sigma}\, dS + \int_{\text{exit}} \mathbf{\Sigma}\, dS + \int_{\text{wall decal}} \mathbf{\Sigma}\, dS + \int_{\text{washer decal}} \mathbf{\Sigma}\, dS$$

On the inlet and exit ports we write the stress vector as $\mathbf{\Sigma} = -p\,\mathbf{n}$. On the two decal surfaces the stress consists of both pressure and shear stress. Thus on these surfaces we use $\mathbf{\Sigma} = -p\,\mathbf{n} + \boldsymbol{\tau}$. The surface force becomes

$$\int_{CS} \mathbf{\Sigma}\, dS = \int_{\text{inlet}} -p\,\mathbf{n}\, dS + \int_{\text{exit}} -p\,\mathbf{n}\, dS + \int_{\text{wall decal}} (-p\,\mathbf{n} + \boldsymbol{\tau})\, dS + \int_{\text{washer decal}} (-p\,\mathbf{n} + \boldsymbol{\tau})\, dS$$

The inlet and exit ports have areas A_1 and A_2. Assuming a uniform pressure on these surfaces, and noting that the effect of pressure on the round decal wall surface cancels for reasons of axisymmetry, we have

$$\int_{CS} \mathbf{\Sigma}\, dS = p_1 A_1 \mathbf{k} - p_2 A_2 \mathbf{k} + \int_{\text{wall decal}} \boldsymbol{\tau}\, dS + \int_{\text{washer decal}} (-p\,\mathbf{n} + \boldsymbol{\tau})\, dS$$

The momentum balance therefore becomes

$$\left(\rho V_2^2 A_2 - \rho V_1^2 A_1\right) \mathbf{k} = p_1 A_1 \mathbf{k} - p_2 A_2 \mathbf{k} + \int_{\text{wall decal}} \boldsymbol{\tau}\, dS + \int_{\text{washer decal}} (-p\,\mathbf{n} + \boldsymbol{\tau})\, dS \tag{A}$$

We see that the momentum transport vector is balanced by the pressure forces on the inlet and exit, the shear force on the wall decal surface, and the pressure and shear force on the washer decal surface adjacent to the shoulder of the expansion.

To evaluate the remaining integrals in (A), we will write the surface force on the wall decal surface as $\int_{\text{wall decal}} \boldsymbol{\tau}\, dS = -\bar{\tau}_{\text{wall}} A_{\text{wall}} \mathbf{k}$, where $\bar{\tau}_{\text{wall}}$ is the average wall shear stress, assumed to be acting on the wall in the flow direction. As in the preceding example, the negative sign is needed because the shear stress acting on the adjacent CV surface acts in the direction opposite to the wall shear stress. Now consider the washer decal surface. The effect of shear stress on this surface cancels for reasons of axisymmetry, so the surface force can be written in terms of an average pressure $\bar{p}_{\text{washer}}$ as

$$\int_{\text{washer decal}} -p\,\mathbf{n}\, dS = -\bar{p}_{\text{washer}} A_{\text{washer}}(-\mathbf{k}) = \bar{p}_{\text{washer}}(A_2 - A_1)\mathbf{k}$$

Note the direction of the outward unit normal on this surface. Empirical data and computational fluid dynamic simulations suggest that the pressure on this surface is uniform and the same as that acting at the inlet. Taking $\bar{p}_{\text{washer}} = p_1$, we find that the surface force on the washer decal surface is $\bar{p}_{\text{washer}}(A_2 - A_1)\mathbf{k} = p_1(A_2 - A_1)\mathbf{k}$.

The completed momentum balance is

$$(\rho V_2^2 A_2 - \rho V_1^2 A_1)\mathbf{k} = p_1 A_1 \mathbf{k} - p_2 A_2 \mathbf{k} - \bar{\tau}_{\text{wall}} A_{\text{wall}} \mathbf{k} + p_1(A_2 - A_1)\mathbf{k}$$

Solving for the pressure change we have

$$(p_2 - p_1)A_2 = -(\rho V_2^2 A_2 - \rho V_1^2 A_1)\mathbf{k} - \bar{\tau}_{\text{wall}} A_{\text{wall}} \quad \text{(B)}$$

which can also be written by using $\dot{M} = \rho A_1 V_1 = \rho A_2 V_2$ (i.e., $V_2 = V_1(A_1/A_2)$) as

$$\frac{p_2 - p_1}{\frac{1}{2}\rho V_1^2} = 2\frac{A_1}{A_2}\left(1 - \frac{A_1}{A_2}\right) - \frac{\bar{\tau}_{\text{wall}} A_{\text{wall}}}{\frac{1}{2}\rho V_1^2 A_2} \quad \text{(C)}$$

Result (C) seems reasonable and tells us that the pressure change consists of an increase in pressure due to the flow slowing down as the area increases and a decrease in pressure due to the effects of the wall shear stress. Note that there is no expansion of the pipe as $A_1/A_2 \to 1$, and we find

$$\frac{p_2 - p_1}{\frac{1}{2}\rho V_1^2} = -\frac{\bar{\tau}_{\text{wall}} A_{\text{wall}}}{\frac{1}{2}\rho V_1^2 A_2} \quad \text{or} \quad p_1 - p_2 = \frac{\bar{\tau}_{\text{wall}} A_{\text{wall}}}{A_2}$$

The only pressure change in this case is a drop in pressure $p_1 > p_2$ due to the effect of friction at the wall. This agrees with (B) of Example 7.7, which considered the flow through a round pipe of constant diameter.

The flow through a sudden expansion at the high Reynolds numbers of engineering interest is accompanied by recirculation zones near the expansion plane, which persist for some distance downstream in the larger pipe, as shown in Figure 7.16A. The wall shear stress is very small and is directed opposite to the flow direction because the wall is exposed to a very low speed flow in the reverse direction. Thus the term in (C) involving the wall shear stress is actually positive but small enough to be neglected, and the pressure change for a flow through a sudden expansion is traditionally written as

$$\frac{p_2 - p_1}{\frac{1}{2}\rho V_1^2} = 2\frac{A_1}{A_2}\left(1 - \frac{A_1}{A_2}\right) \quad \text{(D)}$$

This formula is plotted in Figure 7.16C. It is easy to show that (D) predicts a maximum pressure increase of $(p_2 - p_1)/\frac{1}{2}\rho V_1^2 = \frac{1}{2}$ for $A_1/A_2 = 0.5$, and no change in pressure for $A_1/A_2 \to 0$ and $A_1/A_2 \to 1$.

To determine the loss coefficient, recall that in our case study of Section 3.3.2, on flow through an area change, the empirical formula for the pressure change across a sudden expansion is given by Eq. 3.20 as $p_2 - p_1 = [\frac{1}{2}\rho(\bar{V}_1^2 - \bar{V}_2^2)] - \Delta p_F$, where Δp_F is the frictional pressure loss. This loss, which is due to the effects of turbulence and recirculation, is given by Eq. 3.22 as $\Delta p_F = K_E \frac{1}{2}\rho \bar{V}_1^2$, where the K_E is the expansion loss coefficient. Combining these two equations we find:

$$p_2 - p_1 = \left[\tfrac{1}{2}\rho\left(\bar{V}_1^2 - \bar{V}_2^2\right)\right] - K_E \tfrac{1}{2}\rho \bar{V}_1^2 \tag{E}$$

Our analysis of this problem by means of a momentum balance allows us to find the expansion loss coefficient. Rearranging (D) to put it into the form of (E), we have

$$p_2 - p_1 = 2\frac{A_1}{A_2}\left(1 - \frac{A_1}{A_2}\right)\left(\tfrac{1}{2}\rho V_1^2\right)$$

$$= \left[2\frac{A_1}{A_2} - 2\left(\frac{A_1}{A_2}\right)^2\right]\left(\tfrac{1}{2}\rho V_1^2\right) + \left(\tfrac{1}{2}\rho V_1^2 - \tfrac{1}{2}\rho V_2^2\right) - \left(\tfrac{1}{2}\rho V_1^2 - \tfrac{1}{2}\rho V_2^2\right)$$

$$= \left(\tfrac{1}{2}\rho V_1^2 - \tfrac{1}{2}\rho V_2^2\right) - \left[1 - 2\frac{A_1}{A_2} + 2\left(\frac{A_1}{A_2}\right)^2 - \frac{V_2^2}{V_1^2}\right]\left(\tfrac{1}{2}\rho V_1^2\right)$$

After using the fact that $V_2^2/V_1^2 = (A_1/A_2)^2$ from the mass balance and simplifying, we have

$$p_2 - p_1 = \left(\tfrac{1}{2}\rho V_1^2 - \tfrac{1}{2}\rho V_2^2\right) - \left[1 - 2\frac{A_1}{A_2} + \left(\frac{A_1}{A_2}\right)^2\right]\left(\tfrac{1}{2}\rho V_1^2\right)$$

Since the flow is approximately uniform, $V_1 = \bar{V}_1$ and $V_2 = \bar{V}_2$. Thus we can write $K_E = [1 - 2(A_1/A_2) + (A_1/A_2)^2]$ and conclude that the expansion loss coefficient predicted by the momentum balance is

$$K_E = \left(1 - \frac{A_1}{A_2}\right)^2 \tag{F}$$

This result, which is plotted in Figure 7.16D (and appeared earlier in the case study of Section 3.3.2 as Figure 3.13), is in excellent agreement with experimental measurements.

Students often think of a momentum balance as an approximate method, but in fact there is no approximation involved in the momentum balance itself. However, approximations are often introduced into the momentum balance when different terms are evaluated. If poorly chosen, these approximations do have the potential to affect the accuracy of the final answer. While it is good engineering practice to be aware of the approximations you are making to evaluate terms, well-chosen approximations do not change the fundamental principle that a momentum balance

Although a momentum balance can be used to analyze a number of different aspects of a flow, it is often employed to calculate the force exerted by a fluid on an object or part of a structure. A question of this type may be answered by choosing a CV that has a section of its control surface adjacent to all parts of the object that are in contact with the fluid. This section of the control surface is called a decal surface, as explained earlier. By the law of action–reaction, the force **F** exerted by the fluid inside the control volume on the object is equal and opposite to the reaction force **R** exerted by the object on the fluid inside the CV. Since the decal surface is placed at the interface between the object and the fluid, the integral of the stress on this surface is an accurate expression of Newton's second law. For example, the assumptions of uniform velocities and pressures at ports are two common and well-accepted approximations in applying a momentum balance to problems involving turbulent flow. These comments on the role of approximations also apply to the mass, energy, and angular momentum balances discussed in this chapter.

appears in the momentum balance and defines the reaction force $\mathbf{R}$. The desired force of the fluid on the object $\mathbf{F}$ is the negative of the surface force on the decal surface, thus $\mathbf{F} = -\mathbf{R}$. We can use momentum balance to determine $\mathbf{R}$, then calculate $\mathbf{F}$, since $\mathbf{F} = -\mathbf{R}$.

For example, consider the flow through the sudden expansion in Example 7.8, and suppose we are asked to find the force $\mathbf{F}$ exerted on the expansion by the fluid flowing inside it. The CV we selected earlier has two decal surfaces adjacent to the solid surfaces of the expansion in contact with the fluid inside. Thus according to the foregoing recommendation, this CV is satisfactory. The momentum balance for this CV, (A) of Example 7.8, gives

$$\left(\rho V_2^2 A_2 - \rho V_1^2 A_1\right)\mathbf{k} = p_1 A_1 \mathbf{k} - p_2 A_2 \mathbf{k} + \int_{\text{wall decal}} \boldsymbol{\tau}\, dS + \int_{\text{washer decal}} (-p\mathbf{n} + \boldsymbol{\tau})\, dS$$

The sum of the two integrals over the decal surfaces is the total reaction force $\mathbf{R}$ applied by the solid surfaces of the expansion to the fluid inside the CV. Thus we have

$$\mathbf{R} = \int_{\text{wall decal}} \boldsymbol{\tau}\, dS + \int_{\text{washer decal}} (-p\mathbf{n} + \boldsymbol{\tau})\, dS$$

and the momentum balance can be written as

$$\left(\rho V_2^2 A_2 - \rho V_1^2 A_1\right)\mathbf{k} = p_1 A_1 \mathbf{k} - p_2 A_2 \mathbf{k} + \mathbf{R}$$

Rearranging terms, we find

$$-\mathbf{R} = \left(p_1 + \rho V_1^2\right) A_1 \mathbf{k} - \left(p_2 + \rho V_2^2\right) A_2 \mathbf{k}$$

From the law of action–reaction, the reaction force $\mathbf{R}$ on the CV is equal and opposite to the force of the fluid $\mathbf{F}$ on the wetted surface of the sudden expansion. That is $\mathbf{F} = -\mathbf{R}$, and the momentum balance allows us to write the force applied by the fluid to the expansion as

$$\mathbf{F} = \left(p_1 + \rho V_1^2\right) A_1 \mathbf{k} - \left(p_2 + \rho V_2^2\right) A_2 \mathbf{k}$$

This fundamental idea of determining the reaction force $\mathbf{R}$ and then the force $\mathbf{F}$ exerted by a fluid on an object or part of a structure by using a momentum balance and an appropriately defined decal surface is also valid for a problem involving an external or structural force. Suppose we are to calculate $\mathbf{F}_E$, an external force acting on a structure or object. In this type of problem we first write a force balance on the object which includes the force $\mathbf{F}$ applied by all fluids, the body force acting on the object, and the external force $\mathbf{F}_E$ acting on the object. Next we choose a CV that has a section of its control surface adjacent to all parts of the object that are in contact with a fluid. This section of the control surface is the decal surface, and it allows us to include the reaction force $\mathbf{R}$ in the momentum balance. Finally we use the momentum balance and force balance together to determine the external force $\mathbf{F}_E$. This process is illustrated in detail in Examples 7.9 and 7.10. After solving each problem with a CV containing fluid only, we show how choosing a mixed CV that contains the object or structure greatly simplifies the analysis of a problem involving external forces.

EXAMPLE 7.9

A decorative fountain uses a nozzle that is press-fitted to the end of a high pressure water line as shown in Figure 7.17A. Find the force that must be exerted on the nozzle by the press fitting to keep it in place. Assume uniform velocities and pressures at the inlet and exit of the nozzle.

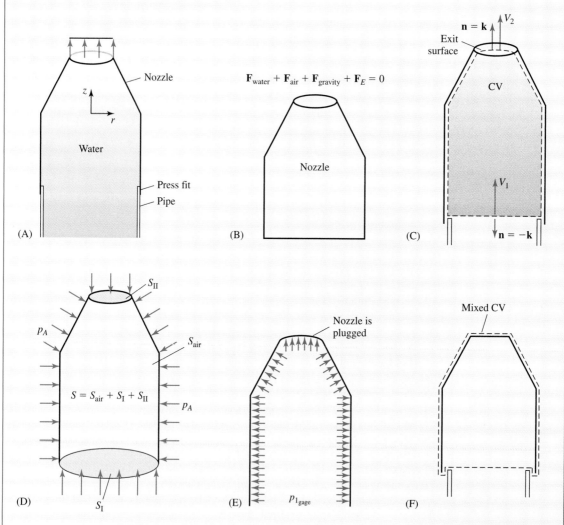

Figure 7.17 Schematic for Example 7.9: (A) geometry, (B) force balance, (C) fluid-only CV, (D) pressure distribution, (E) pressure on plugged nozzle, and (F) mixed CV.

SOLUTION

Figure 7.17A is a sketch of the physical arrangement. We will assume steady flow. The nozzle is the object of interest, and the force applied by the fitting to the nozzle is an external force $\mathbf{F}_E$ on the nozzle. Since we are asked to find this external force, we begin as recommended by writing a vector force balance on the nozzle as shown in Figure 7.17B. The forces acting on the nozzle include the forces due to the water inside and air outside, the force of gravity, and the external force $\mathbf{F}_E$ exerted on the nozzle by the press fit. The force balance in vector form is

$$\mathbf{F}_{\text{water}} + \mathbf{F}_{\text{air}} + \mathbf{F}_{\text{gravity}} + \mathbf{F}_E = 0$$

Thus the external force applied by the fitting to the nozzle is given by

$$\mathbf{F}_E = -\mathbf{F}_{\text{water}} - \mathbf{F}_{\text{air}} - \mathbf{F}_{\text{gravity}} \tag{A}$$

To find the external force, we must find the forces applied to the nozzle by gravity, the water, and the air. The force of gravity on the nozzle is $\mathbf{F}_{\text{gravity}} = -W_{\text{nozzle}}\mathbf{k}$. The force due to the water is found by considering the surface force applied by the water to the interior wetted surface of the nozzle. This surface force is given by integrating the stress vector over this surface, thus we can write

$$\mathbf{F}_{\text{water}} = \int_{\text{nozzle}} \mathbf{\Sigma} \, dS = \int_{\text{nozzle}} (-p\mathbf{n} + \boldsymbol{\tau}) \, dS$$

Since we are not given information about the pressure and shear stress distributions on the nozzle, this integral cannot be evaluated directly. Thus we will find this force by a momentum balance on the CV shown in Figure 7.17C. Note that by placing a decal surface adjacent to the wetted surface of the nozzle, we can use the principle of action–reaction to show that the unknown force $\mathbf{F}_{\text{water}}$ is equal and opposite to the reaction force $\mathbf{R}$ on the decal surface. Thus the force of the water on the nozzle can be found from a momentum balance using the chosen CV.

Now consider the force of the air at atmospheric pressure on the outside of the nozzle. The air is at rest, so the force is given by the integral $\mathbf{F}_{\text{air}} = \int_{S_{\text{air}}} \mathbf{\Sigma} \, dS = \int_{S_{\text{air}}} -p_A \mathbf{n} \, dS$, where the integral is taken over the exterior surface S_{air} of the nozzle in contact with the air. We can use the following trick to evaluate this integral over a curved surface. Consider the surface force due to atmospheric pressure on the surface S shown in Figure 7.17D. This surface duplicates the exterior surface of the nozzle, but with the inlet and exit openings closed. Since a constant pressure acts everywhere on this closed surface, the integral $\int_S -p_A \mathbf{n} \, dS$ over the entire surface must be zero. Thus we can write

$$\int_S -p_A \mathbf{n} \, dS = \int_{S_{\text{air}}} -p_A \mathbf{n} \, dS + \int_I -p_A \mathbf{n} \, dS + \int_{II} -p_A \mathbf{n} \, dS = 0$$

Solving for the desired integral of atmospheric pressure over the exterior surface, which is equal to $\mathbf{F}_{\text{air}}$ we have

$$\mathbf{F}_{\text{air}} = -\int_I -p_A \mathbf{n} \, dS - \int_{II} -p_A \mathbf{n} \, dS$$

If the nozzle wall is thin, the open surfaces I and II are virtually identical to the inlet and exit CV surfaces, and the desired force of atmospheric pressure over the exterior surface is given by

$$\mathbf{F}_{\text{air}} = -\int_{\text{inlet}} -p_A \mathbf{n}\, dS - \int_{\text{exit}} -p_A \mathbf{n}\, dS \qquad \text{(B)}$$

This methodology for evaluating the force of atmospheric pressure on part of the surface of an object should be noted for future use, since it allows us to calculate the surface force due to atmospheric pressure on any surface with openings by considering the standard integral of atmospheric pressure over each opening and changing the sign in front of each integral.

Upon inserting the preceding results into the force balance (A), we see that the external force supplied by the press fitting is given by

$$\mathbf{F}_E = -\int_{\text{nozzle}} \mathbf{\Sigma}\, dS + \int_{\text{inlet}} -p_A \mathbf{n}\, dS + \int_{\text{exit}} -p_A \mathbf{n}\, dS + W_{\text{nozzle}} \mathbf{k} \qquad \text{(C)}$$

To evaluate the integral over the nozzle interior surface in (C), we use the CV with a decal surface adjacent to the interior wall of the nozzle as shown in Figure 7.17C. Applying Eq. 7.19, we have

$$\int_{\text{CS}} (\rho \mathbf{u})(\mathbf{u}\cdot\mathbf{n})\, dS = \int_{\text{CV}} \rho \mathbf{f}\, dV + \int_{\text{CS}} \mathbf{\Sigma}\, dS$$

The momentum transport terms on the inlet and exit are evaluated by inspection. On the inlet we have $\mathbf{u} = V_1\mathbf{k}$, $\mathbf{n} = -\mathbf{k}$, and $\mathbf{u}\cdot\mathbf{n} = -V_1$, so that $\int_{\text{inlet}} (\rho \mathbf{u})(\mathbf{u}\cdot\mathbf{n})\, dS = (\rho V_1 \mathbf{k})(-V_1)A_1$. On the exit, $\mathbf{u} = V_2\mathbf{k}$, $\mathbf{n} = \mathbf{k}$, and $\mathbf{u}\cdot\mathbf{n} = V_2$, so that $\int_{\text{exit}} (\rho \mathbf{u})(\mathbf{u}\cdot\mathbf{n})\, dS = (\rho V_2 \mathbf{k})(V_2)A_2$. The body force integral is simply the weight of the fluid in the CV or $-W_{\text{CV}}\mathbf{k}$. The stress terms are evaluated by noting that the stress on the inlet and exit is due to the pressure. Thus we find

$$\int_{\text{CS}} \mathbf{\Sigma}\, dS = \int_{\text{inlet}} -p_1 \mathbf{n}\, dS + \int_{\text{exit}} -p_2 \mathbf{n}\, dS + \int_{\text{decal}} \mathbf{\Sigma}\, dS$$

Although the stress on the decal surface is unknown, this integral is the reaction force $\mathbf{R}$. By the principle of action–reaction, we know that $\mathbf{F}_{\text{water}} = -\mathbf{R}$. Thus in this problem we can write $\int_{\text{decal}} \mathbf{\Sigma}\, dS = \mathbf{R} = -\mathbf{F}_{\text{water}} = -\int_{\text{nozzle}} \mathbf{\Sigma}\, dS$. The completed momentum balance is then

$$\rho V_2^2 A_2 \mathbf{k} - \rho V_1^2 A_1 \mathbf{k} = -W_{\text{CV}}\mathbf{k} + \int_{\text{inlet}} -p_1 \mathbf{n}\, dS + \int_{\text{exit}} -p_2 \mathbf{n}\, dS - \int_{\text{nozzle}} \mathbf{\Sigma}\, dS \qquad \text{(D)}$$

We now have two equations, (C) and (D), to determine the external force applied to the nozzle by the fitting. Rearranging (D) we have

$$-\int_{\text{nozzle}} \mathbf{\Sigma}\, dS = \left(\rho V_2^2 A_2 \mathbf{k} - \rho V_1^2 A_1 \mathbf{j}\right) + W_{\text{CV}}\mathbf{k} + \int_{\text{inlet}} p_1 \mathbf{n}\, dS + \int_{\text{exit}} p_2 \mathbf{n}\, dS$$

Inserting this into (C) and combining the pressure integrals gives

$$\mathbf{F}_E = \left(\rho V_2^2 A_2 \mathbf{k} - \rho V_1^2 A_1 \mathbf{k}\right) + \int_{\text{inlet}} (p_1 - p_A)\mathbf{n}\,dS + \int_{\text{exit}} (p_2 - p_A)\mathbf{n}\,dS + (W_{\text{CV}} + W_{\text{nozzle}})\mathbf{k}$$

We see that the effect of the air at atmospheric pressure on the outside of the nozzle is equivalent to using gage pressure at the open surfaces of the control volume, i.e., the inlet and exit. Since the pressure is uniform on these surfaces, these integrals give $\int_{\text{inlet}} (p_1 - p_A)\mathbf{n}\,dS = (p_1 - p_A)(-\mathbf{k})A_1$ and $\int_{\text{exit}} (p_2 - p_A)\mathbf{n}\,dS = (p_2 - p_A)(\mathbf{k})A_2$. Thus the external force on the nozzle is given by

$$\mathbf{F}_E = \left(p_{2_{\text{gage}}} + \rho V_2^2\right) A_2 \mathbf{k} - \left(p_{1_{\text{gage}}} + \rho V_1^2\right) A_1 \mathbf{k} + (W_{\text{CV}} + W_{\text{nozzle}})\mathbf{k} \qquad \text{(E)}$$

This is the answer to the problem. To see whether it is reasonable, we can imagine the situation in which the nozzle is plugged so that no water is flowing and a hydrostatic pressure $p_{1_{\text{gage}}}$ acts at the nozzle inlet, as shown in Figure 7.17E. The force of the water on the inside of the nozzle is found by using the indirect method for a curved surface as explained in Section 5.5.3. The result is

$$\mathbf{F}_{\text{water}} = -\left(p_{1_{\text{gage}}}\right)(-\mathbf{k})A_1 - W_{\text{CV}}\mathbf{k} = p_{1_{\text{gage}}} A_1 \mathbf{k} - W_{\text{CV}}\mathbf{k}$$

which is the inlet gage pressure acting on the projected area of the nozzle less the expected small weight contribution, which accounts for the fact that the static pressure on the inside wetted surface of the nozzle is slightly less than $p_{1_{\text{gage}}}$ at the higher elevations. From the force balance (A), the required external force with the nozzle plugged is thus $\mathbf{F}_E = -(p_{1_{\text{gage}}})A_1\mathbf{k} + (W_{\text{CV}} + W_{\text{nozzle}})\mathbf{k}$. By putting $V_1 = V_2 = 0$ and $A_2 = 0$ into (E), we get exactly the same result. In most cases the weight of the fluid in the CV and weight of the nozzle are negligible, so the external force required with the nozzle plugged is usually approximated as

$$\mathbf{F}_E = -\left(p_{1_{\text{gage}}}\right) A_1 \mathbf{k} \qquad \text{(F)}$$

As expected, the press fitting must resist the tendency of the hydrostatic pressure to blow the nozzle off the pipe.

When the water is moving, the jet of liquid exiting the nozzle is at atmospheric pressure. Neglecting the effects of gravity, we have from (E) $\mathbf{F}_E = \rho V_2^2 A_2 \mathbf{k} - (p_{1_{\text{gage}}} + \rho V_1^2)A_1\mathbf{k}$. A mass balance reveals that $\dot{M} = \rho A_1 V_1 = \rho A_2 V_2$, so we can write this as

$$\mathbf{F}_E = -\left(p_{1_{\text{gage}}} A_1\right)\mathbf{k} + \dot{M}(V_2 - V_1)\mathbf{k} \qquad \text{(G)}$$

Once again we see that the press fitting must create a frictional force that resists the tendency of the nozzle to be blown off the pipe. With the water moving, both pressure and shear stress act on the nozzle, but now the pressure distribution inside the nozzle is far from hydrostatic. If the nozzle inlet pressure $p_{1_{\text{gage}}}$ is the same in both cases, which is true for a reasonably small nozzle opening, comparing (G) and (F) shows that the force required to keep the nozzle on the pipe is smaller with the nozzle open than when the nozzle is plugged. Does this agree with your intuition?

We can also observe that the completed momentum balance (D) allows us to write the unknown force of the water on the nozzle as

$$\mathbf{F}_{water} = \int_{nozzle} \mathbf{\Sigma}\, dS = -(\rho V_2^2 A_2 \mathbf{j} - \rho V_1^2 A_1 \mathbf{j}) - W_{CV}\mathbf{k} + \int_{inlet} -p_1 \mathbf{n}\, dS + \int_{exit} -p_2 \mathbf{n}\, dS$$

Evaluating the pressure terms and neglecting the weight of water in the control volume, we have

$$\mathbf{F}_{water} = (p_1 + \rho V_1^2) A_1 \mathbf{k} - (p_A + \rho V_2^2) A_2 \mathbf{k} \qquad \text{(H)}$$

This example problem can be solved more quickly and efficiently by using a mixed control volume that contains both fluid and solid, as shown in Figure 7.17F. The CV surrounds the entire nozzle and has a decal surface along the press fit, where the external force on the nozzle inside the CV occurs. Writing the momentum balance for this CV yields

$$\int_{inlet} (\rho \mathbf{u})(\mathbf{u}\cdot\mathbf{n})\, dS + \int_{exit} (\rho \mathbf{u})(\mathbf{u}\cdot\mathbf{n})\, dS = \int_{CV} \rho \mathbf{f}\, dV + \int_{inlet} \mathbf{\Sigma}\, dS + \int_{exit} \mathbf{\Sigma}\, dS + \int_{decal} \mathbf{\Sigma}\, dS + \int_{air} \mathbf{\Sigma}\, dS$$

where the integral of the surface force over the decal surface adjacent to the press fit defines F_E, the external force acting on the nozzle inside the CV. Solving for this force, we have

$$\mathbf{F}_E = \int_{decal} \mathbf{\Sigma}\, dS = \int_{inlet} (\rho \mathbf{u})(\mathbf{u}\cdot\mathbf{n})\, dS + \int_{exit} (\rho \mathbf{u})(\mathbf{u}\cdot\mathbf{n})\, dS - \int_{CV} \rho \mathbf{f}\, dV - \int_{inlet} \mathbf{\Sigma}\, dS - \int_{exit} \mathbf{\Sigma}\, dS - \int_{air} \mathbf{\Sigma}\, dS$$

Using the procedures outlined earlier to evaluate each term on the right-hand side yields

$$\mathbf{F}_E = (\rho V_1 \mathbf{k})(-V_1)A_1 + (\rho V_2 \mathbf{k})(V_2)A_2 + (W_{CV} + W_{nozzle})\mathbf{k} - p_1 A_1 \mathbf{k} + p_2 A_2 \mathbf{k} - (-p_A A_1 \mathbf{k} + p_A A_2 \mathbf{k})$$

Introducing gage pressures shows that this is identical to (E), the preceding result. Note that the body force term in the momentum balance for a mixed CV must account for the weight of the nozzle and the fluid inside the CV, since both the nozzle and its fluid contents are inside the mixed CV. In this example, using a mixed CV to find the external force is much quicker and easier than using a CV filled only with fluid. However, if the problem had asked for the force of the water on the nozzle, this mixed CV would not work because it does not have a decal surface adjacent to the wetted surface of the nozzle.

EXAMPLE 7.10

A liquid-fueled rocket is in steady operation on a test stand at sea level as shown in Figure 7.18A. What is the force acting on the support pylon at the rocket attachment point? What is the thrust produced by the rocket? Assume the velocity, pressure, and temperature of the exhaust stream are uniform and known at the nozzle exit plane and that the combustion gas obeys the perfect gas law. Neglect the effects of the flows of fuel and oxidizer.

7.6 CONTROL VOLUME ANALYSIS

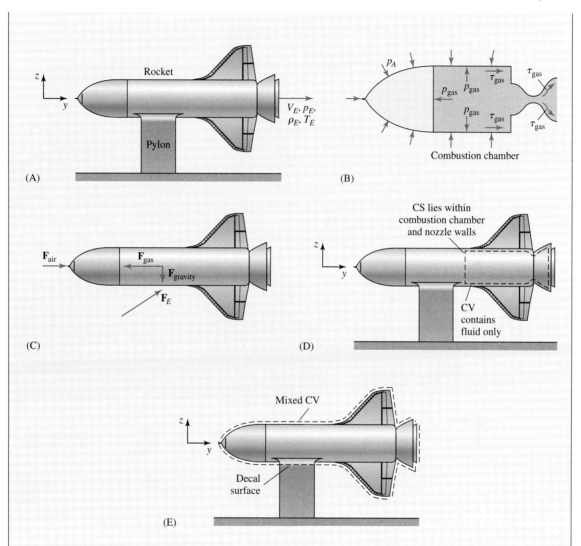

Figure 7.18 Schematic for Example 7.10: (A) geometry, (B) stress on the combustion chamber, (C) free body diagram of the rocket, (D) interior CV, and (E) exterior CV that cuts through the support pylon.

SOLUTION

Figure 7.18A shows the physical arrangement, and Figure 7.18B is an idealized representation of the pressure and shear stress acting inside the combustion chamber and nozzle of the rocket. Both questions relate to the forces acting on the rocket, so we will construct a force balance on the rocket as shown in Figure 7.18C. The relevant forces include the force applied by the high pressure combustion gas to the interior surfaces of the engine, the force of atmospheric air on the outside, the force of gravity, and the external force $\mathbf{F}_E$ applied to the rocket by the support pylon. Thus we have $\mathbf{F}_{\text{gas}} + \mathbf{F}_{\text{air}} + \mathbf{F}_{\text{gravity}} + \mathbf{F}_E = 0$, and the external force acting on the rocket is $\mathbf{F}_E = -\mathbf{F}_{\text{gas}} - \mathbf{F}_{\text{air}} - \mathbf{F}_{\text{gravity}}$. However, in this example we are asked to find the force $\mathbf{F}_{\text{pylon}}$ acting on the support pylon. Since this force is given by $\mathbf{F}_{\text{pylon}} = -\mathbf{F}_E$, the force balance can be written as

$$\mathbf{F}_{\text{pylon}} = \mathbf{F}_{\text{gas}} + \mathbf{F}_{\text{air}} + \mathbf{F}_{\text{gravity}} \tag{A}$$

To find this force, we must find the forces applied to the rocket by the combustion gas, the air, and gravity. The force of gravity on the rocket is $\mathbf{F}_{\text{gravity}} = -W_{\text{rocket}}\mathbf{k}$. The force of the air is found by using the trick involving the negative of the integral of atmospheric pressure over the open surface of the engine. In this case, the only open surface is the exit plane of the nozzle, so assuming that the nozzle wall is thin we have $\mathbf{F}_{\text{air}} = -\int_{\text{exit}} -p_A \mathbf{n}\, dS = p_A A_E \mathbf{j}$. The force acting on the pylon is now

$$\mathbf{F}_{\text{pylon}} = \mathbf{F}_{\text{gas}} + p_A A_E \mathbf{j} - W_{\text{rocket}}\mathbf{k} \tag{B}$$

The force of the gas on the rocket is defined by the integral of the stress vector over the interior surfaces of the engine, $\mathbf{F}_{\text{gas}} = \int_{\text{engine}} \mathbf{\Sigma}\, dS = \int_{\text{engine}} (-p\mathbf{n} + \boldsymbol{\tau})\, dS$. Since we are not given any information about the pressure or shear stress distributions inside the engine, this integral cannot be evaluated directly. Instead we will use a decal surface and a momentum balance to find this force.

To evaluate the force of the combustion gas on the interior surface of the engine, we will define a CV with a decal surface adjacent to the interior wall of the engine as shown in Figure 7.18D. Applying a steady flow momentum balance to this CV, we have

$$\int_{\text{CS}} (\rho \mathbf{u})(\mathbf{u} \cdot \mathbf{n})\, dS = \int_{\text{CV}} \rho \mathbf{f}\, dV + \int_{\text{exit}} \mathbf{\Sigma}\, dS + \int_{\text{decal}} \mathbf{\Sigma}\, dS$$

The terms on the exit port are $\mathbf{u} = V_E \mathbf{j}$, $\mathbf{n} = \mathbf{j}$, and $\mathbf{u} \cdot \mathbf{n} = V_E$. Thus,

$$\int_{\text{exit}} (\rho \mathbf{u})(\mathbf{u} \cdot \mathbf{n})\, dS = (\rho V_E \mathbf{j})(V_E) A_E = \rho V_E^2 A_E \mathbf{j}$$

The body force integral is simply the weight of the gas in the CV or $-W_{\text{CV}}\mathbf{k}$. The stress on the exit port is due to the pressure, so we find that $\int_{\text{exit}} \mathbf{\Sigma}\, dS = \int_{\text{exit}} -p_E \mathbf{n}\, dS = -p_E A_E \mathbf{j}$. The integral over the decal surface is the reaction force. By the principle of action–reaction, we know that $\mathbf{R} = -\mathbf{F}_{\text{gas}}$; thus, $\int_{\text{decal}} \mathbf{\Sigma}\, dS = \mathbf{R} = -\mathbf{F}_{\text{gas}}$. The completed momentum balance is then

$$\rho_E V_E^2 A_E \mathbf{j} = -W_{\text{CV}}\mathbf{k} - p_E A_E \mathbf{j} - \mathbf{F}_{\text{gas}} \tag{C}$$

We now have two equations, (B) and (C), to determine $\mathbf{F}_{\text{pylon}}$. Solving for this force, we find

$$\mathbf{F}_{\text{pylon}} = -\left[\rho_E V_E^2 A_E + (p_E - p_A) A_E\right]\mathbf{j} - (W_{\text{CV}} + W_{\text{rocket}})\mathbf{k} \tag{D}$$

This is our answer. Since we know the pressure and temperature at the exit, we can use the perfect gas law and the specific gas constant of the combustion products to find the gas density. The exit velocity is typically very large (supersonic) for a rocket. We see that our answer is reasonable. It predicts that the total weight of the rocket acts downward on the pylon and that the engine also applies a horizontal force on the pylon to the left, as expected.

The thrust produced by a rocket engine is the force available to accelerate the vehicle containing the engine. In this case we conclude from (D) that the thrust is

$$\mathbf{F}_{\text{thrust}} = -\left[\rho_E V_E^2 A_E + (p_E - p_A) A_E\right]\mathbf{j} \tag{E}$$

Combining (D) and (E) gives

$$\mathbf{F}_{\text{pylon}} = \mathbf{F}_{\text{thrust}}\mathbf{j} - (W_{\text{CV}} + W_{\text{rocket}})\mathbf{k}$$

which sensibly states that the force on the pylon is due to the thrust and the total weight.

Our analysis shows that the pressure at which the rocket exhaust stream exits the nozzle influences the amount of thrust produced. There is a complex relationship between the combustion chamber pressure, exit velocity, exit pressure, and ambient pressure of a rocket nozzle, so the exit velocity and exit pressure are not independent. The analysis of isentropic compressible flow through a converging–diverging nozzle, a topic not covered in this text, shows that the maximum thrust occurs for an exit pressure equal to the ambient pressure or $p_E = p_A$, thus the maximum thrust is given by $\mathbf{F}_{\text{thrust}} = -\rho_E V_E^2 A_E \mathbf{j}$.

We can also solve this problem using a mixed CV with a decal surface to isolate the force applied to the pylon. This CV is shown in Figure 7.18E. Writing a steady flow momentum balance for this CV, we have $\int_{\text{exit}} (\rho \mathbf{u})(\mathbf{u} \cdot \mathbf{n}) \, dS = \int_{\text{CV}} \rho \mathbf{f} \, dV + \int_{\text{decal}} \mathbf{\Sigma} \, dS + \int_{\text{air}} \mathbf{\Sigma} \, dS + \int_{\text{exit}} \mathbf{\Sigma} \, dS$. Evaluating the various terms and noting that by action–reaction $\int_{\text{decal}} \mathbf{\Sigma} \, dS = -\mathbf{F}_{\text{pylon}}$, we find

$$\rho_E V_E^2 A_E \mathbf{j} = -(W_{\text{CV}} + W_{\text{rocket}})\mathbf{k} - \mathbf{F}_{\text{pylon}} + p_A A_E \mathbf{j} + (-p_E A_E \mathbf{j})$$

which is identical to the earlier result (D), and much quicker.

Examples 7.9 and 7.10 suggest that in problems involving the determination of a force, the first piece of control surface to assign is the decal surface, which is the surface across which the desired force acts. Once the decal surface has been assigned, we place sections of the control surface where we have information. This process tends to suggest how the rest of the control surface should be selected, and whether it should contain fluid only or be a mixed CV containing an object. If a problem asks for the force of a fluid on a structure, a CV containing fluid only should be considered. If a problem asks for an external force, a mixed CV usually works best. In most cases, either type of CV will work, but the analysis is easier with the right control volume choice.

For example, consider the jet engine on a test stand shown in Figure 7.19. Suppose you are asked to determine the thrust produced by this engine. Since thrust is the force tending to push the engine forward, it is equal and opposite to the horizontal component of the external force of the pylon on the engine. Thus, we assign a decal surface to cut across the pylon to determine this external force. We are given information about the velocities and flow properties at the engine inlet and exit so it makes sense to put sections of the control surface at these locations. Thus, we arrive at the mixed CV shown in Figure 7.19. This is an excellent choice, as you may have guessed from studying Example 7.10.

Figure 7.19 CV for a jet engine on a test stand.

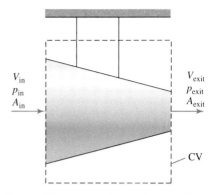

As you gain experience solving force questions with a momentum balance, it becomes possible to combine the ideas we have been discussing into a more streamlined solution technique. This is particularly true in a problem in which all the forces are obvious by inspection and only a single component of the force is to be determined. This streamlined process is illustrated for a case involving a single force component in Example 7.11. As you study it, notice the mixed CV selection and observe how the problem is solved by using only one component of the momentum balance and an understanding of the forces acting on the CV.

EXAMPLE 7.11

What is the horizontal force required to hold the thin sluice gate of height H and length L in the position shown in Figure 7.20A? You may assume 2D steady flow under the gate, with uniform velocities and hydrostatic pressure distributions at the indicated locations. What is the horizontal force required to hold the sluice gate in the closed position as shown in Figure 7.20B if the water levels on both sides remain the same?

SOLUTION

This question asks for the horizontal component of the external force acting on the sluice gate in the open and closed positions. We know that this force component must balance the horizontal force of the air and water on the gate in each case. In the necessary sketches for this problem (Figure 7.20A, 7.20B), the horizontal forces of air and water on the gate are shown in the expected directions. The horizontal external force is shown pointing in the negative y direction. Our horizontal force balance on the gate in each case is

$$F_{\text{water}} - F_{\text{air}} - F_E = 0 \quad \text{(A)}$$

In the closed position the pressure distribution in the water is hydrostatic and it is constant in the air. There is no need for a momentum balance, since we can write the net hydrostatic force applied by both fluids on the gate by considering the pressure distribution on each side as shown in Figure 7.20B. The net horizontal force of the air and water is found to be:

$$F_{\text{water}} - F_{\text{air}} = \left(\frac{\rho g D_1}{2}\right) D_1 L - \left(\frac{\rho g D_2}{2}\right) D_2 L = \left[\frac{1}{2}\rho g D_1^2 L - \frac{1}{2}\rho g D_2^2 L\right]$$

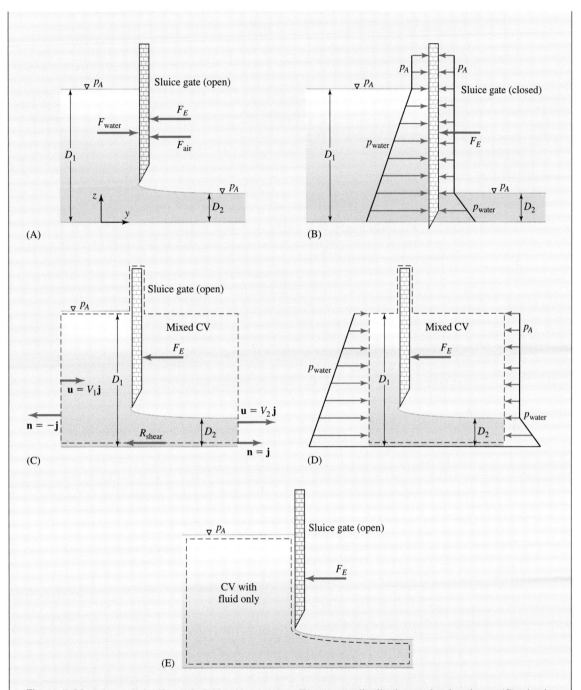

Figure 7.20 Schematic for Example 7.11: (A) geometry, (B) pressure distribution on the closed gate, (C) mixed CV, (D) pressure distribution on open gate and (E) a fluid only CV.

Using (A) the horizontal external force needed to keep the gate stationary in the closed position is found to be

$$F_E = \tfrac{1}{2}\rho g D_1^2 L - \tfrac{1}{2}\rho g D_2^2 L \qquad (B)$$

This positive value confirms that in the closed position the external force points to the left as assumed in Figure 7.20B.

In the open position we choose a mixed CV as shown in Figure 7.20C and use Eq. 7.20b,

$$\int_{CS}(\rho v)(\mathbf{u}\cdot\mathbf{n})\,dS = F_{B_Y} + F_{S_Y}$$

noting that the body force is zero in the y direction. On the inlet we have $\mathbf{u} = V_1\mathbf{j}$, $\mathbf{n} = -\mathbf{j}$, and $\mathbf{u}\cdot\mathbf{n} = -V_1$. On the exit, $\mathbf{u} = V_2\mathbf{j}$, $\mathbf{n} = \mathbf{j}$, and $\mathbf{u}\cdot\mathbf{n} = V_2$. The momentum transport terms are found to be $(\rho V_2^2 A_2 - \rho V_1^2 A_1)$, where $A_1 = D_1 L$ and $A_2 = D_2 L$. We next examine each section of the control surface shown in Figure 7.20C, dropping those on which the surface forces do not act in the y direction. The remaining surfaces are the inlet, the decal surfaces on the left and right sides of the gate exposed to air (on which the surface forces cancel each other), the vertical surfaces at the exit exposed to air and water, respectively, and the decal surface adjacent to the bottom of the channel where we have shown a reaction shear force R_{shear} acting to the left due to friction. We must also include the external horizontal force F_E. Since the gate is completely inside and thus part of the mixed CV, this force is transmitted to the gate across some section of the control surface. The horizontal momentum balance becomes

$$(\rho V_2^2 A_2 - \rho V_1^2 A_1) = \int_{\text{inlet}} p_1(z)\,dS + \int_{\text{air-exit}} -p_A\,dS + \int_{\text{exit}} -p_2(z)\,dS - R_{\text{shear}} - F_E$$

We are told that pressure distribution on the inlet and exit surface is hydrostatic. When we sketch the pressure distributions on the inlet, exit and air–exit surface as shown in Figure 7.20D, we see that the sum of the three integrals containing these pressures yields

$$\int_{\text{inlet}} p_1(z)\,dS + \int_{\text{air-exit}} -p_A\,dS + \int_{\text{exit}} -p_2(z)\,dS$$

$$= \left[p_A + \frac{\rho g D_1}{2}\right]D_1 L - p_A(D_1 - D_2)L - \left[p_A + \frac{\rho g D_2}{2}\right]D_2 L$$

$$-\tfrac{1}{2}\rho g D_1^2 L - \tfrac{1}{2}\rho g D_2^2 L$$

Thus the horizontal component of the external force needed to hold the gate open is

$$F_E = \left(\tfrac{1}{2}\rho g D_1^2 L - \tfrac{1}{2}\rho g D_2^2 L\right) - \left(\rho V_2^2 A_2 - \rho V_1^2 A_1\right) - R_{\text{shear}}$$

In most cases the reaction force due to shear on the bottom is neglected because it is unknown and presumed to be small. Note that it is also possible to solve this problem with the CV containing fluid only shown in Figure 7.20E.

EXAMPLE 7.12

Water is pumped steadily through a 90° reducing elbow welded onto the end of a pipe and exits to the atmosphere as shown in Figure 7.21A. What is the force applied to the elbow at the weld joint? Assume uniform conditions at the elbow inlet and exit and neglect gravity.

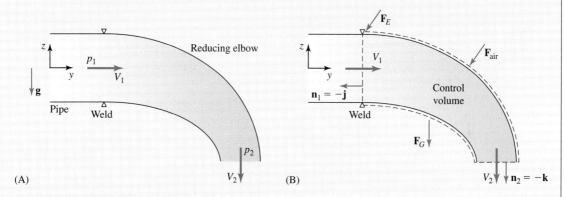

Figure 7.21 Schematic for Example 7.12: (A) geometry and (B) CV.

SOLUTION

We are asked for the structural force applied to the elbow by the pipe to which it is welded. This force holds the elbow in place. From experience we know that this external force must balance the forces of the water inside and air outside acting on the elbow, as well as the weight of the elbow (Figure 7.21A). However, we are told to neglect gravity in this problem. Although we would normally write a force balance on the elbow at this point, we will instead go directly to a momentum balance on the mixed CV shown in Figure 7.21B. Our CV selection is guided by the knowledge that a mixed control volume works best for an external force. We have placed a decal surface through the weld where the external force acts, and sections of control surface at the inlet and the exit of the elbow. The last section of control surface is a decal surface placed on the outside of the elbow where the force applied by the air occurs. A mass balance shows that $\dot{M} = \rho A_1 V_1 = \rho A_2 V_2$.

Since we anticipate momentum transport terms in both the y and z directions and the flow is steady, we will use the vector form of the steady flow momentum balance as given by Eq. 7.19a:

$$\int_{CS} (\rho \mathbf{u})(\mathbf{u} \cdot \mathbf{n}) \, dS = \mathbf{F}_B + \mathbf{F}_S$$

and neglect the body force. We will evaluate the momentum transport at the inlet and exit, surface forces due to the pressure at these locations, the force applied by the air to the decal surface on the outside of the elbow, and the external force $\mathbf{F}_E$ applied by the pipe at the decal surface at the weld. On the inlet we have $\mathbf{u} = V_1 \mathbf{j}$, $\mathbf{n} = -\mathbf{j}$, and $\mathbf{u} \cdot \mathbf{n} = -V_1$. By inspection, the momentum transport vector there is $-\rho V_1^2 A_1 \mathbf{j}$. On the exit, $\mathbf{u} = -V_2\mathbf{k}$, $\mathbf{n} = -\mathbf{k}$, and $\mathbf{u} \cdot \mathbf{n} = V_2$, and the momentum transport is $-\rho V_2^2 A_2 \mathbf{k}$. The net momentum transport is $(-\rho V_1^2 A_1 \mathbf{j} - \rho V_2^2 A_2 \mathbf{k})$.

The surface force on the inlet and exit is due to the pressure at each location. These terms are $(-p_1(-\mathbf{j})A_1) + (-p_2(-\mathbf{k})A_2) = (p_1 A_1 \mathbf{j} + p_2 A_2 \mathbf{k})$. The force of the air on the outside of the elbow can be included by using gage pressure in these terms. This assumes the wall of the elbow is thin, which is not always the case. If the wall is not thin, we can calculate the force of the air by using the trick as $\mathbf{F}_{\text{air}} = -p_A(A_1 + A_{1w})\mathbf{j} + p_A(A_2 + A_{2w})\mathbf{k}$, where we have included the wall areas in the terms. Upon gathering the various terms, our momentum balance for a thin-walled elbow becomes $(-\rho V_1^2 A_1 \mathbf{j} - \rho V_2^2 A_2 \mathbf{k}) = (p_{1_{\text{gage}}} A_1 \mathbf{j} + p_{2_{\text{gage}}} A_2 \mathbf{k}) + \mathbf{F}_E$. Solving for the external force, we have $\mathbf{F}_E = -(p_{1_{\text{gage}}} + \rho V_1^2)A_1 \mathbf{j} - (p_{2_{\text{gage}}} + \rho V_2^2)A_2 \mathbf{k}$. The exit of the elbow is at atmospheric pressure so $p_{2_{\text{gage}}} = 0$, and we have

$$\mathbf{F}_E = -\left(p_{1_{\text{gage}}} + \rho V_1^2\right) A_1 \mathbf{j} - \rho V_2^2 A_2 \mathbf{k}$$

This is the answer. It predicts that the weld must exert a force on the elbow to the left and down. To see if this is sensible, we can imagine that the exit of the elbow is plugged, so that the elbow is subjected to hydrostatic pressure on the inside. This would tend to force the elbow to the right, which requires an external force to the left. With the elbow open the pressure distribution inside will still force the elbow to the right, but also force the elbow up. Think of the elbow exit acting as a jet to produce thrust upward.

7.6.3 Energy Balance

An integral equation expressing conservation of energy may be developed by applying the laws of thermodynamics to a system. From thermodynamics we know that the time rate of change of the internal plus kinetic energy in a system is equal to the rate at which work is done on the system by body and surface forces, plus the rate at which energy is added to the system across its boundary. There are a variety of ways in which energy may be added to a system. Heat conduction occurs in many engineering applications. Chemical reaction and other types of energy input (e.g., radiation and Joule heating), are less frequently encountered, but important in specialized applications. We express conservation of energy for a system by writing

$$\frac{d\Phi_{\text{sys}}}{dt} = \dot{W}_B + \dot{W}_S + \dot{Q}_C + \dot{S}$$

where Φ_{sys} is the sum of the internal plus kinetic energy in the system, $\dot{W}_B$ and $\dot{W}_S$ are the rates at which work is done on the system by body and surface forces, $\dot{Q}_C$ is the rate at which energy is added to the system by heat conduction, and $\dot{S}$ is the net rate of any additional types of energy input. Now consider a fixed CV that coincides with the system. From the Reynolds transport theorem for a control volume, Eq. 7.5, we can write

$$\frac{d\Phi_{\text{sys}}}{dt} = \int_{\text{CV}} \frac{\partial}{\partial t}(\rho \varepsilon) \, dV + \int_{\text{CS}} (\rho \varepsilon)(\mathbf{u} \cdot \mathbf{n}) \, dS$$

where E_{sys} has been replaced by Φ_{sys}, the sum of the internal plus kinetic energy in the system. On a unit volume basis, the internal plus kinetic energy is given by $\rho\varepsilon = \rho(u + \frac{1}{2}\mathbf{u}\cdot\mathbf{u})$. Note the use of similar symbols representing different quantities: the boldface $\mathbf{u}$ is the velocity vector while u is the internal energy per unit mass. The transport theorem thus gives us a second expression for the time rate of change in total energy in the system:

$$\frac{d\Phi_{sys}}{dt} = \int_{CV}\frac{\partial}{\partial t}\left[\rho\left(u + \tfrac{1}{2}\mathbf{u}\cdot\mathbf{u}\right)\right]dV + \int_{CS}\rho\left(u + \tfrac{1}{2}\mathbf{u}\cdot\mathbf{u}\right)(\mathbf{u}\cdot\mathbf{n})\,dS$$

Since the CV and system coincide, the work and energy inputs to the system are identical to the work and energy inputs to the CV. Thus we may combine the preceding expression with that written for the system to obtain

$$\int_{CV}\frac{\partial}{\partial t}\left[\rho\left(u + \tfrac{1}{2}\mathbf{u}\cdot\mathbf{u}\right)\right]dV + \int_{CS}\rho\left(u + \tfrac{1}{2}\mathbf{u}\cdot\mathbf{u}\right)(\mathbf{u}\cdot\mathbf{n})\,dS = \dot{W}_B + \dot{W}_S + \dot{Q}_C + \dot{S} \tag{7.22}$$

This is the integral energy conservation equation, or energy balance, for a fixed CV. Because the equation accounts for the various work and energy flows into and out of a CV, it is useful for analyzing the performance of all types of fluid-handling devices including pumps, blowers, turbines, and compressors. It is valid for all fluids, under all conditions.

The appearance of a thermodynamic state variable such as the internal energy in the energy balance should remind us that the value of a state variable is defined to be that which would be seen by an observer moving with the fluid. Thus, if $u(\mathbf{x}, t)$ is the internal energy in a fluid flow at location $\mathbf{x}$ at time t, then this is the value of internal energy that would be measured by an observer moving with the fluid particle that happens to be at $\mathbf{x}$ at time t. Later we will show that it is possible to write the energy balance in terms of other thermodynamic variables such as enthalpy, entropy, or temperature, as required by the demands of a given problem. We do this by using the appropriate thermodynamic relationships between state variables. The use of these other thermodynamic variables is illustrated in several example problems.

To provide a physical interpretation of the energy balance we rewrite it in symbolic form as

$$\frac{d\Phi_{CV}}{dt} = -\Gamma_C + \dot{W}_B + \dot{W}_S + \dot{Q}_C + \dot{S}$$

where the rate of accumulation of internal plus kinetic energy within the control volume is given by

$$\frac{d\Phi_{CV}}{dt} = \int_{CV}\frac{\partial}{\partial t}\left[\rho\left(u + \tfrac{1}{2}\mathbf{u}\cdot\mathbf{u}\right)\right]dV$$

the net convective transport of this energy into the control volume is

$$-\Gamma_C = -\int_{CV}\rho\left(u + \tfrac{1}{2}\mathbf{u}\cdot\mathbf{u}\right)(\mathbf{u}\cdot\mathbf{n})\,dS$$

and the four remaining terms account for the work and energy inputs already described.

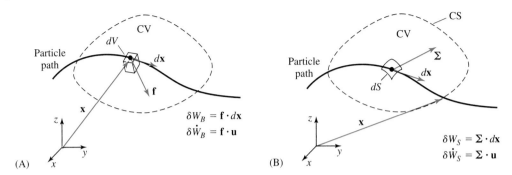

Figure 7.22 (A) Body force acting on a fluid particle within a CV. (B) Surface force acting on a fluid particle at the surface of a CV.

Before we apply the energy balance in an example, it is helpful to have a clear understanding of the work and energy inputs. Consider the rate at which work is done by body forces. As illustrated in Figure 7.22A, the work done by the body force on a moving fluid particle located at a point inside the CV is given by the dot product of the body force vector acting on the particle with a vector element of length in the direction of the particle's path. To calculate the rate at which work is done on this particle, we use the dot product of the body force vector and the velocity vector of the particle. In the Eulerian description, this dot product is given by $\rho \mathbf{f} \cdot \mathbf{u}$. Summing the contributions from all parts of the CV, we find

$$\dot{W}_B = \int_{CV} \rho(\mathbf{f} \cdot \mathbf{u}) dV \tag{7.23}$$

which serves to remind us that the net rate at which work is done by body forces depends on the nature of the body force and the movement of the fluid. For work to be done on the fluid inside the CV, there must be a body force acting in the direction of the fluid velocity field.

Next consider the work done by surface forces. As illustrated in Figure 7.22B, the work done on a moving fluid particle instantaneously located at a point on the control surface is given by the dot product of the stress vector, $\boldsymbol{\Sigma}$, acting on the particle with a vector element of length $d\mathbf{x}$ in the direction of the particle's path. The work done on the particle by the stress vector is given by $\boldsymbol{\Sigma} \cdot d\mathbf{x}$, and the rate at which work is done on this particle is given by the dot product of the stress vector and the particle's velocity, i.e., $\boldsymbol{\Sigma} \cdot \mathbf{u}$. Summing the contributions from all parts of the control surface, we obtain

$$\dot{W}_S = \int_{CS} (\boldsymbol{\Sigma} \cdot \mathbf{u}) dS \tag{7.24}$$

We see that the net rate at which work is done by surface forces depends on the stress on the control surface, and on the velocity of the fluid. There must be a stress acting on the control surface in the direction of the fluid velocity vector for work to be done on the fluid.

The heat conduction term $\dot{Q}_C$ may also be written in terms of an integral over the control surface. The rate at which heat crosses the control surface by conduction may be represented as a diffusive transport of heat. Using Eq. 6.38, and adjusting the sign to get the rate at which heat enters the CV, we have

$$\dot{Q}_C = -\int_{CS} k(\nabla T \cdot \mathbf{n})\,dS \qquad (7.25)$$

where k is the thermal conductivity of the fluid. We see that an energy input across the control surface occurs only in the presence of a temperature gradient normal to the control surface.

Finally, consider $\dot{S}$, which represents the net rate of any additional types of energy input. Many types of energy input such as those created by combustion, chemical reaction, or Joule heating are modeled using a volume integral of the form

$$\dot{S} = \int_{CV} \dot{s}(\mathbf{x}, t)\,dV \qquad (7.26)$$

where $\dot{s}(\mathbf{x}, t)$ represents the energy release per unit volume within the fluid as a function of space and time.

If we replace the work and energy terms in Eq. 7.22 with their integral representations, the energy balance becomes

$$\int_{CV} \frac{\partial}{\partial t}\left[\rho\left(u + \tfrac{1}{2}\mathbf{u}\cdot\mathbf{u}\right)\right]dV + \int_{CS} \rho\left(u + \tfrac{1}{2}\mathbf{u}\cdot\mathbf{u}\right)(\mathbf{u}\cdot\mathbf{n})\,dS$$
$$= \int_{CV} \rho(\mathbf{f}\cdot\mathbf{u})\,dV + \int_{CS}(\boldsymbol{\Sigma}\cdot\mathbf{u})\,dS - \int_{CS} k(\nabla T\cdot\mathbf{n})\,dS + \int_{CV} \dot{s}(\mathbf{x}, t)\,dV \qquad (7.27)$$

Although detailed information is needed to calculate the various work and energy integrals, in the next section we will show that it is usually possible to avoid the calculation of these integrals altogether.

Gravitational Potential Energy: Consider the integral $\dot{W}_B = \int_{CV} \rho(\mathbf{f}\cdot\mathbf{u})\,dV$ representing the rate at which work is done on the fluid inside the CV by body forces. Gravity is a steady conservative body force and by far the most important body force in engineering applications. We can account for the rate at which work is done by gravity by employing an appropriate potential energy term that is included along with the internal and kinetic energy in the accumulation and convective transport terms in the energy equation. That is, we replace the integrand $\rho(u + \tfrac{1}{2}\mathbf{u}\cdot\mathbf{u})$ with $\rho(u + \tfrac{1}{2}\mathbf{u}\cdot\mathbf{u} + gz)$ in the accumulation and convective transport terms in the energy balance, and drop the original term $\dot{W}_B = \int_{CV} \rho(\mathbf{f}\cdot\mathbf{u})\,dV$ from the right-hand side of the energy balance.

Total Heat Transfer Rate: The total heat transfer rate into a CV is illustrated in Figure 7.23. In most engineering problems we do not know the exact temperature distribution over the control surface, so we cannot use the surface integral $\dot{Q}_C = -\int_{CS} k(\nabla T \cdot \mathbf{n})\,dS$ to calculate the total heat transfer rate. However, we can often obtain information about the value of the total heat transfer rate by indirect physical measurements. For example, in a water-cooled compressor or combustion engine, we can calculate the total heat transfer rate by measuring both the coolant flowrate into the compressor or engine and the temperature increase sustained by the coolant in passing through the cooling passages. Thus, if coolant enters the engine at temperature T_1 and

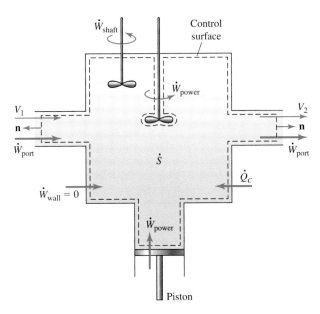

Figure 7.23 Generic CV, illustrating the various energy transfer terms.

leaves at T_2, we can write $\dot{Q}_C = -\dot{M}c(T_2 - T_1)$, where $\dot{M}$ is the coolant mass flowrate and c is the specific heat of the coolant. The negative sign indicates that cooling has occurred, since our convention is that a positive value of $\dot{Q}_C$ is associated with a heat input. In some problems the value of $\dot{Q}_C$ is provided on a unit mass of flowing fluid basis, or is itself the parameter of interest, in which case we evaluate the other terms in the energy balance, and $\dot{Q}_C$ is the unknown.

Total Rate of Energy Addition: The term $\dot{S} = \int_{CV} \dot{s}(\mathbf{x}, t) \, dV$ represents the rate at which energy is added to the fluid in the CV from a variety of sources. We usually do not know the energy release rate $\dot{s}(\mathbf{x}, t)$, but information about the total energy input is often available. For example, gasoline has a heating value of approximately 21,000 Btu/lb$_m$, so in a combustion process involving the burning of gasoline we can write $\dot{S} = \dot{m}(21{,}000 \text{ Btu/lb}_m)$ where $\dot{m}$ is the mass flowrate of the gasoline being burned inside the CV. Thus the associated integral is rarely evaluated, and we simply employ $\dot{S}$ in the energy balance as shown in Figure 7.23.

Work Done by Surface Forces: To further simplify the energy balance, consider the term $\dot{W}_S = \int_{CS} (\mathbf{\Sigma} \cdot \mathbf{u}) \, dS$, which represents the rate at which work is done by surface forces on the fluid inside the CV. The evaluation of this integral depends on the type of control surface involved, as well as whether we have selected a CV containing fluid only or a mixed CV. Here we will consider the four common types of control surface that occur with both fluid only and mixed CVs. These four types of control surface, illustrated in Figure 7.23, will be referred to as stationary wall, port, fluid power surface, and shaft surface. To allow for any combination of these types of control surface in a

7.6 CONTROL VOLUME ANALYSIS

problem, we can consider the surface work term to be the sum of contributions from each type of control surface and write

$$\dot{W}_S = \dot{W}_{\text{wall}} + \dot{W}_{\text{port}} + \dot{W}_{\text{power}} + \dot{W}_{\text{shaft}} \qquad (7.28)$$

Incorporating these results, the energy balance now becomes

$$\int_{\text{CV}} \frac{\partial}{\partial t}\left[\rho\left(u + \tfrac{1}{2}\mathbf{u}\cdot\mathbf{u} + gz\right)\right] dV + \int_{\text{CS}} \rho\left(u + \tfrac{1}{2}\mathbf{u}\cdot\mathbf{u} + gz\right)(\mathbf{u}\cdot\mathbf{n})\, dS \qquad (7.29)$$
$$= \dot{W}_{\text{wall}} + \dot{W}_{\text{port}} + \dot{W}_{\text{power}} + \dot{W}_{\text{shaft}} + \dot{Q}_C + \dot{S}$$

The evaluation of the rate at which work is done by surface forces at each of these types of control surface is explained in the subsections that follow.

Stationary Wall ($\dot{W}_{\text{wall}}$): Consider the decal surfaces shown in Figure 7.23, which are adjacent to the stationary solid walls. The value of the term $\dot{W}_{\text{wall}} = \int_{\text{wall}} (\mathbf{\Sigma}\cdot\mathbf{u})\, dS$ on this type of control surface depends on the stress acting on the wall and the velocity at the wall. Because of the no-slip, no-penetration conditions, the fluid velocity is zero on all stationary surfaces. Thus the rate at which work is done on the fluid is identically zero irrespective of the value of the stress, i.e, $\dot{W}_{\text{wall}} = 0$, so we can ignore the surface work term on stationary solid walls.

Port ($\dot{W}_{\text{port}}$): A port is defined to be any portion of a control surface through which fluid flows. Consider the generic CV and its CS as shown in Figure 7.23. The control surface includes inlet and exit ports. To simplify the term $\dot{W}_{\text{port}} = \int_{\text{port}} (\mathbf{\Sigma}\cdot\mathbf{u})\, dS$ at an inlet or exit port, we must first ensure that the port is properly defined. Recall that at a properly defined port, the fluid velocity vector and unit normal to the control surface are parallel, as shown in Figure 7.23. Since the work rate is defined by the dot product $\mathbf{\Sigma}\cdot\mathbf{u}$, the only component of the stress vector that can do work at a port is the normal stress. A tangential stress may exist at a properly defined port, but it does no work on the fluid because the tangential stress is at a right angle to the velocity vector. As discussed earlier, in most engineering applications the normal stress may be assumed to be equal to the negative of the fluid pressure. Thus at an inlet or exit port we write $\mathbf{\Sigma} = -p\mathbf{n}$ and the dot product with the velocity vector is then $\mathbf{\Sigma}\cdot\mathbf{u} = -p\mathbf{n}\cdot\mathbf{u} = -p(\mathbf{u}\cdot\mathbf{n})$. The rate at which work is done by surface forces at the inlet and exit ports of a fixed CV is therefore given by

$$\int_{\text{port}} (\mathbf{\Sigma}\cdot\mathbf{u})\, dS = -\int_{\text{port}} p(\mathbf{u}\cdot\mathbf{n})\, dS$$

which we can also write as $\int_{\text{port}} (\mathbf{\Sigma}\cdot\mathbf{u})\, dS = -\int_{\text{port}} (p/\rho)\rho(\mathbf{u}\cdot\mathbf{n})\, dS$. This is now seen to be a convective transport integral involving the quantity p/ρ, and we will include this contribution in the convective transport term on the left-hand side of the energy balance. Thus the effects of the surface work term at inlet and exit ports are accounted for by writing the convective transport term in the energy balance as

$$\int_{\text{CS}} \rho\left(u + \frac{p}{\rho} + \frac{1}{2}\mathbf{u}\cdot\mathbf{u} + gz\right)(\mathbf{u}\cdot\mathbf{n})\, dS$$

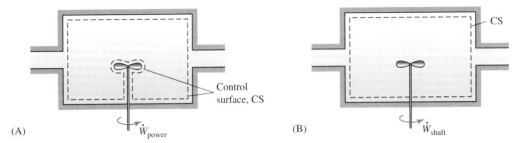

Figure 7.24 (A) Fluid-only CV, where the control surface follows the moving surface of the propeller. (B) Mixed CV, where the control surface cuts the propeller shaft.

We see that the term p/ρ acts like a potential energy. This is the pressure potential energy per unit mass discussed briefly in Section 2.9.3. Notice that this term arises from the rate at which the fluid pressure does work at the inlet and exit ports of the CV. The work term at a port of a CV is therefore normally accounted for by calculating the transport of pressure potential energy at each inlet or exit port.

Fluid Power ($\dot{W}_{\mathbf{power}}$): Work is done on fluid in compressors, pumps, turbines, and other similar devices by moving solid surfaces. The moving surfaces can take the form of blades, vanes, or pistons. If a control surface is placed adjacent to a moving solid surface in contact with fluid, fluid power transmission occurs on this surface and must be accounted for. For such surfaces as a piston in a cylinder, this calculation is possible because the fluid velocity on the surface is equal to the piston velocity, as discussed for the rod in Example 7.3. On the other hand, consider a CV selection that employs the decal surface adjacent to a moving solid surface of the propeller shown in Figure 7.24A. Power is transferred across this decal surface from the propeller to the fluid and this transfer must be accounted for in an energy balance. Since the velocity is nonzero at a moving surface, work is done on the fluid by both normal and shear stresses at a power transmission surface. Although this type of surface is frequently encountered, we rarely have the necessary information about the velocity and stress at each point on the surface. However, recognizing that a moving surface is providing a power input $\dot{W}_{\text{power}}$ to the fluid through the interaction of the force of the moving surface on the fluid, we can write

$$\dot{W}_{\text{power}} = \int_{\text{power}} (\mathbf{\Sigma} \cdot \mathbf{u}) \, dS \tag{7.30}$$

to account for the surface work term on this type of control surface. In some cases the power input (or output) is the quantity to be found from the analysis, in others we are given the power input or output can measure these externally. In both cases the effect is accounted for by using $\dot{W}_{\text{power}}$ to replace the corresponding integral.

Shaft Power ($\dot{W}_{\mathbf{shaft}}$): Finally consider the effect of a shaft that is transmitting power to moving surfaces inside the CV. For example, consider the mixed CV shown in Figure 7.24B, and notice that this is the same physical arrangement discussed in the last section, but with a different CV. The propeller is now inside the CV, so its surface is not part of the control surface. Thus, there is no fluid power transmission term to consider.

While energy integrals are rarely calculated by hand, they are often determined numerically from CFD solutions.

However, notice that the control surface now cuts across the shaft. In this case the power input is given by

$$\dot{W}_{\text{shaft}} = \int_{\text{shaft}} (\mathbf{\Sigma} \cdot \mathbf{u}) \, dS \tag{7.31}$$

where the stress vector now acts on the shaft material. It can be seen that this result and the preceding one, Eq. 7.30, are complementary, for the power transmitted by the shaft must be equal to the power delivered to the fluid inside the CV by the moving surfaces connected to the shaft. The occurrence of a shaft or fluid power input in the energy balance depends on how the CV is defined.

Since we do not usually know the stress distribution in a shaft, we do not evaluate the integral in Eq. 7.31. Instead we take advantage of the fact that the power transmitted by a shaft is the product of the torque and angular velocity. This allows us to write

$$\dot{W}_{\text{shaft}} = T_{\text{shaft}} \, \omega \tag{7.32}$$

where T_{shaft} is the magnitude of the torque transmitted by the shaft and ω is the shaft angular velocity. Torque and angular velocity are easily measured, thus in energy balance problems involving an external power input, it is convenient to use a mixed CV with the fluid power transmission surfaces inside the CV. In such cases, the term $\dot{W}_{\text{power}}$ is dropped, and we use Eq. 7.32 to calculate $\dot{W}_{\text{shaft}}$.

Upon incorporating all the standard simplifications described thus far, the energy balance becomes

$$\int_{CV} \frac{\partial}{\partial t}\left[\rho\left(u + \frac{1}{2}\mathbf{u}\cdot\mathbf{u} + gz\right)\right] dV + \int_{CS} \rho\left(u + \frac{p}{\rho} + \frac{1}{2}\mathbf{u}\cdot\mathbf{u} + gz\right)(\mathbf{u}\cdot\mathbf{n}) \, dS$$
$$= \dot{W}_{\text{power}} + \dot{W}_{\text{shaft}} + \dot{Q}_C + \dot{S} \tag{7.33}$$

Equation 7.33 is the starting point for an energy balance analysis of any type of unsteady process. Recalling the discussion of the various forms of fluid energy in Section 2.9, we see that the left-hand side of this equation contains a term describing the accumulation of internal, kinetic, and gravitational potential energy, and a flux term that describes the transport of the total energy $e = u + p/\rho + \frac{1}{2}\mathbf{u}\cdot\mathbf{u} + gz$ per unit mass. The total energy per unit mass consists of internal energy u, pressure potential energy p/ρ, kinetic energy $\frac{1}{2}\mathbf{u}\cdot\mathbf{u}$, and gravitational potential energy gz. The use of the energy balance to analyze the charging of a tank from a high pressure air line is shown in Example 7.13.

EXAMPLE 7.13

An insulated air tank, initially at a pressure of 0 MPa (gage) and 20°C, is connected by a valve to an air supply that provides air at a constant pressure of 0.5 MPa (gage) and the same temperature. When the valve is opened, the initial mass flowrate into the tank is found to be $\dot{M}_0 = 0.05$ g/s. If the volume of the tank is $V = 0.05 \, \text{m}^3$, and the cross-sectional area of the supply line is 0.5 cm², determine the initial rate of change of density and temperature in the tank.

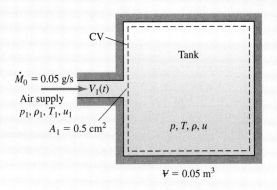

Figure 7.25 Schematic for Example 7.13.

SOLUTION

We are asked to determine the initial rate of change of density and temperature during the filling of an air tank. Figure 7.25 sketches the tank and connection to the air supply as well as an appropriate fixed control volume for use in the solution of this problem. We will assume that the air obeys the perfect gas law and that the values of fluid properties on the section of the control surface defining the port are uniform. We will also assume that the temperature, density, and other fluid properties inside the tank are spatially uniform, that air velocities are small, and that the effects of gravity are negligible.

To find $\partial\rho/\partial t$, the rate of change of density in the tank, recall that the mass balance Eq. 7.11, $\int_{CV}(\partial\rho/\partial t)\,dV + \int_{CS}\rho(\mathbf{u}\cdot\mathbf{n})\,dS = 0$, contains this term. Using our assumptions of a spatially uniform density inside the tank and uniform properties at the port, and applying this equation to the selected control volume, we find $(\partial\rho/\partial t)V\!\!\!\!/ - \rho_1 V_1(t)A_1 = 0$. Upon rearrangement we find

$$\frac{\partial\rho}{\partial t} = \frac{\rho_1 A_1 V_1(t)}{V\!\!\!\!/} = \frac{\dot{M}(t)}{V\!\!\!\!/} \tag{A}$$

which agrees with our mass balance analysis in Example 7.2. We see that the rate of change of density in the tank at any time is given by the instantaneous mass flowrate into the tank divided by the volume of the tank. A check shows that (A) has the expected dimensions of density per unit time. Since the initial mass flowrate is known to be $\dot{M}_0 = 0.05$ g/s, the initial rate of change of density inside the tank is calculated as

$$\left[\frac{\partial\rho}{\partial t}\right]_0 = \frac{\dot{M}_0}{V\!\!\!\!/} = \frac{5\times 10^{-5}\text{ kg/s}}{5\times 10^{-2}\text{ m}^3} = 1\times 10^{-3}\text{ kg/(m}^3\text{-s)}$$

To find an expression for the rate of change of temperature in the tank, we apply an energy balance to the CV, recalling that for a perfect gas internal energy and temperature are related. According to Eq. 7.33 we have

$$\int_{CV} \frac{\partial}{\partial t}\left[\rho\left(u + \frac{1}{2}\mathbf{u}\cdot\mathbf{u} + gz\right)\right]dV + \int_{CS}\rho\left(u + \frac{p}{\rho} + \frac{1}{2}\mathbf{u}\cdot\mathbf{u} + gz\right)(\mathbf{u}\cdot\mathbf{n})\,dS$$
$$= \dot{W}_{\text{power}} + \dot{W}_{\text{shaft}} + \dot{Q}_C + \dot{S}$$

Since there is no fluid power, shaft power, heat, or other energy input to this CV, the right-hand side of this equation is zero. In addition, we will assume velocities are low enough to permit us to neglect the kinetic energy terms. Since we have also decided to neglect the effects of gravity the energy balance reduces to

$$\int_{CV} \frac{\partial}{\partial t}(\rho u)\,dV + \int_{CS}\rho\left(u + \frac{p}{\rho}\right)(\mathbf{u}\cdot\mathbf{n})\,dS = 0$$

With the assumption of uniform properties inside the tank and uniform properties on the control surface defining the port which are the same as in the air supply, we evaluate the integrals to find:

$$\frac{\partial(\rho u)}{\partial t}\mathcal{V} - \rho_1 V_1(t) A_1\left(u_1 + \frac{p_1}{\rho_1}\right) = 0$$

Expanding the time derivative, the above yields $[u(\partial\rho/\partial t) + \rho(\partial u/\partial t)]\mathcal{V} - \rho_1 V_1(t) A_1 (u_1 + p_1/\rho_1) = 0$. Now, $\dot{M}(t) = \rho_1 V_1(t) A_1$, and from (A) we have $\partial\rho/\partial t = \dot{M}(t)/\mathcal{V}$. Thus the preceding expression becomes

$$\dot{M}(t)(u - u_1) + \rho\left(\frac{\partial u}{\partial t}\right)\mathcal{V} - \dot{M}(t)\left(\frac{p_1}{\rho_1}\right) = 0$$

For a perfect gas, the internal energy change in going from state 1 to state 2 is related to the temperature change by Eq. 2.21a as $u_2 - u_1 = c_V(T_2 - T_1)$. We can use this relationship to write $u - u_1 = c_V(T - T_1)$, and $\partial u/\partial t = c_V(\partial T/\partial t)$. We will also make use of the fact that $p = \rho RT$, and write the energy balance as $\dot{M}(t)c_V(T - T_1) + \rho c_V(\partial T/\partial t)\mathcal{V} - \dot{M}(t) RT_1 = 0$. Solving for the rate of change of temperature in the tank, we have

$$\frac{\partial T}{\partial t} = \frac{\dot{M}(t) RT_1}{\rho(t) c_V \mathcal{V}} - \frac{\dot{M}(t)(T(t) - T_1)}{\rho(t)\mathcal{V}} \quad (B)$$

This result is valid at any instant of time. To find the initial rate of change of temperature in the tank, we make use of the fact that at the initial instant the temperature inside the tank is the same as that in the supply line. Thus $T(0) = T_1$, the second term in (B) is zero, and $[\partial T/\partial t]_0 = \dot{M}_0 RT_1/\rho_0 c_V \mathcal{V}$. We are not given the density in the tank at the initial instant, but we can make use of the perfect gas law to write the density inside the tank as $\rho_0 = p_0/RT(0) = p_0/RT_1$, once again making use of the fact that $T(0) = T_1$. We now have

$$\left[\frac{\partial T}{\partial t}\right]_0 = \frac{\dot{M}_0 (RT_1)^2}{p_0 c_V \mathcal{V}} \quad (C)$$

Inserting the given data and values for R and c_V from Table 2.4 into (C) we find

$$\left[\frac{\partial T}{\partial t}\right]_0 = \frac{\dot{M}_0(RT_1)^2}{p_0 c_V \mathcal{V}} = \frac{(5 \times 10^{-5} \text{ kg/s})\{[287(\text{N-m})/(\text{kg-m})](293 \text{ K})\}^2}{(1.01 \times 10^5 \text{ N/m}^2)[717(\text{N-m})/(\text{kg-K})](5 \times 10^{-2} \text{m}^3)}$$
$$= 9.8 \times 10^{-2} \text{ K/s}$$

Note that in using the perfect gas law we are required to use the absolute pressure and temperature.

It is instructive to calculate the magnitude of the neglected kinetic energy flux into the tank at the initial instant. This flux is seen to be

$$\frac{1}{2}\dot{M}_0 V_1(0)^2 = \frac{1}{2}\dot{M}_0 \left(\frac{\dot{M}_0}{\rho_1 A_1}\right)^2 = \frac{1}{2}\dot{M}_0^3 \left(\frac{RT_1}{p_1 A_1}\right)^2$$
$$= \frac{1}{2}(5 \times 10^{-5} \text{ kg/s})^3 \left\{\frac{[287 \,(\text{N-m})/(\text{kg-K})](293 \text{ K})}{[(0.5 + 0.101) \times 10^6 \text{N/m}^2](5 \times 10^{-5} \text{ m}^2)}\right\}^2$$
$$= 4.9 \times 10^{-7} \,(\text{kg-m}^2)/\text{s}^3$$

The corresponding initial flux of pressure potential energy is

$$\dot{M}_0 \frac{p_1}{\rho_1} = \dot{M}_0 R T_1 = (5 \times 10^{-5} \text{ kg/s})[287(\text{N-m})/(\text{kg-K})](293 \text{ K}) = 4.2 \,(\text{kg-m}^2)/\text{s}^3$$

which can be seen to be many orders of magnitude larger. This is typical for gas flows and justifies our neglect of the kinetic energy term in this flow.

As was the case in Example 7.12, in many problems of interest certain terms in the energy balance are identically zero or may be neglected. For example, consider a steady process, i.e., one in which the velocity and fluid and flow properties are independent of time. The accumulation term in the energy balance, Eq. 7.33, is zero, and the total amount of internal, kinetic, and gravitational potential energy inside the control volume is fixed. The energy balance for a steady process is therefore given by

$$\int_{CS} \rho \left(u + \frac{p}{\rho} + \frac{1}{2}\mathbf{u} \cdot \mathbf{u} + gz\right) (\mathbf{u} \cdot \mathbf{n}) \, dS = \dot{W}_{\text{power}} + \dot{W}_{\text{shaft}} + \dot{Q}_C + \dot{S} \quad (7.34)$$

EXAMPLE 7.14

Use an energy balance to analyze the steady, fully developed flow of constant density fluid in a round pipe shown in Figure 7.26. Consider laminar and turbulent flow, and assume that the internal energy and pressure are uniform on any cross section of the pipe.

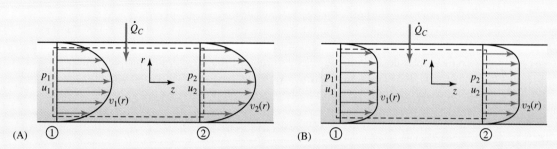

Figure 7.26 Schematic for Example 7.14: (A) laminar and (B) turbulent flow in a pipe.

SOLUTION

As discussed in Example 7.7, the laminar flow velocity profile in the pipe is parabolic (Figure 7.26A), while the turbulent flow velocity profile is nearly uniform (Figure 7.26B). We will analyze each case by assuming that the fully developed, axisymmetric velocity field is given in cylindrical coordinates by $v_r = 0$, $v_\theta = 0$, and $v_z(r)$, and inserting the appropriate function for laminar and turbulent flow. Consider a CV containing all the fluid in a section of pipe of length L. Applying a mass balance, we find that $\dot{M} = \rho_1 A_1 \bar{V}_1 = \rho_2 A_2 \bar{V}_2$. Since the density is constant and the areas are the same, this result shows that the average velocity at each port is the same. The steady process energy balance is given in general by Eq. 7.34 as

$$\int_{CS} \rho \left(u + \frac{p}{\rho} + \frac{1}{2} \mathbf{u} \cdot \mathbf{u} + gz \right) (\mathbf{u} \cdot \mathbf{n}) \, dS = \dot{W}_{\text{power}} + \dot{W}_{\text{shaft}} + \dot{Q}_C + \dot{S}$$

In this problem there is no fluid or shaft power input or energy addition; and because the pipe is horizontal, the effects of gravity cancel. Thus, the energy balance can be written as

$$\int_{CS} \rho \left(u + \frac{p}{\rho} \right) (\mathbf{u} \cdot \mathbf{n}) \, dS + \int_{CS} \rho \left(\frac{1}{2} \mathbf{u} \cdot \mathbf{u} \right) (\mathbf{u} \cdot \mathbf{n}) \, dS = \dot{Q}_C \quad \text{(A)}$$

Note how we have written the flux integral in two parts. Since we are told that the internal energy and pressure are uniform on the port surfaces in both the laminar and turbulent flow, we can evaluate the first integral to obtain

$$\int_{CS} \rho \left(u + \frac{p}{\rho} \right) (\mathbf{u} \cdot \mathbf{n}) \, dS = \dot{M} \left[\left(u_2 + \frac{p_2}{\rho} \right) - \left(u_1 + \frac{p_1}{\rho} \right) \right]$$

$$= \dot{M} \left[(u_2 - u_1) + \frac{p_2 - p_1}{\rho} \right] \quad \text{(B)}$$

The value of the second integral in (A) depends in general on the density and the form of the velocity profile at a port. For turbulent flow, the velocity profile may be assumed to be uniform, and the integral defining the kinetic energy flux at the inlet and exit port gives

$$\int_{CS} \rho \left(\tfrac{1}{2} \mathbf{u} \cdot \mathbf{u} \right) (\mathbf{u} \cdot \mathbf{n}) \, dS = \tfrac{1}{2} \rho_2 A_2 \bar{V}_2^3 - \tfrac{1}{2} \rho_1 A_1 \bar{V}_1^3 = 0$$

Here we have made use of the fact that when the density is constant and the areas are the same, the mass balance shows that the average velocities are also the same. For laminar flow the velocity profile is not uniform but is parabolic. However, since the flow in the pipe is fully developed, the velocity profile is the same at the inlet and exit ports. Furthermore, the density is constant, so we can conclude that for laminar flow the integral is also zero without bothering with the calculation. Thus we have in both cases

$$\int_{CS} \rho(\tfrac{1}{2}\mathbf{u}\cdot\mathbf{u})(\mathbf{u}\cdot\mathbf{n})\,dS = 0 \qquad (C)$$

Inserting (B) and (C) into the energy balance (A), we find $\dot{M}[(u_2 - u_1) + (p_2 - p_1)/\rho] = \dot{Q}_C$, which after rearrangement and dividing by the mass flowrate becomes

$$\frac{p_1 - p_2}{\rho} = (u_2 - u_1) - \frac{\dot{Q}_C}{\dot{M}} \qquad (D)$$

We can interpret this result by noting that the pressure drop down the pipe due to friction results in a loss of pressure potential energy per unit mass in the amount $(p_1 - p_2)/\rho$. The energy balance shows that this energy loss appears as a combination of an increase in the internal energy of the fluid per unit mass $(u_2 - u_1)$, and a heat transfer per unit mass out of the pipe $-\dot{Q}_C/\dot{M}$.

Recall that in the case study of Section 3.3.1 (flow in a round pipe), the pressure drop, $\Delta p = p_1 - p_2$, is given in terms of the friction factor by Eq. 3.15 as $\Delta p = \rho f(L/D)(\bar{V}^2/2)$. Upon using this to substitute for the pressure drop in the energy balance (D), we find

$$f\frac{L}{D}\frac{\bar{V}^2}{2} = (u_2 - u_1) - \frac{\dot{Q}_C}{\dot{M}} \qquad (E)$$

This result can be used to predict the increase in temperature of a liquid flowing in an insulated pipe, i.e., one for which $\dot{Q}_C = 0$, by making use of the fact that the temperature change is related to the change in internal energy by the specific heat: $u_2 - u_1 = c(T_2 - T_1)$. Then according to (E) we have $f(L/D)(\bar{V}^2/2) = c(T_2 - T_1)$. Similarly, the temperature change in a low speed gas flow in an insulated pipe could be estimated using the perfect gas relation Eq. 2.21a to obtain $u_2 - u_1 = c_V(T_2 - T_1)$, with the result $f(L/D)(\bar{V}^2/2) = c_V(T_2 - T_1)$.

Additional simplification of the energy balance is possible if the velocity and other fluid properties (including the gravitational potential energy under certain conditions) are spatially uniform at ports of the CV. This is normally the case in turbulent flow involving pumps, compressors, and comparable devices. In such cases we may write the convective transport at each port in the form of a mass flowrate (with appropriate sign) multiplied by the value of $(u + p/\rho + \tfrac{1}{2}\mathbf{u}\cdot\mathbf{u} + gz)$ at each port, where z is the elevation of the center of the port. That is, at each port of the control surface, the corresponding flux term may be written as

$$\int_{\text{port}} \rho \left(u + \frac{p}{\rho} + \frac{1}{2}\mathbf{u}\cdot\mathbf{u} + gz \right) (\mathbf{u}\cdot\mathbf{n})\, dS = \pm \dot{M}_{\text{port}} \left(u + \frac{p}{\rho} + \frac{1}{2}\bar{V}^2 + gz \right)_{\text{port}} \quad (7.35)$$

where the sign of the mass flowrate is determined by whether the port is an inlet (negative) or exit (positive). This approach is applicable to both the unsteady and steady process energy balances. For example, consider a CV with one inlet and one outlet, and a flow in which velocity and fluid properties are spatially uniform at each port. A mass balance shows that the mass flowrates in and out of the CV are the same, and the steady process energy balance under these conditions takes the following particularly simple form that you may have learned in a thermodynamics course:

$$\dot{M}\left[\left(u + \frac{p}{\rho} + \frac{1}{2}\bar{V}^2 + gz\right)_{\text{out}} - \left(u + \frac{p}{\rho} + \frac{1}{2}\bar{V}^2 + gz\right)_{\text{in}}\right] = \dot{W}_{\text{power}} + \dot{W}_{\text{shaft}} + \dot{Q}_C + \dot{S} \quad (7.36)$$

The use of this form of steady process energy balance to solve practical engineering problems is illustrated in the two examples that follow.

EXAMPLE 7.15

A water pump runs steadily at a flowrate of $Q = 80$ gal/min with a pressure rise of $\Delta p = 40$ psi. The inlet and outlet areas are the same. (1) Determine the minimum power input necessary to run the pump assuming an ideal pumping process. You may assume that this ideal process involves no heat transfer and no increase in the internal energy of the water passing through the pump. (2) If the pump actually requires 2 hp to run, but the heat transfer is negligible, find the temperature increase in the water passing through the pump, and calculate the pump efficiency. (3) What is the useful power input to the water passing through the pump?

SOLUTION

This problem is concerned with the power input to run a pump under ideal and actual conditions. Figure 7.27 shows an appropriate mixed CV containing the pump and the fluid within it and the cut across the shaft connecting the pump to the pump motor. We will assume turbulent flow at the inlet and outlet ports with uniform properties, and that the inlet and outlet elevations are the same. A mass balance shows that $\dot{M} = \rho_1 A_1 \bar{V}_1 = \rho_2 A_2 \bar{V}_2$. Since the density is constant and the areas are the same, the average velocity at each port is also the same, i.e., $\bar{V}_1 = \bar{V}_2$. For this steady constant density flow with uniform properties Eq. 7.36 gives the energy balance. We will drop the gravitational potential energy terms, and since there is no additional energy input, and no fluid power input (all moving parts of the pump in contact with the fluid are

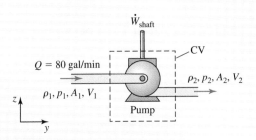

Figure 7.27 Schematic for Example 7.15.

inside the mixed control volume), the corresponding terms in Eq. 7.36 will also be dropped leaving us with

$$\dot{M}\left[\left(u_2 + \frac{p_2}{\rho} + \frac{1}{2}\bar{V}_2^2\right) - \left(u_1 + \frac{p_1}{\rho} + \frac{1}{2}\bar{V}_1^2\right)\right] = \dot{W}_{\text{shaft}} + \dot{Q}_C$$

Since we know $\bar{V}_1 = \bar{V}_2$, the applicable form of the energy balance is

$$\dot{M}\left[\left(u_2 + \frac{p_2}{\rho}\right) - \left(u_1 + \frac{p_1}{\rho}\right)\right] = \dot{W}_{\text{shaft}} + \dot{Q}_C \qquad \text{(A)}$$

We will now apply this equation to analyze the two relevant cases.

1. *Minimum power input:* We are told the minimum power input corresponds to an ideal pumping process in which heat transfer and internal energy changes are absent. Thus the energy balance (A) becomes

$$\dot{M}\left[\left(\frac{p_2}{\rho}\right) - \left(\frac{p_1}{\rho}\right)\right] = \dot{W}_{\text{shaft}}$$

Since the shaft work in this case is the ideal shaft work required to run the pump we will write this as $\dot{M}[(p_2/\rho) - (p_1/\rho)] = \dot{W}_{\text{ideal}}$. Solving for the ideal shaft work, and noting that $\dot{M} = \rho Q$, where Q is the volume flowrate, we find

$$\dot{W}_{\text{ideal}} = Q(p_2 - p_1) \qquad \text{(B)}$$

Equation (B) is often used to provide an estimate of the power required to pump a constant density fluid. It corresponds to a perfectly efficient pump. Substituting the given data into (B) we obtain

$$\dot{W}_{\text{ideal}} = 80 \text{ gal/min} \left(\frac{1 \text{ min}}{60 \text{ s}}\right)\left(\frac{1 \text{ ft}^3}{7.48 \text{ gal}}\right)(40 \text{ lb}_f/\text{in.}^2)\left(\frac{12 \text{ in.}}{1 \text{ ft}}\right)^2$$

$$= [1027(\text{ft-lb}_f)/\text{s}]\left[\frac{1 \text{ hp}}{550(\text{ft-lb}_f)/\text{s}}\right] = 1.87 \text{ hp}$$

Thus under ideal conditions we need $\dot{W}_{\text{ideal}} = 1.87$ hp to run the pump.

2. *Pump efficiency:* To consider an actual pumping process, we start with the energy balance (A), and use the volume flowrate to write

$$\dot{W}_{shaft} = \dot{M}(u_2 - u_1) + Q(p_2 - p_1) - \dot{Q}_C \qquad (C)$$

Since we defined $\dot{W}_{ideal} = Q(p_2 - p_1)$, the energy balance for this process can be written as

$$\dot{W}_{shaft} = \dot{W}_{ideal} + \dot{M}(u_2 - u_1) - \dot{Q}_C \qquad (D)$$

Equation (D) shows that the power required in the actual pumping process is increased over that in an ideal process because of the increase in the internal energy of the fluid caused by viscous effects (friction) and any heat transfer that leaves the CV. The efficiency of the pump can be defined as the ratio of the ideal to actual power requirement or

$$\eta = \frac{\dot{W}_{ideal}}{\dot{W}_{shaft}} \qquad (E)$$

For the pump in this example, the efficiency can be calculated from the data as $\eta = \dot{W}_{ideal}/\dot{W}_{shaft} = 1.87/2 = 0.935$, or approximately 94%.

We are told that the heat transfer is negligible in the operation of the pump. To calculate the temperature increase in the water passing through the pump, we drop the heat transfer term in (D) and solve for the internal energy change. Noting that for a liquid $u_2 - u_1 = c(T_2 - T_1)$, we find

$$T_2 - T_1 = \frac{\dot{W}_{shaft} - \dot{W}_{ideal}}{\dot{M}c} \qquad (F)$$

We can use the data to calculate the mass flowrate as

$$\dot{M} = \rho Q = 1.94 \text{ slugs/ft}^3 (80 \text{ gal/min}) \left(\frac{1 \text{ min}}{60 \text{ s}}\right) \left(\frac{1 \text{ ft}^3}{7.48 \text{ gal}}\right)$$
$$= 3.46 \times 10^{-1} \text{ slug/s}$$

We also know that

$$\dot{W}_{shaft} - \dot{W}_{ideal} = (2 - 1.87) \text{ hp} \left[\frac{550 \text{ (ft-lb}_f)/\text{s}}{1 \text{ hp}}\right] = 71.5 \text{ (ft-lb}_f)/\text{s}$$

Converting the specific heat for water from Table 2.3 to BG, we have

$$c = 4186 \text{ J/(kg-K)} \left(\frac{0.7376 \text{ ft-lb}_f}{1 \text{ J}}\right) \left(\frac{14.59 \text{ kg}}{1 \text{ slug}}\right) \left(\frac{1 \text{ K}}{1.8°\text{F}}\right)$$
$$= 2.5 \times 10^4 \text{ (ft-lbf)/(slugs-°F)}$$

which is the same as the standard value for water of $c = 1$ Btu/(lb$_m$-°F) in EE. The temperature increase is now calculated as

$$T_2 - T_1 = \frac{71.5 \text{ (ft-lb}_f/\text{s})}{(3.46 \times 10^{-1} \text{ slug/s})[25{,}027\text{(ft-lb}_f)/(\text{slugs -°F})]} = 8.3 \times 10^{-3} °\text{F}$$

We see that the temperature increase is imperceptible because of the large heat capacity of the water. The temperature increase in the fluid due to viscous dissipation of energy is responsible for the heat transfer out of the CV. Since the temperature increase is tiny, the heat transfer is negligible.

3. *Useful power input:* To determine the useful power input to the water passing through the pump, note that the total energy in the water per unit mass is given by $(u + p/\rho + \frac{1}{2}\bar{V}^2 + gz)$. For a constant density fluid we can consider the value of the mechanical energy per unit mass $(p/\rho + \frac{1}{2}\bar{V}^2 + gz)$ to represent useful or available energy, since this energy content can be extracted by devices to produce shaft work. In this case the water pressure has increased in traveling through the pump, and thus the pressure potential energy per unit mass has increased. There is no change in the kinetic or potential energy of the water, so the useful power input to the water is $\dot{M}[(p_2/\rho) - (p_1/\rho)] = Q(p_2 - p_1)$. We see that this is the same as the ideal shaft work calculated earlier or $\dot{W}_{\text{ideal}} = 1.87$ hp. This calculation shows that of the 2 hp delivered to the pump, 1.87 hp increases the useful energy of the water passing through the pump. Losses in the pumping process absorb 0.13 hp, and this portion of the total shaft work input causes the internal energy, and hence temperature of the water to rise. We cannot make use of this energy to produce useful work.

EXAMPLE 7.16

A small air compressor is running steadily at an inlet flowrate of $Q_1 = 4$ ft^3/min. The inlet conditions for the air are $p_1 = 14.7$ psia, $T_1 = 70°$F, $A_1 = 2$ in.^{2}; and the outlet conditions are $p_2 = 100$ psig $= 114.7$ psia, $T_2 = 200°$F, $A_2 = 0.05$ in.^{2}. If the power required to run the compressor is 2 hp, what is the heat transfer rate between the compressor and the atmosphere?

SOLUTION

We must determine the heat transfer rate for a compressor operating under the given conditions. Figure 7.28 shows a fixed CV containing the compressor and the air within it. We will assume that all property distributions at the inlet and outlet ports are uniform, that potential energy changes are negligible, and that the air may be treated as a perfect gas. For this steady flow device, a mass balance gives $\dot{M} = \rho_1 A_1 \bar{V}_1 = \rho_2 A_2 \bar{V}_2$, or equivalently $\dot{M} = \rho_1 Q_1 = \rho_2 Q_2$. Upon applying Eq. 7.36, and noting that for this mixed CV we have a shaft power input and heat transfer but no fluid power or other energy input, we write $\dot{M}[(u_2 + p_2/\rho_2 + \frac{1}{2}\bar{V}_2^2) - (u_1 + p_1/\rho_1 + \frac{1}{2}\bar{V}_1^2)] = \dot{W}_{\text{shaft}} + \dot{Q}_C$. For a perfect gas we can introduce the enthalpy $h = u + (p/\rho)$ and use Eq. 2.21b to write $h_2 - h_1 = c_P(T_2 - T_1)$. Thus the energy balance becomes

$$\dot{W}_{\text{shaft}} = \dot{M}\left[c_P(T_2 - T_1) + \tfrac{1}{2}\left(\bar{V}_2^2 - \bar{V}_1^2\right)\right] - \dot{Q}_C \quad \text{(A)}$$

7.6 CONTROL VOLUME ANALYSIS

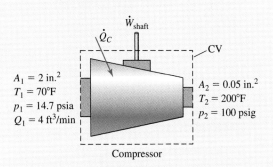

Figure 7.28 Schematic for Example 7.16.

We can also write this result on a per-unit-mass basis of air passing through the machine as

$$\frac{\dot{W}_{\text{shaft}}}{\dot{M}} = \left[c_P(T_2 - T_1) + \frac{1}{2}(\bar{V}_2^2 - \bar{V}_1^2)\right] - \frac{\dot{Q}_C}{\dot{M}} \quad \text{(B)}$$

Using thermodynamic properties for air from Table 2–4, and the given data, we now calculate

$$\rho_1 = \frac{p_1}{RT_1} = \frac{(14.7\ \text{lb}_f/\text{in.}^2)\left(\frac{12\ \text{in.}}{1\ \text{ft}}\right)^2}{[53.3\ (\text{ft-lb}_f)/(\text{lb}_f\text{-}°\text{R})][(70+460)°\text{R}]} = 7.49 \times 10^{-2}\ \text{lb}_m/\text{ft}^3$$

$$\rho_2 = \frac{p_2}{RT_2} = \frac{(114.7\ \text{lb}_f/\text{in.}^2)\left(\frac{12\ \text{in.}}{1\ \text{ft}}\right)^2}{[53.3(\text{ft-lb}_f)/(\text{lb}_m\text{-}°\text{R})][(200+460)°\text{R}]} = 0.470\ \text{lb}_m/\text{ft}^3$$

$$\dot{M} = \rho_1 Q_1 = (7.49 \times 10^{-2}\ \text{lb}_m/\text{ft}^3)\left(\frac{4\ \text{ft}^3}{\text{min}}\right)\left(\frac{1\ \text{min}}{60\ \text{s}}\right) = 5.0 \times 10^{-3}\ \text{lb}_m/\text{s}$$

$$\bar{V}_1 = \frac{\dot{M}}{\rho_1 A_1} = \frac{5.00 \times 10^{-3}\ \text{lb}_m/\text{s}}{(7.49 \times 10^{-2}\ \text{lb}_m/\text{ft}^3)(2\ \text{in.}^2)\left(\frac{1\ \text{ft}}{12\ \text{in.}}\right)^2} = 4.8\ \text{ft/s}$$

$$\bar{V}_2 = \frac{\dot{M}}{\rho_2 A_2} = \frac{5.00 \times 10^{-3}\ \text{lb}_m/\text{s}}{(0.470\ \text{lb}_m/\text{ft}^3)(0.05\ \text{in.}^2)\left(\frac{1\ \text{ft}}{12\ \text{in.}}\right)^2} = 30.6\ \text{ft/s}$$

The individual terms in (B) are now calculated as

$$\frac{\dot{W}_{\text{shaft}}}{\dot{M}} = \frac{(2\ \text{hp})\left[\frac{550(\text{ft-lb}_f)/\text{s}}{1\ \text{hp}}\right]\left(\frac{1\ \text{Btu}}{778\ \text{ft-lb}_f}\right)}{5.00 \times 10^{-3}\ \text{lb}_m/\text{s}} = 283\ \text{Btu/lb}_m$$

$$c_P(T_2 - T_1) = [0.241 \text{ Btu/(lb}_m\text{-}°\text{R)}][(660 - 530)(°\text{R})] = 31.3 \text{ Btu/lb}_m$$

$$\frac{1}{2}\bar{V}_1^2 = \frac{1}{2}(4.8 \text{ ft/s})^2 \left(\frac{1 \text{ lb}_f\text{-s}^2}{32.2 \text{ lb}_m\text{-ft}}\right)\left(\frac{1 \text{ Btu}}{778 \text{ ft-lb}_f}\right) = 4.6 \times 10^{-4} \text{ Btu/lb}_m$$

$$\frac{1}{2}\bar{V}_2^2 = \frac{1}{2}(30.6 \text{ ft/s})^2 \left(\frac{1 \text{ lb}_f\text{-s}^2}{32.2 \text{ lb}_m\text{-ft}}\right)\left(\frac{1 \text{ Btu}}{778 \text{ ft-lb}_f}\right) = 1.9 \times 10^{-2} \text{ Btu/lb}_m$$

$$\frac{1}{2}\left(\bar{V}_2^2 - \bar{V}_1^2\right) = (1.9 \times 10^{-2} \text{ Btu/lb}_m - 4.6 \times 10^{-4} \text{ Btu/lb}_m) = 1.85 \times 10^{-2} \text{ Btu/lb}_m$$

It is evident that the kinetic energy terms are very small compared with the enthalpy and shaft work terms. You will find that in most problems involving gas flows, the kinetic energy terms are negligible unless velocities are very high.

Using (A) to calculate the heat transfer rate, we have

$$\dot{Q}_C = \dot{M}\left[c_P(T_2 - T_1) + \tfrac{1}{2}\left(\bar{V}_2^2 - \bar{V}_1^2\right)\right] - \dot{W}_{\text{shaft}}$$

where

$$\dot{W}_{\text{shaft}} = (2 \text{ hp})\left[\frac{550 \text{ (ft-lb}_f\text{/s)}}{1 \text{ hp}}\right]\left(\frac{1 \text{ Btu}}{778 \text{ ft-lb}_f}\right) = 1.41 \text{ Btu/s}.$$

Thus we find

$$\dot{Q}_C = 5.00 \times 10^{-3} \text{ lb}_m\text{/s}\,[31.3 \text{ Btu/lb}_m + (1.85 \times 10^{-2}) \text{ Btu/lb}_m] - (1.41 \text{ Btu/s})$$
$$= -1.25 \text{ Btu/s}$$

Since $\dot{Q}_C$ is negative, heat is leaving the CV. It is worthwhile to consider a slightly different approach to this problem by rearranging the original energy balance to obtain

$$\dot{M}\left[(u_2 - u_1) + \left(\frac{p_2}{\rho_2} - \frac{p_1}{\rho_1}\right) + \frac{1}{2}\left(\bar{V}_2^2 - \bar{V}_1^2\right)\right] = \dot{W}_{\text{shaft}} + \dot{Q}_C$$

Introducing the perfect gas relationships, we have

$$\dot{W}_{\text{shaft}} = \dot{M}\left[c_V(T_2 - T_1) + R(T_2 - T_1) + \tfrac{1}{2}\left(\bar{V}_2^2 - \bar{V}_1^2\right)\right] - \dot{Q}_C \quad \text{(C)}$$

The first two terms on the right-hand side represent the internal energy and pressure potential energy increase in the gas. We see from this equation that the ratio of the internal energy increase to the pressure potential energy increase is

$$\frac{\dot{M}c_V(T_2 - T_1)}{\dot{M}R(T_2 - T_1)} = \frac{c_V}{R} = \frac{R/(\gamma - 1)}{R} = \frac{1}{\gamma - 1}$$

where we have made use of the perfect gas relation Eq. 2.23c to write $c_V = R/(\gamma - 1)$. Since the specific heat ratio for air is $\gamma = 1.4$, we find $1/(\gamma - 1) = 2.5$. Now recall that of the shaft work input of 1.41 Btu/s, 1.25 Btu/s escapes as heat. Neglecting the negligible kinetic energy increase, of the remaining 0.156 Btu/s, an amount [(2.5/3.5)(0.156 Btu/s) = 0.111 Btu/s] increases the internal energy of the gas, and only the remainder [(1/3.5)(0.156 Btu/s) = 0.045 Btu/s] increases the pressure potential energy of the gas. Unlike the case of a liquid, some of the internal energy of a gas can be extracted by devices to produce useful work.

7.6.4 Angular Momentum Balance

Many fluid-handling devices have rotating parts that change the angular momentum of the fluid passing through them. Examples include pumps, compressors, turbines, and rotating lawn sprinklers. An angular momentum balance on a properly chosen mixed CV that contains the rotating element can be used to analyze flows in which a change in the angular momentum of a fluid is important. By taking the moment of Newton's second law, it can be shown that the time rate of change of the total angular momentum of a system is equal to the sum of the moments created by body and surface forces acting on the fluid within the system. We express conservation of angular momentum for a system by writing

$$\frac{d\mathbf{A}_{\text{sys}}}{dt} = \mathbf{M}_B + \mathbf{M}_S$$

where $\mathbf{M}_B$ and $\mathbf{M}_S$ are the moments created by the body and surface forces, respectively, about a selected point. It is customary to refer to the moments $\mathbf{M}_B$ and $\mathbf{M}_S$ as torques, so we will use moment and torque interchangeably.

Now consider a fixed CV that coincides with the system. Using the Reynolds transport theorem for a control volume, Eq. 7.5, we can write $d\mathbf{A}_{\text{sys}}/dt = \int_{\text{CV}} (\partial/\partial t)(\rho\varepsilon)\, dV + \int_{\text{CS}} (\rho\varepsilon)(\mathbf{u}\cdot\mathbf{n})\, dS$, where E_{sys} has been replaced by $\mathbf{A}_{\text{sys}}$, the total angular momentum in the system, and the integrals are over the coincident CV. The angular momentum per unit mass is $\varepsilon = \mathbf{r} \times \mathbf{u}$, where $\mathbf{r}$ is the moment arm, hence we have $\rho\varepsilon = \rho(\mathbf{r} \times \mathbf{u}) = \mathbf{r} \times \rho\mathbf{u}$. Thus the transport theorem gives an alternate expression for the time rate of change of the total angular momentum in a system as

$$\frac{d\mathbf{A}_{\text{sys}}}{dt} = \int_{\text{CV}} \frac{\partial}{\partial t}(\mathbf{r} \times \rho\mathbf{u})\, dV + \int_{\text{CS}} (\mathbf{r} \times \rho\mathbf{u})(\mathbf{u}\cdot\mathbf{n})\, dS$$

Since the CV and system coincide, the moments (torques) on the system created by the body and surface forces acting on the fluid in the system are identical to those acting on the fluid within the CV. Thus we may combine the preceding expression with that written earlier to obtain

$$\int_{\text{CV}} \frac{\partial}{\partial t}(\mathbf{r} \times \rho\mathbf{u})\, dV + \int_{\text{CS}} (\mathbf{r} \times \rho\mathbf{u})(\mathbf{u}\cdot\mathbf{n})\, dS = \mathbf{M}_B + \mathbf{M}_S \qquad (7.37)$$

This is the integral angular momentum conservation equation, or angular momentum balance for a fixed CV. It is valid for all fluids, under all conditions, and it applies in an inertial reference frame.

We can provide a physical interpretation for the angular momentum balance by writing it as

$$\int_{CV} \frac{\partial}{\partial t}(\mathbf{r} \times \rho\mathbf{u}) \, dV = -\int_{CS} (\mathbf{r} \times \rho\mathbf{u})(\mathbf{u} \cdot \mathbf{n}) \, dS + \mathbf{M}_B + \mathbf{M}_S$$

We see that the rate of accumulation of angular momentum inside the CV

$$\frac{d\mathbf{A}_{CV}}{dt} = \int_{CV} \frac{\partial}{\partial t}(\mathbf{r} \times \rho\mathbf{u}) \, dV$$

equals the net convective transport of angular momentum into the CV

$$-\boldsymbol{\Gamma}_C = -\int_{CS} (\mathbf{r} \times \rho\mathbf{u})(\mathbf{u} \cdot \mathbf{n}) \, dS$$

plus the rate at which angular momentum is created by the sum of the torques applied by body and surface forces: $\mathbf{M}_B + \mathbf{M}_S$. That is, the equation can be written symbolically as

$$\frac{d\mathbf{A}_{CV}}{dt} = -\boldsymbol{\Gamma}_{CV} + \mathbf{M}_B + \mathbf{M}_S$$

We conclude that the angular momentum balance for a fixed CV states that the accumulation of angular momentum within the CV is equal to the net inflow of angular momentum into the CV plus the torques acting on the contents of the CV due to body and surface forces.

A more complete understanding of the body and surface moments acting on a CV can be gained by representing these moments by their corresponding integral expressions. In Chapter 5, we showed that the total body moment is given by a volume integral. Writing this integral for a CV we have $\mathbf{M}_B = \int_{CV} (\mathbf{r} \times \rho\mathbf{f}) \, dV$. Similarly, we can write the total surface moment for a CV as $\mathbf{M}_S = \int_{CS} (\mathbf{r} \times \boldsymbol{\Sigma}) \, dS$. Substituting these integrals for the body and surface moments in Eq. 7.37 yields the following equivalent form of the angular momentum balance for a fixed CV:

$$\int_{CV} \frac{\partial}{\partial t}(\mathbf{r} \times \rho\mathbf{u}) \, dV + \int_{CS} (\mathbf{r} \times \rho\mathbf{u})(\mathbf{u} \cdot \mathbf{n}) \, dS = \int_{CV} (\mathbf{r} \times \rho\mathbf{f}) \, dV + \int_{CS} (\mathbf{r} \times \boldsymbol{\Sigma}) \, dS \tag{7.38}$$

Before applying this equation to a problem, it is worthwhile to examine the various terms to better understand what they represent and how they enter an analysis. In most problems it proves possible to employ a number of approximations and simplifications as described next.

Steady Process: The term $\int_{CV} (\partial/\partial t)(\mathbf{r} \times \rho\mathbf{u}) \, dV$ represents the accumulation of angular momentum in the CV; hence it is zero for a steady process. Most problems involving stationary fluid-handling devices are of this type. The flow in a steadily rotating machine can be considered to be a steady process when analyzed in an appropriate rotating reference frame. This will be explained further in a subsequent section.

Angular Momentum Transport: The term $\int_{CS} (\mathbf{r} \times \rho\mathbf{u})(\mathbf{u} \cdot \mathbf{n}) \, dS$ represents the net transport of angular momentum out of the ports of the CV. In most cases we can assume a uniform, unidirectional velocity profile at each port. After evaluating the integral at each port, we can write the result as:

$$\int_{CS} (\mathbf{r} \times \rho \mathbf{u})(\mathbf{u} \cdot \mathbf{n}) \, dS = \sum \pm \dot{M}_{\text{port}} [\mathbf{r}_{\text{port}} \times \mathbf{u}_{\text{port}}] \qquad (7.39\text{a})$$

where the summation symbol indicates we are to sum the contribution from each port. Here $\dot{M}_{\text{port}}$ is the mass flowrate at a specific port, $\mathbf{r}_{\text{port}}$ is the moment arm to the center of the port, and $\mathbf{u}_{\text{port}}$ is the uniform velocity vector at the port. The sign of $\dot{M}_{\text{port}}$ is negative at an inlet and positive at an exit.

Torque Due to Body Forces: The term $\int_{CV} (\mathbf{r} \times \rho \mathbf{f}) \, dV$ represents the torque applied by body forces to the CV. When the body force involved is gravity, we can write this term as

$$\int_{CV} (\mathbf{r} \times \rho \mathbf{f}) \, dV = \mathbf{r}_G \times (-\rho g \mathcal{V} \mathbf{k}) = \mathbf{r}_G \times (-W_{CV} \mathbf{k}) \qquad (7.39\text{b})$$

where $\mathbf{r}_G$ defines the point of application of the force $(-W_{CV} \mathbf{k})$ acting on the CV.

Torque Due to Surface Forces: The term $\int_{CS} (\mathbf{r} \times \boldsymbol{\Sigma}) \, dS$ represents the total torque applied by surface forces acting on all parts of the control surface. In problems involving a rotating machine, it is best to employ a mixed CV that encloses the rotating parts and cuts across the shaft, providing power to the part. Since the resulting control surface will have segments cutting through a shaft, at other solid surfaces, at ports, and at exterior surfaces, we will write the total torque as the sum of four terms

$$\int_{CS} (\mathbf{r} \times \boldsymbol{\Sigma}) \, dS = \int_{\text{shaft}} (\mathbf{r} \times \boldsymbol{\Sigma}) \, dS + \int_{\text{solid}} (\mathbf{r} \times \boldsymbol{\Sigma}) \, dS + \int_{\text{ports}} (\mathbf{r} \times \boldsymbol{\Sigma}) \, dS$$
$$+ \int_{\text{exterior}} (\mathbf{r} \times \boldsymbol{\Sigma}) \, dS$$

The first integral accounts for the torque created by a rotating shaft, the second for the torque created by a stationary solid structural support of some kind, the third for the effects of surface forces at ports, and the fourth for the torque created by surface forces acting on the exterior of a CV. Since we rarely know the stress distribution in a rotating shaft, we will account for the torque applied by a rotating shaft by writing

$$\mathbf{T}_{\text{shaft}} = \int_{\text{shaft}} (\mathbf{r} \times \boldsymbol{\Sigma}) \, dS \qquad (7.39\text{c})$$

which serves to introduce the shaft torque $\mathbf{T}_{\text{shaft}}$ into the CV analysis. Similarly we will account for the torque applied to the CV by structural supports as

$$\mathbf{T}_{\text{solid}} = \int_{\text{solid}} (\mathbf{r} \times \boldsymbol{\Sigma}) \, dS \qquad (7.39\text{d})$$

At a properly chosen port, the stress vector is due to the pressure. For a uniform pressure at each port, the corresponding integral can be written as a sum of terms:

$$\int_{\text{ports}} (\mathbf{r} \times \boldsymbol{\Sigma}) \, dS = \int_{\text{ports}} (\mathbf{r} \times (-p\mathbf{n})) \, dS = \sum [\mathbf{r}_{\text{port}} \times (-p_{\text{port}} A_{\text{port}} \mathbf{n})] \qquad (7.39\text{e})$$

442 | 7 CONTROL VOLUME ANALYSIS

where $\mathbf{r}_{\text{port}}$ is the moment arm to a port, A_{port} is the area, and $(-p_{\text{port}} A_{\text{port}} \mathbf{n})$ is the pressure force acting at the port in question.

Finally, to account for the torque applied by surface forces to the exterior of a CV we write

$$\mathbf{T}_{\text{exterior}} = \int_{\text{exterior}} (\mathbf{r} \times \mathbf{\Sigma})\, dS \qquad (7.39\text{f})$$

Often the exterior surface of a CV will coincide with the exterior surface of a fluid-handling device. If the machine is at rest in a stationary fluid, we have

$$\mathbf{T}_{\text{exterior}} = \int_{\text{exterior}} [\mathbf{r} \times (-p_A \mathbf{n})]\, dS \qquad (7.39\text{g})$$

which accounts for the effect of an ambient or atmospheric pressure. If the machine is in motion or there is a flow of fluid over the exterior of the machine, then normal and shear stresses may act on the exterior surface of the machine and the integral defining $\mathbf{T}_{\text{exterior}}$ is written as

$$\mathbf{T}_{\text{exterior}} = \int_{\text{exterior}} [\mathbf{r} \times (-p\mathbf{n} + \boldsymbol{\tau})]\, dS \qquad (7.39\text{h})$$

We see that $\mathbf{T}_{\text{exterior}}$ accounts for the retarding torque exerted by aerodynamic forces acting on the exterior of the machine. This torque can be substantial in the case of high speed rotation.

Upon substituting Eqs. 7.39a–7.39f into Eq. 7.38, the angular momentum balance becomes

$$\int_{\text{CV}} \frac{\partial}{\partial t}(\mathbf{r} \times \rho \mathbf{u})\, dV + \sum \pm \dot{M}_{\text{port}} [\mathbf{r}_{\text{port}} \times \mathbf{u}_{\text{port}}]$$
$$= \mathbf{r}_G \times (-W_{\text{CV}} \mathbf{k}) + \mathbf{T}_{\text{shaft}} + \mathbf{T}_{\text{solid}} + \sum [\mathbf{r}_{\text{port}} \times (-p_{\text{port}} A_{\text{port}} \mathbf{n})] + \mathbf{T}_{\text{exterior}}$$
$$(7.40)$$

This equation may be used as the starting point for an angular momentum balance on a fixed CV in an inertial reference frame.

EXAMPLE 7.17

Water flows steadily at 3 m/s through the vertical offset pipe bend shown in Figure 7.29. The pipe diameter is 10 cm, the upstream gage pressure is 275 kPa, and the pressure drop through the bend is estimated to be 50 kPa. Find the forces that must be exerted by each flange on the pipe bend to keep it in place.

SOLUTION

We are asked to determine the forces applied to an offset pipe bend by two flanges under the given conditions. Since the Reynolds number here may be calculated to be $Re = 2.7 \times 10^5$, the flow is turbulent, and we may assume uniform properties at the

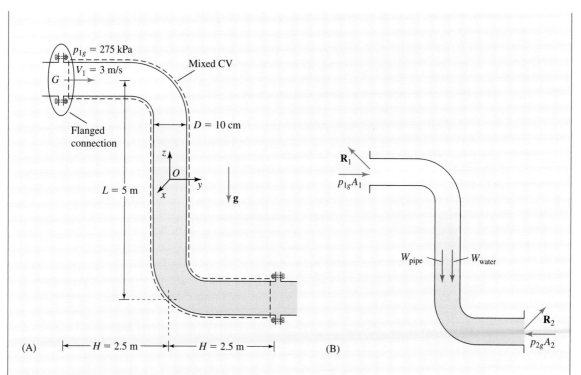

Figure 7.29 Schematic for Example 7.17.

ports. Choosing the mixed CV shown in Figure 7.29A with a decal surface at each flange, and noting that the density and areas are constant, a steady flow mass balance gives $\dot{M} = \rho A \bar{V}_1 = \rho A \bar{V}_2$, or $\bar{V}_1 = \bar{V}_2$. Since we anticipate terms in the momentum balance in both the z and y directions, we will use the vector form of the steady flow momentum balance as given by Eq. 7.19a: $\int_{CS} (\rho \mathbf{u})(\mathbf{u} \cdot \mathbf{n}) \, dS = \mathbf{F}_B + \mathbf{F}_S$. To evaluate the body force we write $\mathbf{F}_B = -(W_{\text{pipe}} + W_{\text{water}})\mathbf{k}$ to account for the weight of the pipe and the water inside the pipe, since these are both inside the CV. To finish the momentum balance, we will evaluate the momentum transport at the inlet and exit, the surface force due to the pressure at these locations, the surface force applied by the air to the decal surface on the outside of the bend, and the reaction forces $\mathbf{R}_1$ and $\mathbf{R}_2$ applied to the CV by the flanges. The forces $\mathbf{R}_1$ and $\mathbf{R}_2$ are the requested forces applied to the pipe bend by each flange. On the inlet we have $\mathbf{u} = \bar{V}_1 \mathbf{j}$, $\mathbf{n} = -\mathbf{j}$, and $\mathbf{u} \cdot \mathbf{n} = -\bar{V}_1$, so that the momentum transport vector there is $-\rho \bar{V}_1^2 A_1 \mathbf{j}$. On the exit, $\mathbf{u} = \bar{V}_2 \mathbf{j}$, $\mathbf{n} = \mathbf{j}$, and $\mathbf{u} \cdot \mathbf{n} = \bar{V}_2$, and the momentum transport is $+\rho \bar{V}_2^2 A_2 \mathbf{j}$. The net momentum transport is $(-\rho \bar{V}_1^2 A_1 \mathbf{j} + \rho \bar{V}_2^2 A_2 \mathbf{j}) = 0$, since $A_1 = A_2 = A$ and $\bar{V}_1 = \bar{V}_2$. The surface force on the inlet and exit is due to the pressure at each location. Thus these terms give $(-p_1(-\mathbf{j})A_1) + (-p_2(\mathbf{j})A_2) = (p_1 - p_2)A\mathbf{j}$. The force of the air on the outside of the elbow will be accounted for by using gage pressure in these terms. Thus the momentum balance for the offset pipe bend gives

$$0 = -(W_{\text{pipe}} + W_{\text{water}})\mathbf{k} + (p_{1_{\text{gage}}} - p_{2_{\text{gage}}})A\mathbf{j} + \mathbf{R}_1 + \mathbf{R}_2$$

Solving for the desired reaction forces we have

$$\mathbf{R}_1 + \mathbf{R}_2 = (p_{2_{\text{gage}}} - p_{1_{\text{gage}}})A\mathbf{j} + (W_{\text{pipe}} + W_{\text{water}})\mathbf{k}$$

This is a vector equation whose components are

$$R_{1x} + R_{2x} = 0 \tag{A}$$

$$R_{1y} + R_{2y} = (p_{2_{\text{gage}}} - p_{1_{\text{gage}}})A \tag{B}$$

$$R_{1z} + R_{2z} = W_{\text{pipe}} + W_{\text{water}} \tag{C}$$

Equation (A) shows that if the flanges apply a force in the x direction, these components must cancel. Since there is no reason to assume that these forces exist, we will assume that $R_{1X} = R_{2X} = 0$. Frictional pressure losses ensure that $p_{1_{\text{gage}}} > p_{2_{\text{gage}}}$, thus we see by (B) that the flanges must apply a reaction force to the left. Equation (C) shows that the weight of the pipe and its contents is supported by the flanges. We will assume that due to symmetry each flange supports half the total weight, i.e.,

$$R_{1z} = R_{2z} = \tfrac{1}{2}(W_{\text{pipe}} + W_{\text{water}}) \tag{D}$$

Note that (B) does not allow us to determine the y component of each reaction force separately, only their sum. Figure 7.29B shows the various forces acting on the CV. To determine the y component of each reaction force, we will use Eq. 7.40 to apply a steady angular momentum balance to this CV. In this problem, the shaft torque term is zero, and the exterior torque is due to the ambient pressure (see Eq. 7.39g). Also, since the torque $\mathbf{T}_{\text{solid}}$ applied by solid supports to the control volume is created by the unknown reaction forces $\mathbf{R}_1$ and $\mathbf{R}_2$, we can write the torque as $\mathbf{T}_{\text{solid}} = \mathbf{r}_1 \times \mathbf{R}_1 + \mathbf{r}_2 \times \mathbf{R}_2$, where $\mathbf{r}_1$ and $\mathbf{r}_2$ are the moment arms of each reaction force. With these substitutions Eq. 7.40 becomes

$$\sum \pm \dot{M}_{\text{port}}[\mathbf{r}_{\text{port}} \times \mathbf{u}_{\text{port}}] = \mathbf{r}_G \times (-W_{\text{CV}}\mathbf{k}) + \mathbf{r}_1 \times \mathbf{R}_1 + \mathbf{r}_2 \times \mathbf{R}_2$$
$$+ \sum [\mathbf{r}_{\text{port}} \times (-p_{\text{port}}A_{\text{port}}\mathbf{n})] + \int_{\text{exterior}} (\mathbf{r} \times (-p_A\mathbf{n})\, dS$$

Since the only forces acting on the CV are in the y and z directions (see (A)), we anticipate that the only component of the angular momentum balance of interest in this problem is the x component. This component describes the tendency of the forces acting on the CV to twist the pipe bend about the x axis.

Consider an origin O at the midpoint of the bend run (Figure 7.29A). To evaluate the transport term in the angular momentum balance, note that on the inlet $\dot{M}_1 = \rho A \bar{V}_1$, $\mathbf{r}_1 = -H\mathbf{j} + (L/2)\mathbf{k}$, $\mathbf{u}_1 = \bar{V}_1\mathbf{j}$, hence $\mathbf{r}_1 \times \mathbf{u}_1 = [-H\mathbf{j} + (L/2)\mathbf{k}] \times \bar{V}_1\mathbf{j} = -(L/2)\bar{V}_1\mathbf{i}$. On the exit $\dot{M}_2 = \rho A \bar{V}_2$, $\mathbf{r}_2 = H\mathbf{j} - (L/2)\mathbf{k}$, $\mathbf{u}_2 = \bar{V}_2\mathbf{j}$, and

$\mathbf{r}_2 \times \mathbf{u}_2 = [H\mathbf{j} - (L/2)\mathbf{k}] \times \bar{V}_2\mathbf{j} = (L/2)\bar{V}_2\mathbf{i}$. Accounting for the signs of $\dot{M}$ at each port, and noting that $\dot{M} = \rho A \bar{V}_1 = \rho A \bar{V}_2$, we find

$$\sum \pm \dot{M}_{\text{port}}[\mathbf{r}_{\text{port}} \times \mathbf{u}_{\text{port}}] = (-\dot{M}_1)\left(-\frac{L}{2}\bar{V}_1\mathbf{i}\right) + (\dot{M}_2)\left(\frac{L}{2}\bar{V}_2\mathbf{i}\right) = \dot{M}L\bar{V}_1\mathbf{i}$$

Consider next the moment $\mathbf{r}_G \times (-W_{\text{CV}}\mathbf{k})$ created by gravity acting on the CV. Owing to the symmetry, the force of gravity $-(W_{\text{pipe}} + W_{\text{water}})\mathbf{k}$ acting on the CV may be considered to act at the point O. Since the moment arm for a force acting at this location is $\mathbf{r}_G = 0$, the moment about O contributed by the body force is zero. The terms $\mathbf{r}_1 \times \mathbf{R}_1$ and $\mathbf{r}_2 \times \mathbf{R}_2$ are the unknowns in the angular momentum balance in this case, so these are left alone.

To evaluate the torque applied by atmospheric pressure on the exterior of the CV, note that if atmospheric pressure acted on the entire control surface (including the ports and small decal surface at each flange), the resulting torque would be zero. We can account for the effects of atmospheric pressure acting on the portion of the control surface in contact with air by using gage pressure at the inlet and exit ports. Thus we evaluate this term and the corresponding term at the ports together by writing

$$\sum [\mathbf{r}_{\text{port}} \times (-p_{\text{port}} A_{\text{port}} \mathbf{n})] + \int_{\text{exterior}} (\mathbf{r} \times (-p_A \mathbf{n})\, dS$$
$$= \mathbf{r}_1 \times (-p_{1_{\text{gage}}} A_1 \mathbf{n}_1) + \mathbf{r}_2 \times (-p_{2_{\text{gage}}} A_2 \mathbf{n}_2)$$

Note that this approach assumes that the pipe wall is thin and the flange areas are negligible. We have $\mathbf{r}_1 = -H\mathbf{j} + (L/2)\mathbf{k}$, $\mathbf{n}_1 = -\mathbf{j}$, and $\mathbf{r}_2 = H\mathbf{j} - (L/2)\mathbf{k}$, $\mathbf{n}_2 = \mathbf{j}$. Thus on the inlet we evaluate the cross product and get $[-H\mathbf{j} + (L/2)\mathbf{k}] \times [-p_{1_{\text{gage}}} A_1(-\mathbf{j})] = -(L/2)p_{1_{\text{gage}}} A_1 \mathbf{i}$. On the exit, the cross product gives $-(L/2)p_{2_{\text{gage}}} A_2 \mathbf{i}$. Since $A = A_1 = A_2$, the two pressure terms together give

$$\mathbf{r}_1 \times (-p_{1_{\text{gage}}} A_1 \mathbf{n}_1) + \mathbf{r}_2 \times (-p_{2_{\text{gage}}} A_2 \mathbf{n}_2) = -\frac{L}{2}(p_{2_{\text{gage}}} + p_{1_{\text{gage}}})A\mathbf{i}$$

When all the terms have been gathered, the angular momentum balance becomes

$$\dot{M}L\bar{V}_1\mathbf{i} = [\mathbf{r}_1 \times \mathbf{R}_1]_O + [\mathbf{r}_2 \times \mathbf{R}_2]_O - \frac{L}{2}(p_{2_{\text{gage}}} + p_{1_{\text{gage}}})A\mathbf{i}$$

Solving for the torques applied by the flanges about point O we find

$$[\mathbf{r}_1 \times \mathbf{R}_1]_O + [\mathbf{r}_2 \times \mathbf{R}_2]_O = \left[\dot{M}L\bar{V}_1 + \frac{L}{2}(p_{2_{\text{gage}}} + p_{1_{\text{gage}}})A\right]\mathbf{i}$$

The left-hand side of this vector equation is evaluated by writing

$$[\mathbf{r}_1 \times \mathbf{R}_1]_O = \left(-H\mathbf{j} + \frac{L}{2}\mathbf{k}\right) \times (R_{1y}\mathbf{j} + R_{1z}\mathbf{k}) = \left(-HR_{1z} - \frac{L}{2}R_{1y}\right)\mathbf{i}$$

$$[\mathbf{r}_2 \times \mathbf{R}_2]_O = \left(H\mathbf{j} - \frac{L}{2}\mathbf{k}\right) \times (R_{2y}\mathbf{j} + R_{2z}\mathbf{k}) = \left(HR_{2z} + \frac{L}{2}R_{2y}\right)\mathbf{i}$$

Thus the angular momentum balance becomes

$$\left(-HR_{1z} - \frac{L}{2}R_{1y}\right)\mathbf{i} + \left(HR_{2z} + \frac{L}{2}R_{2y}\right)\mathbf{i} = \left[\dot{M}L\bar{V}_1 + \frac{L}{2}(p_{2_{\text{gage}}} + p_{1_{\text{gage}}})A\right]\mathbf{i}$$

We see that there is only an x component to this equation, as expected. Earlier we assumed each flange supports half the weight, thus we can put $R_{1z} = R_{2z}$. Solving for R_{1y} we obtain

$$R_{1y} = R_{2y} - 2\dot{M}\bar{V}_1 - (p_{2_{\text{gage}}} + p_{1_{\text{gage}}})A \tag{E}$$

To solve for the y component of each reaction force, we can use (B) to write

$$R_{2y} = -R_{1y} + (p_{2_{\text{gage}}} - p_{1_{\text{gage}}})A$$

Substituting the preceding expression into (E) and solving for R_{1y} we find

$$R_{1y} = -\dot{M}\bar{V}_1 - p_{1_{\text{gage}}}A \tag{F}$$

The remaining component is then found to be

$$R_{2y} = \dot{M}\bar{V}_1 + p_{2_{\text{gage}}}A \tag{G}$$

The forces applied by the two flanges are thus

$$\mathbf{R}_1 = \left[0, -\dot{M}\bar{V}_1 - p_{1_{\text{gage}}}A, \tfrac{1}{2}(W_{\text{pipe}} + W_{\text{water}})\right] \quad \text{and} \quad \mathbf{R}_2 = \left[0, \dot{M}\bar{V}_1 + p_{2_{\text{gage}}}A, \tfrac{1}{2}(W_{\text{pipe}} + W_{\text{water}})\right]$$

By using the data, we find

$$\dot{M} = \rho A \bar{V}_1 = (998 \text{ kg/m}^3)\left[\frac{\pi(0.1 \text{ m})^2}{4}\right](3 \text{ m/s}) = 23.5 \text{ kg/s}$$

and with $p_{1_{\text{gage}}} = 275$ kPa, we know $p_{2_{\text{gage}}} = 275$ kPa $- 50$ kPa $= 225$ kPa. The y components of the reaction force are

$$R_{1y} = -\dot{M}\bar{V}_1 - p_{1_{\text{gage}}}A = -(23.5 \text{ kg/s})(3 \text{ m/s}) - (275{,}000 \text{ N/m}^2)\left[\frac{\pi(0.1 \text{ m})^2}{4}\right]$$

$$= -2.23 \text{ kN}$$

$$R_{2y} = \dot{M}\bar{V}_1 + p_{2_{\text{gage}}}A = (23.5 \text{ kg/s})(3 \text{ m/s}) + (225{,}000 \text{ N/m}^2)\left[\frac{\pi(0.1 \text{ m})^2}{4}\right]$$

$$= 1.84 \text{ kN}$$

We are not given the weight of the pipe bend, so we will write the z components of the reactions in symbolic form as $R_{1z} = R_{2z} = \tfrac{1}{2}(W_{\text{pipe}} + W_{\text{water}})$.

The choice of an origin for the angular momentum balance is arbitrary. We could choose point G in the figure to evaluate the angular momentum balance and obtain the same results.

In problems involving rotating machinery, it is highly advantageous to choose a CV that rotates along with the rotating machine element. The CV is then fixed in a rotating, noninertial reference frame. Since the angular momentum balance derived earlier, Eq. 7.38:

$$\int_{CV} \frac{\partial}{\partial t}(\mathbf{r} \times \rho\mathbf{u})\,dV + \int_{CS}(\mathbf{r} \times \rho\mathbf{u})(\mathbf{u}\cdot\mathbf{n})\,dS = \int_{CV}(\mathbf{r} \times \rho\mathbf{f})\,dV + \int_{CS}(\mathbf{r} \times \boldsymbol{\Sigma})\,dS$$

is valid only in an inertial reference frame, it cannot be used without modification for a rotating CV. The modification needed is straightforward: in a noninertial frame, we must include in the body force term in Eq. 7.38 the additional body forces caused by the motion of the noninertial frame with respect to an inertial frame. The velocities in the angular momentum balance must also be taken as those relative to the rotating CV. As derived in advanced fluid dynamics texts, the additional body forces per unit volume are given as:

$$\rho\left[-\mathbf{A}_O - (2\boldsymbol{\Omega}\times\mathbf{u}) - [\boldsymbol{\Omega}\times(\boldsymbol{\Omega}\times\mathbf{x})] - \left(\frac{d\boldsymbol{\Omega}}{dt}\times\mathbf{x}\right)\right]$$

where $\mathbf{A}_O$ is the rectilinear acceleration of the noninertial frame, and $\boldsymbol{\Omega}$ is the angular velocity vector. The first term accounts for the additional body force due to the rectilinear acceleration of the CV. This term is zero for a CV that is rotating but not translating. The next two terms are the Coriolis and centrifugal forces due to rotation of the CV, and the last is the force created by temporal variation in the rotation rate. For a CV at rest with respect to a rotating machine element (with $\mathbf{A}_O = 0$), the general form of the body force term in the angular momentum balance is

$$\mathbf{r}_G \times (-W_{CV}\mathbf{k}) + \int_{CV}\mathbf{r}\times\rho\left[-(2\boldsymbol{\Omega}\times\mathbf{u}) - (\boldsymbol{\Omega}\times(\boldsymbol{\Omega}\times\mathbf{x})) - \left(\frac{d\boldsymbol{\Omega}}{dt}\times\mathbf{x}\right)\right]dV$$

In most cases we are interested in using an angular momentum balance to analyze the performance of a pump, compressor, or other rotating machine operating at a constant rotation rate. Under these conditions it is appropriate to assume a steady process and drop the accumulation of angular momentum term in the equation. Since the rotation rate is constant, $d\boldsymbol{\Omega}/dt = 0$, the term involving $d\boldsymbol{\Omega}/dt$ may also be dropped. Thus the steady angular momentum balance for a CV in constant rotation is

$$\int_{CS}(\mathbf{r}\times\rho\mathbf{u})(\mathbf{u}\cdot\mathbf{n})\,dS = \int_{CV}(\mathbf{r}\times\rho\mathbf{f})\,dV + \int_{CV}\mathbf{r}\times\rho[-(2\boldsymbol{\Omega}\times\mathbf{u}) - (\boldsymbol{\Omega}\times(\boldsymbol{\Omega}\times\mathbf{x}))]\,dV$$

$$+ \int_{CS}(\mathbf{r}\times\boldsymbol{\Sigma})\,dS$$

Introducing the simplifications discussed earlier, we get

$$\sum \pm\dot{M}_{port}[\mathbf{r}_{port}\times\mathbf{u}_{port}] = \mathbf{r}_G\times(-W_{CV}\mathbf{k}) + \int_{CV}\mathbf{r}\times\rho[-(2\boldsymbol{\Omega}\times\mathbf{u}) - (\boldsymbol{\Omega}\times(\boldsymbol{\Omega}\times\mathbf{x}))]\,dV$$

$$+ \mathbf{T}_{shaft} + \mathbf{T}_{solid} + \sum[\mathbf{r}_{port}\times(-p_{port}A_{port}\mathbf{n})] + \mathbf{T}_{exterior} \quad (7.41)$$

Angular momentum considerations are fundamental to the design and analysis of all types of rotating machinery. For these problems, it is normally possible and convenient to select a coordinate system with the z axis aligned with rotation axis. A typical problem for which an angular momentum balance is performed is to determine the shaft torque needed to run a machine or the torque produced by a machine.

EXAMPLE 7.18

It is proposed to use high pressure water in the design of a power unit for a rotary cleaning brush. As sketched in Figure 7.30, water enters the inlet of the power unit and exits through N jets. If friction in the inlet bearing and aerodynamic drag are negligible, what is the torque available to rotate the brush at a given angular velocity? How many jets should be employed?

Figure 7.30 Schematic for Example 7.18.

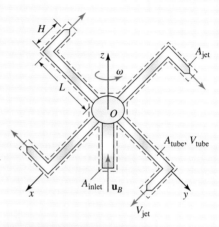

SOLUTION

We will choose a CV enclosing the power unit and rotating with it as shown in Figure 7.30, and assume that the unit has achieved a constant angular velocity. Assuming equal flowrates out of each jet, we can use a mass balance to write $\dot{M} = \rho Q = N\dot{M}_{\text{jet}}$, where $\dot{M}_{\text{jet}} = \rho A_{\text{jet}} \bar{V}_{\text{jet}}$ is the jet mass flowrate, A_{jet} is the jet area, and $\bar{V}_{\text{jet}}$ is the jet exit velocity relative to the device. To apply a steady process angular momentum balance we will employ Eq. 7.41:

$$\sum \pm \dot{M}_{\text{port}}[\mathbf{r}_{\text{port}} \times \mathbf{u}_{\text{port}}] = \mathbf{r}_G \times (-W_{\text{CV}}\mathbf{k}) + \int_{\text{CV}} \mathbf{r} \times \rho[-(2\boldsymbol{\Omega} \times \mathbf{u}) - (\boldsymbol{\Omega} \times (\boldsymbol{\Omega} \times \mathbf{x}))]\,dV$$
$$+ \mathbf{T}_{\text{shaft}} + \mathbf{T}_{\text{solid}} + \sum [\mathbf{r}_{\text{port}} \times (-p_{\text{port}} A_{\text{port}} \mathbf{n})] + \mathbf{T}_{\text{exterior}}$$

and choose the origin on the rotation axis as shown in Figure 7.30. To account for the angular momentum transport at each port of the CV, we write

$$\sum \pm \dot{M}_{\text{port}}[\mathbf{r}_{\text{port}} \times \mathbf{u}_{\text{port}}] = -\dot{M}[\mathbf{r}_B \times \mathbf{u}_B] + N\dot{M}_{\text{jet}}[\mathbf{r}_{\text{jet}} \times \mathbf{u}_{\text{jet}}]$$

The angular momentum transport at the inlet port can be neglected. To understand why, recall that the term $-\dot{M}[\mathbf{r}_B \times \mathbf{u}_B]$ representing the transport at the inlet is actually given by the integral $\int_{\text{inlet}} (\mathbf{r} \times \rho\mathbf{u})(\mathbf{u} \cdot \mathbf{n})\,dS$. The velocity vector in this integral is relative to the CV, and since the CV is rotating, the velocity entering the inlet is actually not unidirectional and uniform but swirling relative to the CV. The angular momentum of this incoming flow is very small, however, since the moment arm is always less than the radius of the inlet. Since this radius is small, we can neglect this term. The total angular momentum transport is thus $N\dot{M}_{jet}[\mathbf{r}_{jet} \times \mathbf{u}_{jet}]$. To evaluate this term, consider the jet leaving the tube aligned with the $+y$ axis. The moment arm to this jet is $\mathbf{r}_{jet} = H\mathbf{i} + L\mathbf{j}$, the jet velocity vector is $\mathbf{u}_{jet} = \bar{V}_{jet}\mathbf{i}$, and we obtain

$$\dot{M}_{jet}[\mathbf{r}_{jet} \times \mathbf{u}_{jet}] = \dot{M}_{jet}[(H\mathbf{i} + L\mathbf{j}) \times \bar{V}_{jet}\mathbf{i}] = -\dot{M}_{jet}L\bar{V}_{jet}\mathbf{k}$$

The angular momentum transport for each of the remaining jets is the same, thus the total angular momentum transport by the jets is

$$\sum \pm \dot{M}_{port}[\mathbf{r}_{port} \times \mathbf{u}_{port}] = N\dot{M}_{jet}[\mathbf{r}_{jet} \times \mathbf{u}_{jet}] = -N\dot{M}_{jet}L\bar{V}_{jet}\mathbf{k} = -\dot{M}L\bar{V}_{jet}\mathbf{k} \quad \text{(A)}$$

Next consider the body force terms. Symmetry causes the term $\mathbf{r}_G \times (-W_{CV}\mathbf{k})$ representing the torque due to gravity to be zero. The term $\int_{CV} \mathbf{r} \times \rho\{-(2\mathbf{\Omega} \times \mathbf{u}) - [\mathbf{\Omega} \times (\mathbf{\Omega} \times \mathbf{x})]\} dV$ is evaluated as follows. Consider the portion of the CV that includes the entire jet tube aligned with the y axis. We must account for the torque exerted by the Coriolis and centrifugal forces acting on both the solid and fluid inside this part of the CV. The centrifugal force does not create a torque about the rotation axis. We can show this by observing that since the jet tube is of small diameter, the moment arm to any point in the control volume is $\mathbf{r} = y\mathbf{j}$. The location of the point is also $\mathbf{x} = y\mathbf{j}$, and $\mathbf{\Omega} = \omega\mathbf{k}$. Thus the centrifugal force $\mathbf{\Omega} \times (\mathbf{\Omega} \times \mathbf{x}) = \omega^2 y\mathbf{j}$ acts along the tube in the y direction, and the cross product $\mathbf{r} \times \rho[-(\mathbf{\Omega} \times (\mathbf{\Omega} \times \mathbf{x}))]$ is zero.

To evaluate the torque created by the Coriolis force on this portion of the CV, we need only consider the fluid, since the solid parts inside the CV are not moving relative to the rotating reference frame. Let the uniform velocity of the fluid inside the portion of the tube of length L lying along the y axis be given by $\mathbf{u}_L = \bar{V}_{tube}\mathbf{j}$ and that in the remaining section of the tube of length H be given by $\mathbf{u}_H = \bar{V}_{tube}\mathbf{i}$. For the tube section of length L, we have $-2\mathbf{\Omega} \times \mathbf{u} = -2\omega\mathbf{k} \times \bar{V}_{tube}\mathbf{j} = 2\omega\bar{V}_{tube}\mathbf{i}$. The moment arm to a point in the fluid inside is $\mathbf{r} = y\mathbf{j}$, and $\mathbf{r} \times \rho[-(2\mathbf{\Omega} \times \mathbf{u})] = \rho[y\mathbf{j} \times 2\omega\bar{V}_{tube}\mathbf{i}] = -2\omega\rho y\bar{V}_{tube}\mathbf{k}$. The integral over this portion of the CV for the N jet tubes can be evaluated for a tube of area A_{tube} as

$$N\int_{L-\text{tube}} \mathbf{r} \times \rho[-(2\mathbf{\Omega} \times \mathbf{u})] dV = NA_{tube}\int_0^L -2\omega\rho y\bar{V}_{tube}\mathbf{k}\, dy$$
$$= -\omega L^2 N(\rho A_{tube}\bar{V}_{tube})\mathbf{k} = -\omega L^2 \dot{M}\mathbf{k} \quad \text{(B)}$$

where we used a mass balance to show that $\rho A_{tube}\bar{V}_{tube} = \rho A_{jet}\bar{V}_{jet} = \dot{M}_{jet}$, and $\dot{M} = N\dot{M}_{jet}$.

For the section of the tube of length H, we have $-2\mathbf{\Omega} \times \mathbf{u} = -2\omega\mathbf{k} \times \bar{V}_{tube}\mathbf{i} = -2\omega\bar{V}_{tube}\mathbf{j}$. The moment arm to a point in this section of the CV can be written as $\mathbf{r} = x\mathbf{i} + y\mathbf{j}$, and thus $\mathbf{r} \times \rho[-(2\mathbf{\Omega} \times \mathbf{u})] = \rho[(x\mathbf{i} + y\mathbf{j}) \times (-2\omega\bar{V}_{tube}\mathbf{j})] = -2\omega\rho\bar{V}_{tube}x\mathbf{k}$. The integral for all N tubes is

$$N\int_{H-\text{tube}} \mathbf{r} \times \rho[-(2\mathbf{\Omega} \times \mathbf{u})] dV = NA_{tube}\int_0^H -2\omega\rho x\bar{V}_{tube}\mathbf{k}\, dx$$
$$= -\omega H^2 N(\rho A_{tube}\bar{V}_{tube})\mathbf{k} = -\omega H^2 \dot{M}\mathbf{k} \quad \text{(C)}$$

We are told to neglect friction in the inlet bearing, so we can set $\mathbf{T}_{solid}$ to zero. Since $\mathbf{T}_{exterior}$ includes the effect of atmospheric pressure on the exterior (which is always present), we will neglect the unknown aerodynamic drag but account for atmospheric pressure by using gage pressures at the ports. Since each jet leaves at atmospheric pressure, the gage pressures are all zero, and the term $\sum [\mathbf{r}_{port} \times (-p_{port} A_{port} \mathbf{n})]$ needs be evaluated only at the inlet. Because of the symmetry and alignment of the rotation axis, the pressure on the inlet does not contribute a torque.

The term $\mathbf{T}_{shaft}$ represents the torque applied to the CV by the shaft on which the brush mounts. By the principle of action–reaction, we can write the torque $T_{brush}\mathbf{k}$ supplied to the brush as

$$-T_{brush}\mathbf{k} = \mathbf{T}_{shaft} \tag{D}$$

Inserting (A) through (D) into the angular momentum balance, we find

$$-\dot{M} L \bar{V}_{jet}\mathbf{k} = -\omega L^2 \dot{M}\mathbf{k} - \omega H^2 \dot{M}\mathbf{k} - T_{brush}\mathbf{k}$$

Solving for the torque available to turn the brush, we have

$$T_{brush} = \dot{M} L \bar{V}_{jet} - \dot{M} \omega (L^2 + H^2) \tag{E}$$

From this result we see that there is no apparent advantage to employing multiple jets, since the torque delivered to the brush is independent of N. Notice also that the maximum torque is delivered to the brush when it is not turning. As the rpm of the brush increases, the torque delivered decreases. The result also shows that it is advantageous to keep the length H as short as possible.

7.7 SUMMARY

A system is a fixed identifiable quantity of fluid. Mass may not cross the system boundary, but other interactions of the system and its surroundings are permitted. For example, there may be a transport of energy across a system boundary by heat conduction, and work may be done on the fluid within a system by body and surface forces. A system is generally characterized by spatially and temporally variable properties. The system approach is useful in those rare instances of system boundaries having shapes that are either independent of time or easily described as a function of time.

A control volume (CV) is a closed region in space through which fluid flows. This region is defined by its boundary or control surface. A segment of the control surface at which fluid enters or leaves a CV is called a port. Ports are usually defined so that the velocity vector at each port has only a normal component. Control volumes may be fixed (stationary) or moving. The size and the shape of a moving CV do not change, thus the velocity of each point on a moving control surface is the same.

7.7 SUMMARY

The Reynolds transport theorem is expressed for a system as

$$\frac{dE_{sys}}{dt} = \int_{R(t)} \frac{\partial}{\partial t}(\rho\varepsilon)\,dV + \int_{S(t)} (\rho\varepsilon)(\mathbf{u}\cdot\mathbf{n})\,dS$$

where $R(t)$ is the region occupied by the system. This theorem states that the time rate of change in the total amount of a property in a system may be calculated as the sum of a volume integral, which accounts for the instantaneous change in the total amount of property within the system due to temporal variations in fluid properties, and a surface integral, which accounts for the contribution from the deformation of the system due to the instantaneous motion of its boundaries.

If we consider a fixed CV that coincides with the system at a particular time, the value of a fluid property at a point in the system is the same as the value of that property at the same point in the CV. Thus, the integrals over the system may be interpreted as applying also to the coincident CV. An equivalent form of the transport theorem is $dE_{sys}/dt = dE_{CV}/dt + \Gamma_{CV}$. This equation tells us that the time rate of change of the total amount of fluid property in a system is equal to the time rate of change of the total amount of this property in a fixed coincident CV (accumulation), plus the instantaneous outflow of the fluid property through the control surface (convective transport).

We can use the transport theorem to obtain the integral form of the mass conservation law (mass balance) for a fixed CV: $\int_{CV} (\partial\rho/\partial t)\,dV + \int_{CS} \rho(\mathbf{u}\cdot\mathbf{n})\,dS = 0$. This expression states that the increase in the total amount of mass in a fixed CV is equal to the instantaneous convective transport of mass into the CV.

Application of the transport theorem gives the integral momentum conservation equation (momentum balance) for a fixed CV: $\int_{CV} (\partial/\partial t)(\rho\mathbf{u})\,dV + \int_{CS} (\rho\mathbf{u})(\mathbf{u}\cdot\mathbf{n})\,dS = \mathbf{F}_B + \mathbf{F}_S$. The interpretation of this equation is that the accumulation of momentum within the CV plus the net outflow of momentum from the CV is equal to the sum of the body and surface forces. By using the definitions of body and surface forces, we obtain an equivalent expression for the law of conservation of momentum:

$$\int_{CV} \frac{\partial}{\partial t}(\rho\mathbf{u})\,dV + \int_{CS} (\rho\mathbf{u})(\mathbf{u}\cdot\mathbf{n})\,dS = \int_{CV} \rho\mathbf{f}\,dV + \int_{CS} \Sigma\,dS$$

The integral energy conservation equation (energy balance) for a fixed CV is

$$\int_{CV} \frac{\partial}{\partial t}\left[\rho\left(u + \tfrac{1}{2}\mathbf{u}\cdot\mathbf{u}\right)\right]dV + \int_{CV} \rho\left(u + \tfrac{1}{2}\mathbf{u}\cdot\mathbf{u}\right)(\mathbf{u}\cdot\mathbf{n})\,dS = \dot{W}_B + \dot{W}_S + \dot{Q}_C + \dot{S}$$

This equation shows that the rate of accumulation of internal plus kinetic energy within the CV plus the net convective transport of this energy out of the CV is equal to the sum of the rate at which work is done on the fluid in the CV by body and surface forces, and the rate at which energy is added to the fluid by heat conduction and other energy inputs.

Finally, the integral angular momentum conservation equation (angular momentum balance) for a fixed CV in an inertial reference frame is

$$\int_{CV} \frac{\partial}{\partial t}(\mathbf{r}\times\rho\mathbf{u})\,dV + \sum \pm\dot{M}_{port}[\mathbf{r}_{port}\times\mathbf{u}_{port}]$$
$$= \mathbf{r}_G \times (-W_{CV}\mathbf{k}) + \mathbf{T}_{shaft} + \mathbf{T}_{solid} + \sum [\mathbf{r}_{port}\times(-p_{port}A_{port}\mathbf{n})] + \mathbf{T}_{exterior}$$

We see that the rate of accumulation of angular momentum inside the CV plus the net convective transport of angular momentum out of the CV equals the rate at which angular momentum is created by body and surface forces. For a CV rotating at constant angular velocity, we can write the angular momentum balance for a steady process as

$$\sum \pm \dot{M}_{port}[\mathbf{r}_{port} \times \mathbf{u}_{port}] = \mathbf{r}_G \times (-W_{CV}\mathbf{k}) + \int_{CV} \mathbf{r} \times \rho[-(2\mathbf{\Omega} \times \mathbf{u}) - (\mathbf{\Omega} \times (\mathbf{\Omega} \times \mathbf{x}))]\,dV$$
$$+ \mathbf{T}_{shaft} + \mathbf{T}_{solid} + \sum [\mathbf{r}_{port} \times (-p_{port}A_{port}\mathbf{n})] + \mathbf{T}_{exterior}$$

PROBLEMS

Section 7.2

7.1 Discuss the similarities and differences between a thermodynamic system and a fluid system.

7.2 Is it possible for a thermodynamic system to be equivalent to a fluid system? If so, give an example. If not, explain why this is impossible.

7.3 Discuss the similarities and differences between a fluid system and a (fluid) control volume.

Section 7.3

7.4 Is the concept of a control volume more compatible with the Lagrangian or Eulerian method of description? Why?

7.5 Explain why the control volume approach was better suited to the description of the single stage piston compressor described in the text. Provide an example of another fluid process that would offer a similar obstacle to the use of a systems approach. In contrast, provide an example of a fluid process that you believe could benefit from a systems approach.

7.6 In the text the concept of mass conservation for an isolated gas compression process was discussed. Did you prefer the systems approach or the control volume approach? Why?

Sections 7.4 and 7.5

7.7 Offer a physical interpretation of the Reynolds transport theorem as applied to a system.

7.8 Offer a physical interpretation of the Reynolds transport theorem as applied to a control volume.

7.9 Explain how the Reynolds transport theorem provides the necessary link between system and control volume analysis.

7.10 Give three examples of fluid properties that could be represented by the variable ε in Eq. 7.4. Also provide two examples of fluid properties that could be represented by the variable λ in the equation developed in Example 7.1.

7.11 Consider the fluid flow that occurs when a fire extinguisher is discharged.
(a) Sketch this device and show an appropriate control volume for use in the analysis of this flow.
(b) Describe an appropriate choice for a system for this flow.
(c) Write an appropriate form of the Reynolds transport theorem for your control volume and offer a physical interpretation for each term in this equation.
(d) Write an appropriate form of the Reynolds transport theorem for your system and offer a physical interpretation for each term in this equation.

(e) Finally, state which approach (CV or system) you would chose to solve this problem. Justify your selection.

7.12 People are routinely walking into and out of the corporate headquarters of Tar Foot Consulting company. Use the Reynolds transport theorem to describe the time rate of change of the number of people in the building.

7.13 Can the Reynolds transport theorem be used to describe the time rate of change of the number of bees in a beehive? How about the time rate of change of the number of cars in a parking garage? How about the time rate of change of the number of people in the maternity ward of a hospital?

Section 7.6

7.14 In the text we list recommendations for choosing a fixed control volume. Discuss the reasons for each of these recommendations.

7.15 After reviewing Example 7.2, find your own example of a similar fluid device and sketch an appropriate choice for a fixed control volume.

Section 7.6.1

7.16 What types of fluid flow problem can be solved by using control volume analysis and a mass balance? Under what conditions will a mass balance be helpful but insufficient to completely solve the problem? Under what conditions is it a bad idea to use a mass balance?

7.17 Calculate the average velocity of fluid in the space between the rod and the cup as shown in Figure P7.1. The rod is moving at a speed of 2 cm/s into the cup, which is filled with water. Does it matter what fluid is in the cup? Why or why not?

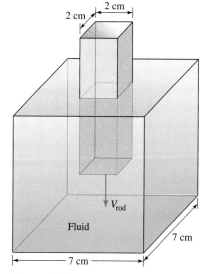

Figure P7.1

7.18 Reconsider the case of water displacing gasoline in a tank as shown in Figure 7.10. Select an appropriate control volume for this problem and develop a quantitative expression for the mass conservation law.

7.19 A fume hood in a chemistry laboratory is being used to store an open container of volatile liquid as shown in Figure P7.2. Molecules of the liquid vapor are leaving the container at a constant rate, and the concentration of the vapor molecules within the hood is

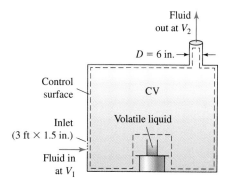

Figure P7.2

maintained at a constant rate by the ventilation fan.

(a) Apply the Reynolds transport theorem to the control volume indicated in Figure P7.2 and offer a physical interpretation for each term in this equation.

(b) The opening at the bottom of the fume hood is 3 ft wide and 1.5 in. high, and the air leaves the hood through a circular duct of diameter 6 in. at a velocity of 400 ft/min. Determine the average air velocity entering the hood through the rectangular port.

(c) Suppose the ventilation fan fails so that the average velocity at both the rectangular and circular ports is zero. Apply the Reynolds transport theorem under these conditions to the control volume indicated in Figure P7.2 and offer a physical interpretation for each term in this equation.

7.20 Water is flowing through the Y-shaped piping system shown in Figure P7.3. The velocities at the three ports are given by $V_1 = 1.5$ m/s (into the device), $V_2 = 2.5$ m/s (into the device), and $V_3 = 3.5$ m/s (out of the device). If the two pipes on the left each have a diameter of 5 cm, determine the diameter of the single pipe on the right.

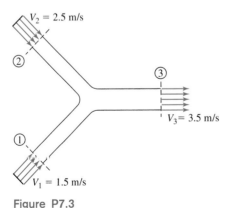

Figure P7.3

7.21 Reconsider the piping flow shown in Figure P7.3. Let the two supply pipes have a diameter of 2.5 in. and assume that over the period $0 < t < 2$ s, the velocities at ports 1 and 2 are given by $V_1 = 10$ (in./s^2)t and $V_2 = 20$ (in./s^2)t. Determine the velocity at port 3 during the period $0 < t < 2$ s if the diameter of the exit pipe is 3.5 in.

7.22 Reconsider the piping flow shown in Figure P7.3. Water is flowing into the device through port 1 with a mass flowrate of 5 slugs/s. Water is flowing out of the device through port 3 with a volume flowrate of 0.8 ft^3/s. Determine the average velocity and direction of water flow through port 2 if the diameter of the pipe at port 2 is 3 in.

7.23 Consider the multiport device shown in Figure P7.4. Water is flowing through the device, and you are given the following information: $A_1 = 100$ cm^2, $A_2 = 200$ cm^2, $A_3 = 300$ cm^2, $A_4 = 400$ cm^2, the average velocity at port one is 3 m/s (into the device), the volume flowrate at port 2 is 0.030 m^3/s (into the device) and the mass flowrate at port 3 is 40 kg/s (out of the device).

(a) Determine the average fluid velocity at port 4.

(b) At which port is the magnitude of the average fluid velocity the greatest?

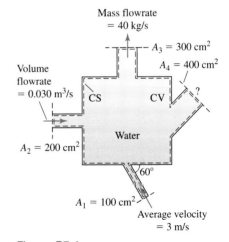

Figure P7.4

7.24 Consider the multiport device shown in Figure P7.4. Water is flowing through the

device, and you are given the following information: $A_1 = 100$ cm^2, $A_2 = 200$ cm^2, $A_3 = 300$ cm^2, the average velocity at port one is 2 m/s (into the device), the mass flowrate at port 2 is 35 kg/s (out of the device), and the volume flowrate at port 3 is 0.040 m^3/s (out of the device).
(a) Determine the mass flowrate at port 4.
(b) Determine the cross-sectional area of port 4 if the magnitude of the average velocity at that port is the same that at port 2.

7.25 An inflatable backyard swimming pool is being filled from a garden hose with a flowrate of 0.12 gal/s.
(a) If the pool is 8 ft in diameter, determine the time rate of change of the depth of the water in the pool.
(b) Suppose that the pool has a drain port on its side. Immediately after the drain is opened, the rate of change of the depth of the water in the pool is observed to decrease to 80% of the value calculated in part a. If the diameter of the drain port is 0.5 in., determine the average fluid velocity at the drain port.

7.26 Consider the incompressible airflow within the boundary layer over a flat plate as shown in Figure P7.5. The velocity profile is given by the equation $\mathbf{u} = U[2(y/h) - (y/h)^2]\mathbf{i}$, where h is a function of x. For the specific value of x in this problem, the value of h is given as 8 mm. If U is 80 m/s and the plate has a width (into the paper) of 0.5 m, determine the mass flowrate of air into or out of the top surface of the control volume.

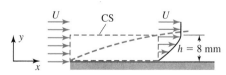

Figure P7.5

7.27 Reconsider the incompressible airflow described in Problem 7.26. At what value of h will the mass flow out the top of the control volume be equal to the mass flow out of the right-hand side of the control volume?

7.28 In Chapter 6 we learned that the velocity profile for fully developed laminar flow through a cylindrical pipe is given by $\mathbf{u} = W_0[1 - (r/R_P)^2]\mathbf{e}_z$. As shown in Figure P7.6, however, when the fluid first enters the pipe, the velocity profile is uniform. Use a mass balance to determine the relationship between the initial uniform velocity, W_I, and the maximum velocity along the centerline of the fully developed flow, W_O.

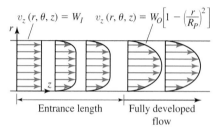

Figure P7.6

7.29 In Chapter 6 we learned that the velocity profile for fully developed laminar flow between flat plates is given by $\mathbf{u} = W_0[1 - (x/h)^2]\mathbf{k}$. As shown in Figure P7.7, however, when the fluid first enters the region between the plates, the velocity profile is uniform. Use a mass balance to determine the relationship between the initial uniform velocity, W_I, and the maximum velocity along the centerline of the fully developed flow, W_O.

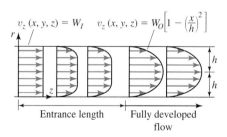

Figure P7.7

7.30 A new jet engine is being tested in a wind tunnel as illustrated in Figure P7.8. Air, with properties characteristic of U.S. Standard Atmosphere at 6000 m enters the engine at a velocity of 275 m/s through a circular intake

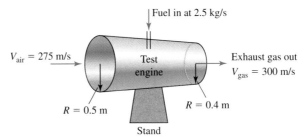

Figure P7.8

port of radius 0.5 m. Fuel enters the engine at a mass flowrate of 2.5 kg/s. If the gas leaves the engine with an average velocity of 300 m/s through an exit port of radius 0.4 m, calculate the density of the exhaust gas. How would your solution to this problem change if the engine had been attached to the wing of an airplane flying through still air at a velocity of 990 km/h?

7.31 Consider the velocity field $\mathbf{u} = C_1 tx\mathbf{i} + C_2 txy\mathbf{j} + C_3 tz\mathbf{k}$, where $C_1 = 1\,\text{s}^{-2}$, $C_2 = 2\,\text{m}^{-1}\text{s}^{-2}$, and $C_3 = -2\,\text{s}^{-2}$. For the region, $0 < x < 1$ m, $0 < y < 2$ m, and $0 < z < 3$ m, use a mass balance to investigate the claim that this flow conserves mass for a constant density fluid.

7.32 The Eulerian velocity field $\mathbf{u} = C[x\mathbf{i} - y\mathbf{j}]$ is characteristic of what is known as "flow in a corner." Do a mass balance over the area $0 < x < 5$ m and $0 < y < 5$ m to investigate the claim that this flow conserves mass for a constant density fluid.

7.33 Reinvestigate Example 7.3 using the mixed control volume shown in Figure 7.11C.

Section 7.6.2

7.34 Provide a physical interpretation for each of the terms of the momentum balance given in Eq. 7.18.

7.35 Show how the integral momentum balance equation can be simplified to yield the integral hydrostatic equation under appropriate conditions. What other types of flow conditions lead to simplification of the momentum balance equations?

7.36 What types of fluid flow problems suggest the use of the momentum balance equation?

7.37 Some of the terms in the momentum balance equation are generally more important in gas flow problems than they are in liquid flow problems. For which terms is this true? Explain your answer.

7.38 Water flow through a horizontal conical nozzle is illustrated in Figure P7.9. The fluid enters the nozzle through a circular port of area 8 in.² and leaves the nozzle through a port of area 1.5 in.². The fluid enters the nozzle at a uniform pressure of 59 psig. The fluid leaves the nozzle at atmospheric pressure with an average velocity of 25 m/s.
(a) Use a mass balance to find the average fluid velocity at the entrance port.
(b) Calculate the horizontal force necessary to hold the nozzle in place.

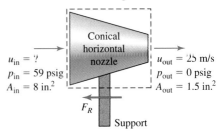

Figure P7.9

7.39 Flow through a vertical conical nozzle is illustrated in Figure P7.10. The fluid

enters the nozzle through a circular port of area 50 cm² and leaves the nozzle through a port of area 10 cm². The average fluid velocity at the entrance port is 3 m/s. The pressure across the entrance port is uniform at 210 kPa, and the fluid leaves the nozzle at atmospheric pressure. The mass of the nozzle is 1.5 kg.
(a) Calculate the total force necessary to hold the nozzle in place. Neglect both the weight of the nozzle and the weight of the fluid in the nozzle.
(b) Calculate the total force necessary to hold the nozzle in place. Include the weight of both the nozzle and the fluid in the nozzle in your calculations.

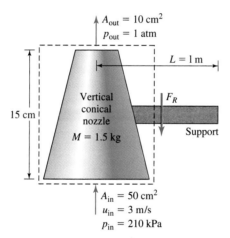

Figure P7.10

7.40 Flow through a 90° reducing elbow is illustrated in Figure P7.11. The diameter of the inlet port is 12 cm, and the diameter of the

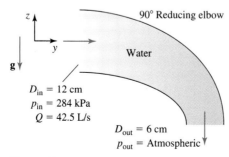

Figure P7.11

exit port is 6 cm. The volume flowrate of water through the elbow is 42.5 L/s. The entrance pressure is 284 kPa, and the fluid exits the elbow at atmospheric pressure. Determine the magnitude and direction of the force necessary to hold the elbow in place. Neglect the weight of the elbow and the fluid in the elbow in your calculations.

7.41 Flow through a horizontal 180° pipe bend is shown in Figure P7.12. The pipe diameter is constant at 15 cm, and the volume flowrate is 125 L/s. The pressure at the entrance port is 183 kPa, and the pressure at the exit port is 148 kPa. Determine the magnitude and direction of the force necessary to hold the elbow in place. Neglect the weight of the elbow and the fluid in the elbow in your calculations.

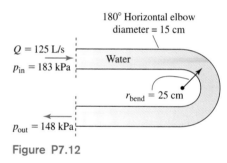

Figure P7.12

7.42 Based on the results of the tests on the prototype jet engine illustrated in Figure P7.8, a second-generation engine has been designed. During this stage of testing it is necessary to determine the thrust generated by the engine using the setup shown in Figure P7.13. Air, with properties characteristic of U.S. Standard Atmosphere at 6000 m enters the engine at a velocity of 275 m/s through a circular intake port of radius 0.5 m. The mass flowrate of the fuel is 2.5 kg/s. The exhaust gas leaves the engine with an average velocity of 300 m/s at atmospheric pressure. Calculate the magnitude and direction of the force exerted on the support structure.

7.43 A cylindrical container of diameter 20 in. is placed on a scale as shown in Figure P7.14. At steady state, the height of the

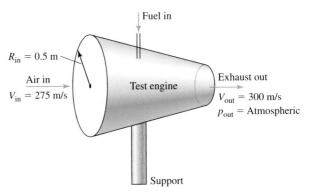

Figure P7.13

water in the tank is 2.5 ft. Water enters the tank through a top port of diameter 6 in. at a velocity of 24 ft/s. The water leaves the tank through two exit ports each with a diameter of 6 in. If the scale shows a reading of 585 lb_f, calculate the weight of the tank when it is empty.

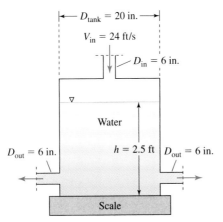

Figure P7.14

7.44 A sluice gate, like the one illustrated schematically in Figure P7.15, is often used to control the flow of a river to prevent or minimize flood damage.
(a) Using the information provided in Figure P7.15, calculate the force per unit width exerted on the gate.
(b) If the gate was closed to stop the flow and the depth of the water remained at 10 ft, would the force on the gate under static conditions be greater than, less than, or equal to the force you calculated in part a?

7.45 Water from the nozzle of a fire hose strikes a flat a stop sign as shown in Fig-

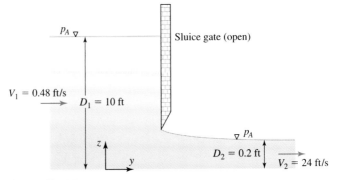

Figure P7.15

ure P7.16. Water exits the nozzle at a velocity of 45 ft/s through a circular exit port of diameter 5 in.
(a) Calculate the horizontal reaction force on the stop sign support.
(b) What, if anything, can you conclude about the fluid flow characteristics in the vertical direction for this situation?

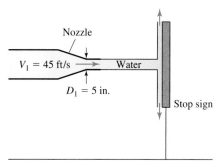

Figure P7.16

7.46 Suppose the water from the fire hose described in Problem 7.45 strikes the device shown in Figure P7.17. Neglecting the change in water elevation as it moves over the device, determine the magnitude and direction of the force exerted by the water stream on the device.

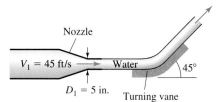

Figure P7.17

7.47 Problems P7.45 and P7.46 dealt with a plate at 90° to the flow and an angled plate that turned the flow 45°, respectively. What happens when the plate is parallel to the flow? Consider the geometry shown in Figure P7.18. The fluid approaches the leading edge of the plate with a uniform velocity $\mathbf{u} = U_0\mathbf{i}$. The no-slip condition at the plate surface forces the fluid velocity to zero at $y = 0$. The slowly moving fluid particles near the surface exert a drag on the adjacent fluid molecules such that a roughly parabolic velocity profile develops in the boundary layer. At any distance y from the leading edge of the plate, the boundary layer has a characteristic thickness δ. The velocity profile at a specific value of x is given by $\mathbf{u} = U_0[2(y/\delta) - (y/\delta)^2]\mathbf{i}$. In this particular problem the fluid is air with a density of 1.22 kg/m³, $U_0 = 140$ km/h, the x coordinate has been selected such that $\delta = 1$ cm, and the plate has a width of 1.3 m.
(a) Calculate the volume flowrate of air through the top surface of the indicated control volume.
(b) Calculate the drag force on the plate in the x direction.

7.48 Reconsider the syringe problem examined in Example 7.8. This time assume that the seal between the piston and syringe wall is imperfect and allows some leakage of fluid out the back end of the syringe. If the volume flowrate of leaking fluid is 5% of the volume flowrate of fluid out of designed exit port of the syringe, calculate the net force necessary to move the piston forward at a velocity of 0.2 cm/s.

Section 7.6.3

7.49 Provide a physical interpretation for each term in Eq. 7.22.

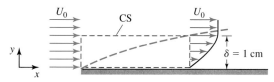

Figure P7.18

7.50 What types of fluid flow problems suggest the use of the integral energy conservation equation for a control volume?

7.51 Discuss the differences between the forms of the energy equation given in Eqs. 7.22 and 7.33. That is, what simplifications and assumptions are built into Eq. 7.33, and under what conditions are these changes appropriate?

7.52 In Example 7.13 we investigated the flow of air into an insulated tank. Reconsider this flow with the initial mass flowrate unknown and the initial rate of temperature change in the tank measured to be 0.04 K/s. Assume that the heat transfer rate is sufficiently small that it can be neglected.

7.53 Compressed nitrogen is stored in a tank of volume 0.3 m³ at 2.0 MPa and 58°C. If when the valve is opened the instantaneous mass flowrate is 45 g/s, calculate the initial rate of temperature change.

7.54 Water is flowing horizontally through a pump as shown in Figure P7.19. The flowrate is 350 gal/min, the entrance port has a cross-sectional area of 10 in.², and the exit port has a cross-sectional area of 1 in.². The pressures at the entrance and exit ports are 4 and 45 psig, respectively. If the temperature increase across the pump is 0.13°F, calculate the power in horsepower required to operate the pump. Assume that the process is adiabatic.

7.55 Air is flowing through a horizontal turbine as shown in Figure P7.20. At the entrance port the conditions are $V = 100$ ft/s, $T = 320°F$, $p = 160$ psi, and area $= 30$ in.². The conditions at the exit port are $V = 300$ ft/s, $T = 50°F$, and $p = 45$ psi. It is also known that there is a heat loss of 600,000 Btu/h at the turbine.
(a) Determine the density of the air at the exit port.
(b) Determine the diameter of the circular exit port.
(c) Determine the power (in hp) produced by this turbine.
(d) Comment on the relative magnitudes of the power generated and the heat loss.

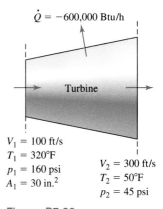

$\dot{Q} = -600{,}000$ Btu/h

Turbine

$V_1 = 100$ ft/s
$T_1 = 320°F$
$p_1 = 160$ psi
$A_1 = 30$ in.²

$V_2 = 300$ ft/s
$T_2 = 50°F$
$p_2 = 45$ psi

Figure P7.20

7.56 Steam is flowing through a horizontal turbine as shown in Figure P7.21. The conditions at the entrance port are $V = 90$ ft/s and $h = 1480$ Btu/lb$_m$. The corresponding conditions at the exit port are $V = 180$ ft/s and $h = 1100$ Btu/lb$_m$. The mass flowrate is 100 lb$_m$/s. The process is occurring adiabatically.

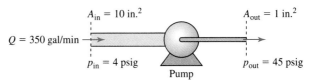

$A_{in} = 10$ in.² $A_{out} = 1$ in.²
$Q = 350$ gal/min
$p_{in} = 4$ psig $p_{out} = 45$ psig
Pump

Figure P7.19

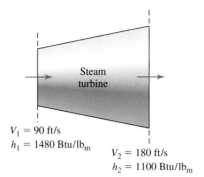

$V_1 = 90$ ft/s
$h_1 = 1480$ Btu/lb$_m$
$V_2 = 180$ ft/s
$h_2 = 1100$ Btu/lb$_m$

Figure P7.21

(a) Determine the change in kinetic energy associated with this process.
(b) Determine the change in enthalpy associated with this process.
(c) Determine the power (in hp) produced by this turbine.
(d) Comment on the relative magnitudes of the quantities calculated in parts a–c.

7.57 A horizontal air compressor requires a power input of 650 hp for a mass flowrate of 0.7 slugs/s. Air enters the device at STP with a negligible velocity and exits the device at $p = 55$ psig, $T = 110°$F through an exit port of diameter 7 in.
(a) Use the ideal gas law to determine the fluid density at the exit port.
(b) Calculate the average velocity at the exit port.
(c) Determine the heat transfer rate for this compressor.
(d) Compare the magnitudes of the power input, the heat transfer rate, and the change in kinetic energy of the fluid for this device.

7.58 The change in elevation of the water passing over Niagara Falls is 52 m. Assuming that the process is adiabatic and using the control volume shown in Figure P7.22, determine the temperature change associated with this flow. Note that $V_1 = V_2 = 0$, since we have selected the control surfaces to be at the surface of each large body of water.

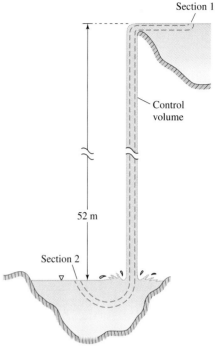

Figure P7.22

7.59 Reconsider the flow illustrated in Figure P7.22. Suppose that you have conclusive evidence that the water temperature at the surface of the lower body of water is slightly lower than that at the surface of the upper body of water. Offer a viable explanation for this observation.

7.60 A steady flow air-handling device experiences a heat transfer rate into the device of 67 kJ/s as illustrated in Figure P7.23. The property values at each of the three ports are as follows: port 1, diameter = 21 cm, flowrate = 3.0 m³/s, $T = 21°$C, $p = 140$ kPa, $z = 0$ m; port 2, diameter = 17 cm, flowrate = 1.5 m³/s, $T = 92°$C, $p = 150$ kPa, $z = 0.5$ m; port 3, diameter = 34 cm, $T = 38°$C, $p = 210$ kPa, $z = 1.2$ m.
(a) Calculate the flowrate at port 3.
(b) Find the power input necessary to run the device.

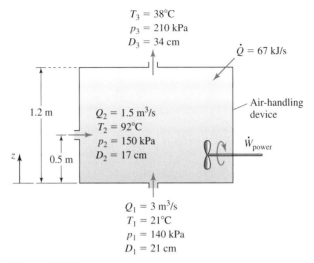

Figure P7.23

Section 7.6.4

7.61 What kinds of fluid flow problems are analyzed using an angular momentum balance equation?

7.62 Provide a physical interpretation of each term in Eq. 7.40. Also discuss the most common simplifications associated with this equation and the conditions under which each simplification is justified.

7.63 What is the torque on the support due to the nozzle shown in Figure P7.10. The moment arm from the base of the support to the centerline of the nozzle is $L = 1$ m.

7.64 Calculate the torque due to the flow on the 180° horizontal pipe bend shown in Figure P7.12. The radius of the bend is $r_{BEND} = 25$ cm. Do not consider body forces.

7.65 Reconsider the rotary cleaning brush analyzed in Example 7.18. In the current design the jets are perpendicular ($\theta = 90°$) to the x or y axis. You are now to analyze the brush with the jets at an arbitrary angle θ. Determine the torque available to turn the brush as in the example, but now also as a function of θ.

7.66 Water flows at 10 ft/s, up for $H = 15$ ft in a 5 in. diameter pipe. The inlet pressure is 30 psi. At the top are two 4 in. diameter, 90° elbows that are orthogonal to each other. From the elbows run $L = 1$ ft pipe stubs that discharge to the atmosphere (see Figure P7.26). The flowrate through each discharge is the same. The mass of the pipe and water is 5 lb_m/ft. Calculate the torque necessary to hold the pipes in place.

Additional, Unclassified Control Volume Problems

7.67 Describe the kinds of fluid problem that suggest the use of each type of balance. That is, under what conditions would you attempt to solve a problem by using a mass balance? When is the use of a momentum balance indicated? When would you use an energy balance? What about an angular momentum balance? Provide simple examples of systems to support your conclusions.

7.68 Explain why a mass balance is frequently used as the first step in solving a great variety of fluid mechanics problems, while an angular momentum balance is used far less commonly.

7.69 What kind of balance (i.e., mass, momentum, energy, or angular momentum,) would you use to answer a question about the torque exerted by a fluid on a rotating mechanical device? Why?

7.70 List and describe three different types of fluid mechanics problem that *cannot* be solved by means of the techniques described in this chapter.

7.71 An incompressible fluid flow is described by the velocity field $\mathbf{U} = C(x^2\mathbf{i} + y^2\mathbf{j} + z^2\mathbf{k})$, where C is a constant with dimensions $\{L^{-1}t^{-1}\}$. Set up a cubic control volume with one corner located at the origin and three edges aligned along the positive x, y, and z axes to investigate whether this flow conserves mass. How could you use the vector concept of a gradient to simplify this problem?

7.72 An incompressible fluid flow is described by the velocity field $\mathbf{U} = C(x^2 y\mathbf{i} + x^3\mathbf{j} - 2xyz\mathbf{k})$, where C is a constant with dimensions $\{L^{-2}t^{-1}\}$. Set up a cubic control volume with one corner located at the origin and three edges aligned along the positive x, y, and z axes to investigate whether this flow conserves mass. How could you use the vector concept of a gradient to simplify this problem?

7.73 Use a control volume analysis to verify that the flow over a cylinder defined in Problem 6.12 conserves mass. How could you use the vector concept of a gradient to simplify this problem?

7.74 Use a control volume analysis to verify that the flow over a sphere defined in Problem 6.13 conserves mass. How could you use the vector concept of a gradient to simplify this problem?

7.75 Use a control volume analysis to verify that the "tornado flow" defined in Problem 6.16 conserves mass. How could you use the vector concept of a gradient to simplify this problem?

7.76 Moist air enters a dehumidifier at the rate of 0.2 lb$_m$/s. Water drains out of the device at the rate of 16 lb$_m$/h. Determine the velocity at which the (less moist) air exits the device if the area of the exit port is 0.5 ft^2.

7.77 Consider the flow situation illustrated in Figure P7.24. The oil has a specific gravity of 0.90.
(a) Suppose that water is flowing into the tank at exactly the same mass flowrate as that of the oil flowing out of the tank. In which direction, if at all, is air moving through the vent?
(b) Suppose that water is flowing into the tank at exactly the same volume flowrate as that of the oil flowing out of the tank. In which direction, if at all, is air moving through the vent?

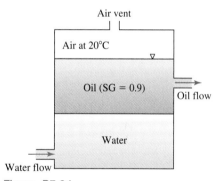

Figure P7.24

7.78 A water jet is turned through an angle of 180° by the device shown in Figure P7.25. The fluid stream has a constant diameter of 2 cm. If the magnitude of the restraining force

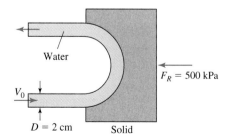

Figure P7.25

is 500 kPa, estimate the average water velocity of the incoming water jet. Do you think this is a realistic set of flow conditions? Why or why not?

7.79 Consider the flow of water through the device shown in Figure P7.26. The supply line is 0.5 in. in diameter and water is flowing through the entrance port at a velocity of 10 ft/s and a pressure of 30 psi. The two exit ports are each 0.4 in. in diameter, and water exits the two ports at atmospheric pressure. If the volume flowrates through the two exit ports are the same, calculate the following.
(a) The average fluid velocity through either exit port.
(b) The magnitude and direction of the restraining force necessary to hold the device in place [the mass of the piping and water in the piping is 90 lb$_m$].

first entrance port has $D = 10$ in. and an average fluid velocity of $V = A + B\cos(\omega t)$, where $A = 20$ ft/s, $B = 10$ ft/s, and $\omega = 2\pi/\text{s}$. The second entrance port has $D = 6$ in. and $V = 15$ ft/s.
(a) Determine an expression for the average velocity through the exit port as a function of time.
(b) Evaluate the volume flowrate through the exit port at $t = 1$ min.
(c) Determine the minimum and maximum mass flowrates through the exit port.

7.81 A circus act involves a dog balancing on a circular plate suspended above a steady water jet as shown in Figure P7.27. The diameter of the water jet is 1.5 in. and the mass of the dog and plate is 25 lb$_m$. What is the average velocity of the fluid in the vertical jet?

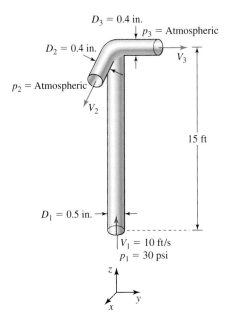

Figure P7.26

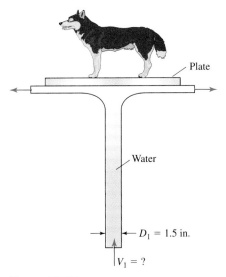

Figure P7.27

7.80 A horizontal fluid-handling device has two entrance ports and one exit port. There is no accumulation of fluid in the device. The exit port has a diameter of 8 in. The

7.82 Air is flowing steadily through a duct of constant cross-sectional area. At a specific upstream cross section, the air properties are $p = 100$ psi, $T = 85°$F, and $V_{\text{average}} = 250$ ft/s. Given that the air properties at a downstream cross section of the duct are $p = 20$ psi and $T = 0°$F, determine the average air velocity at the downstream cross section.

7.83 Two children are attempting to fill a cylindrical inflatable swimming pool by using garden hoses. One hose is supplying water at the rate of 0.5 L/s, while the other hose is supplying water at the rate of 0.4 L/s. Unfortunately, the kids forgot to close the plug on the drain. If the drain, which is on the side of the pool near the bottom, is 1.2 cm in diameter, calculate the average velocity at which fluid is leaving the drain port if the water level in the pool is independent of time. Do you think the flow values in the problem are reasonable? Why or why not?

7.84 One way to calibrate a flow meter is to use a device known as a weigh tank as illustrated in Figure P7.28. The weight of the empty weigh tank is 200 N. Water is entering the tank through a port of diameter 3 cm at an average velocity of 5 m/s.
(a) Determine the reading on the scale after the water has been flowing for 5 seconds.
(b) Does this value represent the true weight of the water in the tank? If so, explain why, if not calculate the true weight of the water in the tank.

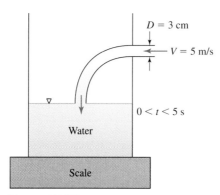

Figure P7.28

7.85 The steady flowrate of water through a hydroelectric turbine is 2.5×10^6 gal/min. The water is supplied to the turbine through a cylindrical pipe of radius 15 ft. Determine the Reynolds number characteristic of the fluid flow through the supply pipe.

7.86 Consider the sluice gate shown in Figure P7.29. The gate has a width of 8 m.
(a) Determine the magnitude of the horizontal restraining force required to hold the gate in place for the indicated flow conditions. [*Hint:* The details of the flow conditions directly under the sluice gate need not be known to solve this problem.]
(b) What is the magnitude of the horizontal restraining force required to hold the gate in place when it is closed and the upstream water depth is 4 m?
(c) Compare your answers to those for parts a and b and explain why they are different.
(d) Your professor believes that the horizontal force on the sluice gate can be used together with knowledge of the upstream depth and velocity to determine the downstream flow conditions. Do you agree with this statement? If so, derive the relevant equation(s). If not, explain why not.

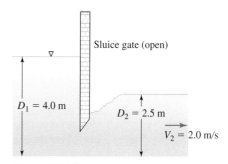

Figure P7.29

7.87 Consider the water tank on a frictionless cart shown in Figure P7.30. The diameter of the exit port is 1.5 in. and the average water velocity through the exit port is 15 ft/s. Water is entering through the tank from above in such a way that the water level in the tank remains constant. After exiting the tank, the water jet is deflected by a vane through an angle of θ, where θ can be varied between 0 and 80°.
(a) Calculate the tension in the cable when $\theta = 30°$.
(b) At what angle will the tension in the cable be a maximum?

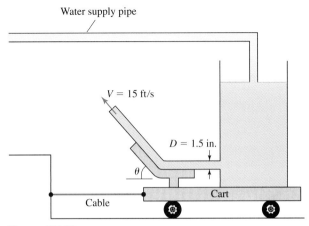

Figure P7.30

7.88 A cone-shaped hole in the ground has a depth of 2 m and a circular opening of diameter 1.5 m. This hole is being filled with water at the rate of 1.5 L/s.
(a) How long will it take to fill the hole with water?
(b) Calculate the rate at which the depth of the water is increasing after the water has been flowing for 1 minute.

7.89 Figure P7.31 shows an illustration of the fluid-handling device known as an expansion chamber. The free surface of the water has a diameter of 2 ft, and the diameter of the entrance and exit ports is 1 ft. If the average fluid velocity through the entrance port is 15 ft/s and the height of the free surface is increasing at the rate of 0.1 in./s, calculate the mass flowrate through the exit port.

7.90 Under steady state conditions in a wind tunnel test, a certain aircraft engine experiences a flowrate of 2 slugs/s of air and 1.4 slugs/min of liquid fuel into the combustion chamber through different ports. The exhaust gases leave the combustion chamber through an exit port of diameter 1 ft with an average velocity of 1000 mph. Estimate the density of the exhaust gases.

7.91 At time $t = 0$, water is flowing through a straight horizontal pipe of diameter 1 ft at the rate of 9.0 ft^3/s and exits the pipe at atmospheric pressure. The flowrate is increasing at the rate of 0.5 ft^3/s^2. What is the

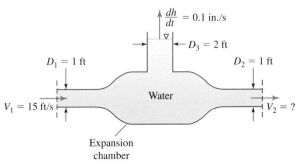

Figure P7.31

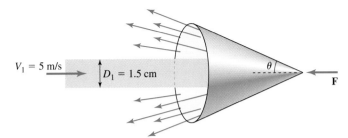

Figure P7.32

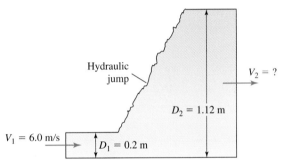

Figure P7.33

pressure in the pipe 200 ft upstream of the exit port? Assume friction is negligible.

7.92 A water jet of velocity 5 m/s and diameter 1.5 cm strikes a fixed cone as shown in Figure P7.32. If the magnitude of the velocity of the deflected fluid is also 5 m/s, determine the magnitude of the horizontal restraining force as a function of the cone angle θ.

7.93 Reconsider the flow geometry shown in Figure P7.32. Let the properties of the incoming water jet remain the same, but rotate the cone 180° so that its axis is still aligned with the water stream but the tip of the cone now points to the left. Determine the magnitude of the horizontal restraining force as a function of the cone angle θ.

7.94 Suppose the cone in Figure P7.32 is replaced with a hemispherical shell whose axis is aligned with the incoming water jet (open end facing jet). Determine the magnitude of the horizontal restraining force acting on the shell. State any assumptions.

7.95 In Chapter 15 we will discuss open channel flow in some detail. It will be shown that under certain conditions, a flow feature known as a hydraulic jump can develop. This is illustrated in Figure P7.33. Upstream of the hydraulic jump the water depth is 0.2 m, and the average fluid velocity is 6.0 m/s. If the depth of the water downstream of the jump is 1.12 m, calculate the average water velocity at the downstream location. Do you think the water heats up or cools down as it goes through the jump? Why?

7.96 Reconsider the flow condition known as a hydraulic jump as illustrated in Figure P7.33. Let the upstream water height be D_1 and the upstream velocity be V_1. Use the continuity and momentum equations to derive expressions for the downstream water height, D_2, and downstream velocity, V_2. Be sure to state any assumptions.

7.97 A rocket has an initial mass of 12,000 kg. After a vertical liftoff it burns fuel at

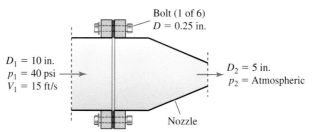

Figure P7.34

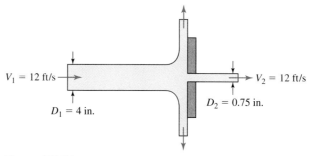

Figure P7.35

the rate of 10 kg/s for 80 s until its fuel is exhausted. During this time a constant thrust of 220 kN is developed. Calculate the final velocity of the rocket. You may neglect air resistance.

7.98 Reconsider the incompressible airflow within the boundary layer over a flat plate as shown in Figure P7.5 and described in Problem 7.26. In this problem all flow conditions are the same except that the horizontal plate is porous and suction beneath the plate is resulting in an average fluid velocity through the plate of magnitude 5 mm/s. Determine the volume flowrate of air into or out of the top surface of the control volume.

7.99 The horizontal nozzle shown in Figure P7.34 is held together at the flange by a set of six bolts. Each bolt has a diameter of 1/4 in. The bolts are substandard, however, and will fail when the stress in any bolt reaches 500 psi. The upstream diameter is 10 in. and the downstream diameter is 5 in. The nozzle is designed to operate with an upstream pressure of 40 psi. and an upstream velocity of 15 ft/s. Will the flawed bolts be able to survive under the design conditions?

7.100 A water jet of diameter 4 in. and velocity 12 ft/s strikes a flat plate with a $\frac{3}{4}$ in. diameter hole drilled through it as shown in Figure P7.35. If the velocity of the downstream jet is the same as that of the upstream jet, calculate the magnitude and direction of the force necessary to hold the plate stationary. State any assumptions.

7.101 A faculty member is about to go from her office to the student center to get some lunch. Unfortunately, it's pouring outside. One of her students suggests that if she runs, she will get less wet than walking. The faculty member isn't sure about the accuracy of this advice so she does a quick calculation. She approximates her body as a control volume in the shape of a rectangular prism. Did she run or walk? Why.

7.102 A cylindrical tank is draining through a circular hole in its base. The tank has a diameter of 0.5 m and the depth of water in the

tank is 0.4 m at the instant in question. If the volume flowrate out of the tank is 0.005 m³/s determine the following.
(a) The instantaneous rate of change of the height of the liquid in the tank.
(b) The diameter of the hole in the bottom of the tank.

7.103 Water flows through a horizontal 180° reducing bend as shown in Figure P7.36. The diameters of the entrance and exit ports are known to be $D_1 = 10$ in. and $D_2 = 3$ in., respectively. The pressure at the entrance port is $p_1 = 50$ psia and the exit port is at atmospheric pressure. The Reynolds number at the upstream cross section is 5.5×10^5.
(a) Calculate the upstream velocity.
(b) Calculate the downstream velocity.
(c) Calculate the magnitude and direction of the restraining force required to hold the reducing bend in place.

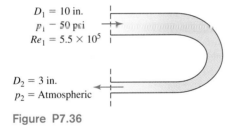

$D_1 = 10$ in.
$p_1 = 50$ psi
$Re_1 = 5.5 \times 10^5$

$D_2 = 3$ in.
$p_2 = $ Atmospheric

Figure P7.36

7.104 A schematic illustration of a room in a home is shown in Figure P7.37. The room has air entering through three windows and exiting through one doorway. The windows have (open) dimensions of 3 ft × 2 ft and the door has (open) dimensions of 3 ft × 6.5 ft. Plot the average air velocity through the door as a function of the angle θ, which describes the wind direction. Let θ vary from 0 to 180°. What is the velocity of the air through the doorway when the wind blows due north?

7.105 One the authors has a basement that leaks during heavy rains. The basement floor has an area of 1200 ft² and the water depth has been known to increase at a rate of 0.5 in./h. What size pump should the author rent to counter this influx of water (i.e., to keep the water level constant)? If the goal is not to keep the level constant, but rather to drain the basement at a rate of 2.0 in./h, even during the flooding, what capacity pump is required? Note that pump capacities at the local Rent-A-Pump store are reported in gallons per minute.

7.106 Consider the sluice gate shown in Figure P7.38. The upstream water depth and velocity are 10 m and 1 m/s, respectively. If the downstream depth is 3 m, what is the force on the gate due to the water?

7.107 In many industrial processes evaporative cooling towers are used to extract heat from circulating water. The process is illustrated in Figure P7.39. Water at $T = 50°C$ enters the evaporator at a flowrate of 30 L/s and exits the device at $T = 32°C$ with a flowrate of 29.3 L/s. The flowrate of the moist air out of the device is 19.8 kg/s.

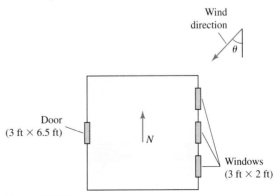

Figure P7.37

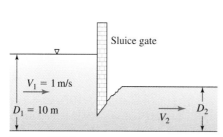

Figure P7.38

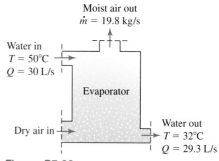

Figure P7.39

(a) Calculate the flowrate of dry air into the evaporator.
(b) Answer this question *without* doing a calculation. If the water enters and exits the device through ports of equal area, is the average fluid velocity higher or lower at the exit port? Why?
(c) Answer this question *without* doing a calculation. If the air enters and exits the device through ports of equal area, is the average fluid velocity higher or lower at the exit port? Why? Are you sure?

7.108 A jet of compressed nitrogen gas has a diameter of 1.5 cm and an average velocity of 50 m/s. This jet strikes a flat plate at an angle, as shown in Figure P7.40. If the gas velocity remains constant, determine the mass flowrates for each of the exit streams. Also calculate the magnitude and direction of the restraining force necessary to hold the flat plate in place.

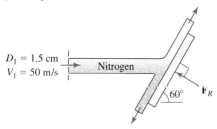

Figure P7.40

7.109 A dehumidifier designed for use in a damp basement takes in moist air (100% relative humidity) at 90°F and atmospheric pressure through a port of area 2.5 in.² at an average velocity of 10 ft/s. The drier air (55% relative humidity) exits the device at 90°F and atmospheric pressure. The water that drains out of the dehumidifier is collected in a reservoir capable of holding one gallon of water. The mass ratio of water vapor to air in the incoming flow is 3.2×10^{-2}. If the reservoir is initially empty, how long can the dehumidifier run under these conditions before the reservoir is filled to 95% of its capacity?

7.110 The tank shown is Figure P7.41 is resting on a frictionless surface. The volume flowrate through the supply line is adjusted so that the water level in the tank remains constant. The upper exit port, which is located 1 ft below the surface of the water, has a diameter of 1 in. The lower exit port is located 2 ft below the water surface. What diameter must the lower exit port have if the tank is to remain motionless? *Hint:* $V_{Exit} = \sqrt{2gh}$ where h is depth below the free surface.

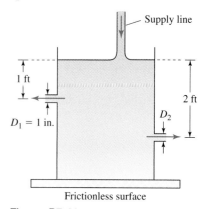

Figure P7.41

7.111 A rocket fueled by solid propellant has no entrance ports and a single exit port. Fuel combustion inside the rocket occurs at $T = 2250°F$ and 140 psi to produce an exhaust gas (with near ideal gas behavior and a molecular weight of 29). The flow conditions at the exit port are area $= 1.75 \text{ ft}^2$, $p = 13$ psi, $V = 3750$ ft/s, and $T = 900°F$. Calculate the mass of propellant required to permit an engine operation time of 1 hour. Does your answer seem reasonable?

7.112 Figure P7.42 shows a device known as a hydraulic accumulator. It is designed to reduce pressure variations in hydraulic systems. The entrance and exit ports each have a diameter of 0.25 in. and the free surface has a diameter of 4 in. If hydraulic fluid is flowing into the device at the rate of 0.25 gal/min and flowing out of the exit port at an average velocity of 0.39 ft/s, calculate the rate at which the free surface is rising or falling.

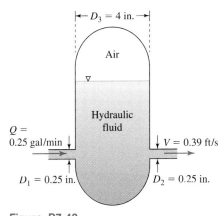

Figure P7.42

7.113 A pop-up sprinkler head for a lawn watering system is illustrated in Figure P7.43. The water flows upward through a supply line of diameter 1 in. at an average velocity of 30 ft/s. The water then moves radially outward between the two parallel solid disks. The disks are 4.0 in. in diameter and are separated by a gap of 0.20 in. Calculate the water velocity as it leaves the sprinkler head.

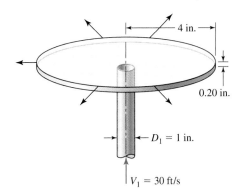

Figure P7.43

7.114 Consider the fluid-handling device shown in Figure P7.44. Each of the three ports has a diameter of 3 cm. Water enters the center port at 100 kPa gage and a volume flowrate of 0.010 m³/s and exits to the atmosphere through the upper and lower port at a volume flowrate of 0.006 m³/s. Calculate the magnitude and direction of the force required to hold the device in place.

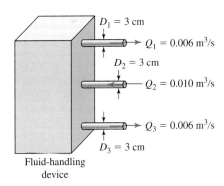

Figure P7.44

7.115 The open channel flow device known as a submerged or "drowned" weir is illustrated in Figure P7.45. The channel has a width of 50 ft and the upstream conditions are $V_u = 5$ ft/s and $h_u = 10$ ft. If the downstream velocity is 15 ft/s, calculate the magnitude of the horizontal force on the weir.

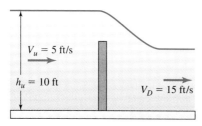

Figure P7.45

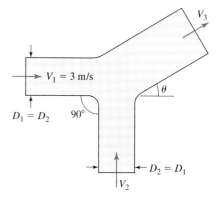

Figure P7.47

7.116 A device known as a jet pump is shown in Figure P7.46. Water flows through the smaller pipe ($D = 7$ cm) with an average velocity of 50 m/s. As this flow exits the supply pipe it entrains a secondary flow of water in the washer-shaped (annular) region between the inner and outer pipe walls. The outer pipe has $D = 25$ cm. At a cross section sufficiently downstream from the exit port of the supply pipe, the two flows are fully mixed and the average fluid velocity is found to be 8.528 m/s. Determine the average velocity of the water in the annular region between the supply pipe and the outer pipe at the plane of the exit of the supply pipe.

7.117 Two fluid jets of equal cross-sectional area collide to produce a single, well-mixed, output stream as shown in Figure P7.47. If the velocity of the horizontal input stream is 3 m/s and the angle θ is found to be 30°, determine the velocity of the vertical input stream.

7.118 Reconsider the flow geometry shown in Figure P7.47. The two input streams are equal in velocity, but while the horizontal stream is water, the vertical stream is SAE 30 W oil. Determine the relative velocity of the exit stream (in terms of V_0, the velocity of the input streams, and the angle θ).

7.119 A compressed gas tank used to inflate party balloons has a volume of 5 ft^3 and is filled with helium. When the valve is opened, He flows out of the tank. The flow conditions at the exit port are $D = 0.25$ in., $V = 1000$ ft/s, $T = 0°$F, and $p = 55$ psi. Calculate $d\rho/dt$ at this instant in time.

7.120 Two immiscible liquids are mixed in the Y-shaped pipe illustrated in Figure P7.48. The first liquid has a specific gravity of 1.05 and enters the device with a flowrate of 1500 gal/min. The second liquid has a specific gravity of 0.85 and enters the device with a flow rate of 1200 gal/min. If the mixture exits the device at a velocity of 30 ft/s, estimate the cross-sectional area of the exit port.

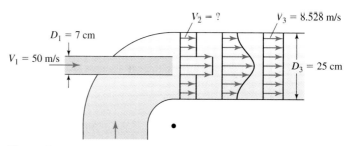

Figure P7.46

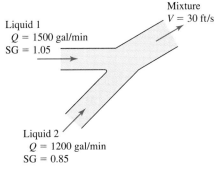

Figure P7.48

7.121 One of the authors lives in a two-story house attached to a single-story garage. During a torrential downpour he wanders outside to see if his gutters and downspouts are working properly. During his examination it appears that the flowrate out of the longer (2-story) downspout is greater than the flowrate from the shorter (1-story) downspout. He goes back inside and does a calculation to determine the expected ratio of these two flowrates. Attempt to reproduce this calculation. Be sure to state any assumptions.

7.122 When a cylindrical tank of radius R drains through a circular hole or radius r in the center of its base, the velocity of the fluid leaving the tank is approximately $V = \sqrt{2gh}$, where h is the instantaneous height of the liquid in the tank. Derive an expression for the time required for the tank to drain from an initial fluid height of h_0 to a final height of $0.5h_0$.

7.123 Figure P7.49 shows a conical tank with a maximum radius R and half-angle θ draining through a circular hole of radius r at its tip. The velocity of the fluid leaving the tank is approximately $V = \sqrt{2gh}$, where h is the instantaneous height of the liquid in the tank. Derive an expression for the time required for the tank to drain from an initial fluid height of h_0 to a final height of $0.5h_0$.

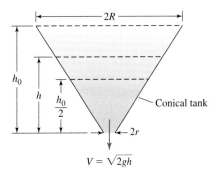

Figure P7.49

8 FLOW OF AN INVISCID FLUID: THE BERNOULLI EQUATION

8.1 Introduction
8.2 Frictionless Flow Along a Streamline
8.3 Bernoulli Equation
    8.3.1 Bernoulli Equation for an Incompressible Fluid
    8.3.2 Cavitation
    8.3.3 Bernoulli Equation for a Compressible Fluid
8.4 Static, Dynamic, Stagnation, and Total Pressure
8.5 Applications of the Bernoulli Equation
    8.5.1 Pitot Tube
    8.5.2 Siphon
    8.5.3 Sluice Gate
    8.5.4 Flow through an Area Change
    8.5.5 Draining of a Tank
8.6 Relationship to the Energy Equation
8.7 Summary
Problems

8.1 INTRODUCTION

One of the oldest approximations in fluid dynamics is that of an inviscid fluid. If such a fluid existed, its viscosity would be exactly zero, it would be incapable of exerting a shear stress, and its flow would be frictionless. The state of stress in an inviscid fluid is characterized solely by the pressure distribution. Since there is no fluid whose viscosity is zero, one might ask why we study the flow of an inviscid fluid, i.e., frictionless flow. One answer, which is particularly relevant in engineering design, is that the highest performance that can be achieved from a fluid-handling device is that which would occur in the absence of friction. Thus the study of frictionless flow through a device may allow us to optimize a preliminary design. A second answer, which is of great importance in aerodynamics, is that the principal phenomenon that arises from the existence of viscosity, i.e. shear stress, is not significant in all flows.

To better understand this comment, consider the flow of the two fluids of greatest engineering interest, air and water. The viscosity of these fluids is not zero, but it is small in comparison to many other fluids. Now according to Eq. 1.2c, the relationship between shear stress, viscosity, and velocity gradient is given by $\tau = \mu(du/dy)$. It is easy to see that if the viscosity is zero, then the shear stress must also be zero. The equation also shows, however, that if the velocity gradient at a point in a flow of air or water is small, then the local shear stress may be negligible (from an engineering point of view) even though the viscosity is nonzero.

A number of important flows in engineering contain regions in which shear stresses are in fact negligible. In these regions the flow is approximately frictionless and indistinguishable from the flow of an inviscid fluid. Thus in this chapter we will consider frictionless flow and the flow of an inviscid fluid to be synonymous, and we will investigate the characteristics of frictionless flow by employing the concept of an inviscid fluid.

In the next section we use mass and momentum balances to investigate the characteristics of frictionless flow along a streamline. The result of this analysis is the Bernoulli equation, which describes the energy content of a flow along a streamline in the absence of friction. After discussing the forms of the Bernoulli equation for incompressible and compressible fluids, we introduce the concepts of static, dynamic, stagnation, and total pressure. These "pressures" are all routinely used in fluid dynamics and their definitions are based on terms in the Bernoulli equation. Next we discuss selected applications of Bernoulli's equation, and conclude the chapter by reviewing the relationship between the Bernoulli equation and the energy balance for steady flow along a streamline. Although an energy balance is not needed to derive the Bernoulli equation, it provides additional insight into the conditions under which the Bernoulli equation is applicable.

 CD/Kinematics/Streamlines and the Streamfunction/Streamlines in 3-D.

8.2 FRICTIONLESS FLOW ALONG A STREAMLINE

Consider a short segment of a streamline passing through a region of frictionless flow as illustrated in Figure 8.1A. To understand how fluid and flow properties vary along this streamline, we will write mass and momentum balances for a differential control volume (CV) of length ds centered at an arbitrary point on this streamline at time t and bounded by nearby streamlines as shown in Figure 8.1B. At any point along the streamline the area is $A = A(s)$, and the velocity is $\mathbf{u} = V(s, t)\mathbf{e}_S$, where the unit vector $\mathbf{e}_S$ lies along the streamline. We will assume that the area A is small enough that the velocity and fluid properties are uniform on any cross section and vary only in the s direction, that the only body force acting is gravity, and that the viscosity of the fluid is zero.

We begin by applying a mass balance to the differential CV shown in Figure 8.1B. Because the side of the CV is defined by other streamlines, the velocity vector on the

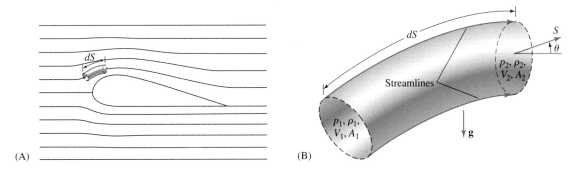

Figure 8.1 (A) Streamlines of flow over an airfoil. (B) CV between streamlines.

side is wholly tangential to the surface. Therefore transport only occurs across the inlet and exit surfaces. We can write a mass balance for this CV using Eq. 7.11,

$$\int_{CV} \frac{\partial \rho}{\partial t} dV + \int_{CS} \rho(\mathbf{u} \cdot \mathbf{n}) dS = 0$$

Making use of the characteristics of the differential CV, we can evaluate the accumulation and flux terms by writing the corresponding integrals in the following respective forms:

$$\int_{CV} \frac{\partial \rho}{\partial t} dV = \int_1^2 \frac{\partial \rho}{\partial t} A \, ds \quad \text{and} \quad \int_{CS} \rho(\mathbf{u} \cdot \mathbf{n}) dS = -\rho_1 A_1 V_1 + \rho_2 A_2 V_2 = \int_1^2 \frac{d}{ds}(\rho A V) \, ds$$

Writing the mass balance as a single integral, we obtain

$$\int_1^2 \left[\frac{\partial \rho}{\partial t} A + \frac{d}{ds}(\rho A V) \right] ds = 0$$

Since points 1 and 2 are arbitrary, we conclude that the following relationship holds at each point along the streamline:

$$\frac{\partial \rho}{\partial t} A + \frac{d}{ds}(\rho A V) = 0 \qquad (8.1)$$

We next write a momentum balance for the differential CV by applying Eq. 7.18:

$$\int_{CV} \frac{\partial}{\partial t}(\rho \mathbf{u}) \, dV + \int_{CS} (\rho \mathbf{u})(\mathbf{u} \cdot \mathbf{n}) \, dS = \int_{CV} \rho \mathbf{f} \, dV + \int_{CS} \mathbf{\Sigma} \, dS$$

Equation 4.19 allows us to write the stress vector in terms of its normal and tangential components as $\mathbf{\Sigma} = -p\mathbf{n} + \boldsymbol{\tau}$. Since the shear stress is zero in the flow of an inviscid fluid, we have $\mathbf{\Sigma} = -p\mathbf{n}$, and the momentum balance becomes

$$\int_{CV} \frac{\partial}{\partial t}(\rho \mathbf{u}) \, dV + \int_{CS} (\rho \mathbf{u})(\mathbf{u} \cdot \mathbf{n}) \, dS = \int_{CV} \rho \mathbf{f} \, dV + \int_{CS} -p\mathbf{n} \, dS$$

We are interested in calculating the component of this equation in the s direction along the streamline. We will again make use of the geometry of the differential CV to evaluate each term in the resulting momentum balance in the s direction. Evaluating the s component of the accumulation term we find

$$\int_{CV} \frac{\partial}{\partial t}(\rho V)\, dV = \int_1^2 \frac{\partial}{\partial t}(\rho V) A\, ds$$

Next we evaluate the s component of the flux term to obtain

$$\int_{CS} (\rho V)(\mathbf{u} \cdot \mathbf{n})\, dS = -\rho_1 A_1 V_1^2 + \rho_2 A_2 V_2^2 = \int_1^2 \frac{d}{ds}(\rho A V^2)\, ds$$

The s component of the body force term is given by

$$\int_{CV} \rho f_S\, dV = \int_1^2 \rho(-g\sin\theta) A\, ds$$

where we have used the fact that f_S, the component of the gravitational body force in the s direction, is $-g\sin\theta$, where $\theta = \theta(s)$, (see Figure 8.1B). For later use we note that $\sin\theta\, ds = dz$, and thus $\sin\theta = dz/ds$.

Before considering the s component of the pressure integral, we use Gauss's theorem to write the surface integral as

$$\int_{CS} -p\mathbf{n}\, dS = \int_{CV} -\nabla p\, dV$$

We are interested only in the component of the volume integral in the s direction. Since the component of the pressure gradient in the s direction is dp/ds, the desired result is $\int_1^2 -(dp/ds) A\, ds$. Gathering terms and writing the s component of the momentum balance as a single integral gives

$$\int_1^2 \left[\frac{\partial(\rho V)}{\partial t} A + \frac{d}{ds}(\rho A V^2) + \rho(g\sin\theta) A + \frac{dp}{ds} A \right] ds = 0$$

Since points 1 and 2 are arbitrary, we conclude that at every point along the streamline we have

$$\frac{\partial(\rho V)}{\partial t} A + \frac{d}{ds}(\rho A V^2) + \rho(g\sin\theta) A + \frac{dp}{ds} A = 0 \qquad (8.2)$$

Equations 8.1 and 8.2 describe frictionless flow along a streamline. We will use them in the next section to derive the Bernoulli equation.

8.3 BERNOULLI EQUATION

In the flow of an inviscid fluid the only forces acting on the fluid are the inertial, body, and pressure forces per unit volume. Viscous (friction) forces are absent. Recalling our discussion in Chapter 2 of the forms of fluid energy, we intuitively expect that in a

HISTORY BOX 8-1

Daniel Bernoulli (1700–1782) was a member of a distinguished family of mathematicians from Basel, Switzerland. His book *Hydrodynamica,* published in 1738, described the physical phenomena predicted by the Bernoulli equation. However, it was several years later that Leonhard Euler first derived the Bernoulli equation as we know it today. It is also interesting to note that Johann Bernoulli, Daniel's father, published a book he called *Hydraulica* in 1743, but dated his work 1728 to take credit from his son for this fundamental physical relationship. Thus, our knowledge of the relationship between pressure and velocity is better understood than the dynamics of this brilliant family.

frictionless flow, energy is conserved. The equation expressing this fact, which is named after Daniel Bernoulli, can be derived using the results of Section 8.2.

Our derivation of Bernoulli's equation begins with Eq. 8.2:

$$\frac{\partial(\rho V)}{\partial t} A + \frac{d}{ds}(\rho A V^2) + \rho(g \sin\theta) A + \frac{dp}{ds} A = 0$$

which expresses conservation of momentum in the flow of an inviscid fluid along a streamline. We can simplify this equation further by expanding the first term as

$$\frac{\partial(\rho V)}{\partial t} A = \rho \frac{\partial V}{\partial t} A + \frac{\partial \rho}{\partial t} A V$$

$$= \rho \frac{\partial V}{\partial t} A - V \frac{d}{ds}(\rho A V)$$

where we have used Eq. 8.1 to write $(\partial \rho/\partial t) A = -(d/ds)(\rho A V)$. We can now write the first and second terms in Eq. 8.2 as:

$$\frac{\partial(\rho V)}{\partial t} A + \frac{d}{ds}(\rho A V^2) = \rho \frac{\partial V}{\partial t} A - V \frac{d}{ds}(\rho A V) + \frac{d}{ds}(\rho A V^2)$$

$$= \rho \frac{\partial V}{\partial t} A + \rho A \frac{d}{ds}\left(\frac{V^2}{2}\right)$$

and write Eq. 8.2 as

$$\rho \frac{\partial V}{\partial t} A + \rho A \frac{d}{ds}\left(\frac{V^2}{2}\right) + \rho (g \sin\theta) A + \frac{dp}{ds} A = 0$$

Dividing through by ρA, and recognizing that $\sin\theta = dz/ds$, we obtain

$$\frac{\partial V}{\partial t} + \frac{d}{ds}\left(\frac{V^2}{2}\right) + g\frac{dz}{ds} + \frac{1}{\rho}\frac{dp}{ds} = 0 \qquad (8.3)$$

Since this equation holds at every point on a streamline, we can integrate it along the streamline from point 1 to point 2 to obtain

$$\int_1^2 \left[\frac{\partial V}{\partial t} + \frac{d}{ds}\left(\frac{V^2}{2}\right) + g\frac{dz}{ds} + \frac{1}{\rho}\frac{dp}{ds}\right] ds = 0$$

Taking advantage of the exact differential form of the middle two terms, we can write them as

$$\int_1^2 \frac{d}{ds}\left(\frac{V^2}{2}\right) ds = \frac{1}{2}\left(V_2^2 - V_1^2\right) \quad \text{and} \quad \int_1^2 g\frac{dz}{ds} ds = \int_1^2 g\, dz = g(z_2 - z_1)$$

8.3 BERNOULLI EQUATION

We also have $\int_1^2 (1/\rho)(dp/ds)\,ds = \int_1^2 dp/\rho$, thus we can write the resulting equation as

$$\int_1^2 \frac{\partial V}{\partial t}\,ds + \int_1^2 \frac{dp}{\rho} + \frac{1}{2}(V_2^2 - V_1^2) + g(z_2 - z_1) = 0 \qquad (8.4)$$

This is the general form of the Bernoulli equation. It applies to the frictionless flow of a compressible or incompressible fluid; the only restriction is that the path connecting the two points must be an instantaneous streamline.

A number of important simplifications to the Bernoulli equation merit additional discussion. These relate to whether the fluid is compressible or incompressible, and also to whether the flow is steady. In most applications, we do not apply the general form of the Bernoulli equation but rather use one of the simplified versions developed in the next sections.

8.3.1 Bernoulli Equation for an Incompressible Fluid

CD/Dynamics/Potential Flow/Incompressibility and Irrotationality.

Since the density of an incompressible fluid is constant, it may come outside the pressure integral in the general form of the Bernoulli equation, Eq. 8.4. This allows us to write the pressure integral as $\int_1^2 (dp)/\rho = (1/\rho)(p_2 - p_1)$. The Bernoulli equation for unsteady flow of a constant density fluid is therefore given by

$$\int_1^2 \frac{\partial V}{\partial t}\,ds + \frac{1}{\rho}(p_2 - p_1) + \frac{1}{2}(V_2^2 - V_1^2) + g(z_2 - z_1) = 0 \qquad (8.5)$$

We can interpret the constant density Bernoulli equation in terms of fluid energy. Recall that the total energy per unit mass in a fluid is defined as $e = u + p/\rho + \frac{1}{2}\mathbf{u} \cdot \mathbf{u} + gz$ and that the sum $p/\rho + \frac{1}{2}\mathbf{u} \cdot \mathbf{u} + gz$ is referred to as the mechanical energy per unit mass since each term relates to a mechanical rather than thermal form of energy. The total energy is the sum of the thermal energy of the fluid (as measured by the internal energy), and the mechanical energy. In the absence of viscous forces, the Bernoulli equation is a statement about the mechanical energy at two points along a streamline in a constant density flow. In an unsteady flow, we see from Eq. 8.5 that the mechanical energy content of the fluid at the two points differs only owing to the effects of the local acceleration, represented here by $\int_1^2 (\partial V/\partial t)\,ds$. This effect is illustrated in the next two examples.

Many flows of engineering interest involving incompressible fluids are steady. In steady flow, $\partial V/\partial t = 0$, and the leading term in Eq. 8.5 vanishes. Thus for a steady flow of an incompressible fluid, the Bernoulli equation becomes

$$\frac{1}{\rho}(p_2 - p_1) + \frac{1}{2}(V_2^2 - V_1^2) + g(z_2 - z_1) = 0$$

We normally write this equation as

$$\frac{p_1}{\rho} + \frac{1}{2}V_1^2 + gz_1 = \frac{p_2}{\rho} + \frac{1}{2}V_2^2 + gz_2 \qquad (8.6)$$

EXAMPLE 8.1

A pipe open at the top and filled with a liquid is closed by a valve at the end as shown in Figure 8.2. When the valve is opened, flow through the pipe begins. Assuming frictionless flow, find an expression for the acceleration of the liquid in the pipe.

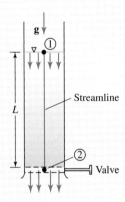

Figure 8.2 Schematic for Example 8.1.

SOLUTION

We will apply the unsteady, constant density Bernoulli equation between points 1 and 2 along a streamline down the center of the pipe at the instant after the valve opens. For this problem Eq. 8.5 takes the form

$$\int_1^2 \frac{\partial V}{\partial t}\, ds + \frac{1}{\rho}(p_2 - p_1) + \frac{1}{2}\left(V_2^2 - V_1^2\right) + g(z_2 - z_1) = 0$$

Since the flow is frictionless, the velocity at any pipe cross section is uniform, and a mass balance yields $V_1 = V_2$. The pressure at each point is atmospheric and the difference in elevation is $g(z_2 - z_1) = -gL$. Thus we have

$$\int_1^2 \frac{\partial V}{\partial t}\, ds - gL = 0 \tag{A}$$

To evaluate the integral along the streamline, we note that a mass balance shows that V is constant along s. Thus $\partial V/\partial t$ is also constant along s, and we can write

$$\int_1^2 \frac{\partial V}{\partial t}\, ds = \frac{dV}{dt}\int_1^2 ds = \frac{dV}{dt}(s_2 - s_1) = \frac{dV}{dt}L$$

Substituting this into (A), we find $(dV/dt)L = gL$, or $dV/dt = g$. That is, the local acceleration of the liquid is equal to g. The liquid is in free fall, and it is interesting to observe that the resulting acceleration is independent of the density of the fluid. Thus mercury and water behave the same in a frictionless free fall out of a tube.

The velocity of the liquid is now easily found to be $V(t) = gt$, and the kinetic energy of the entire column of liquid at any time is $\frac{1}{2}MV^2 = \frac{1}{2}\rho[\pi(d^2/4)L](gt)^2$. In time t the fluid column has dropped a distance $H = \int V(t)\,dt = \frac{1}{2}gt^2$, and the work done on the fluid by gravity is $MgH = \frac{1}{2}\rho[\pi(d^2/4)L](gt)^2$. As expected, the work done equals the increase in kinetic energy. The force applied to the fluid by the gravity field results in a constant acceleration of the liquid column in this frictionless flow. Notice that if the Bernoulli equation is applied just before the valve is opened with the liquid at rest, it predicts $(1/\rho)(p_2 - p_1) - gL = 0$, or $p_2 = p_1 + \rho g L$, which agrees with the hydrostatic equation.

EXAMPLE 8.2

A liquid column of total length L oscillates in a U-tube as shown in Figure 8.3. Assuming frictionless flow, find an expression predicting the frequency of oscillation.

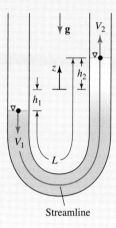

Figure 8.3 Schematic for Example 8.2.

SOLUTION

We will apply the unsteady, constant density Bernoulli equation between points 1 and 2 along a streamline along the center of the U-tube at the instant shown. Since the flow is frictionless, the velocity at any pipe cross section is uniform, and a mass balance shows that at any instant $V = V_1 = V_2$. The pressure at each point is atmospheric, and the difference in elevation is $g(z_2 - z_1) = g[h_2 - (-h_1)] = g(h_2 + h_1)$. Thus writing the Bernoulli equation, Eq. 8.5, under these conditions, we find

$$\int_1^2 \frac{\partial V}{\partial t}\,ds + g(h_2 + h_1) = 0$$

To evaluate the integral, note that V and $\partial V/\partial t$ are independent of s; thus we can write

$$\int_1^2 \frac{\partial V}{\partial t}\,ds = \frac{dV}{dt}\int_1^2 ds = \frac{dV}{dt}(s_2 - s_1) = \frac{dV}{dt}L$$

Thus the differential equation governing the oscillation is

$$\frac{dV}{dt}L + g(h_2 + h_1) = 0 \qquad \text{(A)}$$

To solve this equation we must remember that although L is constant, $h_1 = h_1(t)$, and $h_2 = h_2(t)$. However, from Figure 8.3 we see that $dh_1/dt = -V_1(t) = V$, and $dh_2/dt = V_2(t) = V$. This suggests that we differentiate (A) to obtain

$$\frac{d^2V}{dt^2}L + g\left(\frac{dh_2}{dt} + \frac{dh_1}{dt}\right) = 0$$

Substituting for the first derivatives and rearranging, we get the second-order ordinary differential equation

$$\frac{d^2V}{dt^2} + \frac{2g}{L}V = 0$$

The general solution to this equation is

$$V(t) = A\sin\sqrt{\frac{2g}{L}}\,t + B\cos\sqrt{\frac{2g}{L}}\,t$$

which describes an oscillation at a frequency

$$\omega = \sqrt{\frac{2g}{L}} \qquad \text{(B)}$$

Since the flow is frictionless, this motion is undamped. We see that the frequency depends on gravity and the total length of the liquid column but, perhaps surprisingly, not on the density or diameter of the tube. An experiment performed with a small diameter U-tube filled with water shows that the motion is highly damped owing to viscous friction. Thus we need to be careful about assuming frictionless flow. Even though water has a comparatively low viscosity, the flow in a small diameter tube is not properly modeled as an inviscid flow because the velocity gradients are significant. Thus the shear stresses (and hence friction) are not negligible.

Recall that we discussed Equation 8.6 briefly in Chapter 2 and listed it there in another form as Eq. 2.11. We can interpret the steady flow, constant density Bernoulli equation in terms of the mechanical energy content of the fluid: in the absence of viscous forces, the mechanical energy at two points along a streamline in a steady, constant density flow is the same.

8.3.2 Cavitation

Low pressures routinely occur in regions of high velocity within fluid machinery such as on the inlet (suction) side of a pump and on rapidly rotating propellers and impellers

EXAMPLE 8.3

Find an expression for the velocity and area of the falling jet of liquid shown in Figure 8.4.

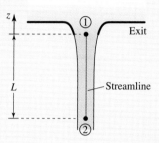

Figure 8.4 Schematic for Example 8.3.

SOLUTION

We will assume steady, constant density, frictionless flow, and apply Bernoulli's equation on the central streamline from a point at the exit where $z = 0$ to a point below at $z = -L$, where the velocity is V. We will also assume that the effects of surface tension are negligible, hence the pressure in the jet is atmospheric. The steady, constant density form of Bernoulli's equation, Eq. 8.6, is

$$\frac{p_1}{\rho} + \frac{1}{2}V_1^2 + gz_1 = \frac{p_2}{\rho} + \frac{1}{2}V_2^2 + gz_2$$

Applying this equation between the exit plane and the point at $z = -L$, and noting that the pressure is atmospheric at both locations, we have

$$\frac{p_A}{\rho} + \frac{1}{2}V_E^2 + g(0) = \frac{p_A}{\rho} + \frac{1}{2}V^2 + g(-L)$$

Canceling terms and solving for the velocity at $z = -L$, we have

$$V(L) = \sqrt{V_E^2 + 2gL} \tag{A}$$

which shows that the velocity increases as a result of the acceleration of gravity. A mass balance on the control volume shown reveals that $\rho V A = \rho V_E A_E$; thus we can write $A(L) = A_E(V_E/V)$ and use (A) to obtain the following formula for the area of the jet at any location below the exit plane:

$$A(L) = \frac{A_E V_E}{\sqrt{V_E^2 + 2gL}}$$

We see that the area of the jet decreases as it falls. This is necessary for mass to be conserved. Observation of a falling jet of liquid shows that the jet breaks into droplets some distance below the exit plane. Prior to the breakup, disturbances appear on the air–liquid interface. These phenomena are not accounted for in the preceding formulas, so once again we see the need for caution in applying the results of a Bernoulli equation analysis.

EXAMPLE 8.4

A decorative fountain employs a small nozzle attached to a high pressure water pipe as shown in Figure 8.5. If the maximum pressure available in the water pipe is 100 psig, how high will the water go? At what velocity does the water exit the nozzle?

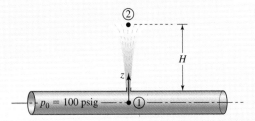

Figure 8.5 Schematic for Example 8.4.

SOLUTION

We will assume steady, constant density, frictionless flow, and apply Bernoulli's equation from a point inside the water pipe where $z = 0$ to a point at the top of the water jet where $z = H$, and the velocity is zero. Since the nozzle area is small in comparison to the cross-sectional area of the water pipe, we will assume that the water in the pipe is nearly at rest at a pressure p_0. The external water jet is at atmospheric pressure. Applying the steady, constant density form of Bernoulli's equation, Eq. 8.6, to this situation yields:

$$\frac{p_0}{\rho} + \frac{1}{2}V_1^2 + g(0) = \frac{p_A}{\rho} + \frac{1}{2}V_2^2 + gH$$

Neglecting the term $\frac{1}{2}V_1^2$, since the velocity in the pipe is very small, we obtain $p_0/\rho = p_A/\rho + gH$. Solving for the height and inserting the data gives:

$$H = \frac{p_0 - p_A}{\rho g} = \frac{(100 \text{ lb}_f/\text{in.}^2)(144 \text{ in.}^2/\text{ft}^2)}{(1.94 \text{ slugs/ft}^3)(32.2 \text{ ft/s}^2)} = 230.5 \text{ ft}$$

To find the exit velocity, we next apply Bernoulli's equation from a point inside the water pipe to the nozzle exit plane where the velocity is V_E, obtaining $p_0/\rho = p_A/\rho + \frac{1}{2}V_E^2 + gh_E$. Ignoring the slight change in gravitational potential energy over the height h_E, we find

$$\frac{p_0}{\rho} = \frac{p_A}{\rho} + \frac{1}{2}V_E^2$$

After solving for the exit velocity and substituting, the data yield

$$V_E = \sqrt{\frac{2(p_0 - p_A)}{\rho}} = \sqrt{\frac{2(100 \text{ lb}_f/\text{ft}^2)(144 \text{ in.}^2/\text{ft}^2)}{1.94 \text{ slugs/ft}^3}} = 121.8 \text{ ft/s}$$

Since $H = (p_0 - p_A)/\rho g$, we can also write the previous formula as $V_E = \sqrt{2gH}$, which illustrates that in the absence of friction, the kinetic energy in the flow as it exits the nozzle is converted to gravitational potential energy as the flow reaches maximum height. As you have undoubtedly experienced, water leaving a nozzle at high speed breaks up into droplets that are quickly slowed by air resistance. Thus this estimate of maximum height is highly optimistic. The result is useful for analyzing carefully designed low speed water jets that maintain a contiguous smooth filament of water as they rise.

(see Figure 8.6A). The result is the formation of water vapor bubbles in certain regions within the flow and on moving surfaces. If bubbles are then swept to regions of higher pressure, the bubbles implode. This phenomenon, called cavitation, can result in shock waves accompanied by pressures on the order of 50,000 psi (345 MPa). You can easily cause mild cavitation in a garden hose by bending the hose back on itself, nearly pinching shut the flow passage. The pinch creates a venturi in which the water moves at high speed, creating a pressure low enough for cavitation to occur. Try this experiment and listen for the sounds of cavitation.

As shown in Figure 8.6B, cavitation can result in significant damage to structural materials. An understanding of cavitation requires a review of the concept of vapor pressure. The vapor pressure of a liquid p_v is the pressure at which a liquid will boil or vaporize at a given temperature. This is why negative pressures in a liquid are never encountered. As the absolute pressure falls toward zero, the liquid boils and maintains a

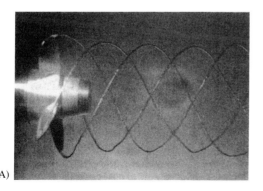
(A)

(B)

Figure 8.6 (A) Cavitation can occur at the lower pressure areas of a propeller. (B) Damage may result from the cavitation as shown on this turbine blade.

small but positive pressure in the vapor cavities that form. Vapor pressures for several liquids are given in Appendix A.

Boiling of a liquid in a vessel can be induced at a given pressure by increasing temperature, or at a given temperature by decreasing pressure. For example, water boils at 212°F at sea level (14.7 psia) but at 203°F in Denver (12.2 psia at 5000 ft). At 40°F the vapor pressure is 0.122 psia, a value that is less than 1% of standard atmospheric pressure. If you lowered the pressure on a container of water at 50°F below its vapor pressure, what would happen? Some fluids, including silicon oils, have very low vapor pressures that make them suitable for use in applications involving high vacuum, where water and common liquids will boil away.

Cavitation may be thought of as boiling or vapor bubble formation at a point in a moving liquid when the local pressure is below the vapor pressure at the point. We can use the Bernoulli equation to predict the occurrence of cavitation, as illustrated in Example 8.5.

EXAMPLE 8.5

Water flows through a round horizontal venturi and exits to the atmosphere as shown in Figure 8.7. Find an expression for the pressure at the throat, i.e., at the minimum area.

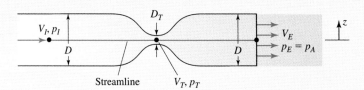

Figure 8.7 Schematic for Example 8.5.

SOLUTION

We will assume steady, constant density, frictionless flow through the venturi and apply the Bernoulli equation along the central streamline. Since the venturi is horizontal, the change in gravitational potential energy is zero, and Eq. 8.6 can be written as

$$\frac{p_1}{\rho} + \frac{1}{2}V_1^2 = \frac{p_2}{\rho} + \frac{1}{2}V_2^2$$

A mass balance shows that $V_I A = V_T A_T$, and also that $V_I A = V_E A$. Thus the velocities at the inlet and exit are the same, $V_I = V_E$, and the velocity at the throat is $V_T = V_I (D/D_T)^2$. Applying the Bernoulli equation between the inlet and the throat, we find

$$\frac{p_I}{\rho} + \frac{1}{2}V_I^2 = \frac{p_T}{\rho} + \frac{1}{2}V_T^2 = \frac{p_T}{\rho} + \frac{1}{2}V_I^2 \left(\frac{D}{D_T}\right)^4$$

Solving for the pressure at the throat, we obtain

$$\frac{p_T}{\rho} = \frac{p_I}{\rho} - \frac{1}{2}V_I^2\left[\left(\frac{D}{D_T}\right)^4 - 1\right] \quad \text{(A)}$$

We see that the pressure is lower at the throat. Applying the Bernoulli equation between the inlet and exit, and noting that the pressure at the exit is atmospheric, i.e., $p_E = p_A$, we obtain

$$\frac{p_I}{\rho} + \frac{1}{2}V_I^2 = \frac{p_A}{\rho} + \frac{1}{2}V_E^2$$

Since $V_I = V_E$, we conclude that

$$p_I = p_A \quad \text{(B)}$$

The pressure is atmospheric at the inlet to the venturi, falls to a minimum at the throat, and returns to atmospheric at the exit. Substituting (B) into (A), we can write the following expression for the pressure in the throat

$$\frac{p_T}{\rho} = \frac{p_A}{\rho} - \frac{1}{2}V_I^2\left[\left(\frac{D}{D_T}\right)^4 - 1\right] \quad \text{(C)}$$

We see that for a given venturi geometry, we can lower the pressure at the throat by increasing the flow speed. This works until we reach the vapor pressure p_V, at which point the liquid cavitates and prevents the pressure from decreasing further. The inlet speed at which cavitation occurs can be estimated by inserting $p_T = p_v$ into (C) and solving for V_I.

A venturi may be used as a vacuum pump by attaching the inlet to a garden hose. The vacuum is created in a second hose attached to a wall tap at the throat. The second hose may then be used to draw a liquid into the throat of the venturi, where it mixes with the main stream, or to lower the pressure on a vessel partially full of water to demonstrate boiling at room temperature.

8.3.3 Bernoulli Equation for a Compressible Fluid

To apply the Bernoulli equation to the frictionless flow of a compressible fluid, we must calculate the pressure integral in the general Bernoulli equation, Eq. 8.4. To do this we need to make some assumptions about the thermodynamic process undergone by the fluid in its passage from point 1 to point 2. Typically the compressible fluid of interest is a gas, and we can employ the perfect gas law $p = \rho RT$ (Eq. 2.20b), to describe the relationship between pressure, density, and temperature. A reversible, adiabatic process (i.e., an isentropic process) is consistent with the frictionless flow assumption underlying Bernoulli's equation. It can be shown that an isentropic process for a perfect gas is described by

$$\frac{p}{p_0} = \left(\frac{\rho}{\rho_0}\right)^\gamma \quad (8.7)$$

where $\gamma = c_P/c_V$ is the ratio of specific heats, and the subscript 0 denotes a datum state. We can also write this relationship as $p = K\rho^\gamma$, where K is a constant. Writing the total derivative of the pressure we have $dp = \gamma K\rho^{\gamma-1}$, and the pressure integral in the general form of the Bernoulli equation, Eq. 8.4, can be integrated to obtain

$$\int_1^2 \frac{dp}{\rho} = \int_1^2 \frac{\gamma K\rho^{\gamma-1}}{\rho} d\rho = \int_1^2 \gamma K\rho^{\gamma-2} d\rho = \frac{\gamma}{\gamma-1} K\rho^{\gamma-1}\bigg|_1^2 = \frac{\gamma}{\gamma-1}\frac{p}{\rho}\bigg|_1^2$$

$$= \left(\frac{\gamma}{\gamma-1}\right)\left(\frac{p_2}{\rho_2} - \frac{p_1}{\rho_1}\right)$$

Thus for an unsteady isentropic flow of a perfect gas the Bernoulli equation is

$$\int_1^2 \frac{\partial V}{\partial t} ds + \left(\frac{\gamma}{\gamma-1}\right)\left(\frac{p_2}{\rho_2} - \frac{p_1}{\rho_1}\right) + \frac{1}{2}(V_2^2 - V_1^2) + g(z_2 - z_1) = 0 \quad (8.8)$$

An alternate way to write this equation is to note that for a perfect gas

$$u + \frac{p}{\rho} = c_V T + \frac{p}{\rho} = \frac{p}{\rho}\left(\frac{c_V}{R} + 1\right)$$

Since $R = c_P - c_V$, we can show that $c_V/R + 1 = \gamma/(\gamma-1)$ and thus write

$$\left(\frac{\gamma}{\gamma-1}\right)\left(\frac{p}{\rho}\right) = u + \frac{p}{\rho}$$

The Bernoulli equation for unsteady, isentropic flow is therefore often written as

$$\int_1^2 \frac{\partial V}{\partial t} ds + (u_2 - u_1) + \left(\frac{p_2}{\rho_2} - \frac{p_1}{\rho_1}\right) + \frac{1}{2}(V_2^2 - V_1^2) + g(z_2 - z_1) = 0 \quad (8.9)$$

We see that in an unsteady isentropic flow, the total energy (i.e., the sum of the internal, pressure potential, kinetic, and gravitational potential energy) differs at two points along a streamline owing to the effects of local acceleration.

The steady flow versions of the Bernoulli equation for an isentropic flow of a perfect gas are developed from Eqs. 8.8 and 8.9 by dropping the unsteady term and rearranging. The results are two equivalent forms of the Bernoulli equation for steady, isentropic flow:

$$\left(\frac{\gamma}{\gamma-1}\right)\left(\frac{p_1}{\rho_1}\right) + \frac{1}{2}V_1^2 + gz_1 = \left(\frac{\gamma}{\gamma-1}\right)\left(\frac{p_2}{\rho_2}\right) + \frac{1}{2}V_2^2 + gz_2 \quad (8.10)$$

and

$$u_1 + \frac{p_1}{\rho_1} + \frac{1}{2}V_1^2 + gz_1 = u_2 + \frac{p_2}{\rho_2} + \frac{1}{2}V_2^2 + gz_2 \quad (8.11)$$

Equation 8.11 shows that the Bernoulli equation may be interpreted in this case as a statement of energy conservation. In a steady isentropic flow of a gas, the total energy (i.e., the sum of the internal, pressure potential, kinetic, and gravitational potential

EXAMPLE 8.6

A gas flows from a large plenum through a passage and exits to the atmosphere as shown in Figure 8.8. If the pressure and temperature of the gas in the plenum and exit plane are p_0, T_0, and p_E, T_E, respectively, find an expression for the exit velocity V_E. Assume steady, isentropic flow.

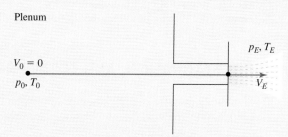

Figure 8.8 Schematic for Example 8.6.

SOLUTION

We can solve this problem by applying the Bernoulli equation for steady isentropic flow along a streamline between the two points shown in the figure. Since the passage is horizontal $g(z_2 - z_1) = 0$ and we can write Eq. 8.10 as

$$\left(\frac{\gamma}{\gamma - 1}\right)\left(\frac{p_1}{\rho_1}\right) + \frac{1}{2}V_1^2 = \left(\frac{\gamma}{\gamma - 1}\right)\left(\frac{p_2}{\rho_2}\right) + \frac{1}{2}V_2^2$$

where point 1 refers to the plenum and point 2 to the exit. We will assume that in the large plenum the kinetic energy of the gas is negligible. Solving for the exit velocity we find

$$V_E = \sqrt{\left(\frac{2\gamma}{\gamma - 1}\right)\left(\frac{p_0}{\rho_0} - \frac{p_E}{\rho_E}\right)} \quad \text{(A)}$$

We see that for given plenum conditions, the exit velocity depends on the pressure and density at the exit. The maximum velocity corresponds to an exit pressure of zero. The geometry of the passage plays a critical role in determining the actual exit conditions for an isentropic flow. These phenomena are covered in books about compressible flow.

By using the perfect gas law, we can also write (A) as

$$V_E = \sqrt{\left(\frac{2\gamma R}{\gamma - 1}\right)(T_0 - T_E)} \quad \text{(B)}$$

We see that for a given plenum temperature, the exit velocity depends solely on the exit temperature. If you have felt high pressure air escaping from a nozzle or hose, you know the escaping air is cold. The maximum velocity that can be obtained corresponds to an exit temperature of zero. This is not possible of course, since the air will liquefy, an effect not accounted for in the preceding analysis.

energy) at two points along a streamline, is the same. Finally we can recall that the enthalpy of a perfect gas is given by Eq. 2.48 as $h = u + p/\rho$ and that the relationship between enthalpy change and temperature change is given by Eq. 2.21b: $h_2 - h_1 = c_P(T_2 - T_1)$. Thus the Bernoulli equation can also be written in terms of enthalpy or temperature if desired.

The use of the steady flow Bernoulli equation is illustrated in Example 8.6. In this particular example the flow is horizontal, so the change in gravitational potential energy $g(z_2 - z_1)$ is zero. In most applications of the Bernoulli equation in compressible flow the change in gravitational potential energy $g(z_2 - z_1)$ is negligible in comparison to the other terms, so this term may be dropped regardless of the orientation of the flow.

8.4 STATIC, DYNAMIC, STAGNATION, AND TOTAL PRESSURE

If we multiply each term in the steady flow, constant density form of Bernoulli's equation (Eq. 8.6) by the fluid density, we obtain an alternate form:

$$p_1 + \tfrac{1}{2}\rho V_1^2 + \rho g z_1 = p_2 + \tfrac{1}{2}\rho V_2^2 + \rho g z_2 \tag{8.12}$$

This can also be written as

$$p + \tfrac{1}{2}\rho V^2 + \rho g z = \text{constant} \tag{8.13}$$

Since pressure occurs alone in this form of Bernoulli's equation, dimensional consistency requires that the remaining terms also have dimensions of pressure. Indeed, examination of those terms shows that they do have dimensions of $\{FL^{-2}\}$, as expected. Fluid mechanics traditionally refers to each term in this equation as a different kind of pressure. The pressure p is often referred to as the static pressure, while the term $\tfrac{1}{2}\rho V^2$ is called the dynamic pressure. The remaining term, $\rho g z$, called the hydrostatic pressure, represents the change in the static pressure that would occur if the fluid moved along the streamline to an elevation of zero. The sum of the static pressure and dynamic pressure, $p + \tfrac{1}{2}\rho V^2$, is called the stagnation pressure, and the sum of all three terms, $p + \tfrac{1}{2}\rho V^2 + \rho g z$, is referred to as the total pressure. Since we potentially have to distinguish between several different "pressures" in a conversation with other engineers, it is worthwhile to take a closer look at these "pressures" and understand where each name comes from.

Static Pressure: The pressure p in fluid mechanics is a measure of the average normal stress existing at a point in a fluid. It is defined to be the pressure that would be measured by an observer or pressure sensor moving with the fluid. To such an observer, the fluid appears to be static or stationary, so this pressure is often called the static pressure. The static pressure might also be called the mechanical pressure, since it is defined by the normal force on a surface exposed to the fluid. In a constant density flow, as well as in nearly all other flows of interest, the mechanical pressure and the thermodynamic pressure defined by an appropriate equation of state are identical. The symbol p in Bernoulli's equation represents the static or mechanical pressure, and this pressure is

8.4 STATIC, DYNAMIC, STAGNATION, AND TOTAL PRESSURE

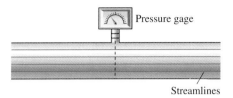

Figure 8.9 The pressure tap in the pipe wall does not disturb the streamlines; thus the gage measures the static pressure.

identical to thermodynamic pressure. Thus the pressure obtained by solving the hydrostatic equation, by performing a CV analysis, or from any analysis using the Bernoulli equation is the static pressure p.

Since it is difficult, if not impossible, to measure pressure by means of a probe moving with the fluid, the measurement of static pressure at a point in a fluid presents a challenge. Fortunately, a number of methods have been developed that do not require a moving probe. If the streamlines of a flow are straight and parallel, it can be shown that there is no variation in pressure normal to the streamlines (other than that due to gravity). This makes it possible to measure static pressure in a region of the flow where the streamlines are straight by using a wall opening or wall tap oriented perpendicular to the flow direction. Figure 8.9 shows a pressure tap located in the wall of the channel through which the fluid is moving. Notice the straight parallel streamlines. By attaching the pressure tap to a manometer or other type of pressure gage, it is possible to measure the static pressure at the wall to determine actual static pressure for all points along the line perpendicular to the flow direction. (A correction for the hydrostatic pressure variation may be added if necessary.)

Dynamic Pressure: The dynamic pressure, $\frac{1}{2}\rho V^2$, represents the pressure increase that would occur if all the kinetic energy of a fluid particle in a frictionless flow were converted into a corresponding increase in pressure potential energy. We can construct an imaginary situation in which this conversion would take place by considering a frictionless stagnation point flow of a constant density fluid and analyzing flow along the stagnation streamline as shown in Figure 8.10.

If we consider the flow along the stagnation streamline shown in Figure 8.10 and apply Eq. 8.12 between the point upstream where the speed is V_∞ and a point on the same streamline where the speed is V_S, the result is

$$p_\infty + \tfrac{1}{2}\rho V_\infty^2 + \rho g z_\infty = p_S + \tfrac{1}{2}\rho V_S^2 + \rho g z_S$$

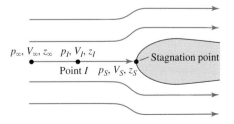

Figure 8.10 Flow along a streamline that ends at a stagnation point.

These points lie on the same elevation, hence $z_\infty = z_S$. Because of the symmetry and no-penetration conditions, $V_S = 0$ at the stagnation point. Thus Bernoulli's equation predicts a pressure at the stagnation point given by

$$p_S = p_\infty + \tfrac{1}{2}\rho V_\infty^2 \tag{8.14}$$

In the process of the fluid coming to rest, the static pressure has increased from its value upstream by an amount equal to the dynamic pressure $\tfrac{1}{2}\rho V_\infty^2$ that exists in the upstream flow.

Each point on the stagnation streamline has a distinct value for its static and dynamic pressure. Consider the intermediate point labeled I in Figure 8.10. Applying Eq. 8.12 between the upstream point and this point we have

$$p_\infty + \tfrac{1}{2}\rho V_\infty^2 = p_I + \tfrac{1}{2}\rho V_I^2 \tag{8.15}$$

Since we can reasonably conclude that the flow is slowing down as it approaches the stagnation point, the dynamic pressure at point I must be less than that upstream, i.e., $\tfrac{1}{2}\rho V_I^2 < \tfrac{1}{2}\rho V_\infty^2$. Thus we conclude that $p_I > p_\infty$: the static pressure increases along the streamline as the dynamic pressure decreases.

Stagnation Pressure: The pressure p_0 at a point in a frictionless flow where the fluid velocity is zero (i.e., the flow is stagnant) is known as the stagnation pressure. The concept of a stagnation pressure can also be understood with reference to the stagnation point flow of Figure 8.10. By definition, the stagnation pressure p_0 occurs where $V_S = 0$; thus the pressure at the stagnation point p_S is equal to the stagnation pressure, i.e., $p_0 = p_S$. Now consider the point upstream. According to Eq. 8.14, we have $p_S = p_\infty + \tfrac{1}{2}\rho V_\infty^2$. Substituting $p_0 = p_S$, we find $p_0 = p_\infty + \tfrac{1}{2}\rho V_\infty^2$, which shows that at the point upstream the sum of the static and dynamic pressures at this point is equal to the stagnation pressure. We can also use Eq. 8.14 to write $p_0 = p_I + \tfrac{1}{2}\rho V_I^2$, which shows that in a frictionless flow the stagnation pressure is the same at every point along the stagnation streamline.

Even if a point is not on a stagnation streamline, we can define the value of the stagnation pressure at the point by adding the static and dynamic pressures. Thus the stagnation pressure at any point in a constant density flow is given by the sum

$$p_0 = p + \tfrac{1}{2}\rho V^2 \tag{8.16}$$

Total Pressure: In a constant density flow, the total pressure p_T is defined as the sum of the static, dynamic, and hydrostatic pressures:

$$p_T = p + \tfrac{1}{2}\rho V^2 + \rho g z \tag{8.17}$$

Comparing this definition of total pressure with Bernoulli's equation (Eq. 8.13)

$$p + \tfrac{1}{2}\rho V^2 + \rho g z = \text{constant}$$

shows that the constant in Bernoulli's equation is the total pressure. Thus in a frictionless, constant density flow, the total pressure is a constant along a streamline.

8.4 STATIC, DYNAMIC, STAGNATION, AND TOTAL PRESSURE

EXAMPLE 8.7

A torpedo travels at 50 mph, 30 ft under the surface, as shown in Figure 8.11A. What are the static, dynamic, stagnation, hydrostatic, and total pressures at the point upstream on the stagnation streamline? What is the pressure at the stagnation point on the nose of the torpedo?

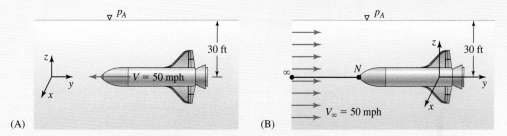

Figure 8.11 Schematic for Example 8.7.

SOLUTION

In a frame of reference fixed to the earth the flow is unsteady. However, by employing a frame of reference fixed to the torpedo, as shown in Figure 8.11B, we see that the flow is steady and is described by the steady, constant density Bernoulli equation.

The static pressure upstream is p_∞. For an observer moving with the fluid (which is at rest), this static pressure is given by $p_\infty = p_A + \rho g D$, where D is the depth. For the given depth of 30 ft, the static pressure upstream is thus

$$p_\infty = 14.7 \text{ psia} + (2.00 \text{ slugs/ft}^3)(32.2 \text{ ft/s}^2)(30 \text{ ft})\left(\frac{1 \text{ ft}^2}{144 \text{ in.}^2}\right) = 28.1 \text{ psia} \quad \text{(A)}$$

The density of seawater is taken from Appendix A in kg/m^3 and converted to slug/ft^3. The dynamic pressure at any location is $\frac{1}{2}\rho V^2$. Upstream the seawater approaches the torpedo with a velocity $V_\infty = 50$ mph. Thus the dynamic pressure upstream is

$$\frac{1}{2}\rho V_\infty^2 = \frac{1}{2}(2.00 \text{ slugs/ft}^3)(50 \text{ mph})^2 \left(\frac{1.467 \text{ ft/s}}{1 \text{ mph}}\right)^2 \left(\frac{1 \text{ ft}^2}{144 \text{ in.}^2}\right) = 37.4 \text{ psia} \quad \text{(B)}$$

The stagnation pressure is defined to be the sum of the static and dynamic pressure; thus the stagnation pressure upstream is

$$p_{0\infty} = p_\infty + \frac{1}{2}\rho V_\infty^2 = 28.1 \text{ psia} + 37.4 \text{ psia} = 65.5 \text{ psia} \quad \text{(C)}$$

Since the stagnation streamline is at an elevation $z = 0$, the hydrostatic pressure term is zero for all points on this streamline. The total pressure upstream is thus equal to the stagnation pressure:

$$p_{T\infty} = p_{0\infty} = 65.5 \text{ psia}$$

To find the pressure on the nose of the torpedo, consider the two points shown on the stagnation streamline. Writing the Bernoulli equation in the form of Eq. 8.12 between these points we have

$$p_\infty + \tfrac{1}{2}\rho V_\infty^2 + \rho g z_\infty = p_N + \tfrac{1}{2}\rho V_N^2 + \rho g z_N$$

On the nose the velocity is zero. Also $z_\infty = z_N$, so the corresponding terms cancel. Solving for the pressure on the nose we have

$$p_N = p_\infty + \tfrac{1}{2}\rho V_\infty^2 \qquad (D)$$

We see that the static pressure on the nose of the torpedo at the stagnation point is the sum of the upstream static pressure and the upstream dynamic pressure. The latter is also equal to the upstream stagnation pressure, thus from (C) the static pressure at the stagnation point on the nose of the torpedo is equal to the stagnation pressure for this streamline or

$$p_N = p_{0\infty} = 65.5 \text{ psia}$$

We see that the motion of the torpedo results in a substantial increase in static pressure in the vicinity of the stagnation point.

A comparison of the dynamic pressure in a low speed flow of air and water is instructive. For example, consider a flow at 50 mph in air and water. The corresponding dynamic pressures using standard density values are

$$\tfrac{1}{2}\rho V_\infty^2 = \tfrac{1}{2}(0.002378 \text{ slug/ft}^3)(50 \text{ mph})^2$$
$$\times \left(\frac{1.467 \text{ ft/s}}{1 \text{ mph}}\right)^2 \left(\frac{1 \text{ ft}^2}{144 \text{ in.}^2}\right)$$
$$= 4.4 \times 10^{-2} \text{ psia for air}$$

$$\tfrac{1}{2}\rho V_\infty^2 = \tfrac{1}{2}(1.94 \text{ slugs/ft}^3)(50 \text{ mph})^2$$
$$\times \left(\frac{1.467 \text{ ft/s}}{1 \text{ mph}}\right)^2 \left(\frac{1 \text{ ft}^2}{144 \text{ in.}^2}\right)$$
$$= 36.2 \text{ psia for water}$$

Looking at this another way, a dynamic pressure of 1 psia occurs at a speed of 348 ft/s in air, but only 12 ft/s in water.

The concepts of static, dynamic, stagnation, and total pressure are not limited to constant density flow. The various "pressures" at a point in a flow of a compressible fluid are defined in a similar way. In a compressible flow, the static pressure is p and the dynamic pressure is $\tfrac{1}{2}\rho V^2$. The stagnation pressure, however, is defined as the pressure that is obtained at a point of zero velocity in a frictionless (isentropic) flow. Considering a steady isentropic flow on a horizontal stagnation streamline and applying Eq. 8.10, we find

$$\left(\frac{\gamma}{\gamma-1}\right)\frac{p}{\rho} + \frac{1}{2}V^2 = \left(\frac{\gamma}{\gamma-1}\right)\left(\frac{p_0}{\rho_0}\right)$$

Thus the stagnation pressure in isentropic flow is given by

$$\frac{p_0}{\rho_0} = \frac{p}{\rho} + \frac{1}{2}\left(\frac{\gamma-1}{\gamma}\right)V^2 \qquad (8.18)$$

By using the perfect gas law and recalling that an isentropic flow obeys Eq. 8.7, $p/p_0 = (\rho/\rho_0)^\gamma$, we can employ the definition of Mach number as $M = V/c$,

8.4 STATIC, DYNAMIC, STAGNATION, AND TOTAL PRESSURE

EXAMPLE 8.8

A cruise missile flies at a Mach number $M = 0.5$ just above the surface of the ocean as shown in Figure 8.12A. What is the pressure at the stagnation point on the nose of the missile? What pressure is predicted by assuming incorrectly that the air is incompressible?

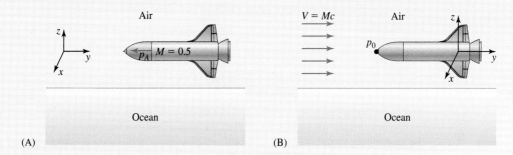

Figure 8.12 Schematic for Example 8.8.

SOLUTION

As in Example 8.7, in a frame of reference fixed to Earth the flow is unsteady. However, by choosing a reference frame fixed to the missile (see Figure 8.12B), the flow is steady and is described by the steady, isentropic Bernoulli equation. We do not need to use this equation, however, since we know that the pressure at the nose of the missile at the stagnation point is the stagnation pressure. Upon applying Eq. 8.19 we find

$$\frac{p_0}{p} = \left[1 + \left(\frac{\gamma - 1}{2}\right) M^2\right]^{\gamma/(\gamma-1)} = \left[1 + \left(\frac{1.4 - 1}{2}\right)(0.5)^2\right]^{1.4/(1.4-1)} = 1.19$$

Thus with the static pressure known to be $p = p_A$ at sea level, the stagnation pressure on the cruise missile at sea level is

$$p_0 = 1.19(14.7 \text{ psia}) = 17.4 \text{ psia}$$

If we incorrectly assume that the air is incompressible and apply Eq. 8.16, a Mach number of 0.5 corresponds to $V = Mc = 0.5(1117 \text{ ft/s}) = 559 \text{ ft/s}$, where the sea-level sound speed is calculated for $T = 59°F$ taken from the Standard Atmosphere at sea level (Appendix B). Then Eq. 8.16 predicts

$$p_0 = p + \frac{1}{2}\rho V^2 = 14.7 \text{ psia} + \frac{1}{2}(0.002378 \text{ slug/ft}^3)(559 \text{ ft/s})^2 \left(\frac{1 \text{ ft}^2}{144 \text{ in.}^2}\right)$$

$$= 17.3 \text{ psia}$$

The error is not great at $M = 0.5$ but is much larger at higher Mach numbers. It is generally accepted that compressibility effects become significant in flows at a Mach number greater than 0.3 in air.

and sound speed as $c = \sqrt{\gamma RT}$ to write the stagnation pressure as

$$\frac{p_0}{p} = \left[1 + \left(\frac{\gamma - 1}{2}\right) M^2\right]^{\gamma/(\gamma-1)} \quad (8.19)$$

The total pressure, p_T, in an isentropic flow is again defined as the sum of the static, dynamic and hydrostatic pressures, but since the gravitational potential energy term is negligible in high speed gas flows the total and stagnation pressures are the same. The terms total pressure and stagnation pressure are thus used interchangeably in gas dynamics and in aerodynamics.

8.5 APPLICATIONS OF THE BERNOULLI EQUATION

Bernoulli's equation is a powerful tool for finding the values of pressure and velocity at two points along a streamline. In applying this equation, it is important to remember that the underlying assumption of frictionless flow means that viscous effects are absent. Since real flows do not perfectly satisfy these underlying assumptions, you should always consider a result obtained with the Bernoulli equation to be an engineering approximation.

In this section we discuss a number of applications of Bernoulli's equation, using practical flow problems to illustrate how the two points at which the equation is applied are selected, and how the analysis is carried out. These applications also show how to use control volume analysis and Bernoulli's equation together to understand a given flow situation.

8.5.1 Pitot Tube

The stagnation pressure can be measured using a device called a pitot (pronounced pē·tō) tube. As shown in Figure 8.13, a pitot tube is inserted into a flow with its open end facing upstream. A suitable gage is used to read the pressure inside the hollow nose

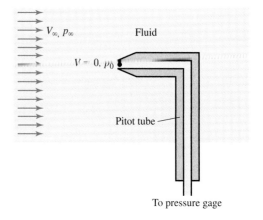

Figure 8.13 Pitot tube used for the measurement of stagnation pressure.

8.5 APPLICATIONS OF THE BERNOULLI EQUATION

of the tube where the fluid has come to rest. It is known that viscous effects on the flow approaching the pitot tube along the stagnation streamline are negligible. Thus as the fluid decelerates to rest just inside the nose of the pitot tube, the flow is frictionless. In a constant density flow, we can apply Bernoulli's equation (Eq. 8.12) from a point upstream to a point just inside the nose of the pitot tube to find

$$p_\infty + \tfrac{1}{2}\rho V_\infty^2 + \rho g z_\infty = p_0 + \tfrac{1}{2}\rho V_0^2 + \rho g z_0$$

For a horizontal streamline, with the fluid inside the pitot tube at rest, the pressure p_0 just inside the nose of the pitot tube is given by $p_0 = p_\infty + \tfrac{1}{2}\rho V_\infty^2$. The right side of this equation is the upstream stagnation pressure, $p_{0\infty}$, confirming our claim that the pitot tube measures the upstream stagnation pressure. Thus the equation governing the pitot tube is

$$p_0 = p_\infty + \tfrac{1}{2}\rho V_\infty^2 \qquad (8.20)$$

where p_0 is the reading of the pitot tube, and the values of static pressure and velocity refer to a location immediately upstream of the nose of the pitot tube. If we happen to know the static pressure at the location of the pitot tube, we can use the stagnation pressure reading of the pitot tube to calculate the velocity from

$$V_\infty = \sqrt{\frac{2(p_0 - p_\infty)}{\rho}} \qquad (8.21)$$

A pitot tube will give an accurate reading of stagnation pressure provided it is aligned with the flow and the probe is small relative to the length scale of the flow field. Equation 8.21 can be used to interpret the reading of a pitot tube in flows of liquids, and also in flows of gases at low Mach numbers, where the assumption of an incompressible fluid is valid.

When a pitot tube is inserted into a gas flow at a low speed, it is reasonable to assume the gas is behaving as an incompressible fluid. Thus Eqs. 8.20 and 8.21 are valid. At higher subsonic speeds, compressibility effects become important, and these equations are not applicable. To develop an accurate formula for interpreting the pressure reading from a pitot tube in a subsonic flow, we must apply the appropriate form of the Bernoulli equation for an isentropic flow from a point upstream to a point just inside the nose of the pitot tube, where the fluid is at rest. Writing Eq. 8.10 between these points we find

$$\left(\frac{\gamma}{\gamma-1}\right)\left(\frac{p_\infty}{\rho_\infty}\right) + \frac{1}{2}V_\infty^2 = \left(\frac{\gamma}{\gamma-1}\right)\left(\frac{p_0}{\rho_0}\right)$$

The fluid is at rest in the pitot tube, so the right side of this equation is the ratio of the stagnation pressure and stagnation density. This expression is the same as that in the last section, where we used it to discuss stagnation pressure in an isentropic flow. Thus we can use Eqs. 8.18 and 8.19 to interpret the reading of the pitot tube. We can use Eq. 8.18 to write the upstream velocity as

$$V_\infty = \sqrt{\frac{2\gamma}{\gamma-1}\left(\frac{p_0}{\rho_0} - \frac{p_\infty}{\rho_\infty}\right)} \qquad (8.22)$$

EXAMPLE 8.9

A pitot tube is mounted on the nose of an aircraft as shown in Figure 8.14. If the aircraft is flying at an altitude of 15,000 ft and the reading of the tube is 8.8 psia, what is the airspeed?

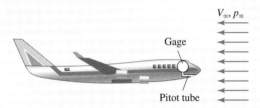

Figure 8.14 Schematic for Example 8.9.

SOLUTION

We will choose a reference frame fixed to the aircraft so that the flow is steady. We will also assume constant density flow and use Eq. 8.21, $V_\infty = \sqrt{2(p_0 - p_\infty)/\rho}$, to predict the airspeed. The air approaching the tube is at a static pressure $p_{15,000} = 8.297$ psia, and density $\rho_{15,000} = 0.001496$ slug/ft^3 as shown in the Standard Atmosphere table in Appendix B. Since the reading of the tube is 8.8 psia, the speed of the approaching air is

$$V_\infty = \sqrt{\frac{2(p_0 - p_\infty)}{\rho}} = \sqrt{\frac{2[(8.8 - 8.297)\,\text{lb}_\text{f}/\text{in.}^2](144\,\text{in.}^2/\text{ft}^2)}{0.001496\,\text{slug/ft}^3}}$$

$$= (311\,\text{ft/s})\left(\frac{1\,\text{mph}}{1.467\,\text{ft/s}}\right) = 212\,\text{mph}$$

Thus the airspeed is 212 mph. To check the Mach number, we can calculate the sound speed at 15,000 ft using data from the Standard Atmosphere table in Appendix B, and calculate the Mach number as $M = V_\infty/c_{15,000} = 311\,\text{ft/s}/1058\,\text{ft/s} = 0.29$. Note that this Mach number is just below the value of 0.3 at which compressibility effects cannot be neglected.

but it is customary in a compressible flow to interpret the reading of the tube in terms of Mach number. According to Eq. 8.19 we have

$$\frac{p_0}{p_\infty} = \left[1 + \left(\frac{\gamma - 1}{2}\right)M_\infty^2\right]^{\gamma/(\gamma-1)} \tag{8.23}$$

To determine the Mach number at which it is necessary to account for the effects of compressibility on the reading of a pitot tube, we can expand Eq. 8.23 in an infinite series to yield

$$\frac{p_0}{p_\infty} = \left[1 + \left(\frac{\gamma - 1}{2}\right) M_\infty^2\right]^{\gamma/(\gamma-1)}$$

$$= 1 + \left(\frac{\gamma}{\gamma - 1}\right)\left(\frac{\gamma - 1}{2}\right) M_\infty^2$$

$$+ \left(\frac{\gamma}{\gamma - 1}\right)\left(\frac{\gamma}{\gamma - 1} - 1\right)\left(\frac{1}{2}\right)$$

$$\times \left(\frac{\gamma - 1}{2}\right)^2 M_\infty^4 + \cdots$$

$$= 1 + \frac{\gamma M_\infty^2}{2} + \frac{\gamma M_\infty^4}{8} + \cdots$$

Upon rearranging and writing this as

$$\frac{p_0 - p_\infty}{p_\infty} = \frac{\gamma M_\infty^2}{2}\left[1 + \frac{M_\infty^2}{4} + \cdots\right]$$

we can make a comparison to the corresponding result for an incompressible fluid obtained from Eq. 8.20: $p_0 = p_\infty + \frac{1}{2}\rho V_\infty^2$. Writing the latter in the same form as the foregoing series, we find

$$\frac{p_0 - p_\infty}{p_\infty} = \frac{1}{2}\frac{\rho}{p_\infty}V_\infty^2 = \frac{\gamma M_\infty^2}{2}$$

For small Mach numbers, which we can define as $M_\infty^2/4 \ll 1$, the incompressible and compressible formulas for $(p_0 - p_\infty)/p_\infty$ agree very well. An acceptable error of 2% in the value of $(p_0 - p_\infty)/p_\infty$ corresponds to $M_\infty^2/4 = 0.02$, which is a Mach number of $M_\infty = 0.28$. As a rule of thumb then, the incompressible pitot tube formulas may be considered to be valid up to Mach numbers of approximately $M_\infty = 0.3$, which corresponds to approximately 230 mph at sea level. This analysis is also used to justify the neglect of the effects of compressibility in the flow of air for $M < 0.3$.

Thus the Mach number of the upstream flow is given by

$$M_\infty = \sqrt{\frac{2}{\gamma - 1}\left[\left(\frac{p_0}{p_\infty}\right)^{(\gamma-1)/\gamma} - 1\right]} \quad (8.24)$$

After determining the Mach number, we can calculate velocity from the sound speed by means of the definition of Mach number, $M_\infty = V_\infty/c_\infty$. The sound speed is calculated using the formula $c_\infty = \sqrt{\gamma R T_\infty} = \sqrt{\gamma p_\infty/\rho_\infty}$, with data for the Standard Atmosphere from Appendix B.

Figure 8.15 shows another type of pressure probe called a pitot-static tube. Since we rarely know the value of the static pressure in a nonuniform flow field, a pitot-static tube combines a static pressure tap on its side with a regular pitot tube nose tap. Thus a pitot-static tube permits the simultaneous measurement of the static and stagnation pressures at a point in a flow. The equations governing the behavior of the pitot-static tube are the same as those governing the pitot tube, namely Eqs. 8.20 and 8.21 for an incompressible fluid and Eqs. 8.22, 8.23, and 8.24 for a compressible fluid. To obtain accurate readings, a pitot-static tube must be aligned with the flow and of the proper size. Because of the sensitivity of the static tap on the side of the pitot-static tube, alignment can be difficult in an flow field whose direction is unknown.

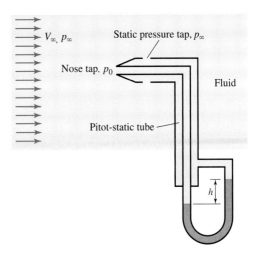

Figure 8.15 A pitot-static tube measures dynamic pressure directly.

EXAMPLE 8.10

A commercial aircraft similar to the one shown earlier in Figure 8.14 cruises at an altitude of 30,000 ft, and a pitot tube mounted on the nose reads 6.65 psia. Find the Mach number of the aircraft and its airspeed.

SOLUTION

We will choose a reference frame fixed to the aircraft so that the flow is steady. Since this is a commercial aircraft, we can assume that it is flying at a relatively large but subsonic Mach number in excess of 0.3. Compressibility effects must therefore be considered. We will apply Eq. 8.24 to determine M. The air approaching the tube is at a static pressure $p_{30,000} = 4.37$ psia, density $\rho_{30,000} = 8.907 \times 10^{-4}$ slug/ft^3, and temperature $T_{30,000} = -47.83°F$ as read from the Standard Atmosphere table in Appendix B. The sound speed at this altitude is calculated as usual using temperature data from Appendix B and is found to be $c_{30,000} = 995$ ft/s. Inserting the data, we find

$$M_\infty = \sqrt{\left(\frac{2}{\gamma - 1}\right)\left[\left(\frac{p_0}{p_\infty}\right)^{(\gamma-1)/\gamma} - 1\right]}$$

$$= \sqrt{\left(\frac{2}{1.4 - 1}\right)\left[\left(\frac{6.65 \text{ psia}}{4.37 \text{ psia}}\right)^{(1.4-1)/1.4} - 1\right]} = 0.8$$

Thus the aircraft is traveling at Mach 0.8, and its airspeed is

$$V = Mc = 0.8 \, (995 \text{ ft/s}) \left(\frac{1 \text{ mph}}{1.467 \text{ ft/s}}\right) = 543 \text{ mph}$$

EXAMPLE 8.11

A pitot-static tube like that shown in Figure 8.15, mounted inside an air duct, is connected to a U tube manometer that reads $h = 0.2$ in. of water. What is the airspeed at the location of the tube?

SOLUTION

We will assume constant density flow and use Eq. 8.21, $V_\infty = \sqrt{2(p_0 - p_\infty)/\rho}$, to calculate the speed, checking the result to ensure that the speed is low enough that the assumption of constant density flow is appropriate. The manometer reading gives the

8.5 APPLICATIONS OF THE BERNOULLI EQUATION

pressure difference $(p_0 - p_\infty)$ in terms of the hydrostatic pressure in a column of water of height h. Thus we can calculate the pressure equivalent of 0.2 in. of water as

$$p_0 - p_\infty = \rho_{H_2O}gh = (1.94 \text{ slugs/ft}^3)(32.2 \text{ ft/s}^2)(0.2 \text{ in.})\left(\frac{1 \text{ ft}}{12 \text{ in.}}\right) = 1.04 \text{ lb}_f/\text{ft}^2$$

Next we calculate the speed as

$$V = \sqrt{\frac{2(p_0 - p_\infty)}{\rho}} = \sqrt{\frac{2(1.04)\text{lb}_f/\text{ft}^2}{2.377 \times 10^{-3} \text{ slug/ft}^3}} = 29.6 \text{ ft/s}$$

which is certainly low enough to allow us to assume constant density flow. This speed, which in practice is usually given in feet per minute (1775 ft/min), is typical for main ducts in heating, ventilating, and air conditioning (HVAC) systems.

EXAMPLE 8.12

Suppose the airplane shown in Figure 8.16 has a ground speed of 150 mph while flying at a constant altitude of 5000 ft into a 50 mph headwind. What values of the static and stagnation pressures will be measured by a pitot-static tube mounted on the aircraft?

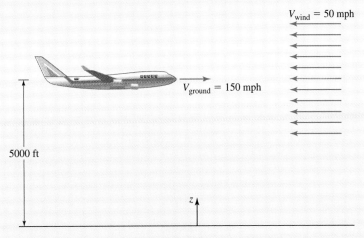

Figure 8.16 Schematic for Example 8.12.

SOLUTION

At this relatively low speed we will assume constant density flow. From the Standard Atmosphere table in Appendix B we find that the atmosphere ahead of the aircraft is at a static pressure $p_{5000} = 12.228$ psia, and density $\rho_{5000} = 2.048 \times 10^{-3}$ slug/ft^3. The ground speed of the aircraft is $V_{\text{ground}} = 150$ mph $= 220$ ft/s, and the wind speed is $V_{\text{wind}} = 50$ mph $= 73.3$ ft/s, where this value refers to the speed of the wind relative to the ground.

It is critical to note that the pitot-static tube responds to the relative velocity of the air with respect to the tube, i.e., to the airspeed, which is the speed of the aircraft with respect to the air. If the air ahead of the aircraft were still, that is, not moving with respect to the ground, the ground speed and the airspeed would be the same. In this case, however, the air is moving at 50 mph toward the plane; thus the airspeed is $V = (150 + 50)$ mph $= 200$ mph $= 293.3$ ft/s, and this is the value we expect the tube to provide based on Eq. 8.21, $V_\infty = \sqrt{2(p_0 - p_\infty)/\rho}$.

Let us check this conclusion. The fact that both the plane and the air are moving does not affect the static pressure upstream of the plane, since static pressure is defined to be the pressure seen by an observer moving with the fluid. We will assume that the air approaching the plane is at the pressure and density of the standard atmosphere at this altitude; thus the static tap on the tube will read p_{5000}. The stagnation pressure reading of the tube is affected by the wind approaching the plane. We can calculate the stagnation pressure seen by the tube by using Eq. 8.20 to write

$$p_0 = p_\infty + \tfrac{1}{2}\rho V_\infty^2 = p_{5000} + \tfrac{1}{2}\rho_{5000} V_\infty^2$$

The tube will therefore output a pressure difference of

$$(p_0 - p) = \left(p_{5000} + \tfrac{1}{2}\rho_{5000} V_\infty^2 - p_{5000}\right) = \tfrac{1}{2}\rho_{5000} V_\infty^2$$

Substituting this pressure difference into Eq. 8.21, we have

$$V_\infty = \sqrt{\frac{2(p_0 - p_\infty)}{\rho}} = \sqrt{\frac{2\left(\tfrac{1}{2}\rho_{5000} V_\infty^2\right)}{\rho_{5000}}} = V_\infty$$

as expected. The actual pressure difference read by the tube under the indicated conditions is

$$p_0 - p_\infty = \tfrac{1}{2}\rho_{5000} V_\infty^2 = \tfrac{1}{2}(2.048 \times 10^{-3}\text{ slug/ft}^3)(293.3\text{ ft/s})^2 \left(\frac{1\text{ ft}^2}{144\text{ in.}^2}\right)$$

$$= 0.612\text{ psia}$$

Thus the tube will show a static pressure of $p_{5000} = 12.228$ psia and a stagnation pressure of

$$p_0 = p_{5000} + \tfrac{1}{2}\rho_{5000} V_\infty^2 = 12.228\text{ psia} + 0.612\text{ psia} = 12.84\text{ psia}$$

8.5 APPLICATIONS OF THE BERNOULLI EQUATION

8.5.2 Siphon

A siphon is a device that allows a liquid to be drawn from a storage vessel without the use of a pump. As illustrated in Figure 8.17, a siphon may be as simple as a length of hose. One end of the hose is inserted into the liquid to depth d, the other is positioned below the level of the free surface at a distance H as shown. Experience shows that nothing happens when the hose is inserted into the liquid. But if suction is applied to the free end of the hose to start the flow, the flow will continue. We can investigate the operation of the siphon after the flow starts by assuming steady, frictionless flow.

The assumption of steady flow is appropriate only if the tank is sufficiently large. To better understand this statement, consider a mass balance on the control volume shown in Figure 8.17. Since the density is constant, we obtain $\rho V_S A_S = \rho V A$, where V_S is the velocity at which the free surface falls, A_S is the cross-sectional area of the free surface (and tank) less the hose area, V is the average velocity in the hose, and A is the cross-sectional area of the hose. The mass balance tells us that the mass flowrate out of the CV (and thus out through the hose) must match the mass flowrate into the CV (due to the motion of the free surface). Upon writing this as

$$V_S = V\left(\frac{A}{A_S}\right) \tag{8.25}$$

we see that if the cross-sectional area of the hose is small in comparison to that of the tank, the velocity at the free surface will be very small in comparison to the velocity in the hose. If so, we may consider the free surface to be stationary, and flow in the tank and hose to be steady. Thus the mass balance suggests that it is valid to assume steady flow when $A/A_S \ll 1$.

To explore the operation of a siphon, we have identified a number of points along a possible streamline that begins at the free surface and follows the hose centerline. Applying the steady, constant density Bernoulli equation, Eq. 8.6, between the free

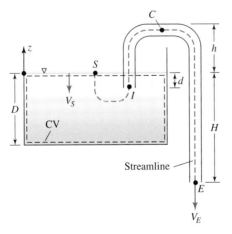

Figure 8.17 Schematic for the analysis of a siphon.

surface and hose exit we have

$$\frac{p_S}{\rho} + \frac{1}{2}V_S^2 + gz_S = \frac{p_E}{\rho} + \frac{1}{2}V_E^2 + gz_E$$

At the free surface the pressure is atmospheric, the kinetic energy is negligible, and the elevation is $z_S = 0$. At the exit the pressure is atmospheric, and the elevation is $z_E = -H$. Thus we find

$$\frac{p_A}{\rho} = \frac{p_A}{\rho} + \frac{1}{2}V_E^2 - gH$$

Solving for the exit velocity we have

$$V_E = \sqrt{2gH} \qquad (8.26)$$

A mass balance on the hose shows that $\rho V A = \rho V_E A$, thus $V = V_E$ and we also have

$$V = \sqrt{2gH} \qquad (8.27)$$

The flowrate leaving the hose is

$$Q = A\sqrt{2gH} \qquad (8.28)$$

We see that the siphon flowrate can be improved by increasing the size of the hose or by positioning the exit end of the hose further below the free surface. The depth of insertion of the hose, d, has no effect on the velocity or flowrate.

To further explore the effect of hose positioning on flowrate, we will apply the Bernoulli equation between point C, which is located at the maximum elevation in the system, and the exit. Applying Eq. 8.6 between these points we obtain

$$\frac{p_C}{\rho} + \frac{1}{2}V_C^2 + gz_C = \frac{p_E}{\rho} + \frac{1}{2}V_E^2 + gz_E$$

Since the velocities at points C and E are the same, these terms cancel. Inserting the known pressure at the exit, $p_E = p_A$, and elevations $z_C = h$ and $z_E = -H$ yields $p_C/\rho + gh = p_A/\rho - gH$. Solving for the pressure at point C we have

$$p_C = p_A - \rho g(h + H) \qquad (8.29)$$

which shows that the pressure at point C is sub atmospheric. If we attempt to increase the flowrate by increasing H with h fixed, the pressure at point C will approach the vapor pressure of the liquid p_v, vapor bubbles will form at point C, and the siphon will cease to operate. The maximum flowrate thus occurs at the value of H for which $p_C = p_v$. When we insert this value into Eq. 8.29 and solve for the value of H at which cavitation first appears, we find

$$H_C = \frac{p_A - p_v}{\rho g} - h \qquad (8.30)$$

8.5 APPLICATIONS OF THE BERNOULLI EQUATION

which suggests that we minimize h to improve the flowrate. Substituting Eq. 8.30 into Eqs. 8.27 and 8.28 gives the velocity and flowrate at the onset of cavitation as

$$V_C = \sqrt{2g\left(\frac{p_A - p_v}{\rho g} - h\right)} \qquad (8.31)$$

and

$$Q_C = A\sqrt{2g\left(\frac{p_A - p_v}{\rho g} - h\right)} \qquad (8.32)$$

We see that the maximum velocity and flowrate will occur with $h = 0$ at the onset of cavitation. From Eq. 8.30 we calculate the maximum H as $H_{\max} = (p_A - p_v)/g$, and Eqs. 8.31 and 8.32 with $h = 0$ give the corresponding velocity and flowrate as

$$V_{\max} = \sqrt{2\left(\frac{p_A - p_v}{\rho}\right)} \qquad (8.33)$$

and

$$Q_{\max} = A\sqrt{2\left(\frac{p_A - p_v}{\rho}\right)} \qquad (8.34)$$

In some applications of a siphon we are more interested in maximizing the height of an obstacle that can be cleared by the hose. This corresponds to the maximum allowable value of h for a given H. The preceding analysis leading up to Eqs. 8.26, 8.27, 8.28, and 8.29 remains valid. From Eq. 8.29 we have $p_C = p_A - g(h + H)$, thus if we attempt to increase h with H fixed, the pressure at point C will again approach the vapor pressure of the liquid p_v, vapor bubbles will form at point C, and the siphon will cease to operate. The maximum h for a given value of H thus also corresponds to $p_C = p_v$. Inserting this value into Eq. 8.29 and solving for the value of h at which cavitation first appears for a given H we find

$$h_C = \frac{p_A - p_v}{\rho g} - H \qquad (8.35)$$

We see that to clear a tall obstacle, we should minimize H. The maximum obstacle is cleared by setting H to zero, which corresponds to

$$h_{\max} = \frac{p_A - p_v}{\rho g} \qquad (8.36)$$

What happens to the velocity and flowrate as h approaches $h_{\max} = (p_A - p_v)/\rho g$? Since this corresponds to $H = 0$, Eqs. 8.27 and 8.28 predict that the velocity and flowrate approach zero as well. We can check this by applying the Bernoulli equation between the free surface and point C to find

$$\frac{p_S}{\rho} + \frac{1}{2}V_S^2 + gz_S = \frac{p_C}{\rho} + \frac{1}{2}V_C^2 + gz_C$$

Inserting known values and noting that $V = V_C$, we have $p_A/\rho = p_C/\rho + \tfrac{1}{2}V^2 + gh$. Solving for h we have

$$h = \frac{p_A - p_C}{\rho g} - \frac{V^2}{2g} \tag{8.37}$$

At $h_{\max} = (p_A - p_v)/\rho g$, we see that the velocity is zero.

What is the pressure at the inlet to the hose? Is the pressure there equal to the hydrostatic value $p_A + \rho g d$? To find out, we apply the Bernoulli equation between the free surface and the inlet to obtain

$$\frac{p_S}{\rho} + \frac{1}{2}V_S^2 + gz_S = \frac{p_I}{\rho} + \frac{1}{2}V_I^2 + gz_I$$

Inserting known values gives $p_A/\rho = p_I/\rho + \tfrac{1}{2}V_I^2 - gd$. Since $V = V_I$, we can solve for the pressure at the inlet to find

$$p_I = (p_A + \rho g d) - \tfrac{1}{2}\rho V^2 \tag{8.38}$$

which shows that the inlet pressure is below the hydrostatic pressure at this elevation.

To understand this as well as the other effects we have discussed in connection with siphons, consider a fluid particle as it flows along the streamline from the free surface to the hose inlet, then through the hose to the exit. The mechanical energy is conserved as the particle moves along the streamline. At the free surface the particle has a mechanical energy content per unit mass of

$$\left(\frac{p}{\rho} + \frac{1}{2}V^2 + gz\right)_S = \frac{p_A}{\rho}$$

which shows that the energy at the free surface is wholly in the form of pressure potential energy.

At the hose inlet the particle has a mechanical energy content of

$$\left(\frac{p}{\rho} + \frac{1}{2}V^2 + gz\right)_I = \frac{p_I}{\rho} + \frac{1}{2}V^2 - gd$$

At this location the energy consists of pressure potential energy, kinetic energy, and a negative gravitational potential energy relative to the starting point. By using Eq. 8.38, however, we can substitute $p_I = (p_A + \rho g d) - \tfrac{1}{2}\rho V^2$ for the inlet pressure and obtain

$$\frac{p_I}{\rho} + \frac{1}{2}V^2 - gd = \frac{(p_A + \rho g d) - \tfrac{1}{2}\rho V^2}{\rho} + \frac{1}{2}V^2 - gd$$

$$= \frac{p_A}{\rho}$$

The statement earlier in this section that the pressure in the exit plane is atmospheric is a part of a general principle we will refer to as the exit rule. This rule may be summarized as follows. A subsonic flow (exit velocity less than sound speed in the fluid under the exit conditions) exits at the ambient pressure. A supersonic flow (exit velocity greater than the sound speed under the exit conditions) may exit at a pressure above or below the ambient depending on a number of other factors. Although we will not discuss the rule further in this text, it does tell us that a jet of liquid or gas at subsonic velocity comes out at the ambient or atmospheric pressure.

which shows that the mechanical energy content of the fluid has been conserved as expected.

At point C the mechanical energy content is

$$\left(\frac{p}{\rho} + \frac{1}{2}V^2 + gz\right)_C = \frac{p_C}{\rho} + \frac{1}{2}V^2 + gh$$

At this location the energy consists of pressure potential energy, kinetic energy, and positive gravitational potential energy. From Eqs. 8.29 [$p_C = p_A - \rho g(h + H)$] and 8.27 ($V = \sqrt{2gH}$), this becomes

$$\frac{p_C}{\rho} + \frac{1}{2}V^2 + gh = \frac{p_A - \rho g(h + H)}{\rho}$$
$$+ \frac{1}{2}\left(\sqrt{2gH}\right)^2 + gh$$
$$= \frac{p_A}{\rho}$$

which shows that the mechanical energy content here is also the same as at the free surface.

Finally at the exit the mechanical energy content is

$$\left(\frac{p}{\rho} + \frac{1}{2}V^2 + gz\right)_E = \frac{p_A}{\rho} + \frac{1}{2}V^2 - gH$$

By means of Eq. 8.27, $V = \sqrt{2gH}$, this simplifies to

$$\frac{p_A}{\rho} + \frac{1}{2}\left(\sqrt{2gH}\right)^2 - gH = \frac{p_A}{\rho}$$

Thus we see that in this steady, frictionless flow, the mechanical energy is redistributed among the different possible energy types but otherwise is conserved in flow along a streamline.

EXAMPLE 8.13

Gasoline is siphoned from a tank with a hose as shown in Figure 8.18. The hose has an inside diameter of 2 cm, the depth of the gasoline in the tank is 1.5 m, the highest point of the tube is 0.5 m above the free surface, and the outlet of the hose is 1.0 m below the free surface. Find the gasoline velocity and flowrate. Will cavitation occur? If a longer hose is used and the outlet of the hose remains 1.0 m below the free surface, how far above the free surface can the bend in the hose be raised?

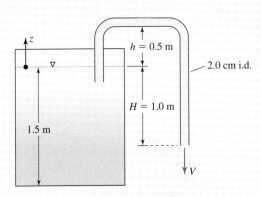

Figure 8.18 Schematic for Example 8.13.

SOLUTION

We will assume steady frictionless flow and apply the results of this section. To determine the velocity in the hose the relevant equation is Eq. 8.27, $V = \sqrt{2gH}$, where H is the distance of the open end of the hose below the free surface. From Figure 8.18, $H = 1.0$ m, and we can calculate the velocity as

$$V = \sqrt{2(9.8 \text{ m/s}^2)(1 \text{ m})} = 4.43 \text{ m/s}$$

For a round hose we can calculate the flowrate as $Q = AV = (\pi D^2/4)V$. Inserting the data, we have

$$Q = \frac{\pi(2 \text{ cm})^2}{4} (4.43 \text{ m/s}) \left(\frac{100 \text{ cm}}{1 \text{ m}}\right) = 1.39 \times 10^3 \text{ cm}^3/\text{s}$$

To express this in more familiar units, we can convert it to liters per minute as follows

$$Q = (1.39 \times 10^3 \text{ cm}^3/\text{s}) \left(\frac{60 \text{ s}}{1 \text{ min}}\right) \left(\frac{1 \text{ L}}{10^3 \text{ cm}^3}\right) = 83.5 \text{ L/min}$$

In gallons per minute, this is flowrate of $Q = (83.5 \text{ L/min}) (1 \text{ gal}/3.784 \text{ L}) = 22$ gal/min. To check for cavitation at the highest point, we can compute the pressure there and compare with to the vapor pressure of gasoline, which is $p_V = 5.51 \times 10^4 \text{ N/m}^2 = 55.1$ kPa from Appendix A. This is point C in the context of this section, and according to Eq. 8.29 the pressure at this point is

$$p_C = p_A - \rho g(h + H)$$

where h is the height of this point above the free surface. Here $h = 0.5$ m and $h + H = 1.5$ m. We can use the density of gasoline from Appendix A, 680 kg/m³, and calculate the pressure at the highest point as

$$p_C = p_A - \rho g(h + H) = 101.3 \text{ kPa} - (680 \text{ kg/m}^3)(9.8 \text{ m/s}^2)(1.5 \text{ m})$$
$$= 91.3 \text{ kPa}$$

The pressure at the highest point is well above the vapor pressure, so cavitation will not occur.

If a longer hose is used and the outlet of the hose remains 1.0 m below the free surface, the bend could be raised to just below the height at which cavitation occurs. This height is given by Eq. 8.35 as $h_C = (p_A - p_V)/\rho g - H$. Inserting the data, we find

$$h_C = \left[\frac{101.3 \text{ kPa} - 55.1 \text{ kPa}}{(680 \text{ kg/m}^3)(9.8 \text{ m/s}^2)}\right] - 1.0 \text{ m} = 5.95 \text{ m}$$

Thus the siphon would fail if the bend approached 6 m above the free surface with the hose end positioned 1.0 m below the free surface.

8.5.3 Sluice Gate

A sluice gate is a simple device that may be used to control and measure the flow of water in an open channel flow such as that in a river, drainage ditch, or irrigation canal. In its simplest form, as shown in Figure 8.19, the movable gate is set at a distance h above the bed of the channel. Once a steady flow condition has been established, the flowrate can be predicted from measurements of the depths of the water upstream and downstream of the sluice gate. To establish the relationship between the velocities and stream depths upstream and downstream, we will assume steady, constant density,

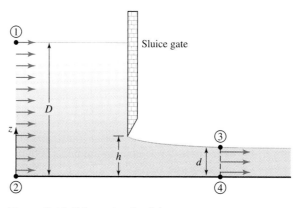

Figure 8.19 Schematic of a sluice gate.

frictionless flow through the sluice gate and apply the Bernoulli equation between points 1 and 3 as shown in Figure 8.19. Note that the assumption of frictionless flow implies that the velocity profiles upstream and downstream are uniform.

Applying the steady, constant density Bernoulli equation, Eq. 8.6, between the points 1 and 3 on the free surface upstream and downstream of the gate yields

$$\frac{p_1}{\rho} + \frac{1}{2}V_1^2 + gz_1 = \frac{p_3}{\rho} + \frac{1}{2}V_3^2 + gz_3$$

The pressures are both atmospheric, and a mass balance shows that the velocities are related by $Q = V_1(DW) = V_3(dW)$, where Q is the volume flowrate and W is the gate width. Thus we can write $V_1 = Q/DW$ and $V_3 = Q/dW$. The elevations at points 1 and 3 are D and d, respectively. Thus the Bernoulli equation becomes

$$\frac{p_A}{\rho} + \frac{1}{2}\left(\frac{Q}{DW}\right)^2 + gD = \frac{p_A}{\rho} + \frac{1}{2}\left(\frac{Q}{dW}\right)^2 + gd$$

Solving for Q yields

$$Q = dW\sqrt{\frac{2gD(1-d/D)}{1-(d/D)^2}} \qquad (8.39)$$

Before discussing this result, it is instructive to ask what would happen in this analysis if we had chosen points 2 and 4 on the channel bottom. Since the velocities are uniform, the mass balance predicts $V_2 = Q/DW$ and $V_4 = Q/dW$. The elevations are now zero for both points, but the pressures at the two points are $p_2 = p_A + \rho g D$ and $p_4 = p_A + \rho g d$. Notice that the pressures are the hydrostatic values at the two locations even though the water is moving along the bottom at the two points. We can assume the pressures are hydrostatic because the streamlines are straight and parallel at these locations. Thus the only variation in pressure across the streamlines is that which occurs from the free surface downward due to gravity. The Bernoulli equation between points 2 and 4 is: $p_2/\rho + \frac{1}{2}V_2^2 + gz_2 = p_4/\rho + \frac{1}{2}V_4^2 + gz_4$, which after inserting the data gives

$$\frac{p_A + \rho g D}{\rho} + \frac{1}{2}\left(\frac{Q}{DW}\right)^2 + g(0) = \frac{p_A + \rho g d}{\rho} + \frac{1}{2}\left(\frac{Q}{dW}\right)^2 + g(0)$$

It can be shown that this leads to Eq. 8.39.

The formula obtained here for the volume flowrate assumes that the flow under the gate exhibits a vena contracta as illustrated in Figure 8.19. It would be desirable to eliminate the need for depth measurements both upstream and downstream by finding a way to introduce the gate opening h in place of the downstream depth d. The ratio of d to h defines the vena contracta; however, the tools we have available do not permit us to predict this ratio theoretically. In some conditions of operation, the downstream flow may recirculate and climb up the back of the sluice gate. The flowrate in this condition is not accurately predicted by Eq. 8.39.

EXAMPLE 8.14

A sluice gate similar to the one shown in Figure 8.19 is used to measure the water flow in the spillway of a dam. The width of the gate is 100 ft, and the water heights are $D = 2$ ft and $d = 0.5$ ft. What is the volume flowrate at the indicated condition of operation? What are the velocities upstream and downstream of the sluice gate?

SOLUTION

We will assume steady frictionless flow and apply the results of this section. The presence of a vena contracta indicates that we can apply Eq. 8.39,

$$Q = dW\sqrt{\frac{2gD(1 - d/D)}{1 - (d/D)^2}}$$

From the data we have $(d/D) = 0.5 \text{ ft}/2 \text{ ft} = 0.25$. Substituting this result into the preceding equation yields

$$Q = (0.5 \text{ ft})(100 \text{ ft})\sqrt{\frac{2(32.2 \text{ ft/s}^2)(2 \text{ ft})(0.75)}{0.9375}} = 508 \text{ ft}^3/\text{s}$$

Recognizing that the volume flowrate is the product of the velocity and flow area allows us to calculate that the upstream and downstream velocities, respectively, are

$$V_1 = \frac{Q}{DW} = \frac{508 \text{ ft}^3/\text{s}}{(2 \text{ ft})(100 \text{ ft})} = 2.54 \text{ ft/s}$$

and

$$V_2 = \frac{Q}{dW} = \frac{508 \text{ ft}^3/\text{s}}{(0.5 \text{ ft})(100 \text{ ft})} = 10.2 \text{ ft/s}$$

8.5.4 Flow through Area Change

In the case study of Section 3.3.2, on flow through area change, we briefly discussed the types of area change encountered in duct and pipe systems and developed a method of calculating the frictional pressure drop in the steady flow of an incompressible fluid through a sudden enlargement or contraction. In this section we will apply Bernoulli's equation to examine the pressure changes that occur in nozzles, diffusers, and sudden area changes in the absence of friction. It is important to keep in mind that in practical applications most of these flows are turbulent, and the flow will experience a frictional pressure drop in passing through the area change. This effect is not accounted for by the Bernoulli equation. More accurate methods to account for the pressure change and flowrate characteristics for turbulent flows through area change can be found in Chapter 13, as well as in Section 3.3.2.

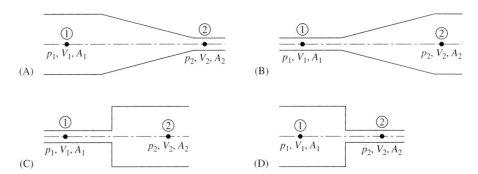

Figure 8.20 Basic types of area change: (A) nozzle, (B) diffuser, (C) sudden enlargement, and (D) sudden contraction.

Consider a frictionless, steady, constant density flow through any of the area changes shown in Figure 8.20. Applying the steady, constant density Bernoulli equation, Eq. 8.6, between points 1 and 2 on the axial streamline upstream and downstream of the area change, and ignoring the effects of gravity, we have $p_1/\rho + \frac{1}{2}V_1^2 = p_2/\rho + \frac{1}{2}V_2^2$, which upon rearrangement gives

$$p_2 - p_1 = \tfrac{1}{2}\rho\left(V_1^2 - V_2^2\right) \tag{8.40}$$

A mass balance allows us to write

$$\frac{V_1}{V_2} = \frac{A_2}{A_1} \tag{8.41}$$

Dividing Eq. 8.40 by $\frac{1}{2}\rho V_1^2$, and using Eq. 8.41, we can also write

$$\frac{p_2 - p_1}{\frac{1}{2}\rho V_1^2} = 1 - \frac{A_1^2}{A_2^2} \tag{8.42}$$

We see that in a steady frictionless flow of a constant density fluid, if the area decreases ($A_2/A_1 < 1$), the velocity increases, and the pressure decreases. If the area increases, the velocity decreases, and the pressure increases. These equations are applicable to any type of area change.

Equation 8.40 for frictionless flow may be compared with Eq. 3.20:

$$p_2 - p_1 = \left[\tfrac{1}{2}\rho\left(\bar{V}_1^2 - \bar{V}_2^2\right)\right] - \Delta p_F$$

which was given in Section 3.3.2 as an empirical model for turbulent flow through a sudden area change. Recall that in our discussion of Eq. 3.20 we noted that the total change in pressure as a flow passes through an area change maybe thought of as the sum of a pressure change $\frac{1}{2}\rho(\bar{V}_1^2 - \bar{V}_2^2)$ associated with the change in average flow velocity (which may be either positive or negative depending on whether the flow slows down or speeds up), and a frictional pressure drop Δp_F (a negative pressure change). We see that Eq. 8.40, derived from the Bernoulli equation, provides a basis for estimating the change in pressure due to the area change. (You might wish to review Example 7.8.)

EXAMPLE 8.15

Water flows from a garden hose having an inner diameter of 0.5 in. through a nozzle as shown in Figure 8.21. The pressure in the hose is 50 psig, and the nozzle exit is 0.125 in. I.D. Find the velocity and flowrate out of the nozzle. What is the velocity in the hose?

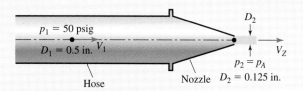

Figure 8.21 Schematic for Example 8.15.

SOLUTION

We can assume a steady, frictionless, constant density flow through the nozzle and apply Eq. 8.42, $(p_2 - p_1)/\frac{1}{2}\rho V_1^2 = (1 - A_1^2/A_2^2)$. Solving first for V_1, the velocity in the hose, we find

$$V_1 = \sqrt{\frac{2(p_2 - p_1)}{\rho(1 - A_1^2/A_2^2)}} \qquad (A)$$

From Eq. 8.41, $V_1/V_2 = A_2/A_1$, the velocity in the nozzle is given by

$$V_2 = \left(\frac{A_1}{A_2}\right) V_1 \qquad (B)$$

The flowrate in the nozzle and hose are the same and can be calculated from

$$Q = A_2 V_2 \qquad (C)$$

From the data we have $A_1/A_2 = D_1^2/D_2^2 = (0.5/0.125)^2 = 16$, $A_1^2/A_2^2 = (16)^2 = 256$, and $(p_2 - p_1) = -50 \text{ lb}_f/\text{in.}^2$. Thus from (A) the velocity in the hose is

$$V_1 = \sqrt{\frac{2(p_2 - p_1)}{\rho(1 - A_1^2/A_2^2)}} = \sqrt{\frac{2(-50) \text{ lb}_f/\text{in.}^2(144 \text{ in.}^2/\text{ft}^2)}{1.94 \text{ slugs/ft}^3(1 - 256)}} = 5.4 \text{ ft/s}$$

By using (B), we find that the velocity at the nozzle exit is

$$V_2 = \left(\frac{A_1}{A_2}\right) V_1 = 16(5.4 \text{ ft/s}) = 86.4 \text{ ft/s}$$

and the flowrate is found from (C) as

$$Q = A_2 V_2 = \left(\frac{\pi D_2^2}{4}\right) V_2 = \left(\frac{\pi}{4}\right)\left(\frac{(0.125 \text{ in.})^2 (1 \text{ ft}^2)}{144 \text{ in.}^2}\right) 86.4 \text{ ft/s} = 7.36 \times 10^{-3} \text{ ft}^3/\text{s}$$

Since 7.48 gal = 1 ft^3, this is a flowrate of

$$Q = 7.36 \times 10^{-3} \text{ ft}^3/\text{s} (7.48 \text{ gal/ft}^3) \left(\frac{60 \text{ s}}{1 \text{ min}}\right) = 3.3 \text{ gal/min}$$

EXAMPLE 8.16

Air at 25°C flows through the diffuser shown in Figure 8.22. At the inlet the velocity and area are 100 m/s and 0.5 m^2. At the exit the area is 1.5 m^2 and the pressure is atmospheric. Find the exit velocity and pressure change through the diffuser.

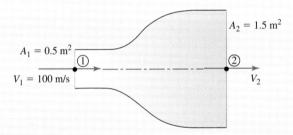

Figure 8.22 Schematic for Example 8.16.

SOLUTION

The velocity here is high enough that the assumption of incompressible flow may be incorrect. We can check this by calculating the Mach number and comparing it with the Mach number that defines the limits of incompressible flow, i.e., $M = 0.3$. Recalling the definition of Mach number as $M = V/c$, the sound speed as $c = \sqrt{\gamma RT}$, and using $R = 287$ (N-m)/(kg-K) as the gas constant for air, we calculate $c = \sqrt{\gamma RT} = \sqrt{1.4[287(\text{N-m})/(\text{kg-K})](298 \text{ K})[(\text{kg-m})/(\text{N-s}^2)]} = 346$ m/s. Thus the Mach number of this flow is $M = V/c = \frac{100}{346} = 0.29$, very close to the Mach number of 0.3, which defines the upper limit of incompressible flow. We can employ Eq. 8.41 to calculate the velocity ratio as $V_1/V_2 = A_2/A_1 = 1.5/0.5 = 3$; thus the exit velocity is $V_2 = V_1/3 = 100 \text{ m/s}/3 = 33$ m/s.

Assuming steady frictionless constant density flow, we can use Eq. 8.42 to calculate the pressure change through the diffuser:

$$\frac{p_2 - p_1}{\frac{1}{2}\rho V_1^2} = 1 - \frac{A_1^2}{A_2^2} \qquad (A)$$

The value of the density of air at 25°C can be calculated by using the perfect gas law:

$$\rho = \frac{p}{RT} = \frac{101{,}325\ \text{N/m}^2}{[287\ (\text{N-m})/(\text{kg-K})](298\ \text{K})} = 1.18\ \text{kg/m}^3$$

where we have assumed the pressure in the diffuser is approximately atmospheric throughout. Rearranging (A) and inserting the data, we have

$$p_2 - p_1 = \frac{1}{2}\rho V_1^2 \left(1 - \frac{A_1^2}{A_2^2}\right) = \frac{1}{2}(1.18\ \text{kg/m}^3)(100\ \text{m/s})^2 \left(1 - \frac{1}{9}\right) = 5244\ \text{N/m}^2$$

Thus the pressure increases by 5.2 kPa in the diffuser because the flow slows down from 100 m/s to 33 m/s. The pressure in the diffuser is 96.1 kPa at the inlet and rises to 101.3 kPa at the exit. We see that the diffuser converts kinetic energy into pressure potential energy. More discussion of diffusers can be found in Chapter 13.

As Example 8.16 illustrates, the low speed gas flow through an area change can be analyzed by assuming the gas to be incompressible. At higher speeds, the compressibility of the gas cannot be neglected. We can analyze a steady, isentropic flow of a gas through the types of area change illustrated in Figure 8.20 by applying Eq. 8.10. Neglecting the effects of gravity we have

$$\left(\frac{\gamma}{\gamma - 1}\right)\left(\frac{p_1}{\rho_1}\right) + \frac{1}{2}V_1^2 = \left(\frac{\gamma}{\gamma - 1}\right)\left(\frac{p_2}{\rho_2}\right) + \frac{1}{2}V_2^2 \qquad (8.43)$$

In this case a mass balance gives us

$$\dot{M} = \rho_1 A_1 V_1 = \rho_2 A_2 V_2 \qquad (8.44)$$

Additional relationships among the flow variables can be developed by using the perfect gas law, the definitions of Mach number and sound speed, and the isentropic relationships between pressure, density, and temperature. The pressure–density relationship in isentropic flow is given by Eq. 8.7, which we can write as

$$\frac{p_1}{p_2} = \left(\frac{\rho_1}{\rho_2}\right)^\gamma \qquad (8.45a)$$

Using the perfect gas law we can write the additional isentropic relationships:

$$\frac{p_1}{p_2} = \left(\frac{T_1}{T_2}\right)^{\gamma/(\gamma-1)} \quad \text{and} \quad \frac{\rho_1}{\rho_2} = \left(\frac{T_1}{T_2}\right)^{1/(\gamma-1)} \qquad (8.45\text{b, c})$$

Mach number plays an important role in the description of compressible flow. If we rearrange Eq. 8.43 to get

$$\left(\frac{\gamma}{\gamma-1}\right)\left(\frac{p_1}{\rho_1}\right)\left[1+\left(\frac{\gamma-1}{2}\right)\left(\frac{\rho_1 V_1^2}{\gamma p_1}\right)\right] = \left(\frac{\gamma}{\gamma-1}\right)\left(\frac{p_2}{\rho_2}\right)\left[1+\left(\frac{\gamma-1}{2}\right)\left(\frac{\rho_2 V_2^2}{\gamma p_2}\right)\right]$$

then since $\rho V^2/\gamma p = V^2/\gamma RT = V^2/c^2 = M^2$, we find

$$\left(\frac{\gamma}{\gamma-1}\right)\left(\frac{p_1}{\rho_1}\right)\left[1+\left(\frac{\gamma-1}{2}\right)M_1^2\right] = \left(\frac{\gamma}{\gamma-1}\right)\left(\frac{p_2}{\rho_2}\right)\left[1+\left(\frac{\gamma-1}{2}\right)M_2^2\right]$$

Dividing out the common factor and using the perfect gas law yields

$$T_1\left[1+\left(\frac{\gamma-1}{2}\right)M_1^2\right] = T_2\left[1+\left(\frac{\gamma-1}{2}\right)M_2^2\right]$$

Writing this as a ratio gives

$$\frac{T_1}{T_2} = \frac{\left[1+\left(\frac{\gamma-1}{2}\right)M_2^2\right]}{\left[1+\left(\frac{\gamma-1}{2}\right)M_1^2\right]} \tag{8.46a}$$

Next we can use the isentropic relationships Eqs. 8.45b and 8.45c to write

$$\frac{p_1}{p_2} = \left[\frac{1+\left(\frac{\gamma-1}{2}\right)M_2^2}{1+\left(\frac{\gamma-1}{2}\right)M_1^2}\right]^{\gamma/(\gamma-1)} \quad \text{and} \quad \frac{\rho_1}{\rho_2} = \left[\frac{1+\left(\frac{\gamma-1}{2}\right)M_2^2}{1+\left(\frac{\gamma-1}{2}\right)M_1^2}\right]^{1/(\gamma-1)}$$

$$\tag{8.46b, c}$$

To develop an expression relating the area ratio to Mach number, we rearrange Eq. 8.44 to yield

$$\frac{A_2}{A_1} = \frac{\rho_1 V_1}{\rho_2 V_2} = \left(\frac{\rho_1}{\rho_2}\right)\left(\frac{M_1\sqrt{\gamma RT_1}}{M_2\sqrt{\gamma RT_2}}\right) = \left(\frac{\rho_1}{\rho_2}\right)\left(\frac{M_1}{M_2}\right)\left(\frac{T_1}{T_2}\right)^{1/2}$$

From Eqs. 8.46a and 8.46c, this becomes

$$\frac{A_2}{A_1} = \left(\frac{M_1}{M_2}\right)\left[\frac{1+\left(\frac{\gamma-1}{2}\right)M_2^2}{1+\left(\frac{\gamma-1}{2}\right)M_1^2}\right]^{(\gamma+1)/[2(\gamma-1)]} \tag{8.46d}$$

These equations can be used to analyze an isentropic flow through an area change. Note, however, that the density of a gas in isentropic flow is an additional variable, and compressible flows have a number of unique features that distinguish them from incompressible flows.

EXAMPLE 8.17

Air flows through a converging–diverging nozzle as shown in Figure 8.23. At section 1, $p = 930$ kPa, $T = 561$ K, $A = 0.2$ m^2, and the Mach number is $M = 1.2$. If the nozzle exit area is 0.6 m^2, find p, T, and M at the nozzle exit. Does the velocity of the gas decrease in the diverging section?

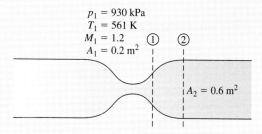

Figure 8.23 Schematic for Example 8.17.

SOLUTION

We will assume steady isentropic flow and apply Eqs. 8.46a–8.46d, as well as the perfect gas law and definitions of M and c. At section 1, the sound speed is calculated as

$$c_1 = \sqrt{\gamma R T_1} = \sqrt{1.4[287 \text{ (N-m)/(kg-K)}](561 \text{ K})} = 475 \text{ m/s}$$

A Mach number of $M_1 = 1.2$ corresponds to a velocity of $V_1 = M_1 c_1 = 1.2\,(475 \text{ m/s}) = 570$ m/s. To use Eqs. 8.46a–8.46c, we need to find M at the nozzle exit. We do this by solving Eq. 8.46d with $M_1 = 1.2$, and $A_2/A_1 = 0.6 \text{ m}^2/0.2 \text{ m}^2 = 3$. One can use either iteration or a symbolic code to obtain the two mathematical solutions $M_2 = 0.19$ and $M_2 = 2.67$. Of these, only $M_2 = 2.67$ is physically possible. To calculate the p and T at the nozzle exit, we use Eqs. 8.46a and 8.46b to find:

$$\frac{T_1}{T_2} = \left[\frac{1 + \left(\frac{\gamma - 1}{2}\right) M_2^2}{1 + \left(\frac{\gamma - 1}{2}\right) M_1^2}\right] = \left[\frac{1 + \left(\frac{1.4 - 1}{2}\right)(2.67)^2}{1 + \left(\frac{1.4 - 1}{2}\right)(1.2)^2}\right] = \frac{2.43}{1.29} = 1.88$$

$$\frac{p_1}{p_2} = \left[\frac{1 + \left(\frac{\gamma - 1}{2}\right) M_2^2}{1 + \left(\frac{\gamma - 1}{2}\right) M_1^2}\right]^{\gamma/(\gamma - 1)} = (1.88)^{3.5} = 9.11$$

Thus the nozzle exit temperature is $T_2 = T_1/1.88 = 561$ K$/1.88 = 298.4$ K, and the exit pressure is $p_2 = p_1/9.11 = 930$ kPa$/9.11 = 102$ kPa. With the exit temperature known, we can compute the sound speed at the exit $c_2 = \sqrt{\gamma R T_2} = \sqrt{1.4[287\text{(N-m)/(kg-K)}](298 \text{ K})} = 346$ m/s, then use the exit Mach number to calculate the exit velocity: $V_2 = M_2 c_2 = 2.67(346 \text{ m/s}) = 924 \text{ m/s}$. We see that the velocity has increased, and the pressure and temperature have decreased, as the flow passed through the diverging section. This is quite different from the behavior we saw in Example 8.16, which featured a constant density flow through a diverging passage. In that case the velocity decreased in the flow through the diverging passage.

8.5.5 Draining of a Tank

Consider the large cylindrical tank of diameter D and height h filled with liquid as shown in Figure 8.24A. The tank is open to the atmosphere and has a well-rounded nozzle of diameter d mounted on the bottom. Suppose the liquid is freely escaping to the atmosphere through the nozzle. We will apply Bernoulli's equation to predict the exit velocity V_E as a function of the fluid and flow parameters, using the streamline connecting points 1 and 2 as shown.

Before applying Bernoulli's equation, challenge yourself with this question: Are there any techniques already studied that might be helpful in understanding this flow? If you thought to answer control volume analysis you are certainly correct. Applying a mass balance to the CV shown in Figure 8.24A and noting that because liquid escapes, the free surface will drop at some velocity V_S, we obtain $\rho V_S (\pi D^2/4) = \rho V_E (\pi d^2/4)$. As expected, the mass flowrate out of the bottom of the CV (and thus out of the nozzle) must match the mass flowrate into the top of the CV (due to the motion of the free surface). Writing this as

$$V_S = V_E \left(\frac{d^2}{D^2}\right) \tag{8.47}$$

we see that if the cross-sectional area of the nozzle is small in comparison to that of the tank, the free surface velocity will be very small in comparison to the nozzle velocity. If so we may consider the flow in the tank and nozzle to be steady. Thus the mass balance suggests that it is valid to employ the steady, constant density form of Bernoulli's equation when $(d^2/D^2) \ll 1$.

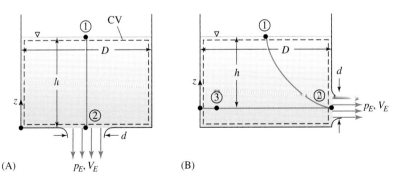

Figure 8.24 Liquid draining from a tank: (A) hole on the bottom and (B) hole on the side.

HISTORY BOX 8-2

Evangelista Torricelli (1608–1647) was a contemporary and close friend of Galileo late in that great man's life. Torricelli invented the barometer, and a unit of pressure equal to one millimeter of mercury is named the torr in his honor. He performed extensive experiments with liquids draining from tanks to develop Torricelli's law that the velocity is proportional to the square root of the height of the liquid, i.e., $V = \sqrt{2gh}$.

The selection of the two points in a Bernoulli analysis is guided by a simple rule: put one of the points at the location for which information is requested and the other at a point for which information is known. This assumes, of course, that these two points are on the same streamline. In this case we will choose the second (downstream) point to be at the exit plane and the first to be at the free surface, assuming that a streamline connects these points. Applying Eq. 8.6, we have

$$\frac{p_S}{\rho} + \frac{1}{2}V_S^2 + gz_S = \frac{p_E}{\rho} + \frac{1}{2}V_E^2 + gz_E$$

The kinetic energy per unit mass, $\frac{1}{2}V_S^2$, can be written by using Eq. 8.47 as $\frac{1}{2}V_S^2 = \frac{1}{2}V_E^2(d^4/D^4)$, which for a large tank is negligible. At the free surface the pressure is atmospheric. The elevation of the free surface is $z_S = h$. At the exit plane, the pressure is also atmospheric (for reasons to be explained in a moment), and the elevation is $z_E = 0$. Thus Bernoulli's equation becomes $\frac{1}{2}V_E^2 = gh$, which is a statement that in the absence of losses, all the gravitational potential energy is converted to kinetic energy. Solving for the exit velocity, we find

$$V_E = \sqrt{2gh} \quad (8.48)$$

a result known as Torricelli's law.

It is also worthwhile to do a Bernoulli analysis for the arrangement with a horizontal nozzle shown in Figure 8.24B. In this case we use points 1 and 2 in the figure and write the Bernoulli equation as $p_S/\rho + \frac{1}{2}V_S^2 + gz_S = p_E/\rho + \frac{1}{2}V_E^2 + gz_E$. Notice that now the origin is at the elevation of the center of the exit, so that $z_E = 0$, while the free surface is at $z_S = h$. The rest of the terms are evaluated exactly the same as before, so the exit velocity is again given by Eq. 8.48.

If we consider a streamline located above or below the central axis of the nozzle, the same analysis predicts a slightly lower or higher velocity. However, since the diameter of the nozzle is assumed to be small, the slight variation of velocity with elevation across the nozzle itself is negligible. One can also analyze this problem by using point 3. As shown in Figure 8.24B, this point lies well to the left of the nozzle, and the velocity here is assumed to be zero. The Bernoulli analysis using this point starts with $p_3/\rho + \frac{1}{2}V_3^2 + gz_3 = p_E/\rho + \frac{1}{2}V_E^2 + gz_E$. Inserting values, we find $p_3/\rho = p_A/\rho + \frac{1}{2}V_E^2$, and the question is, What is the pressure at point 3? The answer is that since the fluid in the tank at point 3 is at rest or nearly so, the pressure at point 3 must be hydrostatic, or $p_3 = p_A + \rho gh$. Inserting this value into Bernoulli's equation yields $p_3/\rho = (p_A + \rho gh)/\rho = p_A/\rho + \frac{1}{2}V_E^2$. It is easy to see that this simplifies to $gh = \frac{1}{2}V_E^2$ and leads to the same result for the exit velocity V_E, Eq. 8.48.

The shape of a nozzle has an effect on the diameter of the jet leaving the nozzle, but not on the velocity, as one might otherwise expect. If the nozzle is well rounded, as shown in Figure 8.25A, a Bernoulli analysis leads to the conclusion that Eq. 8.48 gives the exit velocity right at the end of the nozzle. For a sharp-edged nozzle, as shown in Figure 8.25B, the jet diameter is observed to be less than the diameter of the nozzle

EXAMPLE 8.18

Water escapes from a large storage tank through a small drain hole in the bottom. If the water depth is 2 m. what is the exit velocity? If a similar tank contained gasoline, what would the exit velocity be?

SOLUTION

We will solve this problem using Eq. 8.48, $V_E = \sqrt{2gh}$. Since the density of the liquid does not enter the formula, water and gasoline both exit at the same velocity:

$$V_E = \sqrt{2gh} = \sqrt{2(9.8 \text{ m/s})(2 \text{ m})} = 6.26 \text{ m/s}.$$

orifice, an effect known as vena contracta. Since the streamlines are curved at the orifice, the pressure is atmospheric on the edge of the jet and higher in the middle. In this case, the exit velocity predicted by Torricelli's theorem occurs at the plane where the streamlines have become straight, since this is where the pressure is atmospheric throughout the jet. The degree of contraction is given by a contraction coefficient based on the ratio of the jet to nozzle orifice areas:

$$C_C = \frac{A_J}{A_D} \tag{8.49}$$

Values of this coefficient for different geometries are shown in Figure 8.26.

EXAMPLE 8.19

Consider two nozzle designs for draining a liquid storage tank as shown in Figure 8.25. Both nozzles have the same diameter. If the liquid level is the same, which nozzle provides a larger mass flowrate out the exit?

SOLUTION

The exit velocity in each nozzle is the same as discussed in the text, but the mass flowrate $\rho A V$ in each nozzle differs as a result of the vena contracta effect. The well-rounded nozzle has a contraction coefficient of unity, while the stub nozzle has a coefficient of 0.61 according to Figure 8.26A. Thus the well-rounded nozzle in Figure 8.26B will deliver 39% more mass flow for any given liquid height. Whether this additional flow is worth the added expense of constructing this kind of nozzle is a question of engineering economics.

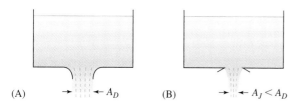

Figure 8.25 (A) Orifice with rounded edges. (B) Orifice with sharp edges results in vena contracta, jet diameter less than orifice diameter.

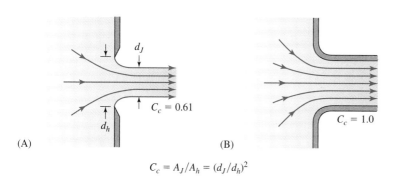

Figure 8.26 Streamline patterns and contraction coefficients for four orifice shapes: (A) sharp edge, (B) well rounded, (C) square, and (D) reentrant.

$$C_c = A_J/A_h = (d_J/d_h)^2$$

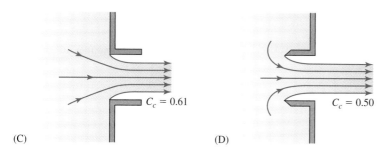

8.6 RELATIONSHIP TO THE ENERGY EQUATION

In deriving the Bernoulli equation, we did not use the energy equation. This seems curious, given our interpretation of the Bernoulli equation as a statement of energy conservation. Perhaps we can learn something by writing an energy balance for flow along a streamline and comparing the result with the Bernoulli equation. To apply an energy balance to the differential CV shown earlier in Figure 8.1B, we use Eq. 7.33:

$$\int_{CV} \frac{\partial}{\partial t} \rho \left(u + \frac{1}{2} \mathbf{u} \cdot \mathbf{u} + gz \right) dV + \int_{CS} \rho \left(u + \frac{p}{\rho} + \frac{1}{2} \mathbf{u} \cdot \mathbf{u} + gz \right) (\mathbf{u} \cdot \mathbf{n}) dS$$
$$= \dot{W}_{\text{power}} + \dot{W}_{\text{shaft}} + \dot{Q}_C + \dot{S}$$

Recalling that the total energy in a fluid is defined as $e = u + p/\rho + \frac{1}{2} \mathbf{u} \cdot \mathbf{u} + gz$, we can write the energy balance as

$$\int_{CV} \frac{\partial}{\partial t} \left[\rho \left(e - \frac{p}{\rho} \right) \right] dV + \int_{CS} \rho(e)(\mathbf{u} \cdot \mathbf{n}) dS = \dot{W}_{\text{power}} + \dot{W}_{\text{shaft}} + \dot{Q}_C + \dot{S}$$

It is possible to account for the unsteady flow in a draining tank in the following approximate way. Referring to Figure 8.24, and noting that at any instant of time the mass balance leading up to Eq. 8.47 is valid, we can write the velocity of the free surface as $V_S = -dh/dt = V_E(d^2/D^2)$. Substituting $V_E = \sqrt{2gh}$ (Eq. 8.48), we find $dh/dt = -(d^2/D^2)\sqrt{2gh}$, which can be integrated as $\int_{h_0}^{h} dh/\sqrt{h} = -(d^2/D^2)\sqrt{2g}\int_{t_0}^{t} dt$, where the height of water in the tank is h_0 at time t_0. The result is the following expression for the height of the liquid in the tank as a function of time

$$2[\sqrt{h(t)} - \sqrt{h_0}] = -\left(\frac{d^2}{D^2}\right)\sqrt{2g}(t - t_0) \tag{8.50}$$

A more accurate analysis of this problem could be made by using the unsteady form of the Bernoulli equation to provide a value for V_E as a function of time.

Evaluating the accumulation and flux terms in a manner similar to that used earlier for the momentum balance, we find

$$\int_{CV} \frac{\partial}{\partial t}\left[\rho\left(e - \frac{p}{\rho}\right)\right] dV = \int_{1}^{2} \frac{\partial}{\partial t}\left[\rho\left(e - \frac{p}{\rho}\right)\right] A\, ds$$

and

$$\int_{CS} \rho e(\mathbf{u}\cdot\mathbf{n})\, dS = -(\rho_1 A_1 V_1)e_1 + (\rho_2 A_2 V_2)e_2$$

$$= \int_{1}^{2} \frac{d}{ds}(\rho A V e)\, ds$$

Since the work and energy terms on the right are the net rates of work and energy additions to the CV between stations 1 and 2, we can write these terms as

$$\dot{W}_{power} + \dot{W}_{shaft} + \dot{Q}_C + \dot{S}$$

$$= \int_{1}^{2} \frac{d}{ds}(\dot{W}_{power} + \dot{W}_{shaft} + \dot{Q}_C + \dot{S})\, ds$$

Writing the energy balance as a single integral, we find

$$\int_{1}^{2}\left\{\frac{\partial}{\partial t}\left[\rho\left(e - \frac{p}{\rho}\right)\right]A + \frac{d}{ds}(\rho A V e) - \frac{d}{ds}(\dot{W}_{power} + \dot{W}_{shaft} + \dot{Q}_C + \dot{S})\right\} ds = 0$$

Since points 1 and 2 are arbitrary, we conclude that

$$\frac{\partial}{\partial t}\left[\rho\left(e - \frac{p}{\rho}\right)\right]A + \frac{d}{ds}(\rho A V e) = \frac{d}{ds}(\dot{W}_{power} + \dot{W}_{shaft} + \dot{Q}_C + \dot{S}) \tag{8.51}$$

along a streamline. In deriving Eq. 8.51 it has not been necessary to introduce any approximations; thus the equation applies to viscous flow.

We can use this equation to discuss the relationship between the energy balance and the Bernoulli equation. For simplicity, we will consider only steady flow, for which the energy balance becomes

$$\frac{d}{ds}(\rho A V e) = \frac{d}{ds}(\dot{W}_{power} + \dot{W}_{shaft} + \dot{Q}_C + \dot{S})$$

Expanding the first term, we find

$$e\frac{d}{ds}(\rho A V) + (\rho A V)\frac{de}{ds} = \frac{d}{ds}(\dot{W}_{power} + \dot{W}_{shaft} + \dot{Q}_C + \dot{S})$$

8.6 RELATIONSHIP TO THE ENERGY EQUATION

For a steady flow the mass balance, Eq. 8.1, is $(d/ds)(\rho A V) = 0$, and the mass flowrate $\dot{m} = \rho A V$ is a constant. The first term in the preceding equation is thus zero, and the steady flow energy balance becomes

$$(\rho A V)\frac{de}{ds} = \frac{d}{ds}(\dot{W}_{\text{power}} + \dot{W}_{\text{shaft}} + \dot{Q}_C + \dot{S}) = 0$$

Dividing this equation by $\dot{m} = \rho A V$ and integrating along a streamline yields

$$\left(u_2 + \frac{p_2}{\rho_2} + \frac{1}{2}V_2^2 + gz_2\right) - \left(u_1 + \frac{p_1}{\rho_1} + \frac{1}{2}V_1^2 + gz_1\right)$$
$$= \int_1^2 \frac{1}{\dot{m}}\frac{d}{ds}(\dot{W}_{\text{power}} + \dot{W}_{\text{shaft}} + \dot{Q}_C + \dot{S})\,ds$$

Completing the integration, we have

$$\left(u_2 + \frac{p_2}{\rho_2} + \frac{1}{2}V_2^2 + gz_2\right) - \left(u_1 + \frac{p_1}{\rho_1} + \frac{1}{2}V_1^2 + gz_1\right)$$
$$= \frac{\Delta \dot{W}_{\text{power}}}{\dot{m}} + \frac{\Delta \dot{W}_{\text{shaft}}}{\dot{m}} + \frac{\Delta \dot{Q}_C}{\dot{m}} + \frac{\Delta \dot{S}}{\dot{m}} \tag{8.52}$$

where the terms on the right-hand side represent the work and energy inputs to the fluid as it moves along the streamline. In developing this equation, the only assumption we have made is steady flow. We did not introduce any assumptions related to frictionless flow or the compressibility of the fluid. The energy balance tells us that in a steady flow along a streamline, the change in total energy between two points is due to the work and energy inputs to the fluid.

The energy balance for the flow of an inviscid fluid along a streamline can be derived by recalling that the viscosity of an inviscid fluid is zero. We can argue that the work terms must be zero because the presence of these terms is inconsistent with the absence of all shear stresses. This means that the Bernoulli equation cannot be used to describe flow through a power-producing or power-absorbing device. With these restrictions, the energy balance reduces to

$$\left(u_2 + \frac{p_2}{\rho_2} + \frac{1}{2}V_2^2 + gz_2\right) - \left(u_1 + \frac{p_1}{\rho_1} + \frac{1}{2}V_1^2 + gz_1\right) = \frac{\dot{Q}_C}{\dot{m}} + \frac{\dot{S}}{\dot{m}} \tag{8.53}$$

Thus, in a steady frictionless flow (compressible or incompressible) along a streamline, a change in total energy can occur only in the presence of a heat addition or an energy input per unit mass.

Suppose we compare this result with the Bernoulli equation for an incompressible fluid. The Bernoulli equation for steady, constant density flow is given by Eq. 8.6 as

$$\frac{p_1}{\rho} + \frac{1}{2}V_1^2 + gz_1 = \frac{p_2}{\rho} + \frac{1}{2}V_2^2 + gz_2$$

Rearranging the energy balance, Eq. 8.53, and noting that for an incompressible fluid, $\rho = \rho_1 = \rho_2$, we can write the energy balance as

$$\frac{p_1}{\rho} + \frac{1}{2}V_1^2 + gz_1 = \left(\frac{p_2}{\rho} + \frac{1}{2}V_2^2 + gz_2\right) + \left[(u_2 - u_1) - \left(\frac{\dot{Q}_C}{\dot{m}} + \frac{\dot{S}}{\dot{m}}\right)\right]$$

Comparing the previous two equations, we see that for the Bernoulli equation to be valid, it is necessary that

$$(u_2 - u_1) - \left(\frac{\dot{Q}_C}{\dot{m}} + \frac{\dot{S}}{\dot{m}}\right) = 0 \tag{8.54}$$

There are two ways in which this equation can be satisfied. Most commonly, heat or energy input of any type is absent, in which case the internal energy of the fluid must be constant. This is the situation in all the examples of this chapter. The second possibility is that the heat added plus energy input raises the internal energy of the fluid but does not affect the mechanical energy. In this case we calculate the change in internal energy from

$$u_2 - u_1 = \frac{\dot{Q}_C}{\dot{m}} + \frac{\dot{S}}{\dot{m}} \tag{8.55}$$

and apply the Bernoulli equation to determine the changes in pressure, and so on. The flow itself is unaffected by the energy input except for the change in internal energy.

Finally, consider a steady isentropic flow for which the Bernoulli equation is given by Eq. 8.11 as

$$u_1 + \frac{p_1}{\rho_1} + \frac{1}{2}V_1^2 + gz_1 = u_2 + \frac{p_2}{\rho_2} + \frac{1}{2}V_2^2 + gz_2$$

Comparing this with the energy balance, Eq. 8.53, we conclude that for the Bernoulli equation to be valid in this type of flow we must have

$$\frac{\dot{Q}_C}{\dot{m}} + \frac{\dot{S}}{\dot{m}} = 0 \tag{8.56}$$

i.e., no net energy addition is allowed.

8.7 SUMMARY

One of the oldest approximations in fluid dynamics is that of an inviscid fluid. If such a fluid existed, its viscosity would be exactly zero, it would be incapable of exerting a shear stress, and its flow would be frictionless. Although all real fluids exhibit nonzero viscosity, there are many important engineering flows for with the inviscid assumption yields reasonable results. For example, many flows of air and water for which the local velocity gradient is small can be adequately modeled by using the inviscid assumption.

Bernoulli's equation, which can be interpreted as a statement of conservation of energy in the flow of an inviscid fluid, can be written as $\int_1^2 (\partial V/\partial t)\, ds + \int_1^2 dp/\rho + \frac{1}{2}(V_2^2 - V_1^2) + g(z_2 - z_1) = 0$. This form of the Bernoulli equation applies to the frictionless flow of a compressible or incompressible fluid, with the only restriction being that the path connecting points 1 and 2 be an instantaneous streamline. Note that if the flow is unsteady, i.e., the temporal derivative is nonzero, this equation shows that the mechanical energy content of the fluid differs from point 1 to point 2 owing to the effects of acceleration. For a steady flow of an incompressible fluid, the Bernoulli equation takes the simplified form $p_1/\rho + \frac{1}{2}V_1^2 + gz_1 = p_2/\rho + \frac{1}{2}V_2^2 + gz_2$, which shows that in the absence of viscous forces, the mechanical energy at two points along a streamline in a steady, constant density flow is the same.

The Bernoulli equation for unsteady, isentropic gas flow is often written in the form

$$\int_1^2 \frac{\partial V}{\partial t} ds + (u_2 - u_1) + \left(\frac{p_2}{\rho_2} - \frac{p_1}{\rho_1}\right) + \frac{1}{2}(V_2^2 - V_1^2) + g(z_2 - z_1) = 0$$

If the isentropic gas flow is steady, then Bernoulli's equation takes the form $u_1 + p_1/\rho_1 + \frac{1}{2}V_1^2 + gz_1 = u_2 + p_2/\rho_2 + \frac{1}{2}V_2^2 + gz_2$. In applications of the Bernoulli equation in compressible flow, the change in gravitational potential energy $g(z_2 - z_1)$ is usually negligible in comparison to the other terms.

If we multiply each term in the steady flow, constant density form of Bernoulli's equation by the fluid density we obtain an alternate form $p + \frac{1}{2}\rho V^2 + \rho gz =$ constant. Fluid mechanics traditionally refers to each term in this equation as a different kind of pressure. The pressure p is referred to as the static, or mechanical, pressure. It is a measure of the average normal stress existing at a point in a fluid and is defined to be the pressure that would be measured by an observer or pressure sensor moving with the fluid. In a constant density flow, as well as in nearly all other flows of interest, the mechanical pressure and the thermodynamic pressure defined by an appropriate equation of state are identical. The term $\frac{1}{2}\rho V^2$ is called the dynamic pressure, and it represents the pressure increase that would occur if all the kinetic energy of a fluid particle in a frictionless flow were converted into a corresponding increase in pressure potential energy. The remaining term, ρgz, called the hydrostatic pressure, represents the change in the static pressure that would occur if the fluid moved along the streamline to an elevation of zero. The sum of the static pressure and dynamic pressure is called the stagnation pressure, and the sum of all three pressure terms is referred to as the total pressure.

Bernoulli's equation is a powerful tool for finding the values of pressure and velocity at two points along a streamline. In applying this equation, it is important to remember that the underlying assumption of frictionless flow means that viscous effects are absent. Real flows do not perfectly satisfy these underlying assumptions, which means that you should consider the results obtained with Bernoulli's equation to be an engineering approximation.

The stagnation pressure can be measured by using a device called a pitot tube. A pitot tube will give an accurate reading of stagnation pressure provided it is aligned with the flow and if the probe is small relative to the length scale of the flow field. A pitot-static tube, which combines a static pressure tap on its side with a regular pitot tube type of nose tap, permits the simultaneous measurement of the static and stagnation pressures at a point in a flow field.

If we compare the Bernoulli equation for steady, constant density flow with the corresponding energy balance, we find that the condition $(u_2 - u_1) - (\dot{Q}_C/\dot{m} + \dot{S}/\dot{m}) = 0$ must be satisfied for the Bernoulli equation to be valid. There are two ways in which this equation can be satisfied. Most commonly, an energy input of any type is absent, in which case the internal energy of the fluid must be constant. This is the situation in all the flows examined in this chapter. The second possibility is that the energy input raises the internal energy of the fluid but does not affect the mechanical energy. In this case we calculate the change in internal energy from $u_2 - u_1 = \dot{Q}_C/\dot{m} + \dot{S}/\dot{m}$ and apply the Bernoulli equation to determine the changes in pressure, and so on. The flow itself is unaffected by the energy input except for the change in internal energy. For an isentropic flow, consideration of the energy balance shows that for the Bernoulli equation to be valid we must have $\dot{Q}_C/\dot{m} + \dot{S}/\dot{m} = 0$; i.e., no net energy addition is allowed.

PROBLEMS

Section 8.3

8.1 Air flows through a nozzle with a centerline velocity of $\mathbf{u} = 50(1+x)\mathbf{i}$ m/s, where x is in meters. Determine the pressure gradient along the center line of the nozzle.

8.2 Water flows through a diffuser with a centerline velocity of $\mathbf{u} = 5(1-x)\sin(0.5t)\mathbf{i}$ m/s, where x is in meters and t is in seconds. Determine the pressure gradient in the diffuser as a function of time.

8.3 Repeat Problem 8.2 with the diffuser in the vertical direction such that $\mathbf{w} = 5(1-z)\sin(0.5t)\mathbf{k}$ m/s.

8.4 The velocity of air in a pipe is $\mathbf{u} = 200(1+x)\mathbf{i}$ m/s, where x is in meters. The pressure and density are related by $p/\rho^n = A$, where n and A are constants. What is the change in pressure in the pipe from $x = 0$ to $x = 1$ m?

8.5 The area of a nozzle varies according to the equation $A = A_{\text{inlet}}(1 - \alpha x^2/L^2)$ where α is a constant and L is the length of the nozzle. The volume flowrate is unsteady and is described by $Q = \beta \sin(\omega t)$ where β and ω are constants. Derive an expression for the pressure change between in the inlet and exit of the nozzle.

8.6 A venturi flume (shown in Figure P8.1) consisting of a bump on the channel floor is used to measure the flowrate. Assuming uniform velocity profiles, derive an expression

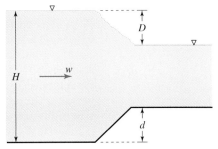

Figure P8.1

for the flowrate as a function of the geometric parameters w, H, D, and d.

8.7 Estimate the maximum height of the flow resulting from a water main break at 80 psig.

8.8 Water flows upward out of the variable area nozzle shown in Figure P8.2. Determine H as a function of D_{exit}.

8.9 The water flow from a garden hose is 10 ft/s. What is the maximum height the water can go above the hose exit? If you place your thumb over the exit to reduce the exit area by half, how high can the water rise?

8.10 For the hose in Problem 8.9, how far can the water shoot horizontally if the exit is at 30°?

8.11 For the water flow system in Figure P8.3, what is the mercury manometer reading if the velocity at the location of the pressure reading is 1 m/s?

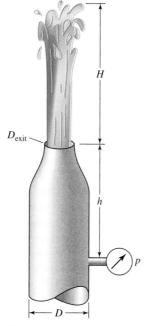

Figure P8.2

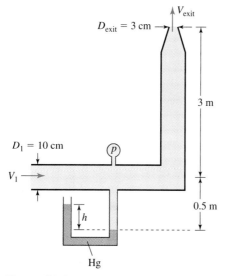

Figure P8.3

8.12 Nitrogen flows isentropically in a 2 in. diameter pipe. Conditions at one point are $v = 400$ ft/s, $p = 80$ psia, and $\rho = 2.03 \times 10^{-3}$ lb$_m$/ft^3. At a nearby point, the pressure has fallen to 76 psia. What is the velocity at the second point?

8.13 At a certain point in an isentropic airflow $v = 30$ m/s, $p = 375$ kPa, and $\rho = 2.85$ kg/m^3. At a second point on the same streamline the velocity is 100 m/s. What is the pressure at the second point?

Section 8.4

8.14 An airplane is flying 150 mph at 10,000 ft. What is the stagnation pressure on the airplane? What is the pressure on top of the wing, where the velocity reaches 195 mph?

8.15 What is the maximum pressure on your hand if you stick it out of the car window traveling at 70 mph through still air at standard conditions?

8.16 A 2 in. ID. pipe carries gasoline at 10 gal/min at a pressure of 50 psig. Determine the stagnation pressure in psi and in feet of gasoline.

8.17 Air flows past an object at 200 m/s. Determine the stagnation pressure in the standard atmosphere at elevations of sea level, 2000 m, and 10,000 m.

8.18 If hurricane winds reach 100 mph, what is the stagnation pressure on a window? What is the force exerted on 6 ft^2 window?

Section 8.5

8.19 Water is siphoned from a tank as shown in Figure P8.4. Calculate the flowrate and the pressure at the upper bend in the tube (point B).

8.20 Fluid is drained from the reservoir by using a plastic tube as shown in Figure P8.5. The tube will collapse if $p < p_c$. Derive an expression for the maximum value of h for which the fluid will flow.

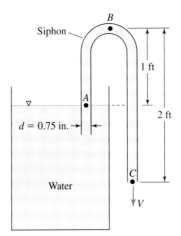

Figure P8.4

8.21 Oil (SG = 0.85) is drained through the siphon ($D = 50$ mm) shown in Figure P8.6. What is the volume flowrate?

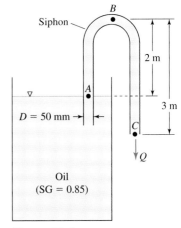

Figure P8.6

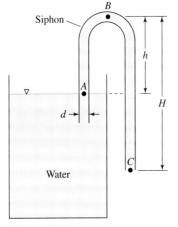

Figure P8.5

8.22 Consider the manometer connected to the pitot-static tube shown in Figure P8.7. What is the velocity in the channel if the flowing fluid is JP-4 fuel (SG = 0.77), water, or air at standard conditions?

8.23 Calculate the flowrate of water through the pipe reduction shown in Figure P8.8.

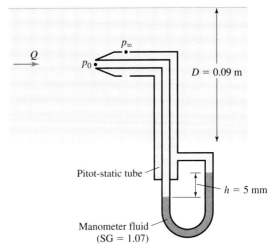

Figure P8.7

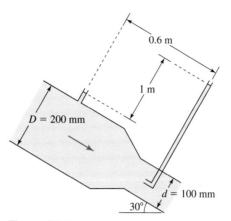

Figure P8.8

8.24 Calculate the pressure and velocity at points 2 and 3 for the water pipe shown in Figure P8.9.

8.25 Repeat Problem 8.24 using air as the fluid. Comment on the effect of the elevation change for air as opposed to water.

8.26 Determine the height of fluid in the manometer, H, for the flow shown in Figure P8.10, if the velocity before the contraction is 1.5 m/s.

8.27 What is the flowrate of water leaving the tank shown in Figure P8.11?

8.28 A hose is added to the exit of Problem 8.27. At what exit height will the flow stop?

8.29 The flow of water from the cylindrical tank is controlled by adjusting the air pressure on top as shown in Figure P8.12. Derive an expression for the volume flowrate of water as a function of p_{air}, H, D_{exit}, and ρ_{water}.

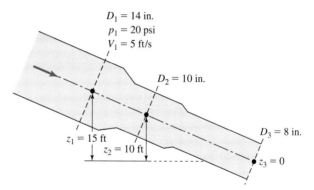

Figure P8.9

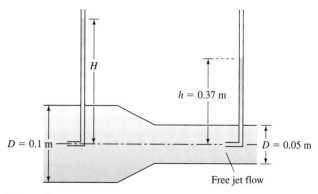

Figure P8.10

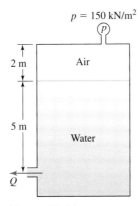

Figure P8.11

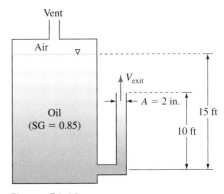

Figure P8.13

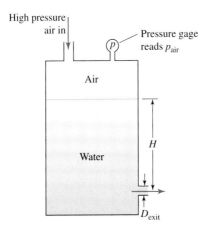

Figure P8.12

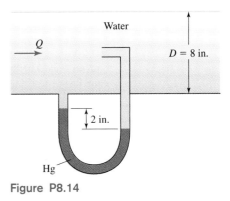

Figure P8.14

8.30 Oil (SG = 0.85) flows from the tank through the pipe shown in Figure P8.13. What is the exit velocity?

8.31 Water flows through an 8 in. pipe as shown in Figure P8.14. The U-tube manometer, filled with mercury, is attached to a stagnation pitot tube centered in the pipe and to a static pressure tube on the wall. Calculate the volume flowrate.

8.32 If the pipe from Problem 8.31 is oriented at 45° and the manometer reading is as shown in the Figure P8.15, calculate the volume flowrate.

8.33 Repeat Problem 8.31 with air flowing through the pipe and water as the manometer fluid.

8.34 Repeat Problem 8.32 with air flowing through the pipe and water as the manometer fluid.

8.35 The venturi shown in Figure P8.16 is used to draw water into the airstream. Derive an expression for the velocity of air in the pipe that will draw water up to the pipe.

8.36 The water tank system shown in Figure P8.17 is in steady state; find the water depth in the second tank H_2.

8.37 Water flows from the tank shown in Figure P8.18. Derive an expression for the exit velocity based on H_1, H_2, ρ_{water}, SG_{oil}, and D_{exit}.

PROBLEMS 531

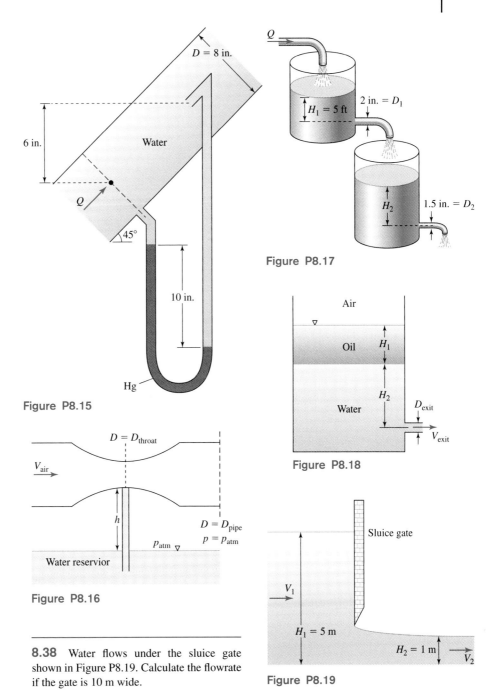

Figure P8.15

Figure P8.16

Figure P8.17

Figure P8.18

Figure P8.19

8.38 Water flows under the sluice gate shown in Figure P8.19. Calculate the flowrate if the gate is 10 m wide.

8.39 Water flows up the ramp in the 10 ft wide channel shown in the Figure P8.20. Calculate the flowrate.

8.40 The spillway shown in Figure P8.21 is 5 m wide. Determine the flowrate in the channel.

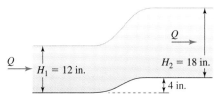

Figure P8.20

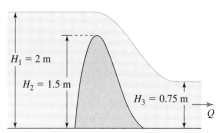

Figure P8.21

8.41 Water flows down the ramp shown in Figure P8.22. Use the Bernoulli equation to determine the downstream depth. Comment on the validity of your solution(s).

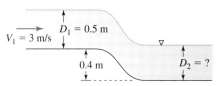

Figure P8.22

8.42 Water flows under the sluice gate shown in Figure P8.23. Calculate the flowrate if the gate is 15 ft wide.

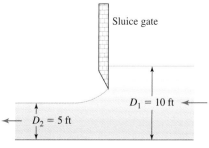

Figure P8.23

8.43 Water flows up the ramp in the 5 m wide channel shown in the Figure P8.24. Calculate the flowrate.

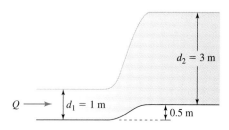

Figure P8.24

8.44 The spillway shown in Figure P8.25 is 10 ft wide. Determine the flowrate in the channel.

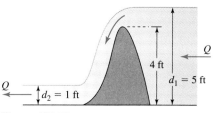

Figure P8.25

8.45 Water flows down the ramp shown in Figure P8.26. Use the Bernoulli equation to determine the downstream depth. Comment on the validity of your solution(s).

8.46 The water level, H, in a tank is kept constant. Determine at what height h a nozzle with exit area A should be placed such that the distance traveled by the resulting horizontal jet is maximized.

8.47 As shown in Figure P8.27, a pitot tube in an air duct is connected to a pressure gage that reads 11.0 psig. An adjacent static pressure gage reads 10.0 psig. What is the velocity in the duct?

PROBLEMS | 533

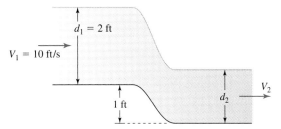

Figure P8.26

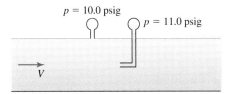

Figure P8.27

8.48 Use Figure P8.28 to derive an expression for the velocity in the duct.

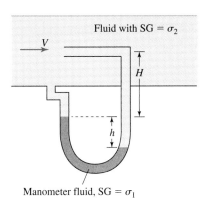

Figure P8.28

8.49 Determine the volume flowrate of water through the venturi meter shown in Figure P8.29 given that the fluid in the manometer is mercury.

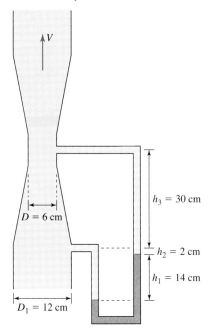

Figure P8.29

9 DIMENSIONAL ANALYSIS AND SIMILITUDE

9.1 Introduction
9.2 Buckingham Pi Theorem
9.3 Repeating Variables Method
9.4 Similitude and Model Development
9.5 Correlation of Experimental Data
9.6 Application to Case Studies
  9.6.1 DA of Flow in a Round Pipe
  9.6.2 DA of Flow through Area Change
  9.6.3 DA of Pump and Fan Laws
  9.6.4 DA of Flat Plate Boundary Layer
  9.6.5 DA of Drag on Cylinders and Spheres
  9.6.6 DA of Lift and Drag on Airfoils
9.7 Summary
Problems

9.1 INTRODUCTION

Fluid mechanics problems can be dealt with by using analytical, computational, and experimental approaches to understand the distribution of fluid and flow properties and the interaction of the fluid with its surroundings. In the preceding chapters you learned how to apply mass, momentum, and energy balances and the Bernoulli equation. In this chapter, we focus our attention on some of the experimental tools engineers use to solve fluid mechanics problems. These tools, known as dimensional analysis, similitude, and modeling, are very powerful but surprisingly easy to apply. Perhaps you remember that the case study results were based on dimensional analysis and experiments. In a later section of this chapter we will show you how dimensional analysis can be used to arrive at the design formulas of those case studies. We will also show you how the tools you learn about in this chapter allow an engineer to efficiently organize and understand the results of experiments.

9.1 INTRODUCTION | 535

Figure 9.1 (A) Model of an aircraft in a wind tunnel. (B) scale model of a section of the Missisipi River. (C) Model of San Antonio, Texas, used for determining wind patterns in this urban environment.

Did you know that most experimental studies in fluid mechanics involve the use of scale models? Examples are shown in Figure 9.1. Perhaps it is obvious that a scale model must be geometrically similar to an actual device or system, but what guarantee do we have that the flow field that occurs with a scale model is similar to the flow field of engineering interest? How do we apply data gathered from an experiment on a model to the design of a full-scale device?

Dimensional analysis (abbreviated DA) is the source of answers to these and many other questions involving experimental research. It is the foundation of the theories of similitude and modeling, and it provides a means to design an efficient experimental program. DA makes use of the principle that all terms of a physical equation must have the same dimensions. Engineers take advantage of this by routinely checking a proposed formula to make sure it is dimensionally consistent. However, this particular application of DA, while helpful, is really of secondary importance compared to using DA to understand the behavior of a physical system without the need for complex mathematics.

Most engineers associate DA with the Buckingham Pi theorem, published in 1914 by E. Buckingham. With the aid of the Pi theorem, DA may be used to determine the number and form of the dimensionless groups describing any fluid system. As discussed in Chapter 3, a dimensionless group is a unitless algebraic combination of several of the physical parameters of a problem. Recall that the most important dimensionless group in fluid mechanics, the Reynolds number, is given by the product of density ρ, a fluid velocity scale V, and a length scale L, all divided by viscosity, μ. Thus we have $Re = \rho V L/\mu$. There are numerous other groups that arise from dimensional analysis. For example, dividing the length of a pipe by its diameter yields the dimensionless group L/D. A given fluid system may have from one to five or more dimensionless groups, depending on its complexity. The value of each dimensionless group is a pure number determined by the specific values of the problem parameters. With experience it is possible to anticipate the behavior of a physical system simply by knowing the values of the dimensionless groups that describe it.

You already know from the case studies that the use of dimensionless groups allows an engineer to classify a fluid mechanics problem, relate it to work done by others, and select an effective solution method. DA is also essential when one is conducting and analyzing experiments. Proper selection of the values of each dimensionless group for a scale model ensures that a condition known as similitude is achieved. Similitude guarantees that a particular experimental model is similar to the true physical system it is intended to simulate and also assures us that experimental data obtained with the model can be scaled and applied to a full-scale prototype. Finally, DA provides an understanding of how to minimize the total number of experiments and enhances the correlation and efficient compilation of experimental data.

The Buckingham Pi theorem is introduced in the following section, followed by a discussion of DA using the repeating variable method of constructing dimensionless groups. Next we demonstrate the use of dimensional groups in similitude and model development, and in the correlation of experimental data. Finally, we revisit the case studies of Chapter 3 and demonstrate the power of DA and empirical correlations in the design of fluid mechanics devices and systems. We do this by showing how the empirical relationships first presented in the case studies without explanation can now be seen to be based on dimensional analysis.

9.2 BUCKINGHAM PI THEOREM

We begin our discussion of the Pi theorem by considering a practical problem in the design of a piping system. What size pump is required to move a particular liquid at a desired flowrate through a pipe? It can be shown that the power needed is a function of the pressure drop down the pipe, (see the case study in Section 3.3.1), which in turn depends on the viscous dissipation of energy in the flow (see Section 2.7.1). The power required also depends on the mass flowrate, defined as the mass of liquid moving through the pipe per unit time. How can the Pi theorem contribute to answering this question?

The first step in a DA of pipe flow is to determine which fluid and flow properties might influence the pressure drop. As illustrated in Figure 9.2, we postulate that the pressure drop Δp may depend upon the pipe length L, diameter D, wall roughness e, average liquid velocity $\bar{V}$, liquid density ρ, and viscosity μ. The wall roughness, defined as the average height of random protuberances, depends on the type of pipe and how long it has been in service. How did we decide on this list of parameters? We know that

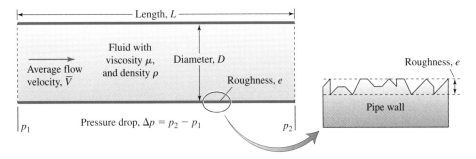

Figure 9.2 The parameters that affect pressure drop in horizontal flow through a round pipe.

Δp is related to frictional losses (viscous dissipation). Since we expect frictional loses to increase with an increase in viscosity, pipe length, and pipe wall roughness, it makes sense to include μ, L, and e in our model. The inclusion of $\bar{V}$ and D should also seem reasonable to you if you remember that the volume flowrate, i.e., the volume of liquid moving through the pipe per unit time, is given by the product of the average liquid velocity and the cross-sectional area of the pipe. Finally, we include density as a variable to be able to relate the volume flowrate to the mass flowrate.

The next step in the DA is to write the proposed functional relationship mathematically as

$$\Delta p = f(L, D, e, \bar{V}, \rho, \mu) \tag{9.1}$$

According to Eq. 9.1, the pressure drop is a function of six independent variables. By the principle of dimension consistency, the unknown function f must combine the independent variables in such a way that it has dimensions of pressure. Suppose we try to determine f by experiment. To explore the effects of the six variables on Δp we systematically vary each independent variable while holding the other independent variables fixed. This can be time-consuming and expensive. Furthermore, the experiments may be difficult to perform. For example, how can we vary viscosity significantly? Is it possible to reduce the number of independent variables in Eq. 9.1?

Buckingham's work established the theoretical basis for a process to reduce the number of independent variables in a functional relationship like Eq. 9.1 to a minimum. The Pi theorem is applicable to a function, f, of k dimensional variables, u_k, in the form

$$u_1 = f(u_2, u_3, \ldots, u_k) \tag{9.2}$$

For example, in Eq. 9.2 the dependent variable u_1 is the pressure drop and there are six independent variables, so the total number of variables is $k = 7$. The Pi theorem states that there exists a functional relationship between at most $k - r$ dimensionless Pi groups, Π_{k-r}, of the form

$$\Pi_1 = g(\Pi_2, \Pi_3, \ldots, \Pi_{k-r}) \tag{9.3}$$

where r is the number of base dimensions needed to describe those parameters. Thus the Pi theorem proves that the number of independent variables in any functional relationship may be reduced from k to $k - r$ if the relationship is expressed in terms of

9 DIMENSIONAL ANALYSIS AND SIMILITUDE

The implications of the Pi theorem are twofold. First, in performing experiments it is necessary to vary only the value of a dimensionless group rather than of each physical parameters it contains. Second, by working with dimensionless groups, there are $k - r$ independent variables rather than k, a substantial reduction.

dimensionless groups. The r base dimensions in most fluid mechanics problems are M, L, t, and T, representing mass, length, time, and temperature. In the absence of thermal effects, a temperature scale T is unnecessary.

What happens when we apply the Pi theorem to our pipe flow problem? The seven physical parameters in pipe flow can be written in terms of base dimensions as

$$\{\Delta p\} = \frac{M}{t^2 L}, \quad \{L\} = \{D\} = \{e\} = L, \quad \{\bar{V}\} = \frac{L}{t}, \quad \{\rho\} = \frac{M}{L^3}, \quad \text{and} \quad \{\mu\} = \frac{M}{Lt}$$

Notice that we did not use the base dimension for temperature, T, and it was not necessary to introduce force, F, as a base dimension, since M, L, and t can be combined to form the dimension of force. Thus, the base dimensions for pipe flow are M, L, and t and we have $r = 3$. Table 9.1, which is a list of common fluid and flow properties and their base dimensions, can be of assistance in this process.

According to the Pi theorem, using dimensionless groups reduces the number of independent variables in any functional relationship from k to $k - r$. For pipe flow, since $k = 7$ and $r = 3$, the theorem suggests replacing Eq. 9.1 with a functional relationship:

$$\Pi_1 = g(\Pi_2, \Pi_3, \Pi_4)$$

where the four Pi groups are dimensionless algebraic combinations of the original set of seven physical parameters. By design, the first Pi group includes the dependent variable,

TABLE 9.1 Base Dimensions for Common Fluid and Flow Properties

Property	Dimensions	Property	Dimensions
Acceleration	Lt^{-2}	Momentum	MLt^{-1}
Angle	Dimensionless	Power	ML^2t^{-3}
Angular momentum	ML^2t^{-2}	Pressure	$ML^{-1}t^{-2}$
Angular velocity	t^{-1}	Specific heat	$L^2t^{-2}T^{-1}$
Area	L^2	Specific weight	$ML^{-2}t^{-2}$
Density	ML^{-3}	Strain	Dimensionless
Energy	ML^2t^{-2}	Stress	$ML^{-1}t^{-2}$
Force	MLt^{-2}	Surface tension	Mt^{-2}
Frequency	t^{-1}	Temperature	T
Heat	ML^2t^{-2}	Time	t
Length	L	Torque	ML^2t^{-2}
Mass	M	Velocity	Lt^{-1}
Modulus of elasticity	$ML^{-1}t^{-2}$	Viscosity (dynamic)	$ML^{-1}t^{-1}$
Moment of a force	ML^2t^{-2}	Viscosity (kinematic)	L^2t^{-1}
Moment of inertia (area)	L^4	Volume	L^3
Moment of inertia (mass)	ML^2	Work	ML^2t^{-2}

so Π_1 includes the pressure drop. Thus in pipe flow the theorem instructs us to think in terms of a relationship between a dimensionless pressure drop and three other dimensionless groups made up of the remaining physical parameters. We conclude that in performing experiments on pipe flow, it is necessary to vary only the value of three dimensionless groups rather than of the six physical parameters they contain.

EXAMPLE 9.1

The drag force F_D on a soccer ball is thought to depend on the velocity of the ball V, diameter D, air density ρ, and viscosity μ. Determine the number of Pi groups that can be formed from these five parameters.

SOLUTION

We begin by writing $F_D = f(V, D, \rho, \mu)$, and establishing that the total number of variables is $k = 5$. The dimensions of these variables are

$$\{F_D\} = \frac{ML}{t^2}, \quad \{V\} = \frac{L}{t}, \quad \{D\} = L, \quad \{\rho\} = \frac{M}{L^3}, \quad \text{and} \quad \{\mu\} = \frac{M}{Lt}$$

Thus, the base dimensions are M, L, and t, and $r = 3$. The expected number of Pi groups is $k - r = 5 - 3 = 2$.

EXAMPLE 9.2

Suppose your company has a policy that engineers investigating the influence of an independent variable on a particular flow must use at least 10 different values of each independent variable during testing. How many experiments would be required to investigate the influence of the six independent variables for pipe flow listed in Eq. 9.1? How many experiments are required if we make use of the Pi theorem with $r = 3$?

SOLUTION

If we require 10 experiments for each independent variable, we will need 10^k experiments to fully describe the influence of k independent variables. In the pipe flow problem, with $k = 6$, we would need to perform a million experiments. Once we understand the power of the Pi theorem, however, we realize that we can collect equivalent data with only 10^{k-r}, or a thousand experiments. Will your knowledge of the Pi theorem save your company time and money?

9.3 REPEATING VARIABLE METHOD

The primary goal of DA is to determine the form of each dimensionless group predicted by the Pi theorem. With experience, this may be done by inspection, since we know that each group is dimensionless and that every valid physical parameter must appear in at least one dimensionless group. In general, however, we recommend the use of the repeating variable method to construct dimensionless groups. It is quick and easy to implement, and it provides the advantage of using a set procedure with less chance of error. The required procedure, stated formally here, is explained in more detail in the subsequent example.

Repeating Variable Method

1. List all physical parameters, assigning one as the dependent variable, and express a relationship between this variable and the others in the form of Eq. 9.2. Let k be the total number of variables, including the dependent variable.
2. Represent each variable in terms of its base dimensions, forming a base dimensions table. Let r be the total number of different base dimensions needed.
3. Choose r independent variables to serve as repeating variables, making sure that all base dimensions are included. There are $(k - r)$ nonrepeating independent variables left, with each appearing in a dimensionless Pi group.
4. Form a Pi group first for the nonrepeating dependent variable by multiplying it by each repeating variable raised to a power. Choose the exponents of each of the r repeating variables to make the overall product and resulting dependent Pi group dimensionless. Repeat for the remaining $(k - r - 1)$ nonrepeating independent variables, forming a Pi group for each.
5. Check each Pi group to be sure it is dimensionless, and rearrange to obtain a standard form if known.
6. Express the $(k - r)$ Pi groups in the functional form of equation 9.3.

Let us now look at each of these steps in more detail by executing the complete procedure for the pipe flow shown earlier in Figure 9.2.

Step 1

We are told that the pressure drop Δp depends on the pipe length L, diameter D, wall roughness e, average velocity $\bar{V}$, liquid density ρ, and viscosity μ. The dependent variable is Δp, so the desired functional relationship between the physical parameters is Eq. 9.1:

$$\Delta p = f(L, D, e, \bar{V}, \rho, \mu)$$

Counting all the physical parameters we see that $k = 7$.

Comments on Step 1: To simply list all the physical parameters that influence a flow problem is deceptively difficult; yet this step is critical and usually a challenge for an inexperienced engineer. We are required to list every fluid and flow property, geometric parameter, and external agent that exerts an influence on the phenomenon of interest. Constructing this list is guided by an understanding of the theoretical and practical aspects of fluid mechanics, and most importantly by experience. A review of published work on the flow of interest is helpful.

If a noninfluential parameter is inadvertently included in a DA, an extra dimensionless group containing that parameter will result. Experiments will, however, show the extra group is superfluous and can be ignored. For example, if we had included gravity in our pipe flow DA, we would have obtained an extra dimensionless group that would have been shown by experiments to have no influence on Δp. On the other hand, if an important parameter is left out, the corresponding dimensionless group will be missing. The effect of a missing group may not be discovered except in hindsight, when mysterious variations in experimental data are finally understood. Therefore, we recommend that you include every parameter in step 1 that can be reasonably expected to influence the flow. This is why wall roughness is included for pipe flow. Does it seem reasonable that the frictional pressure drop in a rough pipe might be different from that in a smooth pipe?

Be careful to avoid including redundant parameters in your parameter list. For example, choosing to include both pipe diameter and cross-sectional area is inappropriate because the effect of a change in the value of one of these parameters completely determines the change in the other. Tradition dictates that diameter be used in pipe flow. The same thinking applies to including both the absolute viscosity μ and the kinematic viscosity ν of a fluid along with the density. The two viscosities are not independent. We recommend using μ rather than ν, but never both.

Step 2

We represented each variable in pipe flow in terms of base dimensions earlier. We therefore use those results to construct the following base dimensions table:

Δp	L	D	e	$\bar{V}$	ρ	μ
$\dfrac{M}{t^2 L}$	L	L	L	$\dfrac{L}{t}$	$\dfrac{M}{L^3}$	$\dfrac{M}{tL}$

By inspection, the number of base dimensions used in the table is $r = 3$. Since there are 7 variables and 3 base dimensions, we anticipate finding a total of $k - r = 7 - 3 = 4$ dimensionless groups.

Comment on Step 2: It is usually straightforward to represent each physical parameter in terms of its base dimensions and determine the total number of base dimensions in a problem. Table 9.1 can be consulted as part of this process. The recommended default set of base dimensions is M, L, t, with T included when needed in thermal problems.

Step 3

We must now choose $r = 3$ independent variables out of the set $L, D, e, \bar{V}, \rho, \mu$ to serve as repeating variables, with only one stated constraint: that all base dimensions be included. From the base dimensions table, it appears that a reasonable choice is $(D, \bar{V}, \rho)$. The remaining nonrepeating variables $\Delta p, L, e, \mu$ will each appear in a dimensionless group.

At this point we pause to insert a word of caution. Normally you can expect the Pi theorem as stated to produce the correct number of dimensionless groups. There are, however, a few exceptions. Consider capillary rise, the well-known phenomenon of a liquid column rising in a tube inserted into liquid as the result of surface tension (see the discussion associated with Eq. 2.41). The height of the column h is known to depend on the tube diameter d, the surface tension σ, and specific weight γ of the liquid. Suppose we attempt to determine the number of Pi groups that can be formed from these four parameters. We begin by writing $h = f(d, \sigma, \gamma)$, and establishing that the total number of variables is $k = 4$. The dimensions of these variables are:

$$\{h\} = L, \ \{d\} = L, \ \{\sigma\} = \frac{M}{t^2}, \text{ and}$$

$$\{\gamma\} = \frac{M}{L^2 t^2}$$

Thus, the base dimensions are M, L, and t, and $r = 3$. The expected number of Pi groups is $k - r = 4 - 3 = 1$. However, practical experience shows that we actually need two Pi groups for this problem rather than only one. What went wrong? The difficulty results from the impossibility, in this case, of picking three variables from which a dimensionless group *cannot* be formed. Thus, the third requirement described in the comments on step 3 cannot be met. Try it yourself. To complete the DA in a case like this, we reduce r by one and use one less repeating variable. In the capillary rise case we reduce r from 3 to 2 and use only two repeating variables, d and σ. These include all the base dimensions but cannot form a dimensionless group. This correctly results in the formation of two Pi groups.

Comments on Step 3: The set of repeating variables shown is not the only possibility, so why pick these three? For example, why not pick $(L, \bar{V}, \rho)$, $(\Delta p, e, D)$, or $(L, \bar{V}, D)$? The choice of repeating variables is restricted by certain requirements. First, each base dimension must be represented among the repeating variables. Note that the $(D, \bar{V}, \rho)$, $(L, \bar{V}, \rho)$, and $(\Delta p, e, D)$ sets satisfy this requirement, but the set $(L, \bar{V}, D)$ does not, since it doesn't contain M. A second requirement is that the dimensions of the repeating variables be independent, meaning that the dimension of one repeating variable should not be equal to that of another raised to a power. The $(\Delta p, e, D)$ set fails this test because e and D have the same dimension. The third requirement is that the repeating variables not be able to form a dimensionless group among themselves. This is checked by inspection.

Both $(D, \bar{V}, \rho)$ and $(L, \bar{V}, \rho)$ satisfy the requirements for a valid set of repeating variables. How do we choose from among several viable options? You should select the set of repeating variables that results in the traditional forms of dimensionless groups commonly used in fluid mechanics (see Chapter 3). To understand this directive, realize that virtually all dimensionless groups generated by DA have already been generated by engineers in one form or another. Thus, it makes sense in communicating results to peers for us to use a traditional form of each dimensionless group whenever possible. For example, in pipe flow we expect to find Reynolds number as one of the dimensionless groups. Reynolds number for pipe flow is defined by using the average velocity and the diameter of a pipe rather than its length. The $(D, \bar{V}, \rho)$ set will produce the traditional Reynolds number, but the $(L, \bar{V}, \rho)$ set will not, since L will appear in the Reynolds number rather than D.

As a final comment, recall that the dependent variable (in our example Δp) may not serve as a repeating variable. The reason is that a dependent variable chosen as a repeating variable will potentially appear in every dimensionless group. When performing DA in the context of the design or interpretation of experiments, we want to isolate the effects of the independent variables on the dependent variable, so the latter must not appear in more than one dimensionless group.

Step 4

The first Pi group always involves the dependent variable. We form this group for pipe flow by writing a product of the pressure drop with each of the repeating variables raised to a power. Since we will pick these exponents to make the Pi group dimensionless, we have:

$$\Pi_1 = (\Delta p)^1 (D)^A (\bar{V})^B (\rho)^C = \left(\frac{M}{t^2 L}\right)^1 (L)^A \left(\frac{L}{t}\right)^B \left(\frac{M}{L^3}\right)^C = M^0 L^0 t^0$$

Note that we have substituted the dimensions of each variable in the Pi group from the base dimensions table. We can ensure that the first Pi group is dimensionless by equating exponents of each base dimension as follows:

$$(M)^{1+C} = M^0, \quad (L)^{-1+A+B-3C} = L^0, \quad \text{and} \quad (t)^{-2-B} = t^0$$

The equations for the exponents are:

$$1 + C = 0, \quad -1 + A + B - 3C = 0, \quad \text{and} \quad -2 - B = 0$$

and these are satisfied by choosing $C = -1$, $B = -2$, $A = 0$. Thus the first Pi group is

$$\Pi_1 = (\Delta p)^1 (D)^0 (\bar{V})^{-2} (\rho)^{-1} = \frac{\Delta p}{\rho \bar{V}^2}$$

The remaining three Pi groups containing the three nonrepeating independent variables (L, e, μ) are constructed following the same procedure. The order in which the remaining groups are created does not matter. The Pi group containing pipe length L is constructed by writing

$$\Pi_2 = (L)^1 (D)^A (\bar{V})^B (\rho)^C = (L)^1 (L)^A \left(\frac{L}{t}\right)^B \left(\frac{M}{L^3}\right)^C = M^0 L^0 t^0$$

so the equations for the exponents are $C = 0$, $1 + A + B - 3C = 0$, and $-B = 0$. By inspection, $A = -1$, $B = 0$, and $C = 0$, so the second Pi group is

$$\Pi_2 = (L)^1 (D)^{-1} (\bar{V})^0 (\rho)^0 = \frac{L}{D}$$

The next Pi group, containing wall roughness e, is constructed as follows:

$$\Pi_3 = (e)^1 (D)^A (\bar{V})^B (\rho)^C = (L)^1 (L)^A \left(\frac{L}{t}\right)^B \left(\frac{M}{L^3}\right)^C = M^0 L^0 t^0$$

The equations determining the required exponents are the same as those for pipe length, since wall roughness and pipe length have the same dimensions. Thus the third Pi group is

$$\Pi_3 = (e)^1 (D)^{-1} (\bar{V})^0 (\rho)^0 = \frac{e}{D}$$

The last Pi group, containing viscosity μ, is constructed as follows:

$$\Pi_4 = (\mu)^1 (D)^A (\bar{V})^B (\rho)^C = \left(\frac{M}{Lt}\right)^1 (L)^A \left(\frac{L}{t}\right)^B \left(\frac{M}{L^3}\right)^C = M^0 L^0 t^0$$

The equations for the exponents are $1 + C = 0$, $-1 + A + B - 3C = 0$, and $-1 - B = 0$, so the resulting exponents are $C = -1$, $B = -1$, and $A = -1$. Thus the final Pi group is

$$\Pi_4 = (\mu)^1 (D)^{-1} (\bar{V})^{-1} (\rho)^{-1} = \frac{\mu}{D\bar{V}\rho}$$

The complete set of four Pi groups for pipe flow are

$$\Pi_1 = \frac{\Delta p}{\rho \bar{V}^2}, \quad \Pi_2 = \frac{L}{D}, \quad \Pi_3 = \frac{e}{D}, \quad \Pi_4 = \frac{\mu}{D\bar{V}\rho}$$

Comment on Step 4: In this particular case, once pipe diameter has been selected as a repeating parameter, it is possible to anticipate that the dimensionless groups containing pipe length and wall roughness must simply involve the division of each by the pipe diameter, since this immediately forms a dimensionless group. In effect, the pipe diameter has been chosen as the length scale for this analysis. Thus you can form these groups by inspection as indicated earlier. With experience, you may find that you can do the entire DA this way.

Step 5

We now check each Pi group to be sure it is dimensionless, referring to the base dimensions table as needed. In this case it is obvious that the second and third Pi groups are dimensionless. Checking the other two, we have

$$\{\Pi_1\} = \left\{\frac{\Delta p}{\rho \bar{V}^2}\right\} = \frac{M/Lt^2}{(M/L^3)(L^2/t^2)} = \frac{M}{Lt^2} \frac{L^3}{M} \frac{t^2}{L^2} = 1$$

$$\{\Pi_4\} = \left\{\frac{\mu}{D\bar{V}\rho}\right\} = \frac{M/Lt}{(L)(L/t)(M/L^3)} = \frac{M}{Lt} \frac{1}{L} \frac{t}{L} \frac{L^3}{M} = 1$$

We conclude that all four Pi groups are dimensionless and that we have carried out the procedure correctly. The next step is to rearrange individual Pi groups to put them into standard form. In this case the fourth Pi group is the inverse of Reynolds number, so we will invert it and write it as

$$\Pi_4 = \frac{\rho \bar{V} D}{\mu}$$

Comment on Step 5: To rearrange a Pi group, it is necessary to know the standard forms of dimensionless groups in fluid mechanics. In Chapter 3 we discussed the important dimensionless groups in fluid mechanics and the context in which you are likely to encounter them. You might wish to reread Section 3.2 at this time.

Step 6

The final step in a DA is to write a relationship between the dependent Pi group and the remaining groups in the form of Eq. 9.3. For pipe flow, the relationship between pressure drop and the other physical parameters was originally expressed by Eq. 9.1 as

$$\Delta p = f(L, D, e, \bar{V}, \rho, \mu)$$

By using DA, we have discovered that this relationship can be expressed more compactly as

$$\frac{\Delta p}{\rho \bar{V}^2} = g\left(\frac{L}{D}, \frac{e}{D}, \frac{\rho \bar{V} D}{\mu}\right) \tag{9.4}$$

Comments on Step 6: DA generally cannot tell us anything about the nature of the functional relationship among Pi groups. The necessary additional information must come from theory, experiment, intuition, or experience. For example, in pipe flow it is known empirically that Δp can be written as:

$$\Delta p = \rho f \frac{L}{D} \frac{\bar{V}^2}{2} \tag{9.5a}$$

where the dimensionless parameter f, called the friction factor, is known from experiments to be a function of the relative roughness e/D, and Reynolds number $\rho \bar{V} D/\mu$. Thus we can rewrite Eq. 9.5a in a dimensionless form as

$$\frac{\Delta p}{\rho \bar{V}^2} = \frac{1}{2} \frac{L}{D} f\left(\frac{e}{D}, \frac{\rho \bar{V} D}{\mu}\right) \tag{9.5b}$$

As you can see, this empirically based formula is consistent with the result of our DA (Eq. 9.4). Notice also that the dependence of the dimensionless pressure drop on the dimensionless group L/D is linear. This is something DA alone could not predict. At this point you might benefit from revisiting Section 3.3.1, where we presented the case study on flow in a round pipe. In that section we simply gave you the required design formulas in the form of equations. Now it should be clear that those equations can be derived by using DA.

The preceding analysis of pipe flow by means of the repeating variable method certainly looks long and unwieldy. This is because we have included comments about all the things an engineer considers at each step. DA is normally a compact and straightforward procedure, as indicated by the following examples.

EXAMPLE 9.3

In the following historical account, DA provided a critical insight into a wartime problem well in advance of a theoretical or intuitive understanding, and five years before a first experiment could be performed. In 1940 an American explosives expert concluded that the destructive effect of the release of energy through nuclear fission would not be nearly as large as expected. G. I. Taylor, a leading British fluid mechanician, was asked to determine the validity of this conclusion. Taylor was able to answer the question rather easily by using DA. He set up the problem by assuming that the fireball in

Figure 9.3 Schematic for Example 9.3.

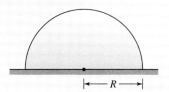

Figure 9.3 has a blast radius R at a time t^* seconds after initiation and that the blast radius depends only on the total energy E released by the bomb and on the initial density ρ_0 of the air in the atmosphere. See if you can recreate Taylor's DA for this problem.

SOLUTION

We apply the repeating variable method to implement the Buckingham Pi theorem by means of the standard six-step procedure.

Step 1. We are told that the radius R of the fireball after the explosion depends on the elapsed time t^*, total energy released E, and initial gas density ρ_0. The desired functional relationship between these physical parameters is $R = f(E, \rho_0, t^*)$. There are four parameters, so $k = 4$.

Step 2. The base dimension table is:

R	E	ρ_0	t^*
L	$\dfrac{ML^2}{t^2}$	$\dfrac{M}{L^3}$	t

The base dimensions are M, L, and t so $r = 3$ and we will have $4 - 3 = 1$ dimensionless groups.

Step 3. The only possible choice for the required set of 3 repeating variables is all three independent variables E, ρ_0, t^*. This selection includes all three base dimensions.

Step 4. The first and only Pi group in this problem is found by writing

$$\Pi_1 = (R)^1 (E)^A (\rho_0)^B (t)^C = (L)^1 \left(\frac{ML^2}{t^2}\right)^A \left(\frac{M}{L^3}\right)^B (t)^C = M^0 L^0 t^0$$

To obtain a dimensionless group we must have

$$(M)^{A+B} = M^0, \quad (L)^{1+2A-3B} = L^0, \quad \text{and} \quad (t)^{-2A-C} = t^0$$

which gives the following equations for the exponents $A + B = 0$, $1 + 2A - 3B = 0$, and $-2A - C = 0$. Solving these we find $A = -\frac{1}{5}$, $B = \frac{1}{5}$, $C = \frac{2}{5}$. The single Pi group in this problem is the dimensionless blast radius: $\Pi_1 = R(\rho_0)^{1/5}/E^{1/5}(t^*)^{2/5}$.

Step 5. Checking this group to see if it is dimensionless, we find

$$\{\Pi_1\} = \left\{\frac{R(\rho_0)^{1/5}}{E^{1/5}(t^*)^{2/5}}\right\} = \frac{L(M/L^3)^{1/5}}{(ML^2/t^2)^{1/5}(t^{2/5})} = \frac{L^{2/5}M^{1/5}}{M^{1/5}L^{2/5}t^0} = 1$$

So the dimensional analysis appears to be correct.

Step 6. There is only one Pi group here. With a little thought perhaps you can convince yourself that in a problem with a single Pi group, dimensional consistency demands that the single Pi group be equal to a dimensionless constant. Thus the relationship between blast radius and the other physical parameters in this problem must be

$$\Pi_1 = \frac{R(\rho_0)^{1/5}}{E^{1/5}(t^*)^{2/5}} = C$$

where C is a dimensionless constant. Through further investigation Taylor determined that this constant is approximately 1.0. The original New Mexico test had a 20 kiloton (8×10^{13} J) yield, so $E = 8 \times 10^{13}$ J. At $t^* = 0.015$ s after detonation, we can calculate that $R = 108$ m by using an air density of $\rho_0 = 1.23$ kg/m^3. Compare this estimate with the photograph in Figure 9.4.

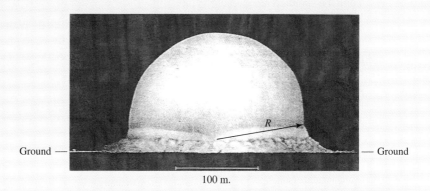

Figure 9.4 Atomic fireball in New Mexico 0.015 s after ignition in 1945. Note the scale and the fact that Taylor's analysis was confirmed.

EXAMPLE 9.4

Consider the open channel flow of water shown in Figure 9.5A. Under flow conditions illustrated, a structure known as a hydraulic jump forms, causing a change of depth from d_1 upstream to d_2 downstream as shown in the schematic Figure 9.5B. The downstream

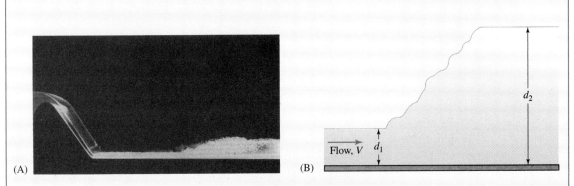

Figure 9.5 (A) Hydraulic jump caused by the flow over the spillway and (B) schematic illustration of the change in depth.

depth d_2 is known to depend on the upstream depth d_1, the upstream velocity V, and the acceleration of gravity g, but (surprisingly) not on density and viscosity. Use DA to find the number of dimensionless groups in this problem, and write a relationship between them.

SOLUTION

We apply the repeating variable method to implement the Buckingham Pi theorem by using the standard six-step procedure.

Step 1. We are told that the downstream depth d_2 in a hydraulic jump depends on the upstream depth d_1, velocity V, and the acceleration of gravity g. The desired functional relationship between these physical parameters is $d_2 = f(d_1, V, g)$. There are four parameters in all so $k = 4$.

Step 2. The base dimension table is:

d_2	d_1	V	g
L	L	$\dfrac{L}{t}$	$\dfrac{L}{t^2}$

The base dimensions are L, t so $r = 2$ and there will be $4 - 2 = 2$ dimensionless groups.

Step 3. There are three choices for the two repeating variables: (d_1, g), (d_1, V), or (V, g). Each pair includes all base dimensions. We choose (d_1, g), leaving V as the non-repeating variable.

Step 4. The first Pi group in this problem contains the dependent variable d_2, and is found by writing

$$\Pi_1 = (d_2)^1 (d_1)^A (g)^B = (L)^1 (L)^A \left(\frac{L}{t^2}\right)^B = L^0 t^0$$

To obtain a dimensionless group, we must have $(L)^{1+A+B} = L^0$ and $(t)^{-2B} = t^0$, which gives the following equations for the exponent: $1 + A + B = 0$ and $-2B = 0$. By inspection we have $A = -1$, and $B = 0$, so the first Pi group is

$$\Pi_1 = \frac{d_2}{d_1}$$

The second Pi group containing V is found from

$$\Pi_2 = (V)^1(d_1)^A(g)^B = \left(\frac{L}{t}\right)^1 (L)^A \left(\frac{L}{t^2}\right)^B = L^0 t^0$$

To obtain a dimensionless group we must have $(L)^{1+A+B} = L^0$ and $(t)^{-1-2B} = t^0$, which yields $1 + A + B = 0$ and $-1 - 2B = 0$. By inspection we have $A = -\frac{1}{2}$, $B = -\frac{1}{2}$, so the second Pi group is

$$\Pi_2 = \frac{V}{\sqrt{gd_1}}$$

Comparing this with Eq. 3.7 we see that it is the Froude number.

Step 5. Checking these groups to see whether they are dimensionless, we see immediately that the first is. To check the second one we write

$$\{\Pi_2\} = \left\{\frac{V}{\sqrt{gd_1}}\right\} = \frac{L/t}{(L^2/t^2)^{1/2}} = 1$$

So the dimensional analysis appears to be correct.

Step 6. There are two Pi groups, so in a hydraulic jump the relationship between downstream depth and the other physical parameters is

$$\frac{d_2}{d_1} = g\left(\frac{V}{\sqrt{gd_1}}\right)$$

Thus we find that the depth ratio is solely determined by the Froude number.

CD/Dynamics/Reynolds Number: Inertia and Viscosity/Dynamic Simulation

CD/Video Library/Flow Past Cars

9.4 SIMILITUDE AND MODEL DEVELOPMENT

Experimental modeling is a fundamental tool in the design of fluid devices and systems, and in the solution of many fluid mechanics problems. Experiments are used extensively to validate the design of airplanes, ships, buildings, bridges, and harbors, where it is important to confirm that the device or system will perform as anticipated before incurring the expense of construction. The large size of such engineering projects makes it impractical and uneconomical to build full-scale prototypes of proposed designs. Thus, the use of models becomes mandatory.

It is critical for an engineer to understand the issues involved in performing experiments with a model rather than the actual device or system of interest. In most cases a model is smaller than the actual device; however, it is sometimes prudent to build a large-scale model of a device that is too small to permit measurements to be taken with conventional sensors. In this section we discuss the design of scale model experiments and the methods for using experimental data obtained from a model study to predict the performance of a full-scale device or system.

There are three similarity conditions that must be met in an experiment using models. The model flow field must be geometrically, kinematically, and dynamically similar to the full-scale prototype it is intended to represent. When all three conditions are satisfied, we achieve complete similarity between the model and full-scale flow. It is then possible to use the experimental results to predict what will occur with the full-scale device. Let us now describe these conditions in more detail and explain how to achieve each one.

By definition, geometric similarity requires that a scale model have the precise shape of the full-scale device or system of interest, with each of the model's physical dimensions in a fixed ratio to the corresponding dimension of the full-scale prototype. For example, a 1/10-scale model has each of its dimensions reduced by a factor of 10. Achieving geometric similarity is the first condition needed to ensure that the fluid dynamic phenomena experienced in the full-scale flow are also present in experiments conducted using the model. Although this looks like a straightforward requirement, it may be impossible to reproduce the surface finish of a full-scale device in a small-scale model. For example, the thousands of rivet heads of an aircraft wing are impossible to incorporate into a wind tunnel model. It is up to the engineer to decide whether this loss of perfect geometric similarity is important. The question to ask is, Does the absence of the feature affect the flow field in a significant way? If it does not, there is no need for concern. If it does, then the engineer must find a way to account for the effect of the missing feature. The missing rivets on a model wing may affect the onset of turbulent flow because the roughness of the model surface is different. Perhaps the model surface could be roughened artificially to produce the missing flow disturbances. This potential problem in using models is called scale effect.

The second condition, called kinematic similarity, is satisfied if the velocity vectors in the model flow field have the same direction as those in the full-scale flow, with the magnitudes of corresponding vectors related by a single velocity scale factor. The third condition, dynamical similarity, is achieved if all forces in the model system have the same direction as those in the full-scale device with the magnitudes of corresponding forces related by a single force scale factor. It is not obvious how to achieve these remaining two conditions, but DA provides the necessary insight.

We are interested in performing experiments on a geometrically similar model in a way that results in the achievement of complete similarity. This means that the set of experimental operating parameters must be picked so that kinematic and dynamic similarity occur in the model flow field. To see how to pick these parameters, suppose a DA has been performed on a full-scale device or system operating under the proposed design conditions. The DA will yield a relationship between all relevant dimensionless groups of the form given by Eq. 9.3:

$$\Pi_1^{FS} = g\left(\Pi_2^{FS}, \Pi_3^{FS}, \ldots, \Pi_{k-r}^{FS}\right) \qquad (9.6)$$

and the proposed design conditions will yield full-scale values for each dimensionless group. A DA performed on an experimental model under the same operating conditions would yield an identical relationship

$$\Pi_1^M = g\left(\Pi_2^M, \Pi_3^M, \ldots, \Pi_{k-r}^M\right) \tag{9.7}$$

but the dimensionless groups would likely take different values for the model and for the full-scale device. Now suppose we adjust the experimental conditions so that each dimensionless group appearing in the function g of Eq. 9.7 has exactly the same value as it would in Eq. 9.6 describing the full-scale device. That is, we design the experiment so that the values of all the independent dimensionless groups are the same for the model and full-scale flow:

$$\Pi_2^{FS} = \Pi_2^M, \quad \Pi_3^{FS} = \Pi_3^M, \quad \text{and} \quad \Pi_{k-r}^{FS} = \Pi_{k-r}^M \tag{9.8}$$

It can be shown that selecting the experimental conditions according to Eq. 9.8 guarantees kinematic and dynamic similarity for a geometrically similar model. Thus, Eq. 9.8 embodies the rules for designing an experiment: use a geometrically similar model, and pick a set of experimental conditions that causes each independent dimensionless group to have the same value as in the proposed full-scale design.

How do we use experimental data from a properly designed experiment to predict the behavior in the full-scale flow? If corresponding dimensionless groups on the right-hand

EXAMPLE 9.5

The scale model of a prototype human-powered airplane is to be tested in a wind tunnel (see Figure 9.6). The design cruising speed of the plane is 20 mph, and it will have a 100 ft wingspan with a 4 ft chord (i.e., wing width in the flow direction). If the model is tested at a tunnel velocity of 160 mph, at what geometric scale should the model be built to maintain dynamic similarity? Assume the conditions of air in the wind tunnel are the same as the atmosphere and that Reynolds number is the single important nondimensional group.

SOLUTION

To maintain dynamic similarity, the Reynolds numbers of the prototype and model must be equal. Thus we can write

$$\left(\frac{\rho V L}{\mu}\right)_p = \left(\frac{\rho V L}{\mu}\right)_m$$

In this case, the length scale L is normally chosen as the chord. Because the density and viscosity of air are the same for the prototype and model, the geometric scale of the model is simply related to the velocity ratio by

$$\text{geometric scale} = \frac{L_m}{L_p} = \frac{V_p}{V_m} = \frac{20 \text{ mph}}{160 \text{ mph}} = \frac{1}{8}$$

Figure 9.6 The *Gossamer Condor,* the first human-powered aircraft to demonstrate sustained, maneuverable flight, won a prize in 1977 for its developer, Paul MacCready.

If the wing span for the human-powered vehicle is 100 ft, the model wing span must be 12.5 ft.

side of the Eqs. 9.6 and 9.7 are identical, then the remaining dimensionless group must have an identical value in the model and full-scale flow. Thus we can write

$$\Pi_1^{FS} = \Pi_1^{M} \tag{9.9}$$

This is the desired relationship between dependent dimensionless groups, which permits the results from an experiment to be applied to the full-scale device.

In many important applications, geometric similarity is achieved, but complete kinematic and dynamic similarity are not possible. This type of scale effect occurs when nongeometric Pi groups cannot be matched between the full-scale flow and a model. The magnitude of influence of scale effects must be considered when one is designing model tests. When scale effects cannot be avoided, care must be taken in interpreting results. This is illustrated in Example 9.6.

EXAMPLE 9.6

The model of a tidal channel in a coastline study is scaled to 1/100 of actual size. Fresh water is to be used in place of seawater in the model. Assuming that the Reynolds number must be matched, what model velocity is needed to ensure dynamic similarity? Will similarity also be achieved for free surface effects related to the Weber and Froude numbers? In your calculations, note that the appropriate velocity and length scales for the actual tidal channel are $V = 0.5$ m/s and $L = 10$ m, respectively.

SOLUTION

To maintain dynamic similarity, the Reynolds number of the model must be the same as that of the actual channel. For seawater (Appendix A) $\rho = 1025$ kg/m³ and $\mu = 1.07 \times 10^{-3}$ kg/(m-s), so

$$Re = \frac{\rho V L}{\mu} = \frac{(1025 \text{ kg/m}^3)(0.5 \text{ m/s})(10 \text{ m})}{1.07 \times 10^{-3} \text{ kg/(m-s)}} = 4.8 \times 10^6$$

The length scale of the model channel is $L_{\text{model}} = L(1/100) = (10 \text{ m}/100) = 0.1$ m. For the fresh water in the model (Appendix A), $\rho = 998$ kg/m³ and $\mu = 1 \times 10^{-3}$ kg/(m-s), so that the Reynolds number for the model is:

$$Re_m = \left(\frac{\rho V L}{\mu}\right)_m = \frac{(998 \text{ kg/m}^3) V_m (0.1 \text{ m})}{1 \times 10^{-3} \text{ kg/(m-s)}} = 4.8 \times 10^6$$

Solving for V_m we obtain

$$V_m = \frac{(4.8 \times 10^6)[1 \times 10^{-3} \text{ kg/(m-s)}]}{(998 \text{ kg/m}^3)(0.1 \text{ m})} = 48 \text{ m/s}$$

To check for similarity of surface effects, we must calculate the Weber and Froude numbers for the full-scale flow and the model. Surface tension for both seawater and fresh water is found in Appendix A to be 7.28×10^{-2} N/m. The Weber numbers for the tidal channel and the model are found by using Eq. 3.5a to be:

$$We = \frac{\rho V^2 L}{\sigma} = \frac{(1025 \text{ kg/m}^3)(0.5 \text{ m/s})^2(10 \text{ m})}{7.28 \times 10^{-2} \text{ N/m}} = 3.5 \times 10^4$$

and

$$We_m = \left(\frac{\rho V^2 L}{\sigma}\right)_m = \frac{(998 \text{ kg/m}^3)(48 \text{ m/s})^2(0.1 \text{ m})}{7.28 \times 10^{-2} \text{ N/m}} = 3.2 \times 10^6$$

respectively. Thus, surface tension effects can be safely neglected for both the tidal channel and model even though the Weber numbers do not exactly match. Next use Eq. 3.3 to calculate Fr:

$$Fr = \frac{V}{\sqrt{gL}} = \frac{0.5 \text{ m/s}}{\sqrt{(9.81 \text{ m/s}^2)(10 \text{ m})}} = 5 \times 10^{-2}$$

$$Fr_m = \left(\frac{V}{\sqrt{gL}}\right)_m = \frac{48 \text{ m/s}}{\sqrt{(9.81 \text{ m/s}^2)(0.1 \text{ m})}} = 48$$

There is a three-order-of-magnitude difference in Fr between the full-scale channel (subcritical), and the model (supercritical). This difference will cause a significant scale effect in the model's ability to mimic the wave propagation characteristics of the full-scale channel. In fact, it is impossible to match both Re and Fr for a scale model when using the same liquid, and in this case we would notice that a velocity of 48 m/s based on Re is impractical. Models of this type are usually used to investigate wave propagation and designed to match Fr rather than Re. Similarity-based on Fr results in $V_m = 5 \times 10^{-2}$ m/s and $Re_m = 5 \times 10^3$. At this lower velocity, the Weber number is $We_m = 3.4$, which does not match the value in the full-scale channel but is still satisfactory. This problem illustrates the care required in the design of experimental fluid mechanics models.

9.5 CORRELATION OF EXPERIMENTAL DATA

Suppose you have been assigned to conduct an experimental program involving liquid flow in a horizontal pipe as discussed in Section 9.1. From your preliminary study of the physical system, you conclude that the pressure drop depends on the pipe length and diameter, wall roughness, average velocity, liquid density, and viscosity. Furthermore, you postulate a relationship among these variables as given by Eq. 9.1:

$$\Delta p = f(L, D, e, \bar{V}, \rho, \mu)$$

Thus, you anticipate that the pressure drop is a function of six independent variables. Next you decide on some reasonable range for each of the independent variables and elect to split that range into 10 equal intervals. As discussed in Example 9.2, the result is, that 10 tests for each of six independent variables will require a total of 10^6 tests! DA shows that the relationship among the variables may be expressed in dimensionless groups by Eq. 9.4:

$$\frac{\Delta p}{\rho \bar{V}^2} = g\left\{\frac{L}{D}, \frac{e}{D}, \frac{\rho \bar{V} D}{\mu}\right\}$$

If we propose 10 tests to explore the effect of each of three dimensionless groups, there will need to be 10^3 tests. This is still a large number of experiments to conduct, but a factor of a thousand less than needed originally.

Another benefit of using dimensionless groups becomes evident if we consider what it means to vary the value of a dimensionless group rather than the value of a specific variable. Consider the group L/D. The influence of this group could be explored by changing pipe diameter, pipe length, or a combination of the two. By electing to vary length while holding diameter constant, we can avoid difficulties in making connections to pumps, control valves, and so on. Similarly, we would have difficulty finding liquids of different densities and viscosities. But by examining Re, we conclude that we can use

9.5 CORRELATION OF EXPERIMENTAL DATA

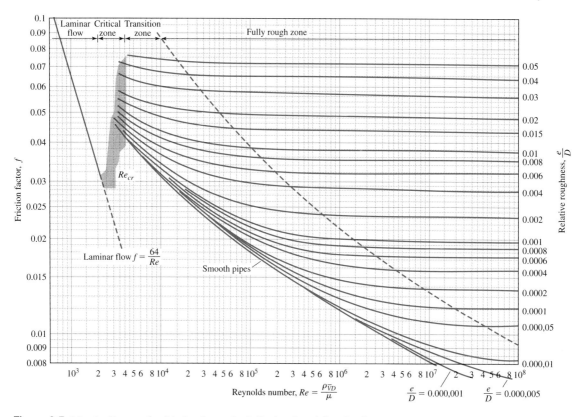

Figure 9.7 Moody diagram for friction factor for fully developed flow in pipes.

water for all tests in a pipe of fixed diameter, exploring the range of Reynolds numbers by changing the average velocity rather than the fluid type. Working with dimensionless groups not only reduces the number of tests but makes them far more convenient and practical.

Finally, note that the use of dimensionless groups greatly enhances our ability to interpret experimental data presented in the form of plots. For example, consider the Moody diagram shown in Figure 9.7. This compact diagram contains the entire spectrum of information for flow in a round horizontal pipe. It is essentially a graphical presentation of Eq. 9.5b

$$\frac{\Delta p}{\rho \bar{V}^2} = \frac{1}{2}\frac{L}{D}f\left(\frac{e}{D}, \frac{\rho \bar{V} D}{\mu}\right)$$

with friction factor, f, plotted as a function of Re, for various values of the relative roughness.

EXAMPLE 9.7

The results of experiments using glycerin and very smooth 36 in. long tubes with diameters of 0.25, 0.5, and 0.75 in. are summarized in the following data table.

D = 0.25 in.		D = 0.5 in.		D = 0.75 in.	
$\bar{V}$ (ft/s)	Δp (lb$_f$/ft^2)	$\bar{V}$ (ft/s)	Δp (lb$_f$/ft^2)	$\bar{V}$ (ft/s)	Δp (lb$_f$/ft^2)
0.44	3046	1.01	1748	2.01	1546
1.02	7062	1.98	3427	4.11	3162
1.51	10454	3.00	5192	6.02	4631
1.97	13638	4.04	6992	7.89	6069
2.50	17308	5.02	8688	9.95	7654
3.03	20977	5.96	10315	12.07	9285
3.54	24508	7.01	12133	14.01	10777
3.99	27623	7.97	13794	15.87	12208
4.44	30738	9.02	15611	17.98	13831
5.03	34823	9.99	17290	20.02	15400

Since flow in a very smooth tube does not depend on relative roughness, a DA suggests that the relationship between dimensionless groups is

$$\frac{\Delta p}{\rho \bar{V}^2} = g\left(\frac{L}{D}, \frac{\rho \bar{V} D}{\mu}\right)$$

Correlate the preceding set of flow data. That is, first plot the data in raw form (Δp vs $\bar{V}$) and then replot the data in nondimensional form.

Correlation of Experimental Data
Glycerin @68°F
Density 2.44 slugs/ft^3
Viscosity 3.13 × 10^{-2} (lb$_f$-s)/ft^2

Re (D = 0.2)	L/(D × Re)	$\Delta p/(\rho \bar{V}^2)$	Re (D = 0.5)	L/(D × Re)	$\Delta p/(\rho \bar{V}^2)$	Re (D = 0.7)	L/(D × Re)	$\Delta p/(\rho \bar{V}^2)$
0.71	201.5142	6448.453	3.28	21.94709	702.3068	9.79	4.901395	156.8446
1.66	86.92768	2781.686	6.43	11.19523	358.2474	20.02	2.397032	76.70504
2.45	58.71936	1879.019	9.74	7.388852	236.4433	29.33	1.636512	52.36839
3.20	45.00824	1440.264	13.12	5.486772	175.5767	38.44	1.248644	39.95662
4.06	35.46649	1134.928	16.31	4.415649	141.3008	48.48	0.990131	31.68419
4.92	29.26278	936.409	19.36	3.719221	119.0151	58.81	0.816222	26.11911
5.75	25.04696	801.5026	22.77	3.162134	101.1883	68.26	0.703198	22.50233
6.48	22.22211	711.1076	25.89	2.781249	88.99998	77.32	0.620782	19.86501
7.21	19.96987	639.0359	29.30	2.45749	78.63967	87.60	0.547931	17.5338
8.17	17.62748	564.0794	32.45	2.218875	71.00399	97.54	0.492098	15.74714

Figure 9.8 Spreadsheet data used for the analysis of Example 9.7.

SOLUTION

Using data from Appendix A and conversion factors from Appendix C we find for glycerin at 68°F, $\rho = 2.44$ slugs/ft^3 and $\mu = 3.13 \times 10^{-2}$ (lb$_f$-s)/ft^2. Then we use a spreadsheet to calculate the product of the inverse Reynolds number, the geometric variable L/D, and $\Delta p/(\rho \bar{V}^2)$ (see Figure 9.8). A plot of the raw data is shown in Figure 9.9A and a similar plot of the correlated data is shown in Figure 9.9B. Note that the data are linear and collapse into a single curve when correlated by means of the dimensionless variables.

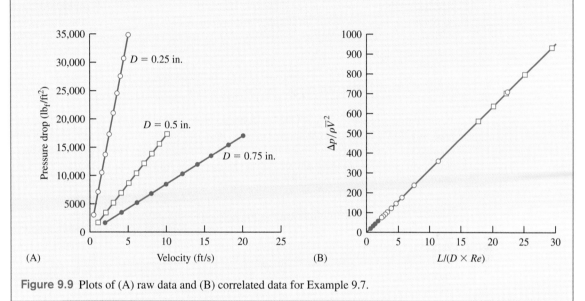

Figure 9.9 Plots of (A) raw data and (B) correlated data for Example 9.7.

9.6 APPLICATION TO CASE STUDIES

An understanding of DA and the role of dimensionless groups enables an engineer to employ the substantial database of previous empirical results describing a fluid flow of interest. The case studies presented earlier in Chapter 3, Sections 3.3.1 to 3.3.6, represent a varied selection of this type of information. Now that we have dimensional analysis tools available, let us briefly revisit those case studies and see what the methods introduced in this chapter can tell us about the sources of the empirical formulas.

9.6.1 DA of Flow in a Round Pipe

In our first case study we considered steady, fully developed incompressible flow in a straight, horizontal, round pipe (refer again to Figure 9.2). Our dimensional analysis of this problem earlier in this chapter led to a relationship between pressure drop and the various parameters describing pipe flow as given by Eq. 9.4: $\Delta p/\rho \bar{V}^2 = g(L/D, e/D, \rho \bar{V} D/\mu)$. After using Eq. 9.5a, i.e., $\Delta p = \rho f (L/D)(\bar{V}^2/2)$, to introduce the concept of a friction factor, we found that the result of the DA could be written as Eq. 9.5b:

$$\frac{\Delta p}{\rho \bar{V}^2} = \frac{1}{2}\frac{L}{D} f\left(\frac{e}{D}, \frac{\rho \bar{V} D}{\mu}\right)$$

If you compare these results with the formulas in the case study (Section 3.3.1), you will see that they are identical. In that case study we noted that to arrive at a useful engineering methodology to analyze pipe flow it is sufficient to provide information on how to obtain the friction factor in laminar and turbulent flow. Thus we see that the combination of dimensional analysis and experimental data allows us to understand how to analyze this important problem.

9.6.2 DA of Flow through Area Change

In presenting this case study in Chapter 3 we began by noting that the total change in pressure as a flow passes through an area change may be thought of as the sum of a pressure change associated with the change in average flow velocity, and a frictional pressure drop. We postulated that a model for this effect in turbulent flow can be written as Eq. 3.20,

$$p_2 - p_1 = \left[\tfrac{1}{2}\rho\left(\bar{V}_1^2 - \bar{V}_2^2\right)\right] - \Delta p_F$$

where p_2 is the downstream pressure, p_1 is the upstream pressure, and Δp_F is the frictional pressure loss. The velocities $\bar{V}_1$ and $\bar{V}_2$ in this formula are the average velocities in the upstream and downstream sections. Next we stated that we can calculate Δp_F, by using DA and empirical results and gave formulas for the cases of sudden expansions and sudden contractions. Suppose we now perform a DA on the case of a sudden expansion, the geometry of which was illustrated in Figure 3.12C. We can begin the DA by assuming that the frictional pressure loss Δp_F is related to the other variables in the problem by a functional relationship of the form $\Delta p_F = f(A_1, A_2, \rho, \bar{V}_1)$, where A_1 is the inlet area, A_2 is the outlet area, ρ is the density, and $\bar{V}_1$ is the average inlet velocity. Note that we do not have to include $\bar{V}_2$ in our analysis, since a mass balance shows that $\bar{V}_1 A_1 = \bar{V}_2 A_2$; thus the inclusion of $\bar{V}_2$ is redundant. According to the Buckingham Pi theorem, there are two dimensionless groups. By using the repeating variable method to define these groups, we can write $\Delta p_F/\rho \bar{V}_1^2 = g(A_1/A_2)$. Next we introduce the customary factor of $\tfrac{1}{2}$ to the ρV_1^2 term (to create a kinetic energy–like term) and define a dimensionless loss coefficient K_E for the enlargement as the ratio $K_E = \Delta p_F/(\tfrac{1}{2}\rho \bar{V}_1^2)$. The DA then shows that K_E is a function of the area ratio of the enlargement. If K_E is known, the frictional pressure drop can be calculated from

$$\Delta p_F = K_E \tfrac{1}{2}\rho \bar{V}_1^2$$

which is Eq. 3.22 as given in the case study. Thus, as noted in Chapter 3, the problem reduces to finding the enlargement loss coefficient, and we presented the necessary data in Figure 3.13.

A dimensional analysis of the sudden contraction leads to a similar expression for the pressure drop. After defining the contraction loss coefficient K_C as $K_C = \Delta p_F/(\frac{1}{2}\rho \bar{V}_2^2)$, we see that the DA leads a formula for calculating the pressure drop for the contraction as given by Eq. 3.23:

$$\Delta p_F = K_C \tfrac{1}{2}\rho \bar{V}_2^2$$

Once again we see that DA and experimental correlations allow us to obtain very useful engineering results without having to apply more complex theoretical methods.

9.6.3 DA of Pump and Fan Laws

In our case study on pump and fan laws, we noted that it is customary to use a parameter called total head H in the design of pipe and duct systems. This total head, with dimensions of energy per unit mass (or equivalently $\{L^2 t^{-2}\}$), is a measure of the total load seen by a pump or fan moving fluid through the system. The power P required by the pump or fan is also an important parameter in the design of these systems. Thus in analyzing the performance of a pump or fan, the head H and power P are both considered to be important dependent variables. A DA of these fluid machines therefore employs H and P as two separate dependent variables that are functions of the other physical parameters.

We begin our DA of pumps and fans with the observation that for geometrically similar machines of a given type, only one length scale is required to specify the machine geometry. This length scale is conveniently taken to be the diameter D of the impeller or other rotating element. We assume that the head and power of a fan or pump depends on ω, the angular speed of the impeller, the volume flowrate, and the density and viscosity of the fluid. Thus we write

$$H = f_1(D, Q, \omega, \rho, \mu) \quad \text{and} \quad P = f_2(D, Q, \omega, \rho, \mu)$$

The head and power are both valid candidates for DA, so we will use the head here as an example, leaving the analysis of power as a homework assignment.

The steps in a DA of the head delivered by a pump or fan are as follows.

1. We are told that the head H depends on the impeller diameter D, angular speed ω, the volume flowrate Q, density ρ, and viscosity μ. The desired functional relationship is: $H = f_1(D, Q, \omega, \rho, \mu)$.

2. The base dimension table is:

H	D	Q	ω	ρ	μ
$\dfrac{L^2}{t^2}$	L	$\dfrac{L^3}{t}$	$\dfrac{1}{t}$	$\dfrac{M}{L^3}$	$\dfrac{M}{tL}$

The base dimensions are M, L, t, so $r = 3$, and there will be $6 - 3 = 3$ dimensionless groups.

3. Choose ρ, ω, D as the three repeating variables. This set includes all base dimensions.

4. Form the Pi groups as usual:

$\Pi_1 = H\rho^A\omega^B D^C$	$\Pi_2 = Q\rho^A\omega^B D^C$	$\Pi_3 = \mu\rho^A\omega^B D^C$
$M^0 L^0 t^0 = \left(\dfrac{L^2}{t^2}\right)\left(\dfrac{M}{L^3}\right)^A \left(\dfrac{1}{t}\right)^B (L)^C$	$M^0 L^0 t^0 = \left(\dfrac{L^3}{t}\right)\left(\dfrac{M}{L^3}\right)^A \left(\dfrac{1}{t}\right)^B (L)^C$	$M^0 L^0 t^0 = \left(\dfrac{M}{Lt}\right)\left(\dfrac{M}{L^3}\right)^A \left(\dfrac{1}{t}\right)^B (L)^C$

with the following result:

$$\Pi_1 = \frac{H}{\omega^2 D^2}, \quad \Pi_2 = \frac{Q}{\omega D^3}, \quad \text{and} \quad \Pi_3 = \frac{\mu}{\rho \omega D^2}$$

5. Checking these groups shows that each is dimensionless.
6. The DA shows that the relationship of dimensionless head to the remaining groups is

$$\frac{H}{\omega^2 D^2} = g_1 \left(\frac{Q}{\omega D^3}, \frac{\rho D^2 \omega}{\mu} \right) \tag{9.10a}$$

where we have inverted and rearranged the third Pi group.

A similar application of DA to the proposed relationship involving power yields

$$\frac{P}{\rho \omega^3 D^5} = g_2 \left(\frac{Q}{\omega D^3}, \frac{\rho D^2 \omega}{\mu} \right) \tag{9.10b}$$

It is easy to see that these are identical to Eqs. 3.24a and 3.24b of the corresponding case study in Section 3.3.3.

In pump and fan engineering, the dependent dimensionless groups $H/\omega^2 D^2$ and $P/\rho\omega^3 D^5$ are known as the head and power coefficients, respectively. The independent dimensionless group $Q/\omega D^3$ is known as the flow coefficient, while the group $\rho D^2 \omega/\mu$ can be considered to be a form of the Reynolds number, since the product $D\omega$ has the dimension of velocity.

From our discussions of the scaling of two geometrically similar systems, we know that all independent dimensionless groups must be the same. However, for typical pumps and fans it is found that the performance is independent of Re as defined earlier. Thus in comparing two pumps or fans in the same family the appropriate scaling law is:

$$\frac{Q_1}{\omega_1 D_1^3} = \frac{Q_2}{\omega_2 D_2^3} \tag{9.11a}$$

When the flow coefficients of two machines are equal, then the head and power coefficients are also equal:

$$\frac{H_1}{\omega_1^2 D_1^2} = \frac{H_2}{\omega_2^2 D_2^2} \quad \text{and} \quad \frac{P_1}{\rho \omega_1^3 D_1^5} = \frac{P_2}{\rho \omega_2^3 D_2^5} \tag{9.11b}$$

These are the equations known as the pump laws or fan laws that we presented in the earlier case study as Eqs. 3.25. Not only do they relate the performance of two differently sized machines in the same family, but they also allow us to determine how a given machine will operate under a new set of operating conditions. We can now see that the form of these equations is predicted by dimensional analysis.

9.6.4 DA of Flat Plate Boundary Layer

Recall that the case study on a flat plate boundary layer (Section 3.3.4) was concerned with the characteristics of a flow over a flat plate. The resulting boundary layer, in which the fluid velocity changes smoothly from zero on the plate to its freestream value, may be laminar or turbulent. We noted that the quantity of greatest interest in the flat plate boundary layer is the wall shear stress. If we know how the wall shear stress varies along the plate, we can calculate the frictional force applied by the fluid to the plate, and use this model to analyze flow over relatively flat surfaces, such as the hulls of ships. The geometry of the boundary layer was illustrated in Figure 3.18. Observations suggest that in an incompressible flow at high Re, the shear stress τ_W on the wall in a flat plate boundary layer depends on the distance from the leading edge x, the freestream velocity V, and the fluid density and viscosity. Thus for the first step in the DA we propose a relationship between these variables of the form:

$$\tau_W = f(x, V, \rho, \mu)$$

A DA using the repeating variable method reveals that this relationship can be expressed as

$$\frac{\tau_W}{\frac{1}{2}\rho V^2} = g\left(\frac{\rho V x}{\mu}\right)$$

As noted in the case study, it is customary in boundary layer analysis to define the skin friction coefficient C_f by means of Eq. 3.27 as $C_f = \tau_W/(\frac{1}{2}\rho V^2)$, and to define a Reynolds number based on the distance x from the leading edge, using Eq. 3.28 ($Re_x = \rho V x/\mu$). From the DA we can immediately conclude that there is a relationship between the skin friction coefficient and the Reynolds number of the form given in Eq. 3.29:

$$C_f = C_f(Re_x)$$

To complete the case study it was only necessary to provide theoretical and empirical results for the skin friction coefficient in laminar and turbulent boundary layers.

 CD/Special Features/Charts & Graphs/Drag Curves on Spheres & Cylinders

9.6.5 DA of Drag on Cylinders and Spheres

As you know, one of the most important problems in fluid mechanics is to determine the drag on a body immersed in a moving fluid. In the case study of Section 3.3.5 we discussed the drag in steady, incompressible flow for two very simple geometries: an infinitely long circular cylinder, and a sphere. Suppose we apply a DA to the steady flow over a cylinder. We are interested in the drag force F_D on a cylinder of diameter D and length L. The drag will depend on these two geometric parameters as well as on the velocity, density, and viscosity of the fluid. We summarize the proposed relationship mathematically as $F_D = f(D, L, V, \rho, \mu)$. According to the Buckingham Pi theorem there are three dimensionless groups. Using the repeating variable method we find that the relationship between these groups is $F_D/D^2V^2\rho = g(Re, L/D)$, where the Re is based on the cylinder diameter. The standard way to present this result is to first write it as $F_D = C_D \frac{1}{2}\rho V^2 DL$, which is Eq. 3.34. The DA predicts that $C_D = C_D(Re, L/D)$. This result was given in the case study as Eq. 3.36. All that remains is to provide the necessary experimental correlation for the drag coefficient for a cylinder and we are immediately able to do calculations in practical problems.

The DA of flow over a sphere is also straightforward. The drag force F_D on a smooth sphere of diameter D will depend on this single geometric parameter as well as on the fluid velocity, density, and viscosity. We summarize the proposed relationship as

$$F_D = f(D, V, \rho, \mu)$$

According to the Buckingham Pi theorem, there are only two dimensionless groups. From the repeating variable method, the relationship between these groups is found to be

$$\frac{F_D}{D^2 V^2 \rho} = g(Re)$$

where the Re is based on the sphere diameter. The drag coefficient for a sphere is defined by Eq. 3.39 as $C_D = F_D/[\frac{1}{2}\rho V^2(\pi D^2/4)]$, and the DA shows that since there is only one length scale for a sphere, the drag coefficient is only a function of Re: $C_D = C_D(Re)$. This result is Eq. 3.40 of the case study.

9.6.6 DA of Lift and Drag on Airfoils

In the case study of Section 3.3.6 we presented results that allow us to calculate the lift and drag forces on airfoil shapes exposed to a uniform incompressible flow. To perform a DA on this problem, we can recall that the standard nomenclature for airfoil geometry is illustrated in Figure 3.23. In steady subsonic flow the lift and drag forces, F_L and F_D, respectively, are each found to depend on the thickness t, span b, chord length c, and angle of attack α. They also depend on the freestream velocity V, and on the fluid density and viscosity. We begin the DA as usual by writing the dependence of lift and drag on the physical parameters as

$$F_L = f(t, b, c, V, \rho, \mu, \alpha)$$
$$F_D = f(t, b, c, V, \rho, \mu, \alpha)$$

The remainder of the DA is straightforward and leads to the following standard relationships among dimensionless groups

$$\frac{F_L}{\frac{1}{2}\rho V^2 bc} = g_1\left(Re_c, \frac{t}{c}, \frac{b}{c}, \alpha\right)$$

$$\frac{F_D}{\frac{1}{2}\rho V^2 bc} = g_2\left(Re_c, \frac{t}{c}, \frac{b}{c}, \alpha\right)$$

where Re_c is the Reynolds number based on chord length. The lift and drag coefficients for an airfoil section are given in the case study as

$$C_L = C_L\left(Re_c, \frac{t}{c}, \frac{b}{c}, \alpha\right) \quad \text{and} \quad C_D = C_D\left(Re_c, \frac{t}{c}, \frac{b}{c}, \alpha\right)$$

which leads to the formulas $F_L = C_L \frac{1}{2}\rho V^2 bc$ and $F_D = C_D \frac{1}{2}\rho V^2 bc$, as given in the case study by Eqs. 3.43a and 3.43b. The DA shows that the dependence of the lift and drag coefficients on the various dimensionless groups. Once again we see that a DA explains why engineering formulas take their precise and customary forms.

In this section we hope we have persuaded you of both the power and the usefulness of the tools you learned in this chapter. As completely new and different flows are encountered by engineers, there is no doubt that dimensional analysis and similitude will continue to be key tools in understanding the relationships between various flow parameters.

9.7 SUMMARY

Dimensional analysis (DA) makes use of the principle that all terms of a physical equation must have the same dimensions. Thus DA can be used to obtain a fundamental understanding of physical systems without the need for complex mathematics. The set of dimensionless groups predicted by the Buckingham Pi theorem gives engineers the ability to classify fluid mechanics problems, apply known design formulas, and choose the most effective solution methods for new problems. DA also helps ensure that experimental models faithfully represent the intended physical system. The correlation and efficient compilation of experimental data is enhanced by DA.

The repeating variable method is used to determine the form of the dimensionless groups required to describe a physical system. This method can be used to analyze any fluid mechanics problem using the following procedure:

1. List the variables that influence the problem as expressed.
2. Represent each variable in terms of its basic dimensions.
3. Determine the total number of basic dimensions, r.
4. Choose r variables from among those listed in step 1 to be the repeating variables.

5. Form each Π group by multiplying one of the nonrepeating variables by all of the repeating variables with each raised to an exponent to make the product dimensionless.

6. Express the Π groups in their common functional form.

Experimental modeling is a critically important approach in the solution of complex fluid mechanics problems. In the design of an experiment, complete similarity should be maintained by imposing the conditions of geometric, kinematic, and dynamic similarity. To be geometrically similar, all the model's linear dimensions must be related to those of the full-scale prototype by a single geometric scale factor. Kinematic similarity requires that all the velocity vectors in the flow field for the model have the same direction as those for the full-scale prototype, and the magnitudes of the vectors must be related by a single velocity scale factor. Dynamic similarity requires that all the forces in the model system have the same direction, and their magnitudes must be related by a single force scale factor to those of the prototype. Kinematic and dynamic similitude requires that the value of each of the dimensionless Π groups of the model be equal to the corresponding value for the full-scale prototype.

DA is a useful tool in the design and execution of experiments and the subsequent correlation of experimental data. For example, in horizontal pipe flow if 10 tests are required for each of six independent variables, a total of a million tests would be needed to completely examine the cross-dependence of each variable. By using DA, the number of independent variables can be reduced to three, thus reducing the number of tests to one thousand. A single diagram by Moody was used to present the experimental data for this problem.

This chapter concludes with the application of DA to each of the six case studies: fully developed flow in pipes and ducts, flow through sudden area change, pump and fan laws, flat plate boundary layer, drag on bluff bodies (cylinders and spheres), and lift and drag on airfoils. These case studies represent a substantial amount of the material with which an engineer could practice design after a single course in fluid mechanics.

PROBLEMS

Section 9.2

9.1 The speed of propagation of a gravity wave in deep water, V, is known to be a function of the wavelength λ, the water depth d, the fluid density ρ, and the local gravitational constant g. Determine the number of Pi groups that can be formed from these five parameters.

9.2 When a valve on a pipe is suddenly closed, a wave is created that results in the phenomenon known as water hammer. The pressure associated with a wave of this type can be substantial and may cause damage to the pipe. The maximum pressure developed, p_{max}, is known to be a function of the fluid density ρ, the bulk modulus E_v, and the initial flow velocity V. Determine the number of Pi groups that can be formed from these four parameters.

9.3 The wall shear stress τ_w in a boundary layer is known to be a function of the fluid density ρ and viscosity μ, the fluid free stream velocity V_{max}, and the distance from the leading edge of the body, x. Determine the number of Pi groups that can be formed from these five parameters.

9.4 The power input, P, to a pump is known to be a function of the volume flowrate Q, density ρ, impeller diameter d, angular velocity ω, and fluid viscosity μ. Determine the number of Pi groups that can be formed from these six parameters.

9.5 The heat flux, q, from a hot body in a stationary fluid is known to be a function of the fluid's thermal conductivity k, and kinematic viscosity, ν, the temperature difference ΔT, the length of the object L, and the product of the local gravitational constant and the thermal expansion coefficient for the fluid, $g\beta$. Determine the number of Pi groups that can be formed from these seven parameters.

9.6 The height of the capillary rise, h, in a crack is known to be a function of the surface tension of the fluid σ, the contact angle θ_c, the width of the gap between the plates w, the fluid density, ρ, and the gravitational constant, g. Determine the number of Pi groups that can be formed from these six parameters.

9.7 The boundary layer thickness, δ, associated with flow over a smooth flat plate of low speeds is known to be a function of the fluid free stream velocity V_{max}, the fluid density and viscosity ρ and μ, and the distance from the leading edge of the plate, x. Determine the number of Pi groups that can be formed from these five parameters.

9.8 The lift force, F, on a rocket is known to be a function of its dimensions L and D, the angle of attack, α, the rocket velocity, V, and the fluid properties density, viscosity, and sound speed, ρ, μ, and c. Determine the number of Pi groups that can be formed from these eight parameters.

9.9 A civil engineering structure known as a weir is shown in Figure P9.1. A weir is an obstruction placed in an open channel that can be used to determine the flowrate in the channel. The flowrate Q is known to vary with the width of the weir (the dimension perpendicular to the view shown in Figure P9.1) w, the upstream water height above the weir h, and the local gravitational constant g. Determine the number of Pi groups that can be formed from these four parameters.

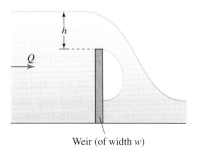

Figure P9.1

9.10 The diameter, D, of fluid droplets produced by a liquid spray nozzle is known to be a function of the nozzle diameter d, the liquid velocity at the nozzle exit V, and the fluid properties density, viscosity, and surface tension, ρ, μ, and σ. Determine the number of Pi groups that can be formed from these six parameters.

9.11 In many sports the spin on a roughly spherical ball can have a huge impact on the game. In the case of a golf ball, the rate of spin is related to the torque experienced by the ball as a result of its motion through the air. The torque, T, in turn is a function of the velocity and angular velocity of the ball, V and ω, the fluid properties ρ and μ, the diameter of the ball D, and the diameter of the "dimples" on the ball, d. Determine the number of Pi groups that can be formed from these seven parameters.

V, the drag D on the plate is found to be a function of L, w, V, and the fluid properties ρ and μ. Application of the Buckingham Pi theorem results in the equation

$$\frac{D}{\rho w V^2} = f\left\{\frac{w}{L}, Re_w\right\}$$

where the subscript on the Reynolds number indicates that the characteristic dimension for Re is w. Assume that L, w, V, ρ, and μ are known for the prototype system and that the width of the model, w_m, has been specified.
(a) Describe the conditions required for there to be similitude between the model and the prototype system.
(b) Derive an expression for the length of the model plate, L_w.
(c) Derive an expression for the fluid velocity in the model system, V_m.
(d) Derive an expression for the drag on the prototype plate, D.

9.33 Your company is planning to build a pipeline to transport gasoline from the refinery to a field of storage tanks. The parameters for the prototype system are a pipe diameter of 1 m, with a flow velocity of 0.5 m/s at 25°C. The model system will use water at STP with a geometric scaling factor of 1 : 20. What fluid velocity is required in the model system to guarantee kinematic similarity in the form of equal Reynolds numbers?

9.34 For wind tunnel testing of new golf ball dimple patterns it is known that similarity of both the Re and St numbers is necessary. Experiments with a representative sampling of members of the PGA tour indicate prototype parameters of $V = 250$ ft/s and $\omega = 900$ s^{-1}. The diameter of a golf ball is 1.68 in. The company president wants to use a video clip from your tests in a commercial and has told you to use a geometric factor of 4 : 1 (the model "ball" will be larger than the prototype). Determine the required model fluid velocity and model angular velocity.

9.35 You have been asked to design a blimp for your company. It is to cruise through air at STP at a speed of 10 mph. Experience suggests that the drag on the blimp created by the pressure distribution is of primary importance.
(a) What dimensionless parameters do you think should be matched to achieve similarity between prototype and model for this problem?
(b) Using a 1 : 20 scale model, determine the velocity you will use in your model. State any assumptions.

9.36 The pressure drop through an oil (SAE 30W at 20°C) supply line at a service station is to be modeled using the flow of water at 20°C through an identical length of tubing. The required oil velocity in the prototype is known to be 35 cm/s.
(a) Use Reynolds number similarity to find the velocity of the model fluid.
(b) Use Euler number similarity to predict the pressure drop in the prototype line given that the pressure drop in the model was 0.05 psi.

9.37 An airplane wing with a 2 m chord is designed to fly through standard atmosphere at an elevation of 8 km and a velocity of 850 km/h. The wing is to be tested in a water tank at STP. What fluid velocity is necessary in the model to ensure Reynolds number similarity? What other dimensionless groups might be important in this flow?

9.38 You are in charge of testing a 1:50 scale model of a flat bottom barge. The prototype fluid is seawater at 20°C, and your analysis tells you that you must obtain similarity of the Reynolds and Froude numbers. What characteristics must your model fluid possess? What fluid would you recommend using for your model?

9.39 In a certain flow situation it is known that both surface tension effects and cavitation may be important. The prototype fluid is gasoline at 20°C and the model fluid is water at 20°C. The prototype operates at 101 kPa and the vapor pressure of gasoline under these conditions is 55.1 kN/m^2. If the model system must operate at pressures between 101 kPa and 400 kPa, determine the appropriate range of scaling factors for the model.

9.40 Analysis suggests that to predict flow characteristics in a canal, it is vital that Froude number similarity be maintained. If the model scale is 1:100, and the fluid velocity in the canal is estimated to be 7 ft/s, what fluid velocity should be used in the model? A colleague points out that because the prototype is only 6 ft deep, the depth of the model will be less than 0.75 in. In this situation the Weber number might be important. If the prototype flow involves water at STP, what characteristics must the model fluid posses to satisfy both Fr and We similarity?

9.41 The wind tunnel testing for the first airplane to break the sound barrier, the Bell X-1, employed a small scale model of about 1/25 scale. What velocity for the model at sea-level conditions is required to maintain dynamic similarity with the actual aircraft at $M = 1$ and 40,000 ft? What is the Mach number for the model?

Section 9.5

9.42 An experiment to investigate the drag force on smooth spheres consists of dropping stainless steel ($\rho = 8010$ kg/m^3) balls of different diameter in glycerin at 20°C. The time for the balls to fall between two lines 25 cm apart is recorded. It is assumed that the balls are falling at constant velocity such that the drag force is $F_D = V_{ball}g(\rho_{ball} - \rho_{glycerin})$. Plot C_D versus Re based on the following data:

D (cm)	Δt (s)
1.0	13,020
1.5	2,545
2.0	805
2.5	329
3.0	159
3.5	86
4.0	51
4.5	33
5.0	21

9.43 The experiment described in Problem 9.42 is repeated in water. Plot C_D versus Re based on the following data:

D (cm)	Δt (s)
1.0	105
1.5	11.2
2.0	8.0
2.5	5.2
3.0	4.0
3.5	3.0
4.0	2.5
4.5	2.1
5.0	1.8

Section 9.6

9.44 Use the repeating variable method to obtain the expression $\Delta p_F/\rho V_1^2 = g(A_1/A_2)$ from the equation $\Delta p_F = f(A_1, A_2, \rho, V_1)$.

9.45 The manufacturer of a room air conditioner for motel rooms has added a sudden contraction of 8 in. to 6 in. in the round internal duct. Each unit provides 200 ft^3/min of supply air. The electric operating power required accommodate this modification is estimated as $1.25Q\Delta p_F$ and costs $0.07/kW-h. How much will this design modification cost a 100-room motel each year? Assume air at standard conditions and that each unit operates 8 h/day.

9.46 Use the repeating variable method to verify Eq. 9.10b.

9.47 Fill in the missing steps of the DA leading to the equation $F_D/\frac{1}{2}\rho V^2 DL = g(Re, L/D)$ for a cylinder.

9.48 Fill in the missing steps of the DA leading to the equation $F_D/D^2 V^2 \rho = g(Re)$ for a sphere.

9.49 Fill in the missing steps of the DA leading to the equation $F_L/\frac{1}{2}\rho V^2 bc = g_1(Re_c, t/c, b/c, \alpha)$ for the lift on an airfoil.

9.50 Fill in the missing steps of the DA leading to the equation $F_D/\frac{1}{2}\rho V^2 bc = g_2(Re_c, t/c, b/c, \alpha)$ for the drag on an airfoil.

PART 2
DIFFERENTIAL ANALYSIS OF FLOW

10 ELEMENTS OF FLOW VISUALIZATION AND FLOW STRUCTURE

10.1 Introduction
10.2 Lagrangian Kinematics
    10.2.1 Particle Path, Velocity, Acceleration
    10.2.2 Lagrangian Fluid Properties
10.3 The Eulerian–Lagrangian Connection
10.4 Material Lines, Surfaces, and Volumes
10.5 Pathlines and Streaklines
10.6 Streamlines and Streamtubes
10.7 Motion and Deformation
10.8 Velocity Gradient
10.9 Rate of Rotation
    10.9.1 Vorticity
    10.9.2 Circulation
    10.9.3 Irrotational Flow and Velocity Potential
10.10 Rate of Expansion
    10.10.1 Dilation
    10.10.2 Incompressible Fluid and Incompressible Flow
    10.10.3 Streamfunction
10.11 Rate of Shear Deformation
10.12 Summary
Problems

10.1 INTRODUCTION

Suppose you are assigned to investigate the possibility of dust, mildew, spores, bacteria, and other debris accumulating in a building's heating, ventilating, and air-conditioning (HVAC) system. This is important because the accumulation of allergens like these may cause respiratory discomfort. Experience suggests that deposits of airborne materials occur in regions of slow or recirculating flow, so you are to review the HVAC design and identify any regions in which allergens might accumulate. Now consider the proposed duct layout for an HVAC return line illustrated in Figure 10.1A. Will regions of slow or recirculating flow occur with this layout? Wouldn't it be nice to be able to "see" the corresponding flow in this duct section so that you could answer the flow question before the ducting was installed? Flow visualization is the tool that can help you with this sort of task.

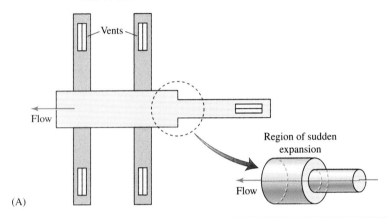

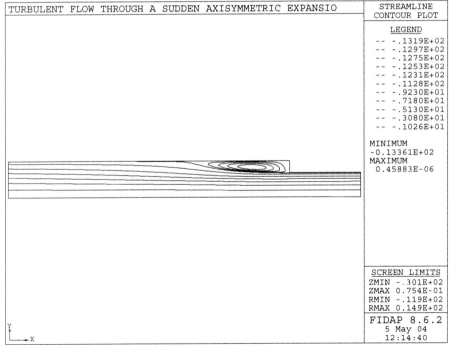

Figure 10.1 (A) HVAC duct layout that includes a sudden expansion. (B) Streamlines in turbulent flow in sudden expansion. Only the top half of the axis-symmetric flow is shown and the streamlines correspond to time-averaged velocities.

Several case studies in Chapter 3 contain empirical results that are applicable to the design and analysis of certain features of an HVAC duct section like the one we are considering, and additional relevant information can be found in the applications chapters of this text. Although we could use those methods to size the fan for our duct system, we cannot answer detailed questions about the flow field or predict particle deposition without flow visualization. This further illustrates why analytical, experimental, and computational approaches to the solution of problems are all important in fluid mechanics.

To determine whether slow or recirculating flow exists in the proposed duct layout, you need to know more about the detailed flow structure at various locations. An experienced engineer is often able to predict flow structure from the geometry and flowrate, or this information may be acquired from a computer flow simulation. For example, consider the sudden area change in the return line in Figure 10.1. Predicting the pressure drop in an area change was discussed in a case study in Section 3.3.2, but nothing was said there about flow structure. A computational fluid dynamics (CFD) analysis of this geometry shows that fluid does tend to recirculate in the region just downstream of the expansion plane. In this recirculation zone, shown in Figure 10.1B, particles carried along by the fluid may settle out and create a site for the growth of mold and mildew. Thus, a CFD visualization of the flow field is helpful in deciding whether it is necessary to modify the design.

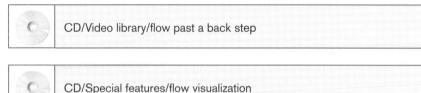

Flow visualization is widely used in a variety of fluid mechanics applications. It is helpful whenever there is a need to understand the detailed structure of a flow field: how fluid moves, if and where boundary layers separate, whether recirculation is present, and many other questions involving the velocity field. Although CFD can often be a cost-effective approach for visualizing a flow, an engineer should also consider obtaining the required information from a physical experiment. Carefully designed flow visualization experiments can produce a wealth of information, with the following advantage: there is no uncertainty about an underlying mathematical model.

The choice between computation and experiment is based on engineering judgment. Some flow problems do not lend themselves to experimentation. Consider what might be involved in modeling lava flow in the lab or the flow of interstellar gas approaching a black hole. Other flow problems are entirely new, with no solutions available in the literature to guide the development of a new computational model. Fortunately, most flow problems are amenable to the use of both approaches. The thought processes involved in making this decision are illustrated in Example 10.1.

The elements of flow visualization presented in this chapter are applicable irrespective of whether a flow visualization is based on a physical experiment, an analytical solution, or a computational simulation. Figure 10.2 illustrates the experimental flow visualization process, in which a fluid entity (a point, line, surface or volume in a flow) is marked with smoke, dye, suspended solid tracer particles, tiny helium or hydrogen

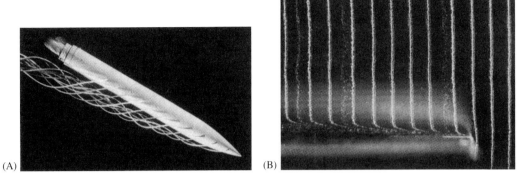

Figure 10.2 Experimental flow visualization using (A) dye and (B) hydrogen bubbles.

EXAMPLE 10.1

Consider each of the flow situations described and decide whether you would use a computational analysis, experiments, or both to predict the flow characteristics.

A. Predict the reentry aerodynamics of a probe entering the atmosphere on Jupiter.
B. Investigate a manufacturer's claim that their new golf ball will add 10 yards to your drive.
C. Predict lift and drag for the wing of a next-generation passenger airplane.
D. Analyze the deposition of an airborne toxin in the human respiratory system.

SOLUTION

A. Owing to the difficulty in reproducing atmospheric conditions on Jupiter and achieving the high reentry velocity, it might be difficult to achieve similitude in an experiment. Thus, it seems reasonable to simulate this flow with a computational model. It is likely, however, that some experiments will be required because of uncertainties in the CFD model and the high cost of failure.

B. An experimental approach seems appropriate, since it is easy to establish the relevant test conditions. Since driving distances vary with any golfer, however, statistical methods will be required to permit confidence in the experimental results. This suggests the use of a device that simulates a golf swing and strikes the ball in a reproducible manner. A golf ball manufacturer might also decide to do some computational modeling to predict how a change in dimple pattern influences drag, since it will be expensive to manufacture golf balls with a range of dimple patterns. Only the patterns that yield favorable computational results would be manufactured and tested.

C. Aerodynamics is a well-established field of fluid dynamics, and many excellent computational codes for predicting flow over a wing are in routine use. The analysis and design of a new wing shape would be carried out first on the computer with selected experiments to verify certain features of the best new design.

D. One cannot learn about the spread of a toxin in the human respiratory system by exposing volunteers to the toxin, and it may be unethical to carry out such a study in animals. This project appears to be a prime candidate for computational simulation or perhaps an experiment that uses a benign tracer material. Intersubject variability in the structure of the lung must be carefully considered.

bubbles, or other materials. The next step is to record an image of the marked fluid entity at one or more time intervals, to permit the investigators to deduce some characteristic of the fluid motion of interest. This creates a visual record of the flow, which an engineer can analyze to better understand the underlying physical processes. Tracking a fluid entity and analyzing the resulting record can also be carried out with an analytical or computational solution.

The first half of this chapter is designed to teach you how to exploit basic features of both the Lagrangian and Eulerian descriptions of fluid mechanics for use in applying and interpreting a variety of flow visualization techniques. In the next section we begin by discussing what can be learned about a flow by applying concepts from the Lagrangian description of a fluid.

After completing our introduction to the subject of flow visualization we turn our attention in Sections 10.7 through 10.11 to an analysis of flow structure. Flow structure can be defined as a coherent organization and relationship between the values of velocity at a number of neighboring spatial points. This structure is best understood by considering the effect of a velocity field on a fluid element, a small volume of fluid of arbitrary shape that always contains the same fluid particles. In a nonuniform flow, different parts of a fluid element have different velocity vectors. A fluid element therefore moves and deforms as fluid particles within the element change their relative positions over time. This behavior is not only the key to understanding flow structure, but also gives rise to the stress–(rate of strain) response of the fluid.

An example of a flow structure that clearly exhibits a coherent relationship between the values of velocity at a number of neighboring points is shown in Figure 10.3A. The distinct oval "spot" is readily recognized as part of Jupiter's "landscape." This atmospheric region is believed to be an enormous vortex or storm that has lasted for hundreds of years. On a smaller scale, the human eye readily identifies the distinct arrangement of a hurricane in the satellite image shown in Figure 10.3B. The flows you encounter every day, from those in the atmosphere or ocean to flows in pumps, pipes, and other engineering devices also exhibit flow structure. In many of these cases the structure is not visible to the eye, but reveals itself when the velocity field is examined in detail.

 CD/Videolibrary/The great red spot of Jupiter & Videolibrary/Everyday flows

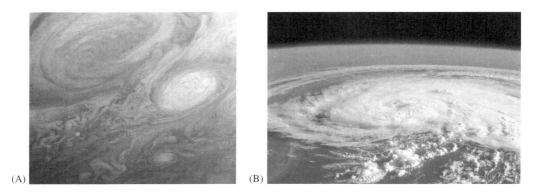

(A) (B)

Figure 10.3 (A) An ancient storm on Jupiter (spot). (B) A storm on Earth.

10.2 LAGRANGIAN KINEMATICS

 CD/Kinematics/Kinematics of points and fluid particles

Kinematics is the study of motion without regard for the forces involved or how the motion is created. From rigid body dynamics you know that the mathematical connection between the position, velocity, and acceleration of a point mass is made with a time derivative. The same is true for a fluid described in the Lagrangian model. Owing to the deformable nature of a fluid, however, there are features of Lagrangian fluid kinematics that you did not encounter in the study of rigid bodies.

The general motion of a rigid body consists of translation of the center of mass and rotation about the center of mass. If we know the translational velocity vector $\mathbf{U}(t)$ and the angular velocity vector $\mathbf{\Omega}(t)$, we can use these two vectors to completely define the motion of every particle in the rigid body. What is the corresponding situation for a moving fluid? Well, there is no overall orientation for a fluid, so there is no need for an angular velocity vector. However, since the fluid is deformable, each fluid particle undergoes a distinct motion that must be tracked separately. Therefore, a complete Lagrangian description of motion consists of knowing the position of all fluid particles as a function of time, plus any additional information needed to establish the thermodynamic state of each fluid particle. The particle velocity and acceleration may be calculated by taking the time derivative of the position similar to a rigid body.

10.2.1 Particle Path, Velocity, Acceleration

 CD/Kinematics/Kinematics of points and fluid particles/Particle kinematics

Consider a Cartesian coordinate system spanning the space occupied by fluid at a particular reference time t_0 as shown in Figure 10.4A. Suppose we focus attention on a

Although the Lagrangian description is used in numerical simulations of certain flow problems, and in flow visualization, we shall not derive or solve the Lagrangian equations of motion in this introductory textbook. One difficulty in the Lagrangian description lies in the need to model the forces of interaction between fluid particles, and between fluid particles and boundaries. Although there is a good understanding of molecular interactions in liquids and gases, a model that directly describes the interactions of fluid particles is lacking. Thus, the most common way to obtain the governing equations of fluid mechanics in the Lagrangian description is to transform the Eulerian forms of the same governing equations; for this task, engineers use relationships developed later in this chapter.

single one of the many fluid particles in this space and assign an initial position vector to this particle:

$$\mathbf{X}_0 = X_0 \mathbf{i} + Y_0 \mathbf{j} + Z_0 \mathbf{k} \qquad (10.1)$$

where (X_0, Y_0, Z_0) are its spatial coordinates at time t_0. At time t, the particle has moved to a new location (Figure 10.4B) given by its position vector:

$$\mathbf{X} = X(t)\mathbf{i} + Y(t)\mathbf{j} + Z(t)\mathbf{k} \qquad (10.2)$$

where $X(t), Y(t), Z(t)$ are functions of time that provide this particle's spatial coordinates at time t. Notice that if we know the position vector of a fluid particle, we know where it is going and where it has been. This is the essential idea behind the Lagrangian description, and it explains why the position vector $\mathbf{X}$ is referred to as the particle path or particle trajectory.

To define the motion of a volume of fluid, we must define the particle path for each of the large number of fluid particles contained in that volume (Figure 10.4B). We can account for every fluid particle without using a separate particle path for each one by noting that two fluid particles cannot physically occupy the same spatial location at the same initial time t_0. The Lagrangian particle path of Eq. 10.2 is therefore modified to account for all fluid particles in a flow by using (X_0, Y_0, Z_0) and time t as independent variables, and writing the particle path in vector form as

$$\mathbf{X} = X(X_0, Y_0, Z_0, t)\mathbf{i} + Y(X_0, Y_0, Z_0, t)\mathbf{j} + Z(X_0, Y_0, Z_0, t)\mathbf{k} \qquad (10.3)$$

Note that the three Cartesian components of this vector are defined by the functions

$$X = X(X_0, Y_0, Z_0, t), \quad Y = Y(X_0, Y_0, Z_0, t), \quad \text{and} \quad Z = Z(X_0, Y_0, Z_0, t) \qquad (10.4\text{a–c})$$

In interpreting Eq. 10.3 we say that $\mathbf{X}$ is the position at time t of the fluid particle with initial position (X_0, Y_0, Z_0) at time t_0. Since the identity of each fluid particle is

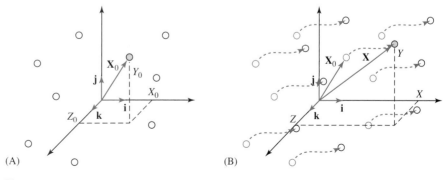

Figure 10.4 (A) Lagrangian fluid particles at time t_0. (B) The same particles at a later time t.

EXAMPLE 10.2

Consider the laminar flow through a channel driven by a pressure gradient as illustrated in Figure 10.5A. The particle paths for this flow are given by

$$X = X_0 + \left[\frac{h^2(p_1 - p_2)}{2\mu L}\right]\left[1 - \left(\frac{Y_0}{h}\right)^2\right](t - t_0), \quad Y = Y_0, \quad \text{and} \quad Z = Z_0$$

where $2h$ is the channel height, and the pressures are measured a distance L apart as shown. Consider the fluid particle initially located at $X_0 = x_1$, $Y_0 = 0$, $Z_0 = 0$ at time t_0 as shown in Figure 10.5B. Where will this particle be at time $t_1 = t_0 + 4\mu L/[h(p_1 - p_2)]$? Where will the particle initially located at $X_0 = x_1$, $Y_0 = h/2$, $Z_0 = 0$ be located at time t_1?

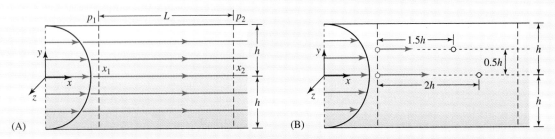

Figure 10.5 Schematic for Example 10.2: (A) laminar flow through a channel driven by a pressure gradient and (B) particle paths for this flow.

SOLUTION

Examining the particle paths, we see that the Y and Z coordinates of a fluid particle cannot change, but the X coordinate changes in time and depends on the initial coordinate Y_0. We are asked first for the future position of a fluid particle that is initially located on the channel centerline at $X_0 = x_1$, $Y_0 = 0$, $Z_0 = 0$. To determine the future position, we enter the specified initial values $X_0 = x_1$, $Y_0 = 0$, $Z_0 = 0$ and desired time $t_1 = t_0 + 4\mu L/[h(p_1 - p_2)]$ into the three components of the particle path to obtain

$$X = (x_1) + \left[\frac{h^2(p_1 - p_2)}{2\mu L}\right]\left[1 - \left(\frac{0}{h}\right)^2\right]\left[\frac{4\mu L}{h(p_1 - p_2)}\right]$$

$$= (x_1) + \left[\frac{h^2(p_1 - p_2)}{2\mu L}\right](1)\left[\frac{4\mu L}{h(p_1 - p_2)}\right] = x_1 + 2h$$

$$Y = 0 \quad \text{and} \quad Z = 0$$

Thus, the location of this fluid particle at time t_1 is given by the vector

$$\mathbf{X} = (x_1 + 2h)\mathbf{i} + 0\mathbf{j} + 0\mathbf{k}$$

We see that the particle has moved the equivalent of one channel height to the right in this amount of time as shown in Figure 10.5B. Note that since this flow is described by $Y = Y_0$ and $Z = Z_0$, the values of these two coordinates never change for any fluid particle.

The location of the second particle at time t_1 is found using the same procedure to be

$$\mathbf{X} = \left(x_1 + \frac{3}{2}h\right)\mathbf{i} + \left(\frac{h}{2}\right)\mathbf{j} + 0\mathbf{k}$$

We see that as expected, this particle remains at its initial elevations $Y_0 = h/2$, $Z_0 = 0$. It has moved in the X direction, but not as far as the first particle. Why is that? From the velocity profile in Figure 10.5B we see that the velocity at the initial location of the second particle is less than that at the initial location of the first. If we check for the future location of fluid particles located initially on either wall ($Y_0 = +h$ or $-h$), we find that these particles do not move at all. This is consistent with the no-slip, no-penetration boundary conditions on the channel walls.

specified by its initial position, we refer to (X_0, Y_0, Z_0) as the identity variables. There is a unique value of these variables for each fluid particle, so Eq. 10.3 (or 10.4a–10.4c) describes the motion of an entire fluid volume.

CD/Kinematics/Kinematics of points and fluid particles/Kinematics of many particles

We can determine the sequence of points in space occupied by a selected fluid particle by advancing the time variable in the Lagrangian position vector. By taking a sufficiently large number of particles and using Eq. 10.3 to track each one through time, we can see how the fluid moves as time evolves. For example, Figure 10.6 shows how an initially vertical line of particles in the parallel plate flow of Example 10.2 would advance in equal time steps. Notice how the velocity of each particle determines its position relative to its neighbors. Can you see why the Lagrangian position vector $\mathbf{X}$ is referred to as the particle path?

To plot each particle position in a sequence at a number of fixed time intervals, we follow the procedure illustrated in Example 10.2 considering each new time in turn for

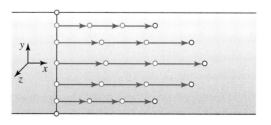
Figure 10.6 Particle paths for laminar channel flow.

a given particle, then doing the same thing for the desired number of different particles. The same process is followed if an analytical solution is available. In that case we might write a short computer code or use one of the popular symbolic codes to produce and plot the points.

 CD/Kinematics/Kinematics of points and fluid particles/fluid particles

It is natural to follow the motion of a rigid body by keeping our eyes focused on the moving body rather than by focusing on a region in space through which the rigid body passes. (Recall in Section 1.3.4 the example of people's heads turning in unison when watching a tennis match.) Our inclination to track moving objects suggests that the Lagrangian particle path is an instinctive and satisfying way to model fluid motion. In a fluid flow, the presence of a large number of fluid particles makes visualization more difficult than observing a single tennis ball; but if you have observed debris blown around by the wind, you have a sense of what a visualization of fluid flow in the Lagrangian description is like.

We saw in Example 10.2 that the particle path allows us to determine the future location of any fluid particle of known identity given by its initial location. In some applications we would like to be able to determine the location at an earlier time $t^* = t_0 - \Delta t$ of a certain fluid particle. This information is also provided by the particle path, Eq. 10.3, for this equation applies at any desired time, even at a time $t^* \leq t_0$. We find the desired past location at time t^* by inserting $t = t^*$ into the particle path.

It is sometimes convenient to have an expression that gives the initial position of a fluid particle as a function of its current position. This representation, called the inverse particle path, is given in vector form as

$$\mathbf{X}_0 = X_0(X, Y, Z, t)\mathbf{i} + Y_0(X, Y, Z, t)\mathbf{j} + Z_0(X, Y, Z, t)\mathbf{k} \tag{10.5}$$

and has three Cartesian components given by the functions

$$X_0 = X_0(X, Y, Z, t), \quad Y_0 = Y_0(X, Y, Z, t), \quad \text{and} \quad Z_0 = Z_0(X, Y, Z, t) \tag{10.6a–c}$$

The inverse particle path (Eq. 10.5) tells us that the particle now at position $\mathbf{X}$ at time t was at initial position $\mathbf{X}_0$ at time t_0. In principle, we can find the inverse particle path analytically if we know the particle path, and vice versa. If the particle paths are complex, however, this inversion may not be possible to carry out analytically.

The inversion process may be illustrated for a uniform flow in the X direction given by

$$X = X_0 + U_\infty(t - t_0), \quad Y = Y_0, \quad \text{and} \quad Z = Z_0$$

The inverse particle paths are found in this case by isolating the identity variables to obtain

$$X_0 = X - U_\infty(t - t_0), \quad Y_0 = Y, \quad \text{and} \quad Z_0 = Z$$

Since the inversion process is deceptively simple in this case, be sure to study Example 10.3, which is slightly more difficult.

EXAMPLE 10.3

Find the inverse particle paths for the channel flow of Example 10.2. Show that the initial position of the fluid particle located at $\mathbf{X} = (2h)\mathbf{i} + 0\mathbf{j} + 0\mathbf{k}$ at time $t = t_0 + 4\mu L/[h(p_1 - p_2)]$ is $\mathbf{X}_0 = 0\mathbf{i} + 0\mathbf{j} + 0\mathbf{k}$.

SOLUTION

The process begins by inspecting the particle paths. As given earlier, these are

$$X = X_0 + \left[\frac{h^2(p_1 - p_2)}{2\mu L}\right]\left[1 - \left(\frac{Y_0}{h}\right)^2\right](t - t_0), \quad Y = Y_0, \quad \text{and} \quad Z = Z_0$$

We immediately note that two of the inverse functions are available by inspection as

$$Y_0 = Y \quad \text{and} \quad Z_0 = Z$$

Isolating the X_0 variable in the remaining function, we have

$$X_0 = X - \left[\frac{h^2(p_1 - p_2)}{2\mu L}\right]\left[1 - \left(\frac{Y_0}{h}\right)^2\right](t - t_0)$$

We are not finished, however, since the identity variable Y_0 occurs in the right-hand side. We must replace this identity variable, since the inverse function is only allowed to be a function of (X, Y, Z, t). In general, the final step in developing an inverse function is to get rid of any identity variables, and this can be difficult if highly complex functions are involved. We do it here by substituting for Y_0 using the previously determined inverse $Y_0 = Y$. The final component of the inverse is then

$$X_0 = X - \left[\frac{h^2(p_1 - p_2)}{2\mu L}\right]\left[1 - \left(\frac{Y}{h}\right)^2\right](t - t_0)$$

To find the initial position of the indicated particle, we substitute the current position $\mathbf{X} = (2h)\mathbf{i} + (0)\mathbf{j} + 0\mathbf{k}$ and current time $t = t_0 + 4\mu L/[h(p_1 - p_2)]$ into the inverse functions to obtain

$$X_0 = 2h - \left[\frac{h^2(p_1 - p_2)}{2\mu L}\right]\left[1 - \left(\frac{0}{h}\right)^2\right]\left[\frac{4\mu L}{h(p_1 - p_2)}\right] = 2h - 2h = 0$$

$$Y_0 = Y = 0 \quad \text{and} \quad Z_0 = Z = 0$$

The initial position of this particle is $\mathbf{X}_0 = 0\mathbf{i} + 0\mathbf{j} + 0\mathbf{k}$ as stated.

The utility of these ideas on tracking fluid particles is limited by the occurrence of turbulence in many flows of interest. Although the idea of fluid particle tracking is sound, and works well in laminar flows, we never have sufficient information in a turbulent flow to track particles for more than a very short period of time. Flow visualization techniques normally use a short time interval to avoid this difficulty. Modern CFD codes attempt to compensate for the effects of turbulence in randomly perturbing the path of a fluid particle (or any other type of particle carried along in the flow) by introducing a statistical model for the randomizing effects of turbulence. How successful this approach actually is remains an open question.

There are some engineering problems for which the Lagrangian particle paths provide invaluable information. For example, consider the automotive situation shown in Figure 10.7. To determine whether there is any possibility of carbon monoxide entering the passenger compartment through a vent or open window, you might want to know where the fluid particles leaving the exhaust pipe go. The particle path provides the required information. On the other hand, if we know carbon monoxide is entering the passenger compartment at a particular location, the inverse particle path allows us to backtrack along the particle trajectories to determine the origin of these fluid particles. Does this explain why a disaster response team trying to warn local residents of a moving toxic gas cloud would prefer to have flow information in the form of particle paths? Suppose another team was responsible for trying to identify the source of the toxic material. Would they prefer to know the inverse particle paths? If you answered yes to both questions, you are correct.

As shown conceptually in Figure 10.8, the Lagrangian fluid velocity $\mathbf{V} = \mathbf{V}(X_0, Y_0, Z_0, t)$ is defined to be the time rate of change of the position of the fluid

Figure 10.7 Car exhaust.

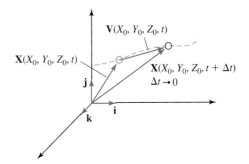

Figure 10.8 Definition of the Lagrangian velocity vector.

particle whose identity is defined by its initial position (X_0, Y_0, Z_0). The velocity vector is represented in vector form as

$$\mathbf{V} = U(X_0, Y_0, Z_0, t)\mathbf{i} + V(X_0, Y_0, Z_0, t)\mathbf{j} + W(X_0, Y_0, Z_0, t)\mathbf{k} \qquad (10.7)$$

where the three Cartesian components are

$$U = U(X_0, Y_0, Z_0, t), \quad V = V(X_0, Y_0, Z_0, t), \quad \text{and} \quad W = W(X_0, Y_0, Z_0, t) \qquad (10.8\text{a–c})$$

By definition, the Lagrangian velocity vector is the time rate of change of position. Thus, it may be obtained from the particle path by taking a time derivative. It is customary to write this relationship in vector form as

$$\mathbf{V} = \frac{d\mathbf{X}}{dt} \qquad (10.9)$$

rather than use a partial time derivative. This does not create confusion provided you remember the time derivative is taken for a specific fluid particle having a fixed set of identity variables. On a component basis we have

$$U = \frac{dX}{dt}, \quad V = \frac{dY}{dt}, \quad \text{and} \quad W = \frac{dZ}{dt} \qquad (10.10\text{a–c})$$

The Lagrangian acceleration vector $\mathbf{A} = \mathbf{A}(X_0, Y_0, Z_0, t)$ is represented in Cartesian coordinates as

$$\mathbf{A} = A_x(X_0, Y_0, Z_0, t)\mathbf{i} + A_y(X_0, Y_0, Z_0, t)\mathbf{j} + A_z(X_0, Y_0, Z_0, t)\mathbf{k} \qquad (10.11)$$

EXAMPLE 10.4

Find the velocity and acceleration in a uniform flow in the X direction for which the particle paths are given by: $X = X_0 + U_\infty(t - t_0)$, $Y = Y_0$, and $Z = Z_0$.

SOLUTION

From Eqs. 10.10a–10.10c, the velocity is defined by $U = dX/dt$, $V = dY/dt$, and $W = dZ/dt$. Taking derivatives of the particle path functions, we obtain:

$$U = \frac{dX}{dt} = \frac{d}{dt}(X_0 + U_\infty(t - t_0)) = U_\infty, \quad V = \frac{dY}{dt} = \frac{d}{dt}(Y_0) = 0$$

and

$$W = \frac{dZ}{dt} = \frac{d}{dt}(Z_0) = 0$$

Thus, the velocity vector is $\mathbf{V} = U_\infty \mathbf{i} + 0\mathbf{j} + 0\mathbf{k}$, and, since this velocity is independent of time, acceleration is zero. The results are consistent with the description of these particle paths as a uniform flow.

EXAMPLE 10.5

Find the velocity and acceleration in the channel flow of Example 10.2. The particle paths are given by

$$X = X_0 + \left[\frac{h^2(p_1 - p_2)}{2\mu L}\right]\left[1 - \left(\frac{Y_0}{h}\right)^2\right](t - t_0), \quad Y = Y_0, \quad \text{and} \quad Z = Z_0$$

SOLUTION

From Eqs. 10.10a–10.10c, the velocity is defined by $U = dX/dt$, $V = dY/dt$, and $W = dZ/dt$. Taking derivatives of the particle path functions, we have

$$U = \frac{dX}{dt} = \frac{d}{dt}\left\{X_0 + \left[\frac{h^2(p_1 - p_2)}{2\mu L}\right]\left[1 - \left(\frac{Y_0}{h}\right)^2\right](t - t_0)\right\}$$

$$= \left[\frac{h^2(p_1 - p_2)}{2\mu L}\right]\left[1 - \left(\frac{Y_0}{h}\right)^2\right]$$

$$V = \frac{dY}{dt} = \frac{d}{dt}(Y_0) = 0 \quad \text{and} \quad W = \frac{dZ}{dt} = \frac{d}{dt}(Z_0) = 0$$

Thus the velocity vector is

$$\mathbf{V} = \left[\frac{h^2(p_1 - p_2)}{2\mu L}\right]\left[1 - \left(\frac{Y_0}{h}\right)^2\right]\mathbf{i} + 0\mathbf{j} + 0\mathbf{k}$$

We note that the velocity is zero on the channel walls as expected, and has a maximum value of $\mathbf{V} = \{[h^2(p_1 - p_2)]/2\mu L\}\mathbf{i} + 0\mathbf{j} + 0\mathbf{k}$ on the channel centerline.

According to Eq. 10.13, the three Cartesian components of the acceleration vector are related to the particle path by second time derivatives:

$$A_x = \frac{d^2X}{dt^2}, \quad A_y = \frac{d^2Y}{dt^2}, \quad \text{and} \quad A_z = \frac{d^2Z}{dt^2}$$

We could also use Eq. 10.12 and write the three components of acceleration in terms of a time derivative of velocity as

$$A_x = \frac{dU}{dt}, \quad A_y = \frac{dV}{dt}, \quad \text{and} \quad A_z = \frac{dW}{dt}$$

Applying the latter approach to the velocity just given, we find by inspection $\mathbf{A} = 0\mathbf{i} + 0\mathbf{j} + 0\mathbf{k}$. The Lagrangian acceleration is zero in this channel flow.

10.2 LAGRANGIAN KINEMATICS

and defined by the first time derivative of velocity

$$\mathbf{A} = \frac{d\mathbf{V}}{dt} \quad (10.12)$$

or, equivalently, by the second time derivative of particle path

$$\mathbf{A} = \frac{d^2\mathbf{X}}{dt^2} \quad (10.13)$$

Notice that the definitions of position, velocity, and acceleration are identical for a fluid particle and for a point mass in dynamics. Identity variables are necessary in the description of a fluid particle, so the position, velocity, and acceleration are functions of time and identity variables, rather than of just time alone. Examples 10.4 and 10.5 illustrate the relationships between position, velocity, and acceleration in the Lagrangian description.

As shown in Figure 10.9, the Lagrangian position vector of a fluid particle in cylindrical coordinates is represented as

$$\mathbf{X} = R(R_0, \theta_0, Z_0, t)\mathbf{e}_r + Z(R_0, \theta_0, Z_0, t)\mathbf{e}_z \quad (10.14a)$$

where $\mathbf{e}_r$ is the unit vector in the radial direction and $\mathbf{e}_z$ is the unit vector in the axial direction. Note that there is no θ component in the position vector, but a complete description of particle position requires that we know the function

$$\theta = \theta(R_0, \theta_0, Z_0, t) \quad (10.14b)$$

The Lagrangian velocity vector, which does have three components, may be written as

$$\mathbf{V} = V_r(R_0, \theta_0, Z_0, t)\mathbf{e}_r + V_\theta(R_0, \theta_0, Z_0, t)\mathbf{e}_\theta + V_z(R_0, \theta_0, Z_0, t)\mathbf{e}_z \quad (10.15)$$

while the Lagrangian acceleration vector is

$$\mathbf{A} = A_r(R_0, \theta_0, Z_0, t)\mathbf{e}_r + A_\theta(R_0, \theta_0, Z_0, t)\mathbf{e}_\theta + A_z(R_0, \theta_0, Z_0, t)\mathbf{e}_z \quad (10.16)$$

The relationships among position, velocity, and acceleration components in cylindrical coordinates are somewhat unusual and worth noting. From dynamics, the relationships between velocity and position are

$$V_r = \frac{dR}{dt} = \dot{R}, \quad V_\theta = R\frac{d\theta}{dt} = R\dot{\theta}, \quad \text{and} \quad V_z = \frac{dZ}{dt} = \dot{Z} \quad (10.17)$$

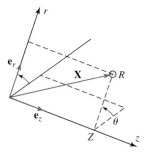

Figure 10.9 The Lagrangian position vector in cylindrical coordinates.

EXAMPLE 10.6

Laminar steady fluid flow through a round pipe, known as Poiseuille flow, is shown in Figure 10.10. In cylindrical coordinates the corresponding particle paths are

$$R = R_0, \quad Z = Z_0 + \left\{\left[\frac{R_P^2(p_1 - p_2)}{4\mu L}\right]\left[1 - \left(\frac{R_0}{R_P}\right)^2\right]\right\}(t - t_0), \quad \text{and} \quad \theta = \theta_0$$

where R_P is the radius of the pipe and the pressures are measured as shown in Figure 10.10. Calculate the Lagrangian velocity and acceleration vectors for this flow.

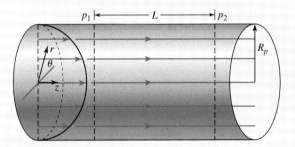

Figure 10.10 Schematic for Example 10.6, Poiseuille flow through a round pipe.

SOLUTION

First note that the particle paths indicate that all fluid particles move straight down the pipe without changing their radial or angular positions. Thus, the particle paths coincide with the streamlines shown in Figure 10.10. The velocity is calculated by using Eqs. 10.17, with the result

$$V_r = \frac{dR}{dt} = \frac{d}{dt}(R_0) = 0, \quad V_\theta = R\frac{d\theta}{dt} = R\frac{d}{dt}(\theta_0) = 0$$

and

$$V_z = \frac{dZ}{dt} = \frac{d}{dt}\left\{Z_0 + \left[\left(\frac{R_P^2(p_1 - p_2)}{4\mu L}\right)\left(1 - \left(\frac{R_0}{R_P}\right)^2\right)\right](t - t_0)\right\}$$

$$= \left(\frac{R_P^2(p_1 - p_2)}{4\mu L}\right)\left(1 - \left(\frac{R_0}{R_P}\right)^2\right)$$

Thus,

$$\mathbf{V} = 0\mathbf{e}_r + 0\mathbf{e}_\theta + \left(\frac{R_P^2(p_1 - p_2)}{4\mu L}\right)\left(1 - \left(\frac{R_0}{R_P}\right)^2\right)\mathbf{e}_z$$

Note that the no-slip, no-penetration conditions are satisfied for a particle at $R_0 = R_P$, i.e., on the pipe wall, and that the single nonzero velocity component points in the flow direction as expected. The maximum velocity, which occurs for a particle on the centerline, is $\mathbf{V}_{\max} = 0\mathbf{e}_r + 0\mathbf{e}_\theta + \{[R_P^2(p_1 - p_2)]/4\mu L\}\mathbf{e}_z$.

The acceleration may be calculated using Eq. 10.18 or 10.19. Applying Eq. 10.18 to the preceding velocity vector, we obtain

$$A_r = \frac{dV_r}{dt} - \frac{V_\theta^2}{R} = \frac{d}{dt}(0) - \frac{(0)^2}{R} = 0, \quad A_\theta = \frac{dV_\theta}{dt} + \frac{V_r V_\theta}{R} = \frac{d}{dt}(0) + \frac{(0)(0)}{R} = 0$$

and

$$A_z = \frac{d}{dt}\left[\left(\frac{R_P^2(p_1 - p_2)}{4\mu L}\right)\left(1 - \left(\frac{R_0}{R_P}\right)^2\right)\right] = 0$$

Thus we have $\mathbf{A} = 0\mathbf{e}_r + 0\mathbf{e}_\theta + 0\mathbf{e}_z$. There is no Lagrangian acceleration in Poiseuille flow.

while those between acceleration and velocity are

$$A_r = \frac{dV_r}{dt} - \frac{V_\theta^2}{R}, \quad A_\theta = \frac{dV_\theta}{dt} + \frac{V_r V_\theta}{R}, \quad \text{and} \quad A_z = \frac{dV_z}{dt} \qquad (10.18)$$

With these results, it is straightforward to show that position and acceleration are related by

$$A_r = \ddot{R} - R\dot{\theta}^2, \quad A_\theta = R\ddot{\theta} + 2\dot{R}\dot{\theta}, \quad \text{and} \quad A_z = \ddot{Z} \qquad (10.19)$$

CD/Videolibrary/Pipe flow

10.2.2 Lagrangian Fluid Properties

CD/Kinematics/Fields, particles, and reference frames/Lagrangian representation: scalar fields

Each fluid particle in a flow field possesses a Lagrangian position, velocity, and acceleration, as well as other fluid and flow properties as discussed in Chapter 2. One of the most important properties of a fluid particle is its density, ρ. We represent the density of a fluid particle in the Lagrangian description as the scalar function

$$\rho = \rho(X_0, Y_0, Z_0, t) \qquad (10.20a)$$

Like position, velocity, and acceleration, density is a function of the identity variables and time. Thus, we interpret Eq. 10.20a by saying that ρ is the density at time t of the fluid particle identified by the identity variables (X_0, Y_0, Z_0). As time evolves and a fluid particle moves, its density may change.

A fluid particle is also characterized by its pressure, internal energy, temperature and other fluid and flow properties. Each of these Lagrangian fluid properties is written as a function of (X_0, Y_0, Z_0, t). For example, pressure in the Lagrangian description is represented as

$$p = p(X_0, Y_0, Z_0, t) \qquad (10.20b)$$

and temperature is

$$T = T(X_0, Y_0, Z_0, t) \qquad (10.20c)$$

10.3 THE EULERIAN–LAGRANGIAN CONNECTION

 CD/Kinematics/Fields, particles, and reference frames/Eulerian and Lagrangian frames: Vector quantities

Regardless of which description one uses to describe a flow, the Eulerian values of velocity, acceleration, and other fluid and flow properties at a given point in space at a given time must be the same as the corresponding Lagrangian values for the fluid particle that is at that spatial location at that instant of time. This is the basis for a connection between the Lagrangian and Eulerian descriptions: a fluid particle whose position is **X** in the Lagrangian description is located at **x** in the Eulerian description (see Figure 10.11). Furthermore, the Eulerian velocity vector **u** at the point **x** at a given time is defined to be the velocity **V** of the fluid particle that happens to be at that location at that instant of time. The Eulerian acceleration vector **a** is likewise defined to be the acceleration **A** of the fluid particle that happens to be located at the point in question. Thus, the formal connection between the two descriptions takes the form

$$\mathbf{X} = \mathbf{x}, \quad \mathbf{V} = \mathbf{u}, \quad \text{and} \quad \mathbf{A} = \mathbf{a} \qquad (10.21)$$

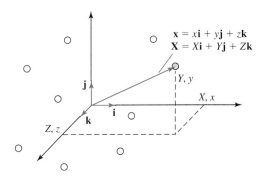

Figure 10.11 The equivalence of Lagrangian and Eulerian position vectors at an instant in time.

10.3 THE EULERIAN–LAGRANGIAN CONNECTION

This connection extends to the components of each vector. Thus, for Cartesian coordinates we find

$$X = x, \quad Y = y, \quad \text{and} \quad Z = z \tag{10.22a}$$

$$U = u, \quad V = v, \quad \text{and} \quad W = w \tag{10.22b}$$

and

$$A_x = a_x, \quad A_y = a_y, \quad \text{and} \quad A_z = a_z \tag{10.22c}$$

Since the particle path is the primary dependent variable in the Lagrangian description, and velocity plays this role in the Eulerian description, the connection between velocities, $\mathbf{V} = \mathbf{u}$, is usually expressed in terms of the Lagrangian particle path as

$$\frac{d\mathbf{X}}{dt} = \mathbf{u} \tag{10.23a}$$

rather than as in Eq. 10.21. Similarly the connection for acceleration is usually expressed as

$$\frac{d\mathbf{V}}{dt} = \mathbf{a} \tag{10.23b}$$

From our discussion in Chapter 6 of the definition of a material derivative and its role in defining acceleration in the Eulerian description, it should be clear that a time derivative in the Lagrangian description is equivalent to a material derivative in the Eulerian description. Thus, we can also write these relationships for velocity and acceleration as

$$\frac{d\mathbf{X}}{dt} = \mathbf{u} = \frac{D\mathbf{x}}{Dt} \quad \text{and} \quad \frac{d\mathbf{V}}{dt} = \mathbf{a} = \frac{D\mathbf{u}}{Dt} \tag{10.24a, b}$$

In analyzing a flow visualization experiment, an engineer would like to be able to use data taken in the Lagrangian description to construct the corresponding quantities in the Eulerian description. Similarly, after performing a flow simulation on the computer in the standard Eulerian description, an engineer may wish to calculate particle paths in the Lagrangian description to visualize the flow field. The connections described in Eqs. 10.21 through 10.24 provide a means of converting results from one description to the other.

In principle, the process of going from the Lagrangian description to the Eulerian description is straightforward. Suppose you have obtained the Lagrangian particle path $\mathbf{X} = \mathbf{X}(\mathbf{X}_0, t)$ for a certain flow. From Eq. 10.24a the Eulerian velocity field may be found by taking the time derivative of the particle path, since $d\mathbf{X}/dt = \mathbf{u}$. The Eulerian acceleration is found by using Eq. 10.24b in the form $d^2\mathbf{X}/dt^2 = d\mathbf{V}/dt = \mathbf{a}$. In both cases it is necessary to eliminate the identity variables after the derivative has been taken, and use $\mathbf{X} = \mathbf{x}$ to complete the conversion. This means we must also have available the inverse particle path.

The conversion of the Eulerian description of a flow to the Lagrangian description also makes use of the relationships developed in Eqs. 10.23 and 10.24, but the process is often more difficult. Recall that we may write the relationship between velocities as $d\mathbf{X}/dt = \mathbf{u}$. In Cartesian coordinates, this gives

$$\frac{dX}{dt} = u, \quad \frac{dY}{dt} = v, \quad \text{and} \quad \frac{dZ}{dt} = w \tag{10.25a}$$

The purpose of introducing the Eulerian–Lagrangian connection in this text is to convince you that it is possible to move back and forth, in a quantitative sense, from one method of description to the other. Those who take additional fluid mechanics classes or go on to work in a related industry (or both) may well make use of the equations in this section either directly or through the use of a CFD code. We will not, however, discuss the quantitative aspects of the Eulerian–Lagrangian connection further in this introductory text.

while in cylindrical coordinates this relationship is

$$\frac{dR}{dt} = v_r, \quad R\frac{d\theta}{dt} = v_\theta, \quad \text{and} \quad \frac{dZ}{dt} = v_z \quad (10.25b)$$

When the Eulerian velocity field is known, the preceding equations provide three coupled, nonlinear, ordinary differential equations describing the Lagrangian particle path in the corresponding coordinate system. These equations are very complex, and it is often necessary to integrate them numerically. Modern computational fluid dynamics packages have this capability, however, so when a flow problem is solved numerically in the Eulerian description, the Lagrangian particle paths may be obtained as well. This can be extremely helpful in visualizing a complex flow field.

Any fluid property written in the Eulerian description can be transformed into the Lagrangian description and vice versa if the particle path and the inverse function are known. The basis for the conversion is simply that the value of a fluid property at some point in space at a given time is the same as the value of the Lagrangian property carried by the fluid particle at the same point in space at that instant.

CD/Demonstrations/Euler vs. Lagrange: What is steady and unsteady

10.4 MATERIAL LINES, SURFACES, AND VOLUMES

CD/Kinematics/Timelines/Definition of timelines and Videolibrary/Demonstration of stirring paint

Material lines, surfaces, and volumes are Lagrangian concepts. The adjective "material" means that the line, surface, or volume in question is always made up of the same fluid particles. It can be shown that to the extent that the continuum hypothesis holds, all lines, surfaces, and volumes in a fluid have this characteristic. In thinking about this statement, keep in mind that molecular diffusion eventually causes fluid molecules initially in a material entity of any type to wander away; but this effect is not accounted for in a continuum model.

Suppose we have made a material line in a stationary fluid visible by marking the fluid particles with dye as shown in Figure 10.12. If we attempt to cut the line with a knife, the line will wrap itself around the edge of the blade. As the blade continues to move, the distance between dyed particles in the material line will increase as the length of the line increases, but the line will not break. This is readily observed in a viscous liquid such as glycerin. In water, the knife may appear to cut the line. Material lines also exist in a moving fluid. As shown in Figure 10.13, in a visualization experiment with a moving fluid, if we instantaneously create a line of dyed fluid particles, then the dyed

The reason the experiments illustrated in Figures 10.12 through 10.14 work so much better in glycerin than in water is due to the effects of diffusion. A dyed material line or surface in water quickly becomes fuzzy and difficult to distinguish because the dye diffuses more rapidly into the surrounding water.

line is a material line. After a material line has been created, it moves with the fluid and thereby undergoes stretching and deformation. If we know the velocity field, it is possible to predict the shape of the line at any later time by tracking each fluid particle over time. Material lines are often referred to as time lines.

Next consider the concept of a material surface. In the continuum hypothesis, every surface in a fluid is a material surface containing the same fluid particles at all times. Imagine marking a horizontal layer within a still liquid with dye and releasing a heavy object just above the layer so that the object falls onto the fluid material surface. We would observe that the object cannot penetrate the material surface. Instead, as shown in Figure 10.14, the initially planar material surface deforms and wraps itself around the object, stretching as needed to remain intact.

An interesting aspect of the process of mixing two miscible fluids by stirring is that the interface between them is a material surface. Mixing takes place by the intricate stretching and layering of the original interface. This is true regardless of whether the mixing is accomplished by a moving object (e.g., stirring with a spoon) or by turbulent fluid motion (e.g., the hot exhaust stream of a jet engine entering the surrounding air).

Figure 10.12 Deformation of a material line.

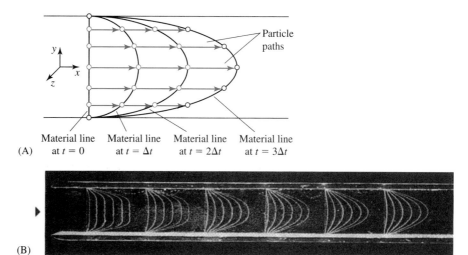

Figure 10.13 Channel flow: (A) material lines and particle paths and (B) flow visualization of material lines with hydrogen bubbles.

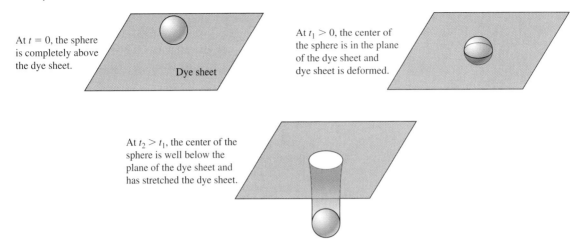

Figure 10.14 Deformation of a material surface.

Ultimately, two fluids are brought into close proximity over a large interfacial area, at which point molecular diffusion is able to complete the mixing process. Turbulence greatly enhances the stretching and layering process thereby reducing the time needed to thoroughly mix. Examine the photograph of the mixing of a plume of cigarette smoke with the surrounding air shown in Figure 10.15. Can you identify the laminar and turbulent mixing regions? In which region is the mixing occurring more rapidly?

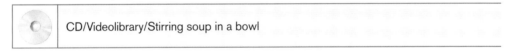

CD/Videolibrary/Stirring soup in a bowl

A material volume is a volume of fluid always made up of the same fluid particles. In a nonuniform flow, a material volume of fluid tends to become distorted over time. This effect is shown in Figure 10.16. The fluid system discussed in Chapter 7 is a

Figure 10.15 Mixing of a plume of cigarette smoke with the surrounding air.

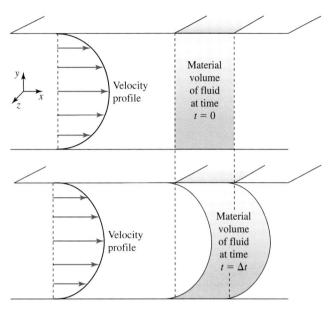

Figure 10.16 Deformation of a material volume at time $t = 0$ and $t = \Delta t$.

material volume. In that chapter we derived the integral governing equations of fluid mechanics by invoking the Reynolds transport theorem to apply the laws of conservation of mass, momentum, and energy to a material volume of fluid. An extension of the same approach leads to the governing differential equations of fluid mechanics as will be shown in Chapter 11.

The inviolability of material entities of all types in the continuum model is a reflection of the fact that the molecular structure of the fluid comes into play only in certain allowable ways. When dye is used to mark fluid particles, the blurring of the dyed line or surface over time is evidence that the random motion of fluid molecules eventually causes molecules initially in a material entity to leave and be replaced by others. The assumed permanence of material lines, surfaces, and volumes in the continuum hypothesis must be kept in mind when one is modeling physical phenomena. Some real phenomena cannot be successfully modeled this way. For example, is the concept of material entities consistent with the breaking up of a liquid jet into droplets?

The Lagrangian description provides a means for tracking material lines, surfaces, and volumes and thus allows these concepts to be profitably employed in the experimental and theoretical analysis of fluid flow by means of observations of marked fluid. If we define the mathematical function that represents a material line, surface, or volume at an initial instant, then the material entity at a later time can be described by using the inverse Lagrangian particle paths to move each fluid particle in the original entity to the correct position at the later time. Tracking material lines, surfaces, and volumes is the basis for a number of techniques in experimental and computational flow visualization. Example 10.7 illustrates the process of using the particle paths to predict the future position and shape of a material entity.

EXAMPLE 10.7

Consider the steady shear flow in the xy plane shown in Figure 10.17A. This flow is described in Cartesian coordinates by the velocity field: $u = \alpha y$, $v = 0$, and $w = 0$. Suppose you have marked three material entities in this flow at time t_0 with dye:

A. A line segment of slope β passing through the origin
B. A circular disk centered about the origin of radius R_c
C. A material volume in the shape of a cube centered at the origin.

Predict the shape of each of these entities at a later time.

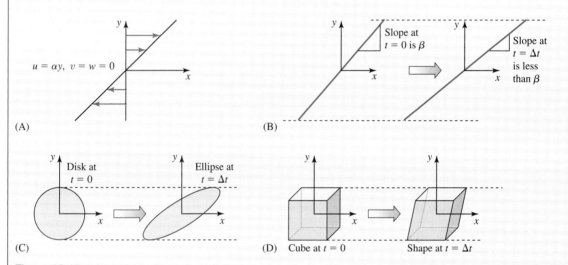

Figure 10.17 Schematic for Example 10.7. Deformation of material entities in a shear flow: (A) shear flow velocity vectors, (B) material line, (C) material surface, and (D) material volume.

SOLUTION

We are asked to predict the shape of three specified material entities of fluid at a future time given the Eulerian velocity field for the flow of interest. Figure 10.17 is the appropriate sketch, and no assumptions are required. To solve this problem qualitatively, we must recognize that in this uniform flow fluid particles do not move in either the y or the z direction, and their velocity in the x direction is proportional to their initial y coordinate. That is, as the initial positive y coordinate for a fluid particle increases, its velocity in the x direction also increases. Fluid particles with an initial y coordinate less than zero will have a velocity in the negative x direction, and particles with an initial y coordinate of zero do not move at all in this flow.

A. Line segment of slope β passing through the origin: The fluid particle initially located at the origin of the chosen coordinate system will remain at the origin for all times. Particles with positive initial y coordinates will move to the right with velocities proportional to their distance from $y = 0$, and particles with negative initial y coordinates will move to the left with velocities proportional to their distance from $y = 0$. The result is that the slope of the material line will decrease in time but never become negative, as shown in Figure 10.17B.

B. Circular disk centered about the origin: Using logic similar to that in case A, we see that the upper half of the material surface will move to the right and the lower half to the left such that the original circular boundary will become elliptical over time (see Figure 10.17C). The marked fluid lies inside this boundary, since fluid particles can never leave a material entity.

C. Material volume in the shape of a cube centered at the origin: The outer boundary of the cube can be thought of as being defined by 8 horizontal lines and 4 vertical lines as shown in Figure 10.17D. The upper and lower sets of 4 horizontal lines each move at the same constant speed to the right and left, respectively, in this shear flow, while the 4 vertical lines change slope but remain straight, as found in case A. The result is that over time, the initially square front and back surfaces of the cube will take on the shape of a parallelogram. Once again the marked fluid particles will remain in the material volume according to continuum theory.

This exercise may also be solved quantitatively by using the concepts and equations introduced in our discussion of the Eulerian–Lagrangian connection.

10.5 PATHLINES AND STREAKLINES

CD/Kinematics/Pathlines

In Section 10.2.1, we stated that the particle path $\mathbf{X} = \mathbf{X}(X_0, Y_0, Z_0, t)$ provides the trajectory through space of the fluid particle identified by (X_0, Y_0, Z_0). We can see a particle path in a fluid flow by introducing a tiny quantity of dye instantaneously at a point in the flow. As our eyes follow the motion of the dyed fluid particle, we are observing its progress along its pathline. A pathline is the trajectory through space of a selected fluid particle during a time interval. If the particle path is known analytically or from a numerical simulation of the flow, a pathline may be constructed by plotting the location of the particle at successive times and connecting these locations to form a line through space. Modern computational fluid dynamics codes have this capability. Figure 10.18 shows the pathlines in laminar flow through a nozzle as calculated by means of the FIDAP CFD code.

The process of marking a pathline to make it visible is different from that used to make a material line visible. A material line is a line in the fluid always made up of the same fluid particles. To make a material line visible we simultaneously dye, mark, or tag

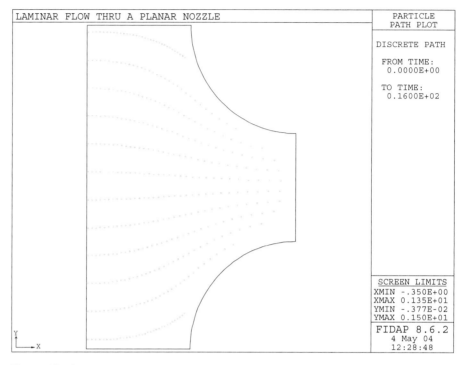

Figure 10.18 CFD visualization of pathlines of laminar flow through a planar nozzle.

its component fluid particles. The dyed entity is therefore a line in the fluid of some chosen shape. Observation of the material line focuses on the movement and deformation of the original dyed line rather than on the path of each particle in the line itself. A comparison of material lines and pathlines for flow over a cylinder is shown in Figure 10.19.

CD/Kinematics/Streaklines

Rather than injecting dye instantaneously at a set of spatial locations, we may instead choose to dye all the fluid particles that pass through a chosen dye injection point during some finite interval of time. Our eyes will see a line of dyed fluid particles extending from the injection point out into the flow field, as shown in Figure 10.20. This line, called a streakline, is the locus of fluid particles that passed through the marking or tagging point during the interval selected. The tagging point used to create a streakline may be stationary or even moving. In either case, the length and shape of a streakline line depends on the motion of the tagging point, how long the marking interval lasts, and how much the flow stretches and deforms the streakline.

CD/Videolibrary/Flow past a blunt object

10.5 PATHLINES AND STREAKLINES

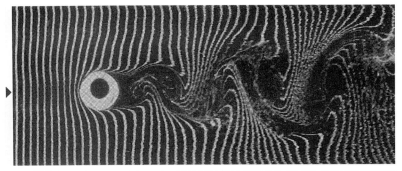

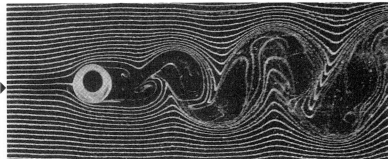

Figure 10.19 The flow upstream of the cylinder is steady. However, in the wake a Karman vortex street develops that is unsteady. (A) Flow visualized with material lines consisting of illuminated hydrogen bubbles. (B) The same flow visualized with pathlines that are coincident with streaklines upstream of the cylinder where the flow is steady. Because the flow downstream of the cylinder is unsteady, these are no longer pathlines but still are considered to be streaklines. This concept is explained in the text.

Figure 10.20 Experimental flow visualization by means of streaklines produced with dye injection.

Figure 10.21 CFD visualization of the flow shown in Figure 10.20. The streaklines are faithfully reproduced in the laminar region, but the simulation cannot resolve the breakup of the lines in the turbulent wake.

It is important to realize that a streakline is not the same as a pathline because the flow might have changed direction or magnitude during the marking interval. Figure 10.22 illustrates this point for a stationary injection point in an unsteady flow. Notice how the particles in the streakline arrived at their locations at the instant of time shown. In an unsteady flow, the paths traveled by individual fluid particles in a streakline can be very different. In a steady flow, it can be shown that a pathline and streakline are identical. Pathlines and streaklines are both Lagrangian concepts because they are defined by the motion of fluid particles. The following example will give you some idea of the logic involved in distinguishing pathlines and streaklines.

Figure 10.22 Schematic illustrating the difference between streaklines and pathlines in unsteady flow.

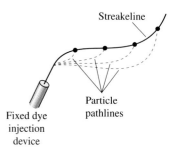

 CD/Kinematics/Flow lines and flow visualization

EXAMPLE 10.8

In each of the following cases, determine whether the visualized line is a pathline, streakline, or material line.

    A. The steam plume released from a vent pipe at a power plant
    B. The vapor trail of a jet airplane
    C. A stream of chum (a chopped bait and seawater mixture) pumped overboard into the wake of a fishing vessel anchored in the Gulf Stream
    D. The smoke released from a source upstream of a model in a wind tunnel experiment
    E. A neutrally buoyant helium balloon carried in the wind

SOLUTION

    A. The steam is being released over a period of time from a fixed tagging point. At any given instant, the plume marks the location of all the fluid particles that passed near the vent during some finite period of time. The plume is a streakline from a fixed tagging point.
    B. In the strict sense, the vapor trail is a streakline from a moving tagging point. However, if the wind speed is small in comparison to the plane's speed, it may be appropriate to think of a short length of the vapor trail as being created instantaneously and then acted on by the velocity field. The short length of vapor trail approximates a material line in the wind in that case.

C. The chum clearly marks a streakline from a fixed tagging point in the water. However, if we can distinguish individual bits of chum and watch their progress in the current, we are observing pathlines.

D. The wind tunnel flow is steady; thus the line of smoke is both a streakline from a fixed tagging point, and a pathline.

E. A neutrally buoyant helium balloon can be thought of as a fluid particle carried along in the flow. Thus the trajectory of the balloon over time is a pathline in the flow.

The use of neutrally buoyant particles of various kinds in flow visualization experiments with both gases and liquids is common. It is important for the particles to be small enough to accurately follow the flow. Micrometer-sized particles are also used in both air and water to reflect laser light in the measuring technique known as laser–Doppler velocimetry (Figure 10.23). This technique relies on measuring the Doppler frequency shift in laser light scattered from seed particles to deduce the velocity of those particles and of the surrounding fluid.

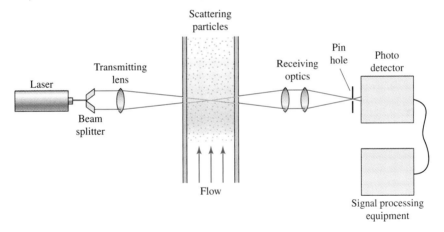

Figure 10.23 Schematic of a typical laser Doppler velocimetry system.

The traditional method for recording images of pathlines and streaklines has been photography. Photographic techniques are excellent for gaining a qualitative understanding of flows. However, they are quite cumbersome when applied to the calculation of particle velocities. A more recently developed technique, particle image velocimetry (PIV), determines the velocity field by recording the light reflected from particles moving through a two-dimensional section of the flow, illuminated by a sheet of laser light (see Figure 10.24). PIV systems provide the capability to experimentally measure a complete velocity field in real time.

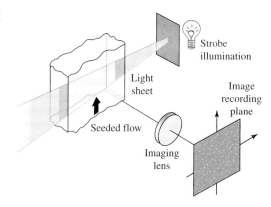

Figure 10.24 Schematic of a typical particle image velocimetry system.

It should be evident that even though the Lagrangian description is not often used to solve flow problems, an understanding of elements of Lagrangian kinematics is important because those concepts provide an easily interpreted visual picture of a flow. In many cases an engineer is able to optimize the design of a vehicle or piece of fluid-handling equipment simply by reshaping contours of solid surfaces to change the way the fluid moves. The combination of flow visualization and model testing, illustrated in Figure 10.25, is quite helpful in this respect.

 CD/Kinematics/Streamlines and streamfunction

Figure 10.25 Flow visualization using a 1/48-scale model of an F-18.

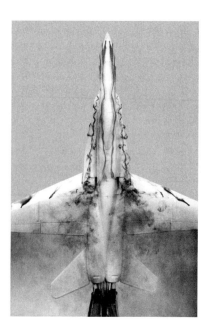

10.6 STREAMLINES AND STREAMTUBES

Material lines, pathlines, and streaklines are Lagrangian concepts that allow us to understand a flow field by observing the behavior of collections of marked fluid particles. In contrast, a streamline is strictly an Eulerian concept. As mentioned in earlier chapters, a streamline is defined to be a line in space that is everywhere tangent to the local velocity vector. In a steady flow, the fluid velocity does not change in time, and it can be shown that streamlines, pathlines, and streaklines are identical. Thus, a streamline may be visualized in a steady flow by marking pathlines or streaklines. An array or rake of continuously operating dye or smoke injection points creates a family of streaklines that, in a steady flow, is also a family of streamlines. A picture of these streaklines provides the direction of the velocity vector at each point along the coincident streamlines and so makes the Eulerian velocity field visible. As shown in Figure 10.26, automobile companies use this technique to improve the aerodynamics of their vehicles. You may have seen commercials designed to demonstrate the "streamlined" nature of automobiles. A CFD visualization of a similar flow is shown in Figure 10.27.

In an unsteady flow the streamlines may change dramatically from one instant to the next. Unsteady streamlines cannot be visualized by using tracers to mark fluid particles but they can be constructed from computational simulations of fluid flow. In an unsteady flow, streamlines, streaklines, and pathlines are generally different, and each line gives us a different picture of the underlying velocity field.

Figure 10.26 Visualization of flow around an automobile by means of smoke. In this steady flow the smoke lines are pathlines, streaklines, and streamlines.

Figure 10.27 CFD visualization of flow around an automobile.

In the design of low drag shapes of all kinds, it is important to reduce or eliminate boundary layer separation, which produces regions of low pressure that contribute to high drag. Boundary layer separation, which will be discussed in Chapter 14, is a complex physical phenomenon related to the balance of inertial, pressure, and viscous forces within the boundary layer. By using smoke to visualize the streamlines in a steady flow, it is usually possible to detect regions of separation. An examples of such a visualization is shown in Figure 10.28.

CD/Boundary Layers/Separation

Now suppose we have a functional representation of a velocity field in the Eulerian description, and we want to plot the streamlines in a certain region. How is this accomplished? How do we identify each streamline and distinguish it from all others? First note that streamlines generally do not cross one another. If they did, a fluid particle at the intersection point would have to be simultaneously moving in two different directions, as shown in Figure 10.29. This is physically impossible, of course, unless the velocity at the crossing point happens to be zero. A streamline is uniquely identified by its passage through a specified point at a specified time.

The process for constructing the streamlines at a given time t of a known 3D velocity field begins by observing that any line passing through space may be represented in Cartesian coordinates by two equations that define the locus of points in the line. These two equations may be of any one of the following forms:

Figure 10.28 Separation zones are clearly visible in the flow over the bluff car model.

$$y = y(x) \quad \text{and} \quad z = z(x) \quad (10.26a)$$
$$x = x(z) \quad \text{and} \quad y = y(z) \quad (10.26b)$$
$$z = z(y) \quad \text{and} \quad x = x(y) \quad (10.26c)$$

The functions corresponding to streamlines of a 3D velocity field can be identified as follows. Let

$$d\mathbf{r} = dx\mathbf{i} + dy\mathbf{j} + dz\mathbf{k} \quad (10.27)$$

be a differential vector element of length along a streamline passing through position (x, y, z) at a given time, as shown in Figure 10.29B. The definition of a streamline as

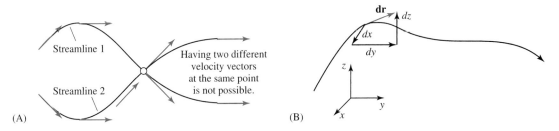

Figure 10.29 (A) Crossing streamlines are physically impossible unless the velocity at the intersection is zero. (B) The vector element of length along a streamline.

tangent to the velocity vector at each point means that $d\mathbf{r}$ and $\mathbf{u}$ are parallel vectors passing through (x, y, z) at the selected instant of time. Parallel vectors have a zero cross product, so a streamline must satisfy the following vector relationship:

$$d\mathbf{r} \times \mathbf{u} = 0 \tag{10.28}$$

This relationship is equivalent to three scalar differential equations:

$$w\,dy - v\,dz = 0, \quad u\,dz - w\,dx = 0, \quad \text{and} \quad v\,dx - u\,dy = 0 \tag{10.29a--c}$$

The solution to these equations provides the two functions that define the streamlines.

The process of finding the streamlines in a 3D flow at specified time requires that we find a set of two functions (like those given in Eqs. 10.26) that describe a line satisfying all three conditions given in Eqs. 10.29. In unsteady flow, the two functions will contain time t in their argument; but t is treated as a fixed parameter during the solution process, since the streamlines found are those for that specific instant of time. In a 2D velocity field, the absence of a third velocity component ensures that a streamline passing through a point in space never leaves the plane defined by the two nonzero velocity components. Finding the streamlines in a 2D flow is therefore relatively straightforward: there is only one function to be found, using the one remaining scalar differential equation of the original set of three.

To see how to find the streamlines in a 2D flow, consider a flow in the xy plane with a velocity field given by $u(x, y, t), v(x, y, t), 0$. The single function $y = y(x)$ in Eq. 10.26a defines the streamlines of this flow, and only needs to satisfy Eq. 10.29c: $v\,dx - u\,dy = 0$. This can be rearranged to obtain

$$\frac{dy}{dx} = \frac{v(x, y, t)}{u(x, y, t)} \tag{10.30}$$

which indicates that the slope of the streamline is such that the streamline is tangent to the velocity vector as shown in Figure 10.30. Equation 10.30 is an ordinary differential equation whose solution $y = y(x)$ describes the streamlines in the flow and contains one constant of integration that can be used to specify the point through which the streamline passes. For an unsteady flow, Eq. 10.30 is solved with t as an independent parameter. The solution in that case also has one constant of integration and describes the streamlines at time t.

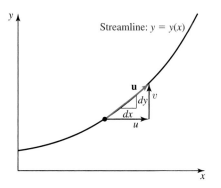

Figure 10.30 The velocity vector is tangent to the streamline.

EXAMPLE 10.9

Consider a steady velocity field as shown in Figure 10.31A, given by $\mathbf{u} = \alpha x \mathbf{i} - \alpha y \mathbf{j} + 0\mathbf{k}$, where α is a constant with units of reciprocal seconds and the spatial coordinates are measured in meters. Suppose the region of interest is defined by $0 < x < 5$ and $0 < y < 5$. Determine the equation for the streamlines of this flow, and find the streamline passing through the point (1, 2). Plot several other streamlines in the region of interest.

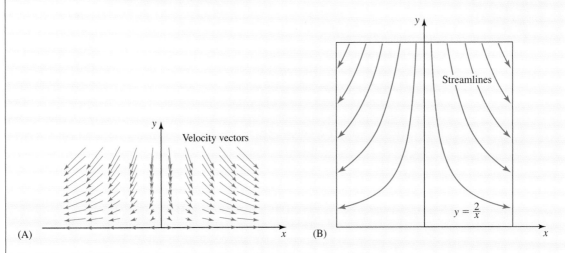

Figure 10.31 Stagnation point flow: (A) velocity field and (B) streamlines.

SOLUTION

We are asked to plot several streamlines for a known velocity field. The appropriate sketch is shown in Figure 10.31, and no assumptions are required. This is a 2D flow in the xy plane, thus the streamlines are described by Eq. 10.30: $dy/dx = [v(x, y, t)]/[u(x, y, t)]$. Substituting the velocity components, we have $dy/dx = -\alpha y/\alpha x = -y/x$. Separating variables and performing an indefinite integration, we obtain $\int dy/y = \int -dx/x$, which yields $\ln y = -(\ln x) + C$. Using the properties of the natural log, we can write this as $xy = C$ or $y = C/x$. The value of the constant C that corresponds to the streamline passing through some point (x_0, y_0) is found by inserting this point into the equation to obtain $x_0 y_0 = C$. Our final equation for the streamline can then be written as

$$y = \frac{x_0 y_0}{x}$$

The specific streamline passing through the point $x_0 = 1$, $y_0 = 2$ is found by writing

$$y = \frac{x_0 y_0}{x} = \frac{(1)(2)}{x} = \frac{2}{x}$$

A number of other streamlines can be plotted by picking the constant C to correspond to various points in the indicated region. A plot of these streamlines is shown in Figure 10.31B.

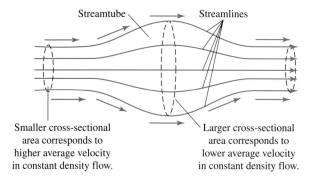

Figure 10.32 A streamtube is made up of a collection of streamlines.

Streamlines also exist in 1D or unidirectional flow. In this case there is only a single nonzero velocity component, and by definition, the streamlines must all be straight lines aligned with the velocity vector. For example consider a 1D flow in which $u = u(y)$. From Eq. 10.30, we can write $dy/dx = 0/u(y) = 0$, so the streamlines are given by $y = C$.

If we draw a closed line in space and enclose a bundle of streamlines, the set of streamlines defines a streamtube, as shown in Figure 10.32. Since the surface of a streamtube is comprised of streamlines, the velocity vector there is wholly tangential. Therefore no fluid may escape a streamtube, and the cross-sectional area of a streamtube anywhere along its length varies in such a way that the mass flow through the streamtube remains constant (see Figure 10.32).

Streamtubes exist in an unsteady flow, but as with unsteady flow streamlines, there is no practical way to make them visible using dye. The streamtube concept is most useful in steady flow, where a selected streamtube may be made visible by releasing a tracer from an injection area rather than from an injection point. If we have an analytical or numerical solution for a steady flow, we can construct a streamtube by plotting the bounding streamlines. Streamtubes are useful for analyzing open turbomachines such as wind turbines and propellers.

 CD/Kinematics/Kinematics of points and fluid particles/motion of rigid bodies

10.7 MOTION AND DEFORMATION

To begin our discussion of flow structure it is helpful to briefly compare the behavior of a nondeformable element of a moving rigid body with a deformable element of a fluid in motion. As shown in Figure 10.33, the nondeformable element of the rigid body translates and rotates as the rigid body itself translates and rotates. A deformable fluid element also translates and rotates in a flow, but as shown in Figure 10.34, it will also experience deformation owing to a change in the relative positions of its constituent fluid particles.

The deformation of a fluid element can be divided into two types: expansion and shear deformation. Expansion is defined to be a relative motion of fluid particles that

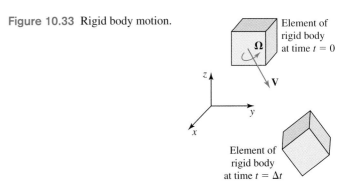

Figure 10.33 Rigid body motion.

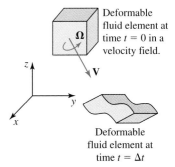

Figure 10.34 Motion of a deforming body such as fluids.

increases (or decreases) their average separation distance without change of shape or rotation. As illustrated in Figure 10.35A, this effect causes a net volume change for a fluid element. If the volume change is positive, a fluid element expands and its density decreases. If the volume change is negative, a fluid element contracts and its density increases. Expansion cannot take place unless the density of the fluid changes. Thus, expansion is absent in all liquid flows and in any other flow that is considered to be incompressible.

As shown in Figure 10.35B, shear deformation is the relative movement of fluid particles in the form of the sliding of one layer of fluid past another. In pure shear deformation, discussed in more detail shortly, the fluid element changes shape, but its volume is unchanged and it does not rotate. Recall from Newton's law of viscosity that shear stress is proportional to the shear strain rate. Therefore, the presence of shear stress indicates shear deformation or strain.

We will now consider four simple example flows that illustrate the effects of the basic types of fluid motion: translation, rotation, expansion, and shear deformation. A

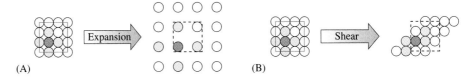

Figure 10.35 Deformation due to (A) expansion and (B) shear.

10.7 MOTION AND DEFORMATION

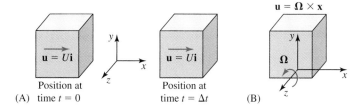

(A) Position at time $t = 0$ Position at time $t = \Delta t$ (B)

Figure 10.36 Rigid body (A) translation and (B) rotation.

uniform and purely translational fluid motion occurs when fluid is placed in a closed container and the entire container is moved at a constant translational velocity U, as shown in Figure 10.36A. After a time, viscosity will cause any initial motion of the fluid relative to its container to cease. The steady velocity field in the fluid will then be given by

$$\mathbf{u} = U\mathbf{i} \quad (10.31)$$

Each fluid element simply moves in the direction of the flow without rotation, change of volume, or change of shape. It can be shown that in a mathematical sense, there is no rotation, expansion, or shear deformation in this uniform translational velocity field because there is no spatial variation in the velocity field. Alternatively phrased, since all partial derivatives of the velocity components with respect to the spatial variables are zero, the types of motion known as rotation, expansion, and shear are absent.

 CD/Kinematics/Two dimensional flow and vorticity/Vorticity and rotation

We can create a velocity field that is purely rotational if we spin the same container at a constant angular velocity $\mathbf{\Omega}_0 = (\Omega_{0x}, \Omega_{0y}, \Omega_{0z})$. Viscosity will damp out any relative motion, and every fluid particle will orbit about the axis of rotation at a speed proportional to its distance from the axis. The corresponding velocity field in the fluid at position $\mathbf{x} = (x, y, z)$ is

$$\mathbf{u} = \mathbf{\Omega} \times \mathbf{x} \quad (10.32a)$$

with velocity components

$$u = \Omega_{0y}z - \Omega_{0z}y, \quad v = \Omega_{0z}x - \Omega_{0x}z, \quad w = \Omega_{0x}y - \Omega_{0y}x \quad (10.32b)$$

This steady velocity field is referred to as rigid body rotation of the fluid. There is no net translation, since every fluid element travels on a circular path, and each element rotates at the same angular velocity without expansion or deformation.

Figure 10.36B illustrates rigid body rotation about the z axis with $\mathbf{\Omega}_0 = (0, 0, \Omega_0)$. This flow has velocity components $u = -\Omega_0 y$, $v = \Omega_0 x$, $w = 0$. It may look more familiar to you when expressed in cylindrical coordinates as

$$v_r = 0, \quad v_\theta = r\Omega_0, \quad v_z = 0 \quad (10.33)$$

This flow, like all others with a rotational component, has spatial derivatives of its velocity components. In fact, we can calculate that $\partial u/\partial y = -\Omega_0$ and $\partial v/\partial x = \Omega_0$.

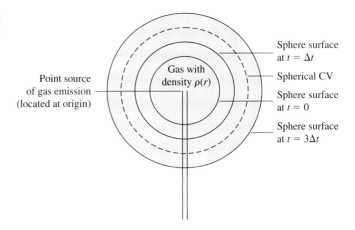

Figure 10.37 Gas released from a point source undergoes expansion and translation as it flows out radially in all directions.

A flow that exhibits expansion (and some translation) results from a steady release of gas from a point source as illustrated in Figure 10.37. As the gas flows out radially from the source, its density will decrease and the volume of a fluid element will increase. Given a spherical CV of radius r surrounding the source, and with the mass flowrate crossing any spherical surface surrounding the emission point defined as $\dot{m}$, a mass balance shows that

$$\rho(r)v_r(r) = \frac{\dot{m}}{4\pi r^2} \qquad (10.34)$$

where $\rho(r)$ and $v_r(r)$ are the density and radial velocity in spherical coordinates. Although the exact form of the functions $\rho(r)$ and $v_r(r)$ depends on the thermodynamics of the expansion process, there is no rotation or shear deformation in this source flow. This flow is not a pure expansion because every fluid element but the one centered on the origin is also translating. The flow is useful, however, in illustrating the concept of expansion and accompanying density variation. Notice also that this flow has a spatial derivative of the radial velocity component, so expansion, like rotation, is connected to the presence of a spatial variation of velocity.

It is difficult to describe a simple flow exhibiting shear deformation unless we limit consideration to 2D flows. The flow of constant density fluid shown in Figure 10.38A is described in Cartesian coordinates by

$$u = Cy, \qquad v - Cx, \qquad w = 0 \qquad (10.35)$$

where C is a constant. As shown in Figure 10.38B, this flow causes a fluid element located at the origin to undergo pure shear deformation. Fluid elements at other locations experience translation and shear deformation but not expansion or rotation. The flow has velocity variations as described by $\partial u/\partial y = C$ and $\partial v/\partial x = C$.

It should be apparent from our four example flows that rotation, expansion, and shear deformation are all connected to spatial variations in a velocity field. However, we can often obtain a good understanding of a flow without a lot of mathematics by applying what we have learned from the four basic flows and an understanding of particle paths. For example, consider the flow created when a fluid is sheared in a narrow gap between two large plates, as shown in Figure 10.39. In this familiar example the lower

10.7 MOTION AND DEFORMATION

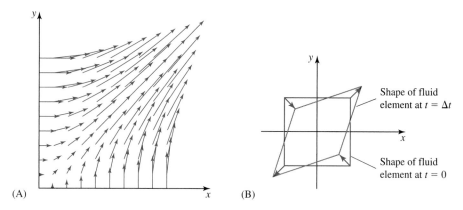

Figure 10.38 (A) Velocity vectors for the flow $u = Cy$, $v = Cx$. (B) the resulting deformation of a fluid element.

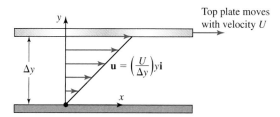

Figure 10.39 Shear flow between parallel plates.

plate is stationary, and the upper plate is moving at a constant velocity U. Fluid dragged in the direction of the top plate by the action of viscosity will create the velocity field $u = (U/\Delta y)\, y$, $v = 0$, $w = 0$, as shown in Figure 10.40. Does it appear to you that the shearing action of the top plate on the fluid will cause translation? (Think about particle paths to see that fluid particles will translate to the right.) Are there expansion, rotation, and shear deformation of a fluid element? This is not an easy question to answer, is it?

The presence of a velocity variation in this flow tells us that expansion, rotation, and shear deformation may be present. Consider rotation first, and examine the cylindrical fluid element shown in Figure 10.40. The sliding motion of the upper plate will be transmitted through successive horizontal layers of the fluid by the action of viscosity.

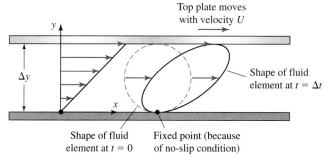

Figure 10.40 Deformation due to a shear flow between parallel plates.

Since the sliding motion disappears at the fixed bottom plate, layers near the top plate slide faster than layers near the bottom plate, and the top of the cylinder will move to the right faster than the bottom of the cylinder. The net effect of this relative motion definitely looks as if it will cause rotation of the cylinder. This same relative motion will also change the shape of the cylinder and drag its "center" to the right. Thus, we conclude that translation, rotation, and shear deformation are all present in parallel plate flow. What about expansion? Although it may not be obvious, a bit of work tracing particle paths shows that the volume of any fluid element is unchanging. Therefore expansion is absent in this flow. In Chapter 11 we will show how to use the continuity equation to determine whether expansion is present in a particular flow.

Flows of engineering interest are generally more complex than those in the preceding examples and do not lend themselves to a visual/graphical analysis. This complexity motivates us to ask the following question: Is there a way to analyze the fluid velocity field to determine the types of motion and underlying flow structure? If so, the mathematical analysis must examine the velocity variations in a flow, since it is spatial variations in the velocity field that give rise to rotation, expansion, and shear deformation.

10.8 VELOCITY GRADIENT

Consider 3D flow in Cartesian coordinates. There are three velocity components, and each of these components may vary in each of the three coordinate directions. To measure this variation completely requires nine partial derivatives, which together make up the velocity gradient, $\nabla \mathbf{u}$. The velocity gradient is a tensor, a doubly directional quantity like the stress tensor; hence it is conveniently represented in Cartesian coordinates by the following array:

$$\nabla \mathbf{u} = \begin{pmatrix} \dfrac{\partial u}{\partial x} & \dfrac{\partial v}{\partial x} & \dfrac{\partial w}{\partial x} \\ \dfrac{\partial u}{\partial y} & \dfrac{\partial v}{\partial y} & \dfrac{\partial w}{\partial y} \\ \dfrac{\partial u}{\partial z} & \dfrac{\partial v}{\partial z} & \dfrac{\partial w}{\partial z} \end{pmatrix} \qquad (10.36)$$

In cylindrical coordinates the velocity gradient is given by

$$\nabla \mathbf{u} = \begin{pmatrix} \dfrac{\partial v_r}{\partial r} & \dfrac{\partial v_\theta}{\partial r} & \dfrac{\partial v_z}{\partial r} \\ \left(\dfrac{1}{r}\dfrac{\partial v_r}{\partial \theta} - \dfrac{v_\theta}{r}\right) & \left(\dfrac{1}{r}\dfrac{\partial v_\theta}{\partial \theta} + \dfrac{v_r}{r}\right) & \dfrac{1}{r}\left(\dfrac{\partial v_z}{\partial \theta}\right) \\ \dfrac{\partial v_r}{\partial z} & \dfrac{\partial v_\theta}{\partial z} & \dfrac{\partial v_z}{\partial z} \end{pmatrix} \qquad (10.37)$$

If we know the velocity field analytically, experimentally, or numerically, we can calculate the velocity gradient by taking the indicated derivatives.

EXAMPLE 10.10

Find the velocity gradient in the flow between parallel plates with the top plate moving as shown in Figure 10.41. The velocity field is given by $\mathbf{u} = (U/2)(y/h + 1)\mathbf{i}$, where U is the velocity of the moving plate and $2h$ is the distance between the plates.

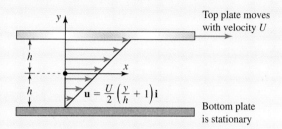

Figure 10.41 Schematic for Example 10.10.

SOLUTION

This exercise is solved by using Eq. 10.36. For this flow we have

$$u = \frac{U}{2}\left(\frac{y}{h} + 1\right), \quad v = 0, \quad \text{and} \quad w = 0$$

Substituting these values into Eq. 10.36 and taking the appropriate derivatives yields

$$\nabla \mathbf{u} = \begin{pmatrix} \frac{\partial u}{\partial x} & \frac{\partial v}{\partial x} & \frac{\partial w}{\partial x} \\ \frac{\partial u}{\partial y} & \frac{\partial v}{\partial y} & \frac{\partial w}{\partial y} \\ \frac{\partial u}{\partial z} & \frac{\partial v}{\partial z} & \frac{\partial w}{\partial z} \end{pmatrix} = \begin{pmatrix} 0 & 0 & 0 \\ \frac{U}{2h} & 0 & 0 \\ 0 & 0 & 0 \end{pmatrix}$$

Thus, the only nonzero term in the velocity gradient is the 2,1 element given by $(U/2h)\mathbf{ji}$.

The physical interpretation of this result is that the x component of velocity is changing in the y direction at the rate of $U/2h$. The remaining elements are zero because the v and w velocity components are zero, and the u velocity component does not vary in the x or z directions.

As suggested earlier, information about the rotation, expansion, and shear deformation in a flow is contained in the partial derivatives making up the velocity gradient. In advanced texts it is shown that $\nabla \mathbf{u}$ itself can be written as the sum of three quantities called the rate of rotation tensor, $\mathbf{R}$, the rate of expansion tensor, $\mathbf{E}$, and the rate of shear deformation tensor, $\mathbf{D}$. Thus we can write

$$\nabla \mathbf{u} = \mathbf{R} + \mathbf{E} + \mathbf{D} \tag{10.38}$$

EXAMPLE 10.11

Poiseuille flow through a steadily rotating circular pipe is illustrated in Figure 10.42. The corresponding velocity field is given by

$$v_r = 0, \quad v_\theta = r\Omega_0, \quad \text{and} \quad v_z = \frac{(p_1 - p_2)R^2}{4\mu L}\left[1 - \left(\frac{r}{R}\right)^2\right]$$

where Ω_0 is the angular velocity of the pipe, R is the pipe radius, and the pressures are measured a distance L apart on the wall. Find the velocity gradient for this flow.

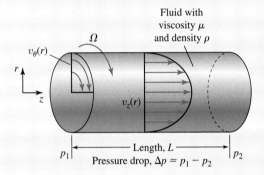

Figure 10.42 Schematic for Example 10.11.

SOLUTION

This exercise is solved by using Eq. 10.37. Substituting the velocity components into Eq. 10.37 and taking the appropriate derivatives yields:

$$\nabla \mathbf{u} = \begin{pmatrix} \dfrac{\partial v_r}{\partial r} & \dfrac{\partial v_\theta}{\partial r} & \dfrac{\partial v_z}{\partial r} \\ \left(\dfrac{1}{r}\dfrac{\partial v_r}{\partial \theta} - \dfrac{v_\theta}{r}\right) & \left(\dfrac{1}{r}\dfrac{\partial v_\theta}{\partial \theta} + \dfrac{v_r}{r}\right) & \dfrac{1}{r}\left(\dfrac{\partial v_z}{\partial \theta}\right) \\ \dfrac{\partial v_r}{\partial z} & \dfrac{\partial v_\theta}{\partial z} & \dfrac{\partial v_z}{\partial z} \end{pmatrix} = \begin{pmatrix} 0 & \Omega_0 & \left[\dfrac{-(p_1 - p_2)r}{2\mu L}\right] \\ -\Omega_0 & 0 & 0 \\ 0 & 0 & 0 \end{pmatrix}$$

There are three nonzero components of the velocity gradient. The 1,2 and 2,1 components arise from the dependence of the θ component of the velocity on r, while the 1,3 component is due to the variation in the axial velocity component in the r direction. The physical interpretation of the latter term is that the z component of velocity is changing in the r direction at the rate of $[-(p_1 - p_2)r/2\mu L]$.

10.8 VELOCITY GRADIENT

In Cartesian coordinates the rate of rotation tensor is given by

$$\mathbf{R} = \begin{pmatrix} 0 & \frac{1}{2}\left(\frac{\partial v}{\partial x} - \frac{\partial u}{\partial y}\right) & \frac{1}{2}\left(\frac{\partial w}{\partial x} - \frac{\partial u}{\partial z}\right) \\ \frac{1}{2}\left(\frac{\partial u}{\partial y} - \frac{\partial v}{\partial x}\right) & 0 & \frac{1}{2}\left(\frac{\partial w}{\partial y} - \frac{\partial v}{\partial z}\right) \\ \frac{1}{2}\left(\frac{\partial u}{\partial z} - \frac{\partial w}{\partial x}\right) & \frac{1}{2}\left(\frac{\partial v}{\partial z} - \frac{\partial w}{\partial y}\right) & 0 \end{pmatrix} \quad (10.39)$$

In the next section we show that the rotation tensor is the part of the velocity gradient that causes a fluid element to rotate without changing shape or expanding.

The rate of expansion tensor $\mathbf{E}$ represents the part of the velocity gradient that causes a fluid element to change its volume without changing its original shape or rotating. This tensor, which is discussed in more detail in Section 10.10, is defined in Cartesian coordinates by

$$\mathbf{E} = \begin{bmatrix} \frac{1}{3}\left(\frac{\partial u}{\partial x} + \frac{\partial v}{\partial y} + \frac{\partial w}{\partial z}\right) & 0 & 0 \\ 0 & \frac{1}{3}\left(\frac{\partial u}{\partial x} + \frac{\partial v}{\partial y} + \frac{\partial w}{\partial z}\right) & 0 \\ 0 & 0 & \frac{1}{3}\left(\frac{\partial u}{\partial x} + \frac{\partial v}{\partial y} + \frac{\partial w}{\partial z}\right) \end{bmatrix} \quad (10.40)$$

Finally, the rate of shear deformation tensor, $\mathbf{D}$, or the shear deformation rate, is the part of the velocity gradient responsible for changing the shape of a fluid element without a change in volume or rotation. This tensor, which is discussed in Section 10.11, is given in Cartesian coordinates by

$$\mathbf{D} = \begin{bmatrix} \frac{\partial u}{\partial x} - \frac{1}{3}\left(\frac{\partial u}{\partial x} + \frac{\partial v}{\partial y} + \frac{\partial w}{\partial z}\right) & \frac{1}{2}\left(\frac{\partial u}{\partial y} + \frac{\partial v}{\partial x}\right) & \frac{1}{2}\left(\frac{\partial u}{\partial z} + \frac{\partial w}{\partial x}\right) \\ \frac{1}{2}\left(\frac{\partial v}{\partial x} + \frac{\partial u}{\partial y}\right) & \frac{\partial v}{\partial y} - \frac{1}{3}\left(\frac{\partial u}{\partial x} + \frac{\partial v}{\partial y} + \frac{\partial w}{\partial z}\right) & \frac{1}{2}\left(\frac{\partial v}{\partial z} + \frac{\partial w}{\partial y}\right) \\ \frac{1}{2}\left(\frac{\partial w}{\partial x} + \frac{\partial u}{\partial z}\right) & \frac{1}{2}\left(\frac{\partial w}{\partial y} + \frac{\partial v}{\partial z}\right) & \frac{\partial w}{\partial z} - \frac{1}{3}\left(\frac{\partial u}{\partial x} + \frac{\partial v}{\partial y} + \frac{\partial w}{\partial z}\right) \end{bmatrix} \quad (10.41)$$

Notice that each of these quantities is an array of combinations of partial derivatives of velocity components. Furthermore, a term by term addition of the elements in the three arrays adds up to the corresponding element in the velocity gradient.

Fluid mechanics also makes use of a quantity known as the rate of strain tensor $\mathbf{S}$. This quantity is responsible for the overall tendency of a fluid element to be strained or deformed by spatial variations in the velocity field. In general, this effect simultaneously alters the shape and volume of a fluid element. By definition, the rate of strain tensor is the sum of the rate of expansion and the rate of shear deformation. Thus we have $\mathbf{S} = \mathbf{E} + \mathbf{D}$, and by using Eqs. 10.40 and 10.41 we see that $\mathbf{S}$ can be written as

$$\mathbf{S} = \begin{bmatrix} \dfrac{\partial u}{\partial x} & \dfrac{1}{2}\left(\dfrac{\partial v}{\partial x} + \dfrac{\partial u}{\partial y}\right) & \dfrac{1}{2}\left(\dfrac{\partial w}{\partial x} + \dfrac{\partial u}{\partial z}\right) \\ \dfrac{1}{2}\left(\dfrac{\partial u}{\partial y} + \dfrac{\partial v}{\partial x}\right) & \dfrac{\partial v}{\partial y} & \dfrac{1}{2}\left(\dfrac{\partial w}{\partial y} + \dfrac{\partial v}{\partial z}\right) \\ \dfrac{1}{2}\left(\dfrac{\partial u}{\partial z} + \dfrac{\partial w}{\partial x}\right) & \dfrac{1}{2}\left(\dfrac{\partial v}{\partial z} + \dfrac{\partial w}{\partial y}\right) & \dfrac{\partial w}{\partial z} \end{bmatrix} \qquad (10.42)$$

Versions of the rotation, expansion, shear deformation, and rate of strain tensors in cylindrical coordinates are not given in this book but can be found in many advanced texts. If we know the velocity field, we can always calculate these three quantities. This is illustrated in the following examples.

EXAMPLE 10.12

The velocity field for pressure-driven flow in the gap between stationary parallel plates is shown in Figure 10.43. This flow is described in Cartesian coordinates by

$$u(y) = \left(\dfrac{h^2(p_1 - p_2)}{2\mu L}\right)\left[1 - \left(\dfrac{y}{h}\right)^2\right], \quad v = 0, \quad \text{and} \quad w = 0$$

where $2h$ is the channel height, and the pressures are measured a distance L apart, as shown in Figure 10.43. Determine the rate of strain and rate of rotation in this flow.

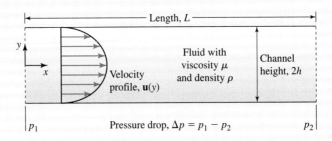

Figure 10.43 Schematic for Example 10.12.

SOLUTION

We can use Eq. 10.42, to calculate the rate of strain by inserting the various partial derivatives into the array describing the rate of strain. The result is

$$\mathbf{S} = \begin{pmatrix} 0 & \dfrac{-(p_1 - p_2)y}{2\mu L} & 0 \\ \dfrac{-(p_1 - p_2)y}{2\mu L} & 0 & 0 \\ 0 & 0 & 0 \end{pmatrix}$$

Next, from Eq. 10.39, the rate of rotation is found to be

$$\mathbf{R} = \begin{pmatrix} 0 & \dfrac{(p_1 - p_2)y}{2\mu L} & 0 \\ \dfrac{-(p_1 - p_2)y}{2\mu L} & 0 & 0 \\ 0 & 0 & 0 \end{pmatrix}$$

We can check this result by using Eq. 10.36 to calculate the velocity gradient. The result is:

$$\nabla \mathbf{u} = \begin{pmatrix} 0 & 0 & 0 \\ \dfrac{-(p_1 - p_2)y}{\mu L} & 0 & 0 \\ 0 & 0 & 0 \end{pmatrix}$$

which we see is also the result obtained by forming the term-by-term sum of $\mathbf{S}$ and $\mathbf{R}$.

EXAMPLE 10.13

A flow field given by $\mathbf{u} = (x^2 y)\mathbf{i} + (-xy^2)\mathbf{j} + 0\mathbf{k}$ is illustrated in Figure 10.44. Find the velocity gradient, rate of strain, rate of rotation, rate of expansion, and rate of shear deformation in this flow.

SOLUTION

The velocity gradient is found by using Eq. 10.36:

$$\nabla \mathbf{u} = \begin{pmatrix} 2xy & -y^2 & 0 \\ x^2 & -2xy & 0 \\ 0 & 0 & 0 \end{pmatrix}$$

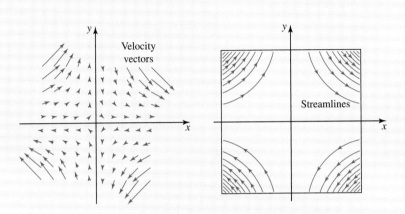

Figure 10.44 Schematic of the velocity field and streamlines for Example 10.13.

Next we use Eq. 10.42 to calculate the rate of strain, and Eq. 10.39 to calculate the rate of rotation to find

$$\mathbf{S} = \begin{pmatrix} 2xy & \frac{1}{2}(x^2 - y^2) & 0 \\ \frac{1}{2}(x^2 - y^2) & -2xy & 0 \\ 0 & 0 & 0 \end{pmatrix} \quad \text{and} \quad \mathbf{R} = \begin{pmatrix} 0 & -\frac{1}{2}(x^2 + y^2) & 0 \\ \frac{1}{2}(x^2 + y^2) & 0 & 0 \\ 0 & 0 & 0 \end{pmatrix}$$

The rate of expansion is calculated by using Eq. 10.40. From the velocity components we find $\frac{1}{3}(\partial u/\partial x + \partial v/\partial y + \partial w/\partial z) = \frac{1}{3}(2xy - 2xy + 0) = 0$. Thus the rate of expansion is

$$\mathbf{E} = \begin{bmatrix} \frac{1}{3}\left(\frac{\partial u}{\partial x} + \frac{\partial v}{\partial y} + \frac{\partial w}{\partial z}\right) & 0 & 0 \\ 0 & \frac{1}{3}\left(\frac{\partial u}{\partial x} + \frac{\partial v}{\partial y} + \frac{\partial w}{\partial z}\right) & 0 \\ 0 & 0 & \frac{1}{3}\left(\frac{\partial u}{\partial x} + \frac{\partial v}{\partial y} + \frac{\partial w}{\partial z}\right) \end{bmatrix} = \begin{pmatrix} 0 & 0 & 0 \\ 0 & 0 & 0 \\ 0 & 0 & 0 \end{pmatrix}$$

Finally, by using Eq. 10.41, we calculate the shear deformation rate as

$$\mathbf{D} = \begin{pmatrix} 2xy & \frac{1}{2}(x^2 - y^2) & 0 \\ \frac{1}{2}(x^2 - y^2) & -2xy & 0 \\ 0 & 0 & 0 \end{pmatrix}$$

We see that the rate of expansion is zero in this flow, while the rate of rotation and the shear deformation rate are nonzero.

10.9 RATE OF ROTATION

In this section we will explore the characteristics of the rate of rotation tensor in more detail. In Cartesian coordinates the rate of rotation is given by Eq. 10.39 as

$$\mathbf{R} = \begin{pmatrix} 0 & \frac{1}{2}\left(\frac{\partial v}{\partial x} - \frac{\partial u}{\partial y}\right) & \frac{1}{2}\left(\frac{\partial w}{\partial x} - \frac{\partial u}{\partial z}\right) \\ \frac{1}{2}\left(\frac{\partial u}{\partial y} - \frac{\partial v}{\partial x}\right) & 0 & \frac{1}{2}\left(\frac{\partial w}{\partial y} - \frac{\partial v}{\partial z}\right) \\ \frac{1}{2}\left(\frac{\partial u}{\partial z} - \frac{\partial w}{\partial x}\right) & \frac{1}{2}\left(\frac{\partial v}{\partial z} - \frac{\partial w}{\partial y}\right) & 0 \end{pmatrix}$$

Note that the diagonal terms are all zero. In addition, each element below the diagonal is simply the negative of the corresponding element above. Thus, there are only three independent elements in this tensor. From vector calculus we know that the curl of the velocity vector is defined to be $\nabla \times \mathbf{u} = (\partial w/\partial y - \partial v/\partial z)\mathbf{i} + (\partial u/\partial z - \partial w/\partial x)\mathbf{j} + (\partial v/\partial x - \partial u/\partial y)\mathbf{k}$. Examining the various elements in the rate of rotation, we see that we can write $\mathbf{R}$ in terms of the three components of $\nabla \times \mathbf{u}$ as

$$\mathbf{R} = \begin{pmatrix} 0 & \frac{1}{2}[\nabla \times \mathbf{u}]_z & -\frac{1}{2}[\nabla \times \mathbf{u}]_y \\ -\frac{1}{2}[\nabla \times \mathbf{u}]_z & 0 & \frac{1}{2}[\nabla \times \mathbf{u}]_x \\ \frac{1}{2}[\nabla \times \mathbf{u}]_y & -\frac{1}{2}[\nabla \times \mathbf{u}]_x & 0 \end{pmatrix} \quad (10.43)$$

This result suggests that the rate of rotation is closely related to the curl of the velocity field.

What characteristic of a flow does the curl measure? Consider the special case of a fluid in rigid body rotation at angular velocity $\mathbf{\Omega}_0$. As discussed in Section 10.7, the angular velocity vector in a rigid body rotation is the same at each point in the flow, and the velocity field is given by Eq. 10.32a as $\mathbf{u} = \mathbf{\Omega}_0 \times \mathbf{x}$, where $\mathbf{\Omega}_0 = (\Omega_{0x}, \Omega_{0y}, \Omega_{0z})$ is a constant vector. Equation 10.32b gives the corresponding velocity components as $u = \Omega_{0y}z - \Omega_{0z}y$, $v = \Omega_{0z}x - \Omega_{0x}z$, $w = \Omega_{0x}y - \Omega_{0y}x$. The corresponding velocity vector is

$$\mathbf{u} = (\Omega_{0y}z - \Omega_{0z}y)\mathbf{i} + (\Omega_{0z}x - \Omega_{0x}z)\mathbf{j} + (\Omega_{0x}y - \Omega_{0y}x)\mathbf{k} \quad (10.44a)$$

and taking the curl of this vector, we find

$$\nabla \times \mathbf{u} = \nabla \times (\mathbf{\Omega}_0 \times \mathbf{x}) = 2\Omega_{0x}\mathbf{i} + 2\Omega_{0y}\mathbf{j} + 2\Omega_{0z}\mathbf{k} = 2\mathbf{\Omega}_0 \quad (10.44b)$$

This important result demonstrates that for a fluid in rigid body rotation at angular velocity $\mathbf{\Omega}_0$, the curl of the velocity field is a vector that points in the same direction as the angular velocity vector and has twice its magnitude.

Because a fluid is a deformable material, different parts of moving fluid generally rotate with different angular velocity vectors. Have you ever observed eddy motions in a river or the chaotic swirls in the gas leaving a smokestack? Each eddy or swirl is a rotating volume of fluid with a distinct rotation rate and axis of rotation (Figure 10.45).

Figure 10.45 Eddys and swirls: (A) in a river and (B) from a smokestack.

To characterize the rotation in a flow, we define the angular velocity vector, or rotation vector, $\mathbf{\Omega}(x, t)$, at a point in a fluid as

$$\mathbf{\Omega} = \tfrac{1}{2} \nabla \times \mathbf{u} \tag{10.45}$$

Notice that this definition of rotation allows the rotation vector to be a function of position and time, but reproduces Eq. 10.44b when applied to the velocity field in the special case of rigid body rotation. We may interpret the value of the rotation vector at position $\mathbf{x}$ in a fluid at time t in terms of the fluid particle that happens to be there. This fluid particle, if thought of as a tiny rigid body, would rotate with an angular velocity vector equal to the rotation vector calculated by using Eq. 10.45. The sign convention for an angular velocity vector is illustrated in Figure 10.46. In general, the resulting rotation is clockwise when looking in the direction of $\mathbf{\Omega}$.

In Cartesian coordinates the three components of the rotation vector are

$$\Omega_x = \frac{1}{2}\left(\frac{\partial w}{\partial y} - \frac{\partial v}{\partial z}\right), \quad \Omega_y = \frac{1}{2}\left(\frac{\partial u}{\partial z} - \frac{\partial w}{\partial x}\right), \quad \text{and} \quad \Omega_z = \frac{1}{2}\left(\frac{\partial v}{\partial x} - \frac{\partial u}{\partial y}\right) \tag{10.46a–c}$$

while in cylindrical coordinates, they are

$$\Omega_r = \frac{1}{2}\left(\frac{1}{r}\frac{\partial v_z}{\partial \theta} - \frac{\partial v_\theta}{\partial z}\right), \quad \Omega_\theta = \frac{1}{2}\left(\frac{\partial v_r}{\partial z} - \frac{\partial v_z}{\partial r}\right), \quad \text{and} \quad \Omega_z = \frac{1}{2}\left[\frac{1}{r}\left(\frac{\partial(rv_\theta)}{\partial r} - \frac{\partial v_r}{\partial \theta}\right)\right] \tag{10.47a–c}$$

Figure 10.46 Sign convention for the angular velocity vector: CW, clockwise; CCW, counterclockwise.

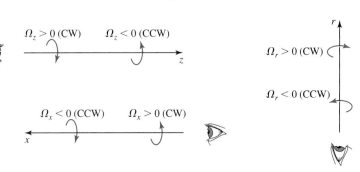

10.9 RATE OF ROTATION

Substituting the components of the rotation vector defined by Eqs. 10.46 into the rate of rotation tensor (Eq. 10.39) shows that we can write the latter in Cartesian coordinates as

$$\mathbf{R} = \begin{pmatrix} 0 & \Omega_z & -\Omega_y \\ -\Omega_z & 0 & \Omega_x \\ \Omega_y & -\Omega_x & 0 \end{pmatrix} \qquad (10.48)$$

We see that the rate of rotation tensor is completely determined by the three components of the rotation vector. A similar relationship holds in cylindrical coordinates.

EXAMPLE 10.14

Find the rotation vector in the flow between parallel plates with the top plate moving as shown in Figure 10.47. Recall that the velocity field for this flow is $\mathbf{u} = (U/2)(y/h + 1)\mathbf{i}$, where U is the velocity of the moving plate and $2h$ is the distance between the plates. Give the rotation and rate of rotation tensor in this flow.

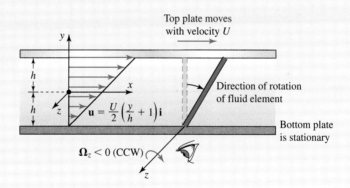

Figure 10.47 Schematic for Example 10.14.

SOLUTION

The first part of this exercise is solved by using Eq. 10.46. For this flow we have

$$u = \frac{U}{2}\left(\frac{y}{h} + 1\right), \quad v = 0, \quad \text{and} \quad w = 0$$

Substituting these values into Eqs. 10.46 and taking the appropriate derivatives yields:

$$\Omega_x = \frac{1}{2}\left(\frac{\partial w}{\partial y} - \frac{\partial v}{\partial z}\right) = \frac{1}{2}(0 - 0) = 0, \quad \Omega_y = \frac{1}{2}\left(\frac{\partial u}{\partial z} - \frac{\partial w}{\partial x}\right) = \frac{1}{2}(0 - 0) = 0$$

and

$$\Omega_z = \frac{1}{2}\left(\frac{\partial v}{\partial x} - \frac{\partial u}{\partial y}\right) = \frac{1}{2}\left(0 - \frac{U}{2h}\right) = -\frac{U}{4h}$$

Thus the rotation vector is $\mathbf{\Omega} = 0\mathbf{i} + 0\mathbf{j} - (U/4h)\mathbf{k}$. To interpret this result, notice that in Figure 10.47 the z axis is pointing out of the paper. Thus, since $\mathbf{\Omega} = 0\mathbf{i} + 0\mathbf{j} - (U/4h)\mathbf{k}$ the z-component of rotation is negative and we should expect to "see" a counter clockwise rotation when looking in the $+z$ direction. If we focus our attention on a thin fluid material volume originally aligned with the y axis, we see that as time passes this fluid element does indeed rotate in a counter clockwise direction as viewed in the $+z$ direction. Thus, our result appears reasonable and consistent with our discussion of this same flow in the final example of Section 10.7. Recall that in that discussion we considered the effect of the velocity field on a small cylinder of fluid.

To calculate the rate of rotation tensor, we can use Eq. 10.48, substituting the components of the rotation vector to obtain

$$\mathbf{R} = \begin{pmatrix} 0 & \Omega_z & -\Omega_y \\ -\Omega_z & 0 & \Omega_x \\ \Omega_y & -\Omega_x & 0 \end{pmatrix} = \begin{pmatrix} 0 & -\dfrac{U}{4h} & 0 \\ \dfrac{U}{4h} & 0 & 0 \\ 0 & 0 & 0 \end{pmatrix}$$

We could also calculate the rate of rotation directly by using Eq. 10.39.

 CD/Kinematics/Pathlines/Examples of complex flows

 CD/Kinematics/Two dimensional flow and vorticity

10.9.1 Vorticity

The term "vortex," designating a region of concentrated rotation in a flow, should be familiar to you. According to the dictionary, a vortex is "an eddy, whirlpool, the depression at the center of a whirling body of air or water." Naturally occurring vortices include hurricanes, tornadoes, waterspouts, and dust devils. Vortices also occur on a smaller scale in the wake of an object—for example, when a paddle is dragged through water or a spoon moved through a cup of tea. It is fair to say that vorticity (and rotation) is almost always present in a moving fluid. This is true even though an organized vortex is not always evident.

EXAMPLE 10.15

Find and interpret the rotation vector in Poiseuille flow through a rotating circular pipe. The velocity field is given in cylindrical coordinates by

$$v_r = 0, \quad v_\theta = r\Omega_0, \quad \text{and} \quad v_z = \frac{(p_1 - p_2)R^2}{4\mu L}\left[1 - \left(\frac{r}{R}\right)^2\right]$$

SOLUTION

Substituting the velocity components into Eqs. 10.47 we find

$$\Omega_r = \frac{1}{2}\left(\frac{1}{r}\frac{\partial v_z}{\partial \theta} - \frac{\partial v_\theta}{\partial z}\right) = \frac{1}{2}\left(\frac{1}{r}(0) - 0\right) = 0$$

$$\Omega_\theta = \frac{1}{2}\left(\frac{\partial v_r}{\partial z} - \frac{\partial v_z}{\partial r}\right) = \frac{1}{2}\left[0 - \left(-\frac{(p_1 - p_2)r}{2\mu L}\right)\right] = \frac{(p_1 - p_2)r}{4\mu L}$$

$$\Omega_z = \frac{1}{2}\left[\frac{1}{r}\left(\frac{\partial(rv_\theta)}{\partial r} - \frac{\partial v_r}{\partial \theta}\right)\right] = \frac{1}{2}\left[\frac{1}{r}\left(\frac{\partial(r^2\Omega_0)}{\partial r} - 0\right)\right] = \Omega_0$$

The rotation vector in this flow is $\mathbf{\Omega} = 0\mathbf{e}_r + [(p_1 - p_2)r/4\mu L]\mathbf{e}_\theta + \Omega_0 \mathbf{e}_z$. To interpret this result, notice that in Figure 10.48 the z axis is pointing to the right along the axis of the pipe. Since the fluid is rotating while simultaneously moving down the axis of the pipe, the rotation vector has a component $\Omega_z = \Omega_0$, as expected. If we focus our attention on the thin fluid material volume originally aligned in the radial direction with $\theta = 0$, we see that as time passes this fluid element rotates in a clockwise direction as viewed in the $+\theta$ direction owing to the nonuniform Poiseuille velocity profile. Thus, the positive value for $\Omega_\theta = (p_1 - p_2)r/4\mu L$ is correct.

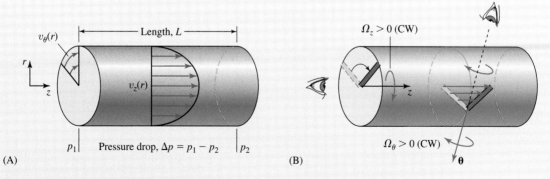

Figure 10.48 Schematic for Example 10.15.

It is customary in fluid mechanics to characterize a velocity field in terms of its vorticity rather than its rotation. The vorticity vector, $\boldsymbol{\omega}$, is defined by

$$\boldsymbol{\omega} = \nabla \times \mathbf{u} \tag{10.49}$$

Comparing this definition of vorticity to that for the rotation vector, Eq. 10.45, we see that these vectors are related by a factor of two:

$$\boldsymbol{\omega} = 2\boldsymbol{\Omega} \tag{10.50}$$

In general, vorticity is a function of position and time. Thus we speak of the vorticity field in the fluid. From Eq. 10.50 and our expanded form for $\boldsymbol{\Omega}$ (Eq. 10.46) we see that the three components of vorticity in Cartesian coordinates are given by

$$\omega_x = \left(\frac{\partial w}{\partial y} - \frac{\partial v}{\partial z}\right), \quad \omega_y = \left(\frac{\partial u}{\partial z} - \frac{\partial w}{\partial x}\right), \quad \text{and} \quad \omega_z = \left(\frac{\partial v}{\partial x} - \frac{\partial u}{\partial y}\right) \tag{10.51a–c}$$

In cylindrical coordinates, the components of vorticity are

$$\omega_r = \left(\frac{1}{r}\frac{\partial v_z}{\partial \theta} - \frac{\partial v_\theta}{\partial z}\right), \quad \omega_\theta = \left(\frac{\partial v_r}{\partial z} - \frac{\partial v_z}{\partial r}\right), \quad \text{and} \quad \omega_z = \frac{1}{r}\left(\frac{\partial (rv_\theta)}{\partial r} - \frac{\partial v_r}{\partial \theta}\right) \tag{10.52a–c}$$

Examination of these equations suggests that a 3D flow will generally have a 3D vorticity field.

In the important case of a planar 2D flow, there is only one nonzero vorticity component. For example, if the Cartesian velocity components are $(u, v, 0)$ and are independent of z, the only component of vorticity in the flow is the z component, given by Eq. 10.51c as $\omega_z = (\partial v/\partial x - \partial u/\partial y)$. This vorticity component represents the tendency of fluid particles to spin about an axis perpendicular to the xy plane of the flow field. Thus, the rotation of a fluid element in a planar 2D flow can only be about an axis perpendicular to the plane of the fluid motion.

EXAMPLE 10.16

Two of your friends are having an argument. They have been asked to find an example of a 2D planar flow in Cartesian coordinates with a nonzero vorticity vector. One claims that the stagnation point flow (see Figure 10.49), with the velocity field $\mathbf{u} = \alpha x \mathbf{i} - \alpha y \mathbf{j} + 0\mathbf{k}$, fits the bill. Another argues that a better example is the shear flow given by the velocity field $\mathbf{u} = Cy\mathbf{i} + Cx\mathbf{j} + 0\mathbf{k}$ (see Eq. 10.35 and Figure 10.38A). Whom will you support in this argument?

SOLUTION

The vorticity vector for a Cartesian flow can be found by using Eqs. 10.51. Examination of these two velocity fields, however, shows that they are both 2D planar flows with u and v independent of z. Therefore, the only component of vorticity that might be nonzero

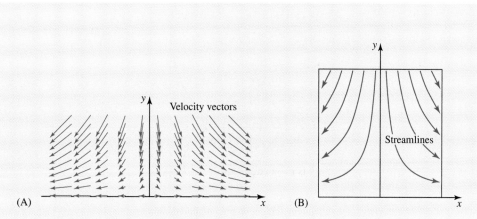

Figure 10.49 Schematic for Example 10.16.

is given by Eq. 10.51c, $\omega_z = (\partial v/\partial x - \partial u/\partial y)$. Evaluating the z components of vorticity for the stagnation point flow and the shear flow gives, respectively,

$$\omega_z = \left(\frac{\partial v}{\partial x} - \frac{\partial u}{\partial y}\right) = (0-0) = 0 \quad \text{and} \quad \omega_z = \left(\frac{\partial v}{\partial x} - \frac{\partial u}{\partial y}\right) = C - C = 0$$

Neither of these 2D flows exhibits a nonzero vorticity vector, so both colleagues are in error. Can you identify an example of a 2D planar flow in Cartesian coordinates with a nonzero vorticity vector? You might want to consider the flows in Examples 10.10 and 10.12.

Earlier we noted that vorticity is almost always present in a fluid in motion. This is because vorticity is generated in a flow field by several mechanisms. The most important of these is the no-slip condition at a solid boundary. To illustrate this effect, consider the constant density 2D flow created by moving a thin flat plate at a constant velocity $-U\mathbf{i}$ through stationary fluid, as shown in Figure 10.50. The vorticity in the fluid far to the left of the moving plate is zero because the fluid velocity is zero there. However, fluid that has passed near the plate has a nonzero vorticity. How is this vorticity created?

To analyze this situation it is helpful to change our reference frame to one fixed to the flat plate. In this frame the flow is steady and the plate is at rest, as shown in Figure 10.51. Do you recognize the flat plate boundary layer discussed in the corresponding case study of Section 3.3.4? The fluid is approaching the plate at a constant velocity $U\mathbf{i}$. In the boundary layer the no-slip condition ensures that there will be a nonzero velocity gradient $\partial u/\partial y$, since the velocity component $u = 0$ on the plate and $u = U$ a short distance away in the y direction outside the boundary layer. The single vorticity component at any point in this 2D flow is given by Eq. 10.51c as $\omega_z = \partial v/\partial x - \partial u/\partial y$. The value of $\partial v/\partial x$ is zero on the wall because v is zero everywhere on the plate by the no-penetration condition, thus the change in v with x must be zero. In the boundary layer, $\partial v/\partial x$ is negligible in comparison to $\partial u/\partial y$, since the boundary is thin and the

Figure 10.50 Velocity field created by a flat plate moving through a stationary fluid.

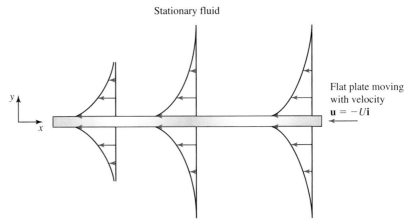

streamlines are nearly parallel to the wall. We therefore anticipate that the maximum vorticity value in the boundary layer will occur at the wall where the velocity gradient $\partial u/\partial y$ is the largest. This maximum value is given by

$$\omega_z|_{\text{wall}} = \left(\frac{\partial v}{\partial x} - \frac{\partial u}{\partial y}\right)\bigg|_{y=0} = -\frac{\partial u}{\partial y}\bigg|_{y=0}$$

The wall shear stress is found for a Newtonian fluid by using Eq. 1.2c to obtain $\tau_{\text{wall}} = \mu(\partial u/\partial y)|_{y=0}$. Comparing this with the vorticity at the wall, we discover that in this flow $\omega_z|_{\text{wall}} = -\tau_{\text{wall}}/\mu$.

An examination of Appendix A shows that in either the SI or BG unit systems, the viscosity is a numerically small number ($\ll 1$) for air, water, and most other fluids. Thus, a small shear stress on the plate is always accompanied by a large vorticity value. We see that when a fluid is moving tangentially relative to a solid boundary, a large value of vorticity will be present at the boundary owing to the no-slip condition. This vorticity simultaneously diffuses into the fluid and is transported downstream. We see that the

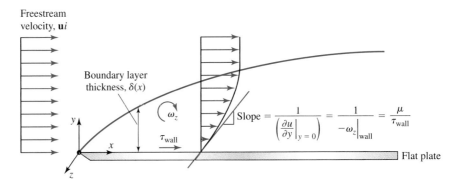

Figure 10.51 Vorticity creation in the boundary layer in flow over a flat plate. The magnitude of the vorticity is dependent on the slope of the velocity field at the wall, $(\partial u/\partial y)|_{y=0}$. In the free stream $(\partial u/\partial y)|_{y>\delta} = 0$ and the vorticity is zero.

EXAMPLE 10.17

Recall that pressure-driven flow between stationary parallel flat plates, illustrated in Figure 10.52A, is described by

$$u(y) = \frac{h^2(p_1 - p_2)}{2\mu L}\left[1 - \left(\frac{y}{h}\right)^2\right], \quad v = 0, \quad \text{and} \quad w = 0$$

Determine the vorticity distribution for this flow and compare the magnitude of the wall shear stress to the vorticity at the top and bottom plates.

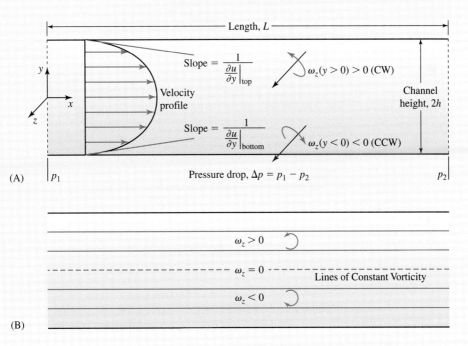

Figure 10.52 Schematic for Example 10.17: Pressure-driven flow between stationary parallel flat plates.

SOLUTION

We will use Eqs. 10.51 to calculate the vorticity, and Eq. 1.2c to calculate the magnitude of the wall shear stress. A straightforward calculation shows that the single component vorticity vector in this flow is $\boldsymbol{\omega} = \omega_z \mathbf{k} = (\partial v/\partial x - \partial u/\partial y)\,\mathbf{k}$ with $\omega_z = (p_1 - p_2)y/\mu L$. Note that the vorticity changes sign in the gap between the plates, meaning that fluid particles in the upper half of the channel rotate in the opposite direction to those in the lower half. Figure 10.52B is a plot of the vorticity contours for this flow generated using the equation above. The contours are horizontal lines.

The vorticity at the top and bottom plates is found to be

$$\omega_z|_{\text{top}} = \left[\frac{(p_1 - p_2)y}{\mu L}\right]\bigg|_{y=h} = \frac{(p_1 - p_2)h}{\mu L}$$

and

$$\omega_z|_{\text{bottom}} = \left[\frac{(p_1 - p_2)y}{\mu L}\right]\bigg|_{y=-h} = \frac{(p_1 - p_2)(-h)}{\mu L} = -\frac{(p_1 - p_2)h}{\mu L}$$

The vorticity has the same magnitude at each plate as expected from the symmetry.

The shear stress at each plate is found by using Eq. 1.2c. By the symmetry the wall shear stress is the same on each plate. We can evaluate it at the bottom plate to obtain:

$$\tau_{\text{wall}} = \mu \frac{\partial u}{\partial y}\bigg|_{y=-h} = \mu \left(\frac{h^2(p_1 - p_2)}{2\mu L}\right)\left(\frac{-2y}{h^2}\right)\bigg|_{y=-h} = \frac{(p_1 - p_2)h}{L}$$

The ratio of the magnitudes of vorticity to the shear stress at each plate is $|\omega_{\text{wall}}|/\tau_{\text{wall}} = 1/\mu$. Since the magnitude of μ is small, the vorticity at each plate is large in comparison to the local shear stress.

layer of fluid that passes near a solid boundary and consequently contains vorticity defines the boundary layer.

It is possible to visualize the vorticity field in a fluid flow by using the same techniques employed to visualize a velocity field. For example, we can use the vector plot of many software packages to represent the vorticity in a 3D flow. A contour plot of the magnitude of vorticity, like the one shown in Figure 10.52B for pressure-driven flow between parallel plates, is especially helpful for sensing where the highest rotation rates occur in a 2D flow. Most commercial CFD packages produce plots of the vorticity.

Velocity and vorticity are examples of vector fields. Recall that streamlines are defined to be lines in a fluid that are everywhere parallel to the velocity vectors. Vortex lines are defined similarly, as lines that are everywhere parallel to the vorticity vectors. Vortex lines generally do not cross one another because that would require that the vorticity field be multivalued at the crossing point, a physical impossibility.

10.9.2 Circulation

Since vorticity is defined by the curl of the velocity field, it is natural to think that Stokes theorem, which involves the curl of a vector field, might provide additional insight into the physical phenomena associated with vorticity. Applying Stokes theorem to the velocity field **u**, we have

$$\oint \mathbf{u} \cdot d\mathbf{r} = \iint_S (\nabla \times \mathbf{u}) \cdot \mathbf{n} \, dS \qquad (10.53a)$$

10.9 RATE OF ROTATION | **629**

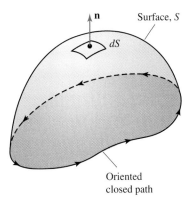

Figure 10.53 The closed path and surface used in the Stokes theorem.

where the line integral is taken around the closed path that bounds the edge of an arbitrarily selected surface drawn in the fluid. The relationship between the orientation of the path and the corresponding unit normal to the surface is shown in Figure 10.53. Introducing the vorticity, we see that the Stokes theorem becomes

$$\oint \mathbf{u} \cdot d\mathbf{r} = \iint_S \boldsymbol{\omega} \cdot \mathbf{n} \, dS \qquad (10.53b)$$

Now consider an arbitrary surface in a fluid in motion and its associated path, oriented as shown in Figure 10.54. If the flow possesses vorticity, then vortex lines may penetrate this surface, and the path will define a vortex tube. A closed path like this in a fluid is said to possess a flow property known as circulation. The circulation, $C(t)$, of a path is defined mathematically by the line integral

$$C(t) = \oint \mathbf{u} \cdot d\mathbf{r} \qquad (10.54)$$

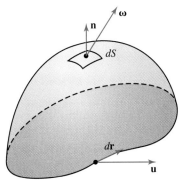

Figure 10.54 The vorticity flux through the surface is related to the circulation of the path.

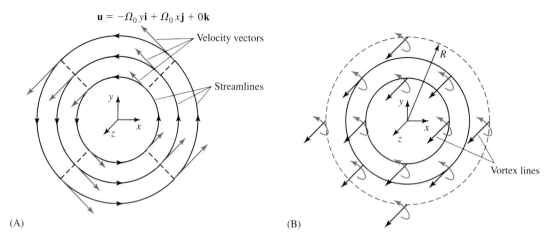

Figure 10.55 (A) Velocity field and streamlines for a fluid in rigid body rotation. (B) The associated vortex lines.

where the integrand is the component of velocity tangent to the path as shown in Figure 10.54. By using the Stokes theorem, we can also write the circulation as

$$C(t) = \iint_S \boldsymbol{\omega} \cdot \mathbf{n}\, dS \qquad (10.55)$$

Thus, the Stokes theorem shows that the circulation around a closed path in a fluid is related to the normal component of vorticity on any surface bounded by the path.

To illustrate the connection between the vorticity flux through a surface and the circulation around the associated path, consider the special case of a fluid in rigid body rotation about the z axis with a constant angular velocity $\boldsymbol{\Omega}_0 = \Omega_0 \mathbf{k}$ and velocity field given by Eq. 10.44a as $\mathbf{u} = (\Omega_{0y}z - \Omega_{0z}y)\mathbf{i} + (\Omega_{0z}x - \Omega_{0x}z)\mathbf{j} + (\Omega_{0x}y - \Omega_{0y}x)\mathbf{k}$. Since $(\Omega_{0x}, \Omega_{0y}, \Omega_{0z}) = (0, 0, \Omega_0)$, we can write the velocity field as $\mathbf{u} = -\Omega_0 y\mathbf{i} + \Omega_0 x\mathbf{j} + 0\mathbf{k}$. (This flow may look more familiar when expressed in cylindrical coordinates as $\mathbf{u} = r\Omega_0 \mathbf{e}_\theta$). The velocity field and streamlines are shown in Figure 10.55A. Applying Eqs. 10.51, the vorticity field is found to be $2\Omega_0 \mathbf{k}$, as expected. Thus the vortex lines are all perpendicular to the plane of motion, as shown in Figure 10.55B. Now consider a circular path of radius R centered on the rotation axis. The surface integral defining the circulation in Eq. 10.55 is easy to evaluate if we select the surface as a circular disk of area πR^2 normal to the z axis and bounded by the selected path. Since the unit normal of the disk is $\mathbf{n} = \mathbf{k}$, we find

$$C(t) = \iint_S \boldsymbol{\omega} \cdot \mathbf{n}\, dS = \iint_S 2\Omega_0 \mathbf{k} \cdot \mathbf{k}\, dS = \iint_S 2\Omega_0\, dS$$

The value of the surface integral is therefore simply $2\Omega_0 \pi R^2$. This value will also be found for a surface of any shape that is bounded by this circular path. Thus, according to Eq. 10.55, in a rigid body rotation, the circulation $C(t)$ on any circular path oriented normal to the z axis and centered on the axis of rotation in a flow in rigid body rotation is a constant equal to $2\Omega_0 \pi R^2$. We can verify this result by calculating the circulation

10.9 RATE OF ROTATION

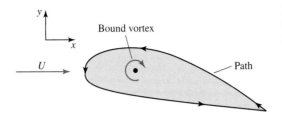

Figure 10.56 A bound vortex can be used to model the circulation and, therefore, the lift of an airfoil.

by means of its definition as the line integral around the path bounding the circular disk (Eq. 10.54). For this path we can write $\mathbf{u} \cdot d\mathbf{r} = v_\theta R \, d\theta$, and since the tangential component of velocity at any point along the path is $v_\theta = R\Omega_0$, the line integral is $\oint \mathbf{u} \cdot d\mathbf{r} = \int_0^{2\pi} (R\Omega_0) R \, d\theta = 2\Omega_0 \pi R^2$, as found with the surface integral.

Circulation plays an important role in the 2D frictionless flow model describing the aerodynamics of a wing. It can be shown that the lift produced by a long wing of length L is related to the circulation around the airfoil by

$$F_L = -\rho U C L \qquad (10.56)$$

Here U is the speed of the fluid approaching the wing, ρ is the fluid density, and C is the circulation on the path illustrated in Figure 10.56.

EXAMPLE 10.18

Calculate the circulation for the rectangular path shown in Figure 10.57 for the pressure-driven flow between stationary parallel plates. The velocity field is given by

$$u(y) = \left[\frac{h^2(p_1 - p_2)}{2\mu L}\right]\left[1 - \left(\frac{y}{h}\right)^2\right], \quad v = 0, \quad \text{and} \quad w = 0$$

Does your answer agree with the value obtained by using Eq. 10.55?

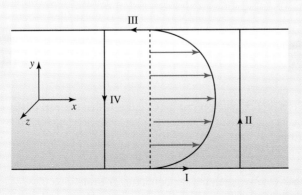

Figure 10.57 Schematic for Example 10.18.

SOLUTION

We will use Eq. 10.54, $C(t) = \oint \mathbf{u} \cdot d\mathbf{r}$, to calculate the circulation on the indicated path, with $\mathbf{u} = u(y)\mathbf{i} + 0\mathbf{j} + 0\mathbf{k}$ and $d\mathbf{r} = dx\,\mathbf{i} + dy\,\mathbf{j} + dz\,\mathbf{k}$. On paths II and IV, we have $d\mathbf{r} = dy\,\mathbf{j}$ and $d\mathbf{r} = -dy\,\mathbf{j}$, respectively, so the dot product $\mathbf{u} \cdot d\mathbf{r}$ on these paths is zero because $\mathbf{i} \cdot \mathbf{j} = 0$. On paths I and III, we have $d\mathbf{r} = dx\,\mathbf{i}$ and $d\mathbf{r} = -dx\,\mathbf{i}$, so the corresponding integrals are $I_I = \int_0^L u(y)|_{y=h}\, dx = 0$ and $I_{III} = \int_0^L -u(y)|_{y=-h}\, dx = 0$. Thus, the circulation on this path is zero. To check this result, recall that we used Eqs. 10.51 to calculate the vorticity in this flow in Example 10.17 and obtained $\boldsymbol{\omega} = \omega_z\mathbf{k}$ with $\omega_z = [(p_1 - p_2)y]/\mu L$. Applying Eq. 10.55, and noting that the unit normal must be pointing out of the paper for this path, we have

$$C(t) = \iint_S \boldsymbol{\omega} \cdot \mathbf{n}\, dS = \iint_S (\omega_z \mathbf{k}) \cdot (+\mathbf{k})\, dS$$

$$= \int_{-h}^{h} \int_0^L \left[\frac{(p_1 - p_2)y}{\mu L}\right] dx\, dy = \frac{p_1 - p_2}{\mu}\int_{-h}^{h} y\, dy = 0$$

Thus, the two methods produce the same answer. Notice what actually occurs here: the flux of vorticity in the top half of the channel cancels that in the bottom half.

 CD/Kinematics/Two dimensional flow and vorticity/Irrotational flows

10.9.3 Irrotational Flow and Velocity Potential

From our discussion of how the no-slip condition creates vorticity in flows past a solid boundary, it appears that we would rarely, if ever, encounter a flow in which vorticity is absent. As engineers, however, we are always alert to the possibility of simplifying our analysis with a valid approximation. For example, in high speed flow over a streamlined body, the boundary layer in which vorticity is present is thin and attached (always adjacent to the body), so most of the flow field is free of vorticity. One can deduce certain aspects of this flow field by using a model that assumes the absence of vorticity. Such a flow is referred to as an irrotational flow, since the fluid elements do not rotate. This is one of the most important approximations in fluid dynamics, and it has a long and continuing history of use in aerodynamics. In an irrotational flow, the velocity field satisfies

$$\nabla \times \mathbf{u} = 0 \tag{10.57}$$

An interesting result from vector calculus is that the curl of the gradient of a scalar is zero. This suggests that we define a scalar velocity potential $\phi(\mathbf{x}, t)$ for irrotational flow by writing

$$\mathbf{u} = \nabla \phi \tag{10.58}$$

By using the velocity potential, the condition of irrotational flow is automatically satisfied, and the three velocity components are replaced by a single scalar potential function.

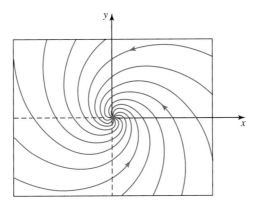

Figure 10.58 Flow streamlines produced by a sink and a vortex.

In Cartesian coordinates, the velocity components are related to the velocity potential by

$$u = \frac{\partial \phi}{\partial x}, \quad v = \frac{\partial \phi}{\partial y}, \quad \text{and} \quad w = \frac{\partial \phi}{\partial z} \qquad (10.59\text{a--c})$$

while in cylindrical coordinates the relationship is

$$v_r = \frac{\partial \phi}{\partial r}, \quad v_\theta = \frac{1}{r}\frac{\partial \phi}{\partial \theta}, \quad \text{and} \quad v_z = \frac{\partial \phi}{\partial z} \qquad (10.60\text{a--c})$$

If we know the velocity potential, we can construct the velocity field by taking the appropriate derivatives. Although the velocity potential will be discussed in more detail in Chapter 12, to understand its power, consider the sink–vortex flow shown in Figure 10.58. This interesting flow appears visually complex but can in fact be completely described by a velocity potential in cylindrical coordinates of the form $\phi = A \ln r - B\theta$, where A and B are constants. Using Eqs. 10.60a–c, we can easily calculate the velocity components in this flow as

$$v_r = \frac{\partial \phi}{\partial r} = \frac{A}{r}, \quad v_\theta = \frac{1}{r}\frac{\partial \phi}{\partial \theta} = -\frac{B}{r}, \quad \text{and} \quad v_z = \frac{\partial \phi}{\partial z} = 0$$

Although the irrotational flow approximation is used in both incompressible and compressible flows, the equation of motion for incompressible, irrotational flow is particularly simple and noteworthy. As shown in Chapter 11, the governing equation expressing mass conservation for incompressible flow reduces to $\nabla \cdot \mathbf{u} = 0$. Substituting for the velocity by means of Eq. 10.58, the velocity potential for irrotational, incompressible flow can be seen to satisfy Laplace's equation,

$$\nabla^2 \phi = 0 \qquad (10.61)$$

one of the most well-understood linear partial differential equations in applied mathematics. Techniques for solving problems involving this type of flow are highly developed. As we shall see in Chapter 12, many flow solutions are available for incompressible, irrotational flow. In fact we have used the velocity fields for a number of these solutions in the examples throughout this book.

An engineer's decision to use the approximation of irrotational flow in analyzing a given problem bears on the type of CFD analysis that would be selected. Commercial CFD packages for irrotational flow analysis are very different from those that do viscous flow analysis.

In Cartesian coordinates, the equation governing the velocity potential is

$$\frac{\partial^2 \phi}{\partial x^2} + \frac{\partial^2 \phi}{\partial y^2} + \frac{\partial^2 \phi}{\partial z^2} = 0 \quad (10.62\text{a})$$

while in cylindrical coordinates the potential satisfies

$$\frac{1}{r}\frac{\partial}{\partial r}\left(r\frac{\partial \phi}{\partial r}\right) + \frac{1}{r^2}\frac{\partial^2 \phi}{\partial \theta^2} + \frac{\partial^2 \phi}{\partial z^2} = 0 \quad (10.62\text{b})$$

EXAMPLE 10.19

The flow illustrated in Figure 10.59A is given by $u = A(x^2 - y^2)$, $v = -2Axy$, $w = 0$, where A is a constant. Is this an irrotational flow? If so, find the velocity potential and demonstrate that it satisfies Laplace's equation.

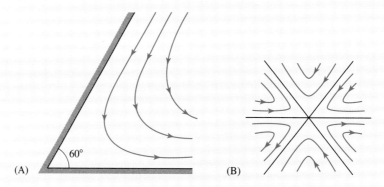

Figure 10.59 Streamlines for Example 10.19.

SOLUTION

An irrotational flow has zero vorticity. We can calculate the vorticity in this flow by using Eqs. 10.51a–c to obtain

$$\omega_x = \left(\frac{\partial w}{\partial y} - \frac{\partial v}{\partial z}\right) = \left[0 - \frac{\partial}{\partial z}(-2Axy)\right] = 0$$

$$\omega_y = \left(\frac{\partial u}{\partial z} - \frac{\partial w}{\partial x}\right) = \left\{\frac{\partial}{\partial z}[A(x^2 - y^2)] - 0\right\} = 0$$

$$\omega_z = \left(\frac{\partial v}{\partial x} - \frac{\partial u}{\partial y}\right) = \left\{\frac{\partial}{\partial x}(-2Axy) - \frac{\partial}{\partial y}[A(x^2 - y^2)]\right\}$$
$$= [(-2Ay) - (-2Ay)] = 0$$

We see that the flow is irrotational. To find the velocity potential, we use Eq. 10.59 to write

$$\frac{\partial \phi}{\partial x} = u = A(x^2 - y^2), \quad \frac{\partial \phi}{\partial y} = v = -2Axy, \quad \text{and} \quad \frac{\partial \phi}{\partial z} = w = 0$$

Thus, the potential is a function of only x and y in this case. To find the potential, we integrate the first equation with respect to x to obtain $\phi = A(x^3/3 - xy^2) + f(y)$ and then take the derivative of ϕ with respect to y to find $\partial \phi/\partial y = -2Axy + df/dy$. Comparing this result with $\partial \phi/\partial y = v = -2Axy$ shows that $df/dy = 0$ and, therefore, f is a constant. Thus, the velocity potential is $\phi = A(x^3/3 - xy^2) + C$. The constant is not significant and may be set to zero. To check whether Laplace's equation is satisfied by this potential, we use Eq. 10.62a to obtain

$$\frac{\partial^2 \phi}{\partial x^2} + \frac{\partial^2 \phi}{\partial y^2} + \frac{\partial^2 \phi}{\partial z^2}$$

$$= \frac{\partial^2}{\partial x^2}\left[A\left(\frac{x^3}{3} - xy^2\right)\right] + \frac{\partial^2}{\partial y^2}\left[A\left(\frac{x^3}{3} - xy^2\right)\right] + \frac{\partial^2}{\partial z^2}\left[A\left(\frac{x^3}{3} - xy^2\right)\right]$$

$$= (2Ax) + (-2Ax) + 0 = 0$$

Thus we see that the potential does indeed satisfy Laplace's equation. The streamlines for this flow are shown in Figure 10.59A. Note that engineers often make use of only part of an irrotational flow. In this case the same velocity field also describes flow in the other five pie-shaped regions, as shown in Figure 10.59B.

10.10 RATE OF EXPANSION

As noted earlier, the rate of expansion tensor is the part of the velocity gradient that causes a change in the volume of a fluid element while leaving its shape unchanged. A fluid must be compressible for the volume of a fluid particle (or element) to change. Hence in discussing the rate of expansion tensor, we anticipate that we will discover a connection to the fluid density. Before exploring this connection, recall that the rate of expansion tensor is defined by Eq. 10.40 as

$$\mathbf{E} = \begin{pmatrix} \frac{1}{3}\left(\frac{\partial u}{\partial x} + \frac{\partial v}{\partial y} + \frac{\partial w}{\partial z}\right) & 0 & 0 \\ 0 & \frac{1}{3}\left(\frac{\partial u}{\partial x} + \frac{\partial v}{\partial y} + \frac{\partial w}{\partial z}\right) & 0 \\ 0 & 0 & \frac{1}{3}\left(\frac{\partial u}{\partial x} + \frac{\partial v}{\partial y} + \frac{\partial w}{\partial z}\right) \end{pmatrix}$$

The rate of expansion is an isotropic tensor, which implies that the deformation of a fluid element is independent of its orientation. This isotropic characteristic can be easily

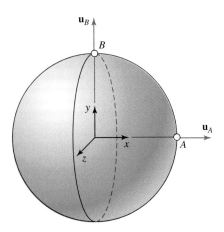

Figure 10.60 Illustration of a spherical fluid element undergoing pure expansion.

identified in any tensor (or square matrix) by noting that such a tensor is the product of a scalar function and the unit matrix. Recall that the unit matrix has "1s" along the diagonal and "0s" elsewhere. In the case of the expansion tensor, the scalar function that multiplies each of the "1s" is $\frac{1}{3}(\partial u/\partial x + \partial v/\partial y + \partial w/\partial z)$.

Consider the spherical fluid element of radius R shown in Figure 10.60, and suppose this element is embedded in a flow with a velocity field having rate of expansion **E**. If we focus on the effect of **E** on the relative velocity of points on the surface of this sphere, we find that a point on the surface has a relative velocity given by $\mathbf{r} \cdot \mathbf{E}$ with respect to the center of the sphere. If we calculate this dot product for the point marked A in Figure 10.60, we find $\mathbf{r} \cdot \mathbf{E}|_A = R[\frac{1}{3}(\partial u/\partial x + \partial v/\partial y + \partial w/\partial z)]\mathbf{i}$. Assuming that $\frac{1}{3}(\partial u/\partial x + \partial v/\partial y + \partial w/\partial z)$ is positive, the relative velocity vector points radially outward at point A. Similarly, at point B we find $\mathbf{r} \cdot \mathbf{E}|_B = R[\frac{1}{3}(\partial u/\partial x + \partial v/\partial y + \partial w/\partial z)]\mathbf{j}$. This relative velocity vector also points radially outward and has the same magnitude as that at point A. In fact, the velocity vector at any point on the surface points radially outward and has the same magnitude. We conclude that the spherical fluid volume will expand (and remain spherical) at a rate proportional to the (positive) value of $\frac{1}{3}(\partial u/\partial x + \partial v/\partial y + \partial w/\partial z)$. Thus, the contribution of **E** to the relative motion of a fluid element is an expansion if $\frac{1}{3}(\partial u/\partial x + \partial v/\partial y + \partial w/\partial z) > 0$, and a contraction if $\frac{1}{3}(\partial u/\partial x + \partial v/\partial y + \partial w/\partial z) < 0$. If $\frac{1}{3}(\partial u/\partial x + \partial v/\partial y + \partial w/\partial z) = 0$ in a given flow, the volume of a fluid element is unchanged in that flow. Does it seem to you that these results must imply something about how the density of the fluid inside this element is changing? The answer to this question is the topic of the next section.

10.10.1 Dilation

In fluid mechanics, we define the dilation, (sometimes called dilatation), $\Delta = \Delta(\mathbf{x}, t)$, as the divergence of the velocity field:

$$\Delta = \nabla \cdot \mathbf{u} \quad (10.63)$$

10.10 RATE OF EXPANSION

The dilation is given in Cartesian and cylindrical coordinates, respectively, by

$$\Delta = \frac{\partial u}{\partial x} + \frac{\partial v}{\partial y} + \frac{\partial w}{\partial z} \quad \text{and} \quad \Delta = \frac{1}{r}\frac{\partial (rv_r)}{\partial r} + \frac{1}{r}\frac{\partial v_\theta}{\partial \theta} + \frac{\partial v_z}{\partial z} \quad (10.64\text{a,b})$$

Using the dilation, the rate of expansion tensor can be written as

$$\mathbf{E} = \begin{pmatrix} \frac{1}{3}\Delta & 0 & 0 \\ 0 & \frac{1}{3}\Delta & 0 \\ 0 & 0 & \frac{1}{3}\Delta \end{pmatrix} = \begin{pmatrix} \frac{1}{3}\nabla \cdot \mathbf{u} & 0 & 0 \\ 0 & \frac{1}{3}\nabla \cdot \mathbf{u} & 0 \\ 0 & 0 & \frac{1}{3}\nabla \cdot \mathbf{u} \end{pmatrix} \quad (10.65\text{a,b})$$

The divergence of a velocity field defines the fractional rate of expansion in a fluid at a point in a flow. If the divergence is positive at a point, the volume of the fluid particle located there is increasing, its density is decreasing, and the particle is said to be dilating because it is expanding. If the divergence is negative, the volume of the fluid particle is decreasing and its density is increasing. A nonzero dilation, therefore, implies a compressible fluid, while an incompressible fluid must always have zero dilation.

To prove these statements, consider a small fluid system and note that an equation for the time rate of change of the volume of this system is given by

$$\frac{d\mathcal{V}(t)}{dt} = \frac{d}{dt}\int_{R(t)} 1\, dV$$

where $\mathcal{V}(t)$ is the volume of the system. Using the Reynolds transport theorem, Eq. 7.4, and noting that in this case we have $\rho\varepsilon = 1$, we obtain

$$\frac{d\mathcal{V}(t)}{dt} = \frac{d}{dt}\int_{R(t)} 1\, dV = \int_{R(t)} \frac{\partial}{\partial t}(1)\, dV + \int_{R(t)}(1)(\mathbf{u}\cdot\mathbf{n})\, dS$$

$$= \int_{R(t)} (1)(\mathbf{u}\cdot\mathbf{n})\, dS$$

Converting the surface integral to a volume integral with Gauss's theorem, we obtain

$$\frac{d\mathcal{V}(t)}{dt} = \int_{R(t)} (\nabla \cdot \mathbf{u})\, dV \quad (10.66)$$

This result shows that the rate at which the volume of a fluid system is increasing is related to the divergence of the velocity field. We can also apply Eq. 10.66 to a fluid particle, since it qualifies as a small fluid system. Suppose the fluid particle's volume at some instant of time is $\delta\mathcal{V}$. According to Eq. 10.66 its volume is changing at a rate given by $d(\delta\mathcal{V})/dt = \int_{\delta\mathcal{V}} (\nabla \cdot \mathbf{u})\, dV$. Since all fluid properties are uniform for a fluid particle, the volume integral is simply the integrand times the volume, so that $d(\delta\mathcal{V})/dt = (\nabla \cdot \mathbf{u})\delta\mathcal{V}$. Dividing both sides by $\delta\mathcal{V}$, we find

$$\frac{1}{\delta\mathcal{V}}\frac{d(\delta\mathcal{V})}{dt} = \nabla \cdot \mathbf{u} \quad (10.67)$$

Thus, at a given instant of time, the fractional rate of increase in the volume of a fluid particle located at a point in the fluid is given by the divergence of velocity at this point, i.e., by the dilation.

10.10.2 Incompressible Fluid and Incompressible Flow

There are two important approximations in fluid mechanics involving the modeling of fluid density. The first approximation, that of an incompressible fluid, assumes a constant fluid density. Thus, in a problem involving an incompressible fluid, the density satisfies the condition

$$\rho(\mathbf{x}, t) = \rho_0 \tag{10.68}$$

where ρ_0 is simply a constant.

The second approximation, called incompressible flow, assumes that the density of a fluid particle does not change as it moves. Since the rate of change of density following the motion of a fluid particle is given by the material derivative of density, the incompressible flow condition is expressed mathematically as

$$\frac{D\rho}{Dt} = 0 \tag{10.69}$$

With the definition of the material derivative (see Eqs. 6.15 and 6.21), we can write the incompressible flow condition in Cartesian coordinates as

$$\frac{D\rho}{Dt} = \frac{\partial \rho}{\partial t} + u\frac{\partial \rho}{\partial x} + v\frac{\partial \rho}{\partial y} + w\frac{\partial \rho}{\partial z} = 0 \tag{10.70a}$$

and in cylindrical coordinates as

$$\frac{D\rho}{Dt} = \frac{\partial \rho}{\partial t} + v_r\frac{\partial \rho}{\partial r} + \frac{v_\theta}{r}\frac{\partial \rho}{\partial \theta} + v_z\frac{\partial \rho}{\partial z} = 0 \tag{10.70b}$$

As discussed in Chapter 6, the rate of change of any quantity following the motion of a fluid particle is given by the sum of the local rate of change (the temporal derivative term) and the convective rate of change (the sum of the three terms involving spatial derivatives). Thus, the density of a fluid particle, like its velocity, may be changing owing to both local and convective effects.

If the density of a fluid is constant (an incompressible fluid), then we see immediately from Eq. 10.69 that the corresponding flow is also an incompressible flow. However, the presence of four terms in Eqs. 10.70a and 10.70b, means that the material derivative of density may be zero in a flow in which the density is not constant. For example, this can occur in a steady process (for which $\partial \rho/\partial t = 0$) if the velocity field takes the specific form needed to make the convective rate of change of density zero. Consider the steady 2D flow shown in Figure 10.61 with a velocity field described by $\mathbf{u} = u(x, y)\mathbf{i} + v(x, y)\mathbf{j} + 0\mathbf{k}$ and a linear density variation in the vertical z direction given by $\rho(z) = \rho_0 - \alpha z$, where α is a constant. This density field could result from a vertical temperature gradient in the atmosphere or a body of water or from a variable salt concentration in seawater. Applying Eq. 10.70a to these velocity and density fields, we find that $\partial \rho/\partial t$, $\partial \rho/\partial x$, and $\partial \rho/\partial y$ are all zero. Thus, $D\rho/Dt = w(\partial \rho/\partial z)$, and since $w = 0$ in this flow, we have $D\rho/Dt = 0$. Therefore, this is an incompressible flow, the rate of change in density following the motion of a fluid particle is zero, and the density of a fluid particle does not change as it moves. Does this make sense in a flow in which the density is spatially variable? Examining the velocity field, we see that a fluid particle remains in the same horizontal layer throughout its motion. Since the density in any horizontal layer is constant, each fluid particle retains its original density value as it moves.

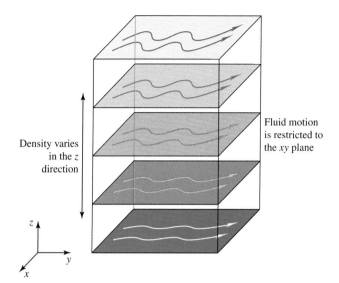

Figure 10.61 A 2D flow with a linear density variation in the z direction.

EXAMPLE 10.20

The velocity and density fields in a flow are given by $u = kx^2y + \alpha/y$, $v = -kxy^2 + \beta/x$, $w = \varepsilon$, and $\rho = \rho_0 \exp[-\gamma(xy - ct)]$, where k, α, β, ε, γ, ρ_0, and c are constants. Is this an incompressible flow? Does it involve an incompressible fluid?

SOLUTION

Since the density is a function of position and time, this is not an incompressible fluid. An incompressible flow must satisfy Eq. 10.70a: $D\rho/Dt = \partial\rho/\partial t + u(\partial\rho/\partial x) + v(\partial\rho/\partial y) + w(\partial\rho/\partial z) = 0$. For the given velocity and density fields, we find:

$$\frac{D\rho}{Dt} = \frac{\partial \rho}{\partial t} + u\frac{\partial \rho}{\partial x} + v\frac{\partial \rho}{\partial y} + w\frac{\partial \rho}{\partial z}$$

$$= \gamma c\{\rho_0 \exp[-\gamma(xy - ct)]\} + \left(kx^2y + \frac{\alpha}{y}\right)(-\gamma y)\{\rho_0 \exp[-\gamma(xy - ct)]\}$$

$$+ \left(-kxy^2 + \frac{\beta}{x}\right)(-\gamma x)\{\rho_0 \exp[-\gamma(xy - ct)]\}$$

Notice that both the local and convective derivatives of the density are nonzero for this flow. Also note that although the flow is steady (i.e., the velocity field is independent of time), it is not a steady process (since the density is a function of time). Simplifying this result by writing $\gamma\rho_0 \exp[-\gamma(xy - ct)] = \gamma\rho$, we have $D\rho/Dt = \gamma\rho(c - kx^2y^2 - \alpha + kx^2y^2 - \beta) = \gamma\rho(c - \alpha - \beta)$. Thus, this could only be an incompressible flow if the constants satisfy $c - \alpha - \beta = 0$.

While as noted earlier, it is easy to confirm that an incompressible fluid satisfies the equation defining incompressible flow, an incompressible flow does not necessarily have a constant fluid density. When analyzing a flow, an engineer must often decide whether to treat the fluid density as a variable (compressible flow) or to apply one of the incompressible approximations. Since every fluid is compressible, varying only in the degree of compressibility as expressed by the bulk modulus, using a compressible flow model is never invalid. Because the resulting governing equations are complex and expensive to solve, however, it is important to avoid using a compressible flow model unnecessarily. Thus, the relevant question is this: When it is permissible to treat the fluid density as a constant?

As you know, a liquid is normally modeled as a constant density or incompressible fluid because there are relatively few circumstances in which the density of a liquid changes significantly. One exception arises in problems involving the deep ocean, where the pressure is so large that seawater is compressed. Another exception is related to the fact that the transmission of sound in liquids is evidence of a change in density, albeit a very small one. Although the pressure change in a sound wave (which causes the density change) is very small, a blast wave in a liquid creates a large pressure change and a correspondingly larger density change. Thus, a sound or blast wave cannot be analyzed in a liquid if constant density is assumed. Compressibility is also significant in problems

EXAMPLE 10.21

What percent density change would be expected for a flow of air with a characteristic velocity equivalent to $M = 0.3$? What is the corresponding percent change in pressure and temperature?

SOLUTION

This exercise is solved by using Eqs. 10.71 to estimate the various changes. Noting that $\gamma = 1.4$ for air and $M^2 = (0.3)^2 = 0.09$, we find:

$$\frac{d\rho}{\rho} = \frac{1}{2}M^2 = \frac{1}{2}(0.09) = 0.045$$

which is a 4.5% density change. The changes in pressure and temperature are calculated as

$$\frac{dp}{p} = \frac{\gamma}{2}M^2 = \frac{1.4}{2}(0.09) = 0.063 = 6.3\%$$

$$\frac{dT}{T} = \frac{\gamma - 1}{2}M^2 = \frac{0.4}{2}(0.09) = 0.018 = 1.8\%$$

A Mach number of 0.3 is generally accepted as the upper limit for assuming that air may be modeled as an incompressible fluid. Nevertheless, since the sound speed in air is about 340 m/s (1130 ft/s), a constant density model for air is appropriate in many practical applications.

10.10 RATE OF EXPANSION

 It is important for an engineer to understand the different types of density approximation available in fluid mechanics. Among other things, the approximation chosen strongly influences the type of CFD code that will be employed and the cost and complexity in the resulting CFD analysis. The interested reader can consult advanced texts and the user manuals for the CFD codes of interest for further study of this important issue.

involving very high liquid pressures and velocities, as in a water jet cutter. In most circumstances, however, it is sensible to apply a constant density (i.e., incompressible fluid) model to a liquid.

Perhaps more remarkable than the need to occasionally model a liquid as a compressible fluid is the frequency with which a gas may be modeled as an incompressible fluid. Considering the highly compressible character of all gases, modeling the density of a gas as a constant seems like a contradiction. However, for a gas flow at low speed in the absence of external heating, the changes in pressure, density, and temperature of a moving gas particle are very small. Compressible flow theory gives us the following estimates of the maximum changes in these variables as function of the Mach number, $M = V/c$, in the isentropic flow of a perfect gas:

$$\frac{dp}{p} = \frac{\gamma}{2}M^2, \quad \frac{d\rho}{\rho} = \frac{1}{2}M^2, \quad \text{and} \quad \frac{dT}{T} = \frac{\gamma-1}{2}M^2 \qquad (10.71\text{a–c})$$

where γ is the specific heat ratio, ($\gamma = 1.4$ for air). We see that if M is small, as it is in low speed flow, the change in density is negligible and may be safely ignored, as indicated by the Example 10.21.

A final contrasting example that illustrates the need for care in the use of the incompressible fluid approximation in fluid dynamics occurs in the analysis of liquid or gas flows created by heating. Natural convection is the term used to describe flows that result from buoyancy forces acting on a differentially heated fluid in a gravity field (see Figure 10.62). For even modest temperature differences, the density change experienced by a fluid particle is large enough for buoyancy forces to set it in motion. The density change caused by heating may be estimated from known hot and cold temperatures by means of the coefficient of thermal expansion. Applying the incompressible fluid approximation in analyzing this type of problem would fail to capture the essence of natural convection in either liquids or gases. With experience, you will recognize when

Figure 10.62 Natural convection from a warm body. The visualization was accomplished with the Schlieren method.

the use of an incompressible fluid model may result in an analysis that fails to predict important physical phenomena.

There is an interesting connection between dilation, density, and incompressible flow. To discuss this connection, it is necessary to use the governing differential equation expressing mass conservation. Although this equation, called the continuity equation, is derived in the next chapter, we will use it here to show how the value of the dilation is related to the fluid density. The continuity equation states that the connection between density and the divergence of velocity, i.e., dilation, in a fluid is given by $D\rho/Dt + \rho \nabla \cdot \mathbf{u} = 0$. Writing this in terms of dilation Δ and rearranging, we have $(1/\rho)(D\rho/Dt) = -\nabla \cdot \mathbf{u} = -\Delta$. The material derivative $D\rho/Dt$ is the time rate of change of density following the motion of a fluid particle at a point in the fluid. When divided by the density, the result, $(1/\rho)(D\rho/Dt)$ is the fractional rate of increase in density of that particle. By Eq. 10.67, the divergence of velocity $\nabla \cdot \mathbf{u}$ is the fractional rate of increase in the volume of the same particle. Thus the continuity equation shows that if the density of a particle is increasing at a certain fractional rate, the volume must be decreasing at the same fractional rate and vice versa, a fact that is recognized as expressing mass conservation.

Since both incompressible fluids and incompressible flows satisfy the equation $D\rho/Dt = 0$, we can use the continuity equation to conclude that when either of these important approximations holds, the corresponding velocity field must satisfy

$$\nabla \cdot \mathbf{u} = \Delta = 0 \tag{10.72}$$

That is, the dilation is zero in an incompressible fluid or in an incompressible flow.

Gauss's theorem, also known as the divergence theorem, provides additional insight into the physical interpretation of dilation, the divergence of velocity, and the concepts of incompressible fluid and incompressible flow. Consider a closed, bounded volume, as shown in Figure 10.63, filled with fluid in motion with a velocity field $\mathbf{u}(\mathbf{x}, t)$. Applying Gauss's theorem to this volume, we find

$$\int_S (\mathbf{u} \cdot \mathbf{n}) \, dS = \int_V (\nabla \cdot \mathbf{u}) \, dV$$

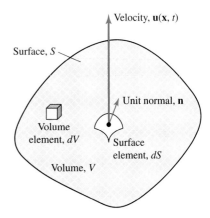

Figure 10.63 A fluid volume that illustrates the terms in Gauss's theorem.

The surface integral is the volume flowrate across the surface of the fluid volume shown in Figure 10.63. Thus, we can replace it with the volume flowrate Q to obtain

$$Q = \int_V (\nabla \cdot \mathbf{u})\, dV$$

The physical interpretation of this result is as follows. If the dilation is positive within the volume, the volume integral is positive, and there is a net outflow of fluid ($Q > 0$). If the dilation is negative, the integral is negative, and there is a net inflow ($Q < 0$). If the dilation is zero, the net outflow through the closed surface is zero. This does not imply that the normal velocity ($\mathbf{u} \cdot \mathbf{n}$) is zero on every element of the surface, only that when integrated over the whole surface, the net volume flow is zero. Fluid may be flowing in on some part of the surface with ($\mathbf{u} \cdot \mathbf{n}$) < 0, and flowing out elsewhere with ($\mathbf{u} \cdot \mathbf{n}$) > 0. It should come as no surprise to discover that for an incompressible fluid or incompressible flow ($\nabla \cdot \mathbf{u} = 0$), the volume flowrate into a control volume equals the volume flowrate out.

CD/Kinematics/Streamlines and streamfunction

10.10.3 Streamfunction

We have seen that the velocity field in an incompressible fluid or in an incompressible flow must be such that the divergence of velocity is zero. This is true for steady and unsteady flow. It is possible to use this result to draw some interesting conclusions about the streamlines in 2D flows. Consider the general case of a 2D velocity field $(u, v, 0)$ whose components are functions of (x, y, t). The continuity equation (Eq. 10.72: $\nabla \cdot \mathbf{u} = \Delta = 0$) reduces in Cartesian coordinates in this case to $\nabla \cdot \mathbf{u} = \partial u/\partial x + \partial v/\partial y = 0$. If we now define a function $\psi = \psi(x, y, t)$ so that

$$u = \frac{\partial \psi}{\partial y} \quad \text{and} \quad v = \frac{-\partial \psi}{\partial x} \quad \text{(10.73a, b)}$$

the continuity equation is automatically satisfied, since

$$\frac{\partial u}{\partial x} + \frac{\partial v}{\partial y} = \frac{\partial}{\partial x}\left(\frac{\partial \psi}{\partial y}\right) + \frac{\partial}{\partial y}\left(\frac{-\partial \psi}{\partial x}\right) = \frac{\partial^2 \psi}{\partial x\, \partial y} - \frac{\partial^2 \psi}{\partial y\, \partial x} = 0$$

The function satisfying Eqs. 10.73 is called the streamfunction.

In a 2D flow described in cylindrical coordinates by velocity components $(v_r, v_\theta, 0)$, the streamfunction $\psi = \psi(r, \theta, t)$ is defined by

$$v_r = \frac{1}{r}\frac{\partial \psi}{\partial \theta} \quad \text{and} \quad v_\theta = -\frac{\partial \psi}{\partial r} \quad \text{(10.74a, b)}$$

In a 2D incompressible or constant density flow, the streamfunction automatically satisfies the governing equation of mass conservation $\nabla \cdot \mathbf{u} = 0$. By defining a streamfunction, we have replaced the two scalar components of the velocity vector by the single scalar streamfunction, a process that reduces the number of unknowns in a flow problem. Many attempts at solving the governing equations of 2D flow have employed the streamfunction to exploit this reduction in the number of unknowns.

The source of the name streamfunction comes from the connection between this function and the streamlines of a 2D flow. Setting the streamfunction equal to a constant defines a streamline in the flow. To prove this statement, let us assume that a certain streamline is defined by the constant ψ_0. The equation of this streamline is then

$$\psi(x, y, t) = \psi_0 \tag{10.75}$$

Consider the pattern of streamlines as they exist at an instant of time. Taking the total derivative of the streamfunction while holding time fixed, we obtain $d\psi = (\partial\psi/\partial x)\,dx + (\partial\psi/\partial y)\,dy$. Substituting for the partial derivatives by using Eqs. 10.73 gives

$$d\psi = -v\,dx + u\,dy \tag{10.76}$$

Recall that a streamline in a 2D flow in the xy plane satisfies Eq. 10.29c, namely, $v\,dx - u\,dy = 0$. Comparing this with Eq. 10.76, we conclude that on a streamline $d\psi = 0$, which shows that the streamfunction has a constant value on a streamline. Thus, if we know the streamfunction, we can construct a family of streamlines by setting the streamfunction equal to different constant values and plotting the resulting curves.

Suppose you are asked to find the equation of the streamline that passes through a certain point (x_P, y_P) at some instant of time. To find the relevant equation, let the unknown constant value of the streamfunction be ψ_P. Since the streamfunction is constant on a streamline, we know that $\psi(x_P, y_P) = \psi_P$. The desired constant can therefore be found by simply inserting the coordinates of the desired point into the streamfunction itself and evaluating the result to obtain the constant. The equation of the streamline through the point is then given by $\psi(x, y, t) = \psi_P$.

There is an interesting relationship between streamlines and the lines of constant potential of a 2D irrotational flow. The slope of a streamline in the xy plane is found by recognizing that along a streamline $d\psi = -v\,dx + u\,dy = 0$. Thus, on a streamline the slope is $(dy/dx)_\psi = v/u$. A line of constant potential, which is usually referred to as a potential line, is defined in a 2D flow by $d\phi = (\partial\phi/\partial x)\,dx + (\partial\phi/\partial y)\,dy = 0$. According to Eqs. 10.59a and 10.59b, we have $u = \partial\phi/\partial x$ and $v = \partial\phi/\partial y$. Thus a potential line is defined by $d\phi = u\,dx + v\,dy = 0$, and the slope of a potential line is $(dy/dx)_\phi = -u/v$. We see that these slopes are the negative reciprocals of each other, thus we conclude that streamlines and potential lines are orthogonal to one another as shown in Figure 10.64.

Figure 10.64 Streamlines and potential lines are orthogonal.

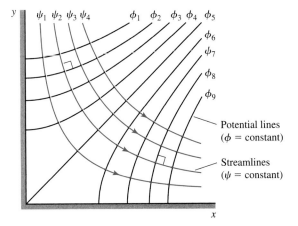

10.10 RATE OF EXPANSION

EXAMPLE 10.22

The flow in a corner can be modeled with the streamfunction $\psi = Axy$, where A is a constant. Sketch the streamlines corresponding to constant values of ψ of 0, A, $2A$, $3A$. Also find the streamline that passes through the point (1, 1).

SOLUTION

To find the streamlines for constant values of ψ, we simply substitute the known value into the streamfunction equation and solve for y. For $\psi = 0$, $x = 0$ and $y = 0$ are solutions that make up the walls of the corner as shown in Figure 10.65. For $\psi = A$, the equation for the streamline is $y = 1/x$. For $\psi = 2A$ and $3A$, the solutions are $y = \frac{1}{2}x$ and $\frac{1}{3}x$, respectively. These streamlines are also shown in Figure 10.65. To find the streamline passing through point (1, 1) we substitute these values into the streamfunction equation to find that $\psi_P = A$. Thus, the desired streamline is the $y = 1/x$, which we have already found.

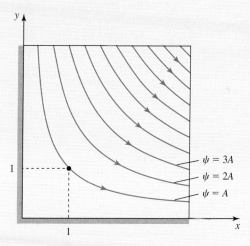

Figure 10.65 Streamlines for Example 10.22.

If we know the streamfunction, we can directly construct the velocity field by taking the indicated derivatives of the streamfunction given in Eqs. 10.73 or 10.74. Conversely, if we know the velocity field, we can construct the streamfunction, since the equations provide two first-order partial differential equations that define the unknown streamfunction. Note that the equations are generally coupled to one another because each velocity component depends on (x, y, t) or (r, θ, t). In steady flow, time does not appear in these equations. In an unsteady flow, t is treated as a constant in solving the equations. It should be remembered that although streamlines exist in 3D flow, the streamfunction applies only to 2D, constant density, or incompressible flow.

EXAMPLE 10.23

Find and plot the velocity field for the flow in the corner. The streamfunction is $\psi = Axy$, where A is a constant.

SOLUTION

This exercise can be solved by using Eqs. 10.73: $u = \partial\psi/\partial y$ and $v = -\partial\psi/\partial x$. For the streamfunction of interest we find

$$u = \frac{\partial \psi}{\partial y} = \frac{\partial}{\partial y}(Axy) = Ax \quad \text{and} \quad v = \frac{-\partial \psi}{\partial x} = \frac{-\partial}{\partial x}(Axy) = -Ay$$

A vector plot of this velocity field is shown in Figure 10.66. Note that this plot is entirely consistent with the streamlines for this flow shown in Figure 10.65.

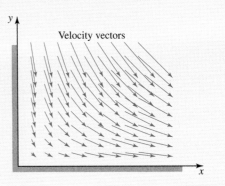

Figure 10.66 Velocity field for Example 10.23.

 CD/Dynamics/Potential Flow/Incompressibility and irrotationality

Although the numerical value ψ_0 defining a certain streamline has no meaning in isolation, there is an important physical interpretation of the difference in the streamfunction values on different streamlines. This interpretation can be established by computing the volume flux of fluid crossing a surface S, of width D into the paper, whose edge is defined by the path C connecting two points on different streamlines as shown in Figure 10.67A. In this 2D flow, the path C lies entirely in the xy plane. Let the two streamlines be defined by the constants ψ_1 and ψ_2. The volume flux crossing this surface is $Q = \int_S (\mathbf{u} \cdot \mathbf{n})\, dS$. To evaluate this integral, note that for a path lying wholly in the xy plane we may write a differential vector element of length $d\mathbf{r}$ along the path as $d\mathbf{r} = dx\,\mathbf{i} + dy\,\mathbf{j}$. An oriented surface element, $d\mathbf{A} = \mathbf{n}\,dS$, on the surface shown in Figure 10.67B, can be described by $d\mathbf{A} = d\mathbf{r} \times dz\,\mathbf{k}$. Completing the cross product, we

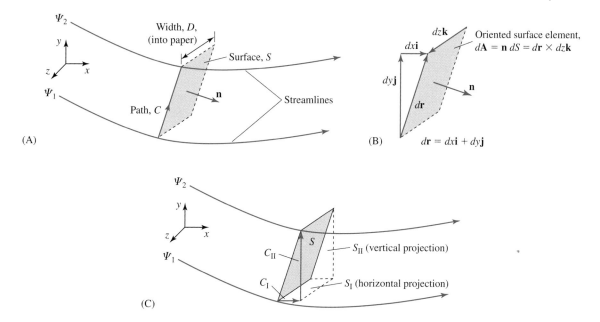

Figure 10.67 The relationship between streamlines and volume flux is illustrated by means of (A) the integral path C in the xy plane, (B) the surface element used in the integration of volume flux, and (C) an alternate integral path for the calculation of volume flux.

find $\mathbf{n}\, dS = dy\, dz\, \mathbf{i} - dx\, dz\, \mathbf{j}$, from which it follows that $(\mathbf{u} \cdot \mathbf{n})\, dS = u\, dy\, dz - v\, dx\, dz$. Thus, the volume flux crossing the surface is given by the double integral $Q = \iint (u\, dy\, dz - v\, dx\, dz)$, with appropriate limits. If we perform the integration over the z variable first, noting that the integrand does not depend on z, the volume flowrate crossing the surface becomes $Q = D \int (u\, dy - v\, dx)$. This result may also be written as the line integral

$$Q = D \int_C (u\, dy - v\, dx) \qquad (10.77)$$

where the path C is defined by the edge of the surface.

From Eq. 10.76, we have $d\psi = -v\, dx + u\, dy$. Thus the line integral of $d\psi$ from point 1 to point 2 along this path is

$$\psi_2 - \psi_1 = \int_C (u\, dy - v\, dx) \qquad (10.78)$$

Comparing Eqs. 10.77 and 10.78, we conclude that the volume flowrate Q crossing the surface formed by projecting the path a depth D in the z direction is

$$Q = D(\psi_2 - \psi_1) \qquad (10.79)$$

Equation 10.79 describes an important result. If D is taken as a unit distance in the z direction, we see that the difference $\psi_2 - \psi_1$ on two streamlines is the volume flowrate per unit depth crossing the surface defined by the path connecting a point on each streamline.

The integrand in Eqs. 10.77 and 10.78 is an exact differential form. Thus, the path chosen to connect the two points on different streamlines is completely arbitrary. This means that the values of Q and $\psi_2 - \psi_1$ will be the same for any path we might select. Consider a new path made up of one segment C_I along the x axis, and a second segment C_II along the y axis, as shown in Figure 10.67C. This path joins the same two points joined by the original path. The two segments define two new surfaces S_I and S_II. The volume flux across these new surfaces are Q_I and Q_II. Since the total volume flux must be the same for any path, the volume flux on the original surface Q must be the sum of the volume fluxes Q_I and Q_II on horizontal and vertical projections of this surface. Your physical intuition should confirm that the volume flowrate of a constant density fluid across the original surface must be the same as that crossing the two orthogonal surfaces.

The preceding interpretation of the values of the streamfunction on different streamlines plays an important role in flow visualization. Consider steady flow over a rotating cylinder placed in a uniform oncoming stream of constant density fluid. The

EXAMPLE 10.24

The streamfunction for a parallel plate flow is $\psi = G(h^2 y - y^3/3)$, where h is half the gap height in centimeters and G is a constant with units of reciprocal centimeter-seconds $(\text{cm-s})^{-1}$. Find the velocity field for this flow, then compare the volume flowrate through the channel as calculated by using the direct volume flux calculation to that found with the streamfunction by using Eq. 10.79. See Figure 10.68 for the geometry and use a depth D into the paper.

SOLUTION

To determine the velocity field use Eqs. 10.73: $u = \partial \psi/\partial y$ and $v = -\partial \psi/\partial x$. By inspection, we see that v must be zero since $\psi = \psi(y)$ only. We take the derivative of the streamfunction to find u

$$u = \frac{\partial \psi}{\partial y} = \frac{\partial [G(h^2 y - y^3/3)]}{\partial y} = G(h^2 - y^2)$$

The flowrate through a surface spanning the channel and having a width D into the paper is given by

$$Q = \int_S (\mathbf{u} \cdot \mathbf{n}) \, dS = \int_{-D/2}^{D/2} \int_{-h}^{h} G(h^2 - y^2) \mathbf{i} \cdot (\mathbf{i}) \, dy \, dz$$

$$= GD \left(h^2 y - \frac{y^3}{3} \right) \bigg|_{-h}^{h} = GD \left[\left(h^3 - \frac{h^3}{3} \right) - \left(-h^3 + \frac{h^3}{3} \right) \right] = \frac{4GDh^3}{3}$$

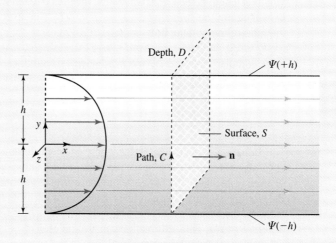

Figure 10.68 Schematic for Example 10.24.

Using Eq. 10.79 and path C of Figure 10.68 we evaluate ψ at $\pm h$ and obtain

$$Q = D[\psi(+h) - \psi(-h)] = GD\left[h^2(+h) - \frac{(+h)^3}{3}\right] - GD\left[h^2(-h) - \frac{(-h)^3}{3}\right]$$

$$= \frac{4GDh^3}{3}$$

Thus, both methods yield the same result, as expected. Note that Q in units of cubic centimeters per second is consistent with D and h in units of centimeters and G in reciprocal centimeter-seconds.

streamlines in this flow are shown in Figure 10.69. Note the equal increments in the streamfunction constant $\Delta\psi_0$ in the upstream region. Each pair of adjacent streamlines forms a streamtube of upstream height h_∞ and depth D in the z direction. By Eq. 10.79, the volume flowrate traveling downstream in any streamtube is $Q = \Delta\psi_0 D$. Since fluid cannot leave a streamtube, the volume flowrate through the streamtube at any point

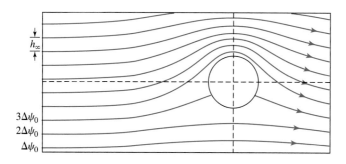

Figure 10.69 Streamlines for flow over a rotating cylinder.

along its length is the same. Upstream, the area of each streamtube there is $A_\infty = h_\infty D$. Downstream, the streamtube area is $A = hD$. The average velocity in a streamtube is the ratio of the volume flowrate to area, so in each streamtube upstream the average velocity is $\bar{V}_\infty = Q/A_\infty = \Delta\psi_0 D/h_\infty D = \Delta\psi_0/h_\infty$. At a point downstream where the perpendicular distance between adjacent streamlines is h, the average velocity is $\bar{V} = Q/A = \Delta\psi_0 D/hD = \Delta\psi_0/h$. We conclude that the ratio of average velocity at the two points is

$$\frac{\bar{V}}{\bar{V}_\infty} = \frac{h_\infty}{h} \tag{10.80}$$

Equation 10.80 indicates that where adjacent streamlines are closer to each other, the streamtube area is smaller, hence the average fluid velocity is larger. If adjacent streamlines diverge from one another, the streamtube area is larger, and the average velocity is smaller. By plotting a family of streamlines upstream of the cylinder that are originally a fixed distance h_∞ apart, we create a flow visualization that tells us how fast the fluid is moving at different points. This information can also be used to estimate the pressure distribution in a flow if viscous losses are minimal. According to Bernoulli's equation, in a flow of constant density fluid, the pressure is higher where the fluid is moving more slowly. Thus the streamline pattern in the flow over the rotating cylinder tells us that there is a higher pressure on the bottom surface of the cylinder than on the top, giving a net upward vertical force or lift on the spinning cylinder. A similar pressure difference explains the sharp, breaking motion of a curve ball in baseball, or the effectiveness of top-spin in tennis.

10.11 RATE OF SHEAR DEFORMATION

The shear deformation rate is given by Eq. 10.41 as

$$\mathbf{D} = \begin{bmatrix} \frac{\partial u}{\partial x} - \frac{1}{3}\left(\frac{\partial u}{\partial x} + \frac{\partial v}{\partial y} + \frac{\partial w}{\partial z}\right) & \frac{1}{2}\left(\frac{\partial u}{\partial y} + \frac{\partial v}{\partial x}\right) & \frac{1}{2}\left(\frac{\partial u}{\partial z} + \frac{\partial w}{\partial x}\right) \\ \frac{1}{2}\left(\frac{\partial v}{\partial x} + \frac{\partial u}{\partial y}\right) & \frac{\partial v}{\partial y} - \frac{1}{3}\left(\frac{\partial u}{\partial x} + \frac{\partial v}{\partial y} + \frac{\partial w}{\partial z}\right) & \frac{1}{2}\left(\frac{\partial v}{\partial z} + \frac{\partial w}{\partial y}\right) \\ \frac{1}{2}\left(\frac{\partial w}{\partial x} + \frac{\partial u}{\partial z}\right) & \frac{1}{2}\left(\frac{\partial w}{\partial y} + \frac{\partial v}{\partial z}\right) & \frac{\partial w}{\partial z} - \frac{1}{3}\left(\frac{\partial u}{\partial x} + \frac{\partial v}{\partial y} + \frac{\partial w}{\partial z}\right) \end{bmatrix}$$

The shear deformation plays a key role in the formulation of a constitutive model. As described in Chapter 1, a constitutive model defines the relationship between the shear stress and the shear deformation rate in a particular fluid. This relationship may be postulated or observed empirically. For a Newtonian fluid, the relationship between shear stress and shear deformation rate is linear, with the absolute viscosity of the fluid serving as the proportionality factor. The complete form of the constitutive model for a Newtonian fluid will be discussed in the next chapter.

Earlier we stated that the tensor $\mathbf{D}$ describes a change of shape of a fluid element exposed to a velocity field (without a volume change or rotation). In general, the rate of shear deformation is a complicated tensor with spatial derivatives of the velocity field defining its various components. To gain further insight into the physical significance of this tensor, reconsider the 2D shear flow shown in Figure 10.38, whose velocity field

10.11 RATE OF SHEAR DEFORMATION

was given by Eq. 10.35 as $u = Cy$, $v = Cx$, $w = 0$. Before calculating the shear deformation rate we will determine the velocity gradient, rate of rotation, and rate of expansion. In Eq. 10.36, the velocity gradient is found to be

$$\nabla \mathbf{u} = \begin{pmatrix} \frac{\partial u}{\partial x} & \frac{\partial v}{\partial x} & \frac{\partial w}{\partial x} \\ \frac{\partial u}{\partial y} & \frac{\partial v}{\partial y} & \frac{\partial w}{\partial y} \\ \frac{\partial u}{\partial z} & \frac{\partial v}{\partial z} & \frac{\partial w}{\partial z} \end{pmatrix} = \begin{pmatrix} 0 & C & 0 \\ C & 0 & 0 \\ 0 & 0 & 0 \end{pmatrix}$$

The rate of rotation **R**, as defined in Eq. 10.43, is zero in this flow. This is readily confirmed by examining the rotation vector or vorticity. The latter is given by the curl of the velocity field, which in this case is

$$\nabla \times \mathbf{u} = \left(\frac{\partial w}{\partial y} - \frac{\partial v}{\partial z}\right)\mathbf{i} + \left(\frac{\partial u}{\partial z} - \frac{\partial w}{\partial x}\right)\mathbf{j} + \left(\frac{\partial v}{\partial x} - \frac{\partial u}{\partial y}\right)\mathbf{k}$$
$$= (0 - 0)\mathbf{i} + (0 - 0)\mathbf{j} + (C - C)\mathbf{k} = 0$$

We conclude that this is an irrotational flow. It is easy to show that the flow can be described by the velocity potential $\phi = Cxy$.

The rate of expansion is given by Eqs. 10.65 as

$$\mathbf{E} = \begin{pmatrix} \frac{1}{3}\Delta & 0 & 0 \\ 0 & \frac{1}{3}\Delta & 0 \\ 0 & 0 & \frac{1}{3}\Delta \end{pmatrix} = \begin{pmatrix} \frac{1}{3}\nabla \cdot \mathbf{u} & 0 & 0 \\ 0 & \frac{1}{3}\nabla \cdot \mathbf{u} & 0 \\ 0 & 0 & \frac{1}{3}\nabla \cdot \mathbf{u} \end{pmatrix}$$

According to Eq. 10.64a, the dilation is given in this case by $\Delta = \partial u/\partial x + \partial v/\partial y + \partial w/\partial z = 0 + 0 + 0$; thus we can conclude that this is an incompressible flow.

The rate of shear deformation for this flow can be determined using Eq. 10.41:

$$\mathbf{D} = \begin{bmatrix} \frac{\partial u}{\partial x} - \frac{1}{3}\left(\frac{\partial u}{\partial x} + \frac{\partial v}{\partial y} + \frac{\partial w}{\partial z}\right) & \frac{1}{2}\left(\frac{\partial u}{\partial y} + \frac{\partial v}{\partial x}\right) & \frac{1}{2}\left(\frac{\partial u}{\partial z} + \frac{\partial w}{\partial x}\right) \\ \frac{1}{2}\left(\frac{\partial v}{\partial x} + \frac{\partial u}{\partial y}\right) & \frac{\partial v}{\partial y} - \frac{1}{3}\left(\frac{\partial u}{\partial x} + \frac{\partial v}{\partial y} + \frac{\partial w}{\partial z}\right) & \frac{1}{2}\left(\frac{\partial v}{\partial z} + \frac{\partial w}{\partial y}\right) \\ \frac{1}{2}\left(\frac{\partial w}{\partial x} + \frac{\partial u}{\partial z}\right) & \frac{1}{2}\left(\frac{\partial w}{\partial y} + \frac{\partial v}{\partial z}\right) & \frac{\partial w}{\partial z} - \frac{1}{3}\left(\frac{\partial u}{\partial x} + \frac{\partial v}{\partial y} + \frac{\partial w}{\partial z}\right) \end{bmatrix}$$

Since $\partial u/\partial x + \partial v/\partial y + \partial w/\partial z = 0$, we find

$$\mathbf{D} = \begin{pmatrix} \frac{\partial u}{\partial x} & \frac{1}{2}\left(\frac{\partial u}{\partial y} + \frac{\partial v}{\partial x}\right) & \frac{1}{2}\left(\frac{\partial u}{\partial z} + \frac{\partial w}{\partial x}\right) \\ \frac{1}{2}\left(\frac{\partial v}{\partial x} + \frac{\partial u}{\partial y}\right) & \frac{\partial v}{\partial y} & \frac{1}{2}\left(\frac{\partial v}{\partial z} + \frac{\partial w}{\partial y}\right) \\ \frac{1}{2}\left(\frac{\partial w}{\partial x} + \frac{\partial u}{\partial z}\right) & \frac{1}{2}\left(\frac{\partial w}{\partial y} + \frac{\partial v}{\partial z}\right) & \frac{\partial w}{\partial z} \end{pmatrix} = \begin{pmatrix} 0 & C & 0 \\ C & 0 & 0 \\ 0 & 0 & 0 \end{pmatrix}$$

10 ELEMENTS OF FLOW VISUALIZATION AND FLOW STRUCTURE

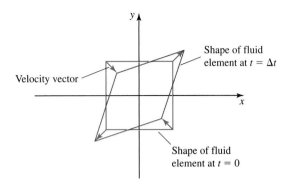

Figure 10.70 Fluid element undergoing shear deformation.

We have established that this incompressible, irrotational flow has a zero rate of rotation and expansion but a nonzero rate of shear deformation. Thus, it can be classified as an example of a 2D shear flow, as suggested earlier when this flow was considered in Section 10.7. Now consider the effect of this flow on a square element of fluid whose side is one unit in length as shown in Figure 10.70; the velocity vectors on the corners

---

EXAMPLE 10.25

Find the rate of shear deformation in the flow described by $u = kx^2y + \alpha/y$, $v = -kxy^2 + \beta/x$, $w = \varepsilon$, where k, α, β, and ε are constants.

SOLUTION

The shear deformation rate is given by Eq. 10.41 as

$$\mathbf{D} = \begin{bmatrix} \frac{\partial u}{\partial x} - \frac{1}{3}\left(\frac{\partial u}{\partial x}+\frac{\partial v}{\partial y}+\frac{\partial w}{\partial z}\right) & \frac{1}{2}\left(\frac{\partial u}{\partial y}+\frac{\partial v}{\partial x}\right) & \frac{1}{2}\left(\frac{\partial u}{\partial z}+\frac{\partial w}{\partial x}\right) \\ \frac{1}{2}\left(\frac{\partial v}{\partial x}+\frac{\partial u}{\partial y}\right) & \frac{\partial v}{\partial y} - \frac{1}{3}\left(\frac{\partial u}{\partial x}+\frac{\partial v}{\partial y}+\frac{\partial w}{\partial z}\right) & \frac{1}{2}\left(\frac{\partial v}{\partial z}+\frac{\partial w}{\partial y}\right) \\ \frac{1}{2}\left(\frac{\partial w}{\partial x}+\frac{\partial u}{\partial z}\right) & \frac{1}{2}\left(\frac{\partial w}{\partial y}+\frac{\partial v}{\partial z}\right) & \frac{\partial w}{\partial z} - \frac{1}{3}\left(\frac{\partial u}{\partial x}+\frac{\partial v}{\partial y}+\frac{\partial w}{\partial z}\right) \end{bmatrix}$$

Because of the many derivatives involved, it is best to solve this problem with a symbolic mathematics code or calculator. The result is

$$\mathbf{D} = \begin{bmatrix} 2kxy & \frac{1}{2}\left[\left(\frac{kx^2y^2-\alpha}{y^2}\right)-\left(\frac{kx^2y^2+\beta}{x^2}\right)\right] & 0 \\ \frac{1}{2}\left[\left(\frac{kx^2y^2-\alpha}{y^2}\right)-\left(\frac{kx^2y^2+\beta}{x^2}\right)\right] & -2kxy & 0 \\ 0 & 0 & 0 \end{bmatrix}$$

We see that in this flow the shear deformation varies significantly with position. This is normally the case for all but the simplest flow fields.

of this element are shown, as are the positions of the fluid particles originally at the corners after one unit of time. Since this is an incompressible, irrotational flow, the fluid element does not rotate or expand as it undergoes a shear deformation that changes its shape from a square to a parallelogram.

10.12 SUMMARY

Flow visualization can be accomplished by means of physical experiment, analytical solution, or computational simulation. The analysis of images generated by flow visualization techniques leads to a visual record of the flow, which engineers can use to enhance their understanding of the physical processes.

The Lagrangian model is useful in flow visualization. A complete Lagrangian description consists of knowing the positions of all fluid particles as a function of time, plus any additional information needed to establish the thermodynamic state of each fluid particle at every instant. The Lagrangian particle path allows us to determine the future location of any fluid particle. The Lagrangian velocity vector may be obtained from the particle path by taking a time derivative. Similarly, the Lagrangian acceleration vector is defined by the first time derivative of the velocity, or equivalently, the second time derivative of the particle path. Identity variables are necessary in the description of a fluid particle, so the position, velocity, and acceleration are functions of time and identity variables.

The connection between the Lagrangian and Eulerian descriptions is based on the observation that a fluid particle whose position is $\mathbf{X}$ in the Lagrangian description is located at $\mathbf{x}$ in the Eulerian description. Thus, the Eulerian velocity vector $\mathbf{u}$ at the point $\mathbf{x}$ at a given time is defined to be the velocity $\mathbf{V}$ of the fluid particle that happens to be at that location at that instant of time. Similarly, the Eulerian acceleration vector $\mathbf{a}$ is defined to be the acceleration $\mathbf{A}$ of the fluid particle that happens to be located at the point in question. In accordance with the concept of a material derivative, the Eulerian–Lagrangian connection is $d\mathbf{X}/dt = D\mathbf{x}/Dt = \mathbf{u}$ and $d\mathbf{V}/dt = D\mathbf{u}/Dt = \mathbf{a}$. These equations provide an engineer with the opportunity to switch between the Eulerian and Lagrangian descriptions in the process of examining a flow field.

In the Lagrangian description, material lines, surfaces, and volumes are always made up of the same fluid particles. To the extent that the continuum hypothesis holds, all lines, surfaces, and volumes in a fluid are of the material type. Although over time, molecular diffusion causes fluid molecules initially in a material entity to wander away, this effect is not accounted for a continuum model.

A pathline is the trajectory through space of a selected fluid particle during a time interval. A streakline is the locus of fluid particles that passed through a specified point during a specified time interval. In general, a streakline is not the same as a pathline because the flow might well have changed direction and magnitude during the marking interval. Material lines, pathlines, and streaklines are Lagrangian concepts that allow us to understand a flow field by observing the behavior of collections of marked fluid particles.

A streamline is an Eulerian concept. It is defined to be a line in space that is everywhere tangent to the local velocity vector. In a steady flow, streamlines, pathlines, and streaklines are identical. A bundle of streamlines defined by a closed material line is called a streamtube. Since the edge of a streamtube comprises streamlines, the fluid velocity vector on the surface of a streamtube is wholly tangential. Therefore no fluid may escape a streamtube, and the cross-sectional area of a streamtube anywhere along its length varies such that the mass flow through the streamtube remains constant.

Information about the rotation, expansion, and shear deformation in a flow is contained in the partial derivatives making up the velocity gradient, $\nabla \mathbf{u}$. The velocity gradient can be written as the sum of the rate of rotation, rate of expansion, and rate of shear deformation: $\nabla \mathbf{u} = \mathbf{R} + \mathbf{E} + \mathbf{D}$.

To characterize the rotation in a flow, we define the angular velocity vector, or rotation vector, at a point in a fluid as $\mathbf{\Omega} = \frac{1}{2} \nabla \times \mathbf{u}$. It is customary, however, to characterize a velocity field in terms of a vorticity vector rather than its rotation vector. The vorticity $\boldsymbol{\omega}$ is defined by $\boldsymbol{\omega} = \nabla \times \mathbf{u}$, so that $\boldsymbol{\omega} = 2\mathbf{\Omega}$. A flow with zero vorticity is called an irrotational flow, and the assumption of irrotational flow is one of the most important approximations in fluid dynamics. Noting that the curl of the gradient of a scalar is zero, we define a scalar velocity potential $\phi(\mathbf{x}, t)$ for irrotational flow by writing $\mathbf{u} = \nabla \phi$. With $\phi(\mathbf{x}, t)$, the condition of irrotational flow is automatically satisfied, and the three velocity components are replaced by a single scalar potential function.

The dilation, Δ, is defined as the divergence of the velocity field, $\Delta = \nabla \cdot \mathbf{u}$. The dilation defines the fractional rate of expansion in a fluid at a point in a flow. If the dilation is positive at a point, the volume of the fluid particle located at that point is increasing and its density is decreasing.

There are two important approximations in fluid mechanics that involve the modeling of fluid density. In the approximation known as an incompressible fluid, the fluid density is assumed to be constant. In the incompressible flow approximation, the density of a fluid particle does not change as it moves; i.e., the material derivative is identically zero. The dilation is zero in an incompressible fluid or in an incompressible flow. Conversely, a nonzero dilation implies a compressible flow.

For an incompressible fluid or flow, a 2D velocity field $(u, v, 0)$ whose components are functions of (x, y, t) must satisfy the relationship $\nabla \cdot \mathbf{u} = \partial u / \partial x + \partial v / \partial y = 0$. If we define a function $\psi = \psi(x, y, t)$ so that $u = \partial \psi / \partial y$ and $v = -\partial \psi / \partial x$, this requirement is automatically satisfied. The function satisfying the preceding pair of equations is called the streamfunction. Setting the streamfunction equal to a constant defines a streamline in the flow. Although the numerical value of the stream function on a given streamline has no meaning by itself, the difference in the streamfunction values on different streamlines, $\psi_2 - \psi_1$, represents the volume flowrate per unit depth crossing the surface defined by the path connecting a point on each streamline. One consequence of this observation is that when adjacent streamlines are closer to each other the average fluid velocity is larger. If adjacent streamlines diverge from one another, the average velocity is smaller. By plotting a family of streamlines, we create a flow visualization that immediately tells us how fast the fluid is moving at different points.

PROBLEMS

Section 10.1

10.1 Consider each of the flow situations and recommend the use of computational and/or experimental analysis.

(a) The interaction of the wind with a bridge
(b) The performance of an artificial heart valve

10.2 Consider each of the flow situations and recommend the use of computational and/or experimental analysis.

(a) The design of the piping systems for a refinery
(b) The performance of a racing yacht

Section 10.2

10.3 The Lagrangian particle paths for a flow are given by

$$X = X_0 e^{C(t-t_0)}, \quad Y = Y_0 e^{-C(t-t_0)}$$

and

$$Z = Z_0$$

where $C = 0.2\ \text{s}^{-1}$ is a constant. Where will a particle located at $X_0 = 1\ \text{m}$, $Y_0 = 1\ \text{m}$, $Z_0 = 0$ at $t = t_0 = 0$ be located at $t = 2\ \text{s}$?

10.4 For the flow given in Problem 10.3, where was the same particle at $t = -2\ \text{s}$?

10.5 Find the inverse of the particle paths given in Problem 10.3. Find the initial position of the particle located at $(2m, 3m, 0)$ at $t = 3\ \text{s}$.

10.6 The Lagrangian particle paths for a flow are given by

$$X = X_0 + U(t - t_0),$$

$$Y = Y_0 + V \sin\left[\omega\left(t_0 - \frac{X_0}{U}\right)\right](t - t_0)$$

and

$$Z = Z_0$$

where U m/s, V m/s, $\omega\ \text{s}^{-1}$ are constants. Find the location of the particle at times $\pi/4\omega$, $\pi/2\omega$, $3\pi/4\omega$, and π/ω initially located at $X_0 = 0$, $Y_0 = 0$, $Z_0 = 0$ at $t = t_0 = 0$. If $t_0 = \pi/4\omega$, where will the particle located at $X_0 = 0$, $Y_0 = 0$, $Z_0 = 0$ be at times $\pi/2\omega$, $3\pi/4\omega$, and π/ω?

10.7 For the flow given in Problem 10.6, where was the particle at $t = -\pi/4\omega$ that is currently located at $X_0 = U\pi/\omega$, $Y_0 = 0$, $Z_0 = 0$ at $t_0 = 0$?

10.8 Find the inverse of the particle paths given in Problem 10.6. Find the initial position of the particle located at $(U\pi/\omega, V\pi/\omega, 0)$ at $t = \pi/\omega$.

10.9 The Lagrangian particle paths for an irrotational vortex are given by

$$R = R_0, \quad \theta = \theta_0 + \frac{K}{2\pi R_0^2}(t - t_0)$$

and

$$Z = Z_0$$

where K is a constant in m²/s. Where will the particle located at $R_0 = 1m$, $\theta_0 = 0$, $Z_0 = 0$ at $t_0 = 0$, be at $t = \pi s$?

10.10 For the flow given in Problem 10.9, where was the same particle at $t = -2\pi$?

10.11 Find the inverse of the particle paths given in Problem 10.9. Find the initial position of the particle located at $(2m, 0, 0)$ at $t = \pi s$.

Section 10.3

10.12 Determine the Lagrangian velocity and acceleration for the flow given in Problem 10.3.

10.13 Determine the Eulerian velocity and acceleration using the Lagrangian particle paths given in Problem 10.3.

10.14 Determine the Lagrangian velocity and acceleration for the flow given in Problem 10.6.

10.15 Determine the Eulerian velocity and acceleration using the Lagrangian particle paths given in Problem 10.6.

10.16 Determine the Lagrangian velocity and acceleration for the flow given in Problem 10.9.

10.17 Determine the Eulerian velocity using the Lagrangian particle paths given in Problem 10.9.

Section 10.4

10.18 Find an expression that predicts the shape of the material line initially located between (0, 0, 0) and (1, 1, 0) for the flow given in Problem 10.3.

10.19 Find an expression that predicts the shape of a disk of radius 0.5 m initially centered at (1, 1, 0) for the flow given in Problem 10.3.

10.20 Find an expression that predicts the shape of a horizontal and a vertical material line for the flow given in Problem 10.6.

10.21 Find an expression that predicts the future shape of a line between the points (0, 0, 0) and (1, 0, 0) at $t = 0$ for the flow given in Problem 10.9.

10.22 Find an expression that predicts the future shape of the horizontal line between the points $(1, \pi/2, 0)$ and $(\sqrt{2}, \pi/4, 0)$ at $t = 0$ for the flow given in Problem 10.9.

Section 10.5

10.23 For the flow described in Problem 10.3, sketch the streakline initiated at (1, 1, 0). Sketch the pathline for the particle at this point at $t = 0$.

10.24 For the flow described in Problem 10.6, sketch the streakline initiated at (0, 0, 0). Sketch the pathline for the particle at this point at $t = 0$.

10.25 As you have seen in many of the figures in this book, hydrogen bubbles are often used as flow visualization technique. The velocity field for a flow is $\mathbf{u} = 2$ ft/s $\mathbf{i} + 1$ ft/s $\mathbf{j}$ for $0 \le t < 1$ and $\mathbf{u} = 1$ ft/s $\mathbf{i} + 2$ ft/s $\mathbf{j}$ for $1 \le t < 2$ s. Plot the pathlines of bubbles that leave (0, 0) at $t = 0, 0.5, 1, 1.5,$ and 2 s. Indicate the streakline at $t = 2$ s.

10.26 Find the equation for the Lagrangian equation for the pathline that passes through a point (X_0, Y_0) at time $t = 0$ for the Eulerian velocity field $\mathbf{u} = C_1 x \mathbf{i} + C_2 y (C_3 + t) \mathbf{j}$. C_1, C_2 and C_3 are constants of dimensions t^{-1}, t^{-2} and t, respectively.

10.27 A fluid particle is instantaneously marked with dye and then a long exposure photograph of the particle's motion is taken. Is the blurred line on the photograph a pathline, a material line, or a streakline? Explain.

10.28 Given the velocity field $\mathbf{u} = C(x^2 - y^2)\mathbf{i} - 2Cxy\mathbf{j}$, where $C = 1 \, \text{m}^{-1}\text{s}^{-1}$, show that the streaklines are given by the equation $x^2 y - y^3/3 = K$ where K is a constant. Sketch the streakline through the point (1,1).

10.29 Sketch the streakline through the origin for the velocity field $\mathbf{u} = Cx\mathbf{i} + C^3 x(x - 1\text{m})(y + 1\text{m})\mathbf{j}$, where $C = 1 \, \text{s}^{-1}$ and the dimensions are in meters.

Section 10.6

10.30 An advertisement for a CFD code states that it can identify where streamlines intersect in complex flows. Is this a good feature for a CFD code?

10.31 For the velocity field given in Problem 10.26, sketch the streamlines at $t = 0$ and 2 s.

10.32 Sketch the streamlines for the velocity field given in Problem 10.28.

10.33 Sketch the streamlines for the velocity field given in Problem 10.29.

10.34 The velocity field for a flow is given by $\mathbf{u} = 3 \, \text{s}^{-1} y \mathbf{i} + 2 \, \text{ft/s} \, \mathbf{j}$, where y is in feet. Determine the equation for the steamlines and make a sketch for the upper half-plane.

10.35 The velocity field for a flow is given by
$$\mathbf{u} = \frac{-Cy}{\sqrt{x^2 + y^2}} \mathbf{i} + \frac{Cx}{\sqrt{x^2 + y^2}} \mathbf{j}$$
where C is a constant. Determine the equations for the streamlines and make a sketch.

10.36 The velocity field for a flow is given by $\mathbf{u} = 2 \, (\text{m-s})^{-1} x^2 \mathbf{i} - 3 \, (\text{m-s})^{-1} xy \mathbf{j}$, where x and y are in meters. Determine the equations for the streamlines and make a sketch.

10.37 The velocity field for a flow is given by $\mathbf{u} = 2 \, (\text{m-s})^{-1} xy \mathbf{i} - 3 \, (\text{m-s})^{-1} y^2 \mathbf{j}$, where x and y are in meters. Determine the equations for the streamlines and make a sketch.

10.38 The velocity field for a flow is given by $\mathbf{u} = 5\,\text{s}^{-1}x\mathbf{i} - 1\,\text{cm}^{1/2}\text{s}^{-1}y^{1/2}\mathbf{j}$, where x and y are in centimeters. Determine the equations for the streamlines and make a sketch.

Section 10.8

10.39 The velocity field $\mathbf{u} = Cx\mathbf{i} - Cy\mathbf{j}$ models flow in a corner for the upper right quadrant. Determine the velocity gradient for this flow. What is the velocity gradient at $(1, 0)$ along the wall?

10.40 Given the velocity field $\mathbf{u} = (x^2 - y^2)\mathbf{i} - 2xy\mathbf{j}$ determine the velocity gradient.

10.41 The velocity field for a flow is given by $\mathbf{u} = 3\,\text{s}^{-1}y\mathbf{i} + 2\,\text{ft/s}\,\mathbf{j}$, where y is in feet. Find the equation for the velocity gradient.

10.42 The velocity field for a flow is given by $\mathbf{u} = 2\,(\text{m-s})^{-1}x^2\mathbf{i} - 3\,(\text{m-s})^{-1}xy\mathbf{j}$, where x and y are in meters. Determine the velocity gradient.

10.43 The velocity field for a flow is given by $\mathbf{u} = 5\,\text{s}^{-1}x\mathbf{i} - 1(\text{cm}^{1/2}/\text{s})y^{1/2}\mathbf{j}$, where x and y are in centimeters. Determine the velocity gradient.

10.44 The velocity field for flow over a cylinder is given by

$$v_r = \frac{-A\cos\theta}{r^2} + U\cos\theta$$

$$v_\theta = \frac{-A\sin\theta}{r^2} - U\sin\theta$$

where A is a constant and U is the oncoming velocity. Determine the velocity gradient at $r = (A/U)^{0.5}$.

Section 10.9

10.45 Determine the vorticity field for the velocity field given in Problem 10.39.

10.46 Determine the vorticity field for the velocity field given in Problem 10.40.

10.47 Determine the vorticity field for the velocity field given in Problem 10.41.

10.48 Determine the vorticity field for the velocity field given in Problem 10.42.

10.49 Determine the vorticity field for the velocity field given in Problem 10.44.

10.50 Determine the circulation around the path shown in Figure P10.1 for the velocity field given in Problem 10.39.

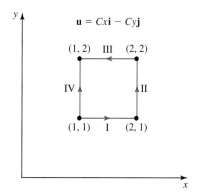

Figure P10.1

10.51 Determine the circulation around the path shown in Figure P10.2 for the velocity field given in Problem 10.44.

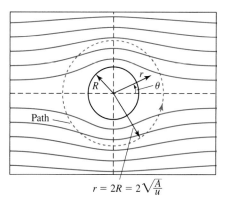

Figure P10.2

10.52 Determine whether the flow given in Problem 10.39 is irrotational. If it is, calculate the velocity potential.

10.53 Determine whether the flow given in Problem 10.44 is irrotational. If it is, calculate the velocity potential.

Section 10.10

10.54 Determine whether the flow given in Problem 10.39 is an incompressible flow.

10.55 Determine whether the flow given in Problem 10.40 is an incompressible flow.

10.56 Determine whether the flow given in Problem 10.41 is an incompressible flow.

10.57 If the density distribution for the flow given in Problem 10.39 is $\rho = Bx$, where B is a constant, is this an incompressible fluid or an incompressible flow?

10.58 If the density distribution for the flow given in Problem 10.44 is $\rho = Br$, where B is a constant, is this an incompressible fluid or an incompressible flow?

Section 10.11

10.59 Determine the shear deformation rate for the flow given in Problem 10.39.

10.60 Determine the shear deformation rate for the flow given in Problem 10.40.

10.61 Determine the shear deformation rate for the flow given in Problem 10.41.

10.62 Determine the shear deformation rate for the flow given in Problem 10.42.

ADDITIONAL PROBLEMS

10.63 For the velocity field given in Problem 10.39, characterize the relative motion in this flow at an arbitrary point (x, y, z) by determining the translation, rotation, expansion, and shear deformation contributions to the velocity at the nearby point $(x + a, y + b, z + c)$.

10.64 For the velocity field given in Problem 10.40, characterize the relative motion in this flow at an arbitrary point (x, y, z) by determining the translation, rotation, expansion, and shear deformation contributions to the velocity at the nearby point $(x + a, y + b, z + c)$.

10.65 For the velocity field given in Problem 10.41, characterize the relative motion in this flow at an arbitrary point (x, y, z) by determining the translation, rotation, expansion, and shear deformation contributions to the velocity at the nearby point $(x + a, y + b, z + c)$.

10.66 For the velocity field given in Problem 10.44, characterize the relative motion in this flow at an arbitrary point (x, y, z) by determining the translation, rotation, expansion, and shear deformation contributions to the velocity at the nearby point $(x + a, y + b, z + c)$.

11 GOVERNING EQUATIONS OF FLUID DYNAMICS

11.1 Introduction
11.2 Continuity Equation
11.3 Momentum Equation
11.4 Constitutive Model for a Newtonian Fluid
11.5 Navier–Stokes Equations
11.6 Euler Equations
    11.6.1 Streamline Coordinates
    11.6.2 Derivation of the Bernoulli Equation
11.7 Energy Equation
11.8 Discussion
    11.8.1 Initial and Boundary Conditions
    11.8.2 Nondimensionalization
    11.8.3 Computational Fluid Dynamics
11.9 Summary
Problems

11.1 INTRODUCTION

In this chapter we develop the differential form of the governing equations of fluid dynamics and describe how they are used to obtain a complete description of a flow. These partial differential equations express conservation of mass, momentum, and energy. The equations equate the time rate of change of mass, momentum, or energy at a point in a fluid to source terms representing various physical mechanisms affecting the rate at which these properties change over time.

11 GOVERNING EQUATIONS OF FLUID DYNAMICS

Figure 11.1 The direct numerical solution of the governing equations was used for this visualization of a supersonic jet. The contours of vorticity magnitude indicate the jet in the foreground; the near acoustic field is visualized with gray levels indicating divergence of velocity, which highlights the weak shocks (dark) and relatively broad expansions (light).

Engineers are normally taught to solve the governing equations that model a problem in engineering science. The attractive feature of a fluid mechanics solution is that it provides a complete description of the flow at every point throughout time. Thus, when a solution is available, there is virtually no limit on an engineer's ability to answer questions concerning the transport of mass, momentum and energy, the kinematics and structure of the fluid velocity field, the distribution of forces within the fluid, and the interaction of the fluid and its surroundings.

Unfortunately, the governing equations of fluid mechanics are a complex set of nonlinear partial differential equations, and the number of known solutions of these equations is small. The practice of fluid dynamics in the past has primarily involved the judicious use of approximations, applying knowledge about the general behavior of a fluid obtained from scarce analytical solutions, and performing a multitude of experiments. Today, the number of known exact analytical solutions remains small, but the likelihood of obtaining a numerical solution of the governing equations is far greater than it once was. Engineers now routinely have access to powerful computational hardware and software that are capable of providing a solution for many important flow problems. We therefore believe that an exposure to the governing equations and their solution, as provided in this chapter, is both an important part of an education in fluid mechanics techniques and serves as the foundation for an understanding of the power of computational fluid dynamics. An example of a CFD solution of a complex flow is shown in Figure 11.1.

 CD/Kinematics/Compressibility/The equation of continuity

11.2 CONTINUITY EQUATION

The partial differential equation expressing conservation of mass is called the continuity equation. The derivation of this scalar equation can be done in a number of ways, but an elegant approach begins by applying the Reynolds transport theorem to calculate the

11.2 CONTINUITY EQUATION

time rate of change of the total amount E_{sys} of some property in a fluid system. We did this earlier, in Chapter 7. The result, Eq. 7.4,

$$\frac{dE_{sys}}{dt} = \int_{R(t)} \frac{\partial}{\partial t}(\rho\varepsilon)\,dV + \int_{S(t)} (\rho\varepsilon)\mathbf{u}\cdot\mathbf{n}\,dS$$

expresses the time rate of change of the total amount E_{sys} of some property in the fluid system or material volume of fluid occupying a region $R(t)$ in terms of a volume and a surface integral over this region. To derive a mass conservation statement, E_{sys} is chosen as M_{sys}, the total mass in the system. The intensive counterpart of the total mass is the mass per unit mass, i.e., $\varepsilon = 1$, so choosing $\rho\varepsilon = \rho$ we have

$$\frac{dM_{sys}}{dt} = \int_{R(t)} \frac{\partial \rho}{\partial t}\,dV + \int_{S(t)} \rho\mathbf{u}\cdot\mathbf{n}\,dS$$

Since mass is conserved for a material volume of fluid, $dM_{sys}/dt = 0$, and the integral form for the law of conservation of mass for a material volume of fluid is

$$\int_{R(t)} \frac{\partial \rho}{\partial t}\,dV + \int_{S(t)} \rho\mathbf{u}\cdot\mathbf{n}\,dS = 0$$

This is identical in form to the integral mass conservation equation for a CV (Eq. 7.11).

To derive the corresponding differential equation, we rewrite the surface integral in this equation as a volume integral using the Gauss theorem. The result is

$$\int_{R(t)} \frac{\partial \rho}{\partial t}\,dV + \int_{R(t)} \nabla\cdot(\rho\mathbf{u})\,dV = 0$$

Next, the two integrals over the same volume can be combined, giving

$$\int_{R(t)} \left(\frac{\partial \rho}{\partial t} + \nabla\cdot\rho\mathbf{u}\right) dV = 0$$

Since the volume is arbitrary, the integrand must be zero. This leads to the following differential equation, known as the continuity equation:

$$\frac{\partial \rho}{\partial t} + \nabla\cdot(\rho\mathbf{u}) = 0 \quad (11.1a)$$

This form of the continuity equation, referred to as the divergence form, is frequently used in computational fluid dynamics. Two other forms of the continuity equation are also in common use. One of these, the expanded form, is obtained by substituting the vector identity $\nabla\cdot(\rho\mathbf{u}) = \mathbf{u}\cdot\nabla\rho + \rho\nabla\cdot\mathbf{u}$ into Eq. 11.1a to obtain

$$\frac{\partial \rho}{\partial t} + \mathbf{u}\cdot\nabla\rho + \rho\nabla\cdot\mathbf{u} = 0 \quad (11.1b)$$

The remaining form is obtained by replacing the first two terms in Eq. 11.1b by the material derivative, $D\rho/Dt = \partial\rho/\partial t + (\mathbf{u}\cdot\nabla)\rho$, (see Eq. 6.9), to obtain:

$$\frac{D\rho}{Dt} + \rho\nabla\cdot\mathbf{u} = 0 \quad (11.1c)$$

The continuity equation involves only the fluid density and the fluid velocity. It applies to all fluids, compressible and incompressible, Newtonian and non-Newtonian, and for the whole range of flow speeds. As one of the three fundamental governing equations of fluid mechanics, it expresses the law of conservation of mass at each point in the fluid. Thus, the continuity equation must be satisfied at every point in the flow field.

When expanded in Cartesian coordinates, the continuity equation is given by

$$\left(\frac{\partial \rho}{\partial t} + u\frac{\partial \rho}{\partial x} + v\frac{\partial \rho}{\partial y} + w\frac{\partial \rho}{\partial z}\right) + \rho\left(\frac{\partial u}{\partial x} + \frac{\partial v}{\partial y} + \frac{\partial w}{\partial z}\right) = 0 \quad (11.2a)$$

In cylindrical coordinates the continuity equation is

$$\left(\frac{\partial \rho}{\partial t} + v_r\frac{\partial \rho}{\partial r} + \frac{v_\theta}{r}\frac{\partial \rho}{\partial \theta} + v_z\frac{\partial \rho}{\partial z}\right) + \rho\left(\frac{1}{r}\frac{\partial(rv_r)}{\partial r} + \frac{1}{r}\frac{\partial v_\theta}{\partial \theta} + \frac{\partial v_z}{\partial z}\right) = 0 \quad (11.2b)$$

EXAMPLE 11.1

Are the flows represented by the following velocity and density functions physically possible?

A. $u = Axy^2$, $v = -Ax^2y$, $w = 0$, and $\rho = Bxy$, where A and B are constants

B. $v_r = U_\infty(1 - R^2/r^2)\cos\theta$, $v_\theta = -U_\infty(1 + R^2/r^2)\sin\theta$, $v_z = 0$, and $\rho = Cz + \rho_0$, where U_∞, R, C, and ρ_0 are constants

SOLUTION

To be physically possible, the continuity equation must be satisfied. For case A, we use Eq. 11.2a, $[\partial \rho/\partial t + u(\partial \rho/\partial x) + v(\partial \rho/\partial y) + w(\partial \rho/\partial z)] + \rho(\partial u/\partial x + \partial v/\partial y + \partial w/\partial z) = 0$, and substitute the given functions for u, v, w, and ρ. Since this a steady flow, the time derivative is zero. Also, there is no z dependence in the density or velocity, and the z velocity component is zero. Thus, after substituting we find:

$$(Axy^2)\frac{\partial(Bxy)}{\partial x} + (-Ax^2y)\frac{\partial(Bxy)}{\partial y} + (Bxy)\left(\frac{\partial(Axy^2)}{\partial x} + \frac{\partial(-Ax^2y)}{\partial y}\right) = 0$$

Simplifying, we obtain:

$$(ABxy^3) + (-ABx^3y) + (ABxy^3) + (-ABx^3y) = (xy^3 - x^3y) \neq 0$$

Since this flow does not satisfy the continuity equation, it cannot be a physically possible flow.

For case B we use Eq. 11.2b:

$$\left(\frac{\partial \rho}{\partial t} + v_r\frac{\partial \rho}{\partial r} + \frac{v_\theta}{r}\frac{\partial \rho}{\partial \theta} + v_z\frac{\partial \rho}{\partial z}\right) + \rho\left(\frac{1}{r}\frac{\partial(rv_r)}{\partial r} + \frac{1}{r}\frac{\partial v_\theta}{\partial \theta} + \frac{\partial v_z}{\partial z}\right) = 0$$

and substitute the given functions, noting that this is a steady flow with $v_z = 0$. Ignoring terms that are zero, the result is

$$(Cz + \rho_0)\left(\frac{1}{r}\frac{\partial\{r[U_\infty(1 - R^2/r^2)\cos\theta]\}}{\partial r} + \frac{1}{r}\frac{\partial[-U_\infty(1 + R^2/r^2)\sin\theta]}{\partial \theta} + \frac{\partial(0)}{\partial z}\right) = 0$$

Evaluating the derivatives we find

$$(Cz + \rho_0)\left[\frac{U_\infty \cos\theta}{r}\left(1 - \frac{R^2}{r^2} - 1 - \frac{R^2}{r^2} + \frac{2R^2}{r^2}\right)\right] = 0$$

Since the continuity equation is satisfied, the flow described in case B is a physically possible flow. In fact it describes inviscid flow over a cylinder.

EXAMPLE 11.2

Consider an infinitesimal CV filled with fluid as shown in Figure 11.2. Apply a mass balance to this CV to derive the continuity equation in Cartesian coordinates.

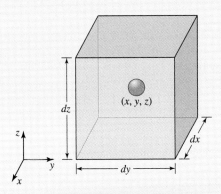

Figure 11.2 Schematic for Example 11.2.

SOLUTION

Figure 11.2 serves as the appropriate sketch. If the value of density at the cube center located at (x, y, z) is ρ, then the value on the near face is given by a Taylor series as

$$\rho_{\text{near}} = \rho + \left(\frac{\partial \rho}{\partial x}\right)\left(\frac{dx}{2}\right) + \frac{1}{2!}\left(\frac{\partial^2 \rho}{\partial x^2}\right)\left(\frac{dx}{2}\right)^2 + \cdots$$

To the first order (since the cube is infinitesimal, we can safely assume that higher order terms are negligible), we have $\rho_{\text{near}} = \rho + (\partial\rho/\partial x)(dx/2)$. Similarly, for velocity we find

$$u_{\text{near}} = u + \left(\frac{\partial u}{\partial x}\right)\left(\frac{dx}{2}\right), \quad v_{\text{near}} = v + \left(\frac{\partial v}{\partial x}\right)\left(\frac{dx}{2}\right)$$

and

$$w_{\text{near}} = w + \left(\frac{\partial w}{\partial x}\right)\left(\frac{dx}{2}\right)$$

The values on the far face are found to be

$$\rho_{\text{far}} = \rho - \left(\frac{\partial \rho}{\partial x}\right)\left(\frac{dx}{2}\right), \quad u_{\text{far}} = u - \left(\frac{\partial u}{\partial x}\right)\left(\frac{dx}{2}\right), \quad v_{\text{far}} = v - \left(\frac{\partial v}{\partial x}\right)\left(\frac{dx}{2}\right),$$

and

$$w_{\text{far}} = w - \left(\frac{\partial w}{\partial x}\right)\left(\frac{dx}{2}\right)$$

The same procedure is applied to the other four faces, using the appropriate spatial derivative and differential element in each case.

To apply a mass balance, we write $\int_{\text{CV}}(\partial\rho/\partial t)\,dV + \int_{\text{CS}} \rho(\mathbf{u}\cdot\mathbf{n})\,dS = 0$ and use the appropriate values of density and velocity to evaluate the integrals. Note that for an infinitesimal cube, the integrand in each volume or surface integral is constant. For example, the volume integral yields $\int_{\text{CV}}(\partial\rho/\partial t)\,dV = (\partial\rho/\partial t)V = (\partial\rho/\partial t)(dx\,dy\,dz)$, while each surface integral provides a term of the form $\int_{\text{CS}} \rho(\mathbf{u}\cdot\mathbf{n})\,dS = \rho(\mathbf{u}\cdot\mathbf{n})A$, where A is the area of the face. The value of $\rho(\mathbf{u}\cdot\mathbf{n})A$ on each face involves only one velocity component, and it is convenient to consider faces in pairs. For example, consider the near and far faces. On the near face we have $(\mathbf{u}\cdot\mathbf{n}) = u_{\text{near}}$, and the mass flux is

$$\rho(\mathbf{u}\cdot\mathbf{n})A = \rho_{\text{near}}(u_{\text{near}})(dy\,dz) = \left[\rho + \left(\frac{\partial\rho}{\partial x}\right)\left(\frac{dx}{2}\right)\right]\left[u + \left(\frac{\partial u}{\partial x}\right)\left(\frac{dx}{2}\right)\right](dy\,dz)$$

$$= (\rho u)\,dy\,dz + \frac{1}{2}\left[\rho\left(\frac{\partial u}{\partial x}\right) + u\left(\frac{\partial\rho}{\partial x}\right)\right]dx\,dy\,dz + \left[\frac{1}{4}\left(\frac{\partial\rho}{\partial x}\right)\left(\frac{\partial u}{\partial x}\right)dx\right]dx\,dy\,dz$$

$$= (\rho u)\,dy\,dz + \frac{1}{2}\left[\rho\left(\frac{\partial u}{\partial x}\right) + u\left(\frac{\partial\rho}{\partial x}\right)\right]dx\,dy\,dz \quad \text{to first order}$$

On the far face we have $(\mathbf{u}\cdot\mathbf{n}) = -u_{\text{far}}$, and the mass flux is

$$\rho(\mathbf{u}\cdot\mathbf{n})A = \rho_{\text{far}}(-u_{\text{far}})(dy\,dz) = \left[\rho - \left(\frac{\partial\rho}{\partial x}\right)\left(\frac{dx}{2}\right)\right]\left[-u + \left(\frac{\partial u}{\partial x}\right)\left(\frac{dx}{2}\right)\right](dy\,dz)$$

$$= (-\rho u)\,dy\,dz + \frac{1}{2}\left[\rho\left(\frac{\partial u}{\partial x}\right) + u\left(\frac{\partial\rho}{\partial x}\right)\right]dx\,dy\,dz \quad \text{to first order}$$

The two faces combine to give a net mass flux of $[\rho(\partial u/\partial x) + u(\partial \rho/\partial x)]\, dx\, dy\, dz$. The other face pairs yield similar net mass fluxes of $[\rho(\partial v/\partial y) + v(\partial \rho/\partial y)]\, dx\, dy\, dz$ and $[\rho(\partial w/\partial z) + w(\partial \rho/\partial z)]\, dx\, dy\, dz$.

Completing the mass balance we find:

$$\frac{\partial \rho}{\partial t}(dx\, dy\, dz) + \left[\rho\left(\frac{\partial u}{\partial x}\right) + u\left(\frac{\partial \rho}{\partial x}\right)\right] dx\, dy\, dz + \left[\rho\left(\frac{\partial v}{\partial y}\right) + v\left(\frac{\partial \rho}{\partial y}\right)\right] dx\, dy\, dz$$

$$+ \left[\rho\left(\frac{\partial w}{\partial z}\right) + w\left(\frac{\partial \rho}{\partial z}\right)\right] dx\, dy\, dz = 0$$

Dividing by $(dx\, dy\, dz)$ and rearranging yields the continuity equation in the form of Eq. 11.2a:

$$\left(\frac{\partial \rho}{\partial t} + u\frac{\partial \rho}{\partial x} + v\frac{\partial \rho}{\partial y} + w\frac{\partial \rho}{\partial z}\right) + \rho\left(\frac{\partial u}{\partial x} + \frac{\partial v}{\partial y} + \frac{\partial w}{\partial z}\right) = 0$$

We see that it is possible to derive this equation in different ways.

In Chapter 10 we discussed two important approximations in fluid mechanics, namely, incompressible fluid and incompressible flow, and we pointed out that in analyzing a flow problem, an engineer must often decide whether to treat the fluid density as a variable or as a constant. Recall that the definition of incompressible fluid is given by Eq. 10.68 as $\rho(\mathbf{x}, t) = \rho_0$, where ρ_0 is constant. In an incompressible flow, the density obeys Eq. 10.69, $D\rho/Dt = 0$. You should be able to easily verify that each term in the material derivative of density is zero if the density is constant. Notice, however, that only the sum of the various derivatives in the material derivative of density must be zero for an incompressible flow. The various derivatives themselves may be nonzero, which means that in an incompressible flow density may actually vary in space but only in a prescribed way.

The continuity equation is considerably simplified by the assumption of either an incompressible fluid or incompressible flow. To see this, note that the general form of the continuity equation, applicable to all fluids and flows, is given by Eq. 11.1c as $D\rho/Dt + \rho \nabla \cdot \mathbf{u} = 0$. For an incompressible fluid or an incompressible flow, we know $D\rho/Dt = 0$, thus the continuity equation reduces to

$$\nabla \cdot \mathbf{u} = 0 \tag{11.3}$$

The velocity field of an incompressible flow, and that of an incompressible fluid, must satisfy this equation, which replaces the general continuity equation. Equation 11.3 says that the dilation, or divergence of velocity, in such a case is zero. Thus in Cartesian coordinates the velocity field must satisfy

$$\frac{\partial u}{\partial x} + \frac{\partial v}{\partial y} + \frac{\partial w}{\partial z} = 0 \tag{11.4a}$$

EXAMPLE 11.3

Which of the following constant density flows are physically possible?

A. $u = \left(\dfrac{h^2(p_1 - p_2)}{2\mu L}\right)\left[1 - \left(\dfrac{y}{h}\right)^2\right]$, $v = 0$, and $w = 0$

B. $v_r = 0$, $v_\theta = 0$, and $v_z(r) = \left(\dfrac{R_P^2(p_1 - p_2)}{4\mu L}\right)\left[1 - \left(\dfrac{r}{R_P}\right)^2\right]$

SOLUTION

The velocity components must satisfy the simplified continuity equation for a constant density fluid. This is Eq. 11.4a or 11.4b, depending on the coordinate system in use.

A. Since we have $u = u(y)$ and $(v = w = 0)$, substituting the three velocity components into Eq. 11.4a gives

$$\frac{\partial u}{\partial x} + \frac{\partial v}{\partial y} + \frac{\partial w}{\partial z} = \frac{\partial}{\partial x}(u(y)) + \frac{\partial}{\partial y}(0) + \frac{\partial}{\partial z}(0) = 0$$

Equation 11.4a is satisfied, so this flow is possible. In fact, you may have recognized it as channel flow.

B. In this case the relevant equation is Eq. 11.4b:

$$\frac{1}{r}\frac{\partial(rv_r)}{\partial r} + \frac{1}{r}\frac{\partial v_\theta}{\partial \theta} + \frac{\partial v_z}{\partial z} = 0$$

Since we are told $v_r = 0$ and $v_\theta = 0$, this reduces to $\partial v_z/\partial z = 0$, which can be seen to be satisfied by inspection. This flow is also possible; in fact, it is Poiseuille flow in a round pipe.

while in cylindrical coordinates the requirement is

$$\frac{1}{r}\frac{\partial(rv_r)}{\partial r} + \frac{1}{r}\frac{\partial v_\theta}{\partial \theta} + \frac{\partial v_z}{\partial z} = 0 \tag{11.4b}$$

11.3 MOMENTUM EQUATION

CD/Dynamics/Newton's second law of motion/The momentum equation

The partial differential equation expressing conservation of momentum for a Newtonian fluid is called the Navier–Stokes equation. The derivation of this vector equation also

11.3 MOMENTUM EQUATION

begins by applying the Reynolds transport theorem to a material volume of fluid in the form:

$$\frac{dE_{sys}}{dt} = \int_{R(t)} \frac{\partial}{\partial t}(\rho\varepsilon)\,dV + \int_{S(t)} (\rho\varepsilon)\mathbf{u}\cdot\mathbf{n}\,dS$$

In this case E_{sys} is chosen to be the total linear momentum in the volume, for which the intensive counterpart is the linear momentum per unit mass $\varepsilon = \mathbf{u}$. By Newton's law, the time rate of change of linear momentum within the material volume equals the sum of the body and surface forces acting on the volume:

$$\int_{R(t)} \frac{\partial}{\partial t}(\rho\mathbf{u})\,dV + \int_{S(t)} (\rho\mathbf{u})\mathbf{u}\cdot\mathbf{n}\,dS = \int_{R(t)} \rho\mathbf{f}\,dV + \int_{S(t)} \mathbf{\Sigma}\,dS$$

Note that the momentum equation for a material volume is identical to Eq. 7.18 for a CV; moreover, we have used the fact that the total body force is given as usual by the volume integral (Eq. 4.7) $\mathbf{F}_B = \int_{R(t)} \rho\mathbf{f}\,dV$, while the total surface force is given by the surface integral (Eq. 4.21) $\mathbf{F}_S = \int_{S(t)} \mathbf{\Sigma}\,dS$. To derive the differential momentum equation, we will write the surface integral in terms of the stress tensor rather than the stress vector, using Eq. 4.32 to write $\mathbf{F}_S = \int_S (\mathbf{n}\cdot\boldsymbol{\sigma})\,dS$. Next we use Gauss's theorem to write the surface integral in terms of the volume integral of the stress divergence $(\nabla\cdot\boldsymbol{\sigma})$, as defined by Eqs. 4.39a–4.39c in Section 4.7. Thus, the surface force is now given by the volume integral $\mathbf{F}_S = \int_{R(t)} (\nabla\cdot\boldsymbol{\sigma})\,dV$. Substituting this result into the momentum equation for a material volume, we have

$$\int_{R(t)} \frac{\partial}{\partial t}(\rho\mathbf{u})\,dV + \int_{S(t)} (\rho\mathbf{u})\mathbf{u}\cdot\mathbf{n}\,dS = \int_{R(t)} \rho\mathbf{f}\,dV + \int_{R(t)} (\nabla\cdot\boldsymbol{\sigma})\,dV$$

To derive a differential equation, we use Gauss's theorem to transform the flux integral into a volume integral, noting that we need the tensor form because $(\rho\mathbf{u})\mathbf{u}$ is a tensor. Next, we combine all the volume integrals into one, obtaining

$$\int_{R(t)} \left(\frac{\partial(\rho\mathbf{u})}{\partial t} + \nabla\cdot(\rho\mathbf{u}\mathbf{u}) - \rho\mathbf{f} - \nabla\cdot\boldsymbol{\sigma}\right)dV = 0$$

Since the volume is arbitrary, the integrand must be zero. Thus the differential equation expressing the law of momentum conservation is given by

$$\frac{\partial(\rho\mathbf{u})}{\partial t} + \nabla\cdot(\rho\mathbf{u}\mathbf{u}) = \rho\mathbf{f} + \nabla\cdot\boldsymbol{\sigma}$$

The preceding equation is referred to as the conservative form of the differential momentum equation. This form serves as the starting point in many numerical algorithms used to solve the governing equations in computational fluid dynamics.

The traditional form of the momentum equation is obtained by expanding the time derivative and divergence terms, then rearranging the remaining terms to obtain

$$\rho\left(\frac{\partial\mathbf{u}}{\partial t} + \mathbf{u}\cdot\nabla\mathbf{u}\right) + \mathbf{u}\left[\frac{\partial\rho}{\partial t} + \mathbf{u}\cdot\nabla\rho + \rho\nabla\cdot\mathbf{u}\right] = \rho\mathbf{f} + \nabla\cdot\boldsymbol{\sigma}$$

The term in the square bracket is the left-hand side of the continuity equation (see Eq. 11.1b). Thus this term is equal to zero, and we can write the differential momentum equation as

$$\rho\left(\frac{\partial \mathbf{u}}{\partial t} + \mathbf{u} \cdot \nabla \mathbf{u}\right) = \rho \mathbf{f} + \nabla \cdot \boldsymbol{\sigma}$$

Using the material derivative, the differential momentum equation takes its traditional form:

$$\rho \frac{D\mathbf{u}}{Dt} = \rho \mathbf{f} + \nabla \cdot \boldsymbol{\sigma}$$

In deriving the momentum equation, we have employed the fluid density, fluid velocity, and stress tensor as variables but have not restricted the discussion to a certain type of fluid. Thus the momentum equation, like the continuity equation, is applicable to all fluids, compressible and incompressible, Newtonian and non-Newtonian, and for the whole range of flow speeds. As one of the three fundamental governing equations of fluid mechanics, it expresses the law of conservation of momentum at each point in the fluid. Although every physically possible fluid flow must satisfy this equation, it cannot be solved unless we introduce a constitutive model that provides relationships between the stress tensor and the velocity field. We will discuss the constitutive model for a Newtonian fluid in the next section.

Let us now consider what the various terms in the momentum equation represent. Recalling that the material derivative of velocity defines the fluid acceleration, we see that the left-hand side of the momentum equation is the product of density and fluid acceleration. Thus we could write the momentum equation as

$$\rho \mathbf{a} = \rho \mathbf{f} + \nabla \cdot \boldsymbol{\sigma}$$

The left-hand side of this equation represents the inertial force per unit volume. The two terms on the right represent the body and surface forces per unit volume (the latter in terms of the stress divergence). Thus, the momentum equation represents a balance of inertial, body, and surface force per unit volume at each point in a fluid. You may find it worthwhile at this point to reread Section 4.7 and review the effects of stress variation in a fluid.

Using the definition of the stress divergence, Eqs. 4.39, we can write the three components of the momentum equation in Cartesian coordinates as

$$\rho\left(\frac{\partial u}{\partial t} + u\frac{\partial u}{\partial x} + v\frac{\partial u}{\partial y} + w\frac{\partial u}{\partial z}\right) = \rho f_x + \left(\frac{\partial \sigma_{xx}}{\partial x} + \frac{\partial \sigma_{yx}}{\partial y} + \frac{\partial \sigma_{zx}}{\partial z}\right) \quad (11.5\text{a})$$

$$\rho\left(\frac{\partial v}{\partial t} + u\frac{\partial v}{\partial x} + v\frac{\partial v}{\partial y} + w\frac{\partial v}{\partial z}\right) = \rho f_y + \left(\frac{\partial \sigma_{xy}}{\partial x} + \frac{\partial \sigma_{yy}}{\partial y} + \frac{\partial \sigma_{zy}}{\partial z}\right) \quad (11.5\text{b})$$

$$\rho\left(\frac{\partial w}{\partial t} + u\frac{\partial w}{\partial x} + v\frac{\partial w}{\partial y} + w\frac{\partial w}{\partial z}\right) = \rho f_z + \left(\frac{\partial \sigma_{xz}}{\partial x} + \frac{\partial \sigma_{yz}}{\partial y} + \frac{\partial \sigma_{zz}}{\partial z}\right) \quad (11.5\text{c})$$

EXAMPLE 11.4

Consider an infinitesimal CV filled with fluid as shown in Figure 11.3A. Apply a momentum balance in the x direction to this CV to derive the x component of the momentum equation in Cartesian coordinates.

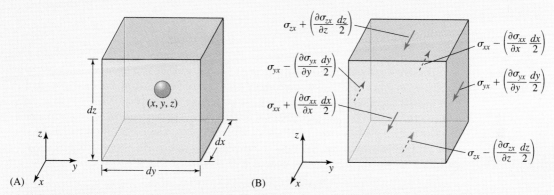

Figure 11.3 Schematic for Example 11.4: (A) infinitesimal fluid volume and (B) stress values.

SOLUTION

We are asked to derive the x component of the momentum equation in Cartesian coordinates for a specified volume of fluid. Figure 11.3 serves as the sketch for this system. Recalling the procedure used to perform a mass balance for this CV in Example 11.2, we will use a Taylor series expansion to relate the values of density, velocity, and stress on each face to the values at the center of this cube. To apply a momentum balance, we write (Eq. 7.18):

$$\int_{CV} \frac{\partial}{\partial t}(\rho \mathbf{u})\, dV + \int_{CS} (\rho \mathbf{u})(\mathbf{u} \cdot \mathbf{n})\, dS = \int_{CV} \rho \mathbf{f}\, dV + \int_{CS} \mathbf{\Sigma}\, dS$$

As discussed earlier, we will write the surface force in terms of the stress tensor rather than the stress vector, using Eq. 4.32, $\mathbf{F}_S = \int_S (\mathbf{n} \cdot \boldsymbol{\sigma})\, dS$. The resulting momentum balance is

$$\int_{CV} \frac{\partial}{\partial t}(\rho \mathbf{u})\, dV + \int_{CS} (\rho \mathbf{u})(\mathbf{u} \cdot \mathbf{n})\, dS = \int_{CV} \rho \mathbf{f}\, dV + \int_{CS} (\mathbf{n} \cdot \boldsymbol{\sigma})\, dS$$

The x component of this equation is $\int_{CV} (\partial/\partial t)(\rho u)\, dV + \int_{CS} (\rho u)(\mathbf{u} \cdot \mathbf{n})\, dS = \int_{CV} \rho f_x\, dV + \int_{CS} (\mathbf{n} \cdot \boldsymbol{\sigma})_x\, dS$, where the integrand of the stress integral, $(\mathbf{n} \cdot \boldsymbol{\sigma})_x$, gives the stresses that act on the faces in the x direction as shown in Figure 11.3B. Notice in this figure that a first order Taylor series expansion has been used to relate the value of a stress on a face to the value at the center of the cube.

Each of the integrals in the x component momentum balance has a constant integrand. We can therefore write the volume integrals in terms of the values of the integrand at the center of the cube multiplied by the volume of the cube $(dx\,dy\,dz)$, to obtain

$$\int_{CV} \frac{\partial}{\partial t}(\rho u)\,dV = \frac{\partial}{\partial t}(\rho u)\,dx\,dy\,dz = \left(\rho\frac{\partial u}{\partial t} + u\frac{\partial \rho}{\partial t}\right)dx\,dy\,dz \quad\text{(A)}$$

and

$$\int_{CV} \rho f_x\,dV = \rho f_x\,dx\,dy\,dz \quad\text{(B)}$$

The stress integral $\int_{CS}(\mathbf{n}\cdot\boldsymbol{\sigma})_x\,dS = (\mathbf{n}\cdot\boldsymbol{\sigma})_x A$ is evaluated by using the stress values shown in Figure 11.3B and considering each pair of faces in turn. For example, on the near and far faces we find, respectively, $(\mathbf{n}\cdot\boldsymbol{\sigma})_x A = [\sigma_{xx} + (\partial\sigma_{xx}/\partial x)(dx/2)]\,dy\,dz$ and $(\mathbf{n}\cdot\boldsymbol{\sigma})_x A = -[\sigma_{xx} - (\partial\sigma_{xx}/\partial x)(dx/2)]\,dy\,dz$. The net surface force on this pair of faces is therefore $(\mathbf{n}\cdot\boldsymbol{\sigma})_x A = (\partial\sigma_{xx}/\partial x)\,dx\,dy\,dz$. The contribution from the remaining two pairs of faces is found to be $(\partial\sigma_{yx}/\partial y)\,dx\,dy\,dz + (\partial\sigma_{zx}/\partial z)\,dx\,dy\,dz$, thus the total surface force on the cube is to first order

$$\left(\frac{\partial\sigma_{xx}}{\partial x} + \frac{\partial\sigma_{yx}}{\partial y} + \frac{\partial\sigma_{zx}}{\partial z}\right)dx\,dy\,dz \quad\text{(C)}$$

The momentum flux integral is of the form $\int_{CS}(\rho u)(\mathbf{u}\cdot\mathbf{n})\,dS = (\rho u)(\mathbf{u}\cdot\mathbf{n})A$ and may be evaluated by using a Taylor series expansion (see Example 11.2) to define the appropriate values of ρ, u, $(\mathbf{u}\cdot\mathbf{n})$ and the area A on the six faces. For example, the momentum flux on the near face, where $\rho_{\text{near}} = \rho + (\partial\rho/\partial x)(dx/2)$ and $(\mathbf{u}\cdot\mathbf{n})$ is given by $+u_{\text{near}} = u + (\partial u/\partial x)(dx/2)$, takes the form

$$(\rho u)(\mathbf{u}\cdot\mathbf{n})A = \left[\rho + \left(\frac{\partial\rho}{\partial x}\right)\left(\frac{dx}{2}\right)\right]\left[u + \left(\frac{\partial u}{\partial x}\right)\left(\frac{dx}{2}\right)\right]\left[u + \left(\frac{\partial u}{\partial x}\right)\left(\frac{dx}{2}\right)\right]dy\,dz$$

$$= \rho uu\,dy\,dz + \left(\rho u\frac{\partial u}{\partial x} + \frac{1}{2}uu\frac{\partial\rho}{\partial x}\right)dx\,dy\,dz \quad\text{to first order}$$

where we have neglected higher order terms as usual. On the far face, where $(\mathbf{u}\cdot\mathbf{n})$ takes the value $-u_{\text{far}} = -u + (\partial u/\partial x)(dx/2)$, we find

$$(\rho u)(\mathbf{u}\cdot\mathbf{n})A = \left[\rho - \left(\frac{\partial\rho}{\partial x}\right)\left(\frac{dx}{2}\right)\right]\left[u - \left(\frac{\partial u}{\partial x}\right)\left(\frac{dx}{2}\right)\right]\left[-u + \left(\frac{\partial u}{\partial x}\right)\left(\frac{dx}{2}\right)\right]dy\,dz$$

$$= -\rho uu\,dy\,dz + \left(\rho u\frac{\partial u}{\partial x} + \frac{1}{2}uu\frac{\partial\rho}{\partial x}\right)dx\,dy\,dz \quad\text{to first order}$$

The sum of these two terms is

$$\left(2\rho u\frac{\partial u}{\partial x} + uu\frac{\partial\rho}{\partial x}\right)dx\,dy\,dz$$

or equivalently

$$\left[\rho u \frac{\partial u}{\partial x} + u\left(\rho \frac{\partial u}{\partial x} + u\frac{\partial \rho}{\partial x}\right)\right] dx\, dy\, dz$$

The two remaining pairs of faces contribute fluxes of

$$\left[\rho v \frac{\partial u}{\partial y} + u\left(\rho \frac{\partial v}{\partial y} + v\frac{\partial \rho}{\partial y}\right)\right] dx\, dy\, dz \quad \text{and} \quad \left[\rho w \frac{\partial u}{\partial z} + u\left(\rho \frac{\partial w}{\partial z} + w\frac{\partial \rho}{\partial z}\right)\right] dx\, dy\, dz$$

Thus, after some rearrangement, the total momentum flux is to first order

$$\left\{\rho\left(u\frac{\partial u}{\partial x} + v\frac{\partial u}{\partial y} + w\frac{\partial u}{\partial z}\right) + u\left[\left(u\frac{\partial \rho}{\partial x} + v\frac{\partial \rho}{\partial y} + w\frac{\partial \rho}{\partial z}\right) + \rho\left(\frac{\partial u}{\partial x} + \frac{\partial v}{\partial y} + \frac{\partial w}{\partial z}\right)\right]\right\} dx\, dy\, dz \tag{D}$$

Gathering terms A–D, rearranging, and dividing by the common factor $dx\, dy\, dz$ yields

$$\rho\left(\frac{\partial u}{\partial t} + u\frac{\partial u}{\partial x} + v\frac{\partial u}{\partial y} + w\frac{\partial u}{\partial z}\right) + u\left[\left(\frac{\partial \rho}{\partial t} + u\frac{\partial \rho}{\partial x} + v\frac{\partial \rho}{\partial y} + w\frac{\partial \rho}{\partial z}\right) + \rho\left(\frac{\partial u}{\partial x} + \frac{\partial v}{\partial y} + \frac{\partial w}{\partial z}\right)\right]$$
$$= \rho f_x + \left(\frac{\partial \sigma_{xx}}{\partial x} + \frac{\partial \sigma_{yx}}{\partial y} + \frac{\partial \sigma_{zx}}{\partial z}\right)$$

The final step is to realize that since the term in square brackets is the continuity equation in the form of Eq. 11.2a, it has a value of zero. Therefore the final result is

$$\rho\left(\frac{\partial u}{\partial t} + u\frac{\partial u}{\partial x} + v\frac{\partial u}{\partial y} + w\frac{\partial u}{\partial z}\right) = \rho f_x + \left(\frac{\partial \sigma_{xx}}{\partial x} + \frac{\partial \sigma_{yx}}{\partial y} + \frac{\partial \sigma_{zx}}{\partial z}\right)$$

which is identical to Eq. 11.5a, as expected.

11.4 CONSTITUTIVE MODEL FOR A NEWTONIAN FLUID

Examination of the three Cartesian components of the momentum equation (11.5a–11.5c) shows that they involve three velocity components, temporal and spatial derivatives of these velocity components, and spatial derivatives of the six independent components of the stress tensor. In their present form these governing equations are incomplete: there are too many unknowns and not enough equations. What is missing is a relationship between stress and rate of strain for the particular fluid involved. This relationship is part of what is known as a constitutive model for a fluid. The key function of a constitutive model is to provide the necessary relationships between the components

It is not necessary to have a constitutive model to describe the dynamics of a rigid body because the rigid body approximation eliminates the state of stress within the object from the governing equation of motion. Since a rigid body is capable of translation and rotation only, the velocity at any point within the rigid body is always known from the position of that point and the angular velocity vector. The situation for a fluid is dramatically different, since fluid particles may move relative to one another. It is the resistance of a fluid to the relative motion of its different parts that gives rise to a state of stress in the fluid. This observation suggests that a constitutive model should relate the components of the stress tensor at a point to the spatial velocity derivatives that describe the relative motion of a fluid at that same point.

of the stress tensor and the fluid velocity field so that the governing equations are mathematically well posed and solvable.

CD/Newton's second law of motion/Flow of particles vs. continuous fluids

The constitutive model employs parameters that reflect the physical and molecular structure of the fluid and its thermodynamic state. The complexity of a constitutive model therefore depends on the molecular structure of the fluid, and on whether the fluid consists of a single component or a mixture of components. There are many different engineering fluids ranging from air, water, and other single component liquids and gases to paints, slurries, colloidal suspensions, and other complex fluid mixtures. Fortunately the entire class of single component liquids and all gases are well described by the Newtonian model. The more complex liquids are challenging to analyze because their constitutive relationships are complicated. Since we focus solely on the Newtonian model in this text, it will be important for you to recognize the limitations of this model, to avoid being tempted to use it in inappropriate situations.

The first assumption of the Newtonian model is that the fluid possesses an isotropic structure that is reflected in its behavior. One manifestation of this isotropic structure is that the stress tensor is symmetric. As a result, there are only six independent components in the (nine-term) stress tensor. In addition, for an isotropic fluid, only two fluid properties are required to define completely the relationship between stress and rate of strain. The absolute viscosity μ defines the shearing behavior of the fluid and relates shear stress to the rate of shear deformation. The bulk viscosity κ defines the compression and expansion behavior of a Newtonian fluid. The second assumption of the Newtonian model is that the shear stress existing at a point in the fluid at time t is linearly related to the velocity gradient that exists at the same point in the fluid at that instant of time. Thus, the fluid does not remember what has happened to it earlier, nor does it exhibit hysteresis or other anomalous behavior. Non-Newtonian fluids that do have a memory are called viscoelastic fluids. Examples include some liquid polymers, and in some circumstances, blood.

In Cartesian coordinates the constitutive relationships defining the Newtonian fluid are:

$$\sigma_{xx} = -p - \frac{2}{3}\mu(\nabla \cdot \mathbf{u}) + 2\mu \frac{\partial u}{\partial x} \tag{11.6a}$$

$$\sigma_{yy} = -p - \frac{2}{3}\mu(\nabla \cdot \mathbf{u}) + 2\mu \frac{\partial v}{\partial y} \tag{11.6b}$$

11.4 CONSTITUTIVE MODEL FOR A NEWTONIAN FLUID

In an incompressible flow, or a flow of an incompressible fluid, the continuity equation reduces to Eq. 11.3, $\nabla \cdot \mathbf{u} = 0$, thus in this case the difference in the mechanical and equilibrium pressures is zero irrespective of the value of the bulk viscosity.

$$\sigma_{zz} = -p - \frac{2}{3}\mu(\nabla \cdot \mathbf{u}) + 2\mu\frac{\partial w}{\partial z} \quad (11.6c)$$

$$\sigma_{xy} = \sigma_{yx} = \mu\left(\frac{\partial u}{\partial y} + \frac{\partial v}{\partial x}\right) \quad (11.6d)$$

$$\sigma_{yz} = \sigma_{zy} = \mu\left(\frac{\partial v}{\partial z} + \frac{\partial w}{\partial y}\right) \quad (11.6e)$$

$$\sigma_{zx} = \sigma_{xz} = \mu\left(\frac{\partial w}{\partial x} + \frac{\partial u}{\partial z}\right) \quad (11.6f)$$

The role of pressure in these constitutive equations requires further discussion. The pressure p is the mechanical or physical pressure. It is the pressure that would be measured by an absolute pressure sensor moving with the fluid at the point in question. We can also define a thermodynamic or equilibrium pressure, p_e. This is the pressure predicted by an equation of state. Since thermodynamic equilibrium may not exist at every point in a flow field, the Newtonian model postulates a difference in the two pressures that is given by $p = p_e - \kappa(\nabla \cdot \mathbf{u})$. The bulk viscosity κ is the fluid property that measures the degree of departure of the mechanical pressure from its equilibrium value.

For monatomic gases, the bulk viscosity is known to be zero. For gases with a complex molecular structure and many liquids, the bulk viscosity is nonzero, and the effect of this viscosity on the flow depends on the rate of expansion. Since we rarely encounter situations in which the difference in the two pressures is significant, we will ignore the effects of bulk viscosity in the remainder of this text, assuming that the constitutive relations are given by Eq. 11.6, with the pressure p (which is equal to both the mechanical and thermodynamic pressure) representing the actual physical pressure in the fluid.

Adding together Eqs. 11.6a–11.6c shows that the pressure is related to the normal stresses by $p = -(1/3)(\sigma_{xx} + \sigma_{yy} + \sigma_{zz})$. We see that the pressure in a Newtonian fluid at a point is defined by the negative of the average normal stress acting in the fluid at that point. The negative sign reflects the common interpretation of pressure as a compressive stress. Although a negative absolute pressure representing a tensile stress is theoretically conceivable, as discussed in Chapter 2 negative pressures do not occur physically. Finally, recall from fluid statics that in a stationary fluid each normal stress is exactly the same and equal to the negative of the pressure; i.e., the static pressure acts equally in all directions. You may confirm that the pressure in the Newtonian model has these same characteristics in a stationary fluid by setting all velocity components and their derivatives to zero in Eqs. 11.6a–11.6c.

There is a tendency to assume that the equality of normal stresses known to occur in a fluid at rest also occurs in a fluid in motion. However, that is not necessarily the case. What is true is that in many flows the differences in the three normal stresses are vanishingly small. For example, in a constant density or incompressible flow, the continuity equation reduces to Eq. 11.3, $\nabla \cdot \mathbf{u} = 0$, and the viscous contributions to the normal stresses in Eqs. 11.6 are $2\mu(\partial u/\partial x)$, $2\mu(\partial v/\partial y)$, and $2\mu(\partial w/\partial z)$. These contributions are usually negligible, so the differences in the three normal stresses are often neglected in analyzing constant density and incompressible flows.

In cylindrical coordinates the constitutive relationships for a Newtonian fluid are:

$$\sigma_{rr} = -p - \frac{2}{3}\mu(\nabla \cdot \mathbf{u}) + 2\mu\frac{\partial v_r}{\partial r} \tag{11.7a}$$

$$\sigma_{\theta\theta} = -p - \frac{2}{3}\mu(\nabla \cdot \mathbf{u}) + 2\mu\left(\frac{1}{r}\frac{\partial v_\theta}{\partial \theta} + \frac{v_r}{r}\right) \tag{11.7b}$$

$$\sigma_{zz} = -p - \frac{2}{3}\mu(\nabla \cdot \mathbf{u}) + 2\mu\frac{\partial v_z}{\partial z} \tag{11.7c}$$

$$\sigma_{r\theta} = \sigma_{\theta r} = \mu\left[r\frac{\partial}{\partial r}\left(\frac{v_\theta}{r}\right) + \frac{1}{r}\frac{\partial v_r}{\partial \theta}\right] \tag{11.7d}$$

$$\sigma_{\theta z} = \sigma_{z\theta} = \mu\left(\frac{\partial v_\theta}{\partial z} + \frac{1}{r}\frac{\partial v_z}{\partial \theta}\right) \tag{11.7e}$$

$$\sigma_{zr} = \sigma_{rz} = \mu\left(\frac{\partial v_z}{\partial r} + \frac{\partial v_r}{\partial z}\right) \tag{11.7f}$$

The next element of the Newtonian constitutive model is a set of equations of state that relate the physical parameters μ and κ (and others that occur in the energy equation to be developed later in this chapter) to the thermodynamic state of the fluid. The state relationship is expressed symbolically for the viscosity μ as

$$\mu = \mu(p, T) \tag{11.8}$$

Similar relationships exist for other fluid properties. We also require an equation of state that relates the pressure to the fluid density and temperature of the form

$$p = p(\rho, T) \tag{11.9}$$

This state equation may be known empirically or through a model such as the perfect gas law.

The final element of the Newtonian model is concerned with fluid transport. In Eq. 6.36, we used Fourier's law to define a diffusive flux vector for a property, relating the flux vector to the gradient of the property. This relationship is valid only for a fluid with an isotropic structure; thus it applies to Newtonian fluids and may be thought of as part of the Newtonian constitutive model.

In summary, the constitutive model for a Newtonian fluid allows us to do two important things. First, it enables us to calculate all components of the stress tensor if we know the velocity and pressure fields. Thus, if we have a solution to a flow problem, the constitutive model allows us to answer any question that might arise in connection with surface forces. Second, by using the constitutive model, we may remove the stress tensor from the governing equations of fluid dynamics by replacing the various stress components that appear in the momentum and energy equations with the corresponding velocity gradients that appear in the constitutive model. In the next section we will use this procedure to produce the Navier–Stokes equation of fluid mechanics.

11.4 CONSTITUTIVE MODEL FOR A NEWTONIAN FLUID

The constitutive relationships for a Newtonian fluid relate stress to rate of strain. We can see this more clearly in Cartesian coordinates by first using the matrix representation of the stress tensor (Eq. 4.30), $\boldsymbol{\sigma} = \begin{pmatrix} \sigma_{xx} & \sigma_{xy} & \sigma_{xz} \\ \sigma_{yx} & \sigma_{yy} & \sigma_{yz} \\ \sigma_{zx} & \sigma_{zy} & \sigma_{zz} \end{pmatrix}$, and writing Eqs. 11.6 in matrix form as

$$\begin{pmatrix} \sigma_{xx} & \sigma_{xy} & \sigma_{xz} \\ \sigma_{yx} & \sigma_{yy} & \sigma_{yz} \\ \sigma_{zx} & \sigma_{zy} & \sigma_{zz} \end{pmatrix} = \begin{pmatrix} -p & 0 & 0 \\ 0 & -p & 0 \\ 0 & 0 & -p \end{pmatrix} + 2\mu \begin{pmatrix} \frac{\partial u}{\partial x} - \frac{1}{3}(\nabla \cdot \mathbf{u}) & \frac{1}{2}\left(\frac{\partial u}{\partial y} + \frac{\partial v}{\partial x}\right) & \frac{1}{2}\left(\frac{\partial u}{\partial z} + \frac{\partial w}{\partial x}\right) \\ \frac{1}{2}\left(\frac{\partial v}{\partial x} + \frac{\partial u}{\partial y}\right) & \frac{\partial v}{\partial y} - \frac{1}{3}(\nabla \cdot \mathbf{u}) & \frac{1}{2}\left(\frac{\partial v}{\partial z} + \frac{\partial w}{\partial y}\right) \\ \frac{1}{2}\left(\frac{\partial w}{\partial x} + \frac{\partial u}{\partial z}\right) & \frac{1}{2}\left(\frac{\partial w}{\partial y} + \frac{\partial v}{\partial z}\right) & \frac{\partial w}{\partial z} - \frac{1}{3}(\nabla \cdot \mathbf{u}) \end{pmatrix}$$

Recalling that the rate of expansion is given by Eq. 10.65b as

$$\mathbf{E} = \begin{pmatrix} \frac{1}{3}\nabla \cdot \mathbf{u} & 0 & 0 \\ 0 & \frac{1}{3}\nabla \cdot \mathbf{u} & 0 \\ 0 & 0 & \frac{1}{3}\nabla \cdot \mathbf{u} \end{pmatrix}$$

and the rate of shear deformation tensor, **D**, or shear deformation rate, is given by Eq. 10.41 as

$$\mathbf{D} = \begin{pmatrix} \frac{\partial u}{\partial x} - \frac{1}{3}\left(\frac{\partial u}{\partial x} + \frac{\partial v}{\partial y} + \frac{\partial w}{\partial z}\right) & \frac{1}{2}\left(\frac{\partial u}{\partial y} + \frac{\partial v}{\partial x}\right) & \frac{1}{2}\left(\frac{\partial u}{\partial z} + \frac{\partial w}{\partial x}\right) \\ \frac{1}{2}\left(\frac{\partial v}{\partial x} + \frac{\partial u}{\partial y}\right) & \frac{\partial v}{\partial y} - \frac{1}{3}\left(\frac{\partial u}{\partial x} + \frac{\partial v}{\partial y} + \frac{\partial w}{\partial z}\right) & \frac{1}{2}\left(\frac{\partial v}{\partial z} + \frac{\partial w}{\partial y}\right) \\ \frac{1}{2}\left(\frac{\partial w}{\partial x} + \frac{\partial u}{\partial z}\right) & \frac{1}{2}\left(\frac{\partial w}{\partial y} + \frac{\partial v}{\partial z}\right) & \frac{\partial w}{\partial z} - \frac{1}{3}\left(\frac{\partial u}{\partial x} + \frac{\partial v}{\partial y} + \frac{\partial w}{\partial z}\right) \end{pmatrix}$$

we can introduce the dilation to write the shear deformation rate as

$$\mathbf{D} = \begin{pmatrix} \frac{\partial u}{\partial x} - \frac{1}{3}(\nabla \cdot \mathbf{u}) & \frac{1}{2}\left(\frac{\partial u}{\partial y} + \frac{\partial v}{\partial x}\right) & \frac{1}{2}\left(\frac{\partial u}{\partial z} + \frac{\partial w}{\partial x}\right) \\ \frac{1}{2}\left(\frac{\partial v}{\partial x} + \frac{\partial u}{\partial y}\right) & \frac{\partial v}{\partial y} - \frac{1}{3}(\nabla \cdot \mathbf{u}) & \frac{1}{2}\left(\frac{\partial v}{\partial z} + \frac{\partial w}{\partial y}\right) \\ \frac{1}{2}\left(\frac{\partial w}{\partial x} + \frac{\partial u}{\partial z}\right) & \frac{1}{2}\left(\frac{\partial w}{\partial y} + \frac{\partial v}{\partial z}\right) & \frac{\partial w}{\partial z} - \frac{1}{3}(\nabla \cdot \mathbf{u}) \end{pmatrix}$$

Using the unit matrix $\mathbf{I} = \begin{pmatrix} 1 & 0 & 0 \\ 0 & 1 & 0 \\ 0 & 0 & 1 \end{pmatrix}$, we can write the stress tensor in vector form as

$$\boldsymbol{\sigma} = -p\mathbf{I} + 2\mu\mathbf{D}$$

11 GOVERNING EQUATIONS OF FLUID DYNAMICS

As a final step, we note that the sum of the rate of strain tensor is $\mathsf{S} = \mathsf{E} + \mathsf{D}$, so the preceding relationship can also be written as

$$\boldsymbol{\sigma} = -p\mathsf{I} + 2\mu\mathsf{S} - 2\mu\mathsf{E}$$

We see that the Newtonian model does indeed relate stress to rate of strain, and also to the rate of expansion. Since the rate of expansion is zero for an incompressible fluid or in an incompressible flow, we can write the stress tensor when using these approximations as

$$\boldsymbol{\sigma} = -p\mathsf{I} + 2\mu\mathsf{S}$$

and recognize that in this case stress is related solely to rate of strain.

EXAMPLE 11.5

Consider the flow of an incompressible Newtonian fluid between parallel plates with the top plate moving as shown in Figure 11.4. The velocity field is $\mathbf{u} = U(y/h)\,\mathbf{i}$, where U is the speed of the moving plate and h is the gap between the plates. Find the stresses in this flow. What can the momentum equation tell us about this flow?

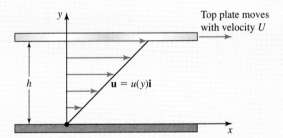

Figure 11.4 Schematic for Example 11.5.

SOLUTION

We are asked to find the stresses in a given flow and to use the momentum equation to gain insight into the nature of the flow. Figure 11.4 serves as the sketch for this fluid system. To find the stresses we will use the Newtonian constitutive relationships, Eqs. 11.6, with the known velocity components. The only nonzero velocity component is u, and the density is constant. Since u is a function of y alone, all x and z spatial derivatives of u are zero. We can begin by confirming that the velocity field satisfies the continuity equation for an incompressible fluid. From Eq. 11.2a we find

$$\frac{\partial u}{\partial x} + \frac{\partial v}{\partial y} + \frac{\partial w}{\partial z} = \frac{\partial}{\partial x}(u(y)) + \frac{\partial}{\partial y}(0) + \frac{\partial}{\partial z}(0) = 0$$

Inserting the velocity components into the Newtonian constitutive relationships, Eqs. 11.6, we find

$$\sigma_{xx} = -p - \frac{2}{3}\mu(\nabla \cdot \mathbf{u}) + 2\mu\frac{\partial u}{\partial x} = -p - \frac{2}{3}\mu(0) + 2\mu(0) = -p$$

$$\sigma_{yy} = -p - \frac{2}{3}\mu(\nabla \cdot \mathbf{u}) + 2\mu\frac{\partial v}{\partial y} = -p - \frac{2}{3}\mu(0) + 2\mu(0) = -p$$

$$\sigma_{zz} = -p - \frac{2}{3}\mu(\nabla \cdot \mathbf{u}) + 2\mu\frac{\partial w}{\partial z} = -p - \frac{2}{3}\mu(0) + 2\mu(0) = -p$$

$$\sigma_{xy} = \sigma_{yx} = \mu\left(\frac{\partial u}{\partial y} + \frac{\partial v}{\partial x}\right) = \mu\left[\frac{\partial}{\partial y}\left(U\frac{y}{h}\right) + \frac{\partial}{\partial x}(0)\right] = \mu\frac{U}{h}$$

$$\sigma_{zy} = \sigma_{yz} = \mu\left(\frac{\partial w}{\partial y} + \frac{\partial v}{\partial z}\right) = \mu\left[\frac{\partial}{\partial y}(0) + \frac{\partial}{\partial z}(0)\right] = 0$$

$$\sigma_{zx} = \sigma_{xz} = \mu\left(\frac{\partial w}{\partial x} + \frac{\partial u}{\partial z}\right) = \mu\left[\frac{\partial}{\partial x}(0) + \frac{\partial}{\partial z}\left(\frac{Uy}{h}\right)\right] = 0$$

We see that the only nonzero shear stress is $\sigma_{xy} = \sigma_{yx} = \mu(U/h)$, a constant. Note, however, that we do not know the pressure distribution because this was not given to us. We can use these results to write the stress tensor in matrix form as

$$\boldsymbol{\sigma} = \begin{pmatrix} -p & \frac{\mu U}{h} & 0 \\ \frac{\mu U}{h} & -p & 0 \\ 0 & 0 & -p \end{pmatrix}$$

realizing that the pressure is an unknown function of the coordinates.

To find out what the momentum equation can tell us about this flow, we will substitute the velocity field into the momentum equations (Eqs. 11.5). The flow is steady, so time derivatives of the velocity components are zero. The only body force is gravity; thus the body force per unit mass is $f = (0, -g, 0)$. The momentum equations become

$$\rho\left[0 + u(0) + (0)\frac{\partial u}{\partial y} + (0)(0)\right] = \rho(0) + \left(\frac{\partial \sigma_{xx}}{\partial x} + \frac{\partial \sigma_{yx}}{\partial y} + \frac{\partial \sigma_{zx}}{\partial z}\right)$$

$$\rho[0 + u(0) + (0)(0) + (0)(0)] = \rho(-g) + \left(\frac{\partial \sigma_{xy}}{\partial x} + \frac{\partial \sigma_{yy}}{\partial y} + \frac{\partial \sigma_{zy}}{\partial z}\right)$$

$$\rho[0 + u(0) + (0)(0) + (0)(0)] = \rho(0) + \left(\frac{\partial \sigma_{xz}}{\partial x} + \frac{\partial \sigma_{yz}}{\partial y} + \frac{\partial \sigma_{zz}}{\partial z}\right)$$

We conclude that the derivatives of the various stresses obey the following equations:

$$\left(\frac{\partial \sigma_{xx}}{\partial x} + \frac{\partial \sigma_{yx}}{\partial y} + \frac{\partial \sigma_{zx}}{\partial z}\right) = 0$$

$$\rho(-g) + \left(\frac{\partial \sigma_{xy}}{\partial x} + \frac{\partial \sigma_{yy}}{\partial y} + \frac{\partial \sigma_{zy}}{\partial z}\right) = 0$$

$$\left(\frac{\partial \sigma_{xz}}{\partial x} + \frac{\partial \sigma_{yz}}{\partial y} + \frac{\partial \sigma_{zz}}{\partial z}\right) = 0$$

Inserting the stresses found earlier into these reduced momentum equations, we find

$$\left(\frac{\partial \sigma_{xx}}{\partial x} + \frac{\partial \sigma_{yx}}{\partial y} + \frac{\partial \sigma_{zx}}{\partial z}\right) = \left[\frac{\partial}{\partial x}(-p) + \frac{\partial}{\partial y}\left(\frac{\mu U}{h}\right) + \frac{\partial}{\partial z}(0)\right] = -\frac{\partial p}{\partial x} = 0$$

$$\rho(-g) + \left(\frac{\partial \sigma_{xy}}{\partial x} + \frac{\partial \sigma_{yy}}{\partial y} + \frac{\partial \sigma_{zy}}{\partial z}\right) = \rho(-g) + \left[\frac{\partial}{\partial x}\left(\frac{\mu U}{h}\right) + \frac{\partial}{\partial y}(-p) + \frac{\partial}{\partial z}(0)\right]$$

$$= -\rho g - \frac{\partial p}{\partial y} = 0$$

$$\left(\frac{\partial \sigma_{xz}}{\partial x} + \frac{\partial \sigma_{yz}}{\partial y} + \frac{\partial \sigma_{zz}}{\partial z}\right) = \left[\frac{\partial}{\partial x}(0) + \frac{\partial}{\partial y}(0) + \frac{\partial}{\partial z}(-p)\right] = -\frac{\partial p}{\partial z} = 0$$

From the first and last of these equations we conclude that the pressure does not vary in the x or z directions. Integrating the remaining equation $\partial p/\partial y = -\rho g$, noting that the density is constant, and evaluating the constant of integration on the top plate at $y = h$, we find $p(y) = p_h - \rho g(y - h)$. Thus, the momentum equations have shown that the pressure distribution in this shear flow is unchanged from the hydrostatic pressure distribution that would exist in the absence of flow. Notice that both the momentum equations and the constitutive relations are needed to solve this (typical) flow problem.

Did you recognize that the flow in this last example is the basis for the definition of viscosity? Notice that the fluid is sheared in the thin gap between parallel plates, and since v and w are zero, we have $\tau = \sigma_{yx} = \mu(\partial u/\partial y + \partial v/\partial x) = \mu(\partial u/\partial y)$, which is the defining equation used in our discussion of Newton's law of viscosity (Eq. 1.2c) in Chapter 1.

11.5 NAVIER–STOKES EQUATIONS

 CD/Dynamics/Navier–Stokes equations

When the constitutive relationships for a Newtonian fluid (Eqs. 11.6) are used to replace the stresses in the differential momentum equations (Eqs. 11.5), the result is the Navier–Stokes equations in Cartesian coordinates. These equations, which describe the behavior of a Newtonian fluid with variable density and viscosity, are applicable to laminar and turbulent flows of liquids and gases throughout the entire range of

11.5 NAVIER–STOKES EQUATIONS

flow speeds. They are the basic equations that all commercial CFD codes employ to solve the most general class of fluid mechanics problems. The Navier–Stokes equations are exceedingly complicated and difficult to solve. Since the flows of interest in this text can be modeled as having constant density and constant viscosity, some simplification is, however, possible. In this case, the continuity equation is given by Eq. 11.3, $\nabla \cdot \mathbf{u} = 0$, and spatial derivatives of the (constant) viscosity are zero. Without going into all the details, we can write the continuity and Navier–Stokes equations for the important case of a constant density, constant viscosity fluid as (Eq. 11.4a) $\partial u/\partial x + \partial v/\partial y + \partial w/\partial z = 0$, and

$$\rho\left(\frac{\partial u}{\partial t} + u\frac{\partial u}{\partial x} + v\frac{\partial u}{\partial y} + w\frac{\partial u}{\partial z}\right) = \rho f_x - \frac{\partial p}{\partial x} + \mu\left(\frac{\partial^2 u}{\partial x^2} + \frac{\partial^2 u}{\partial y^2} + \frac{\partial^2 u}{\partial z^2}\right) \quad (11.10a)$$

$$\rho\left(\frac{\partial v}{\partial t} + u\frac{\partial v}{\partial x} + v\frac{\partial v}{\partial y} + w\frac{\partial v}{\partial z}\right) = \rho f_y - \frac{\partial p}{\partial y} + \mu\left(\frac{\partial^2 v}{\partial x^2} + \frac{\partial^2 v}{\partial y^2} + \frac{\partial^2 v}{\partial z^2}\right) \quad (11.10b)$$

$$\rho\left(\frac{\partial w}{\partial t} + u\frac{\partial w}{\partial x} + v\frac{\partial w}{\partial y} + w\frac{\partial w}{\partial z}\right) = \rho f_z - \frac{\partial p}{\partial z} + \mu\left(\frac{\partial^2 w}{\partial x^2} + \frac{\partial^2 w}{\partial y^2} + \frac{\partial^2 w}{\partial z^2}\right)$$

$$(11.10c)$$

In interpreting these equations, we note that they are actually the three components of a force balance on the fluid. We can write this balance in vector form as

$$\rho\frac{D\mathbf{u}}{Dt} = \rho\mathbf{f} - \nabla p + \mu\nabla^2\mathbf{u} \quad (11.11)$$

You should be able to recognize that the inertial forces per unit volume, given by $\rho\mathbf{a} = \rho(D\mathbf{u}/Dt)$, are balanced by the sum of body forces per unit volume, $\rho\mathbf{f}$, the pressure forces per unit volume as given by $-\nabla p$, and viscous forces per unit volume as given by $\mu\nabla^2\mathbf{u}$. Thus, the vector equation $\rho(D\mathbf{u}/Dt) = \rho\mathbf{f} - \nabla p + \mu\nabla^2\mathbf{u}$ is another way to write the Navier–Stokes equation for a constant density, constant viscosity fluid.

CD/Dynamics/Newton's second law of motion/$F = ma$ for a Newtonian Fluid

In cylindrical coordinates, the continuity and Navier–Stokes equations for a constant density, constant viscosity fluid are $(1/r)\partial(rv_r)/\partial r + (1/r)\partial v_\theta/\partial\theta + \partial v_z/\partial z = 0$, which is Eq. (11.4b), and

$$\rho\left(\frac{\partial v_r}{\partial t} + v_r\frac{\partial v_r}{\partial r} + \frac{v_\theta}{r}\frac{\partial v_r}{\partial \theta} + v_z\frac{\partial v_r}{\partial z} - \frac{v_\theta^2}{r}\right)$$

$$= \rho f_r - \frac{\partial p}{\partial r} + \mu\left(\frac{\partial^2 v_r}{\partial r^2} + \frac{1}{r}\frac{\partial v_r}{\partial r} + \frac{1}{r^2}\frac{\partial^2 v_r}{\partial \theta^2} + \frac{\partial^2 v_r}{\partial z^2} - \frac{v_r}{r^2} - \frac{2}{r^2}\frac{\partial v_\theta}{\partial \theta}\right)$$

$$(11.12a)$$

HISTORY BOX 11.1

Claude Navier (1785–1836) entered the prestigious École Polytechnique in 1802 as a marginal student but emerged at the top of his class. He inherited the role of the leading scholar of mathematics, science, and engineering in France from his teacher and friend Jean Baptiste Fourier. In 1822 he presented a paper that for the first time accurately described the role of friction in the equations of motion for a fluid. His analysis started from a molecular view of fluid. It was left to Jean-Claude de Saint-Venant to explain this result based on the viscous stresses in the fluid and to identify the viscosity as the key material property.

 CD/History/Claude Navier and Sir George Stokes

George Stokes (1819–1903) held the Lucasian chair at Cambridge University, the same position held by Sir Isaac Newton. Thus, he was one of the leading scholars in England. He made many contributions to fluid mechanics and the nature of light. With no knowledge of the work in France, in 1845 Stokes published a derivation of the equations that bear his name, using an analysis based on the internal friction of fluid much like we have presented here.

$$\rho\left(\frac{\partial v_\theta}{\partial t} + v_r\frac{\partial v_\theta}{\partial r} + \frac{v_\theta}{r}\frac{\partial v_\theta}{\partial \theta} + v_z\frac{\partial v_\theta}{\partial z} + \frac{v_r v_\theta}{r}\right)$$

$$= \rho f_\theta - \frac{1}{r}\frac{\partial p}{\partial \theta} + \mu\left(\frac{\partial^2 v_\theta}{\partial r^2} + \frac{1}{r}\frac{\partial v_\theta}{\partial r} + \frac{1}{r^2}\frac{\partial^2 v_\theta}{\partial \theta^2} + \frac{\partial^2 v_\theta}{\partial z^2} - \frac{v_\theta}{r^2} + \frac{2}{r^2}\frac{\partial v_r}{\partial \theta}\right) \quad (11.12b)$$

$$\rho\left(\frac{\partial v_z}{\partial t} + v_r\frac{\partial v_z}{\partial r} + \frac{v_\theta}{r}\frac{\partial v_z}{\partial \theta} + v_z\frac{\partial v_z}{\partial z}\right)$$

$$= \rho f_z - \frac{\partial p}{\partial z} + \mu\left(\frac{\partial^2 v_z}{\partial r^2} + \frac{1}{r}\frac{\partial v_z}{\partial r} + \frac{1}{r^2}\frac{\partial^2 v_z}{\partial \theta^2} + \frac{\partial^2 v_z}{\partial z^2}\right) \quad (11.12c)$$

The continuity and Navier–Stokes equations just given provide a complete set of governing equations to determine the velocity and pressure at every point in a flow. It is not necessary in this case to solve the energy equation to determine the velocity and pressure fields, because we have four equations and four unknowns: three components of velocity and the pressure.

We have often mentioned that solving the governing equations is a difficult task. In Chapter 12 we will demonstrate how to construct analytical solutions for a number of important flows. Here we emphasize that saying we have obtained a solution to the governing equations means one of two things. In an analytical solution, the functions describing the velocity and pressure fields must satisfy all four equations simultaneously, as well as the boundary conditions, when the functions are inserted into the equations

11.5 NAVIER–STOKES EQUATIONS

and the various spatial derivatives evaluated. The next two examples demonstrate this process. In a CFD solution, the set of numerical values for velocity and pressure constituting the solution on some number of spatial points satisfies a massive set of algebraic equations representing the discretized form of the governing equations and boundary conditions to some degree of approximation. This is about all we can say in general about a CFD solution because the details depend to a great degree on the type of approach used by the CFD model.

EXAMPLE 11.6

In the channel flow of a constant density, constant viscosity fluid shown in Figure 11.5, suppose the complete description of the flow is given in Cartesian coordinates by the velocity field $u = \{[h^2(p_1 - p_2)]/2\mu L\}[1 - (y/h)^2]$, $v = 0$, and $w = 0$, and the pressure field $p(x) = p_1 + [(p_2 - p_1)/L](x - x_1)$. Here p_1 and p_2 are the pressures at the indicated locations, and the slight hydrostatic variation in pressure across the channel has been ignored. Show that this flow satisfies the constant density, constant viscosity Navier–Stokes equations in Cartesian coordinates, with the body force neglected.

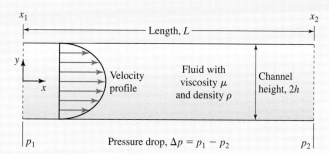

Figure 11.5 Schematic for Example 11.6.

SOLUTION

We will first substitute the three velocity components into the continuity equation for an incompressible fluid, Eq. 11.4a, then substitute the velocity components and pressure into the constant density, constant viscosity forms of the Navier–Stokes equations, Eqs. 11.10a–11.10c. Since we have $u = u(y)$ only, with v and w zero, the continuity equation, $\partial u/\partial x + \partial v/\partial y + \partial w/\partial z = 0$, is satisfied by inspection. Terms in Eqs. 11.10a–11.10c that contain the velocity components v and w are zero, as are time derivatives and spatial derivatives of u with respect to x and z. Writing only the remaining nonzero terms, and setting the body force terms to zero, we find

$$0 = -\frac{\partial p}{\partial x} + \mu \frac{\partial^2 u}{\partial y^2}, \quad 0 = -\frac{\partial p}{\partial y}, \quad \text{and} \quad 0 = -\frac{\partial p}{\partial z}$$

The last two equations are consistent with the given pressure distribution. Substituting the x velocity component and pressure into the first pressure equation yields

$$0 = -\frac{\partial}{\partial x}\left(p_1 + \left(\frac{p_2 - p_1}{L}\right)(x - x_1)\right) + \mu\frac{\partial}{\partial y}\left[\frac{\partial}{\partial y}\left(\left(\frac{h^2(p_1 - p_2)}{2\mu L}\right)\left[1 - \left(\frac{y}{h}\right)^2\right]\right)\right]$$

$$0 = -\left(\frac{p_2 - p_1}{L}\right) + \mu\left(\frac{h^2(p_1 - p_2)}{2\mu L}\right)\frac{\partial}{\partial y}\left(\frac{\partial}{\partial y}\left[1 - \left(\frac{y}{h}\right)^2\right]\right)$$

$$0 = -\left(\frac{p_2 - p_1}{L}\right) + \mu\left(\frac{h^2(p_1 - p_2)}{2\mu L}\right)\left(\frac{-2}{h^2}\right) = 0$$

We see that the velocity and pressure do satisfy the appropriate forms of the continuity and Navier–Stokes equations. It is also straightforward to show that the velocity field satisfies the no-slip, no-penetration conditions at the channel walls.

EXAMPLE 11.7

In the Poiseuille flow of a constant density, constant viscosity fluid in a round pipe, (Figure 11.6), the velocity field is given in cylindrical coordinates by $\mathbf{u} = v_r\mathbf{e}_r + v_\theta\mathbf{e}_\theta + v_z\mathbf{e}_z$ with components $v_r = 0$, $v_\theta = 0$, and $v_z(r) = \{[R_P^2(p_1 - p_2)]/4\mu L\}[1 - (r/R_P)^2]$. Find the pressure distribution in this flow if body forces are neglected.

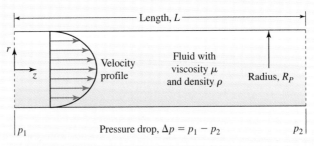

Figure 11.6 Schematic for Example 11.7.

SOLUTION

The velocity field must satisfy the continuity equation, Eq. 11.4b, and the velocity field and pressure distribution must satisfy Eqs. 11.12a–11.12c. We begin by checking the continuity equation:

$$\frac{1}{r}\frac{\partial(rv_r)}{\partial r} + \frac{1}{r}\frac{\partial v_\theta}{\partial \theta} + \frac{\partial v_z}{\partial z} = 0$$

Since we know $v_r = 0$ and $v_\theta = 0$, this reduces to $\partial v_z/\partial z = 0$, which can be seen to be satisfied by inspection, since $v_z = v_z(r)$. Writing only the nonzero terms in Eqs. 11.12a–11.12c, we have

$$0 = -\frac{\partial p}{\partial r}, \quad 0 = -\frac{1}{r}\frac{\partial p}{\partial \theta}, \quad \text{and} \quad 0 = -\frac{\partial p}{\partial z} + \mu\left(\frac{\partial^2 v_z}{\partial^2 r} + \frac{1}{r}\frac{\partial v_z}{\partial r}\right)$$

Thus the pressure is a function of z only. Inserting the given velocity component into the last equation and taking derivatives we find

$$\frac{\partial p}{\partial z} = \mu\left(\frac{R_P^2(p_1 - p_2)}{4\mu L}\right)\left[\left(-\frac{2}{R_P^2}\right) + \frac{1}{r}\left(-\frac{2r}{R_P^2}\right)\right]$$

$$= \mu\left(\frac{R_P^2(p_1 - p_2)}{4\mu L}\right)\left(-\frac{2}{R_P^2} - \frac{2}{R_P^2}\right)$$

which, after simplification, yields $\partial p/\partial z = (p_2 - p_1)/L$. Integrating and evaluating the resulting constant of integration at $z = z_1$, we find $p(z) = p_1 + [(p_2 - p_1)/L](z - z_1)$. This is a linear drop in pressure down the pipe. Can you see by inspection that the no-slip, no-penetration conditions are satisfied on the pipe wall?

CD/Dynamics/Potential Flow

11.6 EULER EQUATIONS

The formidable Navier–Stokes equations have generated many efforts to introduce simplifying approximations. One of the earliest and most valuable of these approximations is that of an inviscid fluid, defined in Chapter 8 to be a fluid whose viscosity is zero. Upon substituting $\mu = 0$ into the constitutive model for a Newtonian fluid (Eqs. 11.6a–11.6f), we find the state of stress in an inviscid fluid in Cartesian coordinates to be

$$\sigma_{xx} = -p, \quad \sigma_{yy} = -p, \quad \sigma_{zz} = -p \quad \text{(11.13a–c)}$$

$$\sigma_{xy} = \sigma_{yx} = 0, \quad \sigma_{zy} = \sigma_{yz} = 0, \quad \sigma_{zx} = \sigma_{xz} = 0 \quad \text{(11.13d–f)}$$

As expected, we see that an inviscid fluid is incapable of exerting a shear stress. This is another way to define an inviscid fluid. The absence of shear stress indicates that an inviscid fluid does not obey the no-slip condition and therefore must slip along a solid surface. The state of stress in an inviscid fluid, with its absence of shear stresses, is given by a pressure distribution alone, just as is the case in a static fluid. However, as we will see in a moment, in an inviscid fluid, the pressure distribution is related to both the body force and the inertial force created by the velocity field, rather than to just the body force as in a static fluid.

HISTORY BOX 11-2

Leonhard Euler (1705–1783) is truly one of the great mathematical physicists in general and fluid dynamicists in particular. He is well known for his contributions to the solution of ordinary differential equations. Of course, for our purposes here he was the first to derive the differential momentum equations for inviscid fluid flow. In fact, he was also the first to write the continuity equation. Euler was born in Basel, Switzerland, where he was a student and colleague of members of the Bernoulli family.

The momentum equation governing the flow of an inviscid fluid was derived by Euler well in advance of the development of the Navier–Stokes equations for viscous fluids. Although we will not describe his derivation of the equation named in his honor, setting $\mu = 0$ in the vector form of Navier–Stokes equation, Eq. 11.11, yields the Euler equation:

$$\rho \frac{D\mathbf{u}}{Dt} = \rho \mathbf{f} - \nabla p \tag{11.14}$$

An alternate way of writing this in terms of acceleration is:

$$\rho \mathbf{a} = \rho \mathbf{f} - \nabla p \tag{11.15}$$

Thus, the flow of an inviscid fluid is governed by a balance of inertial, body, and pressure forces alone. There are no viscous forces of any kind.

 CD/History/Leonhard Euler

The three components of the Euler equation in Cartesian coordinates are:

$$\rho \left(\frac{\partial u}{\partial t} + u\frac{\partial u}{\partial x} + v\frac{\partial u}{\partial y} + w\frac{\partial u}{\partial z} \right) = \rho f_x - \frac{\partial p}{\partial x} \tag{11.16a}$$

$$\rho \left(\frac{\partial v}{\partial t} + u\frac{\partial v}{\partial x} + v\frac{\partial v}{\partial y} + w\frac{\partial v}{\partial z} \right) = \rho f_y - \frac{\partial p}{\partial y} \tag{11.16b}$$

$$\rho \left(\frac{\partial w}{\partial t} + u\frac{\partial w}{\partial x} + v\frac{\partial w}{\partial y} + w\frac{\partial w}{\partial z} \right) = \rho f_z - \frac{\partial p}{\partial z} \tag{11.16c}$$

In cylindrical coordinates they are:

$$\rho \left(\frac{\partial v_r}{\partial t} + v_r \frac{\partial v_r}{\partial r} + \frac{v_\theta}{r}\frac{\partial v_r}{\partial \theta} + v_z \frac{\partial v_r}{\partial z} - \frac{v_\theta^2}{r} \right) = \rho f_r - \frac{\partial p}{\partial r} \tag{11.17a}$$

$$\rho \left(\frac{\partial v_\theta}{\partial t} + v_r \frac{\partial v_\theta}{\partial r} + \frac{v_\theta}{r}\frac{\partial v_\theta}{\partial \theta} + v_z \frac{\partial v_\theta}{\partial z} + \frac{v_r v_\theta}{r} \right) = \rho f_\theta - \frac{1}{r}\frac{\partial p}{\partial \theta} \tag{11.17b}$$

$$\rho \left(\frac{\partial v_z}{\partial t} + v_r \frac{\partial v_z}{\partial r} + \frac{v_\theta}{r}\frac{\partial v_z}{\partial \theta} + v_z \frac{\partial v_z}{\partial z} \right) = \rho f_z - \frac{\partial p}{\partial z} \tag{11.17c}$$

The Euler equations, together with the continuity equation, govern the flow of an inviscid fluid. If we allow the density to be a variable and use the general form of the continuity equation, the Euler equations describe the flow of a compressible, inviscid fluid. If we use the incompressible form of the continuity equation, the Euler equations describe the flow of an incompressible, inviscid fluid. Since an inviscid fluid does not exist, it is tempting to think of the Euler equations as a historical oddity, or of academic interest only. Actually, these equations are widely employed today in the design and analysis of aircraft and turbomachinery! Perhaps you are wondering how this can be. The answer lies in the fact that the Euler equations can also be considered to be the form of the Navier–Stokes equations that apply to an inviscid flow. Notice carefully that we are now talking about an inviscid flow rather than an inviscid fluid.

An inviscid flow is defined to be a one in which the effects of viscosity on the flow are negligible. That is, rather than thinking of a fluid as having no viscosity, we consider instead a flow of a real fluid such as air or water in which the spatial gradients in velocity are sufficiently small that the viscous stresses are negligible. For example, according to Eqs. 11.6d–11.6f, the shear stresses in a Newtonian fluid are given in terms of velocity gradients by

$$\sigma_{xy} = \sigma_{yx} = \mu\left(\frac{\partial u}{\partial y} + \frac{\partial v}{\partial x}\right), \quad \sigma_{yz} = \sigma_{zy} = \mu\left(\frac{\partial v}{\partial z} + \frac{\partial w}{\partial y}\right)$$

and

$$\sigma_{zx} = \sigma_{xz} = \mu\left(\frac{\partial w}{\partial x} + \frac{\partial u}{\partial z}\right)$$

Clearly the shear stresses vanish if the viscosity is zero. But if the values of the velocity gradients are small in a flow of an otherwise low viscosity fluid such as air or water, then the product of these gradients with the viscosity can also result in negligible shear stresses. This is the fundamental idea behind the inviscid flow approximation. A similar conclusion applies to the viscous contributions to the normal stresses. As a model for the flow of a real viscous fluid, rather than substituting $\mu = 0$ into the general version of the Navier–Stokes equation, we can consider the inviscid flow approximation to involve dropping the terms representing viscous effects. Either way, the result is the Euler equations, and these equations do provide an accurate description of many flows of great practical importance.

The Euler equations given here apply to compressible as well as incompressible flow, and also to an incompressible fluid. A solution to a problem in inviscid flow must satisfy the appropriate form of the continuity and Euler equations.

EXAMPLE 11.8

The steady, 2D inviscid flow of an incompressible fluid shown in Figure 11.7 is described by $u = kx$, $v = -ky$, $w = 0$, and $p(x, y) = p_0 - \rho(k^2/2)(x^2 + y^2)$, where k is a constant, p_0 is the pressure at the origin, and body forces have been neglected. This inviscid flow model for a constant density flow approaching a plane wall is referred to as plane stagnation point flow. Show that this flow satisfies the continuity and Euler equations, and comment on whether the no-slip, no-penetration conditions are or are not satisfied.

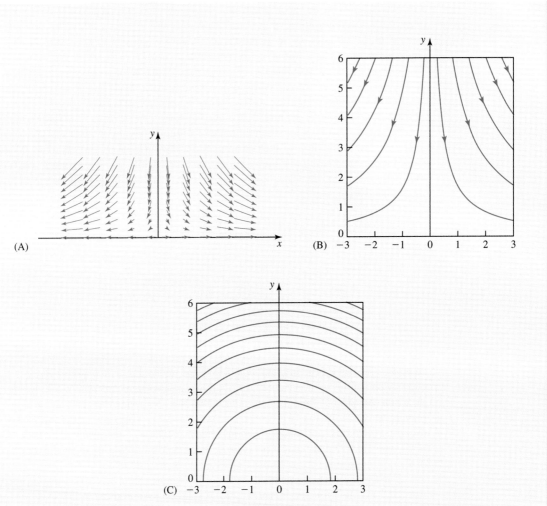

Figure 11.7 Schematic for Example 11.8: (A) velocity vectors, (B) streamlines, and (C) pressure contours for plane stagnation point flow.

SOLUTION

We will check that the continuity equation for an incompressible fluid is satisfied by substituting the velocity components into Eq. 11.4a, $\partial u/\partial x + \partial v/\partial y + \partial w/\partial z = 0$, to obtain

$$\frac{\partial}{\partial x}(kx) + \frac{\partial}{\partial y}(-ky) + \frac{\partial}{\partial z}(0) = k - k = 0$$

Next we substitute the velocity components and pressure into the Euler equations for inviscid flow in Cartesian coordinates, Eqs. 11.16a–c:

$$\rho\left(\frac{\partial u}{\partial t} + u\frac{\partial u}{\partial x} + v\frac{\partial u}{\partial y} + w\frac{\partial u}{\partial z}\right) = \rho f_x - \frac{\partial p}{\partial x}$$

$$\rho\left(\frac{\partial v}{\partial t} + u\frac{\partial v}{\partial x} + v\frac{\partial v}{\partial y} + w\frac{\partial v}{\partial z}\right) = \rho f_y - \frac{\partial p}{\partial y}$$

$$\rho\left(\frac{\partial w}{\partial t} + u\frac{\partial w}{\partial x} + v\frac{\partial w}{\partial y} + w\frac{\partial w}{\partial z}\right) = \rho f_z - \frac{\partial p}{\partial z}$$

To simplify, note that the body forces are zero by assumption, the flow is steady, the velocity components u and v are functions of only one variable and $w = 0$, and the pressure does not depend on z. Inserting the velocity components and pressure, we have

$$\rho[0 + (kx)(k) + (-ky)(0) + 0] = -\frac{\partial}{\partial x}\left[p_0 - \frac{\rho k^2}{2}(x^2 + y^2)\right] = \frac{\rho k^2}{2}(2x)$$

$$\rho[0 + (kx)(0) + (-ky)(-k) + 0] = -\frac{\partial}{\partial y}\left[p_0 - \frac{\rho k^2}{2}(x^2 + y^2)\right] = \frac{\rho k^2}{2}(2y)$$

$$0 = -\frac{\partial}{\partial z}\left[p_0 - \frac{\rho k^2}{2}(x^2 + y^2)\right] = 0$$

After simplifying we find

$$\rho(kx)(k) = \frac{\rho k^2}{2}(2x), \quad \rho(-ky)(-k) = \frac{\rho k^2}{2}(2y), \quad \text{and} \quad 0 = 0$$

Thus, the Euler equations are also satisfied for this flow.

At the wall, $y = 0$; thus we find $u = kx$, $v = 0$, $w = 0$ on the wall. The fluid satisfies the no-penetration condition but slips with $u = kx$ in the x direction.

EXAMPLE 11.9

The streamlines for the 2D, inviscid, constant density flow over a cylinder are shown in Figure 11.8. The streamfunction for this flow is given in cylindrical coordinates by $\psi(r, \theta) = U_\infty r(1 - R^2/r^2)\sin\theta$, where U_∞ is the freestream velocity and R is the cylinder radius. If the body force is neglected, the pressure distribution is given by $p(r, \theta) = p_\infty + \frac{1}{2}\rho U_\infty^2[1 - (1 - R^2/r^2)^2 - 4(R^2/r^2)\sin^2\theta]$. Show that the velocity field in this case is described by

$$v_r = U_\infty\left(1 - \frac{R^2}{r^2}\right)\cos\theta, \quad v_\theta = -U_\infty\left(1 + \frac{R^2}{r^2}\right)\sin\theta, \quad \text{and} \quad v_z = 0$$

Figure 11.8 Schematic for Example 11.9: streamlines for inviscid flow over a cylinder.

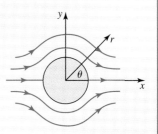

and that the continuity and Euler equations are satisfied. Comment on the boundary conditions.

SOLUTION

The streamfunction for a 2D constant density or incompressible flow that is described in cylindrical coordinates by velocity components $(v_r, v_\theta, 0)$ is defined by Eqs. 10.74a and 10.74b as

$$v_r = \frac{1}{r}\frac{\partial \psi}{\partial \theta} \quad \text{and} \quad v_\theta = -\frac{\partial \psi}{\partial r}$$

Thus we can calculate the velocities from

$$v_r = \frac{1}{r}\frac{\partial \psi}{\partial \theta} = \frac{1}{r}\frac{\partial}{\partial \theta}\left[U_\infty r\left(1 - \frac{R^2}{r^2}\right)\sin\theta\right] = U_\infty\left(1 - \frac{R^2}{r^2}\right)\cos\theta$$

$$v_\theta = -\frac{\partial \psi}{\partial r} = -\frac{\partial}{\partial r}\left[U_\infty r\left(1 - \frac{R^2}{r^2}\right)\sin\theta\right]$$

$$= -U_\infty \sin\theta \left[\left(1 - \frac{R^2}{r^2}\right) + r\left(\frac{2R^2}{r^3}\right)\right]$$

$$= -U_\infty\left(1 + \frac{R^2}{r^2}\right)\sin\theta$$

We see that this agrees with the velocity components given in the problem statement.

Although we know that a streamfunction guarantees that the continuity equation is satisfied, we can check these velocity components by substituting them into that equation in cylindrical coordinates, Eq. 11.4b, $(1/r)[\partial(rv_r)/\partial r] + (1/r)(\partial v_\theta/\partial \theta) + \partial v_z/\partial z = 0$. The left hand side of this equation is

$$\frac{1}{r}\frac{\partial}{\partial r}\left[rU_\infty\left(1 - \frac{R^2}{r^2}\right)\cos\theta\right] + \frac{1}{r}\frac{\partial}{\partial \theta}\left[-U_\infty\left(1 + \frac{R^2}{r^2}\right)\sin\theta\right] + (0)$$

$$= \frac{U_\infty}{r}\left(1 - \frac{R^2}{r^2}\right)\cos\theta + U_\infty\left(\frac{2R^2}{r^3}\right)\cos\theta - \frac{U_\infty}{r}\left(1 + \frac{R^2}{r^2}\right)\cos\theta + (0) = 0$$

To see if the Euler equations are satisfied, we will substitute these velocity components and the pressure into Eqs. 11.17a–11.17c. Writing only the nonzero terms in these equations yields

$$\rho\left(v_r \frac{\partial v_r}{\partial r} + \frac{v_\theta}{r}\frac{\partial v_r}{\partial \theta} - \frac{v_\theta^2}{r}\right) = -\frac{\partial p}{\partial r}, \quad \rho\left(v_r \frac{\partial v_\theta}{\partial r} + \frac{v_\theta}{r}\frac{\partial v_\theta}{\partial \theta} + \frac{v_r v_\theta}{r}\right) = -\frac{1}{r}\frac{\partial p}{\partial \theta}$$

and

$$0 = -\frac{\partial p}{\partial z}$$

Since the pressure given in the problem statement is only a function of r and θ, the last equation is satisfied. Although we leave the details of the final step as an exercise for the interested reader (and we suggest the use of a symbolic mathematics code) substituting the velocities and pressure into the remaining two equations shows that these are also satisfied. On the cylinder, $r = R$, and we find $v_r = 0$, $v_\theta = -2U_\infty \sin\theta$, and $v_z = 0$. So the no-penetration condition is satisfied, but the fluid slips along the surface with a θ velocity of $v_\theta = -2U_\infty \sin\theta$.

11.6.1 Streamline Coordinates

It is possible to gain additional insight into the relationship between pressure and velocity in inviscid flow by making use of the Euler equation in a specialized set of coordinates known as streamline coordinates. Consider a steady 2D flow in the xy plane. The streamlines of the flow and the lines orthogonal to the streamlines form a set of orthogonal curvilinear coordinates, as illustrated in Figure 11.9. The vector form of the Euler equation applies in any coordinate system, including this new set. At an arbitrary point

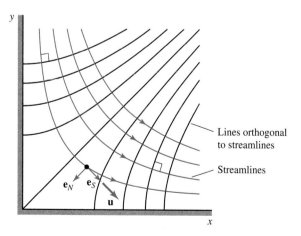

Figure 11.9 Streamline coordinates.

on a streamline, the velocity is tangent to the streamline; thus the velocity vector is $\mathbf{u} = V\mathbf{e}_S$, where V is the speed and $\mathbf{e}_S$ is a unit vector tangent to the streamline at the point. The unit vector normal to the streamline at this point, $\mathbf{e}_N$, is chosen to point away from the center of curvature. Our goal is to write components of the Euler equation in each of these directions.

The Euler equation is written in terms of acceleration, $\mathbf{a}$, using Eq. 11.15 as

$$\rho \mathbf{a} = \rho \mathbf{f} - \nabla p$$

Writing the components of this equation along and normal to the streamline yields

$$\rho a_S = \rho f_S - \frac{\partial p}{\partial s} \quad \text{and} \quad \rho a_N = \rho f_N - \frac{\partial p}{\partial n} \quad (11.18\text{a, b})$$

where we have written the body force per unit mass and pressure gradient in terms of their components along and normal to the streamline.

Now consider the acceleration of a fluid particle at the selected point on the streamline. The local acceleration is zero, since the flow is steady, and the convective acceleration is given by $(\mathbf{u} \cdot \nabla)\mathbf{u}$. Writing the del operator in the streamline coordinates, we find

$$\nabla = \frac{\partial ()}{\partial n}\mathbf{e}_N + \frac{\partial ()}{\partial s}\mathbf{e}_S$$

Thus the dot product $(\mathbf{u} \cdot \nabla)$ is

$$(\mathbf{u} \cdot \nabla) = V\mathbf{e}_S \cdot \left(\frac{\partial ()}{\partial n}\mathbf{e}_N + \frac{\partial ()}{\partial s}\mathbf{e}_S\right) = V\frac{\partial ()}{\partial s}$$

Applying this to the velocity vector $\mathbf{u} = V\mathbf{e}_S$, and noting that the unit vectors are functions of position along the streamline, we see that the convective acceleration is given by

$$(\mathbf{u} \cdot \nabla)\mathbf{u} = V\frac{\partial}{\partial s}(V\mathbf{e}_S) = V\frac{\partial V}{\partial s}\mathbf{e}_S + V^2\frac{\partial \mathbf{e}_s}{\partial s} = V\frac{\partial V}{\partial s}\mathbf{e}_S - \frac{V^2}{\Re}\mathbf{e}_N$$

where $\Re$ is the radius of curvature of the streamline at this point. Note that we have made use of a result you may have encountered in physics or dynamics: $\partial \mathbf{e}_s/\partial s = -\mathbf{e}_N/\Re$.

Substituting these components into Eqs. 11.18a and 11.18b yields the Euler equations in streamline coordinates

$$\rho V\frac{\partial V}{\partial s} = \rho f_S - \frac{\partial p}{\partial s} \quad \text{and} \quad -\rho\left(\frac{V^2}{\Re}\right) = \rho f_N - \frac{\partial p}{\partial n} \quad (11.19\text{a, b})$$

These equations describe the relationship between pressure and speed at a point in an inviscid, 2D, steady flow of a compressible fluid.

To get a sense of how flow speed and the curvature of streamlines cause pressure variations along and across streamlines, consider a constant density flow in the absence of body forces. From Eq. 11.19a, the variation in pressure along a streamline is given by

$$\frac{\partial p}{\partial s} = -\rho V\frac{\partial V}{\partial s} \quad (11.20\text{a})$$

11.6 EULER EQUATIONS

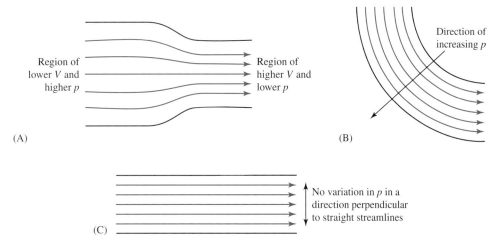

Figure 11.10 Streamlines for 2D inviscid, constant density flows through (A) a contraction, (B) a bend, and (C) a straight section.

indicating that an increase in speed along the streamline is accompanied by a decrease in pressure and vice versa. From Eq. 11.19b, the variation in pressure normal to the streamline is given by

$$\frac{\partial p}{\partial n} = \rho \left(\frac{V^2}{\Re} \right) \tag{11.20b}$$

which shows that at a point on a streamline, the pressure increases along a line normal to the streamline and pointing away from the center of curvature in proportion to the square of the flow speed. If body forces are present, their effect is additive, as can be deduced from Eqs. 11.19a and 11.19b.

To further appreciate the relationship between velocity and pressure in an inviscid flow, consider the three steady, constant density, inviscid 2D flows shown in Figure 11.10. In the contraction, Figure 11.10A, the speed of the fluid increases along the axial streamline, thus $\partial V/\partial s > 0$, and according to Eq. 11.20a, $\partial p/\partial s < 0$. The pressure decreases as the speed increases, a result that is identical to that obtained from the Bernoulli equation. If a streamline is curved, as shown in the 2D flow around a bend in Figure 11.10B, the radius of curvature is finite, and according to Eq. 11.20b, $\partial p/\partial n > 0$. The pressure increases in the direction away from the center of curvature, and the increase is more pronounced for sharper bends having a smaller radius of curvature. The effect is further enhanced for any bend at higher flow speeds, since $\partial p/\partial n$ is proportional to V^2. High speed flow around a sharp bend is accompanied by a relatively large pressure variation across the streamlines from the concave side toward the convex side. Finally, note that if streamlines are straight and parallel as in the flow shown in Figure 11.10C, there is no variation in pressure normal to the streamlines.

These conclusions about pressure variations along and across streamlines are based on the use of the inviscid flow approximation. Nevertheless, they are often used with success to estimate how pressure varies along and across streamlines in viscous flows.

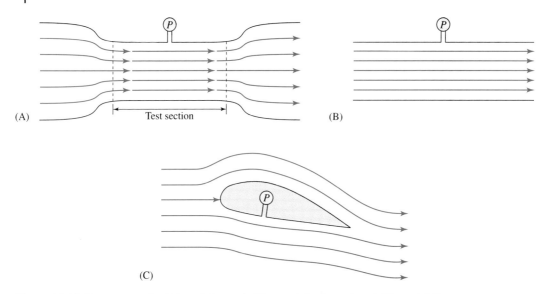

Figure 11.11 Pressure taps for (A) a wind tunnel, (B) a straight channel, and (C) an airfoil.

The accuracy of this approach depends, of course, on whether viscous effects on the pressure are small. Several important cases that satisfy this restriction are shown in Figure 11.11. As shown in Figure 11.11A, the lack of pressure variation across straight parallel streamlines is the basis for measuring the pressure in wind tunnels by means of a wall tap. The value of pressure measured on the tunnel wall is found to be identical to that which occurs everywhere along a line spanning the test section perpendicular to the oncoming flow. Although this line passes through the viscous boundary layer on the wall, the pressure change across the boundary layer on a flat wall is negligible. A wall tap is also the basis for measuring pressure in a pipe or channel flow, as shown in Figure 11.11B. If a wall tap is used on the curved surface of an airfoil, as in Figure 11.11C, it accurately measures the pressure on the wall, but the value of the pressure cannot be used to infer the pressure in the flow away from the wall without accounting for the effects of streamline curvature. Examples 11.10 and 11.11 illustrate these ideas.

11.6.2 Derivation of the Bernoulli Equation

In an inviscid flow the only forces acting on the fluid are inertial, body, and pressure forces per unit volume. Viscous forces, which are frictional, are absent. In Chapter 8 we used a mass and momentum balance to derive the Bernoulli equation for the flow of an inviscid fluid along a streamline. In this section we will derive the Bernoulli equation for an inviscid fluid or inviscid flow from the Euler equation, Eq. 11.14,

$$\rho \frac{D\mathbf{u}}{Dt} = \rho \mathbf{f} - \nabla p$$

EXAMPLE 11.10

Consider a constant density fluid in solid body rotation in the absence of body forces as shown in Figure 11.12. Use the Euler equations in streamline coordinates to analyze the direction of pressure change. Use the Euler equations in cylindrical coordinates to confirm your findings.

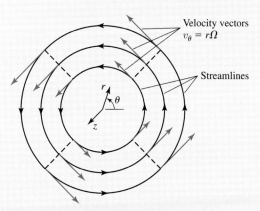

Figure 11.12 Schematic for Example 11.10.

SOLUTION

The velocity field in solid body rotation is described in cylindrical coordinates by

$$av_r = 0, \quad v_\theta = r\Omega, \quad \text{and} \quad v_z = 0$$

where Ω is a constant (see Eq. 10.33). By inspection the streamlines are circles. Consider a point on a streamline of radius r. The velocity at this point is $v_\theta = r\Omega$ and is directed along the streamline. Thus the speed is $V = r\Omega$ and does not change along the streamline. The radius of curvature at this point is $\Re = r$. Applying the Euler equations in streamline coordinates, Eqs. 11.20a and 11.20b, we find

$$\frac{\partial p}{\partial s} = -\rho V \frac{\partial V}{\partial s} = -\rho (r\Omega) \frac{\partial}{\partial s}(r\Omega) = 0$$

$$\frac{\partial p}{\partial n} = \rho \left(\frac{V^2}{\Re}\right) = \rho \left(\frac{(r\Omega)^2}{r}\right) = \rho r \Omega^2$$

Thus there is no change in pressure along a streamline, and the pressure increases with increasing radius.

The Euler equations in cylindrical coordinates, Eqs. 11.17a–11.17c, are

$$\rho\left(\frac{\partial v_r}{\partial t} + v_r\frac{\partial v_r}{\partial r} + \frac{v_\theta}{r}\frac{\partial v_r}{\partial \theta} + v_z\frac{\partial v_r}{\partial z} - \frac{v_\theta^2}{r}\right) = \rho f_r - \frac{\partial p}{\partial r}$$

$$\rho\left(\frac{\partial v_\theta}{\partial t} + v_r\frac{\partial v_\theta}{\partial r} + \frac{v_\theta}{r}\frac{\partial v_\theta}{\partial \theta} + v_z\frac{\partial v_\theta}{\partial z} + \frac{v_r v_\theta}{r}\right) = \rho f_\theta - \frac{1}{r}\frac{\partial p}{\partial \theta}$$

$$\rho\left(\frac{\partial v_z}{\partial t} + v_r\frac{\partial v_z}{\partial r} + \frac{v_\theta}{r}\frac{\partial v_z}{\partial \theta} + v_z\frac{\partial v_z}{\partial z}\right) = \rho f_z - \frac{\partial p}{\partial z}$$

Substituting the velocity components for solid body rotation, and noting the absence of body force, we find

$$\rho\left(-\frac{v_\theta^2}{r}\right) = \rho\left(-\frac{(r\Omega)^2}{r}\right) = -\rho r \Omega^2 = -\frac{\partial p}{\partial r}, \quad 0 = -\frac{1}{r}\frac{\partial p}{\partial \theta}, \quad \text{and} \quad 0 = -\frac{\partial p}{\partial z}$$

Thus we have $\partial p/\partial r = \rho r \Omega^2$, and since n is equivalent to r in this case, this result can be seen to be consistent with that obtained with the Euler equations in streamline coordinates.

EXAMPLE 11.11

Figure 11.13 shows a crude hurricane model in which the flow is circular, and the wind speed increases linearly with radius from 0 at the center of the eye to 150 km/h at $R = 50$ km. Estimate the pressure difference from the eye of the hurricane to the indicated location R.

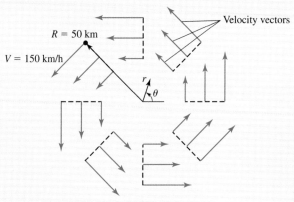

Figure 11.13 Schematic for Example 11.11.

SOLUTION

Circular flow implies circular streamlines with r pointing in the n direction. A linear θ velocity profile fitted to the data is given by

$$v_\theta(r) = \alpha r = \left(\frac{150 \text{ km/h}}{50 \text{ km}}\right) r = \left(3 \frac{\text{km/h}}{\text{km}}\right) r$$

thus $\alpha = 3$ (km/h)/km = 3/h. To determine the pressure distribution, we will apply Eq. 11.20b, using $V = v_\theta \partial/\partial n = \partial/\partial r$ and $\Re = r$ to write the equation as

$$\frac{dp}{dr} = \rho \frac{V^2}{r} = \rho \frac{(\alpha r)^2}{r} = \rho \alpha^2 r$$

Integrating to find the pressure difference, we have $\int_{p_0}^{p_R} dp = \int_0^R \rho \alpha^2 r \, dr$, which yields

$$p_R - p_0 = \rho \frac{\alpha^2 r^2}{2}\bigg|_0^R = \rho \frac{\alpha^2 R^2}{2}$$

Inserting the data, we find that the pressure in the eye of the hurricane is lower than that at $R = 50$ km by the amount

$$p_R - p_0 = 1.225 \text{ kg/m}^3 \frac{(3/\text{h})\left(\frac{1 \text{ h}}{3600 \text{ s}}\right)^2 [(50 \text{ km})(10^3 \text{ m/1 km})]^2}{2} = 1063 \text{ N/m}^2$$

In reality, hurricanes are complex 3D phenomena; therefore, this result of approximately 10 millibars somewhat underestimates recorded pressure differences for hurricanes of this strength. Can you confirm this result using Eq. 11-17a?

If we expand the material derivative and divide by the density, we find

$$\frac{\partial \mathbf{u}}{\partial t} + (\mathbf{u} \cdot \nabla)\mathbf{u} = \mathbf{f} - \frac{\nabla p}{\rho}$$

Next we employ the vector identity $(\mathbf{u} \cdot \nabla)\mathbf{u} = \frac{1}{2}\nabla(\mathbf{u} \cdot \mathbf{u}) - \mathbf{u} \times (\nabla \times \mathbf{u})$ and substitute this identity into the preceding equation. After rearranging slightly, we find

$$\frac{\partial \mathbf{u}}{\partial t} + \frac{\nabla p}{\rho} + \frac{1}{2}\nabla(\mathbf{u} \cdot \mathbf{u}) - \mathbf{f} = \mathbf{u} \times (\nabla \times \mathbf{u})$$

The most common body forces of interest in fluid dynamics are gravity and centrifugal force. Both of these are conservative forces, hence may be represented by an appropriate potential via the relationship $\mathbf{f} = -\nabla \Psi$. Substituting for the body force in the preceding equation, and recalling that vorticity is defined by Eq. 10.49 as $\boldsymbol{\omega} = \nabla \times \mathbf{u}$, we obtain

$$\frac{\partial \mathbf{u}}{\partial t} + \frac{\nabla p}{\rho} + \frac{1}{2}\nabla(\mathbf{u} \cdot \mathbf{u}) + \nabla \Psi = \mathbf{u} \times \boldsymbol{\omega} \tag{11.21}$$

This is an alternate form of the Euler equation for a conservative body force.

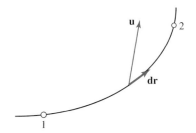

Figure 11.14 Integration path.

The three gradient terms in Eq. 11.21 are a clue to the next step in the derivation of the Bernoulli equation. Using the fact that the line integral of an exact differential form is independent of the path, and bearing in mind that the value of the integral is just the difference in the value of the integrand at the end points, we propose to integrate this equation along a path through a fluid as shown in Figure 11.14. Since this is a vector equation, we will use a dot product of this equation with the vector element of path length $\mathbf{dr}$ to obtain the component of the equation along the path, then integrate from one end of the path to the other. Grouping the three divergence terms together, we find

$$\int_1^2 \frac{\partial \mathbf{u}}{\partial t} \cdot \mathbf{dr} + \int_1^2 \left[\frac{\nabla p}{\rho} + \frac{1}{2}\nabla(\mathbf{u} \cdot \mathbf{u}) + \nabla \Psi \right] \cdot \mathbf{dr} = \int_1^2 (\mathbf{u} \times \boldsymbol{\omega}) \cdot \mathbf{dr}$$

The line integral on the right-hand side of this equation is zero in two important circumstances. First, if the path selected is everywhere parallel to the velocity vector, i.e., if the path is a streamline, the integrand $\mathbf{u} \times \boldsymbol{\omega}$ is a vector that is perpendicular to $\mathbf{u}$, and thus $\mathbf{dr}$, so the dot product $(\mathbf{u} \times \boldsymbol{\omega}) \cdot \mathbf{dr}$ is zero. The second circumstance occurs when $\boldsymbol{\omega} = \nabla \times \mathbf{u} = 0$ everywhere in the flow. In Chapter 10 we noted that this condition defines a special type of flow called an irrotational flow. Thus if the path is a streamline, or the flow is irrotational, the preceding equation becomes

$$\int_1^2 \frac{\partial \mathbf{u}}{\partial t} \cdot \mathbf{dr} + \int_1^2 \left[\frac{\nabla p}{\rho} + \frac{1}{2}\nabla(\mathbf{u} \cdot \mathbf{u}) + \nabla \Psi \right] \cdot \mathbf{dr} = 0$$

The integral $\int_1^2 (\partial \mathbf{u}/\partial t) \cdot \mathbf{dr}$ is not an exact differential form, so we leave it alone. The pressure integral is also not an exact differential form, so we write it as $\int_1^2 (\nabla p/\rho) \cdot \mathbf{dr} = \int_1^2 (dp/\rho)$ to signify that to evaluate this term, we must know the variation in pressure and density along the path. The integrals containing $\frac{1}{2}\nabla(\mathbf{u} \cdot \mathbf{u})$ and $\nabla \Psi$ are in exact differential form; thus they can be evaluated as $\int_1^2 \frac{1}{2}\nabla(\mathbf{u} \cdot \mathbf{u}) \cdot \mathbf{dr} = \frac{1}{2}(V_2^2 - V_1^2)$ and $\int_1^2 \nabla \Psi \cdot \mathbf{dr} = \Psi_2 - \Psi_1$. From these results we obtain

$$\int_1^2 \frac{\partial \mathbf{u}}{\partial t} \cdot \mathbf{dr} + \int_1^2 \frac{dp}{\rho} + \frac{1}{2}\left(V_2^2 - V_1^2\right) + (\Psi_2 - \Psi_1) = 0 \qquad (11.22)$$

This is the general form of the Bernoulli equation for unsteady flow. It applies to either a compressible or an incompressible fluid in a conservative body force field; the only restriction is that the path connecting the two points must be an instantaneous streamline. If the flow is irrotational, the path is arbitrary and the two points do not need to be on the same streamline.

11.6 EULER EQUATIONS

In the case of flow in a gravity field, we have $\Psi_2 - \Psi_1 = g(z_2 - z_1)$, and the Bernoulli equation can be written as

$$\int_1^2 \frac{\partial \mathbf{u}}{\partial t} \cdot d\mathbf{r} + \int_1^2 \frac{dp}{\rho} + \frac{1}{2}\left(V_2^2 - V_1^2\right) + g(z_2 - z_1) = 0 \qquad (11.23)$$

Applying this equation to two points on a streamline, we can write the velocity as $\mathbf{u} = V(s)\mathbf{e}_S$, where the unit vector $\mathbf{e}_S$ lies along the streamline, and since $d\mathbf{r}$ is tangent to the streamline, we have $d\mathbf{r} = ds\,\mathbf{e}_S$. The dot product is then $(\partial \mathbf{u}/\partial t) \cdot d\mathbf{r} = (\partial V/\partial t)\,ds$, and we see that Eq. 11.23 becomes

$$\int_1^2 \frac{\partial V}{\partial t} ds + \int_1^2 \frac{dp}{\rho} + \frac{1}{2}\left(V_2^2 - V_1^2\right) + g(z_2 - z_1) = 0$$

which is identical to the Bernoulli equation (Eq. 8.4) discussed in Chapter 8.

CD/Dynamics/Potential flow/Potential flow builder

Applications of the Bernoulli equation to flow along a streamline were discussed in Chapter 8. In this section we have shown that Bernoulli's equation applies between any two points in an irrotational flow. The use of this equation in an inviscid flow is illustrated in Example 11.12.

EXAMPLE 11.12

A 2D steady, constant density, inviscid flow of air is described by the velocity field $u = Ax$, $v = -Ay$, where $A = 1.5\,\text{s}^{-1}$ and the coordinates are measured in feet. Find the pressure difference between a point at $(1, 1, 0)$ and a point at $(2, 2, 0)$. Are these two points located on the same streamline?

SOLUTION

We are asked to find the pressure difference between two specified points in a flow and to determine whether the points are located on the same streamline. Figure 11.15 serves as a sketch for this flow problem. No additional assumptions are required to solve this problem. Since this is an inviscid flow, we know that the Bernoulli equation is applicable along a streamline. If the flow is also irrotational, then the Bernoulli equation is applicable between any two points in the flow field. We will first check for irrotational flow in this 2D velocity field by using Eq. 10.51c to compute the vorticity. Inserting the known velocity components we find

$$\omega_z = \left(\frac{\partial v}{\partial x} - \frac{\partial u}{\partial y}\right) = \left[\frac{\partial(-Ay)}{\partial x} - \frac{\partial(Ax)}{\partial y}\right] = 0$$

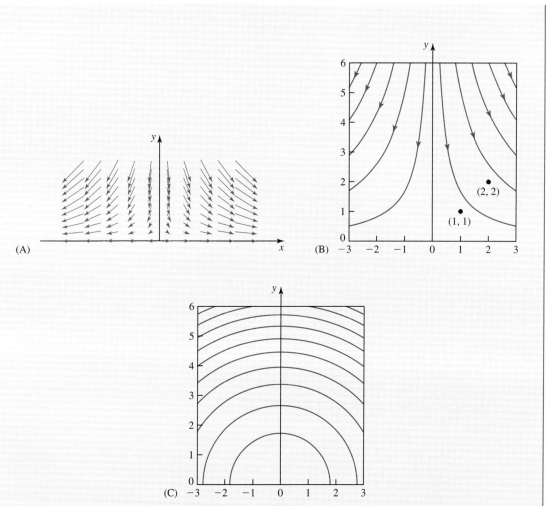

Figure 11.15 Schematic for Example 11.12: (A) velocity vectors, (B) streamlines, and (C) pressure contours for plane stagnation point flow.

Thus, the flow is irrotational and we can apply the Bernoulli equation, Eq. 11.23, between points (1, 1, 0) and (2, 2, 0). For this steady constant density flow, Eq. 11.23 reduces to the traditional form of the Bernoulli equation as given by Eq. 8.6:

$$\frac{p_1}{\rho} + \frac{1}{2}V_1^2 + gz_1 = \frac{p_2}{\rho} + \frac{1}{2}V_2^2 + gz_2$$

Since the points are at the same elevation, $z_1 = z_2$. To calculate the kinetic energy terms we write

$$V^2 = u^2 + v^2 = (Ax)^2 + (-Ay)^2 = A^2(x^2 + y^2)$$

Thus at (1, 1, 0) we can determine that $V_1^2 = A^2(x_1^2 + y_1^2) = A^2(1^2 + 1^2) = 2A^2$ (ft^2), and at the point (2, 2, 0) we have $V_2^2 = A^2(x_2^2 + y_2^2) = A^2(2^2 + 2^2) = 8A^2$ (ft^2). In this case the Bernoulli equation becomes $p_1/\rho + \frac{1}{2}2A^2$ (ft^2) $= p_2/\rho + \frac{1}{2}8A^2$ (ft^2), and the pressure difference is $p_1 - p_2 = 3$ (ft^2)ρA^2. Inserting the data we find

$$p_1 - p_2 = 3 \text{ (ft}^2)\rho A^2 = (3 \text{ ft}^2)(0.002378 \text{ slug/ft}^3)(1.5 \text{ s}^{-1})^2 (\text{lb}_f\text{-s})/(\text{slug-ft})\left(\frac{1 \text{ ft}^2}{144 \text{ in.}^2}\right)$$

$$= 1.1 \times 10^{-4} \text{ psia}$$

To check whether these two points are on the same streamline, we can write the equation for a streamline in this steady flow, using Eq. 10.30 as

$$\frac{dy}{dx} = \frac{v}{u} = \frac{-Ay}{Ax} = -\frac{y}{x}$$

Solving for the streamline we have $dy/y = -dx/x$, or $x\,dy + y\,dx = 0$. Thus the streamlines are given by $xy = C$ (see Example 10.9). The streamline through the point (1, 1, 0) is $xy = 1$, and it is clear that this streamline does not pass through the point (2, 2, 0).

The family of streamlines and the pressure contours for this flow are shown in Figure 11.15. The latter are obtained by realizing that at the origin the velocity is zero; thus the pressure there is the stagnation pressure. Applying the Bernoulli equation between the origin (0, 0, 0) and any point $(x, y, 0)$ we can write $p/\rho + \frac{1}{2}V^2 = p_0/\rho + \frac{1}{2}V_0^2$, where the subscript zero denotes the origin. Since $V^2 = u^2 + v^2 = A^2(x^2 + y^2)$, and $V_0^2 = 0$, we find $p_0 - p = \frac{1}{2}\rho A^2(x^2 + y^2)$, which shows that the pressure is constant on circles centered on the origin.

11.7 THE ENERGY EQUATION

The partial differential equation expressing conservation of energy for a Newtonian fluid is simply referred to as the energy equation. The derivation of this scalar equation, which can be found in advanced texts, is based on the thermodynamic principle that the time rate of change of internal energy plus kinetic energy for a volume of fluid is equal to the rate at which work is done on the fluid plus the rate at which heat is added to the fluid. The resulting differential energy equation is

$$\rho\frac{Du}{Dt} = -p_e \Delta + \kappa \Delta^2 + \rho\Phi + \nabla \cdot (k\nabla T) \tag{11.24}$$

where p_e is the thermodynamic pressure discussed in Section 11.4, the dilation Δ is given by Eq. 10.64a as $\Delta = \partial u/\partial x + \partial v/\partial y + \partial w/\partial z$, and the viscous dissipation per unit mass, Φ, is defined by

$$\Phi = \frac{2\mu}{\rho}\left[\left(\frac{\partial u}{\partial x}\right)^2 + \left(\frac{\partial v}{\partial y}\right)^2 + \left(\frac{\partial w}{\partial z}\right)^2\right]$$
$$+ \frac{\mu}{\rho}\left[\left(\frac{\partial v}{\partial x} + \frac{\partial u}{\partial y}\right)^2 + \left(\frac{\partial w}{\partial y} + \frac{\partial v}{\partial z}\right)^2 + \left(\frac{\partial u}{\partial z} + \frac{\partial w}{\partial x}\right)^2\right] - \frac{2}{3}\frac{\mu}{\rho}\Delta^2 \tag{11.25}$$

Equation 11.24 is a scalar equation expressing the rate at which internal energy is increasing at a point in the fluid. The terms on the right-hand side represent physical processes that affect internal energy. From left to right these terms represent the reversible work of compression by the thermodynamic pressure, viscous dissipation of energy by the bulk viscosity, viscous dissipation of energy by the shear viscosity, and heat transfer by conduction. The dissipation terms, $\kappa \Delta^2$ and $\rho \Phi$, which were briefly discussed in Section 2.7.1, represent an irreversible transformation of fluid mechanical energy into internal energy, i.e., heat.

Thermodynamic properties such as the pressure, density, temperature, internal energy, and entropy may vary in a flowing fluid. As we saw in CV analysis, the values of these quantities in a flow field are critical in understanding and predicting the performance of engines, pumps, compressors, and turbomachinery. Since internal energy is not directly accessible as an experimental variable, thermodynamic relationships are used to express the internal energy in terms of more convenient thermodynamic quantities such as temperature. In the following discussion we show how the material derivative is used to write rate relationships among thermodynamic state variables, and we further discuss some of the important underlying assumptions of the thermodynamic model used in fluid dynamics.

The thermodynamic laws apply to a change from one equilibrium state to another for a fluid system. In fluid mechanics this system is either a finite volume of fluid or an infinitesimal fluid particle. Consider the first law for a reversible process as it applies to a fluid particle: $du = Tds - p_e\, dv$, where u is the internal energy, T is temperature, s is entropy, p_e is thermodynamic pressure, and v is the specific volume. Note that we have used the reversible relationships $dq = Tds$ and $dw = -p_e\, dv$ to replace the work and heat terms in the first law, since the process is assumed to be reversible. Replacing specific volume with density in the preceding equation, we find $du = Tds + p_e(d\rho/\rho^2)$, where we have made use of the fact that $dv = d(1/\rho) = -d\rho/\rho^2$. We can express this relationship as a time rate of change, or rate equation, by writing

$$\frac{du}{dt} = T\frac{ds}{dt} + \frac{p_e}{\rho^2}\frac{d\rho}{dt} \tag{11.26a}$$

This equation, which provides the rate of change in internal energy as a fluid particle moves, is written in the Lagrangian description. To express the equation in the Eulerian description, we replace the time derivatives with material derivatives, obtaining

$$\frac{Du}{Dt} = T\frac{Ds}{Dt} + \frac{p_e}{\rho^2}\frac{D\rho}{Dt} \tag{11.26b}$$

By using the material derivative, any thermodynamic relationship written as a rate equation for a fluid particle may also be expressed in the Eulerian description applied at a point in a fluid. For example, consider the following thermodynamic relationship involving entropy: $ds = (c_p/T)dT - (\beta/\rho)dp_e$, where c_p is the specific heat at constant pressure and β is the coefficient of thermal expansion of the fluid. For a fluid particle, the corresponding rate equation is

$$\frac{ds}{dt} = \frac{c_p}{T}\frac{dT}{dt} - \frac{\beta}{\rho}\frac{dp_e}{dt} \tag{11.27a}$$

11.7 THE ENERGY EQUATION

and the same equation written in the Eulerian description using the material derivative is

$$\frac{Ds}{Dt} = \frac{c_p}{T}\frac{DT}{Dt} - \frac{\beta}{\rho}\frac{Dp_e}{Dt} \qquad (11.27\text{b})$$

We can also combine thermodynamic equations as needed. For example, combining Eqs. 11.26b and 11.27b gives

$$\frac{Du}{Dt} = c_p\frac{DT}{Dt} - \frac{\beta T}{\rho}\frac{Dp_e}{Dt} - \frac{p_e}{\rho^2}\frac{D\rho}{Dt} \qquad (11.28)$$

which relates the change in internal energy to changes in temperature, pressure, and density.

Since virtually all processes involving a moving fluid are irreversible, one might wonder about the utility of the reversible expressions noted earlier in the analysis of real flows. In fact, the power of the reversible expressions stems from the observation that they may be used to relate thermodynamic state variables to one another within the energy equation *even when applied to the description of irreversible processes*. This is a subtle point and must not be taken to mean that changes of state in a moving fluid occur reversibly.

To better understand the irreversible thermodynamic model for a fluid, note that the first law for a fluid undergoing an irreversible process may be written as $(D/Dt)(u + \frac{1}{2}\mathbf{u}\cdot\mathbf{u}) = Dw/Dt + Dq/Dt$, where the terms on the right represent respectively the rate at which work is done on, and heat added to, the fluid. The key issue is that although reversible relationships among state variables are used, expressions for the irreversible work and heat terms must be developed that incorporate the effects of viscosity and heat conduction. We do not use the rate form of the reversible Eulerian relationships $Dq/Dt = T(Ds/Dt)$ and $Dw/Dt = -p_e(Dv/Dt)$, since these approximations are not sufficiently accurate to evaluate the work and heat terms for the irreversible processes of fluid mechanics.

Since internal energy is not directly accessible as an experimental variable, the thermodynamic relationships developed here are transformed to express the energy equation in terms of other more convenient thermodynamic quantities. For example, it is common to write the energy equation in terms of temperature as

$$\rho c_p \frac{DT}{Dt} = \kappa \Delta^2 + \rho \Phi + \beta T \frac{Dp_e}{Dt} + \nabla \cdot (k\nabla T) \qquad (11.29)$$

This form of the differential energy equation for a Newtonian fluid is usually preferred by mechanical engineers. Another form of the energy equation, involving entropy, is often preferred for problems involving high speed gas flows and aerospace applications.

The energy equations given in Eqs. 11.24 and 11.29 are applicable to the general case of a compressible Newtonian fluid with variable absolute and bulk viscosity. The pressure that appears in the energy equation is the thermodynamic or equilibrium pressure, p_e. However, as discussed earlier, in most flow problems the bulk viscosity κ is negligible and the thermodynamic pressure may be assumed to be equal to the mechanical pressure p. In this text we assume this to be the case, and Eqs. 11.24 and 11.29 become

$$\rho\frac{Du}{Dt} = -p\Delta + \rho\Phi + \nabla \cdot (k\nabla T) \qquad (11.30\text{a})$$

$$\rho c_p \frac{DT}{Dt} = \rho\Phi + \beta T\frac{Dp}{Dt} + \nabla \cdot (k\nabla T) \qquad (11.30\text{b})$$

For problems that involve the mixing of different fluids, or the transport of some substance in a fluid, the Reynolds transport theorem may be used to derive additional governing equations that describe the mixing or transport processes involved. This task is beyond the scope of this introductory text, but the process of deriving these equations is very similar to that described in this chapter for mass, momentum, and energy.

Since it is tedious, we will not write out these two forms of the energy equation in Cartesian coordinates, but you will find these equations written in various coordinate systems in many advanced texts.

The continuity, Navier–Stokes, and energy equations, together with the Newtonian constitutive model, constitute a complete mathematical description of fluid flow. CFD codes use these three equations as the starting point for describing fluid flows. As mentioned earlier, however, it is not necessary to solve the energy equation in the flow of a constant density, constant viscosity fluid. In that case, the continuity and Navier–Stokes equations are sufficient to determine the velocity and pressure fields. However, if a flow of this type involves heat transfer, one may find the temperature distribution in the fluid by solving the energy equation after the velocity and pressure fields have been obtained.

11.8 DISCUSSION

We conclude this chapter with a brief discussion of several concepts that have been mentioned in earlier chapters and have direct relevance to solving the governing equations.

 CD/Dynamics/Boundary conditions

11.8.1 Initial and Boundary Conditions

The continuity, Navier–Stokes, and energy equations, together with the appropriate constitutive relationships and state equations, provide a complete mathematical description of the flow of a Newtonian fluid. To obtain a solution of this complex set of governing equations, we must specify an appropriate set of boundary and initial conditions for the flow problem being analyzed. The basic set of unknowns for which a solution is sought in the general case includes the three components of velocity, pressure, density, and the temperature of the fluid.

A complete discussion of the required boundary conditions depends on the exact nature of the problem, the approximations employed, and the set of equations to be solved. Although this discussion is beyond the scope of this text, the boundary conditions associated with the fluid velocity field are generally the no-slip, no-penetration conditions as discussed in Section 6.6. In an unsteady flow problem, the initial conditions take the form of the specification of the spatial distribution of the unknowns (velocity, pressure, etc.) at an initial instant of time. The selection of appropriate boundary and initial conditions will be demonstrated in Chapter 12, where a number of analytical solutions to simplified forms of the governing equations will be discussed.

CD/Dynamics/Reynolds number: Inertia and viscosity/Scaling the Navier–Stokes Equation

11.8.2 Nondimensionalization

In Section 3.2 we described many of the common dimensionless groups in fluid mechanics including the Reynolds number, $Re = \rho V L/\mu$, the Froude number $Fr = V/\sqrt{gL}$, the Euler number $Eu = (p - p_0)/\frac{1}{2}\rho V^2$, and the Strouhal number $St = \omega L/V$. We showed that these and other dimensionless groups naturally occur in applying dimensional analysis (DA) to flow problems. In this section we extend our discussion of DA to describe the process known as nondimensionalization of the governing equations. The value of this process is that complete similitude between two physical systems (e.g., a prototype and the full-scale device of interest) is guaranteed if the dimensionless governing equations and boundary conditions for the two different systems are identical. Another advantage of the dimensionless form of the governing equations is that a solution is applicable over a range of geometric and flow parameters, provided the values of those parameters leave the dimensionless coefficients in the governing equations unchanged.

Nondimensionalization of a governing equation is accomplished by dividing every dependent and independent variable in the equation by an appropriate combination of characteristic dimensions, thereby making each variable dimensionless. A characteristic dimension is a physical dimension that is in some way characteristic of the flow field under investigation. Common examples of characteristic dimensions include a characteristic length scale L, usually derived from the geometry; a characteristic velocity scale U, usually defined as the average fluid velocity; a characteristic pressure P; and a characteristic time scale T.

We will illustrate the process of obtaining nondimensional governing equations for the case of a constant density, constant viscosity, flow of a Newtonian fluid. If we assume that gravity is the only body force, and take the z axis upward as usual, the continuity and Navier–Stokes equations, (Eqs. 11.4a and 11.10a–c), are

$$\frac{\partial u}{\partial x} + \frac{\partial v}{\partial y} + \frac{\partial w}{\partial z} = 0$$

$$\rho\left(\frac{\partial u}{\partial t} + u\frac{\partial u}{\partial x} + v\frac{\partial u}{\partial y} + w\frac{\partial u}{\partial z}\right) = -\frac{\partial p}{\partial x} + \mu\left(\frac{\partial^2 u}{\partial x^2} + \frac{\partial^2 u}{\partial y^2} + \frac{\partial^2 u}{\partial z^2}\right)$$

$$\rho\left(\frac{\partial v}{\partial t} + u\frac{\partial v}{\partial x} + v\frac{\partial v}{\partial y} + w\frac{\partial v}{\partial z}\right) = -\frac{\partial p}{\partial y} + \mu\left(\frac{\partial^2 v}{\partial x^2} + \frac{\partial^2 v}{\partial y^2} + \frac{\partial^2 v}{\partial z^2}\right)$$

$$\rho\left(\frac{\partial w}{\partial t} + u\frac{\partial w}{\partial x} + v\frac{\partial w}{\partial y} + w\frac{\partial w}{\partial z}\right) = -\rho g - \frac{\partial p}{\partial z} + \mu\left(\frac{\partial^2 w}{\partial x^2} + \frac{\partial^2 w}{\partial y^2} + \frac{\partial^2 w}{\partial z^2}\right)$$

Recall that for this case it is not necessary to solve the energy equation to determine the velocity and pressure fields.

The four unknowns in this model are u, v, w, and p, and we have four equations to determine these unknowns. The total number of independent and dependent variables is eight, and their nondimensional counterparts are obtained by dividing each quantity by its corresponding characteristic dimension. Using an asterisk to denote a nondimensional quantity, we write

$$x^* = \frac{x}{L}, \quad y^* = \frac{y}{L}, \quad z^* = \frac{z}{L}, \quad t^* = \frac{t}{T},$$
$$u^* = \frac{u}{U}, \quad v^* = \frac{v}{U}, \quad w^* = \frac{w}{U}, \quad p^* = \frac{p}{P} \tag{11.31}$$

To nondimensionalize the equations, we must derive the nondimensional form of various time and space derivatives. The time derivative with respect to the dimensional variable can be written as

$$\frac{\partial()}{\partial t} = \frac{\partial()}{\partial t^*}\frac{\partial t^*}{\partial t} = \frac{1}{T}\frac{\partial()}{\partial t^*}$$

Similarly, the spatial derivatives are given by

$$\frac{\partial()}{\partial x} = \frac{\partial()}{\partial x^*}\frac{\partial x^*}{\partial x} = \frac{1}{L}\frac{\partial()}{\partial x^*}, \quad \frac{\partial()}{\partial y} = \frac{\partial()}{\partial y^*}\frac{\partial y^*}{\partial y} = \frac{1}{L}\frac{\partial()}{\partial y^*},$$
$$\frac{\partial()}{\partial z} = \frac{\partial()}{\partial z^*}\frac{\partial z^*}{\partial z} = \frac{1}{L}\frac{\partial()}{\partial z^*}$$

and

$$\frac{\partial^2()}{\partial x^2} = \frac{1}{L^2}\frac{\partial^2()}{\partial x^{*2}}, \quad \frac{\partial^2()}{\partial y^2} = \frac{1}{L^2}\frac{\partial^2()}{\partial y^{*2}}, \quad \frac{\partial^2()}{\partial z^2} = \frac{1}{L^2}\frac{\partial^2()}{\partial z^{*2}}$$

Thus, the continuity equation becomes

$$\frac{\partial u}{\partial x} + \frac{\partial v}{\partial y} + \frac{\partial w}{\partial z} = \frac{1}{L}\frac{\partial(u^*U)}{\partial x^*} + \frac{1}{L}\frac{\partial(v^*U)}{\partial y^*} + \frac{1}{L}\frac{\partial(w^*U)}{\partial w^*}$$
$$= \frac{U}{L}\left(\frac{\partial u^*}{\partial x^*} + \frac{\partial v^*}{\partial y^*} + \frac{\partial w^*}{\partial z^*}\right) = 0$$

Upon dividing by U/L, we have the following nondimensionalized form of the continuity equation:

$$\frac{\partial u^*}{\partial x^*} + \frac{\partial v^*}{\partial y^*} + \frac{\partial w^*}{\partial z^*} = 0 \tag{11.32}$$

By using a similar process, the corresponding nondimensionalized Navier–Stokes equations are found to be

$$\left(\frac{L}{UT}\right)\frac{\partial u^*}{\partial t^*} + u^*\frac{\partial u^*}{\partial x^*} + v^*\frac{\partial u^*}{\partial y^*} + w^*\frac{\partial u^*}{\partial z^*}$$
$$= -\left(\frac{P}{\rho U^2}\right)\frac{\partial p^*}{\partial x^*} + \left(\frac{\mu}{\rho UL}\right)\left(\frac{\partial^2 u^*}{\partial x^{*2}} + \frac{\partial^2 u^*}{\partial y^{*2}} + \frac{\partial^2 u^*}{\partial z^{*2}}\right)$$

$$\left(\frac{L}{UT}\right)\frac{\partial v^*}{\partial t^*} + u^*\frac{\partial v^*}{\partial x^*} + v^*\frac{\partial v^*}{\partial y^*} + w^*\frac{\partial v^*}{\partial z^*}$$
$$= -\left(\frac{P}{\rho U^2}\right)\frac{\partial p^*}{\partial y^*} + \left(\frac{\mu}{\rho UL}\right)\left(\frac{\partial^2 v^*}{\partial x^{*2}} + \frac{\partial^2 v^*}{\partial y^{*2}} + \frac{\partial^2 v^*}{\partial z^{*2}}\right)$$

$$\left(\frac{L}{UT}\right)\frac{\partial w^*}{\partial t^*} + u^*\frac{\partial w^*}{\partial x^*} + v^*\frac{\partial w^*}{\partial y^*} + w^*\frac{\partial w^*}{\partial z^*}$$
$$= -\left(\frac{gL}{U^2}\right) - \left(\frac{P}{\rho U^2}\right)\frac{\partial p^*}{\partial z^*} + \left(\frac{\mu}{\rho UL}\right)\left(\frac{\partial^2 w^*}{\partial x^{*2}} + \frac{\partial^2 w^*}{\partial y^{*2}} + \frac{\partial^2 w^*}{\partial z^{*2}}\right)$$

We see that there are four dimensionless groups in the nondimensional Navier–Stokes equations:

$$\frac{L}{UT},\quad \frac{gL}{U^2},\quad \frac{P}{\rho U^2},\quad \frac{\mu}{\rho UL}$$

To identify the first of these, note that in an unsteady flow that involves a forced oscillation of a body immersed in a moving fluid, the oscillation has a characteristic frequency ω. The time scale in such a problem is $T = 1/\omega$, and the group $L/UT = \omega L/U$ can be identified as the Strouhal number, $St = \omega L/U$. The second group gL/U^2 may be seen to be related to the Froude number $Fr = V/\sqrt{gL}$; in fact, this group is $gL/U^2 = 1/Fr^2$. The third group is a form of the Euler number. In Chapter 3 we wrote $Eu = (p - p_0)/\frac{1}{2}\rho V^2$. Choosing the pressure scale to be a characteristic pressure difference $P = \Delta p$, we obtain $P/\rho U^2 = \Delta p/\rho U^2 = Eu/2$. The remaining group is recognized as the inverse of the Reynolds number, i.e., $\mu/\rho UL = 1/Re$.

With the introduction of these dimensionless groups, the Navier–Stokes equations in nondimensional form are

$$(St)\frac{\partial u^*}{\partial t^*} + u^*\frac{\partial u^*}{\partial x^*} + v^*\frac{\partial u^*}{\partial y^*} + w^*\frac{\partial u^*}{\partial z^*}$$
$$= -\left(\frac{Eu}{2}\right)\frac{\partial p^*}{\partial x^*} + \left(\frac{1}{Re}\right)\left(\frac{\partial^2 u^*}{\partial x^{*2}} + \frac{\partial^2 u^*}{\partial y^{*2}} + \frac{\partial^2 u^*}{\partial z^{*2}}\right)$$

$$(St)\frac{\partial v^*}{\partial t^*} + u^*\frac{\partial v^*}{\partial x^*} + v^*\frac{\partial v^*}{\partial y^*} + w^*\frac{\partial v^*}{\partial z^*}$$
$$= -\left(\frac{Eu}{2}\right)\frac{\partial p^*}{\partial y^*} + \left(\frac{1}{Re}\right)\left(\frac{\partial^2 v^*}{\partial x^{*2}} + \frac{\partial^2 v^*}{\partial y^{*2}} + \frac{\partial^2 v^*}{\partial z^{*2}}\right) \quad (11.33)$$

$$(St)\frac{\partial w^*}{\partial t^*} + u^*\frac{\partial w^*}{\partial x^*} + v^*\frac{\partial w^*}{\partial y^*} + w^*\frac{\partial w^*}{\partial z^*}$$
$$= -\left(\frac{1}{Fr^2}\right) - \left(\frac{Eu}{2}\right)\frac{\partial p^*}{\partial z^*} + \left(\frac{1}{Re}\right)\left(\frac{\partial^2 w^*}{\partial x^{*2}} + \frac{\partial^2 w^*}{\partial y^{*2}} + \frac{\partial^2 w^*}{\partial z^{*2}}\right)$$

It is also important to investigate the nondimensional form of the boundary and initial conditions. In most problems the BCs introduce a number of geometric parameters that serve to define the characteristic length L. We see that a solution to a flow problem involving a constant density, constant viscosity fluid depends on the Strouhal, Froude, Euler, and Reynolds numbers defined by the characteristic length, time, velocity, and pressure scales, as well as the density and viscosity of the fluid. As you might suspect, additional dimensionless groups appear in the nondimensional equations describing more complicated flows. For example, the Mach number and the Prandtl number appear in compressible flows, which require the use of the energy equation. The Prandtl number also appears in an incompressible flow involving heat transfer when the energy equation is nondimensionalized. Thus the dimensionless groups discussed in Chapter 3 arise not only from classic dimensional analysis but also from a nondimensionalization of the governing equations.

By nondimensionalizing the governing equations, we conclude that the solution to a flow problem in a specified geometry depends only on the values of the relevant nondimensional groups that appear in the transformed equations and boundary conditions. This means that a given solution to the governing equations applies to any geometrically similar flow that has the same value for the dimensionless groups. Thus, the solution applies to a flow with a different length scale but whose other scales are adjusted in such a way that the dimensionless groups are the same. More importantly, however, we discover that the nondimensional governing equations for two physical systems that are geometrically similar (usually a prototype and the full-scale system) are identical if the values of the dimensionless groups in those governing equations are identical. This proves that the flows are dynamically similar when the values of each dimensionless group are the same, thereby confirming the requirements for complete similitude discussed in Chapter 9.

11.8.3 Computational Fluid Dynamics (CFD)

CD/Dynamics/Navier–Stokes equation/Computational fluid dynamics

In this chapter you have been introduced to the continuity, Navier–Stokes, and energy equations of fluid mechanics. Earlier we discussed the difficulty in finding analytical solutions to this system of nonlinear partial differential equations. Since powerful digital computers are now available in virtually every engineering office, the use of computational fluid dynamics to solve those equations numerically is becoming more common. As computer hardware has improved, so too has the software. There are now many commercial CFD codes available, most of them are based on the finite difference, finite element, or finite volume approaches. These codes usually come complete with advanced graphical user interfaces that make data input and output relatively easy. The choice of topics and emphasis in this book is intended to prepare you to learn more about using commercially available CFD software to analyze practical flow problems. You may encounter this software in elective courses in fluid mechanics or in your professional career.

The usefulness of CFD for an engineer can be appreciated by recalling that we were able to determine much about the relationship between flowrate, pressure rise, power requirements, and other aspects of pump performance by using dimensional analysis (case

Figure 11.16 CFD solution for the flow in a water pump for an automobile.

study in Section 3.3.3) and control volume analysis. But what if you want to improve the efficiency of a certain pump design? In that case a detailed description of the flow field within the pump housing is needed, and this description is not provided by DA or CV analysis. Flow fields of this kind can, however, be calculated by using CFD. An example is shown in Figure 11.16. The use of CFD to investigate the details of the flow fields of fluid machines, vehicles of all types, and virtually every other fluid-related application, including golf balls, has become common.

Although the use of CFD may become widespread, it will never totally replace physical experimentation, which is often needed simply to validate some aspect of a CFD calculation. Figure 11.17A shows a series of streaklines generated around a prolate spheroid as calculated by CFD. Figure 11.17B shows the same flow situation, in this instance photographed from a physical experiment. The general validity of the CFD solution is apparent. But close inspection of the streakline filaments where they break up into turbulence downstream of the spheroid reveals a discrepancy in the CFD solution. The computational resolution required to simulate the small-scale eddies of the turbulence is not adequate here. In fact, only the largest computers have been able to simulate turbulence through solutions of the Navier–Stokes equations. In most applications some sort of turbulence model is employed, as discussed briefly in the next chapter. The difficulty of simulating turbulence is one of the many aspects of CFD that will keep computational researchers active and experimentalists employed for the foreseeable future.

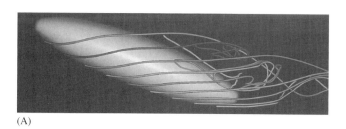
(A)

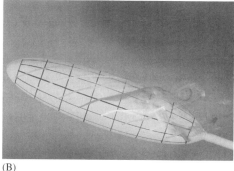
(B)

Figure 11.17 Series of streaklines around a prolate spheroid: (A) calculated by using CFD and (B) created in a water tunnel.

11.9 SUMMARY

The governing equations of fluid dynamics are partial differential equations that apply at every point in the fluid and are completely sufficient to determine the behavior of a fluid under a prescribed set of circumstances.

The differential equation expressing the law of mass conservation, known as the continuity equation, is $D\rho/Dt + \rho \nabla \cdot \mathbf{u} = 0$. For an incompressible flow, defined as one in which the density of a fluid particle does not change as it moves, $D\rho/Dt = 0$. An incompressible fluid, defined as a fluid of constant density, also satisfies this equation. Thus, the continuity equation for either an incompressible fluid or incompressible flow reduces to $\nabla \cdot \mathbf{u} = 0$.

The differential law of momentum conservation is $\rho(D\mathbf{u}/Dt) = \rho \mathbf{f} + \nabla \cdot \boldsymbol{\sigma}$. Every physically possible fluid flow must satisfy this equation. The momentum equation represents a balance of inertial, body, and surface force at each point in a fluid.

A constitutive model provides the link between the stress tensor and the fluid velocity field by using parameters that incorporate the atomic-scale structure of the fluid and its thermodynamic state. The complexity of a constitutive model depends on the molecular structure of the fluid and on whether the fluid consists of a single component or a mixture of components. The constitutive model for a Newtonian fluid, such as air or water, is given by Eqs. 11.6a–11.6f.

When the constitutive model for a Newtonian fluid is used to replace the stresses in the momentum equation, the result is the Navier–Stokes equations. These equations are applicable to laminar and turbulent flows of liquids and gases throughout the entire range of flow speeds. The terms in the Navier–Stokes equations represent the inertial, body, and surface forces acting at a point in the fluid. The surface forces consist of a component of the pressure gradient, which gives the pressure force per unit volume, and terms involving the absolute and bulk viscosity that represent viscous forces. In solving a flow problem for a constant density, constant viscosity fluid, the continuity and Navier–Stokes equations provide a complete set of governing equations to determine the velocity and pressure at every point in the fluid.

One of the most useful approximations in fluid dynamics is that of an inviscid fluid. The momentum equation for the flow of an inviscid fluid, which was derived by Euler, is expressed in vector notation as $\rho(D\mathbf{u}/Dt) = \rho \mathbf{f} - \nabla p$. The Euler equation shows that the flow of an inviscid fluid is governed by a balance of inertial, body, and pressure forces. We used the Euler equations in streamline coordinates to show that an increase in speed along a streamline is accompanied by a decrease in pressure and vice versa. In addition, at a given point the pressure increases along a line normal to the streamline that is pointing away from the center of curvature of the streamline. The magnitude of the pressure increase is proportional to the square of the flow speed and the inverse of the local curvature of the streamline. Thus, if a streamline is highly curved, the pressure increase is significant. There is no variation in pressure perpendicular to straight streamlines.

The Bernoulli equation expressing conservation of energy for inviscid flow can be derived from the Euler equation. For the steady flow of constant density fluid in a gravity field, the Bernoulli equation takes the form $p_2/\rho + \frac{1}{2}V_2^2 + g(z_2 - z_0) = p_1/\rho + \frac{1}{2}V_1^2 + g(z_1 - z_0)$. Thus, in the absence of viscous forces, the sum of the pressure potential, kinetic, and gravitational potential energies at one point along a streamline in the flow of a constant density fluid is equal to the sum of these energies at another point. If the flow is irrotational, the two points need not be on the same streamline.

The differential energy equation, $\rho(Du/Dt) = -p\Delta + \rho\Phi + \nabla \cdot (k\nabla T)$, expresses the rate at which internal energy is increasing at a point in the fluid. The terms on the right-hand side represent, from left to right, the reversible work of compression by the equilibrium pressure, viscous energy dissipation by the bulk viscosity, viscous energy dissipation by the shear viscosity, and heat transfer by conduction. The dissipation terms represent an irreversible transformation of fluid energy into heat.

The presence of spatial and temporal derivatives in the governing equations suggests the need for boundary and initial conditions. If the flow is unsteady, we require an initial condition that specifies the spatial distribution of all flow parameters at $t = 0$. We also require boundary conditions that describe the flow parameters on the physical boundaries defining the flow for all times $t \geq 0$. The most common boundary conditions for fluid velocity are the no-slip, no-penetration conditions introduced in Section 6.6. Although the no-slip condition is widely used, it is not applicable if the fluid is modeled as inviscid or if the flow is modeled as an inviscid flow.

The value of the nondimensional forms of the governing equations is that dynamic and geometric similitude between two physical systems (e.g., prototype and actual) is guaranteed if the dimensionless coefficients of the equations and boundary conditions are matched. The use of nondimensionalized equations also minimizes unit conversion problems.

Commercially available computational fluid dynamics packages offer exciting opportunities for solving the governing equations introduced in this chapter.

PROBLEMS

Section 11.2

11.1 Do the velocity field $\mathbf{u} = x\mathbf{i} + y(1+t)\mathbf{j}$ and density distribution $\rho = z + \rho_0$ represent a flow that is physically possible?

11.2 The velocity field for a flow is given by
$$\mathbf{u} = \frac{-Cy}{\sqrt{x^2 + y^2}}\mathbf{i} + \frac{Cx}{\sqrt{x^2 + y^2}}\mathbf{j}$$
where C is a constant. The density distribution is $\rho = A(x^2 + y^2) + \rho_0$, where A and ρ_0 are constants. Is this flow physically possible?

11.3 You are given the velocity field $\mathbf{u} = (x^2 - y^2)\mathbf{i} - 2xy\mathbf{j}$ for a constant density flow. Is this flow physically possible?

11.4 The velocity field for a constant density flow is $\mathbf{u} = A \ln r \mathbf{e}_r + B\theta \mathbf{e}_\theta$, where A and B are constants. Is this flow physically possible?

11.5 For a two-dimensional incompressible flow, the x component of velocity is $u = 2xy$. What is the y component that will satisfy continuity?

Section 11.4

11.6 The velocity distribution for a Newtonian constant density fluid is given by
$$\mathbf{u} = (6xy^2 - 3x^3)\mathbf{i} + (9x^2y - 2y^3)\mathbf{j}$$
Determine the stress tensor for this flow.

11.7 The velocity distribution for the laminar flow through the rectangular duct shown

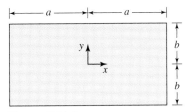

Figure P11.1

in Figure P11.1 with cross section $4ab$ and length L is given by the equation

$$w(x, y) = \frac{7}{4}\alpha \left[1 - \left(\frac{y}{b}\right)^2\right]\left[1 - \left(\frac{x}{a}\right)^6\right]$$

where α is a constant with dimensions of velocity. Find the stress tensor in this flow. If $a = 4b$, which wall has greater stress?

11.8 The velocity distribution for a constant density fluid falling down the outside of a cylindrical tube is given by

$$v_z = \frac{\rho g R^2}{4\mu}\left[1 - \left(\frac{r}{R}\right)^2 + 2a \ln\left(\frac{r}{R}\right)\right]$$

where R is the outside radius of the tube and $r = aR$ corresponds to the free surface of the falling film. Find the stress tensor for this flow and determine the shear stress at the tube wall and at the free surface.

11.9 For the velocity distribution given in Problem 11.6 calculate the deformation rate tensor.

11.10 For the velocity distribution given in Problem 11.7 calculate the deformation rate tensor.

Section 11.5

11.11 Poiseuille flow of a constant density, constant viscosity, Newtonian fluid through a steadily rotating circular pipe was illustrated in Figure 10.42. The corresponding velocity field is given by

$$v_r = 0, \quad v_\theta = r\Omega_0$$

and

$$v_z = \frac{(p_1 - p_2)R^2}{4\mu L}\left[1 - \left(\frac{r}{R}\right)^2\right]$$

where Ω_0 is the angular velocity of the pipe, R is the pipe radius, and the pressures are measured a distance L apart on the wall. The pressure distribution in the z direction is given by

$$p(z) = p_1 + \left(\frac{p_2 - p_1}{L}\right)(z - z_1)$$

Show that this flow satisfies the θ and z components of the Navier–Stokes equations (Eqs. 11.12b and 11.12c) and use the r component of the Navier–Stokes equations (Eq. 11.12a) to investigate the pressure distribution in the radial direction. Ignore the body force due to gravity.

11.12 Imagine taking the channel shown in Figure 11.5 and rotating it 90° clockwise about the z axis (while looking toward the origin) so that the flow is now aligned with the

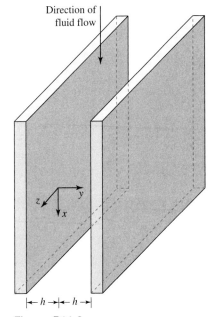

Figure P11.2

direction of the gravitational force (see Figure P11.2). In the absence of an external pressure gradient, the velocity field is found to be

$$u = \left[\frac{h^2\alpha}{2\mu L}\right]\left[1 - \left(\frac{y}{h}\right)^2\right], \quad v = 0$$

and

$$w = 0$$

Use the Navier–Stokes equations for a constant density, constant velocity flow, to investigate the relationship between the constant α and the other parameters in the fluid system.

11.13 Show that the flow given in Problem 11.8 satisfies the constant density, constant viscosity form of the Navier–Stokes equation.

11.14 A thin layer of liquid of constant thickness flow down an inclined plate such that the only velocity component is parallel to the plate. Use the Navier–Stokes equations to determine the relationship between the thickness of the layer and the flowrate per unit width. Assume a steady, laminar, and uniform flow. Also assume that air resistance is negligible.

11.15 An incompressible fluid flows downward through a vertical cylindrical pipe under the action of gravity. The flow is fully developed and laminar. Use the Navier–Stokes equations to derive an expression for the flowrate for the case of zero pressure gradient along the pipe.

Section 11.6

11.16 The streamfunction for an inviscid, incompressible flow is given by $\psi = 2x^2y - \frac{2}{3}y^3$. Find the pressure distribution for this flow.

11.17 The streamfunction for an inviscid, incompressible flow is given by $\psi = Ay^2 - Bx$. Find the pressure distribution for this flow.

11.18 The streamfunction for an inviscid, incompressible flow is given by $\psi = 2x^2y - \frac{2}{3}y^3$, where the stream function ψ is in feet per second. Determine the flowrate passing through the line between (0, 0) and (0, 1).

11.19 The streamfunction for an inviscid, incompressible flow is given by $\psi = -(x - y)$, where ψ the streamfunction has units in meters per second. Find the pressure distribution for this flow.

11.20 The flow near a corner as shown in Figure P11.3 is approximated by $\psi = 2r^{4/3}\sin\frac{4}{3}\theta$. Determine the pressure gradient along the inclined wall.

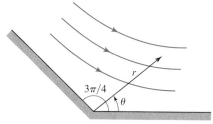

Figure P11.3

11.21 The velocity distribution for an inviscid flow is $\mathbf{u} = (x\,\text{s}^{-1} + 3\,\text{m/s})\mathbf{i} + (y\,\text{s}^{-1} + \frac{4}{3}\,\text{m/s})\mathbf{j}$, where x and y are in meters. The gravitational body force is in the z direction and $\rho = 800$ kg/m³. Find the stagnation points in the flow. Determine the pressure difference between points (0, 0, 0) and (1, 1, 1).

11.22 Find the corresponding velocity field for a flow with the streamfunction $\psi = 2x^2 - 2y^3$.

11.23 Find the corresponding pressure distribution for a steady, inviscid, incompressible flow with the streamfunction $\psi = 6x^2y - 2y^3$.

Section 11.7

11.24 Determine the viscous dissipation for the flow between parallel plates given in Example 11.3, case A.

11.25 Determine the viscous dissipation for the laminar flow through a rectangular duct as given in Problem 11.7.

Section 11.8

11.26 Derive a nondimensional form of the general form of the continuity equation in Cartesian coordinates, Eq. 11.2a.

11.27 Derive a nondimensional form of the z-component of the Euler equations in Cartesian coordinates, Eq. 11.16c.

11.28 Derive a nondimensional form of the z-component of the Euler equations in cylindrical coordinates, Eq. 11.17c.

11.29 Give the nondimensional form of the initial and boundary conditions for the flow between parallel plates given in Example 11.3, case A.

11.30 Give the nondimensional form of the initial and boundary conditions for Poiseuille flow in a round pipe given in Example 11.3, case B.

12 ANALYSIS OF INCOMPRESSIBLE FLOW

12.1 Introduction
12.2 Steady Viscous Flow
    12.2.1 Plane Couette Flow
    12.2.2 Circular Couette Flow
    12.2.3 Poiseuille Flow Between Parallel Plates
    12.2.4 Poiseuille Flow in a Pipe
    12.2.5 Flow over a Cylinder (CFD)
12.3 Unsteady Viscous Flow
    12.3.1 Startup of Plane Couette Flow
    12.3.2 Unsteady Flow over a Cylinder (CFD)
12.4 Turbulent Flow
    12.4.1 Reynolds Equations
    12.4.2 Steady Turbulent Flow Between Parallel Plates (CFD)
12.5 Inviscid Irrotational Flow
    12.5.1 Plane Potential Flow
    12.5.2 Elementary Plane Potential Flows
    12.5.3 Superposition of Elementary Plane Potential Flows
    12.5.4 Flow over a Cylinder with Circulation
12.6 Summary
Problems

12.1 INTRODUCTION

The analytical or computational solution of the governing equations of fluid dynamics can be a challenging task. Not only are these equations nonlinear, but other complicating factors often come into play. Depending on the value of the Reynolds number and other parameters, a flow may be steady or unsteady, laminar or turbulent, and incompressible or compressible. In this chapter we focus on finding analytical and computational solutions to the governing equations for incompressible flows.

714 12 ANALYSIS OF INCOMPRESSIBLE FLOW

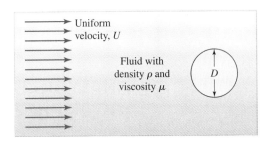

Figure 12.1 Uniform, steady flow over a cylinder.

To construct an analytical solution, we must make engineering approximations that simplify the governing equations to the point that they can be solved. The judgment and decision-making skills you will learn from studying the analytical solutions in this chapter are also relevant to using CFD to solve flow problems. Today's computers, while extraordinarily powerful compared with those available even a few years ago, are still not able to handle a brute force calculation of the majority of engineering flows. This is particularly true for turbulent flow, for which there are no known analytical solutions. Computational solutions for turbulent flow are based on specialized models whose validity and accuracy are critically dependent on the skill of the person employing the CFD code.

 CD/Video Library/Slow flow past a cylinder

The value of experience in solving flow problems can be illustrated by considering the variety of flow fields observed in a uniform flow over a cylinder. Figure 12.1 shows the geometry of this 2D flow. We are interested in understanding this flow over the entire range of Reynolds and Mach numbers, beginning with low speed flows for which Re and M are essentially zero.

In the flow of Figure 12.2, the Reynolds number based on diameter is 0.038 and M is negligibly small. The flow in the vicinity of the cylinder is observed to be steady. Since M is small, it is appropriate to describe this flow as incompressible, i.e., constant density, even if the fluid is a gas. There is no need to solve the energy equation in this case, and the flow is described by the incompressible continuity and Navier–Stokes

Figure 12.2 Flow over a cylinder, $Re = 0.038$. The flow visualization of streamlines is accomplished with aluminum powder in glycerin.

equations. The low value of Re identifies this as a creeping flow. A creeping flow is dominated by the effects of viscosity and is always laminar. Because the fluid velocity is very slow, the Navier–Stokes equations may be simplified by dropping the local and convective acceleration terms, and the resulting governing equations are linear. Although creeping flow over a cylinder looks simple enough to solve analytically, we are only able to construct an approximate solution for a very limited range of Re. The drag coefficient C_D, given in the case study on the drag on cylinders and spheres, Eq. 3.42 of Section 3.3.5, is based on this approximate analytical solution, valid for $Re < 1$ and $M \ll 1$.

 CD/Video library/Flow past a cylinder at moderate Reynolds number

The flow over a cylinder changes character as Re increases. The flow at $Re = 19$ and $M \sim 0$ is shown in Figure 12.3. This flow is also observed to be steady and laminar, and since M is small, it is an incompressible flow. At this higher Re, flow separation has occurred, resulting in symmetric vortices and a well-defined wake. Clearly the velocity field is more complex than in creeping flow at lower Reynolds numbers. Since the flow is steady, the local acceleration terms in the Navier–Stokes equations are zero. The convective acceleration terms are not negligible for $Re = 19$, however, and must be retained. The viscous terms in the Navier–Stokes equations are important in the region close to the cylinder surface (the boundary layer) and in the wake. There is no known analytical solution for a flow at this Re even in an approximate form. The drag coefficient data for this flow, given in Figure 3.20 in the case study of Section 3.3.5, are based on experimental observations, valid for $M \ll 1$.

 CD/Video library/Vortex Shedding

With a further increase in Re vortices are shed and begin to form an oscillating pattern in the wake establishing a structure known as a Karman vortex street (see Figure 12.4 for $Re = 140$). The laminar but unsteady, flow, with M negligible, is described by the incompressible continuity and Navier–Stokes equations with no terms neglected,

Figure 12.3 Flow over a cylinder, $Re = 19$. The flow visualization of streamlines is accomplished with aluminum powder and electrolytic precipitation in water.

HISTORY BOX 12-1

Theodore von Karman (1881–1963) was one of the great aeronautical scientists of the twentieth century. He was born in Budapest, Hungary, but studied under Prandtl in Germany. He immigrated to the United States to take a position at the California Institute of Technology; while there he became a founder of the Jet Propulsion Laboratory. He studied such theoretical subjects as the vortex street that develops behind a bluff body that is named in his honor. He was on the committee organized to study the Tacoma Narrows Bridge collapse because of his work on that subject. In fact for several decades he was at the center of most of the technological advances in aerodynamics facilitated through government research in the U.S.

Figure 12.4 Flow over a cylinder, $Re = 140$ illustrating the Karman vortex street. The flow visualization of streamlines is accomplished by electrolytic precipitation in water.

and no analytical solution is available. The unsteady flow over the cylinder causes a periodic lateral force on the cylinder that is important in many applications. In this case the flow upstream is steady, but the flow over the cylinder is time dependent. Empirical data for this flow are correlated by means of the Reynolds and Strouhal numbers.

At still larger values of Re the boundary layer and wake become turbulent. This is shown in Figure 12.5 for $Re = 2 \times 10^5$. The Mach number here is still negligible, so the flow is described by the incompressible continuity and Navier–Stokes equations. There are no analytical solutions of the governing equations in the turbulent regime, and using computational fluid dynamics to solve this problem requires the Reynolds equations and a turbulence model (see Section 12.4).

As the flow speed increases further, it is no longer possible to assume incompressible flow, and the air must be considered to be a compressible fluid. The problem is then described by the continuity, Navier–Stokes, and energy equations. Figure 12.6 shows the bow shock wave and the wake of a sphere at $M = 1.7$. A shock wave may be considered to be a discontinuity in a continuum flow field; therefore, special techniques are required to treat shock waves analytically or numerically.

The discussion thus far illustrates the challenge you face if asked to solve a practical flow problem. As a first choice, could you use one of the tools you have learned earlier, such as CV analysis or the Bernoulli equation? Should you attempt an analytical or computational solution? Will a single parameter like the drag coefficient, C_D, suffice, or do you need a detailed description of the full flow field? In the case of flow over a cylinder, if C_D alone is needed, how do known solutions and experimental data correlate with the real objects of interest? For instance, can you apply the solutions for 2D flow over a cylinder to the case of the wound wire bridge cables shown in Figure 12.7? Will the solution for an infinitely long cylinder hold for a finite cylinder? Will adjacent objects influence the flow field around the cylinder (see Figure 12.8)? The judgment required to make such choices does not come easily. Even experienced engineers find these questions challenging. We believe that if you encounter these issues in an introductory

12.1 INTRODUCTION | 717

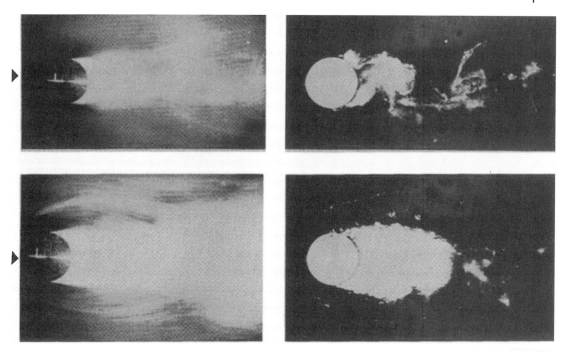

Figure 12.5 Flow over a cylinder, $Re = 2 \times 10^5$. The flow visualization in water illustrates cavitation in the wake.

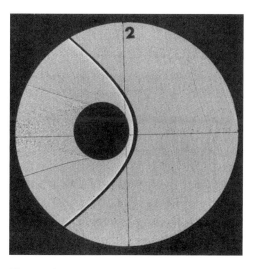

Figure 12.6 Shadowgraph of a sphere at Mach 1.7.

Figure 12.7 Golden Gate Bridge cables.

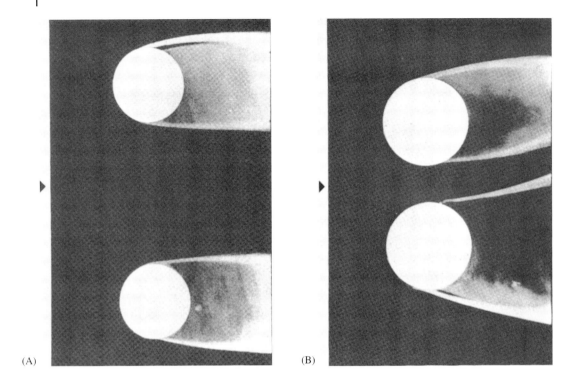

Figure 12.8 (A) Flow over two cylinders, $Re = 900$. (B) Interference between the two wakes as the cylinders are brought together. Flow visualization is by injected streaklines in water.

fluids course, you will begin to acquire good judgment and decision-making skills early in your career. This is essential for avoiding the major mistakes that can be made by those who do not comprehend the full complexity of a fluid mechanics problem.

In the preceding chapters we have been careful to distinguish between an incompressible fluid, i.e., a fluid of constant density, and an incompressible flow. In this chapter our focus is on obtaining solutions of the governing equations for incompressible flows. Although this need not always imply that the fluid density is constant, for simplicity we will assume that all the flows in this chapter are incompressible by reason of constant density. We begin our discussion by solving the governing equations for a number of steady, laminar, viscous flows; then we briefly consider the added complexity of unsteady laminar flow. We close the chapter with a brief discussion of turbulent flows and inviscid potential flows.

12.2 STEADY VISCOUS FLOW

In this section we apply the continuity and Navier–Stokes equations to analyze several steady, laminar flow problems involving a fluid with constant density and viscosity. We are interested in finding the three velocity components and pressure that satisfy the appropriate forms of the governing equations and boundary conditions. The general form of the continuity equation for a constant density fluid is Eq. 11.3, and the Navier–Stokes equations for a constant density, constant viscosity fluid are given in Cartesian and cylindrical coordinates by Eqs. 11.10a–11.10c and 11.12a–11.12c. Since the flow is steady, the time derivatives of velocity in these equations are zero. We will neglect the body force for the sake of concentrating on the relationship between the velocity and pressure fields.

In Cartesian coordinates, the four unknowns, u, v, w, p, are functions of position that satisfy the four governing equations

$$\frac{\partial u}{\partial x} + \frac{\partial v}{\partial y} + \frac{\partial w}{\partial z} = 0 \tag{12.1a}$$

$$\rho\left(u\frac{\partial u}{\partial x} + v\frac{\partial u}{\partial y} + w\frac{\partial u}{\partial z}\right) = -\frac{\partial p}{\partial x} + \mu\left(\frac{\partial^2 u}{\partial x^2} + \frac{\partial^2 u}{\partial y^2} + \frac{\partial^2 u}{\partial z^2}\right) \tag{12.1b}$$

$$\rho\left(u\frac{\partial v}{\partial x} + v\frac{\partial v}{\partial y} + w\frac{\partial v}{\partial z}\right) = -\frac{\partial p}{\partial y} + \mu\left(\frac{\partial^2 v}{\partial x^2} + \frac{\partial^2 v}{\partial y^2} + \frac{\partial^2 v}{\partial z^2}\right) \tag{12.1c}$$

$$\rho\left(u\frac{\partial w}{\partial x} + v\frac{\partial w}{\partial y} + w\frac{\partial w}{\partial z}\right) = -\frac{\partial p}{\partial z} + \mu\left(\frac{\partial^2 w}{\partial x^2} + \frac{\partial^2 w}{\partial y^2} + \frac{\partial^2 w}{\partial z^2}\right) \tag{12.1d}$$

For flows described in cylindrical coordinates, the velocity components and the pressure, v_r, v_θ, v_z, p, satisfy the following governing equations:

$$\frac{1}{r}\frac{\partial (rv_r)}{\partial r} + \frac{1}{r}\frac{\partial v_\theta}{\partial \theta} + \frac{\partial v_z}{\partial z} = 0 \tag{12.2a}$$

$$\rho\left(v_r\frac{\partial v_r}{\partial r} + \frac{v_\theta}{r}\frac{\partial v_r}{\partial \theta} + v_z\frac{\partial v_r}{\partial z} - \frac{v_\theta^2}{r}\right)$$
$$= -\frac{\partial p}{\partial r} + \mu\left(\frac{\partial^2 v_r}{\partial r^2} + \frac{1}{r}\frac{\partial v_r}{\partial r} + \frac{1}{r^2}\frac{\partial^2 v_r}{\partial \theta^2} + \frac{\partial^2 v_r}{\partial z^2} - \frac{v_r}{r^2} - \frac{2}{r^2}\frac{\partial v_\theta}{\partial \theta}\right) \tag{12.2b}$$

$$\rho\left(v_r\frac{\partial v_\theta}{\partial r} + \frac{v_\theta}{r}\frac{\partial v_\theta}{\partial \theta} + v_z\frac{\partial v_\theta}{\partial z} + \frac{v_r v_\theta}{r}\right)$$
$$= -\frac{1}{r}\frac{\partial p}{\partial \theta} + \mu\left(\frac{\partial^2 v_\theta}{\partial r^2} + \frac{1}{r}\frac{\partial v_\theta}{\partial r} + \frac{1}{r^2}\frac{\partial^2 v_\theta}{\partial \theta^2} + \frac{\partial^2 v_\theta}{\partial z^2} - \frac{v_\theta}{r^2} + \frac{2}{r^2}\frac{\partial v_r}{\partial \theta}\right) \tag{12.2c}$$

$$\rho\left(v_r\frac{\partial v_z}{\partial r} + \frac{v_\theta}{r}\frac{\partial v_z}{\partial \theta} + v_z\frac{\partial v_z}{\partial z}\right) = -\frac{\partial p}{\partial z} + \mu\left(\frac{\partial^2 v_z}{\partial r^2} + \frac{1}{r}\frac{\partial v_z}{\partial r} + \frac{1}{r^2}\frac{\partial^2 v_z}{\partial \theta^2} + \frac{\partial^2 v_z}{\partial z^2}\right) \tag{12.2d}$$

To analyze a steady flow problem we must solve the appropriate set of four partial differential equations together with the associated boundary conditions. This process is illustrated in the steady flows discussed in the next several sections.

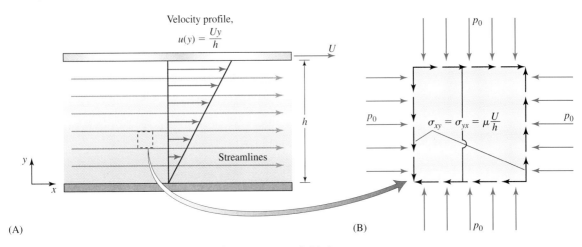

(A) (B)

Figure 12.9 (A) Plane Couette flow and (B) the stresses on a fluid element.

12.2.1 Plane Couette Flow

Consider the flow that results from the steady motion of a flat plate parallel to another stationary plate with the gap between the plates filled with fluid. We are interested in predicting the steady laminar flow in the gap. Flows of this type, which are created by the tangential motion of enclosing walls, are referred to as Couette flows, in honor of M. Couette. The physical arrangement of plane Couette flow is shown in Figure 12.9A. You should recognize this flow as one we have used in a number of examples.

 CD/History/M. Maurice Couette

Our first concern when solving any flow problem is to select a coordinate system. In this case the geometry suggests the use of Cartesian coordinates and the governing equations are Eqs. 12.1a–12.1d. We hope to solve these equations together with the no-slip, no-penetration boundary conditions on the plates. Unfortunately, Eqs. 12.1 cannot be solved without further simplification. To proceed, we ask ourselves what the physical arrangement tells us about each velocity component and its possible dependence on each spatial coordinate. Since we know that a viscous fluid obeys the no-slip condition, it seems likely that the motion of the upper plate will drag fluid in the direction of travel. We conclude that the flow must have a velocity component u in the x direction, and this component must depend on y because the upper plate is moving and the lower plate is fixed. However, u should not depend on x or z if the dimensions of the plates are large in comparison to the gap. Let us assume that we have large plates and a relatively small gap here. Similar reasoning suggests that if there is a velocity component v in the y direction, it should not depend on x or z. What about the velocity component in the remaining direction? Since there is no reason to think that the motion of the

HISTORY BOX 12-2

M. Maurice Couette (1858–1941) lived a happy and productive life as a scientist, teacher, and family man. Couette studied liquid friction under Gabriel Lippman, a Noble Prize winner in physics, at the Sorbonne in Paris. He taught for 43 years at the Catholic University of Angers as a professor of the physical sciences. Couette's careful experimental techniques paved the way for the understanding of the difference between laminar and turbulent flow.

upper plate will cause a velocity component in the z direction, we will assume $w = 0$. Thus we postulate that the flow of interest may have a velocity field of the form $u = u(y)$, $v = v(y)$, $w = 0$.

The next step in developing a solution is to obtain the simplified governing equations by inserting $u = u(y)$, $v = v(y)$, $w = 0$ into Eqs. 12.1a–12.1d. The result is

$$\frac{\partial v}{\partial y} = 0,$$

$$\rho v \frac{\partial u}{\partial y} = -\frac{\partial p}{\partial x} + \mu \frac{\partial^2 u}{\partial y^2},$$

$$\rho v \frac{\partial v}{\partial y} = -\frac{\partial p}{\partial y} + \mu \frac{\partial^2 v}{\partial y^2}, \quad \text{and}$$

$$0 = -\frac{\partial p}{\partial z}$$

Since $v = v(y)$, the first equation is satisfied by $v = C$, a constant. However, the no-penetration condition on each plate is $v = 0$ at $y = 0$ and $y = h$. This can be satisfied only if $C = 0$, which means that the first governing equation is now satisfied by $v = 0$. After inserting $v = 0$ into the other equations, we have $0 = -\partial p/\partial x + \mu(\partial^2 u/\partial y^2)$, $0 = -(\partial p/\partial y)$, and $0 = -(\partial p/\partial z)$. From the second and third equations we conclude that the pressure must be a function of x alone. Thus the solution we are looking for is described by unknown functions $u = u(y)$ and $p = p(x)$, which satisfy the remaining governing equation, $0 = -(\partial p/\partial x) + \mu(\partial^2 u/\partial y^2)$. This equation can be interpreted as expressing a balance between pressure and viscous forces. We must solve this single equation to find the two unknown functions $u = u(y)$ and $p = p(x)$. This appears to be impossible, since we have one equation and two unknowns. However, if we take a partial derivative with respect to x of this equation and interchange derivatives, we obtain

$$0 = -\frac{\partial}{\partial x}\left(\frac{\partial p}{\partial x}\right) + \mu \frac{\partial^2}{\partial y^2}\left(\frac{\partial u}{\partial x}\right)$$

The second term is zero because for $u = u(y)$, the term $\partial u/\partial x = 0$. Thus we know that the unknown pressure $p = p(x)$ satisfies $(\partial/\partial x)(\partial p/\partial x) = 0$. We conclude that the pressure gradient $\partial p/\partial x$ must be a constant. It is customary to write the pressure gradient as $\partial p/\partial x = -G$. The single governing equation describing the velocity profile in the gap can now be written as

$$\frac{\partial^2 u}{\partial y^2} = -\frac{G}{\mu} \quad (12.3)$$

Introducing the constant G does not solve the problem of one equation and two unknowns. However, it is logical to think of Eq. 12.3 as requiring us to determine the

If we include the effect of gravity in this problem, we find that the velocity field is exactly the same as given by Eq. 12.6. The pressure field, rather than being uniform, varies hydrostatically in the y direction. Since the gap in plane Couette flow is typically small, it is permissible to ignore the hydrostatic pressure variation as we have done and simply assume a uniform pressure $p = p_0$ between the plates.

velocity profile $u = u(y)$ for a known imposed value of the pressure gradient. In the present case we are interested in the flow that results from the relative motion of the plates in the absence of any externally imposed pressure gradient. Thus we will assume that pressure between the plates is uniform and equal to a constant. This allows us to write the pressure distribution as

$$p = p_0 \tag{12.4}$$

Plane Couette flow is therefore described by $p = p_0$, where p_0 is the constant ambient pressure in the gap, and by a velocity profile that satisfies Eq. 12.3 with $G = 0$. Since $u = u(y)$ only, we can write Eq. 12.3 with $G = 0$ as the following second-order ordinary differential equation:

$$\frac{d^2 u}{dy^2} = 0 \tag{12.5}$$

This equation expresses the fact that in plane Couette flow the net viscous force is zero.

This second-order differential equation requires two boundary conditions. We can express the no-penetration conditions as $v = 0$ at $y = h$ and at $y = 0$. Since we found $v = 0$ everywhere in this case, the no-penetration conditions are automatically satisfied. The no-slip conditions at the two plates are $u = U$, $w = 0$ at $y = h$, and $u = 0$, $w = 0$ at $y = 0$. Since we have assumed $w = 0$ everywhere at the outset, this part of the no-slip boundary condition is automatically satisfied on both plates. Thus we need be concerned only with the boundary conditions on u. Integrating Eq. 12.5 twice, we find that $u(y) = Ay + B$. Next we evaluate this proposed solution at the two locations $y = h$ and $y = 0$ and insert the values of u prescribed by the no-slip conditions to obtain

$$u(h) = U = Ah + B \quad \text{and} \quad u(0) = 0 = A(0) + B$$

Solving these equations, we find $B = 0$ and $A = U/h$. The velocity profile between the plates is thus given by

$$u(y) = \frac{Uy}{h} \tag{12.6}$$

This linear velocity profile is shown in Figure 12.9A along with the streamlines. The complete solution to the problem is therefore given by $p = p_0$, and $u(y) = Uy/h$, $v = w = 0$. It is interesting to note that neither the density nor the viscosity of the fluid enters the solution. This solution should describe the flow in the gap well away from the edges of the plates, provided the gap is small in comparison to the plate dimensions, a condition we can express for identical plates as $h/L, h/W \ll 1$. Figure 12.10 shows the results of a flow visualization experiment for a Couette flow.

A characteristic feature of plane Couette flow is that the shear stress in the fluid is constant. To see this, recall that the constitutive relationships for a Newtonian fluid are given in Cartesian coordinates by Eqs. 11.6a–11.6f. Since $\nabla \cdot \mathbf{u} = 0$, these relationships reduce for $u(y) = Uy/h$, $v = w = 0$, and $p = p_0$ to

$$\sigma_{xx} = -p_0, \quad \sigma_{yy} = -p_0, \quad \sigma_{zz} = -p_0$$

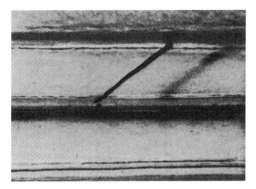

Figure 12.10 Couette flow of glycerin at $Re = 2.7 \times 10^{-2}$. The dyed material line indicates the velocity distribution.

It is also possible to develop solutions for variations on the plane Couette problem in which the lower plate is moving with the upper plate at rest, or both plates are moving. These solutions can be obtained by following the approach to simplifying the equations, then satisfying the boundary conditions that express the appropriate velocities on the plates for each variation.

and

$$\sigma_{xy} = \sigma_{yx} = \mu\left(\frac{\partial u}{\partial y} + \frac{\partial v}{\partial x}\right) = \mu\frac{U}{h},$$

$$\sigma_{zy} = \sigma_{yz} = 0, \quad \sigma_{zx} = \sigma_{xz} = 0$$

Recalling the convention for a stress (the first subscript identifies the plane on which the force acts and the second subscript indicates the direction of the force on this plane), we see that $\sigma_{yx} = \mu(U/h)$ represents the shear stress acting in the x direction on planes in the fluid parallel to the plates, and $\sigma_{xy} = \mu(U/h)$ represents the shear stress acting in the y direction on planes perpendicular to the plates.

We can represent the state of stress in plane Couette flow using the stress tensor as

$$\boldsymbol{\sigma} = \begin{pmatrix} -p_0 & \dfrac{\mu U}{h} & 0 \\ \dfrac{\mu U}{h} & -p_0 & 0 \\ 0 & 0 & -p_0 \end{pmatrix} \quad (12.7)$$

Our solution shows that in plane Couette flow, the normal stresses are each equal to $-p_0$, a constant. The only nonzero shear stresses are σ_{xy} and σ_{yx}, and these are each equal to $\mu(U/h)$, which is also constant. The state of stress in the fluid may be described as one of uniform pressure and uniform shear stress throughout the gap. For an illustration of this state of stress on a small fluid element in the gap between the plates, refer to Figure 12.9B.

12.2.2 Circular Couette Flow

Although the viscosity of a fluid can be measured by using a plane Couette flow between large parallel plates, a more practical viscometer consists of two concentric cylinders

EXAMPLE 12.1

Find the external force and power required to shear a sample of SAE 10W oil between two square plates of edge length 50 cm positioned 0.1 mm apart if the upper plate is moving at 5 cm/s as shown in Figure 12.11.

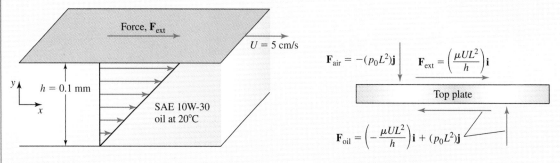

Figure 12.11 Schematic for Example 12.1.

SOLUTION

We are asked to find the external force and power required to shear a sample of oil. Figure 12.11 serves a sketch for this system. It is clear that this is a plane Couette flow, so we can employ the solution to that problem. We will assume that the ambient pressure p_0 outside is the same as the pressure between the plates, and neglect the small shear force applied to the upper moving plate by the surrounding air. A vector force balance on the upper plate will include the external force and the force exerted by the oil and air on the moving plate: $\mathbf{F}_{ext} + \mathbf{F}_{oil} + \mathbf{F}_{air} = 0$. The stress vector applied by the oil on the inside of this plate is obtained from Eq. 4.31B,

$$\Sigma = \mathbf{n} \cdot \boldsymbol{\sigma} = (n_x, n_y, n_z) \begin{pmatrix} \sigma_{xx} & \sigma_{xy} & \sigma_{xz} \\ \sigma_{yx} & \sigma_{yy} & \sigma_{yz} \\ \sigma_{zx} & \sigma_{zy} & \sigma_{zz} \end{pmatrix}$$

where the unit normal is $(0, -1, 0)$. From Eq. 12.7 we find

$$\Sigma = \mathbf{n} \cdot \boldsymbol{\sigma} = (0, -1, 0) \begin{pmatrix} -p_0 & \dfrac{\mu U}{h} & 0 \\ \dfrac{\mu U}{h} & -p_0 & 0 \\ 0 & 0 & -p_0 \end{pmatrix} = \left(-\dfrac{\mu U}{h}\right)\mathbf{i} + p_0 \mathbf{j} + 0 \mathbf{k}$$

The total force applied to the upper plate by the oil is the product of the (constant) stress vector and the plate area, or $\mathbf{F}_{oil} = [(-\mu U/h)\mathbf{i} + p_0 \mathbf{j} + 0 \mathbf{k}]L^2 = (-\mu U L^2/h)\mathbf{i} + (p_0 L^2)\mathbf{j}$. The total force applied by the air to this plate under the stated assumptions is given by $\mathbf{F}_{air} = (-p_0 L^2)\mathbf{j}$. Inserting these results into the force balance and solving for the external force, we obtain $\mathbf{F}_{ext} = (\mu U L^2/h)\mathbf{i}$. We see that the external force acts in the direction of plate motion, as expected. The power required to move the upper plate is given by $P = \mathbf{F}_{ext} \cdot U\mathbf{i} = (\mu U^2 L^2)/h$. Inserting $L = 0.5$ m, $h = 0.0001$ m, and $U = 0.05$ m/s, and obtaining a value of viscosity of $\mu = 0.104$ kg/(m-s) from Appendix A, we find

$$\mathbf{F}_{ext} = \left(\frac{\mu U L^2}{h}\right)\mathbf{i} = \frac{[0.104 \text{ kg/(m-s)}](0.05 \text{ m/s})(0.5 \text{ m})^2}{0.0001 \text{ m}}$$

$$= 13 \text{ (kg-m)/s}^2 \mathbf{i} = 13 \text{ N}\mathbf{i}$$

$$P = \frac{\mu U^2 L^2}{h} = |\mathbf{F}_{ext}| U = (13 \text{ N})(0.05 \text{ m/s})$$

$$= 0.65 \text{ (N-m)/s} = 0.65 \text{ W}$$

Note that in this problem the lower plate is stationary. We can find the retarding force that must act on the lower plate to hold it stationary by writing a force balance and finding the force of the oil and air on the lower plate. The result is $F_R = -(\mu U L^2/h)\mathbf{i} = -13 \text{ N}\mathbf{i}$. Since the movement of oil caused by the upper plate tends to drag the lower plate in the direction of motion, the retarding force necessary to hold the lower plate in place must act opposite to the direction of motion of the upper plate.

EXAMPLE 12.2

A viscometer based on the characteristics of plane Couette flow is shown in Figure 12.12. The viscosity of a fluid sample is to be determined by inserting it into the device and measuring the time Δt it takes the falling weight of mass M to travel a given distance d after reaching terminal velocity. Analyze this arrangement and provide a formula for determining the fluid viscosity.

SOLUTION

We are asked to use our knowledge of Couette flow to obtain a formula for predicting the viscosity of a fluid using the device illustrated in Figure 12.12. We can make use of a result from Example 12.1, namely that the external force required to move the plate is $\mathbf{F}_{ext} = (\mu U L^2/h)\mathbf{i}$. Assuming a frictionless pulley, the magnitude of the external force must be equal to the weight of the mass, Mg, since the plate and the mass are moving at constant velocity. Thus we have $\mu U L^2/h = Mg$, and solving for the viscosity we obtain $\mu = Mgh/UL^2$. Since the velocity of the weight and the plate are the same, we can

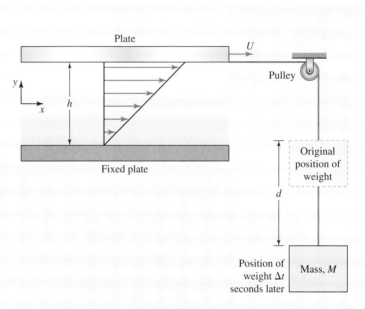

Figure 12.12 Schematic for Example 12.2.

make use of the fact that at constant velocity the weight travels a distance d in time Δt to write $U = d/\Delta t$ and obtain our final result $\mu = Mgh\Delta t/dL^2$.

An alternate approach makes use of the fact that the work done to move the plate a distance d is supplied by the gravitational field. The work done by gravity is Mgd, and the power supplied is $Mgd/\Delta t$. With no power lost to friction in the pulley, this power is equal to the power required to move the plate, which from Example 12.1 is $P = \mu U^2 L^2/h$. Setting the power supplied by gravity equal to this value, we find $Mgd/\Delta t = \mu U^2 L^2/h$, which yields $\mu = Mgdh/\Delta t U^2 L^2$. Inserting $U = d/\Delta t$ and solving for the viscosity, we have $\mu = Mgh\Delta t/dL^2$, which agrees with the earlier result.

with the outer cylinder rotating at a constant angular velocity and the inner cylinder fixed. The fluid whose viscosity is to be determined fills the gap between the two cylinders. The geometry of this circular Couette flow is illustrated in Figure 12.13. Our interest is in finding a solution describing the steady laminar flow in the region well away from the top and bottom of the cylinders for cases in which the gap between the cylinders is small in comparison to their length; i.e., we require $(R_O - R_I)/L \ll 1$.

In this case the geometry suggests the use of cylindrical coordinates. We anticipate that the rotation of the outer cylinder will create a flow with a component of velocity v_θ in the θ direction and that v_θ must depend on r, since the outer cylinder is rotating and the inner cylinder is at rest. Since the gap between the cylinders is small in comparison to their length, v_θ should not depend on z. In other words, end effects are negligible. In

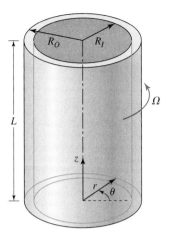

Figure 12.13 Circular Couette flow.

addition, we will assume that the flow is axisymmetric, meaning that there is no variation of any flow property in the θ direction. Thus we conclude that $v_\theta = v_\theta(r)$. Since there is no reason to think that flow will occur in z direction, we will assume $v_z = 0$. If a component of velocity v_r exists in the r direction, it should not depend on θ or z for the reasons just given. Thus we postulate that this flow may have a velocity field of the form $v_r = v_r(r)$, $v_\theta = v_\theta(r)$, $v_z = 0$.

Next we insert $v_r = v_r(r)$, $v_\theta = v_\theta(r)$, $v_z = 0$ into the governing equations, in this case Eqs. 12.2a–12.2d, and make use of the assumption of axisymmetry, $[\partial(\ )/\partial\theta] = 0$, to obtain

$$\frac{1}{r}\frac{\partial(rv_r)}{\partial r} = 0, \quad \rho\left(v_r\frac{\partial v_r}{\partial r} - \frac{v_\theta^2}{r}\right) = -\frac{\partial p}{\partial r} + \mu\left(\frac{\partial^2 v_r}{\partial r^2} + \frac{1}{r}\frac{\partial v_r}{\partial r} - \frac{v_r}{r^2}\right)$$

$$\rho\left(v_r\frac{\partial v_\theta}{\partial r} + \frac{v_r v_\theta}{r}\right) = \mu\left(\frac{\partial^2 v_\theta}{\partial r^2} + \frac{1}{r}\frac{\partial v_\theta}{\partial r} - \frac{v_\theta}{r^2}\right), \quad \text{and} \quad 0 = -\frac{\partial p}{\partial z}$$

We must solve these equations subject to the no-slip, no-penetration boundary conditions on each cylinder. Since $v_r = v_r(r)$ only, the first equation can be written as an ordinary differential equation $(1/r)[d(rv_r)/dr] = 0$, whose solution is $v_r = C/r$. Applying the no-penetration condition $v_r = 0$ at either cylinder shows that $C = 0$. Thus $v_r = 0$, and the remaining equations simplify to

$$\rho\frac{v_\theta^2}{r} = \frac{\partial p}{\partial r}, \quad 0 = \frac{\partial^2 v_\theta}{\partial r^2} + \frac{1}{r}\frac{\partial v_\theta}{\partial r} - \frac{v_\theta}{r^2}, \quad \text{and} \quad 0 = -\frac{\partial p}{\partial z}$$

From the third equation we conclude that the pressure is a function of r only. Since $v_\theta = v_\theta(r)$, and $p = p(r)$, the remaining governing equations can be written as two ordinary differential equations:

$$\frac{dp}{dr} = \rho\frac{v_\theta^2}{r} \quad \text{and} \quad \frac{d}{dr}\left(\frac{1}{r}\frac{d(rv_\theta)}{dr}\right) = 0$$

where we have made use of the identity

$$\frac{d^2 v_\theta}{dr^2} + \frac{1}{r}\frac{dv_\theta}{dr} - \frac{v_\theta}{r^2} = \frac{d}{dr}\left(\frac{1}{r}\frac{d(rv_\theta)}{dr}\right)$$

The equation for dp/dr represents the balance of centrifugal and pressure forces in the r direction, the second equation is a statement that the net viscous force in the θ direction is zero. Integrating the second equation twice, we find

$$v_\theta(r) = \frac{Ar}{2} + \frac{B}{r} \qquad (12.8)$$

where A and B are constants. Now we can find the pressure distribution by inserting this velocity profile into the first equation. The result is

$$\frac{dp}{dr} = \rho\frac{v_\theta^2}{r} = \frac{\rho}{r}\left(\frac{Ar}{2} + \frac{B}{r}\right)^2 = \rho\left(\frac{A^2 r}{4} + \frac{AB}{r} + \frac{B^2}{r^3}\right)$$

Integrating this equation from r to $r = R_O$, we obtain

$$p(R_O) - p(r) = \rho\left[\frac{A^2}{8}(R_O^2 - r^2) + AB\ln\left(\frac{R_O}{r}\right) - \frac{B^2}{2}\left(\frac{1}{R_O^2} - \frac{1}{r^2}\right)\right] \qquad (12.9)$$

We now write the no-slip conditions at each cylinder as $v_\theta = R_O \Omega$ at $r = R_O$ and $v_\theta = 0$ at $r = R_I$ and evaluate the proposed solution for v_θ, Eq. 12.8, at $r = R_O$ and $r = R_I$ to obtain

$$v_\theta(R_O) = R_O\Omega = \frac{AR_O}{2} + \frac{B}{R_O} \qquad \text{and} \qquad v_\theta(R_I) = 0 = \frac{AR_I}{2} + \frac{B}{R_I}$$

Solving these algebraic equations we find $A = 2R_O^2\Omega/(R_O^2 - R_I^2)$ and $B = -R_O^2 R_I^2 \Omega/(R_O^2 - R_I^2)$. We can insert these values of A and B into the velocity profile and pressure distribution to complete the solution. For example, the velocity profile is given by

$$v_\theta(r) = \left(\frac{R_O^2 \Omega}{R_O^2 - R_I^2}\right)r - \left(\frac{R_O^2 R_I^2 \Omega}{R_O^2 - R_I^2}\right)\frac{1}{r} \qquad (12.10)$$

This profile is shown in Figure 12.14. With A and B known, we can now use Eq. 12.9 to evaluate the pressure distribution. Note that neither fluid density nor viscosity enters the solution.

A characteristic feature of circular Couette flow is that the shear stress in the fluid is not constant. To see this, recall that the constitutive relationships for a Newtonian fluid are given in cylindrical coordinates by Eqs. 11.12a–11.12f. Since $\nabla \cdot \mathbf{u} = 0$, these relationships reduce for $v_r = 0$, $v_\theta(r) = Ar/2 + B/r$, $v_z = 0$ and $p = p(r)$, as given by Eq. 12.9 to

$$\sigma_{rr} = -p(r), \quad \sigma_{\theta\theta} = -p(r), \quad \sigma_{rr} = -p(r)$$

$$\sigma_{r\theta} = \sigma_{\theta r} = \mu\left[r\frac{\partial}{\partial r}\left(\frac{B}{r^2}\right)\right] = -2\mu\frac{B}{r^2}, \quad \sigma_{\theta z} = \sigma_{z\theta} = 0, \quad \text{and} \quad \sigma_{zr} = \sigma_{rz} = 0$$

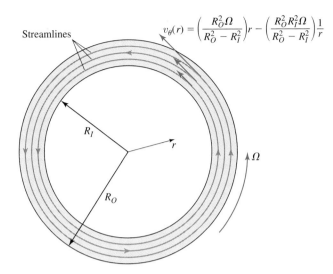

Figure 12.14 Flow distribution for circular Couette flow.

$$v_\theta(r) = \left(\frac{R_O^2 \Omega}{R_O^2 - R_I^2}\right)r - \left(\frac{R_O^2 R_I^2 \Omega}{R_O^2 - R_I^2}\right)\frac{1}{r}$$

If we include the effect of gravity in this problem, we find the velocity field is still given by Eq. 12.10. The pressure field, as given by Eq. 12.9, must be modified to include a hydrostatic variation in the z direction. If the cylinder height is small enough, it is permissible to ignore the hydrostatic pressure variation, as we have done, and use Eq. 12.9 to describe the radial variation in pressure between the cylinders while ignoring the vertical pressure variation.

Recalling the convention for a stress, we see that $\sigma_{r\theta} = -2\mu(B/r^2)$ represents the force acting in the θ direction on a plane in the fluid perpendicular to r, and $\sigma_{\theta r} = -2\mu(B/r^2)$ represents the force acting in the r direction on a plane in the fluid perpendicular to the θ direction.

We can represent the state of stress in circular Couette flow by using the stress tensor as

$$\sigma = \begin{pmatrix} \sigma_{rr} & \sigma_{r\theta} & \sigma_{rz} \\ \sigma_{\theta r} & \sigma_{\theta\theta} & \sigma_{\theta z} \\ \sigma_{zr} & \sigma_{z\theta} & \sigma_{zz} \end{pmatrix}$$

$$= \begin{pmatrix} -p(r) & -2\mu\frac{B}{r^2} & 0 \\ -2\mu\frac{B}{r^2} & -p(r) & 0 \\ 0 & 0 & -p(r) \end{pmatrix} \quad (12.11)$$

Our solution shows that in circular Couette flow, each of the normal stresses is related to the pressure, which is a function of r. The only nonzero shear stresses are $\sigma_{r\theta}$ and $\sigma_{\theta r}$, and each of these is equal to $-2\mu(B/r^2)$, which is also a function of r. The state of stress in the fluid may be described as one of nonuniform pressure and nonuniform shear stress in the gap.

However, for thin gaps the shear stress is nearly uniform and equal to $\sigma_{r\theta} = -(2\mu\Omega R_O^2)/(R_O^2 - R_I^2)$, and the circular Couette flow in a thin gap may be used as a viscometer. This statement can be verified by noting that the ratio of the shear stress at the inner end of the gap to that at the outer end is

$$\frac{\sigma_{r\theta}(\text{inner})}{\sigma_{r\theta}(\text{outer})} = \frac{R_O^2}{R_I^2} = \frac{[R_I + (R_O - R_I)]^2}{R_I^2} = 1 + \frac{2(R_O - R_I)}{R_I} + \left(\frac{R_O - R_I}{R_I}\right)^2$$

EXAMPLE 12.3

Find the external tangential force, torque, and power required to shear a liquid sample in the circular Couette flow arrangement shown in Figure 12.15. Give the tangential force and torque applied by the liquid to the inner cylinder.

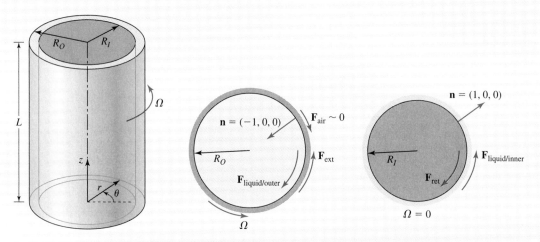

Figure 12.15 Schematic for Example 12.3.

SOLUTION

We are asked to determine the force, torque and power associated with a particular circular Couette flow. Figure 12.15 is an illustration of the geometry. We begin by writing a force balance in the θ direction on the rotating outer cylinder that includes the effect of the external force, the liquid, and the air: $\mathbf{F}_{\text{ext}} + \mathbf{F}_{\text{liquid/outer}} + \mathbf{F}_{\text{air}} = 0$. If we neglect the small tangential force due to the air, our tangential force balance becomes $\mathbf{F}_{\text{ext}} = -\mathbf{F}_{\text{liquid/outer}}$. The stress vector applied by the liquid on the inside of a cylindrical surface of radius r is obtained from the cylindrical equivalent of Eq. 4.31b:

$$\boldsymbol{\Sigma} = \mathbf{n} \cdot \boldsymbol{\sigma} = (n_r, n_\theta, n_z) \begin{pmatrix} \sigma_{rr} & \sigma_{r\theta} & \sigma_{rz} \\ \sigma_{\theta r} & \sigma_{\theta\theta} & \sigma_{\theta z} \\ \sigma_{zr} & \sigma_{z\theta} & \sigma_{zz} \end{pmatrix}$$

where the outward unit normal to the inside is $(-1, 0, 0)$. From Eq. 12.11 we find

$$\boldsymbol{\Sigma} = \mathbf{n} \cdot \boldsymbol{\sigma} = (-1, 0, 0) \begin{pmatrix} -p(r) & -2\mu\dfrac{B}{r^2} & 0 \\ -2\mu\dfrac{B}{r^2} & -p(r) & 0 \\ 0 & 0 & -p(r) \end{pmatrix} = p(r)\mathbf{e}_r + 2\mu\dfrac{B}{r^2}\mathbf{e}_\theta + 0\mathbf{e}_z$$

and the tangential component of the stress is seen to be $\Sigma_\theta = 2\mu(B/r^2)$. The tangential stress acting on the inside of the outer cylinder at $r = R_O$ is thus $\Sigma_\theta = 2\mu(B/R_O^2)$. The tangential force applied to the outer cylinder by the liquid is the product of the constant tangential stress and the cylinder area, or

$$\mathbf{F}_{\text{liquid/outer}} = \left(2\mu\frac{B}{R_O^2}\right)(2\pi R_O L) = -4\pi\mu L\Omega\left(\frac{R_O R_I^2}{R_O^2 - R_I^2}\right)$$

where we have made use of the expression obtained earlier for the constant B (see text preceding Eq. 12.10). Note that the tangential force applied by the liquid acts opposite to the rotation of the cylinder as expected. The force balance on the cylinder, $\mathbf{F}_{\text{ext}} = -\mathbf{F}_{\text{liquid/outer}}$, shows that to keep the cylinder moving at constant velocity, it must be acted upon by a tangential external force in the direction of rotation, given by

$$\mathbf{F}_{\text{ext}} = 4\pi\mu L\Omega\left(\frac{R_O R_I^2}{R_O^2 - R_I^2}\right)$$

The torque required to move the outer cylinder is given by the product of the moment arm and the force, or

$$T_{\text{ext}} = R_O F_{\text{ext}} = 4\pi\mu L\Omega\left(\frac{R_O^2 R_I^2}{R_O^2 - R_I^2}\right)$$

The power required to rotate the outer cylinder is given by the product of the torque and angular velocity or

$$P = T_{\text{ext}}\Omega = 4\pi\mu L\Omega^2\left(\frac{R_O^2 R_I^2}{R_O^2 - R_I^2}\right)$$

To answer the questions concerning the inner cylinder, note that the stress vector acting on the outside of a cylinder of radius r is given by

$$\Sigma = \mathbf{n}\cdot\boldsymbol{\sigma} = (1,0,0)\begin{pmatrix} -p(r) & -2\mu\frac{B}{r^2} & 0 \\ -2\mu\frac{B}{r^2} & -p(r) & 0 \\ 0 & 0 & -p(r) \end{pmatrix} = -p(r)\mathbf{e}_r - 2\mu\frac{B}{r^2}\mathbf{e}_\theta + (0)\mathbf{e}_z$$

where we have been careful to write the outward unit normal as $(1, 0, 0)$. The inner cylinder is located at $r = R_I$, thus the tangential stress is $\Sigma_\theta = -2\mu(B/R_I^2)$, and the tangential force is

$$\mathbf{F}_{\text{liquid/inner}} = \left(-2\mu\frac{B}{R_I^2}\right)(2\pi R_I L) = 4\pi\mu L\Omega\left(\frac{R_O^2 R_I}{R_O^2 - R_I^2}\right)$$

which acts in the direction of rotation (i.e., the viscous liquid tends to drag the inner cylinder in this direction, as expected). The retarding force necessary to hold the inner

cylinder in place is

$$F_{\text{ret}} = -F_{\text{liquid/inner}} = -4\pi\mu L\Omega\left(\frac{R_O^2 R_I}{R_O^2 - R_I^2}\right)$$

and the corresponding torque is

$$T_{\text{ret}} = R_I F_{\text{ret}} = -4\pi\mu L\Omega\left(\frac{R_O^2 R_I^2}{R_O^2 - R_I^2}\right)$$

which is equal and opposite to the torque exerted by the external force on the outer cylinder.

 CD/History/Jean Louis-Marie Poiseuille

Our analysis of circular Couette flow demonstrates that assuming $v_r = 0$ is consistent with the other assumptions but does not exclude the possibility that the axial velocity component v_z is nonzero. In fact, Problem 12.43 asks you to find a solution for circular Couette flow that has an axial velocity component $v_z = v_z(r)$ driven by an axial pressure gradient. It is also possible to develop a more general solution for circular Couette flow in which both cylinders rotate by applying the appropriate boundary conditions on each cylinder.

For a thin gap, $(R_O - R_I)/R_I \ll 1$, and we see that $\sigma_{r\theta(\text{inner})}/\sigma_{r\theta(\text{outer})} \approx 1$.

12.2.3 Poiseuille Flow Between Parallel Plates

Consider the pressure-driven flow in the gap between large parallel plates shown in Figure 12.16. The physical arrangement is identical to that of plane Couette flow, but now both the plates are at rest, and the flow is driven by a pressure difference in the x direction. We are interested in predicting the steady laminar flow in the gap for cases in which the gap is small in comparison to the plate dimensions. Flows of this type, which are created by pressure differences, are referred to as Poiseuille flows, in honor of J. L. M. Poiseuille. Although you are already well acquainted with this flow, we will use the method of the two preceding sections to construct the solution. The geometry suggests the use of Cartesian coordinates so the governing equations are Eqs. 12.1a–12.1d. To simplify these equations, we will neglect body forces and again ask ourselves the following question: What does the physical arrangement tell us about each velocity component and its possible dependence on each spatial coordinate?

In this case, the pressure difference $p_1 - p_2$ suggests that flow will occur in the x direction for $p_1 > p_2$. Thus we expect to have a velocity component u in the x direction, and this component will depend on y, since the no-slip condition demands that $u = 0$ on each plate. Since, however, the plate dimensions are large compared with the gap, u should not depend on x or z. Moreover, since there is no reason to think that this pressure difference will cause a velocity in the z direction, we will assume $w = 0$. Experience with plane Couette flow suggests that the velocity component v in the y direction will be

12.2 STEADY VISCOUS FLOW

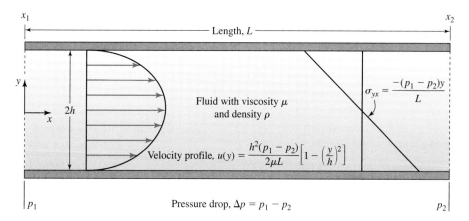

Figure 12.16 Poiseuille flow between parallel plates.

zero, but we can also argue that if v does exist, it should not depend on x or z. Thus, we postulate that this flow may have a velocity field of the form $u = u(y)$, $v = v(y)$, $w = 0$.

The next step in developing a solution is to obtain the simplified governing equations by inserting $u = u(y)$, $v = v(y)$, $w = 0$ into Eqs. 12.1a–12.1d. The result is

$$\frac{\partial v}{\partial y} = 0, \quad \rho v \frac{\partial u}{\partial y} = -\frac{\partial p}{\partial x} + \mu \frac{\partial^2 u}{\partial y^2}, \quad \rho v \frac{\partial v}{\partial y} = -\frac{\partial p}{\partial y} + \mu \frac{\partial^2 v}{\partial y^2}, \quad \text{and} \quad 0 = -\frac{\partial p}{\partial z}$$

Since $v = v(y)$, the first equation is satisfied by $v = C$, a constant. However, the no-penetration condition on each plate is $v = 0$ at $v = -h$ and $v = h$. This can be satisfied only if $C = 0$. The first governing equation is now satisfied. After inserting $v = 0$ into the other equations, we have

$$0 = -\frac{\partial p}{\partial x} + \mu \frac{\partial^2 u}{\partial y^2}, \quad 0 = -\frac{\partial p}{\partial y}, \quad \text{and} \quad 0 = -\frac{\partial p}{\partial z}$$

From the second and third equations we conclude that the pressure is a function of x alone. Thus the solution we are looking for is described by unknown functions $u = u(y)$ and $p = p(x)$, which satisfy the last remaining governing equation, $0 = -\partial p/\partial x + \mu(\partial^2 u/\partial y^2)$. This equation expresses the balance between pressure and viscous forces. In this case, we must solve this single equation to find the two unknown functions $u = u(y)$ and $p = p(x)$, which describe pressure-driven flow between parallel plates. To proceed further, we will again take a partial derivative with respect to x of the equation and interchange derivatives to obtain

$$0 = -\frac{\partial}{\partial x}\left(\frac{\partial p}{\partial x}\right) + \mu \frac{\partial^2}{\partial y^2}\left(\frac{\partial u}{\partial x}\right)$$

The second term is zero because for $u = u(y)$ only, the term $\partial u/\partial x = 0$. Thus the pressure $p = p(x)$ must satisfy $(\partial/\partial x)(\partial p/\partial x) = 0$, which means the pressure gradient must be a constant, i.e., $\partial p/\partial x = -G$. Thus, the governing equation describing the velocity profile in this problem is $\partial^2 u/\partial y^2 = -G/\mu$.

Since $u = u(y)$ only we can write this as an ordinary differential equation and integrate it twice to obtain $u(y) = (-G/2\mu)y^2 + C_1 y + C_2$. The no-slip conditions on the two plates are given by $u = 0$, $w = 0$ at $y = h$, and $u = 0$, $w = 0$ at $y = -h$. We have assumed $w = 0$ everywhere at the outset, so this part of the no-slip boundary condition is automatically satisfied on both plates. Applying the remaining boundary conditions on u, we find:

$$u(h) = 0 = \left(-\frac{G}{2\mu}\right)h^2 + C_1 h + C_2 \quad \text{and} \quad u(-h) = 0 = \left(-\frac{G}{2\mu}\right)(-h)^2 - C_1 h + C_2$$

Solving for the constants we have $C_1 = 0$ and $C_2 = Gh^2/2\mu$, so the velocity profile is given by

$$u(y) = \frac{Gh^2}{2\mu}\left[1 - \left(\frac{y}{h}\right)^2\right] \tag{12.12}$$

The equation governing the pressure distribution is $\partial p/\partial x = -G$. Since $p = p(x)$ only, this is an ordinary differential equation. According to Figure 12.16 we have $p = p_1$ at $x = x_1$, and $p = p_2$ at $x = x_2$. Thus we can integrate from x_1 to x to obtain the linear pressure distribution $p(x) = p_1 - G(x - x_1)$. Since $p = p_2$ at $x = x_2$, we use this solution to write $p_2 = p_1 - G(x_2 - x_1)$. Solving for G, we obtain $G = (p_1 - p_2)/(x_2 - x_1) = (p_1 - p_2)/L$. The pressure distribution is therefore given by

$$p(x) = p_1 - \left(\frac{p_1 - p_2}{L}\right)(x - x_1) \tag{12.13}$$

and the velocity profile can now be written as

$$u(y) = \frac{h^2(p_1 - p_2)}{2\mu L}\left[1 - \left(\frac{y}{h}\right)^2\right] \tag{12.14}$$

This parabolic velocity profile is shown in Figure 12.16. Note that the fluid viscosity occurs in the solution for the velocity profile but does not influence the profile shape. This solution describes the flow in the gap well away from the edges of the plates, provided the gap is small compared with the linear dimensions of the plates, a condition we can express for identical square plates as $h/L \ll 1$.

To determine the state of stress in the fluid, we insert the solution for the velocity and pressure fields into the constitutive relationships for a Newtonian fluid. These are given in Cartesian coordinates by Eqs. 11.6a–11.6f. Since $\nabla \cdot \mathbf{u} = 0$, these relationships reduce in this case to

$$\sigma_{xx} = -p(x), \quad \sigma_{yy} = -p(x), \quad \sigma_{zz} = -p(x)$$

$$\sigma_{xy} = \sigma_{yx} = \mu\left(\frac{\partial u}{\partial y}\right) = \frac{-(p_1 - p_2)y}{L}, \quad \sigma_{zy} = \sigma_{yz} = 0, \quad \text{and} \quad \sigma_{zx} = \sigma_{xz} = 0$$

Recalling the convention for a stress, we see that $\sigma_{yx} = [-(p_1 - p_2)y]/L$ represents the force acting in the x direction on planes in the fluid parallel to the plates, and $\sigma_{xy} = [-(p_1 - p_2)y]/L$ represents the force acting in the y direction on planes

12.2 STEADY VISCOUS FLOW

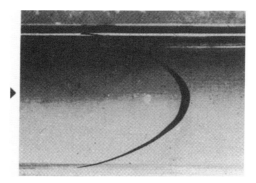

Figure 12.17 Pouiseuille flow of glycerin between parallel plates at $Re = 5.3 \times 10^{-2}$. The dyed material line indicates the velocity distribution.

in the fluid perpendicular to the plates. The distribution of the shear stress $\sigma_{yx} = [-(p_1 - p_2)y]/L$ in the gap is shown in Figure 12.16.

We can represent the state of stress in Poiseuille flow between parallel plates by using the stress tensor:

$$\boldsymbol{\sigma} = \begin{bmatrix} -p(x) & \dfrac{-(p_1 - p_2)y}{L} & 0 \\ \dfrac{-(p_1 - p_2)y}{L} & -p(x) & 0 \\ 0 & 0 & -p(x) \end{bmatrix} \qquad (12.15)$$

Note that each of the normal stresses is related to the pressure, which varies linearly in the x direction. The only nonzero shear stresses are σ_{xy} and σ_{yx}, and these are each equal to $[-(p_1 - p_2)y]/L$; hence the shear stress varies linearly across the gap. The state of stress in the fluid may be described as one of nonuniform pressure and nonuniform shear stress throughout the gap. Figure 12.17 shows the results of a flow visualization experiment for a plane Poiseuille flow.

EXAMPLE 12.4

Find the expressions for volume flowrate, maximum and average velocity, and pressure drop for Poiseuille flow between parallel plates. Use your result to estimate the pressure drop encountered in blowing 60°F air at 2 ft/s through the gap shown in Figure 12.18.

SOLUTION

We will calculate the volume flowrate passing through a cross section perpendicular to the flow direction as shown in Figure 12.18. The volume flowrate is given by $Q = \int_S (\mathbf{u} \cdot \mathbf{n}) \, dS$. We begin by setting $dS = dy \, dz$, $\mathbf{n} = \mathbf{i}$, and $\mathbf{u} = u(y)\mathbf{i}$, with $u(y)$

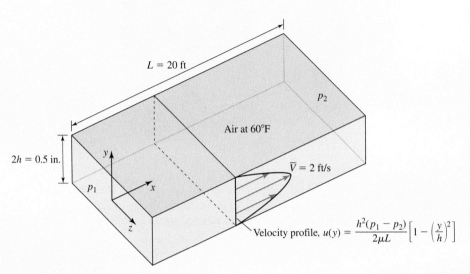

Figure 12.18 Schematic for Example 12.4.

given by Eq. 12.14 as

$$u(y) = \frac{h^2(p_1 - p_2)}{2\mu L}\left[1 - \left(\frac{y}{h}\right)^2\right]$$

Then the integral takes the form

$$Q = \int_{-W/2}^{W/2}\int_{-h}^{h} \frac{h^2(p_1 - p_2)}{2\mu L}\left[1 - \left(\frac{y}{h}\right)^2\right] dy\, dz$$

Performing the integration yields:

$$Q = \frac{2h^3 W}{3\mu}\frac{(p_1 - p_2)}{L} \qquad \text{(A)}$$

The maximum velocity is found by inspection to be

$$V_{max} = \frac{h^2(p_1 - p_2)}{2\mu L} \qquad \text{(B)}$$

The average velocity can now be found by writing $\bar{V} = Q/A = Q/2hW$, and using (A) to obtain

$$\bar{V} = \frac{h^2}{3\mu}\frac{p_1 - p_2}{L} \qquad \text{(C)}$$

Note that $\bar{V} = \frac{2}{3}V_{max}$. Solving (C) for the pressure drop, we find

$$p_1 - p_2 = \frac{3\mu \bar{V} L}{h^2} \qquad \text{(D)}$$

which we can also write in terms of the flowrate as

$$p_1 - p_2 = \frac{3\mu Q L}{2h^3 W} \quad \text{(E)}$$

Inserting the data into (D), we find

$$p_1 - p_2 = \frac{3\mu \bar{V} L}{h^2} = \frac{3[3.75 \times 10^{-7}(\text{lb}_\text{f}\text{-s})/\text{ft}^2](2 \text{ ft/s})(20 \text{ ft})}{(0.25 \text{ in.})^2 \left(\frac{1 \text{ ft}^2}{144 \text{ in.}^2}\right)} = 0.104 \text{ lb}_\text{f}/\text{ft}^2$$

$$= 7.2 \times 10^{-4} \text{ psi}$$

where we have made use of Appendix A to find $\mu = 3.75 \times 10^{-7}$ (lb$_\text{f}$-s)/ft^2 for air at 60°F. We should also check the Reynolds number for this flow noting from Appendix A that for air at 60°F we have $\nu = 1.58 \times 10^{-4}$ ft^2/s^2. Using the gap, $2h$, as the length scale, we have $Re = \bar{V}(2h)/\nu$. Inserting the known values, we find

$$Re = \frac{(2 \text{ ft/s})(0.5 \text{ in.})\left(\frac{1 \text{ ft}}{12 \text{ in.}}\right)}{1.58 \times 10^{-4} \text{ ft}^2/\text{s}^2} = 527$$

which indicates that the flow is laminar (the critical Re for this flow is about 1400).

12.2.4 Poiseuille Flow in a Pipe

Consider the pressure-driven flow in a round pipe as shown in Figure 12.19. We are interested in predicting the steady laminar flow in the pipe for cases in which the pipe diameter is small compared with its length. This is a flow with which you are well acquainted, but it is useful to go through the process of deriving the analytical solution. The geometry suggests the use of cylindrical coordinates. We will neglect the body forces and assume that the flow is axisymmetric; i.e., there is no variation of any flow property in the θ direction. Because the pipe diameter is small in comparison to its length, the flow in the pipe away from the inlet and exit should be fully developed. Thus we will assume that the velocity field is independent of z. We anticipate that the pressure difference will create a flow with a component of velocity v_z, and this component must depend on r because of the no-slip condition on the wall of the pipe. Next we consider the possibility of velocity components in the r and θ directions. Since there is no reason to think that this flow will have a swirl component, we can assume $v_\theta = 0$. If a radial velocity component exists, it cannot depend on θ or z. Thus we postulate a velocity field of the form $v_r = v_r(r)$, $v_\theta = 0$, $v_z = v_z(r)$.

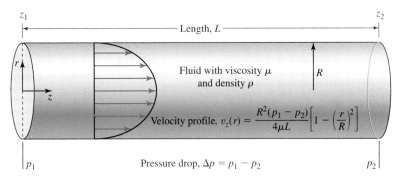

Figure 12.19 Poiseuille flow in a round pipe.

Next we insert $v_r = v_r(r)$, $v_\theta = 0$, $v_z = v_z(r)$ into the governing equations, in this case Eqs. 12.2a–12.2d, and make use of the assumption of axisymmetry, $\partial(\ )/\partial\theta = 0$, to obtain

$$\frac{1}{r}\frac{\partial(rv_r)}{\partial r} = 0, \quad 0 = -\frac{\partial p}{\partial \theta}, \quad \rho v_r \frac{\partial v_r}{\partial r} = -\frac{\partial p}{\partial r} + \mu\left(\frac{\partial^2 v_r}{\partial r^2} + \frac{1}{r}\frac{\partial v_r}{\partial r} - \frac{v_r}{r^2}\right)$$

and

$$\rho v_r \frac{\partial v_z}{\partial r} = -\frac{\partial p}{\partial z} + \mu\left(\frac{\partial^2 v_z}{\partial r^2} + \frac{1}{r}\frac{\partial v_z}{\partial r}\right)$$

We must solve these equations subject to the no-slip, no-penetration boundary conditions (BCs) on the pipe wall. Since $v_r = v_r(r)$ only, the first equation can be written as an ordinary differential equation

$$\frac{1}{r}\frac{d(rv_r)}{dr} = 0$$

whose solution is $v_r = C/r$. Applying the no-penetration condition $v_r = 0$ at $r = R$ shows that $C = 0$. Thus $v_r = 0$, and the remaining equations simplify to

$$0 = -\frac{\partial p}{\partial r}, \quad 0 = -\frac{\partial p}{\partial \theta}, \quad \text{and} \quad 0 = -\frac{\partial p}{\partial z} + \mu\left(\frac{\partial^2 v_z}{\partial r^2} + \frac{1}{r}\frac{\partial v_z}{\partial r}\right)$$

From the first two equations, we conclude that the pressure is only a function of z. Thus the solution we are looking for is described by unknown functions $v_z = v_z(r)$ and $p = p(z)$ that satisfy the third equation. This last equation expresses the balance between pressure and viscous forces. We must solve this equation to find the two unknown functions $v_z = v_z(r)$ and $p = p(z)$, which describe pressure-driven flow in a round pipe. To proceed, we take a partial derivative with respect to z of the equation and interchange derivatives to obtain

$$0 = -\frac{\partial}{\partial z}\left(\frac{\partial p}{\partial z}\right) + \mu\left[\frac{\partial^2}{\partial r^2}\left(\frac{\partial v_z}{\partial z}\right) + \frac{1}{r}\frac{\partial}{\partial r}\left(\frac{\partial v_z}{\partial z}\right)\right]$$

Since $v_z = v_z(r)$, $\partial v_z/\partial z = 0$ and the second term is zero. Thus the pressure $p = p(z)$ must satisfy $\partial/\partial z(\partial p/\partial z) = 0$, and the pressure gradient must be a constant, $\partial p/\partial z = -G$. The governing equation then becomes

$$\frac{\partial^2 v_z}{\partial r^2} + \frac{1}{r}\frac{\partial v_z}{\partial r} = \frac{-G}{\mu}$$

Since $v_z = v_z(r)$ only, we can also write this as

$$\frac{1}{r}\frac{d}{dr}\left(r\frac{dv_z}{dr}\right) = \frac{-G}{\mu}$$

Integrating this ordinary differential equation twice yields $v_z(r) = (-G/4\mu)r^2 + A\ln r + B$.

To evaluate the constants A and B, note that the no-slip condition on the pipe wall is $v_\theta = 0$, $v_z = 0$ at $r = R$. We have assumed $v_\theta = 0$ everywhere at the outset, so this part of the BC is automatically satisfied. Applying the remaining BC, we find $v_z(R) = 0 = (-G/4\mu)R^2 + A\ln R + B$. With no other BC available we seem to be stuck, since we have one equation to determine two constants. The solution to this dilemma is to recognize that if A is nonzero, we will obtain an infinite value of v_z on the axis of the pipe at $r = 0$. Since this is physically impossible, we take $A = 0$ and solve the preceding equation for B to obtain $B = GR^2/4\mu$. Thus, the velocity profile is given by

$$v_z(r) = \frac{GR^2}{4\mu}\left[1 - \left(\frac{r}{R}\right)^2\right]$$

Note that the viscosity of the fluid does not affect the shape of the velocity profile.

The equation governing the pressure distribution is $\partial p/\partial z = -G$. Since $p = p(z)$ only, this is an ordinary differential equation. According to Figure 12.19 we have $p = p_1$ at $z = z_1$, and $p = p_2$ at $z = z_2$. Thus we can integrate from z_1 to z to obtain the linear pressure distribution $p(z) = p_1 - G(z - z_1)$. Since $p = p_2$ at $z = z_2$ we use this solution to write $p_2 = p_1 - G(z_2 - z_1)$. Solving for G, we obtain $G = (p_1 - p_2)/(z_2 - z_1) = (p_1 - p_2)/L$. The pressure distribution is therefore given by

$$p(z) = p_1 - \left(\frac{p_1 - p_2}{L}\right)(z - z_1) \qquad (12.16)$$

and the velocity profile can now be written as

$$v_z(r) = \frac{R^2(p_1 - p_2)}{4\mu L}\left[1 - \left(\frac{r}{R}\right)^2\right] \qquad (12.17)$$

This parabolic velocity profile is shown in Figure 12.19. This solution describes the steady, laminar flow in a round pipe well away from the inlet and exit, provided the pipe diameter is small in comparison to its length, a condition we can express as $D/L \ll 1$. Figure 12.20 shows the results of a flow visualization experiment for Poiseuille flow in a pipe.

 CD/Video library/Pipe flow

740　12　ANALYSIS OF INCOMPRESSIBLE FLOW

Figure 12.20 Material lines at the entrance to the pipe are revealed using the hydrogen bubble method. For this flow $Re = 1.6 \times 10^3$. The fully developed, parabolic profile is evident at the right end of the pipe. The need for the restriction, $D/L \ll 1$, of this solution is obvious. A discussion of the developing flow in a pipe can be found in Chapter 13.

EXAMPLE 12.5

Find expressions for the volume flowrate, maximum and average velocity, and pressure drop for Poiseuille flow in a round pipe. If a pressure drop of 500 kPa is measured in a 10 m long, 5 mm diameter tube when the flowrate of an unknown liquid is 10 cm³/s, what is the fluid viscosity?

SOLUTION

We will calculate the volume flowrate passing through a cross section perpendicular to the flow direction as shown in Figure 12.21. The volume flowrate is given by $Q = \int_S (\mathbf{u} \cdot \mathbf{n}) \, dS$. Using cylindrical coordinates we have $dS = r \, dr \, d\theta$, $\mathbf{n} = \mathbf{e}_z$, and $\mathbf{u} = v_z(r)\mathbf{e}_z$, with $v_z(r)$ given by Eq. 12.17 as $v_z(r) = \{[R^2(p_1 - p_2)]/4\mu L\}[1 - (r/R)^2]$. The integral becomes

$$Q = \int_0^{2\pi} \int_0^R \frac{R^2(p_1 - p_2)}{4\mu L}\left[1 - \left(\frac{r}{R}\right)^2\right] r \, dr \, d\theta$$

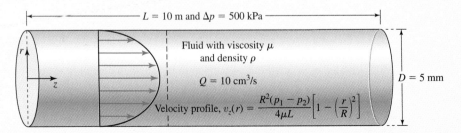

Figure 12.21 Schematic for Example 12.5.

which after substituting $R = D/2$, yields the following expression for the flowrate:

$$Q = \frac{\pi D^4 (p_1 - p_2)}{128 \mu L} \tag{A}$$

The maximum velocity is found by inspection to be

$$V_{max} = \frac{R^2(p_1 - p_2)}{4 \mu L} \tag{B}$$

Since $\bar{V} = Q/A = 4Q/\pi D^2$, we can use (A) to obtain

$$\bar{V} = \frac{D^2(p_1 - p_2)}{32 \mu L} \tag{C}$$

Solving (C) for the pressure drop, we have

$$p_1 - p_2 = \frac{32 \mu \bar{V} L}{D^2} \tag{D}$$

which we can also write in terms of the flowrate as

$$p_1 - p_2 = \frac{128 \mu Q L}{\pi D^4} \tag{E}$$

To find the viscosity of the fluid passing through a tube under known conditions of flowrate and pressure drop, we rearrange (E) to solve for the viscosity:

$$\mu = \frac{\pi D^4 (p_1 - p_2)}{128 QL} \tag{F}$$

Inserting the data, we find

$$\mu = \frac{\pi (0.005 \text{ m})^4 (500 \times 10^3 \text{ N/m}^2)}{128(10 \text{ cm}^3/\text{s}) \left(\frac{1 \text{ m}}{100 \text{ cm}}\right)^3 (10 \text{ m})} = 7.67 \times 10^{-2} \text{ (N-s)/m}^2$$

12.2.5 Flow over a Cylinder (CFD)

As discussed in the introduction to this chapter, the flow over a cylinder changes from a steady, laminar flow at low Reynolds number to an unsteady, laminar flow at some intermediate value of Re and finally becomes turbulent. No analytical solution exists for the low Re steady flow over a cylinder, although approximate solutions are known for

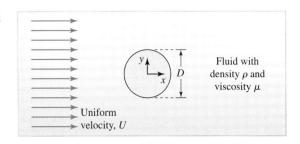

Figure 12.22 Two-dimensional flow over a cylinder.

$Re \ll 1$. In this section we use a commercially available CFD code based on the finite element method to develop a CFD solution for the problem of the 2D laminar steady flow of a uniform stream over a cylinder at low Re. The geometry of this problem, shown in Figure 12.22, suggests the use of cylindrical coordinates. However, it is more convenient to use the Cartesian coordinates shown in the figure with the CFD code. The problem of interest involves a cylinder exposed to a uniform stream with no other nearby boundaries. In effect, the cylinder is inserted into an infinite mass of fluid, all of which is moving at speed U. We are assuming that the flow is steady and 2D; that is, the velocity field is given by $u = u(x, y)$, $v = v(x, y)$, $w = 0$. This assumption is justified by noting that while the flow upstream is in the x direction, the flow must turn to pass around the cylinder. Thus a velocity component in the y direction must be present, and both components must depend on x and y. Neglecting the body force and inserting $u = u(x, y)$, $v = v(x, y)$, $w = 0$ into the continuity and Navier–Stokes equations for steady, constant density, viscous flow (Eqs. 12.1a–12.1d), we obtain

$$\frac{\partial u}{\partial x} + \frac{\partial v}{\partial y} = 0, \quad \rho\left(u\frac{\partial u}{\partial x} + v\frac{\partial u}{\partial y}\right) = -\frac{\partial p}{\partial x} + \mu\left(\frac{\partial^2 u}{\partial x^2} + \frac{\partial^2 u}{\partial y^2}\right)$$

$$\rho\left(u\frac{\partial v}{\partial x} + v\frac{\partial v}{\partial y}\right) = -\frac{\partial p}{\partial y} + \mu\left(\frac{\partial^2 v}{\partial x^2} + \frac{\partial^2 v}{\partial y^2}\right), \quad \text{and} \quad 0 = -\frac{\partial p}{\partial z}$$

The last equation allows us to conclude that the pressure is given by $p = p(x, y)$. We must solve the first three equations for the two unknown velocity components and pressure subject to the no-slip, no-penetration conditions on the surface of the cylinder, and an additional condition that expresses the fact that far from the cylinder in any direction, the flow is uniform. In Cartesian coordinates, the no-slip, no-penetration conditions on the surface of the cylinder are expressed by writing $u = v = 0$ on $x^2 + y^2 = R^2$. This ensures that the velocity of the fluid on the surface of the cylinder matches the velocity of the stationary cylinder, i.e., $\mathbf{u} = 0$. The condition far from the cylinder may be expressed by writing $u = U$, $v = 0$ as $x^2 + y^2 \to \infty$. The governing equations express the balance between inertial, pressure, and viscous forces in the x and y directions.

As part of the computational solution, we must develop a finite element mesh that covers the region in which flow occurs. The CFD solution consists of values of the

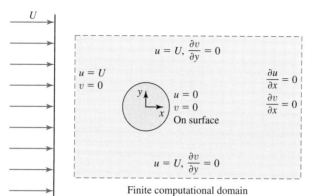

Figure 12.23 Finite computational domain and boundary conditions for flow over a cylinder.

velocity and pressure at a finite number of points in space, namely, at the nodes of the mesh. To ensure the accuracy of the solution, the density of mesh nodes must be high in regions where velocity gradients are large. In this case, a dense mesh is needed near the cylinder, but the mesh density can decrease as we move further away from the cylinder. In theory, the mesh must cover the region outside the cylinder and extend out in all directions to infinity, since this is the region in which flow occurs. Because computer memory is limited, however, we cannot extend the mesh to infinity. We can argue that at some distance upstream, to the sides of the cylinder, and downstream, the disturbance to the uniform stream caused by the presence of the cylinder will vanish. This suggests that we limit the region of interest in our calculation in these directions to some number of multiples of the cylinder diameter. The dashed box in Figure 12.23 shows the computational region that has been selected for further consideration.

In the CFD solution to the problem, the original boundary condition $u = U, v = 0$ as $x^2 + y^2 \rightarrow \infty$ will be replaced by $u = U, v = 0$ on the upstream, $u = U, \partial v/\partial y = 0$ on the sides of the computational domain, and by an outflow condition $\partial u/\partial x = \partial v/\partial x = 0$ on the downstream side. Notice that the outflow condition replaces the condition $u = U, v = 0$ as $x^2 + y^2 \rightarrow \infty$ on the downstream side of the computational domain. These boundary conditions are shown in Figure 12.23. At this point it should be clear that a CFD solution to this problem is approximate and that its accuracy depends on many factors, not least of which is the judgment, skill, and experience of the engineer who is setting up the problem.

With the computation region selected, we can now use the meshing capability of the commercial CFD package to construct a finite element mesh. The mesh used here is shown in Figure 12.24. The final step in constructing the CFD solution is to use the boundary condition and problem statement modules of the commercial package to set the desired boundary conditions, select the appropriate set of governing equations to solve, and provide values for fluid properties such as density and viscosity. In this case we instruct the commercial package that we wish to solve the continuity and Navier–Stokes equations for a steady, 2D, constant density, constant viscosity laminar flow, with BCs on the cylinder and four sides of the computational box as noted earlier.

744 | **12 ANALYSIS OF INCOMPRESSIBLE FLOW**

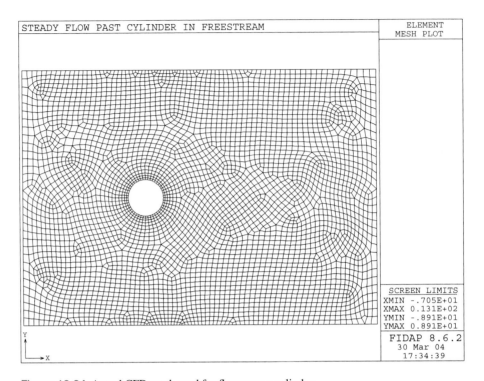

Figure 12.24 Actual CFD mesh used for flow over a cylinder.

After we have given the execution command, the CFD code solves the finite element problem corresponding to the numerical version of the governing equations and returns a set of data describing the distributions of velocity and pressure on all the nodes of the finite element mesh. It is a simple matter to change the values of the BCs and fluid properties if we wish and solve the problem again for a new set of conditions. After obtaining a solution we can invoke the postprocessor capability of the commercial package and examine the flowfield in detail. For example, Figure 12.25 shows the streamlines, and pressure contours for Reynolds numbers of 0.1, 1, 10, and 50. Notice how the flow changes as the Reynolds number changes.

12.3 UNSTEADY VISCOUS FLOW

In this section we will outline the method for using the continuity and Navier–Stokes equations to analyze unsteady flow problems involving a constant density, constant viscosity fluid in the absence of body forces. For the sake of simplicity, we will restrict our discussion to problems best described in Cartesian coordinates. The governing equations in this case are the continuity equation for a constant density fluid as given by Eq. 12.1a, and the Navier–Stokes equations for a constant density, constant viscosity

12.3 UNSTEADY VISCOUS FLOW

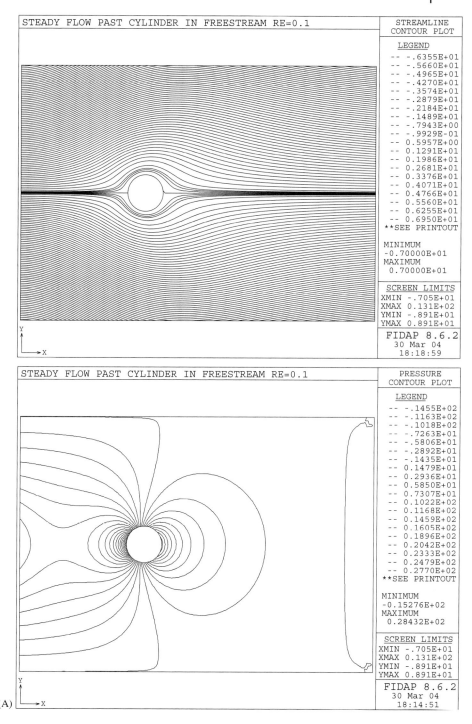

Figure 12.25 CFD solution for flow over a cylinder: streamlines, and pressure contours for (A) $Re = 0.1$, (B) $Re = 1$, (C) $Re = 10$, (D) $Re = 50$.

746 12 ANALYSIS OF INCOMPRESSIBLE FLOW

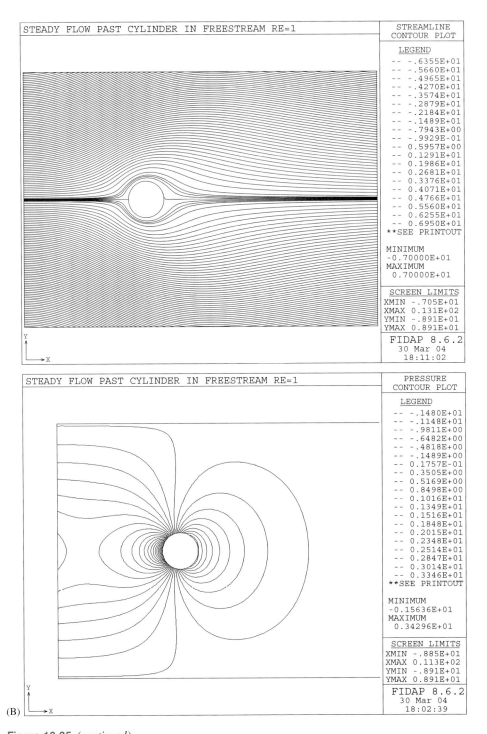

Figure 12.25 (*continued*)

12.3 UNSTEADY VISCOUS FLOW 747

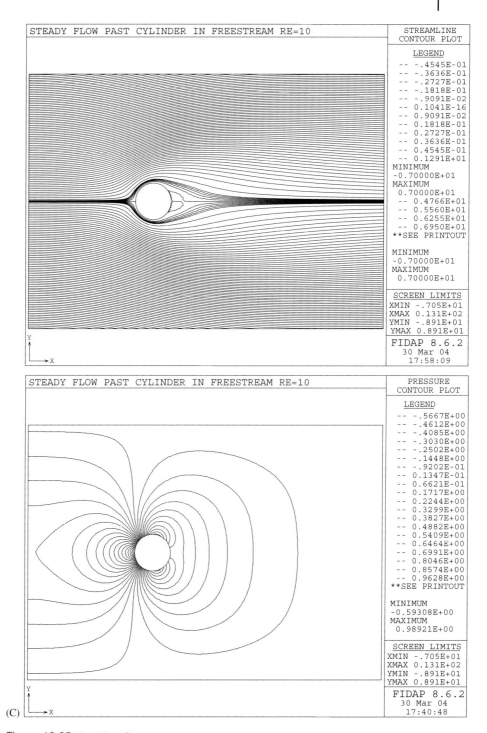

Figure 12.25 (*continued*)

748 12 ANALYSIS OF INCOMPRESSIBLE FLOW

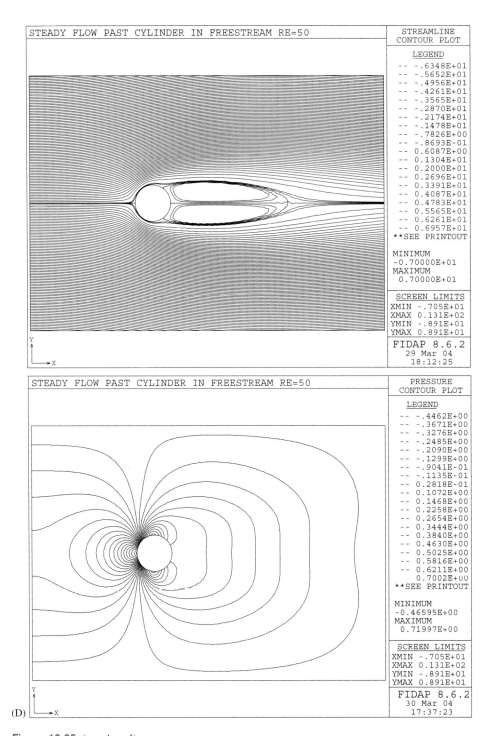

(D)

Figure 12.25 (*continued*)

fluid, (Eqs. 11.10a–11.10c), with the body force terms dropped. For the conditions described in this section the governing equations are

$$\frac{\partial u}{\partial x} + \frac{\partial v}{\partial y} + \frac{\partial w}{\partial z} = 0 \tag{12.18a}$$

$$\rho\left(\frac{\partial u}{\partial t} + u\frac{\partial u}{\partial x} + v\frac{\partial u}{\partial y} + w\frac{\partial u}{\partial z}\right) = -\frac{\partial p}{\partial x} + \mu\left(\frac{\partial^2 u}{\partial x^2} + \frac{\partial^2 u}{\partial y^2} + \frac{\partial^2 u}{\partial z^2}\right) \tag{12.18b}$$

$$\rho\left(\frac{\partial v}{\partial t} + u\frac{\partial v}{\partial x} + v\frac{\partial v}{\partial y} + w\frac{\partial v}{\partial z}\right) = -\frac{\partial p}{\partial y} + \mu\left(\frac{\partial^2 v}{\partial x^2} + \frac{\partial^2 v}{\partial y^2} + \frac{\partial^2 v}{\partial z^2}\right) \tag{12.18c}$$

$$\rho\left(\frac{\partial w}{\partial t} + u\frac{\partial w}{\partial x} + v\frac{\partial w}{\partial y} + w\frac{\partial w}{\partial z}\right) = -\frac{\partial p}{\partial z} + \mu\left(\frac{\partial^2 w}{\partial x^2} + \frac{\partial^2 w}{\partial y^2} + \frac{\partial^2 w}{\partial z^2}\right) \tag{12.18d}$$

A solution to an unsteady flow problem involving a fluid of known density and viscosity consists of finding the three velocity components u, v, w, and pressure p that satisfy these equations as a function of position and time. We must solve this set of four partial differential equations together with the appropriate boundary and initial conditions.

CD/Video library/Impulsively started flow

12.3.1 Startup of Plane Couette Flow

Consider the flow that results if the upper plate in the plane Couette flow arrangement illustrated in Figure 12.26 is suddenly started into motion. Because of the no-slip condition, fluid directly adjacent to the upper plate will begin to move along with it. We are interested in predicting the unsteady laminar flow in the gap between the plates for the case in which the fluid is initially at rest and the upper plate is instantaneously brought to velocity U in the x direction. After a sufficiently long time, we expect the flow to reach steady state, at which point the flow should be described by the solution for plane Couette flow given in Section 12.2.1.

Since the flow is unsteady, the governing equations are Eqs. 12.18a–12.18d. We must solve these equations together with the no-slip, no-penetration boundary conditions

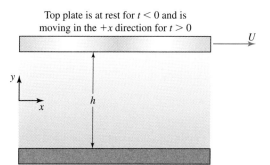

Figure 12.26 Starting flow between parallel plates due to motion of the top plate.

on the two plates, and the initial condition that the fluid is at rest. From the no-slip condition we conclude that at any instant the unsteady flow must have a velocity component u in the x direction, and this component must depend on y because the upper plate is moving and the lower plate is fixed. Since the flow is unsteady, u must depend on t; but since the dimensions of the plates are large in comparison to the gap, we will assume that u does not depend on x or z. There is no reason to think that the impulsive motion of the upper plate will cause a velocity component in the z direction, so we will assume $w = 0$. We will also assume that velocity component v in the y direction is zero. Alternately, the continuity equation and the no-penetration condition can be used to show that $v = 0$. Thus we postulate that the unsteady flow of interest has a velocity field of the form $u = u(y, t)$, $v = 0$, $w = 0$.

Next we insert $u = u(y, t)$, $v = 0$, $w = 0$ into Eqs. 12.18a–12.18d to obtain the simplified governing equations:

$$0 = 0, \quad \rho \frac{\partial u}{\partial t} = -\frac{\partial p}{\partial x} + \mu \frac{\partial^2 u}{\partial y^2}, \quad 0 = -\frac{\partial p}{\partial y}, \quad \text{and} \quad 0 = -\frac{\partial p}{\partial z}$$

We see that the continuity equation is now satisfied and that the pressure may be a function of x. Taking a derivative of the remaining momentum equation with respect to x shows that the pressure gradient $\partial p/\partial x$ is either a constant or a function of time. To focus attention on the flow created solely by the impulsive movement of the upper plate, we will assume that the pressure gradient is zero. Thus the pressure distribution is given by

$$p = p_0 \tag{12.19}$$

The velocity field is described by an unknown function $u = u(y, t)$ that satisfies the single governing equation

$$\rho \frac{\partial u}{\partial t} = \mu \frac{\partial^2 u}{\partial y^2} \tag{12.20}$$

as well as the no-slip conditions on each plate ($u = U$ at $y = h$ for $t > 0$ and $u = 0$ at $y = 0$ for $t \geq 0$), and an initial condition ($u = 0$ at $t = 0$ for $0 \leq y \leq h$). The governing equation expresses the balance between inertial and viscous forces.

To solve this problem we introduce a change of variables, writing the unknown function $u(y, t)$ as the sum of the steady state solution for Couette flow $u(y) = Uy/h$ and a flow transient $w(y, t)$. That is, we define $u(y, t) = Uy/h + w(y, t)$. Then we have

$$\frac{\partial u}{\partial t} = \frac{\partial}{\partial t}\left(\frac{Uy}{h}\right) + \frac{\partial w}{\partial t} = \frac{\partial w}{\partial t} \quad \text{and} \quad \frac{\partial^2 u}{\partial y^2} = \frac{\partial^2}{\partial y^2}\left(\frac{Uy}{h}\right) + \frac{\partial^2 w}{\partial y^2} = \frac{\partial^2 w}{\partial y^2}$$

The governing equation becomes

$$\rho \frac{\partial w}{\partial t} = \mu \frac{\partial^2 w}{\partial y^2} \tag{12.21}$$

The boundary and initial conditions on u can now be used to derive the following boundary and initial conditions on w:

$$w = 0 \text{ at } y = h \text{ for } t > 0, \quad w = 0 \text{ at } y = 0 \text{ for } t \geq 0$$

12.3 UNSTEADY VISCOUS FLOW

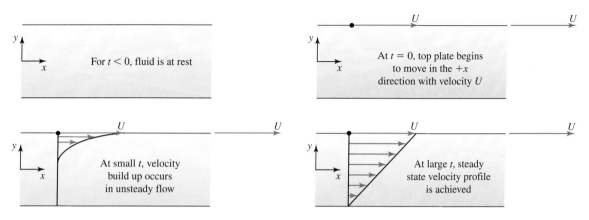

Figure 12.27 Development of the velocity distribution for starting flow between parallel plates.

and

$$w = -\frac{Uy}{h} \text{ at } t = 0$$

The solution of this partial differential equation is usually the topic of advanced mathematics courses not necessarily encountered by students enrolled in a first course in fluid mechanics. Therefore, we will pass over the details of the solution procedure and simply present the results so that we can interpret their significance. The solution to the problem of startup of plane Couette flow is given by

$$u(y,t) = \frac{Uy}{h} + \frac{2U}{\pi} \sum_{n=1}^{\infty} (-1)^n \frac{1}{n} \exp\left(-n^2\pi^2 \frac{\nu t}{h^2}\right) \sin\left(\frac{n\pi y}{h}\right) \quad (12.22)$$

where $n = 1, 2, 3, \ldots$.

This velocity profile is shown at different times after startup in Figure 12.27. Note that the movement of the plate is transmitted first to the layer of fluid directly adjacent to it; then the effect of viscosity is to gradually bring into motion all the fluid in the gap except the fluid on the fixed plate, which must remain at rest owing to the no-slip condition. The time t occurs as part of a dimensionless group $\nu t/h^2$ where $\nu = \mu/\rho$ is the kinematic viscosity. This is not unexpected, since the dimensions of ν are $[\nu] = L^2/t$, which is a diffusivity. The product νt has the dimension of a length squared, or $[\nu t] = L^2$. It is customary to say that in time t, viscous diffusion causes a velocity variation to diffuse a distance $l(t) \approx \sqrt{\nu t}$. The time it takes to diffuse a distance l is then estimated by $t = l^2/\nu$. In the solution for plane Couette flow, each exponential term has a separate time scale that depends on n. Thus we cannot simply estimate the time for the velocity variation to diffuse across the gap as $t = h^2/\nu$. Also note that although neither density nor viscosity of the fluid enters the steady state part of the solution, both enter the transient part of the solution through the kinematic viscosity.

When a constant pressure difference is suddenly applied to a long straight pipe, the fluid will begin to accelerate en masse in the direction of decreasing pressure. After a sufficiently long time the flow will reach steady state and will then be described by the solution for Poiseuille flow given in Section 12.2.4. The initial transient period may be described as the starting flow in a pipe. Use of the procedure outlined in Section 12.2.5, with the obvious change of using the cylindrical coordinate versions of the continuity and Navier–Stokes equations, yields a complex solution for the transient flow velocity field involving Bessel functions of the first and second kind. In Figure 12.28, which shows this velocity field at different times after startup, we see that the suddenly applied pressure difference causes all the fluid to begin to accelerate, except for the fluid on the wall, which must remain at rest because of the no-slip condition. After a sufficiently long time, the acceleration vanishes, and the flow is governed by a balance of pressure and viscous forces alone. This required balance is satisfied by the parabolic profile of steady Poiseuille flow.

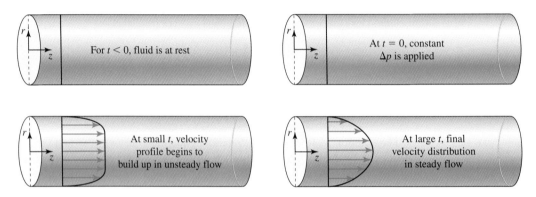

Figure 12.28 Development of the velocity distribution for starting flow in pipe upon imposition of a pressure gradient.

12.3.2 Unsteady Flow over a Cylinder (CFD)

In Section 12.2.5 we discussed a CFD solution for the 2D steady flow over a cylinder. Flows of this type occur for small Reynolds numbers. As Re increases, the flow over a cylinder becomes unsteady, as evidenced by alternately shed vortices in the wake (the Karman vortex street). The unsteadiness occurs even though upstream of the cylinder the oncoming stream remains uniform and steady. In this section we will use a commercially available CFD code based on the finite element method to develop a computational fluid dynamics (CFD) solution for the problem of the 2D unsteady flow of a uniform stream over a cylinder in a wind tunnel. The geometry of this problem is shown in Figure 12.29A.

Using Cartesian coordinates, we employ a computational box that is extended in the downstream direction to allow us to observe the wake region. We assume that the flow is unsteady and 2D; that is, the velocity field is given by $u = u(x, y, t)$, $v = v(x, y, t)$, $w = 0$. Neglecting the body force and inserting $u = u(x, y, t)$, $v = v(x, y, t)$, $w = 0$ into the continuity and Navier–Stokes equations for unsteady, viscous flow (Eqs. 12.18a–12.18d),

12.3 UNSTEADY VISCOUS FLOW

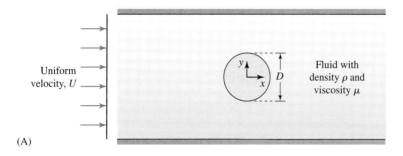

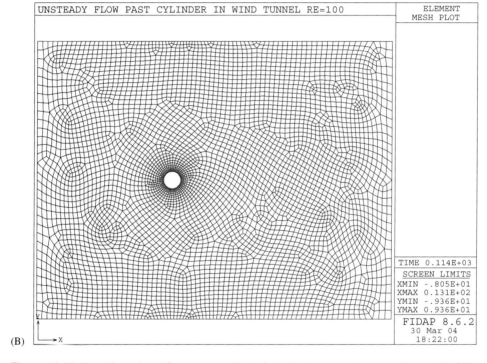

Figure 12.29 Two-dimensional flow over a cylinder in a wind tunnel: (A) schematic, and (B) CFD mesh.

we obtain

$$\frac{\partial u}{\partial x} + \frac{\partial v}{\partial y} = 0, \quad \rho\left(\frac{\partial u}{\partial t} + u\frac{\partial u}{\partial x} + v\frac{\partial u}{\partial y}\right) = -\frac{\partial p}{\partial x} + \mu\left(\frac{\partial^2 u}{\partial x^2} + \frac{\partial^2 u}{\partial y^2}\right)$$

$$\rho\left(\frac{\partial v}{\partial t} + u\frac{\partial v}{\partial x} + v\frac{\partial v}{\partial y}\right) = -\frac{\partial p}{\partial y} + \mu\left(\frac{\partial^2 v}{\partial x^2} + \frac{\partial^2 v}{\partial y^2}\right), \quad \text{and} \quad 0 = -\frac{\partial p}{\partial z}$$

From the last equation and the fact that the flow is unsteady, we conclude that the pressure is given by $p = p(x, y, t)$. We must solve the first three equations for the two unknown velocity components and pressure subject to the no-slip, no-penetration conditions on the surface of the cylinder and side walls, an additional condition indicating that far from the cylinder in the upstream direction the flow is uniform, an outflow condition, and an initial condition.

The boundary condition on the cylinder surface is $u = v = 0$ on $x^2 + y^2 = R^2$. The actual boundary condition far from the cylinder upstream and downstream may be expressed by writing $u \to U$, $v = 0$ as $x^2 + y^2 \to \infty$. On the sides the velocity is zero. In the CFD solution to the problem, the boundary condition is $u = U$, $v = 0$ upstream of the computational domain, $u = 0$, $v = 0$ on the sides, and an outflow condition, $\partial u/\partial x = \partial v/\partial x = 0$, on the downstream side. The initial condition for this calculation, $u = 0$, $v = 0$ at $t = 0$, allows the flow to develop over time and reach the desired periodic state.

As part of the computational solution, we must develop a finite element mesh that covers the region in which flow occurs. The principles that govern the selection of this mesh were discussed in Section 12.2.5. Here we are interested in the flow structure in the wake, so this region is given a reasonably dense mesh as shown in Figure 12.29B. The final step in constructing the CFD solution is to use the boundary condition and problem statement modules of the commercial package to set the desired boundary conditions, select the appropriate set of governing equations to solve, and provide values for fluid properties such as density and viscosity. In this case we instruct the commercial package that we wish to solve the continuity and Navier–Stokes equations for an unsteady, 2D, constant density, constant viscosity laminar flow, with boundary conditions on the cylinder and four sides of the computational box as noted earlier.

After we have given the execution command, the CFD code will solve the finite element problem corresponding to the numerical version of the governing equations and return a data set describing the distributions of velocity and pressure on all the nodes of the finite element mesh at a sequence of time steps. Upon obtaining this solution, we can invoke the postprocessor capability of the commercial package and examine the flow field in detail. For example, Figure 12.30 shows the streamlines for this flow at $Re = 100$. The shed eddies of the Karman vortex street seen earlier in Figure 12.4 are evident as disturbances in the streamlines persisting far downstream.

 CD/Boundary layers/Instability/transition, and turbulence

12.4 TURBULENT FLOW

Most flows in nature and in engineering practice are turbulent. As the Reynolds number of a flow increases, the laminar flow that occurs at low Re undergoes a transition and eventually becomes turbulent. Thus, the Reynolds number is the key nondimensional parameter that determines the transition from laminar to turbulent flow. It is customary to talk of a critical Reynolds number, Re_{cr}, at which transition takes place; but in reality, transition in a given flow is affected by many different phenomena. The Re_{cr} for a certain type of flow should therefore be used only for guidance in deciding whether a flow is likely to be turbulent. For example, in pipe flow transition may occur at Reynolds numbers as low as $Re_{cr} = 2300$, for the flat plate boundary layer at $Re_{cr} = 5 \times 10^5$, and for flow over a cylinder at $Re_{cr} = 3 \times 10^5$.

12.4 TURBULENT FLOW

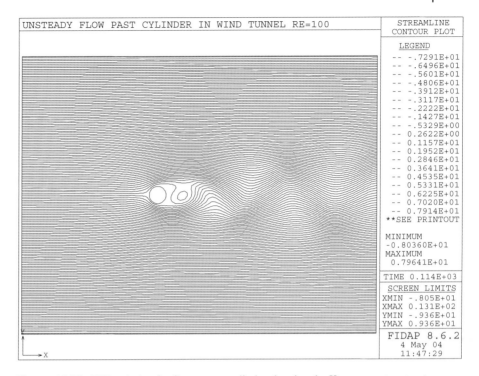

Figure 12.30 CFD solution for flow over a cylinder showing the Karman vortex street.

At this point you might ask, What is turbulence? The answer is that a turbulent flow is characterized by 3D, unsteady velocity and vorticity fields with a wide range of length scales. The complexity of the turbulent flows illustrated in Figure 12.31 has led to the use of the adjective "random" to describe turbulent flows. That is, although we are confident that the governing equations of fluid mechanics describe these flows every bit as well as they do laminar flows, we can never obtain enough information about the precise nature of the disturbances that play a role in causing turbulent flows to consider the problem of turbulence in a given physical system to be completely deterministic. Mathematically, turbulent flows are solutions to the continuity, Navier–Stokes, and energy equations. But they are different from the laminar solutions described in preceding sections because perturbations in initial or boundary conditions, which are damped out in the laminar flow regime at lower Re, gain energy and grow when Re increases.

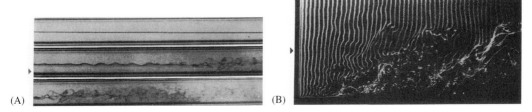

Figure 12.31 (A) Visualization of streak lines with dye in a pipe at $Re = 1.5 \times 10^3$ (top), 2.34×10^3 (middle), and 7.5×10^3 (bottom) indicates laminar, transitional, and turbulent flow, respectively, and (B) visualization of material lines over a flat plate showing the transition to turbulence.

At present there are no known analytical solutions to the Navier–Stokes equations that can be said to describe a turbulent flow. Recently, however, a number of numerical simulations of the Navier–Stokes equations for turbulent flow have been achieved. Although these are interesting, their primary value is in the tuning of turbulence models. The latter are of interest in solving the Reynolds equations, a set of equations describing turbulent flow that are discussed in Section 12.4.1. As you will see, the Reynolds equations are derived from governing equations by means of an averaging process.

In engineering practice empirical methods in forms such as friction factors and drag coefficients have been, and will continue to be, used successfully for analysis and design. You have been exposed to these concepts briefly in the case studies of Chapter 3, but these methods are discussed in greater detail in Chapters 13 and 14.

12.4.1 Reynolds Equations

The random-looking behavior of a turbulent flow suggests that it is useful to consider each of the flow variables to consist of a mean and a fluctuating component. For example, the u component of velocity is written as $u = \bar{u} + u'$, where $\bar{u}$ is called the mean and u' the fluctuation. The determination of the mean is more complicated than it may appear. For our purposes we will consider the mean to be defined by a suitable time average over a period long enough to average out the effects of turbulent fluctuations but not long enough to average out the slower time variations that may occur in the mean values of an unsteady turbulent flow. In general we have

$$u(x, y, z, t) = \bar{u}(x, y, z, t) + u'(x, y, z, t)$$

and the remaining flow variables are written similarly. For example, the pressure is written as

$$p(x, y, z, t) = \bar{p}(x, y, z, t) + p'(x, y, z, t)$$

The rules for the averaging process are discussed in specialized texts on turbulence.

For simplicity, we will limit our discussion to turbulent flows that can be described as steady insofar as the mean flow is concerned and have both constant density and constant viscosity. The governing equations in Cartesian coordinates for such flows are the continuity, and Navier–Stokes equations for unsteady flow given by Eqs. 12.18a–12.18d, plus the necessary boundary and initial conditions. The process of deriving the Reynolds equations for a turbulent flow consists of writing each flow variable u, v, w, p as the sum of a mean and fluctuating component, inserting the sums into each governing equation, then averaging each equation and employing the averaging rules to evaluate each term. The result is the constant density, steady flow Reynolds equations for the mean flow variables:

$$\frac{\partial \bar{u}}{\partial x} + \frac{\partial \bar{v}}{\partial y} + \frac{\partial \bar{w}}{\partial z} = 0 \tag{12.23a}$$

$$\rho\left(\bar{u}\frac{\partial \bar{u}}{\partial x} + \bar{v}\frac{\partial \bar{u}}{\partial y} + \bar{w}\frac{\partial \bar{u}}{\partial z}\right) = -\frac{\partial \bar{p}}{\partial x} + \mu\left(\frac{\partial^2 \bar{u}}{\partial x^2} + \frac{\partial^2 \bar{u}}{\partial y^2} + \frac{\partial^2 \bar{u}}{\partial z^2}\right)$$
$$- \rho\left(\frac{\partial}{\partial x}(\overline{u'^2}) + \frac{\partial}{\partial y}(\overline{u'v'}) + \frac{\partial}{\partial z}(\overline{u'w'})\right) \tag{12.23b}$$

$$\rho\left(\bar{u}\frac{\partial \bar{v}}{\partial x} + \bar{v}\frac{\partial \bar{v}}{\partial y} + \bar{w}\frac{\partial \bar{v}}{\partial z}\right) = -\frac{\partial \bar{p}}{\partial y} + \mu\left(\frac{\partial^2 \bar{v}}{\partial x^2} + \frac{\partial^2 \bar{v}}{\partial y^2} + \frac{\partial^2 \bar{v}}{\partial z^2}\right)$$
$$- \rho\left(\frac{\partial}{\partial x}\overline{(u'v')} + \frac{\partial}{\partial y}\overline{(v'^2)} + \frac{\partial}{\partial z}\overline{(v'w')}\right) \quad (12.23c)$$

$$\rho\left(\bar{u}\frac{\partial \bar{w}}{\partial x} + \bar{v}\frac{\partial \bar{w}}{\partial y} + \bar{w}\frac{\partial \bar{w}}{\partial z}\right) = -\frac{\partial \bar{p}}{\partial z} + \mu\left(\frac{\partial^2 \bar{w}}{\partial x^2} + \frac{\partial^2 \bar{w}}{\partial y^2} + \frac{\partial^2 \bar{w}}{\partial z^2}\right)$$
$$- \rho\left(\frac{\partial}{\partial x}\overline{(u'w')} + \frac{\partial}{\partial y}\overline{(v'w')} + \frac{\partial}{\partial z}\overline{(w'^2)}\right) \quad (12.23d)$$

Applying the same process to the boundary conditions would show that the no-slip, no-penetration conditions apply to both mean and fluctuating velocity components.

Note that equations 12.23b–12.23d resemble the Navier–Stokes equations written in the mean flow variables $\bar{u}, \bar{v}, \bar{w}, \bar{p}$ with additional terms on the right-hand sides. These nine additional terms are referred to as the Reynolds stresses, and their presence as additional unknowns means that Eqs. 12.31a–12.31d are not sufficient to describe the flow. The fact that there are 13 unknowns and only four governing equations constitutes the closure problem in turbulence. To solve this problem, additional equations must be created which relate the Reynolds stresses to the mean flow variables. These additional equations are provided by a turbulence model.

Many different turbulence models have been developed over the years. Most are based on empirical observation of the characteristics of turbulence and require some fine-tuning. Although detailed coverage of these models is beyond the scope of this text, the discussion of turbulent flow between parallel plates in the next section will give you some insight into the application of the Reynolds equations and an appreciation of the closure problem.

12.4.2 Steady Turbulent Flow Between Parallel Plates (CFD)

The analysis of turbulent flow by means of the Reynolds equations can be illustrated by considering the steady turbulent flow between parallel plates. Earlier, in Section 12.2.3, we discussed the laminar flow in this geometry. Using the geometry of Figure 12.16 and reasoning similar to that used earlier, we will assume that the mean flow is described by $\bar{u} = \bar{u}(y)$, $\bar{v} = \bar{w} = 0$, and that the mean pressure is not a function of z. The Reynolds stresses are assumed to be a function of y only. Inserting these velocity components into Eqs. 12.23a–12.23d, we have

$$0 = 0, \quad 0 = -\frac{\partial \bar{p}}{\partial x} + \mu\left(\frac{\partial^2 \bar{u}}{\partial y^2}\right) - \rho\frac{\partial}{\partial y}\overline{(u'v')}, \quad 0 = -\frac{\partial \bar{p}}{\partial y} - \rho\frac{\partial}{\partial y}\overline{(v'^2)}$$

and

$$0 = -\rho\frac{\partial}{\partial y}\overline{(v'w')}$$

Integrating the last equation, we find $\overline{(v'w')} = C$. Since the no-slip, no-penetration conditions require that each of the mean and fluctuating velocity components be zero at the walls, the constant C is zero. Next we integrate the second equation by writing

$$\int_y^h \frac{\partial \bar{p}}{\partial y} dy = -\int_y^h \rho \frac{\partial}{\partial y}(\overline{v'^2}) dy$$

After rearrangement, the result is $\bar{p}(x, y) = \bar{p}(x, h) + \rho(\overline{v'^2})|_{y=h} - \rho(\overline{v'^2})$. The no-penetration condition requires that the fluctuation in the v velocity component disappear at the wall, yielding $\rho(\overline{v'^2})|_{y=h} = 0$, and our final result is

$$\bar{p}(x, y) = \bar{p}(x, h) - \rho(\overline{v'^2}) \tag{12.24}$$

This shows that the fluctuation in the y velocity component creates a pressure variation across the channel. Here $\bar{p}(x, h)$ is the pressure distribution on the upper channel wall, and from the symmetry we know that this must also be the pressure distribution on the lower wall. At this point there is no way to know the variation in $(\overline{v'^2})$ in the y direction across the channel and how this depends on Re. However, experiments suggest that the effect on the pressure is very small.

Now consider the momentum equation in the x direction. Writing this equation as

$$0 = -\frac{\partial \bar{p}}{\partial x} + \frac{\partial}{\partial y}\left[\mu\left(\frac{\partial \bar{u}}{\partial y}\right) - \rho(\overline{u'v'})\right]$$

and substituting for the pressure from Eq. 12.24, we have

$$0 = -\frac{\partial}{\partial x}\left[\bar{p}(x, h) - \rho(\overline{v'^2})\right] + \frac{\partial}{\partial y}\left[\mu\left(\frac{\partial \bar{u}}{\partial y}\right) - \rho(\overline{u'v'})\right]$$

Now recalling that $\bar{u}$, $\overline{v'^2}$, and $\overline{u'v'}$ are assumed to be functions of y only, and since $\bar{p}(x, h)$ is a function of x only, we can take an x derivative to conclude that $(d/dx)[\bar{p}(x, h)]$ is a constant. Thus we know that the wall pressure distribution is linear, i.e., of the form $\bar{p}(x, h) = Ax + B$. We can also write this as $\bar{p}(x, h) = \bar{p}(0, h) + Ax$. If the channel length is L, we have $\bar{p}(L, h) = \bar{p}(0, h) + AL$, so $A = -[\bar{p}(0, h) - \bar{p}(L, h)]/L$. Letting $\bar{p}(0, h) = \bar{p}_1$ and $\bar{p}(L, h) = \bar{p}_2$, we have $A = -(\bar{p}_1 - \bar{p}_2)/L$, and the pressure distribution on the channel walls is given by

$$\bar{p}(x, h) = \bar{p}_1 - \left(\frac{\bar{p}_1 - \bar{p}_2}{L}\right) x \tag{12.25}$$

The mean velocity profile can be seen to obey the equation

$$\frac{\partial}{\partial y}\left[\mu\left(\frac{\partial \bar{u}}{\partial y}\right) - \rho(\overline{u'v'})\right] = -\left(\frac{\bar{p}_1 - \bar{p}_2}{L}\right)$$

Were it not for the presence of the unknown Reynolds stress $\overline{u'v'}$, we would integrate this equation twice to determine the mean velocity profile.

To complete the solution it is necessary to employ a turbulence model that relates $\overline{u'v'}$ to the mean velocity $\bar{u}$. Many different turbulence models have been developed over the years, and most are based on empirical observation of the characteristics of turbulence. Although the discussion of these models is beyond the scope of this text, one of the simplest approaches is to use an eddy viscosity model. In this case the eddy viscosity model allows us to write

$$-\rho(\overline{u'v'}) = \mu_T \frac{\partial \bar{u}}{\partial y} \tag{12.26}$$

where μ_T is called the eddy viscosity. Using Eq. 12.26, the remaining governing equation for turbulent flow between parallel plates becomes

$$\frac{\partial}{\partial y}\left[(\mu + \mu_T)\frac{\partial \bar{u}}{\partial y}\right] = -\left(\frac{\bar{p}_1 - \bar{p}_2}{L}\right)$$

However, we are still unable to solve the problem because the eddy viscosity $\mu_T(y)$ is now an unknown flow property that must be determined.

Modern CFD codes contain a number of built-in turbulence models as well as guidance in their use. A CFD solution of the fully developed turbulent flow between parallel plates at $Re = 12{,}300$ (based on the average velocity) yielded the mean velocity profile shown in Figure 12.32.

CD/Dynamics/Potential flow

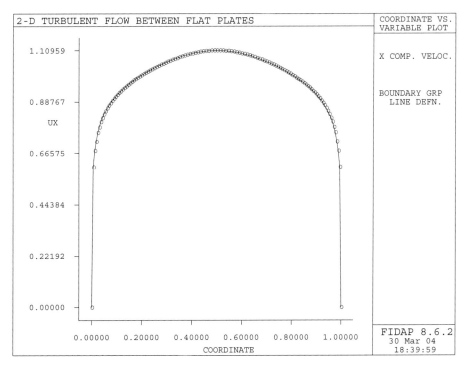

Figure 12.32 CFD solution for the mean velocity profile in turbulent flow between parallel plates at $Re = 12{,}300$.

12.5 INVISCID IRROTATIONAL FLOW

As discussed in Chapter 11, a flow in which the effects of viscosity are negligible is termed an inviscid flow. The governing equations for inviscid flow are the continuity equation, the Euler equations, and an appropriate form of the energy equation. However, for the constant density flows of interest in this chapter, we do not need the energy equation, and the velocity and pressure fields must satisfy only the continuity and Euler equations for a constant density fluid. The appropriate forms of the continuity equation in Cartesian and cylindrical coordinates are Eqs. 12.1a and 12.2a, respectively. The Euler equations in Cartesian and cylindrical coordinates are given in Chapter 11 by Eqs. 11.16a–11.16c and 11.17a–11.17c, respectively.

In Cartesian coordinates, the four unknowns, u, v, w, p, which describe an inviscid, constant density flow, are functions of position and time that satisfy the four governing equations:

$$\frac{\partial u}{\partial x} + \frac{\partial v}{\partial y} + \frac{\partial w}{\partial z} = 0 \qquad (12.27\text{a})$$

$$\rho\left(\frac{\partial u}{\partial t} + u\frac{\partial u}{\partial x} + v\frac{\partial u}{\partial y} + w\frac{\partial u}{\partial z}\right) = \rho f_x - \frac{\partial p}{\partial x} \qquad (12.27\text{b})$$

$$\rho\left(\frac{\partial v}{\partial t} + u\frac{\partial v}{\partial x} + v\frac{\partial v}{\partial y} + w\frac{\partial v}{\partial z}\right) = \rho f_y - \frac{\partial p}{\partial y} \qquad (12.27\text{c})$$

$$\rho\left(\frac{\partial w}{\partial t} + u\frac{\partial w}{\partial x} + v\frac{\partial w}{\partial y} + w\frac{\partial w}{\partial z}\right) = \rho f_z - \frac{\partial p}{\partial z} \qquad (12.27\text{d})$$

In cylindrical coordinates, the four unknowns, v_r, v_θ, v_z, p satisfy the equations

$$\frac{1}{r}\frac{\partial(rv_r)}{\partial r} + \frac{1}{r}\frac{\partial v_\theta}{\partial \theta} + \frac{\partial v_z}{\partial z} = 0 \qquad (12.28\text{a})$$

$$\rho\left(\frac{\partial v_r}{\partial t} + v_r\frac{\partial v_r}{\partial r} + \frac{v_\theta}{r}\frac{\partial v_r}{\partial \theta} + v_z\frac{\partial v_r}{\partial z} - \frac{v_\theta^2}{r}\right) = \rho f_r - \frac{\partial p}{\partial r} \qquad (12.28\text{b})$$

$$\rho\left(\frac{\partial v_\theta}{\partial t} + v_r\frac{\partial v_\theta}{\partial r} + \frac{v_\theta}{r}\frac{\partial v_\theta}{\partial \theta} + v_z\frac{\partial v_\theta}{\partial z} + \frac{v_r v_\theta}{r}\right) = \rho f_\theta - \frac{1}{r}\frac{\partial p}{\partial \theta} \qquad (12.28\text{c})$$

$$\rho\left(\frac{\partial v_z}{\partial t} + v_r\frac{\partial v_z}{\partial r} + \frac{v_\theta}{r}\frac{\partial v_z}{\partial \theta} + v_z\frac{\partial v_z}{\partial z}\right) = \rho f_z - \frac{\partial p}{\partial z} \qquad (12.28\text{d})$$

To analyze an inviscid flow problem, we must solve the appropriate set of four partial differential equations together with the relevant BCs. Because viscous effects are absent, an inviscid flow is able to slip along a solid surface. Thus the no-slip condition is not applicable, and the only BC at a solid surface is the no-penetration condition.

The solution of the governing equations of inviscid flow is a challenging task. In the special case of an irrotational flow, and in particular for 2D flows, methods are available that simplify the problem of finding the velocity and pressure fields that satisfy Eqs. 12.27a–12.27d or 12.28a–12.28d. Irrotational flows are of interest in the analysis of external flows and have a long history of importance in aerodynamics. This is because in aerodynamic flows the effect of viscosity is often confined to thin boundary layers. In

12.5 INVISCID IRROTATIONAL FLOW

the boundary layer the flow is viscous and rotational, but outside the boundary layer the flow is inviscid and irrotational.

In the remainder of this section, we will discuss the important class of flows that are both inviscid and irrotational. For reasons that will become clear in a moment, these flows are also known as potential flows. Recall from Section 10.9.3 that in an irrotational flow, the vorticity $\boldsymbol{\omega} = \nabla \times \mathbf{u}$, is zero, and we can therefore define a scalar velocity potential $\phi(\mathbf{x}, t)$ by using Eq. 10.58 as $\mathbf{u} = \nabla \phi$. By employing the velocity potential, the condition of irrotational flow is automatically satisfied, and the three velocity components are replaced by a single scalar potential function.

In Cartesian coordinates, the velocity components are related to the velocity potential by Eqs. 10.59a–10.59c: $u = \partial\phi/\partial x$, $v = \partial\phi/\partial y$, and $w = \partial\phi/\partial z$, while in cylindrical coordinates the relationship is given by Eqs. 10.60a–10.60c: $v_r = \partial\phi/\partial r$, $v_\theta = (1/r)(\partial\phi/\partial\theta)$, and $v_z = \partial\phi/\partial z$. For a constant density flow, the continuity equation reduces to Eq. 11.3: $\nabla \cdot \mathbf{u} = 0$. Substituting $\mathbf{u} = \nabla \phi$ into this equation, we see that the velocity potential satisfies Laplace's equation (Eq. 10.61): $\nabla^2 \phi = 0$. Writing this equation in Cartesian coordinates, we obtain (Eq. 10.62a): $\partial^2\phi/\partial x^2 + \partial^2\phi/\partial y^2 + \partial^2\phi/\partial z^2 = 0$. In cylindrical coordinates the corresponding result (Eq. 10.62b) is

$$\frac{1}{r}\frac{\partial}{\partial r}\left(r\frac{\partial\phi}{\partial r}\right) + \frac{1}{r^2}\frac{\partial^2\phi}{\partial\theta^2} + \frac{\partial^2\phi}{\partial z^2} = 0$$

The analytical solution to an inviscid flow problem can now be described in general terms. The velocity field is obtained by solving Laplace's equation together with the no-penetration BC on a solid surface to obtain the velocity potential $\phi(\mathbf{x}, t)$. Next we obtain the velocity field by using Eq. 10.58, $\mathbf{u} = \nabla \phi$, where we use Eqs. 10.62a or 10.62b to evaluate this equation in the coordinate system of interest. To obtain the pressure field, recall from Chapter 11 that the Bernoulli equation can be derived from the Euler equations for inviscid flow. The Bernoulli equation holds along a streamline and, of greater importance here, between any two points in an irrotational flow. Thus, the pressure distribution can be obtained from the Bernoulli equation with a known velocity field. Alternately, we can solve the Euler equations for the pressure distribution after inserting a known velocity field, or use one of the many well-developed CFD methods to solve the Euler equations for both the velocity components and pressure. An example of a CFD mesh for use in a potential flow analysis over an aircraft is shown in Figure 12.33.

CD/Dynamics/Potential flow/Potential flow builder

12.5.1 Plane Potential Flow

For simplicity we will now limit consideration to plane potential flow. These are 2D flows that occur in the xy plane with velocity components u and v. In Cartesian coordinates, with a velocity potential $\phi = \phi(x, y, t)$ we have the relations

$$u = \frac{\partial\phi}{\partial x}, \quad \text{and} \quad v = \frac{\partial\phi}{\partial y} \quad (12.29a)$$

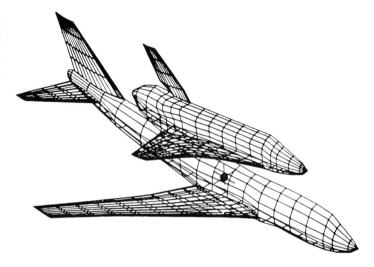

Figure 12.33 CFD implementation of the Euler equations for flow over the Boeing 747 carrying the space shuttle. The mesh represents panels of fundamental flows that will be discussed in the following sections; thus the mesh is used in the panel method.

and since the velocity potential satisfies Laplace's equation, the potential must satisfy

$$\frac{\partial^2 \phi}{\partial x^2} + \frac{\partial^2 \phi}{\partial y^2} = 0 \qquad (12.29\text{b})$$

In a 2D flow we can also make use of the streamfunction. We defined the streamfunction for 2D flows in Section 10.10.3 and showed that streamlines and potential lines are orthogonal to one another. The relevant relationships between the velocity components and streamfunction in Cartesian coordinates are repeated here for convenience:

$$u = \frac{\partial \psi}{\partial y} \quad \text{and} \quad v = \frac{-\partial \psi}{\partial x} \qquad (12.29\text{c})$$

For 2D flow in Cartesian coordinates the condition of irrotationality, $\nabla \times \mathbf{u} = 0$, reduces to $\partial v/\partial x - \partial u/\partial y = 0$. Substituting into this equation from Eq. 12.29c we obtain

$$\frac{\partial^2 \psi}{\partial x^2} + \frac{\partial^2 \psi}{\partial y^2} = 0 \qquad (12.29\text{d})$$

Thus, we see that the streamfunction also satisfies Laplace's equation. Equations 12.29a–12.29d are the basic equations governing plane potential flow in Cartesian coordinates.

In cylindrical coordinates, with the potential given by $\phi = \phi(r, \theta, t)$, the relationships between the velocity components and the potential are

$$v_r = \frac{\partial \phi}{\partial r} \quad \text{and} \quad v_\theta = \frac{1}{r}\frac{\partial \phi}{\partial \theta} \qquad (12.30\text{a})$$

and the potential satisfies Laplace's equation:

$$\frac{1}{r}\frac{\partial}{\partial r}\left(r \frac{\partial \phi}{\partial r}\right) + \frac{1}{r^2}\frac{\partial^2 \phi}{\partial \theta^2} = 0 \qquad (12.30\text{b})$$

12.5 INVISCID IRROTATIONAL FLOW

In addition, we can relate the velocity components to the streamfunction by

$$v_r = \frac{1}{r}\frac{\partial \psi}{\partial \theta} \quad \text{and} \quad v_\theta = -\frac{\partial \psi}{\partial r} \tag{12.30c}$$

The condition of irrotationality in cylindrical coordinates is

$$\frac{1}{r}\left[\frac{\partial}{\partial r}(r v_\theta) - \frac{\partial v_r}{\partial \theta}\right] = 0$$

which means that the streamfunction also satisfies Laplace's equation:

$$\frac{1}{r}\frac{\partial}{\partial r}\left(r\frac{\partial \psi}{\partial r}\right) + \frac{1}{r^2}\frac{\partial^2 \psi}{\partial \theta^2} = 0 \tag{12.30d}$$

Thus a plane potential flow in cylindrical coordinates satisfies equations 12.30a–12.30d.

With the velocity potential for a plane potential flow known, we can construct the pressure distribution using the Bernoulli equation, Eq. 11.22:

$$\int_1^2 \frac{\partial \mathbf{u}}{\partial t}\cdot d\mathbf{r} + \int_1^2 \frac{dp}{\rho} + \frac{1}{2}\left(V_2^2 - V_1^2\right) + (\Psi_2 - \Psi_1) = 0$$

For a steady, constant density flow in the absence of body forces, we can write this equation in Cartesian and cylindrical coordinates as

$$\frac{p_1}{\rho} + \frac{1}{2}\left(u_1^2 + v_1^2\right) = \frac{p_2}{\rho} + \frac{1}{2}\left(u_2^2 + v_2^2\right) \tag{12.31a}$$

$$\frac{p_1}{\rho} + \frac{1}{2}\left(v_{r1}^2 + v_{\theta 1}^2\right) = \frac{p_2}{\rho} + \frac{1}{2}\left(v_{r2}^2 + v_{\theta 2}^2\right) \tag{12.31b}$$

Stagnation points are important concepts related to inviscid flows and are defined to be points in the flow or on the surface of a body where the velocity is zero. Upon applying the Bernoulli equation between the stagnation point and any other point in the flow field, it is evident that the pressure at a stagnation point is a maximum. In inviscid flows over bodies, stagnation points may be found near the nose of the body and near the tail. We will discuss the locations of stagnation points in many of the flows of the following sections.

Consider the problem of finding the velocity potential, streamfunction, and pressure distribution for the uniform flow shown in Figure 12.34A. Uniform flow is an important building block in the study of the irrotational flow created by a body such as a cylinder or airfoil moving through a stationary fluid. In a frame of reference fixed to the moving body, the body is at rest and the undisturbed flow far from the body is uniform. The boundary conditions in such a case are the no-penetration condition on the body surface and a uniform flow far from the body. Thus knowledge of the velocity potential, streamfunction, and pressure distribution for a uniform flow in both Cartesian and cylindrical coordinates is of great interest.

The general approach to solving a problem in plane potential flow is to find a solution of Laplace's equation that satisfies a set of boundary conditions appropriate to the flow of interest. In this case the flow is simple and we can bypass the problem of finding a solution to Laplace's equation altogether. In Cartesian coordinates, we can write the

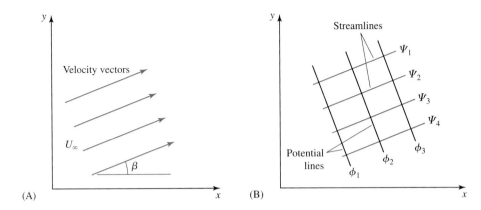

Figure 12.34 Uniform flow.

two velocity components of a uniform flow by inspection (Figure 12.34A) as

$$u = U_\infty \cos\beta \quad \text{and} \quad v = U_\infty \sin\beta \tag{12.32}$$

Applying Eq. 12.29a yields $u = U_\infty \cos\beta = \partial\phi/\partial x$ and $v = U_\infty \sin\beta = \partial\phi/\partial y$. Integrating the first equation with respect to x, we obtain $\phi = U_\infty x \cos\beta + f(y)$. Taking the y derivative of this result and comparing with $\partial\phi/\partial y = U_\infty \sin\beta$ shows that $\partial f/\partial y = U_\infty \sin\beta$. Thus we conclude that $f(y) = U_\infty y \sin\beta$. The velocity potential in Cartesian coordinates for a uniform flow (Figure 12.34B) is thus

$$\phi(x, y) = U_\infty(x\cos\beta + y\sin\beta) \tag{12.33a}$$

Inserting this potential into Eq. 12.29b shows that it satisfies Laplace's equation. Note that if the flow is in the x direction, this reduces to

$$\phi(x, y) = U_\infty x \tag{12.33b}$$

The streamfunction for this flow may be found by applying Eq. 12.29c to obtain

$$u = U_\infty \cos\beta = \frac{\partial\psi}{\partial y} \quad \text{and} \quad v = U_\infty \sin\beta = -\frac{\partial\psi}{\partial x}$$

Integrating the first equation with respect to y gives $\psi = U_\infty y \cos\beta + g(x)$, and after taking the x derivative of this result and comparing with $\partial\psi/\partial x = -U_\infty \sin\beta$, we conclude that $g(x) = -U_\infty x \sin\beta$. Thus as shown in Figure 12.34B, the streamfunction for a uniform flow is

$$\psi(x, y) = U_\infty(y\cos\beta - x\sin\beta) \tag{12.34a}$$

For flow in the x direction we have

$$\psi(x, y) = U_\infty y \tag{12.34b}$$

To find the velocity potential and streamfunction of a uniform flow in cylindrical coordinates, we can make use of the fact that $x = r\cos\theta$, $y = r\sin\theta$. Substituting into Eq. 12.33a we obtain

$$\phi(r, \theta) = U_\infty r(\cos\beta \cos\theta + \sin\beta \sin\theta) \tag{12.35a}$$

For a flow in the x direction we have

$$\phi(r, \theta) = U_\infty r \cos \theta \quad (12.35b)$$

Similarly, the streamfunction in cylindrical coordinates is found to be

$$\psi(r, \theta) = U_\infty r (\cos \beta \sin \theta - \sin \beta \cos \theta) \quad (12.36a)$$

which for flow in the x direction reduces to

$$\psi(r, \theta) = U_\infty r \sin \theta \quad (12.36b)$$

The velocity components in cylindrical coordinates can be obtained by inserting the velocity potential in Eq. 12.30a, or by inserting the streamfunction into Eq. 12.30c. Either method yields

$$v_r = U_\infty (\cos \beta \cos \theta + \sin \beta \sin \theta) \quad \text{and} \quad v_\theta = U_\infty (\sin \beta \cos \theta - \cos \beta \sin \theta) \quad (12.37a)$$

which reduces for flow in the x direction to

$$v_r = U_\infty \cos \theta \quad \text{and} \quad v_\theta = -U_\infty \sin \theta \quad (12.37b)$$

In a uniform inviscid flow we intuitively expect that in the absence of body forces, the pressure is also uniform. Inserting the velocity components in either Cartesian or cylindrical coordinates into the appropriate version of the Bernoulli equation (Eqs. 12.31) shows that this is the case. For example, in Cartesian coordinates applying Eq. 12.31a between any two points in a uniform flow as described by Eq. 12.32 gives

$$\frac{p_1}{\rho} + \frac{1}{2} U_\infty^2 = \frac{p_2}{\rho} + \frac{1}{2} U_\infty^2$$

and we see that $p_1 = p_2$, meaning that the pressure is uniform throughout the flow.

Now consider the problem of finding the velocity potential for uniform flow over a cylinder. The unknown potential satisfies Laplace's equation in cylindrical coordinates, Eq. 12.30b, and the no-penetration boundary condition on the cylinder surface. It must also become identical to the velocity potential for a uniform flow far from the cylinder. The geometry of this flow is shown in Figure 12.35. The use of cylindrical coordinates is indicated here, so we will solve Eq. 12.30b for the velocity potential. The unknown

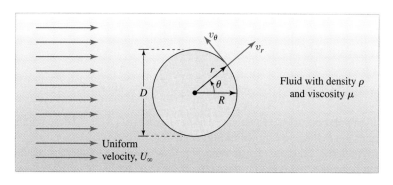

Figure 12.35 Surface velocity vector for inviscid, irrotational flow over a cylinder.

potential is a solution of

$$\frac{1}{r}\frac{\partial}{\partial r}\left(r\frac{\partial \phi}{\partial r}\right) + \frac{1}{r^2}\frac{\partial^2 \phi}{\partial \theta^2} = 0$$

that satisfies the boundary conditions described. The no-penetration condition can be expressed in terms of v_r (see Figure 12.35) by writing $v_r = \partial\phi/\partial r = 0$ at $r = R$. The uniform flow far from the cylinder can be expressed by using Eq. 12.35b to write $\phi = U_\infty r \cos\theta$ at $r = \infty$. At this point we could try a classic separation of variables solution, and indeed this does work. However, by noticing that the unknown potential must have the form of a uniform flow at $r = \infty$, we can try a separable solution of the form $\phi(r, \theta) = f(r)\cos\theta$. The boundary condition on the cylinder now becomes $df/dr = 0$ at $r = R$, and the uniform flow condition requires that $f = U_\infty r$ at $r = \infty$. Substituting into Laplace's equation shows that the function f satisfies the ordinary differential equation $d^2 f/dr^2 + (1/r)(df/dr) - f/r^2 = 0$, whose solution is $f(r) = Ar + B/r$. Since $f = U_\infty r$ at $r = \infty$, we must have $A = U_\infty$. Then to satisfy $df/dr = 0$ at $r = R$, we take $B = U_\infty R^2$. Thus the potential for uniform flow over a cylinder can be written as

$$\phi(r, \theta) = U_\infty r \cos\theta \left(1 + \frac{R^2}{r^2}\right) \tag{12.38}$$

It is straightforward to calculate the velocity components by using Eq. 12.30a:

$$v_r = \frac{\partial \phi}{\partial r} = U_\infty \cos\theta \left(1 - \frac{R^2}{r^2}\right) \tag{12.39a}$$

and

$$v_\theta = \frac{1}{r}\frac{\partial \phi}{\partial \theta} = -U_\infty \sin\theta \left(1 + \frac{R^2}{r^2}\right) \tag{12.39b}$$

With the velocity components known, the streamfunction can be calculated from Eq. 12.30c. The result is

$$\psi(r, \theta) = U_\infty r \sin\theta \left(1 - \frac{R^2}{r^2}\right) \tag{12.40}$$

The streamlines and a velocity profile in this flow are shown in Figure 12.36A. Note the fore and aft symmetry and that the flow is uniform far from the cylinder.

To find the location of any stagnation points, we set the velocity components to zero and solve for the location of points that satisfy the resulting pair of equations. In this case we have

$$v_r = 0 = U_\infty \cos\theta \left(1 - \frac{R^2}{r^2}\right) \quad \text{and} \quad v_\theta = 0 = -U_\infty \sin\theta \left(1 + \frac{R^2}{r^2}\right)$$

Thus, one stagnation point occurs at $(R, 0)$, and a second occurs at (R, π). These points are on the surface of the cylinder as shown in Figure 12.36A.

We can also find the pressure distribution for flow over a cylinder by using Eq. 12.31b and choosing the first point far upstream of the cylinder in the uniform flow, where the pressure is equal to p_∞. Substituting the known velocity components, noting

12.5 INVISCID IRROTATIONAL FLOW

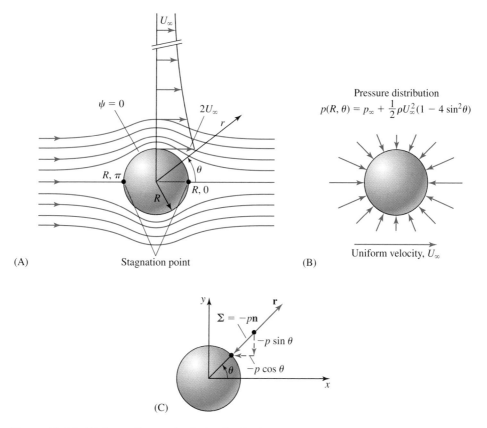

Figure 12.36 (A) Streamlines and velocity distribution. (B) The surface pressure distribution for inviscid, irrotational flow over a cylinder. (C) The force balance on the surface of the cylinder.

that $\frac{1}{2}(v_r^2 + v_\theta^2)|_{r=\infty} = \frac{1}{2}U_\infty^2$, and rearranging, we find

$$p(r, \theta) = p_\infty - \frac{1}{2}\rho U_\infty^2 \left[\frac{R^4}{r^4} + \frac{2R^2}{r^2}(\sin^2 \theta - \cos^2 \theta) \right] \quad (12.41a)$$

On the surface of the cylinder, $r = R$, the pressure distribution is given by

$$p(R, \theta) = p_\infty + \frac{1}{2}\rho U_\infty^2 (1 - 4\sin^2 \theta) \quad (12.41b)$$

The pressure distribution on the surface is shown in Figure 12.36B. To find the pressure at the two stagnation points on the surface of the cylinder, we substitute the location of the points into Eq. 12.41b. The result is that

$$p(R, 0) = p_\infty + \tfrac{1}{2}\rho U_\infty^2 \quad \text{and} \quad p(R, \pi) = p_\infty + \tfrac{1}{2}\rho U_\infty^2$$

We see that the pressure at both stagnation points, $p_\infty + \frac{1}{2}\rho U_\infty^2$, is the stagnation pressure that would be read by means of a pitot tube inserted into the uniform flow upstream of the cylinder.

In this inviscid flow the only stress that may exist is that due to the pressure. The force applied by the fluid to the cylinder can be found by using Eq. 4.21: $\mathbf{F}_S = \int_S \mathbf{\Sigma}\, dS$, where the stress vector acting on the cylinder is $\Sigma = -p\mathbf{n}$, and p is the pressure on the surface. From Figure 12.36C, the components of the stress vector in the x and y directions are given by $\Sigma_x = -p\cos\theta$ and $\Sigma_y = -p\sin\theta$. The surface element is $dS = R\,d\theta\,dz$, thus the components of the force in these directions acting on a cylinder of length L are given by

$$F_{Sx} = \int_S \Sigma_x\,dS = \int_0^L \int_0^{2\pi} -p(R,\theta)\cos\theta\,R\,d\theta\,dz \qquad (12.42a)$$

$$F_{Sy} = \int_S \Sigma_y\,dS = \int_0^L \int_0^{2\pi} -p(R,\theta)\sin\theta\,R\,d\theta\,dz \qquad (12.42b)$$

Performing these integrals with the pressure on the surface given by Eq. 12.41b reveals that the inviscid flow results in no net force on the cylinder. The absence of lift or drag force in this case should not be too surprising because shear stresses are absent and the flow exhibits both fore and aft as well as above and below symmetry in the velocity and pressure fields.

In our discussion of circulation in Chapter 10, we pointed out that circulation plays an important role in the traditional 2D frictionless flow model describing the aerodynamic properties of a wing. According to Eq. 10.56, the lift produced by a long wing of length L is related to the circulation around the airfoil by $F_L = -\rho UCL$, where C is the circulation on a path enclosing the airfoil. The circulation $C(t)$ around the cylinder surface produced by the uniform flow solution can be calculated by applying Eq. 10.54: $C(t) = \oint \mathbf{u} \cdot d\mathbf{r}$ and noting that for a circular path of radius R the line integral is $C(t) = \int_0^{2\pi} v_\theta R\,d\theta$. The value of v_θ on the cylinder surface can be found from Eq. 12.39b to be $v_\theta|_{r=R} = -U_\infty \sin\theta(1 + R^2/r^2)|_{r=R} = -2U_\infty \sin\theta$, so the circulation around the cylinder is

$$C(t) = \int_0^{2\pi} -2U_\infty \sin\theta\,R\,d\theta = 0$$

We see that this results in a prediction of zero lift, which is consistent with our earlier calculation using the pressure in Eq. 12.42b. In Section 12.5.4, where we consider a uniform flow over a cylinder with circulation, we will show that the resulting flow does create lift on the cylinder.

EXAMPLE 12.6

Using the solution for uniform flow over a cylinder, plot the pressure distribution on the stagnation streamline that extends from upstream at infinity and ends on the front stagnation point on the cylinder.

SOLUTION

The pressure distribution is given by Eq. 12.41a as

$$p(r,\theta) = p_\infty - \frac{1}{2}\rho U_\infty^2 \left[\frac{R^4}{r^4} + \frac{2R^2}{r^2}(\sin^2\theta - \cos^2\theta)\right]$$

The stagnation streamline is the line defined by $\theta = \pi$. Inserting this value into the equation and rearranging we obtain

$$\frac{p(r,\pi) - p_\infty}{\frac{1}{2}\rho U_\infty^2} = \left[2\left(\frac{R}{r}\right)^2 - \left(\frac{R}{r}\right)^4\right]$$

This relation is plotted in Figure 12.37. We see that the pressure increases along the stagnation point streamline and reaches its maximum at the front stagnation point, as expected. The pressure on the stagnation streamline emanating from the rear of the cylinder is also shown.

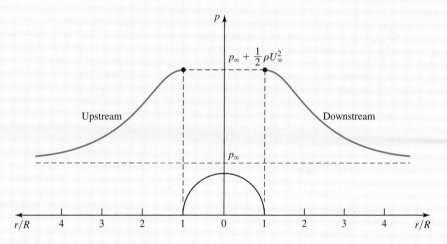

Figure 12.37 Pressure variation along a stagnation point streamline.

A variety of well-developed analytical and computational methods exist for the efficient solution of more complex plane potential flow problems, particularly those that involve a uniform flow over an airfoil or other body. An example of a CFD calculation for flow over an airfoil is shown in Figure 12.38. Here we see the pressure distribution about the airfoil as well as on its surface. With the exception of the superposition methods discussed shortly, most methods for solving plane potential flow problems are beyond the scope of this text. The interested reader should consult advanced fluid mechanics and aerodynamics books.

 CD/Dynamics/Potential flow/Elementary potential flows

12.5.2 Elementary Plane Potential Flows

As a result of the linearity of Laplace's equation, linear combinations of solutions are also valid solutions. Thus, we can combine velocity potentials for known elementary

Figure 12.38 (A) Pressure contours and (B) pressure coefficient for flow over an airfoil obtained using an Euler-based CFD calculation.

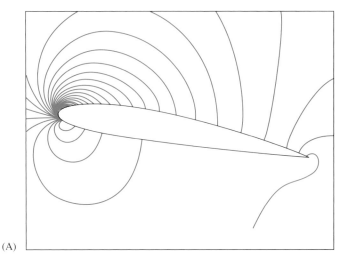

(A)

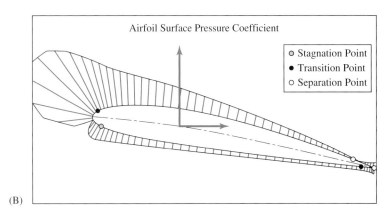

(B)

flows into a potential describing a more complex flow field. One such elementary flow is the uniform flow already discussed. In this section we consider several other elementary flows: the point source and sink, the line vortex, and the doublet. In each case we give the velocity potential, streamfunction, and velocity components in cylindrical coordinates and describe the flow. For homework you may be asked to demonstrate that the velocity potential and streamfunction of these flows satisfy Laplace's equation.

A source or sink of strength m centered on the origin is described by

$$\phi = \frac{m}{2\pi}\ln r, \quad \psi = \frac{m}{2\pi}\theta, \quad v_r = \frac{m}{2\pi r}, \quad \text{and} \quad v_\theta = 0 \quad (12.43)$$

It can be seen that this flow has potential lines that are concentric circles centered on the origin, and radial streamlines. As illustrated in Figure 12.39, the flow is radially outward for a source, $m > 0$, and radially inward for a sink, $m < 0$. Note that the velocity becomes infinite at $r = 0$, an unrealistic feature but one that is not an issue in applications

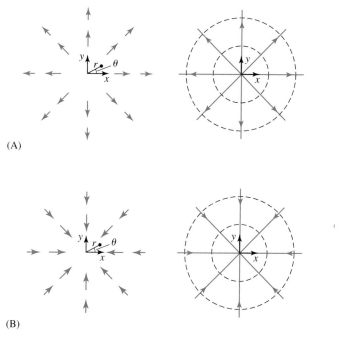

Figure 12.39 Irrotational source (A) and sink (B).

involving the use of a source or sink to construct a more complex flow. If we calculate the volume flowrate (per unit length in the z direction) q, which crosses a circle of radius r centered on the origin of a source or sink, we obtain

$$q = \int_0^{2\pi} v_r r\, d\theta = \int_0^{2\pi} \left(\frac{m}{2\pi r}\right) r\, d\theta = m$$

Thus the strength m is equal to the volume flowrate per unit length leaving the source or entering the sink.

A line vortex of strength K is described by

$$\phi = K\theta, \quad \psi = -K \ln r, \quad v_r = 0, \quad \text{and} \quad v_\theta = \frac{K}{r} \quad (12.44)$$

and illustrated in Figure 12.40. It can be seen that the potential lines are radial lines, and the streamlines are concentric circles centered on the origin. The name "line vortex" is applied to this flow because of the rotary nature of the flow field, evident in the equations $v_r = 0$ and $v_\theta = K/r$. If K is positive, the vortex rotates counterclockwise, if K is negative the rotation is clockwise. We again note the presence of an unrealistic infinite velocity at the origin. If we calculate the circulation, $C(t)$, around a circular path of radius r centered on the origin using Eq. 10.54 we obtain

$$C(t) = \oint \mathbf{u} \cdot d\mathbf{r} = \int_0^{2\pi} v_\theta r\, d\theta = \int_0^{2\pi} \left(\frac{K}{r}\right) r\, d\theta = 2\pi K \quad (12.45)$$

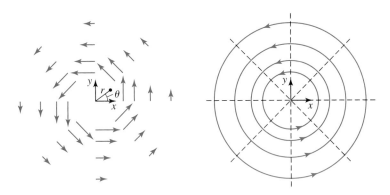

Figure 12.40 Irrotational vortex ($K > 0$, counterclockwise, center at origin).

We see that the circulation of the line vortex is constant and independent of the radius of the circle used to calculate it. The strength of a line vortex is thus related to the circulation about a circular path around the vortex by $K = C/2\pi$. If we calculate the vorticity for the line vortex, it is zero. This requirement for any irrotational flow may come as a surprise to you considering that the flow is a vortex. In this 2D flow there is no variation along the z direction.

A doublet of strength M, centered on the origin, and with its axis aligned with the x axis, is described by

$$\phi = \frac{M}{r}\cos\theta, \quad \psi = -\frac{M}{r}\sin\theta, \quad v_r = -\frac{M}{r^2}\cos\theta, \quad \text{and} \quad v_\theta = -\frac{M}{r^2}\sin\theta \tag{12.46}$$

The doublet is a rather complex and unrealistic looking flow as illustrated in Figure 12.41, but it proves highly useful in constructing more complicated flows. This will be demonstrated in the next section. The doublet may be thought of as the flow that results when a source and a sink are brought infinitely close together, in this case approaching each other (and the origin) along the x axis.

 CD/Dynamics/Potential flow/Superposition of elementary flows

12.5.3 Superposition of Elementary Plane Potential Flows

As mentioned earlier, linear combinations of known solutions of Laplace's equation are also valid solutions. Thus we can combine the solutions for say, uniform flow and one or more of the elementary flows discussed in Section 12.5.2, into new solutions that describe more complex and possibly more useful flow fields. Since the velocity potential and streamfunction each satisfy Laplace's equation, we can use this principle of

12.5 INVISCID IRROTATIONAL FLOW

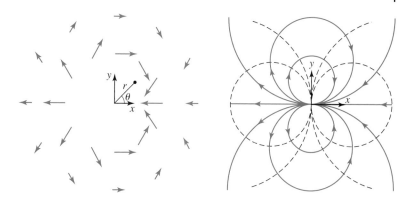

Figure 12.41 Irrotational doublet (center at origin).

superposition to add together potentials as well as the corresponding streamfunctions. The velocity components also add together; in most cases, however, it is best to determine these after the combined potential and streamfunction have been determined. The pressure distribution of the combined flow is not the sum of the pressure distributions of the elementary flows; rather the pressure must always be determined after the new velocity components have been found. This is because Bernoulli's equation involves the sum of the squares of the velocity components; i.e., it is nonlinear.

To illustrate this process, consider the superposition of a sink and vortex at the origin to create the spiral flow shown in Figure 12.42. From Eqs. 12.43 and 12.44, the resulting flow ($m < 0$) is given by

$$\phi = \frac{m}{2\pi} \ln r + K\theta, \quad \psi = -K \ln r + \frac{m}{2\pi}\theta, \quad v_r = \frac{m}{2\pi r}, \quad \text{and} \quad v_\theta = \frac{K}{r} \tag{12.47}$$

Note that we have obtained the velocity components of the combined flow by adding those of the elementary flows together. This is possible because both elementary flows are defined in terms of a common set of cylindrical coordinates.

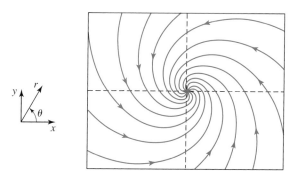

Figure 12.42 Superposition of a sink and a vortex.

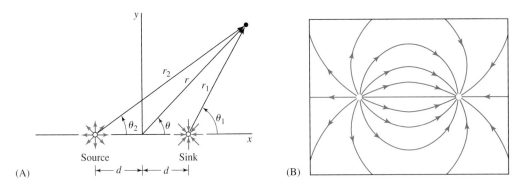

Figure 12.43 (A) Superposition of a source and a sink. (B) Streamlines for a source and a sink.

The use of a set of local coordinates for each elementary flow allows us to include elementary flows that are centered at points away from the origin. For example, suppose we are interested in the flow created by a source and sink of the same strength, located an equal distance d along the x axis on each side of the origin as shown in Figure 12.43A. Using the geometry shown, we can employ Eq. 12.43 to describe the source and sink in two distinct sets of local polar coordinates, one centered on the source and another on the sink. For the source we have $\phi_2 = (m/2\pi)\ln r_2$, $\psi_2 = (m/2\pi)\theta_2$, and for the sink we find $\phi_1 = (-m/2\pi)\ln r_1$, $\psi_1 = (-m/2\pi)\theta_1$. The combined flow is given by the sum of the terms for the elementary flows: $\phi = (m/2\pi)(\ln r_2 - \ln r_1)$, $\psi = (m/2\pi)(\theta_2 - \theta_1)$. Fundamental trigonometric relationships are used to relate the local coordinates to the common coordinates r, θ as follows:

$$\tan\theta_1 = \frac{r\sin\theta}{r\cos\theta - d}, \quad \tan\theta_2 = \frac{r\sin\theta}{r\cos\theta + d} \tag{12.48a}$$

$$r_1^2 = r^2 - 2dr\cos\theta + d^2, \quad r_2^2 = r^2 + 2dr\cos\theta + d^2 \tag{12.48b}$$

Equation 12.48a is used to express the streamfunction as

$$\psi = -\frac{m}{2\pi}\tan^{-1}\left(\frac{2dr\sin\theta}{r^2 - d^2}\right) \tag{12.49}$$

The velocity components may be found using Eq. 12.30c. The result is

$$v_r = -\frac{m}{\pi}\frac{d(r^2 - d^2)\cos\theta}{[(r^2 - d^2)^2 + 4d^2r^2\sin^2\theta]} \quad v_\theta = -\frac{m}{\pi}\frac{d(r^2 + d^2)\sin\theta}{[(r^2 - d^2)^2 + 4d^2r^2\sin^2\theta]} \tag{12.50}$$

The streamlines for this flow are shown in Figure 12.43B.

It is possible to create a flow over a body by superposition of a uniform flow in the x direction and a source located at the origin. The uniform flow is given by Eqs. 12.35b, 12.36b, and 12.37b as

$$\phi = U_\infty r\cos\theta, \quad \psi = U_\infty r\sin\theta, \quad v_r = U_\infty\cos\theta, \quad \text{and} \quad v_\theta = -U_\infty\sin\theta$$

while a source of strength m centered on the origin is described by Eq. 12.43 as

$$\phi = \frac{m}{2\pi}\ln r, \quad \psi = \frac{m}{2\pi}\theta, \quad v_r = \frac{m}{2\pi r}, \quad \text{and} \quad v_\theta = 0$$

12.5 INVISCID IRROTATIONAL FLOW

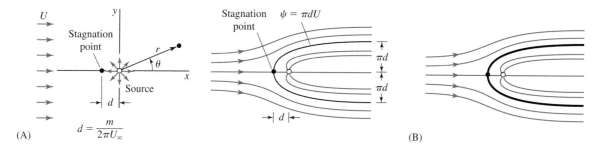

Figure 12.44 Superposition of a uniform flow and a source to form a half-body.

The combined flow is therefore described by

$$\phi = U_\infty r \cos\theta + \frac{m}{2\pi} \ln r, \quad \psi = U_\infty r \sin\theta + \frac{m}{2\pi}\theta$$
$$v_r = U_\infty \cos\theta + \frac{m}{2\pi r}, \quad v_\theta = -U_\infty \sin\theta \tag{12.51}$$

Note that in this case we have obtained the velocity components of the combined flow by adding the velocity components of the elementary flows together. This is possible because both elementary flows are defined in terms of a common set of cylindrical coordinates. The flow corresponding to this combination of a source and a uniform stream is shown in Figure 12.44A.

An interesting characteristic of an inviscid flow is that any streamline may potentially define a solid surface. To understand why, recall that the no-penetration condition is satisfied on any streamline because fluid does not cross a streamline. The motion of fluid along the streamline is of no concern, since fluid is allowed to slip along a surface in an inviscid flow. Thus a streamline in inviscid flow may represent a solid surface. In Figure 12.44B we have highlighted a streamline that can be considered to define the front surface of a body that extends indefinitely far downstream. The pressure distribution on the body surface can be obtained by applying the Bernoulli equation along this streamline.

To obtain a body of finite length in a uniform stream, we must have a closed streamline. Since fluid inside a closed streamline cannot leave, to obtain a closed streamline we must ensure that the flow leaving a source has someplace nearby to go. This reasoning suggests that we add the potentials describing a uniform flow in the x direction and a source and sink located an equal distance up and downstream of the origin on the axis, as shown in Figure 12.45. Can you confirm that the resulting flow over a body known as a Rankine oval is given by

$$\psi = U_\infty r \sin\theta - \frac{m}{2\pi} \tan^{-1}\left(\frac{2dr \sin\theta}{r^2 - d^2}\right) \tag{12.52a}$$

in cylindrical coordinates? Alternately, we can express this in Cartesian coordinates as

$$\psi = U_\infty y - \frac{m}{2\pi} \tan^{-1}\left[\frac{2dy}{(x^2+y^2)-d^2}\right] \tag{12.52b}$$

776 | 12 ANALYSIS OF INCOMPRESSIBLE FLOW

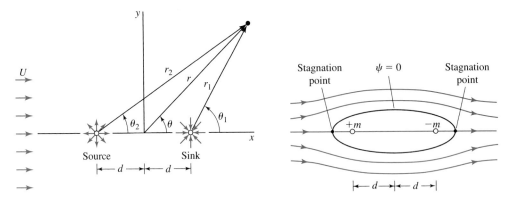

Figure 12.45 Superposition of a uniform flow, a source, and a sink to form the flow over a Rankine oval.

The three parameters in this flow are U_∞, d, and m. When combined into the single parameter $U_\infty d/m$, they control the shape of the Rankine oval. As the value of this parameter increases, the aspect ratio of the Rankine oval increases.

The flow over a Rankine oval at an angle of attack can be created by using a uniform stream that approaches the body at the desired angle. The streamfunction of the resulting flow can be found by replacing the leading terms in Eqs. 12.52a and 12.52b that describe the uniform stream by their counterparts for a uniform stream at an angle β. The result is

$$\psi = U_\infty r(\cos\beta \sin\theta - \sin\beta \cos\theta) - \frac{m}{2\pi}\tan^{-1}\left(\frac{2dr\sin\theta}{r^2 - d^2}\right) \quad (12.52c)$$

in cylindrical coordinates. In Cartesian coordinates the corresponding streamfunction is

$$\psi = U_\infty(y\cos\beta - x\sin\beta) - \frac{m}{2\pi}\tan^{-1}\left[\frac{2dy}{(x^2+y^2)-d^2}\right] \quad (12.52d)$$

EXAMPLE 12.7

Using superposition, combine the streamfunctions for a doublet and a uniform stream and describe the resulting flow.

SOLUTION

The streamfunction for a uniform flow is given by Eq. 12.36b as $\psi(r,\theta) = U_\infty r \sin\theta$, and the streamfunction for a doublet is given by Eq. 12.46 as $\psi = -(M/r)\sin\theta$. Thus, the combined streamfunction is

$$\psi = U_\infty r \sin\theta - \frac{M}{r}\sin\theta = U_\infty r \sin\theta\left(1 - \frac{M}{U_\infty r^2}\right)$$

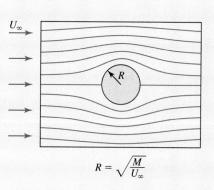

Figure 12.46 Superposition of a uniform flow and a doublet to form the flow over a cylinder. The radius of the cylinder is R.

$$R = \sqrt{\frac{M}{U_\infty}}$$

This streamfunction is plotted for $U_\infty = 5$, $M = 2$ and $r > \sqrt{m/U_\infty}$ in Figure 12.46. Note the closed streamline that forms a circle. Recalling that flow over a cylinder is given by Eq. 12.40:

$$\psi(r, \theta) = U_\infty r \sin\theta \left(1 - \frac{R^2}{r^2}\right)$$

we see that the radius of the cylinder is defined by $R^2 = M/U_\infty$. Thus we have obtained the solution for flow over a cylinder both by solving the potential equation and by using the superposition of a uniform flow and a doublet.

Many interesting flows can be created by the method of superposition. There are several websites that allow one to combine elementary flows and observe the resulting flow fields graphically. Although the superposition of elementary flows may not seem to be a useful approach to solving problems with complicated geometries such as those associated with aircraft, the computational approaches known as panel methods are widely used to model the inviscid flow over virtually any shape by combining a series of source–sink sheets called panels. Vortex panels are used to model lifting bodies. An example of the results obtained with panel models for an airfoil was shown in Figure 12.38. A more complicated panel mesh was shown in Figure 12.33.

12.5.4 Flow over a Cylinder with Circulation

In our analysis of steady uniform flow over a cylinder in Section 12.5.1, we discovered that the force of the fluid on the cylinder is zero. To obtain lift, there must be circulation, and our calculation showed that the circulation created by a doublet and uniform stream is zero. Can we make use of the elementary flow solution for a vortex to add circulation to the uniform flow over a cylinder? The answer is yes. Recall that the strength K of a line vortex is related to the circulation about a circular path around the vortex by $K = C/2\pi$, and that the streamlines of the vortex are concentric circles. Adding the vortex will provide circulation, and since its streamlines are concentric circles, it will not distort the cylindrical closed streamline created by a doublet and uniform stream.

12 ANALYSIS OF INCOMPRESSIBLE FLOW

Suppose we now add a vortex at the center of the cylinder as given by Eq. 12.44:

$$\phi = K\theta, \quad \psi = -K \ln r, \quad v_r = 0 \quad \text{and} \quad v_\theta = \frac{K}{r}$$

to the solution for uniform flow over a cylinder given by Eqs. 12.38, 12.39, and 12.40:

$$\phi = U_\infty r \cos\theta \left(1 + \frac{R^2}{r^2}\right), \quad \psi = U_\infty r \sin\theta \left(1 - \frac{R^2}{r^2}\right)$$

$$v_r = U_\infty \cos\theta \left(1 - \frac{R^2}{r^2}\right), \quad v_\theta = -U_\infty \sin\theta \left(1 + \frac{R^2}{r^2}\right)$$

The combined flow is then described by

$$\phi = U_\infty r \cos\theta \left(1 + \frac{R^2}{r^2}\right) + K\theta, \quad \psi = U_\infty r \sin\theta \left(1 - \frac{R^2}{r^2}\right) - K \ln r$$

$$v_r = U_\infty \cos\theta \left(1 - \frac{R^2}{r^2}\right), \quad v_\theta = -U_\infty \sin\theta \left(1 + \frac{R^2}{r^2}\right) + \frac{K}{r}$$
(12.53)

To check that a cylinder is present after adding the vortex, we look for a constant value of the streamfunction for some value of r. At $r = R$ the streamfunction is $\psi|_{r=R} = -K \ln R$, which is a constant. When the circulation is not present, the value of the streamfunction at $r = R$ is zero. Since the addition of a constant to a streamfunction has no impact on the velocity field, the fact that the streamfunction on the surface of the cylinder is nonzero is not a problem. The streamlines and velocity vectors for this flow are shown in Figure 12.47 for various values of the group $K/2U_\infty R$. This group, which is formed from the three available parameters U_∞, K, and R, can be thought of as measuring the relative contribution of the vortex to the velocity potential. Notice that although the flow has fore and aft symmetry, it is no longer symmetric up and down because of the circulation.

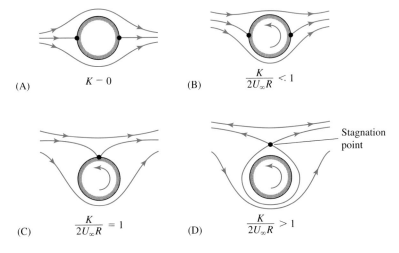

Figure 12.47 Streamlines and stagnation points around a circular cylinder with various levels of circulation.

(A) $K=0$

(B) $\dfrac{K}{2U_\infty R} < 1$

(C) $\dfrac{K}{2U_\infty R} = 1$

(D) $\dfrac{K}{2U_\infty R} > 1$ — Stagnation point

12.5 INVISCID IRROTATIONAL FLOW

To determine the pressure distribution, we use the Bernoulli equation in the form of Eq. 12.31b, with the first point far upstream of the cylinder in the uniform flow where the pressure is equal to p_∞, and $\frac{1}{2}(v_r^2 + v_\theta^2)|_{r=\infty} = \frac{1}{2}U_\infty^2$. Substituting the velocity components from Eq. 12.53 and rearranging, we find the pressure distribution to be

$$p(r,\theta) = p_\infty - \frac{1}{2}\rho U_\infty^2 \left\{ \frac{R^4}{r^4} + \frac{R^2}{r^2}\left[2(\sin^2\theta - \cos^2\theta) + \left(\frac{K^2}{U_\infty^2 r^2}\right)\right] - 2\left(\frac{K}{U_\infty R}\right)\sin\theta\left(\frac{R}{r} + \frac{R^3}{r^3}\right)\right\}$$
(12.54a)

On the surface of the cylinder, $r = R$, the pressure distribution is given by

$$p(R,\theta) = p_\infty + \frac{1}{2}\rho U_\infty^2\left[1 - 4\sin^2\theta - \left(\frac{K^2}{U_\infty^2 R^2}\right) + 4\left(\frac{K}{U_\infty R}\right)\sin\theta\right] \quad (12.54b)$$

The pressure distribution on the surface of the cylinder is shown in Figure 12.48 for $K/U_\infty R = 0.5$. It is evident that the average pressure is higher on the upper half of the cylinder than on the lower half, but the same on the front and the back.

The circulation $C(t)$ around the cylinder surface can be calculated by applying Eq. 10.54: $C(t) = \oint \mathbf{u} \cdot d\mathbf{r}$, and again noting that for a circular path of radius R the line integral is $C(t) = \int_0^{2\pi} v_\theta R\, d\theta$. The value of v_θ on the cylinder surface can be found from Eq. 12.51 to be given by

$$v_\theta|_{r=R} = -U_\infty \sin\theta\left(1 + \frac{R^2}{r^2}\right)\bigg|_{r=R} + \frac{K}{r}\bigg|_{r=R} = -2U_\infty \sin\theta + \frac{K}{R}$$

The circulation around the cylinder is thus found to be

$$C(t) = \int_0^{2\pi}\left(-2U_\infty \sin\theta + \frac{K}{R}\right)R\, d\theta = 2\pi K$$

We see that this is the same as the circulation calculated for the vortex, given earlier by Eq. 12.45. According to Eq. 10.56, the lift produced by a long cylinder of length L is related to the circulation around the cylinder by $F_L = -\rho U C L$. For the cylinder with circulation $C = 2\pi K$ and a uniform stream at speed U_∞, this formula predicts that the lift is

$$F_L = -2\pi \rho U_\infty K L \quad (12.55)$$

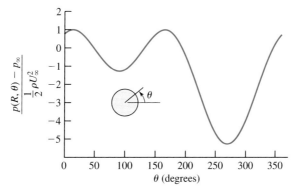

Figure 12.48 Pressure distribution on the surface of a cylinder for $K/U_\infty R = 0.5$. Note that a scaled pressure of 0 corresponds to an actual surface pressure equal to P_∞.

The negative value means the force acts in the negative y direction (for $K > 0$). Note that this is consistent with the action of the pressure distribution on the surface of the cylinder.

We can check this prediction by calculating the force applied by the fluid to the cylinder using Eqs. 12.42 that we derived previously for flow over a cylinder:

$$F_{Sx} = \int_0^L \int_0^{2\pi} -p(R,\theta)\cos\theta\, R\, d\theta\, dz$$

and

$$F_{Sy} = \int_0^L \int_0^{2\pi} -p(R,\theta)\sin\theta\, R\, d\theta\, dz$$

Substituting the pressure on the surface as given by Eq. 12.54b and integrating, we find

$$F_{Sx} = 0$$

and

$$F_{Sy} = LR\int_0^{2\pi} -\left\{p_\infty + \frac{1}{2}\rho U_\infty^2\left[1 - 4\sin^2\theta - \left(\frac{K^2}{U_\infty^2 R^2}\right) + 4\left(\frac{K}{U_\infty R}\right)\sin\theta\right]\right\}\sin\theta\, d\theta$$

Dividing the second integral into three parts and integrating, we have

$$-LR\left\{p_\infty + \frac{1}{2}\rho U_\infty^2\left[1 - \left(\frac{K^2}{U_\infty^2 R^2}\right)\right]\right\}\int_0^{2\pi}\sin\theta\, d\theta = 0$$

$$-LR\left(\frac{1}{2}\rho U_\infty^2\right)4\int_0^{2\pi}\sin^3\theta\, d\theta = 0$$

$$-LR\left(\frac{1}{2}\rho U_\infty^2\right)4\left(\frac{K}{U_\infty R}\right)\int_0^{2\pi}\sin\theta^2\, d\theta = -LR\left(\frac{1}{2}\rho U_\infty^2\right)4\left(\frac{K}{U_\infty R}\right)\pi$$

$$= -2\pi\rho U_\infty KL$$

Thus we see that lift can be obtained either by using the circulation formula or by the direct integration of the pressure on the surface of the cylinder.

12.6 SUMMARY

In this chapter we have developed solutions to the governing equations used to describe incompressible flows. We began by applying the continuity and Navier–Stokes equations to analyze steady, laminar flow problems involving a constant density, constant viscosity fluid. The general procedure to obtain solutions to these flow problems involves the following steps:

1. Select an appropriate coordinate system based on the flow geometry.
2. List the corresponding forms of the continuity and Navier–Stokes equations.

3. Simplify these equations by noting relevant aspects of the velocity and pressure fields.
4. Determine the appropriate form of the no-slip, no-penetration boundary conditions at the relevant solid surfaces.
5. Solve the simplified set of governing equations with the appropriate BCs to obtain expressions for the velocity and pressure fields.

An example of this type of flow is plane Couette flow, which results from the steady motion of a flat plate parallel to another stationary plate with the gap between the plates filled with fluid. When we assumed that the pressure between the plates was uniform, we found that there was a linear velocity profile between the plates and that the shear stress in the fluid was constant. Circular Couette flow is characterized by two concentric cylinders with a thin layer of fluid between them. The resulting velocity field involves only a single component of velocity in the θ direction, but the magnitude of v_θ is a function of the radial coordinate. The pressure and shear stress distributions in this flow are also functions of r.

In Poiseuille flow between parallel plates, the physical arrangement is identical to that of plane Couette flow but the plates are both at rest, and the flow is driven by a pressure difference in the x direction. The result is a parabolic velocity profile, a pressure distribution that depends only on x (with a constant pressure gradient), and a shear stress distribution that varies linearly across the gap. Pressure-driven flow in a round pipe, known as Poiseuille flow, also exhibits a parabolic velocity profile and a constant pressure gradient in the flow direction.

The flow over a cylinder changes from a steady, laminar flow at low Re, to an unsteady, laminar flow at some intermediate value of Re, and finally becomes turbulent. No analytical solution exists for low Re steady flow over a cylinder. Therefore, we developed a computational fluid dynamics (CFD) solution for the problem of the 2D steady flow of a uniform stream over a cylinder using a commercial CFD code. The CFD code solved the finite element problem corresponding to the original governing equations and appropriate BCs and returned a data set describing the distributions of velocity and pressure on all the nodes of the finite element mesh.

To analyze unsteady flows we must solve the appropriate set of four partial differential equations (continuity and three components of Navier–Stokes) along with the associated BCs and an appropriate initial condition. An example of an unsteady flow is the startup phase of plane Couette flow.

The random-looking behavior of a turbulent flow suggests that it is useful to consider each of the flow variables to consist of a mean and fluctuating component. To derive the governing equations for a turbulent flow, one writes each flow variable as the sum of a mean and fluctuating component, inserting the sums into each governing equation and BC, then averaging each equation, and employing the averaging rules to evaluate each term. The result is the Reynolds equations for the mean flow variables. These equations resemble the Navier–Stokes equations written in the mean flow variables with nine additional terms on the right-hand sides. The presence of nine new unknowns and only four governing equations constitutes the closure problem in turbulence. To solve this problem, a turbulence model must be used to generate additional equations that relate the Reynolds stresses to the mean flow variables.

The analytical solution to an inviscid flow problem is developed as follows. First, we solve Laplace's equation with the no-penetration boundary condition on a solid surface to obtain the velocity potential $\phi(\mathbf{x}, t)$. Next we obtain the velocity field using $\mathbf{u} = \nabla\phi$. Finally, we use the known velocity field to obtain the pressure distribution from the Bernoulli equation. An important concept in an inviscid flow is the stagnation point, defined to be a point in the flow or on the surface of a body at which the velocity is zero (and the pressure is a maximum). In flows over bodies, stagnation points are usually found near the nose of the body and near the tail.

Linear combinations of solutions of Laplace's equation are also valid solutions, so we can combine velocity potentials for elementary flows into a potential describing a more complex flow field. Since velocity potential and streamfunction alike satisfy Laplace's equation, we can use this principle of superposition to add potentials as well as the corresponding streamfunctions. The velocity components also add together; however, in most cases it is best to determine these after the overall potential and streamfunction have been determined. The pressure distribution of the combined flow is not the sum of the pressure distributions of the elementary flows, rather the pressure must always be determined after the new velocity components have been found. This is because the Bernoulli equation involves the sum of the squares of the velocity components. An example of the use of superposition is the modeling of a flow over a body by combining a uniform flow and a source. To obtain a body of finite length in a uniform stream we must add the potentials describing a uniform flow, a source, and sink to create what is known as a Rankine oval. If we add a vortex to the model for a Rankine oval, we can obtain a nonzero circulation that in turn permits the incorporation of lift in the model.

PROBLEMS

Section 12.2

12.1 Given laminar, steady flow between parallel plates (total gap width of 4 mm) of oil, $\mu = 0.6$ (N-s)/m², with a pressure gradient of -1000 (N/m²)/m, find the volume flowrate per unit width and the shear stress on the upper plate.

12.2 Redo Problem 12.1 with the top plate moving at 0.5 m/s.

12.3 Redo Problem 12.1 with the top plate moving at 0.5 m/s and the bottom plate moving at –0.5 m/s.

12.4 Redo Problem 12.1 with zero pressure gradient and the top plate moving at 0.5 m/s.

12.5 Redo Problem 12.1 with zero pressure gradient, the top plate moving at 0.5 m/s, and the bottom plate moving at –0.5 m/s.

12.6 Given laminar, steady flow between parallel plates (total gap width of 0.05 in.) of oil, $\mu = 0.015$ (lb$_f$-s)/ft² with a pressure gradient of -10 lb$_f$/ft²/ft, find the volume flowrate per unit width and the shear stress on the upper plate.

12.7 Redo Problem 12.6 with the top plate moving at 0.5 ft/s.

12.8 Redo Problem 12.6 with the top plate moving at 0.5 ft/s and the bottom plate moving at –0.5 ft/s.

12.9 Redo Problem 12.6 with zero pressure gradient and the top plate moving at 0.5 ft/s.

12.10 Redo Problem 12.6 with zero pressure gradient, the top plate moving at 0.5 ft/s, and the bottom plate moving at –0.5 ft/s.

12.11 A laminar, steady flow between parallel plates (total gap width of 10 mm) of oil,

$\mu = 0.5$ (N-s)/m², has an average velocity of 0.5 m/s. Find the pressure drop per unit length in the direction of flow and the maximum velocity.

12.12 Redo Problem 12.11 with the top plate moving at 0.5 m/s.

12.13 Redo Problem 12.11 with the top plate moving at 0.5 m/s and the bottom plate moving at −0.5 m/s.

12.14 The jack at an auto repair shop can support 50,000 N. The hydraulic fluid has a viscosity μ of 0.2 (N-s)/m². The diameter of the piston is 0.25 m, the piston–cylinder gap width is 1 mm, and the piston in the cylinder is 0.25 m long when fully extended. Estimate the leakage of fluid past the piston.

12.15 The jack in Problem 12.14 is overloaded by 50,000 N. Recalculate the leakage rate and determine how fast the jack will descend.

12.16 It is proposed that a 0.15 m diameter piston of a stronger material replace the one described in Problem 12.14. Everything else being the same, what will be the new leakage rate? Compare the two rates.

12.17 The flow in the inlet between parallel plates shown in Figure P12.1 is uniform with $U = 3$ cm/s. Downstream the flow becomes fully developed. If the fluid is glycerin at 20°C and the gap is 1 cm, what is the maximum velocity?

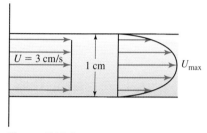

Figure P12.1

12.18 For the flow described in Problem 12.17, find the volume flowrate per unit width and the shear stress on the lower plate.

12.19 Airflows through a heat exchanger made up of 11 plates that form 10 channels as shown in Figure P12.2. The plates are 4 m long and 2 m wide, and each gap is 5 cm. If the total flowrate is 100 m³/min, what is the pressure drop across the heat exchanger and what is the power required to move the air through it?

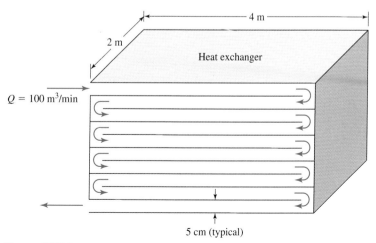

Figure P12.2

12 ANALYSIS OF INCOMPRESSIBLE FLOW

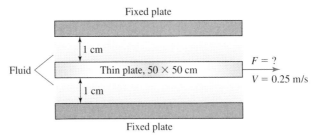

Figure P12.3

12.20 A thin plate is located between two fixed plates as shown in Figure P12.3. The gap is filled with fluid of viscosity $\mu = 0.055$ kg/(m-s). What is the force needed to pull the plate at a steady velocity of 0.25 m/s if the size of the plate is 50 × 50 cm?

12.21 Redo Problem 12.20 assuming a pressure gradient of -10 N/m²/m is present.

12.22 A 50 mm diameter shaft has a 0.20 mm gap from a bearing as shown in Figure P12.4. If the shaft rotates as 100 rpm and the oil has a viscosity μ of 0.18 (N-s)/m², estimate the torque required to overcome viscous friction. What is the power requirement?

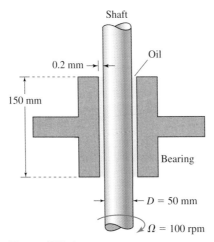

Figure P12.4

12.23 For the shaft and bearing described in Problem 12.22 the angular velocity is doubled. What will now be the torque required to overcome viscous friction? How much more power is required?

12.24 For the shaft and bearing described in Problem 12.22 the length of the bearing must be increased to 300 mm. What will now be the torque required to overcome viscous friction? How much more power is required?

12.25 For the shaft and bearing described in Problem 12.22 estimate the torque required to overcome viscous friction, assuming that this flow can be modeled as flow between infinite parallel plates.

12.26 A journal bearing made up of two concentric cylinders 24 and 25 mm in diameter, and 80 mm long, rotates at 2000 rpm. The torque needed to turn the bearing is 0.15 N-m. What is the viscosity of the lubricating oil in the gap?

12.27 For the bearing described in Problem 12.26 SAE 30 oil is now used as the lubricant. What is the torque needed to turn the bearing?

12.28 A viscometer is designed by using a reservoir and capillary tube as shown in Figure P12.5. Develop an expression for the viscosity as a function of the volume flowrate, the density, and the geometry (H, D, L). Explain any shortcomings of the design.

12.29 A capillary tube 0.35 mm in diameter 0.4 m long is used in the construction of the viscometer described in Problem 12.28. H is maintained at 1 m. What will be the volume flowrate if the fluid is gasoline? Comment on

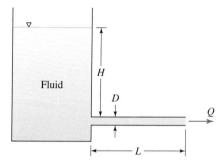

Figure P12.5

any difficulties that might be anticipated if the flowrate is measured by collecting a known volume of fluid leaving the tube.

12.30 Redo Problem 12.29 with SAE 30 oil at 20°C.

12.31 The device described in Problems 12.28 and 12.29 is being used to test a fluid that has a density of 875 kg/m^3. The measured volume flowrate is 75 L/min. What is the viscosity of the fluid? What is the uncertainty in viscosity if the uncertainty of the flowrate measurement is ±0.5 L/min?

12.32 Fuel oil with kinematic viscosity of 4.44×10^{-3} ft^2/s and specific gravity of 0.918 flows through 3000 ft of horizontal 6 in. pipe. If the pressure drop is 150 psi, what is the flowrate?

12.33 A fluid with kinematic viscosity of 2.24×10^{-3} ft^2/s and specific gravity of 0.911 must be carried a horizontal distance of 1000 ft from one side of a factory to another at the rate of 0.7 ft^3/s. If the pressure change is not to exceed 8.0 psi, what size pipe should be used?

12.34 What force is required to push 10.0 mm^3/s of saline solution through a hypodermic needle with diameter 0.1 mm and length of 25 mm? The plunger diameter is 10 mm, and the viscosity of saline is 5 percent greater than that of water.

12.35 For the hypodermic needle described in Problem 12.34 what is the greatest speed that the plunger can be withdrawn to suck saline into the syringe? What force is required?

12.36 Derive an expression between the tube diameter, Reynolds number, kinematic viscosity, and g for a steady, laminar flow in a vertical tube open at each end to the atmosphere.

12.37 Flowrate is measured through the use of the manometer shown in Figure P12.6. Derive an expression for the flowrate as a function of H.

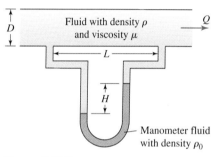

Figure P12.6

12.38 If the liquid flowing in the tube described in Problem 12.37 is SAE 50w oil, the fluid in the manometer is water, $L = 2$ m, $D = 3$ mm, and $H = 2$ mm, what is the flowrate?

12.39 A fluid with viscosity of 9.0×10^{-5}(lb$_f$-s)/ft^2 and density of 1.6 slugs/ft^3 flows steadily with an average velocity of 0.25 ft/s through a 1.5 in. diameter pipe. What is the pressure drop per foot of pipe?

12.40 Water at 20°C flowing in a 5 cm diameter tube at $Re = 50$ is used to model blood flow in a 0.5 cm artery. The flow parameters in the model are selected so that the wall shear stress in the model matches that in the artery. The apparent (blood is non-Newtonian) viscosity of blood is 1.25 times that of water. Derive an expression for the shear stress in the artery based on the model parameters. What is the shear stress in this case? What blood velocity in the artery corresponds to this same shear stress?

12.41 A novelty company is to produce a giant straw. What is the maximum length possible to draw up a drink in laminar flow? Assume the drink has the properties of water and the diameter is 1 cm.

12.42 An air straightening device consists of a hundred 5 mm diameter tubes, each 40 cm long, as shown in Figure P12.7. Estimate the pressure drop across the device for air with an average velocity of 5 m/s.

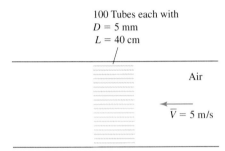

Figure P12.7

12.43 For the annular geometry shown in Figure P12.8 write the differential equation and boundary conditions to solve for the laminar velocity distribution.

Section 12.4

12.44 In a turbulent flow the following velocity data was recorded at 0.5 s intervals:

u (m/s) 25 38 55 10 −6 17 28 −15 28 47

v (m/s)   5 10 −2 −6 19   0 -15   3   7 11

Calculate $\bar{u}, \bar{v}, u', v'$, and $\overline{u'v'}$.

12.45 What is the closure problem in turbulence and how is it typically deal with?

12.46 The following velocity measurements were taken from a traverse in a 20 cm diameter pipe.

r (cm)	$\bar{v}_z$ (m/s)	r (cm)	$\bar{v}_z$ (m/s)
9.95	1.45	8.50	4.91
9.90	2.42	8.00	5.09
9.85	2.99	7.50	5.22
9.80	3.51	7.00	5.43
9.75	3.71	6.50	5.53
9.50	4.12	0.00	6.49
9.00	4.51		

Plot the velocity profile and determine the volume flowrate and mean velocity in the pipe.

12.47 Use the data provided in Problem 12.46 to calculate the wall shear stress in the pipe.

Section 12.5

12.48 The velocity potential for a flow is $\phi = -(x + y/2)$. Find the streamfunction for this flow.

12.49 Find the difference in pressure between points (1, 1, 0) and (2, 2, 0) for the flow described in Problem 12.48. Assume elevation change is zero, the fluid is air at 20°C, the dimensions are in meters, and the units of ϕ are square meters per second.

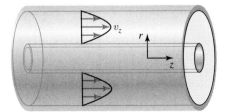

Figure P12.8

12.50 For the flow described in Problem 12.48 the pressure is 100 kPa at point (0, 0, 0). What is the pressure at point (3, 2, 0)? Assume elevation change is zero, the fluid is air at 20°C, the dimensions are in meters, and the units of ϕ are square meters per second.

12.51 The velocity potential for a flow is $\phi = 3x^3 - 9xy^2$. Find the streamfunction for this flow.

12.52 Find the difference in pressure between points (0, 0, 0) and (1, 2, 0) for the flow described in Problem 12.51. Assume elevation change is zero, the fluid is water, the dimensions are in feet, and the units of ϕ are square feet per second.

12.53 For the flow described in Problem 12.51 the pressure is 10 psig at point (1, 0, 0). What is the pressure at point (2, 2, 0)? Assume elevation change is zero, the fluid is water, the dimensions are in feet, and the units of ϕ are square feet per second.

12.54 The streamfunction for a flow is given by $\psi = 2x^2 - 2y^2$. Find the velocity potential for this flow.

12.55 Show that the Bernoulli equation can be applied between any two points in the flow described in Problem 12.54.

12.56 The velocity field for a flow is given by $\mathbf{u} = x(s^{-1})\mathbf{i} - y(s^{-1})\mathbf{j} + 3(m/s)\mathbf{k}$ where the coordinates are measured in meters. Neglecting gravity, find the pressure difference between (0, 0, 0) and (1, 2, 0) if possible.

12.57 Determine whether the Bernoulli equation can be applied between points $(r, \theta) = (1, \pi/6)$, and $(2, \pi/4)$ for the flow $\mathbf{u} = (C/2\pi r)\mathbf{e}_\theta$, where C is a constant.

12.58 Use the velocity potential and streamfunction to show that the velocity distribution for a source or sink is as given in Eq. 12.43.

12.59 Use the velocity potential and streamfunction to show that the velocity distribution for a line vortex is as given in Eq. 12.44.

12.60 Use the velocity potential and streamfunction to show that the velocity distribution for a doublet is as given in Eq. 12.46.

12.61 Combine a sink of strength $m = 8000$ ft^2/s and line vortex of strength $K = 17{,}000$ ft^2/s. Determine the streamfunction and sketch the streamlines. Note that this is a simplistic model of a tornado. What is the pressure at the center?

12.62 A model of a Rankine body is placed in a wind tunnel. If the model is 1 m thick, has its source and sink each located a distance 0.75 m from the origin, and is exposed to an upstream velocity of 25 m/s, what is the equivalent source strength?

12.63 Find the streamfunction and sketch some streamlines for the combination of two sinks of strength $-m$ at $(-d, 0, 0)$ and $-3m$ at $(+d, 0, 0)$.

12.64 Estimate the velocity of water flow over a 10 cm diameter cylinder for which cavitation begins to occur? Assume the upstream pressure is atmosphere.

12.65 Air at 40°C flows at 10 m/s over a porous plate. Air is also injected into the flow through the porous plate. Model this as a uniform flow and with sources of strength $m = 0.1$ m^2/s, 10 cm apart, as shown in Figure P12.9. Use the streamfunction to calculate the volume flowrate through a surface 1 m above, and parallel to the plate and of unit depth.

12.66 For the flow described in Problem 12.65 find the streamfunction, velocity potential, and velocity field.

12.67 A Rankine body is formed with a source and sink, 5 m apart, each of strength 10 m^2/s, and a uniform flow of 5 m/s as shown in Figure P12.10. Find the length L and thickness t of the body.

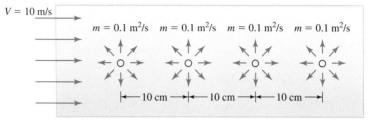

Figure P12.9

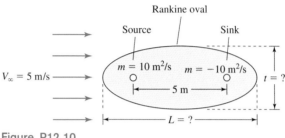

Figure P12.10

12.68 For the flow described in Problem 12.67 find the streamfunction, velocity potential, and velocity field.

12.69 Add the stagnation point flow $\psi = Kxy$ to a source at the origin. Determine the location of the stagnation point for this new flow. This is equivalent to adding a bump to the wall of the stagnation flow.

12.70 Find the streamfunction, velocity potential, and velocity field for a vortex pair located equidistant, d, from the x axis.

12.71 What is the volume flowrate in the y direction per unit depth between $(-d, 0)$ and $(+d, 0)$ for the flow described in Problem 12.70?

12.72 Model the flow into a drain as a line vortex as shown in Figure P12.11. If the velocity at $r = 5$ in. is 10 in./s, what is the velocity at $r = 3$ in.?

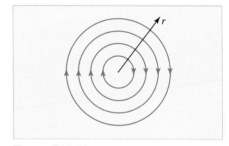

Figure P12.11

12.73 Show that the streamfunction and velocity potential for uniform flow and the doublet satisfy Laplace's equation.

12.74 Show that the streamfunction and velocity potential for the point source flow and line vortex satisfy Laplace's equation.

PART 3
APPLICATIONS

13 FLOW IN PIPES AND DUCTS

13.1 Introduction
13.2 Steady, Fully Developed Flow in a Pipe or Duct
    13.2.1 Major Head Loss
    13.2.2 Friction Factor
    13.2.3 Friction Factors in Laminar Flow
    13.2.4 Friction Factors in Turbulent Flow
13.3 Analysis of Flow in Single Path Pipe and Duct Systems
    13.3.1 Minor Head Loss
    13.3.2 Pump and Turbine Head
    13.3.3 Examples
13.4 Analysis of Flow in Multiple Path Pipe and Duct Systems
13.5 Elements of Pipe and Duct System Design
    13.5.1 Pump and Fan Selection
13.6 Summary
Problems

13.1 INTRODUCTION

 CD/Video library/Flow in the human air duct

One of the challenges for mechanical, civil, and chemical engineers is to design systems in which fluids are pumped through pipes and ducts. The Alaska Pipeline, shown in Figure 13.1, is an example of such a system. Water supply, natural gas, sanitary, and storm sewer systems are part of the critical infrastructure in every city and town, and more complex systems are found in power plants, refineries, and other manufacturing

13 FLOW IN PIPES AND DUCTS

Figure 13.1 The Alaska Pipeline.

environments. The Roman aqueducts and many ancient irrigation systems represent remarkable engineering achievements, as do the modern heating, ventilation, and air-conditioning systems that keep us comfortable year around in our homes, vehicles, and workplaces. Vehicles of all types incorporate pipe and duct systems for the transport of fuel, coolant, refrigerant, hydraulic fluids, and combustion products. Additional examples of successful designs involving flow in pipes and ducts are found in our bodies and those of other animals and plants. Thus, it is safe to say that the material in this chapter is of great practical importance in many fields of engineering, as well as in biology and medicine.

In this text, we describe a round pipe as simply a pipe and employ the word "duct" to describe a fluid passage of any other cross section. Pipe and duct flows are characterized as internal flows because the flow path is completely surrounded by solid surfaces. In the next chapter we consider external flows. These are flows in which the flow path is bounded on one side by a solid surface. Thus, the flows over the external surfaces of airplanes, ships, cars, and sports balls are examples of external flows, as are the flows over buildings and other natural and man-made structures.

Our discussion of flow in pipes and ducts began with three case studies in Chapter 3. Recall that in Section 3.3.1 (flow in a round pipe), we showed how to use a friction factor to calculate the frictional pressure drop in a horizontal pipe. In the next case study, on flow through an area change, we examined the frictional losses that occur when the diameter of a pipe or duct is changed. A key consideration in the design of any pipe or duct system is the selection of a fan or pump of the appropriate size and type. This motivated our discussion of the fan and pump laws in Section 3.3.3. Additional insight was gained when we used control volume analysis to study the flow in a round pipe in Chapter 7, and also when we applied the Bernoulli equation to deduce some characteristics of flow through an area change in Chapter 8. You will also find the analytical solution for laminar flow in a round pipe obtained in Chapter 12 to be valuable background material in your study of flow in pipe and ducts.

In this chapter we provide a comprehensive look at the methods used to analyze steady, constant density flow in a pipe or duct, as well as in more complex systems that incorporate these and other components. In most systems of interest, turbulent flow is the rule, since a laminar flow at the same flowrate would require much larger and more costly components. Thus, this chapter's focus is on turbulent flow, but we do provide sufficient information to allow you to analyze laminar flow. Unsteady and compressible

13.2 STEADY, FULLY DEVELOPED FLOW IN A PIPE OR DUCT

flows in pipes and ducts, although also of engineering interest, are not covered in this text.

Our discussion begins with an analysis by means of mass, momentum, and energy balances of the steady, fully developed flow of a constant density fluid in a single pipe or duct. Our goal is to determine the relationship between pressure drop Δp and volume flowrate Q, since this must be known to predict the power required to move fluid through the pipe or duct. We introduce the concepts of major head loss, friction factor, and hydraulic diameter, and we show how to employ these concepts to analyze laminar and turbulent flow. In the next section we apply mass and energy balances to analyze flow in a single path system. The system may include a pump or fan, any number of pipes or ducts of different diameters and lengths, and other components such as inlets, exits, valves, and elbows. Next, we extend the analysis to multiple path systems, showing how concepts introduced earlier may be applied to solve problems of this type. We conclude with a brief section introducing elements of pipe and duct system design. Our goal throughout this chapter is to show you how to analyze the pipe and duct flows encountered in engineering practice using nothing more than a calculator and charts. Although today's engineer will generally have access to a commercial code for pipe and duct analysis and design, learning to do the analysis by hand will help you develop a sound understanding of the fundamentals of pipe and duct flows.

13.2 STEADY, FULLY DEVELOPED FLOW IN A PIPE OR DUCT

A straight, constant area segment of a pipe or duct is the basic building block of any flow system. Consider the system shown in Figure 13.2 in which a pump supplies fluid to a single flow path consisting of various lengths of pipe joined together by elbows. In Chapter 7 we defined the mechanical energy content of a fluid (per unit mass) as the sum

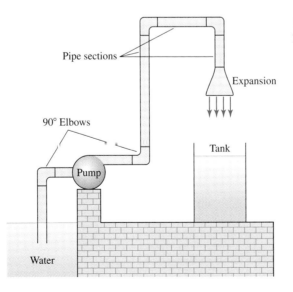

Figure 13.2 Schematic of single path piping system.

of the pressure potential, kinetic, and gravitational potential energies (per unit mass). The mechanical energy represents the useful or available energy content of a fluid, since it is possible to extract this energy by a device to perform useful work. As fluid moves along the flow path shown in Figure 13.2, friction causes a loss of mechanical energy content. As you probably recall, this is referred to as the head loss. The total head loss is further divided into major and minor head loss. The major loss is defined to be the loss incurred in steady, fully developed flow through a straight, constant area segment of pipe or duct. The additional losses from the other components in a system, such as the elbows in Figure 13.2, are called minor losses. In this section our goal is to determine the major head loss and the relationship between pressure drop and volume flowrate. Minor losses are discussed in Section 13.3.

 CD/Video library/Pipe flow

Before we begin our analysis of the major head loss in a steady, fully developed flow, we must recognize that in many instances, the flow in a straight segment of pipe or duct is not fully developed. Consider the situation shown in Figure 13.3, in which a uniform flow enters a pipe and gradually becomes fully developed. For a laminar flow, with $Re < 2300$, the length of pipe L required before a parabolic profile is achieved in a round pipe of diameter D may be estimated as

$$\frac{L}{D} = 0.06\, Re$$

where Re is based on the pipe diameter and average velocity. Thus fully developed flow is achieved in a length of one diameter at $Re = 16$; at $Re = 2000$, however, it requires a length of at least 100 diameters. For a turbulent flow, the length of pipe required to achieve fully developed flow is given by

$$\frac{L}{D} = 4.4\, Re^{1/6}$$

This formula gives the required length to achieve fully developed turbulent flow at the Reynolds numbers normally encountered as 25 to 40 diameters.

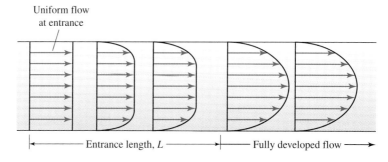

Figure 13.3 Developing flow in a pipe. Compare this sketch to the flow visualization shown in Figure 12.20.

13.2 STEADY, FULLY DEVELOPED FLOW IN A PIPE OR DUCT

The requirement for a certain number of pipe diameters to ensure the reappearance of a fully developed velocity profile after any source of minor head loss can be important. Most fans, pumps, flow meters, heat exchangers, and filters are designed to function optimally with fully developed or uniform velocity profiles at their inlets. If this is not the case, performance of the device may be degraded. Thus, providing a sufficient length of straight pipe in front of devices of these types can be an important part of the design process for a pipe or duct system.

Pipe or duct inlets are not the only locations in a flow system at which we can observe flow that is not fully developed. In fact, the velocity profile changes, and fully developed flow is gradually reestablished after all sources of minor head loss, i.e., downstream of every component in a system. Nevertheless, experience shows that we can ignore this effect in a major head loss calculation and still obtain sufficient accuracy for engineering purposes. Regions of non–fully developed flow that cannot be ignored, such as those at inlets and area change, are accounted for with an appropriate minor loss calculation as will be explained in Section 13.3.

Now consider the steady, fully developed flow of a constant density fluid in a straight, constant area section of a pipe or duct inclined at angle θ as shown in Figure 13.4. Since the flow is fully developed, the velocity profiles at all cross sections along the passage are identical. We can apply mass, momentum, and energy balances to the CV shown in Figure 13.4 to determine the flow characteristics. Since the density is constant, applying a mass balance derived from Eq. 7.15 gives $A_1 \bar{V}_1 = A_2 \bar{V}_2$. The areas at the inlet and exit of the CV are equal and also equal to the area A at any other cross section along the passage. Therefore, the average velocities at each end of the CV are the same (and equal to the average velocity $\bar{V}$ at any section along the passage), i.e.,

$$\bar{V}_1 = \bar{V}_2 \tag{13.1}$$

A momentum balance in the flow direction was given for a horizontal round pipe in Example 7.7. Here we must allow for a noncircular cross section and the fact that the pipe or duct is inclined. Following the approach of Example 7.7, for a steady constant density flow, we apply the momentum balance as given in vector form by Eq. 7.19b:

$$\int_{CS} (\rho \mathbf{u})(\mathbf{u} \cdot \mathbf{n}) \, dS = \int_{CV} \rho \mathbf{f} \, dV + \int_{CS} \mathbf{\Sigma} \, dS$$

Consider the component of this equation in the flow direction. To evaluate the flux terms at the inlet and exit, we note that in a fully developed flow the velocity profiles at the

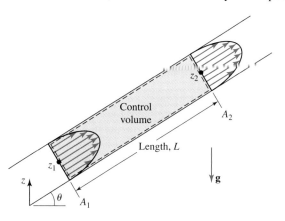

Figure 13.4 Schematic of fully developed flow in a pipe oriented at a nonhorizontal angle.

inlet and exit of the CV are the same. The net momentum transport in the flow direction is therefore zero, since the momentum transport at the inlet is balanced by the momentum transport at the exit.

From Figure 13.4, the gravitational body force acting in the flow direction is seen to be given by

$$\int_{CV} \rho(-g\sin\theta)\,dV = -\rho g\sin\theta \int_{CV} dV = -\rho g\sin\theta(AL)$$

where AL is the volume of the CV. Since $L\sin\theta = z_2 - z_1$, we can write the body force acting in the flow direction as

$$-\rho g\sin\theta(AL) = -(z_2 - z_1)\rho g A$$

The surface force acting in the flow direction consists of a contribution from the pressures acting at each end of the CV, $p_1 A - p_2 A$, and a contribution from the shear force exerted by the wall on the fluid inside the CV. To determine the latter, we note that the shear force exerted by the fluid on the wall is $\bar{\tau}_{wall} A_{wall}$, where $\bar{\tau}_{wall}$ is the average shear stress and A_{wall} is the area of the wall in contact with the fluid. By applying the principle of action–reaction, we find that the shear force exerted by the wall on the fluid inside the CV is $-\bar{\tau}_{wall} A_{wall}$, and the total surface force in the flow direction is thus given by $p_1 A - p_2 A - \bar{\tau}_{wall} A_{wall}$. The momentum balance in the flow direction is thus given by

$$0 = -(z_2 - z_1)\rho g A + p_1 A - p_2 A - \bar{\tau}_{wall} A_{wall}$$

After rearranging we have

$$p_1 - p_2 = \rho g(z_2 - z_1) + \frac{\bar{\tau}_{wall} A_{wall}}{A} \tag{13.2}$$

This result shows that the pressure difference in steady, fully developed flow in an inclined pipe or duct of a specified length and cross section is determined by the difference in elevation of each end of the passage, and viscous friction (in the form of the wall shear stress). Conversely, Eq. 13.2 shows that the average wall shear stress can be determined by measuring the pressure drop and the difference in elevation of the two ends of a pipe.

To continue with our analysis of pipe and duct flow, we can also write the momentum balance, Eq. 13.2, as:

$$\left(\frac{p_1}{\rho} + gz_1\right) - \left(\frac{p_2}{\rho} + gz_2\right) = \frac{\bar{\tau}_{wall} A_{wall}}{\rho A} \tag{13.3}$$

and recall that the mechanical energy content of the fluid at any point along a pipe or duct is given by $p/\rho + \frac{1}{2}\bar{V}^2 + gz$. Thus the change in the mechanical energy content of the fluid as it passes from point 1 to point 2 along the pipe or duct is given by $(p_1/\rho + \frac{1}{2}\bar{V}_1^2 + gz_1) - (p_2/\rho + \frac{1}{2}\bar{V}_2^2 + gz_2)$. In a fully developed flow there cannot be a change in the kinetic energy of the fluid. Thus the change in the mechanical energy of the fluid in a fully developed flow is given solely by $(p_1/\rho + gz_1) - (p_2/\rho + gz_2)$. From the momentum balance, Eq. 13.3, we see that this loss of mechanical energy may be attributed to friction in the form of the wall shear stress.

EXAMPLE 13.1

Water at 80°F flows through a 40 ft long, 6 in. diameter vertical pipe as shown in Figure 13.5. The pressure difference measured between points A and B is found to be $p_A - p_B = 21.8$ psia. Find the direction of flow and average wall shear stress.

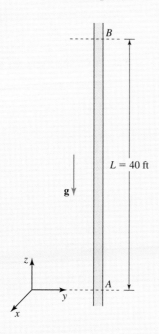

Figure 13.5 Schematic for Example 13.1.

SOLUTION

Before proceeding, it is helpful to observe that if the water in the pipe were stationary, the pressure difference between points A and B could be calculated from the hydrostatic equation ($p_A = p_B + \rho g L$) to be

$$p_A - p_B = \rho g L = (62.4 \text{ lb}_f/\text{ft}^3)(40 \text{ ft})\left(\frac{1 \text{ ft}^2}{144 \text{ in.}^2}\right) = 17.3 \text{ psia}$$

Since the measured pressure difference of 21.8 psia is greater than this, water must be flowing upward in the pipe, since that will create a frictional pressure drop contribution to the measured pressure difference. We conclude that the observed pressure difference is due to a gravitational pressure decrease $\rho g L = 17.3$ psia and a frictional pressure decrease of 4.5 psia due to water flowing upward in the pipe.

To apply Eq. 13.2, we must assign point 1 upstream, which in this case is at point A. Solving Eq. 13.2 for the average wall shear stress, we have

$$\bar{\tau}_\text{wall} = \frac{A}{A_\text{wall}}[(p_1 - p_2) - \rho g(z_2 - z_1)]$$

Since the pipe is vertical $(z_2 - z_1) = L$, and for a round pipe $A/A_{wall} = (\pi D^2/4)/\pi DL = D/4L$, we find:

$$\bar{\tau}_{wall} = \frac{D}{4L}[(p_A - p_B) - \rho g L]$$

Inserting $p_A - p_B = 21.8$ psia and $\rho g L = 17.3$ psia, we find the average wall shear stress to be

$$\bar{\tau}_{wall} = \frac{0.5 \text{ ft}}{160 \text{ ft}}(21.8 \text{ psia} - 17.3 \text{ psia}) = 1.4 \times 10^{-2} \text{ psia} = 2.03 \text{ lb}_f/\text{ft}^2$$

The question of where the lost mechanical energy goes can be answered by applying an energy balance to the CV shown in Figure 13.4 (see Example 7.13). Writing the energy balance for a steady process as given by Eq. 7.34, we have

$$\int_{CS} \rho \left(u + \frac{p}{\rho} + \frac{1}{2}\mathbf{u} \cdot \mathbf{u} + gz \right)(\mathbf{u} \cdot \mathbf{n})\, dS = \dot{W}_{power} + \dot{W}_{shaft} + \dot{Q}_C + \dot{S}$$

In flow in a pipe or duct there is no fluid or shaft power input or energy addition, so for the CV shown in Figure 13.4, we can write the energy balance as

$$\int_{CS} \rho \left(u + \frac{p}{\rho} + \frac{1}{2}\mathbf{u} \cdot \mathbf{u} + gz \right)(\mathbf{u} \cdot \mathbf{n})\, dS = \dot{Q}_C$$

It is convenient to split the flux integral and write this as

$$\int_{CS} \rho \left(u + \frac{p}{\rho} + gz \right)(\mathbf{u} \cdot \mathbf{n})\, dS + \int_{CS} \rho \left(\frac{1}{2}\mathbf{u} \cdot \mathbf{u} \right)(\mathbf{u} \cdot \mathbf{n})\, dS = \dot{Q}_C$$

where the first integral gives the flux of internal, pressure potential, and gravitational potential energy through the control surface and the second describes the flux of kinetic energy through the control surface. These integrals must be evaluated at the inlet and exit of the CV. We will assume that the internal energy and pressure are uniform at each cross section along the flow path, and evaluate the first flux integral at each end of the CV to obtain

$$\int_{CS} \rho \left(u + \frac{p}{\rho} + gz \right)(\mathbf{u} \cdot \mathbf{n})\, dS = \dot{M}\left[(u_2 - u_1) + \left(\frac{p_2}{\rho} + gz_2\right) - \left(\frac{p_1}{\rho} + gz_1\right)\right]$$

The value of the kinetic energy flux depends on the form of the velocity profile. Since this flow is fully developed, however, the velocity profiles at the inlet and exit of the CV are the same, and the kinetic energy fluxes cancel: i.e., the flux integral is identically zero. The energy balance therefore reduces to

$$\dot{M}\left[(u_2 - u_1) + \left(\frac{p_2}{\rho} + gz_2\right) - \left(\frac{p_1}{\rho} + gz_1\right)\right] = \dot{Q}_C$$

13.2 STEADY, FULLY DEVELOPED FLOW IN A PIPE OR DUCT

If the flow through a pipe or duct were frictionless, applying the steady, constant density Bernoulli equation down the axis of the passage between the inlet and exit would yield

$$\left(\frac{p_1}{\rho} + gz_1\right) - \left(\frac{p_2}{\rho} + gz_2\right) = 0$$

showing that there is no loss of mechanical energy in frictionless flow. Equation 13.3 then shows that in a frictionless flow $\bar{\tau}_{wall} A_{wall}/\rho A = 0$; i.e., the wall shear stress is zero as expected. In addition, from Eq. 13.4 we find $(u_2 - u_1) - \dot{Q}_C/\dot{M} = 0$, which shows that in frictionless flow a change in internal energy can occur only if there is heat transfer.

For discussion purposes it is useful to rearrange this result and write the energy equation as

$$\left(\frac{p_1}{\rho} + gz_1\right) - \left(\frac{p_2}{\rho} + gz_2\right) = \left[(u_2 - u_1) - \frac{\dot{Q}_C}{\dot{M}}\right] \quad (13.4)$$

This result shows that in a steady, fully developed flow through a pipe or duct, the lost mechanical energy is converted into a change in the internal energy of the fluid and heat transfer. If the pipe is insulated, $\dot{Q}_C = 0$ and the internal energy (temperature) of the fluid must increase. In most flows, the mechanical energy lost to viscous friction appears as both an increase in the internal energy of the fluid and some heat transfer, usually out through the wall of the pipe or duct. By combining the energy balance, Eq. 13.4, with the momentum balance, Eq. 13.3, we obtain

$$\left[(u_2 - u_1) - \frac{\dot{Q}_C}{\dot{M}}\right] = \frac{\bar{\tau}_{wall} A_{wall}}{\rho A} \quad (13.5)$$

which shows that as a viscous fluid passes through a pipe or duct, there is an irreversible conversion of useful mechanical energy into internal energy and heat transfer due to friction.

13.2.1 Major Head Loss

In most cases involving flow in a pipe or duct we are interested in predicting the pressure drop, but rarely do we know the average wall stress or have sufficient information to evaluate the change in internal energy or heat transfer. To overcome this difficulty, we can make use of the concept of major head loss, h_L, and write the energy balance for fully developed flow, Eq. 13.4, as

$$\left(\frac{p_1}{\rho} + gz_1\right) - \left(\frac{p_2}{\rho} + gz_2\right) = h_L \quad (13.6)$$

It can be seen that the head loss has units of energy per unit mass. If the head loss is known, Eq. 13.6 allows us to calculate the pressure drop. Comparing Eq. 13.6 with Eqs. 13.3 and 13.4, we see that the major head loss is defined in two equivalent ways:

$$h_L = \frac{\bar{\tau}_{wall} A_{wall}}{\rho A} \quad \text{or} \quad h_L = (u_2 - u_1) - \frac{\dot{Q}_C}{\dot{M}} \quad (13.7a, b)$$

If the major head loss can be determined, Eq. 13.7a is useful in determining the average wall shear stress, and Eq. 13.7b provides information about the change in internal energy.

EXAMPLE 13.2

Water at 40°F flows from a supply tank as shown in Figure 13.6 through a buried 65 ft long, 4 in. diameter insulated pipe before entering a building. If the pipe is horizontal, and the measured pressure drop is 3 psia, find the head loss, average wall shear stress, and internal energy change in the chilled water. What is the temperature increase in the water due to friction?

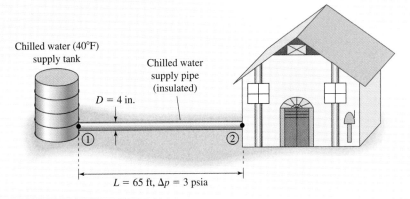

Figure 13.6 Schematic for Example 13.2.

SOLUTION

We first apply Eq. 13.6 to determine the head loss. For a horizontal pipe we have $z_1 = z_2$. Thus upon solving for the head loss we obtain

$$h_L = \frac{p_1 - p_2}{\rho} = \frac{(3 \text{ lb}_f/\text{in.}^2)(144 \text{ in.}^2/\text{ft}^2)}{1.94 \text{ slugs/ft}^3} = 223 \text{ ft}^2/\text{s}^2$$

Next we solve Eq. 13.7a for the average wall shear stress to obtain

$$\bar{\tau}_{wall} = \rho h_L \frac{A}{A_{wall}} = \rho h_L \frac{D}{4L} = (1.94 \text{ slugs/ft}^3)(223 \text{ ft}^2/\text{s}^2) \left[\frac{0.33 \text{ ft}}{4(65 \text{ ft})}\right]$$
$$= 0.55 \text{ lb}_f/\text{ft}^2 = 3.8 \times 10^{-3} \text{ psia}$$

To find the internal energy change in the water, we apply Eq. 13.7b, assuming that the heat transfer through the wall of the insulated pipe is negligible. Solving for the internal energy change, we obtain $u_2 - u_1 = h_L = 223 \text{ ft}^2/\text{s}^2$. Converting this result to familiar thermodynamic units, we find

$$u_2 - u_1 = (223 \text{ ft}^2/\text{s}^2)\left(\frac{\text{lb}_f\text{-s}^2}{\text{slugs-ft}}\right)\left(\frac{1 \text{ slug}}{32.2 \text{ lb}_m}\right)\left(\frac{1 \text{ Btu}}{778 \text{ ft-lb}_f}\right) = 8.9 \times 10^{-3} \text{ Btu/lb}_m$$

To calculate the temperature increase in the chilled water due to the frictional head loss, we note that for liquids, the change in internal energy is directly related to the specific heat of the liquid through the relationship $u_2 - u_1 = c(T_2 - T_1)$. For water, with a specific heat from Table 2.3 of $c = 4816$ J/(kg-K), the temperature change is

$$T_2 - T_1 = \frac{u_2 - u_1}{c} = \frac{8.9 \times 10^{-3} \text{ Btu/lb}_m}{[4816 \text{ J/(kg-K)}]} \left[\frac{(2.3928 \times 10^{-4} \text{ Btu})/(\text{lb}_m\text{-}°\text{R})}{1 \text{ J/(kg-K)}} \right]$$

$$= 7.7 \times 10^{-3} \text{ °R}$$

We see that the temperature increase in the chilled water due to viscous friction is negligible.

13.2.2 Friction Factor

To obtain the major head loss in a given pipe or duct flow, we can make use of results for the pressure drop obtained from experiments in horizontal pipes. Applying Eq. 13.6 to a horizontal pipe, we have $z_1 = z_2$ and thus the head loss is given by $h_L = (p_1/\rho - p_2/\rho) = \Delta p/\rho$. As discussed in Chapters 3 and 9, a dimensional analysis of pipe flow shows that we may write the pressure drop in flow in a horizontal pipe as $\Delta p/\rho \bar{V}^2 = g(L/D, e/D, \rho \bar{V} D/\mu)$, where g is an unknown function that depends on L/D, the relative roughness e/D, and the Reynolds number. Rearranging, and introducing a factor of $1/2$ by convention, we can write this as

$$\frac{\Delta p}{\rho} = \left(\frac{p_1}{\rho} - \frac{p_2}{\rho} \right) = g \left(\frac{L}{D}, \frac{e}{D}, \frac{\rho \bar{V} D}{\mu} \right) \frac{\bar{V}^2}{2}$$

Since the flow is fully developed, the pressure drop should be in direct proportion to the length of the pipe, i.e., the drop over a quarter of the length should be a quarter of the drop over the full length. This means that the unknown function g must depend linearly on L/D, and we can write the preceding equation as

$$\left(\frac{p_1}{\rho} - \frac{p_2}{\rho} \right) = g_1 \left(\frac{e}{D}, \frac{\rho \bar{V} D}{\mu} \right) \frac{L}{D} \frac{\bar{V}^2}{2}$$

We can now define the friction factor to be $f = g_1(e/D, \rho \bar{V} D/\mu)$ and obtain $(p_1/\rho - p_2/\rho) = f(L/D)(\bar{V}^2/2)$. Comparing this with the preceding expression for the head loss in a horizontal pipe, $h_L = (p_1/\rho - p_2/\rho)$, we see that the relationship between major head loss and friction factor is defined for pipe flow by

$$h_L = f \frac{L}{D} \frac{\bar{V}^2}{2} \tag{13.8}$$

Comparing this expression with Eq. 13.7a, we see that

$$f\frac{L}{D}\frac{\bar{V}^2}{2} = \frac{\bar{\tau}_{wall} A_{wall}}{\rho A}$$

Noting that $A_{wall} = \pi D L$ and $A = \pi D^2/4$, we can solve for the friction factor and obtain

$$f = \frac{8\bar{\tau}_{wall}}{\rho \bar{V}^2} \qquad (13.9a)$$

which allows us to write the head loss as

$$h_L = \left(\frac{8\bar{\tau}_{wall}}{\rho \bar{V}^2}\right)\frac{L}{D}\frac{\bar{V}^2}{2} \qquad (13.9b)$$

Since the head loss is due to the effects of friction alone, there is no dependence of head loss (and friction factor) on the orientation of the pipe. Thus for an inclined pipe we can make use of the friction factor as defined by Eq. 13.8 to write the energy balance, Eq. 13.6, as

$$\left(\frac{p_1}{\rho} + gz_1\right) - \left(\frac{p_2}{\rho} + gz_2\right) = f\frac{L}{D}\frac{\bar{V}^2}{2} \qquad (13.10)$$

This is the basic equation describing the steady, fully developed flow in a pipe. If the friction factor is known, we can make use of this equation to determine the pressure change corresponding to a known average velocity (and thus flowrate) in a pipe of known inclination.

EXAMPLE 13.3

In the early stages of the design of the chilled water line of Example 13.2, the friction factor was assumed to be $f = 0.02$ at a design flowrate of 400 gal/min, with a water temperature of 40°F. What is the pressure drop predicted by using this value of f? If the line is installed so that the pipe exit is actually 3 ft above the inlet, what is the predicted pressure drop?

SOLUTION

The chilled water line in Example 13.2 is horizontal, thus by applying Eq. 13.10 we have $(p_1 - p_2)/\rho = f(L/D)(\bar{V}^2/2)$. The volume flowrate is given by $Q = A\bar{V}$, and for a design flowrate of 400 gal/min, the average velocity is calculated as

$$\bar{V} = \frac{Q}{A} = \frac{(400 \text{ gal/min})\left(\frac{1 \text{ ft}^3}{7.48 \text{ gal}}\right)\left(\frac{1 \text{ min}}{60 \text{ s}}\right)}{\left(\frac{\pi(0.33 \text{ ft})^2}{4}\right)} = 10.4 \text{ ft/s}$$

Solving for the pressure drop we have

$$p_1 - p_2 = \rho f \frac{L}{D} \frac{\bar{V}^2}{2} = (1.94 \text{ slugs/ft}^3)(0.02)\left(\frac{65 \text{ ft}}{0.33 \text{ ft}}\right)\left(\frac{1}{2}\right)(10.4 \text{ ft/s})^2$$

$$= 413 \text{ lb}_f/\text{ft}^2 = 2.9 \text{ psia}$$

We see that the predicted pressure drop for a horizontal pipe is about the same as the measured value of 3 psia.

If the pipe is inclined, applying Eq. 13.10 gives $(p_1/\rho + gz_1) - (p_2/\rho + gz_2) = f(L/D)(\bar{V}^2/2)$. Solving for the pressure difference we have $p_1 - p_2 = \rho f(L/D)(\bar{V}^2/2) + \rho g(z_2 - z_1)$. The effect of friction is the same irrespective of the inclination as long as the flowrate is the same, so from the preceding calculation we have

$$\rho f \frac{L}{D} \frac{\bar{V}^2}{2} = 413 \text{ lb}_f/\text{ft}^2 = 2.9 \text{ psia}$$

Since the exit of the pipe is now 3 ft above the inlet, we have

$$\rho g(z_2 - z_1) = 1.94 \text{ slugs/ft}^3 (32.2 \text{ ft/s}^2)(3 \text{ ft}) = 187.4 \text{ lb}_f/\text{ft}^2 = 1.3 \text{ psia}$$

and the predicted pressure drop is

$$p_1 - p_2 = \rho f \frac{L}{D} \frac{\bar{V}^2}{2} + \rho g(z_2 - z_1) = 2.9 \text{ psia} + 1.3 \text{ psia} = 4.2 \text{ psia}$$

The additional pressure required at the inlet is not because of additional friction but because of the elevation change.

We can extend the preceding results to apply to flow in a duct by introducing the concept of a hydraulic diameter D_H. We define the hydraulic diameter of a duct of wetted perimeter P and cross-sectional area A as

$$D_H = \frac{4A}{P} \tag{13.11}$$

and write the head loss in a duct of hydraulic diameter D_H as

$$h_L = f \frac{L}{D_H} \frac{\bar{V}^2}{2} \tag{13.12}$$

The friction factor is related to the average wall shear stress in the duct by

$$f = \frac{8\bar{\tau}_{\text{wall}}}{\rho \bar{V}^2} \tag{13.13a}$$

and the head loss is given by

$$h_L = \left(\frac{8\bar{\tau}_{\text{wall}}}{\rho \bar{V}^2}\right) \frac{L}{D_H} \frac{\bar{V}^2}{2} \tag{13.13b}$$

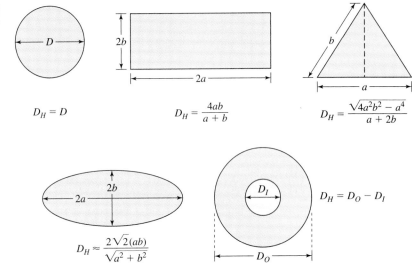

Figure 13.7 The definition of hydraulic diameter $D_H = 4A/P$ for five cross section shapes.

Finally the energy balance is written as

$$\left(\frac{p_1}{\rho} + gz_1\right) - \left(\frac{p_2}{\rho} + gz_2\right) = f \frac{L}{D_H} \frac{\bar{V}^2}{2} \tag{13.14}$$

To analyze a duct flow, the friction factor for a duct of hydraulic diameter D_H and a given value of the Re (based on D_H) is assumed to be the same as the friction factor for a pipe of diameter D_H at the same Re.

The hydraulic diameter model works very well in turbulent flow, but not nearly as well in laminar flow. The failure of this concept in laminar flow is generally not a problem, however, because we have other methods to analyze laminar flow. Care must be taken when one is using the D_H model with turbulent flow in rectangular ducts with aspect ratios greater than 4. In general, the further the cross-sectional geometry of a duct departs from a circle, the greater will be the error in using hydraulic diameter. The hydraulic diameters of several other cross sections are given in Figure 13.7.

EXAMPLE 13.4

Air at 70°F and atmospheric pressure flows in a horizontal 40 ft length of sheet metal duct at a rate of 2000 ft³/min as shown in Figure 13.8. If the duct cross section is rectangular and 1 ft by 2 ft, and the appropriate friction factor is 0.016, what is the pressure drop?

SOLUTION

Applying Eq. 13.14 to this horizontal duct we have $(p_1 - p_2)/\rho = f(L/D_H)(\bar{V}^2/2)$. The hydraulic diameter is given by Eq. 13.11 as $D_H = 4A/P$, and for a rectangular duct of height H and width W, we have $A = HW$, and $P = 2(H + W)$. Thus the

13.2 STEADY, FULLY DEVELOPED FLOW IN A PIPE OR DUCT

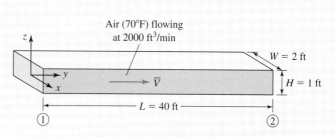

Figure 13.8 Schematic for Example 13.4.

hydraulic diameter is $D_H = 2HW/(H + W)$. Inserting $H = 1$ ft and $W = 2$ ft we find $D_H = 1.33$ ft; and since $L = 40$ ft, $L/D_H = 40$ ft/1.33 ft $= 30$. The volume flowrate is given by $Q = A\bar{V}$, thus we have $\bar{V} = Q/HW$. Inserting the data, we find

$$\bar{V} = \frac{(2000 \text{ ft}^3/\text{min})\left(\dfrac{1 \text{ min}}{60 \text{ s}}\right)}{2 \text{ ft}^2} = 16.7 \text{ ft/s}$$

By using these results to calculate the pressure drop, we obtain

$$p_1 - p_2 = \rho f \frac{L}{D_H} \frac{\bar{V}^2}{2} = (2.329 \times 10^{-3} \text{ slug/ft}^3)(0.016)(30)\left(\frac{1}{2}\right)(16.7 \text{ ft/s})^2$$
$$= 0.16 \text{ lb}_f/\text{ft}^2$$

It is customary to express the pressure drop in a duct flow of air in inches of water. This means that we can write $p_1 - p_2 = 0.16 \text{ lb}_f/\text{ft}^2 = (\rho_{H_2O})gh$. Solving for h we find

$$h = \frac{(0.16 \text{ lb}_f/\text{ft}^2)(12 \text{ in.}/\text{ft})}{(1.94 \text{ slugs/ft}^3)(32.2 \text{ ft/s}^2)} = 0.03 \text{ in.}$$

This is a very modest pressure drop. Most duct systems have significant minor losses.

At this point it should be evident that analyzing the flow in a pipe or duct depends on being able to determine the friction factor. For laminar flows we can often make use of Eq. 13.9a, $f = 8\bar{\tau}_{wall}/\rho\bar{V}^2$, to compute the friction factor directly from a known analytical solution. In turbulent flows the friction factor can be obtained from charts that provide the dependence of the friction factor on the relative roughness of the pipe and Reynolds number.

13.2.3 Friction Factors in Laminar Flow

In this section we will derive the friction factor for laminar flow in a pipe and in a rectangular duct (friction factors can be found in advanced texts for laminar flow in many other geometries). The steady, fully developed, constant density flow in both cases is of the Poiseuille type, meaning that the flow is unidirectional and driven by a pressure

In problems involving duct flows of air, you will often find the units of average velocity given in feet per minute and the flowrate in units of cubic feet per minute. This makes it easy to go back and forth between flowrate and average velocity by means of the duct area in square feet. For example, the 2000 ft³/min flow in Example 13.4 occurred in a duct whose area was given as $A = 2$ ft², so the average velocity is $\bar{V} = Q/A = (2000 \text{ ft}^3/\text{min})/2 \text{ ft}^2 = 1000$ ft/min. Pressure differences in duct flows are generally small, and it is customary to express them in inches of water. Since we can write $(p_1 - p_2) = \rho_{H_2O} gh$, useful conversion factors are 1 in. $H_2O = 5.2$ lb$_f$/ft² $= 0.036$ psia.

gradient. Since the exact analytical solutions for the velocity and pressure fields in both flows are known, it is not necessary to use the friction factor or head loss concept to analyze these flows. However, it is customary to include the friction factor for laminar flow in a discussion of flow in pipes and ducts.

Consider a Poiseuille flow in the horizontal pipe illustrated in Figure 13.9. The velocity field and pressure distribution for this flow were derived in Section 12.2.4. The pressure distribution is given by Eq. 12.16 as $p(z) = p_1 - [(p_1 - p_2)/L](z - z_1)$, and the velocity component in the flow direction is given by Eq. 12.17 as $v_z(r) = [R^2(p_1 - p_2)/4\mu L][1 - (r/R)^2]$. As can be seen from Example 12.5, the volume flowrate through any cross section of the pipe is easily found by integration to be:

$$Q = \int (\mathbf{u} \cdot \mathbf{n}) \, dS = \int_0^{2\pi} \int_0^R \frac{R^2(p_1 - p_2)}{4\mu L} \left[1 - \left(\frac{r}{R}\right)^2\right] r \, dr \, d\theta = \frac{\pi R^4 (p_1 - p_2)}{8\mu L}$$

or in terms of diameter,

$$Q = \frac{\pi D^4 (p_1 - p_2)}{128 \mu L} \quad (13.15)$$

Since $Q = A\bar{V} = (\pi D^2/4)\bar{V}$, the average velocity in this flow is given by

$$\bar{V} = \frac{D^2 (p_1 - p_2)}{32 \mu L} \quad (13.16)$$

To find the dependence of Δp on $\bar{V}$ in flow in a pipe, we can rearrange this equation and obtain

$$p_1 - p_2 = \frac{32 \mu L \bar{V}}{D^2} \quad (13.17)$$

We see that the pressure drop is proportional to the average velocity. If we double the flowrate, the pressure drop doubles. The pressure drop is also proportional to the fluid

Figure 13.9 Cylindrical coordinates used for pipe flows.

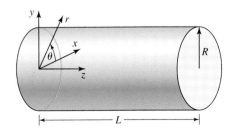

It is worth noting that there is another way to derive the friction factor $f = 64/Re$. Since we know the velocity profile, we can compute the wall shear stress and use Eq. 13.9a to write $f = 8\bar{\tau}_{wall}/\rho \bar{V}^2$. The tangential component of the stress acting in the z direction on a surface in the fluid normal to the r direction is σ_{rz}. The stress on the wall of the pipe is $-\sigma_{rz}$, since the wall is a negative r surface. This stress component is given by Eq. 11.7f as $\sigma_{rz} = \mu(\partial v_z/\partial r + \partial v_r/\partial z) = \mu(\partial v_z/\partial r)$, since there is no velocity component in the radial direction. Using the known velocity field we find

$$\tau_{wall} = -\mu \frac{\partial v_z}{\partial r}\bigg|_{r=R} = -\mu \frac{\partial}{\partial r}$$
$$\times \left\{ \frac{R^2(p_1 - p_2)}{4\mu L}\left[1 - \left(\frac{r}{R}\right)^2\right]\right\}\bigg|_{r=R}$$
$$= \frac{D(p_1 - p_2)}{4L}$$

The wall shear stress is uniform, so the average wall shear stress is $\bar{\tau}_{wall} = [D(p_1 - p_2)]/4L$. From Eq. 13.9a we can now write $f = 8\bar{\tau}_{wall}/\rho \bar{V}^2$ and obtain $f = [2D(p_1 - p_2)]/\rho \bar{V}^2 L$. Writing this in terms of Re, and using the definition of $\bar{V}$ (Eq. 13.16) gives $f = 64/Re$ as expected.

viscosity and to the pipe length, and inversely proportional to the square of the pipe diameter.

To find the friction factor for this flow, first note that to apply Eq. 13.10 to a horizontal pipe, we set $z_1 = z_2$, and obtain

$$\frac{p_1 - p_2}{\rho} = f \frac{L}{D}\frac{\bar{V}^2}{2} \qquad (13.18)$$

To derive the friction factor, we divide Eq. 13.17 by the density, and use $Re = \rho \bar{V} D/\mu$ to write it as

$$\frac{p_1 - p_2}{\rho} = \left(\frac{64}{Re}\right)\frac{L}{D}\frac{\bar{V}^2}{2}$$

Comparing this with Eq. 13.18, we see that the friction factor for laminar flow in a pipe is given by

$$f = \frac{64}{Re} \qquad (13.19)$$

The solution for Poiseuille-type flow through the rectangular duct shown in Figure 13.10 can be found in many advanced texts. Of interest to us here is the velocity component in the flow direction, which for $-a \leq x \leq a$, $-b \leq y \leq b$ with $a > b$, can be expressed as the infinite series:

$$w(x, y) = \frac{(p_1 - p_2)}{2\mu L}$$
$$\times \left[b^2 - y^2 - \frac{4}{b}\sum_{n=0}^{\infty} \frac{(-1)^n}{m^3}\left(\frac{\cosh mx}{\cosh ma}\right)\cos my \right]$$
(13.20)

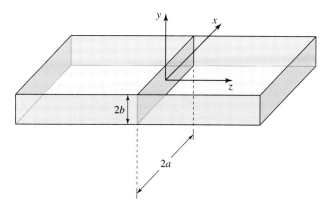

Figure 13.10 Cartesian coordinates used for rectangular duct flows.

where $m = [(2n+1)\pi]/2b$. The volume flowrate is given by

$$Q = \frac{4ab^3(p_1 - p_2)}{3\mu L}\left[1 - \frac{192b}{\pi^5 a}\sum_{n=0}^{\infty}\frac{\tanh ma}{(2n+1)^5}\right]$$

and since the duct area is $A = 4ab$, the average velocity is

$$\bar{V} = \frac{b^2(p_1 - p_2)}{3\mu L}\left[1 - \frac{192b}{\pi^5 a}\sum_{n=0}^{\infty}\frac{\tanh ma}{(2n+1)^5}\right]$$

It is convenient to define a function $F(a/b)$, which depends only on the aspect ratio a/b, as

$$F\left(\frac{a}{b}\right) = \left[1 - \frac{192b}{\pi^5 a}\sum_{n=0}^{\infty}\frac{\tanh ma}{(2n+1)^5}\right] = \left\{1 - \frac{192b}{\pi^5 a}\sum_{n=0}^{\infty}\frac{\tanh[(2n+1)\pi a]/2b}{(2n+1)^5}\right\}$$

This allows us to write the volume flowrate as

$$Q = \frac{4ab^3(p_1 - p_2)}{3\mu L}F\left(\frac{a}{b}\right) \tag{13.21}$$

and the average velocity as

$$\bar{V} = \frac{b^2(p_1 - p_2)}{3\mu L}F\left(\frac{a}{b}\right) \tag{13.22}$$

To find the dependence of Δp on $\bar{V}$, we can rearrange the last equation to obtain

$$p_1 - p_2 = \frac{3\mu L\bar{V}}{b^2 F(a/b)} \tag{13.23}$$

We see that in laminar flow in a rectangular duct the pressure drop is proportional to the average velocity, fluid viscosity, and duct length, and inversely proportional to the height of the duct squared, and the aspect ratio as contained in $F(a/b)$. This behavior is characteristic of all laminar flows. Values of the geometric factor $F(a/b)$ for various aspect ratios can be found in Table 13.1.

By following a procedure similar to that used for the pipe, we can define a friction factor for flow in a rectangular duct to use with the hydraulic diameter model. To apply Eq. 13.14 to a horizontal duct, we set $z_1 = z_2$, and obtain

$$\frac{p_1 - p_2}{\rho} = f\frac{L}{D_H}\frac{\bar{V}^2}{2} \tag{13.24}$$

To derive the friction factor, we divide Eq. 13.23 by the density, and use $Re = \rho\bar{V}D_H/\mu$ and $D_H = 4ab/(a+b)$ to write it as $(p_1 - p_2)/\rho = (96/Re_H)G(a/b)(L/D_H)(\bar{V}^2/2)$, where the geometric factor $G(a/b)$ is defined by $G(a/b) = 1/[(1 + b/a)^2 F(a/b)]$. Comparing our result to Eq. 13.24, we see that the analytical solution shows that the friction factor is

$$f = \left(\frac{96}{Re_H}\right)G\left(\frac{a}{b}\right) \tag{13.25}$$

As the aspect ratio of a rectangular duct gets very large, i.e., $a/b \to \infty$, the flow should become more similar to that between infinite parallel plates as discussed in Section 13.2.3. Employing the solution given there, the velocity component in the flow direction is $u(z) = (p_1 - p_2)/2\mu L \times [b^2 - y^2]$, and we see that this is the leading term in Eq. 13.20 describing the rectangular duct. If the width of the parallel plates is w, the volume flowrate through the channel formed by the plates is given by

$$Q = \int_{-W/2}^{W/2} \int_{-b}^{b} \frac{(p_1 - p_2)}{2\mu L}[b^2 - y^2] dz\, dy$$

$$= w\left[\frac{2b^3(p_1 - p_2)}{3\mu L}\right]$$

If $w = 2a$, we have $Q = [4ab^3(p_1 - p_2)]/3\mu L$. The average velocity is then found by dividing by $4ab$ to obtain $\bar{V} = [b^2(p_1 - p_2)]/3\mu L$, and the pressure drop is related to the average velocity by $p_1 - p_2 = 3\mu L \bar{V}/b^2$. Thus, the rectangular duct results are consistent with these expressions if $F(a/b) \to 1$ as $a/b \to \infty$. Table 13.1 shows that $F(a/b)$ does indeed behave this way.

Values of the geometric factor $G(a/b)$ for various aspect ratios are contained in Table 13.1, along with values of $f Re_H = 96G(a/b)$. The hydraulic diameter model requires that we use the friction factor for flow in a round pipe, which is $f = 64/Re_H$, or $f Re_H = 64$. For this to agree with the exact friction factor for this flow given in Eq. 13.25, the value of $G(a/b)$ would have to be $G(a/b) = 64/96 = 0.667$, or $f Re_H = 64$ for every value of the aspect ratio a/b. It is clear from Table 13.1 that this is not the case. Since the exact solution is available for laminar flow in a rectangular duct, we recommend the use of Eq. 13.23 and values of $F(a/b)$ from Table 13.1 to determine the pressure drop. Alternately, you can employ the exact friction factor as given by Eq. 13.25 within the hydraulic diameter model to obtain an accurate value of the pressure drop.

The analytical friction factors for laminar, fully developed flow in round, rectangular, and several other cross sections are summarized in Table 13.2 and compared with $f = 64/Re_H$. It is evident that in laminar flow, the hydraulic radius approximation is not always accurate. If we know the exact analytical solution this does not matter; but if the solution for a certain cross section is unknown, we can compute the hydraulic diameter, assume the friction factor is $f = 64/Re_H$, and compute an approximate value for the pressure drop.

TABLE 13.1 Values for the Geometric Factors $F(a/b)$ and $G(a/b)$ as a Function of the Aspect Ratio a/b for Rectangular Ducts

(a/b)	$F(a/b)$	$G(a/b)$	$fRe_H = 96G(a/b)$
1	0.421733	0.592794	56.91
2	0.686045	0.647836	62.19
3	0.789951	0.712070	68.36
4	0.842439	0.759699	72.93
5	0.873951	0.794604	76.28
10	0.936975	0.882037	84.68
20	0.968488	0.936242	89.88
50	0.987395	0.973439	93.45
100	0.993698	0.986513	94.71
∞	1	1	96.0

TABLE 13.2 Friction Factors for Laminar Fully Developed Flow for Various Cross Sections

Cross Section Shape	Geometric Parameter	$f Re_H$
Circle ($D_H = D$)	NA	64
Rectangle $D_H = \dfrac{4ab}{a+b}$	(a/b) 1	56.91
	2	62.19
	3	68.36
	4	72.93
	5	76.28
	10	84.68
	20	89.88
	50	93.45
	100	94.71
	∞	96.00
Ellipse $D_H \approx \dfrac{2\sqrt{2}(ab)}{\sqrt{a^2+b^2}}$	NA	64
Right triangle $D_H = \dfrac{2a \sin\theta}{1+\sin\theta+\cos\theta}$	θ (degrees) 0	48.0
	10	49.9
	20	51.2
	30	52.0
	40	52.3
	45	52.5
Equilateral triangle $D_H = \dfrac{a\sqrt{3}}{3}$	NA	53.3
Concentric annulus $D_H = D_2 - D_1$	D_1/D_2 0.0001	71.8
	0.01	80.1
	0.1	89.4
	0.5	95.3
	1.0	96.0

13.2 STEADY, FULLY DEVELOPED FLOW IN A PIPE OR DUCT 811

EXAMPLE 13.5

Microfluidic devices like that in Figure 13.11 are under development for a variety of applications ranging from medicine to environmental sampling. Suppose the dimensions of the horizontal rectangular passage within such a device are $H = 10$ μm, $W = 50$ μm, and $L = 500$ μm. Calculate the pressure drop through this passage for a flowrate of 2.4×10^6 μm³/s. The flow is steady, and the fluid is water at 30°C.

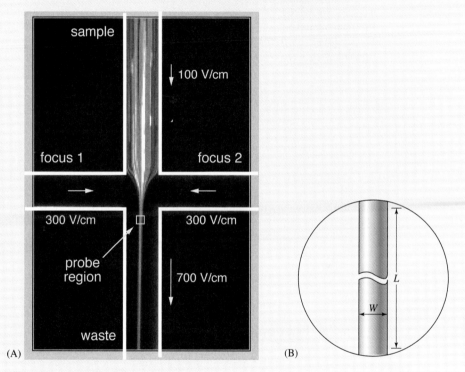

Figure 13.11 (A) Microfluidic channel used in DNA research. (B) Schematic.

SOLUTION

We will first calculate Re to check that this is a laminar flow. Using the definition of hydraulic diameter we find $D_H = 2HW/(H + W) = [2(10\,\mu\text{m})(50\,\mu\text{m})]/(10\,\mu\text{m} + 50\,\mu\text{m}) = 17\,\mu\text{m}$. The average velocity is calculated as

$$\bar{V} = \frac{Q}{A} = \frac{2.4 \times 10^6\,\mu\text{m}^3/\text{s}}{(10\,\mu\text{m})(50\,\mu\text{m})} = 4.8 \times 10^3\,\mu\text{m/s}$$

Then Re is found to be

$$Re_H = \frac{\rho \bar{V} D_H}{\mu} = \frac{(995.7\,\text{kg/m}^3)(4.8 \times 10^3\,\mu\text{m/s})(17\,\mu\text{m})\left(\dfrac{1\,\text{m}}{10^6\,\mu\text{m}}\right)^2}{7.975 \times 10^{-4}\,(\text{N-s})/\text{m}} = 0.1$$

which identifies this as a creeping flow, and hence laminar. To determine the pressure drop we use Eq. 13.23, $p_1 - p_2 = 3\mu L \bar{V}/[b^2 F(a/b)]$, noting in this case that $H = 2b$ and $W = 2a$. Thus, $b = 5\,\mu\text{m}$, $a = 25\,\mu\text{m}$, and $a/b = 25\,\mu\text{m}/5\,\mu\text{m} = 5$. From Table 13.1 we read $F(a/b) = F(5) = 0.873951$, thus

$$p_1 - p_2 = \frac{3\mu L \bar{V}}{b^2 F(a/b)} = \frac{3[7.975 \times 10^{-4}(\text{N-s})/\text{m}^2](500\,\mu\text{m})(4800\,\mu\text{m/s})}{(5\,\mu\text{m})^2(0.873951)} = 263\,\text{N/m}^2$$

For comparison we can use the hydraulic diameter model to calculate Δp. We will use Eq. 13.24: $(p_1 - p_2)/\rho = f(L/D_H)(\bar{V}^2/2)$, and assume $f = 64/Re_H$. Thus Δp is estimated as $p_1 - p_2 = \rho(64/Re_H)(L/D_H)(\bar{V}^2/2)$. Inserting the data and earlier results we have

$$p_1 - p_2 = \rho \frac{64}{Re_H} \frac{L}{D_H} \frac{\bar{V}^2}{2}$$

$$= (995.7\,\text{kg/m}^3)\left(\frac{64}{0.1}\right)\left(\frac{500\,\mu\text{m}}{17\,\mu\text{m}}\right)\left(\frac{1}{2}\right)(4800\,\mu\text{m/s})^2\left(\frac{1\,\text{m}}{10^6\,\mu\text{m}}\right)^2$$

$$= 216\,\text{N/m}^2$$

This is 18% low, and it explains our earlier suggestion to use the exact solution if available for the cross section of interest. If you are wondering whether the flow in microfluidic devices is described by the continuum hypothesis, the length scale of these devices, on the order of 10 μm, is several orders of magnitude larger than the average intermolecular spacing in water.

13.2.4 Friction Factors in Turbulent Flow

There are no analytical solutions for turbulent flow, so we cannot compute the friction factor in a pipe or duct flow from an exact analytical formula for volume flowrate or average velocity. However, we know from dimensional analysis that the friction factor for flow in a round pipe of diameter D is a function of the relative roughness e/D, and Reynolds number $\rho \bar{V} D/\mu$. If we performed a DA on duct flow, we would find that f depends on e/D, the Re, and additional geometric parameters that depend on the shape of the duct. Determining f for every possible duct cross section is an enormous task, but we can avoid this problem by using the hydraulic diameter concept and substituting the friction factor for flow in a round pipe as explained earlier.

To determine the friction factor empirically for steady, fully developed turbulent flow through a pipe, consider a series of experiments in which the pressure drop is measured in a flow through a horizontal section of a pipe. Applying Eq. 13.10 to a horizontal pipe we set $z_1 = z_2$, and obtain $(p_1 - p_2)/\rho = f(L/D)(\bar{V}^2/2)$. This shows that a measurement of the pressure drop is sufficient to determine the friction factor for a

13.2 STEADY, FULLY DEVELOPED FLOW IN A PIPE OR DUCT

given flowrate and type of fluid. Now the friction factor is defined by Eq. 13.9a ($f = 8\bar{\tau}_{\text{wall}}/\rho \bar{V}^2$), and the wall shear stress depends only on the motion of the fluid relative to the wall. We conclude once again that the friction factor does not depend on the angle of inclination of the pipe. Thus there is no need to do experiments in which the pipe is inclined at various angles. Repeating the experiment using various flowrates is sufficient to establish the dependence of the friction factor on Re for the particular value of e/D that applies to the pipe in question.

Through numerous experiments, the data needed to determine the friction factor has been obtained for turbulent flow over the range of values of relative roughness and Reynolds number encountered in engineering practice. This data is organized for convenient use in the form of the chart of relative roughness shown in Figure 13.12, and the Moody chart shown in Figure 13.13. The use of these charts as well as Table 13.3, which contains data on the roughness of various types of pipe, is illustrated in Example 13.6.

EXAMPLE 13.6

A horizontal pipeline will carry crude oil at 60°C in 1 m diameter commercial steel pipe at an average velocity of 3 m/s. If the oil's specific gravity is 0.86 and its viscosity is $\mu = 3.8 \times 10^{-3}$ (N-m)/s², what is the friction factor? What is the pressure drop per kilometer of pipe?

SOLUTION

A specific gravity of 0.86 corresponds to a density of $\rho = 0.86(1000 \text{ kg/m}^3) = 860 \text{ kg/m}^3$. The Reynolds number is then found to be

$$Re = \frac{\rho \bar{V} D}{\mu} = \frac{(860 \text{ kg/m}^3)(3 \text{ m/s})(1 \text{ m})}{3.8 \times 10^{-3} \text{ (N-m)/s}^2}$$

$$= 6.8 \times 10^5$$

From Figure 13.12, we read the value of the relative roughness as $e/D = 4 \times 10^{-5}$ and then use the Moody chart, Figure 13.13, to read the friction factor as $f = 0.012$. The pressure drop is then calculated by using Eq. 13.10 with $z_1 = z_2$ to write

$$p_1 - p_2 = \rho f \frac{L}{D} \frac{\bar{V}^2}{2}$$

$$= (860 \text{ kg/m}^3)(0.012)\frac{1000 \text{ m}}{1 \text{ m}}\left(\frac{1}{2}\right)(3 \text{ m/s})^2$$

$$= 4.6 \times 10^4 \text{ N/m}^2$$

$$= 46 \text{ kPa}$$

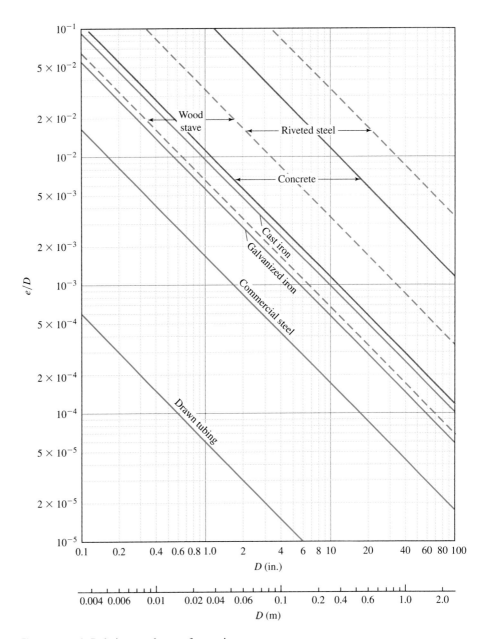

Figure 13.12 Relative roughness of new pipes.

13.2 STEADY, FULLY DEVELOPED FLOW IN A PIPE OR DUCT

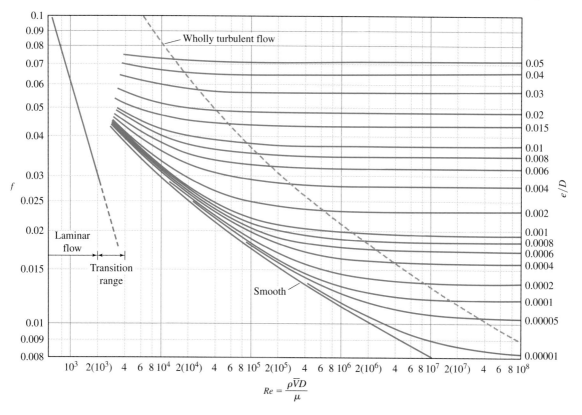

Figure 13.13 The Moody chart, showing the friction factor as a function of Reynolds number and relative roughness.

TABLE 13.3 Equivalent Roughness Values for Various Pipe Materials

Pipe Material	Equivalent Roughness, e	
	ft	mm
Riveted steel	0.003–0.03	0.9–9.0
Concrete	0.001–0.01	0.3–3.0
Wood stave	0.0006–0.003	0.18–0.9
Cast iron	0.00085	0.26
Galvanized iron	0.0005	0.15
Asphalted cast iron	0.0004	0.12
Commercial steel or wrought iron	0.00015	0.045
Drawn tubing	0.000005	0.0015
Plastic or glass	Smooth	Smooth

EXAMPLE 13.7

Find the friction factor for the duct flow described in Example 13.4. Recall that the rectangular duct cross section is 1 ft by 2 ft with a hydraulic diameter $D_H = 1.33$ ft. The flowrate of 2000 ft³/min corresponds to an average velocity $\bar{V} = 16.7$ ft/s, and the air is at 70°F and atmospheric pressure.

SOLUTION

To use the hydraulic diameter model, the friction factor for a duct flow at a certain Re based on D_H is obtained by assuming it to be the same as the friction factor for a flow in a pipe of diameter $D = D_H$ at the same Re. It is also necessary that the pipe have the same relative roughness $e/D = e/D_H$ as the duct. From Appendix A, the kinematic viscosity of air at 70°F is $\nu = 1.64 \times 10^{-4}$ ft²/s. The duct Reynolds number is calculated by using the hydraulic diameter:

$$Re = \frac{\bar{V} D_H}{\nu} = \frac{(16.7 \text{ ft/s})(1.33 \text{ ft})}{1.64 \times 10^{-4} \text{ ft}^2/\text{s}} = 1.35 \times 10^5$$

The duct is described as sheet metal, so we will use the smooth pipe value of relative roughness on the Moody chart, Figure 13.13, to read the friction factor at $Re = 1.35 \times 10^5$ as $f = 0.0165$.

As an alternative to the use of the Moody chart, we can make use of the Colebrook formula on which the chart is based. This formula, given in Chapter 3, Eq. 3.19a, is

$$\frac{1}{\sqrt{f}} = -2.0 \log \left(\frac{e/D}{3.7} + \frac{2.51}{\sqrt{f} Re} \right) \quad (13.26a)$$

As discussed in Chapter 3, this is a transcendental equation requiring iteration to determine the friction factor for known values of relative roughness and Reynolds number.

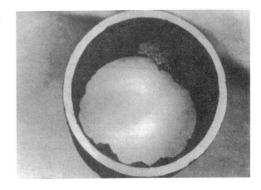

Figure 13.14 Scaling in a water pipe after 40 years of service.

13.3 ANALYSIS OF FLOW IN SINGLE PATH PIPE AND DUCT SYSTEMS

The Chen equation, which was given in Chapter 3, Eq. 3.19b, is also repeated here for convenience:

$$f = \left\{-2.0\log\left[\frac{(e/D)}{3.7065} - \frac{5.0452}{Re}\log\left(\frac{(e/D)^{1.1098}}{2.8257} + \frac{5.8506}{Re^{0.8981}}\right)\right]\right\}^{-2} \quad (13.26b)$$

The advantage of the Chen equation is that it does not require iteration.

It is customary to state that the uncertainty in the value of the friction factor found with any of these methods is on the order of $\pm 10\%$. In arriving at a value for the friction factor, keep in mind that pipes in service tend to degrade over time. The most common forms of degradation are due to biological or corrosion-induced scaling as shown in Figure 13.14. Thus, the relative roughness of a new pipe is likely to be far less than that of a pipe in service for several years.

EXAMPLE 13.8

Using the data of Example 13.6, compare the friction factor obtained with the Colebrook and Chen formulas with that obtained with the Moody chart. How do the friction factors change if the relative roughness increases by an order of magnitude as deposits accumulate in the pipe over time?

SOLUTION

We know that $e/D = 4 \times 10^{-5}$, $Re = 6.8 \times 10^5$, so from the Moody chart, Figure 13.13, we read $f = 0.012$. Using any appropriate computer code, the Colebrook formula, Eq. 13.19a, can be iterated solved to yield $f = 0.0131$, while the Chen equation, Eq. 13.19b, gives $f = 0.01315$. We see that all three methods produce friction factors within 10%. If the relative roughness increases by a factor of 10 to $e/D = 4 \times 10^{-4}$, the value of the friction factor found from the Moody chart is $f = 0.0165$. The Colebrook formula gives $f = 0.0167$, and the Chen equation yields $f = 0.0168$.

As a final note in this section, you should be aware that pipe comes in well-defined standard sizes, as shown in Table 13.4. The schedule of a pipe determines its usage. Higher schedule number pipe has thicker walls and a higher pressure rating. Schedule 40 pipe is normally standard unless otherwise specified.

13.3 ANALYSIS OF FLOW IN SINGLE PATH PIPE AND DUCT SYSTEMS

The piping system shown earlier in Figure 13.2 contains a number of straight constant area sections of pipe joined by elbows. Since all the fluid follows the same path through the system, this is an example of a single path system. A mass balance on any single path

TABLE 13.4 Actual Dimensions of Common Nominal Pipe Sizes

Nominal Diameter (in.)	Schedule	Inside Diameter ft	Inside Diameter cm	Flow Area ft²	Flow Area cm²
1/8	40	0.02242	0.683	0.0003947	0.3664
	80	0.01792	0.547	0.0002522	0.2350
1/4	40	0.03033	0.924	0.0007227	0.6706
	80	0.2517	0.768	0.0004974	0.4632
3/8	40	0.04108	1.252	0.001326	1.233
	80	0.03525	1.074	0.0009759	0.9059
1/2	40	0.05183	1.580	0.002110	1.961
	80	0.04550	1.386	0.001626	1.508
	160	0.03867	1.178	0.001174	1.090
3/4	40	0.06867	2.093	0.003703	3.441
	80	0.06183	1.883	0.003003	2.785
	160	0.03867	1.178	0.002043	1.898
1	40	0.08742	2.664	0.006002	5.574
	80	0.07975	2.430	0.004995	5.083
	160	0.06792	2.070	0.003623	3.365
1¼	40	0.1150	3.504	0.01039	9.643
	80	0.1065	3.246	0.008908	8.275
	160	0.09667	2.946	0.007339	6.816
1½	40	0.1342	4.090	0.01313	13.13
	80	0.1250	3.810	0.01227	11.40
	160	0.1115	3.398	0.009764	9.068
2	40	0.1723	5.252	0.02330	21.66
	80	0.1616	4.926	0.02051	19.06
	160	0.1306	4.286	0.01552	13.43
2½	40	0.2058	6.271	0.03325	30.89
	80	0.1936	5.901	0.02943	27.35
	160	0.1771	5.397	0.02463	22.88
3	40	0.2557	7.792	0.05134	47.69
	80	0.2417	7.366	0.04587	42.61
	160	0.2187	6.664	0.03755	34.88
3½	40	0.2957	9.012	0.06866	63.79
	80	0.2803	8.544	0.06172	57.33
4	40	0.3355	10.23	0.08841	82.19
	80	0.3198	9.718	0.07984	74.17
	120	0.3020	9.204	0.07163	66.54
	160	0.2865	8.732	0.06447	59.88
5	40	0.4206	12.82	0.1389	129.10
	80	0.4801	13.64	0.1810	168.30
	120	0.3803	11.59	0.1136	105.50
	160	0.3594	10.95	0.1015	94.17

13.3 ANALYSIS OF FLOW IN SINGLE PATH PIPE AND DUCT SYSTEMS

TABLE 13.4 (Continued)

Nominal Diameter		Inside Diameter		Flow Area	
(in.)	Schedule	ft	cm	ft²	cm²
6	40	0.5054	15.41	0.2006	186.50
	80	0.4801	13.64	0.1810	168.30
	120	0.4584	13.98	0.1650	153.50
	160	0.4823	13.18	0.1367	136.40
8	20	0.6771	20.64	0.3601	334.60
	30	0.6726	20.50	0.3553	330.10
	40	0.6651	20.27	0.3474	322.70
	60	0.6511	19.85	0.3329	309.50
	80	0.6354	19.37	0.3171	294.70
	100	0.6198	18.89	0.3017	280.30
	120	0.5989	18.26	0.2817	261.90
	130	0.5834	17.79	0.2673	248.60
	160	0.5678	17.31	0.2532	235.30
10	20	0.8542	26.04	0.5730	332.60
	30	0.8446	25.75	0.5604	520.80
	40	0.8350	25.46	0.5476	509.10
	60	0.8125	24.77	0.5185	481.90
	80	0.7968	24.29	0.4987	463.40
	100	0.7760	23.66	0.4730	439.70
	120	0.7552	23.02	0.4470	416.20
	130	0.7292	22.23	0.4176	388.10
	160	0.7083	21.59	0.3941	366.10
12	20	1.021	31.12	0.8185	760.60
	30	1.008	30.71	0.7972	740.71
	40	0.9948	30.33	0.773	722.50
	60	0.9688	29.53	0.7372	684.90
	80	0.9478	28.89	0.7056	655.50
	100	0.9218	28.10	0.6674	620.20
	120	0.8958	27.31	0.6303	585.80
	130	0.8750	26.67	0.6013	558.60
	160	0.8438	25.72	0.5592	519.60
20	20 (std)	1.604	48.89	2.021	1877.00
	160	1.339	40.80	1.407	1307.00

system shows that the volume flowrate through each pipe or duct segment, and through each component, is the same. Multiple path systems, in which the volume flowrate may differ in the various flow paths in the system, will be discussed in the next section.

In both single and multiple path systems, each component along a given flow path contributes to the overall loss of mechanical energy of the fluid passing through the system along that path. In Section 13.2 we discussed the major head loss in the fully

Figure 13.15 Control volume for analysis of a minor loss.

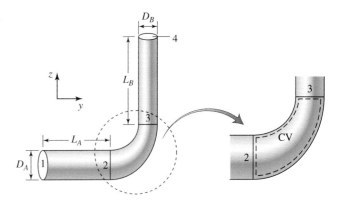

developed flow in a straight section of pipe of constant area and related it to the action of the wall shear stress. In this section we will discuss the use of a minor head loss to account for the loss of mechanical energy due to the presence of various components along a flow path. The loss of the mechanical energy in flows through elbows, tees, valves, filters, and other components is due to not only friction in the form of an average wall shear stress, but also to flow separation and the introduction of flow disturbances.

As a model for the effects of any type of component, consider a steady flow through a system consisting of two straight sections of pipe of different diameters connected by a reducing elbow as shown in Figure 13.15. We would like to predict the relationship between the pressure drop and flowrate through this single path system. To identify the additional loss of mechanical energy created by the reducing elbow, we will apply mass and energy balances to a CV that encloses all of the fluid within the elbow as shown in Figure 13.15. Since the density is constant, applying a mass balance gives $\dot{M} = \rho A_2 \bar{V}_2 = \rho A_3 \bar{V}_3$. Under these conditions the energy balance can be written as

$$\int_{CS} \rho\left(u + \frac{p}{\rho} + gz\right)(\mathbf{u} \cdot \mathbf{n})\,dS + \int_{CS} \rho\left(\frac{1}{2}\mathbf{u} \cdot \mathbf{u}\right)(\mathbf{u} \cdot \mathbf{n})\,dS = \dot{Q}_C$$

We will now assume that the internal energy and pressure are uniform at the inlet and exit of the CV, and evaluate the first flux integral at each end of the CV to obtain

$$\int_{CS} \rho\left(u + \frac{p}{\rho} + gz\right)(\mathbf{u} \cdot \mathbf{n})\,dS = \dot{M}\left[(u_3 - u_2) + \left(\frac{p_3}{\rho} + gz_3\right) - \left(\frac{p_2}{\rho} + gz_2\right)\right]$$

The value of the kinetic energy flux, $\int_{CS} \rho(\frac{1}{2}\mathbf{u} \cdot \mathbf{u})(\mathbf{u} \cdot \mathbf{n})\,dS$, depends on the form of the velocity profile at the inlet and exit of the CV. It is customary to introduce a kinetic energy coefficient α, and write the kinetic energy flux as

$$\int_{CS} \rho(\tfrac{1}{2}\mathbf{u} \cdot \mathbf{u})(\mathbf{u} \cdot \mathbf{n})\,dS = \pm\alpha\left(\tfrac{1}{2}\rho \bar{V}^3 A\right) \tag{13.27}$$

where a positive sign is used for an exit. Since $\dot{M} = \rho A \bar{V}$ and $\alpha(\tfrac{1}{2}\rho \bar{V}^3 A) = \dot{M}(\tfrac{1}{2}\alpha \bar{V}^2)$, the total kinetic energy flux then is $\int_{CS} \rho(\tfrac{1}{2}\mathbf{u} \cdot \mathbf{u})(\mathbf{u} \cdot \mathbf{n})\,dS = \dot{M}(\tfrac{1}{2}\alpha_3 \bar{V}_3^2 - \tfrac{1}{2}\alpha_2 \bar{V}_2^2)$. The energy balance for steady flow through the reducing elbow is thus given by

$$\dot{M}\left[(u_3 - u_2) + \left(\frac{p_3}{\rho} + \tfrac{1}{2}\alpha_3 \bar{V}_3^2 + gz_3\right) - \left(\frac{p_2}{\rho} + \tfrac{1}{2}\alpha_2 \bar{V}_2^2 + gz_2\right)\right] = \dot{Q}_C$$

13.3 ANALYSIS OF FLOW IN SINGLE PATH PIPE AND DUCT SYSTEMS

After rearranging this result we have

$$\left(\frac{p_2}{\rho} + \frac{1}{2}\alpha_2 \bar{V}_2^2 + gz_2\right) - \left(\frac{p_3}{\rho} + \frac{1}{2}\alpha_3 \bar{V}_3^2 + gz_3\right) = (u_3 - u_2) - \frac{\dot{Q}_C}{\dot{M}}$$

In analogy to the major head loss, we now introduce the concept of a minor head loss

$$h_M = (u_3 - u_2) - \frac{\dot{Q}_C}{\dot{M}} \qquad (13.28a)$$

and write the energy balance for the elbow as

$$\left(\frac{p_2}{\rho} + \frac{1}{2}\alpha_2 \bar{V}_2^2 + gz_2\right) - \left(\frac{p_3}{\rho} + \frac{1}{2}\alpha_3 \bar{V}_3^2 + gz_3\right) = h_M \qquad (13.28b)$$

The minor head loss (like the major head loss) has units of energy per unit mass and is determined empirically as explained in the next section. We see from Eq. 13.28a that the minor head loss accounts for the irreversible conversion of mechanical energy to thermal energy as the flow passes through the reducing elbow. In reality, some of the energy loss occurs in the disturbed flow downstream of the elbow. However, since we assume fully developed flow in all pipe or duct segments, in the preceding analysis the loss due to the disturbed flow is accounted for as if it occurred within the elbow.

Assuming steady, fully developed flow in each of the straight pipe segments in Figure 13.15, we can also use an energy balance to analyze each pipe segment, taking into account the elevations, the length and diameter of each pipe, and the friction factor and average velocity in each pipe. By using Eq. 13.10, we obtain

$$\left(\frac{p_1}{\rho} + gz_1\right) - \left(\frac{p_2}{\rho} + gz_2\right) = f_A \frac{L_A}{D_A} \frac{\bar{V}_A^2}{2}$$

and

$$\left(\frac{p_3}{\rho} + gz_3\right) - \left(\frac{p_4}{\rho} + gz_4\right) = f_B \frac{L_B}{D_B} \frac{\bar{V}_B^2}{2}$$

where the subscript on the friction factor indicates that it is to be evaluated appropriately for each pipe. Adding these two equations, we obtain a relationship between conditions at points 1 and 4:

$$\left(\frac{p_1}{\rho} + gz_1\right) - \left(\frac{p_4}{\rho} + gz_4\right) - \left[\left(\frac{p_2}{\rho} + gz_2\right) - \left(\frac{p_3}{\rho} + gz_3\right)\right] = f_A \frac{L_A}{D_A} \frac{\bar{V}_A^2}{2} + f_B \frac{L_B}{D_B} \frac{\bar{V}_B^2}{2}$$

Using Eq. 13.28b to eliminate the term in the square brackets, we can write the result as

$$\left(\frac{p_1}{\rho} + \frac{1}{2}\alpha_2 \bar{V}_2^2 + gz_1\right) - \left(\frac{p_4}{\rho} + \frac{1}{2}\alpha_3 \bar{V}_3^2 + gz_4\right) = f_A \frac{L_A}{D_A} \frac{\bar{V}_A^2}{2} + h_M + f_B \frac{L_B}{D_B} \frac{\bar{V}_B^2}{2}$$

Since we are assuming fully developed flow in each pipe, the kinetic energy flux at the ends of each pipe must be the same. Thus, we can write $\frac{1}{2}\alpha_2 \bar{V}_2^2 = \frac{1}{2}\alpha_1 \bar{V}_1^2$ and

$\frac{1}{2}\alpha_3 \bar{V}_3^2 = \frac{1}{2}\alpha_4 \bar{V}_4^2$. After substituting these relations into the preceding equation, we obtain

$$\left(\frac{p_1}{\rho} + \frac{1}{2}\alpha_1 \bar{V}_1^2 + gz_1\right) - \left(\frac{p_4}{\rho} + \frac{1}{2}\alpha_4 \bar{V}_4^2 + gz_4\right) = f_A \frac{L_A}{D_A} \frac{\bar{V}_A^2}{2} + h_M + f_B \frac{L_B}{D_B} \frac{\bar{V}_B^2}{2}$$

This equation shows that we can analyze flow through a passage made up of two straight pipe segments and a reducing elbow by setting the total change in the mechanical energy content of the fluid equal to the sum of the major loss in each pipe segment and the minor loss due to the elbow. Note carefully that the major loss term for each pipe segment employs the friction factor, length, diameter, and average velocity for the specific segment.

Now consider a much more complex single path pipe or duct system that contains any number of straight pipe or ducts segments plus any combination of components such as elbows, valves, reducers, filters, and traps. From the foregoing analysis we conclude that we can analyze the flow through such a system by setting the total change in the mechanical energy content of the fluid equal to the sum of the major and minor losses for each pipe or duct segment and each component along the flowpath. If point 1 for this system is at the inlet and point 2 is at the exit, we can write an energy balance from point 1 to point 2 as

$$\left(\frac{p_1}{\rho} + \frac{1}{2}\alpha_1 \bar{V}_1^2 + gz_1\right) - \left(\frac{p_2}{\rho} + \frac{1}{2}\alpha_2 \bar{V}_2^2 + gz_2\right) = \sum h_L + \sum h_M \quad (13.29a)$$

where the summation symbols indicate that a major loss term is to be supplied for the major loss in each straight pipe or duct segment, and a minor loss term is to be supplied for each component in the system. In a single path system, a mass balance also shows that

$$A_1 \bar{V}_1 = A_2 \bar{V}_2 \quad (13.29b)$$

thus Eqs. 13.29 are the basic equations governing flow through a single flow path system.

The form of Eq. 13.29a suggests that we define the total head loss h_T for a flow path as

$$h_T = \sum h_L + \sum h_M \quad (13.30)$$

This allows us to write the energy balance from point 1 to point 2 along a flow path as

$$\left(\frac{p_1}{\rho} + \frac{1}{2}\alpha_1 \bar{V}_1^2 + gz_1\right) - \left(\frac{p_2}{\rho} + \frac{1}{2}\alpha_2 \bar{V}_2^2 + gz_2\right) = h_T \quad (13.31)$$

To apply the energy balance in a practical problem, we must be able to determine the total head loss as well as values of the kinetic energy coefficient α that occur at the desired points along the flowpath. Now the major loss may be calculated by means of a friction factor as discussed earlier, and in the next section we discuss methods for obtaining the minor head loss for various components. How do we determine the kinetic energy coefficients?

To answer this question, note that the kinetic energy coefficient is defined by Eq. 13.27 as

$$\int_{CS} \rho\left(\tfrac{1}{2}\mathbf{u}\cdot\mathbf{u}\right)(\mathbf{u}\cdot\mathbf{n})\,dS = \pm\alpha\left(\tfrac{1}{2}\rho\bar{V}^3 A\right)$$

From this we see that the coefficient depends on the velocity profile at the section of interest. The values of this coefficient for steady, fully developed, laminar, and turbulent flow are of particular importance. It is straightforward to show that $\alpha = 1$ for a uniform profile, such as might be used to approximate the velocity profile at an inlet, or at the exit of a well designed nozzle, and that $\alpha = 2$ for the parabolic profile of a fully developed laminar pipe flow. In fully developed turbulent flow in a smooth pipe, we can employ an empirical approximation for the mean velocity profile in the form of the following power-law equation:

$$u(r) = U\left(1 - \frac{r}{R}\right)^{1/n} \tag{13.32}$$

where $u(r)$ is the mean velocity at radius r, U is the centerline mean velocity, and R is the pipe radius. The average velocity corresponding to this velocity profile is given by

$$\frac{\bar{V}}{U} = \frac{2n^2}{(n+1)(2n+1)} \tag{13.33}$$

and from Eq. 13.27, the kinetic energy coefficient is found to be given by

$$\alpha = \frac{(n+1)^3(2n+1)^3}{4n^4(n+3)(2n+3)} \tag{13.34}$$

The exponent n, in turn, depends on Reynolds number as shown in Figure 13.16. It can be seen that n varies from about 6 to 10 over the range of Re usually encountered. The power-law velocity profile in fully developed turbulent pipe flow lies between the two extremes of parabolic and uniform as shown in Figure 13.17. Although the kinetic energy coefficient α is a function n, and hence Re, the functional dependence is weak and it is customary to employ $\alpha = 1$ for fully developed turbulent flow in a pipe irrespective

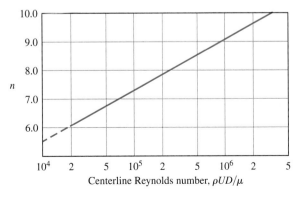

Figure 13.16 The power-law profile exponent.

13 FLOW IN PIPES AND DUCTS

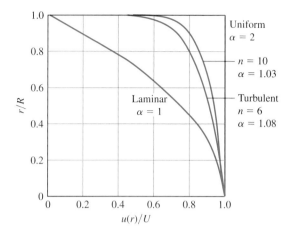

Figure 13.17 Velocity profiles for fully developed pipe flow.

of Reynolds number. The error introduced by using $\alpha = 1$ for turbulent flow is typically less than 8%. Thus, for engineering purposes we use $\alpha = 1$ for uniform or turbulent flow, and $\alpha = 2$ for a laminar pipe flow.

13.3.1 Minor Head Loss

A minor head loss occurs whenever there is a change in direction or geometry along a flow path. Consider the effects on the flow as it passes through the abrupt turn in the 90° miter bend shown in Figure 13.18A. Relatively large eddies form in the corner and the top of the base leg. The formation of these eddies is called flow separation. Within these eddies mechanical energy is converted into heat through viscous dissipation; thus flow separation and eddy formation is a key mechanism in the minor head loss. The eddies also decrease the effective cross-sectional area of the passage because there is no net flow downstream through them. The result is a flow structure known as a vena contracta. The fluid speeds up as it passes through the vena contracta, and some of the additional kinetic energy of this fluid is lost when it later slows down again. If guide vanes are

(A)

(B)

Figure 13.18 Flow ($Re = 2 \times 10^3$) through a 90° right angle bend (A) without and (B) with guide vanes.

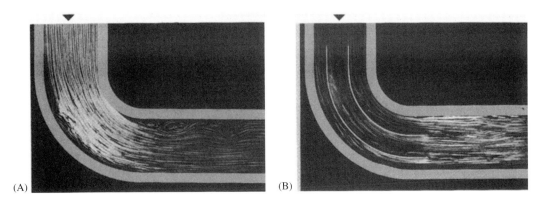

Figure 13.19 Flow ($Re = 2 \times 10^3$) through a 90° gradual bend (A) without and (B) with guide vanes.

added as shown in Figure 13.18B, the eddies in the base leg and vena contracta are eliminated. The result is a smaller loss of mechanical energy than in an abrupt turn without guide vanes. In some applications, such as a closed-circuit wind tunnel, the investment in turning vanes to reduce the minor head loss is justified not only by a lower operating expense, but also to maintain adequate performance (e.g., uniform flow in the test section).

A different approach to reducing head loss in a turn is to employ a gradual bend as shown in Figure 13.19A. In this case there is no corner in which eddies form, and the eddies in the base leg are less intense than those in the elbow. If guide vanes are added to the bend (Figure 13.19B), there is no flow separation at all and the minor head loss is further reduced. We could evaluate the flow in valves, reducers, filters and other flow elements in a similar fashion to obtain a qualitative understanding of the source of the minor head loss.

Values of the minor head loss for common flow system components are tabulated in two forms. The first is in the form of a head loss coefficient K, defined by writing the minor loss as

$$h_M = K \frac{\bar{V}^2}{2} \tag{13.35}$$

The second method is through the use of an equivalent pipe length L_e. A flow element is said to create a loss equivalent to the major loss that would occur at the same flowrate in a length L_e of the same pipe or duct. The minor loss is then written as

$$h_M = f \frac{L_e}{D} \frac{\bar{V}^2}{2} \tag{13.36}$$

where D, f, and $\bar{V}$ refer to the adjacent pipe flow, and L_e is tabulated for various components.

Values of the loss coefficient K, or equivalent length L_e, for a variety of flow elements are given as we develop the subject. It is important when using either form of minor head loss coefficient to take note of its precise definition. In the case of an element in which the inlet and outlet areas differ, it is critical to determine whether the coefficient is based on the inlet or outlet average velocity. You will also find that many manufacturers of products such as valves, filters, and nozzles prefer to give pressure drop as a function of flowrate. To obtain the corresponding minor loss in one of the two forms just described, we can use Eqs. 13.29 to apply a mass and energy balance to the element, dropping the major loss term in the energy balance, and assuming appropriate values for the kinetic energy coefficients.

 CD/Video library/Smoke stack

Inlets and Exits

The minor loss created by an inlet or exit is calculated using a loss coefficient K. There are a variety of possible inlet configurations for connecting a pipe to a reservoir of some type. These range from reentrant to well rounded, and it is possible to decrease the inlet loss significantly if the expense of a well-rounded inlet is justified. Four types of pipe inlet are shown in Figure 13.20, along with the corresponding values of the inlet loss coefficient, K_{in}. According to Eq. 13.35, the minor loss at an inlet is $h_M = K_{in}(\bar{V}^2/2)$, where the average velocity refers to the value in the pipe.

A similar set of exit configurations for joining a pipe to a reservoir is shown in Figure 13.21, along with the corresponding values of the exit loss coefficient, K_{ex}. Since all the kinetic energy contained in the fluid is eventually lost as the exiting fluid decelerates and mixes with the surrounding fluid, there is no advantage whatsoever to employing a certain exit configuration in an attempt to reduce the exit loss. The loss coefficient is $K_{ex} = 1$ for each type of exit, and the minor loss is $h_M = \bar{V}^2/2$, where $\bar{V}$ refers to the value in the pipe. As discussed later, a diffuser, which is a type of gradual expansion, may be used to reduce the kinetic energy loss at an exit.

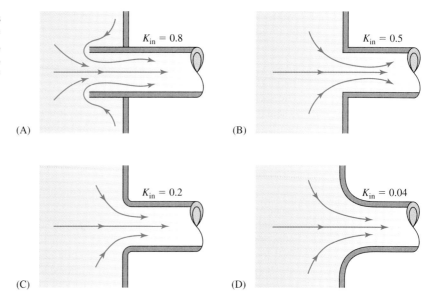

Figure 13.20 Loss coefficients for inlets: (A) reentrant, $K_{in} = 0.8$, (B) sharp edged, $K_{in} = 0.5$, (C) slightly rounded, $K_{in} = 0.2$, and (D) well rounded, $K_{in} = 0.04$.

13.3 ANALYSIS OF FLOW IN SINGLE PATH PIPE AND DUCT SYSTEMS

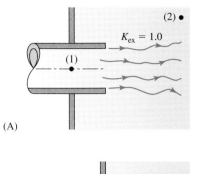
(A)

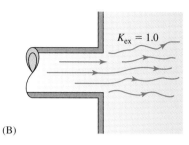
(B)

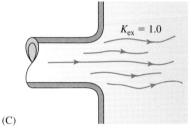
(C)

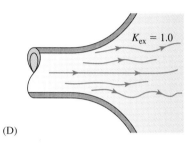
(D)

Figure 13.21 Loss coefficients for exits. (A) reentrant, (B) sharp edged, (C) slightly rounded, and (D) well rounded. $K_{ex} = 1.0$ for all exits.

EXAMPLE 13.9

Find the pressure at the pump inlet shown in Figure 13.22 if the flowrate is 20 gal/min in the 1 in. Schedule 40 galvanized iron inlet pipe. The water is at 60°F.

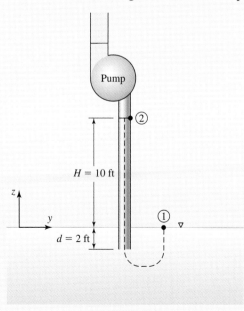

Figure 13.22 Schematic for Example 13.9.

SOLUTION

To compute the change in pressure we will apply Eq. 13.29a

$$\left(\frac{p_1}{\rho} + \frac{1}{2}\alpha_1 \bar{V}_1^2 + gz_1\right) - \left(\frac{p_2}{\rho} + \frac{1}{2}\alpha_2 \bar{V}_2^2 + gz_2\right) = \sum h_L + \sum h_M$$

between a point at the free surface of the reservoir and a point at the pump inlet. We must account for the reentrant inlet and the major loss associated with $(d + H = 12$ ft) of the pipe. From Table 13.4, the inside diameter of the pipe is $D = 1.049$ in., the average velocity corresponding to the 20 gal/min flowrate through this pipe is

$$\bar{V} = \frac{Q}{A} = \frac{(20 \text{ gal/min})\left(\frac{1 \text{ ft}^3}{7.48 \text{ gal}}\right)\left(\frac{1 \text{ min}}{60 \text{ s}}\right)}{\pi \frac{(1.049 \text{ in.})^2}{4}\left(\frac{1 \text{ ft}^2}{144 \text{ in.}^2}\right)} = 7.43 \text{ ft/s}$$

For a density $\rho = 1.938$ slugs/ft^3 and kinematic viscosity of $\nu = 1.21 \times 10^{-5}$ ft^2/s, the following Reynolds number is found:

$$Re = \frac{\bar{V}D}{\nu} = \frac{(7.43 \text{ ft/s})(1.049 \text{ in.})\left(\frac{1 \text{ ft}}{12 \text{ in.}}\right)}{1.21 \times 10^{-5} \text{ ft}^2/\text{s}} = 5.4 \times 10^4$$

The flow is turbulent in the pipe, so we will assume $\alpha_2 = 1$. Writing Eq. 13.29a and noting that the pressure at point 1 is atmospheric, and the kinetic energy is negligible, we have

$$\left(\frac{p_A}{\rho}\right) - \left(\frac{p_{in}}{\rho} + \frac{1}{2}\bar{V}_{in}^2 + gH\right) = f\frac{L}{D}\frac{\bar{V}_{in}^2}{2} + K_{in}\frac{\bar{V}_{in}^2}{2} \quad \text{(A)}$$

Solving for the pressure at the pump inlet, we find

$$p_{in} = p_A - \rho\frac{\bar{V}_{in}^2}{2} - \rho gH - f\frac{L}{D}\rho\frac{\bar{V}_{in}^2}{2} - K_{in}\rho\frac{\bar{V}_{in}^2}{2} \quad \text{(B)}$$

where $L/D = (12 \text{ ft}/1.049 \text{ in.})(12 \text{ in.}/1 \text{ ft}) = 137.3$, the relative roughness $e/D = 0.006$ is read from Figure 13.12, the friction factor is found from the Moody chart, Figure 13.13, to be $f = 0.034$ at $Re = 5.4 \times 10^4$, and K_{in} is taken as $K_{in} = 0.8$. From this equation we conclude that the pressure at the pump inlet is below atmospheric for four reasons: (1) the increase in velocity of the water as it enters the pipe, (2) the higher elevation of the pump inlet, (3) the major loss of 12 ft of pipe, and (4) the loss due to the reentrant inlet. Using the data provided to evaluate each term in (B) yields

$$p_A = 14.7 \text{ lb}_f/\text{in.}^2 \left(\frac{144 \text{ in.}^2}{1 \text{ ft}^2}\right) = 2117 \text{ lb}_f/\text{ft}^2$$

13.3 ANALYSIS OF FLOW IN SINGLE PATH PIPE AND DUCT SYSTEMS

$$\rho \frac{\bar{V}_{in}^2}{2} = (1.938 \text{ slugs/ft}^3)(7.43 \text{ ft/s})^2(0.5) = 53.5 \text{ lb}_f/\text{ft}^2$$

$$\rho g H = (1.938 \text{ slugs/ft}^3)(32.2 \text{ ft/s}^2)(10 \text{ ft}) = 624 \text{ lb}_f/\text{ft}^2$$

$$f\frac{L}{D}\rho\frac{\bar{V}_{in}^2}{2} = (0.034)(137.3)53.5 \text{ lb}_f/\text{ft}^2 = 250 \text{ lb}_f/\text{ft}^2$$

and

$$K_{in}\rho\frac{\bar{V}_{in}^2}{2} = (0.8)53.5 \text{ lb}_f/\text{ft}^2 = 43 \text{ lb}_f/\text{ft}^2$$

Thus the pressure at the pump inlet is

$$p_{in} = (2117 - 53.5 - 624 - 250 - 43) \text{ lb}_f/\text{ft}^2 = 1147 \text{ lb}_f/\text{ft}^2 = 7.97 \text{ psia}$$

which is well below atmospheric pressure.

If the pressure at a pump inlet approaches the vapor pressure of water, cavitation may occur in the pump with a loss of performance and possible damage. As discussed further in Section 13.5.1, pump manufacturers specify a quantity known as the Net Positive Suction Head Required to describe the minimum inlet pressure for the pump. Knowing this value, we can determine whether the pressure calculated at a pump inlet is allowable. In this example it can be seen that the effects of elevation and major loss are of the greatest significance. To increase the pressure at this pump inlet, we could consider a design change that involves a lower elevation and/or a larger inlet pipe.

  CD/Video library/Backward facing step

Sudden Area Change

The effect of a sudden area change should be familiar from the corresponding case study in Chapter 3 (Section 3.3.2). As shown in Figure 13.23, the minor loss associated with a sudden enlargement in a pipe is given by an enlargement loss coefficient K_E. The minor loss with an enlargement is expressed as $h_M = K_E(\bar{V}_1^2/2)$. Note that in this case the average velocity is that in the higher speed flow upstream, i.e., $\bar{V}_1$. In Example 7.8 we showed that the expansion loss coefficient can be obtained by using a momentum balance. The result according to (F) of that example is $K_E = (1 - A_1/A_2)^2$. For a round pipe, the enlargement loss coefficient can be computed from

$$K_E = \left[1 - \left(\frac{D_1}{D_2}\right)^2\right]^2 \qquad (13.37)$$

or simply read from the plot shown in the figure.

Figure 13.23 Loss coefficient for a sudden enlargement.

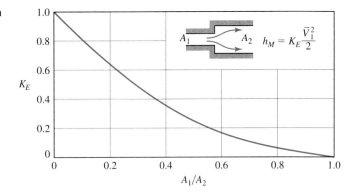

Figure 13.24 Loss coefficient for a sudden contraction.

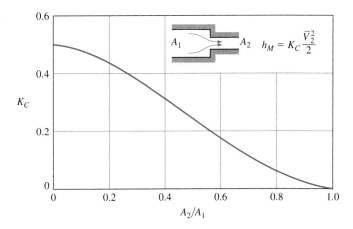

The effects of a sudden contraction are described by a contraction loss coefficient K_C and corresponding minor loss $h_M = K_C(\bar{V}_2^2/2)$ as illustrated in Figure 13.24. Note carefully that the loss is calculated in terms of the higher average velocity $\bar{V}_2$ in the downstream section of pipe. We cannot obtain a formula similar to Eq. 13.35 from a momentum balance in this case, so we must read the contraction loss coefficient from the figure.

EXAMPLE 13.10

An air-conditioning system employs horizontal, round sheet metal ducts of 6 and 12 in. diameter. What is the pressure change in a 400 ft³/min flow of 70°F air through a sudden enlargement from the 6 in. to the 12 in. duct? What is the frictional pressure drop caused by this enlargement? Find the corresponding quantities in a 400 ft³/min flow through a sudden contraction between these duct sizes.

13.3 ANALYSIS OF FLOW IN SINGLE PATH PIPE AND DUCT SYSTEMS

SOLUTION

We are asked to determine the pressure change and frictional pressure drop caused by a specified sudden enlargement in a piping system and then to compare these values with the corresponding quantities caused by a contraction between the same two duct sizes. Figures 13.23 and 13.24 show the appropriate geometries for the expansion and contraction.

First we calculate the average velocity in the 6 in. and 12 in. ducts, respectively, as

$$\bar{V}_6 = \frac{Q}{A} = \frac{400 \text{ ft}^3/\text{min}}{(\pi/4)(0.5 \text{ ft})^2} = 2037 \text{ ft/min}$$

and

$$\bar{V}_{12} = \frac{Q}{A} = \frac{400 \text{ ft}^3/\text{min}}{(\pi/4)(1 \text{ ft})^2} = 509 \text{ ft/min}$$

To analyze the enlargement, we will apply Eq. 13.29a:

$$\left(\frac{p_1}{\rho} + \frac{1}{2}\alpha_1 \bar{V}_1^2 + gz_1\right) - \left(\frac{p_2}{\rho} + \frac{1}{2}\alpha_2 \bar{V}_2^2 + gz_2\right) = \sum h_L + \sum h_M$$

Assuming turbulent flow, and noting that the duct is horizontal with no major loses, we write $(p_1/\rho + \frac{1}{2}\bar{V}_1^2) - (p_2/\rho + \frac{1}{2}\bar{V}_2^2) = h_M$, where for the enlargement point 1 is in the smaller duct. After combining Eqs. 13.35 and 13.37, we can write the minor loss as

$$h_M = K_E \frac{\bar{V}^2}{2} = \left[1 - \left(\frac{D_1}{D_2}\right)^2\right]^2 \frac{\bar{V}^2}{2}$$

Thus the pressure change across the enlargement is given by

$$p_1 - p_2 = \frac{1}{2}\rho\left(\bar{V}_2^2 - \bar{V}_1^2\right) + \rho\left[1 - \left(\frac{D_1}{D_2}\right)^2\right]^2 \frac{\bar{V}_1^2}{2}$$

The first term is the frictionless pressure increase due to the air slowing down; the second term is the pressure drop due to friction. Inserting the data and evaluating each term separately, we have (with $\bar{V}_1 = \bar{V}_6$ and $\bar{V}_2 = \bar{V}_{12}$)

$$\frac{1}{2}\rho(\bar{V}_2^2 - \bar{V}_1^2) = \frac{1}{2}(2.329 \times 10^{-3} \text{ slug/ft}^3)[(509 \text{ ft/min})^2 - (2037 \text{ ft/min})^2]\left(\frac{1 \text{ min}}{60 \text{ s}}\right)^2$$
$$= -1.26 \text{ lb}_f/\text{ft}^2$$

$$\rho\left[1 - \left(\frac{D_1}{D_2}\right)^2\right]^2 \frac{\bar{V}_1^2}{2} = (2.329 \times 10^{-3} \text{ slug/ft}^3)\left[1 - \left(\frac{0.5}{1}\right)^2\right]^2 \left(\frac{1}{2}\right)(2037 \text{ ft/min})^2 \left(\frac{1 \text{ min}}{60 \text{ s}}\right)^2$$
$$= 0.75 \text{ lb}_f/\text{ft}^2$$

We see that that pressure change across the enlargement is

$$p_1 - p_2 = -1.26 \text{ lb}_f/\text{ft}^2 + 0.75 \text{ lb}_f/\text{ft}^2 = -0.51 \text{ lb}_f/\text{ft}^2 \left(\frac{1 \text{ in. H}_2\text{O}}{5.2 \text{ lb}_f/\text{ft}^2}\right)$$
$$= -0.1 \text{ in. H}_2\text{O}$$

which is an increase in pressure. To understand this result, note that the effect of the area increase is larger than that of the minor loss. The enlargement loss coefficient is seen to be

$$K_E = \left[1 - \left(\frac{D_1}{D_2}\right)^2\right]^2 = \left[1 - \left(\frac{0.5}{1}\right)^2\right]^2 = 0.56$$

To analyze the contraction we again apply Eq. 13.29a, and assume turbulent flow to write $(p_1/\rho + \frac{1}{2}\bar{V}_1^2) - (p_2/\rho + \frac{1}{2}\bar{V}_2^2) = h_M$, where for the contraction, point 1 is in the larger duct. The minor loss is $h_M = K_C(\bar{V}_2^2/2)$, where we read the value of K_C from Figure 13.23. Thus the pressure change across the contraction is given by $p_1 - p_2 = \frac{1}{2}\rho(\bar{V}_2^2 - \bar{V}_1^2) + \rho K_C(\bar{V}_2^2/2)$. The first term is the frictionless pressure decrease due to the air speeding up; the second term is the pressure drop due to friction. Note that the area ratio in this case is $A_2/A_1 = D_2^2/D_1^2 = (0.5 \text{ ft})^2/(1 \text{ ft})^2 = 0.25$, hence $K_C = 0.41$. Inserting the data with $\bar{V}_1 = \bar{V}_{12}$ and $\bar{V}_2 = \bar{V}_6$, we have

$$\frac{1}{2}\rho\left(\bar{V}_2^2 - \bar{V}_1^2\right) = \frac{1}{2}(2.329 \times 10^{-3} \text{ slug/ft}^3)[(2037 \text{ ft/min})^2 - (509 \text{ ft/min})^2]\left(\frac{1 \text{ min}}{60 \text{ s}}\right)^2$$
$$= 1.26 \text{ lb}_f/\text{ft}^2$$

$$\rho K_C \frac{\bar{V}_2^2}{2} = (2.329 \times 10^{-3} \text{ slug/ft}^3)(0.41)\left(\frac{1}{2}\right)(2037 \text{ ft/min})^2 \left(\frac{1 \text{ min}}{60 \text{ s}}\right)^2$$
$$= 0.55 \text{ lb}_f/\text{ft}^2$$

We see that that pressure change across the contraction is

$$p_1 - p_2 = 1.26 \text{ lb}_f/\text{ft}^2 + 0.55 \text{ lb}_f/\text{ft}^2 = 1.81 \text{ lb}_f/\text{ft}^2 \left(\frac{1 \text{ in. H}_2\text{O}}{5.2 \text{ lb}_f/\text{ft}^2}\right) = 0.35 \text{ in. H}_2\text{O}$$

which is a decrease due to the flow speeding up, and a decrease due to friction.

Diffusers and Nozzles

Diffusers and nozzles are examples of components in which a gradual area change occurs. The conical diffuser and nozzle shown in Figure 13.25 are characterized respectively by a diffuser loss coefficient K_D and a nozzle loss coefficient K_N. The corresponding losses are given by $h_M = K_D(\bar{V}_{in}^2/2)$ and $h_M = K_N(\bar{V}_{in}^2/2)$. As shown in Table 13.5, both K_D and K_N are functions of the angle θ and relevant area ratio. As shown in Example 13.11, diffusers and nozzles are often used to change the area of a flow passage while avoiding the substantial frictional pressure drop incurred with a sudden expansion or contraction.

13.3 ANALYSIS OF FLOW IN SINGLE PATH PIPE AND DUCT SYSTEMS

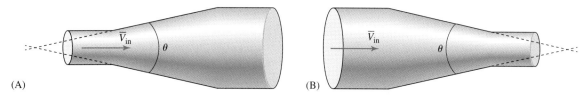

Figure 13.25 Angle definitions for (A) diffusers and (B) nozzles. Loss is based on $\bar{V}_{in}$ as shown.

TABLE 13.5 Loss Coefficients for Diffusers and Nozzles

Area ratio, A_1/A_2	Included Angle (°)									
	10	15	20	30	45	60	90	120	150	180
					Diffuser loss coefficient, K_D					
0.06	0.21	0.29	0.38	0.6	0.84	0.88	0.88	0.88	0.88	0.88
0.1	0.21	0.28	0.38	0.59	0.76	0.8	0.83	0.84	0.83	0.83
0.25	0.16	0.22	0.3	0.46	0.61	0.68	0.64	0.63	0.62	0.62
0.5	0.11	0.13	0.19	0.32	0.33	0.33	0.32	0.31	0.30	0.30
					Nozzle loss coefficient, K_N					
1	0	0	0	0	0	0	0	0	0	0
2	0.2	0.2	0.2	0.2	0.22	0.24	0.48	0.72	0.96	1
4	0.8	0.64	0.64	0.64	0.88	1.1	2.7	4.3	5.6	6.6
6	1.8	1.4	1.4	1.4	2	2.5	6.5	10	13	15
10	5	5	5	5	6.5	8	19	29	37	43

EXAMPLE 13.11

If the sudden enlargement and contraction in the system described in Example 13.10 are replaced with a conical diffuser and conical nozzle, respectively, and the included angle of each is 10°, by how much is the frictional pressure drop reduced?

SOLUTION

To compare the frictional pressure drops calculated for the sudden enlargement and contraction with those for a 10° included angle diffuser and nozzle, we will first compare the loss coefficients for each component. In Example 13.10 we found $K_E = 0.56$, with a frictional pressure drop of $\rho K_E(\bar{V}_1^2/2) = 0.75\ \text{lb}_f/\text{ft}^2$ for the enlargement. For the diffuser with $A_1/A_2 = 0.25$ and $\theta = 10°$ we find from Table 13.5 that $K_D = 0.16$. The ratio of the loss coefficients is $K_D/K_E = 0.16/0.56 = 0.29$, which can be interpreted as a 71% decrease in the loss coefficient realized by replacing the enlargement with a diffuser. Since the loss associated with either device is referenced to the same upstream velocity $\bar{V}_{in}$, the frictional pressure drop associated with the diffuser is simply 29% of that for the enlargement, or 0.22 lb_f/ft^2.

For the contraction we found $K_C = 0.41$, and $\rho K_C(\bar{V}_2^2/2) = 0.55$ lb_f/ft^2. From Table 13.5, we see that a nozzle with $A_1/A_2 = 4.0$ and $\theta = 10°$ has $K_N = 0.80$. It appears contradictory for K_C to be less than K_N but recall that the minor loss for the contraction is based on the higher downstream velocity $\bar{V}_2$, while for the nozzle the table values are based on the lower upstream velocity $\bar{V}_{in}$. The loss for the nozzle is

$$\rho K_N \frac{\bar{V}_1^2}{2} = (2.329 \times 10^{-3} \text{ slug/ft}^3)(0.8)\left(\frac{1}{2}\right)(509 \text{ ft/min})^2 \left(\frac{1 \text{ min}}{60 \text{ s}}\right)^2 = 0.067 \text{ lb}_f/\text{ft}^2$$

which corresponds to a reduction in the frictional pressure drop of $(0.55 \text{ lb}_f/\text{ft}^2 - 0.067 \text{ lb}_f/\text{ft}^2)/0.55 \text{ lb}_f/\text{ft}^2 \times 100\% = 88\%$.

The decrease in pressure drop due to the nozzle and diffuser can be translated into reduced operating cost by recalling that the power needed to move the fluid, $Q\Delta p$, is supplied by a fan typically driven by an electric motor. This savings is offset by increased initial construction costs. Another consideration in duct design is the amount of space an installation will require. In this example each gradual transition requires almost 6 ft of length.

Gradual and Miter Bends

The minor head loss associated with both gradual and miter bends is given in terms of an equivalent length L_e, thus the minor loss is calculated from Eq. 13.36 as $h_M = f(L_e/D)(\bar{V}^2/2)$, where $D, f,$ and $\bar{V}$ refer to the adjacent pipe flow. Data for the equivalent length to be used with both gradual and miter bends are given in Figure 13.26.

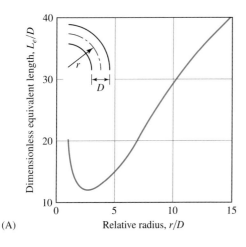
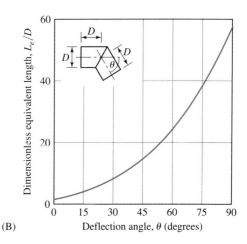

Figure 13.26 Minor loss coefficients (L_e/D) for (A) gradual and (B) miter bends.

13.3 ANALYSIS OF FLOW IN SINGLE PATH PIPE AND DUCT SYSTEMS

TABLE 13.6 Equivalent Lengths for Various Pipe Fittings

Pipe fitting	Description	L_e/D
Ball valve	Fully open	3
Gate valve	Fully open	8
	3/4 open	35
	1/2 open	160
	1/4 open	900
Globe valve	Fully open	340
	1/2 open	500
Elbows	45° standard	16
	90° standard	30
	90° long radius	20
180° bend	Close pattern return	50
90° bends	Bend radius = 1 pipe diameter	20
	Bend radius = 2 pipe diameters	12
	Bend radius = 3 pipe diameters	16
	Bend radius = 4 pipe diameters	24
Standard tee	With flow through run	20
	With flow through branch	60

Pipe Components

There are many different pipe components available for use in systems incorporating threaded, flanged, welded, or glued connections. Values of equivalent lengths for some of these components may be found in Table 13.6.

13.3.2 Pump and Turbine Head

Pumps, fans, blowers, and turbines occur in many pipe and duct systems. A pump, fan, or blower increases the mechanical energy of a fluid flowing through them, while a turbine decreases the mechanical energy of a fluid. To account for the presence of one of these devices along a flow path, we will make use of the concepts of a pump head H_{pump} and turbine head H_{turbine} and write the energy balance from point 1 to point 2 along a flow path like the one shown in Figure 13.27 as

$$\left(\frac{p_1}{\rho} + \frac{1}{2}\alpha_1 \bar{V}_1^2 + gz_1\right) - \left(\frac{p_2}{\rho} + \frac{1}{2}\alpha_2 \bar{V}_2^2 + gz_2\right) = h_T - H_{\text{pump}} + H_{\text{turbine}} \quad (13.38)$$

where h_T is the total head loss. This equation can then be used to analyze a single path system that includes a pump, fan, blower, or turbine along with any additional elements.

To derive an expression for either the pump or the turbine head, consider the mechanical energy change in the fluid passing through a power-producing or power-absorbing device. The energy change is determined as usual by the conditions at the inlet and outlet of the device. At the inlet, the mechanical energy content per unit mass is given by $p_{\text{in}}/\rho + \frac{1}{2}\alpha_{\text{in}} \bar{V}_{\text{in}}^2 + gz_{\text{in}}$. A similar term gives the energy content at the outlet. For devices operating on a constant density fluid, it is usually appropriate to assume

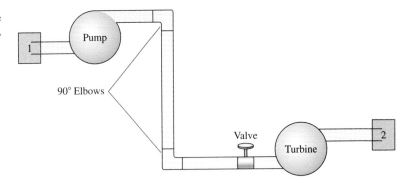

Figure 13.27 Schematic of a single pass flow system including pump, turbine, and minor loss elements.

turbulent flow with $\alpha = 1$, uniform properties at the inlet and outlet, and identical inlet and outlet areas and elevations. The increase in mechanical energy of the fluid passing through the device is then given by $(p_{out}/\rho - p_{in}/\rho)$, so the pump head is defined as

$$H_{pump} = \frac{p_{out} - p_{in}}{\rho} \tag{13.39}$$

In contrast, the turbine head, is defined as

$$H_{turbine} = \frac{p_{in} - p_{out}}{\rho} \tag{13.40}$$

Note carefully that as we have defined them, both pump and turbine heads are always positive quantities. That is, a pump, fan, or blower, which increases the pressure of the fluid, results in a positive value for H_{pump}, and a turbine, which decreases the pressure of the fluid, also results in a positive value for $H_{turbine}$. These definitions are consistent with the fact that on the right-hand side of Eq. 13.38, H_{pump} is subtracted from the total head loss while $H_{turbine}$ is added to h_T.

To relate the pump head H_{pump} to the actual shaft power P_{pump} required to run the pump, we will use the concept of an efficiency. Recall that in Example 7.15, we applied an energy balance to analyze the steady, constant density flow through a pump. The same analysis applies without modification to a fan or blower. In each case we assume turbulent flow with $\alpha = 1$, uniform properties at the inlet and outlet of the device, and identical inlet and outlet elevations and areas. The energy balance on a CV that encloses the device (see Example 7.15) is

$$\dot{M}\left[\left(u_{out} + \frac{p_{out}}{\rho}\right) - \left(u_{in} + \frac{p_{in}}{\rho}\right)\right] = \dot{W}_{shaft} + \dot{Q}_C$$

where $\dot{W}_{shaft} = P_{pump}$. Rearranging, we obtain

$$P_{pump} = Q(p_{out} - p_{in}) + [\dot{M}(u_{out} - u_{in}) - \dot{Q}_C]$$

In the absence of all losses, an energy balance would show that the ideal pump power is

$$P_{ideal} = Q(p_{out} - p_{in}) \tag{13.41}$$

13.3 ANALYSIS OF FLOW IN SINGLE PATH PIPE AND DUCT SYSTEMS

thus we can write the energy balance as

$$P_{\text{pump}} = P_{\text{ideal}} + [\dot{M}(u_{\text{out}} - u_{\text{in}}) - \dot{Q}_C]$$

The term $\dot{M}(u_{\text{out}} - u_{\text{in}}) - \dot{Q}_C$ represents the additional power input required to overcome losses in a real pump, fan, or blower. The efficiency of this type of device is defined as the ratio of the ideal power P_{ideal} to the actual power P_{pump}, or

$$\eta_{\text{pump}} = \frac{P_{\text{ideal}}}{P_{\text{pump}}} \qquad (13.42)$$

We can now write the required pump power by combining Eqs. 13.42 and 13.41 to write

$$P_{\text{pump}} = \frac{Q(p_{\text{out}} - p_{\text{in}})}{\eta_{\text{pump}}} \qquad (13.43)$$

From the definition of the pump head, Eq. 13.39, we can write $(p_{\text{out}} - p_{\text{in}}) = \rho H_{\text{pump}}$, thus from the preceding expression we have the desired expression relating pump head to the efficiency of the pump, shaft power input, and volume flowrate:

$$H_{\text{pump}} = \frac{\eta_{\text{pump}} P_{\text{pump}}}{\rho Q} \qquad (13.44)$$

This equation also applies to a fan or blower. Since the efficiency is always less than unity, the head provided by the pump is always less than the shaft power input per unit of mass flowrate.

Now consider a turbine. To account for the fact that the actual power output of a turbine is always less than the power extracted from the fluid, turbine efficiency is defined as

$$\eta_{\text{turbine}} = \frac{P_{\text{turbine}}}{P_{\text{ideal}}} \qquad (13.45)$$

and an energy balance on a turbine gives

$$\dot{M}\left[\left(u_{\text{out}} + \frac{p_{\text{out}}}{\rho}\right) - \left(u_{\text{in}} + \frac{p_{\text{in}}}{\rho}\right)\right] = \dot{W}_{\text{shaft}} + \dot{Q}_C$$

where now $\dot{W}_{\text{shaft}} = -P_{\text{turbine}}$. (Recall that positive shaft work is done on the CV.) Rearranging the energy balance, we can write

$$P_{\text{turbine}} = Q(p_{\text{in}} - p_{\text{out}}) - [\dot{M}(u_{\text{out}} - u_{\text{in}}) - \dot{Q}_C]$$

In the absence of all losses, an energy balance would show that the ideal turbine power is

$$P_{\text{ideal}} = Q(p_{\text{in}} - p_{\text{out}}) \qquad (13.46)$$

and after combining this with Eq. 13.45, we obtain

$$P_{\text{turbine}} = \eta_{\text{turbine}} Q(p_{\text{in}} - p_{\text{out}}) \qquad (13.47)$$

We can now use Eqs. 13.47 and 13.40 to write

$$H_{\text{turbine}} = \frac{P_{\text{turbine}}}{\rho Q \eta_{\text{turbine}}} \qquad (13.48)$$

EXAMPLE 13.12

A small water turbine rated at 25 hp has an efficiency of 61% and operates at a flowrate of 10 ft³/s. Find the turbine head and pressure drop across the turbine.

SOLUTION

We will apply Eq. 13.48 to determine the turbine head, then use Eq. 13.47 to determine the pressure drop. Inserting the data into Eq. 13.48 we have

$$H_{turbine} = \frac{P_{turbine}}{\rho Q \eta_{turbine}} = \frac{(25 \text{ hp})[550 \text{ (ft-lb}_f)/(\text{hp-s})]}{(1.94 \text{ slugs/ft}^3)(10 \text{ ft}^3/\text{s})(0.61)} = 1162 \text{ ft}^2/\text{s}^2$$

The pressure drop across the turbine is found by rearranging Eq. 13.47 to obtain $p_{in} - p_{out} = P_{turbine}/\eta_{turbine} Q$. Inserting the data yields

$$p_{in} - p_{out} = \frac{P_{turbine}}{\eta_{turbine} Q} = \frac{(25 \text{ hp})[550 \text{ (ft-lb}_f)/(\text{hp-s})]\left(\frac{1 \text{ ft}^2}{144 \text{ in.}^2}\right)}{(0.61)(10 \text{ ft}^3/\text{s})} = 15.7 \text{ psia}$$

The 61% efficiency means that the effective power input to the turbine from the water is 41 hp. To look at the impact of the efficiency of this turbine another way, if the turbine had 100% efficiency, the pressure drop would be only 9.5 psia.

13.3.3 Examples

Problems involving fully developed flow in a single path pipe or duct system are characterized by four parameters: the pressure difference $\Delta p = p_1 - p_2$, volume flowrate Q, pipe diameter D or duct hydraulic diameter D_H, and length L. By asking which of these parameters is unknown, we can identify problems of four distinct types in such flows. In this section we will describe the approaches used to solve each of these four problem types.

The governing equations in the absence of a pump or turbine are the continuity and energy equations as given by Eqs. 13.29a and 13.29b:

$$\left(\frac{p_1}{\rho} + \frac{1}{2}\alpha_1 \bar{V}_1^2 + gz_1\right) - \left(\frac{p_2}{\rho} + \frac{1}{2}\alpha_2 \bar{V}_2^2 + gz_2\right) = \sum h_L + \sum h_M$$

and

$$A_1 \bar{V}_1 = A_2 \bar{V}_2$$

where the major loss for each pipe segment is given by Eq. 13.8 as $h_L = f(L/D)(\bar{V}^2/2)$, and the minor losses are given by either Eq. 13.35 or 13.36 as $h_M = K(\bar{V}^2/2)$ or $h_M = f(L_e/D)(\bar{V}^2/2)$. In a duct flow, of course, we write these equations in terms of hydraulic diameter. If a pump or turbine is present, we employ Eq. 13.38,

13.3 ANALYSIS OF FLOW IN SINGLE PATH PIPE AND DUCT SYSTEMS

which includes the pump or turbine head terms:

$$\left(\frac{p_1}{\rho} + \frac{1}{2}\alpha_1 \bar{V}_1^2 + gz_1\right) - \left(\frac{p_2}{\rho} + \frac{1}{2}\alpha_2 \bar{V}_2^2 + gz_2\right) = h_T - H_{\text{pump}} + H_{\text{turbine}}$$

The elevation difference is normally assumed to be known, and the friction factor is determined by the relative roughness and Moody charts or by using one of the formulas given earlier. A strategy for each type of problem is given, along with a reference to a relevant example. It is helpful to study the indicated examples to learn how to apply the strategy.

1. **Known: Q, D, or D_H, and L; Unknown: Δp** With Q and D or D_H known, we can calculate $\bar{V}$ and Re. The type of pipe or duct determines e/D, and we can read the friction factor from the Moody chart or obtain it from one of the formulas. Since L is also known, we next solve the appropriate energy equation for the unknown Δp. This strategy was demonstrated in Examples 13.6 and 13.9.

2. **Known: Q, D, or D_H, and Δp; Unknown: L** With Q and D or D_H known, we can calculate $\bar{V}$ and Re. The type of pipe or duct determines e/D, and we can read the friction factor from the Moody chart or obtain it from one of the formulas. Since Δp is also known, we can solve the appropriate energy equation for the unknown length. This strategy is demonstrated in Example 13.13.

EXAMPLE 13.13

A new reservoir will use gravity to supply drinking water to a water treatment plant serving several surrounding towns, as shown in Figure 13.28. The required flowrate is 5000 gal/min. The surface of the reservoir is 200 ft above the plain where the water treatment plant is located, and the supply pipe is commercial steel, 3 ft in diameter. If the minimum pressure required at the water treatment plant is 50 psig, how far away can the reservoir be located with this size pipe? Assume that minor losses are negligible and that the water is at 50°F.

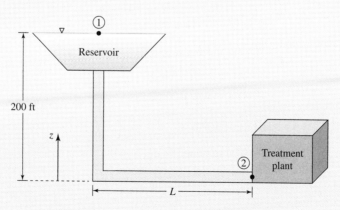

Figure 13.28 Schematic for Example 13.13.

SOLUTION

We are asked to determine the maximum distance between a water treatment plant and a reservoir given a set of piping constraints. Figure 13.28 serves as an appropriate sketch for this flow situation. This is a type 2 problem in that the unknown is a length. We will apply Eq. 13.29a:

$$\left(\frac{p_1}{\rho} + \frac{1}{2}\alpha_1 \bar{V}_1^2 + gz_1\right) - \left(\frac{p_2}{\rho} + \frac{1}{2}\alpha_2 \bar{V}_2^2 + gz_2\right) = \sum h_L + \sum h_M$$

To find the velocity in the pipe, $\bar{V}_2$, we write $\bar{V}_2 = Q/A$ and insert the data to obtain

$$\bar{V}_2 = \frac{(5000 \text{ gal/min})\left(\frac{1 \text{ min}}{60 \text{ s}}\right)\left(\frac{1 \text{ ft}^3}{7.48 \text{ gal}}\right)}{\pi (3 \text{ ft})^2/4} = 1.58 \text{ ft/s}$$

Using properties for water at 50°F, we find $Re = \bar{V}_2 D/\nu = [(1.58 \text{ ft/s})(3 \text{ ft})]/(1.407 \times 10^{-5} \text{ ft}^2/\text{s}) = 3.4 \times 10^5$; thus the flow is turbulent, as expected. From Figure 13.12, the relative roughness for $D = 3$ ft commercial steel pipe is $e/D = 5 \times 10^{-5}$. The friction factor from Figure 13.13, or calculated using the Colebrook or Chen equation, is found to be $f = 0.0147$.

To apply Eq. 13.29a, note that at point 1 on the reservoir surface, $p_1 = 14.7$ psia $= 0$ psig, $\bar{V}_1^2$ is negligible and $z_1 = 200$ ft. At point 2 we have $p_2 = 50$ psig, $\bar{V}_2 = 1.58$ ft/s, and $z_2 = 0$ ft. We will assume $\alpha_2 = 1$ as usual. The major loss is $h_L = f(L/D)(\bar{V}_2^2/2)$, so after neglecting the minor losses, Eq. 13.29a becomes $g(z_1 - z_2) - (p_2/\rho + \frac{1}{2}\alpha_2 \bar{V}_2^2) = f(L/D)(\bar{V}_2^2/2)$. Solving for the length, we obtain

$$L = \frac{2D}{f\bar{V}_2^2}\left[g(z_1 - z_2) - \left(\frac{p_2}{\rho} + \frac{1}{2}\alpha_2 \bar{V}_2^2\right)\right]$$

Inserting the data yields:

$$L = \frac{2(3 \text{ ft})}{(0.0147)(1.58 \text{ ft/s})^2}$$
$$\times \left[(32.2 \text{ ft/s}^2)(200 \text{ ft}) - \frac{(50 \text{ lb}_f/\text{in.}^2)(144 \text{ in.}^2/\text{ft}^2)}{1.94 \text{ slugs/ft}^3} - \frac{1}{2}(1.58 \text{ ft/s})^2\right]$$
$$= 4.46 \times 10^5 \text{ ft}$$

This is a distance of about 84 miles. Note that in this problem we have used gage pressure consistently throughout. In earlier problems we used absolute pressure throughout. Both approaches are correct, but we cannot mix gage and absolute pressure in the head loss equation.

13.3 ANALYSIS OF FLOW IN SINGLE PATH PIPE AND DUCT SYSTEMS

3. Known: Δp, D, or D_H, and L; Unknown: Q With Q unknown, we cannot calculate $\bar{V}$ and Re. The problem requires iteration. Consider a flow in a pipe. One approach is to assume a friction factor on the fully turbulent, i.e., horizontal, portion of the curve for the value of e/D, then solve the appropriate energy equation for the estimated $\bar{V}$. We can now calculate Re for this value of $\bar{V}$ and use it to find a new value of f, followed by a new estimate for $\bar{V}$. The iteration process stops when the change in the estimated average velocity is sufficiently small. The final step is to calculate Q. The same approach is also used for flow in a duct, but in that case we use e/D and Re based on D_H. In Example 13.14, note how this strategy is employed to determine the flowrate through a siphon.

EXAMPLE 13.14

Gasoline is siphoned from a tank with a smooth hose, i.d. 2 cm, as shown in Figure 13.29. For this grade of gasoline $\rho = 719$ kg/m³ and $\mu = 2.92 \times 10^{-4}$ (N-s)/m². Find the flowrate.

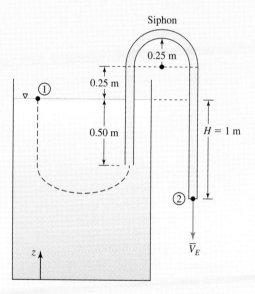

Figure 13.29 Schematic for Example 13.14.

SOLUTION

We are asked to determine the flowrate through a siphon associated with a gasoline tank. Figure 13.29 serves as an appropriate sketch for this flow situation. Since the unknown is Q, this is a type 3 problem. A similar problem was investigated in Example 8.13 by assuming frictionless flow. From the Bernoulli equation, the frictionless exit velocity was found to be $V_E = \sqrt{2gH}$. If we employ that formula to obtain an estimate of the velocity and flowrate in this case the results are:

$$V_E = \sqrt{2(9.8 \text{ m/s}^2)(1 \text{ m})} = 4.43 \text{ m/s}$$

$$Q_{\text{ideal}} = V_E A_E = (4.43 \text{ m/s})\left(\frac{\pi}{4}\right)(0.02 \text{ m})^2 = 1.4 \times 10^{-3} \text{ m}^3/\text{s}$$

Here we intend to consider the effect of friction by including the effects of the major and minor head loss. We will apply Eq. 13.29a:

$$\left(\frac{p_1}{\rho} + \frac{1}{2}\alpha_1 \bar{V}_1^2 + gz_1\right) - \left(\frac{p_2}{\rho} + \frac{1}{2}\alpha_2 \bar{V}_2^2 + gz_2\right) = \sum h_L + \sum h_M$$

with the first point on the free surface of the gasoline and the second point at the siphon exit. At the free surface, the pressure is atmospheric and the kinetic energy is negligible; at the exit, the pressure is atmospheric, the average velocity is $\bar{V}_E$ and the kinetic energy coefficient is α_E. Since $z_1 - z_2 = H$, we can write the resulting equation as $gH = \frac{1}{2}\alpha_E \bar{V}_E^2 + \sum h_L + \sum h_M$. The quantity gH in this equation is the total head available to drive the flow. As fluid moves from the free surface and through the hose to the exit, this total or available head is reduced by the head loss associated with the major and minor losses, leaving a velocity head at the exit of $\frac{1}{2}\alpha_E \bar{V}_E^2 = gH - \sum h_L - \sum h_M$. Solving this equation for the exit velocity we obtain

$$\bar{V}_E = \sqrt{\frac{2(gH - \sum h_L - \sum h_M)}{\alpha_E}} \quad \text{(A)}$$

To check this result, note that in a frictionless flow, the major and minor losses are zero, the velocity profile is uniform, which means that $\alpha_E = 1$, and (A) predicts $\bar{V}_E = \sqrt{2gH}$. For a uniform velocity profile, $V_E = \bar{V}_E$, so that the result from the Bernoulli equation, $V_E = \sqrt{2gH}$, is reproduced. In a flow with friction, where $\alpha_E \geq 1$ and the major and minor losses are nonzero, the average exit velocity is less than $\sqrt{2gH}$; i.e., the velocity and flowrate are reduced by friction.

The total head loss for this problem is due to 2.0 m of hose, a reentrant inlet, and two consecutive 90° bends, each with 0.25 m radius of curvature. Since we are at, but not beyond, the exit, and the kinetic energy of the flow at the exit is unchanged from its value in the hose, there is no exit minor loss in this case. Using Figure 13.20A for the reentrant inlet, we find that $K_{\text{in}} = 0.8$. For the bends, $r/D = 0.25 \text{ m}/0.02 \text{ m} = 12.5$, so from Figure 13.26A we find that $L_e/D = 35$. In this incompressible flow in a constant area hose, the average velocity at any point along the hose is the same as at the exit. The total head loss is therefore given by

$$\sum h_L + \sum h_M = f\frac{L}{D}\frac{\bar{V}_E^2}{2} + K_{\text{in}}\frac{\bar{V}_E^2}{2} + f\frac{L_e}{D}\frac{\bar{V}_E^2}{2} = \frac{\bar{V}_E^2}{2}\left[\frac{f}{D}(L + L_e) + K_{\text{in}}\right]$$

13.3 ANALYSIS OF FLOW IN SINGLE PATH PIPE AND DUCT SYSTEMS

Substituting this result into (A) and solving for $\bar{V}_E$, we obtain

$$\bar{V}_E = \sqrt{\frac{2gH}{\alpha_E + [(f/D)(L + L_e) + K_{\text{in}}]}} \tag{B}$$

It appears that we can solve directly for $\bar{V}_E$; but note that we must be able to calculate Re to find f and α_E. Thus we must first guess $\bar{V}_E$ and iterate. As a first guess we should choose a value less than the frictionless result 4.43 m/s, since friction can be expected to reduce the flowrate. Suppose we choose $\bar{V}_E^{(1)} = 4$ m/s. Using property values given in the problem statement, we find

$$Re^{(1)} = \frac{\rho \bar{V} D}{\mu} = \frac{(719 \text{ kg/m}^3)(4 \text{ m/s})(0.02 \text{ m})}{2.92 \times 10^{-4} \text{ (N-s)/m}^2} = 2 \times 10^5$$

hence the flow is turbulent. We therefore assume $\alpha_E = 1$. From the Moody chart or the Chen equation we can now determine $f^{(1)} = 0.0165$, noting that $e/D = 0$ for smooth tubes. Now we use (B) to determine $\bar{V}_E^{(2)}$ as

$$\bar{V}_E^{(2)} = \sqrt{\frac{2(9.81 \text{ m/s}^2)(1 \text{ m})}{1 + [(0.0165/0.02 \text{ m})(2.0 + 1.4) + 0.8]}} = 2.1 \text{ m/s}$$

then use this result to calculate $Re^{(2)}$ and $f^{(2)}$, repeating the process until the value of $\bar{V}_E$ converges. The results of this iteration are shown in the following table.

Iteration	$\bar{V}$ (m/s)	Re	f
1	4.0	2×10^5	0.0165
2	2.1	1×10^5	0.0178
3	2.0	9×10^4	0.0175
4	2.0	9×10^4	0.0175

We see that the effect of friction on a siphon is significant, reducing the average velocity by over 50% of the ideal frictionless value. The final flowrate is found to be

$$Q = \bar{V}_E A_E = (2.0 \text{ m/s})\left(\frac{\pi}{4}\right)(0.02 \text{ m})^2 = 6.3 \times 10^{-4} \text{ m}^3/\text{s}$$

Notice also that we assumed $\alpha = 1$ for this problem. In fact, Figure 13.16 and Eq. 13.34 show that n is about 7 and $\alpha = 1.06$. Repeating the iteration with this value shows that our use of $\alpha = 1$ results in a 3% error.

844 | 13 FLOW IN PIPES AND DUCTS

4. **Known: Δp, Q, and L; Unknown: D or D_H** With D or D_H unknown, we cannot calculate $\bar{V}$ and Re, even though we know Q. The problem requires iteration. One approach is to assume a value for D or D_H, which allows $\bar{V}$ and Re to be calculated. Since D or D_H is assumed known, e/D and the Moody chart can next be used to find f, after which the appropriate energy equation is solved to find the estimated Δp. The estimated Δp is compared with the known value and used to adjust the diameter upward or downward. If the estimated Δp is larger than the actual value, D or D_H should be increased because this will result in a lower $\bar{V}$ and Re, and ultimately a lower f and estimated Δp. The iteration process stops when the estimated Δp agrees with the known value. At this point the value of D or D_H is known. This strategy is demonstrated in Example 13.15.

EXAMPLE 13.15

A farmer needs to pump 50°F water to a spray irrigation system 100 ft away and 13 ft above the pump as shown in Figure 13.30. The irrigation system requires 30 gal/min at a minimum of 25 psig, and the farmer's pump supplies 30 gal/min at no more than 50 psig. What is the minimum diameter of PVC pipe that will meet the requirements if minor losses are neglected?

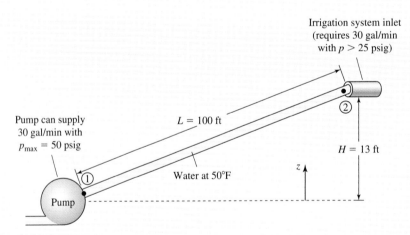

Figure 13.30 Schematic for Example 13.15.

SOLUTION

We are asked to determine the minimum pipe diameter that will meet the requirements of a specified irrigation system. Figure 13.30 serves as an appropriate sketch for this flow situation. Since the unknown is D, this is a type 4 problem. We will apply Eq. 13.29a

$$\left(\frac{p_1}{\rho} + \frac{1}{2}\alpha_1 \bar{V}_1^2 + gz_1\right) - \left(\frac{p_2}{\rho} + \frac{1}{2}\alpha_2 \bar{V}_2^2 + gz_2\right) = \sum h_L + \sum h_M$$

13.3 ANALYSIS OF FLOW IN SINGLE PATH PIPE AND DUCT SYSTEMS

with the first point at the pump exit, and the second point at the inlet of the irrigation system. The elevations of these two points are known, and we also know $\bar{V} = \bar{V}_1 = \bar{V}_2$, and minor losses are negligible. We will assume turbulent flow with $\alpha = 1$. Under these conditions the equation reduces to $(p_1 - p_2)/\rho + g(z_1 - z_2) = f(L/D)(\bar{V}^2/2)$. Since $z_1 - z_2 = -H$, we have:

$$(p_1 - p_2) - \rho g H = \rho f \frac{L}{D}\frac{\bar{V}^2}{2} \qquad (A)$$

Solving for the pressure at the pump we obtain

$$p_1 = p_2 + \rho g H + \rho f \frac{L}{D}\frac{\bar{V}^2}{2} \qquad (B)$$

We see that the pump output pressure p_1, must at a minimum be greater than the ambient hydrostatic pressure to $\rho g H$ at the pump outlet plus the amount needed to overcome the frictional pressure drop in the pipe, $\rho f(L/D)(\bar{V}^2/2)$. In addition, the pump must provide the pressure p_2 required at the irrigation system. To find the maximum frictional pressure drop that can be allowed, we evaluate the left-hand side of (A), using the known values of p_2 and p_1, to obtain

$$(p_1 - p_2) - \rho g H = (50 - 25) \text{ lb}_f/\text{in.}^2 - (1.94 \text{ slugs/ft}^3)(32.2 \text{ ft/s}^2)(13 \text{ ft})\left(\frac{1 \text{ ft}^2}{144 \text{ in.}^2}\right)$$

$$= 19.4 \text{ psi}$$

We must now select a Schedule 40 pipe size whose diameter D results in a frictional pressure drop less than or equal to the maximum allowable pressure drop of 19.4 psi. We can express this requirement by using (A) and the preceding calculation as $\rho f(L/D)(\bar{V}^2/2) \leq 19.4$ psi. We will select an initial value of $D_{40}^{(1)}$, use the inside diameter corresponding to this pipe size and the specified flowrate to calculate $\bar{V}^{(1)}$, $Re^{(1)}$, and $f^{(1)}$, calculate $[\rho f(L/D)(\bar{V}^2/2)]^{(1)}$, then increase or decrease the $D_{40}^{(2)}$ value to the next larger or smaller pipe size as needed to satisfy the inequality. We will assume a relative roughness of zero for smooth plastic.

To illustrate the first iteration, note that for an initial guess of $\frac{3}{4}$ in. pipe, $D = 0.06867$ ft. Using property values of $\rho = 1.94$ slugs/ft^3 and $\nu = 1.407 \times 10^{-5}$ ft^2/s for water at 50°F, we find that the average velocity $\bar{V}^{(1)}$ is given by

$$\bar{V}^{(1)} = \frac{Q}{\pi D^2/4} = \frac{4(30 \text{ gal/min})}{\pi(0.06867 \text{ ft})^2(7.48 \text{ gal/ft}^3)(60 \text{ s/min})} = 18.0 \text{ ft/s}$$

and the remaining parameters are found to be

$$Re^{(1)} = \frac{\bar{V}^{(1)} D^{(1)}}{\nu} = \frac{(18.0 \text{ ft/s})(0.06867 \text{ ft})}{(1.407 \times 10^{-5} \text{ ft}^2/\text{s})} = 87,850$$

$$f^{(1)} = 0.0182$$

$$\rho f \frac{L}{D}\frac{\bar{V}^2}{2} = \frac{(1.94 \text{ slugs/ft}^3)(0.0182)\left(\frac{100 \text{ ft}}{0.06867 \text{ ft}}\right)\frac{1}{2}(18.0 \text{ ft/s})^2}{144 \text{ in.}^2/\text{ft}^2} = 57.8 \text{ psi}$$

Using this procedure we can generate the data shown in the following table:

D_{40} (in.)	D (ft)	$\bar{V}_2$(ft/s)	Re	f	$\rho f \frac{L}{D}\frac{\bar{V}^2}{2}$ (psi)
$\frac{3}{4}$	0.06867	18.0	8.8×10^4	0.0182	57.8
1	0.08742	11.1	6.9×10^4	0.0192	18.2
$1\frac{1}{4}$	0.1150	6.43	5.3×10^4	0.0205	4.98

The table indicates that a nominal 1 in. diameter pipe is satisfactory. If we assume that the pump pressure with this size pipe is 50 psig, then by rearranging (B), the pressure at the irrigation system is found to be

$$p_2 = p_1 - \rho g H - \rho f \frac{L}{D}\frac{\bar{V}^2}{2} = (50 - 5.6 - 18.2) \text{ psig} = 26.2 \text{ psig}$$

which is 1.2 psig above the required pressure of 25 psig. Since there are many uncertainties in pipe flow calculations and the predicted margin is small, it may be prudent to select the $1\frac{1}{4}$ in. pipe if the economics allow.

13.4 ANALYSIS OF FLOW IN MULTIPLE PATH PIPE AND DUCT SYSTEMS

In contrast to the single path piping systems we have studied thus far, most engineering systems have multiple flow paths. In such a system, fluid may travel from inlet to exit via different paths, and there may be multiple inlets and multiple exits. In this section we will analyze multiple path systems by writing mass and energy balances between various points and employing the concepts of major and minor losses. Consider the simple two-branch system shown in Figure 13.31. The points where branches meet are called nodes. In this case node A, located at the upstream entrance of the first tee, is the point at which the two parallel branches of this system begin. The two branches end at node B, located at the downstream exit of the second tee. Branches in a system may be serial (end to end), or parallel, as shown in the system in Figure 13.31.

Suppose fluid enters the system at node A with average velocity $\bar{V}$ at a flowrate Q. Since the entrance and exit pipe diameters are the same, a mass balance shows that fluid exits the system at node B at the same average velocity and flowrate. Some of the fluid

13.4 ANALYSIS OF FLOW IN MULTIPLE PATH PIPE AND DUCT SYSTEMS

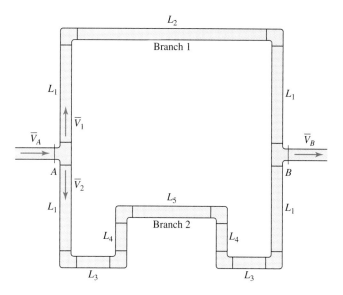

Figure 13.31 Multipath piping system.

Most pipe and duct systems are designed with variable head loss components that are adjusted to cause the flowrate in each path to meet design specifications. This process is known as flow balancing. The variable head loss components are known as control valves in pipe systems and dampers in duct systems. You used one of these devices the last time you washed your hands in warm water in a sink or adjusted the air vent in your automobile.

flows through branch 1 with average velocity $\bar{V}_1$ at flowrate Q_1, and the rest flows through branch 2 at average velocity $\bar{V}_2$ at flowrate Q_2. Since all the fluid entering node A leaves in the two branches, we have $Q = Q_1 + Q_2$. (Since the pipe diameters are the same in this case, we also have $\bar{V} = \bar{V}_1 + \bar{V}_2$.)

Now suppose the pressures and elevations at nodes A and B and flowrate Q are known. How can we determine the unknown flowrate through each branch? To answer this question we will use Eq. 13.31 to write energy balances between the two nodes along each branch. For branch 1, noting that $\bar{V}_A = \bar{V}_B = \bar{V}$, and assuming turbulent flow with $\alpha_A = \alpha_B = 1$, we have $(p_A/\rho + gz_A) - (p_B/\rho + gz_B) = h_{T1}$, where h_{T1} is the total head loss for fluid traveling along branch 1. Similarly, writing the energy balance for branch 2, we obtain $(p_A/\rho + gz_A) - (p_B/\rho + gz_B) = h_{T2}$. Comparing these two equations, we conclude that $h_{T1} = h_{T2}$, meaning that the total head loss must be the same on the parallel branches joining nodes A and B.

These important results can be summarized in general by two statements:

1. The head loss in the parallel branches joining two nodes in a multiple path system is the same.

2. The flowrate into a juncture where one or more branches meet is equal to the sum of the flowrates out of the juncture on each branch.

We can now complete our analysis by writing an energy balance for each branch including both major and minor losses. For branch 1 we obtain

$$\left(\frac{p_A}{\rho} + gz_A\right) - \left(\frac{p_B}{\rho} + gz_B\right) = 2\left(f_1 \frac{L_1}{D} \frac{\bar{V}_1^2}{2}\right) + \left(f_1 \frac{L_2}{D} \frac{\bar{V}_1^2}{2}\right)$$
$$+ 2\left(f_1 \frac{L_{\text{tee}}}{D} \frac{\bar{V}_1^2}{2}\right) + 2\left(f_1 \frac{L_{\text{elbow}}}{D} \frac{\bar{V}_1^2}{2}\right)$$

where L_{tee} is the appropriate equivalent length for the branch run of the identical tees, L_{elbow} is that for the elbow, and f_1 is the friction factor for the relative roughness and the Reynolds number in branch 1. Similarly, writing the energy balance for branch 2, we obtain

$$\left(\frac{p_A}{\rho} + gz_A\right) - \left(\frac{p_B}{\rho} + gz_B\right) = 2\left(f_2 \frac{L_1}{D} \frac{\bar{V}_2^2}{2}\right) + 2\left(f_2 \frac{L_3}{D} \frac{\bar{V}_2^2}{2}\right) + 2\left(f_2 \frac{L_4}{D} \frac{\bar{V}_2^2}{2}\right)$$
$$+ \left(f_2 \frac{L_5}{D} \frac{\bar{V}_2^2}{2}\right) + 2\left(f_2 \frac{L_{\text{tee}}}{D} \frac{\bar{V}_2^2}{2}\right) + 6\left(f_2 \frac{L_{\text{elbow}}}{D} \frac{\bar{V}_2^2}{2}\right)$$

From rule 2 we know $Q = Q_1 + Q_2$. To solve the problem for known pressures and elevations at each node, we must iterate, beginning with a guess for Q (and thus $\bar{V}$) in each branch. The iteration stops when the head loss in each branch is same (rule 1) and equal to $(p_A/\rho + gz_A) - (p_B/\rho + gz_B)$, and the volume flowrates satisfy $Q = Q_1 + Q_2$.

EXAMPLE 13.16

A heating system for a solar house is designed to circulate 120°F hot water from roof-mounted solar collectors through a horizontal heat exchanger embedded in the concrete floor of the living space as shown in Figure 13.32. Control is provided by a thermostatically activated motorized $\frac{3}{4}$ in. globe valve in the bypass line that allows the hot water to return to the solar collectors. The heat exchanger consists of $\frac{3}{4}$ in. Schedule 40 PVC pipe with a total length of 988 ft, and 28 standard 90° elbows, while the bypass line consists of 28 ft of the same pipe. The system is designed for a total flowrate of 80 gal/min. Find the flowrate through the heat exchanger when the globe valve is one-half open.

SOLUTION

We are asked to determine the flowrate through one branch of a specified multiple path flow system for a set of defined conditions. Figure 13.32 serves a sketch of the flow system. We will assign node 1 at the inlet of the tee upstream of the heat exchanger and node 2 at the outlet of the tee just downstream of the heat exchanger. Branch 1 contains the branch run of the upstream tee, the heat exchanger, and the branch run of the downstream tee. Branch 2 contains the line run of the upstream tee, the bypass line and globe valve (GV), and the line run of the downstream tee. Applying the first of the two

13.4 ANALYSIS OF FLOW IN MULTIPLE PATH PIPE AND DUCT SYSTEMS

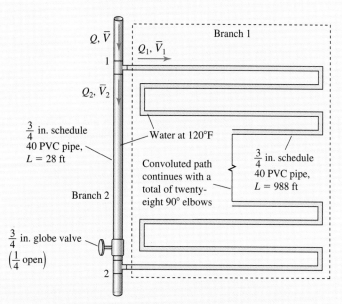

Figure 13.32 Schematic of a heating system for Example 13.16.

principles governing parallel branches, we write $Q = Q_1 + Q_2$, or

$$Q_2 = Q - Q_1 \qquad (A)$$

where $Q = 80$ gal/min $[(1 \text{ min}/60 \text{ s})(1 \text{ ft}^3/7.48 \text{ gal})] = 1.78 \times 10^{-1}$ ft^3/s. Note that in this case the piping is all of the same diameter, thus we also have $\bar{V} = \bar{V}_1 + \bar{V}_2$. We now write the total head loss in each branch, accounting for the major and minor losses to obtain:

$$h_{T1} = 2f_1 \frac{L_{\text{tee-branch}}}{D} \frac{\bar{V}_1^2}{2} + f_1 \frac{L_1}{D} \frac{\bar{V}_1^2}{2} + 28 f_1 \frac{L_{\text{elbow}}}{D} \frac{\bar{V}_1^2}{2}$$

$$h_{T2} = 2f_2 \frac{L_{\text{tee-line}}}{D} \frac{\bar{V}_2^2}{2} + f_2 \frac{L_2}{D} \frac{\bar{V}_2^2}{2} + K_{GV} \frac{\bar{V}_2^2}{2}$$

For convenience, we will write these equations in terms of the flowrates as

$$h_{T1} = \left[2f_1 \frac{L_{\text{tee-branch}}}{D} + f_1 \frac{L_1}{D} + 28 f_1 \frac{L_{\text{elbow}}}{D}\right] \frac{8Q_1^2}{\pi D^2}$$

$$h_{T2} = \left[2f_2 \frac{L_{\text{tee-line}}}{D} + f_2 \frac{L_2}{D} + f_2 \frac{L_{GV}}{D}\right] \frac{8Q_2^2}{\pi D^2}$$

where $L_{\text{tee-branch}}/D = 60$, $L_{\text{tee-line}}/D = 20$, $L_{\text{elbow}}/D = 30$, and $L_{GV}/D = 500$ from Table 13.6. Also, from the problem statement we calculate $L_1/D = (988 \text{ ft}/0.06867 \text{ ft}) = 1.44 \times 10^4$ and $L_2/D = (28 \text{ ft}/0.0687 \text{ ft}) = 408$. Thus these two equations become

$$h_{T1} = (120 f_1 + 1.44 \times 10^4 f_1 + 840 f_1)\frac{8 Q_1^2}{\pi D^2} = 1.54 \times 10^4 f_1 \frac{8 Q_1^2}{\pi D^2} \quad \text{(B)}$$

$$h_{T2} = (40 f_2 + 408 f_2 + 500 f_2)\frac{8 Q_2^2}{\pi D^2} = 948 f_2 \frac{8 Q_2^2}{\pi D^2} \quad \text{(C)}$$

Finally, applying the second principle that the total head loss in parallel branches is the same, we write $h_{T1} = h_{T2}$, and equate (B) and (C) to obtain

$$1.54 \times 10^4 f_1 \frac{8 Q_1^2}{\pi D^2} = 948 f_2 \frac{8 Q_2^2}{\pi D^2}$$

Solving for Q_1, after using (A) to eliminate Q_2, we obtain

$$Q_1 = \frac{Q}{1 + \sqrt{\dfrac{948 f_2}{1.54 \times 10^4 f_1}}} \quad \text{(D)}$$

We will now use (D) to iterate and solve for Q_1, beginning with a guess for the friction factors. To come up with an initial guess for the friction factors, suppose the flowrate in the branches is the same, and equal to half the total flowrate. For $\frac{3}{4}$ in. Schedule 40 pipe, we find $D = 0.06867$ ft, and $A = 3.70 \times 10^{-3}$ ft^2 (Table 13.4). The average velocity in each branch is

$$\bar{V} = \frac{1}{2}\frac{Q}{A} = \frac{1.78 \times 10^{-1} \text{ ft}^3/\text{s}}{2(3.70 \times 10^{-3} \text{ ft}^2)} = 24.0 \text{ ft/s}$$

Using data for water at 120°F, this corresponds to $Re = \bar{V}D/\nu = [(24.0 \text{ ft/s}) \times (0.06867 \text{ ft})]/(6.067 \times 10^{-6} \text{ ft}^2/\text{s}) = 2.7 \times 10^5$, and a friction factor $f = 0.014$, where we have used the Moody chart and assumed a smooth pipe.

Our first iteration now begins by assuming $Q_1^{(1)} = Q_2^{(1)} = 8.9 \times 10^{-2}$ ft^3/s, $\bar{V}_1^{(1)} = \bar{V}_2^{(1)} = 24.0$ ft^2/s, $Re_1^{(1)} = Re_2^{(1)} = 2.7 \times 10^5$ and $f_1^{(1)} = f_2^{(1)} = 0.014$. Inserting the latter into (D), we find

$$Q_1^{(2)} = \frac{Q}{1 + \sqrt{\dfrac{948(0.014)}{1.54 \times 10^4 (0.014)}}} = 0.8 Q$$

and from (A) we have $Q_2^{(2)} = 0.2Q$. The next iteration now proceeds by using $Q_1^{(2)} = 0.8Q = 0.142$ ft^3/s, $Q_2^{(2)} = 0.2Q = 3.56 \times 10^{-2}$ ft^3/s, $\bar{V}_1^{(2)} = 38.5$ ft/s, $\bar{V}_2^{(2)} = 9.6$ ft/s, $Re_1^{(2)} = 4.4 \times 10^5$, $Re_2^{(2)} = 1.1 \times 10^5$, $f_1^{(2)} = 0.013$, and $f_2^{(2)} = 0.018$. Inserting these values into (D) we find

$$Q_1^{(3)} = \frac{Q}{1 + \sqrt{\dfrac{948(0.018)}{1.54 \times 10^4 (0.013)}}} = 0.77 Q$$

and from (A) we have $Q_2^{(3)} = 0.23Q$. These new values turn out to be so close to the earlier values that the friction factors are unchanged. We therefore consider the iteration to have converged, and the final values for the flowrates are $Q_1 = 0.77Q = 61.6$ gal/min through the heat exchanger, and $Q_2 = 0.23Q = 18.4$ gal/min through the bypass line. The values of various parameters at each iteration are contained in the following table.

Iteration	Q_1	Q_2	$\bar{V}_1$ (ft/s)	$\bar{V}_2$ (ft/s)	Re_1	Re_2	f_1	f_2
1	0.5Q	0.5Q	24.0	24.0	2.7×10^5	2.7×10^5	0.014	0.014
2	0.8Q	0.2Q	38.5	9.6	4.4×10^5	1.1×10^5	0.013	0.018
3	0.77Q	0.23Q	37.0	11.1	4.2×10^5	1.3×10^5	0.013	0.018

13.5 ELEMENTS OF PIPE AND DUCT SYSTEM DESIGN

There are several questions to be answered in the design of pipe and duct systems beyond the specification of the size and lengths of the component pipes and ducts. The strength and durability of materials, maintenance issues related to the selection and location of valves, filters, traps, and other components, and space limitations are but a few elements of the overall design specification for a fluid transport system. The key design decisions of relevance to fluid mechanics are generally based on the minimization of the first cost of pipe, pump, and components, as opposed to the operating cost for pumping. Since the frictional pressure drop in turbulent flow is found to be proportional to the average velocity squared, for a given flowrate, the lower first cost of using a smaller diameter pipe and smaller components results in a larger pressure drop. A larger pressure drop results in a higher operating cost of the pump, and in some cases, a higher first cost for a pump of adequate pressure. Thus a successful pipe or duct design involves decisions that require consideration of factors beyond the characteristics of the flow alone. In this section we briefly discuss some of these additional factors.

The keys to understanding the operation of pipe or duct systems are the system and pump (or fan) curves as shown in Figure 13.33. Both curves consist of pressure (or differential) head plotted versus flowrate. First consider the system curve. The pressure or head plotted is that required to drive the flow. Thus, at zero flowrate the curve will indicate the hydrostatic head that must be overcome for the flow to proceed. As the flow begins there is an increase in head required by reason of frictional losses. Thus the system curve is found by using head loss calculations like those performed in the preceding sections. Pump curves like the one in Figure 13.33 are supplied by manufacturers for each of their products, indicating the head provided by the pump as a function of flowrate. This curve is a function of the pump geometry and the rotation speed (e.g., rpm) at which it is run. The intersection of the system and pump curves represents the operating point of the system, where the head provided by the pump equals the head required by the system.

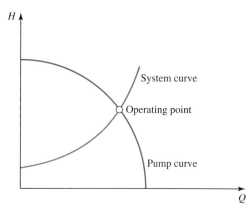

Figure 13.33 System and pump curves.

A typical design involves the flow of a known fluid, at a specified rate, from one location to another. Thus, Q is known. The general layout of the pipe run is usually determined by outside constraints. For example, in designing a building the architect will usually designate the location of the mechanical equipment room and pipe and duct runs. Thus, L is known. The fundamental question is to optimize the overall cost, which consists of the first cost of the pipe and pump and the operating cost, by judicious choice of pipe diameter and pump. This is a type 1 problem, as discussed in Section 13.3.3: to define the system curve, you must usually pick a D to calculate the required Δp or head. Figure 13.34 shows a series of system curves for each possible pipe diameter. Also shown on Figure 13.34 are pump curves that represent different sizes or rpm. Note that there is no intersection of the curves (operating point) exactly at the design flowrate because pipe and pumps have standard sizes. Control valves must be included in the design to adjust the system curve to achieve the proper flowrate.

Figure 13.34 illustrates that there is a choice of system and pump curves such that it is necessary to optimize the design. This approach to optimizing the design consists of assuming a pipe size, performing the head loss calculations, choosing the pump, and determining the total first and operating costs, then assuming a different pipe size and repeating this process until the minimum cost option is found. This can be a time-consuming process. Fortunately, with experience, engineers become adept at choosing

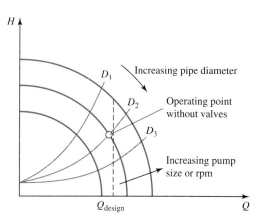

Figure 13.34 System and pump curves with a design flow rate superimposed.

13.5 ELEMENTS OF PIPE AND DUCT SYSTEM DESIGN

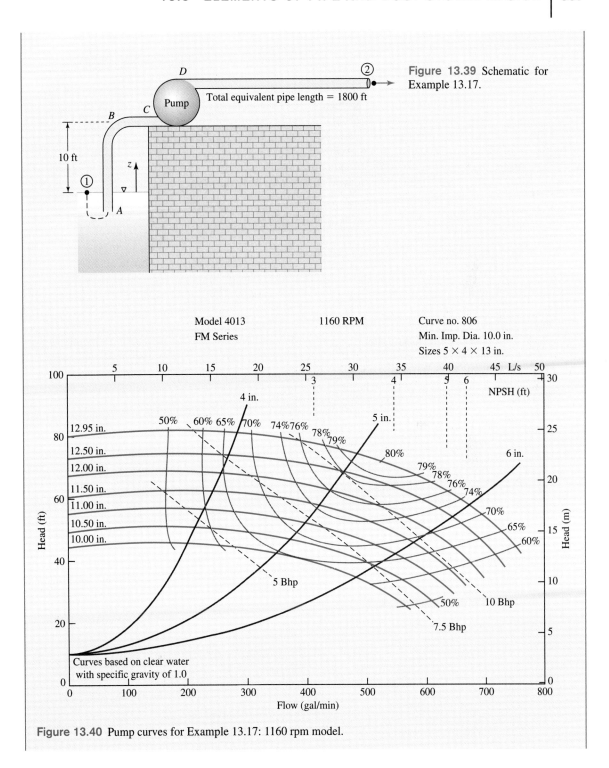

Figure 13.39 Schematic for Example 13.17.

Figure 13.40 Pump curves for Example 13.17: 1160 rpm model.

SOLUTION

We are asked to use a given set of pump curves to select an appropriate pump for a well-defined single path system. Figure 13.39 serves as an appropriate sketch for this problem. There are many approaches one might take to optimize even a simple piping design problem such as this one. We will find an equation for the system curve as a function of the pipe diameter. Then we will use the information provided in the manufacturer's data to choose the most efficient pump by sketching the system curves over the pump curves. Therefore, in this example we will be minimizing operating costs only.

We begin by applying Eq. 13.38 from the reservoir surface to the free exit discharge:

$$\left(\frac{p_1}{\rho} + \frac{1}{2}\alpha_1 \bar{V}_1^2 + gz_1\right) - \left(\frac{p_2}{\rho} + \frac{1}{2}\alpha_2 \bar{V}_2^2 + gz_2\right) = h_T - H_{\text{pump}} + H_{\text{turbine}}$$

Using gage pressures, and noting that $p_1 = 0$ psig, $\bar{V}_1^2 \approx 0$, $z_1 = 0$, and $p_2 = 0$ psig, we have

$$H_{\text{pump}} = \tfrac{1}{2}\bar{V}_2^2 + gz_2 + h_T \qquad (A)$$

where we have assumed $\alpha = 1$ for turbulent flow. We see that the pump must generate an amount of head $H_{\text{pump}} = \tfrac{1}{2}\bar{V}_2^2 + gz_2 + h_T = H_{\text{system}}$. The system head consists of the increase of kinetic energy of the fluid, the elevation potential and the total frictional head loss. The latter is the sum of the major and minor head losses and given by

$$h_T = f\frac{L}{D}\frac{\bar{V}_2^2}{2} + f\frac{L_E}{D}\frac{\bar{V}_2^2}{2} = f\left(\frac{L+L_E}{D}\right)\frac{\bar{V}_2^2}{2}$$

We now want to superimpose the system head curve for each diameter choice. To do so we must transform (A) to units of head in feet and flowrate in gallons per minute, with the understanding that $H_{\text{system}} = H_{\text{pump}}$. By using $Q = \bar{V}_2 A$, $h_T = f[(L+L_E)/D] \times (\bar{V}_2^2/2)$, and $A = \pi D^2/4$, we obtain

$$H_{\text{system}} = \frac{Q^2}{2A^2} + gz_2 + f\left(\frac{L+L_E}{D}\right)\frac{Q^2}{2A^2} = \left[1 + f\left(\frac{L+L_E}{D}\right)\right]\frac{8Q^2}{\pi^2 D^4} + gz_2$$

$$\frac{H_{\text{system}}}{g} = \left[1 + f\left(\frac{L+L_E}{D}\right)\right]\frac{8Q^2}{g\pi^2 D^4} + z_2 \qquad (B)$$

Inserting numerical values into (B) yields

$$\frac{H_{\text{system}}}{g} = \left(1 + f\frac{1800 \text{ ft}}{D}\right)\left(\frac{0.0252(\text{s}^2/\text{ft})Q^2}{D^4}\right)\left(\frac{2.228 \times 10^{-3} \text{ ft}^3/\text{s}}{\text{gal/min}}\right)^2 + 10 \text{ ft}$$

$$\frac{H_{\text{system}}}{g} = \left(1 + f\frac{1800 \text{ ft}}{D}\right)\left(\frac{1.25 \times 10^{-7} \ [\text{ft}^5/(\text{gal/min})^2]Q^2}{D^4}\right) + 10 \text{ ft} \qquad (C)$$

13.5 ELEMENTS OF PIPE AND DUCT SYSTEM DESIGN

TABLE 13.7 System Head (ft) as a Function of Pipe Size and Volume Flowrate

Pipe Diameter: Nominal D (Actual D)	Volume Flowrate, Q (gal/min)								
	0	100	200	300	400	450	500	600	700	800
4 in. (0.3355 ft)	10	18.9	45.7	90.4	152.9	*	*	*	*	*
5 in. (0.4206 ft)	10	12.8	21.0	34.8	54.1	**65.8**	78.9	109.2	*	*
6 in. (0.5054 ft)	10	11.1	13.2	19.6	27.0	**31.5**	36.5	48.2	62.0	77.9

*This combination of flowrate and pipe diameter is not possible.

where D is in feet and Q is gallons per minute. The friction factor can be found by using the Moody chart or the Chen equation. You will find that this system is fully turbulent such that f is independent of flowrate for each diameter. The friction factors are 0.0163, 0.0155, and 0.0139 for the 4, 5, and 6 in. pipes, respectively. Table 13.7 gives system head in feet calculated by using (C) for three choices of pipe diameter and at several flowrates. The entries for the design flow rate of 1 ft³/s = 450 gal/min are given in boldface. These data are superimposed on the pump curves of Figure 13.40.

First of all we can see that this particular pump could not deliver the design flowrate through a 4 in. pipe. A 6 in. pipe could be used, but the pump would be oversized. Thus the obvious choice for the design constraints of this problem is 5 in. diameter pipe. The design point on the 5 in. system curve is not directly on a pump curve. The system curve does intersect the 12.50 in. diameter impeller pump, providing 460 gal/min at 68 ft of head. At this operating point the pump would have an efficiency of 78% and would require a 10 hp motor. To obtain 450 gal/min, an additional 2.2 ft of head loss could be achieved through the use of a control valve. However, we also expect the head loss to increase in the system over time. Another method for moving the operating point to the design point is to change the pump curve by adjusting its rpm. In the case study of Section 3.3.3 we found that the dimensionless group $Q/\omega D^3$, the flow coefficient, applies to pumps and fans. The pump curves in Figure 13.40 are for $\omega = 1160$ rpm. By using the flow coefficient we find that the 12.50 in. diameter impeller operating at 1135 rpm would provide 450 gal/min as desired. However, the head coefficient, $H/\omega^2 D^2$, must also be checked. Taking the head coefficient we find that at 1135 rpm the pump would supply 65 ft of head, which is not enough to achieve the desired flowrate in the system. Thus it should be understood that it is very difficult to make the operating point and design point coincide without control valves.

Before leaving this problem, consider the pressure of a fluid particle moving through this system as indicated in Figure 13.41. At the surface of the reservoir, station 1, the pressure is atmospheric at 0 psig. The pressure increases due to hydrostatic head at the pipe inlet, station A, where it then is drawn into the pipe by the pump. The pressure decreases owing to frictional head losses in the pipe and reduced static head from A to B. From station B to the pump inlet, station C, the reduction in pressure is due to friction alone. Across the pump there is the pressure rise imparted by the pump. From the pump exit, station D, to the pipe exit, station 2, the pressure decreases to atmospheric

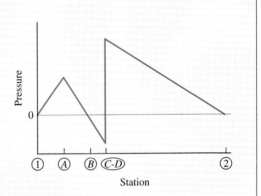

Figure 13.41 Pressure through the piping system analyzed in Example 13.17. The lines connecting each point only give a sense of the process involved and should not be taken literally. For example, pressure changes through a pump, inlet, or exit are typically not linear. However, head loss calculations are always applicable between stations, regardless of the nature of the process between them.

pressure due to frictional losses. However, the fluid at station 2 has greater mechanical energy than at station 1 because it is at a greater elevation and has greater kinetic energy.

The inlet pressure to a pump is called the suction pressure. If the suction pressure falls below the vapor pressure of the liquid, the liquid will boil; i.e., cavitation will result. The vapor bubbles can damage the impeller as they pass through the pump, as shown in Figure 13.42, and cause a decrease in efficiency. All pumping systems should be designed to avoid cavitation.

The potential for cavitation is measured using a concept known as the Net Positive Suction Head (NPSH). The NPSH at any point in a flow system is defined as

$$\text{NPSH} = \frac{p}{\rho g} + \frac{\bar{V}^2}{2g} - \frac{p_v}{\rho g} \tag{13.49}$$

where p is the pressure at the point of interest, $\bar{V}$ is the average velocity, and p_v is the vapor pressure corresponding to the liquid temperature. The dimension of NPSH is length and the units of this quantity are feet or meters. At the inlet of a pump we can calculate the Net Positive Suction Head Available, or NPSHA, by inserting the calculated values of the pump inlet pressure and inlet average velocity, as well as the vapor pressure into Eq. 13.49. Thus we have

$$\text{NPSHA} = \frac{p_{\text{in}}}{\rho g} + \frac{\bar{V}_{\text{in}}^2}{2g} - \frac{p_v}{\rho g} \tag{13.50}$$

Therefore, NPSHA defines the value of the net positive suction head at the pump inlet as determined from a head loss calculation. For cavitation-free operation of a pump, the manufacturer provides a parameter called the Net Positive Suction Head Required, or NPSHR. The value of NPSHR is found experimentally and provided by pump manufacturers; for example, see Figure 13.37. Although NPSHR is specified only as a number in units of feet or meters, NPSHR may be thought of as corresponding to a required combination of pressure and average velocity at the pump inlet, namely, $p_R/\rho g + \bar{V}_R^2/2g$, which exceeds the vapor pressure head $p_v/\rho g$ by a margin sufficient to prevent cavitation from occurring in the high velocity regions inside the pump. For proper operation

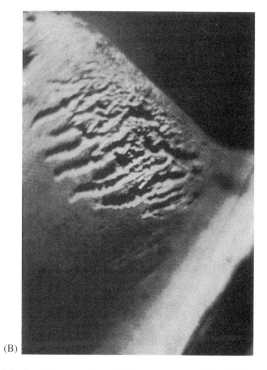

Figure 13.42 (A) Cavitation bubbles forming on a model of a ship propeller. (B) Damage to a turbine blade caused by cavitation.

of a given pump, the available value of the suction head must be greater than or equal to the manufacturer's required value of the suction head. Thus the condition for proper operation of a pump is NPSHA ≥ NPSHR. We can use Eq. 13.50 to write this requirement as

$$\frac{p_{\text{in}}}{\rho g} + \frac{\bar{V}_{\text{in}}^2}{2g} - \frac{p_v}{\rho g} \geq \text{NPSHR} \qquad (13.51)$$

A key question in the design of a system containing a pump is whether the available NPSHA is greater than NPSHR for a given pump and system configuration. To place this discussion in context, consider a system as shown in Figure 13.43 that has both major and minor losses in the flowpath upstream of the pump. Applying Eq. 13.29a between a point upstream and the pump inlet, we have

$$\left(\frac{p_1}{\rho} + \frac{1}{2}\bar{V}_1^2 + gz_1\right) - \left(\frac{p_{\text{in}}}{\rho} + \frac{1}{2}\bar{V}_{\text{in}}^2 + gz_{\text{in}}\right) = \sum h_L + \sum h_M$$

which after dividing by g and rearranging gives

$$\left(\frac{p_1}{\rho g} + \frac{\bar{V}_1^2}{2g} + z_1\right) - \left(\frac{p_{\text{in}}}{\rho g} + \frac{\bar{V}_{\text{in}}^2}{2g} + z_{\text{in}}\right) = \sum \frac{h_L}{g} + \sum \frac{h_M}{g}$$

Figure 13.43 Schematic of pump application for which NPSHA is relevant.

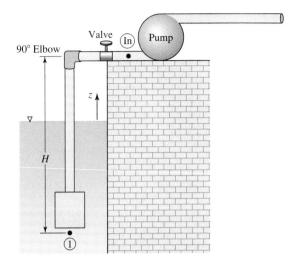

From Eq. 13.50 we can write

$$\left(\frac{p_{in}}{\rho g} + \frac{\bar{V}_{in}^2}{2g}\right) = \text{NPSHA} + \frac{p_v}{\rho g}$$

Thus the preceding equation can be written as

$$\left(\frac{p_1}{\rho g} + \frac{\bar{V}_1^2}{2g}\right) - \left(\text{NPSHA} + \frac{p_v}{\rho g} + H\right) = \sum \frac{h_L}{g} + \sum \frac{h_M}{g}$$

where we have defined $H = z_{in} - z_1$. Then solving for NPSHA we obtain

$$\text{NPSHA} = \left(\frac{p_1}{\rho g} + \frac{\bar{V}_1^2}{2g}\right) - \frac{p_v}{\rho g} - H - \sum \frac{h_L}{g} - \sum \frac{h_M}{g} \quad (13.52)$$

This equation can be used to evaluate NPSHA and to understand the effects of system parameters. We see that NPSHA can be thought of as the value of the suction head at point 1 minus the loss of head due to a pump elevation above point 1 and minus the loss of head due to major and minor losses in the flow path. Example 13.18 illustrates the use of NPSHR and NPSHA in evaluating the design parameters that effect pump performance in a proposed system.

EXAMPLE 13.18

Find the NPSHA for the pump and inlet system described in Example 13.9. This system is shown in Figure 13.44. If the pump manufacturer specifies a value of NPHSR = 20 ft, will the pump operate satisfactorily? If not, can you suggest a design modification that does not change the design flowrate or pipe dimensions?

13.5 ELEMENTS OF PIPE AND DUCT SYSTEM DESIGN

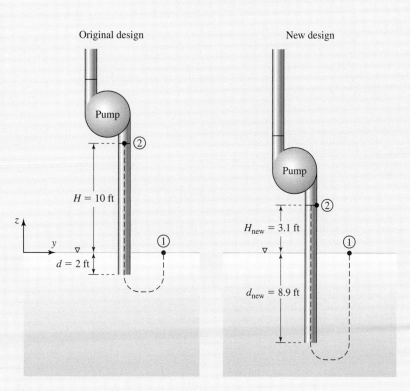

Figure 13.44 Schematic for Example 13.18.

SOLUTION

We are asked to evaluate a specified flow system to determine whether the suggested pump is satisfactory. If our calculations show that the present design is not satisfactory, we are to propose an improved design. Figure 13.44 serves as the appropriate sketch for this flow system. We will use Eq. 13.50 to evaluate NPSHA rather than Eq. 13.52 because the major and minor losses were evaluated in Example 13.9, where we found the following values at the pump inlet: $p_{in} = 800 \text{ lb}_f/\text{ft}^2$, and $\bar{V}_{in} = 7.43$ ft/s. For water at 60°F, Appendix A gives $\rho = 1.938$ slugs/ft^3, so $\rho g = (1.938 \text{ slugs/ft}^3)(32.2 \text{ ft/s}^2) = 62.4 \text{ lb}_f/\text{ft}^3$, and $p_v = 0.2563$ psia $= 36.9 \text{ lb}_f/\text{ft}^2$. Applying Eq. 13.50, we have NPSHA $= p_{in}/\rho g + \bar{V}_{in}^2/2g - p_v/\rho g$, with

$$\frac{p_{in}}{\rho g} = \frac{800 \text{ lb}_f/\text{ft}^2}{62.4 \text{ lb}_f/\text{ft}^3} = 12.8 \text{ ft}, \quad \frac{\bar{V}_{in}^2}{2g} = \frac{(7.43 \text{ ft/s})^2}{2(32.2 \text{ ft/s}^2)} = 0.86 \text{ ft}$$

and

$$-\frac{p_v}{\rho g} = -\frac{36.9 \text{ lb}_f/\text{ft}^2}{62.4 \text{ lb}_f/\text{ft}^3} = -0.59 \text{ ft}$$

Thus the net positive suction head available in this case is

$$\text{NPSHA} = \frac{p_{in}}{\rho g} + \frac{\bar{V}_{in}^2}{2g} - \frac{p_v}{\rho g} = 12.8 \text{ ft} + 0.86 \text{ ft} - 0.59 \text{ ft} = 13.1 \text{ ft}$$

Since we are told NPHSR = 20 ft, the pump will not operate satisfactorily. The design must be changed to increase NPSHA to at least 20 ft.

In many cases, as here, the desired flowrate is fixed, but changes can be made in the elevation of the pump, and in other design parameters that effect the major and minor losses. If we are to keep both the flowrate and the pipe dimensions the same, we must decrease the elevation of the pump above the free surface of the reservoir to a new value H_{new}, as shown in Figure 13.44, while leaving the piping and inlet configuration unchanged. This will increase the pressure at the pump inlet and raise the NPSHA. To find the new conditions at the pump inlet with a decrease in elevation to a value H_{new}, we need to repeat the head loss analysis of Example 13.9. However, in this case we do not actually need the conditions themselves but can make use of Eq. 13.52 to write

$$\text{NPSHA} = \frac{p_A}{\rho g} - H - \frac{p_v}{\rho g} - f\frac{L}{D}\frac{\bar{V}_{in}^2}{2g} - K_{in}\frac{\bar{V}_{in}^2}{2g}$$

This result shows that with the frictional losses the same, the available suction head increases in direct proportion to the decrease in H. Thus to increase NPSHA from 13.1 ft to 20 ft, we must decrease H by 6.9 ft. Thus we conclude that redesigning the system with $H_{new} = 10 \text{ ft} - 6.9 \text{ ft} = 3.1 \text{ ft}$ will allow the pump to operate properly.

13.6 SUMMARY

In this chapter we investigated the methods used to analyze steady, constant density flow in pipes and ducts. The emphasis was on determining the relationship between pressure drop and volume flowrate, since these are the two parameters that must be known to predict the power required to move fluid through the pipe or duct. The concepts of major head loss, friction factor, and hydraulic diameter were introduced and used to analyze laminar and turbulent flow.

For steady, fully developed flow in a pipe or duct, the pressure difference necessary to drive a fluid through a passage is affected by the length and shape of the passage, the difference in elevation of each end of the passage, and viscous friction (in the form of the wall shear stress). In a fully developed flow there is no change in the kinetic energy per unit mass of fluid as it passes through the pipe or duct. Thus, the mechanical energy

13.6 SUMMARY

that is lost in such a flow appears as a combination of a change in the internal energy of the fluid and heat transfer, usually out through the wall of the pipe or duct. In most cases the average wall shear stress is not known, nor do we have sufficient information to evaluate the change in internal energy or heat transfer. Instead we make use of the concept of major head loss, h_L. To determine h_L we depend on empirical results based on the friction factor f. The basic equations describing the steady, fully developed flow in pipes and ducts are $(p_1/\rho + gz_1) - (p_2/\rho + gz_2) = f(L/D)(\bar{V}^2/2)$ and $(p_1/\rho + gz_1) - (p_2/\rho + gz_2) = f(L/D_H)(\bar{V}^2/2)$, respectively. If the friction factor is known, we can make use of these equations to determine the pressure change corresponding to a known average velocity (and thus flowrate) in a pipe or duct.

The friction factor for laminar flow in a pipe and a rectangular duct are given respectively by $f = 64/Re$ and $f = (96/Re_H)G(a/b)$, where $G(a/b)$ is a geometric factor that depends on the aspect ratio of the duct cross section (see Table 13.1). Friction factors for laminar fully developed flow in round, rectangular, and other cross sections are summarized in Table 13.2. In a turbulent flow f depends on the pipe diameter, its physical condition, the density and viscosity of the fluid, and the average velocity. The data needed to determine f were obtained for turbulent flow over the range of values of e/D and Re encountered in engineering practice (see Figure 13.12, and the Moody chart shown in Figure 13.13). Alternatively, one can make use of the Colebrook formula or the Chen equation.

A single path pipe or duct system may contain multiple straight pipe or duct segments plus components such as elbows, valves, reducers, filters, and traps. We analyze the flow through such a system by setting the change in the mechanical energy content of the fluid equal to the sum of the major and minor losses for each pipe or duct segment and each component along the flow path. The energy balance takes the form $(p_1/\rho + \frac{1}{2}\alpha_1 \bar{V}_1^2 + gz_1) - (p_2/\rho + \frac{1}{2}\alpha_2 \bar{V}_2^2 + gz_2) = \sum h_L + \sum h_M$. In a single path system, a mass balance also shows that $A_1 \bar{V}_1 = A_2 \bar{V}_2$. These two equations are the basic equations governing flow through a single flow path system. To apply the energy balance, we must know the values of α that occur at the selected points. For fully developed laminar pipe flow $\alpha = 2$, and it is customary to employ $\alpha = 1$ for fully developed turbulent flow in a pipe even though it is known that α is a weak function of Re in such flows.

Values of the minor head loss for components of flow systems are tabulated in two forms. The first form is as a head loss coefficient, K, defined by $h_M = K(\bar{V}^2/2)$. The second form is through an equivalent pipe length L_e. The minor loss is then defined by $h_M = f(L_e/D)(\bar{V}^2/2)$, where D, f, and $\bar{V}$ refer to the adjacent pipe flow. Values of K and L_e for a variety of flow elements are given in Section 13.3.1. In the case of an element in which the inlet and outlet areas differ, it is critical to determine whether the coefficient is based on the use of inlet or outlet average velocity.

A pump, fan, or blower increases the mechanical energy of a fluid flowing through the device, while a turbine decreases the mechanical energy of a fluid. To account for the presence of one of these devices along a flow path, we make use of the concepts of a pump head H_{pump} and a turbine head H_{turbine}, and write the energy balance from point 1 to point 2 along a flow path as

$$\left(\frac{p_1}{\rho} + \frac{1}{2}\alpha_1 \bar{V}_1^2 + gz_1\right) - \left(\frac{p_2}{\rho} + \frac{1}{2}\alpha_2 \bar{V}_2^2 + gz_2\right) = h_T - H_{\text{pump}} + H_{\text{turbine}}$$

where h_T is the total head loss. For devices operating on a constant density fluid, it is appropriate to assume turbulent flow with $\alpha = 1$, uniform properties at the inlet and outlet, and identical inlet and outlet areas and elevations. The increase in mechanical energy of the fluid passing through the device is then given by $p_{out}/\rho - p_{in}/\rho$, so the pump head is defined as $H_{pump} = (p_{out} - p_{in})/\rho$, while the turbine head, is defined as $H_{turbine} = (p_{in} - p_{out})/\rho$. The relationship between pump head and the pump efficiency, shaft power input, and volume flowrate is $H_{pump} = \eta_{pump} P_{pump}/\rho Q$. This equation also applies to a fan or blower. Since the efficiency is always less than unity, the head provided by the pump is always less than the shaft power input per unit of mass flowrate. For a turbine we find $H_{turbine} = P_{turbine}/\rho Q \eta_{turbine}$.

Problems involving fully developed flow in a single path pipe or duct system are characterized by four parameters: the pressure difference $\Delta p = p_1 - p_2$, volume flowrate Q, pipe diameter D or duct hydraulic diameter D_H, and length L. By asking which of these parameters is unknown, we can identify four distinct types of problems in such flows.

1. Unknown is $\Delta p = p_1 - p_2$: With Q and D known, we can calculate $\bar{V}$ and Re. The type of pipe or duct determines e/D, and we can read f from the Moody chart. Since L is known, we can solve the appropriate energy equation for the unknown Δp.

2. Unknown is L: With Q and D or D_H known, we can calculate $\bar{V}$ and Re. The type of pipe or duct determines e/D, and we can read f from the Moody chart. Since Δp is also known, we can solve the appropriate energy equation for the unknown L.

3. Unknown is Q: With Q unknown, we cannot calculate $\bar{V}$ and Re. The problem requires iteration. One approach is to assume a friction factor on the fully turbulent portion of the curve for the value of e/D, then solve the appropriate energy equation for the estimated $\bar{V}$. For this value of $\bar{V}$ we can now calculate Re and use it to find a new value of f, followed by a new estimate for $\bar{V}$. The iteration process stops when the change in the estimated $\bar{V}$ is sufficiently small. The final step is to calculate Q.

4. Unknown is D or D_H: With D or D_H unknown, we cannot calculate $\bar{V}$ and Re even though we know Q. The problem requires iteration. One approach is to assume a value for D or D_H, which this allows the $\bar{V}$ and Re to be calculated. Since D or D_H is assumed known, e/D and the Moody chart can be used to find f, after which the appropriate energy equation is solved to find the estimated Δp. The estimated Δp is compared to the known value and is used to adjust the diameter upward or downward. The iteration process stops when the estimated pressure difference agrees with the known value. At this point the value of D or D_H is known.

Most engineering systems contain multiple flow paths, defined as those for which fluid may travel from inlet to exit via different paths. The two key facts to keep in mind when analyzing multiple flow paths are (1) the head losses in parallel branches joining two nodes in a multiple path system are the same and (2) the flowrate into a juncture where two or more branches meet is equal to the sum of the flowrates out of the juncture on each branch.

There are several questions to be answered in the design of pipe and duct systems beyond the specification of the size and lengths of the component pipes and ducts. The key design decisions of relevance to fluid mechanics are generally based on the minimization of the first cost of pipe, pump, and components, as opposed to the operating cost for pumping. Since the frictional pressure drop in turbulent flow is proportional to the average velocity squared, for a given flowrate, the lower first cost of using a smaller diameter pipe and smaller components results in a larger pressure drop. A larger pressure drop results in a higher operating cost of the pump, and in some cases, a higher first cost for a pump of adequate pressure. Thus a successful pipe or duct design involves decisions that require consideration of factors beyond the characteristics of the flow alone. Two of the major tools utilized to solve such problems are the system and pump (or fan) curves described in Section 13.5. The pump or fan is usually the single most expensive item in a pipe or duct design. The optimization of pump and fan selection is simplified through the use of charts supplied by manufacturers. An important aspect of pipe and duct design is the uncertainty of the calculations and operating conditions. The uncertainty in head loss calculations is at least 10%. Furthermore, it is common for changes due to interferences to occur during construction, resulting in greater losses than were anticipated in the design. As a consequence, most designers will include a significant safety factor in the pipe and pump selection process.

PROBLEMS

Section 13.2

13.1 A flow-measuring device requires fully developed flow for maximum accuracy. How far downstream of the elbow should the device shown in Figure P13.1 be placed to achieve maximum accuracy if water is flowing in the pipe at 0.1 m/s? At 10 m/s?

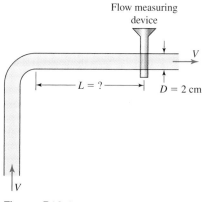

Figure P13.1

13.2 The duct shown in Figure P13.2 carries 1000 ft^3/min of standard air. How far downstream of the transition should the filter be located to maximize its effectiveness? Explain your answer.

13.3 A 1 mm diameter tube, 10 cm in length, is used as a viscometer. Its calibration is based on fully developed laminar flow for its whole length. Develop an equation for the viscosity based on Q and Δp. What is the maximum flowrate for standard air and water at 20°C?

13.4 For the device described in Problem 13.3 what is the viscosity of the fluid if measurements of $Q = 61$ cm^3/s and $\Delta p = 2000$ kPa are taken? What is the potential error of assuming fully developed flow for this case?

13.5 Show that head loss is the viscous dissipation normalized by the volume flowrate for laminar fully developed flow in a pipe. The dissipation in this case is

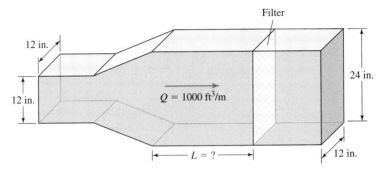

Figure P13.2

$\Phi = (\mu/\rho)(\partial v_z/\partial r)^2$, and the head loss is $h_L = (64/Re)(L/D)(\bar{V}^2/2)$.

13.6 For laminar fully developed flow in a pipe, find the radial location at which the velocity is equal to the average velocity.

13.7 What is the pressure drop in 50 ft of 1 ft diameter pipe for flow of 1000 ft³/min of standard air? What is the wall shear stress for this flow?

13.8 What is the pressure drop in 50 ft of 1 ft × 1 ft duct for flow of 1000 ft³/min of standard air? What is the wall shear stress for this flow?

13.9 Water at 30°C flows upward at the rate of 80 cm/s through the 20 mm diameter pipe shown in Figure P13.3. What is the pressure change in the pipe?

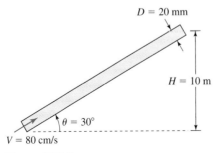

Figure P13.3

13.10 For the pipe described in Problem 13.9, water is now flowing downward. What is the pressure change in this case?

13.11 For the flow described in Problem 13.9, give the wall shear stress and the change in temperature due to friction.

13.12 For the flow described in Problem 13.10, give the wall shear stress and the change in temperature due to friction.

13.13 Air at 50°F is flowing through a 5 in. diameter galvanized steel pipe at 1500 ft/min. If the pipe is 100 ft long and horizontal, what is the pressure change? What is the pressure change if the pipe is vertical?

13.14 Water at 20°C flows in a horizontal 10 cm diameter cast iron pipe with a velocity of 10 cm/s. Use the Moody chart to find the pressure drop in the pipe per meter. What is the power lost to friction per meter?

13.15 Redo Problem 13.14 using the Colebrook equation and the Chen equation.

13.16 Someone designing a piping system without taking a course in fluid mechanics uses Eq. 13.19 for determining the friction factor with $Re = 5000$. Assume that he is using concrete pipe, and approximate the error in his calculation.

13.17 A hypodermic needle is 0.2 mm in diameter and 30 mm long. Determine the flowrate for $Re = 2800$ for a drug with the properties of water at 20°C. This Reynolds number is in the transition zone between laminar and turbulent flow. Calculate the pressure drop assuming laminar flow and the pressure drop calculated through the use of the Chen equation for turbulent flow.

13.18 A 10-mile-long crude oil pipeline is designed to be 1 ft diameter with a flowrate of 200 gal/min. The pipe material is commercial steel. It is expected that over the lifetime of the pipeline, fouling will reduce the effective area by 10% and will double the roughness. Calculate the expected frictional pressure drop for the pipeline when new and at the end of its service.

13.19 A water district has a 30 km pipeline from a reservoir to the treatment plant. The diameter of the pipe is 0.5 m and it carries 1100 m³/day. Without cleaning the commercial steel pipe will foul causing a 2% reduction in diameter and 10% increase in roughness per year? How much money can be spent on cleaning per year if it is paid for by savings in pumping costs and the price of electricity is $0.07/kW-h?

13.20 Air at 30°C flows through a micromachine passage with cross section 0.1 mm × 0.05 mm and length 1 mm. What is the pressure drop if the average velocity is 0.5 mm/s calculated by means of the hydraulic diameter model?

13.21 Redo Problem 13.20 using the rectangular duct solution if possible.

13.22 A horizontal air conditioning duct 50 in. × 10 in. carries 100 ft³/min of 55°F air. What is the pressure drop per foot of duct calculated by using the hydraulic diameter model?

13.23 Redo Problem 13.22 using the rectangular duct solution if possible.

13.24 Consider square duct (of side a) and pipe (of diameter D) of equal area such that $a^2 = \pi D^2/4$. Which will have the greatest frictional losses? Use a calculation based on the hydraulic diameter model to explain your answer.

13.25 Consider equal area rectangular ducts with aspect ratios of 2 and 4. Which will have the greater frictional losses? Use a calculation based on the hydraulic diameter model to explain your answer.

13.26 Air at 70°F flows through a smooth 1 ft diameter pipe. The duct is 75 ft long, and the flowrate is 500 ft³/min. What is the elevation change between the inlet and outlet?

13.27 Water at 20°C flows through 5000 m of 20 cm diameter pipe between two reservoirs whose water surface elevation difference is 75 m. Find the flowrate if $e/D = 0.0015$.

Section 13.3

13.28 Use the power-law velocity profile given in Eq. 13.32 to calculate the wall shear stress for turbulent flow in a smooth pipe. Compare values for $n = 6$ and 10.

13.29 Water flows through a 40 mm pipe with a sudden contraction to 20 mm. If the pressure drop across the contraction is 3.0 kPa, what is the volume flowrate?

13.30 A gradual contraction of 20° is considered to replace the sudden contraction described in Problem 13.29. What will now be the volume flowrate if the pressure drop is unchanged?

13.31 Gasoline flows from one tank to the other as shown in Figure P13.4. The pipe is 1 in., Schedule 40 commercial steel. Calculate the pressure above the gasoline in the first tank that will result in a flowrate of 100 gal/min. The second tank is vented.

13.32 The system described in Problem 13.31 is cleaned with 200°F water. What

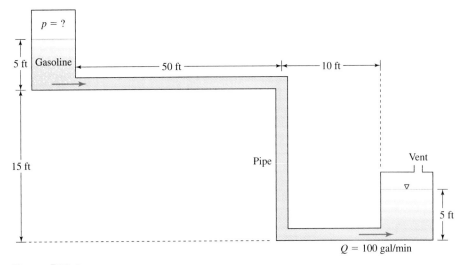

Figure P13.4

pressure is required to move water at the rate of 300 gal/min?

13.33 A 100 m length of 10 cm diameter wrought iron pipe has one open globe valve, one 45° regular elbow, and one pipe bend with a radius of curvature of 1 m. If the fluid is water at 30°C and the velocity is 2 m/s, what is the total head loss?

13.34 How large must a wrought iron pipe be to convey crude oil from one tank to another at a rate of 1 m^3/s if the pipe is 1000 m long and the difference in elevation of the free liquid surfaces that drives the flow is 15 m?

13.35 The 2 ft × 2 ft galvanized steel duct shown in Figure P13.5 carries 1000 ft^3/min of air at 70°F when the filter is new. The filter pressure drop is 0.25 in. H$_2$O when new and should be changed when it is 0.75 in. H$_2$O. What is the volume flowrate when the filter is changed?

13.36 For the system described in Problem 13.35, what total pressure for the system is required to maintain 1000 ft^3/min when the filter is dirty?

13.37 Gasoline is being siphoned through a $\frac{3}{4}$ in. diameter plastic tube, as shown in Figure P13.6. What is the flowrate?

13.38 If the discharge of the siphon described in Problem 13.37 is rotated to 45° to the horizontal, as shown in Figure P13.7, what is the new flowrate?

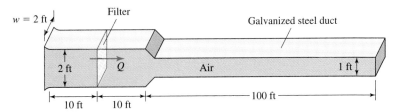

Figure P13.5

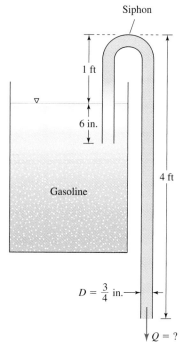

Figure P13.6

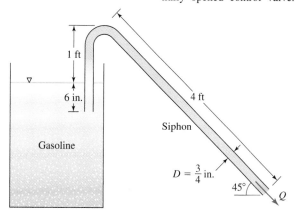

Figure P13.7

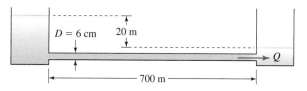

Figure P13.8

13.39 The two reservoirs shown in Figure P13.8 contain water at 20°C. The entrance and exit are both square edged, the length and diameter of the smooth pipe are 700 m and 6 cm, respectively. What will be the flowrate if the difference in surface elevations is 20 m?

13.40 Redo Problem 13.39 to find the flowrate for a diameter of 3 cm.

13.41 Redo Problem 13.39 to find the flowrate for a roughness e/D of 0.0015.

13.42 Find the head loss in a pipeline consisting of 100 ft of 2 in. diameter Schedule 40 steel pipe, four standard 90° elbows, and a gate valve. The flowrate is 2 ft³/s kerosene.

13.43 Reconsider the pipeline described in Problem 13.42. If the head loss remains the same, what diameter pipe is required to double the flowrate to 4 ft³/s?

13.44 Water at 20°C flows through a partially opened control valve. If the average

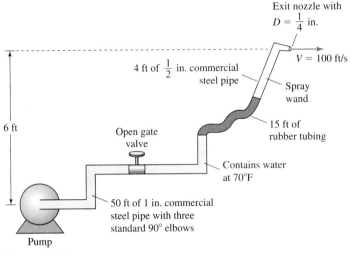

Figure P13.9

velocity is 10 m/s and the pressure drop across the valve is 1×10^5 N/m², what would be the head loss coefficient for the valve in this position?

13.45 The spray wand at a car wash shown in Figure P13.9 has a 1/4 in. exit nozzle and requires 100 ft/s exit velocity. The piping system from the pump consists of 50 ft of 1 in. commercial steel pipe with three standard elbows, one open gate valve, 15 ft of rubber tubing, and 4 ft of $\frac{1}{2}$ in. commercial steel, which makes up the wand. The pump is located about 6 ft below the exit of the wand. What pressure is required to drive water at 70°F through the spray wand?

13.46 A soap solution with a viscosity 10% lower than that of water but having the same density as water is run through the spray wand described in Problem 13.45. What pressure is required for this flow?

13.47 To wash larger vehicles it is proposed to add more rubber tubing to the spray wand described in Problem 13.45. If the pump exit pressure is 100 psig, how much more tubing could be added to maintain 90 ft/s exit velocity?

13.48 The pump at the car wash described in Problem 13.45 has failed. A temporary replacement has a 50 psig exit pressure. What will be the exit velocity?

13.49 An air-handling unit consisting of a supply fan, filter, and heating/cooling coil serve one room as shown in Figure P13.10. The supply ductwork consists of 50 ft of 18 in. × 12 in. galvanized steel duct with three mitered 90° elbows to the supply grille at the exit ($K = 2.5$ combined). The return ductwork consists of 45 ft of 18 in. × 12 in. galvanized steel duct with two elbows from the return grille and entrance ($K = 1.9$ combined). The air-handling unit and the room can be considered to be at the same elevation. The coil and filter have a combined design pressure drop of 1.0 in. H₂O at $Q = 1500$ ft³/min. What is the fan pressure rise required for this system? Assume that the supply air is 55°F.

13.50 To convert the room described in Problem 13.49 to a clean room, a new air-handling unit is installed with a high efficiency particulate air (HEPA) filter. Assume the pressure drop across the filter and

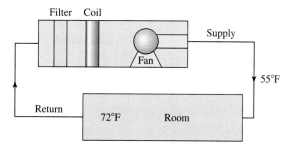

Figure P13.10

coil is now 2.0 in. H_2O. If the fan produces 3.5 in. H_2O of pressure rise, what will be the flowrate?

13.51 It is decided that the clean room described in Problems 13.49 and 13.50 has an inadequate flowrate. What size duct, maintaining one side as 18 in., is required to achieve a flowrate of 3000 ft³/min?

13.52 For the air-handling system described in Problem 13.49, what is the power required by the fan if its efficiency is 75%. Assume the pressure rise across the fan is 2.0 in. H_2O.

13.53 The hydroelectric plant shown in Figure P13.11 uses 200 m of 0.5 m diameter cast iron pipe. If the flowrate is 2 m³/s, what is the pressure drop across the turbine? Neglect minor losses.

13.54 The water from the hydroelectric plant described in Problem 13.53 is to be used for irrigation. If the power output is 200 kW and the turbine is 85% efficient, how far can the pipeline be extended while maintaining a flowrate of 2 m³/s?

13.55 Redo Problem 13.54 but choose the diameter of the pipe extension such that the extended length is 15 km.

13.56 The turbine for the hydroelectric plant described in Problem 13.53 is to be replaced with a more efficient model. If the power output is 300 kW and the turbine efficiency is 90%, what is the flowrate?

13.57 A maintenance check of the hydroelectric plant described in Problem 13.53 indicates a flowrate of 1.95 m³/s and a power

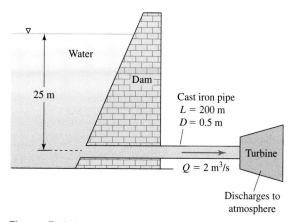

Figure P13.11

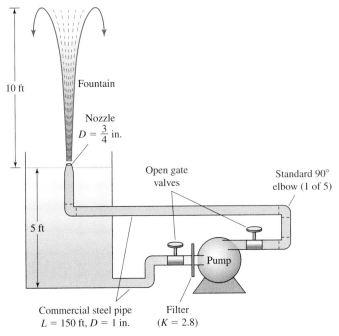

Figure P13.12

output of 250 kW. What is the efficiency of the turbine?

13.58 As shown in Figure P13.12 a fountain is served a pump, 150 ft of 1 in. diameter commercial steel pipe, five standard 90° elbows, a filter, and two gate valves. The fountain nozzle is 3/4 in. diameter. The filter is a minor loss with $K = 2.8$. If the height of the fountain h is 10 ft, how much pressure rise must the pump generate?

13.59 For the fountain described in Problem 13.58 the gate valve on the pump exit can be used to control the fountain height. If the pump provides a 30 psi pressure rise, what will be the height of the fountain if the valve is one-quarter open, half-open, and fully open?

13.60 How much does it cost per hour to operate the fountain described in Problem 13.58 if $h = 15$ ft, the pump efficiency is 80%, and the cost of power is $0.07/kW-h?

13.61 Recirculated air for a cold storage locker is cooled by means of the secondary cooling loop shown in Figure P13.13. The working fluid, ethylene glycol, is cooled from 40°C to 0°C in the heat exchanger located in the chiller. The fluid proceeds to the second heat exchanger in thermal contact with the recirculated air. Here the ethylene glycol absorbs heat and returns to 40°C. The piping system consists of 50 m of 3 cm commercial steel pipe, five standard 90° elbows, and two gates valves, which isolate the pump. The heat exchangers can be considered to be minor losses with a coefficient of $K = 2.0$. Assume no elevation changes. The density and viscosity of ethylene glycol at the working temperatures are $\rho_0 = 1128$ kg/m³, $\rho_{40} = 1100$ kg/m³, $\mu_0 = 5.74$ (kg-m)/s, and $\mu_{40} = 0.95$ (kg-m)/s, respectively. If the flowrate for this system is 1 m³/s, what is the pressure rise required from the pump?

13.62 For the cooling system described in Problem 13.61 new coils are installed which

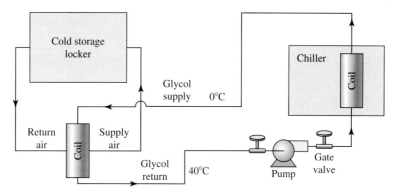

Figure P13.13

have a pressure drop of $15{,}000 \times 10^4$ N/m^2. If the pump is supplying $40{,}000 \times 10^4$ N/m^2 pressure rise, what is the flowrate in the system?

13.63 For the cooling system described in Problem 13.61, what is the power required by the pump if its efficiency is 85%?

13.64 The beer supply system for a bar is shown in Figure P13.14. The beer keg is pressurized to 30 psi by the CO$_2$ tank. The keg is connected to the tap through 1/4 in. plastic tubing. The tap is 5 ft higher than the keg. What distance from the keg can the tap be located to allow a pint glass to be filled in 10 s? Assume that beer has the properties of water at 40°F and that minor losses can be neglected.

13.65 If the tap is located 50 ft from the keg in the beer system described in Problem 13.64, what will be the flowrate in pints per minute?

13.66 For the beer system described in Problem 13.64, what diameter tubing is required to maintain the flowrate at 0.1 pint/s if the tap is 200 ft from the keg?

13.67 The intake and exit of a pump are 6 and 5 cm, respectively. The pressure change across the pump is 20 kPa. The intake and exit are at the same elevation. If the flowrate is 0.75 m^3/s and the power input is 25 kW, what is the efficiency of the pump?

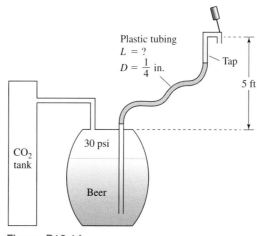

Figure P13.14

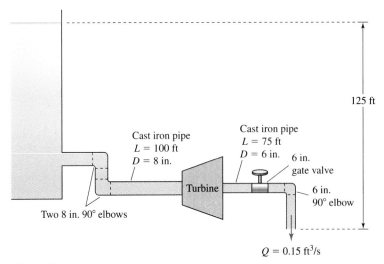

Figure P13.15

13.68 The turbine installation shown in Figure P13.15 consists of 100 ft of 8 in. pipe, 75 ft of 6 in. pipe, all cast iron, two 8 in. 90° elbows, one 6 in. elbow, a 6 in. gate valve. If the surface elevation is 125 ft above the discharge and the flowrate is 0.15 ft³/s of 70°F water, what is the power transferred from the flow?

13.69 All the piping on the turbine installation described in Problem 13.68 is changed to 4 in. diameter. What is the new power output?

Section 13.4

13.70 Three water tanks are connected as shown in Figure P13.16. Determine the flowrate in each pipe. Neglect minor losses.

13.71 Redo Problem 13.70 with cast iron pipe instead of the given friction factors.

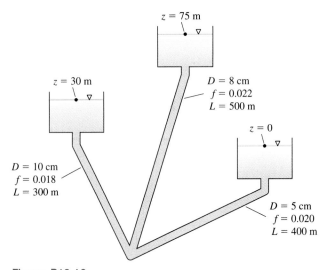

Figure P13.16

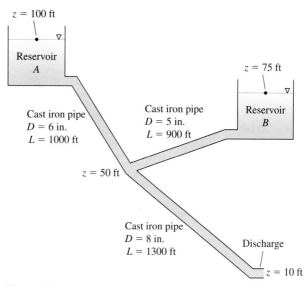

Figure P13.17

13.72 Consider the reservoir system shown in Figure P13.17. The elevations of reservoirs A and B are 100 and 75 ft, respectively. The junction elevation is 50 ft and the discharge is at 10 ft. The pipe from A to the junction is 1000 ft of 6 in. diameter cast iron, from B to the junction is 900 ft of 5 in. diameter cast iron, and from the junction to the discharge is 1300 ft of 8 in. diameter cast iron. Determine the flowrate in each pipe. Neglect minor losses.

13.73 Redo Problem 13.72 assuming that the elevation of reservoir B is unknown and the flow rate leaving it to be 20 ft³/s.

13.74 Consider the schematic of the parallel pipe system in Figure P13.18. The globe valve in branch 2 is fully open. The pressure and elevation at junction A are 1.5×10^5 N/m² and 30 m, respectively. The elevation of junction B is 20 m. The total flowrate is 2 m³/s of 20°C water. Find the

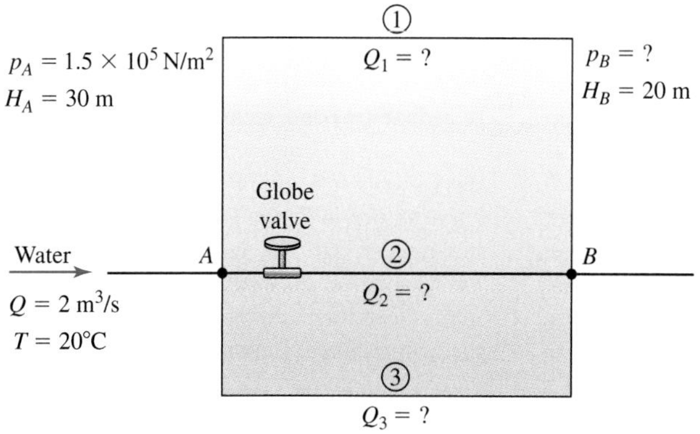

$p_A = 1.5 \times 10^5$ N/m²
$H_A = 30$ m

$Q_1 = ?$

$p_B = ?$
$H_B = 20$ m

Water
$Q = 2$ m³/s
$T = 20°C$

Globe valve

$Q_2 = ?$

$Q_3 = ?$

Figure P13.18

Branch	D (cm)	L (m)	f
1	10	200	0.020
2	8	10	0.018
3	10	150	0.020

flowrate through each branch and the pressure at junction B. Assume that minor losses are included as equivalent length of pipe other than the globe valve.

13.75 Redo Problem 13.74 with the globe valve one-quarter open.

13.76 The commercial steel pipe system shown in Figure P13.19 delivers kerosene at 20°C with a total flow rate of 0.05 m³/s. When the pump is not running it can be considered a minor loss with $K = 1.4$. Determine the flowrate in each pipe and the pressure drop between junctions A and B.

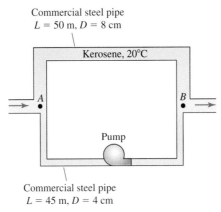

Figure P13.19

13.77 Redo Problem 13.76 with the pump running and delivering 35 kW to the flow.

Section 13.5

13.78 The system shown in Figure P13.20 continuously sprays water on fruit and vegetables moving along the conveyor. The water is recycled in the tank. The pipe is Schedule 40 PVC. The minor losses include a reentrant inlet, two regular 90°, threaded elbows, the filter ($K = 3.0$) and the spray nozzle (10 psi pressure loss at 60 gal/min). Based on the pump curve in Figure 13.37, choose a pipe diameter to maximize the pump efficiency. The water is 70°F.

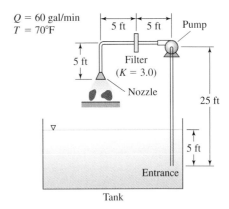

Figure P13.20 System schematic.

13.79 The air velocity must be maintained at 150 ft/min through the open face of the laboratory hood shown in Figure P13.21A. Choose a blower to service the hood (0.1 in. H₂O head loss) from the selection given in Figure P13.21B. Design steel ductwork from the hood to the blower and from the blower to the existing stack. Your choices should be based on minimizing the overall cost. Assume that the ductwork material and fabrication cost is equal to the fan cost. Assume electrical power costs of $0.07/kW-h. Assume a 5-year system life. Neglect the time value of money.

13.80 Redo Problem 13.79 assuming that a more hazardous material, requiring a face velocity of 250 ft/min, is now to be worked inside the hood.

13.81 The design of a nutrient system in a biotechnology lab includes the locations of the nutrient holding tank and the experiment trays shown in Figure P13.22A and the pump curve shown in Figure P13.22B. The required flowrate is 2 gal/min, and the nutrient mix has the properties of water at 70°F. The total minor head losses are $K = 6.0$, not including the control valve. Determine the placement of an existing pump and select the diameter of the plastic tubing.

PROBLEMS

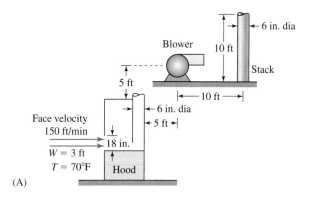

(A)

Shaded Pole Blowers

Flow (ft³/min) at 60 Hz and rpm shown

Free air	0.1 SP	0.2 SP	0.3 SP	0.4 SP	0.5 SP	0.6 SP	0.7 SP	0.8 SP	Cutoff SP
350	340	328	312	296	274	240	202	158	1.00
350	340	328	312	296	274	240	202	158	1.00
465	428	396	352	305	227	125	—	—	0.76
465	428	396	352	305	227	125	—	—	0.76
490	460	425	390	350	305	235	—	—	0.80
495	476	458	437	416	387	360	312	265	1.26
495	476	458	437	416	387	360	312	265	1.26
815	767	716	663	604	537	460	280	—	0.91
980	940	890	843	788	730	655	565	210	1.05

Two-Speed Blowers

									Volts	Hz	Therm. Protect.	rpm	Free Air Watts	Amps	Stock No.	List	Each	Lots 4	shpg. Wt.
435	410	380	345	295	205	100	—	0.78	115	60/50	Auto.	1350	175	2.17	4C565	$131.51	$78.75	$74.80	11.0
296	290	279	265	220	153	90	—	0.75				900	110	1.41					
480	465	445	425	400	370	340	300	245	1.16	115	60/50	Auto.	1525	250	3.55	4C566	169.50	101.55	96.43	13.0
375	365	355	340	320	295	265	230	160	1.12				1165	158	2.21					
760	750	735	710	675	640	610	565	520	1.70	115	60/50	Auto.	1450	335	3.20	5C508	233.16	139.65	132.63	13.0
545	530	525	505	495	480	460	420	380	1.60				1100	215	2.07					

PSC Blowers

									Volts	Hz	Therm. Protect.	rpm	Free Air Watts	Amps	Stock No.	List	Each	Lots 4	shpg. Wt.
382	349	335	319	300	279	248	204	145	1.10	115	60/50	Auto.	1550	125	1.21	4C666	158.11	94.70	89.94	11.0
488	470	450	430	406	377	336	282	217	1.18	115	60/50	Auto.	1580	157	1.43	4C667	168.24	100.75	95.70	12.0
745	715	683	647	605	560	506	—	—	0.92	115/230	60/50	Auto.	1060	249	2.59	4C668	254.86	152.75	144.98	20.0
960	930	890	845	800	745	688	470	253	1.16	115/230	60/50	Auto.	1030	380	4.00	4C830	267.53	160.50	152.19	23.0
1210	1200	1190	1180	1175	1160	1135	1110	1092	2.20	115/230	60/50	Auto.	1400	910	8.60	4C831	300.00	179.75	170.66	22.0

(B)

Figure P13.21 (A) System schematic. (B) Typical manufacturer's chart of blower values.

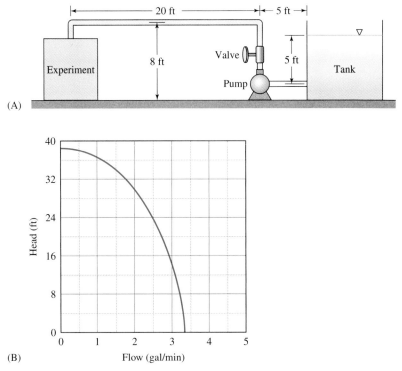

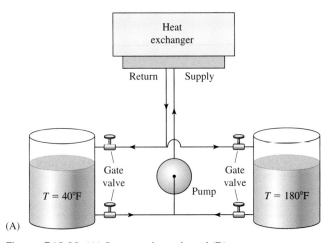

Figure P13.22 (A) System schematic and (B) pump curve.

Figure P13.23 (A) System schematic and (B) pump curves.

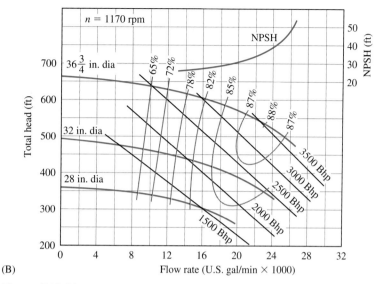

Figure P13.23 (*continued*)

13.82 A heat exchanger can be used for heating or cooling as shown in Figure P13.23A. If the flowrate is 24,000 gal/min, choose the diameter of cast iron pipe and the most efficient pump based on the curves shown in Figure P13.23B for 40 or 180°F water. The total length of pipe, including the equivalent length of minor losses, is 1000 ft.

13.83 The NPSHR for the pump used in Problem 13.78 is 5 ft. Do you recommend that the elevation of the pump be changed?

13.84 Comment on how the location of the pump and the rigidity of the tubing used in Problem 13.81 could affect the performance of the design.

13.85 For the pump system described in Problem 13.82, determine if cavitation will occur if the total head loss from the tanks to the pump is 7 ft. Be sure and check for each temperature.

14 EXTERNAL FLOW

14.1 Introduction
14.2 Boundary Layers: Basic Concepts
    14.2.1 Laminar Boundary Layer on a Flat Plate
    14.2.2 Turbulent Boundary Layer on a Flat Plate
    14.2.3 Boundary Layer on an Airfoil or Other Body
14.3 Drag: Basic Concepts
14.4 Drag Coefficients
    14.4.1 Low Reynolds Number Flow
    14.4.2 Cylinders
    14.4.3 Spheres
    14.4.4 Bluff Bodies
14.5 Lift and Drag of Airfoils
14.6 Summary
Problems

14.1 INTRODUCTION

An external flow occurs whenever an object moves through a fluid or a fluid passes by the surface of a structure. Such flows are present everywhere in the world around us, both in nature and as a result of modern technology. Nearly all living creatures, from the smallest bacterium to the largest mammal, encounter air or water in motion. In fact, some animals, such as pelicans and other diving birds, are equipped to deal with locomotion in both fluids. Humans have often looked to the animal world for inspiration in their desire for enhanced mobility on land, sea, and air. Despite some success in simply copying natural designs, the development of aircraft, ships, and to a lesser extent, land vehicles such as automobiles could not have occurred without a good understanding of external flow. Today, in technical problems ranging from how particles settle onto the surface of the lung to the effects of wind blowing over a building, understanding the interaction between an object moving through a fluid (or equivalently fluid moving over an object or adjacent to a structure), remains of great practical importance.

14.1 INTRODUCTION

Figure 14.1 Aerodynamic testing of a truck with smoke in a wind tunnel.

 CD/Video library/Butterflies & Flow past a larvae & Gold fish & Pine cone & Pine needles & Sperm

Some aspects of external flow have been mentioned in earlier chapters, both because they are fundamentally useful in engineering design and to provide a foundation for this chapter. Before continuing you may find it helpful to revisit the case studies in Sections 3.3.4 (flat plate boundary layer), 3.3.5 (drag on cylinders and spheres), and 3.3.6 (lift and drag on airfoils), as well as those dealing with fluid force exerted on a body in an external flow in Sections 4.5.1 (flow over a flat plate), and 4.5.3 (lift and drag). An experienced engineer can often obtain a good feel for the forces generated by external flows by using flow visualization. Thus you may also wish to review the flow visualization concepts in Chapter 10. Have you seen commercials featuring a car or other vehicle in a wind tunnel with streamlines of smoke passing over the surface, as shown in Figure 14.1? The streamlines reveal regions of flow separation that contribute to drag. A number of other things you have learned are also relevant to your study of external flow. Examples include the analytical and computational fluid dynamics solutions for flow over a cylinder presented in Chapter 12, and the brief discussion of turbulence in that same chapter. As you will learn, turbulence plays a critical role in external flow. The presence or absence of turbulence strongly influences boundary layer development, flow separation, and the forces exerted by a fluid on an immersed object.

 CD/Video library/Flow past cars

Our discussion of external flow begins with the concept of a boundary layer, a thin layer of moving fluid near a solid surface in which the no-slip condition and viscosity combine to create a velocity gradient. That velocity gradient creates a shear stress on the adjacent surface in the direction of the nearby flow. Since the flow in a boundary layer can be laminar or turbulent, we will discuss the characteristics of each type, limiting ourselves at first to the simplest case of flow along a flat plate. We next briefly discuss the boundary layer on airfoils and other objects for which the curvature of the surface and angle of incidence of the freestream are important. We conclude the chapter with sections devoted to the discussion of drag coefficients, and lift and drag of airfoils. Many of the results in these sections are based on empirical observations that have been collapsed by dimensional analysis into lists of the familiar lift and drag coefficients introduced in the case studies. Examples of the use of those coefficients in solving a variety of external flow problems are provided. Throughout this chapter you will also find a qualitative discussion of how lift and drag are generated by the flow field. We include this discussion to help you understand how changes in the flow field about an immersed body can have dramatic effects on the force applied to the body by the fluid.

 CD/Boundary layers/Boundary layer concepts

14.2 BOUNDARY LAYERS: BASIC CONCEPTS

When a body is immersed in a moving fluid, the fluid velocity along a line perpendicular to any point on the body surface is observed to vary from zero on the surface to a maximum value some distance away. At small Reynolds numbers, the distance over which the velocity variation occurs may be of the same magnitude as the dimensions of the body itself. However, at large Re the variation occurs over a relatively small distance, and the body is said to have a boundary layer, meaning that there is a layer of fluid near the surface of the body in which the velocity changes from zero on the surface to the freestream value. Prandtl's insight into this phenomenon and his subsequent development of boundary layer theory are milestones in the development of fluid mechanics.

The characteristics of a boundary layer are affected by the shape of the solid surface of interest, the orientation of the surface relative to the freestream, and many other factors. However, we can illustrate the basic concepts by examining the boundary layer on a thin flat plate aligned with the freestream. Consider the boundary layer on the upper surface of such a plate at large Reynolds number as shown in Figure 14.2. The flow is steady, and we can define a Reynolds number for the flow by using the length L of the plate and freestream velocity U to write $Re_L = UL/\nu$. In a fluid of relatively small kinematic viscosity such as air or water, the requirement of a large Reynolds number means that the freestream velocity U is large. Observation shows that at large Reynolds

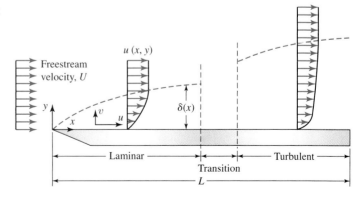

Figure 14.2 Geometry for flow over a flat plate.

numbers the boundary layer is relatively thin, and the thickness of the boundary layer increases in the downstream direction. Moreover, just downstream of the nose of the plate the boundary layer is observed to be laminar, but at some point downstream transition occurs and the boundary layer becomes turbulent.

At any location x along the plate, there is a smooth variation in the streamwise velocity component u along a line perpendicular to the plate surface. Thus, u is a function of x and y inside the boundary layer, but a constant U outside. The transverse velocity component, v, is also observed to be a function of x and y inside the boundary layer, but it is zero outside. The transverse velocity component v in a boundary layer is very small compared to u. Thus one boundary layer characteristic is $v \ll u$. A second characteristic is based on the observation that the velocity changes rapidly in a direction normal to the surface, but slowly in the flow direction. This means that spatial derivatives of the velocity in the flow direction inside the boundary layer are small in comparison to spatial derivatives in the normal direction.

The approach of the streamwise velocity component u to the freestream value is asymptotic. Nevertheless, we can define a boundary layer thickness δ as the height above the plate at which $u = 0.99\,U$, meaning that the streamwise velocity component is within one percent of the freestream value. Observation shows that the boundary layer thickness grows at one rate in the laminar region, another rate in the transition region, and yet another rate in the turbulent region. Since the thickness of a boundary layer depends on the location x along the plate, we write $\delta = \delta(x)$ and recognize that this function is an important characteristic of a boundary layer, since it defines the edge where the boundary layer and freestream meet. Note that this edge (dashed line in Figure 14.2) is not a streamline. In fact, streamlines enter the boundary layer all along its length.

It is customary to define two additional quantities that also characterize the thickness of a boundary layer. The first of these, called the displacement thickness and represented by δ^*, is defined by the following integral

$$\delta^* = \int_0^\infty \left(1 - \frac{u}{U}\right) dy \tag{14.1}$$

This integral takes a different value at each location x along the plate, so we write $\delta^* = \delta^*(x)$. One rationale for defining the displacement thickness in this way is illustrated in Figure 14.3, where we have overlaid the velocity profile in the boundary layer on top of the uniform velocity profile that would exist if the fluid were inviscid and able to slip by the plate. The shaded area can be thought of as the volume flowrate per unit width w into the paper that is missing because of the presence of the boundary layer. The difference in volume flowrate carried by the two velocity profiles is seen to be given by $\Delta Q = w \int_0^\infty U\, dy - w \int_0^\infty u\, dy$. With a little rearrangement we can write this as

$$\Delta Q = w \int_0^\infty (U - u)\, dy = wU \int_0^\infty \left(1 - \frac{u}{U}\right) dy = Uw\delta^*$$

Thus the missing volume flowrate in the boundary layer is seen to be given by $\Delta Q = Uw\delta^*$, and the missing mass flowrate is $\Delta \dot{M} = \rho U w \delta^*$. From this analysis, and recalling that the passage formed by two adjacent streamlines carries a certain volume flowrate in proportion to their distance apart, it is customary to say that in comparison to a fictitious inviscid flow over the plate, a boundary layer displaces streamlines a distance

Figure 14.3 Definitions of boundary layer and displacement thickness.

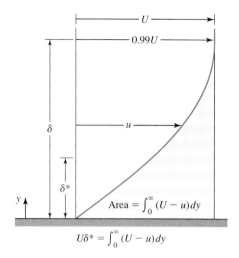

δ^* away owing to viscous effects. Another way of describing δ^* is to say that the boundary layer makes a body appear δ^* thicker owing to the effects of viscosity in slowing down fluid near the body surface. That is, the body surface is effectively defined by the edge of the boundary layer.

The second additional quantity that is used to characterize the thickness of a boundary layer is $\Theta = \Theta(x)$, called the momentum thickness. The momentum thickness is defined by

$$\Theta = \int_0^\infty \frac{u}{U}\left(1 - \frac{u}{U}\right) dy \tag{14.2}$$

A streamwise momentum balance can be used to show that in comparison to an inviscid flow, the missing streamwise momentum flux in the boundary layer is equal to $\rho U^2 w \Theta$. Each of the three thicknesses, $\delta = \delta(x)$, $\delta^* = \delta^*(x)$, and $\Theta = \Theta(x)$ plays an important role in discussions of boundary layers.

Additional quantities of importance in boundary layer theory are the wall shear stress, $\tau_W = \tau_W(x)$, which is a function of position along the plate because of the changing streamwise velocity profile, and the total force exerted by the fluid on the plate.

The total force on the plate can be thought of as consisting of lift and drag. The lift component of this force is defined by Eq. 4.25b as $F_L = \int_S (-p\mathbf{n} + \boldsymbol{\tau}) \cdot \mathbf{n}_L \, dS$, where the unit vector $\mathbf{n}_L$ is normal to the plate. For a flat plate aligned with the freestream, the wall shear stress acts along the plate, i.e., in the streamwise direction, so there can be no contribution to lift from the shear stress. In addition, the symmetry of the flow ensures that the pressure distribution on both sides of the plate is the same. Thus the lift is zero and the total force on a flat plate aligned with the freestream consists solely of drag. The drag force on the plate is defined by Eq. 4.26b as $F_D = \int_S (-p\mathbf{n} + \boldsymbol{\tau}) \cdot \mathbf{n}_\infty \, dS$, where the unit vector $\mathbf{n}_\infty$ points in the flow direction. Since the pressure acts normal to the plate, it cannot contribute to the drag, and the drag on each side of a plate of width w and length L is given by

$$F_D = w \int_0^L \tau_W(x) \, dx \tag{14.3}$$

14.2 BOUNDARY LAYERS: BASIC CONCEPTS

From this we see that the wall shear stress distribution is indeed of great interest in boundary layer analysis.

At this point we can begin to investigate the relationships among these boundary layer parameters by employing dimensional analysis. In doing DA on this problem, it is customary to choose the spatial coordinate x as the length scale, rather than the length of the plate L. The DA is otherwise routine and yields the following relationships:

$$\frac{\delta}{x} = f_1\left(\frac{Ux}{\nu}\right) = f_1(Re_x) \tag{14.4}$$

$$\frac{\tau_W}{\rho U^2} = f_2\left(\frac{Ux}{\nu}\right) = f_2(Re_x) \tag{14.5}$$

where the Reynolds number based on x is given by

$$Re_x = \frac{Ux}{\nu} \tag{14.6}$$

Introducing the skin friction coefficient $C_f = \tau_W / \frac{1}{2}\rho U^2$, we can write Eq. 14.5 as

$$C_f = f_3\left(\frac{Ux}{\nu}\right) = f_3(Re_x) \tag{14.7}$$

DA alone cannot tell us the form of these unknown relationships, but as shown in the next two sections, theory and empirical data can.

14.2.1 Laminar Boundary Layer on a Flat Plate

H. Blasius, one of Prandtl's students, analyzed the steady, laminar boundary layer on a smooth flat plate aligned with the freestream in 1908. We can derive his result, known as the Blasius solution, by using Cartesian coordinates with the plate aligned with the x axis as shown in Figure 14.4. In general, the equations of motion for a steady, constant density, constant viscosity flow are given by Eqs. 12.1a–12.1d. However, there is no reason to expect a cross-stream velocity component w in the flow over a flat plate, nor any variation of a flow property in the z direction. Thus the flow is 2D. Inserting $w = 0$, and

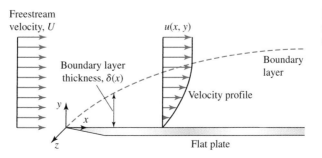

Figure 14.4 The geometry of the boundary layer over a flat plate.

HISTORY BOX 14-1

Heinrich Blasius is famous in fluid mechanics for his boundary layer solution and other contributions he made while studying under Prandtl. However, later in life he left fluid mechanics research to teach at a Hamburg engineering school.

dropping all derivatives with respect to z in Eqs. 12.1a–12.1d, yields the following equations as the starting point for a discussion of the flat plate boundary layer:

$$\frac{\partial u}{\partial x} + \frac{\partial v}{\partial y} = 0$$

$$\rho\left(u\frac{\partial u}{\partial x} + v\frac{\partial u}{\partial y}\right) = -\frac{\partial p}{\partial x} + \mu\left(\frac{\partial^2 u}{\partial x^2} + \frac{\partial^2 u}{\partial y^2}\right)$$

$$\rho\left(u\frac{\partial v}{\partial x} + v\frac{\partial v}{\partial y}\right) = -\frac{\partial p}{\partial y} + \mu\left(\frac{\partial^2 v}{\partial x^2} + \frac{\partial^2 v}{\partial y^2}\right)$$

 CD/Boundary layers/Laminar boundary layers

Further simplification of these equations can be made by recalling that the flow in a boundary layer is predominantly parallel to the surface on which it occurs. Thus we can assume $v \ll u$. Furthermore, the boundary layer is thin, which implies that derivatives of flow variables with respect to x are much smaller than those with respect to y. By using these two assumptions to simplify the preceding set of equations, we obtain

$$\frac{\partial u}{\partial x} + \frac{\partial v}{\partial y} = 0 \tag{14.8a}$$

$$\rho\left(u\frac{\partial u}{\partial x} + v\frac{\partial u}{\partial y}\right) = -\frac{\partial p}{\partial x} + \mu\left(\frac{\partial^2 u}{\partial y^2}\right) \tag{14.8b}$$

$$0 = -\frac{\partial p}{\partial y} \tag{14.8c}$$

These are the Prandtl boundary layer equations. They can be shown to be applicable to boundary layers on moderately curved as well as flat surfaces, an important point to keep in mind throughout the rest of our discussion.

Before we worry about solving these equations, notice that the last equation tells us that the pressure in the boundary layer does not vary across the boundary layer. Thus we conclude that the pressure inside the boundary layer on flat and moderately curved surfaces is the same as it is in the inviscid flow outside the boundary layer. This is a very important aspect of Prandtl's boundary layer equations, for it shows that the pressure gradient in Eq. 14.8b may be considered to be known and determined by finding the pressure distribution in the inviscid flow over the same surface shape.

Blasius was able to solve the Prandtl boundary layer equations for a flat plate by recognizing that since the pressure in an inviscid flow over a flat plate is uniform, the pressure gradient $\partial p/\partial x$ in Eq. 14.8b is zero. Thus the flat plate boundary layer is described by:

$$\frac{\partial u}{\partial x} + \frac{\partial v}{\partial y} = 0 \qquad (14.9a)$$

$$\rho\left(u\frac{\partial u}{\partial x} + v\frac{\partial u}{\partial y}\right) = \mu\left(\frac{\partial^2 u}{\partial y^2}\right) \qquad (14.9b)$$

These two equations, with the associated boundary conditions, must be solved for the unknown velocity components. The no-slip, no-penetration boundary conditions that describe the flow over a flat plate are $u(x, 0) = 0$ and $v(x, 0) = 0$ for $x > 0$. We also want the boundary layer solution to match the inviscid freestream solution $\mathbf{u} = U\mathbf{i}$ that exists upstream of the plate and above the plate outside the boundary layer. These last two conditions can be written as $u \to U$, $v \to 0$ for $x < 0$, and $u \to U$, $v \to 0$ for $y \gg \delta$, i.e., outside the boundary layer. Here U is the magnitude of the freestream velocity.

Although there is no known solution to Eqs. 14.9a and 14.9b that satisfies these conditions exactly, Blasius showed that a similarity solution of these equations can be obtained by introducing a new variable

$$\eta = \left(\frac{U}{\nu x}\right)^{1/2} y \qquad (14.10a)$$

and employing a streamfunction

$$\psi(x, y) = (U\nu x)^{1/2} f(\eta) \qquad (14.10b)$$

The word "similarity" used to describe this solution indicates that when properly scaled in the similarity variable, $\eta = (U/\nu x)^{1/2} y$, the velocity profiles at every location along the flat plate collapse onto a single universal curve. Thus the profiles are similar.

Now by the definition of a streamfunction we have $u = \partial\psi/\partial y$ and $v = -\partial\psi/\partial x$, thus Eq. 14.9a is automatically satisfied. The velocity components are found to be given by

$$u = \frac{\partial}{\partial y}[(U\nu x)^{1/2} f(\eta)] = \frac{\partial f}{\partial \eta}\frac{\partial \eta}{\partial y} = U\frac{\partial f}{\partial \eta}$$

and

$$v = -\frac{\partial \psi}{\partial x} = -\frac{\partial}{\partial x}[(U\nu x)^{1/2} f(\eta)] = \left(\frac{U\nu}{4x}\right)^{1/2}\left(\eta\frac{\partial f}{\partial \eta} - f\right)$$

Substituting these velocity components into Eq. 14.9b, simplifying, and making the key similarity assumption that the function f is not separately a function of x, we obtain the following nonlinear, third-order, ordinary differential equation:

$$2\frac{d^3 f}{d\eta^3} + f\frac{d^2 f}{d\eta^2} = 0 \qquad (14.11)$$

TABLE 14.1 Blasius Solution

η	f	$\dfrac{\partial f}{\partial \eta}$	$\dfrac{\partial^2 f}{\partial \eta^2}$
0	0	0	0.3321
0.5	0.0415	0.1659	0.3309
1.0	0.1656	0.3298	0.3230
1.5	0.3701	0.4868	0.3026
2.0	0.6500	0.6298	0.2668
2.5	0.9963	0.7513	0.2174
3.0	1.3968	0.8460	0.1614
3.5	1.8377	0.9130	0.1078
4.0	2.3057	0.9555	0.0642
4.5	2.7901	0.9795	0.0340
5.0	3.2833	0.9915	0.0159
5.5	3.7806	0.9969	0.0066
6.0	4.2796	0.9990	0.0024
6.5	4.7793	0.9997	0.0008
7.0	5.2792	0.9999	0.0002
7.5	5.7792	1.0000	0.0001
8.0	6.2792	1.0000	0.0000

The boundary conditions for this equation are $f = df/d\eta = 0$ at $\eta = 0$, and $df/d\eta \to 1$ as $\eta \to \infty$.

The function $f(\eta)$ that satisfies this equation and boundary conditions defines the Blasius solution. This function must be obtained numerically. Table 14.1 contains values of f, $df/d\eta$, and $d^2 f/d\eta^2$. These are readily found by using Mathematica or another symbolic code to solve the differential equation. Notice that the edge of the boundary layer, defined as the location at which $df/d\eta = u/U = 0.99$, occurs at $\eta \cong 5.0$.

 CD/Special features/Virtual labs/Blasius Boundary Layer Growth

The streamwise velocity profiles at various locations, shown in Figure 14.5A, exhibit the growth in the thickness of the boundary layer at locations away from the nose of the plate. When properly scaled in the similarity variable, all the profiles in Figure 14.5A collapse onto a single universal curve: the boundary layer velocity profile $u/U = df/d\eta$ as shown in Figure 14.5B. We see the expected boundary layer behavior: a zero velocity on the wall with a gradual approach to the freestream value near $\eta = 5.0$.

Because the governing equations have been simplified to derive the Blasius solution, it is necessary to confirm experimentally that the velocity profiles in the flat plate boundary layer do exhibit similarity. Figure 14.6A shows the similar profiles in the laminar flow region beginning just downstream of the nose of the plate. In the bottom left

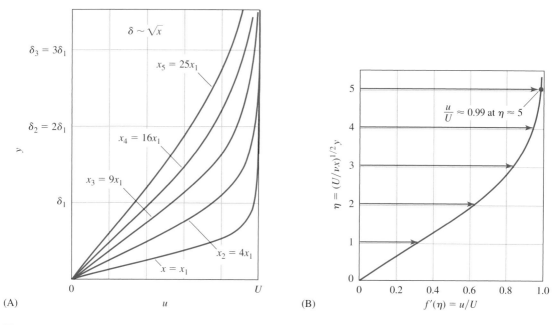

Figure 14.5 Flow over a flat plate: (A) boundary layer velocity profiles along the plate and (B) the equivalent similarity profile.

corner of Figure 14.6B the laminar profile is also present. But after the onset of turbulence the velocity profiles are no longer similar. Thus the Blasius similarity solution, which is valid for the laminar flow boundary layer, confirmed Prandtl's basic ideas about the boundary layer.

> CD/Demonstrations/Blasius and Falkner-Skan solutions

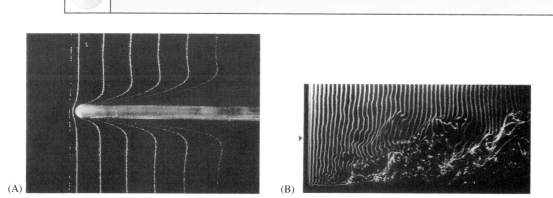

Figure 14.6 Flow visualization of flow over a flat plate using hydrogen bubbles to create material lines. (A) Laminar boundary layer near the nose of the plate, and (B) A wider view showing the transition into a turbulent boundary layer.

The Blasius solution also provides information about the shear stress distribution on the plate. The shear stress is given by Eq. 11.6d as $\sigma_{xy} = \sigma_{yx} = \mu(\partial u/\partial y + \partial v/\partial x)$. Evaluating this expression by using $u = U(\partial f/\partial \eta)$ and $v = (U\nu/4x)^{1/2} \times [\eta(\partial f/\partial \eta) - f]$ results in a complicated expression involving f, $\partial f/\partial \eta$, and $\partial^2 f/\partial \eta^2$. However, at the wall $y = 0$, the expression for the shear stress simplifies to $\sigma_{xy} = \sigma_{yx} = \mu U \sqrt{U/\nu x}(\partial^2 f/\partial \eta^2)|_{y=0}$, so only the value $(\partial^2 f/\partial \eta^2)|_{y=0} = 0.3321$ is needed to calculate τ_W.

Since transition to a turbulent boundary is observed at roughly $Re_x \approx 5 \times 10^5$, the Blasius solution is valid for a value of x greater than zero but smaller than $x \approx 5 \times 10^5(\nu/U)$. However, the precise location of transition in a boundary layer is not as certain as this appears to indicate. Thus this range of validity is approximate. In fact it is possible to force the transition to a turbulent boundary layer near $x = 0$ by various means, including a trip wire or artificial roughness.

Within these constraints, the laminar flat plate boundary problem may be considered solved. The quantities of interest calculated from the Blasius solution may be summarized as follows:

$$Re_x = \frac{Ux}{\nu} \tag{14.12a}$$

$$\frac{\delta(x)}{x} = 5.0(Re_x)^{-1/2} \tag{14.12b}$$

$$\frac{\delta^*(x)}{x} = 1.721(Re_x)^{-1/2} \tag{14.12c}$$

$$\frac{\Theta(x)}{x} = 0.664(Re_x)^{-1/2} \tag{14.12d}$$

$$\tau_W(x) = 0.332\rho U^2 (Re_x)^{-1/2} \tag{14.12e}$$

$$C_f(x) = 0.664(Re_x)^{-1/2} \tag{14.12f}$$

By comparing these results and those obtained by DA in the Section 14.2 (Eqs. 14.4–14.7), we can see that the Blasius solution provides the unknown functional dependence on Re_x. In addition, it is evident that the similarity of the Blasius solution justifies the use of x as the length scale in the dimensional analysis. The important characteristics of the laminar flat plate boundary layer are found in Eqs. 14.12: the boundary layer thickness grows at a rate $\delta(x) \propto x^{1/2}$, and this is also true of the displacement and momentum thicknesses; the wall shear stress decreases at a rate $\tau_W(x) \propto x^{-1/2}$, as does the skin friction coefficient.

It is possible to define a drag coefficient for a flat plate. Consider the drag contributed by the boundary layer on one side of a plate of width w and length L. Inserting Eq. 14.12e into the equation defining the drag force, Eq. 14.3, and substituting $Re_x = Ux/\nu$, we obtain

$$F_D = w\int_0^L \tau_W(x)\,dx = w\int_0^L 0.332\rho U^2 \left(\frac{Ux}{\nu}\right)^{-1/2} dx$$

14.2 BOUNDARY LAYERS: BASIC CONCEPTS

Completing the integration we find

$$F_D = \frac{0.664\rho U^2 wL}{\sqrt{Re_L}} \tag{14.12g}$$

which corresponds to a drag coefficient based on the plate area wL of

$$C_D = \frac{1.328}{\sqrt{Re_L}} \tag{14.12h}$$

Although some of these results appeared in Section 3.3.4, the example that follows further illustrates the characteristics of a laminar flat plate boundary layer.

EXAMPLE 14.1

A thin flat plate, 10 ft tall and 1 ft wide, forms the leading edge of a banner towed by an aircraft at 100 mph on a 70°F day. How far from the leading edge of the plate does the laminar portion of boundary layer extend? What is the boundary layer thickness at the downstream end of the laminar boundary layer? Find the drag force on the plate contributed by the laminar boundary layer, and the corresponding drag coefficient.

SOLUTION

The physical arrangement is shown in Figure 14.7. A laminar boundary layer is expected to transition at $Re_x \leq 5 \times 10^5$. Thus we can solve for the distance x_C to the transition point by using Eq. 14.12a to write: $x_C = (5 \times 10^5 \nu)/U$. Using $U = 100$ mph $= 146.7$ ft/s and, from Appendix A, $\nu = 1.64 \times 10^{-4}$ ft²/s for air, we find

$$x_C = \frac{5 \times 10^5 [1.64 \times 10^{-4} \ (\text{ft}^2/\text{s})]}{146.7 \ \text{ft/s}} = 0.56 \ \text{ft}$$

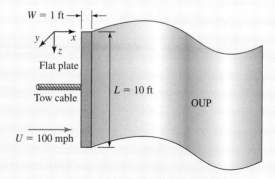

Figure 14.7 Schematic for Example 14.1.

Transition begins to occur just beyond the midpoint of the plate, and the rear portion of the plate has a turbulent boundary layer. To find the laminar boundary layer thickness at $x = x_C$ we use Eq. 14.12b to calculate the thickness as

$$\delta(x_C) = 5.0 x_C (Re_x)^{-1/2} = 5.0(0.56 \text{ ft})(5 \times 10^5)^{-1/2} = 0.00396 \text{ ft} = 0.048 \text{ in.}$$

Thus the laminar boundary layer is only 0.048 in. thick when transition occurs. To find the drag force on both sides of the plate due to the laminar boundary layer, we must multiply Eq. 14.12g, which gives the drag on one side, by 2 and obtain $F_D = 1.328 \rho U^2 w L / \sqrt{Re_L}$. In this case, we must also be careful to insert $L = x_C$, since this defines the portion of the plate covered by the laminar boundary layer. Using $\rho = 2.329 \times 10^{-3}$ slug/ft^3 from Appendix A, inserting the other data, and noting that $Re_L = 5 \times 10^5$, we obtain

$$F_D = \frac{1.328 \rho U^2 w L}{\sqrt{Re_L}}$$

$$= \frac{1.328}{\sqrt{5 \times 10^5}} (2.329 \times 10^{-3} \text{ slug/ft}^3)(146.7 \text{ ft/s})^2 (10 \text{ ft})(0.56 \text{ ft}) = 0.53 \text{ lb}_f$$

The total drag on the plate is actually much larger than this since we have not accounted for the drag of the turbulent boundary layer. The drag coefficient for the laminar portion of the boundary layer, which refers to one side of the plate only, is given by Eq. 14.12h as $C_D = 1.328/\sqrt{Re_L} = 0.0019$.

CD/Boundary layers/Instability. transition, and turbulence

14.2.2 Turbulent Boundary Layer on a Flat Plate

There is no analytical solution available for a turbulent boundary layer on a smooth flat plate, so we are forced in this case to rely on empirical observations. It is customary to model the streamwise velocity profile in the turbulent boundary layer for $0 < y/\delta \leq 1$, by the power law

$$\frac{u}{U} = \left(\frac{y}{\delta}\right)^{1/7} \tag{14.13}$$

with $u = U$ for $y/\delta > 1$. Since the flow is turbulent, u is the average velocity. The boundary layer thickness is a function of x as usual, hence $\delta = \delta(x)$ and must also be determined. Figure 14.8 compares the turbulent velocity profile with the laminar profile. Note that the turbulent profile is fuller and the velocity gradient at the wall is larger than in a laminar flow. The increased mixing due to the turbulence results in a higher streamwise velocity at any given distance from the wall in comparison to the laminar profile.

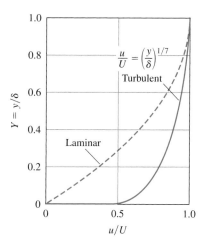

Figure 14.8 Comparison of the flat plate laminar and turbulent boundary layers.

We cannot compute a wall shear stress in turbulent flow because the constitutive relationship between shear stress and average velocity in a turbulent flow is unknown. Instead, we make use of the following empirical result for the wall shear stress

$$\tau_W(x) = 0.0225\rho U^2 \left(\frac{\nu}{U\delta}\right)^{1/4} \tag{14.14a}$$

This allows us to write the skin friction coefficient as

$$C_f(x) = 0.045 \left(\frac{\nu}{U\delta}\right)^{1/4} \tag{14.14b}$$

However, neither formula is useful as is because we do not know the boundary layer thickness $\delta = \delta(x)$. A clever solution to this dilemma consists of making use of a streamwise momentum balance on the boundary layer. Without going into the details here, if the power law given by Eq. 14.13 and the shear stress given by Eq. 14.14a are inserted into the streamwise momentum balance, the following boundary layer thickness is found:

$$\delta(x) = 0.370 \left(\frac{\nu}{U}\right)^{1/5} x^{4/5} \tag{14.15}$$

Comparing this to the corresponding laminar result, Eq. 14.12b, we see that the thickness of the laminar layer grows at the rate $\delta(x) \propto x^{1/2}$, while the turbulent boundary layer grows at the faster rate $\delta(x) \propto x^{4/5}$.

We can now use Eq. 14.15 to evaluate the wall shear stress, and since the velocity profile is known, we can also compute the displacement and momentum thicknesses for the turbulent boundary layer. The important characteristics of a turbulent boundary layer on a flat plate predicted by the power-law velocity profile model are summarized as follows:

$$Re_x = \frac{Ux}{\nu} \tag{14.16a}$$

$$\frac{\delta(x)}{x} = 0.370(Re_x)^{-1/5} \tag{14.16b}$$

$$\frac{\delta^*(x)}{x} = 0.0463(Re_x)^{-1/5} \tag{14.16c}$$

$$\frac{\Theta(x)}{x} = 0.0360(Re_x)^{-1/5} \tag{14.16d}$$

$$\tau_W(x) = 0.0288\rho U^2 (Re_x)^{-1/5} \tag{14.16e}$$

$$C_f(x) = 0.0577(Re_x)^{-1/5} \tag{14.16f}$$

$$F_D = 0.036\rho U^2 w L (Re_L)^{-1/5} \tag{14.16g}$$

$$C_D = 0.072(Re_L)^{-1/5} \tag{14.16h}$$

These results are known to be accurate for Reynolds numbers in the range $5 \times 10^5 < Re_x < 10^7$. We see that in the turbulent flat plate boundary layer, the displacement and momentum thicknesses also grow at a rate proportional to $x^{4/5}$. The wall shear stress decreases at a rate $\tau_W(x) \propto x^{-1/5}$, as does the skin friction coefficient.

Note that a power-law model for the turbulent boundary layer on a flat plate is not the only possible choice. Another model, based on a logarithmic velocity profile, is said to offer the advantage of providing accurate results for the much wider range $10^5 < Re_x < 10^9$. The boundary layer characteristics of this model are:

$$\frac{\delta(x)}{x} = 0.14(Re_x)^{-1/7} \tag{14.17a}$$

$$\tau_W(x) = 0.0125\rho U^2 (Re_x)^{-1/7} \tag{14.17b}$$

$$C_f(x) = 0.025(Re_x)^{-1/7} \tag{14.17c}$$

$$F_D = 0.015\rho U^2 w L (Re_L)^{-1/7} \tag{14.17d}$$

$$C_D = 0.030(Re_L)^{-1/7} \tag{14.17e}$$

EXAMPLE 14.2

A box-shaped truck body 2.5 m wide, 3 m high, and 7 m long (Figure 14.9) is traveling at 20 m/s in 20°C air. Calculate the contributions to the total drag of the truck from the sides and top of the truck body. Assume that a sheet metal seam near the leading edge of each panel causes the boundary layer to be turbulent for the full length of the panel. Also find the wall shear stress and boundary layer thickness along the top panel, and the maximum value of the wall shear stress and boundary layer thickness on this panel.

SOLUTION

We will first use Eq. 14.16a to calculate the maximum Reynolds number at the downstream edge $x = L$ of each panel. With viscosity data from Appendix A, we obtain

$$Re_L = \frac{UL}{\nu} = \frac{(20 \text{ m/s})(7 \text{ m})}{1.51 \times 10^{-5} \text{ m}^2/\text{s}} = 9.3 \times 10^6$$

14.2 BOUNDARY LAYERS: BASIC CONCEPTS

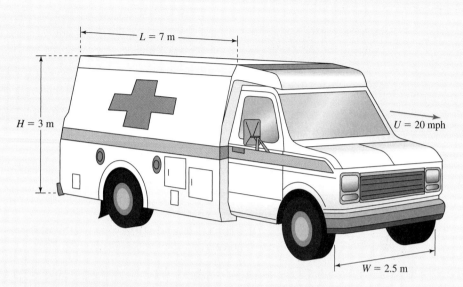

Figure 14.9 Schematic for Example 14.2.

This is just within the applicable range of Eqs. 14.16a–14.16h. The drag force on each panel is calculated by using Eq. 14.16g. For the top we obtain

$$F_{D_{\text{top}}} = 0.036\rho U^2 w L (Re_L)^{-1/5}$$
$$= \frac{0.036(1.204 \text{ kg/m}^3)(20 \text{ m/s})^2(2.5 \text{ m})(7 \text{ m})}{(9.3 \times 10^6)^{1/5}}$$
$$= 12.3 \text{ N}$$

The drag on each side panel is found using the same formula with $w = 3$ m instead of $w = 2.5$ m. Thus we can write $F_{D_{\text{side}}} = (3/2.5)F_{D_{\text{top}}} = 14.8$ N. The drag of all three panels is now calculated as $F_D = 2F_{D_{\text{side}}} + F_{D_{\text{top}}} = 2(14.8 \text{ N}) + 12.3 \text{ N} = 41.9$ N. To find the boundary layer thickness and wall shear stress on the top panel, we use Eqs. 14.16b and 14.16e, respectively. Writing these explicitly in terms of x we have

$$\delta(x) = 0.370x\left(\frac{Ux}{\nu}\right)^{-1/5} = 0.370\left(\frac{20 \text{ m/s}}{1.51 \times 10^{-5} \text{ m}^2/\text{s}}\right)^{-1/5} x^{4/5} = 0.022 x^{4/5} \text{ m}^{1/5}$$

$$\tau_W(x) = 0.0288\rho U^2 \left(\frac{Ux}{\nu}\right)^{-1/5}$$
$$= 0.0288(1.204 \text{ kg/m}^3)(20 \text{ m/s})^2 \left(\frac{20 \text{ m/s}}{1.51 \times 10^{-5} \text{ m}^2/\text{s}}\right)^{-1/5} x^{-1/5}$$
$$= 0.83 x^{-1/5} \text{ (N/m}^2\text{)(m}^{1/5}\text{)}$$

Note that these results for $\delta(x)$ and $\tau_W(x)$ also apply to the side panels. The maximum value of the shear stress and boundary layer thickness on each panel will occur at $x = L$. Inserting the data we find:

$$\delta(L) = 0.022L^{4/5}\text{m}^{1/5} = 0.022(7\text{ m})^{4/5}\text{m}^{1/5} = 0.104\text{ m} = 10.4\text{ cm}$$

$$\tau_W(L) = 0.83L^{-1/5}(\text{N/m}^2)(\text{m}^{1/5}) = (0.83)(7\text{ m})^{-1/5}(\text{N/m}^2)(\text{m}^{1/5}) = 0.56\text{ N/m}^2$$

We can repeat the calculations of the last problem, using the logarithmic model to define the various bounday layer characteristics. We will calculate values at $Re_L = UL/\nu = 9.3 \times 10^6$ and use Eq. 14.17d to calculate the drag, Eq. 14.17a to calculate the boundary layer thickness, and Eq. 14.17b to calculate the wall shear stress. The drag on the top panel is given by

$$F_{D_{top}} = \frac{0.015\rho U^2 wL}{(Re_L)^{1/7}} = \frac{0.015(1.204\text{ kg/m}^3)(20\text{ m/s})^2(2.5\text{ m})(7\text{ m})}{(9.3 \times 10^6)^{1/7}} = 12.8\text{ N}$$

Each side panel contributes $F_{D_{side}} = (3/2.5)F_{D_{top}} = 15.4$ N, yielding a total drag of 43.6 N. This is slightly larger than the 41.9 N drag calculated with the power-law model. For the boundary layer thickness we use Eq. 14.17a to write $\delta(x) = 0.14x(Re_x)^{-1/7}$ and, after inserting the data, we obtain $\delta(x) = 0.14x(Ux/\nu)^{-1/7} = 0.14(20\text{ m/s}/1.51 \times 10^{-5}\text{ m}^2/\text{s})^{-1/7}x^{6/7} = 0.019x^{6/7}\text{ m}^{1/7}$, which yields a thickness at $L = 7$ m of $\delta(L) = 0.019x^{6/7}\text{ m}^{1/7} = 0.019(7\text{ m})^{6/7}\text{ m}^{1/7} = 0.1\text{ m} = 10\text{ cm}$. This is a slightly smaller value than that obtained with the power-law model. From Eq. 14.17b the wall shear stress is given by

$$\tau_W(x) = 0.0125\rho U^2 \left(\frac{Ux}{\nu}\right)^{-1/7} = 0.0125(1.204\text{ kg/m}^3)(20\text{ m/s})^2 \left(\frac{20\text{ m/s}}{1.51 \times 10^{-5}\text{ m}^2/\text{s}}\right)^{-1/7} x^{-1/7}$$

$$= 0.80x^{-1/7}\text{ (N/m}^2\text{)(m}^{1/7}\text{)}$$

and the wall shear stress at $L = 7$ m is found to be

$$\tau_W(L) = 0.80L^{-1/7}\text{ (N/m}^2\text{)(m}^{1/7}\text{)} = 0.80(7\text{ m})^{-1/7}(\text{N/m}^2)(\text{m}^{1/7}) = 0.61\text{ N/m}^2$$

which is slightly larger than that calculated with the power-law model. For engineering purposes these values are equivalent to those found with the power-law model.

14.2.3 Boundary Layer on an Airfoil or Other Body

Consider the high speed flow over an airfoil at a small angle of attack as shown in Figure 14.10A. The upper and lower surfaces of the airfoil are curved, and neither surface is aligned with the freestream. Thus the results obtained earlier for the boundary layer on an aligned flat plate cannot be expected to apply to the boundary layer on this airfoil or, for that matter, to other objects of finite thickness. In fact, observation of boundary layers on airfoils and other bodies show that the shape of an object and its angle of incidence to the freestream have a significant effect on the characteristics of both laminar and turbulent boundary layers. For reasons that will become clear in a moment, this effect is described as the effect of a pressure gradient on the boundary layer.

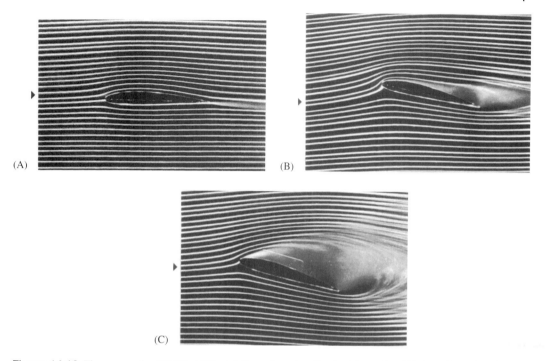

Figure 14.10 Flow around a NACA 4412 airfoil section visualized with smoke. (A) At a 2° angle of attack there is no boundary layer separation. (B) At a 15° angle of attack there is significant boundary layer separation. (C) Slightly increasing the angle of attack over 15° results in stall. In an airplane this would result in a loss of lift such that the aircraft would begin to fall.

Although a complete discussion of laminar and turbulent boundary layers on airfoils and other bodies is beyond the scope of this text, some insight into the effects of body shape and angle of incidence can be gained by considering a laminar boundary layer. Recall that a laminar boundary layer is described by the Prandtl boundary layer equations (Eqs. 14.8a–14.8c):

$$\frac{\partial u}{\partial x} + \frac{\partial v}{\partial y} = 0, \quad \rho\left(u\frac{\partial u}{\partial x} + v\frac{\partial u}{\partial y}\right) = -\frac{\partial p}{\partial x} + \mu\left(\frac{\partial^2 u}{\partial y^2}\right), \quad \text{and} \quad 0 = -\frac{\partial p}{\partial y}$$

These equations apply to the laminar boundary layer on a curved surface, provided the thickness of the boundary layer is small in comparison to the radius of curvature of the surface. It is also necessary for the boundary layer to be attached, meaning that the boundary layer must follow the contour of the surface. This is not the case if flow separation occurs. These constraints are usually met by thin airfoil shapes at a small angle of attack, as is the case in Figure 14.10A. However, if the angle of attack is increased, as shown in Figure 14.10B, the flow begins to separate, the boundary layer is not completely attached, and recirculation occurs. At higher angles of attack the airfoil is said to stall, as illustrated in Figure 14.10C. The lift of the airfoil decreases significantly when an airfoil stalls.

Figure 14.11 Boundary layer flow geometry over a curved surface.

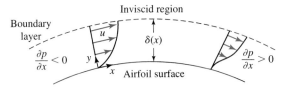

To apply the Prandtl equations to an airfoil or other moderately curved surface, the x coordinate axis is assumed to lie along the surface and the y coordinate axis is then normal to the surface as shown in Figure 14.11. Eq. 14.8c shows that component of the pressure gradient normal to a curved surface, $\partial p/\partial y$, is negligible, just as in the case with the boundary layer on a flat plate. However, the component of the pressure gradient in the streamwise direction, $\partial p/\partial x$, which is zero for a flat plate aligned with the freestream, is nonzero on a curved surface and varies with x. Thus the boundary layer on the curved surface of a body satisfies the following equations:

$$\frac{\partial u}{\partial x} + \frac{\partial v}{\partial y} = 0 \tag{14.18a}$$

$$\rho\left(u\frac{\partial u}{\partial x} + v\frac{\partial u}{\partial y}\right) = -\frac{\partial p}{\partial x} + \mu\left(\frac{\partial^2 u}{\partial y^2}\right) \tag{14.18b}$$

The streamwise pressure gradient, which may be positive or negative, depends on the body shape and angle of incidence, as well as the location x along the body surface. To solve Eqs. 14.18a and 14.18b for the velocity components and extract the desired characteristics of the boundary layer such as its thickness and the wall shear stress, it is necessary to know the pressure gradient.

An essential part of Prandtl's boundary layer theory is the assumption that the streamwise pressure gradient $\partial p/\partial x$ is determined by the inviscid flow just outside the boundary layer. This can be explained as follows. Since the effects of viscosity are confined to the boundary layer, the flow outside is an inviscid flow. There is no pressure gradient $\partial p/\partial y$ normal to the surface in the boundary layer equations; thus the pressure distribution inside the boundary layer is the same as that just outside. Furthermore, in an unseparated flow at the high Reynolds numbers of interest, the boundary layer is very thin and may be considered in a first approximation to have vanishingly small thickness insofar as the inviscid flow is concerned. Thus in solving the inviscid flow problem, we can neglect the presence of a boundary layer and its unknown thickness and simply consider the inviscid flow over the same body at the desired freestream velocity.

After obtaining a solution for the inviscid flow, we can write the inviscid surface pressure distribution on the body using the Bernoulli equation as

$$p_S(x) + \tfrac{1}{2}\rho u_S(x)^2 = C \tag{14.19}$$

where $p_S(x)$ is the pressure on the surface, $u_S(x)$ is the velocity on the surface as predicted by the inviscid flow solution, and C is a constant. The pressure distribution $p_S(x)$ is assumed to be the pressure acting on the boundary layer, and we can use it to determine the streamwise pressure gradient needed to solve the boundary layer equations, writing

$$\frac{\partial p}{\partial x} = \frac{dp_S(x)}{dx} = -\frac{1}{2}\rho\frac{d}{dx}[u_S(x)^2] \tag{14.20}$$

From this discussion, we see that the effects of body shape and angle of incidence are equivalent, insofar as a boundary layer is concerned, to a streamwise pressure gradient. Thus, as mentioned earlier, it is customary to consider the effects of body shape and angle of incidence together as being equivalent to the effect of an imposed pressure gradient.

This approach allows us to understand the effects of a pressure gradient on a laminar flow: we use Eq. 14.20 to replace the pressure gradient in the boundary layer equations to obtain

$$\frac{\partial u}{\partial x} + \frac{\partial v}{\partial y} = 0 \tag{14.21a}$$

$$\rho\left(u\frac{\partial u}{\partial x} + v\frac{\partial u}{\partial y}\right) = \frac{1}{2}\rho\frac{d}{dx}[u_S(x)^2] + \mu\left(\frac{\partial^2 u}{\partial y^2}\right) \tag{14.21b}$$

An interesting solution to these equations, called the Falkner–Skan solution, illustrates the effect of a pressure gradient on a laminar boundary layer by assuming that the surface velocity is of the form $u_S(x) = cx^m$, where c and m are constants with $c > 0$. Since $[du_S(x)]/dx = mcx^{m-1}$, the flow is accelerating for $m > 0$ and decelerating for $m < 0$. From Bernoulli's equation, we know that if the flow is accelerating, the pressure is falling, and vice versa. Thus $m > 0$ implies $dp/dx < 0$, meaning that the pressure is decreasing in the flow direction. This is referred to as a favorable pressure gradient for reasons explained shortly. Similarly, $m < 0$ implies $dp/dx > 0$, which is termed an unfavorable pressure gradient. For $m = 0$, Eqs. 14.21 reduce to those describing the boundary layer on a flat plate.

Although the details of the Falkner–Skan solution are beyond the scope of this text, the qualitative results are important. The favorable pressure gradient of an accelerating freestream tends to thin a laminar boundary layer and bring higher momentum fluid nearer the surface, while the unfavorable pressure gradient of a decelerating freestream tends to do the opposite. Thus, on an airfoil or other moderately curved body, we expect a laminar boundary layer to be thin and to remain attached on portions of the surface where the flow is accelerating, but to become thicker and possibly to separate from portions of the surface where the flow is decelerating. The wall shear stress is zero at the point of separation, and downstream of this point the flow near the wall reverses direction. This also occurs with turbulent boundary layers, although the latter are more resistant to flow separation owing to the increased amount of higher momentum fluid near the surface associated with the more blunt turbulent velocity profile.

These observations are confirmed by flow visualization studies. To illustrate boundary layer separation, consider Figure 14.12 showing velocity profiles over an airfoil at

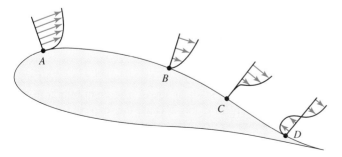

Figure 14.12 Illustration of boundary layer flow over an airfoil: A, a favorable pressure gradient; B, an unfavorable pressure gradient; C, the separation point; D, separation and flow reversal.

Figure 14.13 Flow visualization by the spark tracing method (timelines) of velocity profiles in the boundary layer.

HISTORY BOX 14-2

The boundary layers on the airfoils of early airplanes were turbulent over much of the wing because of the presence of an unfavorable pressure gradient over 90% of that surface. Eastman Jacobs, an engineer for the National Advisory Committee on Aeronautics (NACA), designed an airfoil shape with the intent of producing a favorable pressure gradient over as much as 60% of the wing, thus maintaining laminar flow, reducing the friction drag, and increasing fuel economy. He developed the NACA-66 series of laminar flow airfoils and published his results in 1939. Those results were put to use by North American Aircraft to produce the first airplane with a laminar flow airfoil, the P-51 Mustang.

an angle of attack. At position A, the fluid has accelerated over the front of the airfoil and the velocity profile reflects the favorable pressure gradient. At position B the effects of the unfavorable pressure gradient are apparent in that the flow has slowed in the boundary layer and the velocity profile has become steeper. At point C, the velocity profile clearly shows that $du/dy = 0$, which indicates that this is the point of separation. At position D the flow has reversed near the airfoil surface. Figure 14.13 is a flow visualization of velocity profiles in a boundary layer on an airfoil.

14.3 DRAG: BASIC CONCEPTS

For a stationary object immersed in a moving stream, drag is the component of force exerted on the object by the fluid in the direction of the freestream. An engineer often needs to account for the effect of drag in structural design and stability analysis, since for stationary objects ranging from buildings and trees exposed to wind to bridge piers in a river, the drag exerted by the moving fluid can be significant. For an aircraft or other object moving through a stationary fluid, the drag acts in the direction opposite to the motion of the object. The power required to propel an object through a fluid at constant speed is given by the product of drag and speed. Thus drag not only limits the performance of man-made vehicles of all types and affects the economy of operation but exerts its effects in the natural world as well.

The drag on an object is defined by Eq. 4.26b as

$$F_D = \int_S (-p\mathbf{n} + \boldsymbol{\tau}) \cdot \mathbf{n}_\infty \, dS$$

where the unit vector $\mathbf{n}_\infty$ points in the flow direction. From this we see that the total drag force arises from two mechanisms: pressure and shear stress. The contribution to the total drag due to the pressure is referred to as form drag because the shape or form of the object determines the pressure distribution on its surface. The contribution to the total drag due to the shear stress acting on an object is called friction drag. In a high Reynolds number flow, friction drag can be attributed to the boundary layer and wall shear stress as discussed earlier.

An effective way to illustrate the concepts of friction and form drag is shown in Figure 14.14. For a flat plate aligned with the freestream (Figure 14.14A), the drag is wholly due to the wall shear stress, i.e., skin friction. If the same plate is normal to the freestream (Figure 14.14B), the drag is wholly due to the pressure difference on the front and back surfaces. Note that there is a shear stress distribution on the plate in Figure 14.14B, but the effect of the shear stress cancels owing to symmetry. Even in the absence of this symmetry, the shear stress does not act in the flow direction, hence makes no contribution to drag on a flat plat normal to the freestream. In Figure 14.16C, we see that for a flat plate at an angle of attack, the shear stress and pressure both contribute to the drag.

 CD/Special Features/Demonstrations/Effect of angle of attack on flow structure

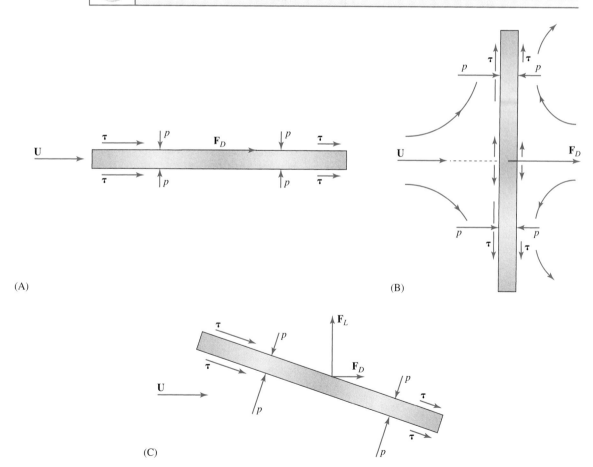

Figure 14.14 Illustration of the types of drag using flow over a flat plate (A) aligned with the flow, friction drag only; (B) normal to the flow, form drag only; and (C) friction and form drag, present for the plate at an angle to the flow.

For any other type of body immersed in a freestream, the total drag will have contributions from both friction and form drag. Although no precise rule can be given, in a high Reynolds number flow the drag on bluff bodies, which share some of the characteristics of a flat plate held normal to the freestream, tends to be dominated by form drag. The drag on long thin bodies, like the flat plate aligned with the freestream, tends to be dominated by skin friction. "Streamlining" is a term used to describe the attempt to design an optimum shape for a bluff body in a high Reynolds number flow by minimizing the total drag. It generally takes the form of elongating the rear of the body. Although this raises the friction drag, it lowers the form drag, and the total drag is reduced. Of course, if the elongation is excessive, the friction drag eventually becomes large and the total drag is not reduced at all. The airfoil at an angle of attack described in the preceding section experiences both friction drag and form drag.

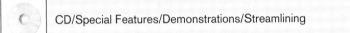
CD/Special Features/Demonstrations/Streamlining

The result of applying the concept of streamlining to a cylinder is illustrated in Figures 14.15 and 14.16. The cylinder and airfoil shape in Figure 14.15 have the same frontal area wL, but the drag on the airfoil shape is a fraction of that on the cylinder. This is emphasized in Figure 14.16, where we see a cylinder and airfoil shape having the same drag. If you are wondering about the relevance of this, early biplanes used wire cables to structurally connect the two wings, but it was eventually recognized that the drag could be lowered by using streamlined airfoil shaped struts instead of cables.

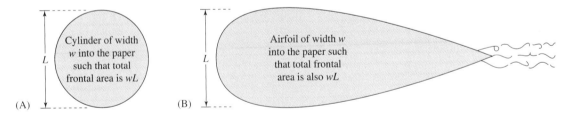

Figure 14.15 The cylinder (A) and the airfoil (B) have the same frontal area, but the drag on the cylinder is much greater.

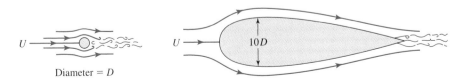

Figure 14.16 The cylinder has much less frontal area than the airfoil, yet the two shapes have the same drag.

14.4 DRAG COEFFICIENTS

Although computational fluid dynamics is increasingly being applied to the problem of determining the drag on objects of engineering interest, much of what is known about drag is the result of experiments. Suppose we apply dimensional analysis to the problem of determining the drag on an object of a specified shape and include a comprehensive set of physical parameters. The dimensionless group containing the drag force is found to be $F_D/\rho U^2 L^2$, where L is a length scale, and U is the freestream velocity. This suggests that the drag coefficient is naturally defined as $C_D = F_D/\rho U^2 L^2$, however it is customary to define the drag coefficient instead as

$$C_D = \frac{F_D}{\frac{1}{2}\rho U^2 A} \tag{14.22a}$$

where the area A normally refers to the frontal area of the object, and the factor of $\frac{1}{2}$ is introduced to produce a denominator that is the product of the dynamic pressure of the upstream flow $\frac{1}{2}\rho U^2$ times the frontal area. (In defining the drag coefficient for a wing, the area is said to be the planform area, i.e., the area of the wing as seen from above.)

The DA further shows that the drag coefficient may be written as

$$C_D = C_D\left(Re, M, Fr, St, We, \frac{e}{L}\right) \tag{14.22b}$$

which shows that the drag coefficient for an object of a given shape may depend on Reynolds number, Mach number, Froude number, Strouhal number, Weber number, and relative roughness. In some cases the drag coefficient may even depend on additional dimensionless groups. For example, if the object is rotating, the drag coefficient will depend on a dimensionless rotation group. From your study of DA, you know that the mere presence of a group in Eq. 14.22b does not mean that all groups are equally important. For example, it is difficult to see how the effect of surface tension, which is characterized by the value of the Weber number, would be significant in engineering applications involving ships and other large objects moving on or through an air–water interface. On the other hand, if we were asked to estimate the drag experienced by a water strider or other aquatic insect, the effect of surface tension would likely be quite important. In the next four sections we present empirical results for the drag coefficient, beginning with low Reynolds number flows.

CD/Dynamics/Low Reynolds number flows

14.4.1 Low Reynolds Number Flow

Low Reynolds number flows of engineering interest include creeping flows, for which $Re \ll 1$, as well as flows in which the Reynolds number is not large enough for a distinct boundary layer to be observed. The Reynolds number range over which the boundary layer is indistinct might be roughly estimated as $0.1 < Re < 100$. In low Reynolds number flows involving air and water, the Reynolds number is usually small because the

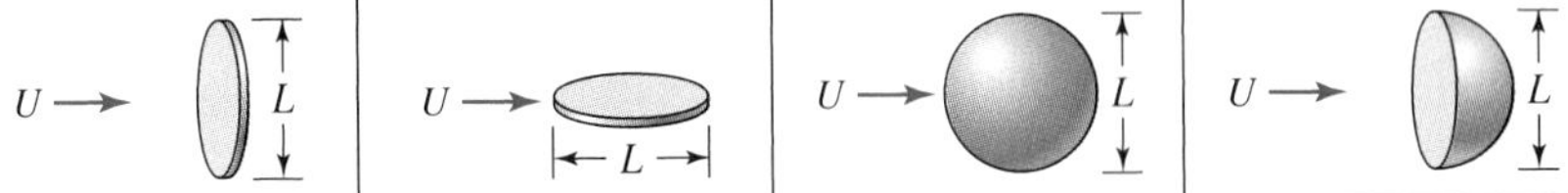

Object	Circular disk normal to flow	Circular disk parallel to flow	Sphere	Hemisphere
$C_D = \dfrac{F_D}{\frac{1}{2}\rho U^2 \left(\frac{\pi L^2}{4}\right)}$ (for $Re \leqslant 1$)	$20.4/Re$	$13.6/Re$	$24.0/Re$	$22.2/Re$

Figure 14.17 Drag coefficients for creeping flow. ($Re = UL/\nu \ll 1$).

length scale of the object is small. Viscous effects in low Reynolds number flows are not confined near the body surface, and both friction and form drag contribute to the total drag. Drag coefficients derived from analytical solutions can be found for many objects of simple shape in creeping flows. Experimental data for slightly larger Reynolds numbers can also be found for cylinders and spheres. If analytical or experimental results are not available for the shape of interest, it is also possible to use computational fluid dynamics to determine the flow over an object in both creeping flow and at slightly larger Reynolds numbers. The drag coefficient can then be calculated directly from the solution.

Drag coefficients are shown in Figure 14.17 for a several shapes in creeping flow. The dependence of these drag coefficients on the inverse of the Reynolds number can be explained by recalling that inertial forces, which depend on the fluid density, are negligible in creeping flows. This means that a dimensional analysis applicable only to creeping flow would not include density; rather, it would assume $F_D = f(L, U, \mu)$, where L is the length scale defining a particular smooth object. Choosing L as a repeating parameter leads to $F_D/\mu U L = C$, where C is a constant. This correctly indicates that a drag coefficient for creeping flow should not contain density. However, since $F_D = C\mu UL$, forming the drag coefficient in the customary way gives

$$C_D = \frac{F_D}{\frac{1}{2}\rho U^2 L^2} = C\frac{\mu UL}{\frac{1}{2}\rho U^2 L^2} = \frac{2C}{Re}$$

We see that all objects in creeping flow have drag coefficients that are proportional to Re^{-1}. Note carefully, however, that this decrease in drag coefficient with increasing Reynolds number applies only to creeping flow, $Re \ll 1$, not to larger Reynolds numbers. The drag on an object in a creeping flow increases linearly with velocity.

EXAMPLE 14.3

A playful child left alone has run a vacuum cleaner in reverse, creating a dust cloud. If the cloud consists of 0.001, 0.01, and 0.1 mm diameter particles, and the particle density is 700 kg/m³, will the child be able to clean up the mess by dusting the furniture before her mother returns an hour later? Assume that the particles near the ceiling must settle 2.5 m before depositing on various surfaces, and that the air temperature is 20°C.

SOLUTION

For a particle settling at terminal velocity, a vertical force balance shows that

$$W = F_{\text{air}}$$

where $W = \rho_P g V$ is the weight of a particle of density ρ_P, and F_{air} is the total force applied by the air to the particle. The force applied by the air consists of a drag force F_D that accounts for the relative motion of the particle through the stationary air and a buoyancy force F_B that accounts for the effects of the hydrostatic pressure variation in the air. Because the buoyancy force is not included when the drag force is calculated by using a drag coefficient, we must write

$$W = F_D + F_B \tag{A}$$

(Another way to think about this problem is revealed by rearranging this equation as $W - F_B = F_D$. Since $W - F_B$ is the weight of the particle as measured in air, we see that the force balance equates this weight, which causes the particle to settle, to the drag force that resists the settling motion.) Substituting for each term in the force balance (A) gives

$$\rho_P g V = C_D \tfrac{1}{2}\rho_{\text{air}} U^2 A + \rho_{\text{air}} g V$$

Solving for the terminal velocity we obtain

$$U = \sqrt{\frac{2(\rho_P - \rho_{\text{air}})g V}{C_D \rho_{\text{air}} A}} \tag{B}$$

We will assume a spherical particle and a creeping flow drag coefficient given by $C_D = 24/Re$. Note that since $C_D = 24/Re = 24\mu/(\rho_{\text{air}} U D)$ in this case, the terminal velocity also occurs in the drag coefficient in (B). The area and volume are $A = \pi D^2/4$ and $V = \pi D^3/6$, respectively; hence $V/A = 2D/3$. Inserting these values into (B) shows that the terminal velocity is given by

$$U = \frac{(\rho_P - \rho_{\text{air}})g D^2}{18\mu} \tag{C}$$

The time needed for a particle to settle from a height H is $t = H/U$. Thus the settling time is

$$t = \frac{18\mu H}{(\rho_P - \rho_{\text{air}})g D^2} \tag{D}$$

Since the diameter occurs in the denominator, the smallest particles take the longest time to settle. The maximum settling time $t_{\max}$ is found using $H = 2.5$ m. Inserting data for air $\rho_{\text{air}} = 1.2$ kg/m^3, $\mu = 1.81 \times 10^{-5}$ (N-s)/m^2, and other values into (B)–(D), we can construct the following table showing $t_{\max}$ for each particle size:

D (mm)	U (m/s)	$t_{\max}$ (minutes)	Re
0.001	2.1×10^{-5}	2000	1.4×10^{-6}
0.01	2.1×10^{-3}	20	1.4×10^{-3}
0.1	2.1×10^{-1}	0.2	1.4

The calculated Reynolds numbers confirm the validity of the creeping flow assumption. It is evident that the smallest particles, which take ~33 h to settle, will pose a problem. Note that since $\rho_{air} = 0.7\% \, \rho_P$, buoyancy is negligible in this example.

  CD/Video library/Flow past a cylinder

14.4.2 Cylinders

Figure 14.18 provides drag coefficient data for a smooth cylinder over a large range of Reynolds number. The frontal area of the cylinder enters the drag coefficient as $A = DL$, where D is the diameter of the cylinder and L its length. Although the data in Figure 14.18 apply only in principle to a cylinder of infinite length, the information is used to estimate the drag on finite length cylinders. The error made will increase as the aspect ratio L/D decreases. In applications for which $L/D < 4$, it is better to use the drag coefficient given later (Section 14.4.4, Table 14.2).

Examination of Figure 14.18 shows that the drag coefficient is a complex function of Reynolds number. To understand the influence of Re on C_D, consider Figure 14.19, a flow visualization of the velocity profiles on a cylinder. As you follow the flow around the cylinder, notice that the laminar boundary layer velocity profile is gradually deformed until the flow reverses direction. The corresponding pressure distributions, both the inviscid approximation and the empirically observed distribution, are shown in

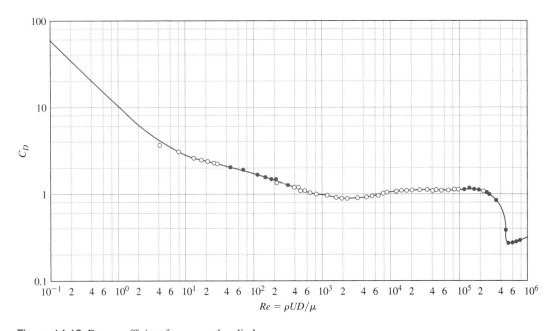

Figure 14.18 Drag coefficient for a smooth cylinder.

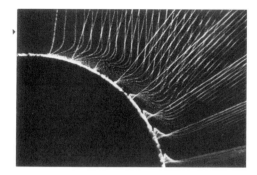

Figure 14.19 Flow visualization of the boundary layer on a cylinder created with hydrogen bubbles.

The visual aspects of the flow field over a cylinder are a strong function of the Reynolds number. Consider the obvious difference in the flows illustrated in Figures 12.2 and 12.3. In the creeping flow shown in Figure 12.2 ($Re = 0.038$), there is no wake behind the cylinder. However, in Figure 12.3 ($Re = 19$), a wake has formed because flow separation has occurred even though there is no boundary layer present. At much larger Reynolds numbers the size of the wake is determined by boundary layer separation, and as noted earlier, this influences the pressure drag.

Figure 14.20. Focus your attention on the true pressure distribution and note that as the fluid in the boundary layer moves from the front of the cylinder to the top, it is accelerated by a favorable pressure gradient. However, as the fluid passes the top, the pressure gradient becomes unfavorable, tending to slow the fluid down. As long as the fluid has enough momentum, it can still move forward; but viscous effects and the pressure gradient tend to exert a decelerating effect. Eventually all the forward momentum is dissipated and the flow reverses direction (see Figure 14.19). This is the phenomenon that was evident in the Falkner–Skan solution discussed earlier.

If the pressure distribution around the cylinder is integrated for the inviscid case, the net force is zero. That is, the pressure drag in the inviscid model is identically zero. In Figure 14.20, however, it can be seen that the actual pressure recovery is much less than that for the inviscid flow case. Thus, for the real viscous flow we find a net force retarding the motion of the cylinder. This is the source of the pressure drag (or form drag) on a cylinder.

We are now in a position to explain the dependence of C_D on Re for a cylinder, as illustrated in Figure 14.18. In the low Reynolds number regime, the drag coefficient is proportional to the inverse of Reynolds number as already explained. For $1 < Re < 10^3$ the friction drag of the laminar boundary layer tends to dominate the form drag, so C_D varies with Re in much the same way that it did for laminar flow over an aligned flat plate (i.e., $C_D \propto Re^{-1/2}$). In the range $10^3 < Re < 10^5$, C_D has only a weak dependence on Re and the total drag is dominated by the pressure or form drag. This behavior is similar to that displayed by "bluff bodies," as described later (Section 14.4.4).

Over the range $10^5 < Re < 10^6$ the cylinder drag coefficient falls dramatically by about 80%. At this critical point the drag actually decreases with increasing speed. Imagine increasing the speed of your car while letting up on the gas. What accounts for this extraordinary behavior? You might attribute it to the laminar-to-turbulent transition of the boundary layer. Recall Figure 14.8, which shows laminar and turbulent velocity profiles. The higher velocity and momentum flux near the surface for turbulent flow causes an increase of friction drag, but results in a substantial decrease in pressure drag. How is this possible? Well, remember that the boundary layer separates when the streamwise momentum of the flow is insufficient to overcome the adverse pressure

Figure 14.20 Comparison of the pressure distribution around a cylinder based on inviscid theory and empirical observations.

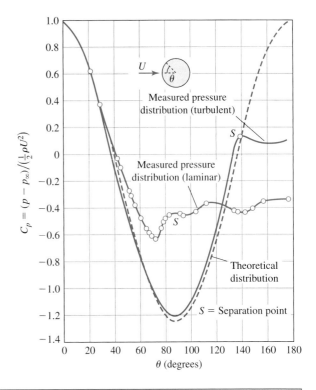

EXAMPLE 14.4

A modern sculpture includes a wind gage in the form of a circular cylinder suspended from two fine wires as shown in Figure 14.21. If the presence of the wires is assumed to have a negligible influence on the flow field, at what angle will the cylinder hang in a wind of 25 km/h? The cylinder weighs 6 N and its dimensions are $D = 10$ cm and $L = 1$ m.

SOLUTION

Since the wind applies a lift and drag to the cylinder, after neglecting the tiny buoyancy force on the cylinder, and writing a force balance on the cylinder in the x and y directions we obtain

$$(F_D \cos\theta + F_L \sin\theta) - W \sin\theta = 0 \quad \text{and} \quad 2T - W\cos\theta + (F_L \cos\theta - F_D \sin\theta) = 0$$

Noting that the x component of force of the wind $(F_D \cos\theta + F_L \sin\theta) = C_D \frac{1}{2}\rho_{air}(U\cos\theta)^2 A$, the force balance in the x direction shows that

$$\sin\theta = \frac{\left(C_D \frac{1}{2}\rho_{air} U^2 A\right)(\cos^2\theta)}{W} \quad \text{(A)}$$

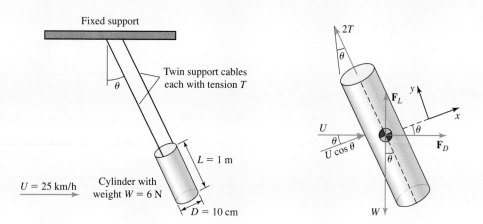

Figure 14.21 Schematic for Example 14.4.

We expect that the angle will be larger for a lighter cylinder at any given wind speed but can never exceed 90°. For small angles, $\cos^2\theta = 1$, and since $A = DL$, we obtain the approximate result

$$\theta = \sin^{-1}\left(\frac{C_D \rho_{air} U^2 DL}{2W}\right) \tag{B}$$

Assuming 20°C air for which $\rho = 1.2$ kg/m³ and $\nu = 1.51 \times 10^{-5}$ m²/s, the Reynolds number is

$$Re = \frac{UD}{\nu} = \frac{(25 \text{ km/h})(1000 \text{ m/km})(\text{h}/3600 \text{ s})(0.1 \text{ m})}{1.51 \times 10^{-5} \text{ m}^2/\text{s}} = 4.6 \times 10^4$$

From Figure 14.18, at this Reynolds number $C_D = 1.2$. Inserting the data into (A) we must iterate or use a symbolic code to solve

$$\sin\theta = \left[\frac{(1.2)(1.2 \text{ kg/m}^3)[(25 \text{ km/h})(1000 \text{ m/km})(\text{h}/3600 \text{ s})]^2(0.1 \text{ m})(1 \text{ m})}{2(6 \text{ N})}\right]\cos^2\theta$$

The result is $\theta = 27.2°$. It is easy to confirm that (B) does not deliver good accuracy in this case. Depending on the range of wind speeds expected at the sculpture site, it may be best to employ a lighter cylinder to obtain a larger angle of deflection. We should also be aware that the aspect ratio of the cylinder affects the drag coefficient so it may be best to validate a design based on (A) or (B) by careful calibration experiments.

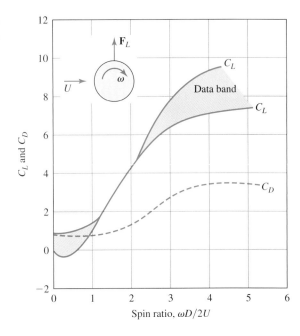

Figure 14.22 Lift and drag coefficients for a spinning cylinder.

gradient. The increased momentum in the boundary layer in turbulent flow causes separation to be delayed, and the resulting wake is smaller. The pressure on the downstream side of the cylinder is therefore not quite as low as it is with a large wake, and the result is a much lower form drag. This is evident in the changed pressure distribution on the cylinder surface as shown in Figure 14.20.

Spinning a cylinder in a freestream results in an increase in drag if the rotation rate is sufficiently large. The rotation also creates a side force or lift on the cylinder. This is known as the Magnus effect. The lift and drag coefficients for a spinning cylinder are shown in Figure 14.22. In this case the lift coefficient is defined as $C_L = F_L/\frac{1}{2}\rho U^2 DL$. Over the years a number of interesting uses for spinning cylinders have been proposed, including a rotor-based wind-powered ship (Figure 14.23).

(A)

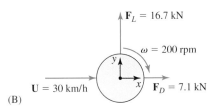

(B)

Figure 14.23 (A) The *Bruckau*, designed by Anton Flettner. (B) Schematic for Example 14.5.

EXAMPLE 14.5

The Flettner rotor-powered ship shown in Figure 14.23A had two rotors, each 3 m diameter and 15 m tall. If $\omega = 200$ rpm and the speed of the wind relative to the rotor is 30 km/h, find the force applied to each rotor by the wind.

SOLUTION

To find the force generated by each rotor, we will use the results for a spinning cylinder as shown in Figure 14.22 to determine the lift and drag coefficients. First we calculate the rotational velocity as

$$V_\theta = R\omega = (0.0015 \text{ km})(200 \text{ rpm})(2\pi \text{ rad/rev})(60 \text{ min/h}) = 113 \text{ km/h}$$

Dividing this value by the wind speed gives us the spin ratio $WD/2U$. Thus we have

$$\frac{V_\theta}{U} = \frac{113 \text{ km/h}}{30 \text{ km/h}} = 3.75$$

From Figure 14.22 we find $C_L = 8.9$ and $C_D = 3.8$. Thus the lift and drag forces are

$$F_L = C_L \tfrac{1}{2}\rho U^2 DL = (8.9)\left(\tfrac{1}{2}\right)(1.2 \text{ kg/m}^3)(8.33 \text{ m/s})^2(3 \text{ m})(15 \text{ m}) = 16.7 \text{ kN}$$

$$F_D = C_D \tfrac{1}{2}\rho U^2 DL = (3.8)\left(\tfrac{1}{2}\right)(1.2 \text{ kg/m}^3)(8.33 \text{ m/s})^2(3 \text{ m})(15 \text{ m}) = 7.1 \text{ kN}$$

where we have assumed air at 20°C in calculating the density. The force applied by the wind to each rotor is thus given by $\mathbf{F}_{\text{wind}} = 7.1 \text{ kN}\,\mathbf{i} + 16.7 \text{ kN}\,\mathbf{j}$ as shown in Figure 14.23B. This force acts at an angle of $\theta = \tan^{-1}(16.7/7.1) = 67°$ to the left of the relative wind direction.

CD/Video library/Flow past a sphere

14.4.3 Spheres

Figure 14.24 shows drag coefficient data for a smooth sphere over a broad range of Re. The frontal area of the sphere enters the drag coefficient as $A = \pi D^2/4$. Spheres exhibit drag coefficient behavior with Reynolds number that is similar to that of cylinders, for much the same reasons. The change in separation point due to the transition from a laminar to turbulent boundary layer is evident in Figure 14.25.

It is interesting to note that golf balls in flight have Reynolds numbers near the point at which the laminar-to-turbulent boundary layer transition occurs. To ensure that the boundary layer is turbulent, roughness is added to the surface of the ball in the form of dimples. These dimples reduce flow separation, thereby lowering the drag and increasing the flight distance. The effect of roughness on C_D for spheres near the turbulent transition is shown in Figure 14.26.

14 EXTERNAL FLOW

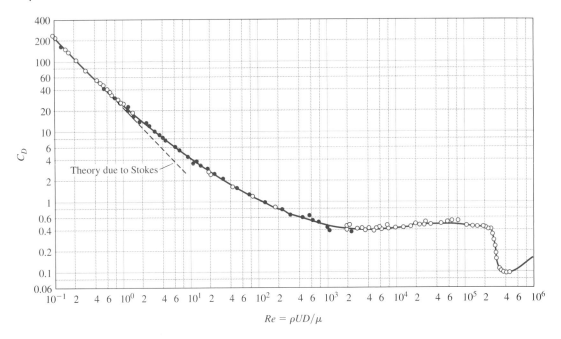

Figure 14.24 Drag coefficient for a smooth sphere.

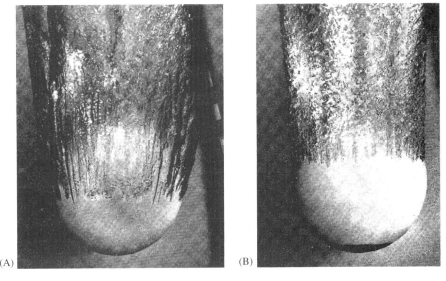

Figure 14.25 Boundary layer separation on a sphere for (A) laminar flow and (B) turbulent flow caused by roughing the nose.

14.4 DRAG COEFFICIENTS

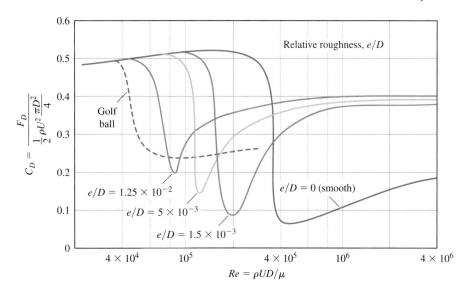

Figure 14.26 The effect of roughness on the drag on a sphere.

EXAMPLE 14.6

To make a car easier to find in crowded parking lots, a colorful 2 in. diameter smooth plastic ball is attached to the end of the vehicle's 3 ft antenna as shown in Figure 14.27. What is the bending moment on the antenna due to the ball if the car is moving at 50 mph?

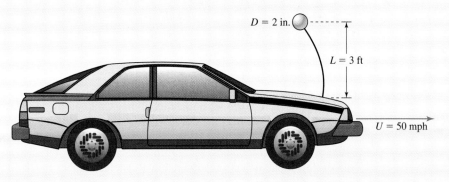

Figure 14.27 Schematic for Example 14.6.

SOLUTION

We will calculate the drag force on the ball, then find the resulting bending moment. We will assume that the flow over the ball is the same as it would be without having the

antenna nearby. Assuming air at 70°F,

$$Re = \frac{UD}{\nu} = \frac{(50 \text{ mph})[1.47 \text{ ft/(mph-s)}](2 \text{ in.})(\text{ft}/12 \text{ in.})}{1.64 \times 10^{-4} \text{ ft}^2/\text{s}} = 7.5 \times 10^4$$

From Figure 14.24 we find $C_D = 0.5$. The drag force on the ball is calculated next from $F_D = C_D \frac{1}{2}\rho U^2 A$, where $A = \pi D^2/4$. Inserting the data, we find

$$F_D = (0.5)\left(\frac{1}{2}\right)(2.329 \times 10^{-3} \text{ slug/ft}^3)(73.3 \text{ ft/s})^2 \frac{\pi (0.1667 \text{ ft})^2}{4}$$

$$= 6.83 \times 10^{-2} \text{ lb}_f$$

Ignoring any curvature of the antenna, the bending moment is

$$M = F_D L = (6.83 \times 10^{-2} \text{ lb}_f)(3 \text{ ft}) = 0.2 \text{ ft-lb}_f$$

Would you recommend adding roughness to the ball?

CD/Boundary layers/Separation

No discussion of the external flow over a sphere would be complete without including the effect of rotation, which plays a prominent role in the flight of sport balls of all types. As was the case with a cylinder, rotation of a sphere not only affects the drag but also produces a sideforce or lift. The lift and drag coefficients for rotating spheres are shown in Figure 14.28. The lift coefficient for a sphere is defined by $C_L = F_L/\frac{1}{2}\rho U^2 A$, where $A = \pi D^2/4$.

14.4.4 Bluff Bodies

Suppose you were asked what feature buildings, billboards, and beams have in common that might strongly affect their drag? If you recognized that each of these objects has a relatively flat face with sharp edges, you are correct. These and other nonstreamlined objects are called bluff bodies. More formally, "bluff body" refers to an object that experiences flow separation at a relatively low Reynolds number and has a flow field after separation occurs that is relatively unchanged as Re increases. As a result, the drag coefficient for a bluff body after separation is nearly independent of Reynolds number (over a large range of Re). The separation process on a bluff body is often, but not always, associated with a sharp corner or other change in geometry.

From our discussions, you know that the onset of flow separation generally corresponds to an increase in total drag resulting from a substantial increase in the form drag, and that form drag can be reduced by streamlining. Consider the tractor trailer truck

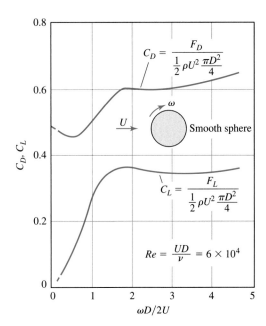

Figure 14.28 Lift and drag coefficients for a spinning sphere.

EXAMPLE 14.7

A baseball pitcher throws his curve ball at 80 mph with a rotational speed of 1800 rpm. The ball has a mass of 5 oz and a 9 in. circumference. Estimate how much this pitch will break as it travels a distance of 55 ft in a spring game in New York, when the air temperature is 50°F. What is the break in a summer game when the temperature is 90°F? In your calculation, assume that the rotation axis of the ball is vertical.

SOLUTION

In the coordinates shown in Figure 14.29, the rotation of the ball will cause a force tending to move the ball in the y direction. The equation of motion for the ball in the y direction of break is $Ma_y = \sum F_y$. The drag on the ball acts in the x direction and tends to slow the ball down slightly during its travel to the plate. We will neglect this effect and assume for the ball a constant velocity of 80 mph. The time of flight of the ball is therefore given by

$$T = \frac{S}{V_{ball}} \quad \text{(A)}$$

where S is the distance to the plate and V_{ball} is the speed of the ball. Inserting the data, we calculate a flight time

$$T = \frac{S}{V_{ball}} = \left(\frac{55 \text{ ft}}{80 \text{ mph}}\right)\left(\frac{1 \text{ mph}}{1.467 \text{ ft/s}}\right) = 0.47 \text{ s}$$

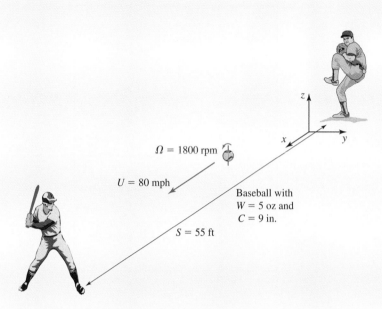

Figure 14.29 Schematic for Example 14.7.

Gravity acts on the ball in the negative z direction and causes the ball to drop as it travels to the plate. We will assume this effect acts independently of the lift force created by the rotation of the ball; hence the drop can be calculated as approximately 2.9 ft. The lift force for a rotating smooth sphere is given by

$$F_L = \frac{1}{2}\rho V_{\text{ball}}^2 \frac{\pi D^2}{4} C_L \tag{B}$$

and acts in the y direction. A baseball has raised stitches that are known to affect the trajectory of a pitch. Since we have lift and drag data only for a smooth sphere as given in Figure 14.29, we will use the smooth sphere data in our calculation. Finally, note that if the rotational speed of the ball is assumed constant, then the lift force is constant during the flight.

We can write the equation of motion for the ball in the y direction as $ma_y = m(d^2y/dt^2) = F_L$. Integrating this twice we obtain $y(t) = (F_L/m)(t^2/2) + C_0 t + C_1$. To evaluate the constants, note that at time $t = 0$, the ball is at an initial location y_0, thus $C_1 = y_0$, and we can write $y(t) - y_0 = (F_L/m)(t^2/2) + C_0 t$. The remaining constant is found by assuming that the ball has no velocity component in the y direction when released by the pitcher. This allows us to set $C_0 = 0$ and obtain $y(t) - y_0 = (F_L/m)(t^2/2)$. At time T the break or distance traveled in the y direction is given by

$$\Delta y = \frac{F_L}{m}\frac{T^2}{2} \tag{C}$$

Inserting (A) and (B) we have

$$\Delta y = \frac{\pi D^2 \rho C_L S^2}{16m} \quad \text{(D)}$$

To determine the lift coefficient, we first calculate the speed ratio

$$\frac{U}{V_{\text{ball}}} = \frac{D\omega}{2V_{\text{ball}}} = \frac{(2.865 \text{ in.})(\text{ft}/12 \text{ in.})(1800 \text{ rpm})(2\pi)(60 \text{ min/h})}{2(80 \text{ mph})(5280 \text{ ft/mile})}$$
$$= 0.19$$

then use the chart in Figure 14.28 to find that $C_L \approx 0.05$. To calculate the break at the different air temperatures, note that $\rho_{50} = 2.420 \times 10^{-3}$ slug/ft^3, and $\rho_{90} = 2.244 \times 10^{-3}$ slug/ft^3. The diameter of the ball is found to be $D = (9 \text{ in.}/\pi)(1 \text{ ft}/12 \text{ in.}) = 0.2387$ ft. Thus from (D) we find the break at 50°F is

$$\Delta y_{50} = \frac{\pi D^2 \rho C_L S^2}{16m}$$

$$\Delta y_{50} = \left(\frac{\pi}{16}\right)\left(\frac{1}{5 \text{ oz}}\right)\left(\frac{1 \text{ oz}}{1.943 \times 10^{-3} \text{ slug}}\right)(0.2387 \text{ ft})^2 (2.420 \times 10^{-3} \text{ slug/ft}^3)(0.05)(55 \text{ ft})^2$$

$$\Delta y_{50} = 0.42 \text{ ft}$$

At 90°F the break is

$$\Delta y_{90} = \left(\frac{\rho_{90}}{\rho_{50}}\right)\Delta y_{90} = \left(\frac{2.244 \times 10^{-3} \text{ slug/ft}^3}{2.420 \times 10^{-3} \text{ slug/ft}^3}\right) 0.42 \text{ ft}$$
$$= 0.39 \text{ ft}$$

or about 8% less in the "lighter" summer air. This break is not sufficient to fool a batter if the curve ball is thrown with the rotation axis vertical, as indicated in Figure 14.29. Instead the pitcher throws the ball so that the break is down, i.e., the rotation axis is nearly horizontal. Pitchers use a variety of spins to induce movement of the ball. A knuckle ball is thrown with no spin and darts erratically owing to flow separation. A discussion of the physics of sports balls can be found in an article by R. D. Mehta entitled "Aerodynamics of Sports Balls," in *Annual Review of Fluid Mechanics*, volume 17, pages 151–189, 1985.

shown in Figure 14.30A. The large flat section of the trailer exposed to the air above the cab is a good example of a bluff body (the cab itself is somewhat streamlined); as such, it causes substantial drag, which reduces gas mileage. Figure 14.30B shows a similar rig with a wind deflector mounted on the roof of the cab in front of the bluff body. This simple and inexpensive streamlining device substantially reduces the drag on the trailer and,

TABLE 14.2 Drag Coefficients for selected 3D objects.

Geometry	Reference Area, A	Drag Coefficient, C_D, and Remarks
Sphere	$\dfrac{\pi D^2}{4}$	**Re**: 10^2, 10^3, 10^4, 10^5, 10^6, 5×10^6 **C_D**: 1.0, 0.41, 0.39, 0.52, 0.12, 0.18 For $Re < 1$, $C_D \approx 24/\{Re[1 + (3/16)Re]\}$.
Hemisphere	$\dfrac{\pi D^2}{4}$	$C_D = 0.42$ (Sphere side facing upstream) $C_D = 1.17$ (Flat side facing upstream)
Ellipsoid of Revolution	$\dfrac{\pi D^2}{4}$	$C_D = 0.44(D/L) + 0.016(L/D) + 0.016(D/L)^{1/2}$ $1 < L/D < 10$. $Re < 2 \times 10^5$, laminar flow.
Sphere in a Circular Duct	$\dfrac{\pi D^2}{4}$	$C_D = \left[1 + 1.45\left(\dfrac{D}{D_0}\right)^{4.5}\right] C_D\big|_{D_0/D=\infty}$ $0 < D/D_0 < 0.92$, $C_D(D_0/D = \infty)$ is that of sphere above.
Thin Circular Disk	$\dfrac{\pi D^2}{4}$	**Re**: 1, 2, 5, 10, 10^2, 10^3, 10^4, 10^5 **C_D**: 25, 15, 6, 3.6, 1.5, 1.1, 1.1, 1.15
Circular Rod Parallel to Flow	$\dfrac{\pi D^2}{4}$	**L/D** **C_D** ~0 1.15 0.5 1.10 1.0 0.93 1.5 0.85 $R_e \geq 10^4$ 2.0 0.83 3.0 0.85 4.0 0.85 5.0 0.85

(continued)

TABLE 14.2 (continued)

Geometry	Reference Area, A	Drag Coefficient, C_D, and Remarks
Cylindrical Rod Perpendicular to Flow	LD	L/D C_D 1.0 0.64 1.98 0.68 2.96 0.74 5.0 0.74 $R_e \geq 10^4$ 10. 0.82 20. 0.91 40. 0.98 ∞ 1.20
Cone	$\dfrac{\pi D^2}{4}$	θ (deg) C_D 10 0.30 20 0.40 30 0.55 40 0.65 $R_e \geq 10^4$ 60 0.80 75 1.05 90 1.15 180 1.40
Thin Rectangular Plate Perpendicular to Flow	LD	L/D C_D 1.0 1.05 2.0 1.10 4.0 1.12 8.0 1.20 $R_e \geq 10^4$ 10.0 1.22 12.0 1.22 17.8 1.33 ∞ 1.90
Square Rod Parallel to Flow	D^2	L/D C_D ~ 0 1.25 0.5 1.25 1.0 1.15 1.5 0.97 $R_e \geq 10^4$ 2.0 0.87 2.5 0.90 3.0 0.93 4.0 0.95 5.0 0.95
Average Man	See data at right. For $C_D A$ product appropriate to different flow directions and posture.	$\rightarrow C_D A = 9$ ft^2 (0.84 m^2) $\uparrow C_D A = 1.2$ ft^2 (0.11 m^2) $\bullet C_D A = 5$ ft^2 (0.46 m^2) Sitting $\rightarrow C_D A = 6$ ft^2 (0.56 m^2) Crouching $\rightarrow C_D A = 2$ to 3 ft^2 (0.19 m^2 to 0.28 m^2)

Figure 14.30 (A) Bluff body truck design. (B) Streamlined truck design.

therefore, increases fuel economy. A person riding a bicycle is another example of a bluff body. Have you noticed that riders in the Tour de France generally wear helmets designed to provide a more streamlined shape and reduce pressure drag?

Table 14.2 includes drag coefficients for a few common bluff bodies as adapted from a variety of sources. The data are for $Re > 10^4$ with accuracy of $\pm 5\%$. The interested reader is referred to *Applied Fluid Dynamics Handbook*, by Robert Blevins for a more complete listing of drag coefficients for bluff bodies. Notice that, as expected,

EXAMPLE 14.8

As shown schematically in Figure 14.31, square columns 4 in. × 4 in. and 10 ft tall are to be used in the construction of a porch in south Florida. If the columns are exposed to hurricane force winds of 100 mph (= 147 ft/s), what force must each column withstand?

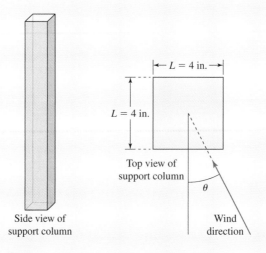

Figure 14.31 Schematic for Example 14.8.

SOLUTION

The Reynolds number of the flow is found to be

$$Re = \frac{UD}{\nu} = \frac{(147 \text{ ft/s})(4 \text{ in.})(\text{ft}/12 \text{ in.})}{1.64 \times 10^{-4} \text{ ft}^2/\text{s}} = 3 \times 10^5$$

which is above the value $Re > 10^4$ for which the data for a square section in Table 14.3 are valid. Thus it is appropriate to use the drag coefficient in this table for our analysis. In Table 14.3 we see that for a square section, the maximum $C_D = 2.4$ is at 45° angle to the wind, thus the maximum force will occur for a wind that comes from this direction. This is the force the column must potentially withstand in the worst case. To calculate it, we will assume 70°F air and use $A = (4/12 \text{ ft})(10 \text{ ft}) = 3.33 \text{ ft}^2$ and $U = 147$ ft/s. The drag force is then found to be

$$F_D = \tfrac{1}{2}\rho U^2 A C_D = \tfrac{1}{2}(2.329 \times 10^{-3} \text{ slug/ft}^3)[147 \text{ ft/s}]^2(3.33 \text{ ft}^2)(2.4) = 200 \text{ lb}_f$$

the reported C_D values for sharp edged bluff bodies are either independent of, or only a weak function of, Re over the range indicated.

Table 14.3 provides similar data for 2D bluff sections. A bluff section is defined to be the (constant) cross section of an object that has infinite depth. The C_D for the sections is based on force per unit span. The reported drag coefficients are also for $Re > 10^4$ with accuracy of ±5%.

You may have noticed that the outside mirrors on modern automobiles are highly streamlined in appearance, and noticeably different in shape from the disk-shaped mirrors seen on early automobiles. Door handles on automobiles have also undergone changes over the years and are now almost always flush with the surface of the door rather than projecting. Mirrors and door handles on automobiles are examples of protuberances, objects that are partly immersed in a freestream and capable of creating a considerable amount of drag. The streamlining of mirrors and door handles reflects a concern for fuel economy, as does the overall lowering of drag coefficients on vehicles of all types. The effect of a protuberance is enhanced if the nearby flow has been accelerated to a high speed by the shape of the body to which the protuberance is attached. Thus if a protuberance is necessary to the function of a vehicle or device, it is wise to locate it in a region of retarded airflow rather than where the airflow is moving at or above the freestream value. In the early days of automotive design, windshield wipers on many models moved from the top of the windshield to the base, then were tucked under the hood, getting them completely out of the airflow.

CD/Boundary layers/Separation/Airfoil separation

CD/Dynamics/Dependence of forces on Reynolds number and geometry/Effect of Re and geometry on flow

14 EXTERNAL FLOW

TABLE 14.3 Drag Coefficients for selected 2D sections.

Geometry	Drag Coefficient, C_D, and Remarks
Circular Cylinder	Re: 10^2, 10^3, 10^4, 10^5, 10^6, 10^7 C_D: 1.4, 1.0, 1.1, 1.2, 0.4, 0.8 For $Re < 1$, $C_D \approx 8\pi/[Re \log_e (7.4/Re)]$.
Cylinder Near a Wall	E/D C_D C_L 0 0.8 0.6 0.25 1.1 0.25 0.5 1.2 0.15 1.0 1.3 0.05 1.5 1.2 0.02 2.0 1.2 0 4.0 1.2 0 6.0 1.2 0 $10^4 < Re < 10^5$ Lift force is away from wall.
Cylinder Downstream of Another Cylinder Drag on Downstream Cylinder	$T/D = 0$ $T/D = 0.5$ L/D C_D L/D C_D 1.0 −0.4 1.0 0.65 1.5 −0.2 1.5 0.50 2.0 0.0 2.0 0.45 2.5 0.2 2.5 0.45 3.0 0.2 3.0 0.40 4.0 0.3 4.0 0.40 $T/D = 1.0$ $T/D = 2$ L/D C_D L/D C_D 1.0 1.1 1.0 1.1 1.5 1.0 1.5 1.0 2.0 0.70 2.0 1.0 2.5 0.70 2.5 1.0 3.0 0.65 3.0 1.0 4.0 0.65 4.0 1.0 $10^4 < Re < 10^5$
Rectangle	L/D C_D L/D C_D 0.1 ≤ 1.9 1.0 2.2 0.2 2.1 1.2 2.1 0.4 2.35 1.5 1.8 $Re \geq 10^4$ 0.5 2.5 2.0 1.6 0.65 2.9 2.5 1.4 0.8 2.3 3.0 1.3 6.0 0.89

14.4 DRAG COEFFICIENTS

TABLE 14.3 (*continued*)

Geometry	Drag Coefficient, C_D, and Remarks
Two Cylinders Side by Side 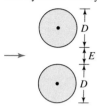	E/D $\quad$ C_D $\quad$ C_L 0 $\quad\quad$ 1.6 $\quad$ 0.8 0.25 $\quad$ 1.0 $\quad$ 0.6 0.5 $\quad\,$ 0.9 $\quad$ 0.4 1.0 $\quad\,$ 1.1 $\quad$ 0.2 1.5 $\quad\,$ 1.3 $\quad$ 0.1 2.0 $\quad\,$ 1.2 $\quad$ 0.05 4.0 $\quad\,$ 1.2 $\quad$ 0.0 6.0 $\quad\,$ 1.2 $\quad$ 0.0 C_D, C_L for each cylinder. $10^4 < Re < 10^5$. Lift force is repulsive.
Inclined Square 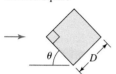	θ (Deg) $\|$ 0 $\;$ 5 $\;$ 10 $\;$ 15 $\;$ 20 $\;$ 25 $\;$ 30 $\;$ 35 $\;$ 40 $\;$ 45 C_D $\quad\quad\;$ 2.2 $\;$ 2.1 $\;$ 1.8 $\;$ 1.3 $\;$ 1.9 $\;$ 2.1 $\;$ 2.2 $\;$ 2.3 $\;$ 2.4 $\;$ 2.4 $Re \geq 10^4$
Rounded Nose Section 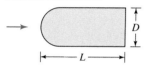	L/D $\quad$ C_D 0.5 $\quad\;$ 1.16 1.0 $\quad\;$ 0.90 2.0 $\quad\;$ 0.70 $\quad$ $Re \geq 10^4$ 4.0 $\quad\;$ 0.68 6.0 $\quad\;$ 0.64
Thin Flat Plate Inclined to Flow 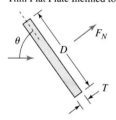 $T < 0.1D$	$C_N \approx \begin{cases} 2\pi \tan\theta, \theta < 8° \\ \dfrac{1}{0.222 + 0.283/\sin\theta}, \quad 90° \geq \theta > 12° \end{cases}$ $\quad Re \geq 10^4$ $C_L = C_N \cos\theta$ $C_D = C_N \sin\theta$ There is a discontinuity in the range $8° < \theta < 12°$ with $C_N \approx 0.8$ as flow separates from upper surface. See Table 14.2 for $\theta = 0°$.
Thin Plate Extending from a Wall 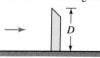	$C_D = 1.4$ $\quad\quad\quad\quad Re \geq 10^4$
Ellipse 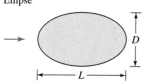	D/L $\quad$ C_D 0.125 $\;$ 0.22 0.25 $\;\;$ 0.3 0.50 $\;\;$ 0.6 $\quad Re \geq 10^4$ 1.0 $\quad\;$ 1.0 2.0 $\quad\;$ 1.6 Laminar flow only.

(*continued*)

TABLE 14.3 (continued)

Geometry	Drag Coefficient, C_D, and Remarks			
I Shape			C_D	
			L/D	
	Flow direction	0.5	1.0	$Re \geq 10^4$
	$\rightarrow$	2.05	1.6	
	$\uparrow$	0.9	1.9	

14.5 LIFT AND DRAG OF AIRFOILS

As discussed briefly in the case study of Section 3.3.6 (lift and drag on airfoils), a wing is a specially shaped body designed to produce lift when exposed to a stream of fluid. Lift is defined to be the component of fluid force acting on a body at a right angle to the oncoming stream. Thus lift is a vertical force for a vehicle or object in level flight. The total lift developed by a wing supports the weight of an aircraft. The spoiler, or upside-down wing, on a racing car produces negative lift, a downward force intended to keep the car on the track.

The cross section at any given point along the span of a wing has the form known as an airfoil. This airfoil shape is carefully designed to maximize lift and minimize drag. There are many different airfoil shapes for different applications. Before discussing airfoil shapes and some of the characteristics of flow over an airfoil further, consider the distribution of pressure and shear stress on a typical airfoil shape as shown in Figure 14.32.

The lift applied by the fluid to this airfoil is defined by Eq. 4.25b as

$$F_L = \int_S (-p\mathbf{n} + \boldsymbol{\tau}) \cdot \mathbf{n}_L \, dS$$

where the unit vector $\mathbf{n}_L$ is normal to the flow direction. The drag component of this same force is

$$F_D = \int_S (-p\mathbf{n} + \boldsymbol{\tau}) \cdot \mathbf{n}_\infty \, dS$$

where the unit vector $\mathbf{n}_\infty$ points in the flow direction. We conclude that in principle both the pressure and the shear stress contribute to the lift and drag of an airfoil. In practice,

Figure 14.32 Typical pressure (normal stress) and shear stress distributions on an airfoil.

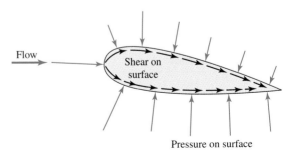

14.5 LIFT AND DRAG OF AIRFOILS 931

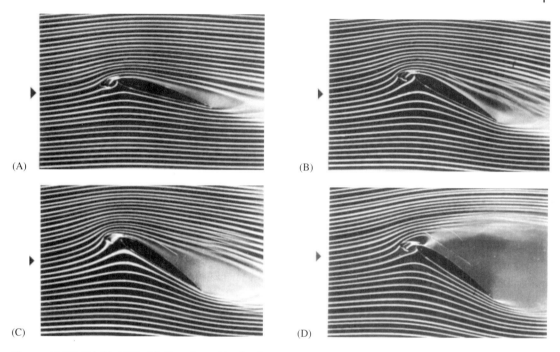

Figure 14.37 NACA 4412 airfoil section with a leading edge flap that delays flow separation from about 15° to 30°.

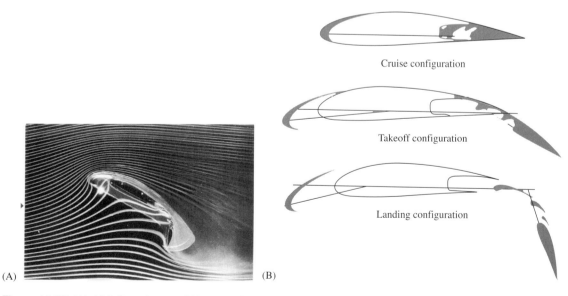

Figure 14.38 (A) Airfoil section at a 25° angle of attack. (B) Sophisticated mechanical high lift devices for an airfoil section.

14 EXTERNAL FLOW

Figure 14.39 Vortex generators on a commercial aircraft.

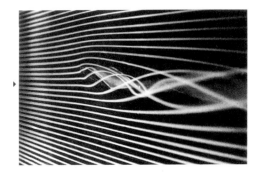

Figure 14.40 Wing tip vortex.

EXAMPLE 14.9

A fully loaded aircraft weighing 900 kN has a wing area of 230 m². If the wing has the characteristics of a NACA 23012 airfoil and during takeoff operates at a 6° angle of attack, what is the required takeoff speed at sea level? What is the takeoff speed at an elevation of 2000 m?

SOLUTION

To take off, the lift force must overcome the weight, so $F_L = W$,

$$\tfrac{1}{2}\rho U^2 A C_L = W$$

$$U = \sqrt{\frac{2W}{\rho A C_L}}$$

From Appendix B for the U.S. Standard Atmosphere $\rho(0 \text{ m}) = 1.225 \text{ kg/m}^3$ and $\rho(2000 \text{ m}) = 1.007 \text{ kg/m}^3$. From Figure 14.36 $C_L = 0.79$. Substituting the data into the preceding expression yields:

$$U = \sqrt{\frac{2(900 \times 10^3 \text{ N})}{(1.225 \text{ kg/m}^3)(230 \text{ m}^2)(0.79)}} = 90 \text{ m/s} = 324 \text{ km/h}$$

at sea level and at 2000 m

$$U = \sqrt{\frac{2(900 \times 10^3 \text{ N})}{(1.007 \text{ kg/m}^3)(230 \text{ m}^2)(0.79)}} = 99 \text{ m/s} = 356 \text{ km/h}$$

Thus an increase of almost 10% in takeoff speed is required at the higher elevation.

14.6 SUMMARY

"External flow" is the term used to describe either an object moving through a fluid or fluid moving over an object. The study of external flows is important in a wide variety of fields including transportation (lift and drag) and structural design (fluid forces).

When a body is immersed in a moving fluid, the fluid velocity along a line perpendicular to any point on its surface is observed to vary from zero on the surface to a maximum value some distance away. At large Re the variation occurs over a relatively small distance, and the body is said to have a boundary layer. The boundary layer characteristics are affected by several factors including the shape of the surface and its orientation relative to the freestream.

The Blasius solution applies to the steady, laminar boundary layer on a smooth flat plate aligned with the freestream and yields the following important results: (1) the pressure inside the boundary layer is the same as it is in the inviscid flow outside the boundary layer; (2) the boundary layer thickness grows at a rate $\delta(x) \propto x^{1/2}$; and (3) the wall shear stress and skin friction coefficient decreases at a rate $\tau_W(x) \propto x^{-1/2}$. The quantitative results from the Blasius solution are provided in Eqs. 14.12a–14.12h. Since it is observed that the transition to turbulence occurs at $Re_x \approx 5 \times 10^5$ for a flat plate, the Blasius solution is valid for $0 < x < 5 \times 10^5 (\nu/U)$.

There is no analytical solution for a turbulent boundary layer on a flat plate, so we are forced to rely on empirical observations. The results of the corresponding power-law model are given in Eqs. 14.16a–14.16h and are valid for the range $5 \times 10^5 < Re_x < 10^7$. The boundary layer thickness grows at a rate proportional to $x^{4/5}$, while the wall shear stress and skin friction coefficient decreases at a rate proportional to $x^{-1/5}$.

Observation of boundary layers on airfoils and other curved bodies show that an object's shape and angle of attack have a significant effect on the characteristics of both laminar and turbulent boundary layers. For a boundary layer, the effects of shape and angle of attack are equivalent to a streamwise pressure gradient. A pressure gradient of the form $dp/dx < 0$, meaning that the pressure is decreasing (and the fluid is accelerating) in the flow direction, is referred to as a favorable pressure gradient, while

$dp/dx > 0$ is termed an unfavorable pressure gradient. The favorable pressure gradient of an accelerating freestream tends to thin a laminar boundary layer and bring higher momentum fluid nearer the surface, while the unfavorable pressure gradient of a decelerating freestream tends to do the opposite. Thus, on an airfoil a laminar boundary layer is relatively thin and remains attached on portions of the surface where the flow is accelerating, but becomes thicker and may separate from portions of the surface where the flow is decelerating. At the point of separation the wall shear stress is zero, and downstream of this point the flow near the wall reverses direction. This also occurs with turbulent boundary layers, although they are more resistant to flow separation owing to the increased amount of higher momentum fluid near the surface associated with the more blunt turbulent velocity profile.

For a stationary object immersed in a moving stream, drag is the component of force exerted on the object by the fluid in the direction of the freestream. For an object moving through a stationary fluid, the drag acts in the direction opposite to the object's motion. The power required to propel an object through a fluid at constant speed is given by the product of the drag and the speed. Thus, drag limits the performance of vehicles of all types and affects their fuel economy. The total drag force arises from two mechanisms: pressure and shear stress. The pressure contribution to the total drag is referred to as form drag, and the contribution due to the shear stress acting on an object's surface is called friction drag. In a high Re flows the drag on bluff bodies tends to be dominated by form drag. The drag on long thin bodies tends to be dominated by skin friction. "Streamlining" is a term used to describe the attempt to design an optimum shape for a bluff body in a high Re flow by minimizing the total drag. It generally takes the form of elongating the rear of the body. Although this raises the friction drag, it lowers the form drag, and the total drag is reduced.

Although CFD is increasingly being used to determine the drag on objects of engineering interest, much of what is known about drag is the result of experiments. Dimensional analysis shows that the drag coefficient may depend on Re, M, Fr, St, We, and e/D.

Drag coefficients are shown in Figure 14.17 for a number of shapes in creeping flow. All objects in creeping flow have drag coefficients that are proportional to Re^{-1}. Figure 14.18 provides C_D data for a smooth cylinder. For $Re < 10^3$ the friction drag of the laminar boundary layer dominates, so C_D varies with Re as it did for laminar flow over an aligned flat plate (i.e., $C_D \propto Re^{-1/2}$). In the range $10^3 < Re < 10^5$, C_D has only a weak dependence on Re and the total drag is dominated by pressure drag. Over the range $10^5 < Re < 10^6$, C_D falls dramatically by about 80%. The drag actually decreases with increasing speed as a result of the laminar-to-turbulent transition of the boundary layer. Spinning a cylinder can result in an increase in drag and also creates a sideforce or lift on the cylinder. Figure 14.24 provides C_D data for a smooth sphere. The trends for the dependence of C_D on Re for a sphere are similar to those for a cylinder, for many of the same reasons. As was the case with a cylinder, rotation of a sphere not only affects the drag but also produces lift.

The term "bluff body" refers to an object that experiences flow separation at a relatively low Re and for which the point of separation is essentially independent of Re. The fixed point of separation is often, but not always, associated with a sharp corner or change in geometry. Since the flow separation point is independent of Re, the drag coefficients for a bluff body is also nearly independent of Reynolds number (over a large range of Re). Tables 14.2 gives the relevant geometry, characteristic area, and drag coefficients for a few common bluff bodies.

PROBLEMS

Section 14.2

14.1 Assuming that the wing on an airplane is behaving like a flat plate, what is the length of the laminar boundary layer if it is flying at a speed 150 mph at an altitude of 5000 ft.

14.2 Calculate δ, δ^*, and Θ for the boundary layer flow described in Problem 14.1.

14.3 Air, at 20°C, with incoming velocity of $U = 18$ m/s, flows over a horizontal flat plate. The velocity profile in the boundary layer is modeled by

$$\frac{u}{U} = \sin\left(\frac{\pi}{2}\frac{y}{\delta}\right) + C$$

where C is a constant. At $x = 0.15$ m the boundary layer thickness is $\delta = 5.0$ mm. What are the boundary conditions that this profile must satisfy? What is the value of C? Is the boundary layer laminar or turbulent at this point? Why?

14.4 Determine δ^*, Θ, and τ_w at $x = 0.15$ m for the flow described in Problem 14.3.

14.5 The laminar, and turbulent velocity profiles in Figure P14.1 have the same boundary layer thickness. The laminar profile is parabolic

$$\frac{u}{U} = 2\left(\frac{y}{\delta}\right) - \left(\frac{y}{\delta}\right)^2$$

and the turbulent profile is the 1/7-power law equation

$$\frac{u}{U} = \left(\frac{y}{\delta}\right)^{1/7}$$

Calculate the moment flux for each profile.

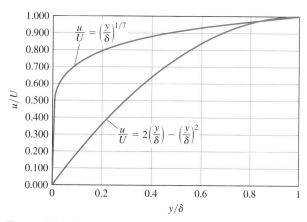

Figure P14.1

14.6 Calculate the kinetic energy flux for each of the profiles given in Problem 14.5.

14.7 Determine δ^*/δ and Θ/δ for each of the profiles given in Problem 14.5.

14.8 A hydrofoil 1.4 ft long and 6 ft wide is put in 50°F water flowing at 30 ft/s. Estimate the boundary layer thickness at the end of the plate.

14.9 Estimate the drag force on the hydrofoil described in Problem 14.8 assuming turbulent flow from the leading edge.

14.10 Estimate the drag force on the hydrofoil described in Problem 14.8 assuming laminar-to-turbulent transition at $Re = 5 \times 10^5$.

14.11 Redo Problem 14.8 in air.

14.12 Redo Problem 14.9 in air.

14.13 Redo Problem 14.10 in air.

14.14 The velocity profile of the atmospheric boundary layer profile over a city is estimated to be $u = y^{0.40}$. If the velocity is 5 mph on the second floor at 25 ft, what is the velocity on the 75th floor? Assume 12 ft for each additional floor.

14.15 For the building described in Problem 14.14, what is the velocity pressure felt on the windows on the second and 75th floors?

14.16 Which of these equal area plates will have the most drag: $L \times L$, $L/2 \times 2L$ short side into the flow, $2L \times L/2$ long side into the flow? Explain your answer.

14.17 Would the object shown in Figure P14.2 have more drag moving toward the left or toward the right? Explain your answer.

Figure P14.2

14.18 For the control volume shown in Figure P14.3, determine the volume flowrate leaving through surface 3-4. The entering profile at 1-4 is uniform. The exiting profile at 2-3 is parabolic. The free stream velocity is $U = 3$ m/s and $\delta = 6.0$ mm at surface 2-3.

14.19 Use the numerical solution obtained by Blasius (Table 14.1) to determine the drag force on a 100 cm long, 50 cm wide plate immersed in SAE 30 oil at 20°C and moving at a speed of 5 m/s. What is the boundary layer thickness at the end of the plate?

14.20 For the flow described in Problem 14.3, determine the displacement thickness by using numerical integration of the Blasius solution data in Table 14.1. Compare your answer to the one found using Eq. 14.12c.

14.21 For the flow described in Problem 14.3, determine the momentum thickness by using numerical integration of the Blasius solution data in Table 14.1. Compare your answer to the one found using Eq. 14.12d.

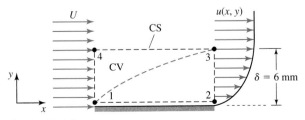

Figure P14.3

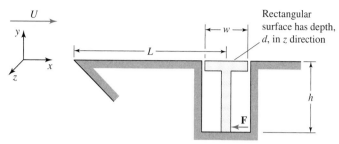

Figure P14.4

14.22 The instrument shown in Figure P14.4 determines the wall shear stress by measuring the force at the base that holds the surface element in place. Determine a formula for τ_w as a function of F and the geometric parameters L, w, d, and h.

14.23 For the device described in Problem 14.22, if water is flowing at 1 m/s, $L = 4$ m, $w = 5$ mm, $d = 2$ mm, and $h = 12$ mm, determine the value of F.

14.24 A fan consists of four blades, 18 in. long and 6 in. wide, acting like flat plates. If the fan rotates at 120 rpm, estimate the torque required to overcome air friction.

14.25 Explain why shape is more critical to the efficient operation of a diffuser than of a nozzle.

Section 14.4

14.26 What is the drag force on pollen falling in 20°C air? Assume the pollen are 10 μm diameter spheres with a density of 800 kg/m³.

14.27 Dirt has become suspended in the water of a garden pool with a depth of 1.5 m. If the smallest particles are 50 μm and the density of the particles is 1250 kg/m³, how long will it take for the pool to clear?

14.28 For the pool described in Problem 14.27, what is the size of the smallest particles if the pool clears in 2 h?

14.29 Sedimentation is one mechanism for deposition of particles in the lung. Determine the terminal velocity of fibers with a density of 1100 kg/m³ in 37°C air. Model them as 2 μm diameter by 10 μm long rods settling both normally and parallel.

14.30 Derive an expression for the terminal velocity of a sphere of diameter D and density ρ, in a fluid of density ρ_f assuming low Re flow.

14.31 Derive an expression for the settling time of a distance h for a sphere of diameter D and density ρ, in a fluid of density ρ_f assuming low Re flow.

14.32 Derive an expression for the viscosity of a fluid μ if a sphere of diameter D and density ρ, falls at terminal velocity V_t, through a fluid with density ρ_f at $Re < 1$.

14.33 A 100 ft tall, 2 ft diameter smokestack must withstand 100 mph winds during a hurricane. What is the bending moment on the stack? Assume a uniform velocity profile.

14.34 Redo Problem 14.33 assuming a 1/7 power-law profile.

14.35 An elevated spherical storage tank 20 m in diameter is supported by a 5 m diameter cylindrical base that is 30 m tall. Assuming that the drag on the tank and the support can be calculated separately and summed, what is the total bending moment on the structure in a 50 km/h wind? Assume a uniform velocity profile.

14.36 What is the power per foot of length required for a 0.5 in. diameter support wire on an antique biplane at 80 mph at an altitude of 500 ft?

14.37 A spherical ($d = 2$ cm) lead ($\rho = 11{,}340$ kg/m^3) weight hangs from a fishing boat. The line is initially let out to a depth of 10 m in still water. The boat then begins to troll at 10 km/h. At what angle to the vertical will be the line? At what depth will be the weight? Ignore the weight and drag on the line.

14.38 Repeat Problem 14.37 considering the drag on the line.

14.39 Bubbles released from the air tank of a scuba diver are 2 cm in diameter at a depth of 15 m. Calculate the terminal velocity of a bubble as a function of depth. Use this result to estimate the time needed for a bubble to rise to the surface.

14.40 Estimate the terminal speed of hailstones the size of peas. State all assumptions.

14.41 A machine ejects a tennis ball horizontally at a velocity of 100 km/h. How far will the ball go before it drops 0.5 m? 2 m? The ball has a mass of 57 g and diameter of 64 mm.

14.42 A home run ball hit into San Francisco Bay is retrieved by your dog. However you drop the exchange. How fast will the ball be falling into the bay? Assume the water temperature is 50°F. The ball has a mass of 5 oz and a circumference of 9 in.

14.43 It was once thought that a baseball dropped from the top of the Washington Monument would be impossible to catch. Estimate the velocity of the ball dropped from the 550 ft level. The ball has a mass of 5 oz and a circumference of 9 in. Explain all your assumptions and suggest more accurate methods. This experiment was attempted and eventually the ball was caught.

14.44 A golf ball with a mass of 1.62 oz and a diameter of 1.68 in. is hit with an initial speed of 240 ft/s at an angle of 18° above horizontal. Determine the distance traveled by the ball at sea level.

14.45 Redo Problem 14.44 at an elevation of 5000 ft.

14.46 Redo Problem 14.44 assuming the ball also has underspin of 1000 rpm.

14.47 A fastball pitcher releases the pitch horizontally at 95 mph with an underspin of 250 rpm. How much will the elevation have changed when the ball reaches the plate approximately 55 ft away? The ball has a mass of 5 oz and a circumference of 9 in.

14.48 Redo Problem 14.47 with no underspin.

14.49 Many curve ball pitchers have had difficulty pitching in Denver. Investigate why this is the case by redoing Example 14.7 at an elevation of 5000 ft.

14.50 For the Flettner rotor described in Example 14.5, estimate the power requirement if the wind speed is 40 km/h and the rotation speed is 500 rpm.

14.51 In Example 14.6 the drag on the antenna itself was not considered. Compare the bending moment on the antenna if it is $\frac{1}{4}$ in. in diameter with that created by the drag on the ball.

14.52 For a laboratory demonstration, the drag on a cylinder is to be investigated. If the wind tunnel has a 1 m square test section, what is the maximum size you would recommend for the cylinder?

14.53 An underwater pipeline is 1 ft diameter and 1 ft up from the bottom. If the water flowrate over the pipe is 4 ft/s, what is the force pulling the pipe off the bottom per unit foot of pipe? What is the drag force on the pipe?

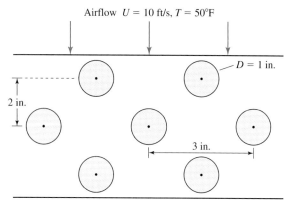

Figure P14.5

14.54 A heat exchanger consists of the tube arrangement as shown in Figure P14.5. What is the drag per unit width on a tube in the third row if the flowrate of air at 50°F is 10 ft/s? Use the drag data supplied in Table 14.2.

14.55 Air at 100°C flows at 10 m/s over the heat exchanger consisting of tubes as shown in Figure P14.6. What is the drag force on the heat exchanger per unit width?

14.56 A large rectangular building, 50 m × 75 m, is 250 m tall. What is the total force and bending moment on the building in a 50 km/h wind directed on the short side of the building? On the long side? Assume a uniform velocity profile.

14.57 Redo Problem 14.56 assuming a 1/7 power-law profile.

14.58 A bridge pier is a canal is 1 m × 2 m, as shown in Figure P14.7. If the flowrate is 10 m/s, what is the bending

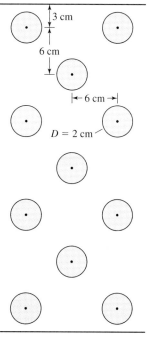

Figure P14.6

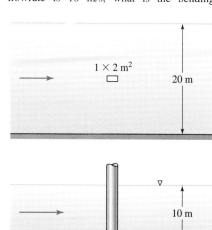

Figure P14.7

moment on the pier? Assume a uniform velocity profile.

14.59 Redo Problem 14.58 with a rounded nose section.

14.60 A 10 cm square piling in 10 m of water is acted on by flow of 2 m/s. Estimate the maximum bending moment on the base of the piling. State all assumptions.

14.61 What is the drag force on 15 ft × 9 ft billboard in a 50 mph wind?

14.62 A kite with a surface area of 1 m^2 is at an angle of 10° with the horizontal. Model the kite as a flat plate to determine the resultant force due to a wind of 25 km/h.

14.63 The scorekeeping sign carried at a golf tournament is 4 ft wide by 3 ft high and is mounted on a 3 ft pole. What is the moment felt by the person walking at 5 mph with the sign?

14.64 A 10 cm × 5 cm elliptical wing strut, 2 m in length, moves through the air at sea level on takeoff at 130 km/h. What is the drag on the strut?

14.65 A 3 m tall pylon with a 1 m equilateral triangular cross section is used to display announcements. Determine the moment on the pylon in a 30 km/h wind.

14.66 Modeling your cupped hand as a hemispherical cup and your arm as a cylinder, estimate the drag created when you stick your arm out the window of an automobile traveling at 50 mph. State assumptions.

14.67 The four-cup anemometer shown in Figure P14.8 starts to rotate with a breeze of 2 km/h. What is the starting torque due to the breeze?

14.68 Estimate the power required to cruise a mini-submarine at constant depth at 10 m/s. Model the submarine as an ellipsoid of revolution of length 5 m and diameter 2 m.

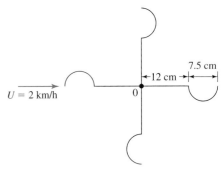

Figure P14.8

14.69 To determine the viscosity, the time is measured for a 1 cm diameter ball bearing to fall 1 m inside a 2 cm diameter cylindrical tube. The bearing is made of stainless steel, $\rho_{ss} = 7800$ kg/m^3, and the density of the fluid is measured to be $\rho_f = 820$ kg/m^3. What is the viscosity of the fluid in the tube if the time measured is 2.4 s? If the effect of the cylinder walls was not taken into account, what would be the error in the viscosity?

14.70 A 1 m per edge cubic crate weighing 9×10^4 N falls into the water from the deck of a ship. How long will it take for the crate to hit the bottom if the depth is 150 m?

14.71 How much drag is on a 5 cm diameter disk oriented normal to the fluid with a density of 900 kg/m^3 and a viscosity of 2×10^{-3} (N-s)/m^2?

14.72 Estimate the drag force on yourself when you are standing in a 40 mph wind.

14.73 Estimate the reduction in drag force obtained by a bicycle rider who was crouching rather than sitting up while peddling at 20 mph.

Section 14.5

14.74 An aircraft uses a NACA 23012 airfoil section at a 2° angle of attack for its wing. The craft travels at 275 km/h at level flight through standard conditions. Its mass is

900 kg. What is the effective area for lift for this craft?

14.75 An aircraft weighing 4800 lb_f with a wing area of 340 ft^2 takes off with a horizontal velocity of 105 mph. What is the necessary lift coefficient of the wing? Assume sea level conditions.

14.76 An aircraft weighing 400 kN empty has a wing area of 220 m^2. It takes off at velocity of 310 km/h at an angle of attack of 12°. Assume that the wing has the characteristics of the NACA 0018 and the air density is 1.19 kg/m^3. Find the allowable weight of the cargo.

14.77 Do you think it takes more power to fly at an altitude of 1000 m or 10,000 m in horizontal flight at the same altitude? Explain your answer by using a calculation based on an aircraft that has the same lift and drag characteristics at each altitude.

14.78 A race car setup for the Indianapolis 500 has a $C_D = 0.669$ based on $A = 12$ ft^2. At 220 mph, what is the drag force and how much power is required to overcome the drag? Assume that the air temperature is 70°F.

14.79 For the Indy car described in Problem 14.78, what is the downward force if the lift-to-drag ratio is 2.92?

14.80 The glide slope angle θ is defined in Figure P14.9 for an aircraft in unpowered flight such that the lift, drag, and weight are in equilibrium. Show that $\theta = \tan^{-1}(C_D/C_L)$.

Figure P14.9

14.81 Assume that an aircraft has lift and drag characteristics of the NACA 23012 airfoil at an angle of attack of 5°. How far can it glide from an altitude of 10,000 ft? This problem is related to Problem 14.80.

14.82 An aircraft has a glide slope angle of 3°. Ten miles from the airport its engine fails. What is the minimum elevation required for it to make it to the airport? This problem is related to Problem 14.80.

14.83 A hydrofoil-based water craft travels at 20 km/h on its foils. The mass of the craft is 2000 kg. If the foils have lift and drag coefficients of 1.65 and 0.58, respectively, what is their effective area? What is the power required at this speed?

14.84 If the hydrofoil water craft described in Problem 14.83 has a power plant that provides 120 kW to overcome the drag, what is its top speed?

14.85 Assume that the lift coefficient for an aircraft wing varies linearly between 0.1 at 0° angle of attack to 0.9 at 8°. At what angle of attack must the wing be for the plane to fly horizontally at 250 km/h at an altitude of 2000 m? The weight is 18 kN and the wing area is 30 m^2.

14.86 What wing area is required to support a 5500 lb_f plane when flying at an angle of attack of 4° at a speed of 90 ft/s. Use the lift coefficient data given in Problem 14.85.

15 OPEN CHANNEL FLOW

15.1 Introduction
15.2 Basic Concepts in Open Channel Flow
15.3 The Importance of the Froude Number
    15.3.1 Flow over a Bump or Depression
    15.3.2 Flow in a Horizontal Channel of Varying Width
    15.3.3 Propagation of Surface Waves
    15.3.4 Hydraulic Jump
15.4 Energy Conservation in Open Channel Flow
    15.4.1 Specific Energy
    15.4.2 Specific Energy Diagrams
15.5 Flow in a Channel with Uniform Depth
    15.5.1 Uniform Flow Examples
    15.5.2 Optimum Channel Cross Section
15.6 Flow in a Channel with Gradually Varying Depth
15.7 Flow under a Sluice Gate
15.8 Flow over a Weir
15.9 Summary
Problems

15.1 INTRODUCTION

 CD/Video library/River Flow

Open channel flow can be defined as a flow of liquid that occurs in a sloped channel having a solid bottom and sidewalls and open to the atmosphere at the top. Thus a river qualifies, and if you have been white water rafting or kayaking you have undoubtedly

15.1 INTRODUCTION

Figure 15.1 (A) Tellico River rapids. (B) Roman aqueduct. (C) Satellite photo of the Mississippi River. (D) Flood waters.

observed the incredible variety of waves and depressions on the water surface (see Figure 15.1A). These and other flow features that make a white water experience so much fun are but one example of the interesting phenomena associated with open channel flow.

Predicting the flow of water in rivers, canals, culverts, flumes, and sewers is of significant engineering interest. The well-defined geometries of man-made channels allow us to use a relatively simple engineering analysis in the investigation of the resulting flows. As the Roman aqueduct in Figure 15.1B attests, engineers have been designing open channel flow structures for many centuries. The flow in rivers, streams, and tidal channels, however, defies simple analysis and is beyond the scope of this text. The reason of course is that their channels typically have very complex geometry. Consider, for example, the satellite image of the river shown in Figure 15.1C. Imagine trying to describe the geometry of each cross section along its length. Although we will not discuss the flow in natural channels further, many of the basic concepts introduced in this chapter do apply to these natural open channel flows. This is fortunate because understanding these flows is of great importance in civil and environmental engineering. In fact, the control of flows in natural channels is a vital topic in disaster abatement, as indicated by the flood damage shown in Figure 15.1D.

 CD/Video library/Glen Canyon Dam Outflow

The study of open channel flow traditionally includes some discussion of the flow in weirs, sluice gates, and spillways. Examples of these structures are shown in Figure 15.2. Their purpose is to measure or control the flowrate in both natural and man-made channels. Another interesting open channel flow phenomenon, which you may have witnessed in miniature when rainwater drains down a steeply sloped parking lot or gutter, is called a hydraulic jump. In a hydraulic jump, fast-moving water abruptly slows

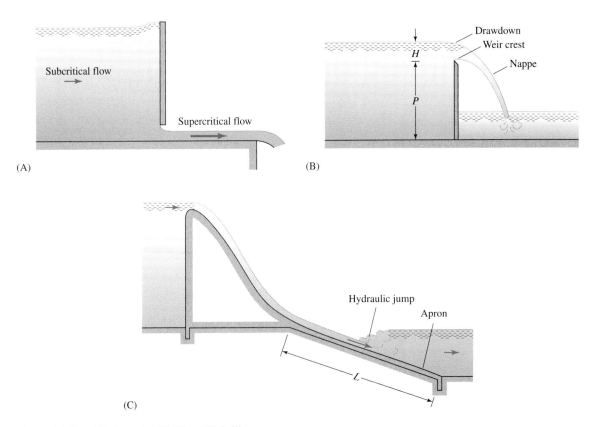

Figure 15.2 (A) Sluice gate. (B) Weir. (C) Spillway.

Figure 15.3 Spillway of the Itaipu Dam in Brazil.

down and its depth increases. An example of a full-size hydraulic jump, deliberately designed to occur on a dam spillway, is shown in Figure 15.3.

In this chapter we discuss the basic concepts used to describe and analyze open channel flow. Among the important considerations are the effects of depth and slope, the shape of the channel, and the behavior of surface waves. Dimensional analysis shows that the Reynolds and Froude numbers are the relevant parameters describing an open channel flow. In most flows these numbers are very large, and the flow is governed solely by value of the Froude number. We demonstrate the importance of the Froude number in open channel flow by analyzing frictionless flow over a bump or depression, flow through a channel of varying width, the behavior of surface waves, and the hydraulic jump. In each case we discover that the presence or absence of phenomena depends on whether the Froude number is greater than or less than one. Next we use an energy balance, first to understand the role of gravity and friction in open channel flow, and then to discuss the concept of specific energy. We continue our discussion by analyzing two important cases: flow in a channel of uniform depth and flow in a channel in which the depth varies slowly along the channel. Mass, momentum, and energy balances prove valuable in developing a theory for predicting the nature of these two types of open channel flow. The chapter concludes with a discussion of methods to analyze flow in weirs and sluice gates. We begin Section 15.2 by describing the basic concepts and terminology of this important area in civil and environmental engineering.

 CD/Demonstrations/Boundary conditions between liquids, solids, and gases

15.2 BASIC CONCEPTS IN OPEN CHANNEL FLOW

In an open channel, liquid flows in a partially filled, sloped channel owing to the action of gravity. As illustrated in Figure 15.4A, the liquid moving down the channel is in contact with solid walls on part of its boundary, and a gas on the rest of its boundary. Because of the no-slip, no-penetration condition, the liquid in contact with the solid walls has zero velocity. Thus the moving liquid must exert a shear stress on the wall, and vice versa. The magnitude of this shear stress is unknown; hence it must be determined as part of the solution to the flow problem.

15 OPEN CHANNEL FLOW

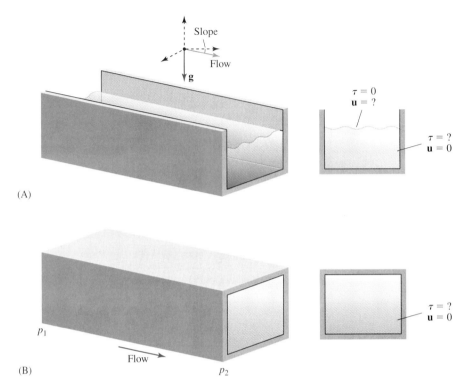

Figure 15.4 (A) Open channel flow. (B) Internal flow.

At the liquid–gas interface, which is referred to as a free surface, the liquid is free to move. The exact location of the free surface is unknown, however, and must be determined. At the free surface the no-slip, no-penetration condition ensures that the liquid velocity matches the gas velocity, but the shear stress and liquid velocity at this interface are unknown. However, by observing that the gas is easily dragged along by the moving liquid, we can conclude that the shear stress at this interface is negligible. Moreover, by noting that the velocity at the free surface is not large, we can use the Bernoulli equation to deduce that the pressure, i.e., normal stress, on this surface is atmospheric.

It should be evident that open channel flows are quite different from the internal flows covered in Chapter 13. For example, consider the balance of forces acting on the fluid in each type of flow. In the steady internal flow shown in Figure 15.4B an externally imposed pressure gradient drives the fluid through the passage and is balanced by the shear force exerted on the fluid by the enclosing solid boundaries. In contrast, in the steady open channel flow shown in Figure 15.4A, the moving fluid is only partially enclosed and is bounded on one side by a free surface. Although there is a hydrostatic pressure variation in the liquid from the free surface to the channel bottom, there can be no imposed pressure gradient in the flow direction under these conditions. What drives the flow then? The answer is the down-slope component of the gravitational body force. In an open channel flow, gravity drives the liquid down the channel and is opposed by the shear force exerted on the liquid by the solid walls. The depth of an open channel flow

15.2 BASIC CONCEPTS IN OPEN CHANNEL FLOW | 947

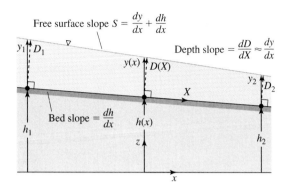

Figure 15.5 Channel coordinates and slope definitions.

is thus an important unknown, since it determines the amount of wetted solid surface over which shear forces act.

Consider now the geometry of an open channel flow as illustrated in Figure 15.5, and notice the two distinct sets of coordinates (X, D) and (x, y). Here h is the elevation of the bottom of the channel (also called the bed) above a horizontal datum, X is the actual distance as measured along the bed, and D is the actual depth of the flow, i.e., the distance from the channel bed to the free surface as measured along a line perpendicular to the bed. The coordinate x is the distance along the horizontal, and y is defined as the distance from the bottom of the channel to the free surface as measured vertically. Because of the very small slopes encountered in open channel flow, we can assume $y = D$ and $x = X$ as an engineering approximation. It also proves convenient to use (x, y) in analyzing the flow rather than the natural coordinates (X, D).

Now consider the three slopes illustrated in Figure 15.5. We see that the slope of the channel bed, called the bed slope, is given by dh/dx, and the slope of the free surface is given by $S = (d/dx)(y + h) = dy/dx + dh/dx$. The variation in depth, D, along the channel, or depth slope, is given by dD/dX. However, since we are using the small slope approximation ($y = D$ and $x = X$), we can write $dy/dx \approx dD/dX$. Thus in discussing open channel flow there are three physically distinct slopes of interest: the bed slope dh/dx, the depth slope dy/dx, and the free surface slope $S = dy/dx + dh/dx$. All three slopes are potentially different, and along with the depth y and distance along the channel x, they play an important role in open channel flow.

Since the shear stress in open channel flow acts at the solid walls defining the channel cross section, the shape of the channel cross section is very important. A number of cross sections of engineering interest are shown in Figure 15.6. Notice that the cross-sectional area of a channel is easily determined if the geometry is known. However, the area of greatest interest in open channel flow is the flow area A, defined to be the area perpendicular to the flow direction through which liquid flows. It can be seen that the flow area depends on the depth y.

There are a number of other geometric parameters of interest. The channel perimeter is defined as the length of the edge of the cross section as measured along the channel walls and bed but excluding the open side. This parameter may be calculated from the known geometry. However, the wetted perimeter P, which is defined as the length of the channel perimeter in contact with liquid, depends on the depth y. The hydraulic

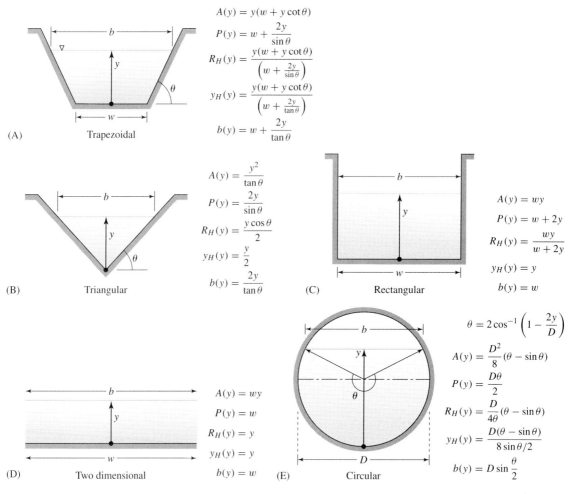

Figure 15.6 Open channel flow cross sections of engineering interest: flow area, A; wetted perimeter, P; hydraulic radius R_H; hydraulic depth y_H; free surface width, $b(y)$.

You probably recall that in internal flow the hydraulic diameter is defined as four times the cross section of the passage divided by the perimeter. Since the fluid completely fills the passage in an internal flow, the hydraulic diameter is defined entirely by the geometry of the passage and does not depend on any characteristic of the flow. It is possible to define a hydraulic diameter for open channel flow as $D_H = 4A/P = 4R_H$, where A is the flow area and P the wetted perimeter, but it is customary to use the hydraulic radius.

radius R_H of an open channel flow is defined as the ratio of the flow area to the wetted perimeter, i.e., as

$$R_H = \frac{A}{P} \quad (15.1)$$

The hydraulic radius of an open channel flow depends on both the channel cross section and the depth, since together they determine the flow area A. We shall see in a later section that an important length scale in open channel flow is the hydraulic depth, defined as

$$y_H = \frac{A}{b} \quad (15.2)$$

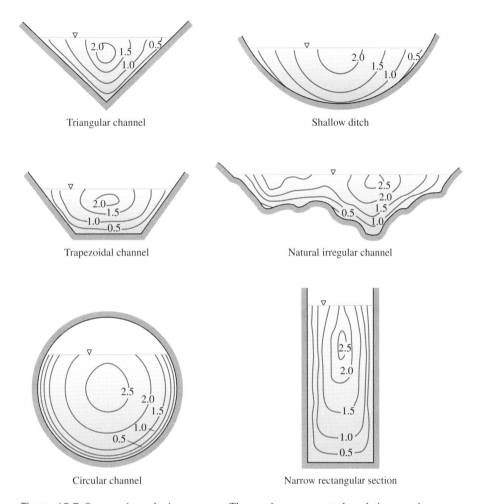

Figure 15.7 Streamwise velocity contours. The numbers represent the relative speed.

where b is the width of the free surface. Values of these different geometric characteristics for channel cross sections of engineering interest are also included in Figure 15.6.

It is tempting to think that in an open channel flow the streamwise velocity at a cross section is approximately uniform. However, this is not the case. Measured streamwise velocity contours are shown for several channel shapes in Figure 15.7. Notice that there is wide variation in speed at different points in the cross sections and that the maximum streamwise velocity occurs some depth below the free surface. Because there is negligible shear stress at the free surface, we would expect the maximum velocity to occur on the free surface, with viscosity causing a reduction in velocity as the walls and bottom are approached. This is not what is observed, however, because of the occurrence of secondary flows. Indeed, the location of the point of maximum velocity below the free surface is an indication of the presence of secondary flows.

The use of the term "secondary flow" means that the velocity components in the vertical and cross-stream directions are nonzero. The nonuniform flow in natural streams and rivers is often cited as the driving force behind the development of meander

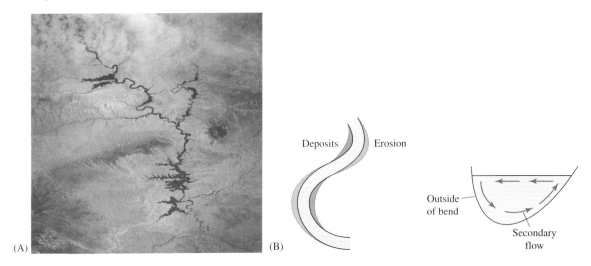

Figure 15.8 (A) The Colorado River. (B) The erosion process in a meandering river.

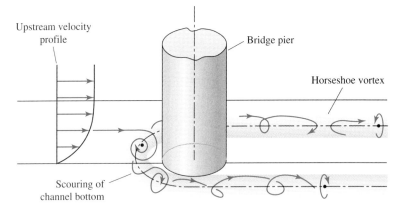

Figure 15.9 Erosion near a bridge pier.

in a streambed. A river with a prominent meander is shown in Figure 15.8A. The presence of a nonuniform flow tends to erode material at the outside of the bend and deposit material at the inside, as shown in Figure 15.8B. Nonuniform flow can also be responsible for erosion at the base of man-made obstructions such as piers and pilings for bridges and other structures. The secondary flow responsible for this process is sketched in Figure 15.9.

Although we know that the velocity field in an open channel flow is actually three dimensional, an open channel flow in a man-made channel is analyzed by assuming uniform, 1D flow, with the average streamwise velocity $V(x)$ aligned parallel to the bed. This assumption is illustrated in Figure 15.10 and used in the rest of this chapter. Since $V(x)$ is the average streamwise velocity in a cross section located at x, it is related to the constant volume flowrate Q, and flow area $A(x)$ by

$$V(x) = \frac{Q}{A(x)} \tag{15.3}$$

15.2 BASIC CONCEPTS IN OPEN CHANNEL FLOW

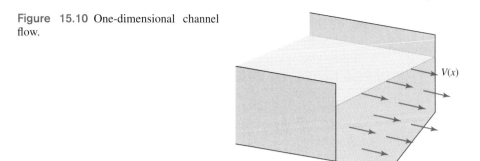

Figure 15.10 One-dimensional channel flow.

An open channel may well incorporate changes in cross section along the channel as well as changes in slope. In this text, our discussion of open channel flow is primarily limited to those in prismatic channels, i.e., channels in which the cross section and slope of the channel bed are fixed along the length of the channel. Open channel flow in a prismatic channel is classified according to the manner in which the depth varies along the flow direction. To understand these classifications, consider the depth slope dy/dx at various locations in the flow illustrated in Figure 15.11.

If a flow occurs in a section of a prismatic channel with constant liquid depth, then $dy/dx = 0$, the slope of the free surface is $S = dy/dx + dh/dx = dh/dx$, and the flow in that section is referred to as a uniform flow (UF). The free surface in a uniform flow is parallel to the bed, the flow is fully developed, and the velocity does not vary along the channel.

A flow in a prismatic channel section with varying liquid depth has $dy/dx \neq 0$ and is referred to as nonuniform flow or varying flow (VF). In varying flow we can further distinguish between a rapidly varying flow (RVF) for which $dy/dx \approx 1$, and a gradually varying flow (GVF), where $dy/dx \ll 1$. The slope of the free surface, $S = dy/dx + dh/dx$, may be greater or less than the slope of the channel bed dh/dx,

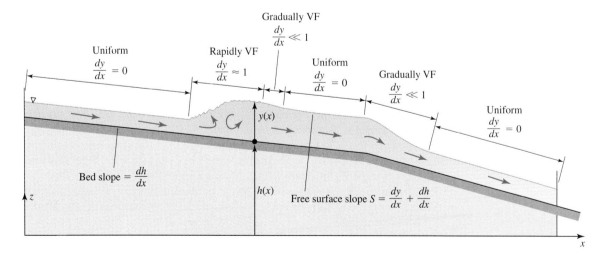

Figure 15.11 Classification of flow with depth along a channel: VF, varying flow.

and since the depth y varies along the channel, the flow is not fully developed. Thus, the velocity varies along the channel in nonuniform flow.

15.3 THE IMPORTANCE OF THE FROUDE NUMBER

An open channel flow may be steady or unsteady, laminar or turbulent, and it may involve any liquid. For example, if you topple a can of motor oil on a sloped pavement, a sheet of oil will flow downhill and create a laminar, unconfined open channel flow. On the other hand, the flow in a river is a turbulent, confined open channel flow. Either of these flows may be steady or unsteady depending on the circumstances.

The important parameters governing an open channel flow are revealed by a dimensional analysis. The DA of open channel flow begins by recognizing that for a given channel geometry, the important factors are the density and viscosity of the fluid, the speed of the flow V, gravity, a length scale L, and surface tension σ. After a dimensional analysis has been performed, the dimensionless parameters are found to be the Reynolds number $Re = \rho V L/\mu$, the Froude number $Fr = V/\sqrt{gL}$, and the Weber number $We = \rho V^2 L/\sigma$.

Consider first the role of surface tension in open channel flow. The Weber number may be thought of as the ratio of flow kinetic energy $\rho V^2 L^3$ to the surface energy σL^2 associated with the surface tension. With the exception of a thin sheet of liquid flowing downhill, as in the case of spilled motor oil mentioned earlier, the Weber numbers of open channel flows are very large. For example, for water at 60°F flowing at 1 ft/s in a river with a hydraulic radius of 20 ft we find:

$$We = \frac{\rho V^2 R_H}{\sigma} = \frac{(1.938 \text{ slugs/ft}^3)(1.00 \text{ ft/s})^2(20.0 \text{ ft})}{5.03 \times 10^{-3} \text{ lb}_f/\text{ft}} = 7710$$

Thus the surface energy is negligible in comparison to the flow kinetic energy, and we can safely ignore surface tension effects in large-scale open channel flows.

The value of the Reynolds number is important in predicting whether an open channel flow is laminar or turbulent. In most cases of interest, Re is relatively large and the flow is turbulent. As a rule of thumb, an open channel flow may be assumed to be laminar for Re based on hydraulic radius of less than 500 and turbulent if Re exceeds 12,500. Using the data just given for the river, Re is found to be

$$Re_R = \frac{\rho V R_H}{\mu} = \frac{(1.938 \text{ slugs/ft}^3)(1 \text{ ft/s})(20 \text{ ft})}{2.344 \times 10^{-5} \text{ (lb}_f\text{-s)/ft}^2} = 1.7 \times 10^6$$

This is well into the turbulent flow range. Observation suggests that a flow in a culvert or flood control channel is also likely to be turbulent. As will be discussed in more detail later, the frictional forces in large Re open channel flows are found to depend primarily on the roughness of the channel walls; they are nearly independent of Re. Thus, Reynolds number is not as important in open channel flow as in other flows you have studied.

The Froude number proves to be the single significant dimensionless parameter in open channel flow. Using the river data given earlier, the Froude number is calculated as

$$Fr = \frac{V}{\sqrt{gR_H}} = \frac{1 \text{ ft/s}}{\sqrt{(32.2 \text{ ft/s}^2)(20 \text{ ft})}} = 0.04$$

15.3 THE IMPORTANCE OF THE FROUDE NUMBER

As will be explained further in the next section, this value corresponds to what is called subcritical flow. Critical flow occurs at $Fr = 1$, and supercritical flow for $Fr > 1$. The value of the Froude number is implicated in a number of interesting phenomena in open channel flow, including flow over a bump or depression, the behavior of waves on a free surface, the response of the flow to a change in channel area, and the phenomenon known as hydraulic jump. This will be demonstrated in our discussion of these topics in Sections 15.3.1 through 15.3.4. We begin by showing that the response of an open channel flow to a bump or depression in the bed of the channel depends on whether the flow is subcritical or supercritical.

15.3.1 Flow over a Bump or Depression

Consider steady open channel flow in a horizontal rectangular channel of width w that has a bump in the channel bed. To focus solely on the effects of the bump on the flow, we will ignore friction. The geometry of the channel bed and coordinates used to describe the flow over a bump are shown in Figure 15.12A. The liquid depth at any location x along the channel is given by $y(x)$, and the height of the channel bed above an arbitrarily chosen datum elevation is given by $h(x)$. The channel bed and its bump are described by $z = h(x)$, and the free surface is given by $z = y(x) + h(x)$. Upstream of the bump at station x_1 in the horizontal section, the depth of the stream is y_1, the bed height is $z_1 = h(x_1)$, the uniform velocity is V_1, and the flow area is $A_1 = y_1 w$. At a downstream station located at x, the depth of the stream is $y(x)$, the bed height is $z = h(x)$, the uniform velocity is $V(x)$, and the flow area is $A(x) = y(x)w$. Our goal is to predict the depth, knowing the shape of the channel bed and the upstream conditions. Thus the problem is to find the function $y(x)$ with the function $h(x)$ and the upstream flow conditions known.

Applying a steady flow mass balance to the control volume shown in Figure 15.12A, we have $\dot{M} = \rho V_1 A_1 = \rho V(x) A(x)$, which after substituting for the flow areas and dividing by the density gives $Q = V_1 y_1 w = V(x) y(x) w$. Thus we can write

$$V(x) = \frac{V_1 y_1}{y(x)} \tag{15.4}$$

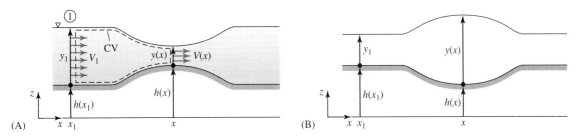

Figure 15.12 (A) Flow over a bump. (B) Flow over a depression.

and also express the velocities in terms of flowrate as

$$V_1 = \frac{Q}{wy_1} \quad (15.5a)$$

$$V(x) = \frac{Q}{wy(x)} \quad (15.5b)$$

Next we apply Bernoulli's equation for steady, constant density flow along the streamline that defines the free surface. Writing this equation from the point on the free surface upstream at location 1 to the point on the free surface at the downstream location x, we obtain

$$\frac{p_1}{\rho} + \frac{V_1^2}{2} + g[y_1 + h(x_1)] = \frac{p}{\rho} + \frac{V(x)^2}{2} + g[y(x) + h(x)]$$

The pressure everywhere on the free surface is atmospheric, so after dividing by g the Bernoulli equation becomes

$$\frac{V_1^2}{2g} + y_1 + h(x_1) = \frac{V(x)^2}{2g} + y(x) + h(x) \quad (15.6)$$

Substituting Eq. 15.4 into Eq. 15.6 and rearranging yields the following equation for the unknown function $y(x)$:

$$\frac{1}{2g}\left[\frac{V_1 y_1}{y(x)}\right]^2 + y(x) + h(x) = \frac{V_1^2}{2g} + y_1 + h(x_1) \quad (15.7)$$

We can use the same approach to analyze flow over a depression in a channel bed. As shown in Figure 15.12B, the shape of the depression is also described by $z = h(x)$, and applying a mass balance and Bernoulli's equation along the free surface leads to exactly the same equations as found earlier for the bump. Thus if the function $h(x)$ describing the complete shape of a channel bed with its bump or depression is known, Eq. 15.7 can be used to find the water depth $y(x)$ at all points along the channel for known values of the upstream velocity V_1, depth y_1, and bed height $h(x_1)$. An example of such a calculation is shown in Example 15.1.

EXAMPLE 15.1

Water flows through a horizontal rectangular channel as shown in Figure 15.13. The flow depth and speed upstream are 2 ft and 2 ft/s, respectively. If the flow encounters the ramp shown, determine the depth and free surface height at the downstream end of the ramp.

SOLUTION

We can use Eq. 15.7 to find the depth $y(x)$ because we know the geometry of the bed and thus $h(x)$. Next we can obtain the free surface height, which is given by the

15.3 THE IMPORTANCE OF THE FROUDE NUMBER

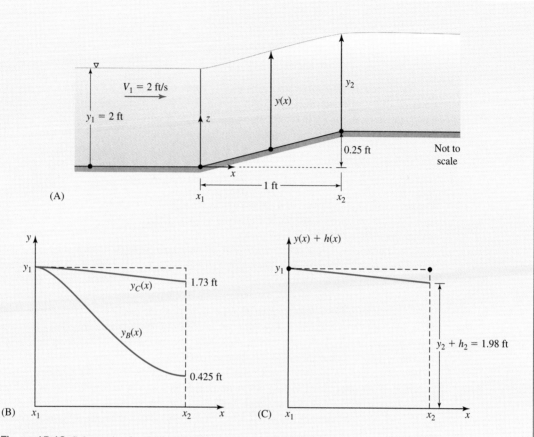

Figure 15.13 Schematics for Example 15.1.

function $y(x) + h(x)$, and construct the requested plots. It is convenient to choose the datum elevation to coincide with the channel bed upstream and place the coordinate origin at the beginning of the ramp as shown. This allows us to represent the shape of the ramp section as $h(x) = 0.25x$ where the units of h and x are feet.

Multiplying Eq. 15.7 by y^2 and rearranging we have

$$y^3 + y^2 \left[h(x) - \frac{V_1^2}{2g} - y_1 - h(x_1) \right] + \frac{1}{2g}(V_1 y_1)^2 = 0$$

After substituting the known values $h(x) = 0.25x$ ft, $y_1 = 2$ ft, $h(x_1) = 0$ ft, as well as the values

$$\frac{V_1^2}{2g} = \frac{(2 \text{ ft/s})^2}{2(32.2 \text{ ft/s}^2)} = 0.0621 \text{ ft}, \quad \frac{1}{2g}(V_1 y_1)^2 = \frac{[(2 \text{ ft/s})(2 \text{ ft})]^2}{2(32.2 \text{ ft/s}^2)} = 0.248 \text{ ft}^3,$$

we obtain

$$y^3 + y^2[0.25x \text{ ft} - 0.0621 \text{ ft} - 2 \text{ ft} - 0 \text{ ft}] + 0.248 \text{ ft}^3 = 0 \quad \text{(A)}$$

We expect to find three solutions to this cubic equation: $y_A(x)$, $y_B(x)$, and $y_C(x)$. Each of these solutions for the depth distribution $y(x)$ along the ramp will correctly match the known upstream depth, $y_1 = 2$ ft, and each will yield a different downstream depth. The three downstream depths can be found by evaluating (A) at $x = 1$ ft to obtain

$$y_2^3 + y_2^2[0.25 \text{ ft} - 0.0621 \text{ ft} - 2 \text{ ft} - 0 \text{ ft}] + 0.248 \text{ ft}^3 = 0$$
$$y_2^3 + y_2^2[-1.812 \text{ ft}] + 0.248 \text{ ft}^3 = 0 \quad \text{(B)}$$

The three solutions to this cubic equation are $y_2 = -0.34$ ft, 0.425 ft, and 1.73 ft, corresponding to the unknown depth distributions $y_A(x)$, $y_B(x)$, and $y_C(x)$. Which downstream depth is correct?

The negative depth can be dismissed on physical grounds, thus $y_A(x)$ can be ignored, but how do we choose between 0.425 ft, and 1.73 ft, and the corresponding solutions $y_B(x)$ and $y_C(x)$ to (A) as shown in Figure 15.13B? One way is to argue that for a small change in bed height, we expect a small change in depth, thus we might guess that $y_C(x)$, which has a downstream depth of 1.73 ft, is the depth profile along the ramp. This turns out to be a lucky guess here but the argument cannot be relied upon in general. The proper way to decide which of the two possible solutions is correct is to calculate the Froude number downstream for each solution and compare it to that upstream. In this case, the upstream Froude number is found to be $Fr_1 = V_1/\sqrt{gy_1} = 0.249$. The Froude numbers for the two downstreams depths can be calculated by noting that since $V_1 y_1 = V_2 y_2$, the downstream velocities for depths 0.425 ft and 1.73 ft are $V_2 = 9.41$ ft/s and 2.31 ft/s respectively. This allows us to calculate the corresponding downstream Froude numbers as $Fr_2 = 2.54$ and 0.31. As will be discussed in the next section, this flow cannot go from a Froude number upstream less than 1 to a Froude number downstream greater than 1. Thus solution $y_C(x)$ as shown in Figure 15.13B is indeed the correct distribution of depth along the channel. The free surface height $y(x) + h(x)$ is shown in Figure 15.13C. Although not illustrated, note that the depth and free surface height are each uniform upstream of the ramp leading edge and also uniform downstream of the ramp trailing edge.

If the upstream speed in Example 15.1 is made large enough (or upstream depth made small enough), the depth over the ramp will actually increase rather than decrease as previously calculated. This difference in behavior is a result of differences in the value of the upstream Froude number, which may be subcritical or supercritical depending on the values of the speed and depth.

We can arrive at some general conclusions about flows over a bump or depression if we differentiate Eq. 15.7 to obtain

$$-\frac{1}{g}\left[\frac{(V_1 y_1)^2}{y(x)^3}\right]\frac{dy}{dx} + \frac{dy}{dx} + \frac{dh}{dx} = 0$$

15.3 THE IMPORTANCE OF THE FROUDE NUMBER

Using Eq. 15.4 and rearranging, we have

$$\frac{dy}{dx} = \frac{\dfrac{dh}{dx}}{\left[\dfrac{V(x)^2}{gy(x)} - 1\right]}$$

where dy/dx is the depth slope (i.e., the rate of change of depth along the channel at station x) and dh/dx is the bed slope. If we now define a local Froude number at the station located at x as

$$Fr = \frac{V(x)}{\sqrt{gy(x)}} \qquad (15.8)$$

the previous relationship between depth slope and bed slope becomes

$$\frac{dy}{dx} = \frac{-dh/dx}{[1 - Fr^2]} \qquad (15.9a)$$

We can now use this result to write the free surface slope $S = dy/dx + dh/dx$ as

$$S = \frac{-Fr^2(dh/dx)}{[1 - Fr^2]} \qquad (15.9b)$$

and after combining the two preceding equations we obtain

$$S = Fr^2 \frac{dy}{dx} \qquad (15.9c)$$

The Froude number defined by Eq. 15.8 is said to be local because it varies along the channel depending on the local values of velocity and depth. Thus Eqs. 15.9 are local relationships, meaning that they apply at a location x, and use the values of Froude number, depth slope, and bed slope at that location.

It is significant that the term $[1 - Fr^2]$ in Eqs. 15.9a and 15.9b changes sign at $Fr = 1$. Indeed you may have noticed that the right-hand sides of the two equations appear to "blow up" if $Fr = 1$. Although we will discuss this possibility in more detail in a moment, a local Froude number of unity, i.e., $Fr = 1$, is termed the critical value. The flow at a location where the value of the local Froude number is less than unity, or $Fr < 1$, is called subcritical flow, while a flow at a location where the value of the local Froude number is greater than unity, or $Fr > 1$, is called supercritical flow. The flow where $Fr = 1$ is said to be critical.

Now consider how the value of the Froude number determines the characteristics of flow over a bump or depression. Starting with Eq. 15.9c, we see first that the free surface slope and the depth slope always have the same sign irrespective of the value of Fr. Next, Eqs. 15.9a and 15.9b reveal that as we approach the leading edge of a bump (for which $dh/dx > 0$), the depth slope and free surface slope will decrease for $Fr < 1$ but increase for $Fr > 1$. Thus, as shown in Figure 15.14A, the liquid depth decreases and the free surface elevation drops over a bump in a subcritical flow, but the depth increases

Figure 15.14 Flow over a bump that is (A) subcritical and (B) supercritical. Flow over a depression that is (C) subcritical and (D) supercritical.

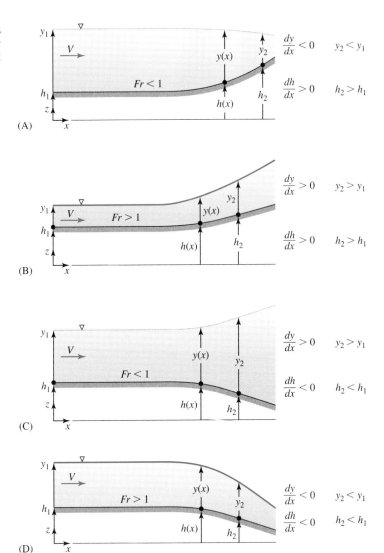

and free surface elevation rises over the same bump in a supercritical flow, as shown in Figure 15.14B.

Furthermore, if the flow over a bump is critical at some location (i.e., if $Fr = 1$) Eqs. 15.9a and 15.9b predict that the depth slope and free surface slope are infinite. Since this is physically impossible, we conclude that critical flow must occur at a location at which the bed slope dh/dx is zero. For the bump shown earlier in Figure 15.12A, this means that critical flow can occur only at the peak of the bump. Later in this chapter we will show that if critical flow does occur at the peak of a bump, then a subcritical flow upstream of the bump may become a supercritical flow downstream if conditions are right. However, if critical flow is not reached at the peak, the flow downstream of the bump will remain subcritical.

15.3 THE IMPORTANCE OF THE FROUDE NUMBER

Equations 15.9a and 15.9b also allow us to predict what will happen to the depth slope and the free surface slope as we approach the leading edge of a depression (for which $dh/dx < 0$). In this case these slopes will increase for $Fr < 1$ but decrease for $Fr > 1$. That is, the liquid depth increases and the free surface elevation rises over a depression in a subcritical flow (Figure 15.14C), but the depth decreases and free surface elevation drops over the same depression in a supercritical flow (Figure 15.14D). As was the case with a bump, critical flow can occur only where $dh/dx = 0$. Thus critical flow can occur only if a depression has a trough or deepest point.

In Example 15.1 we showed how to find the water depth $y(x)$ at all points along a channel when the values of the upstream velocity V_1, depth y_1, and bed height $h(x)$ are known. In some cases we may know the value of the bed height only at a few points along a channel bed. In that event we can make use of Eq. 15.7 and our new understanding of the behavior of subcritical and supercritical flow over a bump or depression to calculate flow properties at these points. Example 15.2 illustrates this type of problem.

EXAMPLE 15.2

Water flows through a horizontal rectangular channel as shown in Figure 15.15A before encountering a depression 1 ft deep. If the flow depth and speed upstream are 4 ft and 1 ft/s, respectively, find the water depth and flow speed over the depression.

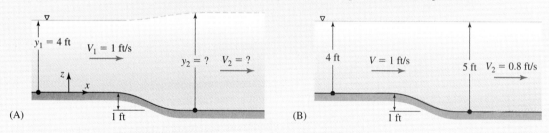

Figure 15.15 Schematics for Example 15.2.

SOLUTION

We can solve this problem by using equation 15.7:

$$\frac{1}{2g}\left[\frac{V_1 y_1}{y(x)}\right]^2 + y(x) + h(x) = \frac{V_1^2}{2g} + y_1 + h(x_1)$$

where point 1 is upstream and x is the location of the arbitrary second point. At a point anywhere along the depression where the bed height is $z = h_2$, the water depth is y_2. Upstream the bed height is $z = h_1$ and the water depth is y_1. Thus we have the following relationship between variables at these two points:

$$\frac{1}{2g}\left[\frac{V_1 y_1}{y_2}\right]^2 + y_2 + h_2 = \frac{V_1^2}{2g} + y_1 + h_1$$

Multiplying this equation by y_2^2 and rearranging, we have

$$y_2^3 + y_2^2\left[h_2 - \frac{V_1^2}{2g} - y_1 - h_1\right] + \frac{1}{2g}(V_1 y_1)^2 = 0$$

It is convenient to choose the coordinate origin and datum elevation at the level of the upstream bed as shown in Figure 15.15A. Then taking point 2 in the depression, we have $h_1 = 0$ ft, $h_2 = -1$ ft, and $h_2 - h_1 = -1$ ft. The remaining known values are $V_1 = 1$ ft/s, $y_1 = 4$ ft. Inserting these values into the preceding equation, we obtain

$$y_2^3 + y_2^2\left[-1\text{ ft} - \frac{(1\text{ ft/s})^2}{2(32.2\text{ ft/s}^2)} - 4\text{ ft} - 0\text{ ft}\right] + \frac{1}{2(32.2\text{ ft/s}^2)}[(1\text{ ft/s})(4\text{ ft})]^2 = 0$$

as the equation to be solved to determine the value of y_2. This equation simplifies to

$$y_2^3 - (5.015\text{ ft})y_2^2 + 0.2484\text{ ft}^3 = 0$$

Solving this cubic equation yields three possible values of the water depth: -0.217 ft, $+0.228$ ft, and $+5.00$ ft. The negative value is not physically possible so can be ignored. But how do we decide whether the water depth is $+0.228$ ft, or $+5.00$ ft? The answer can be found by recalling that the behaviors of the depth slope and the free surface slope depend on the value of the local Froude number. We can use Eq. 15.8 to calculate the local Froude number upstream of the depression:

$$Fr_1 = \frac{V_1}{\sqrt{gy_1}} = \frac{1\text{ ft/s}}{\sqrt{(32.2\text{ ft/s}^2)(4\text{ ft})}} = 0.088$$

Since the flow is subcritical the water depth must increase (see Eq. 15.9a or Figure 15.14C). We conclude that the correct answer in this case is a water depth over the depression of 5.00 ft, as illustrated in Figure 15.15B. The flow speed over the depression can be determined by using Eq. 15.4: $V(x) = V_1 y_1 / y(x)$. Inserting the data, we have

$$V_2 = \frac{(1\text{ ft/s})(4\text{ ft})}{5.00\text{ ft}} = 0.8\text{ ft/s}$$

Before concluding this example, notice carefully that if you choose to use a different datum from which to measure bed height, say one such that the upstream bed height is $z = h_0$, then you would write $h_1 = h_0$, $h_2 = h_0 - 1$ ft, and get $h_2 - h_1 = (h_0 - 1\text{ ft}) - h_0 = -1$ ft. This leads to the same equation for y_2, and thus the same answer. This is to be expected, since the choice of datum should not affect the answer.

In the last example did it seem odd to you that the depth increased by almost exactly 1 foot over the 1-foot-deep depression? There is an explanation for this, and by thinking about it you can learn something interesting about the response of a subcritical flow to a change in bed height. Take a fresh look at Eq. 15.9a, which relates the depth slope to the

15.3 THE IMPORTANCE OF THE FROUDE NUMBER

bed slope: $dy/dx = -(dh/dx)/[1 - Fr^2]$. If the Froude number is very small, then Fr^2 is effectively zero and this equation becomes $dy/dx = -dh/dx$. Integrating this result along the channel gives $y_2 - y_1 = -(h_2 - h_1)$, or $\Delta y = -\Delta h$. Thus in the limit of $Fr \sim 0$ the depth increases by exactly the decrease in the bed height. Another way of grasping this is to note that under these conditions the free surface slope is $S = dy/dx + dh/dx = (-dh/dx) + dh/dx = 0$. In Example 15.2 the $Fr = 0.088$ qualifies as very small, and we did indeed find that a 1 foot decrease in bed height resulted in a 1 foot increase in depth, leaving the free surface at the same elevation. This analysis also predicts that for very small Fr if the bed height increases by some amount, the depth will drop by the same amount.

By studying frictionless flow over a bump or depression in a channel bed, we conclude that the value of Fr allows us to predict what will happen to the depth of the liquid and shape of the free surface in response to a change in the elevation of the channel bed. Next we show that the value of the Froude number also plays a role in determining the response of a flow in a horizontal channel to a change in the channel width.

15.3.2 Flow in a Horizontal Channel of Varying Width

As a second example illustrating the importance Fr in open channel flow, consider the steady frictionless flow through a horizontal rectangular channel whose width $w(x)$ gradually changes downstream as shown in Figure 15.16. The depth of liquid at any location along the channel is given by $y(x)$. Since the channel bed is horizontal, it is convenient to locate the coordinate origin on the bed, hence $h(x) = 0$, and the free surface is given by $z = y(x)$. Upstream of the venturi section at station x_1, the stream depth is y_1, the width is $w_1 = w(x_1)$, the uniform velocity is V_1, and the flow area is $A_1 = y_1 w_1$. At a downstream location at x the stream depth is $y(x)$, the width is $w = w(x)$, the uniform velocity is $V(x)$, and the flow area is $A(x) = y(x)w(x)$. Our goal is to predict the response of the flow to the change in channel width given the upstream conditions. Thus the problem is to find the function $y(x)$ knowing $w(x)$ and the upstream flow conditions.

Applying a steady flow mass balance to the control volume shown in Figure 15.16, we have $\dot{M} = \rho V_1 A_1 = \rho V(x)A(x)$. Substituting for the flow areas and dividing by

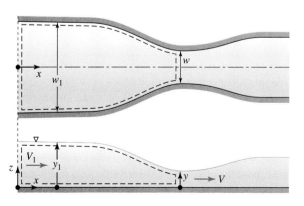

Figure 15.16 Flow through a horizontal channel with changing width.

the density gives $Q = V_1 y_1 w_1 = V(x)y(x)w(x)$. Thus we can write

$$V(x) = \frac{V_1 y_1 w_1}{y(x)w(x)} \tag{15.10}$$

and express the velocities in terms of flowrate as

$$V_1 = \frac{Q}{w_1 y_1} \tag{15.11a}$$

$$V(x) = \frac{Q}{w(x)y(x)} \tag{15.11b}$$

Writing the Bernoulli equation along the free surface from the point upstream at location 1 to a point at any downstream location x, we obtain $p_1/\rho + V_1^2/2 + gy_1 = p/\rho + V(x)^2/2 + gy(x)$. The pressure everywhere on the free surface is atmospheric. Thus the pressure terms cancel and after dividing by g the Bernoulli equation becomes

$$\frac{V_1^2}{2g} + y_1 = \frac{V(x)^2}{2g} + y(x) \tag{15.12}$$

Substituting Eq. 15.10 into Eq. 15.12, and rearranging yields:

$$\frac{1}{2g}\left[\frac{V_1 y_1 w_1}{y(x)w(x)}\right]^2 + y(x) = \frac{V_1^2}{2g} + y_1 \tag{15.13}$$

If the function $w(x)$ describing the width of the channel is known, Eq. 15.13 can be solved to find the water depth $y(x)$ for known values of the upstream velocity V_1, water depth y_1, and width w_1.

We can arrive at some general conclusions about the effect of a change in width on the depth of the liquid if we differentiate Eq. 15.13 with respect to x to obtain

$$-\frac{1}{g}\left[\frac{(V_1 y_1 w_1)^2}{y(x)^3 w(x)^3}\right]\left[w\left(\frac{dy}{dx}\right) + y\left(\frac{dw}{dx}\right)\right] + \frac{dy}{dx} = 0$$

Using Eq. 15.10, rearranging, and introducing the Froude number, we obtain

$$\frac{dy}{dx} = \frac{(V^2/gw)(dw/dx)}{[1 - Fr^2]} \tag{15.14}$$

We see that for subcritical flow through a contraction, for which $[1 - Fr^2] > 0$ and $dw/dx < 0$, Eq. 15.14 predicts $dy/dx < 0$. Thus, the depth will decrease. In supercritical flow through a contraction, we have $dw/dx < 0$ and $[1 - Fr^2] < 0$, so $dy/dx > 0$, and the depth will increase. Critical flow requires that $dw/dx = 0$.

For flow through an expansion, we have $dw/dx > 0$. Thus, in subcritical flow through an expansion the depth will increase, while in supercritical flow the depth will decrease. These general conclusions about the effect of a change in channel width are illustrated in Figure 15.17.

15.3 THE IMPORTANCE OF THE FROUDE NUMBER

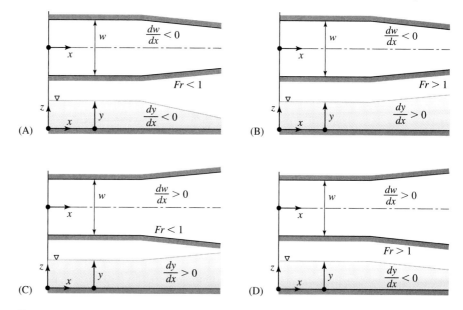

Figure 15.17 Contraction and expansion of channel width: (A) contraction with $Fr < 1$, (B) contraction with $Fr > 1$, (C) expansion with $Fr < 1$, and (D) expansion with $Fr > 1$.

EXAMPLE 15.3

Water flows through a 4 ft wide rectangular irrigation channel that gradually contracts to a width of 3 ft as shown in Figure 15.18. If the water depth upstream of the contraction is 1 ft, and the flow velocity there is 2 ft/s, find the water depth and flow speed downstream of the contraction.

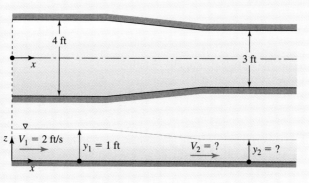

Figure 15.18 Schematic for Example 15.3.

SOLUTION

We can solve this problem by applying Eq. 15.13 between a point upstream where the velocity, width, and depth are V_1, w_1, and y_1, and a point downstream where the velocity, width, and depth are V_2, w_2, and y_2 as shown in Figure 15.18. Writing Eq. 15.13 between these points gives

$$\frac{1}{2g}\left[\frac{V_1 y_1 w_1}{y_2 w_2}\right]^2 + y_2 = \frac{V_1^2}{2g} + y_1$$

Multiplying by y_2^2 and rearranging we obtain the cubic equation

$$y_2^3 - y_2^2\left[\frac{V_1^2}{2g} + y_1\right] + \frac{1}{2g}\left[\frac{V_1 y_1 w_1}{w_2}\right]^2 = 0$$

After inserting the data, we have $y_2^3 - y_2^2[1.062 \text{ ft}] + 0.1104 \text{ ft} = 0$, which can be solved to obtain the three solutions: -0.286 ft, 0.412 ft, and 0.936 ft. The negative root can be immediately discarded on physical grounds. Next we calculate the downstream velocities and corresponding Froude numbers for each of the positive roots. For $y_2 = 0.412$ ft we find:

$$V_2 = \frac{V_1 y_1 w_1}{y_2 w_2} = \frac{(2 \text{ ft/s})(1 \text{ ft})(4 \text{ ft})}{(0.412 \text{ ft})(3 \text{ ft})} = 6.47 \text{ ft/s}$$

and

$$Fr_2 = \frac{V_2}{\sqrt{g y_1}} = \frac{6.47 \text{ ft/s}}{\sqrt{(32.2 \text{ ft/s}^2)(1 \text{ ft})}} = 1.14$$

For $y_2 = 0.936$ ft we find:

$$V_2 = \frac{V_1 y_1 w_1}{y_2 w_2} = \frac{(2 \text{ ft/s})(1 \text{ ft})(4 \text{ ft})}{(0.936 \text{ ft})(3 \text{ ft})} = 2.85 \text{ ft/s}$$

and

$$F_2 = \frac{V_2}{\sqrt{g y_1}} = \frac{2.85 \text{ ft/s}}{\sqrt{(32.2 \text{ ft/s}^2)(1 \text{ ft})}} = 0.50$$

Calculating the upstream Froude number, we find

$$Fr_1 = \frac{V_1}{\sqrt{g y_1}} = \frac{2 \text{ ft/s}}{\sqrt{(32.2 \text{ ft/s}^2)(1 \text{ ft})}} = 0.352$$

hence the flow is subcritical. Since the depth of a subcritical for flow must decrease in a contraction, the correct depth downstream of the contraction is 0.936 ft and the corresponding flow speed is 2.85 ft/s.

15.3 THE IMPORTANCE OF THE FROUDE NUMBER

It is clear that the behavior of a flow through a channel of varying width is similar in many respects to the flow over a bump or depression. Both involve a change in flow area. If we compare the corresponding equations for the depth slope in flow through an open channel of varying width as given by Eq. 15.14,

$$\frac{dy}{dx} = \frac{(V^2/gw)(dw/dx)}{1 - Fr^2}$$

with the corresponding result for the depth slope in flow over a bump or depression, Eq. 15.9a,

$$\frac{dy}{dx} = \frac{-dh/dx}{1 - Fr^2}$$

we see that a contraction ($dw/dx < 0$) is similar in its effect on depth slope to a bump ($dh/dx > 0$), and an expansion ($dw/dx > 0$) is similar in effect to a depression ($dh/dx < 0$). In both types of flow, knowing the value of the Froude number allows us to make a qualitative prediction of the change in depth and behavior of the free surface.

15.3.3 Propagation of Surface Waves

The behavior of surface waves in open channel flow is also governed by the value of the local Froude number. We can demonstrate this important result by considering a horizontal rectangular channel of width, w, filled with liquid at rest to a uniform depth, y. A vertical wall at the left end of this channel is suddenly given a small constant velocity $V_W \mathbf{i}$ to the right, creating a surface wave of height Δy that propagates down the channel as shown in Figure 15.19A. Note that the depth of the liquid behind the wave is $y + \Delta y$ and that the wave propagates at a constant speed c. The fluid ahead of the wave is at rest, and the fluid behind the wave must be all moving to the right at speed V_W as shown in Figure 15.19A. If you are wondering about this last statement, think about how the moving wall is pushing the liquid to the right. Since the liquid cannot be compressed, and the depth behind the wave is uniform, the liquid between the wall and the wave front must be moving to the right at the same speed as the wall.

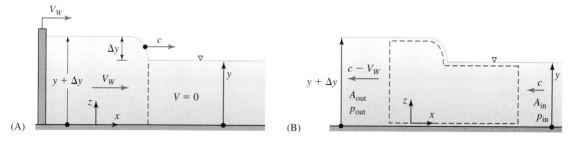

Figure 15.19 The moving end wall with (A) a fixed reference frame and (B) a moving reference frame.

We wish to analyze this model of wave motion and predict the relationships between wave speed c, wave height Δy, and the other physical parameters of the problem. Since the wave is propagating along the channel, the flow is unsteady for an observer in a fixed reference frame attached to the channel. Rather than analyze this unsteady flow, we can use a reference frame that is moving with the wave as shown in Figure 15.19B. Since this reference frame is moving at constant velocity, it is inertial. In the moving frame liquid appears to be approaching the stationary wave from the right at speed c, and moving at speed $c - V_W$ to the left after passing through the wave. Thus the uniform velocity vector on the upstream side of the stationary wave is $\mathbf{u} = -c\mathbf{i}$ and on the downstream side it is $\mathbf{u} = -(c - V_W)\mathbf{i}$ as shown in Figure 15.19B. The flow is steady in this frame.

Applying a steady flow mass balance to the control volume shown in Figure 15.19B, we obtain $-\rho c A_{in} + \rho(c - V_W) A_{out} = 0$, where the inlet and outlet flow areas can be seen to be given by $A_{in} = wy$ and $A_{out} = w(y + \Delta y)$. Thus the mass balance can be written as

$$-\rho c y + \rho(c - V_W)(y + \Delta y) = 0 \tag{15.15}$$

and we can solve for the wave speed to find

$$c = \frac{V_W(y + \Delta y)}{\Delta y} \tag{15.16}$$

At this point, the wave height Δy is unknown, so this single equation is not sufficient to determine the wave speed. Alternately we can use Eq. 15.16 to write

$$V_W = \frac{c \Delta y}{y + \Delta y} \tag{15.17}$$

We can obtain a second equation by applying a steady flow momentum balance in the x direction to the control volume shown in Figure 15.19B. We will neglect friction on the bed surface and assume that the pressure distributions on the inlet and exit surfaces are hydrostatic. The resulting surface forces can then be calculated to be $-wy(p_A + \rho g y/2)\mathbf{i}$ on the inlet flow area, $w(y + \Delta y)\{p_A + [\rho g(y + \Delta y)]/2\}\mathbf{i}$ on the outlet flow area, and a contribution $-p_A w \Delta y \mathbf{i}$ from the free surface at atmospheric pressure. The momentum balance in the x direction is

$$\rho c^2 w y - \rho(c - V_W)^2 w(y + \Delta y)$$
$$= -wy\left[p_A + \frac{\rho g y}{2}\right] + w(y + \Delta y)\left[p_A + \frac{\rho g (y + \Delta y)}{2}\right] - p_A w \Delta y$$

We can write the flux term at the outlet as

$$-\rho(c - V_W)^2 w(y + \Delta y) = -\rho w(c - V_W)[(c - V_W)(y + \Delta y)]$$

then use the mass balance, Eq. 15.15, to replace the term in the square brackets by cy getting

$$-\rho(c - V_W)^2 w(y + \Delta y) = -\rho w c y (c - V_W)$$

15.3 THE IMPORTANCE OF THE FROUDE NUMBER

Thus the two flux terms give $\rho c^2 wy - \rho wcy(c - V_W) = \rho wcy V_W$. Adding and simplifying the three surface force terms, we obtain $w\Delta y(\rho g y + \rho g \Delta y/2)$; thus the momentum balance is

$$\rho wcy V_W = w\Delta y \left[\rho g y + \frac{\rho g \Delta y}{2}\right]$$

Solving for the wave speed we have

$$c = \frac{g\Delta y}{V_W}\left[1 + \frac{\Delta y}{2y}\right] \quad (15.18)$$

Equations 15.16 and 15.18 can now be used to determine the wave speed c and wave height Δy that describe the wave produced when a wall moves at velocity V_W.

The moving wall concept was introduced at the beginning of our analysis as simply the easiest way to visualize a wave propagating along a free surface. At this point we can dispense with the moving wall. That is, we can now simply imagine a solitary wave propagating at some wave speed c, with a wave height Δy, with the liquid behind the wave moving in the same direction at speed V_W. The preceding analysis holds for this isolated wave. Thus, using Eq. 15.17: $V_W = c\Delta y/(y + \Delta y)$, to eliminate V_W from Eq. 15.18, we find that the wave speed obeys the equation:

$$c = \sqrt{gy\left[1 + \frac{\Delta y}{2y}\right]\left[1 + \frac{\Delta y}{y}\right]} \quad (15.19)$$

This equation applies to a wave moving in either direction. Equation 15.19 shows that wave speed increases as the wave height, Δy, increases. Also, by Eq. 15.17, a wave traveling at speed c in stationary water causes the water to move at a speed $V_W = c\Delta y/(y + \Delta y)$ in the direction of wave propagation.

If the amplitude of a wave is sufficiently small, i.e., $\Delta y/y \ll 1$, Eq. 15.19 simplifies to

$$c = \sqrt{gy} \quad (15.20)$$

We see that a small amplitude wave travels at a speed that is determined by the water depth in which it is propagating. In water of uniform depth, a small amplitude wave propagates at a constant speed, and since the wave speed does not depend on Δy, the wave travels without change of shape. Introducing the small amplitude approximation $\Delta y/y \ll 1$ into Eq. 15.17 shows that such a wave causes a velocity in the direction of propagation of $V_W = c\Delta y/y$. The wave created in a still pond by tossing a rock into the water is likely to satisfy the small amplitude assumption. Can you think of a way of determining the water depth in pond or lake by observing the propagation of ripples?

We can also analyze a wave propagating on the surface of a moving stream, i.e., a wave traveling in an open channel flow. Consider first a wave traveling upstream at speed c in a stream moving at speed V. This wave can be analyzed by making use of a change

The speed of a finite amplitude wave is always greater than $c = \sqrt{gy}$. This is easily seen by comparing Eqs. 15.19 and 15.20. If you were using $c = \sqrt{gy}$ to estimate the time of arrival of a large amplitude tidal wave moving in shallow water, you might not have allowed yourself enough time to escape.

EXAMPLE 15.4

Figure 15.20 represents a wave-making machine at a water park. The machine works by moving the wall into a pool 1.25 m deep. The wave amplitude is 0.3 m. Find the wave speed and velocity of the water behind the wave.

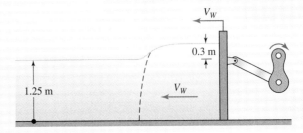

Figure 15.20 Schematic for Example 15.4.

SOLUTION

The wave speed is given by Eq. 15.19 as

$$c = \sqrt{gy\left[1 + \frac{\Delta y}{2y}\right]\left[1 + \frac{\Delta y}{y}\right]}$$

Inserting the data we obtain

$$c = \sqrt{(9.81 \text{ m/s}^2)(1.25 \text{ m})[1 + 0.3 \text{ m}/2(1.25 \text{ m})][1 + 0.3 \text{ m}/1.25 \text{ m}]} = 4.13 \text{ m/s}$$

If we had incorrectly assumed that the wave is of small amplitude and used Eq. 15.20 ($c = \sqrt{gy}$), we would have obtained $c = \sqrt{gy} = \sqrt{(9.8 \text{ m/s}^2)1.25 \text{ m}} = 3.5$ m/s. By checking the ratio of wave amplitude to depth, i.e., $\Delta y/y = 0.3 \text{ m}/1.25 \text{ m} = 0.24$, we would have recognized that the small amplitude approximation is invalid and likely to produce erroneous results.

The speed of the water behind the wave, which is the same as the wall speed, can be found by using Eq. 15.17 and the correct wave speed $c = 4.13$ m/s. The result is

$$V_W = \frac{c\Delta y}{y + \Delta y} = \frac{(4.13 \text{ m/s})(0.3 \text{ m})}{1.25 \text{ m} + 0.3 \text{ m}} = 0.80 \text{ m/s}$$

EXAMPLE 15.5

Most of the waves we observe are generated by the action of the wind over the surface of the water. Tsunamis are waves generated by seismic activity such as earthquakes and volcanoes, or by catastrophic events such as asteroids impacting in the ocean. A tsunami would seem quite harmless if observed on the open ocean, where its amplitude might be as small as 10 cm. However these waves transmit a tremendous amount of energy. What is the wave speed of a tsunami traveling across the Pacific, where a typical depth is 4000 m? Compare this with the speed of a wave of the same amplitude in 1 m of water.

SOLUTION

The wave is sketched in Figure 15.21A. Clearly the small amplitude approximation, Eq. 15.20, is appropriate for the tsunami in the open ocean. Inserting the data, we have

$$c = \sqrt{gy} = \sqrt{(9.81 \text{ m/s}^2)(4000 \text{ m})} = 198 \text{ m/s}$$

which is over 400 mph! When this wave approaches a coastline, it will slow down because the water is shallower, but its amplitude will grow because the energy flux, which is a function of the speed and amplitude, is constant. Tsunamis are very destructive because their amplitude can easily exceed several meters. Figure 15.21B shows the effect of a tsunami that struck Hawaii in 1960.

For a 10 cm amplitude wave in 1 m deep water, we can calculate the wave speed from

$$c = \sqrt{gy\left[1 + \frac{\Delta y}{2y}\right]\left[1 + \frac{\Delta y}{y}\right]} = \sqrt{(9.81 \text{ m/s}^2)(1 \text{ m})\left[1 + \frac{0.1 \text{ m}}{2(1 \text{ m})}\right]\left[1 + \frac{0.1 \text{ m}}{1 \text{ m}}\right]}$$
$$= 3.37 \text{ m/s}$$

Using the small amplitude approximation $c = \sqrt{gy} = \sqrt{(9.81 \text{ m/s}^2)(1 \text{ m})}$ for this wave yields a wave speed of 3.13 m/s, which is in error by only about 7%.

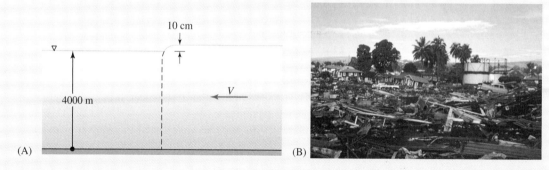

Figure 15.21 (A) Schematic for Example 15.5. (B) Damage due to a tsunami that struck Hilo, Hawaii, in 1960.

of reference frame. Figure 15.22A shows the wave as seen by an observer fixed as usual in the frame of reference attached to the channel. In Figure 15.22B, however, we have changed to a frame of reference fixed to the undisturbed stream moving at speed V. In this moving reference frame, the wave speed is now $c + V$, and the water behind the wave is moving in the direction of wave motion at speed $V - V_R$. The wave now looks exactly like the preceding case of a wave propagating into stationary water, provided we write $V_W = V - V_R$ and recognize that the wave speed in the stationary water case is

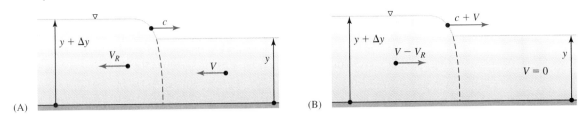

Figure 15.22 (A) Wave in fixed reference frame. (B) Wave in moving reference frame.

now replaced by the sum $c + V$. We conclude that applying a mass and momentum balance to a wave propagating upstream results in the same equations found for propagation of a wave into stationary water. No additional analysis is required to solve this new problem.

To modify the preceding results for a wave propagating upstream, we make the substitutions $V_W = V - V_R$, and $c = c + V$ into the equations obtained earlier. For example, the wave speed is found by substituting $c + V$ for c in Eq. 15.19 to obtain

$$c = \sqrt{gy\left[1 + \frac{\Delta y}{2y}\right]\left[1 + \frac{\Delta y}{y}\right]} - V \quad (15.21a)$$

To analyze a wave propagating downstream, we again use a reference frame moving with the undisturbed stream. The wave speed in the new frame is now $c - V$, and the water behind the wave is moving in the direction of wave motion at speed $V + V_R$. Thus the wave speed for a wave propagating downstream may be obtained by substituting $c - V$ for c in Eq. 15.19 to obtain

$$c = \sqrt{gy\left[1 + \frac{\Delta y}{2y}\right]\left[1 + \frac{\Delta y}{y}\right]} + V \quad (15.21b)$$

Notice that a wave that would travel at a speed $c_0 = \sqrt{gy[1 + \Delta y/2y][1 + \Delta y/y]}$ in stationary liquid will propagate upstream in open channel flow at the reduced wave speed $c = c_0 - V$. That is, the wave's rate of progress upstream relative to the channel bed is reduced because of the motion of the liquid. However, if this wave travels downstream, its wave speed is increased to $c = c_0 + V$.

We can summarize these results for the wave speed in an open channel flow at speed V as

$$c = c_0 \pm V \quad (15.22)$$

using the plus sign for propagation with the stream and the minus sign for propagation against the stream. Note that a wave propagating upstream whose strength is such that its wave speed in stationary water would be c_0 will be stationary in a stream whose speed is $V = c_0$. Such stationary waves can often be observed in river rapids.

Now suppose the wave propagating in an open channel flow is of small amplitude. Then the value of c_0 is given by Eq. 15.20 as $c_0 = \sqrt{gy}$, and using Eq. 15.22, the speed of this wave moving on the surface of an open channel flow is given by

$$c = \sqrt{gy} \pm V \quad (15.23)$$

where the minus sign is used for a wave propagating upstream. The significance of this result is revealed if we think of the small amplitude wave as having been created by some mechanism, perhaps an actual disturbance of the free surface or an obstruction of some type. The resulting small amplitude wave carries information about the disturbance or obstruction to more distant parts of the stream both upstream and downstream. The wave speed for travel upstream is $c = \sqrt{gy} - V$. Now as the value of the channel speed V is increased, the wave speed decreases. At a value $V = \sqrt{gy}$, the wave speed is zero and the wave is unable to propagate upstream at all. Thus information about a surface disturbance or obstruction is not transmitted upstream when $V = \sqrt{gy}$, and if $V > \sqrt{gy}$, the wave is actually washed downstream. We conclude that the characteristic of wave motion (and information flow) in an open channel flow changes dramatically when $V = \sqrt{gy}$. Rearranging this equation, we can write it as $Fr = V/\sqrt{gy} = 1$. Once again we see that the critical Froude number plays a key role in a phenomenon of interest in open channel flow. A small amplitude wave is able to propagate upstream in subcritical flow but not in supercritical flow. Thus the flow of information about surface disturbances and obstructions, which is carried by small amplitude waves, is strongly affected by the value of the local Froude number.

Keep in mind that our analysis of wave propagation on the surface of an open channel flow was based on a simple model. Wave phenomena are complex, and other factors

EXAMPLE 15.6

If a person throws a pebble into a 10 ft deep river flowing at 1 ft/s (Figure 15.23), will the resulting ripples travel upstream? At what speed will the ripples travel in each direction?

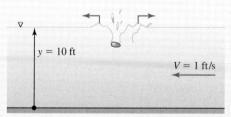

Figure 15.23 Schematic for Example 15.6.

SOLUTION

For a small pebble, we can assume the waves (i.e., ripples) are of small amplitude. If that is the case, the value of the Froude number allows us to determine if a ripple can propagate upstream. A simple calculation yields $Fr = V/\sqrt{gy} = 1 \text{ ft/s}/\sqrt{(32.2 \text{ ft/s}^2)(10 \text{ ft})} = 0.056$. Since $Fr < 1$, the ripples can travel upstream. For small amplitude waves, the wave speed in stationary fluid of this depth is

$$c_0 = \sqrt{gy} = \sqrt{(32.2 \text{ ft/s}^2)(10 \text{ ft})} = 17.9 \text{ ft/s}$$

Thus, the ripples will propagate upstream on this river at $c = c_0 - V = 17.9 - 1 = 16.9$ ft/s and downstream at $c = c_+ - V = 17.9 + 1 = 18.9$ ft/s.

such as wavelength may come into play in determining wave speed. For example, as you have seen at the beach, waves break when they reach shallow water. Nevertheless, we once again see that the value of the Froude number is of paramount importance in predicting the behavior of isolated surface waves and understanding the flow of information in open channel flow.

The hydraulic jump discussed in the next section can be thought of as a special case of a surface wave. However, its importance in open channel flow merits a separate treatment and discussion. As you will learn, a hydraulic jump can occur only when the flow is supercritical, that is, when the Froude number is greater than one.

15.3.4 Hydraulic Jump

In Section 15.3.3 we showed that if a flow is supercritical, disturbances cannot propagate upstream. Thus if downstream conditions require that a supercritical flow become subcritical, a smooth transition is impossible, and the flow goes through a phenomenon known as a hydraulic jump. In a hydraulic jump, the flow changes from supercritical to subcritical in a relatively short distance, with an abrupt decrease in velocity, increase in depth, and a substantial head loss. An example of a hydraulic jump on a dam spillway was shown in Figure 15.3, but you can also create a hydraulic jump yourself by running water from the kitchen faucet onto a flat or slightly sloped surface. As the flow moves radially outward on the surface, a hydraulic jump will occur if the flowrate is large enough.

The general characteristics of a hydraulic jump are shown in Figure 15.24. A hydraulic jump may occur on an inclined or horizontal bed, and in a channel of any shape. For simplicity, we will assume a rectangular horizontal channel. Observations have determined that the maximum length of a jump does not exceed seven downstream depths, thus we will neglect the shear stress applied to the flow by the bed. Note, however, that we are not assuming frictionless flow. In fact, our analysis must allow for a head loss due to viscous dissipation in the jump. Although the turbulent velocity field in a hydraulic jump is generally 3D, we can arrive at the important characteristics by assuming steady, uniform flow, and applying a mass, momentum, and energy balance to the control volume as shown in Figure 15.24.

A mass balance on this CV yields $\dot{M} = \rho Q = \rho A_1 V_1 = \rho A_2 V_2$, where $A_1 = wy_1$ and $A_2 = wy_2$. Thus we can write

$$V_1 y_1 = V_2 y_2 \tag{15.24}$$

Figure 15.24 Control volume for the hydraulic jump.

15.3 THE IMPORTANCE OF THE FROUDE NUMBER

To apply a momentum balance, we use Eq. 7.19b:

$$\int_{CS}(\rho \mathbf{u})(\mathbf{u}\cdot\mathbf{n})\,dS = \int_{CV}\rho \mathbf{f}\,dV + \int_{CS}\Sigma\,dS$$

and consider the component of this equation in the x (flow) direction. The flux terms on the inlet and outlet give $[-\rho V_1^2 A_1 + \rho V_2^2 A_2]\mathbf{i}$, which we can write as $[-\rho V_1^2 w y_1 + \rho V_2^2 w y_2]\mathbf{i}$. There is no component of the body force in the x direction. To evaluate the stress term, we will assume that the pressure distribution on the inlet and exit are hydrostatic and given by $p(y) = p_A - \rho g(y - y_1)$ and $p(y) = p_A - \rho g(y - y_2)$, respectively. The pressure on the free surface is atmospheric, and we can account for the net force applied on this surface in the x direction by using gage pressure in the inlet and outlet terms. On the bottom, the hydrostatic pressure varies from inlet to outlet but contributes no net force in the x direction. The effect of the shear stress here is neglected as explained earlier. The surface force terms involving pressure are

$$\int_{inlet}-p\mathbf{n}\,dS + \int_{outlet}-p\mathbf{n}\,dS = \int_0^{y_1}-(-\rho g(y-y_1))(-\mathbf{i})w\,dy$$
$$+\int_0^{y_2}-(-\rho g(y-y_2))(\mathbf{i})w\,dy$$

which yield $[\rho g w(y_1^2/2) - \rho g w(y_2^2/2)]\mathbf{i}$. Combining terms, we see that the momentum balance yields $-\rho V_1^2 w y_1 + \rho V_2^2 w y_2 = \rho g w(y_1^2/2) - \rho g w(y_2^2/2)$. Our final result after rearranging is

$$\frac{V_1^2}{g}y_1 + \frac{y_1^2}{2} = \frac{V_2^2}{g}y_2 + \frac{y_2^2}{2} \qquad (15.25)$$

We can write an energy balance for this CV by employing Eq. 7.33:

$$\int_{CV}\frac{\partial}{\partial t}\left(\rho\left(u + \frac{1}{2}\mathbf{u}\cdot\mathbf{u} + gz\right)\right)dV + \int_{CS}\rho\left(u + \frac{p}{\rho} + \frac{1}{2}\mathbf{u}\cdot\mathbf{u} + gz\right)(\mathbf{u}\cdot\mathbf{n})\,dS$$
$$= \dot{W}_{power} + \dot{W}_{shaft} + \dot{Q}_C + \dot{S}$$

There is no fluid power, shaft power, or other energy input in this case, and the flow is steady. Thus after the inlet and outlet surfaces of the CV have been identified, the energy balance is given by

$$\int_{inlet}\rho\left(u + \frac{p}{\rho} + \frac{1}{2}\mathbf{u}\cdot\mathbf{u} + gz\right)(\mathbf{u}\cdot\mathbf{n})\,dS + \int_{outlet}\rho\left(u + \frac{p}{\rho} + \frac{1}{2}\mathbf{u}\cdot\mathbf{u} + gz\right)(\mathbf{u}\cdot\mathbf{n})\,dS = \dot{Q}_C$$

On the inlet surface, the streamwise velocity is V_1, the uniform internal energy is u_1, the pressure is $p(y) = p_A - \rho g(y - y_1)$, and the gravitational potential energy is given by $gz = gy$. Thus on the inlet we can write

$$\frac{p}{\rho} + gz = \frac{p_A}{\rho} - g(y-y_1) + gy = \frac{p_A}{\rho} + gy_1$$

which is a constant. At the exit we have V_2 and u_2, with $gz = gy$. Thus on this surface we have

$$\frac{p}{\rho} + gz = \frac{p_A}{\rho} + gy_2$$

Gathering terms and dividing by the mass flowrate $\dot{M}$ the energy balance yields

$$-\left(u_1 + \frac{p_A}{\rho} + gy_1 + \frac{V_1^2}{2}\right) + \left(u_2 + \frac{p_A}{\rho} + gy_2 + \frac{V_2^2}{2}\right) = \frac{\dot{Q}_C}{\dot{M}}$$

The atmospheric pressure terms cancel, so after rearranging we have

$$gy_1 + \frac{V_1^2}{2} = gy_2 + \frac{V_2^2}{2} + (u_2 - u_1) - \frac{\dot{Q}_C}{\dot{M}}$$

In open channel flow, it is customary to define the head loss as

$$h_L = \frac{1}{g}\left[(u_2 - u_1) - \frac{\dot{Q}_C}{\dot{M}}\right] \tag{15.26a}$$

Notice that this head loss has units of energy per unit weight, i.e., units of length. The corresponding head loss (and energy loss) per unit mass is then gh_L, the head loss per unit volume is $\rho g h_L$, and the power consumed in the jump is

$$P = \rho g h_L Q \tag{15.26b}$$

where Q is the volume flowrate through the jump. Dividing the energy balance by g, we obtain

$$\frac{V_1^2}{2g} + y_1 = \frac{V_2^2}{2g} + y_2 + h_L \tag{15.27}$$

A hydraulic jump is described by Eqs. 15.24, 15.25, and 15.27. A trivial solution of the first two equations is $y_2 = y_1$, and $V_2 = V_1$, which using the energy equation corresponds to $h_L = 0$. This corresponds to no jump at all. The solution of interest can be found by using Eq. 15.24 to write $V_2 = (y_1/y_2)V_1$, substituting into Eq. 15.25, and rearranging to obtain $(y_2/y_1)^2 + y_2/y_1 - 2(V_1^2/gy_1) = 0$. Writing the upstream Froude number as $Fr_1 = V_1^2/gy_1$, we have $(y_2/y_1)^2 + y_2/y_1 - 2Fr_1^2 = 0$. By using the quadratic formula, we find the two solutions to this equation for the depth ratio to be

$$\frac{y_2}{y_1} = -\frac{1}{2} \pm \frac{1}{2}\sqrt{1 + 8Fr_1^2}$$

First consider the solution with the negative sign: $y_2/y_1 = -\frac{1}{2} - \frac{1}{2}\sqrt{1 + 8Fr_1^2}$. Since this predicts a negative depth ratio, it is physically meaningless. Thus we conclude that the depth ratio across a hydraulic jump is given by taking the positive sign and writing the depth as

$$\frac{y_2}{y_1} = \frac{1}{2}\left(\sqrt{1 + 8Fr_1^2} - 1\right) \tag{15.28}$$

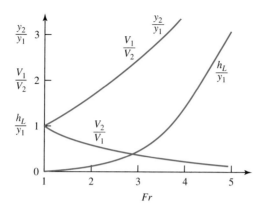

Figure 15.25 Depth ratio, velocity, and head loss across the hydraulic jump as a function of Froude number.

To find the velocity ratio V_1/V_2, we can write the mass balance, Eq. 15.24, as $y_2/y_1 = V_1/V_2$ and substitute this into Eq. 15.28 to obtain

$$\frac{V_1}{V_2} = \frac{1}{2}\left(\sqrt{1 + 8Fr_1^2} - 1\right) \tag{15.29}$$

The curve $y_2/y_1 = V_1/V_2 = \frac{1}{2}(\sqrt{1 + 8Fr_1^2} - 1)$ is plotted in Figure 15.25. Note that depth ratios and Froude numbers less than one are not plotted.

The reason for excluding depth ratios and Froude numbers less than one can be explained as follows. First notice that we can rearrange Eq. 15.28 to obtain

$$Fr_1^2 = \frac{1}{2}\left[\left(\frac{y_2}{y_1}\right)^2 + \frac{y_2}{y_1}\right] \tag{15.30a}$$

Next by substituting $V_2 = (y_1/y_2)V_1$ into the energy balance, Eq. 15.27, we obtain the dimensionless head loss in terms of the depth ratio and upstream Froude number as

$$\frac{h_L}{y_1} = \frac{Fr_1^2}{2}\left[1 - \left(\frac{y_1}{y_2}\right)^2\right] + \left(1 - \frac{y_2}{y_1}\right) \tag{15.30b}$$

Finally by using Eq. 15.30a to eliminate Fr_1^2 in Eq. 15.30b, we obtain

$$\frac{h_L}{y_1} = \frac{1}{4}\frac{(y_2/y_1 - 1)^3}{y_2/y_1} \tag{15.31}$$

Since the head loss must be positive, Eq. 15.31 shows that the depth ratio y_2/y_1 across a hydraulic jump must be greater than one. Equation 15.30a then shows that the Froude number must be greater than one. As a convenience, Figure 15.25 also includes plots of the dimensionless head loss h_L/y_1 and velocity ratio V_2/V_1 as functions of Froude number.

By using a mass, momentum, and energy balance, we have shown that a hydraulic jump can only occur for $Fr > 1$. The hydraulic jump causes an increase in depth, a decrease in velocity, and an accompanying head loss.

Although we know the internal structure of a hydraulic jump is more complicated than a 1D model suggests, the formulas derived here for depth ratio and head loss are

Figure 15.26 Types of hydraulic jump.

Fr	y_2/y_1	Classification	Sketch
<1	1	Jump impossible	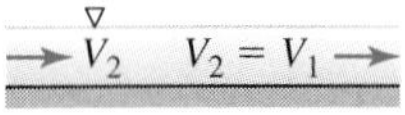
1–1.7	1–2.0	Standing wave or undulant jump	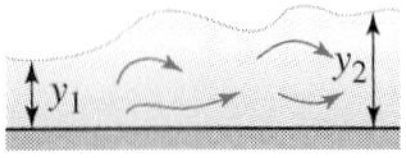
1.7–2.5	2.0–3.1	Weak jump	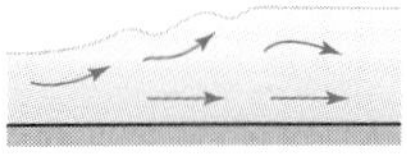
2.5–4.5	3.1–5.9	Oscillating jump	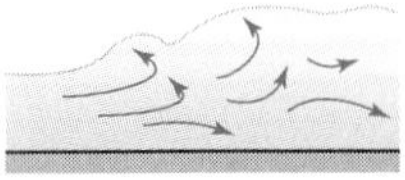
4.5–9.0	5.9–12	Stable, well-balanced steady jump; insensitive to downstream conditions	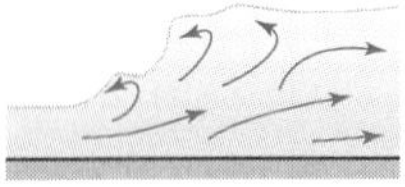
>9.0	>12	Rough, somewhat intermittent strong jump	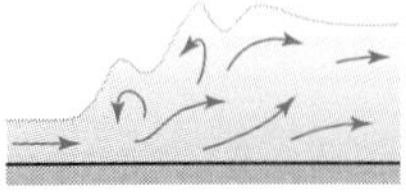

found to be quite accurate. Other features of the hydraulic jump have been determined from experiments. For example, the classification of hydraulic jumps according to the value of the upstream Fr and jump characteristics is illustrated in Figure 15.26. Figure 15.27 shows the experimentally determined jump length L as a function of upstream Fr. Since the jump occurs over such a short length, our analysis, which ignores the slope of the channel (and thus the body force acting in the flow direction), is also applicable to inclined rectangular channels. In an inclined channel the error made in ignoring the body force is proportional to the length of the CV and thus small. Hydraulic jumps also occur in channels of other shapes, and results in such cases may be found in advanced texts.

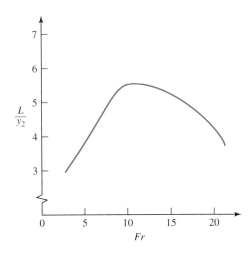

Figure 15.27 Length of hydraulic jump as a function of Froude number.

Figure 15.28 Spillway with obstructions and stilling basin to create a controlled hydraulic jump to dissipate energy as recommended by the U.S. Bureau of Reclamation.

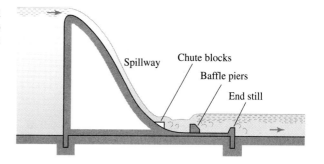

Hydraulic jumps are used to safely dissipate energy in a high speed flow and to prevent streambed erosion. An example of a model hydraulic jump is shown in Figure 15.28. There can be very large normal and shear stresses exerted on the bed underneath the jump itself, so it is important that the material of the bed underneath the jump be able to withstand the severe scouring effects.

EXAMPLE 15.7

The flow of a stream from a steep slope to a horizontal section results in a hydraulic jump. The depth in the horizontal section is 1 m and the velocity is 16 m/s. The stream is 10 m wide. Determine the change in depth, velocity, and Froude number across the jump, and the power dissipated in the jump.

SOLUTION

Figure 15.24 can serve as a sketch for this problem. The first step in solving any hydraulic jump problem is to find the upstream Froude number:

$$Fr_1 = \frac{V_1}{\sqrt{gy_1}} = \frac{16 \text{ m/s}}{\sqrt{(9.81 \text{ m/s}^2)(1 \text{ m})}} = 5.1$$

Then from Eq. 15.28 the depth ratio across the jump is determined to be

$$\frac{y_2}{y_1} = \frac{1}{2}\left(\sqrt{1 + 8Fr_1^2} - 1\right) = \frac{1}{2}\left(\sqrt{1 + 8(5.1)^2} - 1\right) = 6.7$$

Thus, since $y_1 = 1$ m we find $y_2 = 6.7$ m. The velocity ratio is given by $V_1/V_2 = y_2/y_1 = 6.7$, so $V_2 = (16 \text{ m/s})/(6.7) = 2.4$ m/s. The downstream Froude number is now easily calculated to be

$$Fr_2 = \frac{V_2}{\sqrt{gy_2}} = \frac{2.4 \text{ m/s}}{\sqrt{(9.81 \text{ m/s}^2)(6.7 \text{ m})}} = 0.3$$

The power dissipated in the jump is found from Eq. 15.26b: $P = \rho g h_L Q$. We can obtain the head loss by using Eq. 15.30 as follows:

$$\frac{h_L}{y_1} = \frac{1}{4}\frac{[(y_2/y_1) - 1]^3}{y_2/y_1} = \frac{1}{4}\frac{[6.7 - 1]^3}{6.7} = 6.9$$

so $h_L = 6.9 y_1 = 6.9(1\text{ m}) = 6.9\text{ m}$. The power dissipated in the jump is then

$$P = \rho g h_L (w y_1 V_1) = (1000\text{ kg/m}^3)(9.81\text{ m/s}^2)(6.9\text{ m})[(10\text{ m})(1\text{ m})(16\text{ m/s})]$$
$$= 1.1 \times 10^7 \text{ (N-m)/s}$$

This is over 10 megawatts of power. If you are wondering where this enormous amount of energy goes, it is converted into heat. The heat capacity of water is so large however, that the temperature rise is minimal. Perhaps you can estimate the temperature rise?

It has long been noted that open channel flow and high speed compressible flow share some remarkable similarities. Although we will not develop the underlying theoretical basis for this observation here, we can point out that the Mach number of compressible flow is analogous to the Froude number of open channel flow. The shock wave observed in supersonic flow has its counterpart in the hydraulic jump of open channel flow.

15.4 ENERGY CONSERVATION IN OPEN CHANNEL FLOW

In open channel flow the solid walls of the channel apply a shear stress on the moving liquid. The retarding effect of this frictional force is revealed if we perform an energy balance on a steady flow in an inclined prismatic channel. For simplicity, we consider a long, rectangular channel inclined at a small slope as shown in Figure 15.29. In our analysis we will assume that the streamwise velocity is uniform at any given cross section and that the pressure distribution at a cross section is hydrostatic. The depth of the liquid may vary slowly along the channel, thus our analysis applies only to a gradually varying flow (GVF).

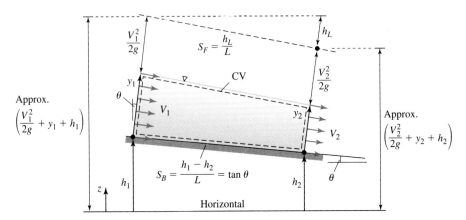

Figure 15.29 A control volume for open channel flow.

15.4 ENERGY CONSERVATION IN OPEN CHANNEL FLOW

Applying a mass balance to the CV shown in Figure 15.29 yields $\dot{M} = \rho Q = \rho A_1 V_1 = \rho A_2 V_2$, where $A_1 = w y_1$ and $A_2 = w y_2$. Thus we can write

$$\frac{Q}{w} = V_1 y_1 = V_2 y_2 \tag{15.32}$$

Next we write an energy balance for this CV by employing Eq. 7.33:

$$\int_{CV} \frac{\partial}{\partial t}\left(\rho\left(u + \frac{1}{2}\mathbf{u}\cdot\mathbf{u} + gz\right)\right) dV + \int_{CS} \rho\left(u + \frac{p}{\rho} + \frac{1}{2}\mathbf{u}\cdot\mathbf{u} + gz\right)(\mathbf{u}\cdot\mathbf{n})\, dS$$
$$= \dot{W}_{\text{power}} + \dot{W}_{\text{shaft}} + \dot{Q}_C + \dot{S}$$

and noting that there is no fluid power, shaft power, or other energy input in this case. The flow is steady, so after the inlet and outlet surfaces of the CV have been identified, the energy balance is given by

$$\int_{\text{inlet}} \rho\left(u + \frac{p}{\rho} + \frac{1}{2}\mathbf{u}\cdot\mathbf{u} + gz\right)(\mathbf{u}\cdot\mathbf{n})\, dS + \int_{\text{outlet}} \rho\left(u + \frac{p}{\rho} + \frac{1}{2}\mathbf{u}\cdot\mathbf{u} + gz\right)(\mathbf{u}\cdot\mathbf{n})\, dS = \dot{Q}_C$$

On the inlet surface, the velocity is V_1, and the uniform internal energy is u_1. The value of z on this surface is $z = h_1 + y\cos\theta$, and the hydrostatic pressure distribution is $p(y) = p_A - \rho g \cos\theta(y - y_1)$. Thus on the inlet we can write

$$\frac{p}{\rho} + gz = \frac{p_A}{\rho} - g\cos\theta(y - y_1) + g(h_1 + y\cos\theta) = \frac{p_A}{\rho} + g(y_1\cos\theta + h_1)$$

which is a constant. At the exit we have V_2, u_2, $z = h_2 + y\cos\theta$ and $p(y) = p_A - \rho g \cos\theta(y - y_2)$. Thus on this surface we obtain $p/\rho + gz = p_A/\rho + g(y_2\cos\theta + h_2)$, also a constant. After we have performed the flux integrals on the two surfaces and divided by the mass flowrate $\dot{M}$, the energy balance becomes

$$-\left[u_1 + \frac{p_A}{\rho} + g(y_1\cos\theta + h_1) + \frac{V_1^2}{2}\right] + \left[u_2 + \frac{p_A}{\rho} + g(y_2\cos\theta + h_2) + \frac{V_2^2}{2}\right] = \frac{\dot{Q}_C}{\dot{M}}$$

The atmospheric pressure terms cancel, and after dividing by g, rearranging, and identifying the head loss per unit weight, we obtain

$$y_1\cos\theta + h_1 + \frac{V_1^2}{2g} = y_2\cos\theta + h_2 + \frac{V_2^2}{2g} + h_L$$

The final step is to note that for the small slopes normally encountered in open channel flow we can use the approximation $\cos\theta \approx 1$. Thus the energy balance is

$$\frac{V_1^2}{2g} + y_1 + h_1 = \frac{V_2^2}{2g} + y_2 + h_2 + h_L \tag{15.33}$$

In open channel flow we define a quantity referred to as the bed slope S_B as

$$S_B = \tan\theta = \frac{h_1 - h_2}{L} \tag{15.34}$$

The value of S_B may be positive or negative, but note that as shown in Figure 15.29, S_B is defined so that if the channel bed elevation decreases in the flow direction, its value is positive. An alternate form of the energy balance may be developed that includes the bed slope by first noting that $h_1 - h_2 = S_B L$. Since the head loss in Eq. 15.33 has units of length, we can also define a dimensionless quantity $S_F = h_L/L$, called the friction slope or slope of the energy line. The value of the friction slope S_F is always positive, since it involves the ratio of two positive-definite quantities. Using these two slopes, we can write the energy balance Eq. 15.33 as

$$\frac{V_1^2}{2g} + y_1 = \frac{V_2^2}{2g} + y_2 + (S_F - S_B)L \tag{15.35}$$

Now consider a rectangular channel of differential length $L = dx$, with the inlet at x and the outlet at $x + dx$. Equation 15.35 can be written for this channel as

$$\left[\frac{V^2}{2g} + y\right]_x = \left[\frac{V^2}{2g} + y\right]_{x+dx} + (S_F - S_B)\,dx$$

Rearranging, dividing by dx, and taking the limit as $dx \to 0$, we obtain the following differential statement of the energy balance for gradually varying flow in a rectangular channel

$$\frac{d}{dx}\left[\frac{V^2}{2g} + y\right] = -(S_F - S_B) \tag{15.36}$$

Here S_F and S_B are interpreted as local values of the friction slope and bed slope, respectively. The complexity of gradually varying open channel flow is revealed if we evaluate the derivative as

$$\frac{d}{dx}\left[\frac{V^2}{2g} + y\right] = \frac{V}{g}\frac{dV}{dx} + \frac{dy}{dx}$$

and write Eq. 15.36 as

$$\frac{V}{g}\frac{dV}{dx} + \frac{dy}{dx} = -(S_F - S_B)$$

Noting that the volume flowrate for a rectangular channel is given by $Q = Vwy$, where w is the constant width of the channel, we have $dV/dx = (d/dx)(Q/wy) = -(Q/wy^2)(dy/dx) = -(V/y)(dy/dx)$. Thus the preceding equation becomes $-(V^2/gy)(dy/dx) + (dy/dx) = -(S_F - S_B)$. Rearranging and introducing the local Froude number we obtain

$$\frac{dy}{dx} = \frac{S_B - S_F}{[1 - Fr^2]} \tag{15.37}$$

We see that the depth slope dy/dx at a given point along a gradually varying flow in a rectangular channel is determined by the local values of the friction slope (always positive), the local bed slope (positive or negative), and whether the local Froude number is less than unity (subcritical) or greater than unity (supercritical).

For a frictionless flow, $S_F = 0$ and Eq. 15.37 becomes $dy/dx = S_B/[1 - Fr^2]$. With a bed described by $z = h(x)$, the value of the bed slope is given by $S_B = -dh/dx$, and Eq. 15.37 becomes $dy/dx = (-dh/dx)/[1 - Fr^2]$. This is identical to Eq. 15.9a, obtained by applying the Bernoulli equation to frictionless flow over a bump or depression in a rectangular channel in Section 15.3.1.

15.4.1 Specific Energy

The energy balance for gradually varying flow in a rectangular channel is given by Eq. 15.35 as $V_1^2/2g + y_1 = V_2^2/2g + y_2 + (S_F - S_B)L$. We see that a change in the value of $[V^2/2g + y]$ at two points along a channel is related to the difference between the friction and bed slopes. It is customary to define the specific energy E in an open channel flow as

$$E = \frac{V^2}{2g} + y \tag{15.38}$$

where E has units of length. Using specific energy, we can write the energy balance as

$$E_2 - E_1 = (S_B - S_F)L \tag{15.39}$$

Note that a positive bed slope (i.e., a bed sloping downward in the flow direction) increases the specific energy, while a negative bed slope decreases the specific energy. Friction always decreases the specific energy.

The specific energy in a rectangular channel flow can be written in terms of the depth by recognizing that since the flowrate Q and width w of the channel are constant, the mass balance as given by Eq. 15.32 can be written as $Q/w = V_1 y_1 = V_2 y_2 = V y$. Thus we have $V = Q/wy$, and using Eq. 15.38 the specific energy is given by

$$E = \frac{Q^2}{2gw^2 y^2} + y \tag{15.40}$$

Since this is a cubic equation in y, for given values of Q and E there are three solutions. We will designate these solutions as y_{neg}, y_{super}, and y_{sub}, where the subscripts are determined by ordering the solutions so that $y_{neg} < y_{super} < y_{sub}$. The solutions y_{super} and y_{sub} are positive, and thus are physically meaningful. The remaining solution y_{neg} is negative, which represents a mathematically valid but physically meaningless result.

Now consider the specific energy diagram shown in Figure 15.30. For zero flowrate, Eq. 15.40 reduces to $E = y$, which can be recognized as the straight line labeled $Q = 0$ in the figure. Each positive value of Q yields a distinct curve, and as shown, each of these curves exhibits a minimum specific energy E_{min}. We can find the depth at which this minimum occurs for each flowrate by setting $dE(y)/dy = (d/dy)[Q^2/2gw^2 y^2 + y] = 0$ and solving for y. The result is called the critical depth y_C for that particular flowrate. It is straightforward to show that the critical depth for a flowrate Q is given by

$$y_C = \left[\frac{Q^2}{gw^2}\right]^{1/3} \tag{15.41a}$$

and that the corresponding minimum specific energy for that flowrate is

$$E_{min} = \tfrac{3}{2} y_C \tag{15.41b}$$

Figure 15.30 Specific energy diagram.

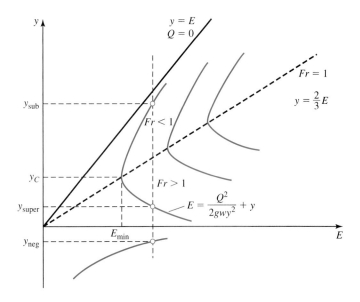

To find the velocity at critical depth we can evaluate Eq. 15.38 at the minimum and obtain $E_{min} = V_C^2/2g + y_C$. Then we can use Eq. 15.41b to write this as $\frac{3}{2}y_C = V_C^2/2g + y_C$ and obtain the critical velocity V_C as

$$V_C = \sqrt{gy_C} \qquad (15.41c)$$

Examining Eq. 15.41c, and recognizing that $Fr = V/\sqrt{gy}$, we conclude that the minimum specific energy for a given flowrate occurs at $Fr = 1$. This explains why we have used the adjective critical to describe the depth and velocity at the point of minimum specific energy for a given flowrate. As Example 15.8 illustrates, the critical depth and critical velocity can be calculated for any open channel flow for which the flowrate and geometric data are known.

EXAMPLE 15.8

Determine the critical depth and flow velocity in a 10 m wide rectangular channel carrying 50 m³/s.

SOLUTION

No sketch is necessary here. We know the flowrate and channel width; thus we can use Eq. 15.41a to calculate the critical depth as

$$y_C = \left[\frac{Q^2}{gw^2}\right]^{1/3} = \left[\frac{(50 \text{ m}^3/\text{s})^2}{(9.81 \text{ m/s}^2)(10 \text{ m})^2}\right]^{1/3} = 1.37 \text{ m}$$

The critical velocity is found from Eq. 15.41c as

$$V_C = \sqrt{gy_C} = \sqrt{(9.81 \text{ m/s}^2)(1.37 \text{ m})} = 3.67 \text{ m/s}$$

15.4 ENERGY CONSERVATION IN OPEN CHANNEL FLOW

Additional features of the specific energy diagram shown in Figure 15.30 are worthy of mention. First we can observe that a line connecting the minimum energy point on each curve must correspond to $Fr = 1$. To find the equation of this line, note that since $V^2/2g = \frac{1}{2}Fr^2 y$, we can write the specific energy as defined by Eq. 15.38 as $E = y(\frac{1}{2}Fr^2 + 1)$. For $Fr = 1$, we find $y = \frac{2}{3}E$. The straight line $y = \frac{2}{3}E$ must therefore pass through the minimum energy point on each curve. This is the dashed curve labeled $Fr = 1$ in Figure 15.30.

Also note that the specific energy curve for each flowrate has an upper and a lower branch. For a certain flowrate, a given specific energy greater than the minimum corresponds to a vertical line that intersects these upper and lower branches. Consider the vertical dashed line in Figure 15.30. The two intersection points correspond to two allowable depths for the flow at this specific energy. The larger depth y_{sub} lies on the upper branch, and the smaller depth y_{super} lies on the lower branch. The velocities corresponding to these two depths may be calculated by noting that since $V = Q/wy$, we can write $V_{\text{sub}} y_{\text{sub}} = Q/w$ and $V_{\text{super}} y_{\text{super}} = Q/w$. Since Q and w are constant, we conclude that $V_{\text{super}}/V_{\text{sub}} = y_{\text{sub}}/y_{\text{super}}$. Then since $y_{\text{super}} < y_{\text{sub}}$, we know $V_{\text{super}} > V_{\text{sub}}$. Thus the flow speed is faster on the lower branch corresponding to a smaller depth. We see that an open channel flow with a given flowrate and specific energy will manifest itself in only two possible ways: at a high speed and shallow depth, or at a lower speed and greater depth. At its minimum specific energy, the only possible manifestation of a flow is at its critical speed and depth.

The value of the Froude number on each branch can be deduced by noting that at depth y, $Fr = V/\sqrt{gy}$. Thus, at the critical point $Fr_C = 1 = V_C/\sqrt{gy_C}$. Forming the ratio of these Froude numbers we obtain $Fr = (V/V_C)(y_C/y)^{1/2}$. Since the flowrate is fixed on each curve, we also know that $Q = Vwy = V_C w y_C$, and thus $V/V_C = y_C/y$. Substituting this into the preceding expression gives

$$Fr = \left(\frac{y_C}{y}\right)^{3/2} \qquad (15.42)$$

Thus if we know the depth at any point on a specific energy curve, we can determine the Froude number at that depth from this equation. On the upper branch we have $y > y_C$, so $Fr < 1$, while on the lower branch we have $y < y_C$, so $Fr > 1$. Thus the flow on the upper branch is subcritical and the flow on the lower branch is supercritical. The two solutions for depth correspond to the possibilities of a subcritical or supercritical flow at a given flowrate and specific energy greater than the minimum.

With this background, we can employ the specific energy diagram shown in Figure 15.30 to make predictions about a flow in a rectangular channel. Our conclusions can be summarized as follows.

1. In subcritical flow, an increase/decrease in specific energy is accompanied by an increase/decrease in depth.

2. In supercritical flow, an increase/decrease in specific energy is accompanied by a decrease/increase in depth.

3. If the flow is critical (i.e., at the point of minimum specific energy), a small increase in specific energy causes a relatively large change in depth.

4. For a subcritical flow to become a supercritical flow, the specific energy must first decrease. The flow must then go through the critical point and along the

The preceding predictions, which were obtained by graphical analysis of Figure 15.30, may be confirmed mathematically. By using Eq. 15.40 to write $dE = d(Q^2/2gw^2y^2 + y)$, completing the derivative, and using the definition of Froude number, we obtain $dE = y[1 - Fr^2]\,dy$. After rearranging, we have $dy = dE/y[1 - Fr^2]$, from which the preceding conclusions can also be derived.

supercritical branch. This process may occur, for example, via a bump or decrease in channel width.

Although our discussion of specific energy has been limited to flow through a rectangular channel, we can also develop results for flow through a prismatic channel whose flow area is A. For this nonrectangular channel, the specific energy is defined as

$$E = \frac{Q^2}{2gA^2} + y \qquad (15.43)$$

In this case A is a known function of y (and the width at the free surface b). The specific energy curve for a given flowrate can now be plotted, and the point of minimum specific energy for this flowrate located by setting $dE/dy = (d/dy)(Q^2/2gA^2 + y) = 0$. Taking the derivative shows that at the critical point, the equation $-(Q^2/gA^3)(dA/dy) + 1 = 0$ is satisfied. Thus, this equation allows us to determine the critical depth y_C. For a rectangular channel we can solve for y_C analytically, since $A = wy = by$. For a nonrectangular channel, we may be able to solve the equation analytically for a regular shape such as those shown in Figure 15.6, but we can always solve the equation numerically if necessary.

Now as shown in Figure 15.31, for a prismatic channel we can write $dA = b\,dy$, thus $dA/dy = b$, and the critical point is defined by $Q^2b_C/gA_C^3 = 1$, where b_C is the width of the free surface at the critical depth. Thus the critical area is given by

$$A_C = \left(\frac{Q^2 b_C}{g}\right)^{1/3} \qquad (15.44a)$$

and, since $Q = AV$, the critical velocity is given by

$$V_C = \sqrt{\frac{gA_C}{b_C}} \qquad (15.44b)$$

Recalling that the hydraulic depth is defined by Eq. 15.2 as $y_H = A/b$, we can write the hydraulic depth at critical conditions, or critical hydraulic depth, as

$$y_{HC} = \frac{A_C}{b_C} \qquad (15.45)$$

Using the critical hydraulic depth, Eq. 15.44b becomes

$$V_C = \sqrt{gy_{HC}} \qquad (15.46)$$

The Froude number in a nonrectangular channel is based on hydraulic depth and is given by $Fr = V/\sqrt{gy_H}$. Thus, at the point of minimum specific energy, where

Figure 15.31 Prismatic channel with area $A(y)$.

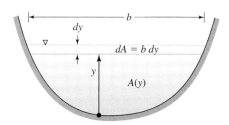

15.4 ENERGY CONSERVATION IN OPEN CHANNEL FLOW

$y_H = y_{HC}$ and $V = V_C = \sqrt{gy_{HC}}$, we have $Fr_C = V_C/\sqrt{gy_{HC}} = 1$. We see for the nonrectangular channel that the Froude number based on hydraulic depth is unity at the minimum on each curve. Further analysis shows that the flow in a nonrectangular channel has the same four characteristics discussed earlier for flow in a rectangular channel. Thus when flow in a prismatic channel is discussed in terms of hydraulic depth, and Fr is based on hydraulic depth, it is similar to flow in a rectangular channel. It is straightforward to show that the all these new formulas reduce to their equivalents for rectangular flow. To prove this result, note that for the rectangular channel $y_{HC} = A_C/b_C = wy_C/w = y_C$.

Finding the critical depth and other critical parameters for a nonrectangular channel is not always easy. Note however, that since A is a known function of w and y (as shown in Figure 15.6 for regular shapes), or known numerically (as in Figure 15.31), we can always determine the critical depth from the fundamental geometry of the channel and the flowrate. Once the critical depth has been found, we can use the formulas to determine other critical parameters, doing this numerically if necessary.

EXAMPLE 15.9

Find the critical depth and critical flow velocity in a 20 ft wide, 45° trapezoidal channel carrying water at the rate of 640 ft³/s. What are the critical hydraulic depth, critical area, and critical free surface width?

SOLUTION

The geometry of the flow channel is shown in Figure 15.32. Using the formula for flow area given in Figure 15.6 for a trapezoidal channel, $A(y) = y(w + y \cot\theta)$, and evaluating this formula for 45°, we have $A(y) = y(w + y)$. We also know $dA/dy = b$, thus we find $b = w + 2y$. At the critical point the depth satisfies $-(Q^2/gA^3)(dA/dy) + 1 = 0$, so y_C is determined by solving

$$\frac{Q^2(w + 2y_C)}{g[y_C(w + y_C)]^3} = 1 \qquad (A)$$

Generally speaking, this equation will need to be solved iteratively or with a symbolic math code. To demonstrate the iterative approach, first note that (A) can be written as

$$\frac{Q^2(w + 2y_C)}{g[y_C(w + y_C)]^3} = \frac{Q^2 w \left(1 + 2\frac{y_C}{w}\right)}{g\left[y_C w \left(1 + \frac{y_C}{w}\right)\right]^3} = 1$$

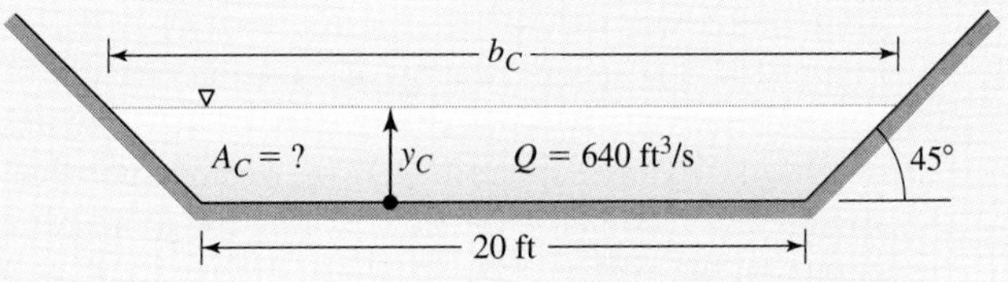

Figure 15.32 Schematic for Example 15.9.

Rearranging, and solving for y_C^3 we have

$$y_C^3 = \frac{Q^2\left(1+2\frac{y_C}{w}\right)}{gw^2\left(1+\frac{y_C}{w}\right)^3} = \frac{Q^2}{gw^2}\left(1+2\frac{y_C}{w}\right)\left(1+\frac{y_C}{w}\right)^{-3} \tag{B}$$

For $y_C \ll w$, this equation can be approximated as $y_C^3 \approx Q^2/gw^2$. Thus we can obtain a first estimate for y_C by writing

$$y_{C,1} = \left(\frac{Q^2}{gw^2}\right)^{1/3}$$

The iteration converges when (A) or equivalently (B) is satisfied. In either case the result for the critical depth is $y_C = 3.0$ ft.
Since we know that $A(y) = y(w+y)$ and $b = w+2y$, the critical flow area and critical free surface width are found to be

$$A_C = y_C(w+y_C) = 3.0 \text{ ft}(20 \text{ ft} + 3.0 \text{ ft}) = 69 \text{ ft}^2$$

and

$$b_C = w + 2y_C = 20 \text{ ft} + 2(3.0 \text{ ft}) = 26 \text{ ft}$$

Also, since $Q = AV$, the critical velocity is given by $V_C = Q/A_C = 640 \text{ ft}^3/\text{s}/69 \text{ ft}^2 = 9.28$ ft/s. Finally, the critical hydraulic depth is found by inserting the data into Eq. 15.45: $y_{HC} = A_C/b_C = 69 \text{ ft}^2/26 \text{ ft} = 2.65$ ft.

15.4.2 Specific Energy Diagrams

The concepts of specific energy and supercritical and subcritical flow can be helpful in understanding and solving a variety of open channel flow problems. For example, suppose we revisit the hydraulic jump illustrated in Figure 15.24. The energy balance for a hydraulic jump was given by Eq. 15.27 as $V_1^2/2g + y_1 = V_2^2/2g + y_2 + h_L$. Writing this in terms of specific energy $E = V^2/2g + y$, we have $E_1 = E_2 + h_L$. Thus a supercritical flow enters the jump with a specific energy E_1 and leaves the jump at a lower specific energy $E_2 = E_1 - h_L$. When written in terms of volume flowrate, $Q = Vwy$, the energy balance for a hydraulic jump is given by $Q^2/2gw^2y_1^2 + y_1 = Q^2/2gw^2y_2^2 + y_2 + h_L$, where the specific energy is $E = Q^2/2gw^2y^2 + y$. Thus we can draw the specific energy diagram for a hydraulic jump as illustrated in Figure 15.33.

We know that the flow approaches the jump with specific energy E_1 and leaves the jump at a lower specific energy E_2; i.e., $E_2 = E_1 - h_L$. If we draw a vertical line in Figure 15.33 at specific energy E_1, the intersection of this line with the specific energy curve defines the two possible depths of the flow upstream. Since a hydraulic jump is observed only in a supercritical flow, we conclude that point 1 on this line defines the upstream flow. Similarly, the flow exiting the jump can be at only one of the two possible points on the vertical line shown in Figure 15.33 at specific energy E_2. In a hydraulic jump the flow downstream is known to be subcritical; hence the downstream flow must be at point 2. The remaining question is, How did the flow get from point 1 to point 2?

15.4 ENERGY CONSERVATION IN OPEN CHANNEL FLOW

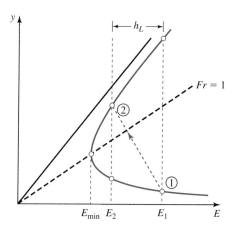

Figure 15.33 Specific energy curve for a hydraulic jump.

EXAMPLE 15.10

Water flows over a 0.25 m bump in a constant width, horizontal rectangular channel as shown in Figure 15.34A. The upstream depth and velocity are 0.5 m and 0.2 m/s, respectively. Draw the specific energy diagram for this flow and find the depth and velocity over the bump at point B. Then find the depth and velocity downstream of the bump at point 2. Neglect the effect of friction.

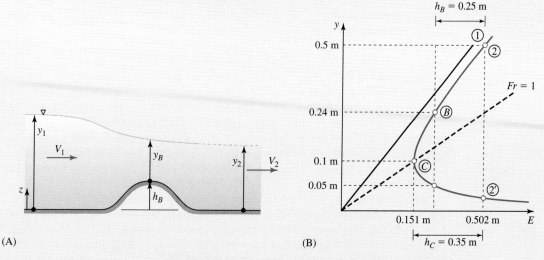

Figure 15.34 (A) Schematic for Example 15.10. (B) Specific energy diagram.

SOLUTION

We first calculate the upstream Froude number for the specified conditions as $Fr_1 = V_1/\sqrt{gy_1} = 0.2 \text{ m/s}/\sqrt{9.81 \text{ m/s}^2(0.5 \text{ m})} = 0.09$. Thus the flow is subcritical and we expect the depth to decrease over the bump (see Figure 15.14A). The specific energy for flow in a rectangular channel is given by Eq. 15.38 as $E = V^2/2g + y$. Substituting $V = V_1 y_1/y$, we find $E = V_1^2 y_1^2/2gy^2 + y$, inserting the data yields $E = (5.09 \times 10^{-4} \text{ m}^3)/y^2 + y$. This equation forms the basis for the specific energy diagram as shown in Figure 15.34B. Various points on this curve can now be identified. Upstream at point 1 we have $V_1 = 0.2$ m/s and $y_1 = 0.5$ m, thus we can calculate $E_1 = V_1^2/2g + y_1 = (0.2 \text{ m/s})^2/2(9.81 \text{ m/s}^2) + (0.5 \text{ m}) = 0.502$ m. The critical point C can be located by substituting $Q = V_1 y_1 w$ into Eq. 15.41a, $y_C = [Q^2/gw^2]^{1/3}$, to obtain $y_C = [V_1^2 y_1^2/g]^{1/3}$. Inserting the data, we find

$$y_C = \left[\frac{(0.2 \text{ m/s})^2 (0.5 \text{ m})^2}{9.81 \text{ m/s}^2}\right]^{1/3} = 0.1006 \text{ m}.$$

Next we use Eq. 15.41c to compute the velocity at the critical point:

$$V_C = \sqrt{gy_C} = \sqrt{(9.81 \text{ m/s}^2)(0.1006 \text{ m})} = 0.993 \text{ m/s}$$

Finally we use Eq. 15.41b to write $E_{\min} = \frac{3}{2} y_C = \frac{3}{2}(0.1006 \text{ m}) = 0.151$ m.

To calculate the depth and velocity over the bump and downstream, recall that in Section 15.3.1 we analyzed frictionless flow over a bump and obtained Eq. 15.6:

$$\frac{V_1^2}{2g} + y_1 + h(x_1) = \frac{V(x)^2}{2g} + y(x) + h(x)$$

Applying this equation between point 1 and any point downstream, we have $V_1^2/2g + y_1 + h_1 = V^2/2g + y + h$. Introducing the specific energy and noting that with the choice of horizontal datum in Figure 15.34A $h_1 = 0$, this equation becomes $E_1 = E + h$. For calculations we can write this as

$$E = E_1 - h \tag{A}$$

For the bump we know $h_B = 0.25$ m, thus $E_B = E_1 - h_B = 0.502$ m $- 0.25$ m $= 0.252$ m. Drawing a vertical line at this value of specific energy gives us two possible depths over the bump, as expected. Before we find these two depths, note that we can also use (A) to calculate the critical bump height h_C, the bump height needed to make the flow critical. If we write (A) as $h_C = E_1 - E_{\min}$, we find $h_C = E_1 - E_{\min} = 0.502$ m $- 0.151$ m $= 0.35$ m. We see immediately that the flow must remain subcritical, since the bump is only 0.25 m high, a value less than the 0.35 m needed for the flow to become critical.

To find the depth y_B over our 0.25 m bump, we make use of the equation of the specific energy diagram $E = (5.09 \times 10^{-4} \text{ m}^3)/y^2 + y$ and write $E_B = 0.252$ m $= (5.09 \times 10^{-4} \text{ m}^3)/y_B^2 + y_B$. This yields the cubic equation $y_B^3 - 0.252$ m $y_B^2 + 5.09 \times 10^{-4}$ m$^3 = 0$. The resulting three solutions are $y_B = -0.04$ m, $+0.05$ m, and $+0.24$ m. However, we already know from the specific energy diagram that the only one of interest to us must lie between $y_1 = 0.5$ m and $y_C = 0.1$ m. Thus the solution of interest is $y_B = 0.24$ m. The velocity over the bump can now be calculated as

$$V_B = \frac{V_1 y_1}{y_B} = \frac{(0.2 \text{ m/s})(0.5 \text{ m})}{(0.24 \text{ m})} = 0.42 \text{ m/s}$$

To find the depth and velocity at point 2 downstream, we use (A) to write $E_2 = E_1 - h_2$. Since $h_2 = 0$ we have $E_2 = E_1 = 0.502$ m. The specific energy at point 2 is given by $E_2 = 0.502 \text{ m} = (5.09 \times 10^{-4} \text{ m}^3)/y_2^2 + y_2$, which we can write as the cubic equation $y_2^3 - 0.502 \text{ m } y_2^2 + 5.09 \times 10^{-4} \text{ m}^3 = 0$. One solution is known immediately: $y_2 = y_1 = 0.5$ m, corresponding to a subcritical flow identical to that upstream of the bump. The other two solutions are $+0.032$ m, corresponding to a supercritical flow, and -0.030 m, which is meaningless. The flow cannot become supercritical after the bump unless it has gone through the critical point. Since the bump height of 0.25 m is less than the critical bump height of 0.35 m calculated earlier, this could not have happened. Thus we conclude that at point 2, the depth and velocity are the same as that upstream.

Before leaving this example, we could ask ourselves what we would observe if the upstream conditions were fixed and the bump height was increased to exactly the critical height $h_C = 0.35$ m. The answer is found by looking at the specific energy diagram of Figure 15.34B and realizing that we would move along the upper branch of the specific energy curve from point 1 to point C. The flow would speed up going up the bump and become critical at the crest of the bump. What would happen as the flow went down the bump? The specific energy diagram reveals we could move back along the upper branch to the initial subcritical state (points 1 and 2) or move along the supercritical branch to point 2′. This point corresponds to a supercritical flow. Conditions far downstream determine which of these two possibilities occurs.

We could also ask what would happen if the upstream conditions were fixed and the bump height increased beyond the critical height of 0.35 m. The answer to this question is interesting, since now point B would not be on the specific energy curve corresponding to the upstream conditions. Thus this flow is impossible with the upstream depth and flowrate fixed. If the flowrate is fixed, and the flow at the crest is critical when the bump height is slowly increased, then the flow at the crest will remain critical and the upstream depth will increase as needed.

The control volume analysis presented here cannot give us the specific path because of the complexity of this 3D flow field (we would need CFD to model this flow more completely). Therefore, we represent the path schematically by a dashed line in Figure 15.33.

15.5 FLOW IN A CHANNEL OF UNIFORM DEPTH

In a steady, fully developed flow in a prismatic channel, the liquid depth is constant, and depth slope is $dy/dx = 0$. Thus the free surface is parallel to the bed as shown in Figure 15.35. This type of open channel flow is referred to as uniform flow or flow at normal depth. The latter name reflects the fact that the depth of a uniform flow is referred to as the normal depth y_N. Channels are often designed to have uniform flow, so understanding their characteristics is important.

We can analyze a uniform flow by using mass, momentum, and energy balances, and noting that the flow is fully developed and steady. Applying a mass balance to the

Figure 15.35 Uniform depth channel and control volume.

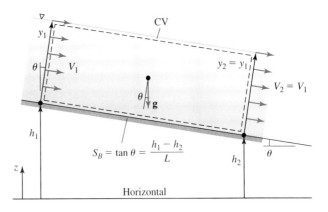

CV shown in Figure 15.35, we have $\dot{M} = \rho Q = \rho A_1 V_1 = \rho A_2 V_2$. For a prismatic channel the flow area A is constant; thus

$$V_1 = V_2 \tag{15.47}$$

The energy balance for flow of gradually varying depth in a rectangular channel was given by Eq. 15.35 as $V_1^2/2g + y_1 = V_2^2/2g + y_2 + (S_F - S_B)L$. This result also applies to gradually varying or uniform flow in any prismatic channel. Since in uniform flow $y_1 = y_2 = y_N$ and $V_1 = V_2$, the energy balance reduces to

$$S_F = S_B \tag{15.48a}$$

We see that in a flow at uniform depth, the friction slope and bed slope are the same. According to Eq. 15.34 we have $S_B = \tan\theta = (h_1 - h_2)/L$, and we know that the head loss is defined by $h_L = LS_F$. Thus we can write $S_F = h_L/L$ and $S_B = (h_1 - h_2)/L$, and substitute these into Eq. 15.48a to obtain

$$h_L = S_F L = S_B L = h_1 - h_2 \tag{15.48b}$$

We see that since the head loss is positive, the friction slope and bed slope are also positive. Thus flow at uniform depth can occur only in a downward-sloping channel. Finally, note that in a uniform flow, gravitational potential energy is continuously lost as a result of the action of viscous shear stresses. The energy loss per unit volume is $\rho g h_L = \rho g S_B L = \rho g (h_1 - h_2)$, and the rate at which energy is dissipated in a uniform flow is then

$$P = \rho g(h_1 - h_2)Q \tag{15.48c}$$

To determine the momentum balance for the CV shown in Figure 15.35, note that since the flow is fully developed and the momentum flux at the inlet is equal in magnitude and opposite in sign to that at the outlet, the momentum balance reduces to $\mathbf{F}_B + \mathbf{F}_S = 0$. That is, the sum of the body and surface forces acting on the CV equals zero. The body force acting on the CV due to gravity is seen by inspection to be $\rho g \mathdsVol$ and acts vertically downward. Writing $\mathbf{F}_B$ in terms of the flow area A and length L of the CV, we see that the body force component acting in the flow direction is $\rho g A L \sin\theta$. The momentum balance shows that this force must be balanced by the sum of pressure acting on each end of the CV and the shear force exerted on the CV at its wetted perimeter. Since the bed slope is assumed to be small, the pressure distribution may be assumed to be hydrostatic throughout the CV. Thus for constant depth, the pressure forces on each

15.5 FLOW IN A CHANNEL OF UNIFORM DEPTH

end of the CV cancel one another. The shear force, which acts opposite to the flow direction, can be written as $\bar{\tau}_W PL$, where $\bar{\tau}_W$ is the average shear stress acting at the channel walls, and P is the wetted perimeter. Thus the momentum balance gives $\bar{\tau}_W PL = \rho g AL \sin\theta$, and we may solve for the average wall shear stress to obtain $\bar{\tau}_W = \rho g (A/P) \sin\theta$. Next we use the hydraulic radius, given by Eq. 15.1 as $R_H = A/P$, to write $\bar{\tau}_W = \rho g R_H \sin\theta$. The final step in this analysis is to write the bed slope as $S_B = \tan\theta = \sin\theta/\cos\theta$, and approximate this for the small slopes of interest as $S_B \approx \sin\theta$. Thus the momentum balance for uniform flow can be written as

$$\bar{\tau}_W = \rho g R_H S_B \tag{15.49}$$

EXAMPLE 15.11

The normal depth in a 6 m wide rectangular irrigation channel is 0.5 m. If the volume flowrate is 0.75 m³/s, and the bed slope is 0.002, as shown in Figure 15.36, find the head loss and power dissipation per kilometer of channel. What are the average velocity, Froude number, and average wall shear stress in the channel?

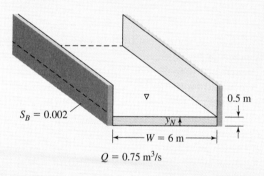

Figure 15.36 Schematic for Example 15.11.

SOLUTION

From $S_B = \tan\theta = (h_1 - h_2)/L$, which is Eq. 15.34, the drop in elevation per kilometer of channel is found to be $h_1 - h_2 = S_B L = 0.002(1 \text{ km})(1000 \text{ m}/1 \text{ km}) = 2$ m. The head loss is given by Eq. 15.48b as $h_L = h_1 - h_2$, so the head loss is $h_L = 2$ m. The power dissipation in a kilometer of channel is calculated by using Eq. 15.48c and found to be

$$P = \rho g (h_1 - h_2) Q = 1000 \text{ kg/m}^3 (9.81 \text{ m/s}^2)(2 \text{ m})(0.75 \text{ m}^3/\text{s})$$
$$= 1.47 \times 10^4 \text{ (N-m)/s} \approx 15 \text{ kW}$$

Next we calculate the average velocity as $V = Q/A = (0.75 \text{ m}^3/\text{s})/[(6 \text{ m})(0.5 \text{ m})] = 0.25$ m/s, and $Fr = V/\sqrt{gy_N} = (0.25 \text{ m/s})/\sqrt{(9.81 \text{ m/s}^2)(0.5 \text{ m})} = 0.11$. Note that this flow is subcritical. The average wall shear stress is calculated by Eq. 15.49: $\bar{\tau}_W = \rho g R_H S_B$. Using $R_H = A/P = [(6 \text{ m})(0.5 \text{ m})]/(6 \text{ m} + 1 \text{ m}) = 0.429$ m, we find

$$\bar{\tau}_W = \rho g R_H S_B = (1000 \text{ kg/m}^3)(9.81 \text{ m/s}^2)(0.429 \text{ m})(0.002) = 8.42 \text{ N/m}^2$$

You may have noticed that the preceding results are not sufficient to allow us to design a uniform flow to carry a certain flowrate. The reason is that we do not know the head loss as a function of the various parameters. We encountered a similar situation in analyzing pipe flow, where it proved necessary to introduce an empirical model for the head loss in the turbulent flows of engineering interest via a friction factor. The same basic approach is employed in open channel flow. Thus using f as the friction factor, and recalling that head loss in open channel flow has units of feet, we write the head loss for open channel flow as

$$h_L = f \frac{L}{4R_H} \frac{V^2}{2g} \tag{15.50}$$

Note the use of hydraulic radius rather than hydraulic diameter in this formula.

Now according to Eq. 15.48b, $h_L = LS_B$. Inserting this into Eq. 15.50 and solving for the velocity we obtain

$$V = \sqrt{\frac{8g}{f}} \sqrt{R_H S_B}$$

The friction factor for open channel flow depends on the roughness of the channel walls, but normally it is independent of Reynolds number. If we let $C = \sqrt{8g/f}$, the preceding equation becomes

$$V = C\sqrt{R_H S_B} \tag{15.51}$$

This is known as the Chezy equation, and the unknown term C that now contains the friction factor is called the Chezy coefficient. This coefficient has the dimensions of $\sqrt{g}$. Manning performed a series of experiments to determine empirical values for C for channels of various types and proposed that C be written as $C = R_H^{1/6}/n$, where n is called the Manning roughness coefficient. Thus the velocity in uniform open channel flow is given by the Manning equation

$$V = \frac{R_H^{2/3} S_B^{1/2}}{n} \tag{15.52a}$$

where values for n may be obtained from Table 15.1. The flowrate is then given by

$$Q = \frac{A R_H^{2/3} S_B^{1/2}}{n} \tag{15.52b}$$

Notice that the values of n shown in Table 15.1 do not have a unit indicated despite the ease with which we could demonstrate that the unit of n is time divided by length to the one-third power. This is because the Manning equation was developed exclusively with SI units. To avoid any confusion and to allow us to determine velocity and flowrate in both SI and BG units, we will use the following equations instead:

$$V = \frac{C_0}{n} R_H^{2/3} S_B^{1/2} \quad \text{and} \quad Q = \frac{C_0}{n} A R_H^{2/3} S_B^{1/2} \tag{15.53a, b}$$

The values of n to be used with Eqs. 15.53 are the original Manning coefficients given in Table 15.1, but we will use the factor $C_0 = 1$ for SI, and $C_0 = 1.49$ for BG.

In the uniform flow described in Example 15.11 $Fr = 0.11$, hence the flow is subcritical. Depending on conditions, including the wall roughness and bed slope, a

The values of the Manning roughness coefficient in Table 15.1 were determined from experiments conducted in SI units. For a BG calculation it is convenient to be able to use the same coefficients, introducing the factor $C_0 = 1.49$ as a correction for the change in units. But where did the value 1.49 come from? Note that $3.281 \text{ ft} = 1 \text{ m}$ and $(3.281)^{1/3} = 1.49$. Thus, if we have a value of n from the Table 15.1 in SI units of $s/m^{\frac{1}{3}}$ and we wish to convert it into BG units of $s/ft^{\frac{1}{3}}$ we would write $n_{BG} = n_{SI}(1 \text{ m}/3.281 \text{ ft})^{1/3}$. This is the justification for Eqs. 15.53.

uniform open channel flow can be subcritical, critical, or supercritical. If the flow is critical, then the normal depth y_N must be equal to the critical hydraulic depth $y_{HC} = A_C/b_C$, and we also know $V_C = \sqrt{gy_{HC}}$ (Eq. 15.46). Solving Eq. 15.53b for the critical bed slope needed to produce this velocity we have

$$S_{BC} = \frac{V_C^2 n^2}{C_0 R_{HC}^{4/3}} \quad (15.54a)$$

which after substituting $V_C = \sqrt{gy_{HC}}$, we can write as

$$S_{BC} = \frac{gy_{HC} n^2}{C_0 R_{HC}^{4/3}} \quad (15.54b)$$

TABLE 15.1 Values of the Manning Roughness Coefficient, n

Wetted Perimeter	n
A. Natural channels	
Clean and straight	0.030
Sluggish with deep pools	0.040
Major rivers	0.035
B. Floodplains	
Pasture, farmland	0.035
Light brush	0.050
Heavy brush	0.075
Trees	0.15
C. Excavated earth channels	
Clean	0.022
Gravelly	0.025
Weedy	0.030
Stony, cobbles	0.035
D. Artificially lined channels	
Glass	0.010
Brass	0.011
Steel, smooth	0.012
Steel, painted	0.014
Steel, riveted	0.015
Cast iron	0.013
Concrete, finished	0.012
Concrete, unfinished	0.014
Planed wood	0.012
Clay tile	0.014
Brick work	0.015
Asphalt	0.016
Corrugated metal	0.022
Rubble masonry	0.025

HISTORY BOX 15-1

Antoine Chezy (1718–1798) was a French engineer who developed these formulas based on experiments in the Seine River and the Courpalet canal. He was a student at L'École des Ponts et Chaussés (School of Bridges and Roads), considered to be the first engineering college, where much of the fluid dynamics we use today was first developed in the eighteenth and nineteenth centuries. Robert Manning (1816–1897) made his contribution based on data he obtained as the chief engineer at the Office of Public Works in Ireland.

or its equivalent

$$S_{BC} = \frac{gA_C n^2}{C_0 b_c R_{HC}^{4/3}} \tag{15.54c}$$

The bed slope of a particular uniform flow is classified according to the following criteria:

1. For $y_N > y_{HC}$, the slope must be $S_B < S_{BC}$, and the slope is called mild or subcritical.
2. For $y_N = y_{HC}$, the slope must be $S_B = S_{BC}$, and the slope is called critical.
3. For $y_N < y_{HC}$, the slope must be $S_B > S_{BC}$, and the slope is called steep or supercritical.

For the important case of a very wide 2D rectangular channel of free surface width $w = b$, Figure 15.6 shows that $R_H = y_H = y$. Thus $R_{HC} = y_{HC} = y_C$, and from Eq. 15.54b the critical slope is

$$S_{BC} = \frac{gn^2}{C_0 y_C^{1/3}} \tag{15.55}$$

For example, consider a critical uniform flow in a wide rectangular channel of unfinished concrete at a normal depth of 0.5 ft. We have $n = 0.014$ from Table 15.1, and in BG we take $C_0 = 1.49$. The critical slope is found to be

$$S_{BC} = \frac{(32.2 \text{ ft/s}^2)(0.014)^2}{1.49(0.5 \text{ ft})^{1/3}} = 0.0053$$

To put this bed slope into perspective, note that this would be a drop of 28 ft per mile, or 5.3 m per kilometer. A bed slope less than this is mild, and one greater than this is steep. However, if the same channel had walls of unfinished earth, then $n = 0.022$, and the critical slope is found to be

$$S_{BC} = \frac{(32.2 \text{ ft/s}^2)(0.022)^2}{1.49(0.5 \text{ ft})^{1/3}} = 0.013$$

which is steeper by a factor of 2.47 owing to the greater frictional resistance of the earthen walls.

15.5.1 Uniform Flow Examples

A uniform flow may be characterized by the size and shape of the channel, the normal depth, the slope of the channel bottom, the roughness of the walls, and either the average velocity or flowrate. Thus a problem may require that we find any of these variables. In addition, we may think of a problem as involving analysis or design. In an analysis problem, we would typically be given information regarding the channel size, shape, material, and slope, and we would be asked to find flowrate for a certain normal depth or vice versa. In a design problem, we might be asked to determine the slope of a

15.5 FLOW IN A CHANNEL OF UNIFORM DEPTH

The slope of a channel can be described in various ways. According to Eq. 15.34, the bed slope is given by $S_B = \tan\theta = (h_1 - h_2)/L$. Thus a channel with a bed slope of $S_B = 0.003$ could be described as having a drop of 3 m/km or 16 ft/mile. But this channel could also be said to have a slope of $0.17°$ or as having a slope of 0.003 radian.

channel for a case in which the desired shape, flowrate, and normal depth are specified. The examples in this section include problems of both types. We begin with several analysis problems, the first of which asks us to determine flowrate.

In the problem discussed in Example 15.12 the normal depth is given and no iteration is required to determine the flowrate. As Example 15.13 illustrates, iteration is required if an analysis problem asks for the normal depth.

EXAMPLE 15.12

A 12 in. diameter, round cast iron drain pipe is laid on a 0.2° slope. Assuming uniform flow, find the flowrate if the pipe is one-quarter, one-half, three-quarters, and completely full. What are the velocity and Froude number for each of these flows?

SOLUTION

The pipe is sketched in Figure 15.37. From Table 15.1 the Manning coefficient for cast iron is $n = 0.013$. Since we know that the slope of the channel is 0.2°, we can use Eq. 15.34 to calculate the bed slope as $S_B = \tan\theta = \tan 0.2° = 0.00349$. Using the formulas in Figure 15.6 for a round pipe we can first calculate $\theta = 2\cos^{-1}(1 - 2y/D)$, where y is the depth, and $D = 1$ ft. Next we calculate the flow area $A(y) = (D^2/8)(\theta - \sin\theta)$, wetted perimeter $P(y) = D\theta/2$, and hydraulic radius $R_H(y) = (D/4)(1 - \sin\theta/\theta)$. The results are shown in Table 15.2. The volume flowrate is calculated next, from Eq. 15.53b: $Q = (C_0/n)AR_H^{2/3}S_B^{1/2}$, with $C_0 = 1.49$, since we are working in BG units. Velocity can then be obtained from $V = Q/A$, and we can calculate $Fr = V/\sqrt{gy}$. The results are included in Table 15.2.

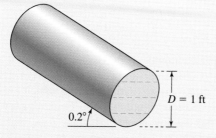

Figure 15.37 Schematic for Example 15.12.

TABLE 15.2 Solution Table for Example 15.12

y_N (ft)	θ_N^{rad}	$A(y)$ ft²	$P(y)$ ft	$R_H(y)$ ft	$R_H^{2/3}$	Q ft³/s	V ft/s	Fr
0.25	2.094	0.153	1.05	0.146	0.277	0.29	1.90	0.67
0.5	3.14	0.393	1.57	0.250	0.397	1.06	2.70	0.67
0.75	4.19	0.632	2.10	0.302	0.450	1.93	3.05	0.62
1.0	6.28	0.785	3.14	0.250	0.397	2.11	2.69	0.47

EXAMPLE 15.13

A culvert made of unfinished concrete and measuring 2 m × 2 m, as shown in Figure 15.38, carries a small stream. If the culvert drops 0.5 m over its 100 m length, and is designed for uniform flow, what is the expected normal depth when the culvert is carrying a flow of 4.7 m³/s?

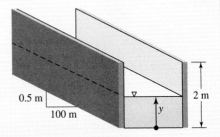

Figure 15.38 Schematic for Example 15.13.

SOLUTION

The culvert is sketched in Figure 15.38. From Table 15.1 the Manning coefficient for unfinished concrete is $n = 0.014$. Since we know that the slope of the channel is described in terms of an elevation change over its length, we can use Eq. 15.34 to calculate the bed slope as $S_B = (h_1 - h_2)/L = 0.5/100 = 0.005$. The volume flowrate is given by Eq. 15.53b, $Q = (C_0/n)AR_H^{2/3}S_B^{1/2}$, with $C_0 = 1$, since we are working in SI units. Next we use Figure 15.6 to write the formulas $A(y) = wy$ and $R_H(y) = wy/(w+2y)$. Inserting these into the formula for Q and rearranging yields

$$(wy)\left(\frac{wy}{w+2y}\right)^{2/3} = \frac{nQ}{S_B^{1/2}} \quad (A)$$

The desired depth satisfies this equation. Inserting $Q = 4.7$ m³/s, $n = 0.014$, $S_B = 0.005$, and $w = 2$ m, we obtain

$$2y\left(\frac{2y}{2+2y}\right)^{2/3} = 0.931$$

as the equation to be solved. Using a symbolic math code, we find $y = 0.8$ m.

We see that a flow of 4.7 m³/s in this culvert occurs at a depth of 0.8 m. The flow area in that case is $A = wy = 2$ m$(0.8$ m$) = 1.6$ m², and the velocity is $V = Q/A = 4.7$ m³/s/1.6 m² $= 2.94$ m/s. The Froude number is $Fr = V/\sqrt{gy} = 2.94$ m/s/$\sqrt{(9.81 \text{ m/s}^2)(0.8 \text{ m})} = 1.05$. Thus the flow is supercritical.

Note that to iterate this problem by hand we can recognize that the answer must be a depth of 2 m or less. Several iterations are shown:

for $y = 0.6$:

$$2y\left(\frac{2y}{2+2y}\right)^{2/3} = 0.624$$

for $y = 0.9$:

$$2y\left(\frac{2y}{2+2y}\right)^{2/3} = 1.33$$

for $y = 0.7$:
$$2y\left(\frac{2y}{2+2y}\right)^{2/3} = 0.77$$

for $y = 0.8$:
$$2y\left(\frac{2y}{2+2y}\right)^{2/3} = 0.932$$

At times we may encounter a channel shape that is not standard or a channel whose roughness is different on different walls. Example 15.14 illustrates how we can solve a problem of this type.

In our final example, the slope of the channel is unknown, but the size and shape of the channel is specified. This could be considered to be a design problem.

EXAMPLE 15.14

A trapezoidal flood control channel has a relatively clean bottom but weedy sides, as shown in Figure 15.39A. Find an expression for the equivalent Manning coefficient n_E for this channel.

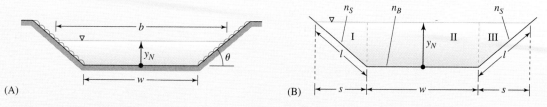

Figure 15.39 (A) Schematic for Example 15.14. (B) Decomposition of channel for Example 15.14.

SOLUTION

We can think of this channel first as a whole and suppose that it has an equivalent Manning coefficient n_E, a flow area A_E, and a hydraulic radius R_{HE}. From the geometry in Figure 15.39A and the formulas in Figure 15.6 for a trapezoidal channel at normal depth y_N, we have

$$A_E = y_N(w + y_N \cot\theta) \quad \text{and} \quad R_{HE} = \frac{y_N(w + y_N \cot\theta)}{w + 2y_N/\sin\theta}$$

By using Eq. 15.53b, $Q = (C_0/n)AR_H^{2/3}S_B^{1/2}$, we can write the volume flowrate for the whole channel as

$$Q_E = \frac{C_0}{n_E}A_E R_{HE}^{2/3} S_B^{1/2} \tag{A}$$

The velocity in the channel is $V_E = Q_E/A_E$.

Next we can think of the channel as if it were three separate channels, as indicated in Figure 15.39B, with the same velocity V_E in each of them. The flowrates in the three channels are then seen to be $Q_I = V_E A_I$, $Q_{II} = V_E A_{II}$, and $Q_{III} = V_E A_{III}$. The total volume flowrate of the whole channel must then be given by the sum of the flows occurring in the three subsections. We can write this as

$$Q_E = Q_I + Q_{II} + Q_{III} \tag{B}$$

Finally we will assume that the volume flowrate in each of the subsections may be calculated separately as a uniform open channel flow in each subsection, with the flow corresponding to the characteristics of the particular subsection.

Two of these channels are triangular and can be seen to have a single wetted side of length $l = y_N/\sin\theta$ and a free surface width $s = y_N/\tan\theta$. Thus they have a flow area $A_I = A_{III} = y_N^2/(2\tan\theta)$ and a wetted perimeter of $P_I = P_{III} = l = y_N/\sin\theta$. The hydraulic radius of each of the triangular channels is $R_{HI} = R_{HIII} = \frac{1}{2}y_N\cos\theta$. By using Eq. 15.53b, and writing the Manning coefficient for the weedy sides as n_S, we can give the following volume flowrate in each of these channels:

$$Q_I = Q_{III} = \frac{C_0}{n_S}A_I R_{HI}^{2/3} S_B^{1/2} \tag{C}$$

The central channel can be seen to have a flow area $A_{II} = wy_N$, wetted perimeter $P_{II} = w$, and thus a hydraulic radius $R_{HII} = y_N$. Again from Eq. 15.53b, and writing the Manning coefficient for the clean bottom as n_B, we find the volume flowrate in the central section:

$$Q_{II} = \frac{C_0}{n_B}A_{II} R_{HII}^{2/3} S_B^{1/2} \tag{D}$$

Inserting (A), (C), and (D) into (B) and simplifying, we obtain

$$\frac{1}{n_E}\left(A_E R_{HE}^{2/3}\right) = \frac{2}{n_S}\left(A_I R_{HI}^{2/3}\right) + \frac{1}{n_B}\left(A_{II} R_{HII}^{2/3}\right) \tag{E}$$

This equation allows us to calculate the equivalent Manning coefficient for the entire channel if we know the parameters describing the three subsections. After finding n_E, we can use (A) to calculate the flowrate in the whole channel. It is interesting to note that (E) is similar to the formula for the equivalent electrical resistance of three resistors in parallel. We see that in uniform open channel flow, Eq. 15.53b, $Q = (C_0/n)AR_H^{2/3}S_B^{1/2}$, can be thought of as equivalent to Ohm's law $I = V/R$, with the volume flowrate, Q, equivalent to the current I, the square root of the bed slope $S_B^{1/2}$, acting as the driving voltage V, and the resistance R equivalent to the combination $n/(AR_H^{2/3})$. Does it make sense to you that the resistance of a channel to flow would increase with Manning coefficient?

EXAMPLE 15.15

The slope of the culvert in Example 15.13 is to be adjusted so that normal depth will be at least 1.5 m when the culvert is carrying its design flowrate of 4.7 m³/s (Figure 15.40). What slope would you recommend?

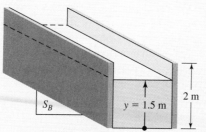

Figure 15.40 Schematic for Example 15.15.

SOLUTION

We know that the volume flowrate is given by Eq. 15.53b, $Q = (C_0/n)AR_H^{2/3}S_B^{1/2}$, with $C_0 = 1$, since we are working in SI units. Solving for the bed slope we obtain $S_B = n^2Q^2/C_0^2A^2R_H^{4/3}$. We can again use Figure 15.6 to write $A(y) = wy$ and $R_H(y) = wy/(w+2y)$. Inserting these into the formula yields

$$S_B = \frac{n^2 Q^2}{(wy)^2 \left(\dfrac{wy}{w+2y}\right)^{4/3}} \tag{A}$$

Inserting the known values as well as a Manning coefficient $n = 0.014$ s/m$^{1/3}$ (notice the units here, which are often omitted in calculations) and width $w = 2$ m, we obtain

$$S_B = \frac{(0.014 \text{ s/m}^{1/3})^2 (4.7 \text{ m}^3/\text{s})^2}{[(2 \text{ m})(1.5 \text{ m})]^2 \left[\dfrac{2 \text{ m}(1.5 \text{ m})}{2 \text{ m} + 2(1.5 \text{ m})}\right]^{\frac{4}{3}}} = 9.5 \times 10^{-4}$$

Thus the bed slope needed is about one-fifth of the slope that produced the slightly supercritical flow in Example 15.13.

15.5.2 Optimum Channel Cross Section

In designing an open channel for uniform flow, an engineer must consider the cost of excavation, the available slope, and the choice of liner materials to calculate the channel shape and size. Now suppose the flowrate, slope, and liner material have been selected. The question then becomes, What is the optimum channel cross section? Using the Manning formula of Eq. 15.53b, $Q = (C_0/n)AR_H^{2/3}S_B^{1/2}$, and $R_H = A/P$, and noting that the type of liner determines the Manning coefficient, we can write the flowrate as $Q = (C_0/n)(A^{5/3}/P^{2/3})S_B^{1/2}$. Solving for the flow area yields

$$A = \left[\frac{nQ}{C_0 S_B^{1/2}}\right]^{3/5} P^{2/5} \tag{15.56}$$

This shows that the optimum channel shape is one that minimizes the wetted perimeter for a given area, which is the same as saying the shape that minimizes the hydraulic radius. Taking the shapes from Figure 15.6 (except for the 2-D flow), we can optimize each geometry by writing the wetted perimeter as a function of flow area and depth. Next we set $dP/dy = 0$ with A held constant and solve for the optimal depth, which for uniform flow is the normal depth y_N. This is the optimal normal depth for that channel geometry and maximizes the flowrate for that particular cross section. The results of this process, including expressions for the optimal normal depth and corresponding flow area, are summarized in Figure 15.41. It can be seen that when optimized, each channel shape has a factor in its expression for flow area that multiplies the common term $[nQ/C_0 S_B^{1/2}]^{3/4}$. Since the smallest factor gives the minimum flow area for the desired flowrate, liner, and bed slope, it is evident that the optimal shape of any channel is circular and filled to the halfway point.

EXAMPLE 15.16

A highway drainage ditch is to carry 10 ft³/s of water in a finished concrete channel during the worst expected rainstorm. If the slope is $S_B = 0.001$, find the dimensions of the optimum rectangular channel and compare with those of an optimum circular channel. Assume uniform flow and that the water fills each channel to the brim at peak flow.

SOLUTION

No sketch is necessary here. From Table 15.1 the Manning coefficient for finished concrete is $n = 0.012$, and we can use the appropriate formulas in Figure 15.41 with $C_0 = 1.49$, since we are working in BG units. For the rectangular channel we have

$$y_N = 2^{-1/8}\left[\frac{nQ}{C_0 S_B^{1/2}}\right]^{3/8} = 2^{-1/8}\left[\frac{(0.012)(10)}{(1.49)(0.001)^{1/2}}\right]^{3/8} = 1.30 \text{ ft}$$

$$A = 2^{3/4}\left[\frac{nQ}{C_0 S_B^{1/2}}\right]^{3/4} = 2^{3/4}\left[\frac{(0.012)(10)}{(1.49)(0.001)^{1/2}}\right]^{3/4} = 3.39 \text{ ft}^2$$

Since the channel is filled to the brim and $y_N = w/2 = b/2$, the rectangular channel is 1.30 ft deep and $w = A/y_N = 3.39$ ft²/1.30 ft = 2.6 ft wide. For the optimum circular channel, the calculations yield

$$y_N = 2^{5/8}\pi^{-3/8}\left[\frac{nQ}{C_0 S_B^{1/2}}\right]^{3/8} = 2^{5/8}\pi^{-3/8}\left[\frac{(0.012)(10)}{(1.49)(0.001)^{1/2}}\right]^{3/8} = 1.42 \text{ ft}$$

15.7 FLOW UNDER A SLUICE GATE

Writing the Bernoulli equation along a surface streamline that connects points 1 and 2 and travels along the inside surface of the gate, and noting that the pressure is atmospheric at both end points, we obtain

$$\frac{V_1^2}{2g} + y_1 = \frac{V_2^2}{2g} + y_2 \tag{15.58b}$$

Using Eq. 15.58a, we can write $V_2 = V_1 y_1 / y_2$, and eliminate V_2 to obtain the cubic equation

$$y_2^3 - \left(\frac{V_1^2}{2g} + y_1\right) y_2^2 + \frac{V_1^2 y_1^2}{2g} = 0 \tag{15.58c}$$

We can solve this equation for the downstream depth given the upstream depth y_1 and velocity V_1. This equation has one real root for subcritical upstream flow. Also, if we write Eq. 15.58b in terms of flowrate, we obtain

$$\frac{Q^2}{2gw^2 y_1^2} + y_1 = \frac{Q^2}{2gw^2 y_2^2} + y_2 \tag{15.58d}$$

Hence solving for flowrate yields

$$Q = \left[\frac{2gw^2 y_1^2 y_2^2}{y_1 + y_2}\right]^{1/2} \tag{15.58e}$$

Since the specific energy of a flow at depth y is given by

$$E = \frac{Q^2}{2gw^2 y^2} + y$$

we see from Eq. 15.58d that the flow through a sluice gate occurs at constant specific energy, i.e., $E_1 = E_2$. Recalling the characteristics of a specific energy curve as discussed in Section 15.4.1, we conclude that the upstream and downstream flows correspond to the alternate depths on the specific energy curve for the given flowrate. This is illustrated by the sluice gate, and its specific energy curve as shown in Figure 15.44A. The upstream state is subcritical, and since specific energy is constant, as the flow approaches and passes under the gate it travels along the specific energy curve reaching the critical state at a minimum specific energy, then becoming supercritical at the downstream state.

At a given flowrate there are generally two operating conditions for a sluice gate, corresponding to two different sluice gate openings. These two operating conditions are illustrated in Figure 15.44. To prove this statement, we can imagine that far upstream of the gate, there is a reservoir at which the depth is y_0 and the velocity is $V_0 = 0$. Writing the Bernoulli equation between this imaginary reservoir location and point 2 we have

$$y_0 = \frac{V_2^2}{2g} + y_2 \tag{15.59a}$$

and using Eq. 15.58b we can also write this as

$$y_0 = \frac{V_1^2}{2g} + y_1 \tag{15.59b}$$

Figure 15.44 Specific energy diagrams for slow under a sluice gate at two different operating conditions.

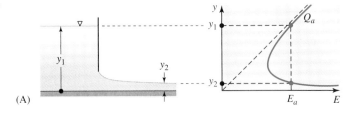

(A)

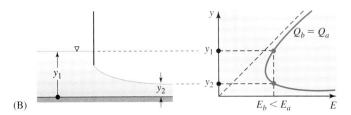

(B)

Rearranging and writing Eq. 15.59a in terms of flowrate, we obtain $y_0 = Q^2/2gw^2y_2^2 + y_2$, or

$$\left(\frac{y_2}{y_0}\right)^2 \left(1 - \frac{y_2}{y_0}\right) = \frac{Q^2}{2gw^2y_0^3} \qquad (15.59c)$$

This equation is plotted in Figure 15.45. We see that there are two depth ratios for any given flowrate less than the maximum flowrate. These two operating conditions and the corresponding gate openings are those illustrated earlier (Figure 15.44).

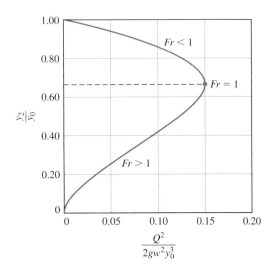

Figure 15.45 Relationship between downstream depth and flowrate for flow through a sluice gate.

15.7 FLOW UNDER A SLUICE GATE

The labeling of the upper and lower branches in Figure 15.45 can be explained by noting that to find the downstream depth at maximum flowrate, we write $(d/dy_2)[Q^2/2gw^2] = 0$, and use Eq. 15.59b to solve $(d/dy)[y_0^3(y_2/y_0)^2(1 - y_2/y_0)] = 0 = 2y_0y_2 - 3y_2^2$. The solution is

$$\frac{y_2}{y_0} = \frac{2}{3} \tag{15.60a}$$

Thus the maximum flowrate with a free discharge occurs when the gate opening is adjusted to produce this depth ratio. The value of the maximum flowrate for a given reservoir depth is then found from Eq. 15.59c to be given by

$$\frac{Q_{max}^2}{gw^2} = \frac{8}{27}y_0^3 \tag{15.60b}$$

The downstream Froude number at the maximum flowrate can be found by writing $Fr_2^2 = V_2^2/gy_2$ and using the preceding results to obtain

$$Fr_2^2 = \frac{V_2^2}{gy_2} = \frac{Q_{max}^2}{gw^2y_2^3} = \frac{8}{27}\left(\frac{y_0}{y_2}\right)^3 = \frac{8}{27}\left(\frac{3}{2}\right)^3 = 1$$

Thus, at the maximum flowrate the downstream Froude number is unity, and with some effort we can determine that the upper branch of the curve in Figure 15.45 corresponds to subcritical downstream flow and the lower branch to supercritical downstream flow.

The final step in our analysis of the free discharge from a sluice gate is to determine the sluice gate opening needed to produce a desired operating condition. Because the flow passing through the gate undergoes a vena contracta (see Figure 15.43A), it is not possible to establish an analytical relationship between the downstream depth y_2 and the height of the gate opening y_G. However, it has been suggested that by using an empirical discharge coefficient C_G, the free discharge flowrate can be modeled as

$$Q = C_G w y_G \sqrt{2gy_1} \tag{15.61a}$$

where values for the discharge coefficient are given by

$$C_G = \frac{0.61}{\sqrt{1 + 0.61(y_G/y_1)}} \tag{15.61b}$$

provided $y_G/y_1 < 0.5$.

EXAMPLE 15.17

A sluice gate in a 24 ft wide rectangular channel is observed to be operating with an upstream depth $y_1 = 6$ ft and a downstream depth $y_2 = 1$ ft (Figure 15.46). Find the volume flowrate passing through the channel and all other geometric and flow parameters.

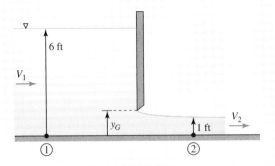

Figure 15.46 Schematic for Example 15.17.

SOLUTION

Applying Eq. 15.58e, $Q = [2gw^2 y_1^2 y_2^2/(y_1 + y_2)]^{1/2}$, the flowrate is found to be

$$Q = \left[\frac{2(32.2 \text{ ft/s}^2)(24 \text{ ft})^2(6 \text{ ft})^2(1 \text{ ft})^2}{6 \text{ ft} + 1 \text{ ft}}\right]^{1/2} = 436.8 \text{ ft}^3/\text{s}$$

The velocities upstream and downstream can now be determined by using Eq. 15.58a, $Q/w = V_1 y_1 = V_2 y_2$, to write $V_1 = Q/wy_1$ and $V_2 = Q/wy_2$. Inserting the data, we find $V_1 = Q/wy_1 = 436.8 \text{ ft}^3/\text{s}/24 \text{ ft}(6 \text{ ft}) = 3.0 \text{ ft/s}$ and $V_2 = Q/wy_2 = 436.8 \text{ ft}^3/\text{s}/24 \text{ ft}(1 \text{ ft}) = 18.2 \text{ ft/s}$. The corresponding Froude numbers are found to be $Fr_1 = V_1/\sqrt{gy_1} = 3.0 \text{ ft/s}/\sqrt{(32.2 \text{ ft/s}^2)(6 \text{ ft})} = 0.22$ and $Fr_2 = 18.2 \text{ ft/s}/\sqrt{(32.2 \text{ ft/s}^2)(1 \text{ ft})} = 3.21$. The reservoir depth y_0 is found by using Eq. 15.59a to write $y_0 = V_2^2/2g + y_2 = (18.2 \text{ ft/s})^2/2(32.2 \text{ ft/s}^2) + 1 \text{ ft} = 6.14 \text{ ft}$. Note that we could also have used Eq. 15.59b to write $y_0 = V_1^2/2g + y_1 = (3.0 \text{ ft/s})^2/2(32.2 \text{ ft/s}^2) + 6 \text{ ft} = 6.14 \text{ ft}$. To find the gate opening we rearrange Eq. 15.61a to obtain $y_G = Q/C_G w \sqrt{2gy_1}$ and note that the discharge coefficient is given by Eq. 15.61b:

$$C_G = \frac{0.61}{\sqrt{1 + 0.61(y_G/y_1)}}$$

Assuming $C_G = 0.61$, a first iteration yields

$$y_G = \frac{Q}{C_G w \sqrt{2gy_1}} = \frac{436.8 \text{ ft}^3/\text{s}}{0.61(24 \text{ ft})\sqrt{2(32.2 \text{ ft/s}^2)(6 \text{ ft})}} = 1.52 \text{ ft}$$

However the 1.52 ft corresponds to a discharge coefficient of

$$C_G = \frac{0.61}{\sqrt{1 + 0.61(y_G/y_1)}} = \frac{0.61}{\sqrt{1 + 0.61(1.52 \text{ ft}/6 \text{ ft})}} = 0.57$$

so further iteration is required. The final results are $y_G = 1.13$ ft and $C_G = 0.58$.

15.8 FLOW OVER A WEIR

A weir is a device used to measure and control open channel flows. A sharp crested weir with a ventilated free nappe is illustrated in Figure 15.47A. This is the configuration of greatest interest in measuring flowrate, hence it will be analyzed in this section. Operation is also possible with a partially or fully submerged nappe (Figure 15.47B, 15.47C). The analysis of these flows is not covered here but may be found in advanced texts.

Although friction can be neglected in analyzing flow over a sharp crested weir, the use of the Bernoulli equation is not straightforward because of streamline curvature and nonuniform flow in the vicinity of the nappe. Additional complications arise when we consider the three principal types of sharp crested weir illustrated in Figure 15.48 and recognize the possibility of end effects. Thus we are forced to rely on empirical results.

In the case of either type of rectangular weir, empirical measurements show that the flowrate can be written as

$$Q = C_W w \left(\frac{L}{w}\right) \sqrt{g}(y_1 - y_W)^{3/2} \qquad (15.62a)$$

where C_W is the rectangular weir discharge coefficient, w is the width of the weir, L is the width of the nappe, y_W is the weir height, and y_1 is the upstream depth. The empirically determined discharge coefficient[1] is given by

$$C_W = 0.59 + 0.08 \left(\frac{y_1 - y_W}{y_W}\right) \qquad (15.62b)$$

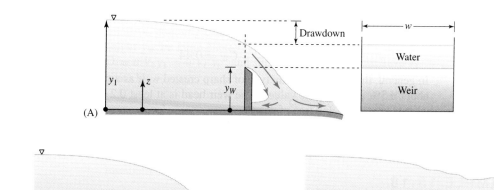

Figure 15.47 Operating conditions for sharp crested weirs: (A) ventilated free nappe, (B) partially submerged nappe, and (C) fully submerged nappe.

---

[1] For more details on weirs see P. Ackers et al., *Weirs and Flumes for Flow Measurement*, Wiley, New York, 1978.

EXAMPLE 15.19

A broad crested weir rises 0.3 m high above the bottom of a stream channel. If the measured weir head is $y_1 - y_W = 0.65$ m, and the weir is 10 m wide, find the volume flowrate crossing the weir and the critical water depth, y_C.

SOLUTION

Figure 15.50 can serve as a sketch. To find the volume flowrate, we insert the data into Eq. 15.65, obtaining $Q = w\sqrt{g}\,(\frac{2}{3})^{3/2}(y_1 - y_W)^{3/2} = 10 \text{ m}\sqrt{9.81 \text{ m/s}^2}\,(\frac{2}{3})^{3/2} \times (0.65 \text{ m})^{3/2} = 8.9 \text{ m}^3/\text{s}$. To find the crtical water depth, we note that $y_C \approx \frac{2}{3}(y_1 - y_W)$. Since we know $y_1 - y_W = 0.65$ m, we obtain $y_C \approx \frac{2}{3}(y_1 - y_W) = \frac{2}{3}(0.65 \text{ m}) = 0.43$ m.

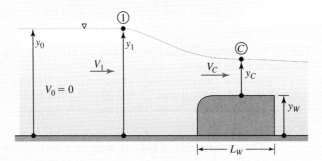

Figure 15.50 Schematic for Example 15.19.

Thus we can use Eq. 15.65 to calculate the flowrate for a broad crested weir, provided the downstream conditions permit critical flow over the weir block. For proper operation, the length of the weir block L_W should be roughly in the range $2 < L_W/(y_1 - y_W) < 10$.

15.9 SUMMARY

The most important parameter in the analysis of open channel flows of engineering interest is the Froude number, $Fr = V/\sqrt{gL}$. For example, as the flow approaches the leading edge of a bump, the depth slope and free surface slope will decrease for $Fr < 1$ (subcritical flow) but increase for $Fr > 1$ (supercritical flow). Thus, the liquid depth decreases and the free surface elevation drops over a bump in a subcritical flow, but the depth increases and free surface elevation rises over the same bump in a supercritical flow. Critical flow can occur only at the peak of the bump. If critical flow does occur at the peak of a bump, then a subcritical flow upstream of the bump may become a supercritical flow downstream if conditions are right.

Similarly, as the flow approaches the leading edge of a depression, the depth slope and free surface slope will increase for $Fr < 1$ but decrease for $Fr > 1$. That is, the liquid depth increases and the free surface elevation rises over a depression in a subcritical flow, but the depth decreases and free surface elevation drops over the same depression in a supercritical flow. As was the case with a bump, critical flow can occur only where the bed slope is horizontal, i.e., if a depression has a trough.

The value of Fr also plays a role in determining the response of a flow in a horizontal channel to a change in the channel width. For subcritical flow through a contraction, the depth will decrease, while for supercritical flow through the same contraction the depth will increase. For subcritical flow through an expansion the depth will increase, while in supercritical flow the depth will decrease.

The behavior of surface waves in an open channel flow is also governed by the value of the local Froude number. A small amplitude wave is able to propagate upstream in subcritical flow but is unable to propagate upstream in supercritical flow. Thus the flow of information about surface disturbances and obstructions, which is carried by small amplitude waves, is strongly affected by the value of the local Froude number.

Since disturbances cannot propagate upstream in a supercritical flow, if downstream conditions require that a supercritical flow become subcritical, a smooth transition is impossible, and a phenomenon known as a hydraulic jump is observed in the flow. In a hydraulic jump, the flow changes from supercritical to subcritical in a relatively short distance, with an abrupt decrease in velocity, increase in depth, and a substantial head loss. Designers use hydraulic jumps to safely dissipate energy in a high speed flow and thus prevent erosion in a streambed downstream.

Specific energy concepts can be used to make predictions of the following types about a flow in a rectangular channel:

1. In subcritical flow, an increase/decrease in specific energy is accompanied by an increase/decrease in depth.

2. In supercritical flow, an increase/decrease in specific energy is accompanied by a decrease/increase in depth.

3. If the flow is critical, i.e., at the point of minimum specific energy, a small increase in specific energy causes a relatively large change in depth.

4. For a subcritical flow to become a supercritical flow, the specific energy must first decrease. The flow must then go through the critical point and along the supercritical branch. For example, this process may occur via a bump or decrease in channel width.

In a steady, fully developed flow in a prismatic channel, the liquid depth is constant, and depth slope is zero. Thus the free surface is parallel to the bed. This type of open channel flow is referred to as uniform flow or flow at normal depth. The latter name reflects the fact that the depth of a uniform flow is referred to as the normal depth y_N. Channels are often designed to have uniform flow, so understanding their characteristics is of some importance. The velocity in uniform open channel flow is given by the Manning equation, $V = R_H^{2/3} S_B^{1/2}/n$, and the volume flowrate is given by $Q = A R_H^{2/3} S_B^{1/2}/n$. The bed slope of a particular uniform flow is classified according to the following criteria:

1. For $y_N > y_{HC}$, the slope must be $S_B < S_{BC}$, and the slope is called mild or subcritical.

2. For $y_N = y_{HC}$, the slope must be $S_B = S_{BC}$, and the slope is called critical.

3. For $y_N < y_{HC}$, the slope must be $S_B > S_{BC}$, and the slope is called steep or supercritical.

A uniform flow may be characterized by the size and shape of the channel, the normal depth, the slope of the channel bottom, the roughness of the wall, and either the average velocity or the flowrate. Thus a problem may require that we find any of these variables. In addition, we may think of a problem as involving analysis or design. In an analysis problem, we would typically be given information regarding the channel size, shape, material, and slope and be asked to find flowrate for a certain normal depth or vice versa. In a design problem, we might be asked to determine the slope of a channel for a case in which the desired shape, flowrate, and normal depth are specified.

Sluice gates are structures used to control and measure the flowrate in systems such as irrigation channels. Given the depths upstream and downstream of the gate and the other physical parameters of the flow, one can predict the flowrate through the gate.

A weir is a device used to measure and control open channel flows. Although friction can be neglected in analyzing flow over a sharp crested weir, the use of the Bernoulli equation is not straightforward because of streamline curvature and nonuniform flow in the vicinity of the nappe. Thus we are forced to rely on empirical results. In general, the accuracy of any type of sharp crested weir as a flow-measuring device is about ±5%, provided the value of the weir head is at least 0.2 ft (0.06 m) and the weir edge is kept sharp and free of debris.

PROBLEMS

Section 15.1

15.1 Find four photographs of open channel flow on the Internet. Include both natural and man-made examples.

15.2 Find the geometry and flow characteristics of a river or stream near your school. What is the Reynolds number?

Section 15.2

15.3 Water flows in a rectangular, open channel that is 36 ft wide and 8 ft deep. The flow rate is $Q = 2 \times 10^6$ gal/min. Determine the average velocity.

15.4 Determine the expression for the hydraulic radius R_H for a trapezoidal channel of width w, depth d, and slope angle θ.

Section 15.3

15.5 For the data you found in Problem 15.2, determine the Froude number.

15.6 For the flow described in Problem 15.3, determine the Froude number.

15.7 Water flows at 1.5 m/s and a depth of 0.8 m in a horizontal rectangular channel. If this flow encounters a bump of 0.1 m, what is the depth above the bump?

15.8 In Problem 15.7, what is the Froude number upstream?

15.9 Redo Problem 15.7 with a flow 4.5 m/s.

15.10 In Problem 15.9, what is the Froude number upstream?

15.11 A flow in a rectangular channel 1 m deep with a velocity of 0.3 m/s approaches a smooth channel rise of 0.3 m. What is the depth after the rise?

15.12 A flow in a 3 ft wide rectangular channel is 3 ft deep and has a flow rate of 0.2 ft^3/s approaching a smooth drop of 2.5 in. What is the depth after the drop?

15.13 Water flows through a 4 ft wide rectangular channel at a velocity of 3 ft/s and a depth of 2 ft. If the channel width contracts to 3 ft what is downstream depth?

15.14 Redo Problem 15.13 if the upstream depth is 0.5 ft and the volume flow rate is 24 ft^3/s.

15.15 A width contraction in a horizontal channel is called a venturi flume. If the upstream channel width and depth are 0.6 and 0.3 m, respectively, and the flow rate is 0.1 m^3/s, find the downstream critical depth, velocity, and width.

15.16 The width of a horizontal, rectangular channel expands smoothly from 6 ft to 7 ft. The upstream depth is 3 ft and the velocity is 5 ft/s. What is the downstream depth?

15.17 Redo Problem 15.16 with an upstream velocity of 10 ft/s.

15.18 Two objects are dropped in a stream two seconds apart; the stream is 4 m deep and is flowing at 1 m/s. Sketch two circular ripples one second after the second object hits the surface of the water.

15.19 Redo Problem 15.18 with the stream flowing at 10 m/s.

15.20 An object is dropped in a stream of uniform depth. The resulting wave travels downstream 15 ft and upstream 6 ft in one second. Determine the velocity and depth of the stream.

15.21 A submerged object breaks the surface of a stream causing a wave as shown in Figure P15.1. If the stream depth is 1 ft and the half-angle of the wedgelike wave pattern is 40°, what are the flow speed and Froude number?

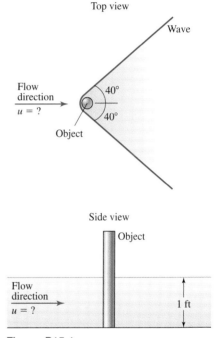

Figure P15.1

15.22 Redo Problem 15.21 with a half-angle of 20°.

15.23 The flowrate into a hydraulic jump is 0.3 m^3/s; the square channel is 0.5 m wide and the depth downstream of the jump is 0.4 m. Determine the depth upstream of the jump, the Froude numbers upstream and downstream of the jump, the headloss, and the power dissipated in the jump.

15.24 A hydraulic jump is formed in a 150 ft wide channel. If the upstream and downstream depths are 1 and 3 ft, respectively, determine the upstream velocity, headloss, and the power dissipated.

15.25 Downstream from a sluice gate a hydraulic jump occurs. Leaving the sluice gate the flow depth is 2 in. and the velocity is 16 ft/s. Determine the downstream headloss and power dissipation per foot of width.

15.26 A hydraulic jump occurs in a rectangular channel that carries a flow with a depth of 0.3 m and a flow rate of 0.6 m^3/s per meter of width. Determine the downstream headloss and power dissipation per meter of width.

15.27 For the hydraulic jump described in Problem 15.24, determine the increase in temperature of the water if all the power dissipated is converted to heat that is absorbed by the water. Comment on whether the temperature change could easily be measured.

15.28 A moving hydraulic jump, as shown in Figure P15.2, is possible naturally in the form of a tidal bore, or can be created by suddenly opening a sluice gate. The conditions can be calculated by using the same procedure as for a stationary hydraulic jump if we consider the velocities relative to the wave. Obtain an expression for the velocity of the wave in terms of the depth on either side of the wave.

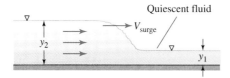

Figure P15.2

Section 15.4

15.29 Determine the critical depth and flow velocity in a 40 ft wide rectangular channel carrying water at 1800 ft^3/s.

15.30 For the flow described in Problem 15.29, if the actual depth is 20 ft, what is the Froude number?

15.31 A 60° trapezoidal channel (similar in shape to the one shown in Figure 15.32) is 10 m wide and carries 20 m^3/s of water. Determine the critical depth, critical hydraulic depth, critical area, critical velocity, and critical free surface width.

15.32 Draw the specific energy diagram for the flow described in Problem 15.7, and use it to solve this problem.

15.33 Draw the specific energy diagram for the flow described in Problem 15.9, and use it to solve this problem.

15.34 Draw the specific energy diagram for the flow described in Problem 15.11, and use it to solve this problem.

15.35 Draw the specific energy diagram for the flow described in Problem 15.12, and use it to solve this problem.

15.36 Draw the specific energy diagram for the flow described in Problem 15.24.

15.37 Draw the specific energy diagram for the flow described in Problem 15.25.

15.38 Draw the specific energy diagram for the flow described in Problem 15.26.

Section 15.5

15.39 Water flows in a rectangular channel 25 ft wide, with a Manning coefficient of 0.022. Plot a graph of flow rate as a function of slope for several depths.

15.40 A creek has area, wetted perimeter, and slope of 130 ft^2, 46 ft, and 0.04 ft/100 ft, respectively. Determine the velocity and the average shear stress on the wetted perimeter of the creek channel.

15.41 A 5 m wide rectangular channel has a slope of 0.004 m/m and carries 20 m^3/s of water at a depth of 4 m. Determine the Manning coefficient and the average shear stress on the channel walls and bottom.

15.42 Water flows in a 35° trapezoidal channel that is 10 ft wide. The depth is 6 ft and the slope is 0.0015. Determine the flowrate if this is a clean excavated earth channel. Also determine what the flowrate would be if weeds covered the perimeter.

15.43 Water flows in a 45° trapezoidal channel that is 10 m wide. The depth is 5 m, and the slope is 0.0012. Determine the flowrate if the channel is lined with finished concrete. Also determine what the flowrate would be for a lining of unfinished concrete.

15.44 At a mine, a rectangular channel lined with wood carries 200 ft^3/s of water at a slope of 1 ft per 100 ft of length. What is the normal depth of the flow?

15.45 A flooded stream can be modeled as shown in Figure P15.3. What is the flowrate if the channel is considered to be clean and straight and the floodplain is farmland? Assume that the slope is 0.001.

15.46 Redo Problem 15.45 with the floodplain covered by trees.

15.47 A 1 m diameter clay tile sewer pipe runs one-third full on a slope of 0.25°. Determine the flowrate.

15.48 Redo Problem 15.47 for a completely full pipe.

15.49 Redo Problem 15.48 using the pipe flow methods of Chapter 13.

15.50 Two sewer pipes like the one described in Problem 15.47 flow into a single finished concrete pipe at the same slope. What diameter should it be such that it is one-third full?

15.51 Derive the expression for optimal depth for a rectangular channel given in Figure 15.6.

15.52 What are the best dimensions for a rectangular channel lined with finished concrete to carry 10 m^3/s of water with a slope of 0.0015?

15.53 What are the best dimensions for a wooden rectangular channel to carry 100 ft^3/s of water with a slope of 0.0012?

15.54 Derive the expression for optimum depth for a trapezoidal channel given in Figure 15.6.

15.55 What is the optimum depth for a 60° trapezoidal channel lined with finished concrete to carry 10 m^3/s of water with a slope of 0.0015?

15.56 What is the optimum depth for a 60° trapezoidal earth channel to carry 100 ft^3/s of water with a slope of 0.0012?

15.57 Derive the expression for optimal depth for a triangular channel given in Figure 15.6.

15.58 What is the optimum depth for a 45° triangular channel lined with finished concrete to carry 10 m^3/s of water with a slope of 0.0015?

15.59 What is the optimum depth for a 45° triangular channel lined with rough concrete to carry 100 ft^3/s of water with a slope of 0.0012?

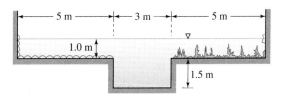

Figure P15.3

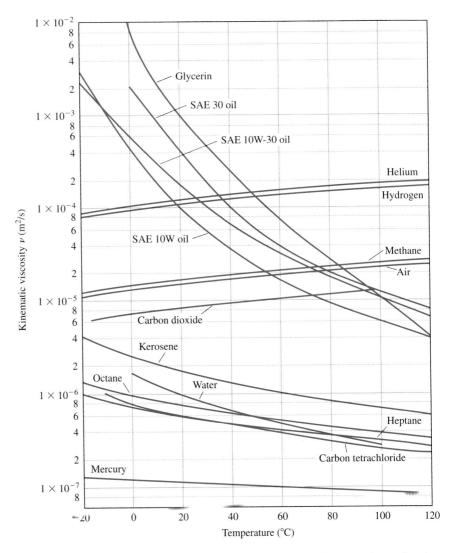

Figure A.2 Kinematic viscosity of common fluids (at atmospheric pressure) as a function of temperature.

APPENDIX A FLUID PROPERTY DATA FOR VARIOUS FLUIDS

TABLE A.3 Physical Properties of Water in BG Units*

Temperature (°F)	Density ρ (slug/ft^3)	Specific Weight γ (lb/ft^3)	Dynamic Viscosity μ [(lb·s)/ft^2]	Kinematic Viscosity ν (ft^2/s)	Surface Tension** σ (lb/ft)	Vapor Pressure p_v [lb/in.^{2} (abs)]	Speed of Sound c (ft/s)
32	1.940	62.42	3.732 E−5	1.924 E−5	5.18 E−3	8.854 E−2	4603
40	1.940	62.43	3.228 E−5	1.664 E−5	5.13 E−3	1.217 E−1	4672
50	1.940	62.41	2.730 E−5	1.407 E−5	5.09 E−3	1.781 E−1	4748
60	1.938	62.37	2.344 E−5	1.210 E−5	5.03 E−3	2.563 E−1	4814
70	1.936	62.30	2.037 E−5	1.052 E−5	4.97 E−3	3.631 E−1	4871
80	1.934	62.22	1.791 E−5	9.262 E−6	4.91 E−3	5.069 E−1	4819
90	1.931	62.11	1.500 E−5	8.233 E−6	4.86 E−3	6.979 E−1	4960
100	1.927	62.00	1.423 E−5	7.383 E−6	4.79 E−3	9.493 E−1	4995
120	1.918	61.71	1.164 E−5	6.067 E−6	4.67 E−3	1.692 E+0	5049
140	1.908	61.38	9.743 E−6	5.106 E−6	4.53 E−3	2.888 E+0	5091
160	1.896	61.00	8.315 E−6	4.385 E−6	4.40 E−3	4.736 E+0	5101
180	1.883	60.58	7.207 E−6	3.827 E−6	4.26 E−3	7.507 E+0	5195
200	1.869	60.12	6.342 E−6	3.393 E−6	4.12 E−3	1.152 E+1	5089
212	1.860	59.83	5.886 E−6	3.165 E−6	4.04 E−3	1.469 E+1	5062

*Adapted from various sources.
**In contact with air.

TABLE A.4 Physical Properties of Water in SI Units*

Temperature (°C)	Density ρ (kg/m^3)	Specific Weight γ (kN/m^3)	Dynamic Viscosity μ [(N·s)/m^2]	Kinematic Viscosity ν (m^2/s)	Surface Tension** σ (N/m)	Vapor Pressure p_v [N/m^2 (abs)]	Speed of Sound c (m/s)
0	999.9	9.806	1.787 E−3	1.787 E−6	7.56 E−2	6.105 E+2	1403
5	1000.0	9.807	1.519 E−3	1.519 E−6	7.49 E−2	8.722 E+2	1427
10	999.7	9.804	1.307 E−3	1.307 E−6	7.42 E−2	1.228 E+3	1447
20	998.2	9.789	1.002 E−3	1.004 E−6	7.28 E−2	2.338 E+3	1481
30	995.7	9.765	7.975 E−4	8.009 E−7	7.12 E−2	4.243 E+3	1507
40	992.2	9.731	6.529 E−4	6.580 E−7	6.96 E−2	7.376 E+3	1526
50	988.1	9.690	5.468 E−4	5.534 E−7	6.79 E−2	1.233 E+4	1541
60	983.2	9.642	4.665 E−4	4.745 E−7	6.62 E−2	1.992 E+4	1552
70	977.8	9.589	4.042 E−4	4.134 E−7	6.44 E−2	3.116 E+4	1555
80	971.8	9.530	3.547 E−4	3.650 E−7	6.26 E−2	4.734 E+4	1555
90	965.3	9.467	3.147 E−4	3.260 E−7	6.08 E−2	7.010 E+4	1550
100	958.4	9.399	2.818 E−4	2.940 E−7	5.89 E−2	1.013 E+5	1543

*Adapted from various sources.
**In contact with air.

APPENDIX A FLUID PROPERTY DATA FOR VARIOUS FLUIDS

TABLE A.5 Physical Properties of Air at Standard Atmospheric Pressure in BG Units*

Temperature (°F)	Density ρ (slug/ft^3)	Specific Weight γ (lb/ft^3)	Dynamic Viscosity μ [(lb·s)/ft^2]	Kinematic Viscosity ν (ft^2/s)	Specific Heat Ratio k	Speed of Sound c (ft/s)
−40	2.939 E−3	9.456 E−2	3.29 E−7	1.12 E−4	1.401	1004
−20	2.805 E−3	9.026 E−2	3.34 E−7	1.19 E−4	1.401	1028
0	2.683 E−3	8.633 E−2	3.38 E−7	1.26 E−4	1.401	1051
10	2.626 E−3	8.449 E−2	3.44 E−7	1.31 E−4	1.401	1062
20	2.571 E−3	8.273 E−2	3.50 E−7	1.36 E−4	1.401	1074
30	2.519 E−3	8.104 E−2	3.58 E−7	1.42 E−4	1.401	1085
40	2.469 E−3	7.942 E−2	3.60 E−7	1.46 E−4	1.401	1096
50	2.420 E−3	7.786 E−2	3.68 E−7	1.52 E−4	1.401	1106
60	2.373 E−3	7.636 E−2	3.75 E−7	1.58 E−4	1.401	1117
70	2.329 E−3	7.492 E−2	3.82 E−7	1.64 E−4	1.401	1128
80	2.286 E−3	7.353 E−2	3.86 E−7	1.69 E−4	1.400	1138
90	2.244 E−3	7.219 E−2	3.90 E−7	1.74 E−4	1.400	1149
100	2.204 E−3	7.090 E−2	3.94 E−7	1.79 E−4	1.400	1159
120	2.128 E−3	6.846 E−2	4.02 E−7	1.89 E−4	1.400	1180
140	2.057 E−3	6.617 E−2	4.13 E−7	2.01 E−4	1.399	1200
160	1.990 E−3	6.404 E−2	4.22 E−7	2.12 E−4	1.399	1220
180	1.928 E−3	6.204 E−2	4.34 E−7	2.25 E−4	1.399	1239
200	1.870 E−3	6.016 E−2	4.49 E−7	2.40 E−4	1.398	1258
300	1.624 E−3	5.224 E−2	4.97 E−7	3.06 E−4	1.394	1348
400	1.435 E−3	4.616 E−2	5.24 E−7	3.65 E−4	1.389	1431
500	1.285 E−3	4.135 E−2	5.80 E−7	4.51 E−4	1.383	1509
750	1.020 E−3	3.280 E−2	6.81 E−7	6.68 E−4	1.367	1685
1000	8.445 E−4	2.717 E−2	7.85 E−7	9.30 E−4	1.351	1839
1500	6.291 E−4	2.024 E−2	9.50 E−7	1.51 E−3	1.329	2114

*Adapted from various sources.

TABLE A.6 Physical Properties of Air at Standard Atmospheric Pressure in SI Units*

Temperature (°C)	Density ρ (kg/m^3)	Specific Weight γ (N/m^3)	Dynamic Viscosity μ [(N·s)/m^2]	Kinematic Viscosity ν (m^2/s)	Specific Heat Ratio k	Speed of Sound c (m/s)
−40	1.514	14.85	1.57 E−5	1.04 E−5	1.401	306.2
−20	1.395	13.68	1.63 E−5	1.17 E−5	1.401	319.1
0	1.292	12.67	1.71 E−5	1.32 E−5	1.401	331.4
5	1.269	12.45	1.73 E−5	1.36 E−5	1.401	334.4
10	1.247	12.23	1.76 E−5	1.41 E−5	1.401	337.4
15	1.225	12.01	1.80 E−5	1.47 E−5	1.401	340.4

APPENDIX A FLUID PROPERTY DATA FOR VARIOUS FLUIDS

TABLE A.6 (*continued*)

Temperature (°C)	Density ρ (kg/m³)	Specific Weight γ (N/m³)	Dynamic Viscosity μ [(N·s)/m²]	Kinematic Viscosity ν (m²/s)	Specific Heat Ratio k	Speed of Sound c (m/s)
20	1.204	11.81	1.82 E−5	1.51 E−5	1.401	343.3
25	1.184	11.61	1.85 E−5	1.56 E−5	1.401	346.3
30	1.165	11.43	1.86 E−5	1.60 E−5	1.400	349.1
40	1.127	11.05	1.87 E−5	1.66 E−5	1.400	354.7
50	1.109	10.88	1.95 E−5	1.76 E−5	1.400	360.3
60	1.060	10.40	1.97 E−5	1.86 E−5	1.399	365.7
70	1.029	10.09	2.03 E−5	1.97 E−5	1.399	371.2
80	0.9996	9.803	2.07 E−5	2.07 E−5	1.399	376.6
90	0.9721	9.533	2.14 E−5	2.20 E−5	1.398	381.7
100	0.9461	9.278	2.17 E−5	2.29 E−5	1.397	386.9
200	0.7461	7.317	2.53 E−5	3.39 E−5	1.390	434.5
300	0.6159	6.040	2.98 E−5	4.84 E−5	1.379	476.3
400	0.5243	5.142	3.32 E−5	6.34 E−5	1.368	514.1
500	0.4565	4.477	3.64 E−5	7.97 E−5	1.357	548.8
1000	0.2772	2.719	5.04 E−5	1.82 E−4	1.321	694.8

*Adapted from various sources.

TABLE A.7 Properties of Common Liquids at Standard Atmospheric Pressure and 20°C in SI Units

Liquid	Density ρ (kg/m³)	Dynamic Viscosity μ [kg/(m·s)]	Surface Tension* σ (N/m)	Vapor Pressure p_v (N/m²)	Bulk Modulus k (N/m²)	Viscosity Parameter $C^\dagger$
Ammonia	608	2.20 E−4	2.13 E−2	9.10 E+5	—	1.05
Benzene	881	6.51 E−4	2.88 E−2	1.01 E+4	1.4 E+9	4.34
Carbon tetrachloride	1,590	9.67 E−4	2.70 E−2	1.20 E+4	9.65 E+8	4.45
Ethanol	789	1.20 E−3	2.28 E−2	5.7 E+3	9.0 E+8	5.72
Ethylene glycol	1,117	2.14 E−2	4.84 E−2	1.2 E+1	—	11.7
Freon 12	1,327	2.62 E−4	—	—	—	1.76
Gasoline	680	2.92 E−4	2.16 E−2	5.51 E+4	9.58 E+8	3.68
Glycerin	1,260	1.49	6.33 E−2	1.4 E−2	4.34 E+9	28.0
Kerosene	804	1.92 E−3	2.8 E−2	3.11 E+3	1.6 E+9	5.56
Mercury	13,550	1.56 E−3	4.84 E−1	1.1 E−3	2.55 E+10	1.07
Methanol	791	5.98 E−4	2.25 E−2	1.34 E+4	8.3 E+8	4.63
SAE 10W oil	870	1.04 E−1‡	3.6 E−2	—	1.31 E+9	15.7
SAE 10W-30 oil	876	1.7 E−1‡	—	—	—	14.0

(*continued*)

TABLE A.7 (continued)

Liquid	Density ρ (kg/m³)	Dynamic Viscosity μ [kg/(m-s)]	Surface Tension* σ (N/m)	Vapor Pressure p_v (N/m²)	Bulk Modulus k (N/m²)	Viscosity Parameter C†
SAE 30W oil	891	2.9 E−1‡	3.5 E−2	—	1.38 E+9	18.3
SAE 50W oil	902	8.6 E−1‡	—	—	—	20.2
Water	998	1.00 E−3	7.28 E−2	2.34 E+3	2.19 E+9	Table A.4
Seawater (30%)	1,025	1.07 E−3	7.28 E−2	2.34 E+3	2.33 E+9	7.28

*In contact with air.
†The viscosity parameter C can be used to estimate the liquid viscosity as a function of temperature using the empirical relationship $\mu/\mu_{20°C} = \exp[C(293\,K/T\,K - 1)]$ with an accuracy of ±6% in the range of $0 \leq T \leq 100°C$.
‡Representative values.

TABLE A.8 Physical Properties of Common Gases at Standard Atmospheric Pressure and 20°C in SI Units*

Gas	Molecular Weight	R [m²/(s²·K)]	ρg (N/m³)	Dynamic Viscosity μ [(N·s)/m²]	Specific Heat Ratio	Power-law Exponent n**
H₂	2.016	4124	0.822	9.05 E−6	1.41	0.68
He	4.003	2077	1.63	1.97 E−5	1.66	0.67
H₂O	18.02	461	7.35	1.02 E−5	1.33	1.15
Ar	39.944	208	16.3	2.24 E−5	1.67	0.72
Dry air	28.96	287	11.8	1.80 E−5	1.40	0.67
CO₂	44.01	189	17.9	1.48 E−5	1.30	0.79
CO	28.01	297	11.4	1.82 E−5	1.40	0.71
N₂	28.02	297	11.4	1.76 E−5	1.40	0.67
O₂	32.00	260	13.1	2.00 E−5	1.40	0.69
NO	30.01	277	12.1	1.90 E−5	1.40	0.78
N₂O	44.02	189	17.9	1.45 E−5	1.31	0.89
Cl₂	70.91	117	28.9	1.03 E−5	1.34	1.00
CH₄	16.04	518	6.54	1.34 E−5	1.32	0.87

*Adapted from various sources.
**The power law exponent, n, can be used to estimate the liquid viscosity as a function of temperature by using the relationship $\mu/\mu_{293K} \approx (T\,K/293\,K)^n$ with an accuracy of ±6% in the range of $250 \leq T \leq 1000$ K.

APPENDIX B PROPERTIES OF THE U.S. STANDARD ATMOSPHERE

TABLE B.1 Properties of the U.S. Standard Atmosphere in BG Units

Altitude (ft)	Temperature (°F)	Acceleration of Gravity g (ft/s²)	Pressure p [lb/in.² (abs)]	Density ρ (slug/ft³)	Dynamic Viscosity μ [(lb·s)/ft²]
−5,000	76.84	32.189	17.554	2.745 E−3	3.836 E−7
0	59.00	32.174	14.696	2.377 E−3	3.737 E−7
5,000	41.17	32.159	12.228	2.048 E−3	3.637 E−7
10,000	23.36	32.143	10.108	1.756 E−3	3.534 E−7
15,000	5.55	32.128	8.297	1.496 E−3	3.430 E−7
20,000	−12.26	32.112	6.759	1.267 E−3	3.324 E−7
25,000	−30.05	32.097	5.461	1.066 E−3	3.217 E−7
30,000	−47.83	32.082	4.373	8.907 E−4	3.107 E−7
35,000	−65.61	32.066	3.468	7.382 E−4	2.995 E−7
40,000	−69.70	32.051	2.730	5.873 E−4	2.969 E−7
45,000	−69.70	32.036	2.149	4.623 E−4	2.969 E−7
50,000	−69.70	32.020	1.692	3.639 E−4	2.969 E−7
60,000	−69.70	31.990	1.049	2.256 E−4	2.969 E−7
70,000	−67.42	31.959	0.651	1.392 E−4	2.984 E−7
80,000	−61.98	31.929	0.406	8.571 E−5	3.018 E−7
90,000	−56.54	31.897	0.255	5.610 E−5	3.052 E−7
100,000	−51.10	31.868	0.162	3.318 E−5	3.087 E−7
150,000	19.40	31.717	0.020	3.658 E−6	3.511 E−7
200,000	−19.78	31.566	0.003	5.328 E−7	3.279 E−7
250,000	−88.77	31.415	0.000	6.458 E−8	2.846 E−7

APPENDIX B PROPERTIES OF THE U.S. STANDARD ATMOSPHERE

TABLE B.2 Properties of the U.S. Standard Atmosphere in SI Units

Altitude (m)	Temperature (°C)	Acceleration of Gravity g (m/s^2)	Pressure p [N/m^2 (abs)]	Density ρ (kg/m^3)	Dynamic Viscosity μ [(N·s)/m^2]
−1,000	21.50	9.810	1.139 E+5	1.347 E+0	1.821 E−5
0	15.00	9.807	1.013 E+5	1.225 E+0	1.789 E−5
1,000	8.50	9.804	8.988 E+4	1.112 E+0	1.758 E−5
2,000	2.00	9.801	7.950 E+4	1.007 E+0	1.726 E−5
3,000	−4.49	9.797	7.012 E+4	9.093 E−1	1.694 E−5
4,000	−10.98	9.794	6.166 E+4	8.194 E−1	1.661 E−5
5,000	−17.47	9.791	5.405 E+4	7.364 E−1	1.628 E−5
6,000	−23.96	9.788	4.722 E+4	6.601 E−1	1.595 E−5
7,000	−30.45	9.785	4.111 E+4	5.900 E−1	1.561 E−5
8,000	−36.94	9.782	3.565 E+4	5.258 E−1	1.527 E−5
9,000	−43.42	9.779	3.080 E+4	4.671 E−1	1.493 E−5
10,000	−49.90	9.776	2.650 E+4	4.135 E−1	1.458 E−5
15,000	−56.50	9.761	1.211 E+4	1.948 E−1	1.422 E−5
20,000	−56.50	9.745	5.529 E+3	8.891 E−2	1.422 E−5
25,000	−51.60	9.730	2.549 E+3	4.008 E−2	1.448 E−5
30,000	−46.64	9.715	1.197 E+3	1.841 E−2	1.475 E−5
40,000	−22.80	9.684	2.871 E+2	3.996 E−3	1.601 E−5
50,000	−2.50	9.654	7.978 E+1	1.027 E−3	1.704 E−5
60,000	−26.13	9.624	2.196 E+1	3.097 E−4	1.584 E−5
70,000	−53.57	9.594	5.221 E+0	8.283 E−5	1.438 E−5
80,000	−74.51	9.564	1.052 E+0	1.846 E−5	1.321 E−5

APPENDIX C UNIT CONVERSION FACTORS

Length

	cm	m	km	inch	foot	mile
1 cm (cgs) =	1	10^{-2}	10^{-5}	3.937×10^{-1}	3.281×10^2	6.214×10^{-6}
1 m (SI) =	10^2	1	10^{-3}	3.937×10^1	3.281	6.214×10^{-4}
1 km =	10^5	10^3	1	3.937×10^4	3.281×10^3	6.214×10^{-1}
1 inch =	2.540	2.540×10^{-2}	2.540×10^{-5}	1	8.333×10^{-2}	1.578×10^{-5}
1 foot (EE, BG) =	3.048×10^1	3.048×10^{-1}	3.048×10^{-4}	1.200×10^1	1	1.894×10^{-4}
1 mile =	1.609×10^5	1.609×10^3	1.609	6.336×10^4	5.280×10^3	1

nautical mile = 1.151 miles

Area

	cm^2	m^2	km^2	$inch^2$	$foot^2$	1 $mile^2$	acre
1 cm^2 (cgs) =	1	10^{-4}	10^{-10}	1.550×10^{-1}	1.076×10^{-3}	3.863×10^{-11}	1.681×10^{14}
1 m^2 (SI) =	10^4	1	10^{-6}	1.550×10^3	1.076×10^1	3.863×10^{-7}	1.681×10^{10}
1 km^2 =	10^{10}	10^6	1	1.550×10^9	1.076×10^7	3.863×10^{-1}	1.681×10^4
1 $inch^2$ =	6.452	6.452×10^{-4}	6.452×10^{-10}	1	6.944×10^{-3}	2.491×10^{-10}	1.593×10^{-7}
1 $foot^2$ (EE, BG) =	9.290×10^2	9.290×10^{-2}	9.290×10^{-8}	1.440×10^2	1	3.587×10^{-9}	2.294×10^{-5}
1 $mile^2$ =	2.589×10^{10}	2.589×10^6	2.589	4.014×10^9	2.788×10^8	1	6.40×10^2
1 acre =	5.948×10^5	5.948×10^1	5.948×10^{-5}	6.278×10^6	4.360×10^4	1.563×10^{-3}	1

Volume

	cm^3	m^3	l	$inch^3$	$foot^3$	gallon
1 cm^3 (cgs) =	1	10^{-6}	10^{-3}	6.102×10^{-2}	3.531×10^{-5}	2.641×10^{-4}
1 m^3 (SI) =	10^6	1	10^3	6.102×10^4	3.531×10^1	2.641×10^2
1 L =	10^3	10^{-3}	1	6.102×10^1	3.531×10^{-2}	2.641×10^{-1}
1 $inch^3$ =	1.639×10^1	1.639×10^{-5}	1.639×10^{-2}	1	5.787×10^{-4}	4.329×10^{-3}
1 $foot^3$ (EE, BG) =	2.832×10^4	2.832×10^{-2}	2.832×10^1	1.728×10^3	1	7.481
1 gallon =	3.786×10^3	3.786×10^{-3}	3.786	2.310×10^2	1.337×10^{-1}	1

1 cm^3 = mL

APPENDIX C UNIT CONVERSION FACTORS

Mass

	g	kg	slug	lb_m
1 g (cgs) =	1	10^{-3}	6.852×10^{-5}	2.204×10^{-3}
1 kg (SI) =	10^3	1	6.852×10^{-2}	2.204
1 slug =	1.459×10^4	1.459×10^1	1	3.217×10^1
1 lb_m (EE) =	4.537×10^2	4.537×10^{-1}	3.108×10^{-2}	1

Density

	g/cm^3	kg/m^3	$slug/ft^3$	$lb_m/in.^3$	lb_m/ft^3
1 g/cm^3 (cgs) =	1	10^3	1.940	3.612×10^{-2}	6.241×10^1
1 kg/m^3 (SI) =	10^{-3}	1	1.940×10^{-3}	3.612×10^{-5}	6.241×10^{-2}
1 $slug/ft^3$ (BG) =	5.154×10^{-1}	5.154×10^2	1	1.862×10^{-2}	3.217×10^1
1 $lb_m/in.^3$ =	2.764×10^1	2.764×10^4	5.371×10^1	1	1.728×10^3
1 lb_m/ft^3 (EE) =	1.601×10^{-2}	1.601×10^1	3.108×10^{-2}	5.787×10^{-4}	1

Time

	second	minute	hour	day	year
second (cgs, SI, EE, BG) =	1	1.667×10^{-2}	2.778×10^{-4}	1.157×10^{-5}	3.169×10^{-8}
minute =	6.000×10^1	1	1.667×10^{-2}	6.944×10^{-4}	1.901×10^{-6}
hour =	3.600×10^3	6.000×10^1	1	4.167×10^{-2}	1.141×10^{-4}
day =	8.640×10^4	1.440×10^3	2.400×10^1	1	2.738×10^{-3}
year =	3.156×10^7	5.259×10^5	8.766×10^3	3.652×10^2	1

Force

	dyne	N	lb_f	poundal
1 g cm s^{-2} (dyne), (cgs) =	1	10^{-5}	2.248×10^{-6}	7.233×10^{-5}
1 kg m s^{-2} = N (SI) =	10^5	1	2.248×10^{-1}	7.233
1 slug ft s^{-2} (lb_f), (BG, EE) =	4.448×10^5	4.448	1	3.217×10^1
1 lb_m ft s^{-2} (poundal) =	1.383×10^4	1.383×10^{-1}	3.108×10^{-2}	1

Velocity

	cm s^{-1}	m s^{-1}	km h^{-1}	ft s^{-1}	mile h^{-1}	knot
1 cm s^{-1} (cgs) =	1	10^{-2}	3.6×10^{-2}	3.281×10^{-2}	2.237×10^{-2}	1.944×10^{-2}
1 m s^{-1} (SI) =	10^2	1	3.600	3.281	2.237	1.944
1 km h^{-1} =	2.778×10^1	2.778×10^{-1}	1	9.113×10^{-1}	6.214×10^{-1}	5.400×10^{-1}
1 ft s^{-1} (EE, BC) =	3.048×10^1	3.048×10^{-1}	1.097	1	6.818×10^1	5.925×10^1
1 mile h^{-1} =	4.470×10^1	4.470×10^{-1}	1.6009	1.467	1	8.689×10^{-1}
1 knot =	5.144×10^1	5.144×10^{-1}	1.852	1.688	1.151	1

1 knot = nautical mile h^{-1}.

APPENDIX C UNIT CONVERSION FACTORS | C-3

Pressure

	dyne cm^{-2}	N m^{-2}	bar	atm	psi	psf	mmHg	in H$_2$O
1 dyne cm^{-2} (cgs) =	1	10^{-1}	10^{-6}	9.869×10^{-7}	1.450×10^{-5}	2.089×10^{-3}	7.501×10^{-4}	4.015×10^{-4}
1 N m^{-2} (Pa), (SI) =	10^1	1	10^{-5}	9.869×10^{-6}	1.450×10^{-4}	2.089×10^{-2}	7.501×10^{-3}	4.015×10^{-3}
1 bar	10^6	10^5	1	9.869×10^{-1}	1.450×10^1	2.089×10^3	7.501×10^2	4.015×10^2
1 atmosphere	1.013×10^6	1.013×10^5	1.013	1	1.470×10^1	2.116×10^3	7.600×10^2	4.068×10^2
1 lb$_f$ in.$^{-2}$ (psi) =	6.895×10^4	6.895×10^3	6.895×10^{-2}	6.805×10^{-2}	1	1.440×10^2	5.171×10^1	2.768×10^1
1 lb$_f$ ft^{-2} (psf), (EE, BG) =	4.788×10^2	4.788×10^1	4.788×10^{-4}	4.725×10^{-4}	6.944×10^{-3}	1	3.591×10^{-1}	1.922×10^{-1}
1 mmHg	1.333×10^3	1.333×10^2	1.333×10^{-3}	1.316×10^{-3}	1.934×10^{-2}	2.785	1	5.353×10^{-1}
1 in. H$_2$O =	2.491×10^3	2.491×10^2	2.491×10^{-3}	2.458×10^{-3}	3.613×10^{-2}	5.202	1.868	1

H$_2$O @ 4°C, Hg @ 0°C

Absolute Viscosity

	centipoise	poise	kg m^{-1} s^{-1}	lb$_f$ s ft^{-2}	lb$_m$ ft^{-1} s^{-1}
1 centipoise =	1	10^{-2}	10^{-3}	2.089×10^{-5}	6.720×10^{-4}
1 g cm^{-1} s^{-1} (poise), (cgs) =	10^2	1	10^{-1}	2.089×10^{-3}	6.720×10^{-2}
1 kg m^{-1} s^{-1} (SI) =	10^3	10	1	2.089×10^{-2}	6.720×10^{-1}
1 lb$_f$ s ft^{-2} (EE, BG) =	4.788×10^4	4.788×10^2	4.788×10^1	1	3.217×10^1
1 lb$_m$ ft^{-1} s^{-1} =	1.488×10^3	1.488×10^1	1.488	3.108×10^{-2}	1

Kinematic Viscosity

	centistoke	stoke	m^2 s^{-1}	ft^2 s^{-1}	ft^2 h^{-1}
1 centistoke =	1	10^{-2}	10^{-6}	1.395×10^2	3.875×10^{-2}
1 cm^2 s^{-1} (stoke), (cgs) =	10^2	1	10^{-4}	1.395×10^4	3.875
1 m^2 s^{-1} (SI) =	10^6	10^4	1	1.395×10^8	3.875×10^4
1 ft^2 s^{-1} (EE, BG) =	7.169×10^{-3}	7.169×10^{-5}	7.169×10^{-9}	1	2.778×10^{-4}
1 ft^2 h^{-1} =	2.581×10^1	2.581×10^{-1}	2.581×10^{-5}	3.600×10^3	1

Volume Flowrate

	cm^3 s^{-1}	m^3 s^{-1}	l min^{-1}	in.^{3} s^{-1}	ft^3 s^{-1}	ft min^{-3}	gal min^{-1}	gal h^{-1}
1 cm^3 s^{-1} (cgs) =	1	10^{-6}	6.000×10^{-2}	6.102×10^{-2}	3.531×10^{-5}	2.119×10^{-3}	1.585×10^{-2}	9.510×10^{-1}
1 m^3 s^{-1} (SI) =	10^6	1	6.000×10^4	6.102×10^4	3.531×10^1	2.119×10^3	1.585×10^4	9.510×10^5
1 l min^{-1}	1.667×10^1	1.667×10^{-5}	1	1.017	5.885×10^{-4}	3.531×10^{-1}	2.641×10^{-1}	1.585×10^1
1 in.^{3} s^{-1}	1.639×10^1	1.639×10^{-5}	9.833×10^{-1}	1	5.787×10^{-4}	3.472×10^{-2}	2.598×10^1	1.559×10^1
1 ft^3 s^{-1} (EE, BG) =	2.832×10^4	2.832×10^{-2}	1.699×10^3	1.728×10^3	1	6.000×10^1	4.489×10^2	2.693×10^4
1 ft^3 min^{-1} =	4.719×10^2	4.719×10^{-4}	2.832×10^1	2.880×10^1	1.667×10^{-2}	1	7.481	4.489×10^2
1 gal min^{-1} =	6.309×10^1	6.309×10^{-5}	3.786	3.849×10^{-2}	2.228×10^{-3}	1.337×10^{-1}	1	1.667×10^{-2}
1 gal h^{-1} =	1.052	1.052×10^{-6}	6.310×10^{-2}	6.416×10^{-2}	3.713×10^{-5}	2.223×10^{-3}	6.000×10^1	1

C-4 APPENDIX C UNIT CONVERSION FACTORS

Energy, Work, Heat

	erg	joule	kW-h	calorie	$lb_m\ ft^2\ s^{-2}$	$ft\ lb_f$	Btu	$hp\ h^{-1}$
1 g cm² s⁻² (erg), (cgs) =	1	10^{-7}	2.778×10^{-14}	2.389×10^{-8}	2.373×10^{-6}	7.376×10^{-8}	9.481×10^{-11}	3.725×10^{-14}
1 kg m² s⁻² (joule), (SI) =	10^7	1	2.778×10^{-7}	2.389×10^{-1}	2.373×10^{1}	7.376×10^{-1}	9.481×10^{-4}	3.725×10^{-7}
1 kW-h =	3.600×10^{13}	3.600×10^{6}	1	8.601×10^{5}	8.543×10^{7}	2.655×10^{6}	3.413×10^{3}	1.341
1 calorie =	4.186×10^{7}	4.186	1.163×10^{-6}	1	9.922×10^{1}	3.087	3.968×10^{-3}	1.559×10^{-6}
1 $lb_m\ ft^2\ s^{-2}$ (EG, BG) =	4.214×10^{5}	4.214×10^{-2}	1.171×10^{-8}	1.008×10^{-2}	1	3.108×10^{-2}	3.994×10^{-5}	1.570×10^{-8}
1 $ft\ lb_f$ =	1.356×10^{7}	1.356	3.766×10^{-7}	3.239×10^{-1}	3.217×10^{1}	1	1.285×10^{-3}	5.051×10^{-7}
1 Btu =	1.055×10^{10}	1.055×10^{3}	2.930×10^{-4}	2.520×10^{2}	2.504×10^{4}	7.779×10^{2}	1	3.929×10^{-4}
1 $hp\ h^{-1}$ =	2.685×10^{13}	2.685×10^{6}	7.547×10^{-1}	6.414×10^{5}	6.371×10^{7}	1.980×10^{6}	2.545×10^{3}	1

Power

	$erg\ s^{-1}$	watt	kilowatt	$ft\ lb_f\ s^{-1}$	$cal\ s^{-1}$	$Btu\ h^{-1}$	hp
1 $erg\ s^{-1}$ (cgs) =	1	10^{-7}	10^{-10}	7.376×10^{-8}	2.389×10^{-8}	3.413×10^{-7}	1.341×10^{-10}
1 $joule\ s^{-1}$ (watt), (SI) =	10^7	1	10^{-3}	7.376×10^{-1}	2.389×10^{-1}	3.413	1.341×10^{-3}
1 kilowatt =	10^{10}	10^{3}	1	7.376×10^{2}	2.389×10^{2}	3.413×10^{3}	1.341
1 $ft\ lb_f\ s^{-1}$ (EE, BG) =	1.356×10^{7}	1.356	1.356×10^{-3}	1	3.239×10^{-1}	4.628	1.818×10^{-3}
1 $cal\ s^{-1}$ =	4.186×10^{7}	4.186	4.186×10^{-3}	3.087	1	1.429×10^{1}	5.613×10^{-3}
1 $Btu\ h^{-1}$ =	2.930×10^{6}	2.930×10^{-1}	2.930×10^{-4}	2.161×10^{-1}	7.000×10^{-2}	1	3.929×10^{-4}
1 horsepower =	7.457×10^{9}	7.457×10^{2}	7.457×10^{-1}	5.500×10^{2}	1.782×10^{2}	2.545×10^{3}	1

Temperature

	°C	K	°F	°R
Celsius =	1	K − 273.15	(5/9)°F − 17.78	(5/9)°R − 273.15
Kelvin =	°C + 273.15	1	(5/9)°F + 255.37	(5/9)°R
Fahrenheit =	(9/5)°C + 32	(9/5)K − 459.7	1	°R − 459.7
Rankine =	(9/5)°C + 491.7	(9/5)K	°F + 459.7	1

CREDITS

Chapter 1, opening figure courtesy of NASA; figure HB1-1courtesy of Julianne Corkery; figure HB1-2a courtesy of Bettmann/CORBIS; figure HB1-2b courtesy of Bettmann/CORBIS; figure 1-1a courtesy of Marc Garanger/CORBIS; figure 1-1b courtesy of NASA; figure HB1-3a courtesy of Bettmann/CORBIS; figure 1-3 is supplied courtesy of MIRA, adapted with permission from Fluent Inc.; figure HB1-5a courtesy of Underwood & Underwood/CORBIS; figure HB 1-5b courtesy of Smithsonian Institution; figure 1-4 courtesy of NASA; figure HB1-6a courtesy of AIP Emilio Segrè Visual Archives, Landé Collection; figure HB1-6b is adapted from Batchelor, G.K., *An Introduction to Fluid Dynamics,* Cambridge University Press, Cambridge, 1967, plate 4; figure 2-1b courtesy of Prof. Andrew Davidhazy, RIT; figure 2-4 courtesy of Bettmann/CORBIS; figure 2-6 courtesy of Jet Edge, Inc., St. Michael, Minnesota; figure 2-10a courtesy of NASA; figure 3-4 courtesy of NASA; figure 3-7 courtesy of University of Washington Libraries, Special Collection; figure 3-11 courtesy of University of Nevada, Las Vegas Libraries; figure 4-1 courtesy of Lyn Topinka/USGS/CVO; figure 5-20 courtesy of NASA Johnson Space Center, Orbital Debris Program Office; figure 5-51a courtesy of Peter Turnley/CORBIS; figure 8-6a is adapted from Munson, B.R., Young, D.F., Okiishi, T.H., *Fundamentals of Fluid Mechanics, 4e,* John Wiley & Sons, Inc., New York, 1999, used by permission of John Wiley & Sons, Inc.; figure 9-1a courtesy of NASA; figure 9-1b courtesy of Sultan Alam/DNR; figure 9-4 is adapted from Taylor, Sir Geoffrey I.; Batchelor, G.K. (ed.) "The Scientific Papers of Sir Geoffrey Ingram Taylor", Vol. 3, Cambridge University Press, Cambridge, U.K., 1963, reprinted with the permission of Cambridge University Press; figure 9-6 courtesy of Bettmann/CORBIS; figure 10-2a is adapted from Merzkirch, W., *Flow Visualization,* Academic Press, New York, 1974, reprinted with permission from Elsevier; figure 10-2b is adapted from Merzkirch, W., *Flow Visualization,* Academic Press, New York, 1974, reprinted with permission from Elsevier; figure 10-3a courtesy of NASA; figure 10-3b courtesy of NASA; figure 10-7 courtesy of Jose Luis Pelaez, Inc./CORBIS; figure 10-15 courtesy of Paul Cherfils/Stone/Gettyimages; figure 10-20 courtesy of Steve Harrington/Flometrics; figure 10-21 courtesy of Steve Harrington/Flometrics; figure 10-25 courtesy of NASA; figure 10-27 image supplied courtesy of Daimler Chrysler Corporation, adapted with permission from Fluent Inc.; figure 10-45a courtesy of Dean Conger/CORBIS; figure 10-45b courtesy of Lester Lefkowitz/CORBIS; figure 10-62 courtesy of Prof. Andrew Davidhazy, RIT; figure 11-1 courtesy of Jonathan Freund/Physics of Fluids, Gallery of Fluid Motion; figure 11-16 courtesy of ITT Fygt Corporation, adapted with permission from FLUENT Inc.; 11-17a courtesy of Steve Harrington/Flometrics; figure 11-17b courtesy of Steve Harrington/Flometrics; figure 12-6 is adapted from Merzkirch, W., *Flow Visualization,* Academic Press, New York, 1974, reprinted with permission from Elsevier; figure 12-7 courtesy of Craig Lovell/CORBIS; figure 13-1 courtesy of Galen Rowell/CORBIS; table 13-5 information from *ASHRAE Fundamentals,* American Society of Heating, Refrigeration, and Air-Conditioning, Engineers, Washington, DC, 1989; figure 13-11 courtesy of Oak Ridge National Lab; figure 13-14 is adapted from Fox, R.W., McDonald, A.T., Pritchard, P.J., *Introduction to Fluid Mechanics, 6e,* John Wiley & Sons, Inc., New York, 2004, used by permission of John Wiley & Sons, Inc.; figure 13-38 is adapted from Fox, R.W., McDonald, A.T., Pritchard, P.J., *Introduction to Fluid Mechanics, 6e,* John Wiley & Sons, Inc., New York, 2004, used by permission of John Wiley & Sons, Inc.; figure 14-1 courtesy of Volvo Truck Corporation; figure 14-23a courtesy of Hulton-Deutsch Collection/CORBIS; figure 14-25 a,b is adapted from Batchelor, G.K.: "An Introduction to Fluid Dynamics", Cambridge University Press, Cambridge, U.K., 1967, reprinted with the permission of Cambridge University Press; figure14-30a courtesy of Danny Lehman/CORBIS; figure 14-30b courtesy of Picture Arts/CORBIS; figure 14-39 courtesy of Richard Cutter; figure 15-1a courtesy of Julie Keller Photography; figure 15-1b courtesy of Julianne Corkery; figure 15-1c courtesy of NASA; figure 15-1d courtesy of G. Kopelow/firstlight.ca; figure 15-3 courtesy of Nehemias Lacerda; figure 15-8a courtesy of NASA; figure 15-21b courtesy of Pacific Tsunami Museum Archives—Martin Polhemus Collection; table 15-3 and figure 15-42 adapted from Gerhart, Gross, and Hochstein, *Fundamentals of Fluid Mechanics,* 2nd edition; figures 1-2a, 2-1, 3-1, 3-3 a,b,c, 3-6, 6-20a,b, 6-24, 8-6b, 9-1c, 9-5a, 10-13b, 10-19a, 10-19b, 10-26, 10-28, 12-2, 12-3, 12-4, 12-5, 12-8a, 12-8b, 12-10, 12-17, 12-20, 12-31a, 12-31b, 13-18 left, 13-18 right, 13-19 left, 13-19 right, 13-42a, 13-42b, 14-6a, 14-6b, 14-10a, 14-10b, 14-10c, 14-13, 14-19, 14-37a,b,c,d, 14-38b, 14-40, 14-10a,b,c, 14-13, 14-19, 14-37 a,b,c,d, 14-38b, 14-40 were adapted from *Visualized Flow: Fluid Motion in Basic and Engineering Situations revealed by Flow Visualization,* compiled by The Japan Society of Mechanical Engineers, Pergamon Press, New York, 1988, reprinted with permission from Elsevier.

D-1

INDEX

A
Absolute pressure, 53–54
Absolute viscosity, 16, 701–702, 1029
   conversion factors for, 1029
   vs. dynamic viscosity, 16
   introduction to, 701–702
Acceleration, fluid, 229–233, 312–319, 578–589, 690
   convective acceleration, 316–318, 690
   coordinates for, 316–319
   introduction to, 312–314
   local acceleration, 316–318, 690
   particle paths for, 314–316
   pathlines for, 314–316
   rectilinear acceleration, 229–233
   substantial/material/co-moving derivatives of, 315–316
   Taylor series expansions for, 315
Addition rates, total energy, 424
Air, 1019–1020, 1022–1024
Airfoils, 926–933. *See also under individual topics*
   attack angles and, 928
   Bleriot airfoils, 138
   boundary layers of, 898–902
   camber and, 928–931
   chords and, 928
   Clark Y airfoils, 138
   drag and, 137–140, 177, 562–563, 926–933
   Göttingen airfoils, 138
   historical perspectives of, 138
   introduction to, 926–932
   Joukowsky airfoils, 138
   lift and, 137–140, 562–563, 926–933
   M-6 airfoils, 138
   National Advisory Committee on Aeronautics (NACA) airfoils, 899–902, 928–933
      NACA 0018, 928
      NACA 66 series, 902
      NACA 4412, 899
      NACA 23012, 928–933
   nomenclature of, 138
   of P-51 Mustangs, 902
   planforms and, 928
   protuberances and, 930–931
   spans and, 928
   stall and, 929
   thicknesses and, 928
   vortices and, 932
Aligned surfaces, planar, 234–238, 256–260
Ammonia, 1023
Amplitude waves, small, 971–972
   flow analyses, 817–851
      multipath analyses, 846–851
      single path analyses, 817–846
   incompressible flow analyses, 713–788
      introduction to, 713–718, 780–782
      of inviscid irrotational flow, 760–780
      of problems for, 782–788

   of steady viscous flow, 718–744
   of turbulent flow, 754–759
   of unsteady viscous flow, 744–754
   system volume analyses, 376–382
Angle valves, conventional, 853
Angles, 90, 138
   attack angles, 138
   contact angles, 90
Angular momentum, 341, 439–450
   balance, 439–450
      examples for, 442–446, 448–450
      introduction to, 439–440
      reaction forces and, 443
      steady processes and, 440
      torque and, 439–440
         body forces and, 441
         introduction to, 439–440
         surface forces and, 441–442
   introduction to, 341
   transport, 440–441
Apparent forces, 148–149
Applied Fluid Dynamics Handbook, 922–923
Aqueducts, Roman, 4, 934
Archimedes, 5, 273
Archimedes' principle, 58–60, 269–275
Area change flow, 120–124, 511–518, 558–559
Area conversion factors, 1027
Atmosphere models, 217–218
Atmosphere properties, standard U.S., 45–46, 222–224, 1025–1026
Attack angles, 138, 903, 928
Attractive forces, intermolecular, 80–83
Average densities, 363
Average velocities and flow rates, 358–363
   average densities for, 363
   examples for, 360–362
   introduction to, 358–363
Axial flow compressors and pumps, 854
Axial stress, 252
Axisymmetric flow, 327–330

B
Balance, 187–190, 386–450. *See also under individual topics*
   angular momentum balance, 439–450
   energy balance, 420–439
   force balance, 187–190
   mass balance, 386–397
   momentum balance, 397–420
Ball valves, 853
Barometric pressure, 213–215
Bed slopes, 947
Bends, gradual *vs.* miter, 834–835
Benzene, 1023
Bernoulli, Daniel, 7, 478
Bernoulli, Johann, 478
Bernoulli equation, 477–521
   applications of, 496–521

   area change flow and, 511–518
   Bernoulli, Daniel and, 7, 478
   Bernoulli, Johann and, 478
   bubbles and, 485
   case study applications of, 121
   cavitation and, 482–487
   compressible fluids and, 487–490
   contraction coefficients and, 520–521
   diffusers and, 512
   examples for, 480–487, 489, 500–502, 507–509, 511, 513–515, 517–518, 520
   historical perspectives of, 7, 478
   Hydraulica and, 478
   incompressible fluids and, 479–482
   introduction to, 477–479
   inviscid fluid flow and, 477–490, 496–521. *See also* Inviscid fluid flow and Bernoulli equation
   isentropic processes and, 487–488
   Mach numbers and, 499
   nozzles and, 512
   oscillation frequencies and, 481
   perfect gas law and, 487–488, 515
   pitot-static tubes, 499
   pitot tubes, 496–503
   pressure variations and, 60
   siphons and, 503–509
   sluice gates, 509–511
   steady fully-developed flow and, 799
   subsonic flow, 507
   sudden contraction and, 512
   sudden enlargement and, 512
   supersonic flow, 507
   tank draining, 518–521
   Torricelli, E. and, 519
   Torricelli's law and, 519–521
   U-tube manometers, 500–501
   vapor bubbles and, 504–505
   vapor pressure and, 485–487, 504
   vena contracta effects and, 510, 520–521
   venturi and, 486–487
Bessel functions, 752
BG (British Gravitational) Unit System, 31
Blasius, H., 130–131, 887–888
Blasius solution, 887–894
Bleriot airfoils, 138
Blevins, Robert, 922–923
Blow fans, radial, 854
Bluff bodies, 177, 904
Body forces, 148–160, 177, 200–201
   examples for, 152–154
   gravitational body force per unit mass, 152–153
   hydrostatic stress and, 200–201
   introduction to, 148–160
   origins of, 149–151
   streamlined body drag, 177

I-1

Bond numbers, 112
Boundaries. *See also under individual topics*
  boundary layers, 128–132, 171–172, 561, 603–605, 884–916
    of airfoils, 898–902
    Blasius, H. and, 887–888
    Blasius solution and, 887–894
    of curved surfaces, 900
    displacement thicknesses of, 885–886
    drag and, 886–887
    external flow and, 884–902
    Falkner-Skan solution and, 901
    of flat plates, 128–132, 171–172, 561, 887–898
    historical perspectives of, 902
    introduction to, 884–887
    Jacobs, Eastman and, 902
    kinematic viscosity and, 884–885
    laminar boundary layers, 130–131, 887–894
    lift and, 886–887
    momentum thicknesses of, 886–887
    National Advisory Committee on Aeronautics (NACA) and, 902
    power-law velocity profile model and, 895–896
    Prandtl equations for, 888–894
    Reynolds numbers for, 884–887
    separation of, 604–605, 916
    streamfunctions of, 889
    streamlines and streamtubes of, 604–605
    thicknesses of, 885–887
    turbulent boundary layers, 894–898
    wall shear stress of, 886–887
  bounding surfaces, 339
  conditions of, 336–340, 702–703
    equations for, 702–703
    Navier-Stokes equations and, 702–703
    no-slip *vs.* no-penetration conditions, 336–337
    permeable *vs.* impermeable conditions, 339–340
Bourdon tube pressure gages, 54
British Gravitational (BG) Unit System, 31
Broad crested weirs, 1012
Brukau, 914
Bubbles, 88–89, 485, 504–505, 593, 599
  hydrogen bubbles, 593, 599
  introduction to, 485
  surface tension and, 88–89
  vapor bubbles, 504–505
Buckingham Pi theorem, 536–540
Bulk compressibility modulus, 73–80
  examples of, 74–79
  introduction to, 73–74
  isentropic processes and, 74–78
  isothermal processes and, 74
  Mach numbers and, 79–80
  speed of sound and, 76–80
  values of, 73–74
Bulk viscosity, 83–85, 673, 700–702
Bumps, flow over, 953–961
Buoyancy, 269–275
  Archimedes' principle and, 58–60, 269–275
  center of gravity *vs.* center of buoyancy, 277–278
  examples for, 271–272, 274–275
  fluid statics and, 269–275
  introduction to, 269
  Newton's second law and, 269–270
  pressure and, 58–60

C
Calderas, 218–219
Camber, 928–931
Capillary actions, 90–93, 111

Carbon compounds, 1019–1024
  carbon dioxide, 1019–1020, 1024
  carbon monoxide, 1024
  carbon tetrachloride, 1020, 1023
Cartesian coordinates, 301–302, 316, 397–398, 678–679, 684
Case study applications, 103–145.
    See also under individual topics
  for Bernoulli equation, 121
  for common dimensionless groups, 103–114
  for dimensional analyses (DA) and similitude, 557–563
  for flow classifications, 332
  introduction to, 103–105, 140–141
  problems for, 141–145
  specific examples of, 114–140
    airfoil lift and drag, 137–140
    area change flow, 120–122
    attack angles, 138
    Bernoulli equation and, 121
    Blasius, H. and, 130–131
    chord lengths, 138
    cylinder drag, 132–137
    drag coefficients, 133–137
    examples for, 118–120, 123–124, 127–128, 131–132, 136–137, 139–140
    fan laws, 124–128
    flat plate boundary areas, 128–132
    head coefficients, 127–128
    introduction to, 114–115
    laminar boundary layers, 130–131
    laminar flow, 117
    laminar-to-turbulent transitions, 129
    National Advisory Committee on Aeronautics (NACA) and, 138–140
    planforms, 138
    power coefficients, 127–128
    Prandtl, Ludwig and, 128–129
    pump and fan laws, 124–128
    round pipe flow, 115–120
    skin fiction coefficients, 130
    span, 138
    sphere drag, 132–137
    sudden contraction, 124
    sudden expansion, 122–124
    total head parameters, 125
    turbulent boundary layers, 131–132
    turbulent flow, 117–118
    Wright brothers and, 138
  for velocity fields, 306
Cavitation, 482–487, 860–862
  Bernoulli equation and, 482–487
  system design elements and, 860–862
Celsius temperature scale, 64–65
Center of gravity *vs.* center of buoyancy, 277–278
Centrifugal compressors and pumps, 854
Centrifugal forces, 148–149
CFD (computational fluid dynamics), 14, 706–707, 752–759
  cylinder flow and, 752–754
  equations for, 706–707
  historical perspectives of, 14
  parallel plates and, 757–759
Channels, 942–1018. *See also under individual topics*
  open channel flow applications, 942–1018
  optimal channel cross sections, 999–1002
  perimeters of, 947–948
  prismatic channels, 951
  varying width horizontal channels, 961–965
Chen equation, 817
Chezy, Antoine, 994
Chezy coefficient, 992
Chezy equation, 992
Chords and chord lengths, 138, 928

Circular Couette flow, 723–732
Circular disks, 906, 909
Circular rods, 909
Circulation, 628–632
Clark Y airfoils, 138
Classifications, 148–149, 320–336.
    See also under individual topics
  of flow, 320–336
  of fluid forces, 148–149
Clermont, 52
Closed *vs.* open surfaces, 341
Co-moving derivatives, 315–316
Coefficients. *See also under individual topics*
  Chezy coefficient, 992
  contraction coefficients, 520–521
  diffusion coefficients, 351–352
  drag coefficients, 133–134, 905–926
  head coefficients, 127–128, 560
  head loss coefficients, 825–826
  inlet coefficients, 826–829
  kinetic energy coefficients, 822–823
  Manning roughness coefficient, 992–993
  power coefficients, 127–128, 560
  skin friction coefficients, 130, 887
  thermal expansion coefficients, 67–70
Colebrook formula, 816–817
Common dimensionless groups, 103–114.
    See also Dimensionless groups
Common gas parameters, 72
Common nominal pipe size dimensions, 818–819
Compressibility, 18, 73–80, 487–490
  bulk compressibility modulus, 73–80
  compressible fluids, 487–490
  introduction to, 18
Compressors, axial flow and centrifugal, 854
Computational fluid dynamics (CFD), 14, 706–707, 752–759
  cylinder flow and, 752–754
  equations for, 706–707
  historical perspectives of, 14
  parallel plates and, 757–759
Concentrations, 341
Conductivity, thermal, 65–66
Cones, 910
Conservation, energy, 978–989
Constant density fluids, 211–233
  in gravity fields, 211–218
  in rectilinear acceleration, 229–233
  in rigid rotations, 222–229
  vs. variable density fluids, 218–222
Constitutive model, Newtonian fluids, 671–678
  equations for, 671–678
  examples for, 676–678
  introduction to, 671–672
Contact angles, 90
Continuity equation, 660–666
  examples for, 662–666
  Gauss's theorem and, 661
  introduction to, 660–662
  Reynolds transport theorem and, 660–662
Continuum hypothesis, 22–24
Continuum models, 301, 348–350, 595
Continuum *vs.* noncontinuum descriptions, 24–26
Contour plots, 305–308
  density contour plots, 308
  speed contour plots, 305–306
  streamline contour plots, 307
  temperature contour plots, 309, 322–323
Contraction, 120, 510–521
  coefficients for, 520–521
  sudden contraction, 124, 512
  vena contracta effects and, 510, 520–521

INDEX

Flat plate boundary layers, 128–132, 171–172, 561, 887–898
  flow over, 171–172
  introduction to, 128–132, 171–172
  laminar boundary layers, 887–894
  turbulent boundary layers, 894–898
Flettner, Anton, 914
Flow-related topics. *See also under individual topics*
  applications, 789–1018
    external flow applications, 882–941
    open channel flow applications, 942–1018
    pipe and duct flow applications, 791–881
  differential analyses, 571–788
    fluid dynamics governing equations, 659–712
    incompressible flow analyses, 713–788
    visualization and structure elements, 573–658
  flow classifications, 320–336
    axisymmetric flow, 327–330
    case study applications for, 332
    examples for, 324–327, 335
    fully developed flow, 330–332
    introduction to, 320–321
    Karman vortices and, 334
    one-dimensional flow, 321–327
    spatially periodic flow, 327–330
    steady flow, 332–336
    steady processes and, 332–336
    Strouhal numbers and, 335
    temperature contour plots and, 322–323
    temporally periodic flow, 332–336
    three-dimensional flow, 321–327
    two-dimensional flow, 321–327
    uniform flow, 327–330
    velocity vector plots and, 322–323, 328–329
{FLtT} systems, 31–32
Fluid acceleration, 312–319
  acceleration, convective and local, 316–318
  coordinates for, 316–319
    Cartesian coordinates, 316
    cylindrical coordinates, 317–319
  introduction to, 312–314
  particle paths for, 314–316
  pathlines for, 314–316
  substantial/material/co-moving derivatives of, 315–316
  Taylor series expansions for, 315
Fluid dynamics governing equations, 659–712. *See also* Equations
  Bernoulli equation, 692–699
  computational fluid dynamics (CFD) and, 706–707
  constitutive model, Newtonian fluids, 671–678
  continuity equation, 660–666
  energy equation, 699–702
  Euler equations, 683–692
  initial and boundary conditions and, 702–703
  introduction to, 659–660, 708–709
  momentum equation, 666–671
  Naiver-Stokes equations, 678–683
  nondimesionalization and, 703–706
  problems for, 709–712
Fluid energy, 93–97
  enthalpy and, 96
  internal energy, 93–94
  introduction to, 93
  kinetic energy, 94–95
  potential energy, 95
  total energy, 95–96
Fluid forces, 146–196
  body forces, 148–160
  classifications of, 148–149
    apparent forces, 148–149

    body forces, 148–149
    centrifugal forces, 148–149
    Coriolis forces, 148–149
    introduction to, 148
    surface forces, 148–149
  fluid stress, 178–187
  force balance, 187–190
  introduction to, 146–148, 190–191
  Mount St. Helens volcano and, 147
  origins of, 149–151
    electromagnetic forces and, 150–151
    electrostatic precipitators and, 150
    gravitational forces and, 149–150
  problems for, 191–196
  surface forces, 148–151
Fluid properties, 42–102, 1019–1026
  Archimedes' principle and, 58–60
  bulk compressibility modulus, 73–80
  comparison data for, 1019–1024
  fluid energy, 93–97
  introduction to, 42–43, 97–98
  Langrangian fluid properties, 589–590
  mass, weight, and density, 43–51
  origins of, 149–151
  pressure, 51–64
  problems for, 99–102
  surface tension, 85–93
  temperature and thermal properties, 64–70
  viscosity, 80–85
Fluid statics, 197–298
  buoyancy and, 269–275
  hydrostatic equation and, 201–210
  hydrostatic forces and, 233–252
  hydrostatic moments and, 252–267
  hydrostatic pressure distribution and, 198–199, 210–233
  hydrostatic stress and, 199–201
  immersed bodies, equilibrium and stability of, 275–278
  introduction to, 197–199, 278–280
  problems for, 280–298
  resultant forces and points of application (POAs), 267–269
Fluid stress, 178–187
  examples for, 184–187
  introduction to, 178–179
  stress tensors and, 178–187
  stress vectors and, 178–187
Fluid transport, 299–374. *See also* Velocity fields and fluid transport
  average velocities and flow rates and, 358–363
  description and components of, 337–358
  flow classifications and, 320–336
  fluid acceleration, 312–319
  fluid velocity fields, 300–312
  introduction to, 299–300, 363–365
  no-slip *vs.* no-penetration boundary conditions and, 336–337
  problems for, 365–374
  substantial derivatives and, 319–320
Fluid velocity fields, 299–374. *See also* Velocity fields and fluid transport
Fluids, basic terminology, 14–22
Flux, 341, 351
  convective flux integrals, 341
  diffusive flux vectors, 351
{FMLtT} systems, 31–32
Forces. *See also under individual topics*
  apparent forces, 148–149
  centrifugal forces, 148–149
  conversion factors for, 1028
  Coriolis forces, 148–149
  electromagnetic forces, 150–151
  fluid forces, 146–196

  gravitational body forces and, 149–150, 210–211
  hydrostatic forces, 233–252
  reaction forces, 407–410, 443
  resultant forces and points of application (POAs), 267–269
  surface forces, 148–151, 160–177
  work done by surface forces, 424–425
Form/pressure drag, 177, 903, 911–912
Fourier, Jean Baptiste, 680
Fourier's law, 65–67, 351
Free surface slopes, 946–947
Freon 12, 1023
Frequencies, oscillation, 481
Friction factors, 475–477, 793, 801–817, 903, 980
  Chen equation and, 817
  Colebrook formula and, 816–817
  examples for, 802–805, 813, 816–817
  friction drag, 903
  friction slopes, 980
  frictionless flow, 475–477
  hydraulic diameters and, 803–804
  introduction to, 793, 801–805
  in laminar flow, 805
  Moody charts and, 813–815
  pipe schedules and, 817
  Poiseuille flow and, 806–807
  relative roughness and, 813–815
  turbulent flow and, 812–817
Froude, William, 109
Froude numbers, 109–110, 553, 703, 952–978
Full-width weirs, 1010
Fully-developed flow, 173–174, 330–332, 793–817. *See also* Steady fully-developed flow
Fulton, Robert, 52

G

Gages, 53–54, 411
  Bourdon tube pressure gages, 54
  pressure and, 53–54, 411
Gas turbines, 854
Gases, liquids, and solids (basic terminology), 14–22
Gasoline, 1023
Gates, 416, 509–511, 853, 944, 1003–1008
  sluice gates, 416, 509–511, 944, 1003–1008
  wedge gate valves, 853
Gauss's theorem, 205–210, 477, 637, 642–643, 661–662, 667
Geometric similarity, 550
Globe valves, conventional, 853
Glycerin, 1019–1020, 1023
Gossamer Condor, 552
Göttingen airfoils, 138
Governing equations, 659–712. *See also* Equations
Gradients, velocity, 17, 475, 612–618
Gradual *vs.* miter bends, 834–835
Gradually varying flow (GVF), 951, 978–979
Gravity, 49–51, 149–153, 210–211, 277–278, 423
  center of gravity *vs.* center of buoyancy, 277–278
  gravitational body forces, 149–153, 210–211
  gravitational potential energy, 423
  gravity fields, 211–222
  hydrostatic pressure distribution and, 211–222
  specific gravity, 49–51
Groups, dimensionless, 103–114. *See also* Dimensionless groups
Guide vanes, 824
GVF (gradually varying flow), 951, 978–979

H

Heads and head loss, 125–128, 560, 793–846, 860–862
  coefficients for, 127–128, 560, 825–826
  head loss, 793, 799–801, 824–835
    coefficients for, 825–826
    major head loss, 793, 799–801
    minor head loss, 824–835
  Net Positive Suction Heads (NPSHs), 860–862
  pump and turbine heads, 835–846
  total head parameters, 125
Heat, 222, 341, 423–439, 573–578, 1030
  conversion factors for, 1030
  fluid transport and, 341
  heating, ventilating, and air-conditioning (HVAC) systems, 573–578
  specific heat, 222, 439
  total heat transfer rates, 423–424
Helium, 1019–1020, 1024
Hemispheres, 906, 909
Heptane, 1019–1020
Historical perspectives, 4–14. *See also under individual topics*
  of airfoils, 138
  of books, 7–8
    Mathematical Principles of Natural Philosophy, 7–8
    Principia, 7–8
  of boundary layers, 902
  of computational fluid dynamics (CFD), 14
  of equations, 478, 680–684
    Bernoulli equation, 478
    Euler equations, 684
    Navier-Stokes equations, 680
  of incompressible flow analyses, 716
  introduction to, 4–14
  of persons, 4–14
    Archimedes, 5, 273
    Bernoulli, Daniel, 7
    da Vinci, Leonardo, 5
    d'Alembert, Jean le Rond, 7–8, 12
    Euler, Leonhard, 7–8
    Navier, Claude, 8
    Newton, Isaac, 7, 14
    Pascal, Blaise, 5
    Pitot, Henri, 8–9
    Poiseuille, J. L. M., 8–9
    Prandtl, Ludwig, 12–13
    Stokes, George, 8, 14
    Torricelli, E., 5
    von Karman, T., 12
    Wright, Wilbur, 12
    Wright brothers, 12
  Roman aqueducts, 4
  of steady viscous flow, 721–722
  of uniform depths, 994
Hoop stress, 252
Horizontal channels, varying width, 961–965
Hydraulic diameters, 793
Hydraulic jumps, 972–978
  classifications of, 976–978
    impossible jumps, 976
    rough, intermittent, strong jumps, 976
    stable, balanced, steady jumps, 976
    standing wave/undulant jumps, 976
    weak jumps, 976
  examples for, 977–978
  introduction to, 548, 972–978
  spillways and, 977
Hydraulic radius, 948
Hydrogen, 1019
Hydrogen bubbles, 593, 599
Hydrostatic equation, 201–210
  differential hydrostatic equation, 205–210
  examples for, 204–210
  Gauss's theorem and, 205–210
  integral hydrostatic equation, 202–205, 242–248
  introduction to, 201–202
  Newton's second law and, 202
Hydrostatic forces, 233–252
  examples for, 236–237, 240–241, 244–252
  integral hydrostatic equation and, 202–205, 242–248. *See also* Hydrostatic equation
  introduction to, 233–234
  outward unit normal for, 235
  stress and, 233–234, 251–253
    axial stress, 252
    hoop stress, 252
    stress vectors, 233–234
  surfaces and, 234–252
    curved surfaces, 248–252
    planar aligned surfaces, 234–238
    planar nonaligned surfaces, 238–248
Hydrostatic moments, 252–267
  examples for, 255, 258–259, 261–263, 265–267
  introduction to, 252–256
  planar nonaligned surfaces and, 260–267
  surfaces and, 256–267
    planar aligned surfaces, 256–260
    planar nonaligned surfaces, 260–267
Hydrostatic pressure distribution, 198–199, 210–233
  barometric pressure and, 213–215
  calderas and, 218–219
  constant density fluids and, 211–233
  control volume (CV) analyses and, 411
  Earth's atmosphere models and, 217–218
  gravitational body forces and, 210–211
  gravity fields and, 211–222
  introduction to, 210–211
  isothermal perfect gasses and, 221–222
  Lake Nyos and, 218–219
  polytropic law and, 221–222
  pump intakes and, 213
  rectilinear acceleration and, 229–233
  rigid rotations and, 222–229
  specific heat and, 222
  U-tube manometers and, 215–218
  U.S. standard atmosphere properties and, 222–224
  vapor pressure, mercury, 214–215
  variable density fluids and, 218–222
Hydrostatic stress, 199–201
  body forces and, 200–201
  introduction to, 199–200
  Newton's second law and, 200–201
  normal stress and, 199–200
  static pressure and, 201
  stress vectors and, 199–200
Hypotheses, continuum, 22–24

I

Immersed bodies, equilibrium and stability, 275–278
Impermeable *vs.* permeable boundaries, 339–340
Incompressible flow analyses, 713–788
  historical perspectives of, 716
  introduction to, 713–718, 780–782
  of inviscid irrotational flow, 760–780
  Karman vortices and, 715–717
  Newtonian fluid constitutive model and, 672–673
  of problems for, 782–788
  of steady viscous flow, 718–744
  of turbulent flow, 754–759
  of unsteady viscous flow, 744–754
  von Karman, Theodore and, 716
Incompressible fluids, 479–482
Initial conditions, 702–703
Inlets, 826–829
Intakes, pumps, 213
Integral hydrostatic equation, 202–205, 242–248
Integrals, convective flux, 341
Interfaces, curved, 87–90
Intermolecular attractive forces, 80–83
Internal energy, 70–71, 93–94, 341, 799
Internal surfaces, 339
Inviscid fluid flow and Bernoulli equation, 474–533
  Bernoulli equation and applications, 477–490, 496–521
  energy equation and, 521–524
  introduction to, 474–475, 524–526
  problems for, 526–533
  shear stress and, 475
  static, dynamic, stagnation, and total pressure and, 490–496
    dynamic pressure, 491–492
    examples for, 493–494
    introduction to, 490
    isentropic flow and, 494–496
    perfect gas law and, 494–496
    speed of sound and, 496
    stagnation pressure, 492
    static pressure, 490–491
    total pressure, 492–496
  streamline frictionless flow and, 475–477
    Gauss's theorem and, 477
    introduction to, 475–477
    velocity gradients and, 475
    viscosity and, 475
Inviscid irrotational flow, 760–780
  cylinder flow and, 777–780
  elementary plane potential flow and, 761–777
  examples for, 768–769, 776–777
  introduction to, 760–761
  panel methods and, 777
  superimposition and, 772–777
Irrotational flow, 632–635, 760–780. *See also under individual topics*
  doublets, 775–778
  inviscid flow, 760–780
  inviscid irrotational flow, 760–780
  vorticity, 632–635
Isentropic processes, 74–78, 487–488
Isothermal perfect gasses, 221–222
Isothermal processes, 74

J

Jacobs, Eastman, 902
Joukowsky airfoils, 138
Jumps, 87–90, 548, 972–978
  hydraulic jumps, 548, 972–978. *See also* Hydraulic jumps
  impossible jumps, 976
  pressure jumps, 87–90
  rough, intermittent, strong jumps, 976
  stable, balanced, steady jumps, 976
  standing wave/undulant jumps, 976
  weak jumps, 976

K

Karman, Theodore von, 12, 716
Karman vortices, 113, 334, 599, 715–717
Kerosene, 1019–1020, 1023
Kinematics. *See also under individual topics*
  kinematic similarity, 550
  kinematic viscosity, 82–83, 114, 884–885, 1020, 1029

Kinematics (*continued*)
 Langrangian kinematics, 576–590
  acceleration and, 578–589
  Euler-Langrangian connection and, 590–592
  examples for, 576–578
  introduction to, 578
  Langrangian fluid properties and, 589–590
  particle paths and, 578–589
  rigid body dynamics and, 578
  velocity and, 578–589
Kinetic energy, 94–95, 341, 822–823
 coefficient for, 822–823
 introduction to, 94–95, 341
Kinetic theory, 71

L

Lake Nyos, 218–219
Laminar boundary layers, 130–131, 887–894
Laminar flow, 104, 117, 805, 793
Langrangian descriptions, 27
Langrangian fluid properties, 589–590
Langrangian kinematics, 576–590
 acceleration and, 578–589
 Euler-Langrangian connection and, 590–592
 examples for, 576–578
 introduction to, 578
 Langrangian fluid properties and, 589–590
 particle paths and, 578–589
 rigid body dynamics and, 578
 velocity and, 578–589
Laplace's equation, 633–635
Laws and principles. *See also under individual topics*
 Archimedes' principle, 58–60, 269–275
 Fan laws, 124–128, 559–561
 Fourier's law, 65–67, 351
 Newton's laws, 17, 67, 80–83, 200–201, 269–270, 397, 407, 666–671
  Newton's law of viscosity, 17, 67, 80–83
  Newton's second law, 200–201, 269–270, 397, 407, 666–671
 perfect gas law, 70–73, 221–222, 412, 487–488, 515, 674–675
 polytropic law, 221–222
 power-law equation, 823
 power-law velocity profile model, 895–896
 pump laws, 124–128, 559–561
 Torricelli's law, 519–521
Length conversion factors, 1027
Lift, 137–140, 175–177, 562–563, 866–867, 926–933
 airfoil lift, 137–140, 562–563, 926–933
 boundary layers and, 886–887
 introduction to, 175–177
Lines, material, 592–597
Liquids, solids, and gases (basic terminology), 14–22
Local acceleration, 316–318, 690
Low Reynolds numbers flow, 905–908

M

M-6 airfoils, 138
MacCready, Paul, 552
Mach, Ernst, 107
Mach numbers, 79–80, 107–109, 499
Magnus effect, 914
Major head loss, 793, 799–801
Manning, Robert, 994
Manning equation, 992–993
Manning roughness coefficient, 992–993
Manometers, 57–58, 215–218, 500–501
 readings of, 57–58
 U-tube manometers, 57–58, 215–218, 500–501

Mass, weight, and density, 43–51. *See also under individual topics*
 density, 44–45, 48
 examples for, 44, 49–51
 introduction to, 43–45
 mass, 43–51, 343, 381, 386–397, 1028
  convective mass transport rates, 381
  conversion factors for, 1028
  examples for, 44, 49–51
  flowrates of, 343
  fluid transport and, 341
  introduction to, 43
  mass balance, 386–397
 molecular weight, 47–48
 Schlieren method and, 45
 solar ponds and, 47
 specific gravity, 49–51
 specific volume, 44–45
 specific weight, 48–49
 standard temperature and pressure (STP), 46–51
 U.S. standard atmosphere properties and, 45–46, 1025–1026
Material derivatives, 315–316
Material lines, surfaces, and volumes, 592–597
 continuum models and, 595
 diffusion and, 593–594
 examples for, 596–597
 hydrogen bubbles and, 593
 introduction to, 592–597
Material surfaces, 592–597
Material volumes, 592–597
Mathematical Principles of Natural Philosophy, 7–8
Mechanical energy content, 796
Mercury, 214–215, 1020, 1023
Methane, 1019–1020
Methanol, 1023
Methods of description, 22–34. *See also* Description methods
Minor head loss, 794, 824–835
Mississippi River, 934
Miter *vs.* gradual bends, 834–835
Mixed control volume (CV) analyses, 377–381, 415–420. *See also* Control volume (CV) analyses
{*MLtT*} systems, 30–32
Models. *See also under individual topics*
 constitutive model, Newtonian fluids, 671–678
  examples for, 676–678
  introduction to, 671–678
 continuum hypotheses, 22–24
 continuum models, 301, 348–350, 595
 Earth's atmosphere models, 217–218
 Newtonian fluid constitutive model, 671–678
 power-law velocity profile model, 895–896
 similitude and model development, 549–554
Modulus. *See also under individual topics*
 bulk compressibility modulus, 73–80
 shear modulus, 16
Molecular descriptions, 26
Molecular weight, 47–48, 70
Moments, hydrostatic, 252–267. *See also* Hydrostatic moments
Momentum. *See also under individual topics*
 angular momentum, 341, 439–450
 momentum equation, 666–671
  examples for, 669–671
  Gauss's theorem and, 667
  introduction to, 666–671
  Newton's second law and, 666–671
 momentum thicknesses, 886–887
Moody charts and diagrams, 555, 813–815
Motion and deformation, 607–612. *See also* Deformation
 expansion, 607–608

 introduction to, 607–612
 rigid body motion, 608
 rigid body rotation, 609–612
 shear deformation, 608–609
Mount St. Helens volcano, 147
Multiple path flow analyses, 846–851
 examples for, 848–851
 introduction to, 846

N

National Advisory Committee for Aeronautics (NACA), 138–140, 899–933
 airfoils of, 899–933. *See also* Airfoils
 introduction to, 138–140
Navier, Claude, 8, 680
Navier-Stokes equations, 6–12, 667–668, 678–683, 702–703
 conditions and, 702–703
  boundary conditions, 702–703
  initial conditions, 702–703
 coordinates for, 678–680
  Cartesian coordinates, 678–679
  cylindrical coordinates, 679–680
 examples for, 681–683
 Fourier, Jean Baptiste and, 680
 historical perspectives of, 680
 introduction to, 6–12, 667–668, 678–683
 Navier, Claude and, 680
 Newton, Isaac and, 680
 Stokes, George and, 680
Net Positive Suction Heads (NPSHs), 860–862
Newton, Isaac, 7, 14, 680
Newtonian fluids, 16–17, 21, 667–668
 constitutive model of, 668–678
  bulk viscosity and, 673
  cylindrical coordinates of, 674
  as equation set, 671–678
  equilibrium pressure and, 673
  examples for, 676–678
  incompressible flow and, 672–673
  introduction to, 671–678
  vs. non-Newtonian fluids, 662–671
  perfect gas law and, 674–675
  shear deformation and, 675–676
  strain tensors and, 676
  stress tensors and, 672–676
 vs. non-Newtonian fluids, 21, 662–671
Newton's laws, 17, 67, 80–83, 200–202, 269–270, 397, 407, 666–671
 Newton's law of viscosity, 17, 67, 80–83
 Newton's second law, 200–202, 269–270, 397, 407, 666–671
Nitrogen compounds, 1024
 nitrogen dioxide, 1024
 nitrogen oxide, 1024
No-penetration boundary conditions, 336–337
No-slip *vs.* no-penetration boundary conditions, 336–337
Nominal pipe size dimensions, 818–819
Non-Newtonian fluids, 21, 662–671
Nonaligned surfaces, planar, 238–248, 260–267
Noncontinuum *vs.* continuum descriptions, 24–26
Nondimesionalization, 703–706
 equations for, 703–706
 introduction to, 703–706
 numbers for, 703
 similitude and, 706
Normal stress, 163–171, 199–200
Nozzles, 408, 512, 832–834
NPSHs (Net Positive Suction Heads), 860–862
Numbers. *See also under individual topics*
 Bond numbers, 112
 Euler numbers, 112–113, 703
 Froude numbers, 109–110, 553, 703, 952–978

INDEX I-8

Mach numbers, 79–80, 107–109, 499
Reynolds numbers, 104–107, 117–118, 129, 135, 555, 703, 884–887, 905–908
Strouhal numbers, 113–114, 335, 703
Weber numbers, 110–112, 553

O

Octane, 1019–1020
Oil, 1019–1020, 1023–1024
One-dimensional flow, 321–327
Open channel flow applications, 942–1018
  channel perimeters and, 947
  Colorado River and, 950
  energy conservation and, 978–989
  Froude numbers and, 952–978
  of gradually varying depths, 1003
  gradually varying flow (GVF) and, 951
  hydraulic radius and, 947–948
  introduction to, 942–951, 1012–1014
  Mississippi River and, 934
  prismatic channels and, 951
  problems for, 1014–1018
  rapidly varying flow (RVF) and, 951
  Roman aqueducts and, 934
  secondary flow and, 949–950
  slopes and, 946–947
  sluice gates, 944, 1003–1008
  spillways, 944
  Tellico River and, 934
  of uniform depths, 989–1002
  uniform flow (UF) and, 951
  weirs, 944, 1009–1012
Open vs. closed surfaces, 341
Optimal channel cross sections, 999–1002
Oscillation frequencies, 481
Outward unit normal, 161–163, 235

P

P-51 Mustangs, 902
Panel methods, 777
Parallel plates, 732–737, 757–759
Partial-width weirs, 1010
Particle image velocimetry (PIV), 601–602
Particle paths, 314–316, 578–589
Pascal, Blaise, 5
Pathlines vs. streaklines, 597–603
  examples for, 600–601
  F-18 Hornets and, 602
  hydrogen bubbles and, 599
  introduction to, 314–316, 597–603
  Karman vortices and, 599
  particle image velocimetry (PIV) and, 601–602
Paths, 314–316, 578–589, 817–851
  multiple path flow analyses, 846–851
  particle paths, 314–316, 578–589
  single path flow analyses, 817–846
Penetration, crack, 92–93
Perfect gas law, 70–73, 221–222, 412, 487–488, 515, 674–675
  applicability limits of, 71–73
  Bernoulli equation and, 487–488, 515
  common gas parameters and, 72
  control volume (CV) analyses and, 412
  enthalpy and, 70–71
  internal energy and, 70–71
  introduction to, 70
  isothermal perfect gasses and, 221–222
  kinetic theory of gasses and, 71
  molecular weight and, 70
  specific gas constant, 70
  specific heats of, 70–71
  universal gas constant, 70

Perimeters, 947–948, 993
  channel perimeters, 947–948
  wetted perimeters, 993
Periodic flow, 327–336
  spatially periodic flow, 327–330
  temporally periodic flow, 332–336
Permeable vs. impermeable boundaries, 339–340
Pi theorem, Buckingham, 536–540. See also Buckingham Pi theorem
Pipe and duct flow applications, 791–881
  introduction to, 791–793, 864–867
  multiple path flow analyses, 846–851
  pipe size dimensions, 818–819
  problems for, 867–881
  round pipe flow, 115–120, 173–174, 557–558
  single path flow analyses, 817–846
  steady fully developed flow, 793–817
  steady viscous flow in, 737–741
  system design elements, 851–864
Pitot, Henri, 8–9
Pitot-static tubes, 499
Pitot tubes, 496–503
PIV (particle image velocimetry), 601–602
Planar surfaces, 234–267
  aligned surfaces, 234–238, 256–260
  nonaligned surfaces, 238–248, 260–267
Plane Couette flow, 720–723, 749–752
Plane potential flow, elementary, 761–777
Planforms, 138, 928
Points of application (POAs), 267–269
Poise, 83
Poiseuille, J. L. M., 8–9, 732
Poiseuille flow, 394, 732–741, 806–807
Polar coordinates, 301–302
Polytropic law, 221–222
Ponds, solar, 47
Ports, 425–426
Potential energy, 95
Power, 127–128, 426–427, 560, 823, 895–896, 1030
  coefficients for, 127–128, 560
  conversion factors for, 1030
  fluid power, 426
  power-law equation, 823
  power-law velocity profile model, 895–896
  shaft power, 426–427
Prandtl, Ludwig, 12–13, 128–129, 337, 884–885
Prandtl boundary layer equations, 888–894
Precipitators, electrostatic, 150
Preferred unit systems, 32
Press-fitted nozzles, 408
Pressure, 51–64. See also under individual topics
  absolute pressure, 53–54
  barometric pressure, 213–215
  Bourdon tube pressure gages and, 54
  buoyancy and, 58–60
  conversion factors for, 1029
  dynamic pressure, 490–496
  equilibrium pressure, 673
  examples for, 52, 55–63
  form/pressure drag and, 177, 903, 911–912
  Fulton, Robert and, 52
  gage pressure, 53–54, 411
  hydrostatic pressure distribution, 198–199, 210–233, 411. See also Hydrostatic pressure distribution
  introduction to, 51–52
  jumps for, 87–90
  manometer readings of, 57–58
  pressure stress, 399
  pumps and, 64
  stagnation pressure, 490–496
  standard temperature and pressure (STP), 46–51
  static pressure, 201, 490–496
  surface tension and, 87–90

total pressure, 490–496
vapor pressure, 214–215, 485–487, 504
variations of, 54–57, 60–64, 691–692
  Bernoulli equation and, 60
  in moving fluids, 60–64
  in stationary fluids, 54–57
  venturi and, 60–62
Principia, 7–8
Prismatic channels, 950–951
Problem solving approaches, 34–35
  introduction to, 34–35
  procedures for, 35
Processes, 332–336, 430, 440, 487–488
  isentropic processes, 487–488
  steady processes, 332–336, 430, 440
Propeller turbines, 854
Properties. See also under individual topics
  fluid properties, 42–102, 149–151, 1019–1024. See also Fluid properties
  standard U.S. atmosphere properties, 45–46, 1025–1026
  thermal properties, 64–70
Protuberances, 930–931
Pumps, 64, 124–128, 213, 559–561, 835–864
  axial flow pumps, 854
  centrifugal/radial flow pumps, 854
  heads of, 835–846
  intakes of, 213
  performance of, 855
  pressure and, 64
  pump curves, 852
  pump laws, 124–128, 559–561
  selection of, 853–864

R

Radial pumps, 854
R.A.F. (Royal Air Force) airfoils, 138
Rapidly varying flow (RVF), 951
Reaction forces, 407–410, 443
Rectilinear acceleration, 229–233
Reducing elbows, 821–822
Repeating variable method, 540–549
  examples for, 545–549
  hydraulic jumps and, 548
  introduction to, 540
  steps of, 540–545
  surface tensions and, 542
  Taylor, G. I. and, 545–546
Resultant forces, 267–269. See also Points of application (POAs)
Reynolds, Osborne, 104
Reynolds equations, 756–757
Reynolds numbers, 104–107, 117–118, 129, 135, 555, 703, 884–887, 905–908
Reynolds transport theorem, 376, 381–385, 637, 660–662, 702
  for control volumes (CVs), 382–385
  examples for, 384
  introduction to, 376, 381, 637, 660–662, 702
  for system volumes, 381–382
Rigid bodies, 222–229, 578, 609–630
  dynamics of, 578
  motion of, 608
  rotation of, 222–229, 609–612, 619–620, 630
Roman aqueducts, 4, 934
Rotation and rotation rates, 619–622
  eddys and, 620
  examples for, 621–622
  introduction to, 619–622
  rigid body rotation, 222–229, 609–612, 619–620, 630
  solid body rotation, 693
  swirls and, 620
Round pipe flow, 115–120, 173–174, 557–558

Royal Air Force (R.A.F.) airfoils, 138
RVF (rapidly varying flow), 951

S

SAE oil, 1019–1020, 1023–1024
Schedules, pipe, 817
Schlieren method, 45, 641
Seawater, 1024
Secondary flow, 949–950
Secondary walls, 425
Separation, flow, 820–821
Shaft power, 426–427
Sharp crested weirs, 1010
Shear. *See also under individual topics*
 deformation and deformation rates, 608–609, 650–653, 675–676
 modulus, 16
 strain and strain rates, 15–16
 strain rates of, 16
 stress, 15, 399–405, 413–415, 475, 886–887
 viscosity, 83, 700
 wall shear stress, 886–887
SI (Système International d'Unités), 30–31
Similitude, 534–570. *See also* Dimensional analyses (DA) and similitude
 Gossamer Condor and, 552
 introduction to, 549–552
 MacCready, Paul and, 552
 model development and, 549–554
 nondimesionalization and, 706
 numbers for, 553–554
 similarities, 550–551
Single-dimensional flow, 321–327
Single path flow analyses, 817–846
 coefficients for, 822–829
 common nominal pipe size dimensions, 818–819
 examples for, 827–846
 exits and, 826–829
 flow separation and, 820–821
 gradual *vs.* miter bends and, 834–835
 guide vanes and, 824
 inlets and, 826–829
 introduction to, 817–824
 minor head loss and, 824–835
 nozzles and, 832–834
 pipe fitting lengths and, 835
 power-law equation and, 823
 pump and turbine heads and, 835–846
 reducing elbows and, 821–822
 sudden area changes and, 829–832
Single path piping systems, 793
Siphons, 503–509
Skin friction coefficients, 130, 887
Slopes, 947–980
 bed slopes, 947
 depth slopes, 947
 energy line slopes, 980
 free surface slopes, 946–947
 friction slopes, 980
 open channel flow applications and, 946–947
Sluice gates, 416, 509–511, 944, 1003–1008
Small amplitude waves, 971–972
Solar ponds, 47
Solid body rotation, 693
Solid tubes, 90–93
Solids, liquids, and gases (basic terminology), 14–22
Sound, speed of, 76–80, 107–109, 499
Span, 138, 928
Spatially periodic flow, 327–330
Specific energy, 981–989
Specific gas constant, 70
Specific gravity, 49–51
Specific heat, 65–66, 70–71, 222
Specific volume, 44–45

Speed contour plots, 305–306
Speed of sound, 76–80, 107–109, 499
Spheres, drag, 132–137, 562
Spillways, 944
Square rods, 910
Stability, immersed bodies, 275–278
Stagnation pressure, 490–496
Stall, 929
Standard atmosphere properties, U.S., 45–46, 222–224, 1025–1026
Standard temperature and pressure (STP), 46–51
Statics, fluid, 197–298. *See also* Fluid statics
 buoyancy and, 269–275
 hydrostatic equation and, 201–210
 hydrostatic forces and, 233–252
 hydrostatic moments and, 252–267
 hydrostatic pressure distribution and, 210–233
 hydrostatic stress and, 199–201
 immersed bodies, equilibrium and stability of, 275–278
 introduction to, 197–199, 278–280
 problems for, 280–298
 resultant forces and points of application (POAs), 267–269
 static pressure and, 201, 490–496
Stationary fluids, pressure variations, 54–57
Steady fully developed flow, 332–336, 793–817
 Bernoulli equation and, 799
 entrance lengths and, 794
 examples for, 797–798, 800–801, 804–805, 813, 816–817
 friction factors and, 793
 hydraulic diameters and, 793
 internal energy and, 799
 introduction to, 332–336, 793–799
 laminar flow and, 793
 major head loss and, 793–794, 799–801
 mechanical energy content and, 796
 single path piping systems and, 793
 turbulent flow and, 793
Steady processes, 332–336, 430, 440
Steady viscous flow, 718–744
 Couette, Maurice and, 720–722
 Couette flow, 720–732
 cylinder flow and, 741–744
 examples for, 724–726, 730–732, 735–737, 740–741
 historical perspectives of, 721–722
 introduction to, 718–720
 parallel plates and, 732–737
 in pipes, 737–741
 Poiseuille, J. L. M. and, 732
 Poiseuille flow and, 732–741
Steam turbines, 854
Stokes, George, 8, 14, 83, 680
Stokes theorem, 628–632, 916
STP (standard temperature and pressure), 46–51
Strain, 15–16, 676
 shear, 15–16
 tensors, 676
Streaklines *vs.* pathlines, 597–603
 examples for, 600–601
 F-18 Hornets and, 602
 hydrogen bubbles and, 599
 introduction to, 314–316, 597–603
 Karman vortices and, 599
 particle image velocimetry (PIV) and, 601–602
Streamfunctions, 643–650, 889
Streamlines and streamtubes, 177, 603–607, 923–924
 body drag and, 177
 boundary layer separation and, 604–605
 contour plots for, 307

coordinates for, 689–692
 examples for, 606–607
 introduction to, 603–607, 923–924
 streamline frictionless flow, 475–477
  Gauss's theorem and, 477
  introduction to, 475–477
Stress. *See also under individual topics*
 axial stress, 252
 fluid stress, 178–187
 hoop stress, 252
 hydrostatic stress, 199–201
 normal stress, 163–171, 199–200
 pressure stress, 399
 shear stress, 15, 399–405, 413–415, 475, 886–887
 tensors, 178–187, 350–351, 672–676
 vectors, 160–161, 178–187, 199–200, 233–234
 wall shear stress, 886–887
Strouhal, Vincenz, 113
Strouhal numbers, 113–114, 335, 703
Structure elements, 573–658. *See also* Visualization and structure elements
 Eulerian-Langrangian connections of, 590–592
 expansion rates, 635–650
 introduction to, 573–578, 653–654
 Langrangian kinematics, 578–590
 material lines, surfaces, and volumes, 592–597
 motion and deformation, 607–612
 pathlines *vs.* streaklines, 597–603
 problems for, 654–658
 rotation rates, 619–622
 shear deformation rates, 650–653
 streamlines and streamtubes, 603–607
 velocity gradients, 612–618
 vorticity, 622–635
Subsonic flow, 507
Substantial derivatives, 315–320
Sudden area changes, 829–832
Sudden contraction, 124, 512
Sudden enlargement, 512
Sudden expansion, 122–124, 395, 402–403
Supercritical *vs.* critical flow, 953
Superimposition, 772–777
Supersonic flow, 507
Surface forces, 86–87, 148–177. *See also under individual topics*
 drag, 175–177
  airfoil drag, 177
  bluff body drag, 177
  introduction to, 175–177
  pressure/form drag, 177
  streamlined body drag, 177
 flat plates and, 171–172
  boundary layers of, 171–172
  flow over, 171–172
 fully developed flow and, 173–174
 introduction to, 148–151, 160–161
 lift, 175–177
 origins of, 149–151
 outward unit normal of, 161–163
 round pipe flow, 173–174
 stress, 160–161
Surface tension, 85–93
 bubbles and, 88–89
 capillary actions and, 90–93
 contact angles and, 90
 crack penetration and, 92–93
 curved interfaces and, 87–90
 dimensionless groups and, 111–112
 examples for, 89, 92
 introduction to, 85–86
 pressure jumps and, 87–90
 in solid tubes, 90–93

INDEX | I-10

surface energy and, 86–87
  values for, 87
  wetting and, 90
Surfaces. *See also under individual topics*
  bounding surfaces, 339
  closed *vs.* open surfaces, 341
  curved surfaces, 248–252, 900
  decal surfaces, 400–420
  free surface slopes, 946–947
  internal surfaces, 339
  material surfaces, 592–597
  planar surfaces, 234–267
  surface energy, 86–87
  surface forces, 148–177
  surface tension, 85–93, 542
  surface wave propagation, 965–972
  work done by surface forces, 424–425
Swirls, 620
System curves, 852
System design elements, 851–864
  cavitation and, 860–862
  compressors, 854–855
  examples for, 856–860, 862–864
  fans, 853–864
  introduction to, 851–853
  Net Positive Suction Heads (NPSHs), 860–862
  pumps, 853–864
  system curves and, 852
  turbines, 854–855
  valves, 852–853
Système International d'Unités (SI), 30–31

T

Tacoma Narrows Bridge, 113
Tangents, vector, 163
Tanks, draining of, 518–521
Taylor, G. I., 545–546
Taylor series expansions, 315
Tellico River, 934
Temperature and thermal properties, 64–70, 114, 309, 322–323, 1030
  conductivity, thermal, 65–66
  contour plots for, 309, 322–323
  conversion factors for, 1030
  diffusity, thermal, 114
  examples for, 66–69, 72
  expansion, thermal, 65–70
  introduction to, 64–65
  laws for, 65–73
  specific heat, 65–66
  temperature scales, 64–65
Temporally periodic flow, 332–336
Tension, surface, 85–93. *See also* Surface tension
Tensors, 178–187, 350–351, 672–676
  strain tensors, 676
  stress tensors, 178–187, 350–351, 672–676
Theorems. *See also under individual topics*
  Buckingham Pi theorem, 536–540
  Gauss's theorem, 205–210, 477, 637, 642–643, 661–662, 667
  Reynolds transport theorem, 376, 381–385, 637, 660–662, 702
  Stokes theorem, 628–632, 916
Thermal properties, 64–70. *See also* Temperature and thermal properties
Thicknesses, 885–887, 928
  of airfoils, 928
  of boundary layers, 885–887
  displacement thicknesses, 885–886
  momentum thicknesses, 886–887
Three-dimensional flow, 321–327
Thrust, 412
Time, conversion factors, 1028
Torque, 427, 439–440

Torricelli, E., 5, 519
Torricelli's law, 519–521
Total energy, 95–96, 424
Total head parameters, 125
Total heat transfer rates, 423–424
Total pressure, 490–496
Total transport, 352–358
Transitions, 104, 129
  laminar-to-turbulent transitions, 129
  turbulent transitions, 104
Transport. *See also under individual topics*
  angular momentum transport, 440–441
  convective mass transport rates, 381
  fluid transport, 299–374
  Reynolds transport theorem, 376, 381–385, 637, 660–662, 702
Tsunamis, 968–969
Tubes. *See also under individual topics*
  pitot-static tubes, 499
  pitot tubes, 496–503
  solid tubes, 90–93
  streamlines and streamtubes, 177, 603–607, 923–924
  U-tube manometers, 500–501
Turbines, 835–854
  gas turbines, 854
  heads of, 835–846
  propeller turbines, 854
  steam turbines, 854
Turbulent boundary layers, 894–898
Turbulent flow, 104, 117–118, 754–756, 793, 812–817
  friction factors for, 812–817
  incompressible analyses and, 754–756
  vs. laminar flow, 104
  steady fully developed flow and, 793
Turbulent transitions, 104
Two-dimensional flow, 321–327

U

U-tube manometers, 57–58, 215–218, 500–501
UF (uniform flow), 327–330, 951
Uniform depths, 992–1002
  Chezy, Antoine, 994
  Chezy coefficient and, 992
  Chezy equation and, 992
  introduction to, 989–994
  Manning, Robert and, 994
  Manning equation and, 992–993
  Manning roughness coefficient and, 992–993
  open channel flow applications and, 989–1002
  optimal channel cross sections and, 999–1002
  wetted perimeters and, 993
Uniform flow (UF), 327–330, 951
Unit conversion factors, 33–34, 1027–1030
Unit system description methods, 22–34
  British Gravitational (BG) Unit System, 31
  continuum hypothesis, 22–24
  for continuum *vs.* noncontinuum descriptions, 24–26
  for conversions, 33–34
  decision-making criteria for, 28–29
  for dimension descriptions, 29–30
  English Engineering (EE) Unit System, 32
  examples for, 24–25, 29, 31, 33–34
  introduction to, 22, 29–30
  for specific descriptions, 26–31
  Système International d'Unités (SI), 30–31
  for systems, 30–32
Universal gas constant, 70
Unsteady viscous flow, 744–754
  Bessel functions and, 752
  cylinder flow and, 752–754

  introduction to, 744–749
  plane Couette flow and, 749–752
U.S. standard atmosphere properties, 45–46, 222–224, 1025–1026

V

V-notch weirs, 1010
Valves, 852–853
  ball valves, 853
  conventional valves, angle and globe, 853
  types of, 852–853
  wedge gate valves, 853
Vanes, guide, 824
Vapor, 214–215, 485–487, 504–505
  bubbles, 504–505
  mercury vapor pressure, 214–215
  pressure, 214–215, 485–487, 504
Variations. *See also under individual topics*
  pressure variations, 54–57, 60–64
  repeating variable method, 540–549
  variable density fluids, 218–222
  varying width horizontal channels, 961–965
Vectors. *See also under individual topics*
  diffusive flux vectors, 351
  plots of, 302
  stress vectors, 160–161, 178–187, 199–200, 233–234
  tangents for, 163
  velocity plots of, 322–323, 328–329
Velocimetry, particle image, 601–602
Velocity fields and fluid transport, 299–374
  average velocities and flow rates, 358–363
  conversion factors for, 1028
  flow classifications and, 320–336
  fluid acceleration, 312–319
  fluid transport, 337–358
  introduction to, 299–300, 363–365
  Langrangian kinematics and, 578–589
  no-slip *vs.* no-penetration boundary conditions and, 336–337
  power-law velocity profile model, 895–896
  problems for, 365–374
  substantial derivatives of, 319–320
  velocity fields, 300–312
    Cartesian coordinates for, 301–302
    case study applications for, 306
    continuum models of, 301
    cylindrical coordinates for, 301–302
    density contour plots for, 308
    Eulerian description of, 300–312
    examples for, 302–304, 309–312
    introduction to, 300–301
    polar coordinates for, 301–302
    speed contour plots for, 305–306
    streamline contour plots for, 307
    temperature contour plots for, 309
    vector plots of, 302
  velocity gradients, 17, 475, 612–618
  velocity potential, 632–635
  velocity vector plots, 322–323, 328–329
  vorticity and, 632–635
Vena contracta effects, 510, 520–521
Venturi, 60–62, 486–487
Viscosity, 80–85. *See also under individual topics*
  absolute viscosity, 16, 701–702, 1029
  bulk viscosity, 83–85, 673, 700–702
  conversion factors for, 1029
  dynamic viscosity, 16
  energy dissipation rates and, 83–85
  examples for, 81–82, 84–85
  intermolecular attractive forces and, 80–83
  introduction to, 80–85
  inviscid fluid flow and, 475

INDEX

Viscosity (*continued*)
  kinematic viscosity, 82–83, 114, 884–885, 1020, 1029
  Newton's law of viscosity, 17, 67, 80–83
  poise and, 83
  shear viscosity, 83–85, 700
  stoke and, 83
  viscous dissipation, 83, 699–700
  viscous flow, 718–754
Visualization and structure elements, 573–658
  Eulerian-Langrangian connections of, 590–592
  expansion rates, 635–650
  heating, ventilating, and air-conditioning (HVAC) systems and, 573–578
  introduction to, 573–578, 653–654
  Langrangian kinematics, 576–590
  material lines, surfaces, and volumes, 592–597
  motion and deformation, 607–612
  pathlines *vs.* streaklines, 597–603
  problems for, 654–658
  rotation rates, 619–622
  shear deformation rates, 650–653
  streamlines and streamtubes, 603–607
  velocity gradients, 612–618
  vorticity, 622–635
    circulation and, 628–632
    examples for, 623–625, 627–628, 631–632, 634–635
    introduction to, 622–628
    irrotational flow and, 632–635
    Laplace's equation and, 633–635
    rigid body rotation and, 630
    Stokes theorem and, 628–632
    velocity potential and, 632–635
Volumes. *See also under individual topics*
  control volume (CV) analyses, 375–473
  conversion factors for, 1027
  flowrates for, 343, 1029
  fluid transport and, 341
  material volumes, 592–597
  specific volume, 44–45
  system volume analyses, 376–382
Von Karman, Theodore, 12, 716
Vortices and vorticity, 622–635. *See also under individual topics*
  airfoils and, 932
  circulation and, 628–632
  Euler equations and, 695
  examples for, 623–625, 627–628, 631–632, 634–635
  introduction to, 622–628
  irrotational flow and, 632–635
  Karman vortices, 113, 334, 599, 715–717
  Laplace's equation and, 633–635
  rigid body rotation and, 630
  Stokes theorem and, 628–632
  velocity potential and, 632–635
  visualization and structure elements and, 622–635

W

Walls, 425, 886–887
  secondary walls, 425
  shear stress of, 886–887

Water, 1019–1021, 1024
Waves, 965–972
  small amplitude waves, 971–972
  surface wave propagation, 965–972
Weber, Mortiz, 110–111
Weber numbers, 110–112, 553
Wedge gate valves, 853
Weight, 43–51. *See also* Mass, weight, and density
  examples for, 44, 49–51
  introduction to, 43–45
  molecular weight, 47–48
  specific weight, 48–49
Weirs, 944, 1009–1012
  broad crested weirs, 1012
  full-width weirs, 1010
  introduction to, 944, 1009–1012
  open channel flow applications and, 944, 1009–1012
  partial-width weirs, 1010
  sharp crested weirs, 1010
  V-notch weirs, 1010
Wetted perimeters, 993
Wetting and surface tension, 90
Work, 424–425, 1030
  conversion factors for, 1030
  surface force work, 424–425
Wright, Wilbur, 12
Wright brothers, 12, 138